Teacher, Stude

One-Stop Internet Resources

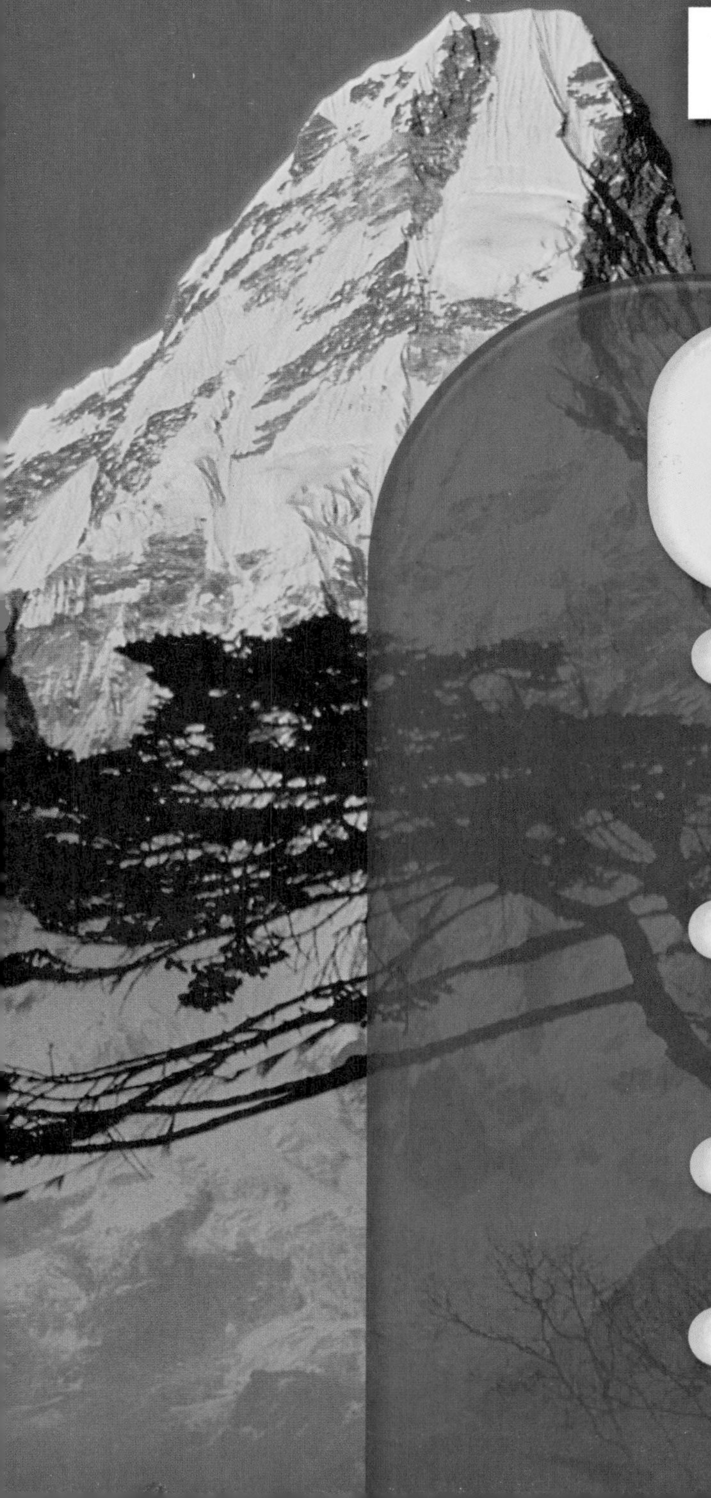

Earth Science Online

Log on to earthgeu.com

ONLINE STUDY TOOLS

- Section Self-Check Quizzes
- Chapter Tests
- Standardized Test Practice
- Vocabulary PuzzleMaker
- Multilingual Science Glossary

ONLINE RESEARCH

- National Geographic Society
- NASA's Picture of the Day
- Prescreened Web Links
- Internet GeoLabs
- Extending the Content
- Problem of the Week
- Science Fair Ideas

INTERACTIVE ONLINE STUDENT EDITION

- Complete Interactive Student Edition
- Textbook Updates

FOR TEACHERS

- Teacher Bulletin Board
- Teaching Today—Professional Development

Topographic Map Symbols

Roads and Railroads

Primary highway, hard surface

Secondary highway, hard surface

Light-duty road, hard or improved surface

Unimproved road

Railroad: single track and multiple track

Railroads in juxtaposition

Buildings and Structures

Buildings

School, church, and cemetery — cem

Barn and warehouse

Wells, not water (with labels) — oil gas

Tanks: oil, water, etc. (labeled if water) — water

Open-pit mine, quarry, or prospect

Tunnel

Benchmark — BM Δ 293

Bridge

Campsite

Habitats

Marsh (swamp)

Wooded marsh

Woods or brushwood

Vineyard

Submerged marsh

Mangrove

Coral reef, rocks

Orchard

Urban area

Perennial streams

Elevated aqueduct

Water well and spring

Small rapids

Large rapids

Intermittent lake

Intermittent stream

Glacier

Large falls

Dry lake bed

Surface Elevations

Spot elevation — **x 7369**

Water elevation — **670**

Index contour — 100

Intermediate contour

Depression contour

Boundaries

National

State

County, parish, municipal

Civil township, precinct, town, barrio

Incorporated city, village, town, hamlet

Reservation, national or state

Small park, cemetery, airport, etc.

Land grant

Township or range line, United States land survey

Township or range line, approximate location

CALIFORNIA

Earth Science: Geology, the Environment, and the Universe

Table of Contents

Glencoe

New York, New York   Columbus, Ohio   Chicago, Illinois   Peoria, Illinois   Woodland Hills, California

California Student Edition: ISBN-10: 0-07-877269-9 ISBN-13: 978-0-07-877269-6

Focus on California

Focus on California lists photos in this textbook that depict things or places in California.

Larry Ulrich

Baron Wolman/Stone

Tom Dean/DRK Photo

Focus on California

Larry Ulrich

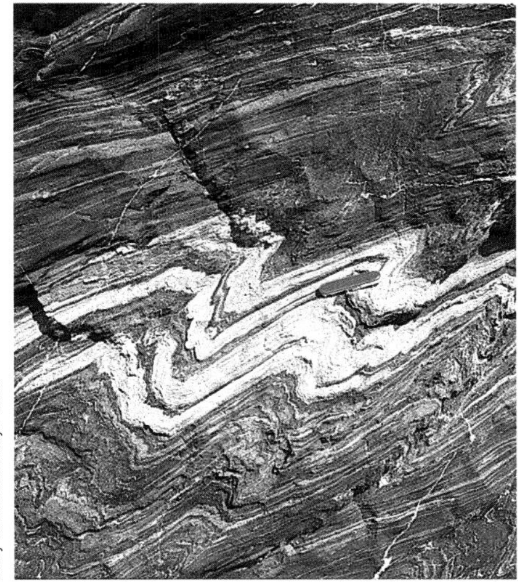

Betty Crowell/Faraway Places

Anthony Cook/Griffith Observatory

GeoCalifornia

Earth Sciences 9.a Students know the resources of major economic importance in California and their relation to California's geology. **9.c** Students know the importance of water to society, the origins of California's fresh water, and the relationship between supply and need.

Economic Geology of California

The formation and distribution of almost all of California's valuable natural resources resulted from geologic processes. These processes include plate tectonic activity that is driven by Earth's internal thermal energy and surface processes that are driven by energy from the Sun. California's abundant resources—minerals, energy, water, and living things—support our lifestyles.

Mineral and Energy Resources

California is a major supplier of mineral and energy resources. It ranks first in the United States for mineral production and fourth in oil production. This abundance of resources provides a foundation for California's economic growth.

Mineral Resources

Deposits of rocks or minerals that can be extracted economically for use are **mineral resources.** Almost all minerals are found dispersed throughout Earth's crust. A mineral deposit is a resource when it is found in a concentration greater than its average crustal concentration. Geological processes above and below Earth's surface concentrate these materials.

Sand and gravel is California's most valuable mineral resources product, as shown below. Its value per ton is not as great as other mineral resource products, but more of it is produced than other mineral resources. The deposition of sand and gravel sediments in river valleys and basins formed this resource. California's second most valuable mineral resource product is cement. Cement is processed from sedimentary limestone deposits.

California is the only state that produces boron. Boron is also found in sedimentary rock deposits. Boron is formed in arid basins containing volcanic ash.

Tectonic forces associated with subduction and mountain building produce molten rock called magma. Magma intrudes into older rocks where precious metals crystallize as it cools. This forms concentrated ore deposits of valuable gold and silver.

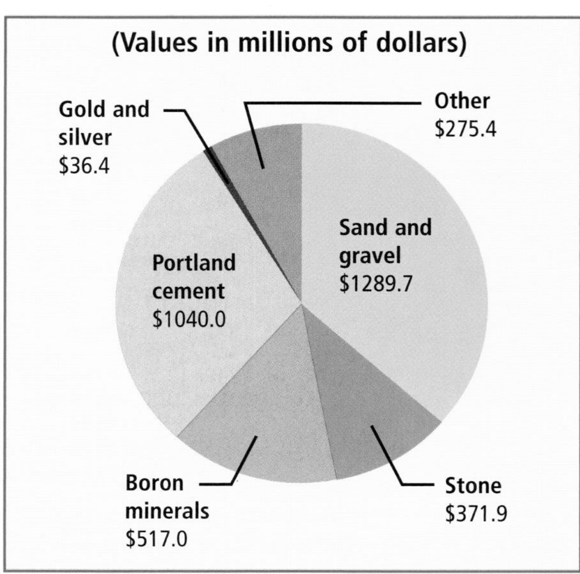

Non-fuel mineral production in California, 2004

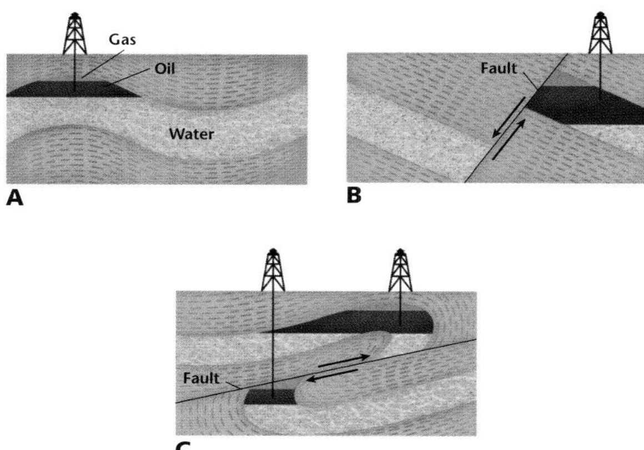

Structural traps of oil and gas deposits

Energy Resources

California has a variety of energy resources including fossil fuels, such as crude oil and natural gas, and geothermal energy. These, like mineral resources, are valuable because they have been concentrated in specific locations by geologic processes.

Fossil Fuels The Sun is the ultimate energy source of fossil fuels. Photosynthetic organisms called producers transform energy from the Sun, which is stored in their cells. As other organisms consume producers, and other organisms consume these organisms, this stored energy is transformed repeatedly and is distributed throughout land and aquatic environments.

In ancient saltwater environments, the remains of microscopic, energy-storing organisms accumulated on the seafloor. Over time, oxygen-poor sediments buried them. After millions of years, pressure from overlying rocks and sediments, and thermal energy from within Earth turned this energy-rich organic material into hydrocarbon molecules. These molecules are the sources of both crude oil and natural gas—California's most important fossil-fuel resources.

The Central Valley is a major source of California's fossil fuels. It was once the location of an ancient sea that formed as the result of tectonic activity. This area sank down and filled with water. The Sierra Nevada was pushed up by folding and faulting. Organic material accumulated on the seafloor and turned into fossil fuels over time. Subsequent folding and faulting concentrated these fossil fuels in geologic structures called traps, as shown in the figure to the left. Fossil fuels in traps can be pumped out when wells are drilled into them.

California is the fourth largest producer of oil and a major producer of natural gas in the United States, but the use of these resources exceeds production. California imports 84 percent of the natural gas and 58 percent of the oil used in the state.

Geothermal Energy Earth's internal thermal energy is released to the surface all around the world. This energy is called geothermal energy, but, just like mineral resources, it can be concentrated in certain locations.

California has active tectonic activity that concentrates the release of Earth's internal thermal energy. In northern California, an active subduction zone releases thermal energy near Earth's surface. In southeastern California, continental spreading brings thermal energy to the surface. In both cases, thermal energy flows to the surface as hot water or steam and is used to generate electricity, as shown below.

Although geothermal energy supplies California with only one percent of its energy needs, it is an important energy resource. Using geothermal energy to produce electricity is considered more environmentally friendly than burning fossil fuels because it produces fewer emissions and is renewable.

David R. Frazier/Photo Researchers

Geothermal power plant, Imperial Valley, California

Cliff Leight

Sierra Nevada snow pack

Water Resources

Most of Earth's water is salt water found in oceans and seas. Only about four percent of Earth's water is freshwater, and most of it is frozen in glaciers and polar ice sheets. However, freshwater is replenished when water evaporates from the oceans and is released from plants, and then returns to land as freshwater precipitation—rain, snow, sleet, and hail. California has abundant freshwater resources supplied by this precipitation.

Sources

Most of the precipitation that supplies California's freshwater falls in autumn and winter. There is little precipitation in spring and summer when water demands are highest. However, California's mountain ranges play a role in the redistribution of this precipitation.

Mountain Watersheds California depends heavily on high elevation snow packs, shown in the photo above, to supply its rivers and other freshwater resources. Watersheds in the Sierra Nevada and other mountain ranges collect snow that melts in the spring. This runoff slowly replenishes rivers, reservoirs, and groundwater aquifers.

Surface and Groundwater Most of California's water is supplied by surface water. Small streams flowing from the mountains join and form rivers.

California's rivers have been dammed, channeled, and diverted to control floods and supply water for hydroelectric power. The Sacramento-San Joaquin River system forms an integral part of California's water resources. It drains a large part of the Central Valley and serves as the water supply for 20 million people. Even a portion of the Colorado River has been diverted to satisfy southern California's water needs.

Groundwater supplies are much larger than surface-water supplies, but they meet a smaller portion of California's water needs. This is because of the poor quality of some groundwater and the expense of pumping it to the surface.

Why not the ocean? When freshwater supplies become scarce, humans sometimes rely on the oceans for water. Desalinization is the process of removing salt from saltwater sources to make freshwater. Forcing salt water through a very fine filter can remove the salt. This method is called reverse osmosis. Salt water also can be evaporated and then condensed, which removes the salt. The expense of both methods limits their use. In response to a severe drought during the early 1990s, the city of Santa Barbara built the largest reverse osmosis desalinization facility in the United States. The facility is no longer used because it is too expensive to operate, but the plant is maintained for use in case of another water shortage.

Distribution

Water in California is not evenly distributed. Northern California receives most of the rain and snow. Southern California is arid to semiarid, and more people live there than in northern California. The water needs of southern California exceed its supply.

Vital Statistics

California's Usable Surface Water

Flows to the ocean in rivers	36%
Other aquatic environments	28%
Used for agriculture	28%
Used by cities and industry	8%

John Livsey/Stone

Los Angeles aqueduct

Distribution Systems

The imbalanced natural distribution of water in California is adjusted using aqueducts and canals that transport water from water-rich locations to the southern part of the state. The Los Angeles aqueduct, as shown in the photo above, diverts water from the Owens River Valley on the eastern side of the Sierra Nevada to the Los Angeles area. It supplies about 70 percent of Los Angeles's water. The remainder comes from the Colorado River aqueduct and the California aqueduct. The California aqueduct transports water from the northern part of the state to the southern part.

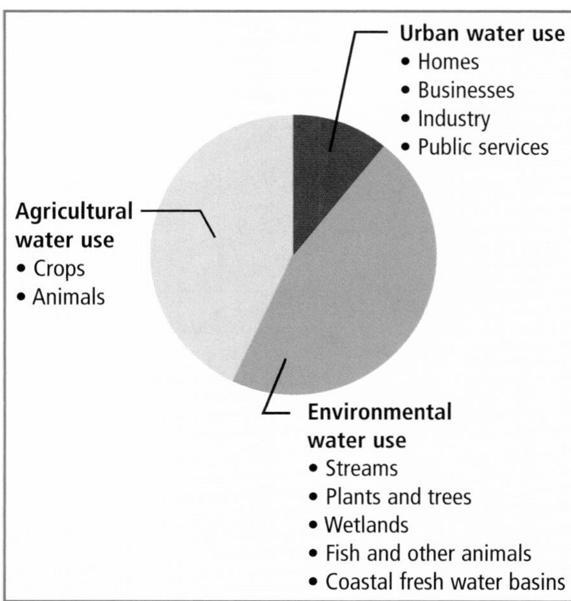

Uses of California's average annual total water supply

Uses

All of California's freshwater supply once was available for use by native plants, animals, and other organisms. Humans now use part of the freshwater supply for residential, industrial, and agricultural needs, as shown below. Water is important because agriculture and industry require large quantities of water and form the basis of California's economy.

Agriculture California's agricultural productivity is the highest in the nation. In addition, agriculture is California's largest industry; farm receipts totaled over $27 billion in 2001. Because much of the state has an arid or semiarid climate, water is essential for crop irrigation. Agricultural processes use over 30 percent of California's freshwater resources. This is about equal to the amount of freshwater that is remaining for natural ecosystems.

Industry Although less water is used for industry than agriculture, maintenance of this supply is crucial for California's growing economy. As industry grows, increased water needs are met by using water from natural ecosystems or from agricultural supplies.

FOCUS ON CAREERS

Mining Engineer

Mining engineers design and manage the development of mines. They locate mines and work to extract minerals and energy resources. They conduct surveys of mineral deposits to determine the size and quality of the resources. Mining engineers must have the ability to visualize three-dimensional objects from two-dimensional drawings. They must be able to adapt to a variety of work situations.

EARTH SCIENCES ASSESSMENT

Use the table below to answer question 1.

Average Seasonal Rainfall (cm)				
City	Winter	Spring	Summer	Fall
Yreka	7.6	3.0	1.8	4.1
Sacramento	7.6	3.6	0.3	3.6
Blythe	1.0	0.5	0.8	0.8
Los Angeles	6.9	3.3	0.0	0.8

1 When do these cities receive the most rainfall?

A spring and summer
B spring and fall
C fall and winter
D winter and spring

2 Where did California's gold and silver deposits form?

A limestone outcrops
B magma intrusions
C sedimentary rock deposits
D volcanic ash deposits

3 What is the ultimate energy source of fossil fuels?

A geothermal energy
B hydroelectric power
C the Sun
D tectonic activity

4 What energy resources are found in the Central Valley?

A boron and oil
B geothermal energy and boron
C natural gas and oil
D oil and cement

5 What concentrates geothermal energy?

A tectonic activity
B weathering and erosion
C photosynthesis
D volcanic eruptions

6 What process allows mountain watersheds to release water slowly?

A heavy summer rains
B accumulation of winter rainfall
C clear cutting of vegetation
D gradual spring snowmelt

7 Why isn't desalination a major source of freshwater for California?

A It is too expensive.
B California lacks abundant salt water.
C The attempt in Santa Barbara failed.
D The technology is not yet developed.

Use the table below to answer question 8.

Mineral Resource	Percent of California's Mineral Economy
Sand and gravel	36
Clays	1
Crushed stone	10
Boron minerals	14
Gold	1
Portland cement	29
Diatomite	2
Other	7

8 What is California's most valuable mineral resource?

A boron minerals
B sand and gravel
C gold
D portland cement

9 Which has the greatest demand on California's freshwater resources?

A agriculture
B businesses
C homes
D industry

Earth Sciences 7.a Students know the carbon cycle of photosynthesis and respiration and the nitrogen cycle. *Also covers* 7.b, 7.c, 7.d

Biogeochemical Cycles

Biogeochemical cycles describe how chemicals move within and between Earth's major spheres. While many elements are involved in the biogeochemical cycles, some are more important than others. **Biogeochemistry** examines the interaction between biological and geochemical processes during biogeochemical cycles.

Earth's Spheres

Earth has four major systems or spheres—the lithosphere, the hydrosphere, the atmosphere, and the biosphere. The lithosphere is Earth's solid, outermost layer. It includes soils, sediments, and rocks and is found beneath the oceans and covering continents. The hydrosphere is the part of Earth composed of water and includes oceans, lakes, rivers, groundwater, ice caps, glaciers, and clouds. The atmosphere is the air that surrounds Earth. The biosphere consists of all organisms, both living and dead.

Sometimes the four spheres overlap. For example, organisms of the biosphere can be found in the other three spheres and clouds are part of the hydrosphere and the atmosphere. Also, chemicals can move among the four spheres.

The Global Carbon Cycle

Carbon is part of many different compounds on Earth and continuously moves throughout the four spheres, as shown in the figure on the following page. The movement of carbon can happen quickly or it can take millions of years. This continuous recycling of Earth's carbon is called the **global carbon cycle** and is essential to the existence of life on Earth.

Carbon Reservoirs

We refer to the major locations on Earth where carbon is found as **carbon reservoirs.** The amount of carbon on Earth is constant, and it continuously cycles through these reservoirs and supports new life. Scientists divide some of these reservoirs into smaller units in order to examine better the ways that various forms of carbon are held in and released from these reservoirs.

Vital Statistics

The Amount of Carbon in Earth Reservoirs

Reservoir	Amount of Carbon (billions of tons)
Atmosphere	750
Hydrosphere	
upper ocean	1000
deep ocean	38,000
Lithosphere	
fossil fuels	10,000
soil	1500
sea-floor methane	20,000
carbonate rocks and sediment	750,000
deep mantle	>750,000
Biosphere	
land plants	500
dead organic material	1500
marine organisms	1000

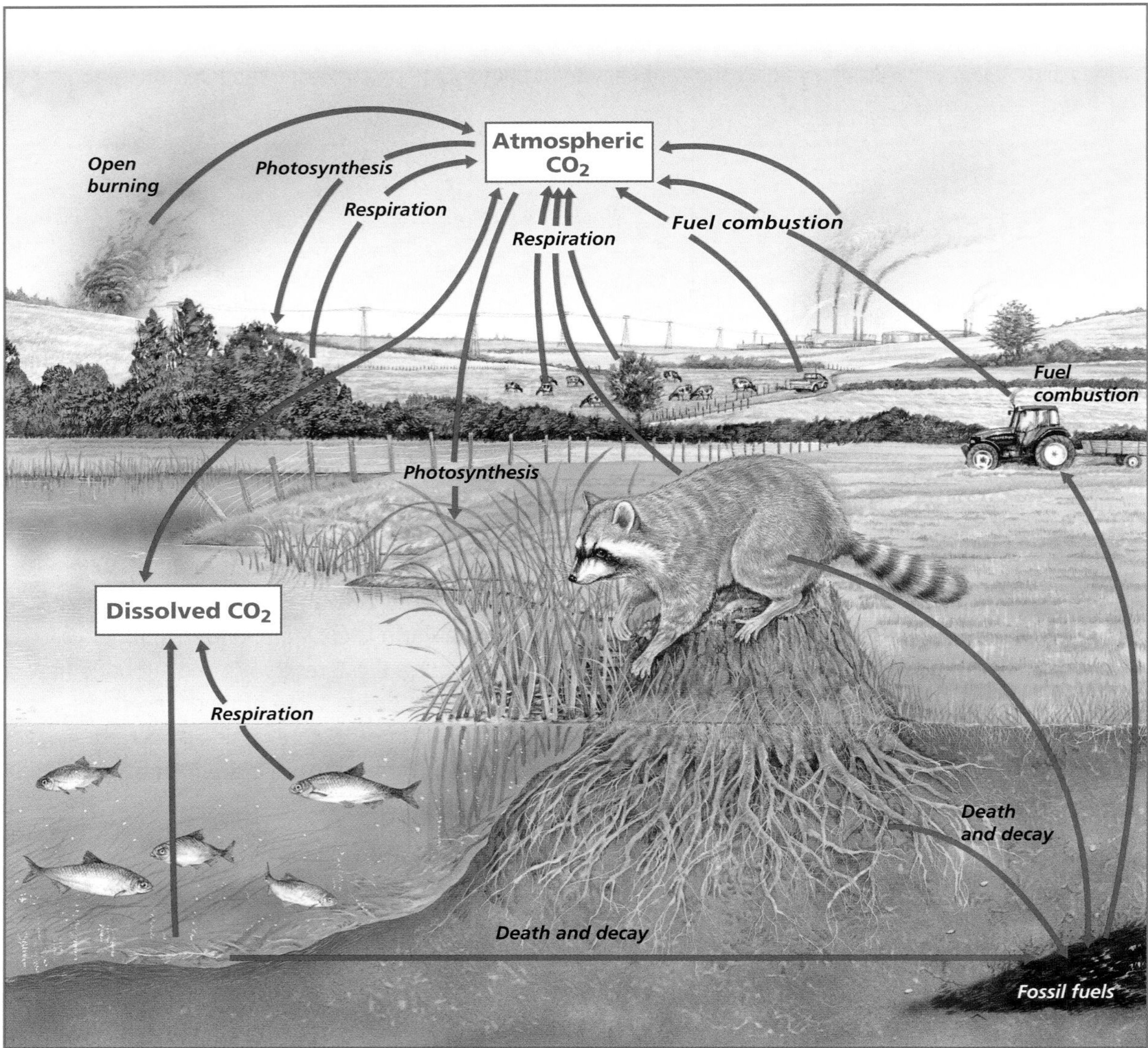

The carbon cycle

Atmosphere Because gases mix throughout the atmosphere, it is considered to be a carbon reservoir. Only about 0.1 percent of carbon is in the atmosphere and its primary form is carbon dioxide (CO_2). It is known that carbon dioxide greatly affects the climate and life on Earth because it contributes to the warming of Earth's atmosphere.

Hydrosphere Most of the hydrosphere's carbon is in the oceans as dissolved carbon dioxide, carbonate ions $(CO_3)^{2-}$, and bicarbonate ions $(HCO_3)^-$. Scientists recognize two distinct reservoirs within the oceans—the surface ocean and the deeper ocean. The surface-ocean reservoir is the smaller of the two ocean reservoirs and is the layer of water that is mixed by winds and currents. This layer of water exchanges carbon with the atmosphere and the deeper waters below. Deep ocean waters hold more carbon than the surface ocean but are isolated from the carbon in the atmosphere.

The amount of carbon that ocean reservoirs can hold depends on the water's temperature. Warm salt water holds less dissolved carbon dioxide than cold salt water. Therefore, an increase in ocean temperature decreases the storage capacity of the ocean's carbon reservoir, and vice versa.

Lithosphere The lithosphere is more solid than the atmosphere or hydrosphere, which results in distinct reservoirs within it. One reservoir is composed of fossil-fuel deposits, including coal and crude oil. This carbon is in the form of complex hydrocarbon molecules that we use for fuel. Earth's soil layer is another carbon reservoir. It is made from a combination of minerals and decayed organic matter. Some metamorphic rocks contain carbon as the mineral graphite. These carbon-rich rocks form a huge reservoir, holding more carbon than the atmosphere, hydrosphere, and biosphere combined.

The most visible reservoir is found in carbonate rocks, such as those shown below, and sediments. The carbon is in the form of minerals, such as calcium carbonate ($CaCO_3$), which make up limestone, dolomite, and marble. The calcium carbonate came from the shells of dead marine organisms deposited on the ocean floor over time, or it precipitated directly from saturated ocean water.

CORBIS (uc)

Carbonate rocks—a carbon reservoir

USGS

Burning methane hydrate

The deep mantle is probably Earth's largest carbon reservoir. Though scientists do not know exactly how much carbon is stored deep in the mantle, most estimate that it probably holds more carbon than carbonate rocks and sediment. Some carbon enters the mantle during subduction of carbonate rocks. However, much of the mantle's carbon has been trapped there since the formation of Earth.

Located deep within seafloor sediments is additional carbon reservoir composed of methane hydrate deposits. Methane gas is a by-product of the anaerobic decomposition of organic material. When methane gas is compressed beneath deep, cold, seafloor sediments, it combines with water molecules and forms a crystalline solid called methane hydrate. Though methane can be a fuel source, as shown above, scientists have not developed a practical method for extracting the gas from the methane hydrate deposits.

Biosphere The biosphere has three major carbon reservoirs—land plants, dead and decaying organic matter, and the marine environment. The carbon in all three is in the form of sugars and other complex organic molecules. The source of carbon is carbon dioxide in the atmosphere or dissolved in water. The biosphere reservoirs are located within the atmosphere, hydrosphere, and lithosphere. In this way, they play a major role in the transport of carbon between reservoirs.

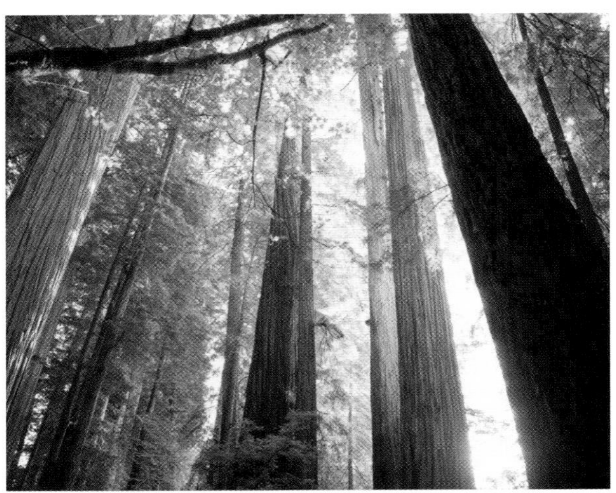
CORBIS

Residence time in a redwood tree can be hundreds of years.

Transport Among Reservoirs

Carbon constantly passes from one reservoir to another in a continuous flow but is temporarily stored in each reservoir. There is great variability in the amount of time it takes for the carbon to pass through each of these reservoirs. Scientists measure this time using the concept of residence time.

Residence Time The average amount of time that carbon remains in any particular reservoir is known as its **residence time.** The residence time (*R*) is calculated by dividing the capacity of the reservoir (*C*) by the rate of inflow (*i*) or outflow. This can be represented by the formula:

$$R = C / i$$

Some carbon reservoirs have residence times measured in hours, and others have residence times measured in millions of years. For example, carbon can remain in a plant for just a few days or up to hundreds of years, such as in a redwood tree, as shown above. Residence times for carbon in rocks can be hundreds of thousands to millions of years. Because of these long residence times, carbon within lithosphere reservoirs does not always directly affect humans like carbon in other reservoirs. Some pathways between carbon reservoirs are driven by energy from the Sun, and others are driven by Earth's internal store of geothermal heat.

Solar Driven Transport Solar energy powers the vast majority of transport processes through the carbon cycle. The process that moves the greatest amount of carbon is photosynthesis. Through photosynthesis, plants and other photosynthetic organisms remove carbon dioxide from the atmosphere, combine carbon dioxide and water using light energy, produce sugars, and release oxygen. This process moves 120 billion tons of carbon from the atmosphere to the biosphere annually. Half of this carbon quickly returns to the atmosphere through the process of cellular respiration. The other half of the carbon becomes incorporated into the bodies of photosynthetic organisms, but some is excreted in their wastes. Photosynthetic organisms are eventually eaten by other organisms or decay through the action of microorganisms.

Solar energy also powers the geological process of weathering. This occurs after carbonate rocks are uplifted to Earth's surface. Acidic rainwater that contains dissolved carbon dioxide helps to break down the rocks. This removes carbon from the atmosphere and eventually transports it to the ocean reservoir as dissolved bicarbonate ions. Weathering also returns some carbon to the atmosphere as carbon dioxide. In the ocean, some organisms incorporate carbon into their shells, which eventually settle on the seafloor and dissolve back into the water or are incorporated into the seafloor sediment. The Sun also indirectly drives the exchange of carbon dioxide between the atmosphere and ocean waters when winds stir the ocean surface, as shown below, increasing carbon dioxide transport.

Paul Berger/Stone

Carbon dioxide transport by wave action

FOCUS ON CAREERS

Environmental Chemist

Environmental chemists study the effects of both natural and industrial chemicals on our environment. Since the environment is very complex, environmental chemists must be familiar with other disciplines including biology, geology, and ecology. They often work for governmental agencies, public health organizations, environmental groups, or industrial enterprises.

Geothermal Driven Transport The geologic processes driven by geothermal energy progress more slowly than biological processes driven by the Sun. Thermal energy from deep in Earth can melt rocks, producing magma that moves toward the surface and carries carbon dioxide with it. When this molten rock flows to the surface during volcanic activity, it releases carbon dioxide. Sometimes this carbon has a deep mantle source as described on page CA11. Volcanoes located near subduction zones can also produce carbon dioxide, which is released from subducted carbonate rocks. As the rocks are subducted, they are heated, which decomposes the calcium carbonate. This process returns carbon dioxide to the atmosphere.

The Nitrogen Cycle

Nitrogen is essential for life because it is a key component of proteins. Therefore, the nitrogen cycle, as shown in the figure below, is another key biogeochemical cycle. Elemental nitrogen gas makes up most of Earth's atmosphere. However, plants and animals cannot use nitrogen in this form. Certain soil bacteria can convert atmospheric nitrogen to a form of nitrogen that can be used by plants. When an organism eats a plant, this form of nitrogen passes into the organism and becomes part of the compounds and structures in its body. After an organism dies, decomposers, such as bacteria and fungi, convert the nitrogen compounds within the organism into elemental nitrogen, and the cycle repeats.

The Nitrogen Cycle

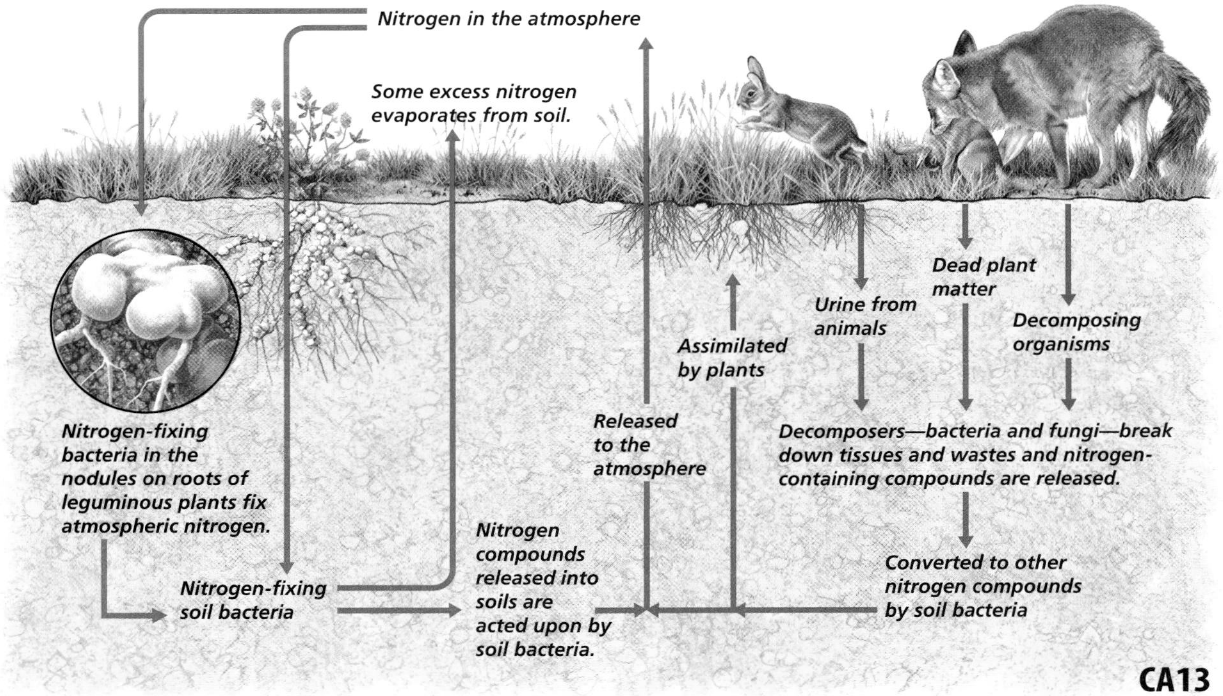

EARTH SCIENCES ASSESSMENT

1 **Which term best describes the average amount of time that carbon remains in any reservoir?**

A carbonate time
B residence time
C storage time
D transport time

2 **Which elements undergo biogeochemical cycles?**

A boron and oxygen
B manganese and sulfur
C nitrogen and carbon
D potassium and lead

Use the table below to answer questions 3 and 4.

Atmospheric Elements	Percent in Atmosphere
Nitrogen (N_2)	78
Oxygen (O_2)	21
Water vapor (H_2O)	0–4
Carbon dioxide (CO_2)	.04
Methane (CH_4)	>.0002

3 **What makes up most of Earth's atmosphere?**

A carbon dioxide
B methane
C nitrogen
D oxygen

4 **What is the most common carbon compound in the atmosphere?**

A carbon dioxide
B methane
C nitrogen
D water vapor

Use the diagram below to answer questions 5–6.

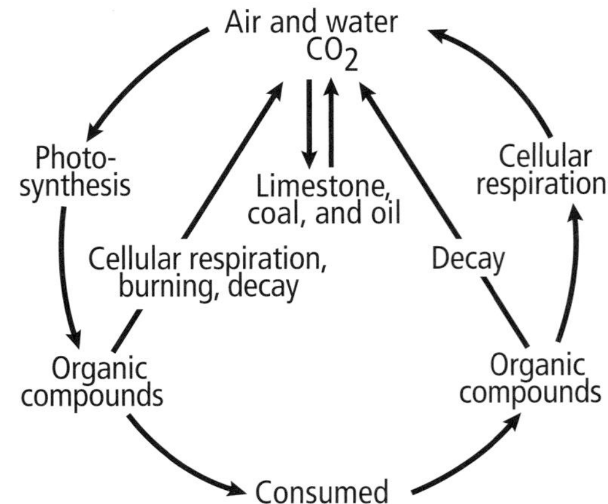

5 **What forms of carbon are common in solid Earth?**

A carbonate rocks and hydrocarbons
B carbon dioxide and carbonate ions
C organic molecules and carbon dioxide
D nitrogen gas and carbonate ions

6 **What are two biological processes that transfer carbon?**

A photosynthesis and respiration
B photosynthesis and subduction
C subduction and weathering
D uplift and respiration

7 **Which term best describes the rate of movement of carbon between reservoirs?**

A constant
B variable
C very fast
D very slow

GeoCalifornia

Earth Sciences: 6.d Students know how computer models are used to predict the effects of the increase in greenhouse gases on climate for the planet as a whole and for specific regions. *Also covers:* **7.d, 8.c**

Climate Modeling

Earth's climate system involves a complex set of components and interactions among these components. Climate models can simplify and organize this complexity, test proposed hypotheses, and quantify global climate change. The use of computers has dramatically increased the sophistication of climate models.

Computer Modeling

Computers have been used for weather forecasting since the late 1940s. Meteorologists developed computer models of atmospheric circulation for this purpose. These weather models became the precursors of modern climate models. Computer modeling uses a computer to analyze data sets for predicting a value of a phenomenon.

Since climate is the average weather of a specific area over a long time, long-term changes in weather can be used to describe climate. Scientists now know enough about what controls atmospheric circulation to construct computer-generated climate models on both global and regional scales. The models provide useful predictions about changes in global and regional mean temperatures.

Global Scale

Earth's global climate system includes many components and processes. Computer models use mathematical equations to represent this complex system. The equations are based on the laws of physics and experimental data from climate research. The calculations for solving these equations require greater computer power than weather models. The increase in computing speed has contributed to the success of many global climate models, such as the one shown at right.

Modeling Parameters In order to use climate models, scientists must collect numerical values for the properties that determine the behavior of a climate system called the model's parameters. Some examples of parameters are temperature, atmospheric pressure, and wind speed. Climate models describe how the parameters change as the result of processes that drive climate variability, referred to as climate-forcing mechanisms. Three natural kinds of climate-forcing mechanisms are tectonic processes, changes in Earth's orbit, and the change of the Sun's strength.

Gary Strand/NCAR

Lawrence Berkeley National Laboratory global climate model

Tectonic processes include the uplift of mountains, eruption of volcanoes, and shifting of continents. Changes in Earth's orbit are changes in the shape of Earth's orbit around the Sun or changes in Earth's tilt on its axis. Orbital changes alter the amount of solar energy that different parts of Earth's surface receive. Lastly, the Sun has been gaining strength since the formation of the solar system. Recently, a fourth mechanism of climate forcing has been recognized. **Anthropogenic climate forcing** is the effect that humans have on climate change.

Human Influences on Greenhouse Gases The most significant way humans affect climate is by adding greenhouse gases to the atmosphere. Carbon dioxide is a primary greenhouse gas. Human activities have accelerated the transfer of carbon to the atmosphere from both the lithosphere and the biosphere. Without human interference, fossil fuels store carbon within the lithosphere for millions of years. Even after they are uplifted to the surface, weathering releases carbon to the atmosphere very slowly. In the last few hundred years, humans have been burning fossil fuels to transform the energy stored in them into usable forms of energy. This releases stored carbon into the atmosphere rapidly as carbon dioxide.

Humans accelerate the transfer of carbon stored in the biosphere when they clear trees and other plants from the land for development or agriculture. If forests are cleared by burning, the carbon stored in the plant tissues is quickly converted to carbon dioxide and enters the atmosphere.

Humans also have increased the concentrations of another primary greenhouse gas—methane. While we have released far less methane than carbon dioxide, methane is a more potent greenhouse gas. Burning fossil fuels and biomass releases some methane into the atmosphere. However, agricultural practices, including raising cattle and rice, produce more methane than burning fuels. Humans now produce 70 percent of the methane entering the atmosphere.

Nitrous oxide is another greenhouse gas produced by human activities. Agricultural practices and the burning of biomass and fossil fuels produce nitrous oxide. Other greenhouse gases, such as chlorofluorocarbons (CFCs) and halocarbons, do not form naturally. These are manufactured chemicals used for refrigeration and in solvents. Like methane, they are present in the atmosphere in smaller quantities than carbon dioxide but are potent greenhouse gases. Atmospheric circulation distributes greenhouse gases, which means that local input can change the climate on a global scale. Humans rarely experience global changes directly, but we can experience regional changes, which result from global climate changes.

Vital Statistics

Increased Greenhouse Gas Concentrations

Gas	Avg. 100 Years Ago (parts/billion)	Approx. Current (parts/billion)
Carbon dioxide	288,000	370,000
Methane	848	1800
Nitrous oxide	285	312
CFC's and halocarbons	0	1.2

Regional Scale

Models that show climate change on the global scale do not provide detailed regional information. At this time, it is not practical to run high-resolution global models for long periods of time because the process requires too much computer time. Regional models have higher resolution, cover smaller areas, and run for shorter periods of time. As computer power increases, so will our ability to increase model resolution.

Model Types

If a model reproduces the present-day climate reasonably well, then it might also be used to model past climates. The past-climate model is successful if its climate prediction matches independent geological data representing that past time. A model that successfully reproduces both past and present climates can be used to predict future climate change.

Models can be one-, two- or three-dimensional. A one-dimensional model could be a single column of air representing the average atmosphere over a location. A two-dimensional model could represent variations in the atmosphere with changes in latitude. With these models, differences between the poles and the equator can be included. Three-dimensional models can reproduce complex climatic variations.

Simpler models can represent climate change over longer periods of time. They are less expensive to operate because they require less computer time. Complicated three-dimensional models incorporate greater detail but are slow and expensive to operate.

General Circulation Models

The most complete three-dimensional representation of global climate is a general circulation model (GCM). A grid is constructed around and above the globe. Values that represent the initial description of the climate for the model parameters are entered for each grid point, along with equations for the physics of atmospheric processes. As the model runs in the computer, the values change at each grid point over time. Atmosphere GCMs are now well developed. Some climatologists are working to produce ocean GCMs.

Coupled Atmosphere/Ocean Models

Coupled models, like the one shown below, link an atmospheric GCM to an oceanic GCM. Because of continuous interaction between the ocean surface and the atmosphere, these combined models greatly improve results. To create a coupled model, the atmospheric model and the oceanic models run separately. After the atmospheric model runs for a short time period, the results are entered into the oceanic model, which then runs for a short time period. Data continues to be exchanged after short-time steps until the model is completed.

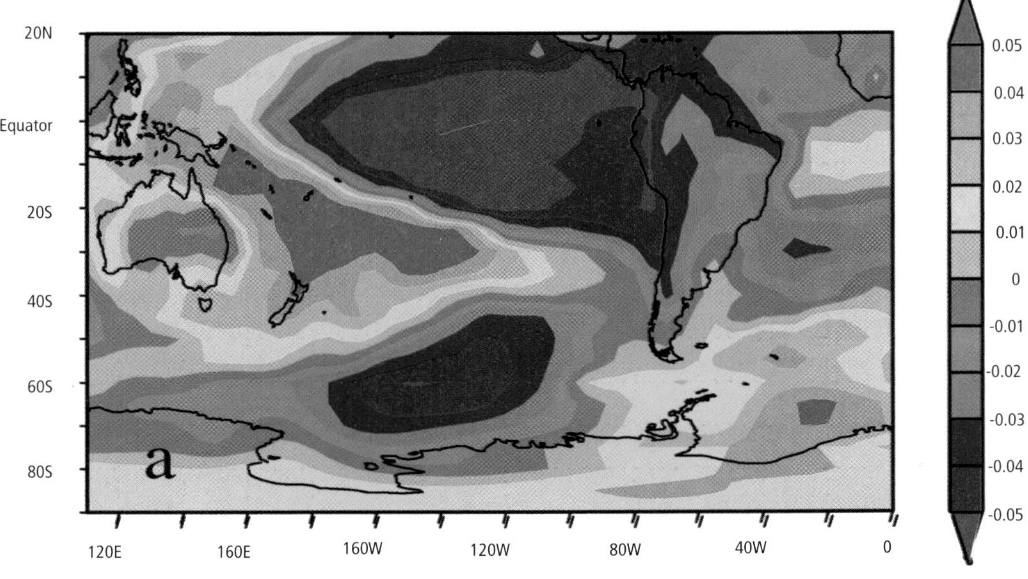

Leading EOF mode eigenvector from SAT

Climate Model Predictions

Computer-generated models now can make useful predictions about the consequences of greenhouse gases, such as the potential for changes in global and regional mean temperatures. While models consistently show increased temperatures on a global scale, regional results are more variable. For example, global warming causes warming in some locations and cooling in others.

Modeling Ozone Depletion

Computer models have predicted depletion of ozone in the stratosphere. These predictions have been less controversial than global-warming predictions for two reasons. First, the depletion of ozone follows an annual cycle. This makes testing models and hypotheses possible every year. Global warming, on the other hand, takes place on a longer time scale. This makes testing hypotheses more difficult. Second, ozone depletion can cause harm to all organisms. This is not true for climate change. Some regions might benefit by the regional changes that global warming produces.

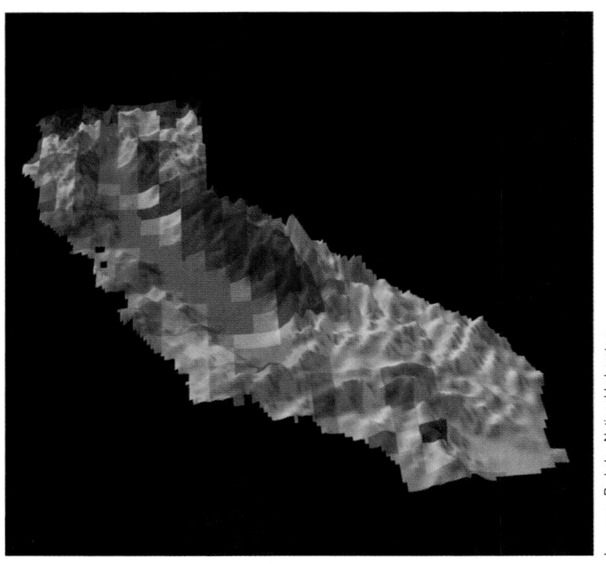

Lawrence Berkeley National Laboratory

Projected decreased precipitation in April-August is shown in orange and dark red and increased precipitation is purple.

California Climate Predictions

Coupling portions of a global model with a more detailed regional model can produce regional climate predictions. Using such coupled models, scientists at the Lawrence Berkeley National Laboratory produced climate projections for California. They modeled climate changes that could take place between the years 2040 and 2049. Their projections are based on the condition that carbon dioxide concentrations in the atmosphere will continue to rise by a rate of one percent each year. Like many other climate models, they predict warmer overall temperatures with projected rainfall increases for the Sierra Nevada and the northern Coast Range.

Scientists at the University of California, Santa Cruz, also modeled future precipitation for the California. Their results differed from the results of the Lawrence Berkeley study. They modeled the years 2080–2099 and found precipitation decreases in the Sierra Nevada between April and August, as shown above. In either case, these precipitation changes could produce significant effects on California water resources. The frequency of both floods and droughts could be altered. Recall that California depends on high-elevation snow packs to supply its rivers and reservoirs with water during the dry months of the year. Warming produces earlier snowmelt. This, along with increased rainfall, means more flooding in the spring and a reduced water supply in the summer. Climate models suggest that California's water-resource managers might need to release water from reservoirs to avoid flooding during the wet season. This too would decrease the amount of water available for the dry season.

It is important to remember that Earth's climate system includes a complex set of feedback mechanisms that are not fully understood. Therefore, predictions of climate changes contain some uncertainty.

EARTH SCIENCES ASSESSMENT

1 How do the details shown on global climate models compare with the details of regional climate models?

A The details are similar.
B Global models are more detailed.
C Global models are less detailed.
D There is no difference.

2 Which gases are primary greenhouse gases?

A carbon dioxide and methane
B carbon dioxide and nitrogen
C methane and nitrogen
D oxygen and methane

Use the table below to answer question 3.

U.S. Methane Emissions in 2003	
Source	**Percentage of Total**
Landfills	24
Natural gas systems	23
Enteric fermentation	21
Coal mining	10
Manure management	7
Wastewater treatment	7
Petroleum systems	3
Rice cultivation	1
Other	4

3 What emissions source produces the most methane?

A coal mining
B enteric fermentation
C landfills
D petroleum systems

4 What types of computer models were used to develop climate models?

A carbon dioxide models
B ocean-circulation models
C ozone models
D weather models

5 What type of model produces the least complex climatic variations?

A circulation model
B coupled model
C three-dimensional model
D two-dimensional model

6 Which can be predicted using a climate model?

A changes in the Sun's strength
B global temperatures
C human population
D ocean depths

Use the graph below to answer question 7.

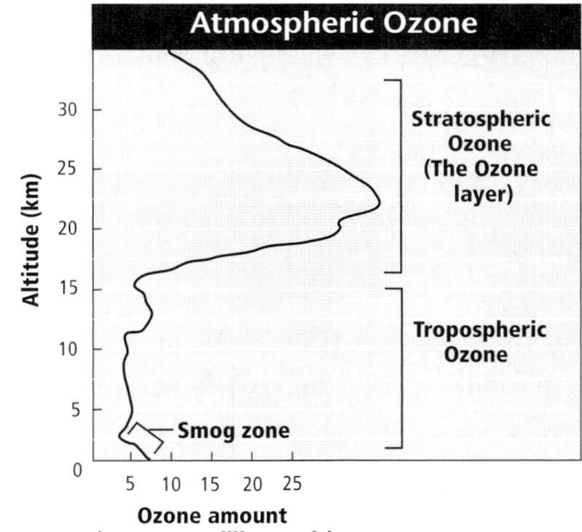

7 Which describes the ozone amount at an altitude of 20 km?

A greatest amount in the atmosphere
B less than the amount at 30 km
C equal to the amount at 10 km
D greater than the amount at 35 km

GeoCalifornia English-Spanish Glossary

English	Español
anthropogenic climate forcing (p. CA16) effect that humans have on climate change	**anthropogenic clima de fuerzo** (p. CA16) el efecto que los seres humanos tienen en cambio del clima
biogeochemistry (p. CA9) scientific discipline that examines the interaction between biological and geochemical processes	**biogeochemistry** (p. CA9) disciplina científica que examina la interacción entre los procesos biológicos y geoquímicos
carbon reservoir (p. CA9) any material that stores carbon	**depósito carbon** (p. CA9) las esferas principales de la tierra donde se almacena el carbón
computer model (p. CA15) computer program that uses available data to predict a value for some phenomenon, such as weather or climate	**modelo de lo computadora** (p. CA15) programa de computadora que utiliza datos disponibles para predecir un valor para un cierto fenómeno, tal como tiempo o clima
global carbon cycle (p. CA9) natural movement of carbon into and out of the various carbon reservoirs	**ciclo carbón global** (p. CA9) el movimiento natural del carbón en y fuera de los varios depósitos del carbón
mineral resource (p. CA4) deposits of rocks or minerals that can be economically extracted for use	**recursos minerals** (p. CA4) depósitos de las rocas o de los minerales que se pueden extraer económicamente para el uso
residence time (p. CA12) average amount of time that carbon remains in any particular reservoir; is calculated by dividing the capacity of the reservoir (*C*) by the rate of inflow (*i*) or outflow, $R = C / i$	**tiempo residencia** (p. CA12) la cantidad de tiempo media que el carbón sigue siendo en cualquier depósito particular:es calculado dividiendo la capacidad del depósito (*c*) por el índice de la afluencia (*i*) o de la salida, $R = C / i$

EARTH SCIENCE: Geology, the Environment, and the Universe correlated to *The Science Content Standards for Grades Nine through Twelve: Earth Sciences* and *Investigation and Experimentation*

Earth Sciences	Pages
Standard Set 1. Earth's Place in the Universe (Solar System)	
1. Astronomy and planetary exploration reveal the solar system's structure, scale, and change over time. As a basis for understanding this concept:	
1.a. Students know how the differences and similarities among the sun, the terrestrial planets, and the gas planets may have been established during the formation of the solar system.	578, 793–797
1.b. Students know the evidence from Earth and moon rocks indicates that the solar system was formed from a nebular cloud of dust and gas approximately 4.6 billion years ago.	793–797
1.c. Students know the evidence from geological studies of Earth and other planets suggest that the early Earth was very different from Earth today.	387–388, 553–556, 570–571, 577–579, 580–583, 584–588, 589–593
1.d. Students know the evidence indicating that the planets are much closer to Earth than the stars are.	775–779, 798–799, 815–820, 921
1.e. Students know the Sun is a typical star and is powered by nuclear reactions, primarily the fusion of hydrogen to form helium.	805, 809–810, 821–825, 859
1.f. Students know the evidence for the dramatic effects that asteroid impacts have had in shaping the surface of planets and their moons and in mass extinctions of life on Earth.	633–634, 754, 780–781
1.g.* Students know the evidence for the existence of planets orbiting other stars.	800

* Not assessed on the California Standards Tests in Science

George Post/Science Photo Library/Photo Researchers.

Earth Sciences	Pages
Standard Set 2. Earth's Place in the Universe (Stars, Galaxies, and the Universe)	
2. Earth-based and space-based astronomy reveal the structure, scale, and changes in stars, galaxies, and the universe over time. As a basis for understanding this concept:	
2.a. Students know the solar system is located in an outer edge of the disc-shaped Milky Way galaxy, which spans 100,000 light years.	833–838
2.b. Students know galaxies are made of billions of stars and comprise most of the visible mass of the universe.	833–838, 839–846
2.c. Students know the evidence indicating that all elements with an atomic number greater than that of lithium have been formed by nuclear fusion in stars.	821–825, 859
2.d. Students know that stars differ in their life cycles and that visual, radio, and X-ray telescopes may be used to collect data that reveal those differences.	813–820, 821–825
2.e.* Students know accelerators boost subatomic particles to energy levels that simulate conditions in the stars and in the early history of the universe before stars formed.	854
2.f.* Students know the evidence indicating that the color, brightness, and evolution of a star are determined by a balance between gravitational collapse and nuclear fusion.	821–825
2.g.* Students know how the red-shift from distant galaxies and the cosmic background radiation provide evidence for the "big bang" model that suggests that the universe has been expanding for 10 to 20 billion years.	842–846, 847–851, 856–857,860
Standard Set 3. Dynamic Earth Processes	
3. Plate tectonics operating over geologic time has changed the patterns of land, sea, and mountains on Earth's surface. As the basis for understanding this concept:	
3.a. Students know features of the ocean floor (magnetic patterns, age, and sea-floor topography) provide evidence of plate tectonics.	448–454, 455–459, 464–465, 469, 546
3.b. Students know the principal structures that form at the three different kinds of plate boundaries.	455–459, 460–463, 478–479, 480–487, 528–534, 535–539
3.c. Students know how to explain the properties of rocks based on the physical and chemical conditions in which they formed, including plate tectonic processes.	99–106, 107–115, 118–119, 121–127, 128–141, 144–145, 147–148, 602, 606, 613, 626
3.d. Students know why and how earthquakes occur and the scales used to measure their intensity and magnitude.	455–459, 495–499, 500–504, 505–510
3.e. Students know there are two kinds of volcanoes: one kind with violent eruptions producing steep slopes and the other kind with voluminous lava flows producing gentle slopes.	471–475, 480–487, 488–489, 490, 492–493, 547
3.f.* Students know the explanation for the location and properties of volcanoes that are due to hot spots and the explanation for those that are due to subduction.	457, 484–487, 492, 636

* Not assessed on the California Standards Tests in Science

Earth Sciences	Pages
Standard Set 4. Energy in the Earth System (Solar Energy Enters, Heat Escapes)	
4. Energy enters the Earth system primarily as solar radiation and eventually escapes as heat. As a basis for understanding this concept:	
4.a. Students know the relative amount of incoming solar energy compared with Earth's internal energy and the energy used by society.	275–277, 578–579, 683–689, 690–697, 698
4.b. Students know the fate of incoming solar radiation in terms of reflection, absorption, and photosynthesis.	275–277, 296–297, 380, 683–684
4.c. Students know the different atmospheric gases that absorb the Earth's thermal radiation and the mechanism and significance of the greenhouse effect.	272, 375–377, 380, 725–726
4.d.* Students know the differing greenhouse conditions on Earth, Mars, and Venus; the origins of those conditions; and the climatic consequences of each.	375–377, 380, 725–726, 729, 782–785, 803
Standard Set 5. Energy in the Earth System (Ocean and Atmospheric Convection)	
5. Heating of Earth's surface and atmosphere by the sun drives convection within the atmosphere and oceans, producing winds and ocean currents. As a basis for understanding this concept:	
5.a. Students know how differential heating of Earth results in circulation patterns in the atmosphere and oceans that globally distribute the heat.	299–301, 305–307, 403–405, 411
5.b. Students know the relationship between the rotation of Earth and the circular motions of ocean currents and air in pressure centers.	305–311, 341–343, 404
5.c. Students know the origin and effects of temperature inversions.	281, 284, 296
5.d. Students know properties of ocean water, such as tempera-ture and salinity, can be used to explain the layered structure of the oceans, the generation of horizontal and vertical ocean currents, and the geographic distribution of marine organisms.	392–398, 404–407
5.e. Students know rain forests and deserts on Earth are distributed in bands at specific latitudes.	364–365
5.f.* Students know the interaction of wind patterns, ocean currents, and mountain ranges results in the global pattern of latitudinal bands of rain forests and deserts.	361–363
5.g.* Students know features of the ENSO (El Niño southern oscillation) cycle in terms of sea-surface and air temperature varia-tions across the Pacific and some climatic results of this cycle.	370, 376, 383, 411

* Not assessed on the California Standards Tests in Science

Earth Sciences	Pages
Standard Set 6. Energy in the Earth System (Climate and Weather)	
6. Climate is the long-term average of a region's weather and depends on many factors. As a basis for understanding this concept:	
6.a. Students know weather (in the short run) and climate (in the long run) involve the transfer of energy into and out of the atmosphere.	275–277, 299–304, 305–311, 359–363
6.b. Students know the effects on climate of latitude, elevation, topography, and proximity to large bodies of water and cold or warm ocean currents.	361–363, 364–368, 411
6.c. Students know how Earth's climate has changed over time, corresponding to changes in Earth's geography, atmospheric composition, and other factors, such as solar radiation and plate movement.	369–374, 375–377, 380, 382
6.d.* Students know how computer models are used to predict the effects of the increase in greenhouse gases on climate for the planet as a whole and for specific regions.	CA15–CA19
Standard Set 7. Biogeochemical Cycles	
7. Each element on Earth moves among reservoirs, which exist in the solid earth, in oceans, in the atmosphere, and within and among organisms as part of biogeochemical cycles. As a basis for understanding this concept:	
7.a. Students know the carbon cycle of photosynthesis and respiration and the nitrogen cycle.	CA9–CA14, 377, 664–665, 668
7.b. Students know the global carbon cycle: the different physical and chemical forms of carbon in the atmosphere, oceans, biomass, fossil fuels, and the movement of carbon among these reservoirs.	CA9–CA14, 272, 375–377, 380, 686–688, 696–697, 725–726
7.c. Students know the movement of matter among reservoirs is driven by Earth's internal and external sources of energy.	CA9–CA14, 285, 290–291
7.d.* Students know the relative residence times and flow characteristics of carbon in and out of its different reservoirs.	CA9–CA14, 380

* Not assessed on the California Standards Tests in Science

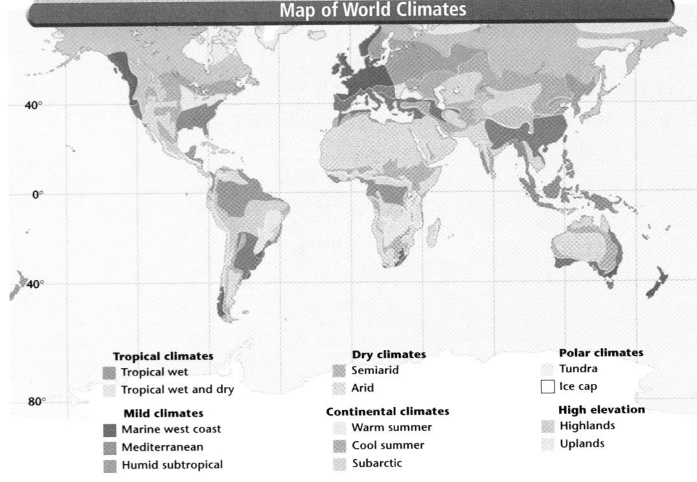

Earth Sciences	Pages
Standard Set 8. Structure and Composition of the Atmosphere	
8. Life has changed Earth's atmosphere, and changes in the atmosphere affect conditions for life. As a basis for understanding this concept:	
8.a. Students know the thermal structure and chemical composition of the atmosphere.	271–274, 278–284, 285, 297
8.b. Students know how the composition of Earth's atmosphere has evolved over geologic time and know the effect of outgassing, the variations of carbon dioxide concentration, and the origin of atmospheric oxygen.	375–377, 380, 584–588, 725–726
8.c. Students know the location of the ozone layer in the upper atmosphere, its role in absorbing ultraviolet radiation, and the way in which this layer varies both naturally and in response to human activities.	CA15–CA19, 273–274, 285, 294, 297, 726, 729
Standard Set 9. California Geology	
9. The geology of California underlies the state's wealth of natural resources as well as its natural hazards. As a basis for understanding this concept:	
9.a. Students know the resources of major economic importance in California and their relation to California's geology.	CA4–CA8, 89–91, 111–113, 659–663, 673, 690–697
9.b. Students know the principal natural hazards in different California regions and the geologic basis of those hazards.	162–166, 181–190, 220–221, 495–499, 511–515
9.c. Students know the importance of water to society, the origins of California's fresh water, and the relationship between supply and need.	CA4–CA8, 228–231, 234, 239–243, 249–257, 260, 669–675, 734
9.d.* Students know how to analyze published geologic hazard maps of California and know how to use the map's information to identify evidence of geologic events of the past and predict geologic changes in the future.	33–36, 198, 204–205, 430–431, 509–510, 513–514

* Not assessed on the California Standards Tests in Science

Percentage of Gases That Make Up Earth's Atmosphere

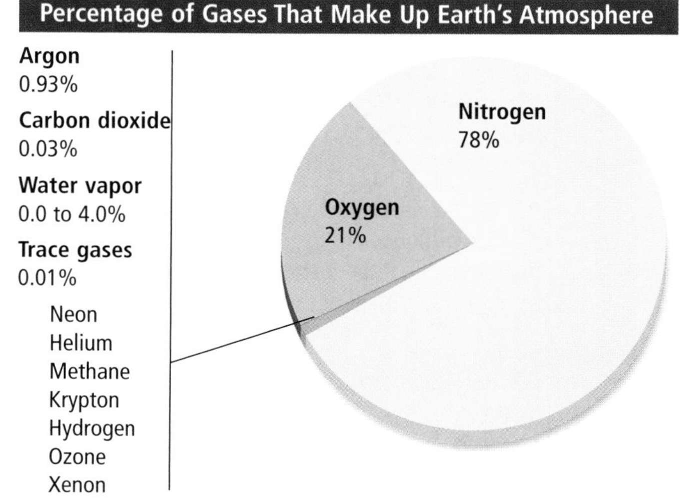

Earth Sciences	Pages
Investigation and Experimentation	
1. Scientific progress is made by asking meaningful questions and conducting careful investigations. As a basis for understanding this concept and addressing the content in the other four strands, students should develop their own questions and perform investigations. Students will:	
1.a. Select and use appropriate tools and technology (such as computer-linked probes, spreadsheets, and graphing calculators) to perform tests, collect data, analyze relationships, and display data. 20–21, 67, 70–71, 92–93,	101, 114–115, 232–233, 352–353, 378–379, 488–489, 642–643, 676–677, 704–705
1.b. Identify and communicate sources of unavoidable experimental error.	110
1.c. Identify possible reasons for inconsistent results, such as sources of error or uncontrolled conditions.	141, 163, 175, 826–827
1.d. Formulate explanations by using logic and evidence.	63, 126, 140–141, 217, 317, 348, 378–379, 474, 587, 618–619, 795
1.e. Solve scientific problems by using quadratic equations and simple trigonometric, exponential, and logarithmic functions.	20–21, 232–233, 644, 770, 777, 778-779, 791, 805, 810, 843
1.f. Distinguish between hypothesis and theory as scientific terms.	11, 19, 25, 754, 780
1.g. Recognize the usefulness and limitations of models and theories as scientific representations of reality.	18, 121, 174–175, 229, 285, 290, 299, 359, 376, 406–407, 456, 471, 719, 747, 798-799, 800
1.h. Read and interpret topographic and geologic maps.	32–36, 42–43, 258–259, 430–431, 540–541
1.i. Analyze the locations, sequences, or time intervals that are characteristic of natural phenomena (e.g., relative ages of rocks, locations of planets over time, and succession of species in an ecosystem).	12, 329–333, 338–339, 396–398, 553–556, 557–561, 564–565, 570–571, 793–797
1.j. Recognize the issues of statistical variability and the need for controlled tests.	11–12, 16, 25, 232–233, 378–379, 704–705, 930–931
1.k. Recognize the cumulative nature of scientific evidence.	17, 443–447, 448–454, 455–460, 463, 570, 775–779
1.l. Analyze situations and solve problems that require combining and applying concepts from more than one area of science	44, 380, 570, 620, 642–643, 678–679, 704–705
1.m. Investigate a science-based societal issue by researching the literature, analyzing data, and communicating the findings. Examples of issues include irradiation of food, cloning of animals by somatic cell nuclear transfer, choice of energy sources, and land and water use decisions in California.	142, 176, 234, 258, 260, 294, 324, 596, 706, 736
1.n. Know that when an observation does not agree with an accepted scientific theory, the observation is sometimes mistaken or fraudulent (e.g., the Piltdown Man fossil or unidentified flying objects) and that the theory is sometimes wrong (e.g., the Ptolemaic model of the movement of the Sun, Moon, and planets).	13, 443–447, 775–779

* Not assessed on the California Standards Tests in Science

Your textbook is arranged by broad topics called units. Each unit consists of chapters that provide detailed information about that topic.

Each chapter begins with a chapter opener page that has the following:

What You'll Learn

This bulleted list introduces the big ideas of a chapter.

Why It's Important

This short paragraph explains the reason for learning about the topic of a chapter.

Earth Science Online

The Glencoe Earth Science Web site provides links to resources with more information about the topic of a chapter.

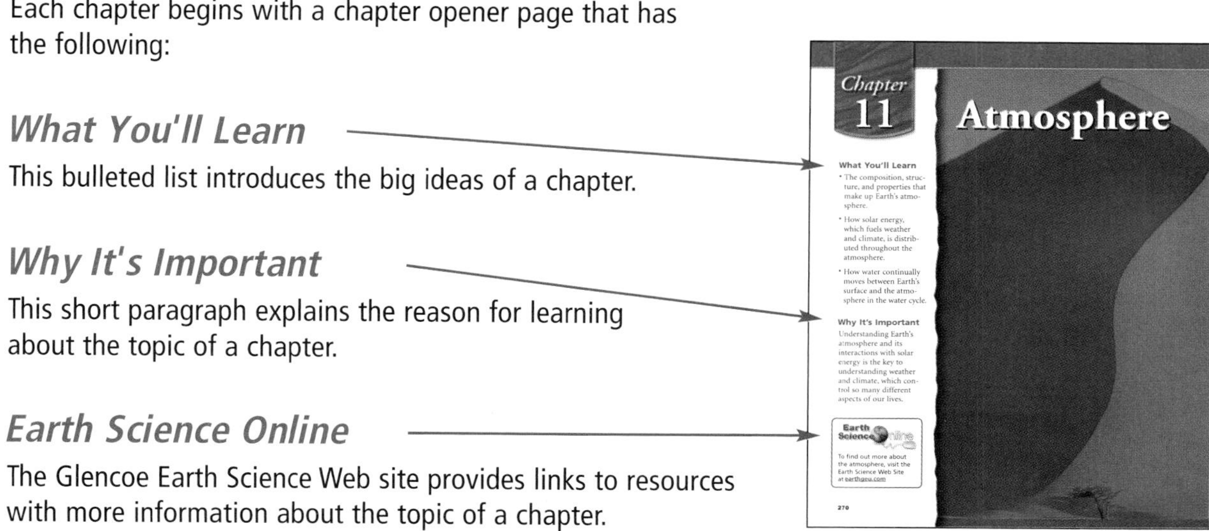

Chapter 11

Atmosphere

What You'll Learn

- The composition, structure, and properties that make up Earth's atmosphere.
- How solar energy, which fuels weather and climate, is distributed throughout the atmosphere.
- How water continually moves between Earth's surface and the atmosphere in the water cycle.

Why It's Important

Understanding Earth's atmosphere and its interactions with solar energy is the key to understanding weather and climate, which control so many different aspects of our lives.

Earth Science Online

To find out more about the atmosphere, visit the Earth Science Web Site at earthgeu.com

270

Discovery Lab

This activity is at the beginning of each chapter. It introduces, reviews, or reinforces science skills that you will use as you study Earth science.

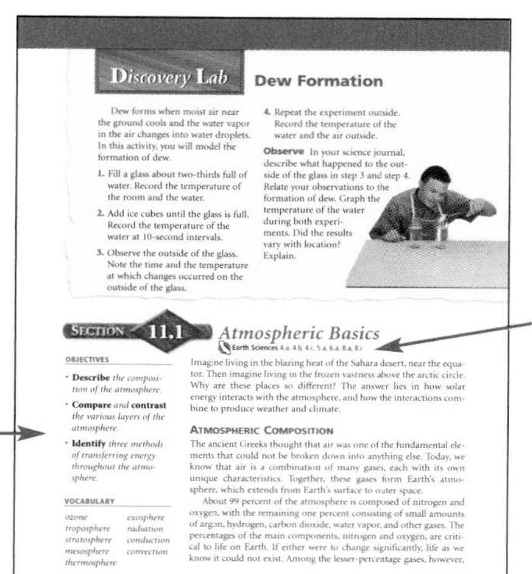

Discovery Lab **Dew Formation**

Dew forms when moist air near the ground cools and the water vapor in the air changes into water droplets. In this activity, you will model the formation of dew.

1. Fill a glass about two-thirds full of water. Record the temperature of the room and the water.
2. Add ice cubes until the glass is full. Record the temperature of the water at 10-second intervals.
3. Observe the outside of the glass. Note the time and the temperature at which changes occurred on the outside of the glass.
4. Repeat the experiment outside. Record the temperature of the water and the air outside.

Observe In your science journal, describe what happened to the outside of the glass in step 3 and step 4. Relate your observations to the formation of dew. Graph the temperature of the water during both experiments. Did the results vary with location? Explain.

Section 11.1 *Atmospheric Basics*

Earth Sciences 4.a, 4.b, 4.c, 5.a, 6.a, 8.a, 8.c

OBJECTIVES

- **Describe** *the composition of the atmosphere.*
- **Compare** *and* **contrast** *the various layers of the atmosphere.*
- **Identify** *three methods of transferring energy throughout the atmosphere.*

VOCABULARY

ozone, troposphere, stratosphere, mesosphere, thermosphere, exosphere, radiation, conduction, convection

Imagine living in the blazing heat of the Sahara desert, near the equator. Then imagine living in the frozen vastness above the arctic circle. Why are these places so different? The answer lies in how solar energy interacts with the atmosphere, and how the interactions combine to produce weather and climate.

ATMOSPHERIC COMPOSITION

The ancient Greeks thought that air was one of the fundamental elements that could not be broken down into anything else. Today, we know that air is a combination of many gases, each with its own unique characteristics. Together, these gases form Earth's atmosphere, which extends from Earth's surface to outer space.

About 99 percent of the atmosphere is composed of nitrogen and oxygen, with the remaining one percent consisting of small amounts of argon, hydrogen, carbon dioxide, water vapor, and other gases. The percentages of the main components, nitrogen and oxygen, are critical to life on Earth. If either were to change significantly, life as we know it could not exist. Among the lesser-percentage gases, however,

11.1 *Atmospheric Basics* 271

On the first page of each section are learning objectives—the main ideas of that section—and important vocabulary terms. Each vocabulary term is boldfaced where it is defined in the section.

Also, on the first page of each section is a list of ***The Science Content Standards for Grades Nine through Twelve: Earth Sciences*** and ***Investigation and Experimentation*** covered in that section.

Other labs . . .

MiniLabs

Reinforce and review the content of every chapter and practice skills with these quick hands-on-science or pencil-and-paper activities.

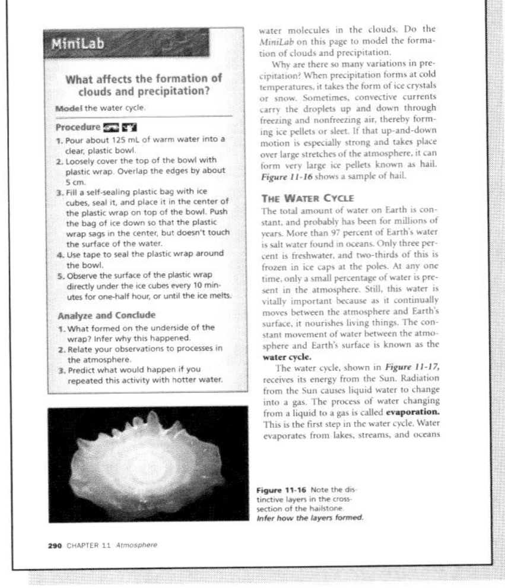

MiniLab

What affects the formation of clouds and precipitation?

Model the water cycle.

Procedure

1. Pour about 125 mL of warm water into a clear, plastic bowl.
2. Loosely cover the top of the bowl with plastic wrap. Overlap the edges by about 5 cm.
3. Fill a self-sealing plastic bag with ice cubes, seal it, and place it in the center of the plastic wrap on top of the bowl. Push the bag of ice down so that the plastic wrap sags in the center, but doesn't touch the surface of the water.
4. Use tape to seal the plastic wrap around the bowl.
5. Observe the surface of the plastic wrap directly under the ice cubes every 10 minutes for one-half hour, or until the ice melts.

Analyze and Conclude

1. What formed on the underside of the wrap? Infer why this happened.
2. Relate your observations to processes in the atmosphere.
3. Predict what would happen if you repeated this activity with hotter water.

water molecules in the clouds. Do the *MiniLab* on this page to model the formation of clouds and precipitation.

Why are there so many variations in precipitation? When precipitation forms at cold temperatures, it takes the form of ice crystals or snow. Sometimes, convective currents carry the droplets up and down through freezing and nonfreezing air, thereby forming ice pellets or sleet. If that up-and-down motion is especially strong and takes place over large stretches of the atmosphere, it can form very large ice pellets known as hail. ***Figure 11-16*** shows a sample of hail.

The Water Cycle

The total amount of water on Earth is constant, and probably has been for millions of years. More than 97 percent of Earth's water is salt water found in oceans. Only three percent is freshwater, and two-thirds of this is frozen in ice caps at the poles. At any one time, only a small percentage of water is present in the atmosphere. Still, this water is vitally important because as it continually moves between the atmosphere and Earth's surface, it nourishes living things. The constant movement of water between the atmosphere and Earth's surface is known as the **water cycle.**

The water cycle, shown in ***Figure 11-17,*** receives its energy from the Sun. Radiation from the Sun causes liquid water to change into a gas. The process of water changing from a liquid to a gas is called **evaporation.** This is the first step in the water cycle. Water evaporates from lakes, streams, and oceans

Figure 11-16 Note the distinctive layers in the cross-section of the hailstone. ***Infer how the layers formed.***

290 CHAPTER 11 *Atmosphere*

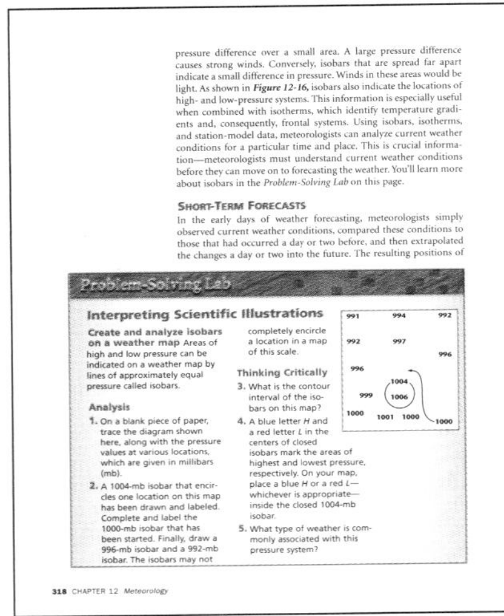

pressure difference over a small area. A large pressure difference causes strong winds. Conversely, isobars that are spread far apart indicate a small difference in pressure. Winds in these areas would be light. As shown in ***Figure 12-16,*** isobars also indicate the locations of high- and low-pressure systems. This information is especially useful when combined with isotherms, which identify temperature gradients and, consequently, frontal systems. Using isobars, isotherms, and station-model data, meteorologists can analyze current weather conditions for a particular time and place. This is crucial information—meteorologists must understand current weather conditions before they can move on to forecasting the weather. You'll learn more about isobars in the *Problem-Solving Lab* on this page.

Short-Term Forecasts

In the early days of weather forecasting, meteorologists simply observed current weather conditions, compared these conditions to those that had occurred a day or two before, and then extrapolated the changes a day or two into the future. The resulting positions of

Problem-Solving Lab

Interpreting Scientific Illustrations

Create and analyze isobars on a weather map Areas of high and low pressure can be indicated on a weather map by lines of approximately equal pressure called isobars.

Analysis

1. On a blank piece of paper, trace the diagram shown here, along with the pressure values at various locations, which are given in millibars (mb).
2. A 1004-mb isobar that encircles one location on this map has been drawn and labeled. Complete and label the 1000-mb isobar that has been started. Finally, draw a 996-mb isobar and a 992-mb isobar. The isobars may not completely encircle a location in a map of this scale.

Thinking Critically

3. What is the contour interval of the isobars on this map?
4. A blue letter *H* and a red letter *L* in the centers of closed isobars mark the areas of highest and lowest pressure, respectively. On your map, place a blue *H* or a red *L*—whichever is appropriate—inside the closed 1004-mb isobar.
5. What type of weather is commonly associated with this pressure system?

318 CHAPTER 12 *Meteorology*

Problem-Solving Labs

In every chapter, you can apply your reasoning skills and practice a science skill as you answer questions or find solutions to problems related to Earth science.

GeoLabs

After studying each chapter, you can perform one of these labs and

- practice scientific methods;
- investigate or model Earth science phenomena;
- analyze maps;
- search the Internet for data and information.

Earth Sciences standards fulfilled by performing a lab are listed on the first page of the lab.

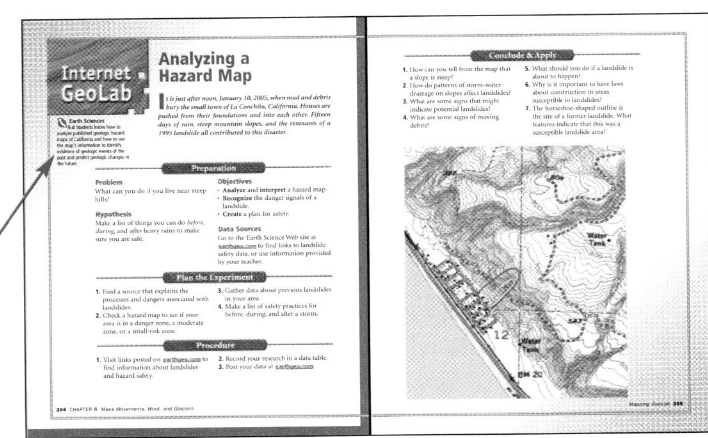

Internet GeoLab

Analyzing a Hazard Map

Earth Sciences 9.d Students know how to analyze published geologic hazard maps of California and how to use the map's information to identify evidence of geologic events of the past and predict geologic changes in the future.

It is just after noon, January 10, 2005, when mud and debris bury the small town of La Conchita, California. Houses are pushed from their foundations and into each other. Fifteen days of rain, steep mountain slopes, and the remnants of a 1995 landslide all contributed to this disaster.

Preparation

Problem

What can you do if you live near steep hills?

Hypothesis

Make a list of things you can do *before*, *during*, and *after* heavy rains to make sure you are safe.

Objectives

- **Analyze** and **interpret** a hazard map.
- **Recognize** the danger signals of a landslide.
- **Create** a plan for safety.

Data Sources

Go to the Earth Science Web site at earthgeu.com to find links to landslide safety data, or use information provided by your teacher.

Plan the Experiment

1. Find a source that explains the processes and dangers associated with landslides.
2. Check a hazard map to see if your area is in a danger zone, a moderate zone, or a small-risk zone.
3. Gather data about previous landslides in your area.
4. Make a list of safety practices for before, during, and after a storm.

Procedure

1. Visit links posted on earthgeu.com to find information about landslides and hazard safety.
2. Record your research in a data table.
3. Post your data at earthgeu.com

204 CHAPTER 8 *Mass Movements, Wind, and Glaciers*

Conclude & Apply

1. How can you tell from the map that a slope is steep?
2. How do patterns of storm-water drainage on slopes affect landslides?
3. What are some signs that might indicate potential landslides?
4. What are some signs of moving debris?
5. What should you do if a landslide is about to happen?
6. Why is it important to have laws about construction in areas susceptible to landslides?
7. The horseshoe-shaped outline is the site of a former landslide. What features indicate that this was a susceptible landslide area?

Mapping GeoLab 205

Textbook preview of *Earth Science: Geology, the Environment, and the Universe*

In the margins . . .

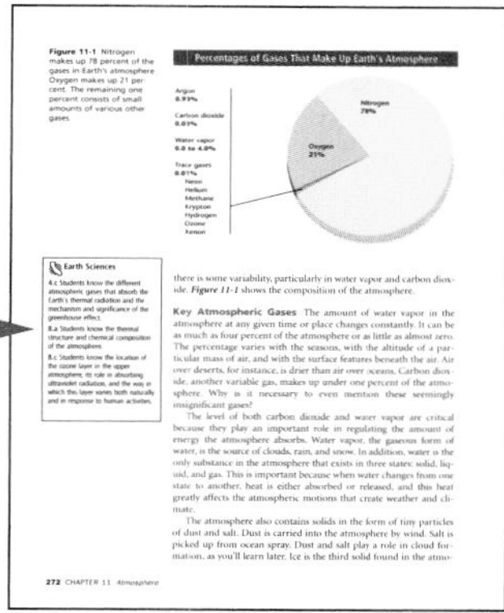

You can read the Earth Sciences standards that correlate with the section's content.

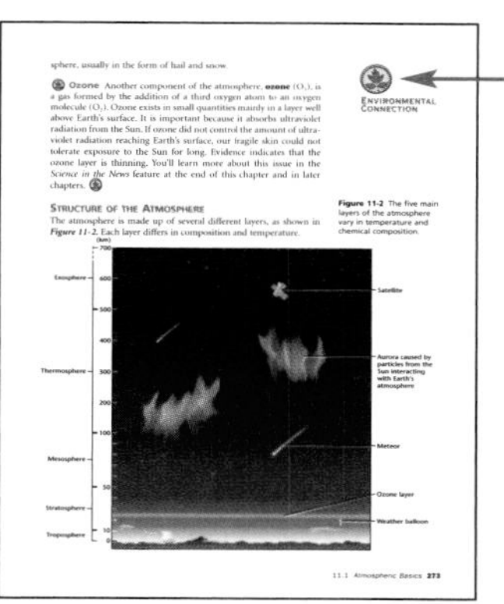

This logo indicates text connections among the environment, Earth sciences, and the real world.

Using Math

At least once in every chapter, you can practice different math skills and solve Earth-science-related problems.

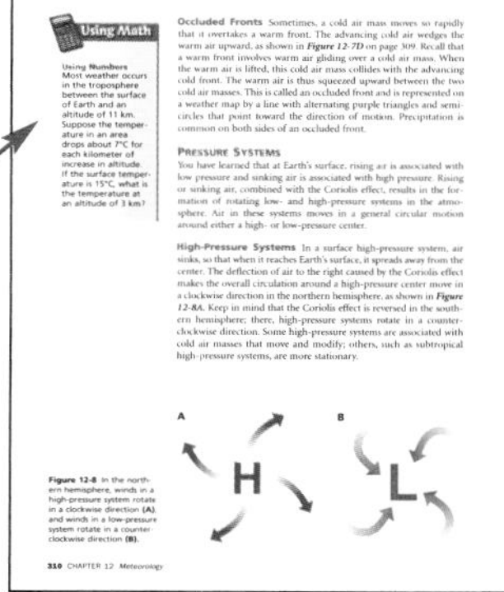

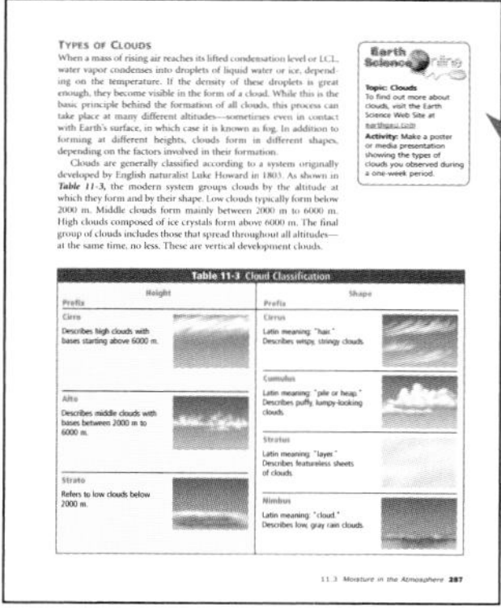

Earth Science Online

Explore various Earth science topics in depth using the Glencoe Earth Science Web site. You can complete the activity using information found at links on the Web site.

Textbook preview of *Earth Science: Geology, the Environment, and the Universe*

. . . Features

Explore Earth science topics that relate to technology, math, the environment, or current issues when you read one of these features.

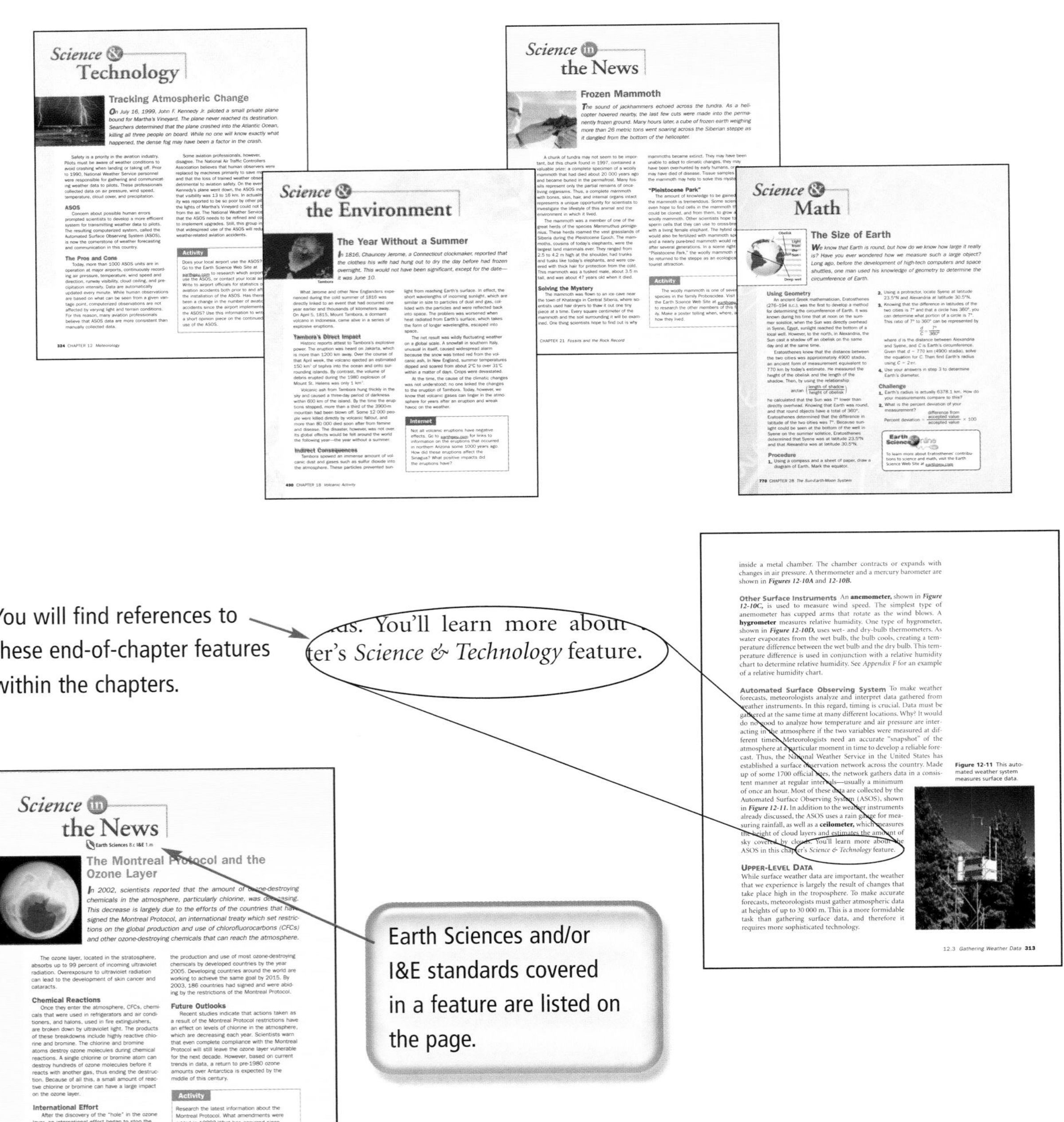

You will find references to these end-of-chapter features within the chapters.

Earth Sciences and/or I&E standards covered in a feature are listed on the page.

Assessments . . .

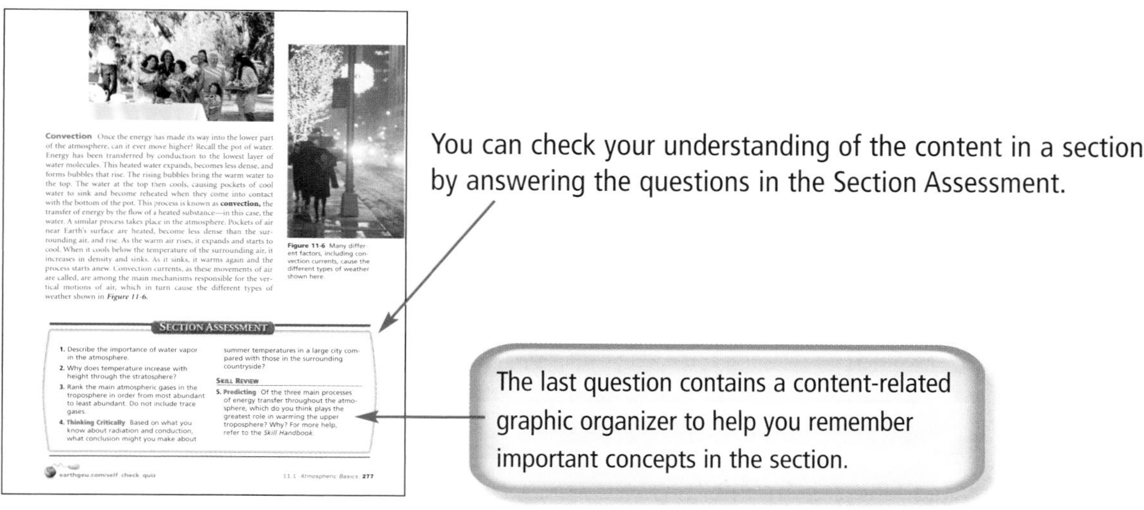

Convection Once the energy has made its way into the lower part of the atmosphere, can it ever move higher? Recall the pot of water. Energy has been transferred by conduction to the lowest layer of water molecules. This heated water expands, becomes less dense, and forms bubbles that rise. The rising bubbles bring the warm water to the top. The water at the top then cools, causing pockets of cool water to sink and become reheated when they come into contact with the bottom of the pot. This process is known as **convection,** the transfer of energy by the flow of a heated substance—in this case, the water. A similar process takes place in the atmosphere. Pockets of air near Earth's surface are heated, become less dense than the surrounding air, and rise. As the warm air rises, it expands and starts to cool. When it cools below the temperature of the surrounding air, it increases in density and sinks. As it sinks, it warms again and the process starts anew. Convection currents, as these movements of air are called, are among the main mechanisms responsible for the vertical motions of air, which in turn cause the different types of weather shown in ***Figure 11-6.***

Figure 11-6 Many different factors, including convection currents, cause the different types of weather shown here.

SECTION ASSESSMENT

1. Describe the importance of water vapor in the atmosphere.
2. Why does temperature increase with height through the stratosphere?
3. Rank the main atmospheric gases in the troposphere in order from most abundant to least abundant. Do not include trace gases.
4. **Thinking Critically** Based on what you know about radiation and conduction, what conclusion might you make about summer temperatures in a large city compared with those in the surrounding countryside?

SKILL REVIEW

5. **Predicting** Of the three main processes of energy transfer throughout the atmosphere, which do you think plays the greatest role in warming the upper troposphere? Why? For more help, refer to the *Skill Handbook*.

earthgeu.com/self check quiz

11.1 *Atmospheric Basics* **277**

You can check your understanding of the content in a section by answering the questions in the Section Assessment.

The last question contains a content-related graphic organizer to help you remember important concepts in the section.

Each chapter ends with two pages of assessment questions to help you check your understanding of the chapter's important concepts.

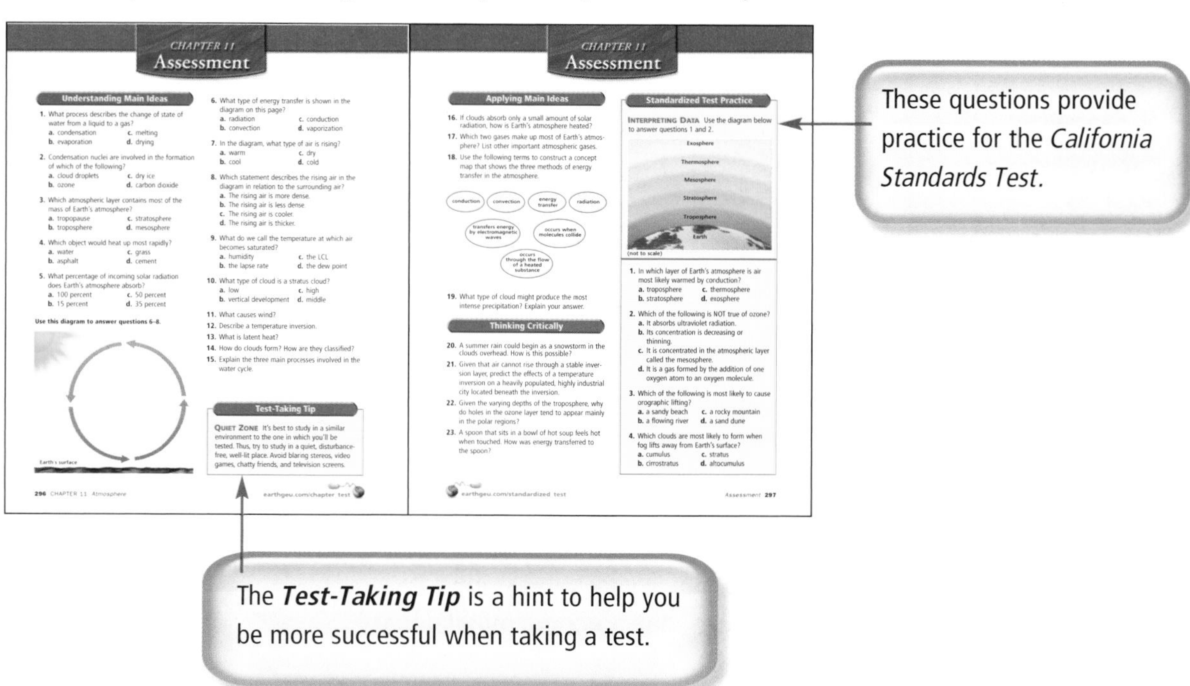

CHAPTER 11
Assessment

Understanding Main Ideas

1. What process describes the change of state of water from a liquid to a gas?
   a. condensation c. melting
   b. evaporation d. drying
2. Condensation nuclei are involved in the formation of which of the following?
   a. cloud droplets c. dry ice
   b. ozone d. carbon dioxide
3. Which atmospheric layer contains most of the mass of Earth's atmosphere?
   a. tropopause c. stratosphere
   b. troposphere d. mesosphere
4. Which object would heat up most rapidly?
   a. water c. grass
   b. asphalt d. cement
5. What percentage of incoming solar radiation does Earth's atmosphere absorb?
   a. 100 percent c. 50 percent
   b. 15 percent d. 35 percent

Use this diagram to answer questions 6–8.

6. What type of energy transfer is shown in the diagram on this page?
   a. radiation c. conduction
   b. convection d. vaporization
7. In the diagram, what type of air is rising?
   a. warm c. dry
   b. cool d. cold
8. Which statement describes the rising air in the diagram in relation to the surrounding air?
   a. The rising air is more dense.
   b. The rising air is less dense.
   c. The rising air is cooler.
   d. The rising air is thicker.
9. What do we call the temperature at which air becomes saturated?
   a. humidity c. the LCL
   b. the lapse rate d. the dew point
10. What type of cloud is a stratus cloud?
   a. low c. high
   b. vertical development d. middle
11. What causes wind?
12. Describe a temperature inversion.
13. What is latent heat?
14. How do clouds form? How are they classified?
15. Explain the three main processes involved in the water cycle.

Test-Taking Tip

QUIET ZONE It's best to study in a similar environment to the one in which you'll be tested. Thus, try to study in a quiet, disturbance-free, well-lit place. Avoid blaring stereos, video games, chatty friends, and television screens.

296 CHAPTER 11 *Atmosphere*

earthgeu.com/chapter test

CHAPTER 11
Assessment

Applying Main Ideas

16. If clouds absorb only a small amount of solar radiation, how is Earth's atmosphere heated?
17. Which two gases make up most of Earth's atmosphere? List other important atmospheric gases.
18. Use the following terms to construct a concept map that shows the three methods of energy transfer in the atmosphere.

19. What type of cloud might produce the most intense precipitation? Explain your answer.

Thinking Critically

20. A summer rain could begin as a snowstorm in the clouds overhead. How is this possible?
21. Given that air cannot rise through a stable inversion layer, predict the effects of a temperature inversion on a heavily populated, highly industrial city located beneath the inversion.
22. Given the varying depths of the troposphere, why do holes in the ozone layer tend to appear mainly in the polar regions?
23. A spoon that sits in a bowl of hot soup feels hot when touched. How was energy transferred to the spoon?

Standardized Test Practice

INTERPRETING DATA Use the diagram below to answer questions 1 and 2.

1. In which layer of Earth's atmosphere is air most likely warmed by conduction?
   a. troposphere c. thermosphere
   b. stratosphere d. exosphere
2. Which of the following is NOT true of ozone?
   a. It absorbs ultraviolet radiation.
   b. Its concentration is decreasing or thinning.
   c. It is concentrated in the atmospheric layer called the mesosphere.
   d. It is a gas formed by the addition of one oxygen atom to an oxygen molecule.
3. Which of the following is most likely to cause orographic lifting?
   a. a sandy beach c. a rocky mountain
   b. a flowing river d. a sand dune
4. Which clouds are most likely to form when fog lifts away from Earth's surface?
   a. cumulus c. stratus
   b. cirrostratus d. altocumulus

earthgeu.com/standardized test

Assessment **297**

These questions provide practice for the *California Standards Test.*

The ***Test-Taking Tip*** is a hint to help you be more successful when taking a test.

Textbook preview of *Earth Science: Geology, the Environment, and the Universe*

At the back of your book . . .

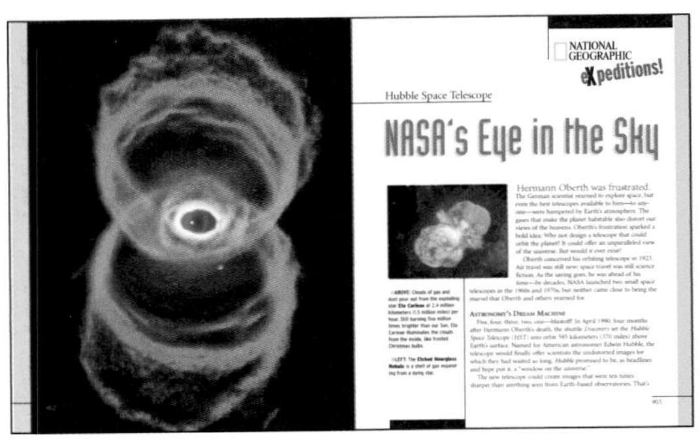

Each *eXpeditions!* allows you to share in the excitement, adventures, and journeys of explorers, scientists, and naturalists as they delve into natural phenomena and the unknown.

Appendices contain useful and concise information and data used frequently when studying Earth science.

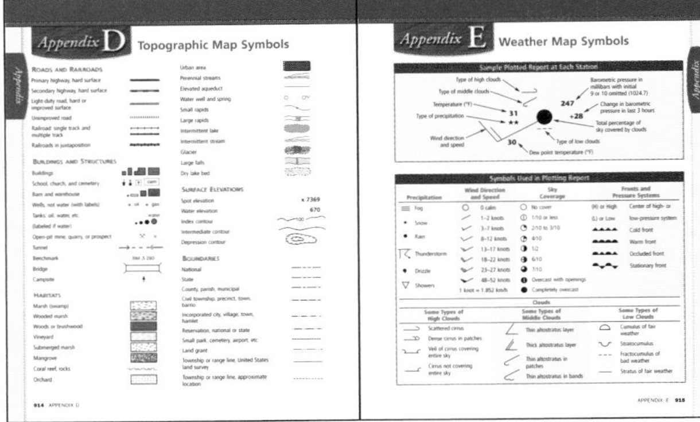

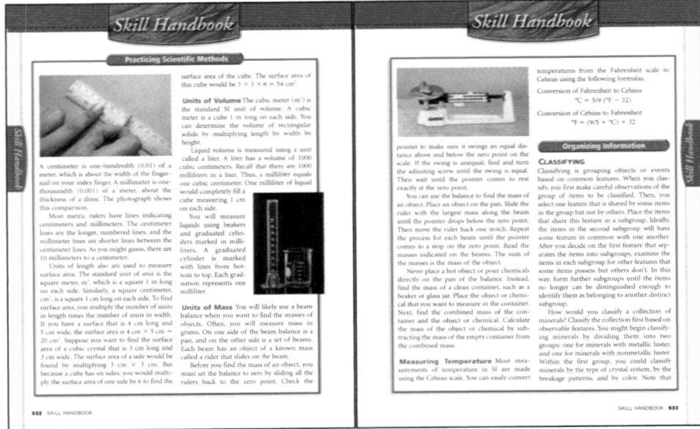

The *Skill Handbook* provides explanations and descriptions of essential science skills—*Thinking Critically, Practicing Scientific Methods,* and *Organizing Information*.

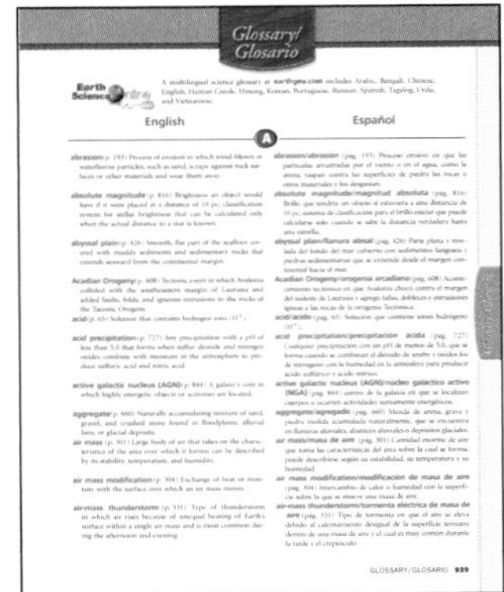

The English-Spanish, side-by-side glossary helps you quickly discover the English and Spanish definitions of the textbook's vocabulary terms. If English or Spanish is not your first language, you can find a multilingual science glossary at earthgeu.com.

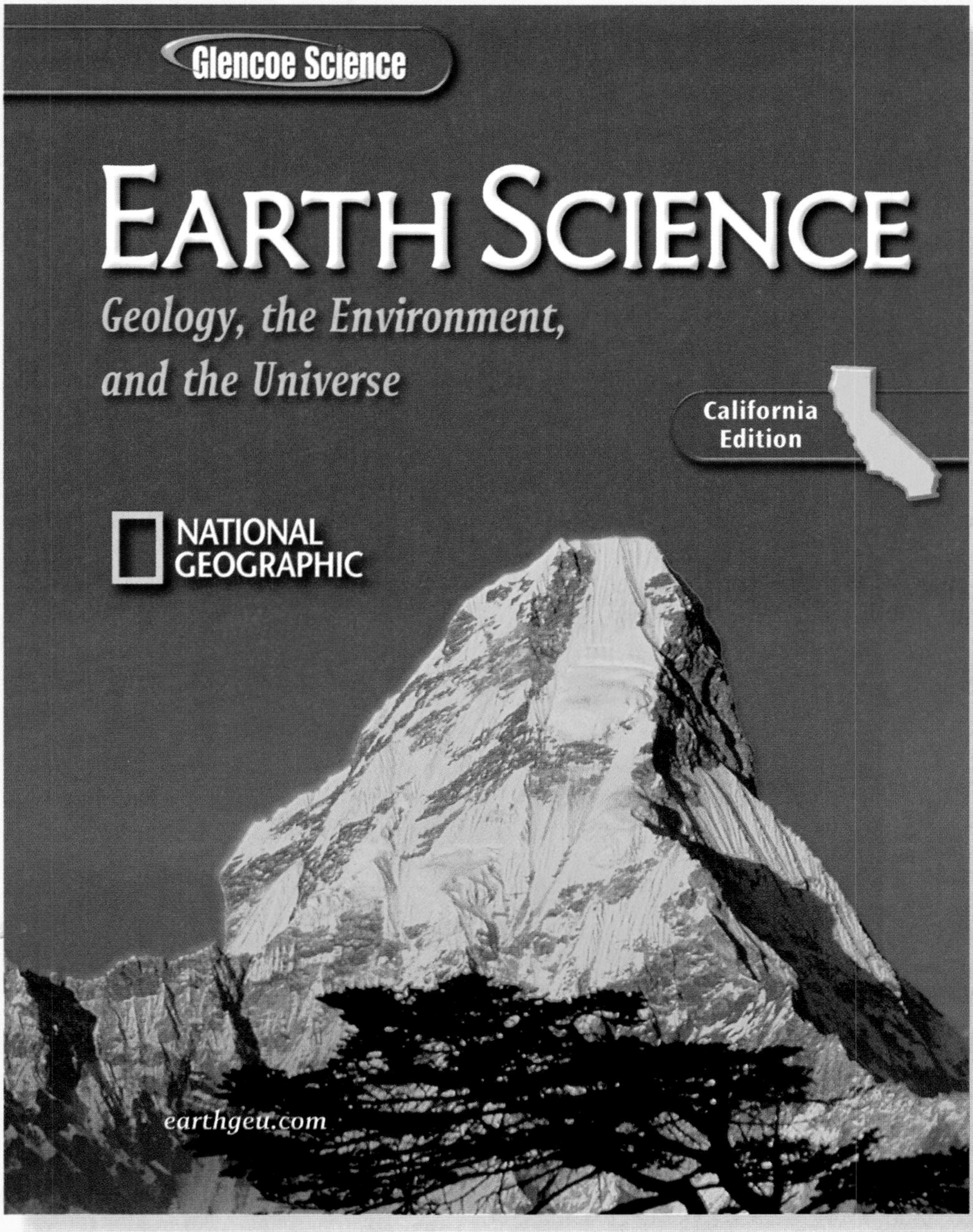

Glencoe Science

EARTH SCIENCE

Geology, the Environment, and the Universe

California Edition

NATIONAL GEOGRAPHIC

earthgeu.com

Glencoe

New York, New York Columbus, Ohio Chicago, Illinois Peoria, Illinois Woodland Hills, California

A Glencoe Program

Earth Science: Geology, the Environment, and the Universe

Visit the Earth Science Web site at earthgeu.com

You'll find:

Online Student Edition, Online Study Tools, Interactive Tutor, Online Quizzes, Internet Geolabs, National Geographic, Extending the Content, Problem of the Week, Web Links, Science Fair Ideas, NASA's Picture of the Day **and much more!**

Cover photo is the peak Ama Dablan (6747.9 m) in Nepal.

Send all inquiries to:
Glencoe/McGraw-Hill
8787 Orion Place
Columbus, OH 43240

ISBN-13: 978-0-07-877269-6
ISBN-10: 0-07-877269-9
Printed in the United States of America.

2 3 4 5 6 7 071/043 10 09 08 07

About the Authors

Frances Scelsi Hess teaches Earth science at Cooperstown High School in New York. She received her B.S. and M.S. in Science Education from the State University at Oneonta, and her Ed.D from Columbia University. Dr. Hess is a Fellow of the Science Teachers Association of New York State, and has received numerous teaching awards, including the Phi Delta Kappa Reed Travel Scholarship to Australia and New Zealand.

Gerhard Kunze is professor emeritus of Geology at the University of Akron in Ohio. He has a B.S. in Science and a Ph.D in Geophysics from Penn State University. He was an NRC research associate at Johnson Space Center, Houston, Texas from 1973-1974. In 1990, Dr. Kunze was awarded a senior Fulbright scholarship to teach geophysics at the Institute of Geophysics, a department of the University of Kiel in Germany.

Stephen A. Leslie is an associate professor of Geology in the Department of Earth Sciences at the University of Arkansas in Little Rock. His areas of research include paleontology, stratigraphy, and the evolution of early life on Earth. He has a B.S. in Geology from Bowling Green State University, an M.S. in Geology from the University of Idaho, and a Ph.D. in Geology from The Ohio State University.

Steve Letro has been a meteorologist for the National Weather Service, the media, and private industry since 1971. He currently serves as the Meteorologist-in-Charge of the National Weather Service office in Jacksonville, Florida. He received his B.S. in Meteorology from Florida State University with an emphasis in tropical meteorology. He is a member of the National Hurricane Center's Hurricane Liaison Team, and has received numerous awards, including an award for his role in restructuring the National Weather Service.

Clayton Millage has been teaching general science to eighth-graders at Lynden Middle School in Lynden, Washington, for six years. Prior to that, he taught marine science and astronomy at Cypress Creek High School in Houston, Texas. Mr. Millage has a B.A. in Geology from Occidental College, an M.S. in Geology from Stanford University, and an M.Ed. in Science Education from the University of Houston.

Len Sharp has taught Earth science at Liverpool High School, New York, for 26 years. He has a B.S. in Secondary Education and an M.S. in Science Education from Syracuse University. Mr. Sharp was president of the Science Teachers Association of New York in 1991-1992, and president of the National Earth Science Teachers Association from 1992-1994. He was a Presidential Awardee in 1995, and received the Fulbright Memorial Fund Fellowship to study in Japan in 1999.

Theodore Snow is a professor of Astronomy at the University of Colorado. He has a B.A. from Yale University, and an M.S. and Ph.D from the University of Washington. Dr. Snow is a founder and former director of the Center for Astrophysics and Space Astronomy at the University of Colorado, which conducts research on planets, stars, and galaxies. He has published numerous papers and several textbooks, and is a Fellow of the Royal Astronomical Society.

NATIONAL GEOGRAPHIC

National Geographic Society, founded in 1888 for the increase and diffusion of geographic knowledge, is the world's largest nonprofit scientific and educational organization. The School Publishing Division supports the Society's mission by developing innovative educational programs—ranging from print materials to multimedia programs, including CD-ROMS, DVDs, and software.

Contributing Writers

Joy Dickerson
McArthur, Ohio

Dale Gnidovec
Columbus, Ohio

Tracey Smeltzer
Derry, Pennsylvania

Jenipher Willoughby
Forest, Virginia

Content Consultants

William Ausich, Ph.D.
Professor
Department of Geological Sciences
The Ohio State University
Columbus, Ohio

Steven Dorobek, Ph.D.
Professor
Department of Geology and Geophysics
Texas A&M University
College Station, Texas

Janet Herman, Ph.D.
Professor and Director
Program of Interdisciplinary Research in Contaminant Hydrogeology
Environmental Sciences Department
University of Virginia
Charlottesville, Virginia

Timothy Horner, Ph.D.
Associate Professor
Department of Geology
California State University, Sacramento
Sacramento, California

Raymond Ingersoll, Ph.D.
Professor
Department of Earth and Space Sciences
University of California, Los Angeles
Los Angeles, California

William Keel, Ph.D.
Professor
Department of Physics and Astronomy
University of Alabama
Tuscaloosa, Alabama

Kirsten Menking, Ph.D.
Assistant Professor
Department of Geology and Geography
Vassar College
Poughkeepsie, New York

Burrell Montz, Ph.D.
Environmental Studies Director
Department of Geological Sciences and Environmental Studies
State University of New York at Binghamton
Binghamton, New York

Richard Pfeffer, Ph.D.
Distinguished Research Professor of Meteorology
Department of Meteorology
Florida State University
Tallahassee, Florida

Amy Leventer Reed, Ph.D.
Visiting Assistant Professor
Department of Geology
Colgate University
Hamilton, New York

Kristen St. John, Ph.D.
Assistant Professor
Department of Geology
Appalachian State University
Boone, North Carolina

John Vidale, Ph.D.
Professor
Department of Earth and Space Sciences
University of California, Los Angeles
Los Angeles, California

Safety Consultants

Anne B. Davidson, Ph.D.
Assistant Principal for Instruction
Madison City Schools
Madison, Alabama

John Longo
Coordinator of Laboratory Instruction
St. Joseph's University
Philadelphia, Pennsylvania

Kenneth Russell Roy, Ph.D.
Director of Science and Safety
Glastonbury Public Schools
Glastonbury, Connecticut

Teacher Reviewers

Frank Blatnik, M.S.
Science Facilitator
Swanton High School
Swanton, OH

Martha Boyd Buchanan
Freedom High School
Morganton, NC

Robert Ellyson, M.S.
Great Bridge High School
Chesapeake, VA

Kimberly Harmelink
Commerce High School
Commerce, GA

Regina Huffman, M. A.
Uniondale High School
Uniondale, NY

Al Janulaw
Creekside Middle School
Rohnert Park, CA

Bill Martin, M.Ed.
Fort Payne Middle School
Fort Payne. AL

Heather Monteleone
Page County High School
Shenandoah, VA

La Moine Motz, Ph.D.
Coordinator of Science Education
Oakland County Schools
Waterford, MI

Teresa Potter, M.A.
Science Department Chairperson
Rio Rico High School
Rio Rico, AZ

Katherine Richter
Athens Drive High School
Raleigh, NC

RevaBeth Russell
Science Department Chairperson
Lehi High School
Lehi, UT

Elsie Santiago, M.S.
Benjamin Cardoza High School
Bayside, NY

Robert Smith
Science Department Chairperson
Terry Parker High School
Jacksonville, FL

Gina Watkiss, M.A., M.S.
Science Coordinator
The Heritage School
Newnan, GA

Activity Testers

Gregory Beckway
Glenbrook North High School
Northbrook, IL

Paul Craft
Upper Arlington High School
Upper Arlington, OH

Georgia O'Hara
Centennial High School
Columbus, OH

Contents in Brief

UNIT 5 *The Dynamic Earth* 440

UNIT 6 *Geologic Time* 550

UNIT 7 *Resources and the Environment* 652

UNIT 8 *Beyond Earth* 744

NATIONAL GEOGRAPHIC

Table of Contents

(icon) = Environmental Connection

UNIT 3 Surface Processes on Earth 150

= ENVIRONMENTAL CONNECTION

UNIT 4 The Atmosphere and the Oceans 268

UNIT 5 The Dynamic Earth 440

UNIT 6 Geologic Time 550

UNIT 7 Resources and the Environment 652

(icon) = ENVIRONMENTAL CONNECTION

100
ml
90
80
70
60
50
40
30
20
10
300ml
250
200
150
100
50
APPROXIMATE VOLUMES

MiniLabs

Each chapter has one MiniLab. These short activities use simple materials and can be tried on your own at home or with help from a teacher at school.

GeoLabs

Each chapter contains a two-page GeoLab. There are four types of labs: GeoLab, Internet GeoLab, Design Your Own GeoLab, and Mapping GeoLab.

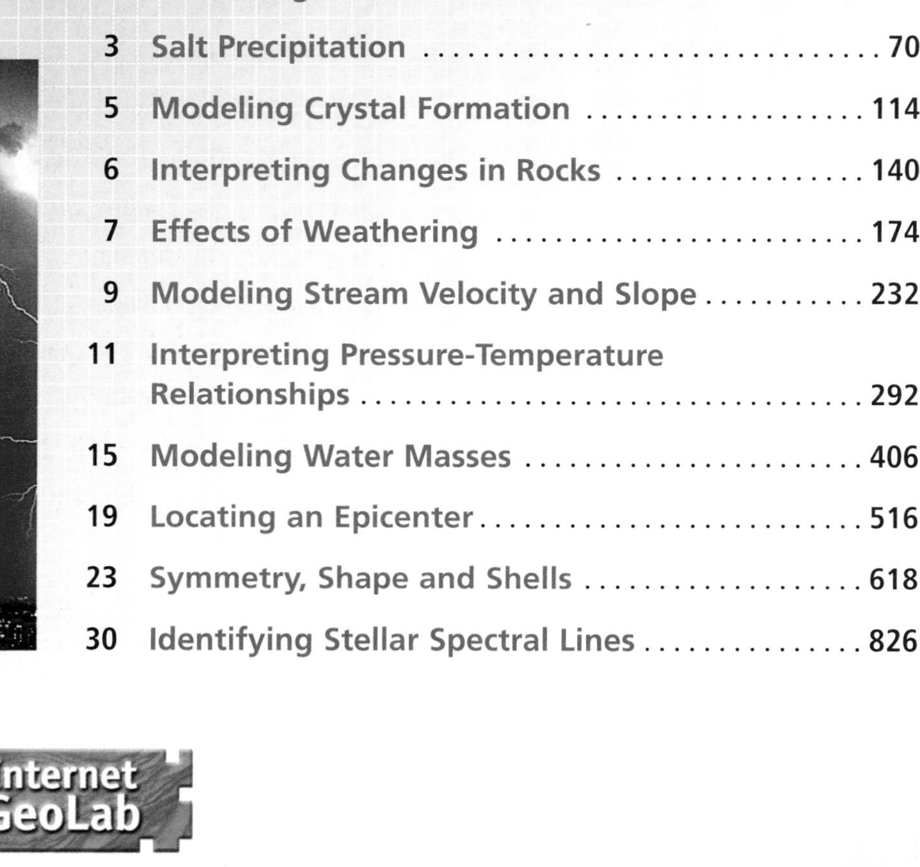

GeoLab

Internet GeoLab

Mapping GeoLab

DESIGN YOUR OWN GeoLab

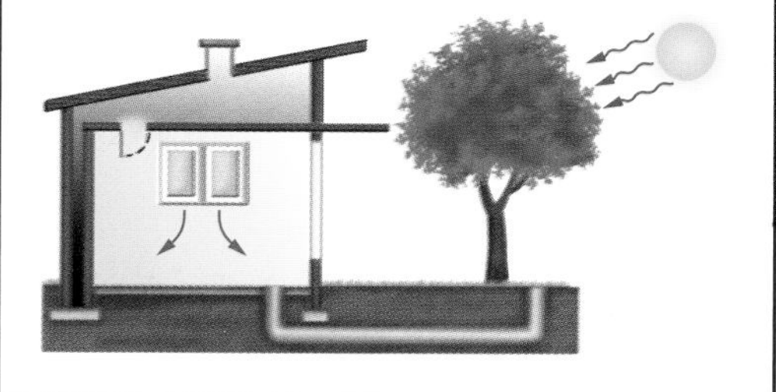

Discovery Labs

Each chapter has one Discovery Lab. These activities are used to introduce the students to the chapter content.

Problem-Solving Labs

Each chapter has one Problem-Solving Activity. These activities are math-based skill activities, and often require data interpretation and graphing.

Features

Each chapter has one Feature. These features expand and extend the chapter content.

Science & Technology

Science & Math

Science in the News

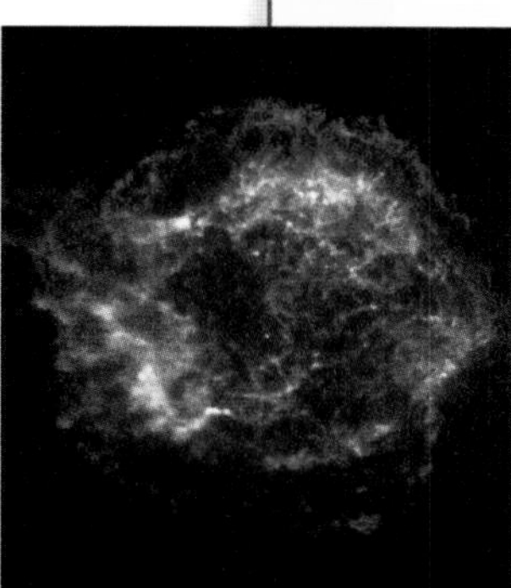

Science & the Environment

Explore the *Titanic*...

Tame the mighty Colorado River...

Dig for Dinosaurs...

What is it like to map the highest mountain on Earth? Or to explore the ghostly wreck of the unsinkable *Titanic?* The **National Geographic Expeditions** allow you to share in the excitement and adventures of explorers, scientists, and lovers of the environment as they delve into the unknown. You can explore the sparkling beauty of a cave in a river of ice. Or watch the explosion of a dying star. Each Expedition will take you on a journey that reaches from the distant galaxies to the depths of the oceans.

As you learn about our dynamic planet, you can use the **National Geographic Expeditions** to extend your knowledge and challenge yourself. To learn more about the Expeditions, go to the **Earth Science Web Site** (earthgeu.com) and click on the **National Geographic Expeditions** link. Or read the original **National Geographic Society** magazine articles to find out how geologists work to discover how a volcano erupts and try to predict future eruptions. The goal is to save lives! It's explorers and scientists like these that discovered the fascinating topics you are studying in your Glencoe **Earth Science: Geology, the Environment, and the Universe** textbook.

National Geographic Expeditions! are referenced within the chapters at point of use, to support or extend chapter content.

Table of Contents

Unit 1

Earth Science

Earth science is a blend of many different sciences, including geology, meteorology, oceanography, and astronomy. Earth scientists in these different specialties study and model the processes that change our planet. Some of these changes take place in a matter of seconds; others take millions of years to occur. The rocks and structures shown here formed millions of years ago as a result of many interactions among some of Earth's systems. In this unit, you'll learn about some of the methods used by Earth scientists, how various parts of Earth interact to produce changes, and how our planet can be represented by models known as maps.

Unit Contents

Go to the **National Geographic Expedition** on page 864 to learn more about topics that are connected to this unit.

Sunset Arch, Grand Staircase—Escalante Wilderness, Utah

Chapter 1

The Nature of Science

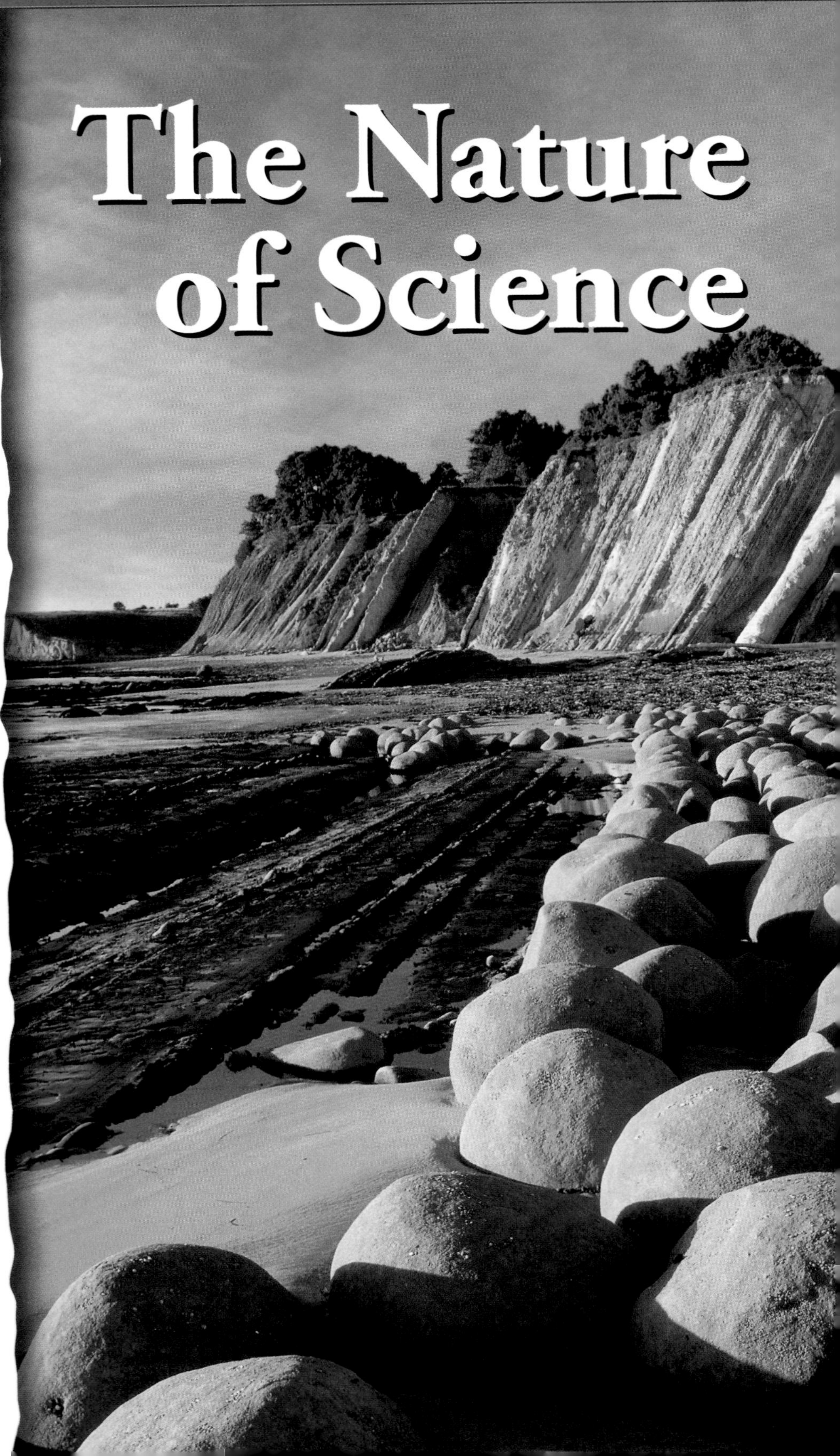

What You'll Learn

- How Earth science is a blend of sciences.
- How Earth's four major systems interact.
- What is involved in carrying out scientific experiments.
- Why it is important to communicate scientific methods and results accurately.

Why It's Important

In order to better understand Earth and how its processes affect our lives and the environment, it is necessary to learn about its major systems, the methods used by Earth scientists, and how scientific work is done.

Earth Science Online

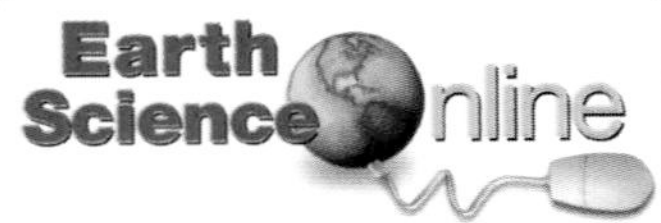

To find out more about the planet on which you live, visit the Earth Science Web Site at earthgeu.com

Discovery Lab Scientific Communication

Have you ever explained something to someone only to later find out that what you thought was a crystal-clear explanation was confusing, misleading, or even incorrect? Precise communication in a manner that is not influenced by your expectations or beliefs is a very important skill. In this activity, you will describe objects provided by your teacher.

1. Obtain an object from your teacher. Don't show it to your partner.
2. Write only one sentence that accurately describes the object in detail without actually saying what the object is.
3. Give your partner the description and allow him or her a few minutes to try to determine what your object is.
4. Now use your partner's description to determine what his or her object is.

Communicate Work together to rewrite each description in your science journals to make them as accurate as possible. Trade the new descriptions with another pair of students. Did this pair of students have an easier time at determining the objects than you and your partner did? Why or why not?

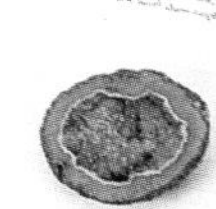

Section 1.1 Earth Science

OBJECTIVES

- **Differentiate** *among the four major branches of Earth science.*
- **Contrast** *the four systems of Earth.*
- **Discuss** *how Earth science affects your daily life.*

VOCABULARY

astronomy
meteorology
geology
oceanography
lithosphere
asthenosphere
hydrosphere
atmosphere
biosphere

It is easy to see from the photograph on page 4 why this strip of the California coast is called Bowling Ball Beach. These round structures are concretions—masses of rock that form as the result of processes at work on Earth's surface. In this book, you'll learn about Earth and the processes and forces that change it, the materials from which it is made, its long history, and its place in the universe.

The Scope of Earth Science

The scope of Earth science is vast. Dinosaur bones on display at museums were once embedded in the rocks that make up some of Earth's cliffs and canyons. Mining certain rocks produces some of the gold used by jewelers and dentists. Computer models simulate the flow of the blanket of air that surrounds Earth so that scientists better understand stormy weather. Ocean-floor exploration has led to the discovery of bizarre creatures that never see the light of day, while the study of objects in space has revealed much about our own planet.

Figure 1-1 Astronomy includes the study of Earth, its neighbors, and distant stars. The Keck Telescope in Hawaii, shown here, is used to study stars trillions of kilometers from Earth.

As you can see, there are many different areas of Earth science. This broad field can be broken into four major areas of specialization: astronomy, meteorology, geology, and oceanography.

Astronomy **Astronomy** is the study of objects beyond Earth's atmosphere. Prior to the invention of sophisticated instruments, such as the telescope shown in ***Figure 1-1,*** many astronomers merely described the locations of objects in space in relation to one another. Today, these Earth scientists study the universe and everything in it, including Earth, its neighbors, and other bodies in the universe.

Meteorology The branch of Earth science that studies the air that surrounds our planet is called **meteorology.** Meteorologists study the forces and processes that cause the atmosphere to change to produce weather. These Earth scientists also try to predict the weather and how changes in weather might affect Earth's climate.

Geology The study of the materials that make up Earth and the processes that form and change these materials is the branch of Earth science known as **geology.** Geologists identify rocks, study glacial movements, interpret clues to Earth's 4.6 billion-year history, and determine how forces change our planet, among many other things.

Figure 1-2 Some oceanographers study how human activities affect Earth's oceans. This oil spill occurred off the coast of Wales in 1996.

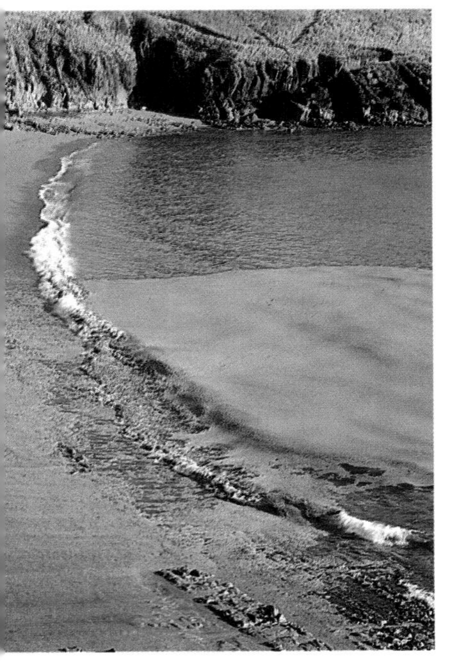

Oceanography The study of Earth's oceans, which cover nearly three-fourths of the planet, is called **oceanography.** Oceanographers study the creatures that inhabit salty water, measure different physical and chemical properties of the oceans, and observe various processes in these bodies of water. Some oceanographers study the effects of human activities on Earth's saltwater bodies. The oil shown in ***Figure 1-2*** is just a very small portion of the 70 000 tonnes that were spilled off the coast of Wales in 1996.

The study of our planet is a broad endeavor, and thus it requires a variety of subspecialties of the four major areas of Earth science. Some of these subspecialties are listed in ***Table 1-1.*** What kinds of things does a paleontologist study? Which subspecialty is concerned with the environment? What types of things are studied by scientists specializing in tectonics? What might a hydrologist study?

EARTH'S SYSTEMS

Scientists who study Earth have identified four main Earth systems: the lithosphere, the hydrosphere, the atmosphere, and the biosphere. Each system is unique, yet each interacts with the others. None of Earth's systems is independent of the others, nor of the global system of Earth itself.

Table 1-1 Some Subspecialties of Earth Science

Subspecialty	Subjects Studied	Subspecialty	Subjects Studied
Climatology	Patterns of weather over a long period of time; effects of human activities on weather and climate	Ecology	Habitats of organisms and how organisms interact with each other and their environments
Paleontology	Remains of organisims that once lived on Earth; ancient environments	Geochemistry	Earth's composition and the processes that change it
Hydrology	Water flow on and below Earth's surface; sources of and solutions to water pollution	Tectonics	Effects of internal processes on Earth's surface, including earthquakes and mountain building

The Lithosphere Earth's **lithosphere** is the rigid outer shell of the planet and includes the crust and the solid, uppermost part of the layer below the crust, the mantle. There are two kinds of crust: continental crust and oceanic crust. Earth's continental crust is made mostly of a rock called granite. Oceanic crust is mainly basalt, a rock that is denser than granite. Earth's mantle is mainly composed of a rock called peridotite. Some of Earth's upper mantle behaves like a rigid solid while other parts of this layer are partially molten and flow like a soft plastic. This partially molten layer is the **asthenosphere.**

Beneath Earth's mantle is the core, which can be divided into two parts: an outer, liquid part and a solid, inner part. Earth's core is thought to be made of iron and nickel. While Earth's core and asthenosphere are not parts of the lithosphere, they do interact with this system of Earth to produce many of the features at the planet's surface. You'll learn how the lithosphere and asthenosphere interact to produce volcanoes, mountains, and earthquakes in Unit 5.

Topic: Earth's Systems
To learn more about Earth's four major systems, visit the Earth Science Web Site at earthgeu.com

Activity: Write a study question for each of the Earth's four major systems based on your new knowledge.

The Hydrosphere The water in Earth's oceans, seas, lakes, rivers, and glaciers, as well as the water in the atmosphere, makes up the **hydrosphere.** About 97 percent of Earth's water exists as salt water; the remaining 3 percent is freshwater contained in glaciers, in lakes and rivers, and beneath Earth's surface as groundwater. About three fourths of all freshwater is contained in glaciers and icebergs, such as the one shown in ***Figure 1-3;*** most of the rest of this freshwater is groundwater. On a fraction of Earth's total amount of freshwater is in lakes and rivers. You'll find out more about Earth's hydrosphere in Units 3, 4, and 7.

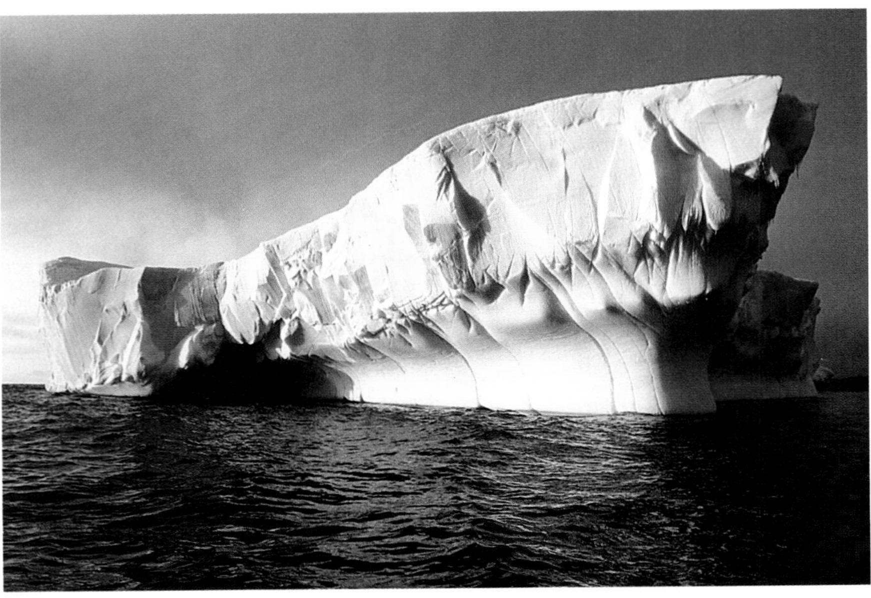

Figure 1-3 Most of Earth's freshwater is contained in glaciers. The iceberg shown here broke off of one of the glaciers that covers nearly all of the continent of Antarctica.

Figure 1-4 The biosphere, hydrosphere, atmosphere, and lithosphere are interdependent systems of Earth.

The Atmosphere The blanket of gases that surrounds our planet is called the **atmosphere.** Among other things, Earth's atmosphere is necessary for respiration by most living things, protects Earth's inhabitants from harmful radiation from the Sun, and helps to keep the planet at a temperature suitable for life. Earth's atmosphere contains about 78 percent nitrogen and 21 percent oxygen. The remaining 1 percent of gases in the atmosphere include water vapor, argon, carbon dioxide, and other trace gases. You will learn more about Earth's atmosphere and how parts of this system interact to produce weather in Unit 4.

The Biosphere The **biosphere** includes all organisms on Earth as well as the environments in which they live. Most organisms exist within a few meters of Earth's surface, but some live deep beneath the ocean's surface, and others live high atop Earth's mountains. Earth's biosphere appears to be unique in that scientists have not yet found any confirmed evidence of life on other planets in our solar system or elsewhere in the galaxy.

As you can see in ***Figure 1-4,*** Earth's biosphere, lithosphere, hydrosphere, and atmosphere are interdependent systems. Earth's present atmosphere, for example, formed millions of years ago as a result of volcanic activity, respiration and transpiration by ancient organisms, and photosynthesis. Today's organisms, including humans, continue to change the atmosphere through their life processes and activities. You'll explore interactions among Earth's biosphere and other systems, both past and present, in Units 3, 4, 6, and 7.

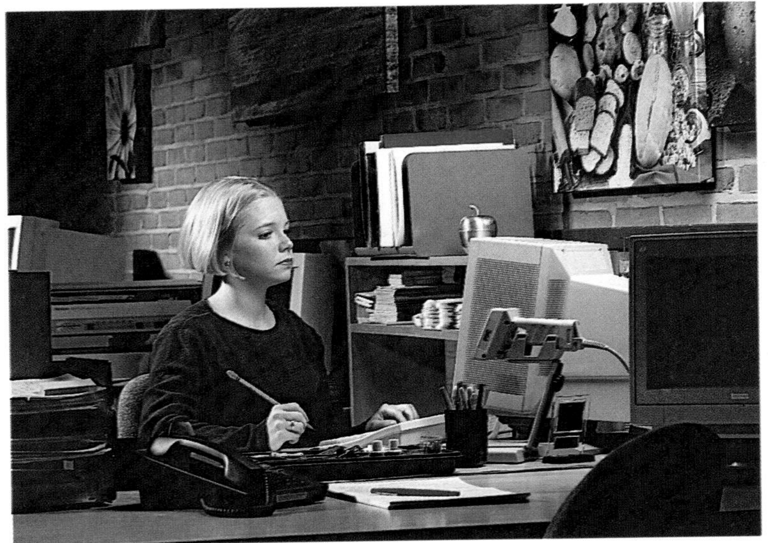

Figure 1-5 Computers, calculators, telephones, and electricity are technological advances used by many people in their daily lives.

EARTH SCIENCE IN YOUR EVERYDAY LIFE

You and the billions of other lifeforms that live on Earth are part of the biosphere. Together with many of these creatures, you live on Earth's crust, which is part of the lithosphere, and breathe the gases in Earth's atmosphere. You also depend in many ways on the substance that covers nearly three-fourths of Earth—water, which makes up the hydrosphere. In what other ways is Earth science a part of your everyday life?

Technology While you might not realize it, the study of science, including Earth science, has led to the discovery of many things that you use every day. This application of scientific discoveries is called technology. Freeze-dried foods, ski goggles, micro-fabrics, and the ultra-light materials used to make many pieces of sports equipment are just a few examples of technological advances developed as a result of scientific study. Today, these items, along with those shown in ***Figure 1-5,*** are common.

Technology is transferable, which means that it can be applied to new situations. The technological developments just described were first developed for use in space. Later, they were modified for use here on Earth. In the *Science & Technology* feature at the end of this chapter, you'll find out how medical technology has been used to study dinosaurs!

SECTION ASSESSMENT

1. Name and briefly describe the four branches of Earth science.
2. What does a geologist study?
3. What does a geochemist study?
4. Compare and contrast Earth's lithosphere and asthenosphere.
5. Describe the subdivisions of Earth's hydrosphere.
6. **Thinking Critically** What kinds of interactions do you think occur between Earth's hydrosphere and atmosphere?

SKILL REVIEW

7. **Outlining** Outline the main ideas of this section. For more help, refer to the *Skill Handbook.*

earthgeu.com/self_check_quiz

SECTION 1.2 Methods of Scientists

I&E 1.f, 1.j, 1.n

Have you ever picked up a rock and peered at it for clues about its origin? Have you ever made a decision about what to wear after having observed the clouds in the sky? Have you wondered where soil comes from? If you answered *yes* to any of these or similar questions, you have thought like a scientist. Scientists use scientific methods to discover the rules—from simple to extremely complex—that govern our vast universe.

OBJECTIVES

- **List** *the steps used in a scientific method.*
- **Compare** *and* **contrast** *experimental variables and controls.*
- **Identify** *basic SI units.*
- **Explain** *how to write numbers using scientific notation.*

VOCABULARY

hypothesis
independent variable
dependent variable
control
Le Système International d'Unités (SI)
scientific notation

THE NATURE OF SCIENTIFIC INVESTIGATIONS

A scientific method is a planned, organized approach to solving a problem. While the steps taken to solve the problem can vary, the first step involved in scientific problem solving, as shown in ***Figure 1-6,*** is usually identifying the problem, or determining what it is you want to know. Often, scientific problem solving involves researching the problem. Once the problem is defined and research is complete, a **hypothesis,** or suggested explanation for an observation, is made. Often, a hypothesis is stated in the form of a question that can be answered by the results of a test or an experiment.

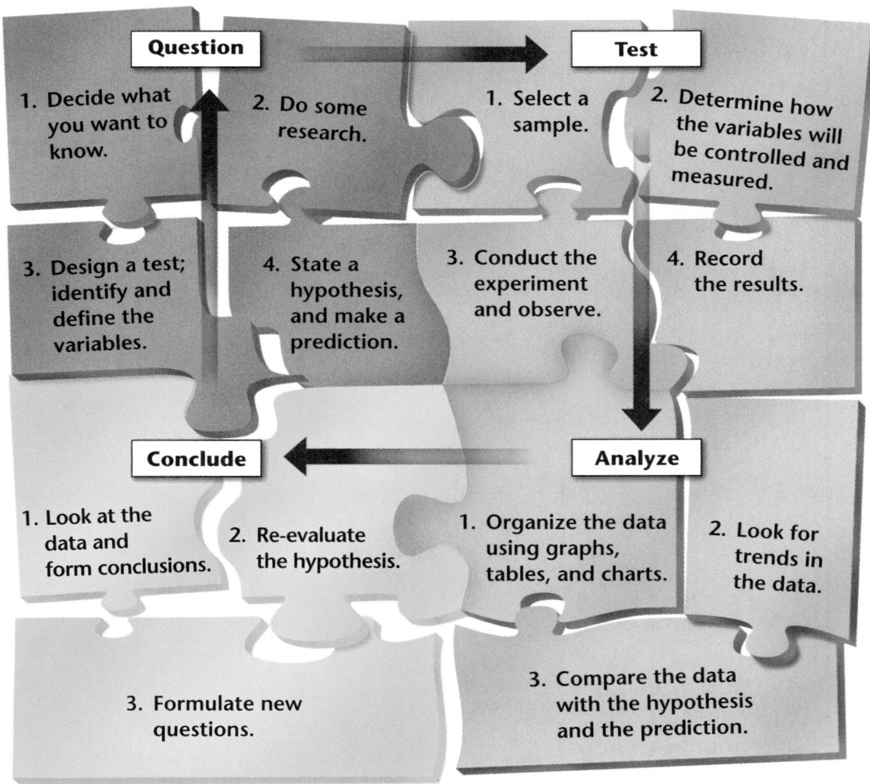

Figure 1-6 Various steps and processes are involved in a scientific approach to problem solving.

I&E

1.f Distinguish between hypothesis and theory as scientific terms.

1.j Recognize the issues of statistical variability and the need for controlled tests.

MiniLab

How do soil and water absorb and release heat?

Experiment to determine the relationship between variables.

Procedure

1. Obtain the materials for this lab from your teacher.
2. Put soil into one container until it is half full. Put water into other container until it is half full.
3. Place one thermometer in the soil so that the bulb is barely covered. Use masking tape to secure another thermometer about 1 cm from the top of the soil.
4. Repeat step 3 with the container of water.
5. Put the containers on a very sunny windowsill. Record the temperature shown on each thermometer. Write these values in a table. Then record temperature readings every 5 minutes for half an hour.
6. Remove the containers from the windowsill and immediately record the temperature on each thermometer every 5 minutes for half an hour.

Analyze and Conclude

1. Which substance absorbed heat faster?
2. Which substance lost heat faster?
3. What was your independent variable? Your dependent variable?

Experimentation A hypothesis is tested by conducting an experiment, which is an organized procedure that involves making measurements and observations. A good scientific experiment tests only one variable, or changeable factor, at a time. The **independent variable** in an experiment is the factor that is manipulated by the experimenter. A **dependent variable** is a factor that can change if the independent variable is changed. Constants are factors that do not change during an experiment. A **control** is used in an experiment to show that the results of an experiment are actually a result of the condition being tested. Refer to the *Skill Handbook* for more information on variables. You will experiment with variables in the *MiniLab* on this page and in many other activities throughout this book.

Safety in the Science Classroom Some of the labs and activities in this book will require that you handle various materials and equipment, including those shown in ***Figure 1-7.*** When conducting any scientific investigation, it is important to use all materials and equipment only as instructed. Follow the safety rules listed in ***Table 1-2*** to

Figure 1-7 Safety goggles and a lab apron should be worn during any activity or experiment in the science lab.

Table 1-2 Some Important Safety Rules for the Science Lab

1. Before beginning any investigation, understand the safety symbols noted by referring to the symbols and their meanings in *Appendix B.*
2. Wear safety goggles and a safety apron during all investigations that involve heating, pouring, or using chemicals.
3. Tie back long hair and loose clothing before you begin any investigation.
4. Always slant test tubes away from yourself and others when heating the tubes. Keep all materials away from open flames.
5. Never eat or drink in the lab and never use laboratory glassware as food or drink containers.
6. Never inhale chemicals, and never taste any substance used in the lab. Also, don't draw any material into a tube with your mouth.
7. Know what to do in case of fire. Also, know the location and proper use of the fire extinguisher, safety shower, fire blanket, first-aid kit, and fire alarm.
8. Report any spill, accident, or injury to your teacher immediately.
9. When cleaning up, dispose of chemicals and other materials only as directed by your teacher.
10. Always wash your hands thoroughly with soap after working in the lab.

help prevent injury to you and others in the lab as well as make you aware of possible hazards in a science lab. Refer to *Appendix B* for additional safety information and a table of symbols.

I&E

1.n Know that when an observation does not agree with an accepted scientific theory, the observation is sometimes mistaken or fraudulent (e.g., the Piltdown Man fossil or unidentified flying objects) and that the theory is sometimes wrong (e.g., the Ptolemaic model of the movement of the Sun, Moon, and planets).

Analysis and Conclusions New ideas in science are limited by the context in which they are conceived. Processes, data, and conclusions must be examined to eliminate bias—influence by expectations or beliefs. During a scientific experiment, all data are carefully recorded. Once an experiment is complete, graphs, tables, and charts are commonly used to format and display data, which are then analyzed so that a conclusion can be drawn. Sometimes, a conclusion is contrary to the original hypothesis. In such a case, because the conclusion is supported by the data, the hypothesis, *not* the conclusion or the data, must be re-evaluated.

It is important to note, as shown in ***Figure 1-6,*** that scientific methods are not rigid, step-by-step outlines to solve problems. Scientists can take many different approaches to solving a scientific problem. In many scientific investigations, for example, scientists form new ideas after observing unexpected results. Sometimes, an experimenter might encounter a different problem and choose to pursue the new problem rather than the original hypothesis.

Topic: SI Measurements
To learn more about SI, visit the Earth Science Web Site at earthgeu.com

Activity: List three facts about the history of SI.

Measuring Use a metric ruler to measure the length of the following objects in both millimeters and centimeters: this book, a small paper clip, and your pencil. Would you use these same units to measure the length of your classroom? Why or why not?

MEASUREMENT

Scientific experiments often involve making measurements. A measurement, as you already know, includes both a number that identifies how many units there are and a unit of measure. Most scientific studies and experiments use a standard system of units called **Le Système International d'Unités,** or **SI** for short. This system is a modern version of the metric system. SI is based on a decimal system that uses the number 10 as the base unit. You will make various measurements in SI in the *GeoLab* at the end of this chapter.

Length The standard SI unit to measure length is the meter (m). The distance from a doorknob to the floor is about 1 m. A guitar is also about 1 m long. The meter is divided into 100 equal parts called centimeters (cm). Thus, 1 cm is 1/100 of 1 m. One millimeter (mm) is smaller than 1 cm. There are 10 mm in 1 cm. How many millimeters are in 1 m? Long distances are measured in kilometers (km). There are 1000 m in 1 km. How many centimeters are in 1 km?

Weight and Mass Weight is a measure of the gravitational force on an object. Weight is typically measured with some type of scale. Unlike mass, weight varies with location. For example, the weight of the astronaut shown in ***Figure 1-8*** while on the Moon is about one-sixth the astronaut's weight on Earth. This is because the gravitational force exerted by the Moon on the astronaut is one-sixth the force exerted by Earth on the astronaut. Weight is a force, and the SI unit for force is the newton (N). The SI unit of mass is the kilogram (kg). A half-cup of water with a mass of 4 ounces weighs about 1 N, and a person with a mass of 60 kg weighs about 600 N.

Mass is the amount of matter in an object and depends on the number and kinds of atoms that make up the object. The mass of an

Figure 1-8 The gravitational force exerted by the Moon is less than the gravitational force exerted by Earth. Thus, when this astronaut visited the Moon in 1969, he weighed less than he did on Earth.

object, unlike weight, does not change with an object's position. Mass can be measured with a balance like the one shown in ***Figure 1-9.***

Figure 1-9 Mass is measured with a balance.

Area and Volume Some measurements, such as area, require a combination of SI units. Area is the amount of surface included within a set of boundaries and is expressed in square units of length, such as square meters (m^2) or square centimeters (cm^2). Determine the area, in square centimeters, of this book by multiplying the length of the book by its width.

The amount of space occupied by an object is the object's volume. The SI units for volume, like those of area, are derived from the SI units used to measure length. The basic SI unit of volume for a regularly shaped, solid object is the cubic meter (m^3). SI measurements for fluid volumes are usually made in milliliters (mL) or liters (L). Volume can also be expressed in cubic centimeters (cm^3); 1 cm^3 equals 1 mL.

Density Density is a measure of the amount of matter that occupies a given space. Density is calculated by dividing the mass of the matter by its volume. Density is often expressed in grams per cubic centimeter (g/cm^3), grams per milliliter (g/mL), or kilograms per cubic meter (kg/m^3).

Time Time is the interval between two events and is usually measured with a watch or clock. The clock shown in ***Figure 1-10*** is an atomic clock, which provides the most precise measure of time. The SI unit of time is the second (s). In the activities in this book, you will generally measure time in seconds or minutes.

Figure 1-10 Atomic clocks provide much more precise measures of time than ordinary clocks.

Temperature Temperature is a measure of the average vibrations of the particles that make up a material. A mass made up of particles that vibrate quickly has a higher temperature than a mass whose particles vibrate more slowly. Temperature is measured in degrees with a thermometer. In science, temperature is often measured on the Celsius (C) scale. On the Celsius scale, a comfortable room temperature is about 25°C, and the normal temperature of the human body is about 37°C. In SI, temperature is measured on the Kelvin scale. On this scale, the coldest possible temperature is absolute zero, or 0 K, which is equal to –273°C.

Figure 1-11 The Sombrero Galaxy, shown here, is just one of many groups of stars in the universe.

SCIENTIFIC NOTATION

In many branches of science, some numbers are very small, while others are quite large. To conveniently express these numbers, scientists use a type of shorthand called **scientific notation** to express the number as a multiplier and a power of 10.

In scientific notation, a number is expressed as a value between 1 and 10 multiplied by a power of 10. The power of 10 is the number of places the decimal point must be shifted so that only a single digit remains either to the left or right of the decimal point. If the decimal point must be shifted to the left, the exponent of 10 is positive. For example, the approximate number of stars in the Sombrero Galaxy, some of which are shown in ***Figure 1-11,*** is 90 000 000 000. In scientific notation, this number is written as 9×10^{10}. The mass of Earth, which is 5 974 200 000 000 000 000 000 000 kg, is written as 5.9742×10^{24} kg in scientific notation. If the decimal point in a number must be shifted to the right, then the exponent of 10 is negative. The diameter of an atom in meters, for example, which is approximately 0.0000000001 m, is written as 1×10^{-10} m.

All of the quantities discussed in this section and the units used to measure them are summarized in *Appendix A*. You can also refer to *Appendix A* for explanations of how to convert between the units you are familiar with, such as feet and pounds, and SI units. In the next section, you will learn about other ways in which scientific information is communicated.

SECTION ASSESSMENT

1. Describe the steps used in a scientific method.
2. Contrast dependent and independent variables.
3. What is the purpose of a control in a scientific experiment?
4. Explain how to write a large number using scientific notation.
5. **Thinking Critically** An increase in the temperature of matter generally results in a decrease in its density. Water is an exception to this rule. What do you think happens to water when it freezes?

SKILL REVIEW

6. **Concept Mapping** Use the terms below to make a concept map that summarizes the units used to measure each quantity discussed in this section. For help, refer to the *Skill Handbook*.

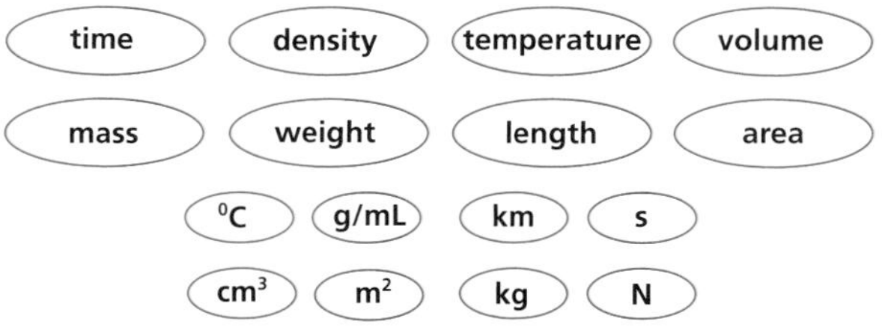

SECTION 1.3 Communicating in Science

I&E 1.a, 1.e, 1.f, 1.g, 1.k

There are many ways to communicate the same information, such as newspapers and magazines, TV and the Internet, and scientific journals. Scientists have the responsibility to truthfully report their methods and results to other scientists and the public. Often, scientists propose models to try to explain ideas or systems. When an explanation withstands the test of repeated experiments, a theory might be proposed. Scientific models and theories can be modified when new observations and data are collected.

OBJECTIVES

- **List** *several ways in which scientific information is communicated.*
- **Differentiate** *between a scientific theory and a scientific law.*

VOCABULARY

theory
law

COMMUNICATING RESULTS

One important goal of science is to make results available to others. Communicating scientific data and results allows others to learn of new discoveries, to possibly verify what has been reported and examine it for bias, and to conduct new experiments using the information. From the laboratory reports that you will generate as you use this book to scientific papers published in professional journals, scientific results are communicated in many ways.

I&E

1.k Recognize the cumulative nature of scientific evidence.

Lab Reports Throughout this book, you will conduct many Earth science experiments and activities. During and after each activity or experiment, you will be asked to record and analyze the information that you collected and to draw conclusions based on your data. Your resulting lab report, similar to the one shown in ***Figure 1-12,*** will be used by your teacher to assess your understanding of the activity or experiment. You might also be asked to compare your results with the results of other students to help you find both similarities and differences among the results.

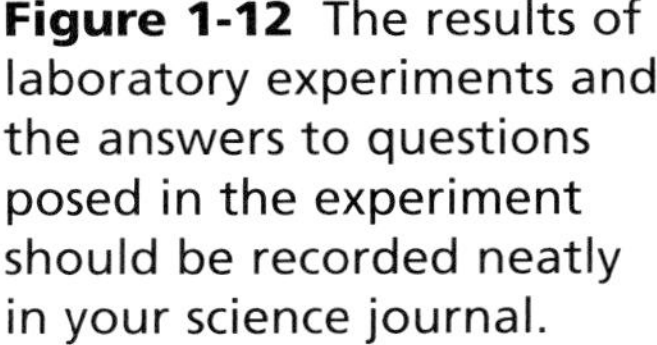

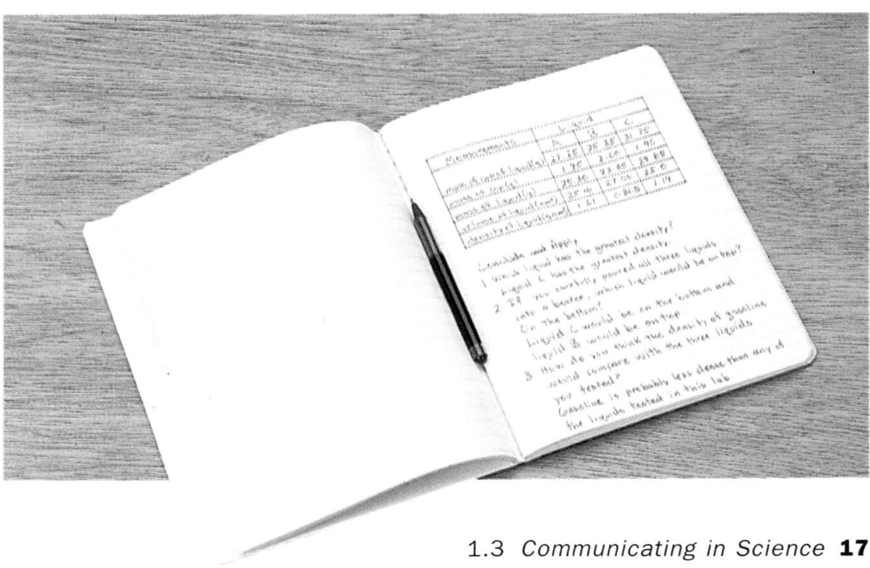

Figure 1-12 The results of laboratory experiments and the answers to questions posed in the experiment should be recorded neatly in your science journal.

Graphs You will be asked to graph the results of many experiments and activities in this book. As you will find out in the *Problem-Solving Lab* on page 18, a line graph is a visual display that shows how two variables are related. On a line graph, the independent variable is plotted on the horizontal (x) axis, and the dependent

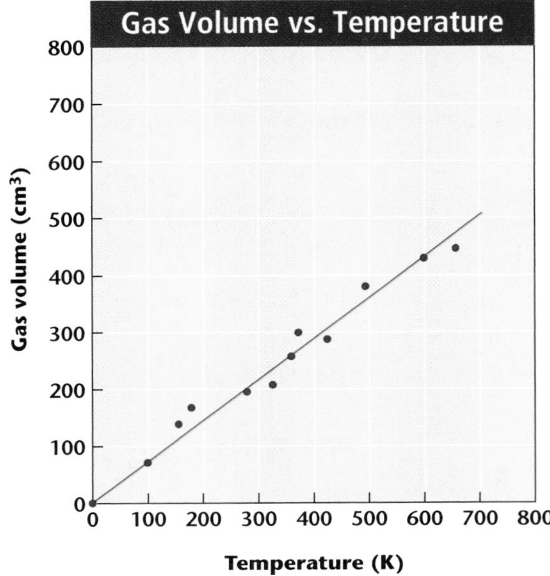

Figure 1-13 A line graph shows the relationship between two variables. Refer to the *Skill Handbook* for other types of graphs.

variable is plotted on the vertical (y) axis. Refer to the line graph in ***Figure 1-13.*** What is the independent variable? The dependent variable?

Models

In some of the activities and experiments in this book, you will be making and using models. A scientific model is an idea, a system, or a mathematical expression that is similar to the idea being explained. While a model might not have all of the components of a given idea, it should be a fairly accurate representation. Models can change when more data are gathered. Early astronomers, for example, thought that Earth was the center of the solar system, as shown in ***Figure 1-14A.*** This model was changed as the result of careful observations of the motions of the Sun and the planets in the night sky that showed that the planets in our solar system orbit the Sun, as shown in ***Figure 1-14B.***

Problem-Solving Lab

Making and Using Graphs

Make and use a graph that shows how the annual, average surface temperature of Earth has varied over the past 500 years. The data in the table are global, average surface temperatures, in Kelvins, starting in the year 1500.

Average Surface Temperatures	
Year	Average Surface Temperature (K)
1500	285.8
1600	285.9
1700	286.0
1800	286.2
1900	286.5
2000	286.9

Procedure

1. Convert each temperature from kelvins to degrees Celsius by subtracting 273 from each value.
2. Determine appropriate scales for your graph. Plot the year on the x-axis and temperature, in degrees Celsius, on the y-axis.

Analysis

3. Describe the general trend shown by the data.
4. How has Earth's average surface temperature changed with time?

Thinking Critically

5. Use the graph to determine the average surface temperature for 1980.
6. Extrapolate the data to predict what the average surface temperature will be in the year 2100.

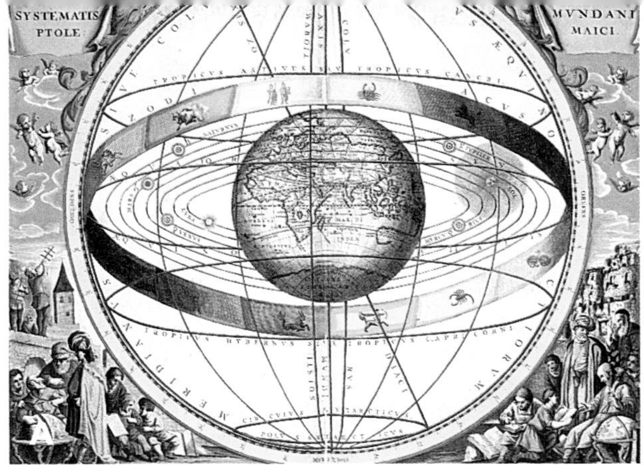

Figure 1-14 Some early astronomers thought that the Sun and other planets orbited Earth **(A).** It is now known that Earth and its eight neighbors orbit a star we call the Sun **(B).**

THEORIES AND LAWS

A scientific **theory** is an explanation based on many observations during repeated experiments. A scientific theory is valid only if it is consistent with observations, makes predictions that can be tested, and is the simplest explanation of observations. Like a scientific model, a theory can be changed or modified with the discovery of new data.

A scientific **law** is a basic principle that describes the behavior of a natural phenomenon. A scientific law can be thought of as a "rule of nature," even though the cause of the law may not be known. The events described by a law are observed to be the same every time. An example of a scientific law is Sir Isaac Newton's first law of motion, which states that an object at rest or in motion stays at rest or in motion unless it is acted upon by an outside force. This law explains why Earth and eight other planets remain in orbit around the Sun. Theories are often used to explain scientific laws.

In this book, you will communicate your observations and draw conclusions based on scientific data. You also will find out that many of the models, theories, and laws used by Earth scientists to explain various processes and phenomena grow from the work of many scientists and sometimes spring from unexpected findings.

I&E

1.f Distinguish between hypothesis and theory as scientific terms.

1.g Recognize the usefulness and limitations of models and theories as scientific representations of reality.

SECTION ASSESSMENT

1. What is the purpose of communicating scientific information?
2. What is the purpose of writing lab reports for experiments and activities in this book?
3. How are data plotted on a line graph?
4. What is a scientific model?
5. Contrast scientific theories and laws.
6. **Critical Thinking** When ice is heated, it melts. Is this a theory or a law? Explain.

SKILL REVIEW

7. **Recognizing Cause and Effect** Refer to *Figure 1-13.* Explain the relationship between the independent and dependent variables. For more help, refer to the *Skill Handbook.*

earthgeu.com/self_check_quiz

GeoLab Measuring in SI

I&E

1.a Select and use appropriate tools and technology (such as computer-linked probes, spreadsheets, and graphing calculators) to perform tests, collect data, analyze relationships, and display data.

1.e Solve scientific problems by using quadratic equations and simple trigonometric, exponential, and logarithmic functions.

Suppose someone asked you to measure the area of your classroom in square cubits. What would you use? A cubit is an ancient unit of length equal to the distance from the elbow to the tip of the middle finger. Since this length varies among individuals, the cubit is not a standard unit of measure. SI units are standard units, which means that they are exact quantities that have been agreed upon to use for comparison. In this GeoLab, you will use SI units to measure various properties of rock samples.

Preparation

Problem

Measure various properties of rocks and use the measurements to explain the relationships among the properties.

Materials

water
250-mL beaker
graph paper
balance
pieces of string
spring scale
rock samples

Objectives

In this GeoLab, you will:

- **Measure** the area, volume, mass, and weight of several rock samples.
- **Calculate** the density of each sample.
- **Explain** the relationships among the quantities.

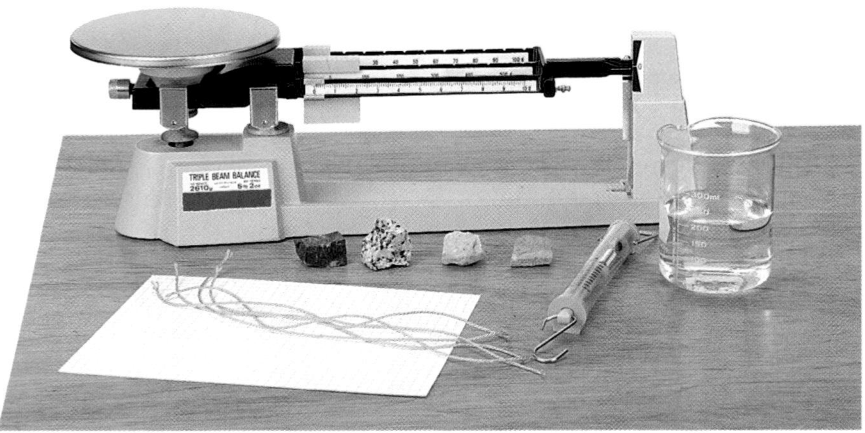

Procedure

1. Use the information in the *Skill Handbook* to design a data table in which to record the following measurements for each sample: area, volume, mass, weight, and density.
2. Obtain rock samples from your teacher. Carefully trace the outline of each rock onto the graph paper. Determine the area of each sample and record the values in your data table.
3. Pour water into the beaker until it is half full. Record this volume in the table. Tie a piece of string securely around one rock sample. Slowly lower the sample into the beaker. Record the volume of the water. Subtract the two values to determine the volume of the rock sample.
4. Repeat step 3 for the other rocks. Make sure the original volume of water is the same as when you measured your first sample.
5. Follow your teacher's instructions about how to use the balance to determine the mass of each rock. Record the measurements in your table.
6. Again, secure each rock with a piece of dry string. Make a small loop in the other end of the string. Place the loop over the hook of the spring scale to determine the weight of each rock sample. Record the values in your data table.

Analyze

1. Compare the area of each of your samples with the areas determined by other students for the same samples. Explain any differences.
2. Compare the volume of each of your samples with the volumes determined by other students for the same samples. Explain any differences.
3. Compare the weight and mass of each of your samples with the values for these quantities determined by other students. Again, explain any differences.
4. Use your measurements to calculate the density of each sample using this formula: *density = mass/volume.* Record these values in your data table.

Conclude & Apply

1. How accurate do you think your measurement of the area of each sample is? Explain.
2. What were the variables you used to determine the volume of each sample?
3. How could you find the volume of a rock such as pumice, which floats in water?
4. Does mass depend on the size or shape of a rock? Explain.

Science & Technology

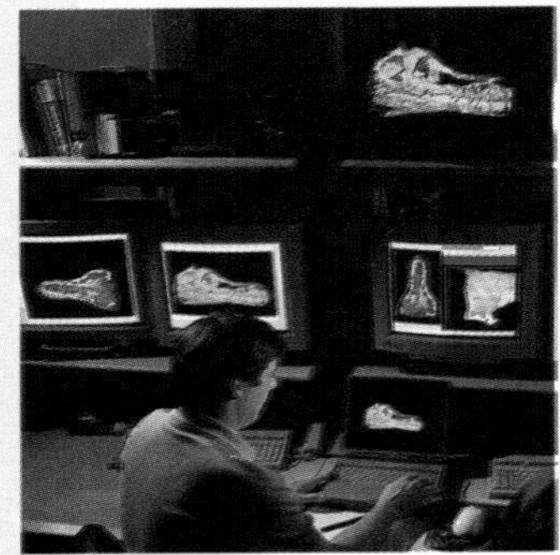

Willo, Sue, and Technology, Too

Paleontology, the area of Earth science that studies ancient life forms, has long been associated with hands-on work—digging, cleaning, and handling fossils. However, technology is playing an increasing role in this area. Technologies borrowed from medicine, manufacturing, and the aerospace industry are leading to new discoveries about dinosaurs—especially Willo and Sue.

Willo, the fossil remains of a 66-million year old *Thescelosaurus,* and Sue, the nearly complete skeleton of a 66-million year old *Tyrannosaurus rex,* were both studied using computerized tomographic scanning, or CT scans. A CT scan is a type of scan in which X rays move through a specimen at different rates depending on the density of the tissues encountered. A CT scan produces a picture of a very thin slice of a portion of a specimen. A computer is used to record and process the rates to produce an image on the screen. Multiple CT slices can be stacked to generate a three-dimensional image of the complete specimen.

A Dinosaur with Heart

Willo was found by paleontologist Michael Hammer in Harding County, South Dakota, in 1993. Willo was a plant-eater about the size of a pony. A CT scan of a dark mass of rock found in Willo's chest cavity revealed a structure that appears to be a heart—a four-chambered heart. A four-chambered heart would strongly support a relatively new hypothesis that dinosaurs were warm-blooded rather than cold-blooded animals.

Sue's Sniffer

Sue, the most complete skeleton of a *T. rex* ever recovered, was put on display at the Field Museum in Chicago, Illinois, in May 2000. Sue was found and excavated by amateur paleontologist Sue Hendrickson in the South Dakota Badlands in 1990. The skull of Sue was sent for CT scanning to a company that makes jet airplane engines because medical scanners couldn't accommodate the skull's 1.6-meter length! The scan, which is shown above and is on display with a cast of the skull, revealed that the ferocious carnivore had a much more acute sense of smell than had been expected. The scan of Sue's skull showed olfactory bulbs the size of grapefruits! The discovery of these scent-sensing organs would never have been made without the use of CT technology.

Technology is changing how discoveries are made in Earth science, but also how they are shared. The images from CT scans are digital. They can be e-mailed and downloaded by scientists and students all over the world. The ease of access to important data will, in turn, lead to more exciting discoveries.

Internet

For more information on the use of CT scans in the area of paleontology, visit the Earth Science Web Site at earthgeu.com. Compare and contrast scanned skulls of meat-eating dinosaurs with those of plant-eating dinosaurs. Present your findings in a table.

CHAPTER 1
Study Guide

Summary

SECTION 1.1 Earth Science

Main Ideas

- There are four major areas in Earth science. Astronomy is the study of objects beyond Earth's atmosphere. Meteorology is the branch of Earth science that deals with Earth's atmosphere. The study of the materials that make up Earth and the processes that form and change these materials is known as geology. The study of Earth's oceans is called oceanography.
- Earth can be divided into four main systems. The lithosphere includes the rocks that make up the crust and rigid, upper mantle. The atmosphere is the blanket of gases that surrounds Earth. Earth's hydrosphere is the system of all of the water on the planet. The biosphere is Earth's inhabitants and their environments.
- All of Earth's systems interact. You are part of the biosphere and you live on the crust, which is part of the lithosphere. You breathe the gases in that atmosphere and depend in many ways on the water in the hydrosphere.

Vocabulary

asthenosphere (p. 8)
astronomy (p. 6)
atmosphere (p. 9)
biosphere (p. 9)
geology (p. 6)
hydrosphere (p. 8)
lithosphere (p. 8)
meteorology (p. 6)
oceanography (p. 6)

SECTION 1.2 Methods of Scientists

Main Ideas

- The order of steps in a scientific method can vary. Most scientific methods to solving a problem, however, include defining the problem, stating a hypothesis, testing the hypothesis, analyzing the results of the test, and drawing conclusions.
- Variables are factors that change in an experiment. A dependent variable can change in response to changes in the independent variable. A control is a standard for comparison.
- Basic units used in SI include the liter, the meter, the second, the kilogram, the Newton, and degrees Celsius.
- In scientific notation, a number is expressed as a multiplier and a power of 10.

Vocabulary

control (p. 12)
dependent variable (p. 12)
hypothesis (p. 11)
independent variable (p. 12)
Le Système International d'Unités (SI) (p. 14)
scientific notation (p. 16)

SECTION 1.3 Communicating in Science

Main Ideas

- Scientific information is communicated through lab reports, professional papers, tables and graphs, and models.
- A scientific theory is an explanation based on many observations during repeated experiments. A scientific theory is valid only if it is consistent with observations, makes predictions that can be tested, and is the simplest explanation of observations. A theory can be changed or modified if it is found to be incorrect.
- A scientific law is a basic fact that describes the behavior of a natural phenomenon. A scientific law can be thought of as a "rule of nature," even though the cause of the law may not be known.

Vocabulary

law (p. 19)
theory (p. 19)

CHAPTER 1

Assessment

Understanding Main Ideas

1. Which area of Earth science includes the study of ancient organisms?
a. astronomy
b. meteorology
c. paleontology
d. geology

2. Which area of Earth science includes the study of stars?
a. meteorology
b. hydrology
c. geology
d. astronomy

3. What is geology?
a. the study of the processes that form and change Earth
b. the study of Earth's oceans
c. the study of objects beyond Earth's atmosphere
d. the systematic study of weather and climate

4. Which of the following is NOT a part of Earth's lithosphere?
a. the inner core
b. the crust
c. the upper mantle
d. rocks on the surface

5. What is Earth's hydrosphere?
a. the gases in the air
b. the solid, rocky part of Earth
c. all of the water on the planet
d. the study of Earth's atmosphere

6. What are the two most common gases in the atmosphere?
a. hydrogen and oxygen
b. nitrogen and water vapor
c. oxygen and nitrogen
d. hydrogen and nitrogen

7. Which of the following scientists would most likely study Earth's past biosphere?
a. hydrologist
b. geochemist
c. meteorologist
d. paleontologist

8. What is technology, and how is it different from science?

9. List the steps involved in a scientific approach to solving a problem.

10. What is a hypothesis, and how is it different from a scientific theory?

11. What is a dependent variable in a scientific investigation? How does it differ from an independent variable?

12. List the SI units that would be used to measure the following quantities: the mass of an apple, the length of a beetle, the weight of the planet Jupiter, the volume of a medium-sized soft drink, and the volume of a cube of sugar.

13. How are area and volume alike? How do they differ?

14. Complete the table below. Once you have made the conversions, express each answer in scientific notation.

Some SI Conversions		
1 m	____ mm	____ km
1 g	____ mg	____ kg
1 cm^3	____ m^3	____ mL
3.5 km	____ m	____ cm
18.6 cm	____ km	____ m

15. Refer to *Appendix B.* What are the safety symbols for a biological hazard, an electrical hazard, an open flame, and the need to wear safety goggles?

Test-Taking Tip

PREPARING FOR A TEST As soon as you find out about an upcoming test, ask which concepts and topics will be tested. When you study for the test, make sure you cover all of the material on which you may be tested.

Assessment

16. Compare and contrast scientific theories and laws. Give an example of each.
17. Explain how variables are plotted on a line graph.
18. What is a scientific model?

Applying Main Ideas

19. Which of the safety symbols in *Appendix B* would be shown in an activity in which you were asked to test the acidity of several liquids?
20. Explain how you might test which of three paper towels is most absorbent.
21. Suppose you were testing the effects of the amount of fertilizer needed to produce tall grass. What would be your independent variable? Your dependent variable? Your control?
22. A doctor is testing a new cancer drug. She chooses 50 patients who have the particular cancer to take part in the study. She gives 25 patients the new drug and the other 25 patients a placebo, which is a substance that contains no active ingredients. What is the purpose of this second group in the doctor's study?

Thinking Critically

23. Suppose you want to find out whether doubling the amount of potassium in a soil will increase the yield of tomato plants. Describe how you would test this hypothesis. What would be your variables? What would you use as a control?
24. How might elements in Earth's hydrosphere interact with Earth's lithosphere?
25. Explain your dependence on each of Earth's four systems.
26. When air or helium is added to a balloon, the balloon expands. Suggest a model that could be used to explain why this happens.

Standardized Test Practice

1. Which of the following lists Earth's layers from the inside out?
   a. inner core, outer core, mantle, crust
   b. crust, mantle, outer core, inner core
   c. crust, inner core, outer core, mantle
   d. mantle, outer core, inner core, crust

2. A block is 2 cm wide, 5.4 cm deep, and 3.1 cm long. The density of the block is 8.5 g/cm^3. What is the mass of the block?
   a. 33.48 g　　**c.** 399.3 g
   b. 85.10 g　　**d.** 284.58 g

USING GRAPHS Use the graph below to answer questions 3 and 4.

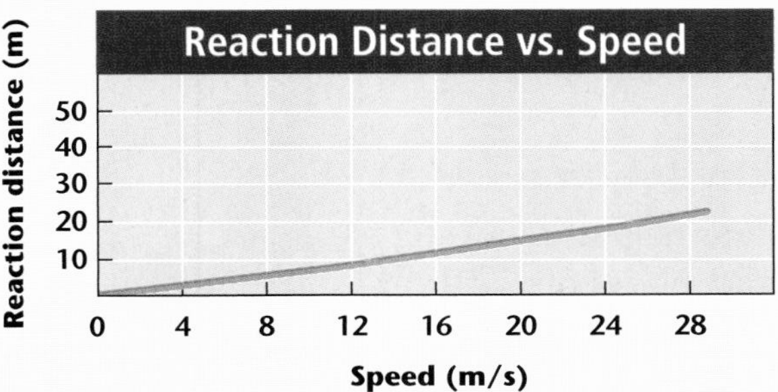

3. The distance a car travels between the time the driver decides to stop the car and the time the driver puts on the brakes is called the reaction distance. How does the reaction distance change with speed?
   a. Reaction distance decreases with speed.
   b. Reaction distance is the same as speed.
   c. Reaction distance increases with speed.
   d. You cannot tell from this graph.

4. What is the reaction distance of a driver traveling 20 m/s?
   a. 3 m　　**c.** 20 m
   b. 15 m　　**d.** 28 m

Chapter 2

Mapping Our World

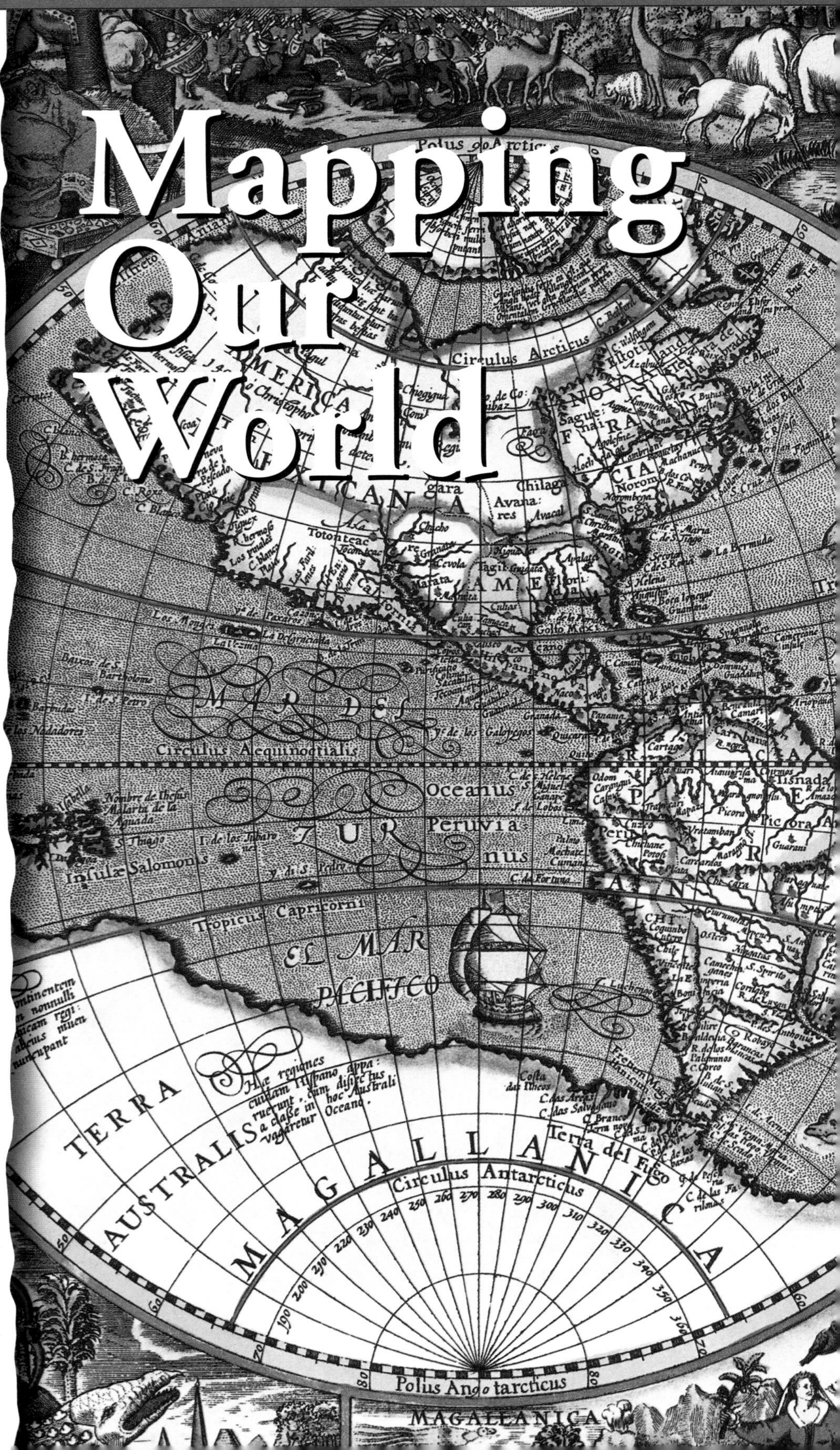

What You'll Learn

- How latitude and longitude are used to locate places on Earth.
- How maps are made, and what types of maps are best suited to particular purposes.
- What technology is used to map Earth from space.

Why It's Important

Maps help us to locate exact places on Earth. All forms of transportation, including ships, planes, cars, and trucks, rely on accurate maps for guidance.

To find out more about maps, visit the Earth Science Web Site at earthgeu.com

Discovery Lab — Make and Use a Map

Have you ever been asked for directions? If so, you know that it's important to include as much detail as possible so that the person asking for directions will not get lost. You also may have realized that it helps to draw a detailed map of the destination in question.

1. Give verbal directions from your school to your home to a classmate who does not know where you live. Include as much detail as possible in your description.
2. Use a sheet of graph paper and colored pencils to draw a map from your school to your home. Include landmarks and other details. Share this map with your classmate.
3. Have your classmate also give you a description of where his or her home is located in relation to your school. Your classmate should then draw a map to his or her home for you to examine.

Observe Which did you find more helpful, the verbal directions or the map? Explain your answer. What kind of information did you include in your map? With your classmate, discuss how you could improve your maps. What details would you add?

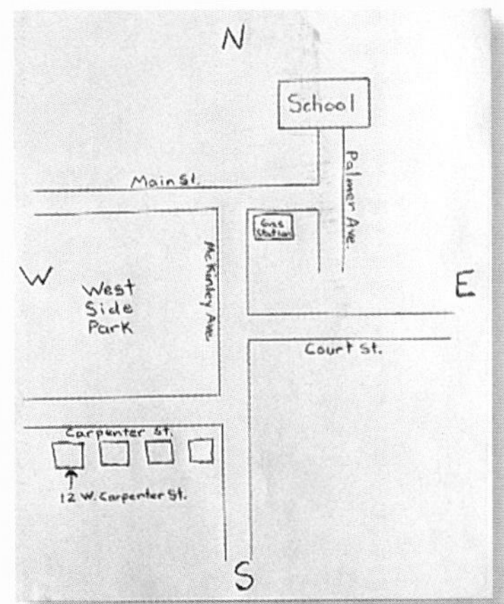

Section 2.1 Latitude and Longitude

OBJECTIVES

- **Compare** *and* **contrast** *latitude and longitude.*
- **Describe** *how time zones vary.*

VOCABULARY

cartography
equator
latitude
longitude
prime meridian
International Date Line

For thousands of years people have used maps such as the one shown at left to define borders and to find places. We still rely on maps for a variety of purposes. The science of mapmaking is called **cartography.** Cartographers use an imaginary grid of parallel lines and vertical lines to locate points on Earth exactly. In this grid, the **equator** circles Earth halfway between the north and south poles. The equator separates Earth into two equal halves called the northern hemisphere and the southern hemisphere.

LATITUDE

Lines running parallel to the equator are called lines of latitude. **Latitude** is the distance in degrees north or south of the equator. The equator, which serves as the reference point for latitude, is numbered 0° latitude. The poles are each numbered 90° latitude. Latitude is thus measured from 0° at the equator to 90° at the poles. Locations

Figure 2-1 Lines of latitude are parallel to the equator **(A).** The value in degrees of each line of latitude is determined by measuring the imaginary angle created between the equator, the center of Earth, and the line of latitude **(B).**

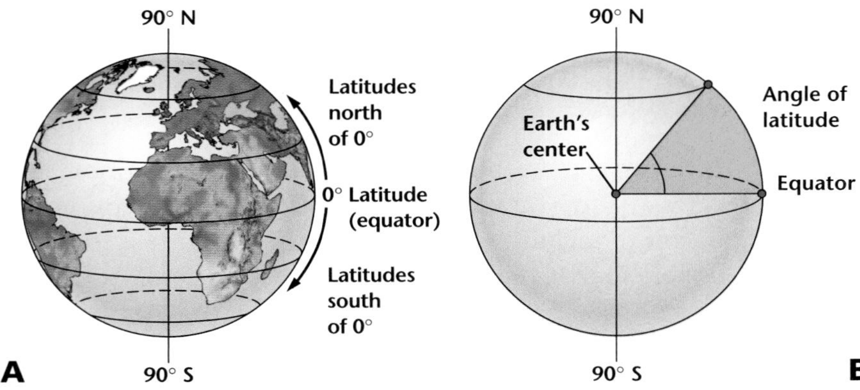

north of the equator are referred to by degrees north latitude (N). Locations south of the equator are referred to by degrees south latitude (S). For example, Syracuse, New York, is located at 43° north latitude, and Christchurch, New Zealand, is located at 43° south latitude. Lines of latitude are illustrated in ***Figure 2-1.***

Using Math

Using Numbers Your plane has flown from 30° north latitude to 42° north latitude. Approximately how many kilometers have you traveled?

Degrees of Latitude Each degree of latitude is equivalent to about 111 km on Earth's surface. How did cartographers determine this distance? Earth is a sphere, and can be divided into 360 degrees. The circumference of Earth is about 40 000 km. To find the distance of each degree of latitude, cartographers divide 40 000 km by 360°. To locate positions on Earth more precisely, cartographers break down degrees of latitude into 60 smaller units, called minutes. The symbol for a minute is ′. The actual distance on Earth's surface of each minute of latitude is 1.85 km, which is obtained by dividing 111 km by 60′. A minute of latitude can be further divided into seconds, which are represented by the symbol ″. Longitude, which is discussed next, is also divided into degrees, minutes, and seconds.

Figure 2-2 The reference line for longitude is the prime meridian **(A).** The degree value of each line of longitude is determined by measuring the imaginary angle created between the prime meridian, the center of Earth, and the line of longitude **(B).**

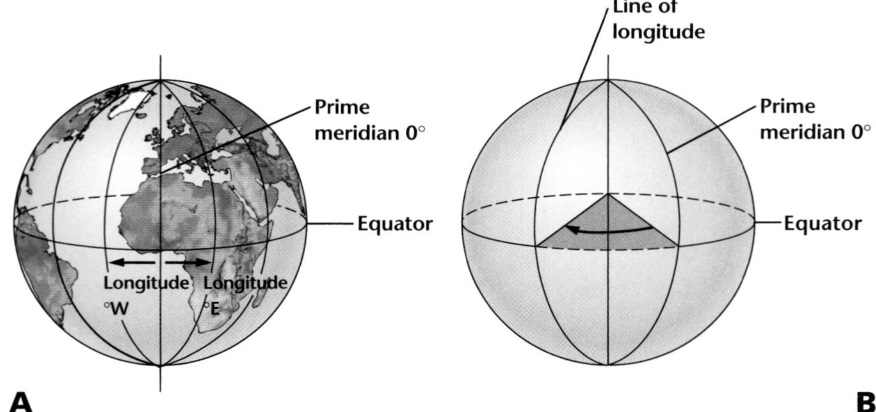

LONGITUDE

To locate positions in east and west directions, cartographers use lines of longitude, also known as meridians. As shown in ***Figure 2-2,*** **longitude** is the distance in degrees east or west of the prime meridian, which is the reference point for longitude. The **prime meridian** represents 0° longitude. In 1884, astronomers decided that the prime meridian should go through Greenwich, England, home of the Royal Naval Observatory. Points west of the prime meridian are numbered from 0° to 180° west longitude (W); points east of the prime meridian are numbered from 0° to 180° east longitude (E).

Semicircles Unlike lines of latitude, lines of longitude are not parallel. Instead, they are large semicircles that extend vertically from pole to pole. For instance, the prime meridian runs from the north pole through Greenwich, England, to the south pole. The line of longitude on the opposite side of Earth from the prime meridian is the 180° meridian. There, east lines of longitude meet west lines of longitude. This meridian is also known as the International Date Line, as you'll learn later in this section.

Degrees of Longitude Degrees of latitude cover relatively consistent distances. The distances covered by degrees of longitude, however, vary with location. Refer back to ***Figure 2-2.*** As you can see, lines of longitude converge at the poles into a point. Thus, one degree of longitude varies from about 111 km at the equator to essentially the distance covered by a point at the poles.

Locating Places with Coordinates Both latitude and longitude are needed to precisely locate positions on Earth, as you'll see in the *MiniLab* on this page. For example, it is not sufficient to say that New Orleans,

MiniLab

How can you locate places on Earth?

Determine latitude and longitude for specific places.

Procedure

1. Use a world map or globe to locate the prime meridian and the equator.
2. Take a few moments to become familiar with the grid system. Examine lines of latitude and longitude on the map or globe.

Analyze and Conclude

1. Use a map to find the latitude and longitude of the following places.
   Mount St. Helens, Washington
   Niagara Falls, New York
   Mt. Everest, Nepal
   Great Barrier Reef, Australia
2. Use the map to find the name of the places with the following coordinates.
   0°03′S, 90°30′W
   27°07′S, 109°22′W
   41°10′N, 112°30′W
   35°02′N, 111°02′W
   3°04′S, 37°22′E
3. Find the latitude and longitude of your hometown, the nearest national or state park, and your state capital.

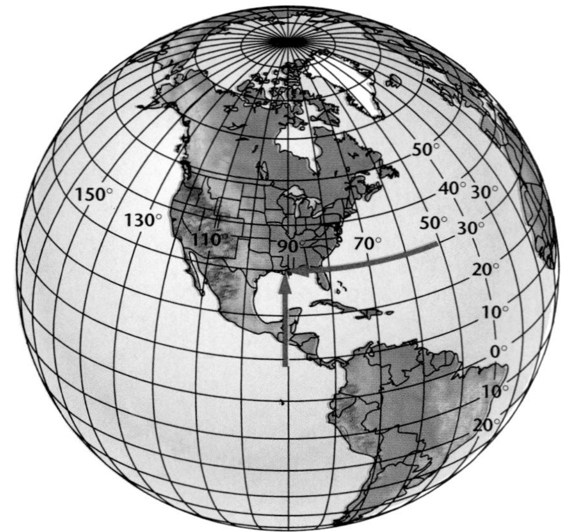

Figure 2-3 The precise location of New Orleans is 29°57′N, 90°04′W.

Louisiana, is located at 29°57′ north latitude because that measurement includes any place on Earth located along the 29°57′ line of north latitude. The same is true of the longitude of New Orleans—90°04′ west longitude could be any point along that longitude from pole to pole. To precisely locate New Orleans, we use its complete coordinates, latitude and longitude, as shown in ***Figure 2-3.*** Note that latitude comes first in reference to the coordinates of a particular location.

TIME ZONES

As ***Figure 2-4*** shows, Earth is divided into 24 time zones. Why 24? Earth takes about 24 hours to rotate once on its axis. Thus, there are 24 times zones, each representing a different hour. Because Earth is constantly spinning, time is always changing. Each time zone is 15° wide, corresponding roughly to lines of longitude. For convenience's sake, however, time zone boundaries have been adjusted in local areas. For example, if a city were split by a time zone, confusion would result. In such a situation, the time zone boundary is moved outside of the city. Large countries, however, often have several times zones. There are six different time zones in the United States, as shown in ***Figure 2-5.*** When it's 10 A.M. in Atlanta, Georgia, it's 7 A.M. in Los Angeles, California. What time is it in Chicago, Illinois?

Figure 2-4 Earth is divided into 24 time zones. Each zone represents a different hour.

Source: Time Almanac 2001

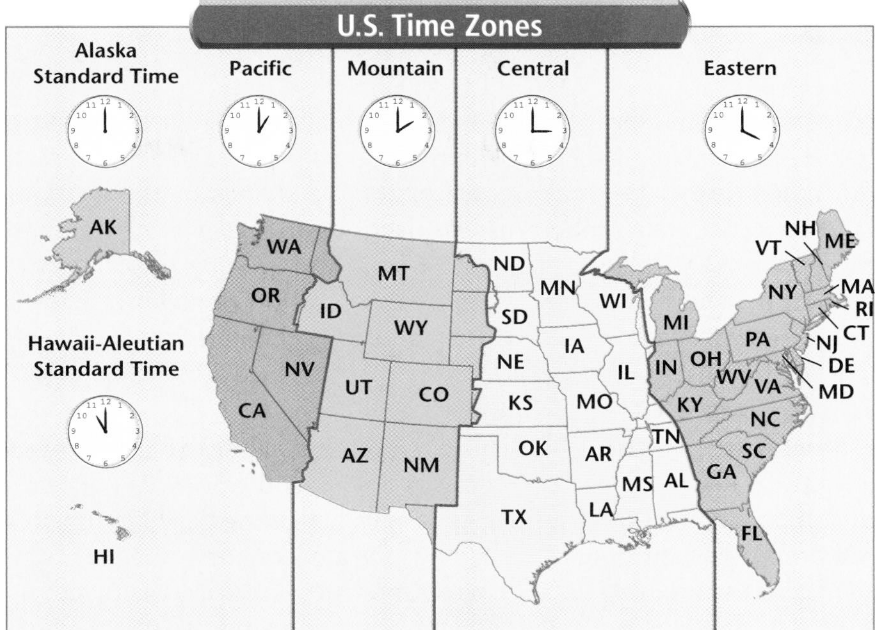

Figure 2-5 Large countries such as the United States are often split into multiple time zones. The United States has six time zones, including Alaska and Hawaii.

Calendar Dates Each day ends and the next day begins at the stroke of midnight. Every time zone experiences this transition from one day to the next, with the calendar advancing to the next day at midnight. Each time you travel through a time zone, you gain or lose time until, at some point, you gain or lose an entire day. The **International Date Line,** or 180° meridian, serves as the transition line for calendar days. If you were traveling west across the International Date Line, you would advance your calendar one day. If you were traveling east, you would move your calendar back one day.

SECTION ASSESSMENT

1. What is cartography?
2. Compare and contrast latitude and longitude. What is the reference point for lines of latitude? What is the reference point for lines of longitude?
3. What is the International Date Line? If it is 3 P.M. on Thursday, July 4, in Salt Lake City, Utah, what time and day is it in Tokyo, Japan? Use ***Figure 2-4*** for help.
4. Estimate the time difference between your home and places that are 60° east and west longitude of your home.
5. **Critical Thinking** If you were flying directly south from the north pole and reached 70° north latitude, how many more degrees of latitude would be left to pass over before you reached the south pole?

SKILL REVIEW

6. **Comparing and Contrasting** Describe how the distance of a degree of longitude varies from the equator to the poles. For more help, refer to the *Skill Handbook.*

earthgeu.com/self_check_quiz

SECTION 2.2 Types of Maps

Earth Sciences 9.d **I&E** 1.h

OBJECTIVES

- **Compare** *and* **contrast** *different map projections.*
- **Analyze** *topographic maps.*
- **Describe** *map characteristics, such as map scales and map legends.*

VOCABULARY

Mercator projection
conic projection
gnomonic projection
topographic map
contour line
contour interval
map legend
map scale

Maps are flat models of a three-dimensional object, Earth. Because Earth is curved, it's difficult to represent on a piece of paper. Thus, all flat maps distort to some degree either the shapes or the areas of landmasses. Cartographers use projections to make maps. A map projection is made by transferring points and lines on a globe's surface onto a sheet of paper. You'll use a projection of a world map in the *Science & Math* feature at the end of this chapter.

MERCATOR PROJECTIONS

A **Mercator projection** is a map that has parallel lines of latitude and longitude. Recall that lines of longitude meet at the poles. When lines of longitude are projected as being parallel on a map, landmasses near the poles are exaggerated. Thus, in a Mercator projection, the shapes of the landmasses are correct, but their areas are distorted. As shown in ***Figure 2-6,*** Greenland appears much larger than Australia. In reality, Greenland is much smaller than Australia. Because Mercator projections show the correct shapes of landmasses and also clearly indicate direction in straight lines, they are used for the navigation of planes and ships.

CONIC PROJECTIONS

A **conic projection** is made by projecting points and lines from a globe onto a cone, as shown in ***Figure 2-7.*** The cone touches the globe at a particular line of latitude. There is very little distortion in the areas or shapes of landmasses that fall along this line of latitude. Distortion is evident, however, near the top and bottom of the projection. Because conic projections have a high degree of accuracy for limited areas, they are excellent for mapping small areas. Hence, they are used to make road maps and weather maps.

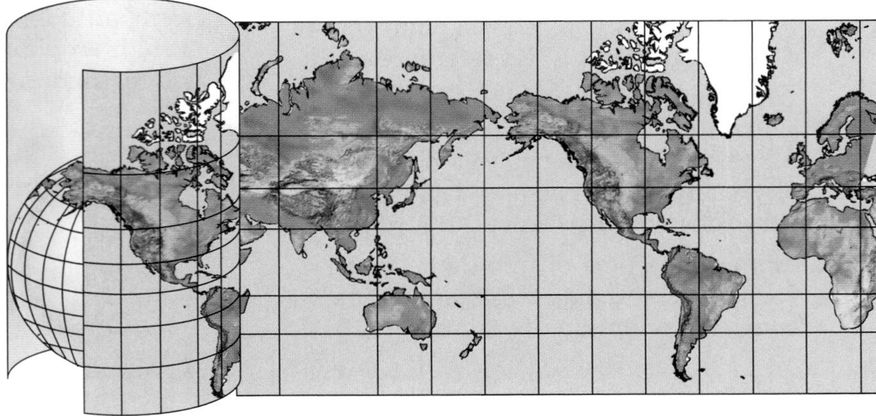

Figure 2-6 In a Mercator projection, points and lines on a globe are transferred onto a cylinder-shaped paper. Mercator projections show true direction but distort areas near the poles.

Gnomonic Projections

A **gnomonic projection** is made by projecting points and lines from a globe onto a piece of paper that touches the globe at a single point. As shown in ***Figure 2-8,*** gnomonic projections distort direction and distance between landmasses. However, they are useful in plotting long-distance trips by air and by sea. To understand why, you must understand the concept of a great circle. Great circles are imaginary lines that divide Earth into two equal halves. The equator is a great circle, as are any two lines of longitude that connect at the poles to form a complete circle. On a sphere such as Earth, the shortest distance between two points lies along a great circle. Navigators connect points on gnomonic projections to plot great-circle routes.

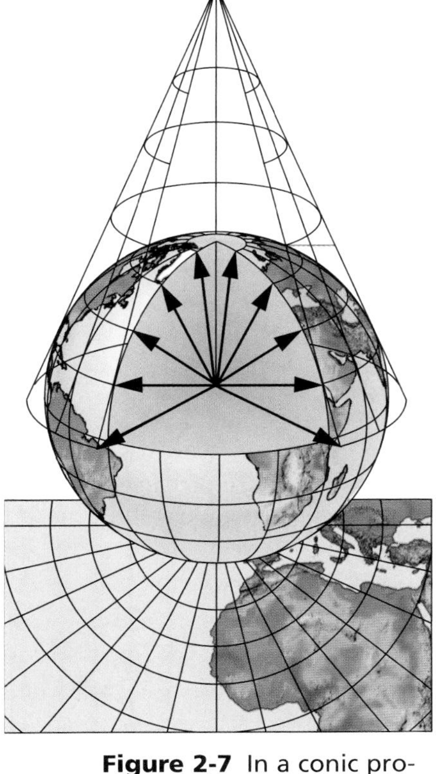

Figure 2-7 In a conic projection, points and lines on a globe are projected onto a cone-shaped paper. Along the line of latitude touched by the paper, there is little distortion.

Topographic Maps

Detailed maps showing the hills and valleys of an area are called topographic maps. **Topographic maps** show changes in elevation of Earth's surface. They also show mountains, rivers, forests, and bridges, among other features. Topographic maps use lines, symbols, and colors to represent changes in elevation and features on Earth's surface.

Contour Lines Elevation on a topographic map is represented by a contour line. A **contour line** connects points of equal elevation. Elevation refers to the distance of a location above or below sea level. Because contour lines connect points of equal elevation, they never cross. If they did, it would mean that the point where they crossed had two different elevations, which would be impossible.

 **Earth Sciences**

9.d Students know how to analyze published geologic hazard maps of California and know how to use the map's information to identify evidence of geologic events of the past and predict geologic changes in the future.

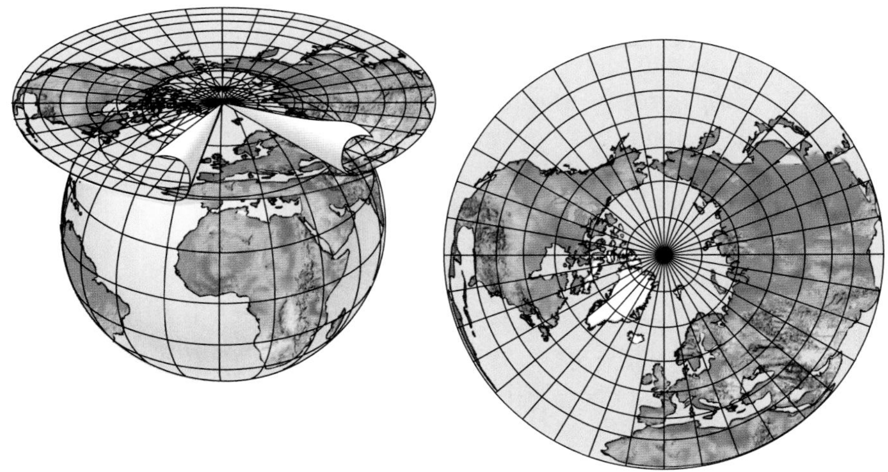

Figure 2-8 In a gnomonic projection, points and lines from a globe are projected onto paper that touches the globe at a single point.

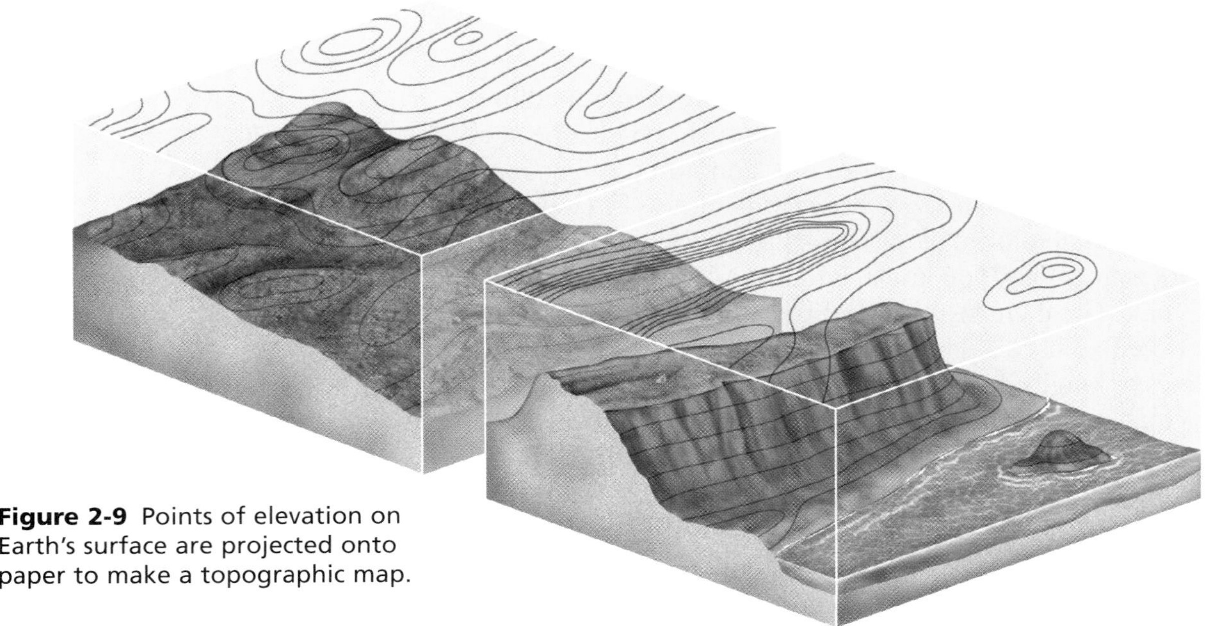

Figure 2-9 Points of elevation on Earth's surface are projected onto paper to make a topographic map.

 I&E

1.h Read and interpret topographic and geologic maps.

Contour Intervals As ***Figure 2-9*** shows, topographic maps use contour lines to show changes in elevation. The difference in elevation between two side-by-side contour lines is called the **contour interval.** The contour interval is dependent on the terrain. For mountains, the contour lines might be very close together, and the contour interval might be as great as 100 m. This would indicate that the land is quite steep because there is a large change in elevation between lines. You'll learn more about topographic maps in the *Problem-Solving Lab* on the next page and in the *Mapping GeoLab* at the end of this chapter.

Index Contours To aid in the interpretation of topographic maps, some contour lines are marked by numbers representing their elevations. These are index contours, and they are used hand-in-hand with contour intervals. If a contour interval on a map is 5 m, you can determine the elevations represented by other lines around the index contour by adding or subtracting 5 m from the elevation indicated on the index contour.

Figure 2-10 The depression contour lines shown here indicate that the center of the area has a lower elevation than the outer portion of the area.

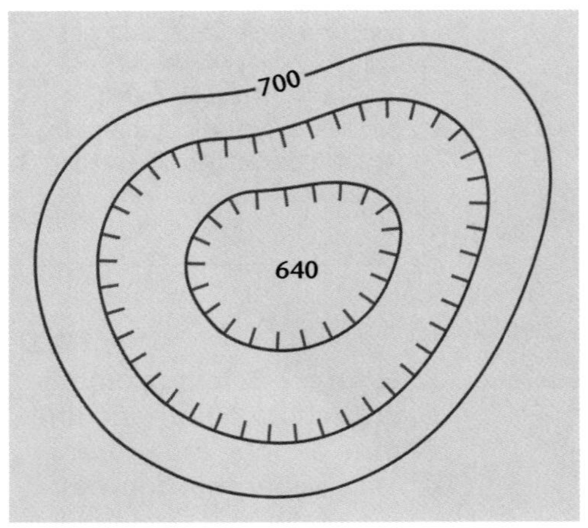

Depression Contour Lines The elevations of some features such as volcanic craters and mines are lower than that of the surrounding landscape. Depression contour lines are used to represent such features. On a map, depression contour lines have *hachures,* or short lines at right angles to the contour line, to indicate depressions. The hachures point toward lower elevations, as shown in ***Figure 2-10.***

Map Legends

Topographic maps and most other maps include both human-made and natural features that are located on Earth's surface. These features are represented by symbols, such as black dotted lines for trails, solid red lines for highways, and small black squares and rectangles for buildings. A **map legend,** such as the one shown in ***Figure 2-11,*** explains what the symbols represent. For more information about the symbols in map legends, see *Appendix D.*

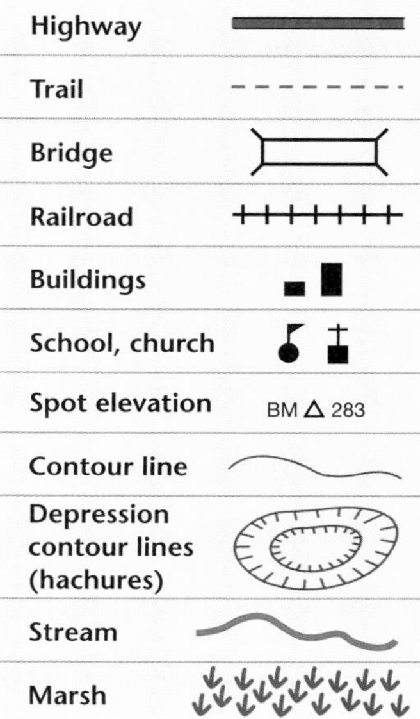

Figure 2-11 Map legends explain what the symbols on maps represent.

Map Scales

When using a map, you need to know how to measure distances. This is accomplished by using a map scale. A **map scale** is the ratio between distances on a map and actual distances on the surface of Earth. There are three types of map scales: verbal scales, graphic scales, and fractional scales. A verbal scale expresses distance as a statement, such as "One centimeter is equal to one kilometer." This means that one centimeter on the map represents one kilometer on Earth's surface. A graphic scale consists of a line that represents a certain distance, such as 5 km or 5 miles. The line is broken down into sections, with each section representing a distance on Earth's surface. For instance, a graphic scale of 5 km may be broken down into five sections, with each section representing 1 km.

Problem-Solving Lab

Calculating Gradients

Analyze changes in elevation

Gradient refers to the steepness of a slope. To measure gradient, divide the change in elevation between two points on a map by the distance between the points. Use the map to answer the questions; convert your answers to SI.

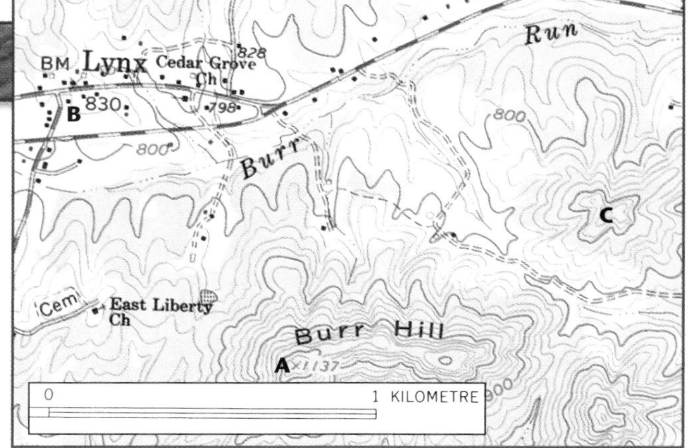

Analysis

1. Use the map scale and a ruler to determine the distance from point A to point B. Record the change in elevation between the two points.
2. If you were to hike this distance, what would be the gradient of your climb?

Thinking Critically

3. Calculate the gradient from point B to point C. Would it be more difficult to hike from point A to point B, or from point B to point C? Explain.
4. Between point A and point C, where is the steepest part of the hike? How do you know?

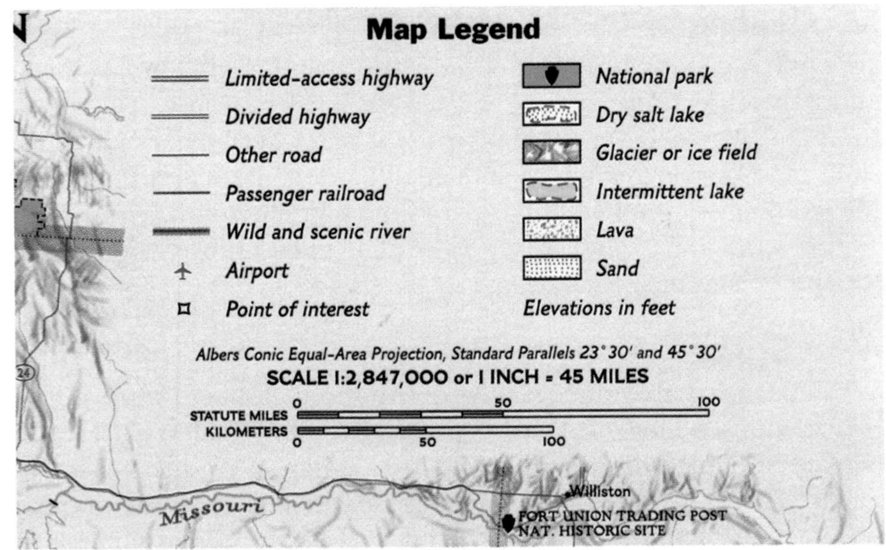

Figure 2-12 The map scale and legend shown here are from a map of the Rocky Mountain area in Montana.

A fractional scale expresses distance as a ratio, such as 1:63 500. This means that one unit on the map represents 63 500 units on Earth's surface. One centimeter on a map, for instance, would be equivalent to 63 500 cm on Earth's surface. The unit of distance may be feet or meters or any other measure of distance. However, the units on each side of the ratio must always be the same. A large ratio indicates that the map represents a large area, while a small ratio indicates that the map represents a small area. A map with a large fractional scale such as 1:100 000 would therefore show less detail than a map with a small fractional scale such as 1:1000. ***Figure 2-12*** shows the map scale and legend found on a typical map.

Section Assessment

1. Compare and contrast Mercator and gnomonic projections. What are these projections most commonly used for?
2. How is a conic projection made? Why is this type of projection best suited for mapping small areas?
3. What is a contour line? How are areas of depression represented on a topographic map?
4. A topographic map has a fractional scale of 1:80 000. The units are in centimeters. If two cities are 3 km apart, how far apart would they be on the map?
5. **Thinking Critically** The equator is the only line of latitude that is a great circle. Why?

Skill Review

6. **Interpreting Scientific Illustrations** Use *Appendix D* to draw symbols in their appropriate colors for the following features: barn, school, church, orchard, woods, perennial stream, marsh, and primary highway. For more help, refer to the *Skill Handbook.*

earthgeu.com/self_check_quiz

SECTION 2.3 Remote Sensing

I&E 1.h, 1.l

Until recently, mapmakers had to go on-site to collect the data needed to make maps. Today, advanced technology has changed the way maps are made. The process of collecting data about Earth from far above Earth's surface is called **remote sensing.** Let's examine how satellites, which use remote sensing, gather information about Earth's surface.

OBJECTIVES

- **Compare** *and* **contrast** *the different forms of radiation in the electromagnetic spectrum.*
- **Discuss** *how satellites and sonar are used to map Earth's surface and its oceans.*
- **Describe** *the Global Positioning System.*

VOCABULARY

remote sensing
electromagnetic spectrum
frequency
Landsat satellite
Topex/Poseidon *satellite*
Global Positioning System
sonar

THE ELECTROMAGNETIC SPECTRUM

Satellites, such as the one being launched in ***Figure 2-13,*** detect different wavelengths of energy reflected or emitted from Earth's surface. This energy has both electric and magnetic properties. Thus, it is referred to as electromagnetic radiation. Visible light is a form of electromagnetic radiation. Other types include gamma rays, X rays, ultraviolet waves, infrared waves, radio waves, and microwaves.

Wave Characteristics All electromagnetic waves travel at the speed of 300 000 km/s in a vacuum, a value commonly referred to as the speed of light. In addition, electromagnetic waves have distinct

Figure 2-13 *Landsat 7,* launched in 1999, is equipped to measure differences in thermal energy emitted by features on Earth's surface.

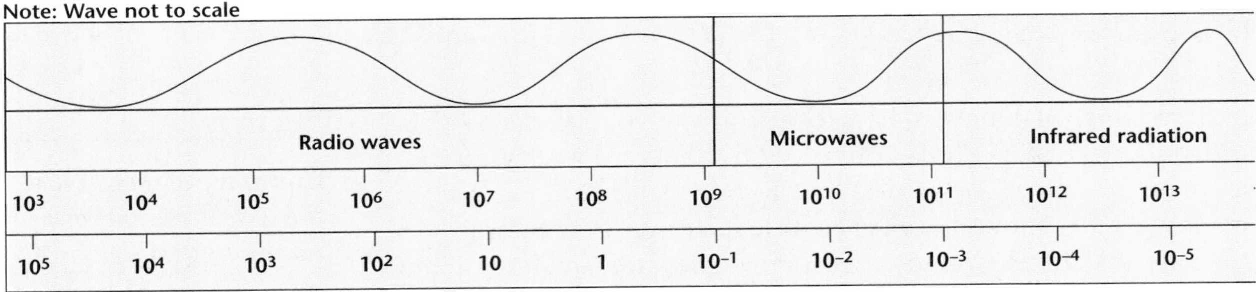

Figure 2-14 In the electromagnetic spectrum, the waves with the longest wavelengths have the lowest frequencies.

wavelengths. The arrangement of electromagnetic radiation according to wavelengths is called the **electromagnetic spectrum,** as shown in ***Figure 2-14.*** Gamma rays have wavelengths of less than 0.000 000 000 01 m, while radio waves have wavelengths of 100 000 m. An electromagnetic wave also can be described according to its **frequency,** which refers to the number of waves that pass a particular point each second. Gamma rays have the highest frequencies and radio waves have the lowest. The wavelengths, speeds, and frequencies of electromagnetic waves help determine how the energy is used by different satellites to map Earth.

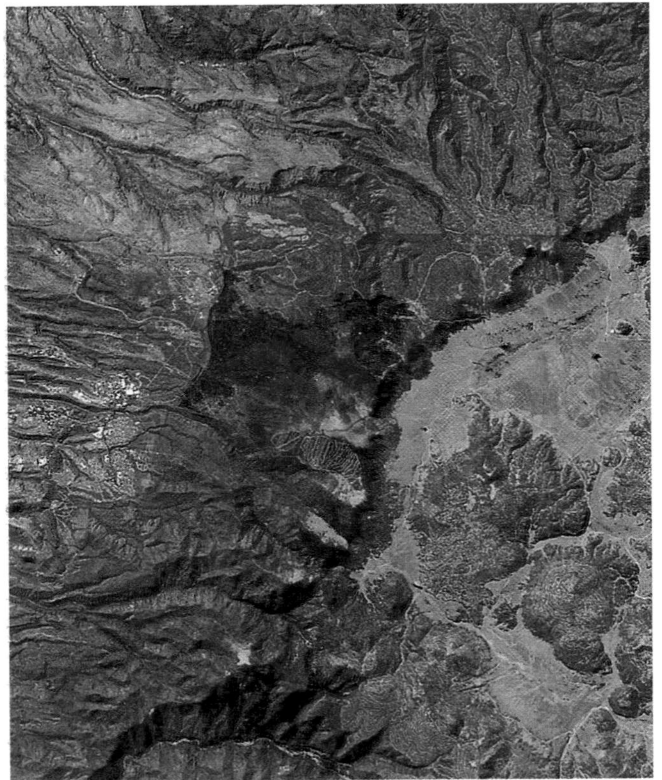

Figure 2-15 The blue area in this *Landsat 7* image shows the range of a fire that occurred in Los Alamos, New Mexico, in May 2000.

Landsat Satellites

A **Landsat satellite** receives reflected wavelengths of energy emitted by Earth's surface, including some wavelengths of visible light and infrared radiation. Features on Earth's surface, such as rivers and forests, radiate warmth at slightly different frequencies. Thus, these features show up as different colors in images such as the one in ***Figure 2-15.*** To obtain such images, each Landsat satellite is equipped with a moving mirror that scans Earth's surface. This mirror has rows of detectors that measure the intensity of energy received from Earth. This information is then converted by computers into digital images that show landforms in great detail. *Landsat 7,* launched in 1999, maps 185 km at a time and scans the entire surface of the planet in 16 days. Landsat data also are used to study the movements of Earth's plates, rivers, earthquakes, and pollution.

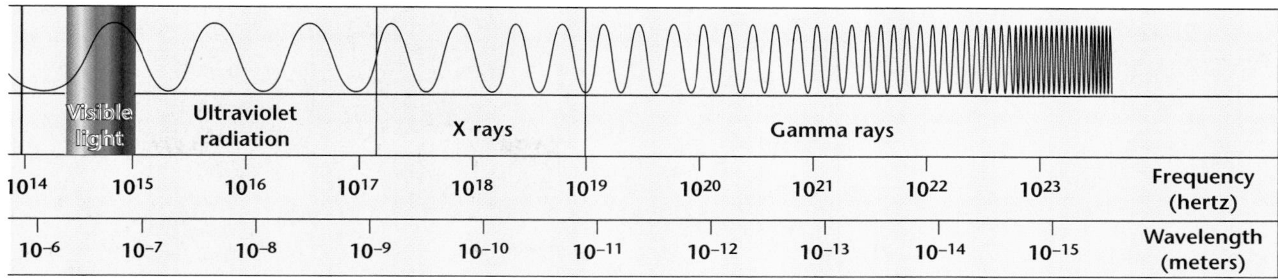

Topex/Poseidon Satellite

Other satellites, such as the ***Topex/Poseidon* satellite,** shown in ***Figure 2-16,*** use radar to map features on the ocean floor. *Topex* stands for "topography experiment." Radar uses high-frequency signals that are transmitted from the satellite to the surface of the ocean. A receiving device then picks up the returning echo as it is reflected off the water. The distance to the water's surface is calculated using the known speed of light and the time it takes for the signal to be reflected. Variations in time indicate the presence of certain features on the ocean floor. For instance, ocean water bulges over seafloor mountains and forms depressions over seafloor valleys. These changes are reflected in satellite-to-sea measurements. Based on these data, computers create maps of ocean-floor features. The *Topex/Poseidon* satellite also has been used to study tidal changes and global ocean currents.

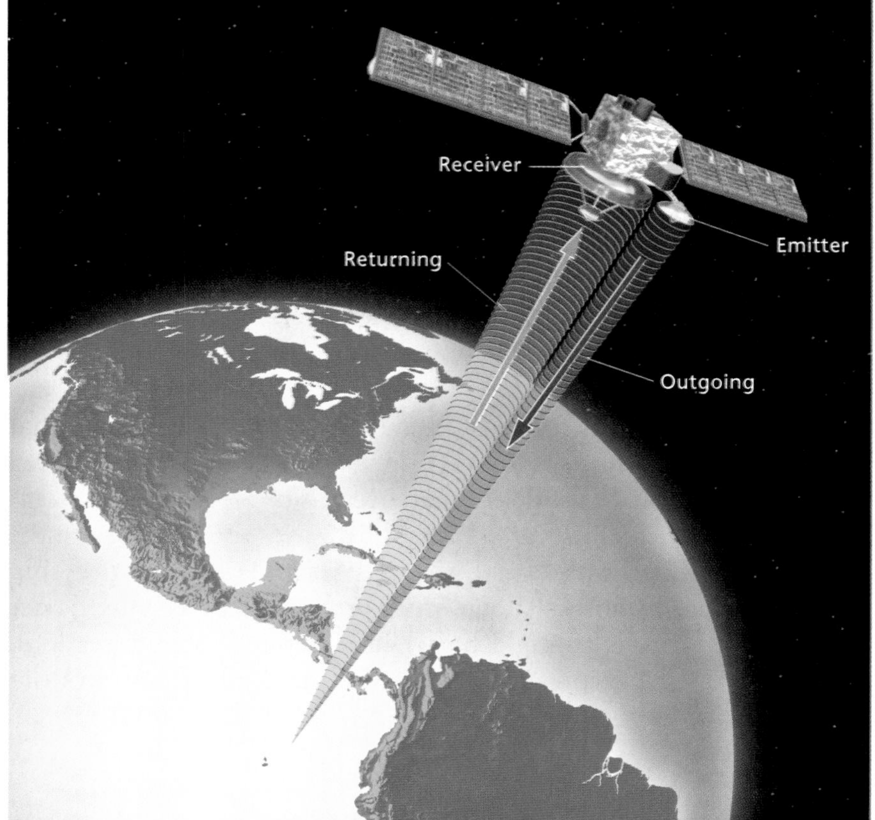

Figure 2-16 In the *Topex/Poseidon* satellite, an emitter sends an outgoing signal to the surface of the ocean. A receiver times the returning signal. The distance to the ocean's surface is calculated using the known speed of light and the return time.

Figure 2-17 This hiker is using a hand-held, GPS receiver.

To learn more about mapping, go to the **National Geographic Expedition** on page 864.

The Global Positioning System

The **Global Positioning System** (GPS) is a radio-navigation system of at least 24 satellites that allows its users to determine their exact position on Earth. Each satellite orbits Earth and transmits high-frequency microwaves that contain information about the satellite's position and the time of transmission. The orbits of the satellites are arranged so that signals from several satellites can be picked up at any given moment by a GPS user equipped with a hand-held receiver, as shown in ***Figure 2-17.*** The receiver calculates the user's precise latitude and longitude by processing the signals emitted by multiple satellites. The satellites also can relay information about elevation, direction, and speed. GPS technology is used extensively for navigation by airplanes and ships. However, it is also used to detect earthquakes, create maps, and track wildlife. Lately, it has become increasingly popular among hikers, backpackers, and other travelers.

Sea Beam

Sea Beam technology is similar to the *Topex/Poseidon* satellite in that it is used to map the ocean floor. However, Sea Beam is located on a ship rather than on a satellite. To map ocean-floor features, Sea Beam relies on **sonar,** which is the use of sound waves to detect and measure objects underwater. First, a sound wave is sent from a ship toward the ocean floor, as shown in ***Figure 2-18.*** A receiving device then picks up the returning echo when it bounces off the seafloor. Computers on the ship calculate the distance to the ocean bottom

Figure 2-18 In a ship equipped with Sea Beam, a sound wave is sent to the ocean floor. The wave bounces off the seafloor and its returning echo is recorded by a receiver on the ship. The distance to the ocean floor is then calculated using the known speed of sound in water and the return time of the sound wave.

using the speed of sound in water and the time it takes for the sound to be reflected. A ship equipped with Sea Beam has more than a dozen sonar devices aimed at different parts of the sea. Sea Beam technology is used by fishing fleets, deep-sea drilling operations, and scientists such as oceanographers, volcanologists, and archaeologists.

Section Assessment

1. What is the electromagnetic spectrum? Sequence the forms of electromagnetic radiation from longest wavelength to shortest wavelength.
2. How do Landsat satellites collect and analyze data to map Earth's surface?
3. What features are mapped by the *Topex/Poseidon* satellite? Describe the mapping process.
4. Describe the Global Positioning System.
5. **Thinking Critically** Explain why electromagnetic waves with short wavelengths have higher frequencies than electromagnetic waves with long wavelengths.

Skill Review

6. **Concept Mapping** Use the following words and phrases to complete a concept map about remote sensing. For more help, refer to the *Skill Handbook.*

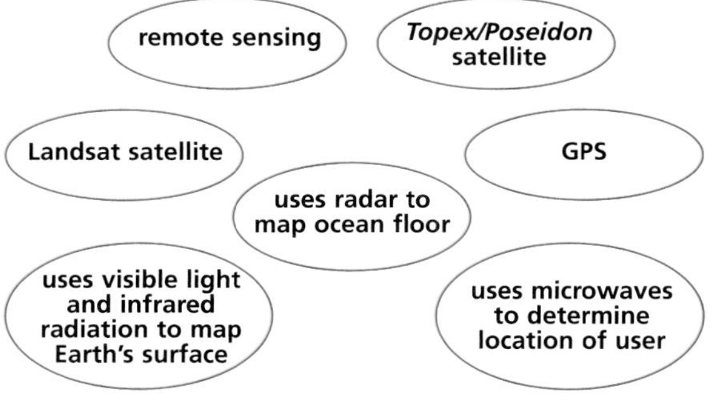

Mapping GeoLab

Using a Topographic Map

Topographic maps show two-dimensional representations of Earth's surface. With these maps, you can determine how steep a hill is, what direction streams flow, and where mines, wetlands, and other features are located.

I&E

1.h Read and interpret topographic and geologic maps.

Preparation

Problem

How can you use a topographic map to interpret information about an area?

Materials

ruler
pencil
string

Procedure

1. Use the map to answer the following questions. Be sure to check the map's scale.
2. Use the string to measure distances between two points that are not in a straight line. Lay the string along the curves, and then measure the distance by laying the string along the ruler.
3. Remember that elevations on United States Geological Survey maps are given in feet.

Analyze

1. What is the contour interval?
2. Calculate the stream gradient of Big Wildhorse Creek from the Gravel Pit in section 21 to where the creek crosses the road in section 34.
3. What is the highest elevation of the jeep trail? If you followed the jeep trail from the highest point to where it intersects an unimproved road, what would be your change in elevation?
4. If you started at the bench mark (BM) on the jeep trail and hiked along the trail and the road to the Gravel Pit in section 21, how far would you have hiked?
5. What is the straight line distance between the two points in question 4? What is the change in elevation?

Conclude & Apply

1. Does Big Wildhorse Creek flow all year round? Explain your answer.
2. What is the shortest distance along roads from the Gravel Pit in section 21 to the secondary highway?
3. Draw a profile of the land surface from the bench mark in section 22 to the Gravel Pit in section 33.

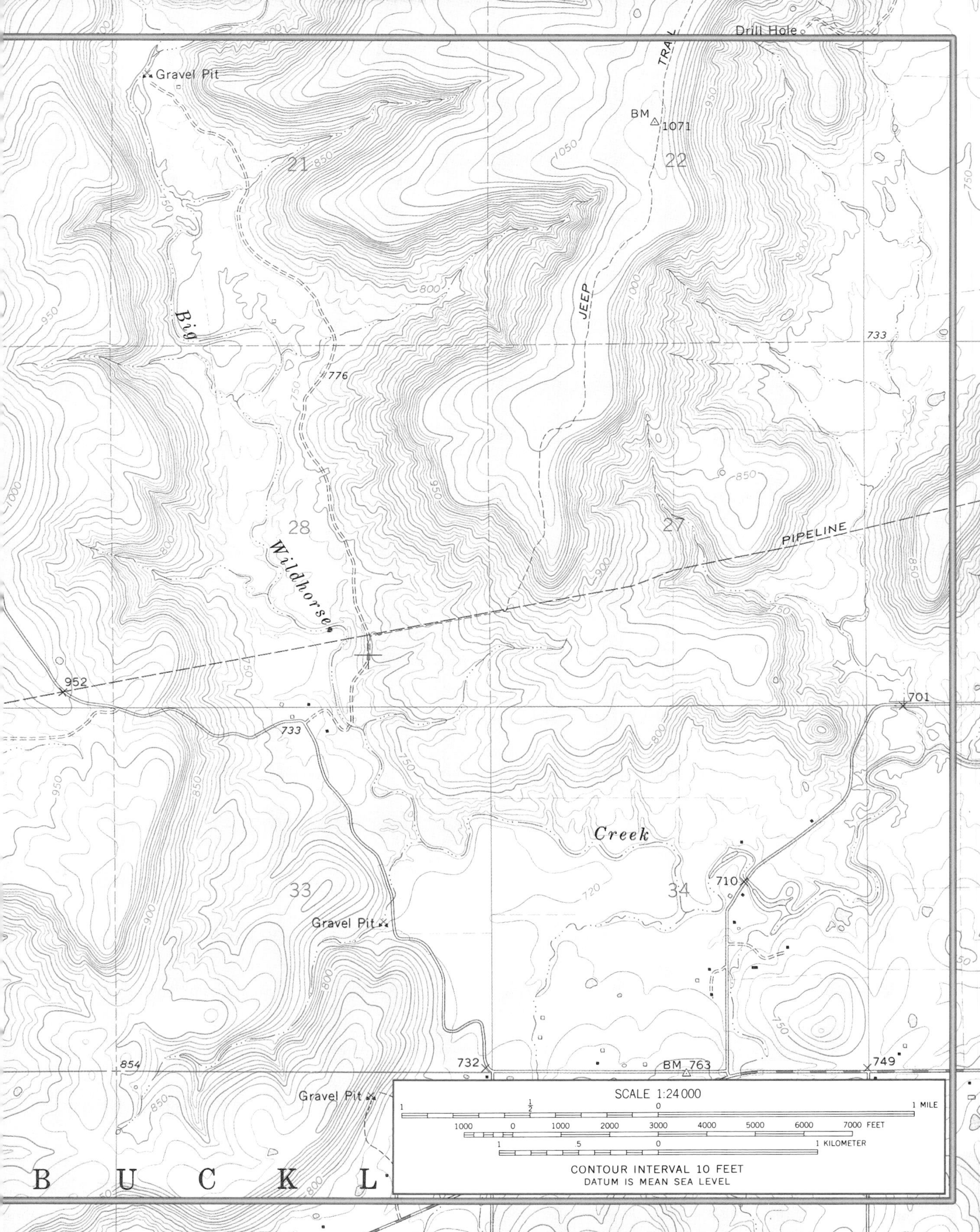

Drill Hole
TRAIL
Gravel Pit
BM 1071
21
850
1050
22
950
750
800
JEEP
1000
Big
733
776
850
1000
28
27
Wildhorse
PIPELINE
900
952
701
733
Creek
33
34
710
720
Gravel Pit
854
732
BM 763
749
Gravel Pit
B U C K L
SCALE 1:24 000
1 MILE
1000 0 1000 2000 3000 4000 5000 6000 7000 FEET
1 KILOMETER
CONTOUR INTERVAL 10 FEET
DATUM IS MEAN SEA LEVEL

Science & Math

I&E 1.l

Thriving in the Arctic

Polar bear

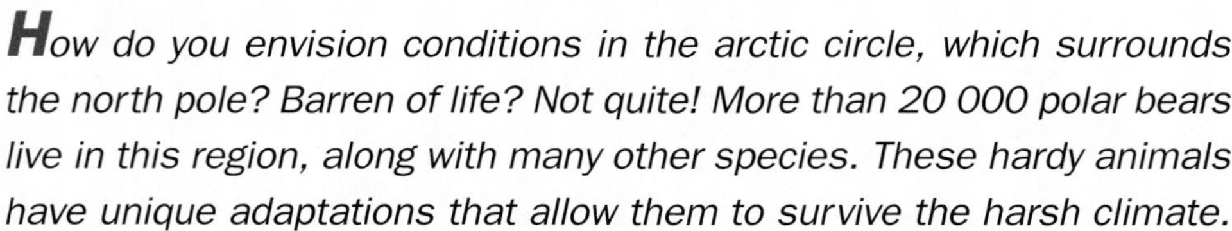
***H**ow do you envision conditions in the arctic circle, which surrounds the north pole? Barren of life? Not quite! More than 20 000 polar bears live in this region, along with many other species. These hardy animals have unique adaptations that allow them to survive the harsh climate.*

The Ring of Life

The borders of five countries—Russia, Norway, Greenland, Canada, and the United States—meet in a rough U-shape around the Arctic Ocean. The vast majority of this region is covered with ice some 2 m thick. In a climate where average winter temperatures hover around –35°C, survival is tenuous. The southern boundaries of this region, however, teem with life. Polar bears, walruses, beluga whales, fish, birds, and seals make the arctic circle their home.

Animal Adaptations

Polar bears in particular thrive where the ocean meets the shoreline, an area of constant freezing and thawing. Supremely adapted to this environment, they have long necks that help them keep their heads above water and huge forepaws that act as paddles. Light-colored fur provides camouflage to help them hunt, and an outer coat of hollow hairs makes the half-ton bears fairly buoyant in the water.

Traveling Bears

Polar bears can swim for an average of approximately 96.5 km without stopping for a rest. They have been tracked on land traveling 30 km a day for several days in a row. A polar bear's home range—the area in which it hunts, mates, and cares for its young—may be around 259 000 km^2. The home ranges of polar bears vary in size from 50 000 km^2 to as much as 350 000 km^2. Polar bear ranges are much greater than those of other mammals because the sea ice on which they live changes from season to season and year to year.

Procedure

1. Calculate the range of a polar bear that travels for six hours a day for seven days at a speed of 5.5 km/h.
2. Calculate how far a polar bear could swim in six hours at a speed of 10 km/h.
3. Convert your answers for questions 1 and 2 into U.S. units.

Challenge

1. Assume that polar bears do equal amounts of swimming and walking, and that they travel an average of four hours a day. Use your calculations and a world map or globe to determine whether a polar bear could travel around the circumference of Greenland in a year.

To find out more about polar bears, visit the Earth Science Web Site at earthgeu.com

CHAPTER 2

Study Guide

Summary

Section 2.1

Latitude and Longitude

Main Ideas

- Cartographers use a grid system to locate exact positions on Earth. Lines of latitude refer to distances north and south of the equator. Lines of longitude refer to distances east and west of the prime meridian.
- Earth is divided into 24 time zones. Each zone represents a different hour. The International Date Line, or 180° meridian, is the transition line for calendar days. The calendar advances to the next day in each time zone at midnight.

Vocabulary

cartography (p. 27)
equator (p. 27)
International Date Line (p. 31)
latitude (p. 27)
longitude (p. 29)
prime meridian (p. 29)

Section 2.2

Types of Maps

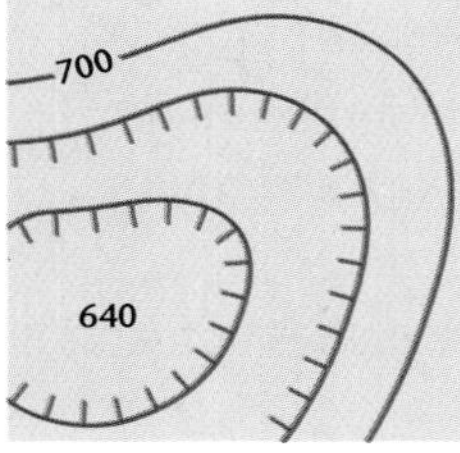

Main Ideas

- Maps are flat models of Earth's surface. All maps contain some sort of distortion in the shapes or areas of landmasses.
- Maps are made by transferring points and lines on a globe onto paper. Mercator projections and gnomonic projections are commonly used for navigation by ships and planes. Conic projections are best suited for mapping small areas.
- Topographic maps show changes in elevation of Earth's surface. Contour lines connect points of equal elevation. A map legend explains the symbols on a map. A map scale shows the relationship between distances on a map and actual distances on Earth.

Vocabulary

conic projection (p. 32)
contour interval (p. 34)
contour line (p. 33)
gnomonic projection (p. 33)
map legend (p. 35)
map scale (p. 35)
Mercator projection (p. 32)
topographic map (p. 33)

Section 2.3

Remote Sensing

Main Ideas

- The process of gathering data about Earth from far above the planet is called remote sensing. The electromagnetic spectrum shows the arrangement of electromagnetic radiation, which is often used by remote-sensing devices to map Earth.
- Landsat satellites use visible light and infrared radiation to map Earth's surface. The *Topex/Poseidon* satellite uses radar to map features on the ocean floor.
- The Global Positioning System is a satellite-based navigation system that allows a user to pinpoint his or her exact location on Earth.

Vocabulary

electromagnetic spectrum (p. 38)
frequency (p. 38)
Global Positioning System (p. 40)
Landsat satellite (p. 38)
remote sensing (p. 37)
sonar (p. 40)
Topex/Poseidon satellite (p. 39)

earthgeu.com/vocabulary_puzzlemaker

CHAPTER 2
Assessment

Understanding Main Ideas

1. What feature on a map shows the ratio of map distance to actual distance on Earth?
   a. map legend
   b. map scale
   c. map symbol
   d. contour line

2. What type of map shows changes in elevation on Earth's surface?
   a. Mercator projection
   b. gnomonic projection
   c. topographic map
   d. GPS

3. Which of the following is NOT true of lines of longitude?
   a. They are semicircles.
   b. They measure distances east and west of the prime meridian.
   c. They run from pole to pole.
   d. They are parallel lines.

4. What technology is used to map seafloor features?
   a. conic projections
   b. *Topex/Poseidon* satellite
   c. the Global Positioning System
   d. Landsat satellite

5. What is the main disadvantage of a Mercator projection?
   a. It distorts areas near the equator.
   b. It distorts the shapes of landmasses.
   c. It distorts areas near the poles.
   d. It does not show true direction.

6. What is the reference point for lines of latitude?
   a. the equator
   b. the prime meridian
   c. the International Date Line
   d. the 180° meridian

7. What is the distance of one degree of latitude?
   a. 11 km
   b. 111 km
   c. 40 000 km
   d. 1.85 km

8. Some areas have lower elevations than the surrounding land. Which of the following represents these areas on a topographic map?
   a. index contours
   b. contour intervals
   c. depression contour lines
   d. map legends

9. What is the Global Positioning System? Describe how it might be used by a hiker lost in the woods.

10. Compare and contrast a verbal scale, a graphic scale, and a fractional scale.

11. Would a topographic map of the Great Plains have a large or small contour interval? Explain.

12. Why can't two contour lines overlap?

13. How could you leave home on Monday to go sailing, sail for an hour on Sunday, and return home on Monday?

14. What is a map legend? Give examples of features found in a map legend.

Applying Main Ideas

15. What type of map would best show true direction?

16. Do closely spaced contour lines indicate a steep slope or a gradual slope? Explain.

Test-Taking Tip

WHERE HAVE I HEARD THAT BEFORE?
If you don't know the definition of a word, you can usually work through the question by thinking about how you've heard the word used before. Think about the context in which the word was used. This will narrow its meaning.

Assessment

17. Approximately how many kilometers separate Orlando, Florida, at 29° north latitude and Cleveland, Ohio, at 41° north latitude?
18. If it is 10 A.M. in Syracuse, New York, at 76° west longitude, what time is it in Athens, Greece, at 24° east longitude?

Use the map to answer questions 19–21.

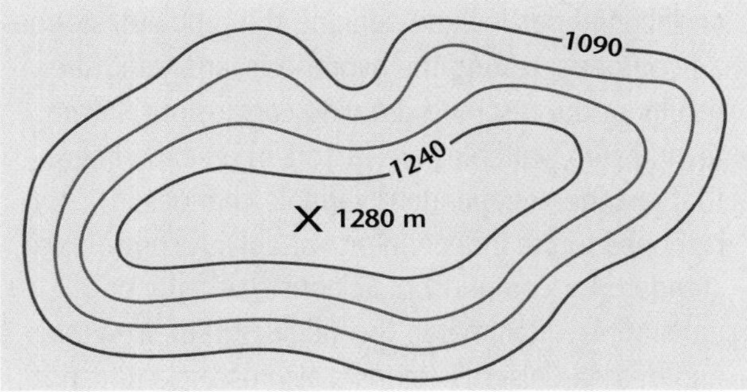

19. Copy the map shown here. What is its contour interval?
20. Based on the contour interval, label the elevations of all the contour lines.
21. Does the map represent a flat or hilly terrain? Explain.

Thinking Critically

22. Would a person flying from Virginia to California have to set his or her watch backward or forward? Explain.
23. If you wanted to study detailed features of a volcano on the island of Hawaii, would you use a map with a scale of 1:150 or 1:150 000? Why?
24. Based on what you have learned in this chapter, infer how astronomers map objects in the night sky.
25. Which direction would you travel along Earth's surface so that your longitude would not change? Explain your answer.

Standardized Test Practice

1. What is the reference point for lines of longitude?
   a. the equator
   b. the prime meridian
   c. the International Date Line
   d. the 360th meridian
2. Which would be most useful if you were lost in the Sahara desert?
   a. Landsat satellite
   b. *Topex/Poseidon* satellite
   c. Global Positioning System
   d. topographic map of Africa

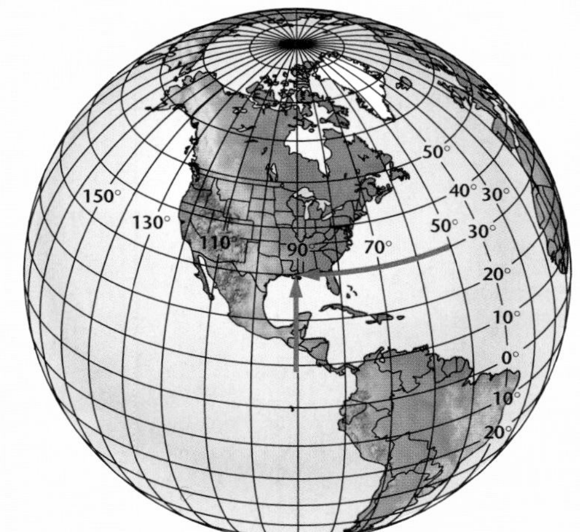

USING MAPS Use the map to answer questions 3 and 4.

3. Roughly how many degrees of longitude does the United States cover?
   a. 10° **b.** 20° **c.** 30° **d.** 40°
4. Roughly how many degrees of latitude does the United States cover?
   a. 10° **b.** 15° **c.** 20° **d.** 25°

UNIT 1

GeoDigest

For a **preview** of Earth science, study this GeoDigest before you read the chapters. After you have studied the unit, you can use the GeoDigest to **review.**

Earth Science

The Nature of Science

Earth Science Earth science is divided into four areas of specialization. Astronomy studies objects beyond Earth's atmosphere. Meteorology studies the atmosphere. Geology studies the materials of Earth and the processes that form them. Oceanography studies the oceans. The application of scientific discoveries is technology. Earth is made up of interacting systems. The lithosphere includes the rocks that make up the crust and upper mantle. The atmosphere is the gas layer that surrounds Earth. The hydrosphere is Earth's water. The biosphere is all of the life and habitats on Earth.

Methods and Communication Most scientific methods include defining the problem, stating a hypothesis, testing the hypothesis, analyzing the results of the test, and drawing conclusions. In the testing step, variables are factors in an experiment that change. A dependent variable changes in response to the independent variable. A control is a standard for comparison. Scientists use standard units of SI—liter, meter, second, kilogram, Newton, and degree Celsius. Scientists also use scientific notation, in which a number is expressed as a multiplier and a power of ten. Scientists communicate in reports and papers, and use tables, graphs, and models. A scientific theory is an explanation based on observations from repeated experiments. It is valid only if it is consistent with observations, leads to testable predictions, and is the simplest explanation. Scientific theories are changed if they are found to be incorrect. A scientific law is a basic fact that describes the behavior of a natural phenomenon.

FOCUS ON CAREERS

Science Teacher

Science teachers often provide a student's first exposure to science and may spark a life-long interest in a particular topic. High school science teachers must have at least a bachelor's degree, often from a five-year program, with an emphasis in their area of interest, such as Earth science.

GeoDigest

Mapping Our World

Latitude, Longitude, and Maps

Cartographers use a grid system of latitude and longitude to locate exact positions on Earth. Latitude refers to distances north and south of the equator. Longitude refers to distances east and west of the prime meridian. Earth is divided into 24 time zones, with each zone representing a different hour. The International Date Line, or the 180° meridian, is the transition line for calendar days. Maps are flat models of Earth's round surface, thus all maps contain some sort of distortion. Maps are made by transferring points and lines on a globe onto paper. A map legend explains map symbols. A map scale shows how distances on a map and actual distances on Earth are related. Mercator and gnomonic projections are used for aircraft and ship navigation. Conic projections are suited to mapping small areas. Topographic maps show changes in elevation of Earth's surface. Gathering data about Earth from far above is called remote sensing. Examples of remote-sensing devices include *Landsat* satellites, the *Topex-Poseidon* satellite, and the Global Positioning System. These different types of technology can be used to map Earth's surface and oceans, and to locate places on Earth.

Vital Statistics

Earth's Land Area

Continent	Area in km²
Asia, Middle East	44 579 000
Africa	30 065 000
North America	24 256 000
South America, Central America, and Caribbean	17 819 000
Antarctica	13 209 000
Europe	9 938 000
Australia and Oceania	7 687 000
Earth Total	**148 429 000**

ASSESSMENT

Understanding Main Ideas

1. Which of the following is an area of specialization in Earth science?
   a. hydrosphere **c.** meteorology
   b. Mercator projection **d.** remote sensing

2. What happens if a scientific theory is found to be incorrect?
   a. It is published.
   b. It is changed.
   c. It becomes a scientific law.
   d. It becomes a control.

3. Which type of map shows changes in elevation of Earth's surface?
   a. conic projection **c.** topographic map
   b. gnomonic projection **d.** latitude map

4. What does a map legend contain?
   a. contour lines
   b. longitude lines
   c. latitude lines
   d. the symbols used in a map

5. What is the application of science called?
   a. technology **c.** scientific law
   b. latitude **d.** theory

Thinking Critically

1. Describe the steps commonly used in scientific methods.
2. Why isn't a conic projection used to navigate a ship or aircraft?

Unit 2

Composition of Earth

In the setting Sun, the giant monolith called Uluru by the Aborigines of the Northern Territory of Australia glows a fiery red. Uluru, also known as Ayers Rock, is 2.5 km long, 1.6 km wide, and oval in shape. This rock is a conglomerate, a type of sedimentary rock composed of large, rounded chunks of rocks and minerals. Uluru is a solitary rock that rises nearly 350 m above the surrounding desert plain. How did such a huge sedimentary rock form in a desert? In this unit, you will explore the geologic forces that formed Uluru and also shape rocks and minerals.

Unit Contents

Go to the **National Geographic Expedition** on page 870 to learn more about topics that are connected to this unit.

Ayers Rock, Australia

Chapter 3

Matter and Atomic Structure

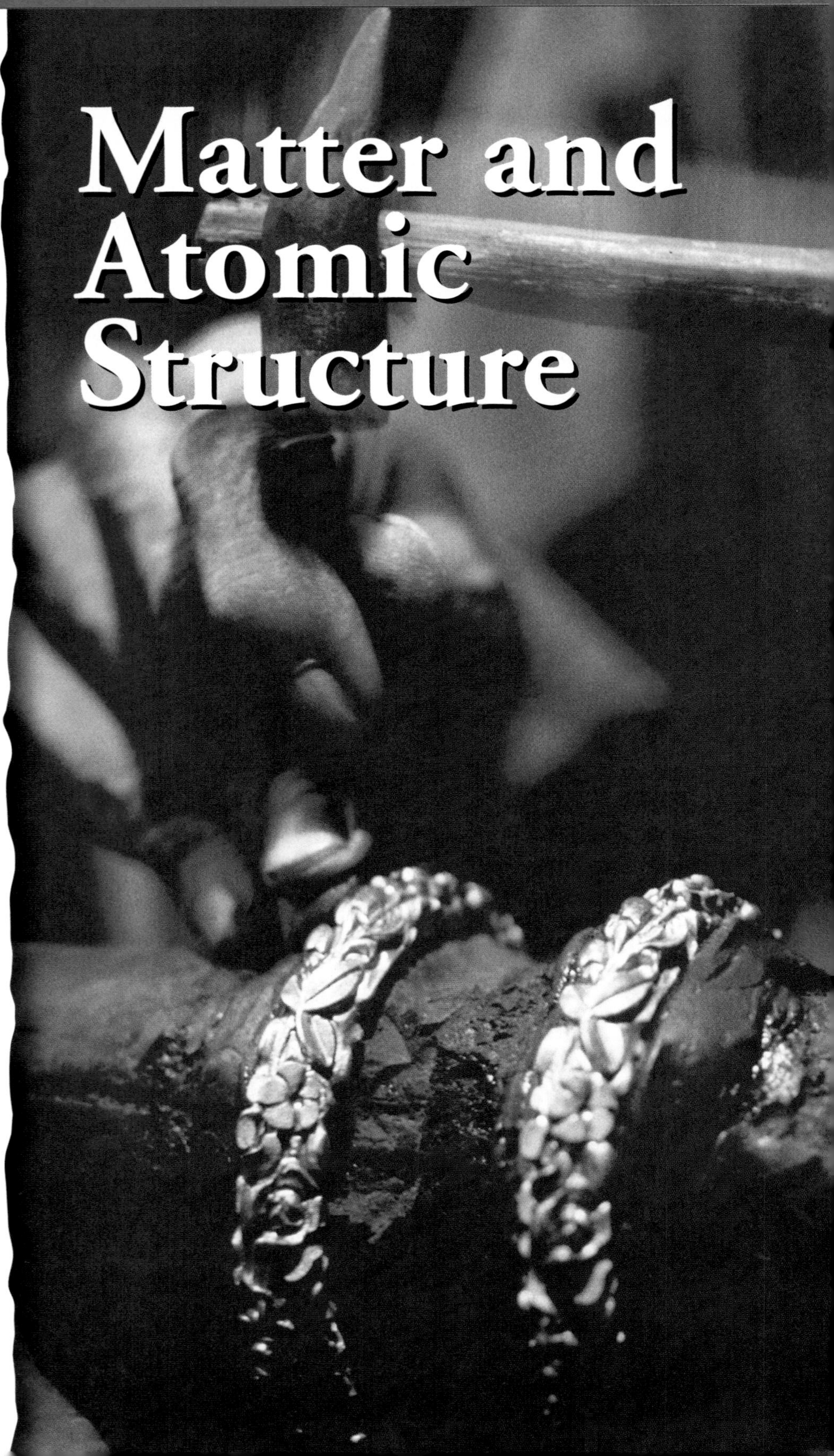

What You'll Learn

- What the basic structures are of the elements that make up Earth.
- How atoms interact to form compounds.
- What states of matter occur on Earth.

Why It's Important

Earth consists of many elements and compounds. Understanding how rocks and minerals form requires a basic knowledge of chemistry, the science of matter.

To find out more about matter and atomic structure, visit the Earth Science Web Site at earthgeu.com

Discovery Lab Fortified Cereals

Advertisements for breakfast cereals often indicate that they are fortified with substances that increase their nutritional value. In this activity, you will identify one substance that is added to cereals.

1. Tape a small strong magnet to the eraser end of a pencil.
2. Pour a sample of dry, fortified cereal into a small plastic bag. Smooth the bag as you close it.
3. Using a rolling pin, thoroughly crush the cereal in the plastic bag.
4. Pour the crushed cereal into a 250-mL glass beaker. Add 150 mL of tap water to the beaker.
5. Using the pencil-magnet stirrer, stir the cereal/water mixture for 10 minutes, stirring slowly for the last minute.
6. Remove the stirrer from the mixture and examine the magnet end of the stirrer with a hand lens.

Observe In your science journal, describe what you see on the end of the pencil stirrer. Study the cereal box to determine what the substance on the magnet might be.

Section 3.1 *What are elements?*

OBJECTIVES

- **Describe** *the particles within atoms and the structure of atoms.*
- **Relate** *the energy levels of atoms to the chemical properties of elements.*
- **Define** *the concept of isotopes.*

VOCABULARY

element
atom
nucleus
proton
neutron
atomic number
mass number
electron
energy level
valence electron
isotope
atomic mass
radioactivity

When a jewelry designer plans a new piece, he or she often chooses to work in gold. Gold is soft and easy to work with. It can be molded, hammered, sculpted, or made into wire. But whatever shape the jewelry takes, the gold remains the same. Gold is a type of matter. The jewelry designer also is made up of matter. The physical world that surrounds you and all living things are composed of matter. What exactly is matter? Matter is anything that has volume and mass. On Earth, matter usually can be found as a solid, liquid, or gas.

ELEMENTS

All matter—that is, everything on Earth and beyond—is made of substances called elements. An **element** is a substance that cannot be broken down into simpler substances by physical or chemical means. For example, gold is still gold whether it has been melted, pulled into wire, hammered into a thin sheet, or divided into small particles.

Ninety-two elements occur naturally on Earth and in the stars. Other elements have been produced in laboratory experiments. Each element is identified by a one-, two-, or three-letter abbreviation known as a chemical symbol. For example, the symbol H represents

Table 3-1 Chemical Symbols of Some Elements

Element	Symbol	Element	Symbol	Element	Symbol
Hydrogen	H	Helium	He	Lithium	Li
Beryllium	Be	Boron	B	Carbon	C
Nitrogen	N	Oxygen	O	Fluorine	F
Neon	Ne	Sodium	Na	Magnesium	Mg
Aluminum	Al	Silicon	Si	Phosphorus	P
Sulfur	S	Chlorine	Cl	Argon	Ar
Potassium	K	Calcium	Ca	Gold	Au
Silver	Ag	Mercury	Hg	Copper	Cu

Sulfur

Mercury

Copper

the element hydrogen, C represents carbon, and O represents oxygen. Elements known in ancient times, such as gold and mercury, have symbols that reflect their Latin origins. For example, gold is identified by the symbol Au, for its Latin name, *aurum.* The chemical symbols of some elements are shown in ***Table 3-1.***

Topic: Elements
To find out more about the elements, visit the Earth Science Web Site at earthgeu.com

Activity: Choose three elements. Design a chart that describes how each element was discovered, and its common uses.

ELEMENTS ARE MADE OF ATOMS

Each element has distinct characteristics. You've already learned some of the characteristics of the element gold. Aluminum has different characteristics from gold, but both aluminum and gold are elements that are made up of atoms. An **atom** is the smallest particle of an element that has all of the characteristics of that element.

All atoms consist of even smaller particles: protons, neutrons, and electrons. The center of an atom is called the nucleus (*plural,* nuclei). A **nucleus** of an atom is made up of protons and neutrons. A **proton** (p^+) is a tiny particle that has mass and a positive electrical charge. A **neutron** (n^0) is a particle with about the same mass as a proton, but it is electrically neutral; that is, it has no electrical charge. All atomic nuclei have a positive charge because they are composed of protons with positive electrical charges and neutrons that have no electrical charges.

The number of protons and neutrons in different atoms varies widely. The lightest of all atoms is the hydrogen atom, which has only one proton in its nucleus. The heaviest naturally occurring atoms are those of uranium. Uranium-238 has 92 protons and 146 neutrons in its nucleus. The number of protons in an atom's nucleus is its **atomic number.** The combined number of protons and neutrons is its **mass number.** For example, the atomic number of uranium is 92 and its mass number is 238 (92 + 146). The atomic numbers and mass numbers of 14 naturally occurring elements are shown in ***Table 3-2.*** You can explore the elements that you can find in your classroom in the *MiniLab* on this page. For a complete list of the elements arranged according to their chemical properties, see the Periodic Table of the Elements in *Appendix G* on page 917.

Surrounding the nucleus of an atom are smaller particles called electrons. An **electron** (e^-) has little mass, but it has a negative electrical charge that is exactly the same magnitude as the positive charge of a proton. An atom has an equal number of protons and electrons; thus, the electrical charge of an electron cancels the positive charge of a proton to produce an atom that has no overall charge.

Have you ever let a jawbreaker melt in your mouth? It may have started out as a large, red sphere, but as it melted, you may have observed layers of yellow, green, blue, and so on, until you reached the candy center. If you think of a jawbreaker's candy center as the nucleus of an atom, you can think of the various colored layers as the energy levels where electrons can be found. An **energy level** represents the area in an atom where an electron is most likely to be found.

MiniLab

Identifying Elements

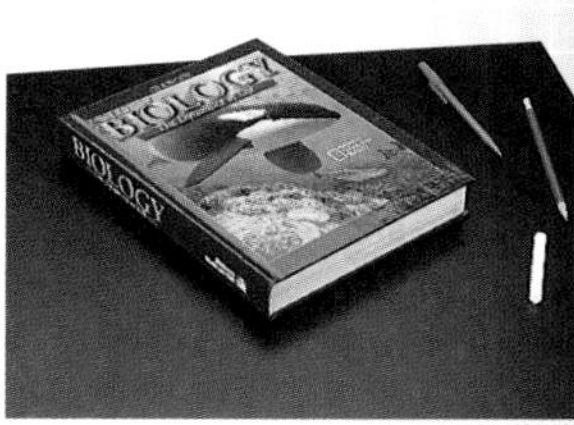

Describe Most substances on Earth occur in the form of chemical compounds. Around your classroom, there are numerous objects or substances that consist mostly of a single element.

Procedure

1. Name three of these objects and the three different elements of which they are made.
2. List the atomic numbers of these elements and describe some of their properties.

Article	Element	Atomic Number	Properties

Analyze and Conclude

1. Matter can be solid, liquid, or gaseous. Give one example of a solid, liquid, and gaseous object or substance.
2. How does a liquid differ from a solid? How does a gas differ from a liquid?

Table 3-2 Atomic Structure of 14 Elements

Element	Symbol	Atomic Number	Mass Number	Element	Symbol	Atomic Number	Mass Number
Hydrogen	H	1	1	Calcium	Ca	20	40
Helium	He	2	4	Iron	Fe	26	56
Oxygen	O	8	16	Sulfur	S	16	32
Carbon	C	6	12	Sodium	Na	11	23
Neon	Ne	10	20	Chlorine	Cl	17	35
Nitrogen	N	7	14	Potassium	K	19	39
Magnesium	Mg	12	24	Argon	Ar	18	40

Figure 3-1 shows models of several atoms with their energy levels represented as cloudlike regions. Notice that the volume of an atom is mostly empty space. Because the electrons have little mass, the *mass* of an atom depends mostly upon the number of protons and neutrons in its nucleus. However, the *size* of an atom depends upon the number and arrangement of its electrons. You can explore how electrons can be used to produce images of objects in the *Science & Technology* feature at the end of this chapter.

Figure 3-1 Electrons move around the nucleus of an atom.

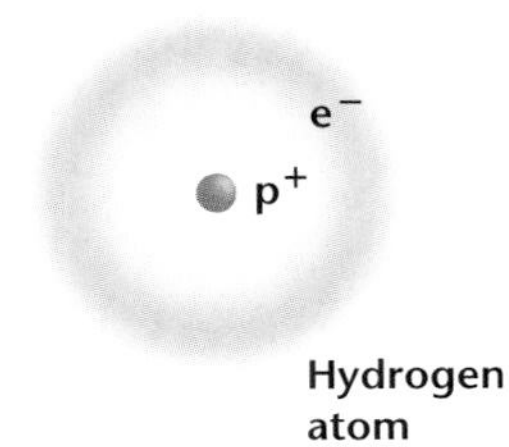

A Hydrogen has just one proton in its nucleus and one electron in its innermost energy level.

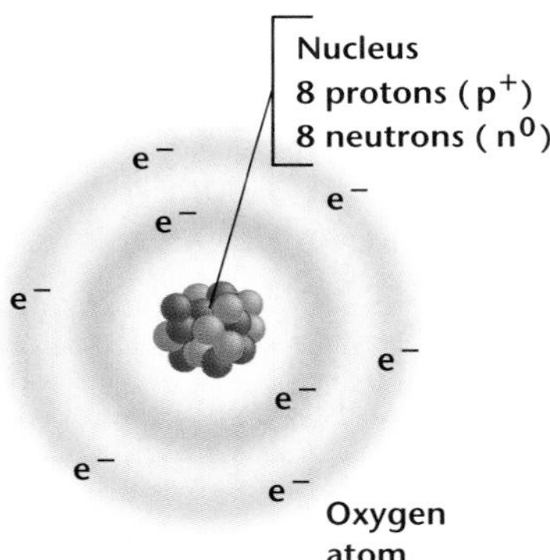

B Oxygen has eight protons and eight neutrons in its nucleus. Two electrons fill the innermost energy level; six electrons are found in the second energy level.

Electrons In Energy Levels

Study ***Figure 3-1.*** Note that electrons are distributed over one or more energy levels in a predictable pattern. Each energy level can hold only a limited number of electrons. For example, the smallest, innermost energy level can hold only two electrons, as illustrated by the oxygen atom shown in ***Figure 3-1B.*** The second energy level is larger, and it can hold up to eight electrons. The third energy level is larger still; it can hold up to 18 electrons. The fourth energy level can hold up to 32 electrons. Depending upon the element, an atom may have electrons in as many as seven energy levels surrounding its nucleus.

Electrons tend to occupy the lowest available energy level. For example, the aluminum (Al) atom in ***Figure 3-1C*** has 13 protons in its nucleus and 13 electrons in its energy levels. The first energy level in an aluminum atom is filled by two electrons. The second energy level is also filled, by eight electrons. The third energy level has only three electrons, so it is not filled.

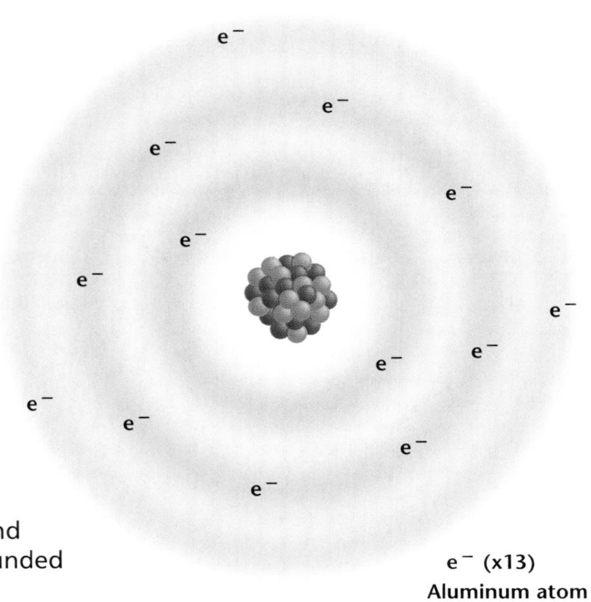

C An aluminum atom has a nucleus that is composed of 13 protons and 13 neutrons. The nucleus is surrounded by 13 electrons.

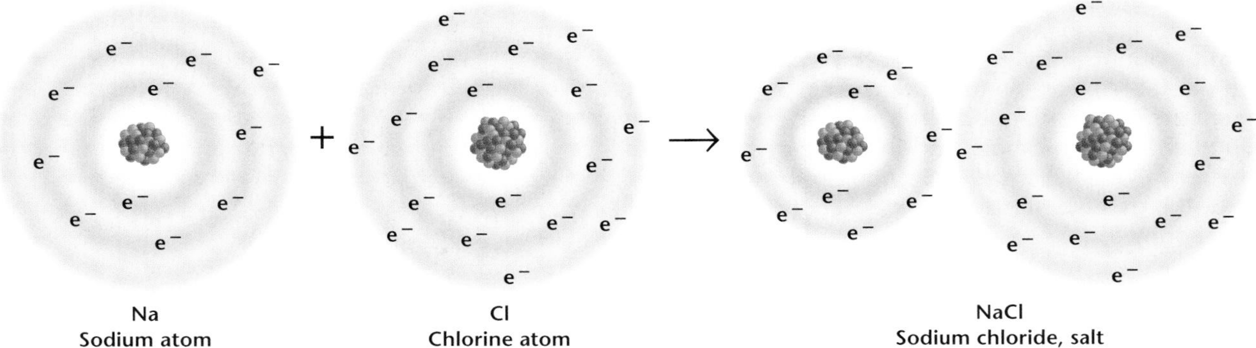

Figure 3-2 A sodium (Na) atom, with one valence electron, combines with a chlorine (Cl) atom, with seven valence electrons, to form a common substance, table salt (NaCl). The new substance forms when the sodium atom loses its valence electron to the chlorine atom.

The electrons in the outermost energy level determine the chemical behavior of the different elements. These outermost electrons are called **valence electrons.** Elements with the same number of valence electrons have similar chemical properties. For example, a sodium (Na) atom, with the atomic number 11, and a potassium (K) atom, with the atomic number 19, both have just one valence electron. Thus, both sodium and potassium are highly reactive metals, which means that they combine easily with other elements. ***Figure 3-2*** illustrates how the common substance table salt is formed when sodium combines with the element chlorine.

Elements such as helium (He), neon (Ne), and argon (Ar) are inert, which means that they do not easily combine with other elements. This is because they have full outermost energy levels. For example, a neon atom has ten electrons in its energy levels. The innermost energy level is filled with two electrons, and the second energy level, which is the outermost energy level, also is filled, with eight electrons. With a filled outermost energy level, neon is unlikely to combine chemically with other elements.

Isotopes

You have learned that all atoms of an element have the same number of protons. However, the number of neutrons in the nuclei of an element's atoms can vary. For example, all chlorine atoms have 17 protons in their nuclei, but they may have either 18 or 20 neutrons. This means that there are two types of chlorine atoms: one with a mass number of 35 (17 protons + 18 neutrons) and one with a mass number of 37 (17 protons + 20 neutrons). When atoms of the same

Figure 3-3 The Periodic Table of the Elements in *Appendix G* provides information about every element. Each block on the table gives the name of the element, its chemical symbol, its atomic number, its atomic mass, and the state of matter in which it is usually found. In this text, the gaseous state is represented by a balloon.

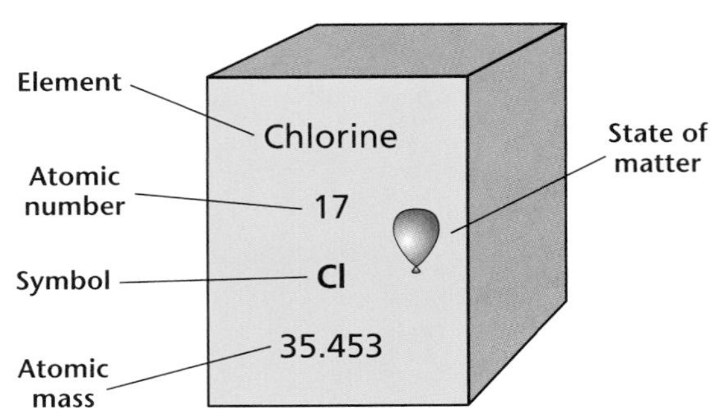

element have different mass numbers, they are known as **isotopes** of that element. The element chlorine has two isotopes: chlorine-35 and chlorine-37. Naturally occurring chlorine is a mixture of these two isotopes. Many elements are mixtures of isotopes. Because the number of electrons in an atom equals the number of protons, isotopes of an element have the same chemical properties.

If many elements are mixtures of isotopes, how do scientists know how many neutrons are found in an element's atoms? Scientists have measured the mass of atoms of elements and found an average atomic mass for each element. The **atomic mass** of an element is the average of the mass numbers of the isotopes of an element. For example, in ***Figure 3-3,*** note that the atomic mass of chlorine is 35.453. This number is the average of the mass numbers of the naturally occurring isotopes of chlorine-35 and chlorine-37.

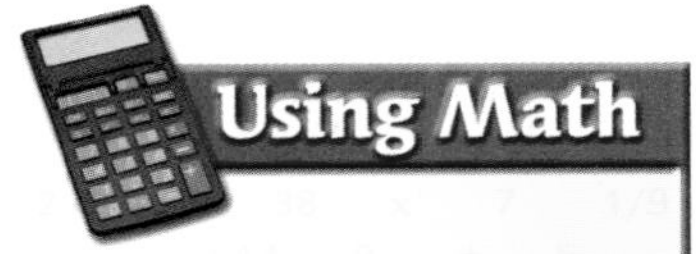

Using Numbers As the radioactive isotope uranium-235 (atomic number 92) decays, it emits two protons and two neutrons. How many protons and neutrons are left in the nucleus after these neutrons and protons have been ejected? What is the atom's new atomic number? What is the name of this element?

The nuclei of some isotopes are unstable and release radiation. **Radioactivity** is the spontaneous process through which unstable nuclei emit radiation. During radioactive decay, a nucleus can lose protons and neutrons, change a proton to a neutron, or change a neutron to a proton. Because the number of protons in a nucleus identifies an element, decay changes the identity of an element. For example, the isotope uranium-238 decays over time into lead-206, so uranium originally present in a rock gradually and predictably is replaced by lead. By measuring the amount of uranium and lead in rocks, scientists can calculate their age. You will find out more about the radioactive dating of rocks in Chapter 21.

What Elements are Most Abundant?

Astronomers have identified the two most abundant elements in the universe as hydrogen and helium. All other elements account for less than one percent of all atoms in the universe, as you can see in ***Figure 3-4A.*** Analyses of the composition of rocks and minerals on Earth indicate that the percentages of elements in Earth's crust

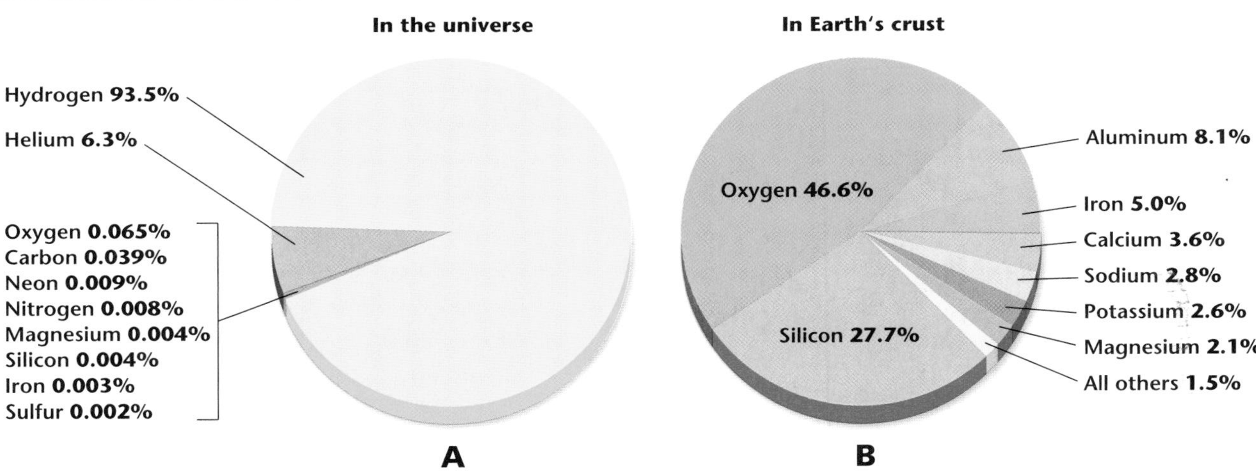

Figure 3-4 The most abundant elements in the universe are hydrogen and helium **(A).** The most abundant elements in Earth's crust are oxygen, silicon, and aluminum **(B).**

differ from the percentages in the universe. The circle graph in ***Figure 3-4B*** shows the percentages of elements in Earth's crust. Note that 98.5 percent of Earth's crust is made up of only eight elements, and that two elements, oxygen and silicon, account for almost 75 percent of the crust. This means that most of the rocks and minerals on Earth contain oxygen and silicon. You might wonder how rocks can contain oxygen, as you usually think of oxygen as a gas in the atmosphere. Oxygen is a reactive element that is mostly found in chemical combinations with other elements. In the next section, you'll learn how elements combine to form compounds.

Section Assessment

1. Name the three particles that make up an atom of an element and discuss their relative masses.
2. The elements magnesium and calcium have similar chemical properties. Explain why.
3. The atomic mass for the element carbon (C) is 12.011. Explain how this number indicates that carbon is a mixture of isotopes. What is the mass number of the most common, naturally occurring isotope of carbon?
4. **Thinking Critically** Oxygen is often found in chemical combinations with other elements, such as magnesium. Using the concepts of valence electrons and energy levels, explain why oxygen might combine easily with magnesium.

Skill Review

5. **Applying Concepts** The element copper (Cu) has 29 electrons. Draw a diagram of an atom of copper that shows the placement of its electrons in the correct energy levels and the number of protons it has. For more help, refer to the *Skill Handbook.*

SECTION 3.2

How Atoms Combine

I&E 1.d

OBJECTIVES

- **Describe** *the chemical bonds that unite atoms to form compounds.*
- **Relate** *the nature of chemical bonds that hold compounds together to the physical structures of compounds.*
- **Distinguish** *among different types of mixtures and solutions.*

VOCABULARY

compound
chemical bond
covalent bond
molecule
ion
ionic bond
chemical reaction
solution
acid
base

Can you identify the materials in ***Figure 3-5?*** The greenish gas in the flask is the element chlorine, which is poisonous. The solid, silvery metal is the element sodium, which also is toxic. Yet these two elements combine chemically to form the third material in the photograph: table salt. How can two toxic elements combine to form a material that you sprinkle on your french fries?

COMPOUNDS

Table salt is a common substance. However, table salt is not an element, but a compound. A **compound** is a substance that is composed of atoms of two or more different elements that are chemically combined. Water is another example of a compound, because it is composed of two elements, hydrogen and oxygen. Most compounds have totally different properties from the elements of which they are composed. For example, both oxygen and hydrogen are gases at room temperature, but in combination they form water, a liquid.

For most elements, an atom is chemically stable when its outermost energy level is full. We know this is true because the most stable elements are the gases helium, neon, and argon. A state of stability is achieved by other elements through **chemical bonds,** which are the forces that hold the elements together in a compound.

Covalent Bonds

One way in which atoms fill their outermost energy levels is by sharing electrons. For example, two hydrogen atoms can combine with each other by sharing electrons. Individual atoms of hydrogen each have just one electron. Each atom becomes more stable when it shares its electron with another hydrogen atom so that each atom has two electrons in its outermost energy level. How do these two atoms stay together? The nucleus of each atom has one proton with a positive charge, and the two positively charged protons attract the two negatively charged electrons. This attraction of two atoms for a shared pair of electrons that holds the atoms together is called a **covalent bond.**

Figure 3-5 Two elements, sodium and chlorine, combine chemically to form table salt, a compound also known as halite.

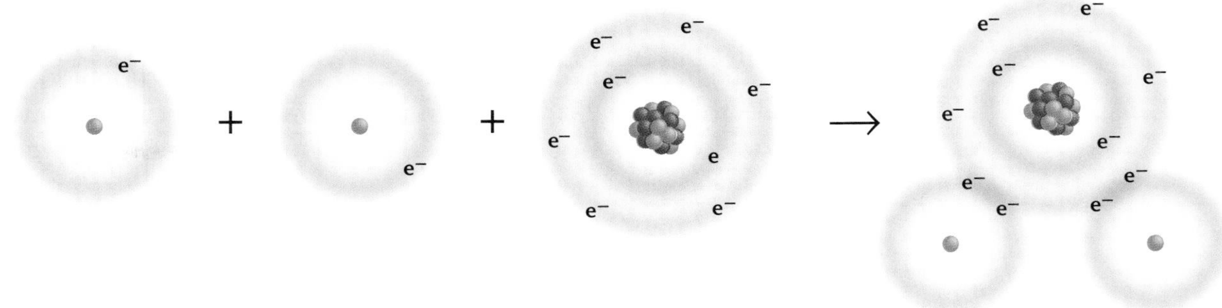

Figure 3-6 A water molecule forms when two hydrogen atoms share electrons with one oxygen atom by covalent bonding.

A **molecule** is composed of two or more atoms held together by covalent bonds. Molecules have no overall electrical charge because the total number of electrons equals the total number of protons. Molecules are represented in chemistry by chemical formulas that include the symbol for each element followed by a subscript number that stands for the number of atoms of that element in the molecule. If there is only one atom of an element, no subscript number follows the symbol. The chemical formula for hydrogen gas is written H_2 because two atoms of hydrogen make up one molecule of hydrogen gas. Water is an example of a compound whose atoms are held together by covalent bonds, as illustrated in ***Figure 3-6.*** The chemical formula for a water molecule is H_2O because, in this molecule, two atoms of hydrogen are combined with one atom of oxygen. A compound comprised of molecules is called a molecular compound.

Polar Molecules Although water molecules are held together by covalent bonds, the atoms do not share electrons equally. As shown in ***Figure 3-7,*** the shared electrons in a water molecule are attracted

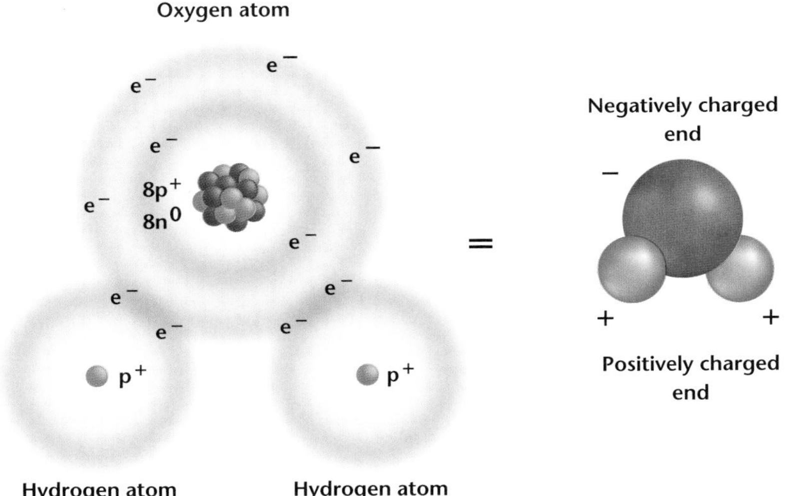

Figure 3-7 At one end of a water molecule, the hydrogen atoms have a positive charge, while at the opposite end, the oxygen atom has a negative charge.

more strongly by the oxygen atom than by the hydrogen atoms. As a result, the electrons spend more time near the oxygen atom than they do near the hydrogen atoms. When atoms in a covalent bond do not share electrons equally, they form polar bonds. A polar bond has a positive end and a negative end. The overall shape of a molecule indicates whether it is polar.

IONS

Not all atoms bond by sharing electrons. Sometimes, atoms gain or lose electrons from their outermost energy levels. An atom that gains or loses an electron is a charged particle called an **ion.** In general, an atom in which the outermost energy level is less than half-full—that is, it has fewer than four valence electrons—tends to lose its valence electrons. When an atom loses its valence electrons, it becomes positively charged. In chemistry, a positive ion is indicated by a superscript plus sign. For example, a sodium ion is represented by Na^{+}. If an ion results from the loss of more than one electron, the number of electrons lost is placed before the plus sign. A magnesium ion, which forms when a magnesium atom has lost two electrons, is thus represented by Mg^{2+}.

An atom in which the outermost energy level is more than half-full—that is, it has more than four valence electrons—tends to fill its outermost energy level by adding one or more needed electrons. Such an atom forms a negative ion. By including additional electrons, negative ions, such as O^{2-} and Cl^{-}, tend to be larger than positive ions. If the outermost energy level is exactly half-full, an atom may form either a positive or negative ion.

Some compounds contain ions made up of covalently bonded atoms. Two such compounds that are important in forming the materials at Earth's surface are silicate ions (SiO_4^{4-}) and carbonate ions (CO_3^{2-}).

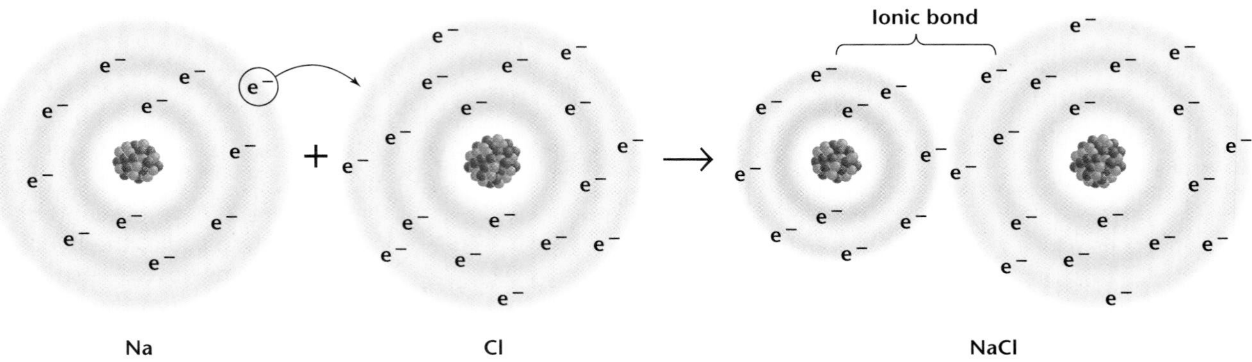

Figure 3-8 The positive charge of a sodium ion attracts the negative charge of a chlorine ion. The two ions are held together by an ionic bond. Note that the negative chlorine ion is slightly larger than the positive sodium ion.

Ionic Bonds As you might expect, positive and negative ions attract each other. The attractive force between two ions of opposite charge is known as an **ionic bond.** ***Figure 3-8*** illustrates an ionic bond between a positive ion of sodium and a negative ion of chlorine. Common table salt (NaCl) consists of equal numbers of sodium (Na^+) ions and chlorine (Cl^-) ions. Note that positive ions are always written first in chemical formulas.

Within the compound NaCl, there are as many positive as negative ions; therefore, the positive charge on the sodium ion cancels the negative charge on the chloride ion, and the net electrical charge of the compound NaCl is zero. Magnesium and oxygen ions combine in a similar manner to form the compound magnesium oxide (MgO), one of the most common compounds on Earth. Compounds formed by ionic bonding are known as ionic compounds. You can determine if elements form ionic compounds in the *Problem-Solving Lab* on this page.

Other ionic compounds have different proportions of ions. For example, oxygen and sodium ions combine in the ratio shown by the chemical formula for sodium monoxide, Na_2O, in which there are two sodium ions to each oxygen ion. With any other ratio of sodium ions and oxygen ions, the electrical charges wouldn't cancel, as you can prove by adding up the ionic charges.

Problem-Solving Lab

Interpreting Scientific Illustrations

Forming compounds Many atoms gain or lose electrons in order to have eight electrons in the outermost energy level. In the diagram, energy levels are indicated by the circles around the nucleus of each element. The colored spheres in the energy levels represent electrons, and the spheres in the nucleus represent protons and neutrons.

Analysis

1. What is the name and symbol of element A?
2. What is the name and symbol of element B?
3. How many electrons are present in atoms of element A? Element B?

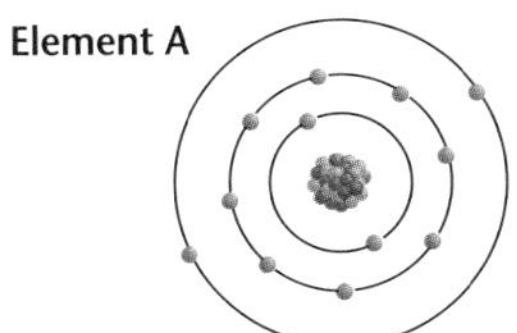

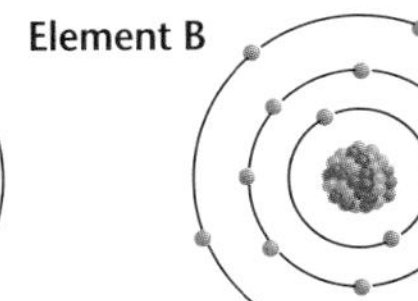

Thinking Critically

4. Can these elements form ions? If so, how many protons would be present in the nuclei of these ions? What would be the electrical charges (magnitude and sign) and chemical symbols of these ions?
5. Can these two elements form a compound? If so, what is the chemical formula of the compound?

Metallic Bonds

Most compounds on Earth are held together by ionic or covalent bonds, or by a combination of these two types of bonds. However, there are other types of bonds. In metals, for example, the valence electrons are shared by all the atoms, not just by adjacent atoms, as in covalent compounds. You could think of a metal as a group of positive ions floating in a sea of negative electrons. The positive ions of the metal are held together by the negative electrons between them. This type of bond, known as a metallic bond, allows metals to conduct electricity because the electrons can move freely throughout the entire solid metal.

Chemical Reactions

You have learned that atoms gain, lose, or share electrons to become more stable, and that these atoms form compounds. Sometimes, compounds break down into simpler substances. The change of one or more substances into other substances is called a **chemical reaction.** Chemical reactions are described by chemical equations. For example, water is formed by the chemical reaction between hydrogen gas (H_2) and oxygen gas (O_2). The formation of water can be described by the following equation.

$$2H_2 + O_2 \rightarrow 2H_2O$$

A chemist reads this chemical equation as "Two molecules of hydrogen and one molecule of oxygen react to yield two molecules of water." In this reaction, hydrogen and oxygen are the reactants, and water is the product. When you write a chemical equation, you must balance the equation by showing an equal number of atoms for each element on each side of the equation. This is because the same amount of matter is present both before and after the reaction. Note that there are four hydrogen atoms on each side of the above equation (2 + 2 = 4). There are also two oxygen atoms on each side of the reaction.

Another example of a chemical reaction, one that takes place between iron (Fe) and oxygen (O), is represented by the following chemical equation.

$$4Fe + 3O_2 \rightarrow 2Fe_2O_3$$

This reaction forms the mineral hematite, an important iron ore. You may be more familiar with another form of this compound shown in ***Figure 3-9***—rust!

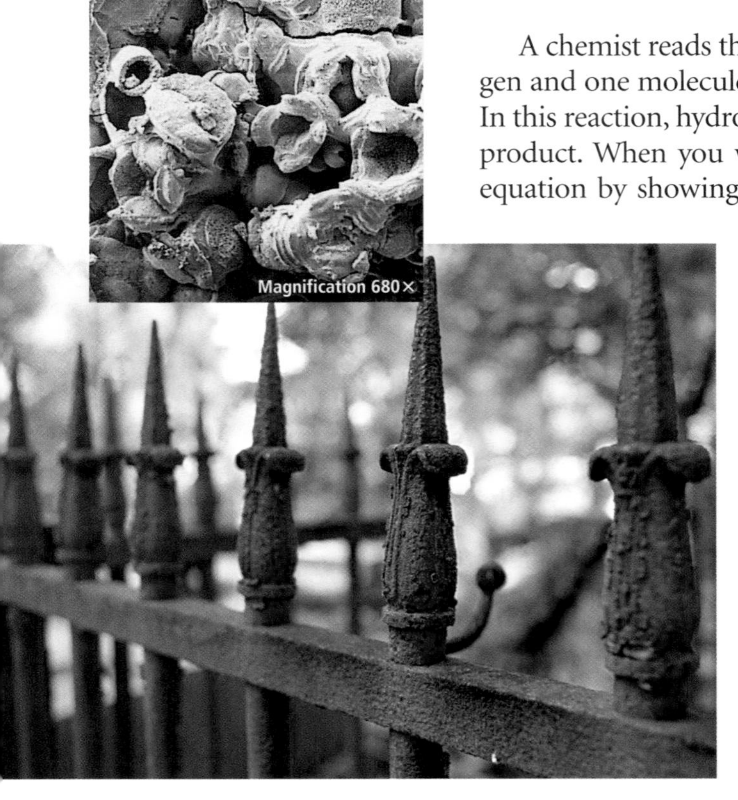

Figure 3-9 Rust, shown magnified in the top photo, forms on metals that contain iron when they are exposed to moist air.

MIXTURES AND SOLUTIONS

Unlike a compound, in which the constituent atoms combine and lose their identities, a *mixture* is a combination of two or more components that retain their identities. When a mixture's components are easily recognizable, it is called a heterogeneous mixture. For example, soil, as shown in ***Figure 3-10A,*** is a heterogeneous mixture because its components are still recognizable: bits of minerals such as quartz and feldspar, clay particles, fragments of plants, and so on. In contrast, in a homogeneous mixture, the component particles cannot be distinguished, even though they still retain their original properties. Brewed coffee is an example of a homogeneous mixture, which is also called a **solution.**

A solution may be liquid, gaseous, or solid. Seawater is a liquid solution consisting of water molecules and ions of many elements that exist on Earth. You will investigate liquid solutions in the *GeoLab* at the end of this chapter. Magma is also a liquid solution; it is composed of ions representing all atoms that were present in the crystals of the rock before it melted. Air is a solution of gases, mostly nitrogen and oxygen molecules together with other atoms and molecules. Metal alloys, such as bronze and brass, are also solutions. Bronze is a homogeneous mixture of copper and tin atoms; brass is a similar mixture of copper and zinc atoms. Such solid homogeneous mixtures are called *solid solutions.* You will learn more about solid solutions in Chapters 4 and 5.

Figure 3-10 Soil is a heterogeneous mixture in which the component parts are easily recognizable **(A).** Coffee is a homogeneous mixture, called a solution, in which the component parts cannot be distinguished **(B).**

Acids and Bases Many chemical reactions that occur on Earth involve solutions called acids and bases. An **acid** is a solution containing a substance that produces hydrogen ions (H^+) in water. Recall that a hydrogen atom consists of one proton and one electron. When a hydrogen atom loses its electron, it becomes simply a proton. The most common acid in our environment is carbonic acid, which is produced when carbon dioxide is dissolved in water by the following reaction.

$$H_2O + CO_2 \longrightarrow H_2CO_3$$

Some of the carbonic acid molecules in the water dissociate, or break apart, into hydrogen ions and bicarbonate ions, as represented by the following equation.

$$H_2CO_3 \longrightarrow H^+ + HCO_3^-$$

These two equations play a major role in the dissolution and precipitation of limestone and the formation of caves, discussed in Chapter 10. Many of the reaction rates involved in geological

ENVIRONMENTAL CONNECTION

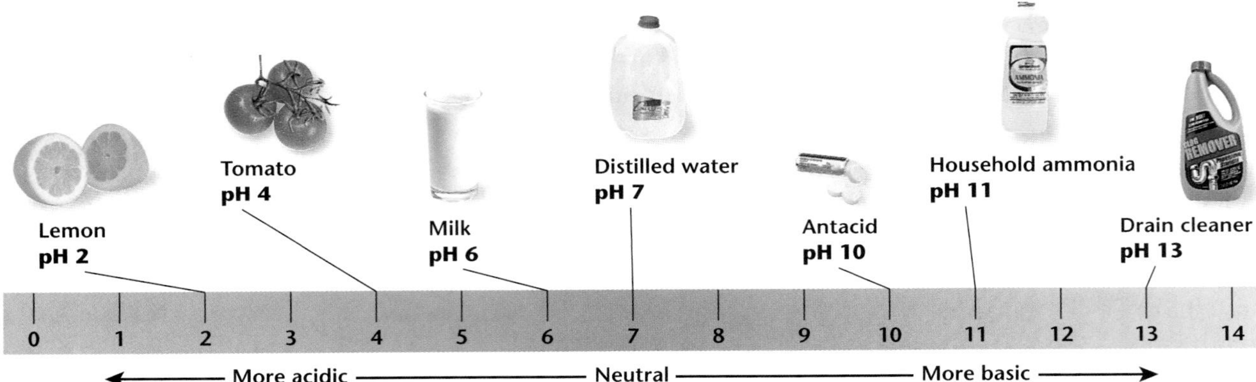

Figure 3-11 The pH values of some common substances are shown on this scale. Treated pH paper can be used to determine the acidity of a solution.

processes are exceedingly slow. For example, it may take thousands of years for the carbonic acid in groundwater to break down limestone to form a cave.

Bases produce hydroxide ions (OH^-) in solution. A base can neutralize an acid by combining with hydrogen ions of the acid to form water through the following reaction.

$$H^+ + OH^- \longrightarrow H_2O$$

The pH scale measures the hydrogen and hydroxide ions in solutions, with 7 being neutral. A solution with a pH reading below 7 is considered to be acidic. The lower the number, the more acidic the solution is. A solution with a reading above 7 is considered to be basic. The higher the number, the more basic the solution is. Distilled water usually has a pH of 7, but rainwater is slightly acidic, having a pH of 5.0 to 5.6. The pH values of some common substances are shown in ***Figure 3-11.***

Section Assessment

1. What is the smallest unit of a molecular compound with the properties of that compound?
2. Why are negative ions usually larger than positive ions?
3. Explain why molecules held together by covalent bonds don't have electrical charges.
4. **Thinking Critically** Whole milk consists of microscopic fat globules suspended in a solution of nutrients. Is milk a homogeneous or a heterogeneous mixture? Explain.

Skill Review

5. **Predicting** What kind of bond forms between the nitrogen atoms and hydrogen atoms in ammonia (NH_3)? For more help, refer to the *Skill Handbook.*

earthgeu.com/self_check_quiz

SECTION 3.3 States of Matter

I&E 1.a

An iceberg floating in the ocean beneath a blue sky not only captures the beauty of nature, but it also illustrates three states of matter found on Earth. Matter may be solid, like the iceberg; liquid, like the ocean; or gaseous, like the water vapor in the air. But are these the only states in which matter can exist? At room temperature and standard atmospheric pressure, matter normally exists in one of these three states. However, there is another state of matter found on Earth, called plasma, which you will learn more about later in this section.

OBJECTIVES

- **Describe** *the states of matter on Earth.*
- **Explain** *the reasons that matter exists in these states.*
- **Relate** *the role of thermal energy to changes of state in matter.*

VOCABULARY

crystalline structure
glass
evaporation
sublimation
plasma
condensation

SOLIDS

What do ice crystals, table-salt crystals, and diamonds have in common? All of these substances are solids, and all of them form crystals. Solids are substances with densely packed particles, which may be ions, atoms, or molecules, depending upon the substance. The particles of a solid are arranged in a definite pattern; thus, a solid has both a definite shape and a definite volume. Most solids have a **crystalline structure,** in which the particles are arranged in regular geometric patterns, as illustrated in ***Figure 3-12.*** Crystals form symmetrical solid objects with flat faces and straight edges between faces. The angles between the faces depend upon the internal arrangement of the particles. For example, vanadium and quartz crystals look different from table-salt crystals because of the different internal arrangements of their particles. Magnesium, quartz, and many other substances have hexagonal crystals.

Figure 3-12 Crystals of table salt are cubic, whereas vanadium crystals are hexagonal.

Figure 3-13 Granite, which is abundant in Earth's crust, is a mass of intergrown crystals.

Well-formed crystals are rare. When many crystals form in the same space at the same time, mutual interference prevents the formation of regular crystals with smooth boundaries. The result is a mass of intergrown crystals, called a polycrystalline solid. Most solid substances on Earth, including rocks, are polycrystalline materials. ***Figure 3-13*** shows the polycrystalline nature of the rock granite.

Some solid materials have no regular internal patterns. **Glasses** are solids that consist of densely packed atoms arranged at random. Glasses form when molten material is chilled so rapidly that atoms don't have enough time to arrange themselves into a regular pattern. These solids do not form crystals, or their crystals are so small that they cannot be seen. Window glass consists mostly of disordered silicon and oxygen with the chemical composition SiO_2.

Liquids

At any temperature above absolute zero (−273°C), the atoms in solids vibrate. Because these vibrations increase with increasing temperature, they are called thermal vibrations. At the melting point of the material, these vibrations become sufficiently vigorous to break the forces holding the solid together. The particles can then slide past each other, and the substance becomes liquid. Liquids do not have their own shape; they take the shape of the container they are placed in, as you can see in ***Figure 3-14A.*** However, liquids do have definite volume.

Figure 3-14 Because the particles in liquids slide past each other, liquids have no definite shape **(A).** Dry ice, which is solid carbon dioxide (CO_2), sublimates from a solid directly into a gas **(B).**

A

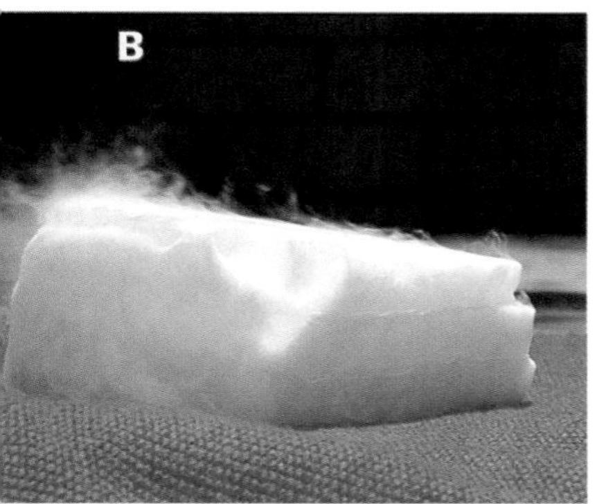

Gases

The particles in liquids vibrate vigorously, and individual particles may gain sufficient energy to escape the liquid. This process of change from a liquid to a gas is called **evaporation,** or vaporization. When any liquid reaches its boiling point, it vaporizes quickly and becomes a gas. However, some evaporation takes place even below the boiling point. In fact, thermal vibrations can enable individual atoms or molecules to escape even from a solid. You may have noticed that even on winter days with temperatures below freezing, snow gradually disappears. This slow change of state from a solid, ice crystals, to a gas, water vapor, without an intermediate liquid state is called **sublimation,** as illustrated in ***Figure 3-14B.***

In gases, the particles are separated by relatively large distances and move about at extremely high speeds. Gas particles move independently of each other and travel randomly. They travel in one direction until they bump into another gas particle or the walls of a container. Gases, like liquids, have no definite shape. Gases also have no definite volume and can expand into any space available, unless they are restrained by a container or a force such as gravity. Earth's gravity keeps the gases in the atmosphere from escaping into space.

Plasma

When matter is heated to temperatures greater than 5000°C, the collisions between particles are so violent that electrons are knocked away from atoms. Such extremely high temperatures exist in stars, and, as a result, the gases of stars consist entirely of positive ions and free electrons. These hot, highly ionized, electrically conducting gases are called **plasmas.** ***Figure 3-15*** shows the plasma that forms the Sun's corona. You have seen matter in the plasma state if you have ever seen lightning or a neon sign. Both lightning and the matter inside a neon tube are in the plasma state.

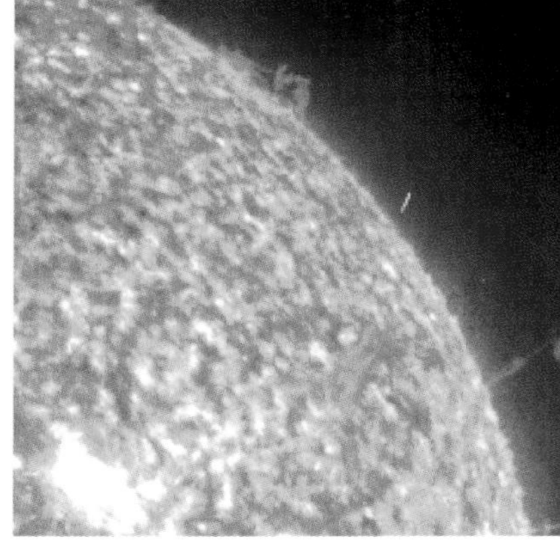

Figure 3-15 The corona around the Sun is formed from hot, glowing plasma.

Changes of State

Solids melt when they absorb thermal energy and their temperatures rise. When a liquid absorbs thermal energy from the environment, it evaporates. This actually has a cooling effect on the surrounding environment. What do you suppose happens when a liquid freezes? The same thermal energy is then released back into the environment. Finally, when a gas is cooled, it releases thermal energy in the process of **condensation,** the change from a gas to a liquid.

Conservation of Matter and Energy

The identity of matter can be changed through chemical reactions and nuclear processes, and its state can be changed under different thermal conditions. However, matter cannot be created or destroyed but can change from one form to another. This fundamental fact is called the law of conservation of matter. Like matter, energy cannot be created or destroyed but it can be changed from one form to another. This law of the conservation of energy is also called the first law of thermodynamics.

Section Assessment

1. What shape do salt crystals have? What determines the shape of a crystal?
2. Why is the puddle underneath a melting ice cube as cold as the ice cube itself?
3. Contrast what happens to thermal energy in evaporation and condensation.
4. **Thinking Critically** Water boils at 100°C at sea level. What do you think its boiling point would be if water molecules were not polar molecules?

Skill Review

5. **Concept Mapping** Use the following terms to construct a concept map to compare and contrast the three common states of matter. You may use some terms more than once. For more help, refer to the *Skill Handbook.*

definite shape | states of matter | solid | no definite shape
definite volume | liquid | gas | no definite volume

GeoLab: Salt Precipitation

I&E

1.a Select and use appropriate tools and technology (such as computer-linked probes, spreadsheets, and graphing calculators) to perform tests, collect data, analyze relationships, and display data.

Many rocks on Earth form from salts precipitating out of seawater. Salt ions precipitate when a salt solution becomes saturated. Solubility is the ability of a substance to dissolve in a solution. When a solution is saturated, no more of that substance can be dissolved. What is the effect of temperature and evaporation on salt precipitation? How do precipitation rates affect the size of crystals?

Preparation

Problem

Under what conditions do salt solutions become saturated and under what conditions does salt precipitate out of solution?

Materials

halite (sodium chloride)
250-mL glass beakers (2)
distilled water
plastic wrap
laboratory scale
hot plate
shallow glass baking dish
refrigerator
glass stirring rod

Objectives

In this GeoLab, you will:

- **Observe** salt dissolving and precipitating from a saturated salt solution.
- **Identify** the precipitated salt crystals.
- **Compare** the salt crystals that precipitate out under different conditions.
- **Hypothesize** why different conditions produce different results.

Safety Precautions

Always wear safety goggles and an apron in the lab. Wash your hands after handling salt solutions. Use care in handling hot solutions. Use protection handling hot glassware.

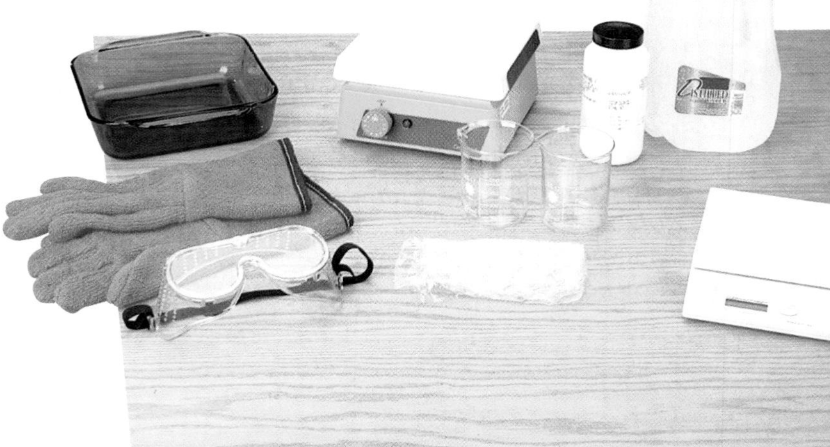

Procedure

1. Pour 150 mL of distilled water into a 250-mL glass beaker.
2. Measure 54 g of sodium chloride. Add the sodium chloride to the distilled water in the beaker and stir until only a few grains remain on the bottom of the beaker.
3. Place the beaker on the hot plate and turn the hot plate on. As the solution inside the beaker heats up, stir it until the last few grains of sodium chloride dissolve. The salt solution will then be saturated.
4. Pour 50 mL of the warm, saturated solution into the second 250-mL glass beaker. Cover this beaker with plastic wrap so that it forms a good seal. Put this beaker in the refrigerator.
5. Pour 50 mL of the saturated solution into the shallow glass baking dish. Place the dish on the hot plate and heat the salt solution until all the liquid evaporates. ***CAUTION:*** *The baking dish will be hot. Handle with care.*
6. Place the original beaker with 50 mL of the remaining solution on a shelf or windowsill. Do not cover the beaker.
7. Observe both beakers one day later. If crystals have not formed, wait another day to make your observations and conclusions.
8. Once crystals have formed in all three containers, observe the size and shape of the precipitated crystals. Describe your observations in your science journal.

Analyze

1. What is the shape of the precipitated crystals in the three containers? Does the shape of the crystals alone identify them as sodium chloride?
2. Why didn't all of the salt solution dissolve in step 2 above? How did heating affect the solubility of sodium chloride? Why did heating have the observed effect? Explain.
3. What effect does cooling have on the solubility of salt?
4. What happens when a salt solution evaporates? What effect does evaporation have on the solubility of salt?
5. Suppose you have two samples of volcanic rock of identical chemical composition but different crystal sizes. What conclusions can you make about the conditions under which each rock sample cooled?

Conclude & Apply

1. What are the sizes of the crystals in the different containers? Which container has the smallest crystals? Which crystals formed in the shortest time interval?
2. Why does salt precipitate from solution? How is crystal size related to precipitation rate?
3. Design an experiment to separate a heterogeneous mixture of different salts, such as NaCl and $MgCl_2$, into its components, by dissolving and precipitation.

Science & Technology

Extreme Magnification

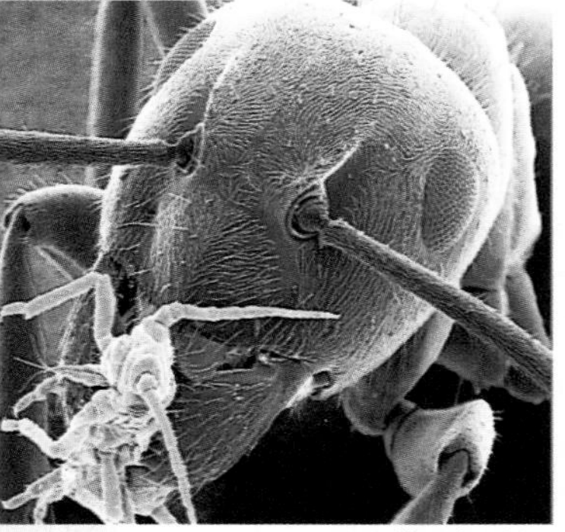
Magnification 33×

Students from Malibu, California, to Atlanta, Georgia, are collecting bugs with more enthusiasm than usual. They are sending insects to a project called Bugscope. Bugscope workers use a scanning electron microscope (SEM) to take close-up photographs of these amazing animals. The photographs are then posted on the Internet. The photos provide detailed information about the structure and composition of matter in our world.

Electron Microscopy

Microscopes have been used by scientists to magnify objects since the 1600s. Traditional light microscopes, however, have limitations: they can magnify objects only up to 2000 times their size, and there is a limit to how clear these microscopes can make objects appear. In the early 1930s, German scientists developed a new type of microscope that could magnify objects 10 000 times their size. This microscope did not use light for the examination of objects; instead, it used a beam of high-energy electrons. Today, electron microscopes scan material to create detailed, three-dimensional, black-and-white images of objects that are magnified up to 100 000 times.

Applications

When a scanning electron microscope is used to examine an object, a beam of electrons is directed onto the object's surface. As scanning coils move the electron beam over the object in a gridlike pattern, the beam causes electrons to be knocked off the object's surface. An image is created on a fluorescent screen by the calculation of the number of electrons that are bounced off each spot on the object. Electron microscopes provide information about the properties of matter being examined, including its reactivity, strength, and reflectivity. Currently, scientists are using electron microscopes to further research in areas ranging from combating leukemia to searching for evidence of product tampering to developing better ways to process and store food.

Magnification 18×

Magnification 140×

Activity

The photographs on this page are SEM images. What does each image represent? Go to earthgeu.com to find links to SEM images of solids and create an image gallery. Present your images, along with clues about their identities, to your classmates, and see how many they can identify.

CHAPTER 3
Study Guide

Summary

SECTION 3.1

What are elements?

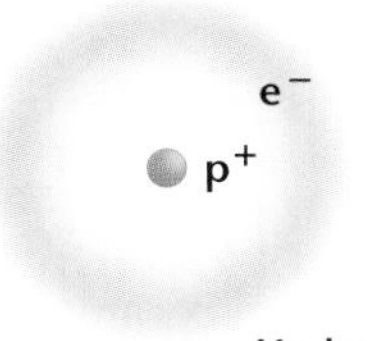

Hydrogen atom

Main Ideas

- The basic building blocks of matter are atoms. Atoms consist of protons, neutrons, and electrons.
- Protons have a positive electrical charge, electrons have a negative electrical charge, and neutrons are electrically neutral. Protons and neutrons make up the nucleus of an atom; electrons surround the nucleus in energy levels.
- An element is a substance consisting of atoms with a specific number of protons in their nuclei. Isotopes of an element differ by the number of neutrons in their nuclei. Many elements are mixtures of isotopes.
- The number of electrons in the outermost energy levels of atoms determines their chemical behavior. Elements with the same number of electrons in their outermost energy levels have similar chemical properties.

Vocabulary

atom (p. 54)
atomic mass (p. 58)
atomic number (p. 54)
electron (p. 55)
element (p. 53)
energy level (p. 55)
isotope (p. 58)
mass number (p. 54)
neutron (p. 54)
nucleus (p. 54)
proton (p. 54)
radioactivity (p. 58)
valence electron (p. 57)

SECTION 3.2

How Atoms Combine

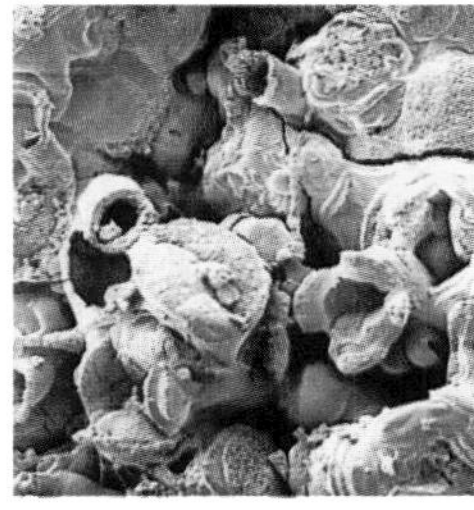

Main Ideas

- Atoms of different elements combine to form compounds.
- Atoms held together by the sharing of electrons in covalent bonds form molecular compounds.
- Ions are electrically charged atoms or groups of atoms. Positive and negative ions attract each other and form ionic compounds.
- Acids are solutions containing hydrogen ions. Bases are solutions containing hydroxide ions. Acids and bases can neutralize each other.
- A mixture is a combination of components that retain their identities. A solution is a mixture in which the components can no longer be distinguished as separate. Solutions can be liquid, solid, gaseous, or combinations.

Vocabulary

acid (p. 65)
base (p. 66)
chemical bond (p. 60)
chemical reaction (p. 64)
compound (p. 60)
covalent bond (p. 60)
ion (p. 62)
ionic bond (p. 63)
molecule (p. 61)
solution (p. 65)

SECTION 3.3

States of Matter

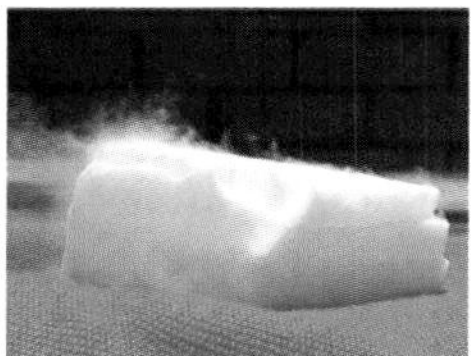

Main Ideas

- Matter on Earth exists in three common physical states: solid, liquid, or gaseous. Matter in the universe includes plasma.
- Most solids have a crystalline structure.
- Liquids are densely packed arrangements of particles.
- Gases consist of widely separated, individual particles. Plasmas are hot, highly ionized, electrically conducting gases.
- Changes of state involve thermal energy.

Vocabulary

condensation (p. 69)
crystalline structure (p. 67)
evaporation (p. 68)
glass (p. 68)
plasma (p. 69)
sublimation (p. 68)

earthgeu.com/vocabulary_puzzlemaker

CHAPTER 3
Assessment

Understanding Main Ideas

1. What particles make up the nucleus of an atom?
   a. protons only
   b. neutrons only
   c. neutrons and electrons
   d. protons and neutrons

2. Which of these makes up an atom's mass number?
   a. number of protons
   b. number of neutrons
   c. neutrons and protons
   d. protons and electrons

3. Which is the average of the mass numbers of an element's isotopes?
   a. atomic number
   b. energy levels
   c. atomic mass
   d. valence electrons

4. What is the lightest of all the elements?
   a. helium
   b. lithium
   c. hydrogen
   d. magnesium

5. One of the isotopes of chlorine (atomic number 17) has a mass number of 35. How many neutrons does this isotope have in its nucleus?
   a. 17 b. 18 c. 35 d. 53

6. Which is NOT an element?
   a. hydrogen
   b. water
   c. argon
   d. uranium

7. What element is the final (nonradioactive) decay product of uranium?
   a. lead
   b. neon
   c. plutonium
   d. hydrogen

8. Many musical instruments are made of brass, which consists of copper and zinc atoms. What is brass an example of?
   a. an ionic compound
   b. a solid solution
   c. a chemical reaction
   d. a base

9. What are formed when sodium ions and chlorine ions combine to form NaCl?
   a. ionic bonds
   b. solid solutions
   c. isotopes
   d. covalent bonds

Use the following diagram to answer questions 10 and 11.

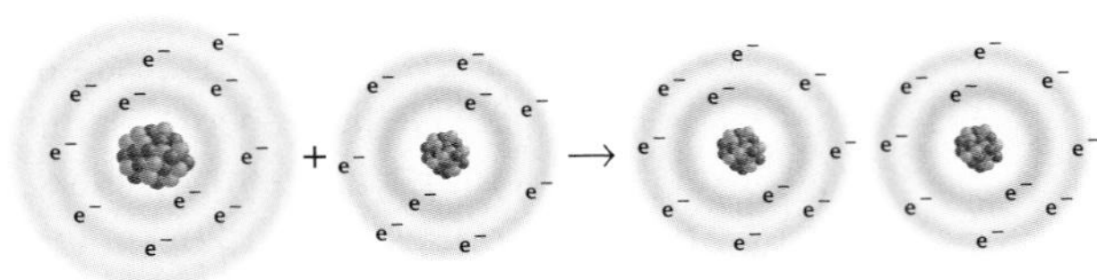

10. Write the chemical equation that the diagram represents.

11. Is the compound formed in the diagram a molecular compound or an ionic compound? Explain your answer.

12. What is the most abundant element in Earth's crust?
    a. hydrogen
    b. uranium
    c. silicon
    d. oxygen

13. Which chemical formula represents a polar molecule?
    a. MgO b. H_2O c. SiO_2 d. NaCl

14. If an atom gains electrons, what does it become?
    a. a positive ion
    b. a negative ion
    c. a different element
    d. a heavier isotope

15. What kind of ions characterize an acid?
    a. hydroxide ions
    b. hydrogen ions
    c. oxygen ions
    d. negative ions

16. Why do metals conduct electricity?

Test-Taking Tip

TERMS If a test question involves a term that you don't remember, see if you can figure out its meaning from the question. Sometimes standardized tests will give a definition or example right in the question.

earthgeu.com/chapter_test

CHAPTER 3

Assessment

17. What happens to the thermal energy of a gas when it condenses and forms a liquid?
18. How are coffee and air chemically alike?
19. Explain the differences between a molecular compound and an ionic compound.
20. What is a molecule?

Applying Main Ideas

21. How many valence electrons do beryllium atoms (atomic number 4) have? Explain your answer.
22. Name two elements with chemical properties that are similar to those of sodium.
23. What is the charge on ions of fluorine (atomic number 9)?
24. List two elements that have ions with a 2+ charge.
25. What compound do the ions Al^{3+} and O^{2-} form?
26. Explain why the mass numbers of elements in the Periodic Table of Elements rarely are whole numbers.

Thinking Critically

27. Why don't gases such as neon and argon combine chemically with other elements?
28. Suppose you want to find out if the elements copper (atomic number 29) and sulfur (atomic number 16) could combine into a compound. How could you check this before you proceeded to do an experiment in the laboratory?

Standardized Test Practice

INTERPRETING DATA Use the table below to answer the following questions.

Atomic Structure

Element	Atomic Number	Atomic Mass
Beryllium	4	9.01
Calcium	20	40.08
Silicon	14	28.09
Scandium	21	44.96
Titanium	22	47.88
Zirconium	40	91.22

1. If titanium has 22 protons in its nucleus, how many neutrons are present in the nucleus of its most common isotope?
   a. 48 **c.** 60
   b. 26 **d.** 28

2. If the most common isotope of scandium has 24 neutrons in its nucleus, how many protons does scandium have?
   a. 66 **c.** 21
   b. 45 **d.** 13

3. If calcium's most common isotope has 20 neutrons in its nucleus, how many neutrons can be found in another naturally occurring isotope of calcium?
   a. 60 **c.** 30
   b. 41 **d.** 21

4. How many valence electrons does oxygen have?
   a. 2 **c.** 6
   b. 4 **d.** 9

5. Does silicon have any isotopes? Explain your answer.

Chapter 4

Minerals

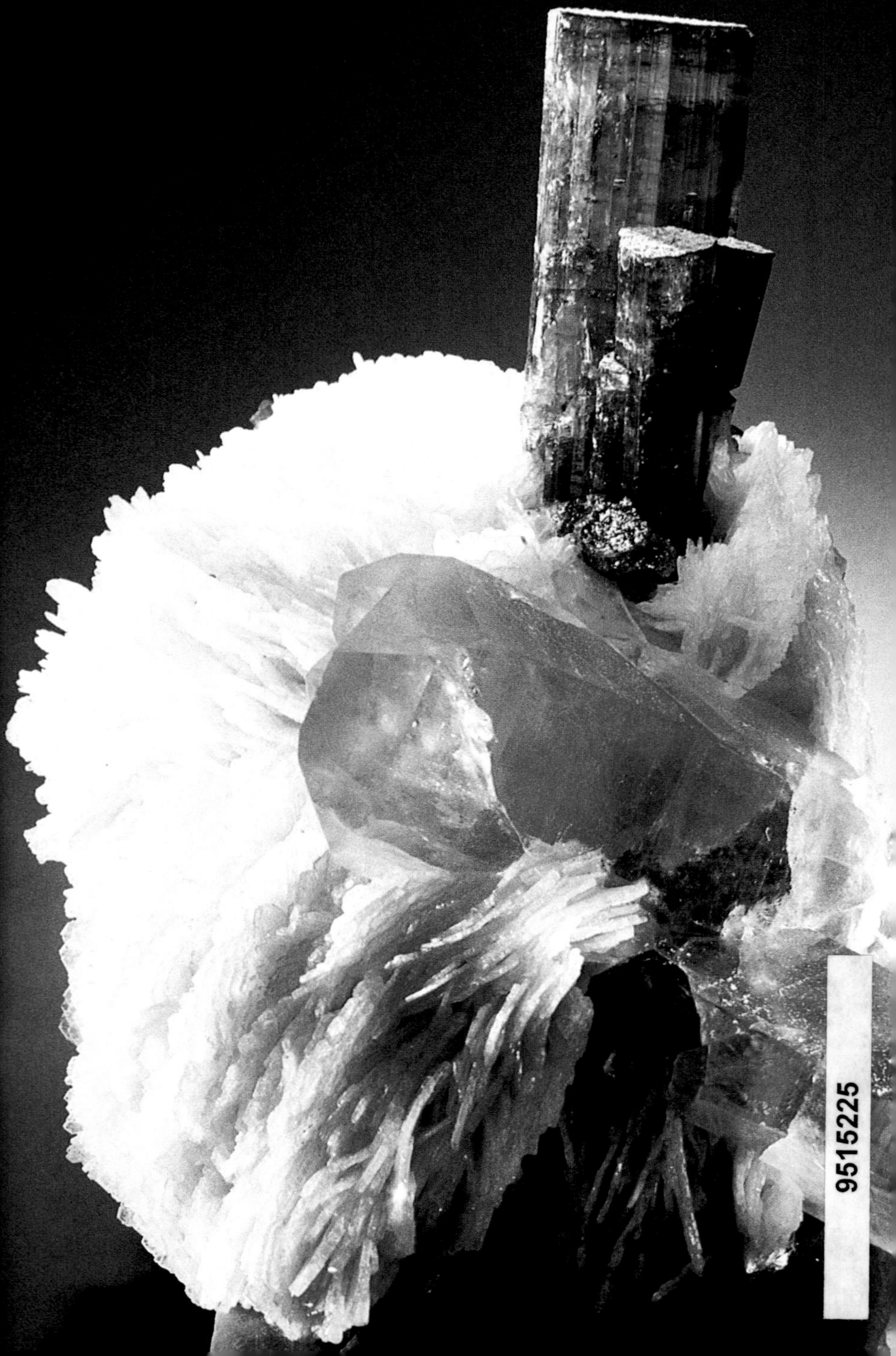

What You'll Learn

- How minerals form, and which are most common in Earth's crust.
- Which properties can be used to identify and classify minerals.
- Why certain minerals are ores and gems.

Why It's Important

Many products used in daily life are made directly or indirectly from minerals. Minerals also play a vital role in the processes that shape Earth. Some minerals form crystals that are valued for their beauty. The crystals shown here consist of albite, watermelon tourmaline, and smoky quartz.

Earth Science Online

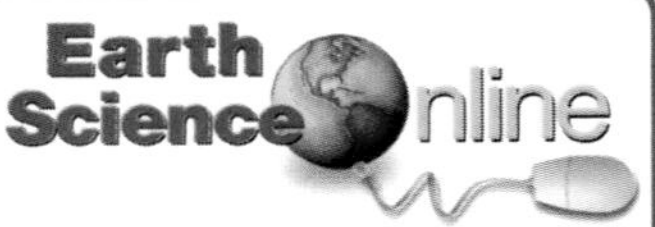

To find out more about minerals, visit the Earth Science Web Site at earthgeu.com

Discovery Lab Observing Mineral Shapes

Although there are thousands of minerals in Earth's crust, each type of mineral has unique characteristics. These characteristics are clues to a mineral's chemical composition and to the way it formed. Physical properties can also be used to distinguish one type of mineral from another.

1. Place a few grains of table salt—the mineral halite—on a microscope slide. Place the slide on the microscope stage and separate the grains. Or, simply observe the grains with a magnifying glass.
2. Focus on one grain at a time. Count the number of sides each grain has. Make sketches of the grains.
3. Next, examine a sample of quartz with the microscope or magnifying glass. Count the number of sides in the quartz sample. Sketch the shape of the quartz sample.

Observe Compare and contrast the shapes of the samples of halite and quartz. What might account for the differences you observed? In your science journal, describe some other physical properties of your mineral samples.

SECTION 4.1 *What is a mineral?*

Earth Sciences 3.c

OBJECTIVES

- **Define** *a mineral.*
- **Describe** *how minerals form.*
- **Identify** *the most common elements in Earth's crust.*

VOCABULARY

mineral
crystal
magma
silicate

Earth's crust is composed of about 3000 minerals. Minerals play important roles in forming rocks and in shaping Earth's surface, and a select few have played—and continue to play—a role in shaping civilization. For example, great leaps in prehistory were made when early humans began making tools from iron. Calcite is the mineral that forms the 2 million limestone blocks that make up the Great Pyramid in Egypt. It is also the primary mineral in the marble found in the Parthenon in Greece. Throughout history, wars have been fought and empires have crumbled over minerals such as gold and silver.

Mineral Characteristics

Look around your classroom. The metal in your desk, the graphite in your pencil, and the glass in the windows are just three examples of how modern humans use products made from minerals. But what exactly is a mineral? A **mineral** is a naturally occurring, inorganic solid with a specific chemical composition and a definite crystalline structure. Let's examine each part of this definition in turn.

Figure 4-1 The chemical composition of olivine varies within a limited range.

Naturally Occurring and Inorganic To say that minerals are naturally occurring simply means that they are formed by natural processes, which you'll learn about later in this section. Thus, synthetic diamonds and other substances developed in labs are not minerals. Secondly, all minerals are inorganic. That is, they aren't alive and never were alive during any part of their existence. Based on this criterion, salt is a mineral, but sugar, which is harvested from plants, is not. What about coal? According to the scientific definition of minerals, coal is not a mineral because hundreds of millions of years ago, it formed from organic processes.

Solids with Specific Compositions The third characteristic of minerals is that they all are solids. Solids have definite shapes and volumes. Liquids and gases do not. Thus, no gas or liquid can be considered a mineral. Next, each type of mineral has a chemical composition unique to that mineral. A few minerals, such as copper, silver, and sulfur, are composed of single elements. The vast majority, however, are made from compounds. The mineral quartz, for instance, is a combination of two atoms of oxygen and one atom of silicon. Although other minerals may contain silicon and oxygen, the arrangement and proportion of these elements in quartz are unique to quartz.

Table 4-1 Crystal Systems

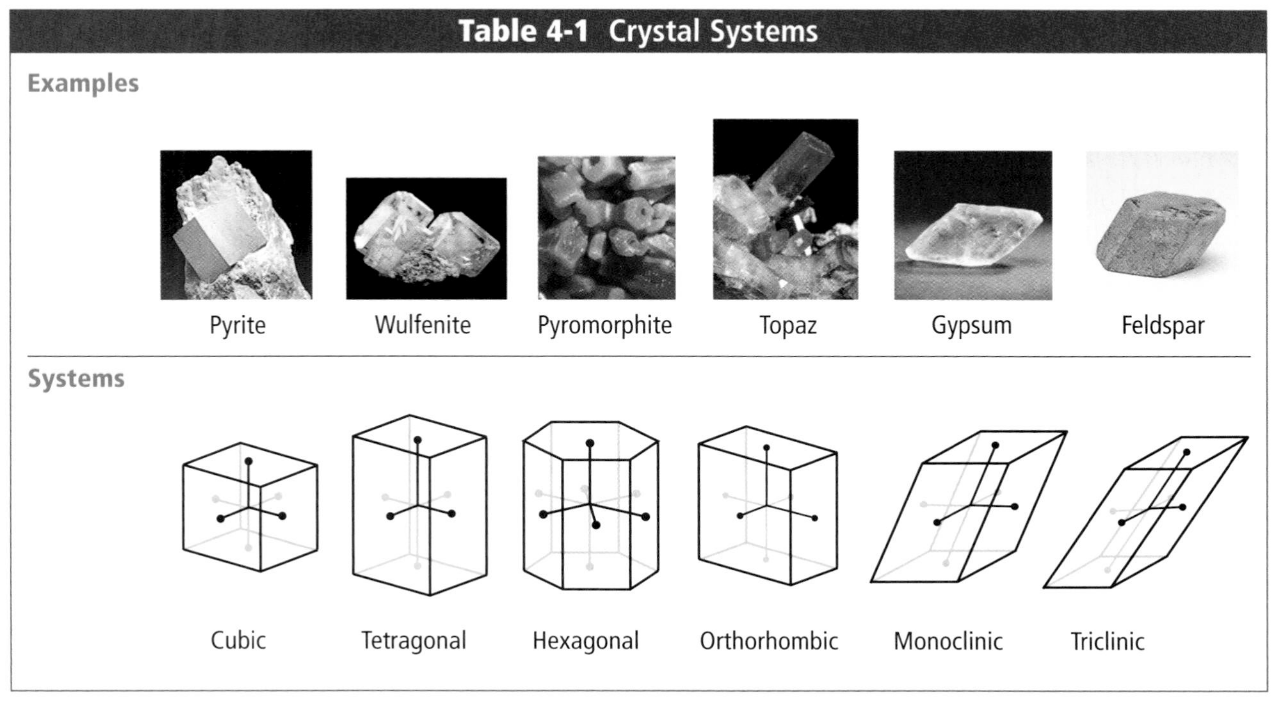

Examples	Pyrite	Wulfenite	Pyromorphite	Topaz	Gypsum	Feldspar
Systems	Cubic	Tetragonal	Hexagonal	Orthorhombic	Monoclinic	Triclinic

In some minerals, such as the one shown in ***Figure 4-1,*** chemical composition may vary within a certain range. For instance, the amount of individual iron and magnesium atoms in the mineral olivine may vary, with some forms of olivine containing more iron than others. But the ratio of the total amount of iron and magnesium atoms to the amount of silicon atoms in olivine is always the same. Thus, the chemical composition of this mineral varies, but only within a well-defined range.

Definite Crystalline Structure The last part of the definition of minerals relates to crystalline structures. The atoms in minerals are arranged in regular geometric patterns that are repeated again and again. A **crystal** is a solid in which the atoms are arranged in repeating patterns. At times, a mineral will form in an open space and grow into one large crystal. The resulting mineral crystal may take the shape of one of the six major crystal systems shown in ***Table 4-1.*** You'll model crystal systems in the *MiniLab* on this page. The well-defined crystal shapes shown in the table are fairly rare. More commonly, the internal atomic arrangement of a mineral is not so readily apparent because the mineral formed in a restricted space. ***Figure 4-2*** compares a crystal that grew in an open space with one that grew in a restricted space.

MiniLab

How can crystal systems be modeled?

Model the six major crystal systems, then classify mineral samples according to these systems.

Procedure

1. Using ***Table 4-1*** for guidance, cut pieces of foam board into geometric shapes. Your largest geometric shape should be no more than about 8 cm in length. Your group will need about 38 various shapes.
2. Tape or glue the geometric shapes into models of the six major crystal systems. Again, use ***Table 4-1*** for guidance.
3. Use the mineral samples provided by your teacher to classify minerals according to their crystal shapes.

Analyze and Conclude

1. What geometric shapes did you use to model the crystal systems?
2. Was the crystal structure readily apparent in all mineral samples? Infer why or why not.
3. Use *Appendix H* to identify your minerals. Besides crystal shape, what properties did you use for identification purposes?

Figure 4-2 The well-shaped crystals of this sample of watermelon tourmaline indicate that it grew in an open space **(A).** This sample of watermelon tourmaline does not have well-defined crystals and thus grew in a restricted space **(B).**

Figure 4-3 Gypsum, shown in the foreground, forms when elements evaporate from a supersaturated solution.

Using Numbers Of the 3000 known minerals, ten make up about 90 percent of the rocks in Earth's crust. What percentage of the total number of minerals do these ten minerals represent?

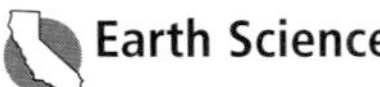

Earth Sciences

3.c Students know how to explain the properties of rocks based on the physical and chemical conditions in which they formed, including plate tectonic processes.

Minerals from Magma

Minerals can form from the cooling of magma. **Magma** is molten material found beneath Earth's surface. Density differences can force magma upward into cooler layers of Earth's interior, where the magma cools. The compounds in the magma no longer move freely in the cooling material, and they may begin to interact chemically to form minerals. The type and amount of elements present in the magma help determine which minerals will form, while the rate at which the magma cools determines the size of the mineral crystals. If the magma cools slowly within Earth's heated interior, the atoms have time to arrange themselves into large crystals. If the magma reaches Earth's surface, comes in contact with air or water, and cools quickly, the atoms don't have time to arrange themselves into large crystals. Thus, small crystals form from rapidly cooling magma and large crystals form from slowly cooling magma. You'll learn more about crystal size in Chapter 5.

Minerals from Solution

A given volume of water in a solution can dissolve only so much of a solid before the water becomes saturated. At that point, the saturated water cannot dissolve any more of the solid. In nature, if a solution becomes supersaturated, or overfilled, with another substance, mineral crystals may begin to precipitate, or drop out of solution. This is one way that minerals can form from a supersaturated solution.

Minerals can also form when elements dissolve in a supersaturated solution. When liquid evaporates from the solution, the elements remain behind and may begin to arrange into crystals. ***Figure 4-3*** shows gypsum deposits that were formed from the evaporation of water. This is the second way that minerals form from a supersaturated solution.

Mineral Groups

Earlier, we said that 3000 minerals are found in Earth's crust. However, only about 30 of these minerals are common. The most common minerals are often referred to as rock-forming minerals

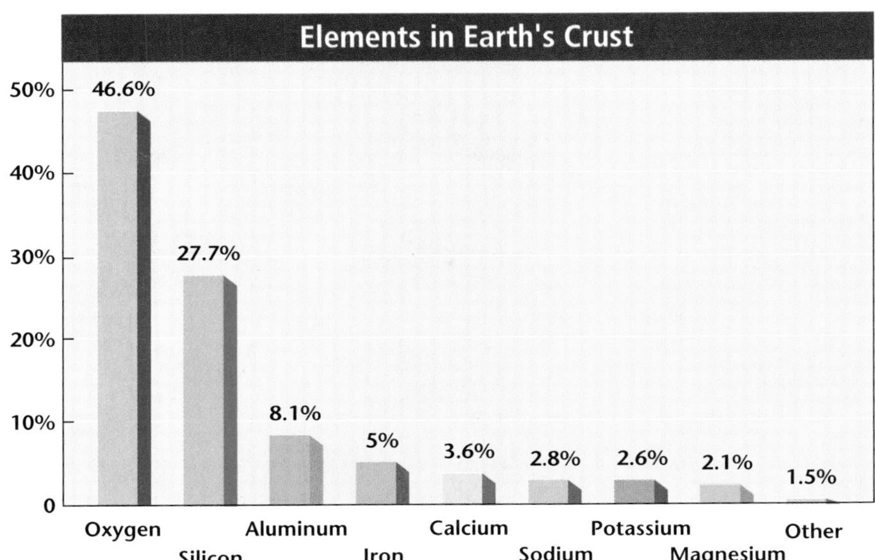

Figure 4-4 Oxygen is the most common element in Earth's crust, followed by silicon. The eight most common elements make up most minerals.

because they make up most of the rocks found in Earth's crust. Elements also are present in Earth's crust. About 90 known elements occur naturally in the crust. The vast majority of minerals are made up of the eight most common elements. ***Figure 4-4*** shows the percentages by weight of the common elements in Earth's crust.

Silicates Oxygen (O) is the most abundant element in Earth's crust, followed by silicon (Si). Minerals that contain silicon and oxygen, and usually one or more other elements, are known as **silicates.** Silicates make up approximately 96 percent of the minerals found in Earth's crust. The most common minerals, feldspar and quartz, are silicates.

Figure 4-5 shows how one silicon atom attaches to four oxygen atoms to form a silica tetrahedron. A tetrahedron is a three-dimensional shape structured like a pyramid. The basic silica tetrahedron has the ability to share oxygen atoms with other tetrahedron molecules. This unique structure allows molecules to combine chemically and

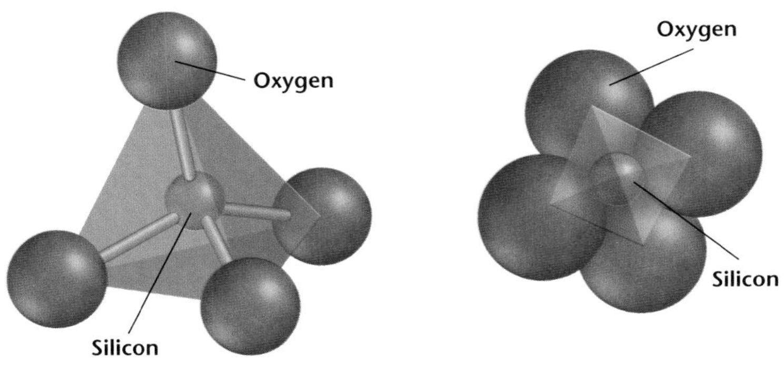

Figure 4-5 A silica tetrahedron is made up of one silicon atom bonded to four oxygen atoms.

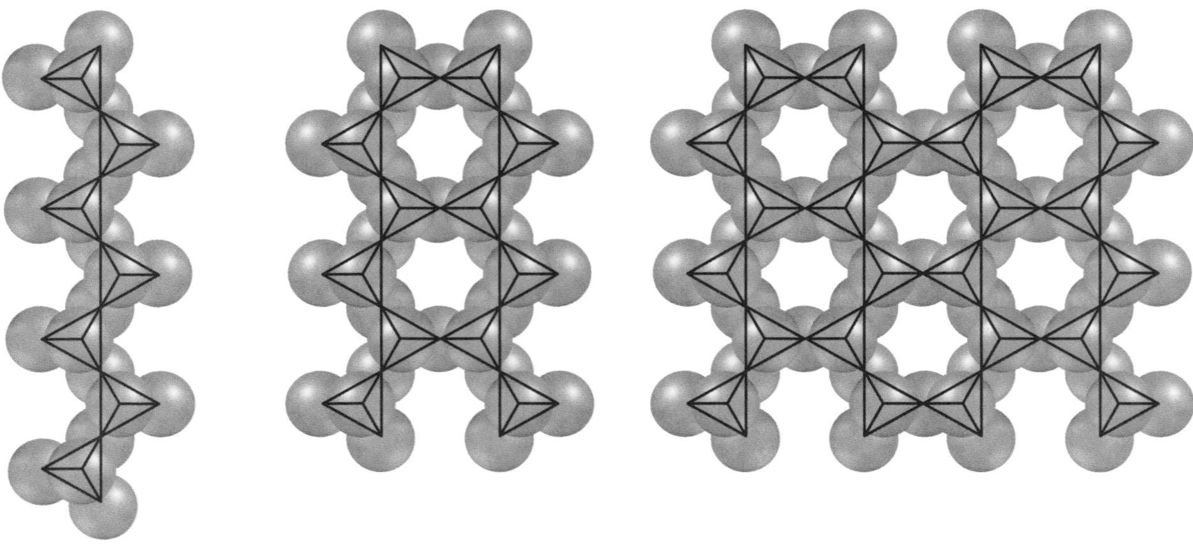

Figure 4-6 Silica tetrahedrons can combine in many ways, including single chains, double chains, and sheets. The different structural combinations account for the diversity of silicates.

structurally in a vast number of ways, which accounts for the diversity of silicates. ***Figure 4-6*** shows some possible arrangements formed by silica tetrahedrons, including single chains, double chains, and sheets. The bonds between the atoms help determine several mineral properties, including the way a mineral splits. Minerals generally split along planes of weak bonds. For instance, mica is an example of a sheet tetrahedron, wherein an atom of aluminum (Al) or potassium (K) bonds sheets together. Mica separates easily into sheets because the attraction between the tetrahedrons and the atom of aluminum or potassium is weak. Quartz, on the other hand, has an intricate network. Quartz is highly resistant to weathering and does not break easily along any planes because its atoms are strongly bonded together.

Table 4-2 Mineral Groups

Group	Example
Native elements	Copper metal (Cu)
Oxides and hydroxides	Hematite (Fe_2O_3) Brucite ($Mg[OH]_2$)
Halides	Halite (NaCl)
Carbonates	Calcite ($CaCO_3$)
Sulfates	Anhydrite ($CaSO_4$)
Silicates	Olivine (Mg_2SiO_4)
Sulfides	Pyrite (FeS_2)

Carbonates Oxygen easily combines with many other elements and thus forms other mineral groups, such as the carbonates and the oxides. Carbonates are minerals composed of one or more metallic elements with the carbonate compound CO_3. Examples of carbonates are calcite, dolomite, and rhodochrosite. Carbonates are the primary minerals found in rocks such as limestone, coquina, and marble. Some carbonates have distinctive colorations, such as the greenish hue of malachite and the blue of azurite, shown in ***Figure 4-7.***

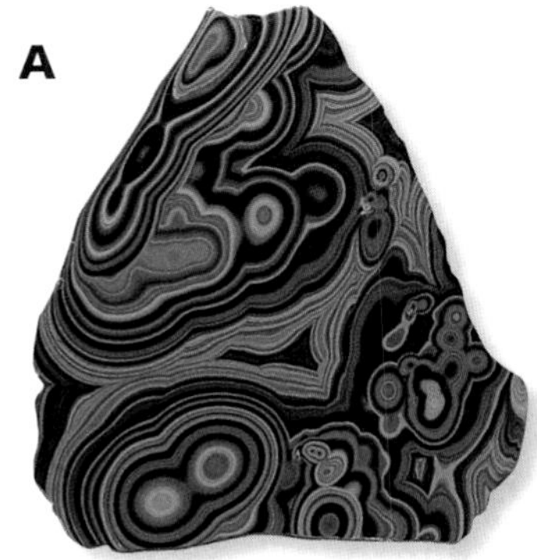

Figure 4-7 The carbonates malachite **(A)** and azurite **(B)** have distinct colorations.

Oxides Oxides are compounds of oxygen and a metal. Hematite (Fe_2O_3) and magnetite (Fe_3O_4) are common iron oxides and good sources of iron. The mineral uraninite is valuable because it is the major source of uranium, which is used to generate nuclear power.

Other major mineral groups are sulfides, sulfates, halides, and native elements. Sulfides such as pyrite (FeS_2) are compounds of sulfur and one or more elements. Sulfates such as anhydrite ($CaSO_4$) are composed of elements with the sulfate compound SO_4. Halides such as halite (NaCl) are made up of chloride or fluoride along with calcium, sodium, or potassium. A native element such as silver (Ag) or copper (Cu) is made up of one element only. ***Table 4-2*** on the previous page summarizes the mineral groups. *Appendix H* contains further information about individual minerals. In the next section, you'll learn how to identify some of the minerals discussed thus far.

Section Assessment

1. Define a mineral. Give two reasons why petroleum is not a mineral.
2. How do minerals form from solution? How do they form from magma?
3. What are the two most abundant elements in Earth's crust? What mineral group do these elements form?
4. Identify the other major mineral groups.
5. Describe a crystal. What determines the size of a mineral crystal formed from magma?
6. **Thinking Critically** Water is an inorganic substance formed by natural processes on Earth. It has a unique chemical composition. Under what conditions, if any, could water be considered a mineral?

Skill Review

7. **Concept Mapping** Use the following terms and *Appendix H* to construct a concept map of the six major crystal systems. For more help, refer to the *Skill Handbook.*

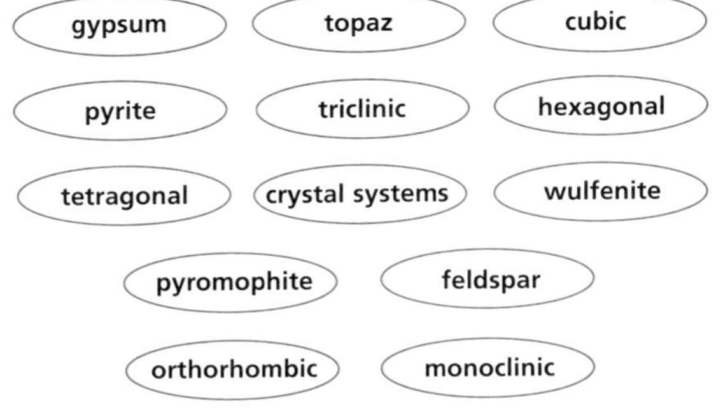

SECTION 4.2 Identifying Minerals

Earth Sciences 9.a **I&E** 1.a

OBJECTIVES

- **Classify** *minerals according to their physical and chemical properties.*
- **Identify** *different types of minerals.*
- **Discuss** *how minerals are used.*

VOCABULARY

luster
streak
hardness
cleavage
fracture
specific gravity
ore
gem

At the beginning of this chapter, we discussed just a few of the many ways in which humans use minerals. Before a mineral can be used, it must first be identified. With more than 3000 minerals in Earth's crust, this presents a problem. How does one go about identifying an unknown mineral?

MINERAL IDENTIFICATION

Geologists rely on several relatively simple tests to identify minerals. These tests are based upon a mineral's physical and chemical properties. As you'll see in the *Design Your Own GeoLab* at the end of this chapter, it's usually best to use a combination of tests rather than just one to identify minerals.

Color One of the most noticeable characteristics of a mineral is its color. Color is sometimes caused by the presence of trace elements or compounds within a mineral. For example, quartz can be found in a variety of colors, as shown in ***Figure 4-8,*** and these different colors are the result of different trace elements in the quartz samples. Red jasper has trace elements of iron oxides, purple amethyst contains ferric iron, orange citrine contains iron hydrates, and rose quartz contains manganese or titanium. The appearance of milky quartz, on the other hand, is caused by the numerous bubbles of gas and liquid trapped within the crystal. In general, color is one of the least reliable clues to a mineral's identity.

Luster The way that a mineral reflects light from its surface is called **luster.** Luster is described as being either metallic or nonmetallic. Silver, gold, copper, and galena have shiny surfaces that reflect light like the chrome trim on cars. Thus, they are said to have a metallic luster.

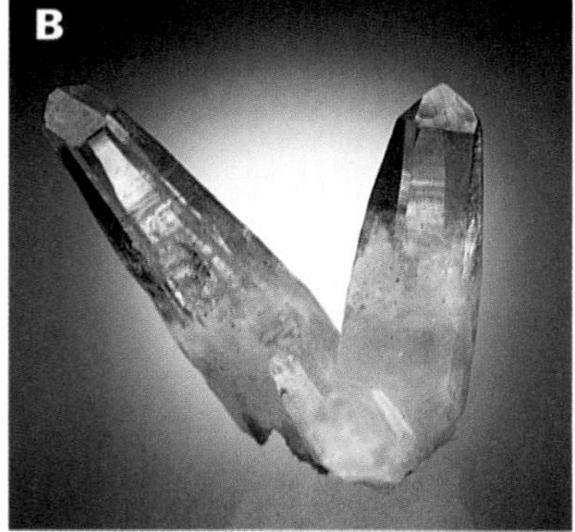

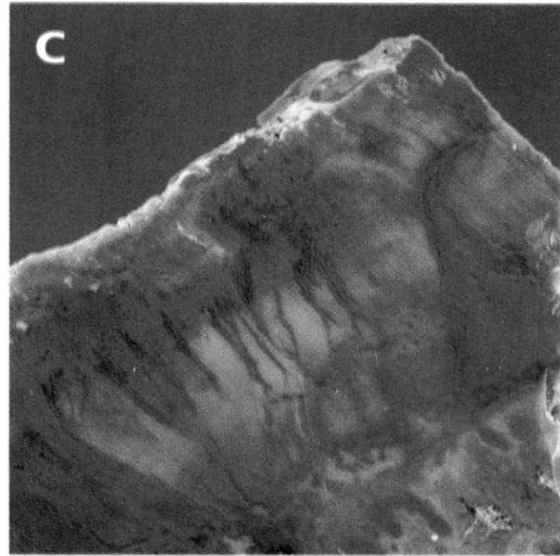

Figure 4-8 Orange citrine **(A),** purple amethyst **(B),** and red jasper **(C)** are all varieties of quartz. The different colors are caused by trace elements.

Figure 4-9 Galena has a metallic luster **(A).** Gypsum has a nonmetallic luster **(B).**

Nonmetallic minerals, such as calcite, gypsum, sulfur, and quartz, do not shine like metals. Their lusters might be described as dull, pearly, waxy, or silky. Differences in luster, shown in ***Figure 4-9,*** are caused by differences in the chemical compositions of minerals.

Topic: Minerals
To find out more about mineral identification, visit the Earth Science Web Site at earthgeu.com

Activity: Choose three minerals. Describe their color, luster, texture, and hardness.

Texture Texture describes how a mineral feels to the touch. Like luster and color, texture is often used in combination with other tests to identify a mineral. The texture of a mineral might be described as smooth, rough, ragged, greasy, soapy, or glassy. For example, fluorite has a smooth texture, while the texture of talc is greasy.

Streak A mineral rubbed across an unglazed porcelain plate will sometimes leave a colored powdered streak on the surface of the plate. **Streak** is the color of a mineral when it is broken up and powdered. Sometimes, a mineral's streak does not match the mineral's external color, as shown in ***Figure 4-10.*** For example, pyrite, which is also known as fool's gold because it looks like gold, leaves a greenish-black streak. Gold, on the other hand, leaves a yellow streak. Thus, streak is one of the main tests used to distinguish pyrite from gold.

A mineral's streak rarely changes, even if it is weathered or its external color varies slightly. For example, fluorite can be purple, yellow, green, or blue, but its streak is always white. The streak test can be used only on minerals that are softer than a porcelain plate. Thus, this test cannot be used to identify all minerals.

Figure 4-10 Both gray and black samples of hematite leave rust-colored streaks on a porcelain plate.

Hardness One of the most useful tests for identifying minerals is hardness. **Hardness** is a measure of how easily a mineral can be scratched. German geologist Friedrich Mohs developed a scale in which an unknown mineral's hardness can be compared to the known hardnesses of ten minerals. The minerals in the Mohs scale of mineral hardness were selected because they are easily recognized and —with the exception of diamond—readily found in nature. Talc is one of the softest minerals and can be scratched by a fingernail; thus, talc represents 1 on the Mohs scale of hardness. In contrast, diamond is so hard that it can be used as a sharpener and cutting tool; diamond represents 10 on the Mohs scale of hardness. The scale, shown in ***Table 4-3,*** works like this: any mineral with a greater hardness than another mineral will scratch that softer mineral. For example, topaz will scratch quartz but not corundum. Hardness, which is one of the most reliable tests of mineral identification, is determined by the arrangement of a mineral's atoms.

Cleavage and Fracture Atomic arrangement also determines how a mineral will break. Minerals break along planes where atomic bonding is weak. A mineral that splits relatively easily and evenly along one or more flat planes is said to have **cleavage.** To identify a mineral by cleavage, geologists count the number of cleaved planes and study the angle or angles between them. For instance, mica, shown in ***Figure 4-11A,*** has perfect cleavage in one direction. It breaks in sheets because of weak atomic bonds, as you learned in the earlier discussion about silica tetrahedrons. Halite has a cubic cleavage,

Table 4-3 Mohs Hardness Scale

	Hardness		Hardness of Common Objects
Talc	1	(softest)	
Gypsum	2		fingernail (2.5)
Calcite	3		piece of copper (3.5)
Fluorite	4		iron nail (4.5)
Apatite	5		glass (5.5)
Feldspar	6		steel file (6.5)
Quartz	7		streak plate (7)
Topaz	8		scratches quartz
Corundum	9		scratches topaz
Diamond	10	(hardest)	scratches all common materials

Figure 4-11 Mica has perfect cleavage in one direction **(A).** Quartz breaks along rough or jagged edges **(B).** The rock obsidian fractures into arc-like patterns **(C).**

which means that it breaks in three directions along planes of weak atomic attraction. Quartz, shown in ***Figure 4-11B,*** breaks unevenly along jagged edges because of its tightly bonded atoms. Minerals that break with rough or jagged edges are said to have **fracture.** Flint, chalcedony, and the rock obsidian share a unique fracture with arc-like patterns resembling clam shells, as shown in ***Figure 4-11C.*** This is called conchoidal fracture.

Density and Specific Gravity Sometimes, two minerals of the same size may feel quite different when they are lifted—one is much heavier than the other. Differences in weight are the result of differences in density, which is defined as mass per unit of volume. Density is expressed as a ratio of the mass of a substance divided by its volume, or $D = M/V$. Pyrite, for instance, has a density of 5.2 g/cm^3, and gold has a density of 19.0 g/cm^3.

Density reflects the atomic weight and structure of a mineral. Because density is not dependent on the size or shape of a mineral, it is a particularly useful identification tool. Often, however, differences in density are too small to be distinguished by simply lifting different minerals and estimating their perceived weights. Thus, for accurate mineral identification, density must be measured. The most common measure of density used by geologists is **specific gravity,** which is the ratio of the weight of a substance to the weight

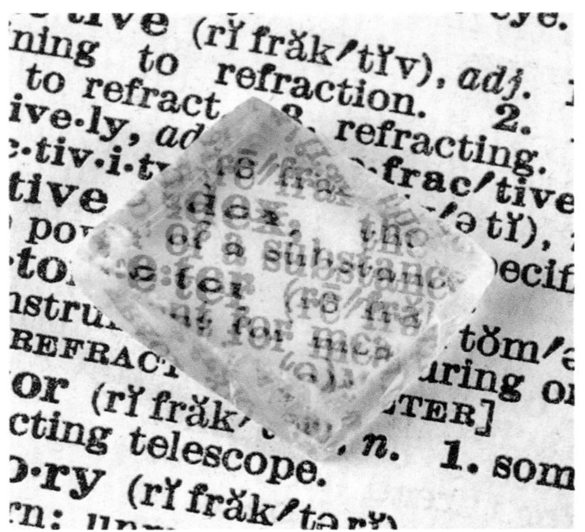

Figure 4-12 Light is bent in two directions when it passes through a sample of Iceland spar. The refraction creates the appearance of two images.

of an equal volume of water at 4°C. Do the *Problem-Solving Lab* to learn how specific gravity and other mineral properties are used to identify minerals.

SPECIAL PROPERTIES

Several special properties of minerals also can be used for identification purposes. For instance, the arrangement of atoms in a type of calcite called Iceland spar causes light to be bent in two directions when it passes through the mineral. The refraction of the single ray of light into two rays creates the appearance of two images, as shown in ***Figure 4-12.*** This process is known as double refraction. Double refraction is also a property of the mineral zircon.

Calcite exhibits another special property as a result of its chemical composition. Calcite ($CaCO_3$) fizzes when it comes into contact with hydrochloric acid (HCl). HCl reacts with calcite to release CO_2 in the form of bubbling gas. In this reaction, shown below, calcium chloride ($CaCl_2$) also forms.

$$CaCO_3 + 2HCl \longrightarrow CaCl_2 + H_2O + CO_2$$

Problem-Solving Lab

Making and Using Tables

Complete a mineral identification table Minerals can be identified by their physical and chemical properties. Common properties include color, streak, hardness, specific gravity, and crystalline structure.

Analysis

1. Copy the data table shown here. Use *Appendix H* to complete the table.
2. Expand the table to include the names of the minerals, information about breakage patterns, and mineral uses.

Thinking Critically

3. Which of these minerals will scratch quartz? How do you know?
4. Which of these minerals might be found in a painting? Which might be found in your desk?
5. What other information could you have included in the table?

Mineral Identification Chart

Mineral Color	Streak	Hardness	Specific Gravity	Crystal System
copper red		3	8.5–9	cubic
	red or reddish brown	6	5.3	hexagonal
pale to golden yellow	yellow			cubic
	colorless	7.5	3.5	
gray, green or white			2.5	triclinic
	colorless		4.0	hexagonal

Other special properties are exhibited by magnetite, an iron ore. Magnetite is naturally magnetic. Lodestone, a form of magnetite, will pick up nails like a magnet, as shown in *Figure 4-13.* The mineral sphalerite produces a distinctive rotten-egg odor when it is rubbed vigorously across a streak plate. The smell is the result of the presence of sulfide in the mineral.

Figure 4-13 Magnetite can pick up iron nails because it is naturally magnetic.

Mineral Uses

There's a good chance that as you read these words, you're sitting on minerals, wearing minerals, and perhaps even eating minerals. Minerals are virtually everywhere. They are used to make computers, cars, televisions, desks, roads, buildings, jewelry, beds, paints, sports equipment, and medicines. This list is by no means exhaustive. In fact, it barely touches upon the many ways in which people use minerals.

Environmental Connection

Earth Sciences

9.a Students know the resources of major economic importance in California and their relation to California's geology.

Ores Many of the items mentioned above are made from ores. A mineral is an **ore** if it contains a useful substance that can be mined at a profit. Hematite, for instance, is a useful ore that contains the element iron. Look around the room. Are any items made of iron? If so, their original source may have been the mineral hematite. Are any items in the room made of aluminum? The element aluminum is found in the ore bauxite. Another mineral and its use are shown in *Figure 4-14.*

Figure 4-14 The ore rutile **(A)** contains the element titanium, a durable, lightweight metal often used in sports equipment **(B).**

Figure 4-15 The open-pit mine in the background, located in Carajas, Brazil, is the world's largest iron mine.

Mines Ores that are located deep within Earth's crust are removed by underground mining. Ores that are near Earth's surface are obtained from large, open-pit mines, such as the one shown in ***Figure 4-15.*** When a mine is excavated, unwanted rock and dirt, known as waste material, are dug up along with the valuable ore. The waste material must be separated from the ore before the ore can be used. Removing the waste material can be expensive and, in some cases, harmful to the environment, as you'll learn in later chapters. If the cost of removing the waste material becomes higher than the value of the ore itself, then the mineral will no longer be classified as an ore. It would no longer be economical to mine it.

The classification of a mineral as an ore may also change if the supply of or demand for that mineral changes. Consider a mineral that is used to make computers. Engineers might develop a more efficient design or a less costly alternative material. In either of these cases, the mineral would no longer be used in computers. Demand for the mineral would drop substantially, and the mineral would no longer be considered an ore.

GEMS

What makes a ruby more valuable than mica? Rubies are much rarer and more visually pleasing than mica. Rubies are thus considered gems. **Gems** are valuable minerals that are prized for their rarity and beauty. Gems such as rubies, emeralds, and diamonds are cut, polished, and used for jewelry. ***Figure 4-16*** shows a raw

Figure 4-16 A raw emerald can be cut, polished, and used in jewelry. This emerald ring is in the Smithsonian Museum collection.

emerald and a polished emerald. Because of their rareness, rubies and emeralds are actually more valuable than diamonds. You'll learn more about diamonds in the *Science and Math* feature at the end of this chapter.

In some cases, the presence of trace elements can make one variety of a mineral more colorful and thus more prized than other varieties of the same mineral. Amethyst, for instance, is the gem form of quartz. Amethyst contains trace minerals, which give the gem a lovely purple color. The mineral corundum, which is often used as an abrasive, also can be found as rubies and sapphires. Rubies contain trace amounts of chromium; sapphires contain trace amounts of cobalt or titanium.

SECTION ASSESSMENT

1. Explain why color is not a good test for distinguishing between pyrite and gold. What test is most reliable for identifying these two minerals?
2. What is a mineral's texture? List several words that are used to describe texture.
3. Compare and contrast cleavage and fracture. Give an example of a mineral with cleavage and a mineral with fracture.
4. Describe the chemical reaction that takes place when hydrochloric acid comes in contact with calcite.
5. What is the hardness of a mineral if it scratches a penny but will not scratch glass?
6. What is an ore?
7. Why are some minerals classified as gems? List several gems.
8. **Critical Thinking** The mineral fluorite can be found in a variety of colors, yet its streak is always white. Why?

SKILL REVIEW

9. **Making and Using Tables** Use *Appendix H* and reference materials to make a data table that includes examples of minerals with the following special properties: magnetism, double refraction, reaction to acid, and smell. For more help, refer to the *Skill Handbook.*

DESIGN YOUR OWN GeoLab

Making a Field Guide to Minerals

I&E

1.a Select and use appropriate tools and technology (such as computer-linked probes, spreadsheets, and graphing calculators) to perform tests, collect data, analyze relationships, and display data.

Have you ever used a field guide to identify a bird, flower, rock, or insect? If so, you know that field guides include far more than simply photographs. A typical field guide for minerals might include background information about minerals in general, plus specific information about the formation, properties, and uses of each mineral. In this activity, you'll create a field guide to minerals.

Preparation

Problem

How would you go about identifying minerals? What physical and chemical properties would you test? Which of these properties should be included in a field guide to help others to identify unknown minerals?

Possible Materials

mineral samples
hand lens
glass plate
streak plate
Mohs scale of mineral hardness
5 percent hydrochloric acid (HCl) with dropper
Appendix H
steel file or nail
piece of copper
paper clip
magnet

Hypothesis

As a group, form a hypothesis about which property or properties might be most useful in identifying minerals.

Objectives

In this GeoLab, you will:

- **Conduct** tests on unknown minerals to determine their physical and chemical properties.
- **Identify** minerals based on the results of your tests.
- **Design** a field guide for minerals.

Safety Precautions

Review the safe use of acids. HCl may cause burns. If a spill occurs, rinse your skin with water and notify your teacher immediately.

Plan the Experiment

1. As a group, list the steps that you will take to test your hypothesis. Keep the available materials in mind as you plan your procedure. Be specific, describing exactly what you will do at each step. Properties that you may want to test include luster, color, reaction to HCl, magnetism, cleavage, fracture, texture, hardness, streak, double refraction, and density.
2. Should you test any of the properties more than once for any of the minerals? How will you determine whether certain properties indicate a specific mineral?
3. Design a data table to summarize your results. You can use this table as the basis for your field guide.
4. Read over your entire experiment to make sure that all steps are in a logical order.
5. Have you included a step for additional research? You may have to use the library or the Earth Science Web Site to gather all the necessary information for your field guide.
6. What information will be included in the field guide? Possible data include how each mineral formed, its uses, its chemical formula, and a labeled photograph or drawing of the mineral.
7. Make sure your teacher approves your plan before you proceed with your experiment.

Analyze

1. **Interpreting Results** Which properties were most reliable for identifying minerals? Which properties were least reliable? Discuss reasons why one property is more useful than others.
2. **Defending Your Hypothesis** Was your hypothesis supported? Why or why not?
3. **Thinking Critically** How could you use a piece of paper, a steel knife, and a glass bottle to distinguish between Iceland spar and quartz?
4. **Observing and Inferring** What mineral reacted with the HCl? Why did the mineral bubble? Write the balanced equation that describes the chemical reaction that took place between the mineral and the acid.
5. **Conducting Research** What information did you include in the field guide? What resources did you use to gather your data? Describe the layout of your field guide.

Conclude & Apply

1. Compare and contrast your field guide with those of other groups. How could you improve your field guide?
2. What are the advantages and disadvantages of field guides?
3. Based on your results, is there any one definitive test that can always be used to identify a mineral? Explain.

Science & Math

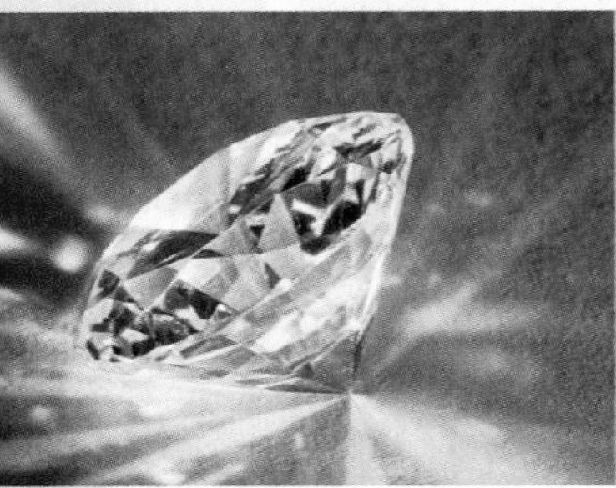

The Price of Diamonds

For centuries people have valued diamonds for their beauty, sparkle, and hardness. The cost of a diamond in a jewelry store depends mainly on four things, often called the four Cs.

Color and Cut

Diamonds come in many colors. Colorless diamonds cost more than those with a slight yellowish tinge, but stones with a deep, rich color, such as the famous blue Hope Diamond, are the most valued of all. The different colors are caused by minor impurities.

Diamonds are cut into many shapes. Some of the most common shapes are the round brilliant; the rose, which is round with a flat-bottom and pointed top; the marquis, which is an oval with pointed ends; and the emerald cut, which is rectangular with rounded corners. The cut of a diamond affects how it reflects light and thus how much it sparkles. An uncut diamond crystal looks like a greasy piece of glass.

Clarity and Carat Weight

Clarity refers to the presence or absence of visible flaws and impurities in the diamond. A diamond is considered flawless if no such defects are visible under a magnification power of ten. Dozens of minerals occur in diamonds, the most common being olivine, garnet, and clinopyroxene. Bubbles of liquid or gas are also found within diamond crystals.

The weight of a diamond is represented by an ancient unit of measurement called the carat. A carat is 1/5 of a gram, or 1/142 of an ounce. The name probably came from the tropical carob tree, which has quite uniform seeds.

Comparing the Costs

Because the combination of the four *Cs* varies among diamonds, there is no set price for a diamond. A 1-carat diamond may cost $7500. How does that compare with the cost of other items you buy?

Procedure

1. The data table lists some items you may frequently buy, along with their estimated price and mass. Copy the table, then calculate and record the weight in carats of each item.

Challenge

1. Calculate and record the cost per carat of each item in the data table.
2. The price of a 1-carat diamond may be $7500. How much would the items in the data table cost if they were priced at $7500 per carat?

Cost Comparison

Item	Price	Mass (g)	Weight (carats)	Cost/ Carat	Cost at $7500/ carat
music CD	$14.00	16			
magazine	$3.99	148			
can of soda	.75¢	355			
gallon of milk	$1.89	3629			

CHAPTER 4
Study Guide

Summary

SECTION 4.1

What is a mineral?

Main Ideas

- A mineral is a naturally occurring, inorganic solid with a specific chemical composition and a definite crystalline structure. There are at least 3000 known minerals in Earth's crust.
- A crystal is a solid in which the atoms are arranged in repeating patterns. The six main crystal systems are cubic, tetragonal, hexagonal, orthorhombic, monoclinic, and triclinic.
- Minerals form from magma or from supersaturated solution. Most minerals are formed from the eight most common elements in Earth's crust.
- Oxygen readily combines with other elements to form a diverse group of minerals, including silicates, carbonates, and oxides. A silica tetrahedron is a three-dimensional shape structured like a pyramid. In a silica tetrahedron one silicon atom attaches to four oxygen atoms.
- Other major mineral groups include sulfides, sulfates, halides, and native elements. Native elements such as silver or copper are made of one element only.

Vocabulary

crystal (p. 79)
magma (p. 80)
mineral (p. 77)
silicate (p. 81)

SECTION 4.2

Identifying Minerals

Main Ideas

- Minerals can be identified based on their physical and chemical properties. The most reliable way to identify a mineral is by using a combination of several tests.
- A mineral's color is generally the result of trace elements within the mineral. Texture describes how a mineral feels, and luster describes how a mineral reflects light. Cleavage and fracture describe how minerals break.
- A mineral's streak, hardness, and density are reliable methods of identification. Special properties of minerals such as magnetism also can be used for identification purposes.
- An ore contains a useful substance that can be mined at a profit. If the cost of mining the ore becomes higher than the value of the ore, then the mineral is no longer classified as an ore. The classification of a mineral as an ore may also change if the supply of or demand for the mineral changes.
- Gems are valuable minerals that are prized for their rarity and beauty. Trace elements can make one variety of a mineral more valuable than other varieties of the same mineral.

Vocabulary

cleavage (p. 86)
fracture (p. 87)
gem (p. 90)
hardness (p. 86)
luster (p. 84)
ore (p. 89)
specific gravity (p. 87)
streak (p. 85)

CHAPTER 4

Assessment

Understanding Main Ideas

1. How many minerals are found in Earth's crust?
   a. 1000
   b. 2000
   c. 3000
   d. 4000

2. Which of the following is part of the definition of a mineral?
   a. liquid
   b. organic
   c. synthetic
   d. inorganic

3. What element is the most abundant in Earth's crust?
   a. oxygen
   b. aluminum
   c. silicon
   d. potassium

4. What property causes the mineral galena to break into tiny cubes?
   a. its density
   b. the internal arrangement of its atoms
   c. its hardness
   d. its luster

5. What mineral fizzes when it comes in contact with hydrochloric acid?
   a. quartz
   b. calcite
   c. gypsum
   d. fluorite

6. A student rubs a mineral across an unglazed porcelain plate. What mineral property is the student testing?
   a. hardness
   b. luster
   c. color
   d. streak

7. A mineral has a mass of 100 g and a volume of 50 cm^3. What is its density?
   a. 5000 g
   b. 2 g/cm^3
   c. 5 g/cm^3
   d. 150 g/cm^3

8. *Dull, silky, waxy,* and *pearly* are descriptive terms that best describe which property of minerals?
   a. color
   b. luster
   c. streak
   d. cleavage

9. What would you use the Mohs scale of hardness for?
   a. to identify a mineral
   b. to find the mass of a mineral
   c. to calculate the density of a mineral
   d. to determine a mineral's fracture

10. The streak of which mineral is different from its external color?
    a. pyrite
    b. gold
    c. copper
    d. magnetite

11. What is an ore?
    a. a mineral that contains a useful substance
    b. a mineral found in food
    c. a mineral that has streak, but no color
    d. a mineral for which there is no demand

12. Why do some minerals attract magnets? What are some other special properties of minerals?

13. What property do the minerals copper, galena, magnetite, and pyrite have in common? Use *Appendix H* for help.

14. Why do minerals have different crystal shapes? Give examples of minerals that exhibit each of the six main crystal systems.

15. Sapphires and rubies are both forms of the mineral corundum, but they are different colors. Why?

Test-Taking Tip

STOCK UP ON SUPPLIES Bring all your test-taking tools: pencils, pens, erasers, correction fluid, a sharpener, a ruler, a calculator, and a protractor. Bring munchies, too. You might not be able to eat in the testing room, but healthy snacks come in handy during outside breaks.

CHAPTER 4

Assessment

Applying Main Ideas

16. A student places a clean, transparent sample of Iceland spar on top of the word *geology* in a textbook. How will the word *geology* appear to the student? Explain.

17. Use *Appendix H* to identify the mineral in the photo below. It contains iron, has a metallic luster, and has a streak that is the same color as the mineral itself.

18. Make a data table that compares and contrasts the chemical and physical properties of graphite and diamond. Use *Appendix H* and reference materials for help.

19. Calculate the density of copper if the volume of a sample is 30 cm^3 and the mass is 267 g.

Thinking Critically

20. Topaz will not leave a streak on an unglazed porcelain plate. Why? What method could you use to observe the streak of topaz?

21. Other than diamond, what mineral would be best for making a sandpaper product? Why? Use *Appendix H* for help.

22. Infer how early prospectors used density to determine whether they had found gold or pyrite.

23. When would a mineral no longer be an ore? Explain.

Standardized Test Practice

1. What is the second most abundant element in Earth's crust?

a. magma
b. oxygen
c. silicon
d. carbon

Use the table below to answer questions 2–3.

Mineral Characteristics

Mineral	Hardness	Specific Gravity	Luster/ Color
Feldspar	6–6.5	2.5–2.8	nonmetallic/ colorless or white
Fluorite	4	3–3.3	nonmetallic/ yellow, blue, purple, rose, green, or brown
Galena	2.5–2.75	7.4–7.6	metallic/ grayish black
Quartz	7	2.65	nonmetallic/ colorless in pure form

2. What is the hardest mineral in the table?

a. feldspar
b. fluorite
c. galena
d. quartz

3. Which mineral most likely has a shiny appearance?

a. feldspar
b. fluorite
c. galena
d. quartz

4. Which is the most reliable clue to a mineral's identity?

a. color
b. streak
c. hardness
d. luster

Chapter 5

Igneous Rocks

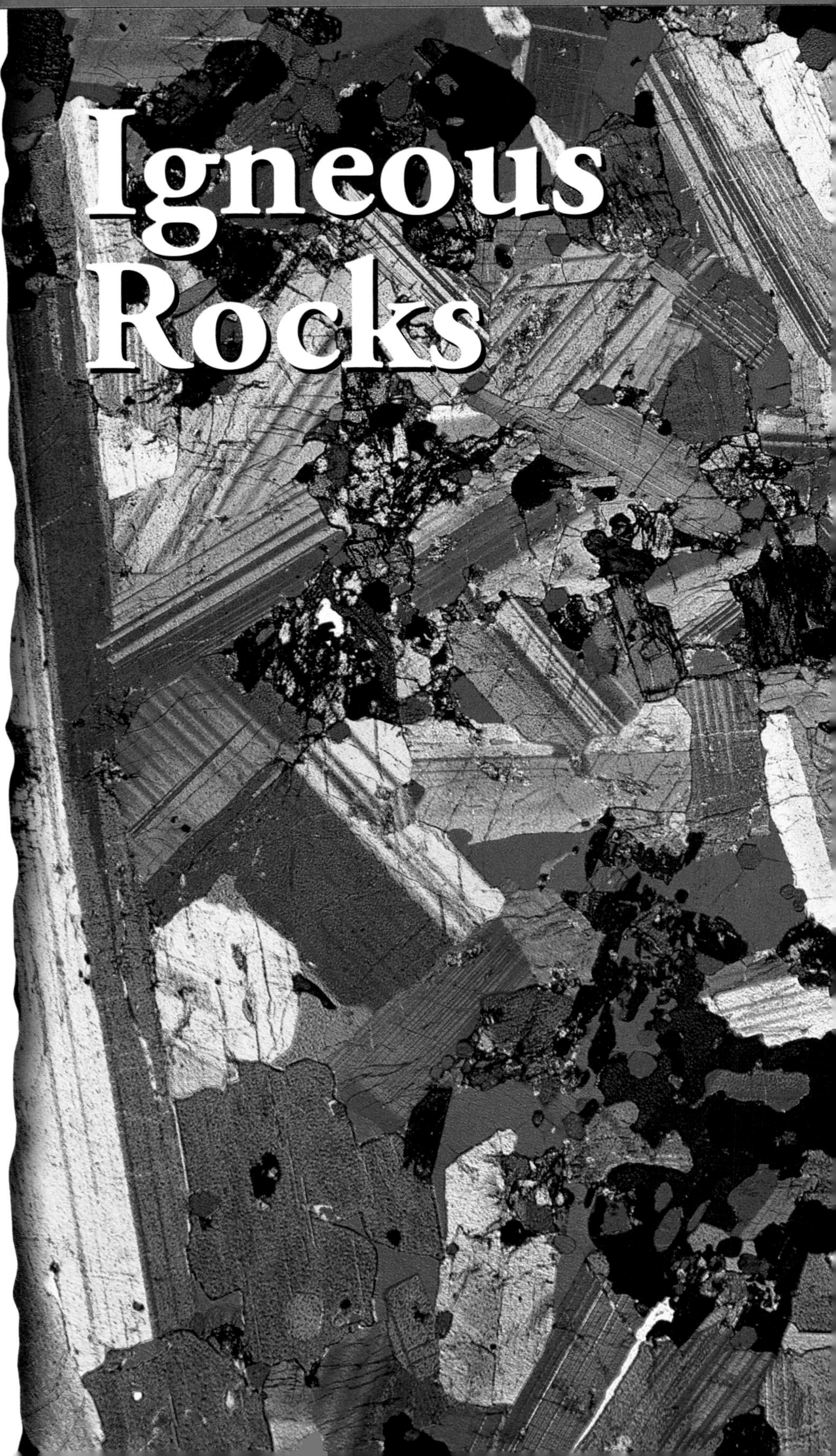

What You'll Learn

- How magma melts and crystallizes to form igneous rocks.
- How igneous rocks are classified.
- How igneous rocks are used.

Why It's Important

This photograph shows a microscopic view of gabbro, a type of igneous rock. Igneous rocks are the most abundant rocks in Earth's crust. Many important mineral and metal deposits are associated with igneous rocks.

Earth Science Online

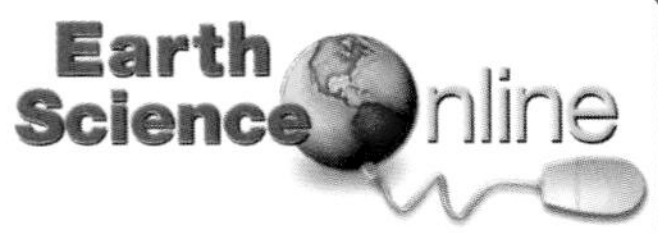

To find out more about igneous rocks, visit the Earth Science Web Site at earthgeu.com

Discovery Lab Identifying Minerals

Rocks are mixtures of minerals, organic matter, and other materials. Sometimes, it's possible to identify the different minerals in a sample of rock.

1. Examine a sample of granite from a distance of about 1 m. Record your observations.
2. Use a magnifying glass or stereomicroscope to observe the granite sample. Record your observations.

Observe In your science journal, draw a picture of what you saw through the magnifying glass or stereomicroscope. Include a scale for your drawing. How many different minerals did you observe in the rock? What minerals can you identify? Describe the sizes and shapes of the minerals. Do you see any evidence that these minerals crystallized from molten rock? Explain.

SECTION 5.1 What are igneous rocks?

Earth Sciences 3.c

OBJECTIVES

- **Compare** *and* **contrast** *intrusive and extrusive igneous rocks.*
- **Describe** *the composition of magma.*
- **Discuss** *the factors that affect how rocks melt and crystallize.*

VOCABULARY

igneous rock
lava
extrusive
intrusive
partial melting
fractional crystallization
Bowen's reaction series

In the *Discovery Lab,* you examined the minerals in a piece of granite. Granite is an igneous rock formed from magma, which, as you learned in Chapter 4, is molten rock below Earth's surface. **Igneous rocks** are formed from the crystallization of magma. The term *igneous* comes from the Latin word *ignis,* which means "fire" because early geologists often associated igneous rocks with fiery lava flows. **Lava** is magma that flows out onto Earth's surface.

TYPES OF IGNEOUS ROCKS

If you live near an active volcano, you can literally watch igneous rocks form. A hot, molten mass of rock may solidify into solid rock overnight. Fine-grained igneous rocks that cool quickly on Earth's surface are called **extrusive** igneous rocks. Coarse-grained igneous rocks that cool slowly beneath Earth's surface are called **intrusive** igneous rocks. Granite is the most common intrusive igneous rock. Initially, scientists did not believe that granite was igneous in origin because it was coarse grained and thus unlike the fine-grained surface rocks that formed from lava. In the late 1700s, however, careful study of granite rock formations revealed that they cut across other rock formations.

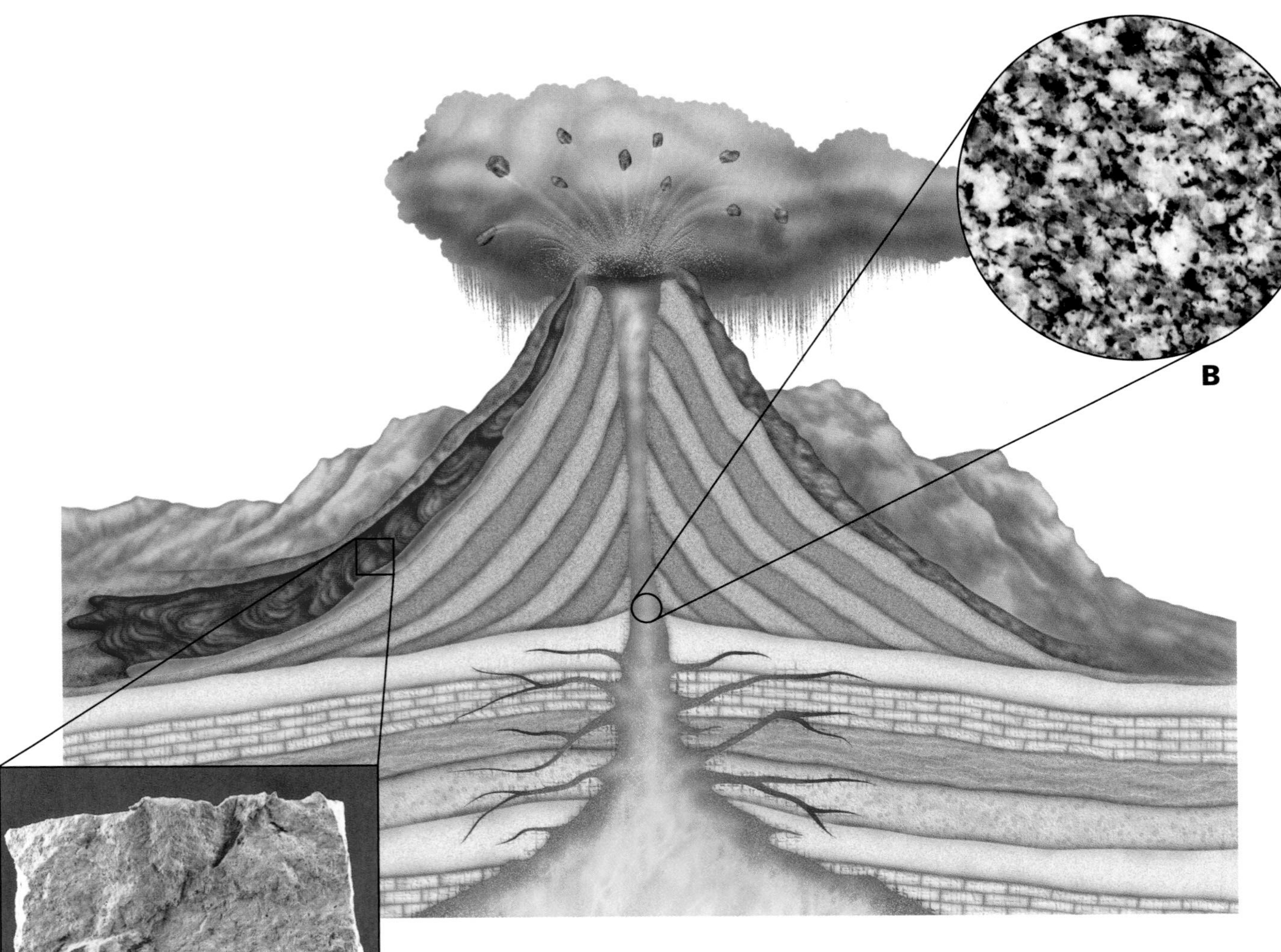

Figure 5-1 Lava cools quickly on Earth's surface and forms fine-grained igneous rocks such as rhyolite **(A).** Magma cools slowly beneath Earth's surface and forms coarse-grained igneous rocks such as granite. **(B).**

These cross-cutting relationships are evidence that the granite was intruded, or forced into, existing rocks. Cross-cutting relationships and other geologic clues provided direct evidence that granites were igneous rocks that formed underground. ***Figure 5-1*** shows the conditions under which granite and other igneous rocks form.

Composition of Magma

Magma is often a slushy mix of molten rock, gases, and mineral crystals. The elements found in magma are the same major elements found in Earth's crust: oxygen (O), silicon (Si), aluminum (Al), iron (Fe), magnesium (Mg), calcium (Ca), potassium (K), and sodium (Na). Of all the compounds found in magma, silica (SiO_2) is the most abundant and has the greatest effect on magma characteristics. As summarized in ***Table 5-1,*** magmas are classified as basaltic, andesitic, and rhyolitic, based on the amount of SiO_2 they contain. Silica content affects melting temperature and also impacts how quickly magma flows.

Origins of Magma

In the laboratory, most rocks must be heated to temperatures of 800°C to 1200°C before they melt. In nature, these temperatures are found in the upper mantle and lower crust. Where does this heat come from? Scientists theorize that the remaining energy from Earth's molten formation and the heat generated from the decay of radioactive elements are the sources of Earth's thermal energy.

Table 5-1
Types of Magma

Group	SiO_2 content
Rhyolitic	70 percent
Andesitic	60 percent
Basaltic	50 percent

Factors That Affect Magma Formation The main factors involved in the formation of magma are temperature, pressure, water content, and mineral composition. Temperature generally increases with depth in Earth's crust. This temperature increase, known as the geothermal gradient, is plotted in ***Figure 5-2A.*** Oil-well drillers and miners, such as those shown in ***Figure 5-2B,*** have firsthand experience with the geothermal gradient. Temperatures encountered when drilling deep oil wells can exceed 200°C.

Pressure also increases with depth. This is a result of the weight of overlying rock. Laboratory experiments show that as pressure on a rock increases, its melting point also increases. Thus, a rock may melt at 1100°C at Earth's surface, but the same rock will melt at 1400°C under the intense pressure found at a depth of 100 km.

The third factor that affects the formation of magma is water content. Rocks and minerals often contain small percentages of water, which changes the melting point of the rocks. As water content increases, the melting point decreases.

 **Earth Sciences**

3.c Students know how to explain the properties of rocks based on the physical and chemical conditions in which they formed, including plate tectonic processes.

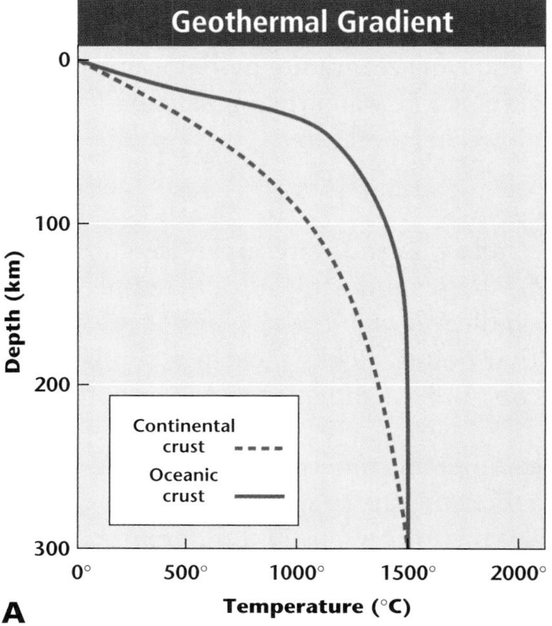

Figure 5-2 Differences in mineral composition cause the geothermal gradient to be higher in oceanic crust than in continental crust **(A),** as you'll learn on the next page. Also, the geothermal gradient causes temperatures in deep mines to be quite high **(B).**

Figure 5-3 Granite's higher water content and mineral composition cause it to melt at lower temperatures than basalt.

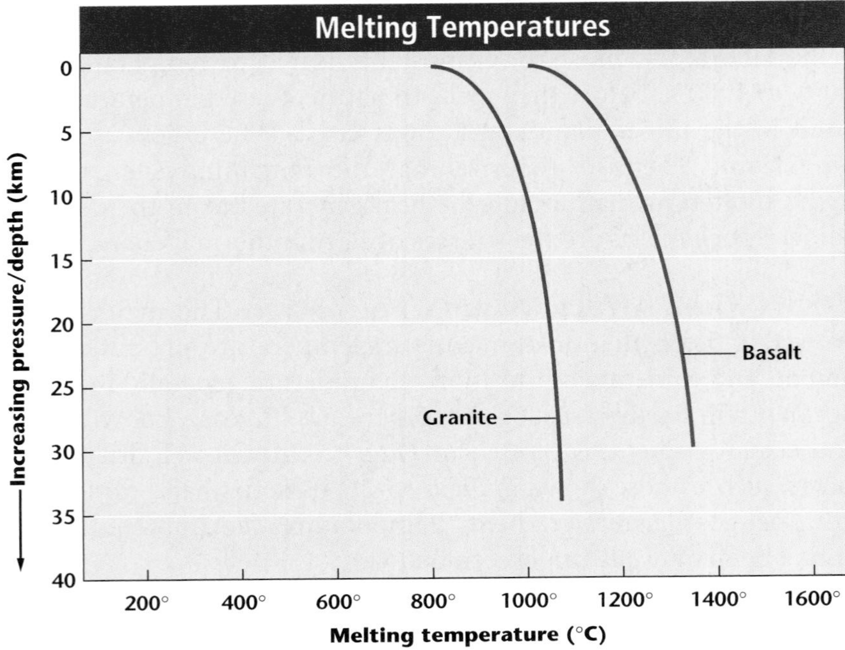

Mineral content also impacts how magma is formed. Different minerals have different melting points. For example, rocks formed of olivine, calcium feldspar, and pyroxene melt at higher temperatures than rocks containing quartz and potassium feldspar. In general, oceanic crust is rich in iron and magnesium and therefore melts at higher temperatures than continental crust, which contains higher levels of silicon and aluminum. Rocks melt only under certain conditions—the right combination of temperature, pressure, and composition must be present. ***Figure 5-3*** shows the melting curves of both granite and basalt. As you can see, granite has a lower melting point. This is because it contains more water than basalt and is made up of minerals that melt at lower temperatures.

How Rocks Melt

Suppose you froze bits of candle wax and water in an ice-cube tray. If you took the tray out of the freezer and left it at room temperature, the ice would melt but the candle wax would not. Why? The two substances have different melting points. Rocks melt in a similar way because the minerals they contain have different melting points.

Partial Melting Because different minerals have different melting points, not all parts of a rock melt at the same time. This explains why magma is often a slushy mix of crystals and molten rock. The process whereby some minerals melt at low temperatures while other minerals remain solid is called **partial melting.** Partial melting is

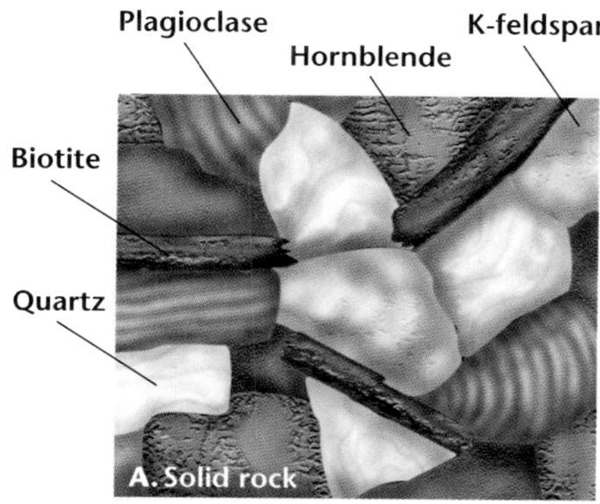

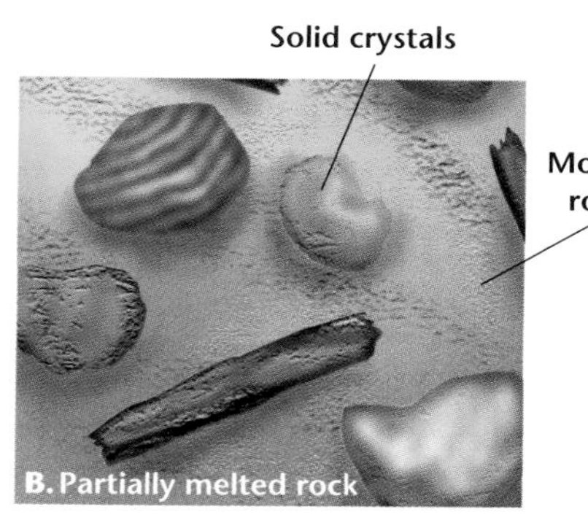

Figure 5-4 A rock is made up of different minerals that melt at different temperatures **(A).** Thus, during the melting process, some minerals are molten while others remain solid **(B).**

illustrated in ***Figure 5-4.*** As each group of minerals melts, different elements are added to the magma "stew," thereby changing its composition. If temperatures are not great enough to melt the entire rock, the resulting magma will have a different chemistry from that of the original rock. This is one way in which different types of igneous rocks form.

Fractional Crystallization When magma cools, it crystallizes in the reverse order of partial melting—the first minerals to crystallize from magma are the last minerals to melt during partial melting. The process wherein different minerals form at different temperatures is called **fractional crystallization.** This process, which is illustrated in ***Figure 5-5,*** is similar to partial melting in that the composition of magma may change. However, during fractional crystallization, the changes occur because as each group of minerals crystallizes, it removes elements from the remaining magma instead of adding new elements.

Bowen's Reaction Series

In the early 1900s, Canadian geologist N. L. Bowen demonstrated that as magma cools, minerals form in predictable patterns. **Bowen's reaction series** illustrates this relationship between cooling magma and mineral formation. Bowen discovered two main patterns, or branches, of crystallization. The first pattern is characterized by a continuous, gradual change of mineral compositions in the feldspar group. The second pattern is characterized by an abrupt change of mineral type in the iron-magnesium groups. ***Figure 5-6*** on page 104 illustrates Bowen's reaction series.

Feldspars In Bowen's reaction series, the right branch represents the feldspar minerals, which undergo a continuous change of composition.

Figure 5-5 Magma is made up of different minerals that crystallize at different temperatures **(A).** Thus, during the crystallization process, some minerals become solid while others remain molten **(B).**

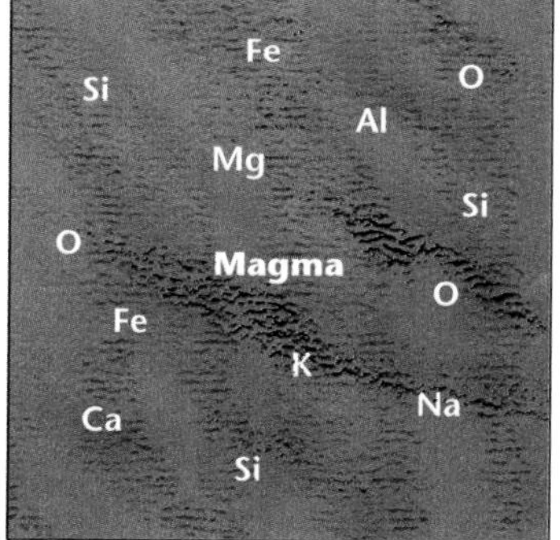

A

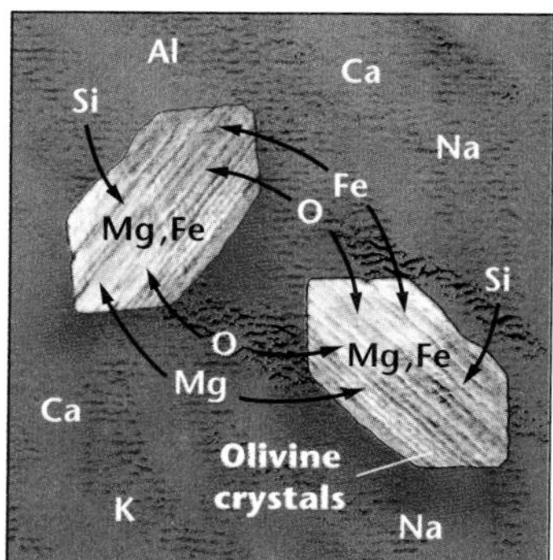

B

Figure 5-6 In the left branch of Bowen's reaction series, Fe-Mg minerals change to different minerals during the crystallization process. In the right branch, calcium-rich feldspars change gradually to sodium-rich feldspars.

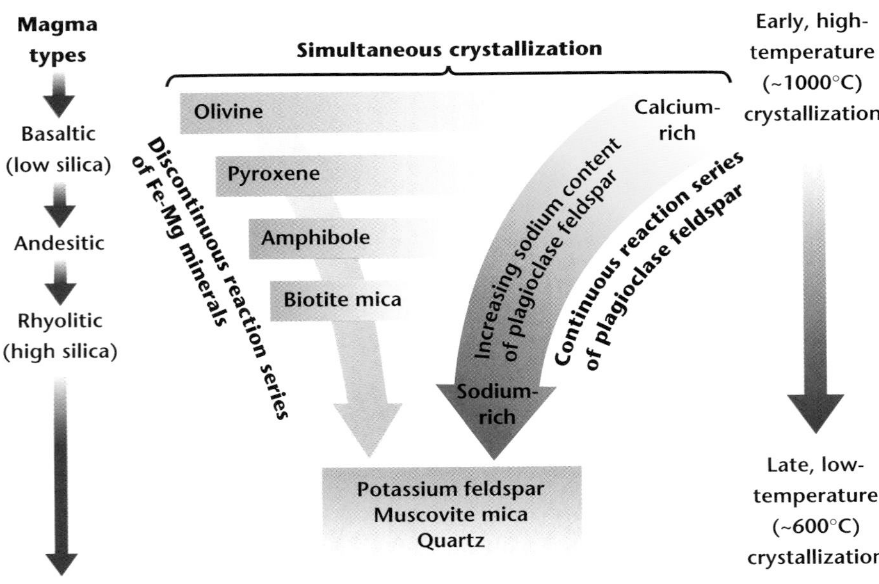

As magma cools, the first feldspars to form are rich in calcium. As cooling continues, these feldspars react with magma, and their calcium-rich compositions change to sodium-rich compositions. In some instances, as when magma cools rapidly, the calcium-rich cores are unable to react completely with the magma. The result is a zoned crystal that has sodium-rich outer layers and calcium-rich cores, as shown in ***Figure 5-7.***

Iron-Rich Minerals The left branch of Bowen's reaction series represents the iron-rich minerals. These minerals undergo abrupt changes during fractional crystallization. For example, when a

Figure 5-7 When magma cools quickly, a feldspar crystal may not have time to react completely with the magma and it retains a calcium-rich core **(A).** The result is a crystal with distinct zones **(B).**

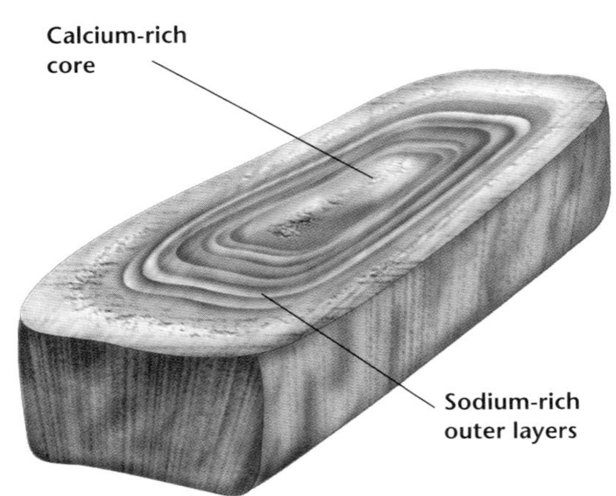

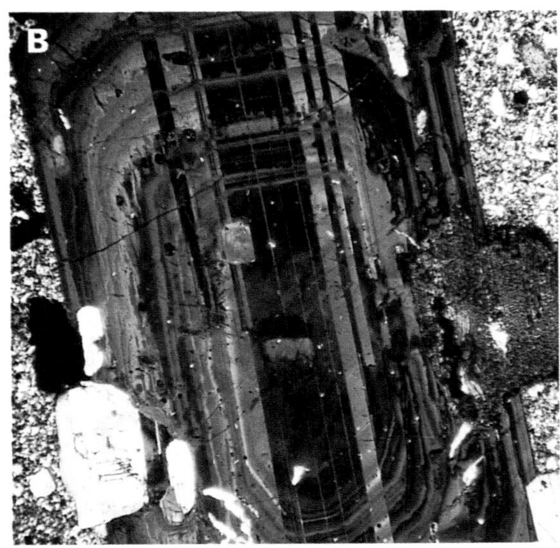

Figure 5-8 Quartz, the last mineral to crystallize, often forms in rock veins when the remaining magma is squeezed into rock fractures and cools.

magma rich in iron and magnesium cools to around 1800°C, olivine begins to crystallize. Olivine continues to form until the temperature drops to 1557°C. At that temperature, a completely new mineral, pyroxene, begins to form. All the olivine that previously formed reacts with the magma and is converted to pyroxene. Similar mineral changes have been observed in amphiboles and biotite.

As minerals form in the order shown in Bowen's reaction series, elements are removed from the remaining magma. Silica and oxygen, the most abundant elements in magma, are left over at the end of the reaction series. When the remaining melt, enriched with silica and oxygen, finally crystallizes, quartz is formed. Quartz often occurs in veins, as shown in ***Figure 5-8,*** because it crystallizes as the last liquid portion of magma is squeezed into rock fractures.

Topic: Bowen's Reaction
To find out more about Bowen's reaction series, visit the Earth Science Web Site at earthgeu.com

Activity: Explain which types of minerals weather more quickly than others. What unique rocks demonstrate this weathering process?

Crystal Separation As is often the case with scientific inquiry, Bowen's reaction series led to more questions. For example, if olivine converts to pyroxene during cooling, why is olivine found in rock? Geologists hypothesize that under certain conditions, newly formed crystals are separated from magma, and the chemical reactions between the magma and the minerals stop. Crystal separation can occur when crystals settle to the bottom of the magma body, and when liquid magma is squeezed from the crystal mush to form two distinct igneous bodies with different compositions.

Layered Intrusions In some magma bodies, the minerals form into distinct bands in the order shown in Bowen's reaction series. The result

Figure 5-9 The settling of crystals, flowing currents in magma, and temperature differences may cause the formation of a layered intrusion, which sometimes has metal-rich layers.

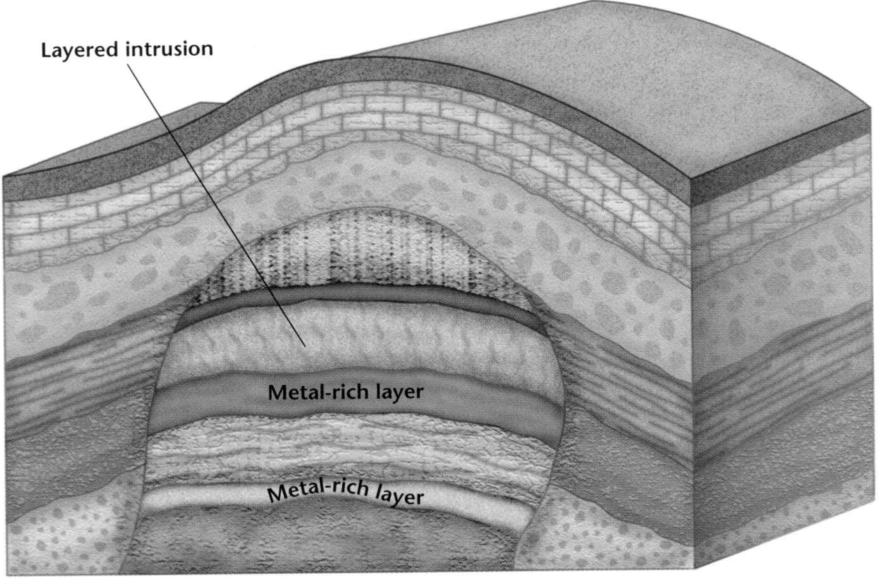

is a layered intrusion, as shown in ***Figure 5-9.*** Geologists are uncertain how these layers form. The settling of crystals, flowing currents in the magma, and temperature gradients within the magma chamber may all play a role. Layered igneous intrusions can be valuable sources of rare metals. Some have very high concentrations of elements such as platinum, chromium, nickel, or gold. For instance, a layered intrusion in Montana called the Stillwater Complex is the only source of platinum in the United States. Platinum is a critical component in catalytic converters, which are used to reduce the amount of pollutants that vehicles emit.

SECTION ASSESSMENT

1. Compare and contrast magma and lava. What two types of igneous rock are formed as each cools?
2. Make a data table that lists the eight major elements found in most magma. Include the chemical symbol of each element.
3. What are the factors that affect the formation of magma?
4. Compare the ways in which iron-magnesium minerals and feldspars crystallize from magma.
5. **Thinking Critically** Geologists have found zoned pyroxene crystals that have magnesium-rich cores and iron-rich outer layers. Which has a higher melting temperature, magnesium-rich pyroxene or iron-rich pyroxene? Explain your reasoning.

SKILL REVIEW

6. **Comparing and Contrasting** Compare and contrast how partial melting and fractional crystallization can change the composition of magma. For more help, refer to the *Skill Handbook.*

earthgeu.com/self_check_quiz

SECTION 5.2 Classifying Igneous Rocks

Earth Sciences 3.c, 9.a **I&E** 1.a

Igneous rocks are broadly classified as intrusive or extrusive. However, geologists further classify these rocks by their mineral compositions. In addition, physical properties such as grain size and texture serve as clues for the identification of various igneous rocks.

OBJECTIVES

- **Classify** *different types and textures of igneous rocks.*
- **Recognize** *the effects of cooling rates on the grain sizes of igneous rocks.*
- **Describe** *some uses of igneous rocks.*

VOCABULARY

felsic
mafic
ultramafic
porphyritic
pegmatite
kimberlite

MINERAL COMPOSITION

As shown in ***Table 5-2,*** the three main groups of igneous rocks—felsic, mafic, and intermediate—are classified according to their mineral compositions. **Felsic** rocks such as granite are light-colored, have high silica contents, and contain quartz and the feldspars orthoclase and plagioclase. **Mafic** rocks such as gabbro are dark-colored, have lower silica contents, and are rich in iron and magnesium. Mafic rocks contain plagioclase, biotite, amphibole, pyroxene, and olivine. Diorite is a good example of an intermediate rock with moderate amounts of biotite, amphibole, and pyroxene.

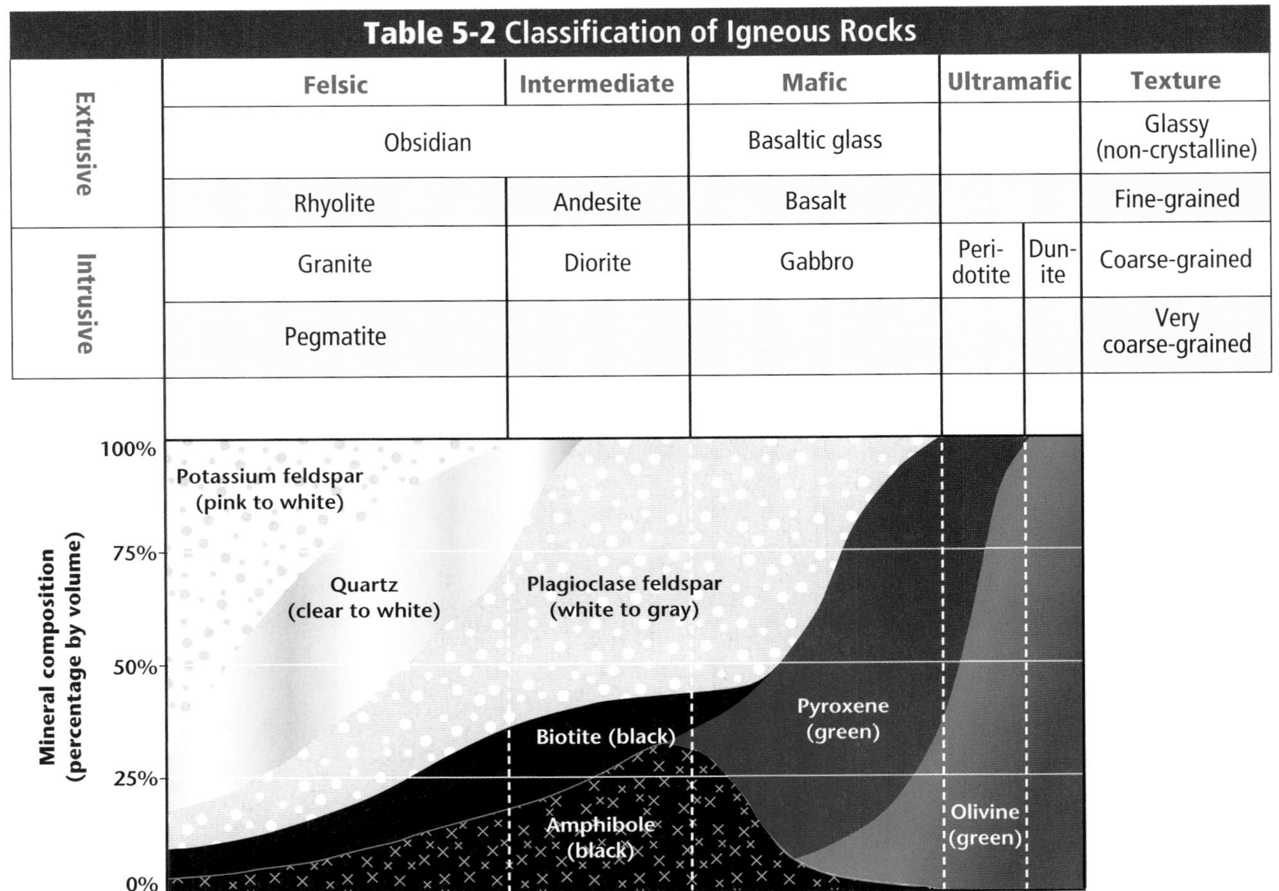

Table 5-2 Classification of Igneous Rocks

	Felsic	Intermediate	Mafic	Ultramafic		Texture
Extrusive	Obsidian		Basaltic glass			Glassy (non-crystalline)
Extrusive	Rhyolite	Andesite	Basalt			Fine-grained
Intrusive	Granite	Diorite	Gabbro	Peridotite	Dunite	Coarse-grained
Intrusive	Pegmatite					Very coarse-grained

MiniLab

How do igneous rocks differ?

Compare and **contrast** the different characteristics of igneous rocks.

Procedure

1. Using the igneous rock samples provided by your teacher, carefully observe the following characteristics of each rock: color, grain size, texture, and, if possible, mineral composition.
2. Design a data table in your science journal to record your observations.

Analyze and Conclude

1. Classify your rock samples as extrusive or intrusive rocks.
2. What characteristics do the extrusive rocks share? How do they differ? What characteristics do the intrusive rocks share? How do they differ?
3. Classify your rock samples as felsic, intermediate, mafic, or ultramafic.

Ultramafic Rocks Two unusual igneous rocks, peridotite and dunite, have low silica contents and very high levels of iron and magnesium, and thus, they are classified as **ultramafic** rocks. Some scientists theorize these ultramafic rocks, shown in ***Figure 5-10,*** are formed by the fractional crystallization of olivine and pyroxene. The minerals may have been separated from magma and did not convert to another mineral upon reaching a particular temperature. Another hypothesis is that ultramafic rocks represent pieces of the upper mantle that have been brought close to Earth's surface. In the *MiniLab* on this page, you'll analyze the mineral compositions of various igneous rocks.

Grain Size

In addition to differences in their mineral compositions, igneous rocks differ in the sizes of their grains. ***Figure 5-11*** compares obsidian, a glassy extrusive rock, and gabbro, a coarse-grained intrusive rock. What might account for the lack of visible crystals in obsidian and the large crystals of gabbro?

Figure 5-10 Dunite **(A)** and peridotite **(B)** are ultramafic rocks. They have low silica contents and high levels of iron and magnesium.

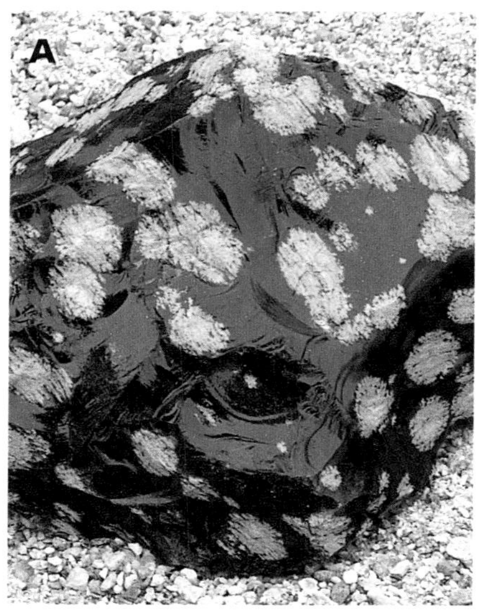

Figure 5-11 Obsidian cools quickly and has no visible mineral grains **(A).** The white patches that characterize this snowflake obsidian are mineral crystals that formed from impurities as the obsidian cooled. Gabbro cools slowly and thus large mineral grains form **(B).**

Earth Sciences

3.c Students know how to explain the properties of rocks based on the physical and chemical conditions in which they formed, including plate tectonic processes.

Cooling Rates When lava flows on Earth's surface, it is exposed to air and moisture. Under these conditions, the lava cools quickly, and there is not enough time for large crystals to form. Thus, extrusive igneous rocks such as obsidian have no visible mineral grains. In contrast, when magma cools slowly beneath Earth's surface, there is sufficient time for large crystals to form. Thus, intrusive igneous rocks such as gabbro may have crystals larger than 1 cm. You'll investigate the effects of cooling rate on crystal size in the *GeoLab* at the end of this chapter.

Texture

Often, it's easier to observe the sizes of mineral grains than it is to observe their shapes. Geologists solve this problem by making thin sections, which are slices of rock so thin that light can pass through them. As shown in the thin section in ***Figure 5-12,*** many mineral grains have interlocking edges. As the grains crystallize from magma, they grow together and form these irregular edges. Although irregular crystal shapes are characteristic of many igneous rocks, well-shaped crystals can form under certain conditions. During fractional crystallization, the minerals that form early in the process float in a liquid and have space in which to grow distinct crystal shapes.

Figure 5-12 The interlocking edges of mineral grains are evident in this thin section of diorite.

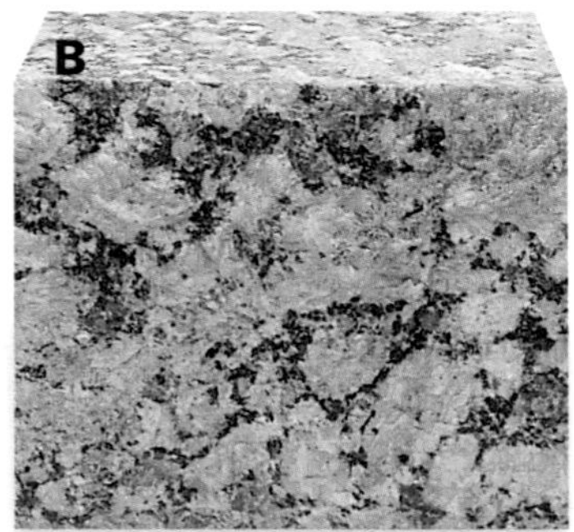

Figure 5-13 The sample of granite has crystals of the same size **(A).** In contrast, the other granite sample has a porphyritic texture with crystals of different sizes **(B).**

Porphyritic Texture Compare the crystal textures of the rocks shown in ***Figure 5-13.*** One of the rocks has grains of two different sizes. This rock has a **porphyritic** texture, which is characterized by large, well-formed crystals surrounded by finer-grained crystals of the same mineral or different minerals.

What causes minerals to form both large and small crystals in the same rock? Porphyritic textures indicate a complex cooling history wherein a slowly cooling magma suddenly began cooling rapidly. Imagine a magma body cooling slowly deep in Earth's crust. As it cools, the resulting crystals grow large. If this magma were to be

Problem-Solving Lab

Interpreting Scientific Illustrations

Estimate mineral composition
Igneous rocks are classified by their mineral compositions. In this activity, you'll estimate the different percentages of minerals in an igneous sample, then use your results to classify the rock.

Analysis

1. Using the diagram of the thin section shown here, design a method to estimate the percentages of the minerals in the rock sample.
2. Make a data table that lists the minerals and their estimated percentages.

Thinking Critically

3. Use ***Table 5-2*** to determine which type of igneous rock the thin section represents.
4. Compare your estimates of the percentages of minerals in the rock with those of your classmates. Hypothesize why the estimates vary. What are some possible sources of error?
5. What could you do to improve the accuracy of your estimate?

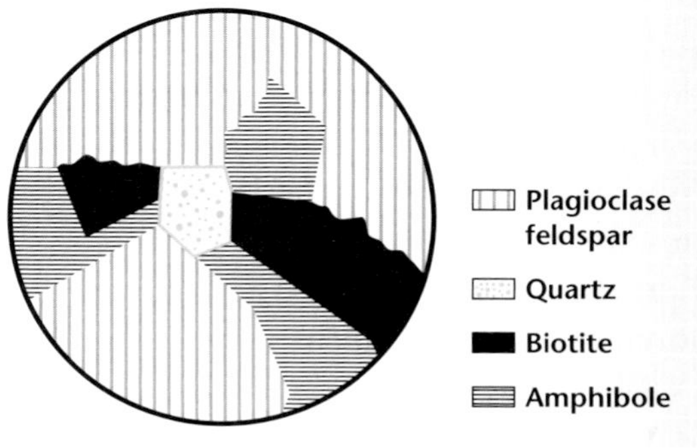

suddenly intruded higher in the crust, or if it erupted onto Earth's surface, the remaining magma would cool quickly and form smaller crystals. You'll explore other characteristics of igneous rocks in the *Problem-Solving Lab* on the previous page.

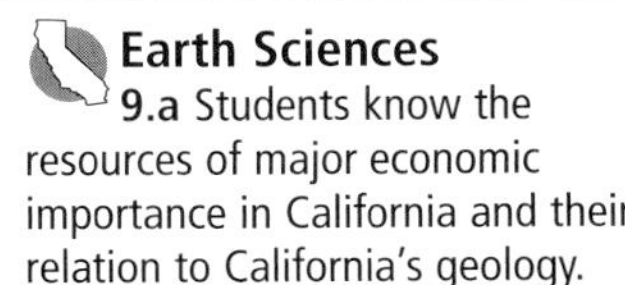

Earth Sciences
9.a Students know the resources of major economic importance in California and their relation to California's geology.

IGNEOUS ROCKS AS RESOURCES

Igneous rocks have several characteristics that make them especially useful as building materials. The interlocking grain textures of igneous rocks help to give them strength. In addition, many of the minerals found in igneous rocks are resistant to weathering. Granite is among the most durable of igneous rocks. Some common construction uses of granite are shown in ***Figure 5-14.*** You'll learn more about uses of other igneous rocks in the *Science & Technology* feature at the end of this chapter.

ENVIRONMENTAL CONNECTION

ORE DEPOSITS

As you learned in Chapter 4, ores are minerals that contain a useful substance that can be mined at a profit. Valuable ore deposits are often associated with igneous intrusions. Sometimes, these ore deposits are found within igneous rock, such as the layered intrusions mentioned earlier. Other times, ore minerals are found in the rocks surrounding intrusions. These type of deposits sometimes occur as veins.

Veins Recall from Bowen's reaction series that the fluid left during magma crystallization contains high levels of silica and water. This fluid also contains any leftover elements that

Figure 5-14 The columns in the Rhodes Memorial in Cape Town, South Africa **(A);** the kitchen tiles **(B);** and the Vietnam Memorial in Washington, D.C. **(C)** are all made of granite.

Figure 5-15 This sample of gold-bearing quartz came from El Dorado County, California.

Using Numbers
A granite slab has a density of 2.7 g/cm^3. What is the mass of a 2-cm thick countertop that is 0.6 m x 2.5 m? How many pounds is this?

were not incorporated into the common igneous minerals. Some important metallic elements that are not included in common minerals are gold, silver, lead, and copper. These elements, along with the dissolved silica, are released at the end of magma crystallization in a hot, mineral-rich fluid that fills cracks and voids in the surrounding rock. This fluid solidifies to form metal-rich quartz veins, such as the gold-bearing veins found in the Sierra Nevada mountains of California. An example of gold-bearing quartz is shown in ***Figure 5-15.***

Pegmatites Vein deposits may contain other valuable resources in addition to metals. Veins of extremely large-grained minerals, such as the one shown in ***Figure 5-16A,*** are called **pegmatites.** Ores of rare elements such as lithium and beryllium are found in pegmatites. In addition to ores, pegmatites can produce beautiful crystals. Because these veins fill cavities and fractures in rock, minerals grow into voids and retain their shapes. Some of the world's most beautiful minerals have been found in pegmatites. An example is shown in ***Figure 5-16B.***

Figure 5-16 Pegmatites are veins of extremely large-grained minerals **(A).** Stunning crystals such as this garnet are often found in pegmatites **(B).**

Figure 5-17 Diamonds are found in kimberlites **(A),** such as those in the Kimberly Diamond Mine in South Africa **(B).**

Kimberlites Diamond is a valuable mineral found in rare, ultramafic rocks known as **kimberlites,** named after Kimberly, South Africa, where the intrusions were first identified. These unusual rocks are a variety of peridotite. They likely form deep in the crust at depths of 150 to 300 km or in the mantle because diamond and other minerals found in kimberlites can form only under very high pressures.

Geologists hypothesize that kimberlite magma is intruded rapidly upwards towards Earth's surface, where it forms long, narrow, pipelike structures. These structures extend several kilometers into the crust, but they are only 100 to 300 m in diameter. Most of the world's diamonds come from South African mines, such as the one shown in ***Figure 5-17.***

Section Assessment

1. Describe the three major groups of igneous rocks. What are ultramafic rocks?
2. Why does rhyolite have smaller crystals than granite?
3. What chemical property is most commonly used to classify igneous rocks? List two physical properties that you could use to identify igneous rocks.
4. Why is gold often found in veins of quartz that are in and around igneous intrusions?
5. **Thinking Critically** Would quartz or plagioclase be more likely to form a well-shaped crystal in an igneous rock? Explain.

Skill Review

6. **Concept Mapping** Use the following terms to construct a concept map about igneous rock classification. For more help, refer to the *Skill Handbook.*

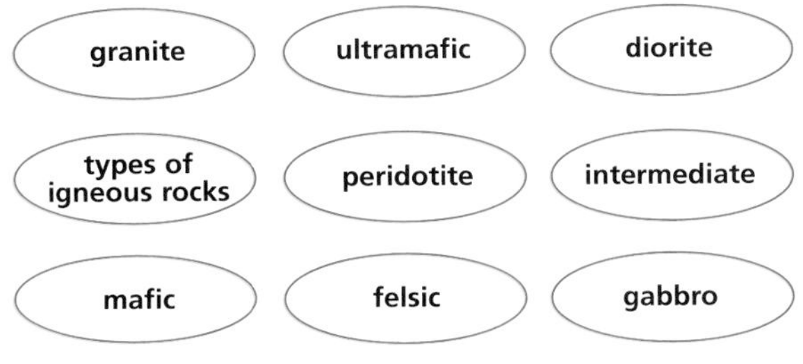

Modeling Crystal Formation

 I&E

1.a Select and use appropriate tools and technology (such as computer-linked probes, spreadsheets, and graphing calculators) to perform tests, collect data, analyze relationships, and display data.

The rate at which magma cools has an effect on the grain size of the resulting igneous rock. Observing the crystallization of magma is difficult because molten rock is very hot and the crystallization process is sometimes very slow. Other materials, however, crystallize at lower temperatures. These materials can be used to model crystal formation.

Preparation

Problem

Model the crystallization of minerals from magma.

Materials

clean, plastic petri dishes
saturated alum solution
200-mL glass beaker
magnifying glass
piece of dark-colored construction paper
thermometer
paper towels
water
hot plate

Objectives

In this GeoLab, you will:

- **Determine** the relationship between cooling rate and crystal size.
- **Compare** and **contrast** different crystal shapes.

Safety Precautions

The alum mixture can cause skin irritation and will be hot when it is first poured into the petri dishes. If splattering occurs, wash skin with cold water. Always wear safety goggles and an apron in the lab.

Procedure

1. As a group, plan how you could change the cooling rate of a hot solution poured into a petri dish. For instance, you may want to put one sample in a freezer or refrigerator for a designated period of time. Assign each group member a petri dish to observe during the experiment.
2. Place a piece of dark-colored construction paper on a level surface where it won't be disturbed. Place the petri dishes on top of the paper.

3. Carefully pour a saturated alum solution that is about 95°C to 98°C, or just below boiling temperature, into each petri dish so that it is half-full. Use caution when pouring the hot liquid to avoid splatters and burns.
4. Observe the petri dishes. Record your observations in your science journal. Draw what you observe happening in the petri dish assigned to you.
5. Every 5 minutes for 30 minutes, record your observations of your petri dish. Make accurate, full-sized drawings of any crystals that begin to form.

Analyze

1. How did you vary the cooling rate of the solutions in the petri dishes? Compare your methods with those of other groups. Did one method appear to work better than others? Explain.
2. Use a magnifying glass or binocular microscope to observe your alum crystals. What do the crystals look like? Are all the crystals the same size?
3. Compare your drawings and petri dish with those of other students in your group. Which petri dish had the smallest average crystal size? Describe the conditions under which that petri dish cooled.
4. Do all the crystals have the same shape? Draw the most common shape. Share your drawings with other groups. Describe any patterns that you see.

Conclude & Apply

1. What factors affected the size of the crystals in the different petri dishes? How do you know?
2. Infer why the crystals changed shape as they grew.
3. How is this experiment different from magma crystallization? How is it the same?
4. Describe the relationship between cooling rate and crystal formation.

Science & Technology

Cutting-Edge Surgery

***W**hen we hear the word* technology, *we often envision complicated gadgets. But one of the earliest forms of technology centered around common rocks. To better hunt their prey, for instance, our early ancestors created razor-sharp arrowheads and spears from the igneous rock obsidian. Today, there's a new use for this old rock: plastic surgery.*

Knapping

Knapping is the process of shaping a rock by using a mallet-like instrument to break off pieces of the rock. For thousands of years, this technique has been used to shape obsidian into tools and decorative pieces. This fast-cooling, extrusive rock has a conchoidal fracture, which allows the rock to break in predictable ways. Knappers use three techniques, sometimes in combination, to shape obsidian. Percussion flaking involves using a hammer or mallet to shape the rock. Pressure flaking involves using specially designed tools to pry off flakes of the rock. Lastly, in indirect percussion, a tool called a punch is placed on the edge of the rock. Flaking results when the punch is struck by a hammer or mallet.

Obsidian Scalpels

Knapping is carried on today by skilled artisans. Some of these modern knappers have taken their craft into the medical field, and are creating scalpels made from obsidian. Because obsidian scalpels are handmade, their surfaces look somewhat rough compared with traditional stainless steel scalpels. However, obsidian scalpels actually have a much sharper, smoother edge than stainless steel scalpels. When viewed under an electron microscope, the edges of an obsidian scalpel meet at a single point, which gives the scalpel its fine, sharp edge. This sharpness allows the scalpel to "divide" rather than tear flesh. Unlike the stainless steel scalpel, the obsidian scalpel creates such a small incision it barely leaves a scar. For this reason, obsidian scalpels are particularly well-suited to plastic surgery.

Disadvantages

At present, only a few doctors use obsidian scalpels, largely because these handmade-tools are relatively expensive. The price of the scalpels is high because only a few knappers are producing the scalpels, and each scalpel takes days or even weeks to complete. The average obsidian scalpel may cost $20. In contrast, stainless steel scalpels, which can be mass-produced, cost approximately $2 each.

Reading Analysis

In addition to scalpels, obsidian is used to make knives. Other igneous rocks, such as granite, are used in the construction industry. Go to the Earth Science Web Site at earthgeu.com to research and write a report about some common uses for igneous rocks.

Study Guide

Summary

SECTION 5.1

What are igneous rocks?

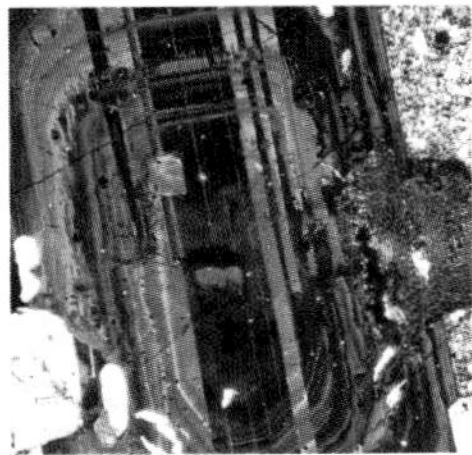

Main Ideas

- Igneous rocks are formed by the cooling and crystallization of magma. Intrusive rocks form inside Earth's crust, and extrusive rocks form on Earth's surface. Extrusive rocks, which cool more rapidly than intrusive rocks, are generally more fine grained.
- Magma is a slushy mix of molten rock, gases, and mineral crystals. The elements found in magma are the same major elements found in Earth's crust: oxygen (O), silicon (Si), aluminum (Al), iron (Fe), magnesium (Mg), calcium (Ca), potassium (K), and sodium (Na).
- Silica (SiO_2) is the most abundant compound in magma. Magmas are classified as basaltic, andesitic, and rhyolitic, based on the amount of SiO_2 they contain.
- Different minerals melt and crystallize at different temperatures in the processes of partial melting and fractional crystallization. Minerals crystallize from magma in a sequential pattern known as Bowen's reaction series.

Vocabulary

Bowen's reaction series (p. 103)
extrusive (p. 99)
fractional crystallization (p. 103)
igneous rock (p. 99)
intrusive (p. 99)
lava (p. 99)
partial melting (p. 102)

SECTION 5.2

Classifying Igneous Rocks

Main Ideas

- Igneous rocks are classified as felsic, mafic, intermediate, and ultramafic, depending upon their mineral compositions. Felsic rocks such as granite are light-colored, have high silica contents, and contain quartz and feldspars. Mafic rocks such as gabbro are dark-colored, have lower silica contents, and are rich in iron and magnesium. Intermediate rocks have moderate silica levels. Ultramafic rocks have low silica contents and very high levels of iron and magnesium. Igneous groups can be further identified by crystal size and texture.
- Early forming minerals may have well-shaped crystals, while later-forming minerals have irregular shapes. Porphyritic textures contain both large and small crystals.
- Igneous rocks such as granite are often used as building materials because of their strength, durability, and beauty.
- Valuable ore deposits and gems are often associated with igneous intrusions. Ores of rare elements such as lithium and beryllium are found in veins of extremely large-grained minerals called pegmatites. Diamonds are found in rare types of igneous intrusions known as kimberlites.

Vocabulary

felsic (p. 107)
kimberlite (p. 113)
mafic (p. 107)
pegmatite (p. 112)
porphyritic (p. 110)
ultramafic (p. 108)

CHAPTER 5

Assessment

Understanding Main Ideas

1. What term describes igneous rocks that crystallize inside Earth?
   a. magma **b.** intrusive **c.** lava **d.** extrusive

2. What magma type contains the greatest amount of SiO_2?
   a. basaltic **b.** andesitic **c.** rhyolitic **d.** peridotitic

3. What igneous rock has no visible crystals as a result of rapid cooling?
   a. gabbro b. andesite c. obsidian d. pegmatite

4. What type of ultramafic rock sometimes contains diamond?
   a. pegmatite **b.** kimberlite **c.** granite **d.** rhyolite

5. What minerals are associated with the right branch of Bowen's reaction series?
   a. olivine and pyroxene **b.** feldspars **c.** mica and feldspars **d.** quartz and biotite

6. What is the last mineral to crystallize from magma?
   a. biotite **b.** plagioclase **c.** olivine **d.** quartz

7. What are veins of extremely coarse-grained igneous rocks called?
   a. gabbros **b.** layered intrusions **c.** pegmatites d. crystals

8. What effect does a fast cooling rate have on grain size in igneous rocks?
   a. It forms fine-grained crystals.
   b. It forms large-grained crystals.
   c. It forms light crystals.
   d. It forms dark crystals.

9. What term describes magma that flows out onto Earth's surface?
   a. layered intrusion **b.** lava **c.** crystallization **d.** ultramafic

10. Which of the following affects the melting temperature of magma?
    a. ore deposits **b.** silica content **c.** oxygen content **d.** potassium content

11. Which of the following does not affect the formation of magma?
    a. temperature **b.** pressure **c.** volume **d.** mineral composition

12. What are some uses of igneous rocks in the construction industry?

13. Why do scientists theorize that kimberlites originate deep within Earth's crust or mantle?

14. What are the three main types of magma? What factor determines these classifications?

15. Describe Bowen's reaction series. Be sure to discuss the two branches of the series.

16. Why are olivine and calcium-rich plagioclase often found together in igneous rocks?

Test-Taking Tip

GET TO THE ROOT OF THINGS. If you don't know the definition of a word, you can infer its meaning by examining its roots, prefixes, and suffixes. For instance, words that start with *non-*, *un-*, *a-*, *dis-*, and *in-* generally reverse what the rest of the word means. Words that end in *-ly* are usually adverbs, and thus, are descriptive terms.

earthgeu.com/chapter_test

CHAPTER 5
Assessment

Applying Main Ideas

17. Why is magma usually a slushy mixture of crystals and molten rock?

18. What is unusual about peridotite and dunite?

Use the table below to answer questions 19-21.

Rock Composition				
Mineral	**Mineral Percentage**			
	Rock 1	**Rock 2**	**Rock 3**	**Rock 4**
Quartz	5	35	0	0
Potassium feldspar	0	15	0	0
Plagioclase feldspar	55	25	0	55
Biotite	15	15	0	10
Amphibole	25	10	0	30
Pyroxene	0	0	40	5
Olivine	0	0	60	0

19. Which rock is most likely granite?

20. Which rock is an ultramafic rock?

21. Rock 4 is fine grained. What type of rock is it?

22. What characteristics of igneous rocks make them good building materials?

Thinking Critically

23. How is it possible for magma to have a higher silica content than the rock from which it formed?

24. Would you expect to find plagioclase feldspar or biotite in a greater variety of igneous rocks? Explain.

25. Which would make a lighter-colored kitchen counter, granite or gabbro? Why?

26. Why are mineral deposits often found around the perimeter of igneous intrusions?

Standardized Test Practice

1. Which of the following is most abundant in magma and has the greatest effect on its characteristics?
- **a.** O
- **b.** Ca
- **c.** Al
- **d.** SiO_2

2. Which process describes how different minerals form at different rates?
- **a.** partial melting
- **b.** Bowen's reaction series
- **c.** fractional crystallization
- **d.** geothermal gradient

USING TABLES Use the table to answer questions 3 and 4.

Characteristics of Rocks			
	Color	**Silica Content**	**Composition**
Rock A	light	high	quartz and feldspars
Rock B	dark	low	iron and magnesium

3. Rock A is most likely what kind of rock?
- **a.** felsic
- **b.** mafic
- **c.** ultramafic
- **d.** intermediate

4. Which type of rock is rock B?
- **a.** granite
- **b.** diorite
- **c.** gabbro
- **d.** pegmatite

Chapter 6

Sedimentary and Metamorphic Rocks

What You'll Learn

- How sedimentary rocks are formed.
- How metamorphic rocks are formed.
- How rocks continuously change from one type to another in the rock cycle.

Why It's Important

Sedimentary rocks provide information about surface conditions and organisms that existed in Earth's past. In addition, mineral resources are found in sedimentary and metamorphic rocks. The rock cycle further provides evidence that Earth is a dynamic planet, constantly evolving and changing.

Earth Science Online

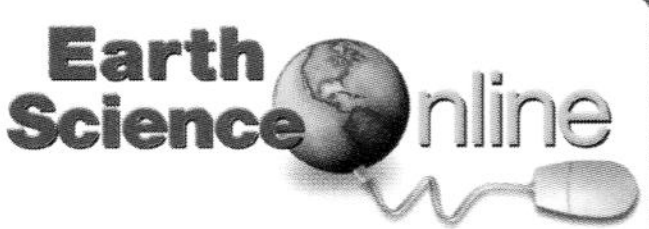

To find out more about sedimentary and metamorphic rocks, visit the Earth Science Web Site at earthgeu.com

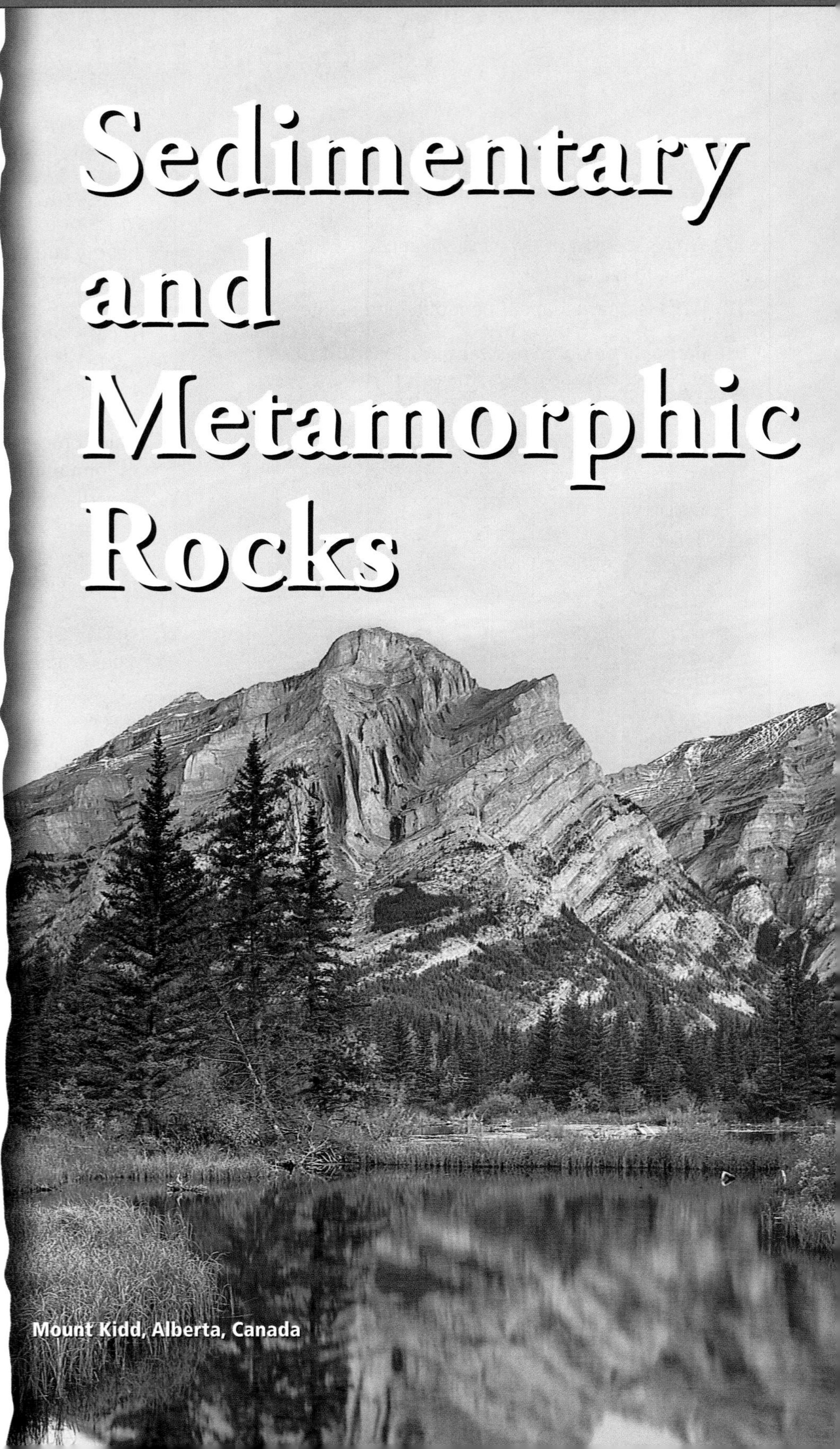

Mount Kidd, Alberta, Canada

Discovery Lab Model Sediment Layering

Sedimentary rocks are usually found in layers. How do these layers form? In this activity, you will investigate how layers form from particles that settle in water.

1. Obtain 100 mL of soil from a location specified by your teacher. Place the soil in a tall, narrow, jar.
2. Add water to the jar until it is three-fourths full. Put the lid on the jar so that it is tightly sealed.
3. Pick up the jar with both hands and turn it upside down several times to mix the water and soil.
4. Quickly turn the jar upright and set it on a flat surface.

Observe In your science journal, draw a diagram of what you observe. What type of particles settled out first? What type of particles form the topmost layers? How is this activity related to the layering that occurs in sedimentary rocks?

SECTION 6.1 Formation of Sedimentary Rocks

OBJECTIVES

- **Sequence** *the formation of sedimentary rocks.*
- **Explain** *the formation and classification of clastic sediments.*
- **Describe** *features of sedimentary rocks.*

VOCABULARY

sediment
clastic
deposition
lithification
cementation
bedding
graded bedding
cross-bedding

Earth Sciences 3.c

You learned in Chapter 5 that igneous rocks are the most common rocks in Earth's crust, yet when you look at the ground, you may not see igneous rocks. In fact, you usually don't see any solid rock at all. Why is this? Much of Earth's surface is covered with sediments. **Sediments** are pieces of solid material that have been deposited on Earth's surface by wind, water, ice, gravity, or chemical precipitation. When sediments become cemented together, they form sedimentary rocks. The formation of sedimentary rocks begins when weathering and erosion produce sediments.

WEATHERING

Wherever Earth's crust is exposed at the surface, it is continuously being worn away by weathering, a set of physical and chemical processes that break rock into smaller pieces. Chemical weathering occurs when the minerals in a rock are dissolved or otherwise

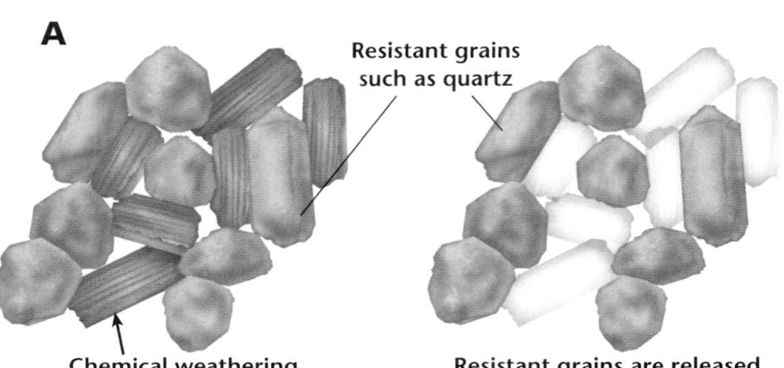

Figure 6-1 The process of chemical weathering is illustrated in **(A).** Granite breaks apart easily under the effects of chemical weathering **(B).**

chemically changed. Study ***Figure 6-1.*** What happens to more-resistant minerals during weathering? While the less-stable minerals are chemically broken down, the more-resistant grains are broken off of the rock as smaller grains. During physical weathering, on the other hand, minerals remain chemically unchanged. Rock fragments simply break off of the solid rock along fractures or grain boundaries.

Weathering produces rock and mineral fragments known as **clastic** sediments. The word *clastic* comes from the Greek word *klastos,* meaning "broken." Clastic sediments range in size from huge boulders to microscopic particles. ***Table 6-1*** summarizes the classification of clastic sediments based on size. Clastic sediment particles usually have worn surfaces and rounded corners caused by physical abrasion during erosion and transport.

 **Earth Sciences**

3.c Students know how to explain the properties of rocks based on the physical and chemical conditions in which they formed, including plate tectonic processes.

Erosion and Transport

After rock fragments have been weathered out of outcrops, they are transported to new locations. The removal and movement of surface materials from one location to another is called erosion.

Table 6-1 Classification of Clastic Sediments

Particle Size	Sediment		Rock
> 256 mm	Gravel	Boulder	Conglomerate
256–64 mm		Cobble	
64–2 mm		Pebble	
2–0.062 mm	Sand		Sandstone
0.062–0.0039 mm	Silt		Siltstone
<0.0039 mm	Clay		Mudstone or shale

Figure 6-2 shows the four main agents of erosion: wind, moving water, gravity, and glaciers. Visible signs of erosion are all around you. For example, water in streams becomes muddy after a storm because silt and clay particles have been added to it. The dust that collects on shelves in your home is another indication of erosion. Where do you think this dust comes from? How is it carried and how does it eventually settle on the shelves?

Eroded materials are almost always carried downhill. Although wind can sometimes carry fine sand and dust to higher elevations, particles transported by water are almost always moved downward. Eventually, even wind-blown dust and fine sand are pulled downhill by gravity. You will learn more about this in the next chapter.

Figure 6-2 Winds blow sand into dunes on the Navajo Indian Reservation in Arizona **(A).** The river Lethe cuts through the Valley of Ten Thousand Smokes in Katmai National Park, Alaska **(B).** This landslide in Papua, New Guinea, carried the entire hillside 300 m into the canyon **(C).** This terminal moraine was built up by the Athabasca Glacier in Jasper National Park, Alberta, Canada **(D).**

Deposition When sediments are laid down on the ground or sink to the bottoms of bodies of water **deposition** occurs. During the *Discovery Lab* at the beginning of this chapter, what happened when you stopped moving the jar full of sediment and water? The sediment sank to the bottom and was deposited. Similarly, in nature, sediments are

Figure 6-3 This large sand dune **(A)** in Algeria, Africa, is made up of fine sand such as this from Kalahari, South Africa **(B).**

deposited when transport stops. Perhaps the wind stops blowing or a river enters a quiet lake or the ocean. In each case, the particles being carried will settle out, forming layers of sediment. You observed this when you completed the *Discovery Lab* at the beginning of this chapter, that the sediment formed layers as it settled to the bottom of the jar. The sediment formed a layered deposit with the largest, grains at the bottom and the smallest particles at the top. As the water in the jar slowed down, the largest particles settled out first. Why? Faster-moving water can transport larger particles. As water slows down, the largest particles settle out first, then the next-largest, and so on, so that different-sized particles are sorted into layers. Such deposits are characteristic of sediment transported by water and wind. Wind, however, can move only small grains. For this reason, sand dunes, such as the ones shown in ***Figure 6-3A,*** are commonly made of fine, well-sorted sand like the sand in ***Figure 6-3B.***

Not all sediment deposits are sorted. Glaciers, for example, move all materials with equal ease. Large boulders, sand, and mud are all carried along by the ice and dumped in an unsorted pile at the end of the glacier. Landslides create similar deposits when sediment moves downhill in a jumbled mass.

Burial Most sediments are ultimately deposited on Earth in depressions called sedimentary basins. These basins may contain layers of sediment that together are more than 8 km thick. As more and more sediment is deposited in an area, the bottom layers are subjected to increasing pressure and temperature. These conditions cause **lithification,** the physical and chemical processes that transform sediments into sedimentary rocks. *Lithify* comes from the Greek word *lithos,* which means "stone."

LITHIFICATION

Lithification begins with compaction. The weight of overlying sediments forces the sediment grains closer together, causing the physical changes shown in ***Figure 6-4.*** Layers of mud may contain up to 60 percent water, and these shrink as excess water is squeezed out. Sand, however, is usually well compacted during deposition, and resists additional compaction during burial. Grain-to-grain contacts in sand form a supporting framework that helps maintain open spaces between the grains. Groundwater, oil, and natural gas are commonly found in these spaces in sedimentary rocks.

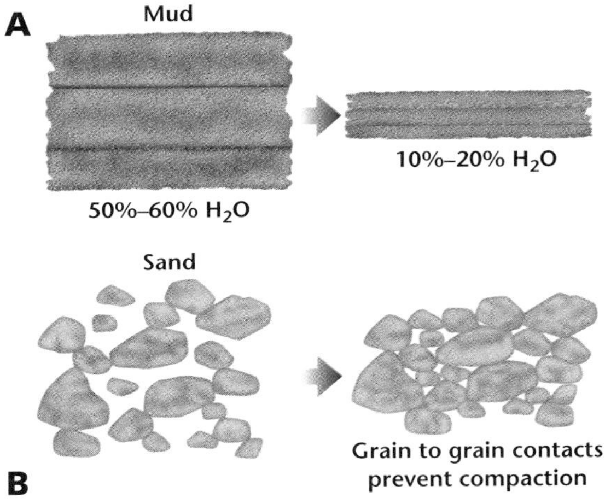

Figure 6-4 Pressure and weight from overlying sediments causes flat clay particles to compact **(A).** The irregular shape of sand grains prevents similar amounts of compaction **(B).**

The temperature in Earth's crust increases with depth by about 30°C per kilometer. Sediments that are buried 3 to 4 km deep experience temperatures that are high enough to start the chemical and mineral changes that cause cementation. **Cementation** occurs when mineral growth cements sediment grains together into solid rock. There are two common types of cementation. The first type occurs when a new mineral, such as calcite ($CaCO_3$) or iron oxide (Fe_2O_3) grows between sediment grains as dissolved minerals precipitate out of groundwater. The second type occurs when existing mineral grains grow larger as more of the same mineral precipitates from groundwater and crystallizes around them. These two types of cementation are shown in ***Figure 6-5.***

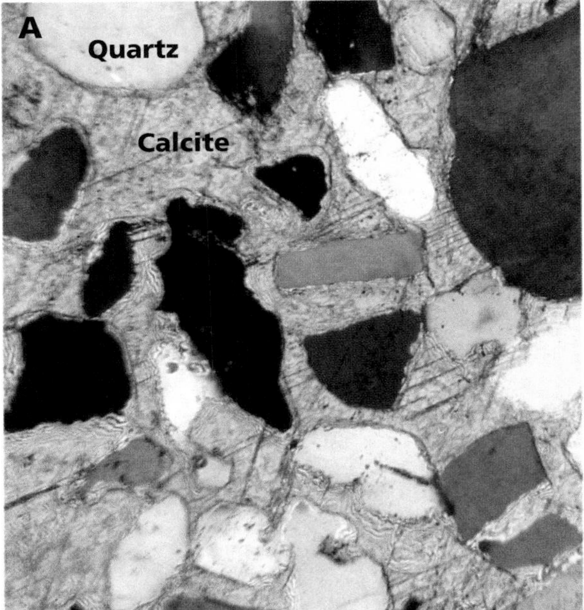

Figure 6-5 Cementation occurs in one of two ways. Either a new mineral, such as the calcite shown in **A,** grows between the grains **(B)** or, the same mineral grows between and over the grains in a process called overgrowth **(C).**

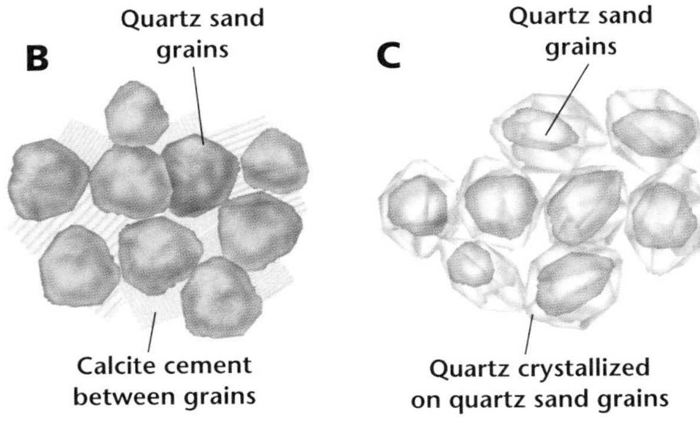

MiniLab

What happened here?

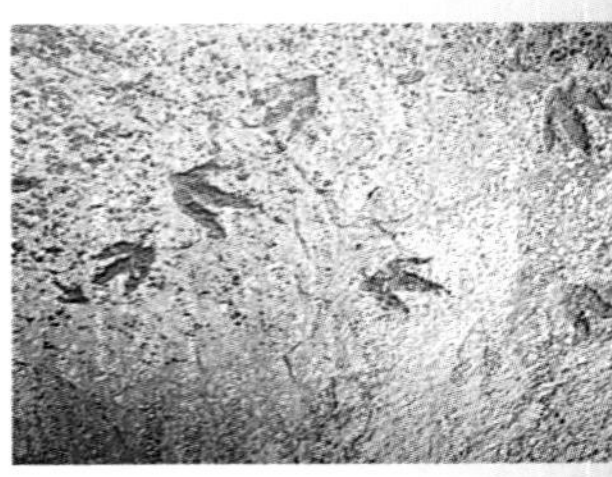

Interpret animal activity from patterns of fossil footprints.

Procedure

1. Study the photograph of a set of footprints that has been preserved in sedimentary rocks.
2. Write a description of how these tracks might have been made.
3. Draw your own diagram of a set of fossilized footprints that record the interactions of organisms in the environment.
4. Give your diagram to another student and have them interpret what happened.

Analyze and Conclude

1. How many animals made the tracks shown?
2. What types of information can be inferred from a set of fossil footprints?
3. Did other students interpret your diagram the same way? What might have caused any differences?

FEATURES OF SEDIMENTARY ROCKS

The primary feature of sedimentary rocks is horizontal layering, called **bedding.** Bedding can range from a millimeter-thick layer of shale to sandstone deposits several meters thick. The type of bedding depends upon the method of transport, while the size of the grains and the material within the bedding depend upon many factors.

Bedding in which the particle sizes become progressively heavier and coarser towards the bottom layers is called **graded bedding.** Graded bedding is often observed in marine sedimentary rocks that were deposited by underwater landslides. As the sliding material slowly came to rest underwater, the largest and heaviest material settled out first and was followed by progressively finer material.

Another characteristic feature of sedimentary rocks is cross-bedding. As you can see in ***Figure 6-6A,*** **cross-bedding** is formed as inclined layers of sediment move forward across a horizontal surface. Small-scale cross-bedding can be observed at sandy beaches and along sandbars in streams and rivers. Most large-scale cross-bedding, such as that shown in ***Figure 6-6B,*** is formed by migrating sand dunes. Small sedimentary features, such as the ripple marks shown

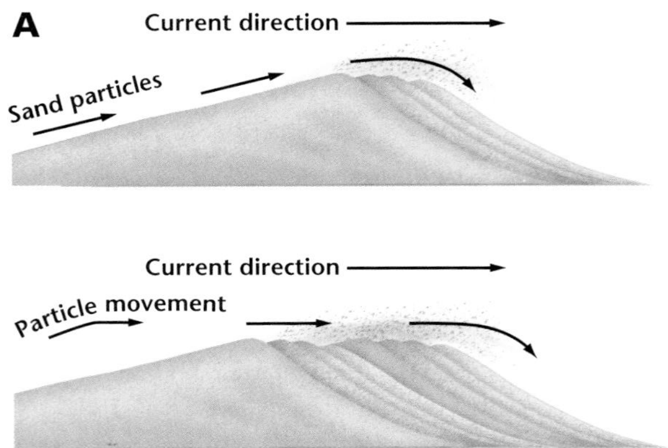

Figure 6-6 Cross-bedding is formed as sediment is carried forward across a layer of sediment, and cascades down the front face of the layer **(A).** Large-scale cross-bedded sandstones are common in Zion National Park **(B).**

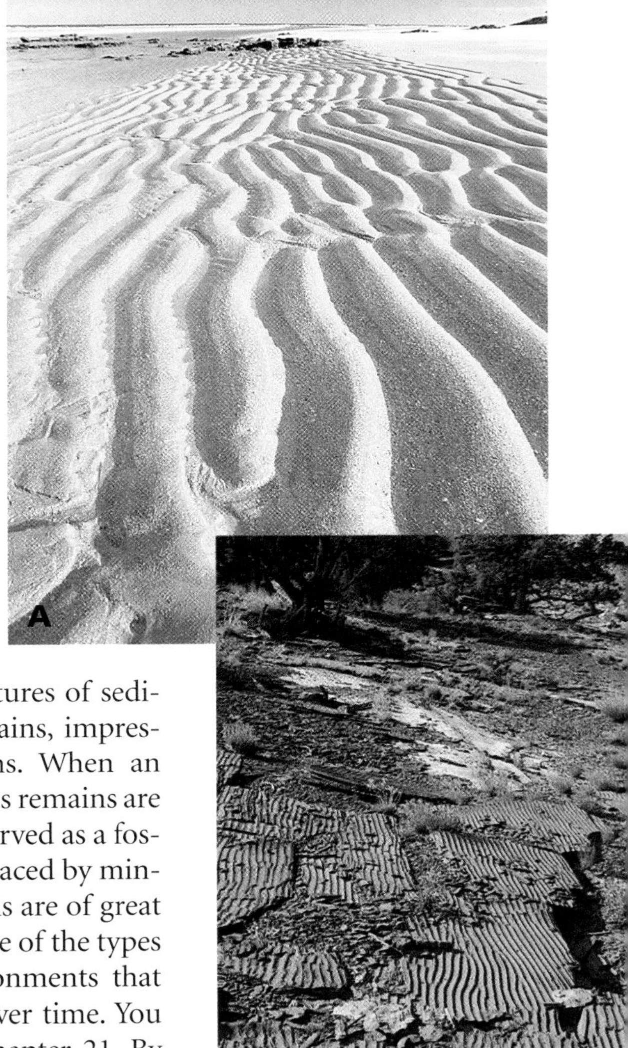

Figure 6-7 Given the right sedimentary conditions, these modern sand ripples **(A)** could become preserved like these ripple marks in Capitol Reef National Park, Utah **(B).**

in ***Figure 6-7,*** are also preserved in sedimentary rocks. Ripple marks form when sediment is moved into small ridges by wind or wave action, or by a river current. The back-and-forth movement of waves creates ripples that are symmetrical, while a current flowing in one direction, such as in a river or stream, produces asymmetrical ripples. If a rippled surface is buried gently by more sediment without being disturbed, it might later be preserved in solid rock.

Evidence of Past Life Probably the best-known features of sedimentary rocks are fossils. Fossils are the preserved remains, impressions, or any other evidence of once-living organisms. When an organism dies, it may be buried before it decomposes. If its remains are further buried without being disturbed, it might be preserved as a fossil. During lithification, parts of the organism can be replaced by minerals and turned into rock, such as fossilized shells. Fossils are of great interest to Earth scientists because fossils provide evidence of the types of organisms that lived in the distant past, the environments that existed in the past, and how organisms have changed over time. You will learn more about fossils and how they form in Chapter 21. By doing the *MiniLab* on the previous page, you can learn first-hand how fossils can be used to interpret past events.

Section Assessment

1. How are clastic sediments formed, and how do scientists classify them?
2. Why do sediment deposits tend to form layers?
3. As sediments are buried, what two factors increase with depth? How do these factors cause lithification?
4. Compare and contrast graded bedding and cross-bedding.
5. **Thinking Critically** Is it possible for a layer of cross-bedded strata to show graded bedding as well? Explain.

Skill Review

6. **Sequencing** Sequence the processes by which a sedimentary rock is formed from clastic sediments. For more help, refer to the *Skill Handbook.*

SECTION 6.2 Types of Sedimentary Rocks

Earth Sciences 3.c

OBJECTIVES

- **Describe** *the types of clastic sedimentary rocks.*
- **Explain** *how chemical sedimentary rocks form.*
- **Describe** *organic sedimentary rocks.*
- **Recognize** *the importance of sedimentary rocks.*

VOCABULARY

clastic sedimentary rock
porosity
evaporite

The classification of sedimentary rocks is based on how they were formed. There are three main groups of sedimentary rocks: clastic, chemical, and organic. ***Table 6-2*** summarizes the classification system for sedimentary rocks.

CLASTIC SEDIMENTARY ROCKS

The most common type of sedimentary rocks, **clastic sedimentary rocks,** are formed from the abundant deposits of loose sediments found on Earth's surface. Clastic sedimentary rocks are further classified according to the sizes of their particles. This classification system was shown in ***Table 6-1*** on page 122.

Coarse-Grained Clastics Sedimentary rocks consisting of gravel-sized rock and mineral fragments are classified as coarse-grained clastics, as shown in ***Figure 6-8.*** What differences between these two rocks do you notice? Conglomerates are coarse-grained sedimentary rocks that have rounded particles, whereas breccias contain angular fragments. How are these different-shaped particles formed? Because of its relatively large mass, gravel is transported by high-energy flows of water, such as those generated by mountain streams, flooding rivers, some ocean waves, and glacial meltwater. During transport, gravel becomes abraded and rounded as the particles scrape against one another. This is why beach and river gravels are often well rounded. Conglomerates provide evidence that this type of transport occurred in the past. In contrast, the angularity of

Earth Sciences

3.c Students know how to explain the properties of rocks based on the physical and chemical conditions in which they formed, including plate tectonic processes.

Table 6-2 Classification of Sedimentary Rocks

Rock Type	Rock Name	Method of Formation
Clastic		
Coarse-grained	Conglomerate or breccia	Lithification of clastic sediments
Medium-grained	Sandstone	
Fine-grained	Shale	
Chemical		Precipitation of dissolved minerals from water
Calcite	Limestone	
Halite	Rock salt	
Gypsum	Rock gypsum	
Organic		Accumulation and lithification of remains of living things
Calcium carbonate–shells	Limestone	
plant matter	Coal	

particles in breccias indicates that the sediments from which they formed did not have time to become rounded. This suggests that the particles were transported only a short distance and deposited close to their source. Under what kinds of circumstances might this type of transport occur?

Figure 6-8 This coarse-grained breccia **(A)** has angular fragments, and the conglomerate **(B)** has rounded fragments.

Medium-Grained Clastics In what types of environments is sand found? Stream and river channels, beaches, and deserts often contain abundant sand-sized sediments. Sedimentary rocks that contain sand-sized rock and mineral fragments are classified as medium-grained clastic rocks. When these medium-sized sediments are buried and lithified, sandstone is formed. Sandstone usually contains several features of interest to scientists. For example, because ripple marks and cross-bedding indicate the direction of current flow, geologists find sandstone layers particularly useful in mapping old stream and river channels.

Another important feature of sandstone is its relatively high porosity. **Porosity** is the percentage of open spaces between grains in a rock. Loose sand can have a porosity of up to 40 percent; some of its open spaces are maintained during the formation of sandstone. The incomplete cementation of mineral grains can result in porosities as high as 30 percent. When pore spaces are connected to one another, fluids can move through sandstone. This feature makes sandstone layers valuable as underground reservoirs of oil, natural gas, and groundwater.

Using Percentages Assume that the volume of a layer of mud will decrease by 35 percent during burial and compaction. If the original sediment layer is 30 cm thick, what will be the thickness of the shale layer after compaction and lithification?

Fine-Grained Clastics Sedimentary rocks consisting of silt and mud are called siltstone and mudstone. Siltstone is mostly composed of silt-sized grains, while shale is composed mostly of silt and clay-sized particles. Shale often breaks along thin layers, as shown in ***Figure 6-9.*** Unlike sandstone, this fine-grained sedimentary rock has very low porosity. It often forms barriers that hinder the movement of groundwater and oil.

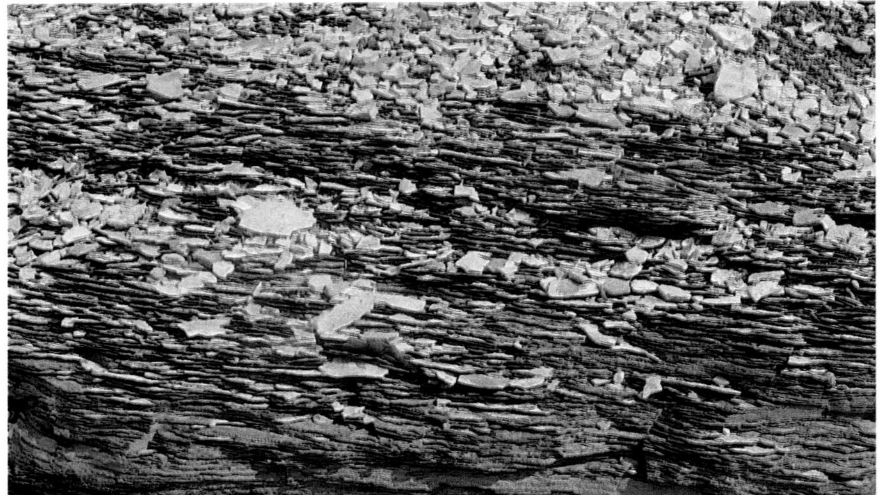

Figure 6-9 The thin bedding that is characteristic of shale is clearly seen in this outcrop from Ontario, Canada.

Chemical Sedimentary Rocks

What happens when you allow a glass of saltwater to evaporate? Eventually, the water disappears and a layer of salt accumulates on the bottom of the glass. A similar process occurs in nature when chemical sedimentary rocks are formed. During chemical weathering, minerals can be dissolved and carried into lakes and oceans. As water evaporates from the lakes and oceans, the dissolved minerals are left behind. In arid regions, high evaporation rates can increase the concentration of dissolved minerals in bodies of water. The Great Salt Lake, shown in ***Figure 6-10A,*** is a well-known example of a lake that has high concentrations of dissolved minerals.

Rocks Formed from Evaporation When the concentration of dissolved minerals in a body of water reaches saturation, which is the point at which no more minerals can be dissolved in the water, crystal grains precipitate out of solution and settle to the bottom. The layers of chemical sedimentary rocks that form as a result are called **evaporites**. Evaporites most commonly form in arid regions, in oceans and in drainage basins on continents that have low water flow. Because little freshwater flows into these areas, the concentration of dissolved minerals remains high.

As more dissolved minerals are carried into the basins, evaporation continues to remove freshwater and maintain high mineral concentrations. Over time, thick layers of evaporite minerals can accumulate on the basin floor, as illustrated in ***Figure 6-10B.***

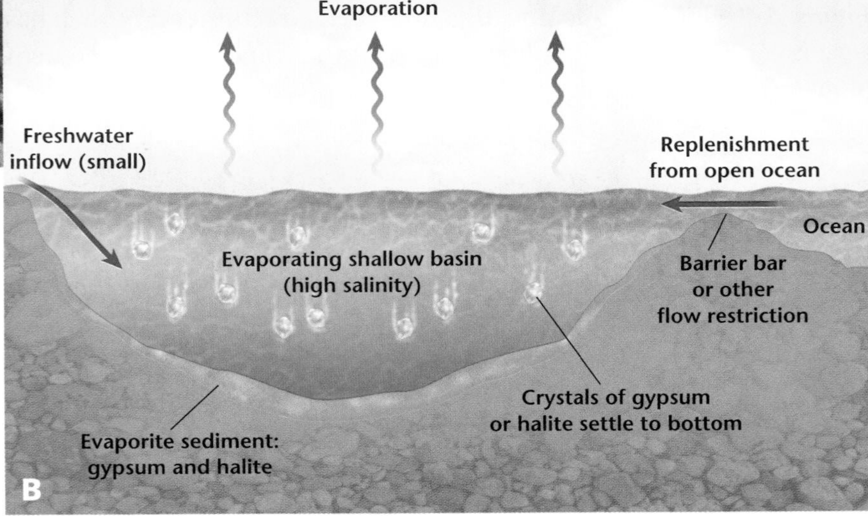

Figure 6-10 Evaporation of water from the Great Salt Lake, Utah, has resulted in salt precipitation on these boulders **(A).** The process of evaporite formation is illustrated in **B.**

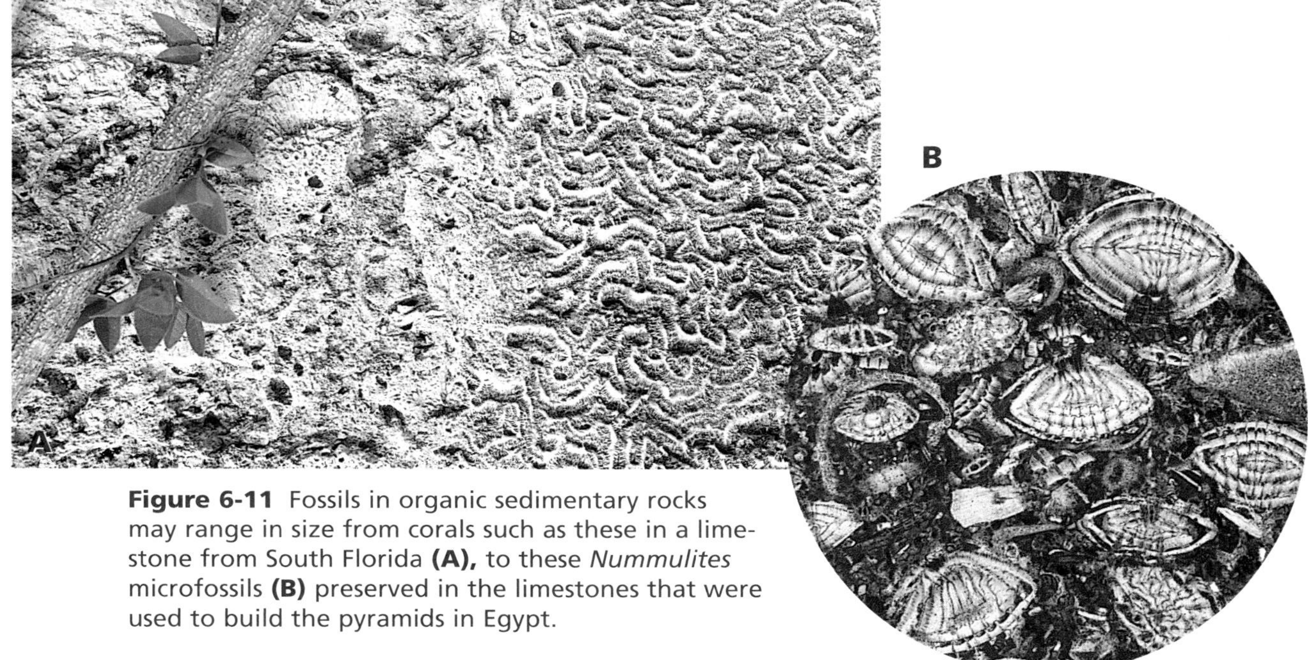

Figure 6-11 Fossils in organic sedimentary rocks may range in size from corals such as these in a limestone from South Florida **(A),** to these *Nummulites* microfossils **(B)** preserved in the limestones that were used to build the pyramids in Egypt.

The three most common evaporite minerals are calcite ($CaCO_3$), halite (NaCl), and gypsum ($CaSO_4$). Layers of these minerals are often mined for their chemical content.

Organic Sedimentary Rocks Organic sedimentary rocks are formed from the remains of once-living things. The most abundant organic sedimentary rock is limestone, which is composed primarily of calcite. Some organisms that live in the ocean use the calcium carbonate dissolved in seawater to make their shells. When these organisms die, their shells settle to the bottom of the ocean and can form thick layers of carbonate sediment. During burial and lithification, calcium carbonate precipitates out of the water, crystallizes between the grains of carbonate sediment, and forms limestone. Limestone is common in shallow water environments such as those in the Bahamas, where coral reefs thrive in 15 to 20 m of water just offshore. The skeleton and shell materials that are currently accumulating there will someday become limestone as well. Many types of limestone contain evidence of their biologic origin in the form of abundant fossils. As shown in ***Figure 6-11,*** these fossils range from large corals to microscopic unicellular organisms. Other organisms use silica to make their shells. These shells form sediment that is often referred to as siliceous ooze because it is rich in silica.

Another type of organic sedimentary rock, coal, forms from the remains of plant material. Over long periods of time, thick layers of vegetation slowly accumulate in swamps and coastal areas. When these layers are buried and compressed, they are slowly lithified into coal. Coal is composed almost entirely of carbon and can be burned for fuel. You will learn more about coal as an energy source in Chapter 25.

Topic: Coral Reefs
For an online update of coral reefs in the Bahamas, visit the Earth Science Web Site at earthgeu.com

Activity: Discuss the cause and significance of the major bleaching event of 1997-1998 in coral reefs around the world.

IMPORTANCE OF SEDIMENTARY ROCKS

The characteristic textures and features of sedimentary rocks, such as cross-bedding, ripple marks, layering, and fossils, provide a geologic "snapshot" of surface conditions in Earth's past. Fossils, for example, provide information about animals and plants that existed in the past. Other sedimentary features indicate the location and direction of flow of ancient rivers, the wave or wind direction over lakes and deserts, and ancient shoreline positions. Rock fragments found in conglomerates and breccias are large enough to easily identify what types of bedrock they were eroded from. By considering all of this information, geologists can reconstruct the nature of Earth's surface at various times in the past. Thus, they can better understand how geologic changes occur over time.

ENVIRONMENTAL CONNECTION

Energy Resources The study of sedimentary rocks provides information about Earth's past, but it also has great practical value. Many of the natural resources used by humans come from sedimentary rocks. For example oil, natural gas, and coal are found in sedimentary rocks. Uranium, which is used for nuclear power, is often mined from sandstone. Large deposits of phosphate, which is used for fertilizer, and iron, which is used to make steel, are also found in sedimentary rocks. Limestone is processed to make cement for the construction industry. Sandstone and limestone are often cut into blocks for use in walls and buildings. Were any sedimentary rocks used to construct your school? What sedimentary rocks were used in the construction of your home?

SECTION ASSESSMENT

1. Compare and contrast the main types of clastic sedimentary rocks.
2. Why do chemical sedimentary rocks form primarily in areas that have high rates of evaporation?
3. Why is coal an organic sedimentary rock?
4. What are some of the commercial values of sedimentary rocks?
5. **Thinking Critically** The original concentration of dissolved minerals in a restricted ocean basin was enough to form only a thin evaporite layer. How, then, is it possible that thick evaporite layers formed there?

SKILL REVIEW

6. **Comparing and Contrasting** Make a data table to compare and contrast the formation of the three types of sedimentary rock. For more help, refer to the *Skill Handbook*.

earthgeu.com/self_check_quiz

SECTION 6.3 Metamorphic Rocks

Earth Sciences 3.c **I&E** 1.c, 1.d, 1.m

OBJECTIVES

- **Compare** *and* **Contrast** *the different types and causes of metamorphism.*
- **Distinguish** *among metamorphic textures.*
- **Explain** *how mineral and compositional changes occur during metamorphism.*
- **Understand** *how rocks continuously change from one type to another in the rock cycle.*

VOCABULARY

regional metamorphism
contact metamorphism
hydrothermal metamorphism
foliated
nonfoliated
porphyroblast
rock cycle

You have learned that increasing pressure and temperature during burial cause recrystallization and cementation of sediments. What happens when rocks are buried at even greater depths?

CAUSES OF METAMORPHISM

Pressure and temperature increase with depth. When temperature or pressure becomes high enough, rocks melt and form magma. But what happens if the rocks do not quite reach the melting point? When high temperature and pressure combine to alter the texture, mineralogy, or chemical composition of a rock without melting it, a metamorphic rock forms. The word *metamorphism* is derived from the Greek words *meta,* meaning "change," and *morphē* meaning "form." During metamorphism, a rock changes form while remaining solid.

The high temperatures required for metamorphism ultimately are derived from Earth's internal heat, either through deep burial or from nearby igneous intrusions. The high pressures required for metamorphism can be generated in two ways: from vertical pressure caused by the weight of overlying rock, or from the compressive forces generated as rocks are deformed during mountain building.

TYPES OF METAMORPHISM

Different combinations of temperature and pressure result in different types of metamorphism, shown in ***Figure 6-12.*** Each combination produces a different group of metamorphic minerals and

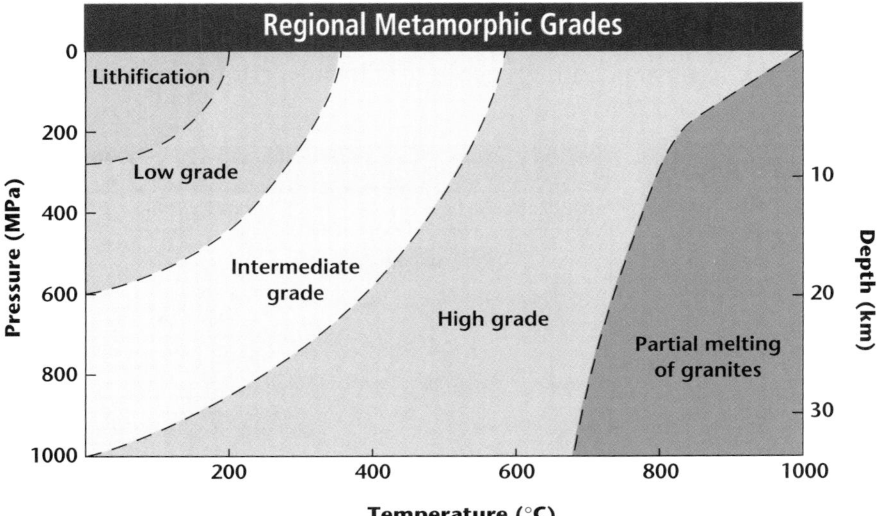

Figure 6-12 The grade of metamorphism, whether it is low, medium or high, is dependent upon the pressure on the rocks, the temperature and the depth below the surface.

Figure 6-13 The northeast portion of North America has undergone several episodes of regional metamorphism. The results can be seen in the distribution of metamorphic rocks.

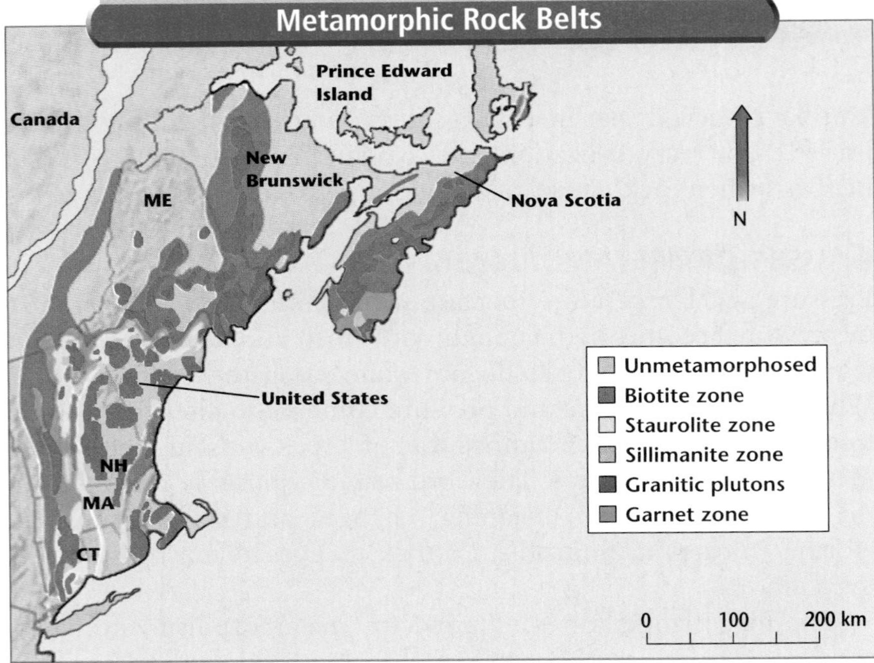

Earth Sciences

3.c Students know how to explain the properties of rocks based on the physical and chemical conditions in which they formed, including plate tectonic processes.

textures. When high temperature and pressure affect large regions of Earth's crust, they produce large belts of **regional metamorphism.** Regional metamorphism can be low grade, intermediate grade, and high grade. The grade of regional metamorphism reflects the relative intensity of temperature and pressure, with low-grade metamorphism reflecting the lowest temperature and pressure. ***Figure 6-13*** shows the regional metamorphic belt that has been mapped in the northeastern United States. Geologists have divided the belt into zones based upon the mineral groups found in the rocks. Some of the key minerals used to map metamorphic zones are listed in ***Figure 6-14.*** Knowing the temperatures that certain areas experienced when rocks were forming can help geologists locate economically valuable

Figure 6-14 Mineral changes in shale **(A)** and basalt **(B),** as a result of metamorphism follow a specific path. The grade of metamorphism (low, intermediate, or high) determines which minerals will form.

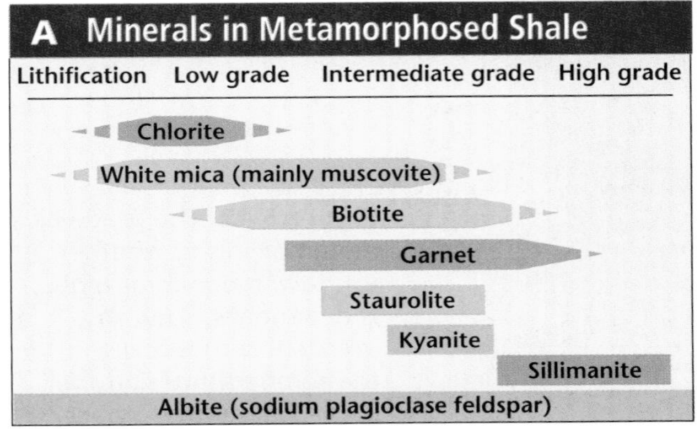

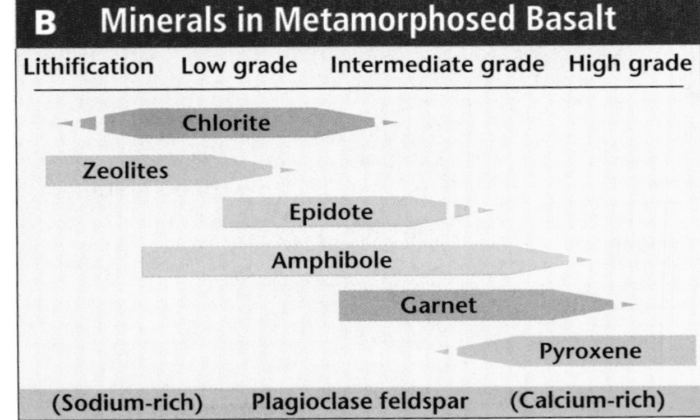

metamorphic minerals such as garnet and talc. An interesting connection between talc and asbestos, another metamorphic mineral, is described in the *Science in the News* feature at the end of this chapter.

When molten rocks, such as those in an igneous intrusion, come in contact with solid rock, a local effect called **contact metamorphism** occurs. High temperature and moderate-to-low pressure form the mineral assemblages that are characteristic of contact metamorphism. ***Figure 6-15*** shows zones of different minerals surrounding an intrusion. Why do you think these zones occur? Because temperature decreases with distance from an intrusion, metamorphic effects also decrease with distance. Recall from Chapter 4 that minerals crystallize at specific temperatures. Minerals that crystallize at high temperatures are found closest to the intrusion, where it is hottest. Contact metamorphism from extrusive igneous rocks is limited to thin zones. Normally, lava cools too quickly for the heat to penetrate very far into surface rocks.

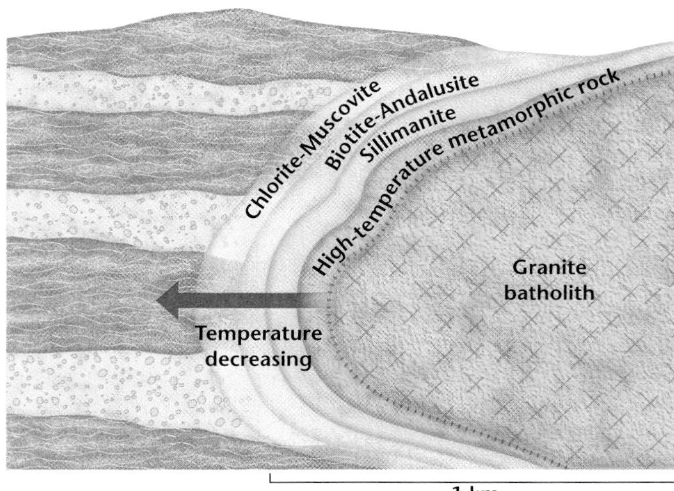

Figure 6-15 Contact metamorphism results in large-scale mineral changes with little deformation. Geologists can follow the occurrence of metamorphic minerals to locate the igneous intrusion.

When very hot water reacts with rock and alters its chemistry and mineralogy **hydrothermal metamorphism** occurs. The word *hydrothermal* is derived from the Greek words *hydro,* meaning "water," and *thermal,* meaning "heat." Hydrothermal fluids can dissolve some minerals, break down others, and deposit new minerals. These types of changes caused the yellow color of the cliffs shown in ***Figure 6-16.*** Hydrothermal metamorphism is common around igneous intrusions and near active volcanoes.

Figure 6-16 Yellowstone National Park's name comes from the hydrothermally changed rocks of the area. These rocks are beautifully exposed in the Grand Canyon of the Yellowstone, Wyoming.

A

Metamorphic Textures

Metamorphic rocks are classified into two textural groups: foliated and nonfoliated. Wavy layers and bands of minerals characterize **foliated** metamorphic rocks. High pressure during metamorphism causes minerals with flat or needlelike crystals to form with their long axes perpendicular to the pressure, as shown in ***Figure 6-17.*** This parallel alignment of minerals creates the layers observed in foliated metamorphic rocks. The two most common types of foliated metamorphic rock are schist, which is derived from shale, and gneiss, which is derived from granite. Some common foliated metamorphic rocks are compared in ***Figure 6-18.***

Unlike foliated rocks, **nonfoliated** metamorphic rocks lack mineral grains with long axes in one direction. Nonfoliated rocks are composed mainly of minerals that form with blocky crystal shapes. Two common examples of nonfoliated rocks, shown in ***Figure 6-19,*** are quartzite and marble. Quartzite is a hard, light-colored rock formed by the metamorphism of quartz-rich sandstone. Marble is formed by the metamorphism of limestone. Some marbles have very smooth textures that are formed by interlocking grains of calcite. Such marbles are sought by artists for sculptures.

Porphyroblasts Under certain conditions, new metamorphic minerals can grow quite large while the surrounding minerals remain small. The large crystals, which can range in size from a few millimeters to a few centimeters, are called **porphyroblasts.** Porphyroblasts are found in areas of both contact and regional metamorphism. These crystals resemble the very large crystals found in porphyritic igneous rocks but form not from magma but

B

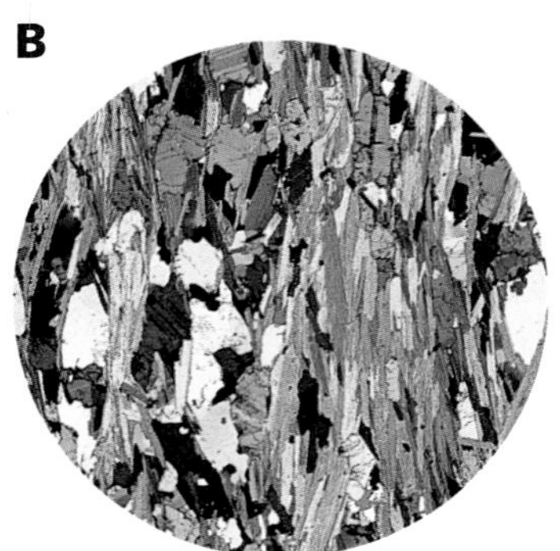

Magnification: 7×

Figure 6-17 Compressional pressure causes elongate minerals to line up perpendicular to the pressure direction **(A).** This photomicrograph of a mica schist **(B)** shows the resulting foliation.

Figure 6-18 These common foliated rocks are arranged in order of increasing metamorphic grade: slate **(A),** phyllite **(B),** gneiss **(C),** and schist **(D).**

A

Figure 6-19 Despite the extreme temperature and pressures, cross-bedding and ripple marks are often preserved in quartzite **(A).** Fossils, on the other hand, are never preserved in marble **(B).**

in solid rock by the reorganization of atoms during metamorphism. Garnet, shown in ***Figure 6-20,*** is a mineral that commonly forms porphyroblasts.

MINERAL CHANGES

How do minerals change without melting? Think back to the concept of fractional crystallization, discussed in Chapter 5. Minerals are stable at certain temperatures and crystallize from magma at different temperatures. Scientists have discovered that these stability ranges also apply to minerals in solid rock. During metamorphism, the minerals in a rock change into new minerals that are stable under the new temperature and pressure conditions. Minerals that change in this way are said to undergo solid-state alterations. Scientists have conducted experiments to identify the metamorphic conditions that create specific minerals. When these same minerals are found in rocks, scientists are able to interpret the conditions inside the crust during the rocks' metamorphism. Recall from page 134 that these conditions are temperature- and pressure-related, with low temperatures and pressures resulting in low-grade metamorphism. You will compare the changes in mineralogy as a result of high- and low-grade metamorphism in the *Problem-Solving Lab* on the next page.

Figure 6-20 This garnet mica schist comes from an exposure in Roxbury Falls, Connecticut.

COMPOSITIONAL CHANGES

Most metamorphic rocks reflect the original chemical composition of the parent rock. Gneiss, for example, has the same general chemical composition as granite. In some instances, however, the chemistry of a rock can be altered along with its minerals and texture. This occurs because hot fluids migrate in and out of the rock during metamorphism, which can change the original composition of the rock. Chemical changes are especially common during contact metamorphism near igneous intrusions. Hydrothermal fluids invade the surrounding rocks and change their mineralogy, textures, and chemistry. Valuable ore deposits of gold, copper, zinc, tungsten, and lead are formed in this manner.

The Rock Cycle

Metamorphic rocks are formed by the changing of other rocks. What types of rocks can metamorphic rocks be changed into? The three types of rock—igneous, sedimentary, and metamorphic—are grouped according to how they form. Igneous rocks crystallize from magma; sedimentary rocks form from cemented sediments; and metamorphic rocks form by changes in temperature and pressure. Once a rock forms, does it remain the same type of rock forever? Maybe, but probably not. Heat and pressure may change an igneous rock into a metamorphic rock. A metamorphic rock may be changed into another metamorphic rock or melted to form an igneous rock. Or, the metamorphic rock may be weathered and eroded into sediments that may become cemented into a sedimentary rock. In fact, any rock can be changed into any other type of rock. This continuous changing and remaking of rocks is called the **rock cycle.** The rock cycle is summarized in ***Figure 6-21.*** The arrows represent the different processes that change rocks into different types. Essentially, rocks are recycled into different rock types much like glass is recycled into different types of jars and bottles. You will learn more about the similarities and differences between rock types when you complete the *GeoLab* at the end of this chapter.

Problem-Solving Lab

Interpreting Scientific Illustrations

Determine which metamorphic minerals will form The types of minerals found in metamorphic rocks depends on metamorphic grade and composition of the original rock. In this activity, you will compare how these factors affect metamorphic minerals.

Analysis

1. Study ***Figure 6-14*** on page 134, showing the different mineral groups that are created under different metamorphic conditions. What mineral is formed when shale and basalt are exposed to low-grade metamorphism?
2. What mineral is formed when shale is exposed to high-grade metamorphism that is not found in basalts under the same conditions?

Thinking Critically

3. Compare the mineral groups you would expect to form from intermediate metamorphism of shale, basalt and limestone.
4. What are the major compositional differences between shale and basalt? How are these differences reflected in the minerals formed during metamorphism?
5. When limestones are metamorphosed, there is very little change in mineralogy, and calcite is still the dominant mineral. Explain why this happens.

Other Possible Paths

There is more than one path in the rock cycle. Sandstone might just as easily become uplifted and weathered back into sediments, thereby bypassing the metamorphic and igneous stages. Another possibility is that sandstone could become intruded by magma and melted, thereby directly becoming an igneous rock and bypassing metamorphism. Can you think of other possible paths? How might a metamorphic rock become a sedimentary rock?

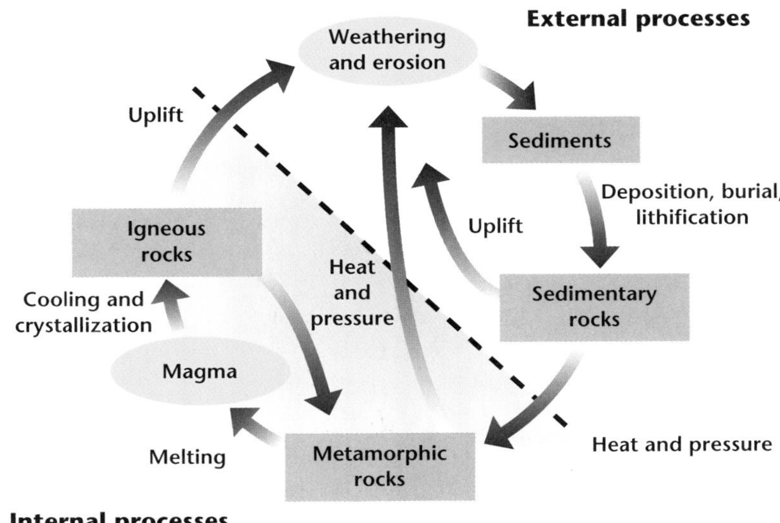

Figure 6-21 The rocks of Earth, whether at the surface or below the crust, are always positioned somewhere on the rock cycle.

The rocks of Earth's crust are constantly being recycled from one type to another. At any given time, magma is crystallizing, sediments are being cemented, and deeply buried rocks are metamorphosing. These processes take place underground, where they cannot be easily observed. However, as you have learned, not all phases of the rock cycle occur beneath Earth's surface. The processes that help shape Earth's landscapes are also part of the rock cycle. You will learn more about these surface processes in the next few chapters.

Section Assessment

1. How can the chemical composition of a rock be changed during metamorphism?
2. What are the three main types of metamorphism? Compare and contrast the factors that cause each type.
3. How does quartzite differ from schist?
4. What causes foliated metamorphic textures?
5. How are the three types of rocks classified?
6. What parts of the rock cycle occur deep in Earth's crust?
7. **Thinking Critically** Which would you expect to cause the greatest amount of contact metamorphism: an intrusion of basaltic magma or an intrusion of rhyolitic magma?

Skill Review

8. **Interpreting Photos** Study *Figure 6-18C.* If this sample is in the exact position it was when it formed, from which direction did the compressional forces originate? What group of metamorphic minerals would you expect during high-grade metamorphism of shale? For more help, refer to the *Skill Handbook.*

earthgeu.com/self_check_quiz

Interpreting Changes in Rocks

Earth Sciences

3.c Students know how to explain the properties of rocks based on the physical and chemical conditions in which they formed, including plate tectonic processes.

Also covers: I&E 1.c, 1.d, 1.m

As the rock cycle continues, and rocks change from one type to another, more changes occur than meet the eye. Color, grain size, texture and mineral composition are easily observed and described visually. Yet, with mineral changes come changes in crystal structure and density. How can these be accounted for and described? Studying pairs of sedimentary and metamorphic rocks can show you how.

Preparation

Problem

How do the characteristics of sedimentary and metamorphic rocks compare?

Materials

samples of sedimentary rocks and their metamorphic equivalents
magnifying glass or hand lens
paper
pencil
beam balance
100-mL graduated cylinder or beaker large enough to hold the rock samples
water

Objectives

In this GeoLab, you will:

- **Describe** the characteristics of sedimentary and metamorphic rocks.
- **Determine** the density of different rock types.
- **Infer** how metamorphism changes the structure of rocks.

Safety Precautions

Always wear safety goggles and an apron in the lab.

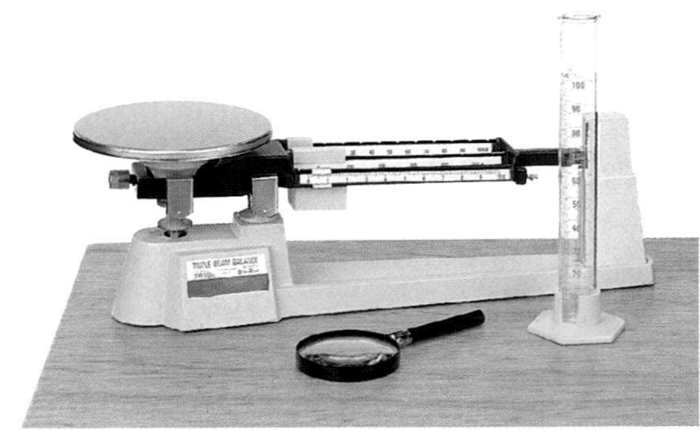

Procedure

1. Prepare a data table similar to the one shown below.
2. Observe each rock sample. Record your observations in the data table.
3. Recall that density = mass/volume. Make a plan that will allow you to measure the mass and volume of a rock sample.
4. Determine the density of each rock sample and record this information in the data table.

Analyze

1. Compare and contrast a shale and a sandstone.
2. How does the grain size of a sandstone change during metamorphism?
3. What textural differences do you observe between a shale and a slate?
4. Compare the densities you calculated with other students. Does everybody have the same answer? What are some of the reasons that answers may vary?

Conclude & Apply

1. Why does the color of a sedimentary rock change during metamorphism?
2. Compare the density of a slate and a quartzite. Which rock has a greater density? Explain.
3. Compare the densities of shale and slate, sandstone and quartzite, and limestone and marble. Does density always change in the same way? Explain the results that you observed.

Data Table

Sample number	Rock Type	Specific characteristics	Mass	Volume	Density
1					
2					
3					
4					

Science in the News

Good News—Crayons Safe

***The** Consumer Products Safety Commission, after extensive testing, declared that crayons are safe for children to use. Although trace amounts of asbestos and asbestos-related fibers were found in many of the crayons tested, the amounts were not considered to be dangerous. No recall was issued, but crayon manufacturers were urged to create a new "recipe" for crayons to exclude the asbestos fibers.*

What do crayons have to do with rocks? A metamorphic mineral—talc—is used in the manufacture of crayons. In fact, it's the properties of talc that led to the asbestos scare.

Talc and Asbestos

Talc is a soft white mineral with a hardness of 1. It is used in many cosmetics and art supplies. Talc is often found in association with serpentine, which is the parent rock of asbestos. Asbestos comes from the mineral chrysotile, which breaks into tiny, hairlike fibers. Asbestos has been used in insulation and in fireproof fabrics and building products. However, in the 1960s and 1970s, it was discovered that inhaling asbestos fibers leads to lung cancer, asbestosis, and mesothelioma—each of which can be fatal. Beginning in the 1980s, virtually all uses of asbestos were banned in the United States and much of Europe.

Blue, green, and asbestos?

Ground up talc is used as a hardener in crayons, which are basically a mixture of pigments, hardeners, and wax. Without talc, crayons would get too sticky to handle. However, when talc is ground, fibers that resemble asbestos are created. Testers aren't sure whether the tiny amount of asbestos found in crayons comes from these asbestos-like fibers, from chrysotile contamination of the talc, or from both. They point out, though, that if the talc fibers look so much like asbestos fibers, they are likely to have the same affect in the lungs.

The risk to children from the asbestos and asbestos-like fibers in crayons is extremely small. The amount of fibers present is very low, and, because the talc is embedded in wax, there is little likelihood that the fibers will be inhaled. The Consumer Product Safety Commission did a test that simulated one half-hour of hard coloring and found no airborne fibers. Still, crayon manufacturers understand that, when it comes to children, any slight risk is too much. All of the major crayon makers have decided to change their crayon formulas to eliminate all use of talc. The use of talc in children's chalk, clay, and sand is also being investigated. The use of asbestos was phased out in the United States in 2001.

Activity

Research the use of talc in other products. Why is talcum powder no longer recommended for infants? Does the use of talc in make-up expose teens to the same asbestos risk that crayons posed for younger children? Present your findings in a written report.

CHAPTER 6

Study Guide

Summary

SECTION 6.1

Formation of Sedimentary Rocks

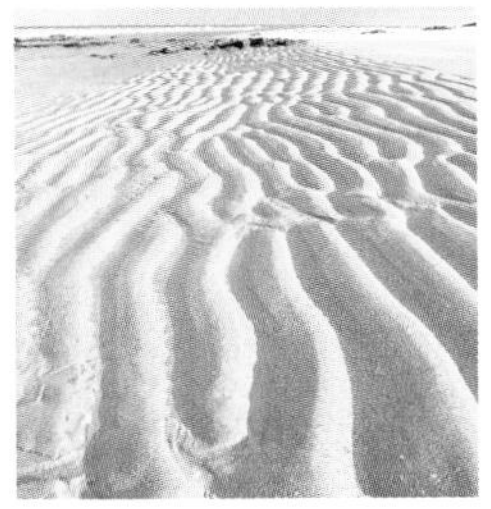

Main Ideas

- The processes of weathering, erosion, deposition, burial, and lithification form sedimentary rocks.
- Clastic sediments are rock and mineral fragments produced by weathering and erosion. They are classified based on particle size.
- Sediments are lithified into rock by the processes of compaction and cementation.
- Sedimentary rocks can contain depositional features such as horizontal bedding, cross-bedding, and ripple marks.
- Fossils are the remains or other evidence of once-living things that are preserved in sedimentary rocks.

Vocabulary

bedding (p. 126)
cementation (p. 125)
clastic (p. 122)
cross-bedding (p. 126)
deposition (p. 123)
graded bedding (p. 126)
lithification (p. 124)
sediment (p. 121)

SECTION 6.2

Types of Sedimentary Rocks

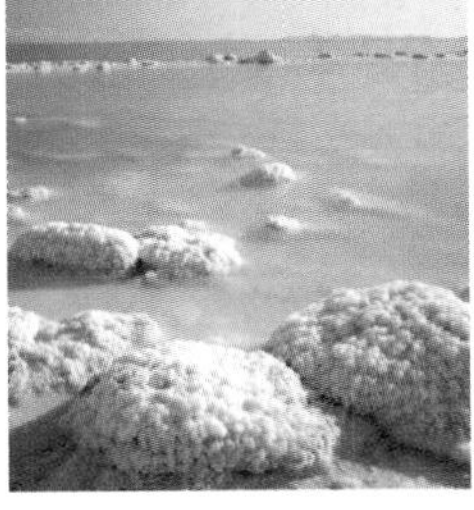

Main Ideas

- There are three main classes of sedimentary rocks: clastic, which are formed from clastic sediments; chemical, which are formed from minerals precipitated from water; and organic, which are formed from the remains of once-living things.
- Clastic sedimentary rocks are classified by particle size and shape.
- Evaporites are chemical sedimentary rocks that form primarily in restricted ocean basins in regions with high evaporation rates.
- Limestone, composed primarily of calcite, is the most abundant organic sedimentary rock. Coal is another organic sedimentary rock.
- Sedimentary rocks provide geologists with information about surface conditions that existed in Earth's past.

Vocabulary

clastic sedimentary rock (p. 128)
evaporite (p. 130)
porosity (p. 129)

SECTION 6.3

Metamorphic Rocks

Main Ideas

- Metamorphic rocks are formed when existing rocks are subjected to high temperature and pressure, which cause changes in the rocks' textures, mineralogy, and composition.
- The three main types of metamorphism are regional, contact, and hydrothermal.
- Metamorphic rocks are divided into two textural groups: foliated and nonfoliated.
- During metamorphism, minerals change into new minerals that are stable under the conditions of temperature and pressure at which they formed.
- The rock cycle is the set of processes whereby rocks continuously change into other types of rock.

Vocabulary

contact metamorphism (p. 135)
foliated (p. 136)
hydrothermal metamorphism (p. 135)
nonfoliated (p. 136)
porphyroblast (p. 136)
regional metamorphism (p. 134)
rock cycle (p. 138)

earthgeu.com/vocabulary_puzzlemaker

CHAPTER 6

Assessment

Understanding Main Ideas

1. What are solid particles that have been deposited on Earth's surface called?
   a. porphyroblasts **c.** schists
   b. sediments **d.** quartzites

2. What process breaks solid rock into smaller pieces?
   a. deposition **c.** weathering
   b. cementation **d.** metamorphism

3. What agent of erosion can usually move only sand-sized or smaller particles?
   a. landslides **c.** water
   b. glaciers **d.** wind

4. Which of the following is an example of a medium-grained clastic sedimentary rock?
   a. conglomerate **c.** evaporite
   b. breccia **d.** sandstone

5. Which of the following are formed by the chemical precipitation of minerals from water?
   a. sandstones **c.** salt beds
   b. coal beds **d.** shale

6. Which of the following would you expect to have the greatest porosity?
   a. sandstone **c.** shale
   b. gneiss **d.** quartzite

7. Which of the following is a common mineral found in both organic and chemical sedimentary rocks?
   a. calcite **c.** garnet
   b. quartz **d.** biotite

8. By what process are surface materials removed and transported from one location to another?
   a. weathering **c.** deposition
   b. erosion **d.** cementation

9. What mineral commonly forms porphyroblasts?
   a. quartz **c.** talc
   b. garnet **d.** calcite

10. What are the two primary causes of lithification?

11. Why is the term *clastic* appropriate for particles weathered from solid rock?

12. Describe the two main types of cementation.

13. How are the three types of sedimentary rocks classified?

14. Rearrange the terms below to create a concept map of the rock cycle.

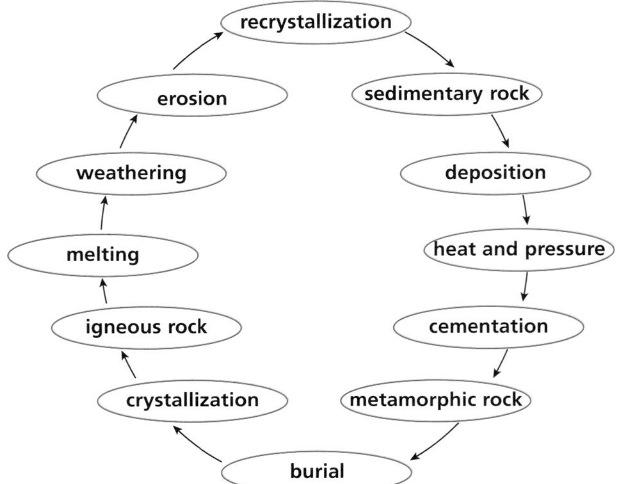

15. What are the two most common types of foliated metamorphic rocks?

16. What are porphyroblasts, and how do they form?

17. What parts of the rock cycle occur at Earth's surface?

Test-Taking Tip

WORDS ARE EASY TO LEARN Whenever you hear or read a word that you cannot define, jot it down on an index card. Then look it up in the dictionary and write the definition on the back of the card. Try to write a sentence or draw a picture using the word, too. Practice saying the word aloud until you are comfortable with it.

CHAPTER 6
Assessment

Applying Main Ideas

18. How can chemical weathering assist physical weathering?
19. Glaciers move very slowly, yet they are able to carry large particles with ease. Why?
20. How does graded bedding form?
21. Geologists have uncovered a mudstone layer containing mud cracks and ripple marks. This layer lies beneath a layer of sandstone. Explain how these structures and layers might have formed during the deposition of the sediments.
22. What information about sedimentary environments can be interpreted from a breccia?
23. What is the source of the calcite found in organically formed limestone?
24. What type of rock is marble? What characteristics make it well suited for sculptures?
25. How might a sedimentary rock become another sedimentary rock without first changing into another rock type?

Thinking Critically

26. Why are sand dunes commonly composed of fine, well-sorted sand?
27. Why do muds lose more volume during compaction than sands do?
28. How could you tell the difference between sedimentary rocks formed from an underwater landslide and sedimentary rocks formed from a landslide on Earth's surface?
29. Sand is often found between the larger grains of conglomerates, but large particles are seldom found in sandstone. Why is this?
30. Would you expect foliated metamorphic textures in rocks that have undergone contact metamorphism? Why or why not?

Standardized Test Practice

1. What initiates the process that turns sediments into sedimentary rocks?
   - **a.** bedding
   - **b.** burial
   - **c.** cementation
   - **d.** compaction

2. Which sedimentary rock is used to make cement for the construction industry?
   - **a.** shale
   - **b.** sandstone
   - **c.** phosphate
   - **d.** limestone

3. Which of the following are rocks composed of minerals that form with blocky crystal shapes?
   - **a.** foliated
   - **b.** nonfoliated
   - **c.** porphyroblasts
   - **d.** phenocrysts

INTERPRETING SCIENTIFIC ILLUSTRATIONS
Use the illustration below to answer question 4.

4. Which rocks are most likely to metamorphose from a lava flow?
   - **a.** only the rocks in the crater of the volcano, where the lava is hottest
   - **b.** rocks in the crater and rocks along the top half of the mountain
   - **c.** all the rocks on the mountain
   - **d.** all the rocks reached by the lava flow

earthgeu.com/standardized_test

UNIT 2

GeoDigest

For a **preview** of Earth's composition, study this GeoDigest before you read the chapters. After you have studied the chapters, you can use the GeoDigest to **review** the unit.

Composition of Earth

Matter and Atomic Structure

Elements Atoms are the basic building blocks of matter. They are made of protons, which have positive electrical charges; electrons, which have negative electrical charges; and neutrons, which are neutral. Protons and neutrons make up the nucleus of an atom; electrons surround the nucleus in energy levels. An element is a substance consisting of atoms with a specific number of protons in their nuclei. Examples of elements include hydrogen, neon, gold, carbon, and uranium. Isotopes of an element differ by the number of neutrons in their nuclei. All elements are mixtures of isotopes. The number of electrons in the outermost energy levels of atoms determines their chemical behavior. Elements with the same number of electrons in their outermost energy levels have similar chemical properties.

How Atoms Combine Atoms of different elements combine to form compounds. Molecular compounds are formed when atoms are held together by the sharing of electrons in covalent bonds. Atoms also combine ionically. Ions are electrically charged atoms or groups of atoms. Positive and negative ions attract each other and form ionic compounds. Acids are solutions containing hydrogen ions. Bases are solutions containing hydroxide ions. Acids and bases can neutralize each other. A mixture is a combination of components that retain their identities and can still be distinguished. A solution is a mixture in which the components can no longer be distinguished as separate. Solutions can be liquid, solid, or gaseous.

States of Matter On Earth, matter exists in three physical states: solid, liquid, and gas. The universe also contains a fourth state of matter: plasma. Most solids have crystalline structures. Atoms, ions, or molecules in crystals are arranged in regular geometric patterns. Most rocks are polycrystalline materials. Liquids are densely packed arrangements of mobile particles. Gases are widely separated, individual particles. Plasmas are hot, highly ionized, electrically conductive gases. Changes of state involve thermal energy. Melting and evaporation absorb thermal energy, whereas freezing and condensation release thermal energy.

Minerals

A mineral is a naturally occurring, inorganic solid with a specific chemical composition and a definite crystalline structure. There are at least 3000 known minerals in Earth's crust. Minerals form from

Sodium and chlorine reaction

Earth Sciences 3.c

magma or from solution. Most minerals are formed from the eight most common elements in Earth's crust. Oxygen readily combines with other elements to form a diverse group of minerals, including silicates, carbonates, and oxides. Minerals are virtually everywhere. Your body needs many of the elements found in minerals to survive, such as iron, calcium, and sodium. Some minerals are found as ores. A mineral is an ore if it contains a useful substance that can be mined at a profit. Ores from deep within Earth are removed by underground mining. Ores close to Earth's surface are obtained from open-pit mines. If responsible procedures are not followed, mining can cause environmental damage. Gems, such as diamonds and rubies, are valuable minerals that are prized for their rarity and beauty. The presence of trace elements can make one variety of a mineral more colorful and thus more prized than other varieties of the same mineral. For example, amethyst which contains traces of manganese, is a gem form of quartz.

Identifying Minerals Minerals can be identified based on their physical and chemical properties. The most reliable way to identify a mineral is to use a combination of tests of color, hardness, and density, among other characteristics. A mineral's color is generally the result of trace elements within the mineral. Texture describes how a mineral feels. Luster describes how a mineral reflects light. Cleavage and fracture describe how a mineral breaks. A mineral's streak, its color in powdered form, its hardness, and its density are also methods of identification. Special properties of minerals, such as magnetism, can also be used for identification purposes.

Igneous Rocks

Formation and Types Igneous rocks, formed by the cooling and crystallization of magma, may be intrusive or extrusive. Intrusive rocks form inside Earth's crust; extrusive rocks form at or near Earth's surface. Minerals crystallize from magma in a sequential pattern known as Bowen's reaction series. Different minerals melt and crystallize at different temperatures in the processes of partial melting and fractional crystallization. Igneous rocks are classified as felsic, intermediate, mafic, and ultramafic, depending upon their mineral compositions. Igneous groups are further identified by crystal size, also called texture. For example, extrusive rocks, which cool more rapidly than intrusive rocks, are generally more fine grained meaning they have small crystals. Early forming minerals may have well-shaped crystals, while later-forming minerals have irregular shapes. Porphyritic textures contain both large and small crystals.

Igneous Rock Resources Igneous rocks are often used as building materials because of their strength, durability, and beauty. Valuable ore deposits and gem crystals are often associated with igneous intrusions. For example, diamonds are found in rare types of igneous intrusions known as kimberlite pipes.

Dioptase on panchéite

Vital Statistics

Top Ten Diamond-Mining Countries

Country	Mine Production (in carats)
Botswana	15 000 000
Australia	13 400 000
Russia	11 500 000
South Africa	4 000 000
Kinshasa	3 500 000
Canada	2 000 000
Namibia	1 990 000
Angola	1 080 000
Ghana	649 000
Liberia	600 000

GeoDigest

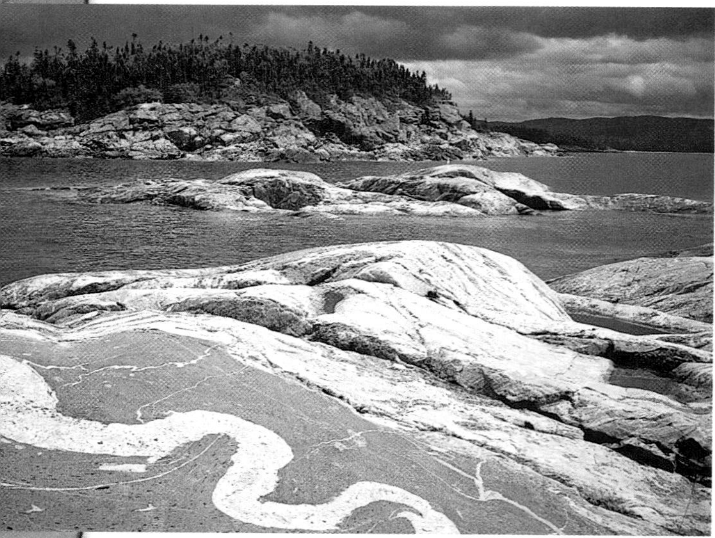

Metamorphic rock, Ontario, Canada

Sedimentary and Metamorphic Rocks

Formation and Types Sedimentary rocks are formed by weathering, erosion, deposition, burial, and lithification. Clastic sediments are rock and mineral fragments produced by weathering and erosion. Lithification occurs through the processes of compaction and cementation. Sedimentary rocks can be identified by depositional features such as horizontal bedding, cross-bedding, and ripple marks. Sedimentary rocks often contain the remains or evidence of once-living things: fossils. Sedimentary rocks also provide geologists with information about surface conditions that existed in Earth's past. Clastic sedimentary rocks form from sediments and are classified by particle size and shape. Chemical sedimentary rocks are formed from minerals precipitated from water. Such rocks include evaporites, which form primarily in restricted ocean basins in regions of high evaporation. Organic sedimentary rocks are formed from the remains of once-living things. Limestone and coal are organic sedimentary rocks.

Metamorphism and the Rock Cycle

Metamorphic rocks are formed when existing rocks are subjected to high temperature and pressure, which cause changes in the rocks' texture, mineralogy, and composition. The three main types of metamorphism are regional, contact, and hydrothermal. The two textural groups of metamorphic rocks are foliated and nonfoliated. During metamorphism, minerals change into new minerals that are stable for the temperature and pressure conditions under which they formed. Geologists use the stability ranges for these minerals to infer the history of Earth's crust. Metamorphism is part of the rock cycle, whereby rocks continuously change into other types of rock. Any type of rock can be changed into any other type of rock.

FOCUS ON CAREERS

Sculptor

Sculptors use rocks, minerals, and other Earth materials to create works of art. Many sculptors cast in bronze, an alloy of copper and tin. Others carve the metamorphic rock marble. Sculptors usually refine their talents in art schools or in the art departments of universities. A good understanding of the materials used is critical to creating a sculpture. The sculptor must know which tools to use on rocks of different hardnesses, how a material will fracture, and how a material in an outdoor sculpture will hold up in weather.

ASSESSMENT

Understanding Main Ideas

1. How are atoms best described?
   a. negatively charged
   b. the building blocks of matter
   c. isotopes
   d. energy levels surrounded by nuclei
2. Hydrogen, neon, gold, carbon, and uranium are examples of what?
   a. elements
   b. energy levels
   c. protons
   d. nuclei
3. What is a combination of components that retain their identities called?
   a. an ionic solution
   b. acids and bases
   c. hydroxide ions
   d. a mixture
4. How are atoms, ions, or molecules in crystals arranged?
   a. as widely separated particles
   b. as densely packed mobile particles
   c. in regular geometric patterns
   d. in solution
5. What is a useful substance that can be mined at a profit called?
   a. calcium
   b. hematite
   c. an ore
   d. a mineral

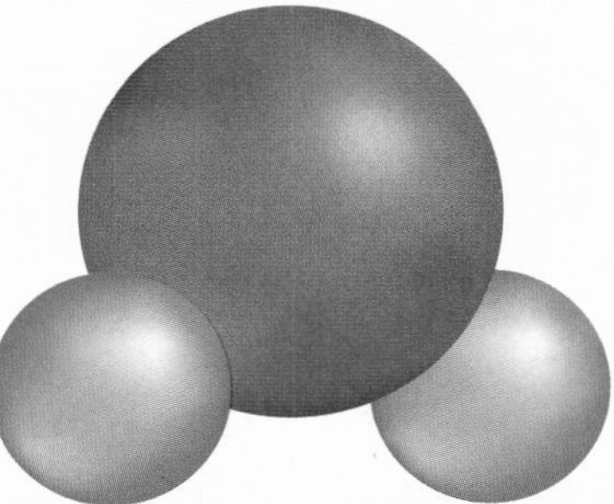

Model of a water molecule

6. Which of the following tests is the most reliable means of identifying a mineral?
   a. hardness
   b. streak
   c. density
   d. a combination of tests
7. Where do intrusive igneous rocks form?
   a. on Earth's surface
   b. inside Earth's crust
   c. in the ocean
   d. in Bowen's reaction series
8. What is a kimberlite pipe?
   a. an igneous intrusion
   b. an igneous extrusion
   c. a durable gem
   d. an extrusive igneous rock
9. How are clastic sedimentary rocks classified?
   a. by particle color
   b. by ripple marks
   c. by the presence of fossils
   d. by particle size and shape
10. What is the process whereby rocks continuously change into other types of rocks?
    a. the rock cycle
    b. metamorphism
    c. porphyry
    d. erosion

Thinking Critically

1. Why would a tossed salad be classified as a mixture?
2. Compare and contrast texture and luster.
3. Describe how clastic sedimentary rocks are formed.

Unit 3

Surface Processes on Earth

Earth has a system of external processes that shape its surface. For example, weathering and erosion change landforms and form soil, an important natural resource. Other external processes such as landslides, glaciers, and avalanches change the landscape. Earth's external processes also have impacts on human populations. Every year, flooding and mudflows, as well as landslides and avalanches cause the loss of life and property in many regions in the world. The photo shows the Emerald Pools in Zion Canyon in Zion National Park in Utah.

Unit Contents

Go to the National Geographic Expedition on page 870 to learn more about topics that are connected to this unit.

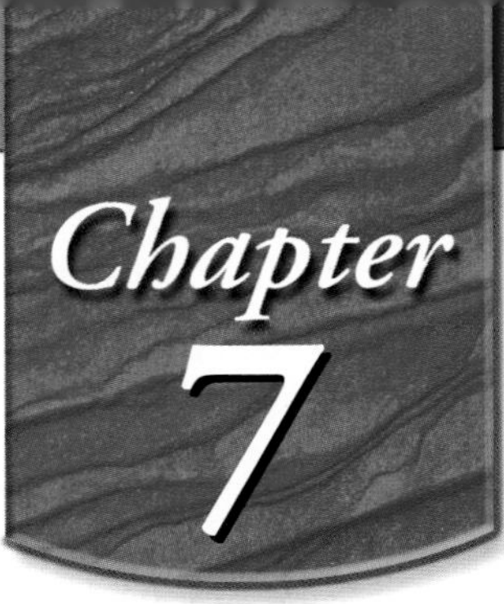

Weathering, Erosion, and Soil

What You'll Learn

- How the process of weathering breaks down rocks and how erosion transports weathered materials from one place to another.
- How soil is formed and why soil is an important natural resource.

Why It's Important

The processes of weathering and erosion change Earth's landforms and form soil, an important natural resource.

To find out more about weathering, erosion, and soil, visit the Earth Science Web Site at earthgeu.com

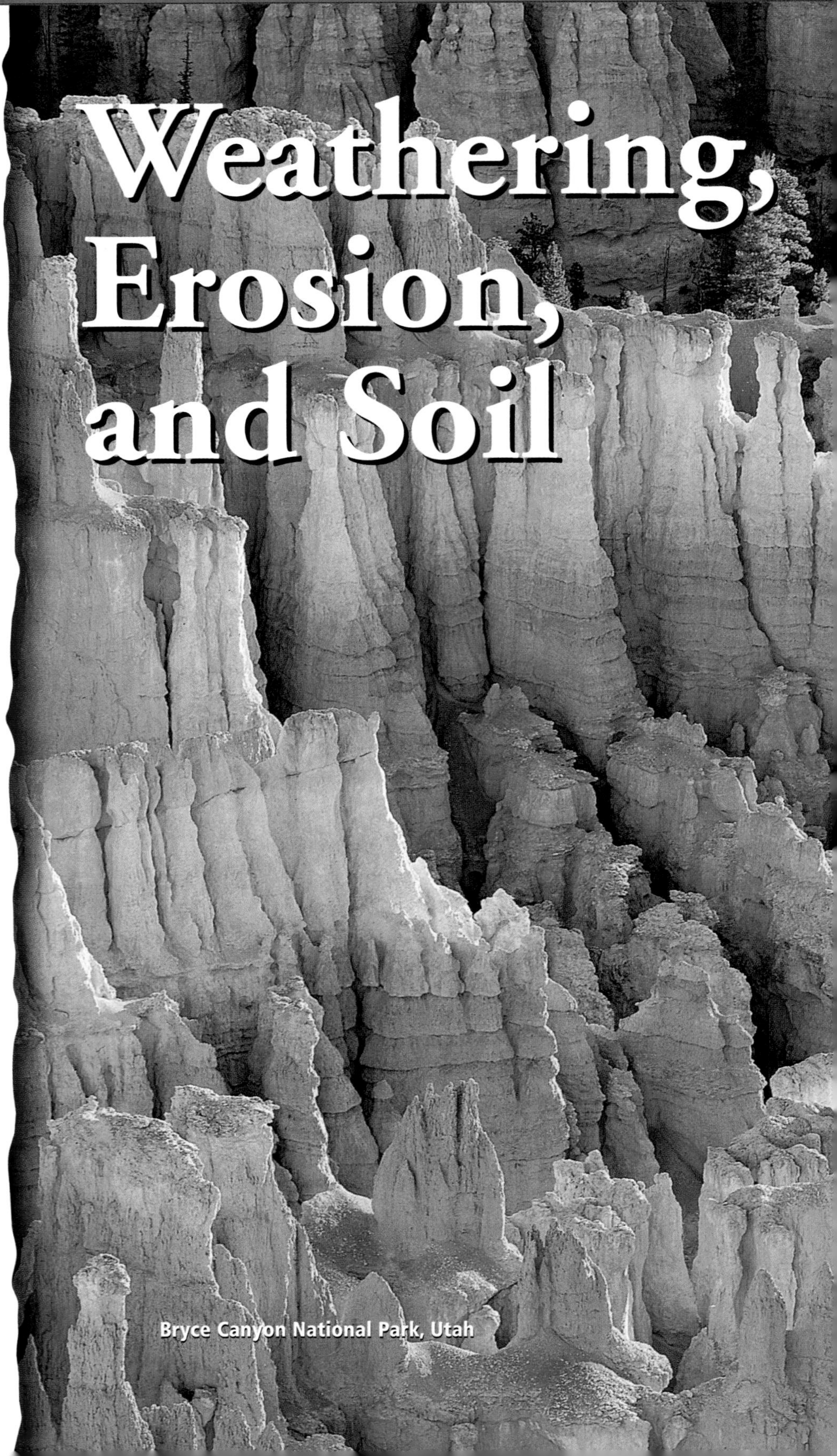

Bryce Canyon National Park, Utah

Discovery Lab Model Interfaces

Changes can take place at the interface between substances. An interface is where a substance comes in contact with another substance or condition. For example, the surface of a rock is the interface where the rock comes in contact with its environment. The more surface area that is exposed to environmental conditions, the more changes that can take place.

1. Fill two 250-mL beakers with water at room temperature.
2. Drop a sugar cube in one beaker and 5 mL of granulated sugar in the other beaker at the same time. Record the time.
3. Slowly and continuously stir the solution in each beaker. Use caution in stirring so as not to crack or break the beaker.
4. Observe the sugar in both beakers. Record the amount of time it takes for the sugar to completely dissolve in each beaker of water.

CAUTION: Always wear safety goggles and an apron in the lab.

Observe In your science journal, describe what happened to the sugar cube and the granulated sugar. Explain why one form of sugar dissolved faster than the other. Infer how you could decrease the time required to dissolve the other form of sugar.

SECTION 7.1 Weathering

Earth Sciences 7.c

OBJECTIVES

- **Distinguish** *between weathering and erosion.*
- **Identify** *variables that affect the rate of weathering.*

VOCABULARY

weathering
erosion
mechanical weathering
frost wedging
exfoliation
chemical weathering
hydrolysis
oxidation

In 1880, an impressive granite monument, Cleopatra's Needle, was moved from Egypt to Central Park, in New York City. Although the monument had existed in Egypt for more than 3500 years, in less than 75 years in New York City's climate, the monument had become dramatically worn and damaged. Today, markings on the surface can barely be read.

Why do you think this has happened? Changes occur every day to Earth's rocks and surface features. Just as the granite of Cleopatra's Needle has undergone changes, so, too, does granite in Earth's crust. The process by which rocks on or near Earth's surface break down and change is called **weathering.** The removal and transport of weathered material from one location to another is known as **erosion.** The processes of weathering and erosion have been going on since the crust of Earth formed, billions of years ago.

MECHANICAL WEATHERING

Before the agents of erosion can pick up and transport Earth materials, these materials must undergo some form of weathering. Weathering caused the result seen in ***Figure 7-1.*** Mechanical and chemical weathering are the two processes that can wear down rocks and minerals. Both types of weathering occur at the same time on Earth's landforms.

Figure 7-1 This 3600-year-old monument, Cleopatra's Needle was moved from Egypt to Central Park, New York City. After many years in New York City's climate, some markings on the monument have all but disappeared as a result of weathering.

The process by which rocks and minerals break down into smaller pieces is **mechanical weathering,** also called physical weathering. Mechanical weathering does not involve any change in a rock's composition, only changes in the size and sometimes the shape of the rock, as shown in ***Figure 7-2.*** A variety of factors are involved in mechanical weathering.

Temperature Temperature plays a significant role in mechanical weathering. When water freezes, it expands and increases in volume by approximately nine percent. Thus, ice takes up approximately nine percent more space than liquid water does. You have observed this increase in volume if you have ever made ice in an ice-cube tray in a freezer. In many places on Earth's surface, water collects in the cracks of rocks and rock layers. If the temperature drops to the freezing point of water, it freezes, expands, exerts pressure on the rocks, and may cause them to split, as shown in ***Figure 7-3A.*** When the temperature then increases, the ice in the cracks of rocks and rock layers melts. The repeated thawing and freezing of water in the cracks of rocks is called **frost wedging.** Frost wedging is also responsible for the formation of potholes in many roads in the northern United States in early spring, as shown in ***Figure 7-3B.***

Earth Sciences

7.c Students know the movement of matter among reservoirs is driven by Earth's internal and external sources of energy.

Pressure Pressure is another factor in mechanical weathering. Bedrock at great depths is under pressure from the overlying rock layers. When the overlying rock layers are removed, the pressure on the bedrock below is reduced. The bedrock surface, formerly

Figure 7-2 The mechanical weathering of these rocks occurred in Monument Valley in Arizona. ***How do you know that these formations occurred as a result of weathering?***

Figure 7-3 Frost wedging has split this granite boulder in the Sierra Nevada, California **(A).** Frost wedging also causes potholes to develop in early spring in Washington State **(B).** This boulder in Baja, California, is undergoing the process of exfoliation **(C).** This birch tree is growing out of a crack in bedrock in New Jersey **(D).**

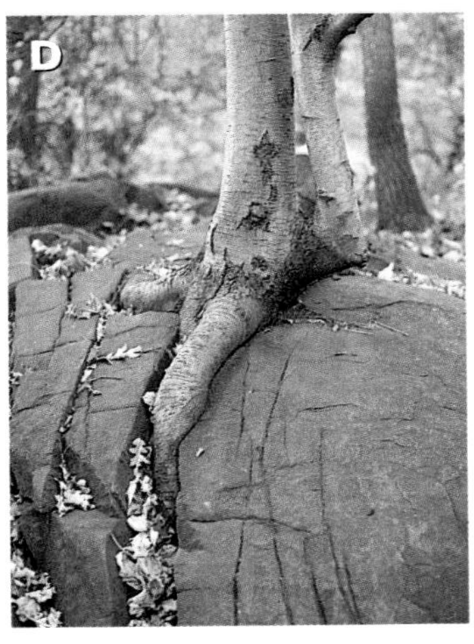

buried, is then able to expand, and long, curved cracks can form. These cracks, also known as joints, occur parallel to the surface of the rock. Reduction of pressure also allows existing cracks in the bedrock to widen.

Over time, the outer layers of rock are stripped away in succession, similar to an onion's layers being peeled off one by one. The process by which outer rock layers are stripped away is called **exfoliation,** shown in ***Figure 7-3C.*** Exfoliation often results in dome-shaped formations such as Liberty Cap and Half Dome in Yosemite National Park and Stone Mountain in Georgia. Sometimes, the effects of reduced pressure on rock layers are dramatic. For example, when several layers of overlying rocks are removed from a deep mine, the sudden decrease of pressure can cause large pieces of rock to explode off the walls of the mine tunnels.

The roots of trees and other plants can wedge themselves into cracks in rocks. As the roots grow and expand, they can exert pressure on the rocks to split, as shown in ***Figure 7-3D.***

Using Numbers
When water freezes, it expands and increases in volume by nine percent. What is the volume of ice that will form from 100 cm^3 of water?

Chemical Weathering

The process by which rocks and minerals undergo changes in their composition as the result of chemical reactions is called **chemical weathering.** Significant agents of chemical weathering include water, oxygen, carbon dioxide, and acids. Chemical reactions between rocks and water result in the formation of new minerals and the release of dissolved substances. The new minerals have different properties from those of the original rocks. For example, rust on an iron chain has a different chemical composition from that of the iron on which it formed. To some extent, the composition of rocks determines the effects that chemical weathering will have on them.

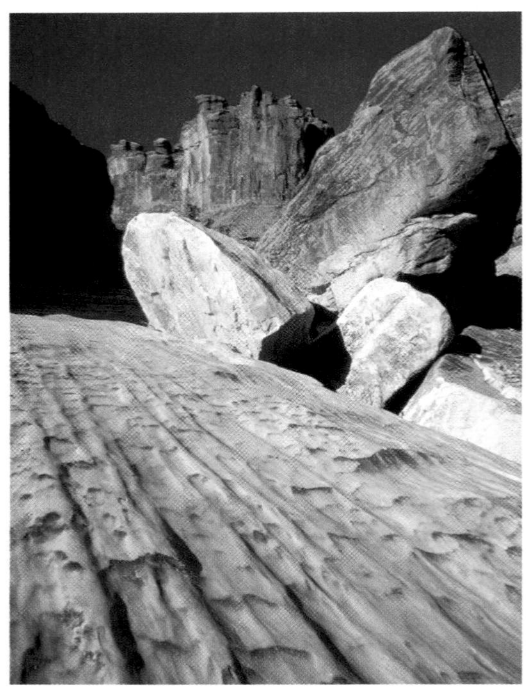

Figure 7-4 The surface of this limestone rock was chemically weathered by the activities of water. The rock is located in Slickhorn Canyon, Glen Canyon National Recreation Area, in Arizona.

Some minerals, such as calcite, may dissolve completely. Rocks that contain calcite, such as limestone and marble, are also greatly affected by chemical weathering. Buildings and monuments made of these rocks readily show signs of wear resulting from weathering.

Temperature is another significant factor in chemical weathering because it influences the rate at which chemical reactions occur. Usually, chemical reaction rates increase as temperature increases. With all other factors being equal, the rate of chemical weathering reactions doubles with each 10°C increase in temperature.

Water Water is an important agent in chemical weathering because it can dissolve many kinds of minerals and rocks, as shown in ***Figure 7-4.*** Water has an active role in some reactions, while it simply serves as a medium through which other reactions occur. The reaction of water with other substances is known as **hydrolysis.** Hydrolysis occurs in the decomposition of silicate minerals, such as the decomposition of potassium feldspar into kaolinite, a fine-grained clay mineral common in soils.

Figure 7-5 This limestone cave in Guatemala was formed when carbonic acid dissolved the calcite in the limestone rock.

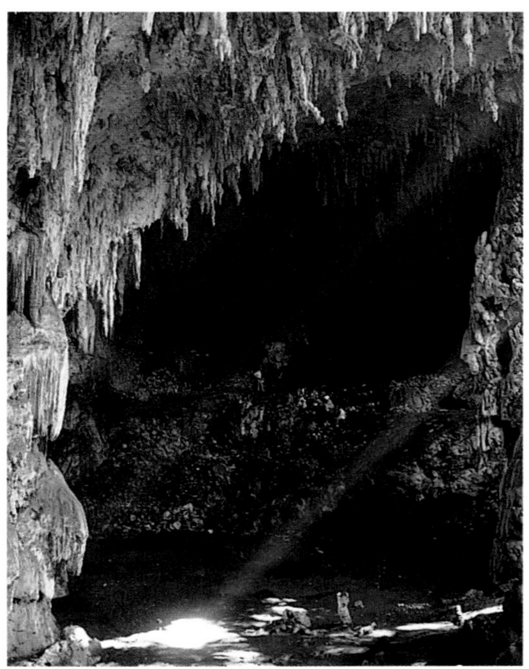

Oxygen Like water, oxygen can combine with other substances. The chemical reaction of oxygen with other substances is called **oxidation.** Approximately 21 percent of Earth's atmosphere is oxygen gas. Iron in rocks and minerals readily combines with this atmospheric oxygen to form minerals with the oxidized form of iron as shown in the following reaction.

$$\underset{\text{magnetite}}{2Fe_3O_4} + \frac{1}{2}O_2 \rightarrow \underset{\text{hematite}}{3Fe_2O_3}$$

Common minerals that contain the reduced form of iron include magnetite, hornblende, biotite, and pyrite.

Carbon Dioxide Another atmospheric gas that contributes to the chemical weathering process is carbon dioxide, which is produced by living organisms during the process of respiration. When carbon dioxide combines with water in the atmosphere, it forms a weak carbonic acid that falls to Earth's surface as precipitation. The formation of carbonic acid is shown in the following reaction.

$$\underset{\text{water}}{H_2O} + \underset{\text{carbon dioxide}}{CO_2} \rightarrow \underset{\text{carbonic acid}}{H_2CO_3}$$

Carbonic acid reacts with minerals such as calcite in limestone and marble to dissolve rocks. For example, limestone caverns, as shown in ***Figure 7-5,*** can form when carbonic acid dissolves the calcite in limestone rocks. Carbonic acid can also affect silicate minerals such as mica, and feldspar by reacting with elements in the minerals, such as magnesium and calcium. This chemical weathering process results in the formation of clay minerals. High concentrations of carbonic acid accumulate in soil, where decaying organic matter and plant respiration produce high levels of carbon dioxide. When water from precipitation seeps into the ground and combines with carbon dioxide, large amounts of carbonic acid become available for the process of chemical weathering.

Topic: Weathering
To find out more about weathering, visit the Earth Science Web Site at earthgeu.com

Activity: Identify an example of weathering in your state. Is it the result of mechanical weathering, chemical weathering, or both?

Acid Precipitation Another agent of chemical weathering is acid precipitation, which is caused mainly by the oxidation of sulfur dioxide and nitrogen oxides that are released into the atmosphere by human activities. Sulfur dioxide forms from the industrial burning of fossil fuels, while nitrogen oxides are emitted from motor-vehicle exhausts. These two gases combine with oxygen and water in the atmosphere to form sulfuric and nitric acids.

ENVIRONMENTAL CONNECTION

We describe how acidic a solution is by using the pH scale. ***Figure 7-6*** illustrates the pH scale. The lower the pH number, the greater the acidity of a substance. Acid precipitation is precipitation that has a

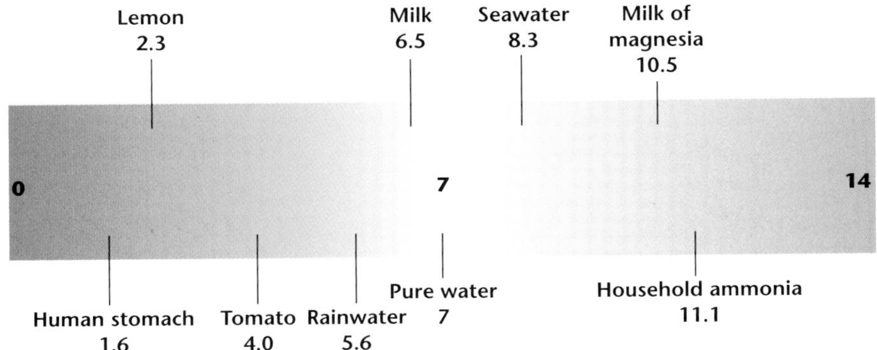

Figure 7-6 The pH scale is used to determine the acidity of substances.

pH value below 5.6, the pH of normal rainfall. Because acids can be harmful to many organisms and destructive to nonliving things, acid precipitation creates problems. It adversely affects fish and aquatic plant populations in lakes. Most freshwater lakes have a natural pH in the range of 6 to 8. These lakes can support many kinds of amphibians, aquatic invertebrates, and fish. However, when the lake water becomes too acidic, the species diversity decreases as shown in ***Table 7-1.*** The table indicates which organisms can survive at a particular pH. As you can see, wood frogs are able to survive at a lower pH than other species.

WHAT AFFECTS THE RATE OF WEATHERING?

The natural weathering of Earth materials occurs very slowly. For example, it may take 2000 years to weather 1 cm of limestone, and yet most rocks weather at even slower rates. Certain conditions and interactions can accelerate or slow the weathering process as demonstrated in the *GeoLab* at the end of the chapter.

Climate The climate of an area is a major influence on the rate of chemical weathering of Earth materials. Variables of climate include precipitation, temperature, and evaporation. The interaction between temperature and precipitation has the greatest effect on a region's rate of weathering. Chemical weathering occurs readily in

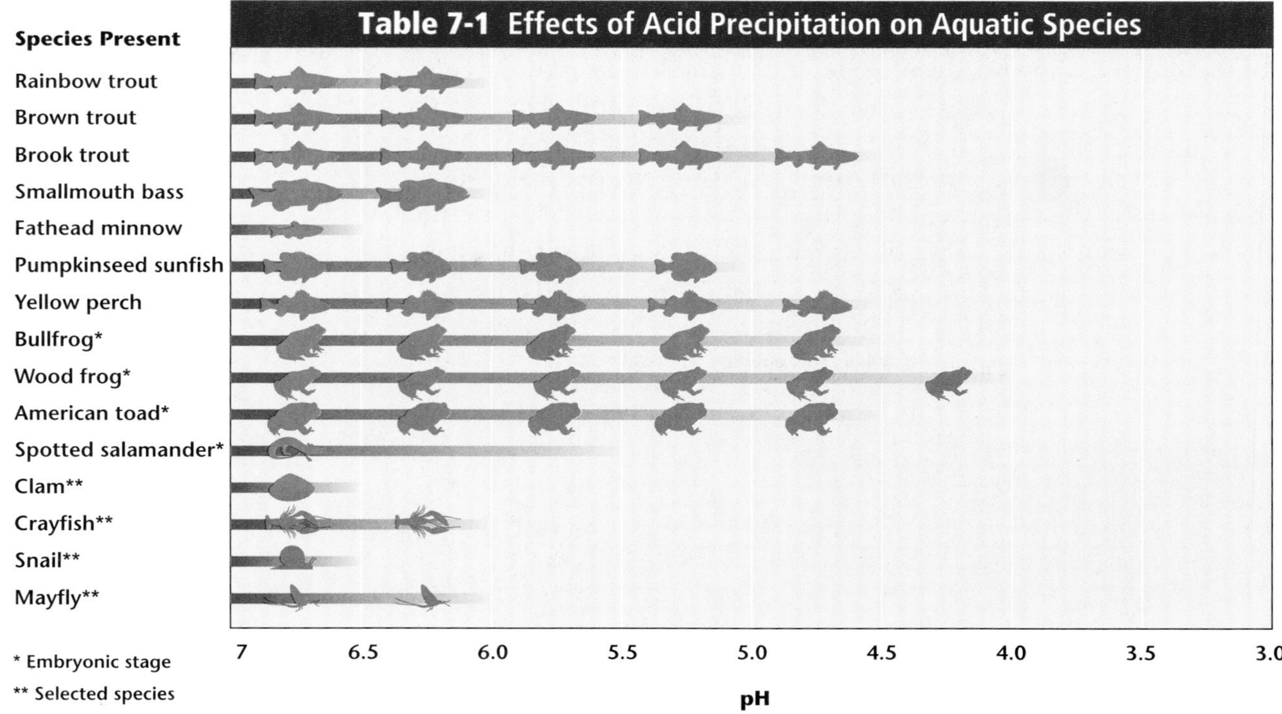

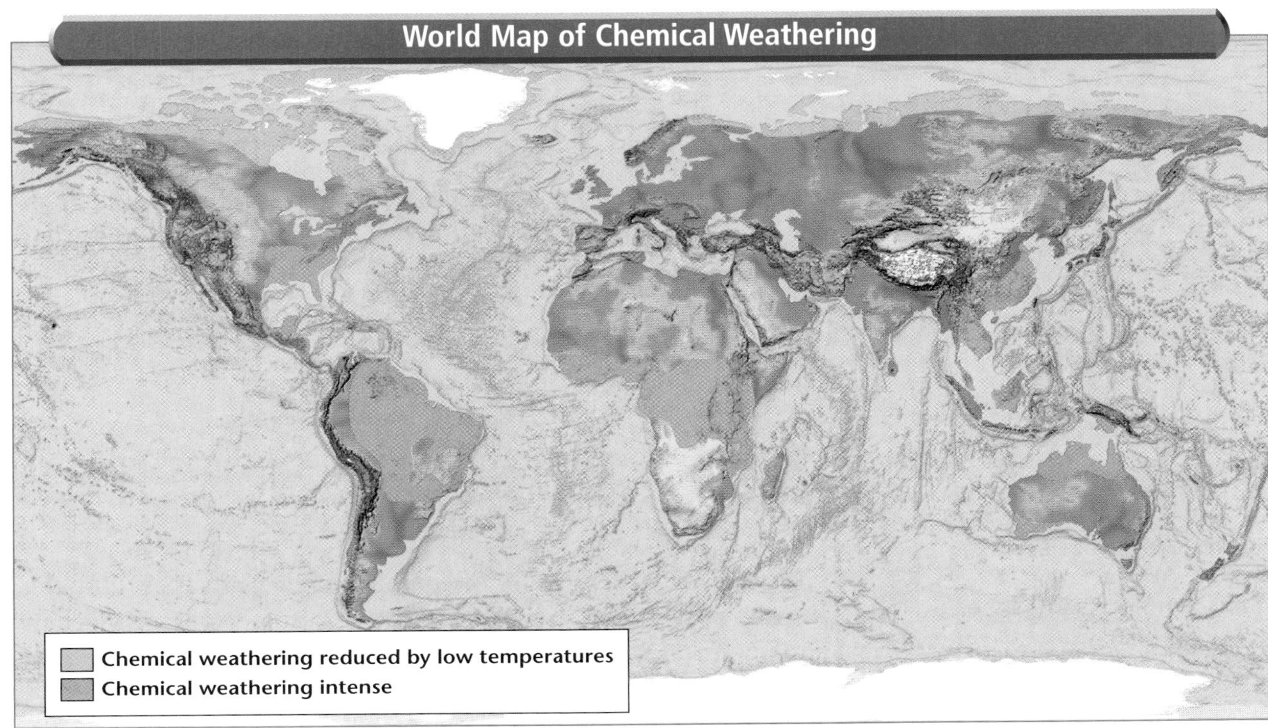

Figure 7-7 This world map shows areas where chemical weathering occurs. ***What areas in the world are subject to the most intense chemical weathering?***

climates with warm temperatures, abundant rainfall, and lush vegetation. These climatic conditions produce thick soils that are rich in organic matter. When water from heavy rainfalls combines with the carbon dioxide in this organic matter to produce high levels of carbonic acid, the weathering process is accelerated. Chemical weathering is evident in tropical Central America, Southeast Asia and other areas as shown in ***Figure 7-7.***

Conversely, physical weathering occurs readily in cool, dry climates. Physical weathering rates are highest in areas where water undergoes repeated freezing and thawing. Conditions in such climates do not favor chemical weathering because cool temperatures slow or inhibit chemical reactions. Little or no chemical weathering occurs in areas that are frigid year-round.

The different rates of weathering caused by different climatic conditions can be illustrated by a comparison of Asheville, North Carolina, and Phoenix, Arizona. Phoenix has dry, warm, conditions; temperatures do not drop below the freezing point of water, and humidity is low. In Asheville, temperatures sometimes drop below the freezing point during the colder months. Asheville has more monthly rainfall and higher levels of humidity than Phoenix does, as

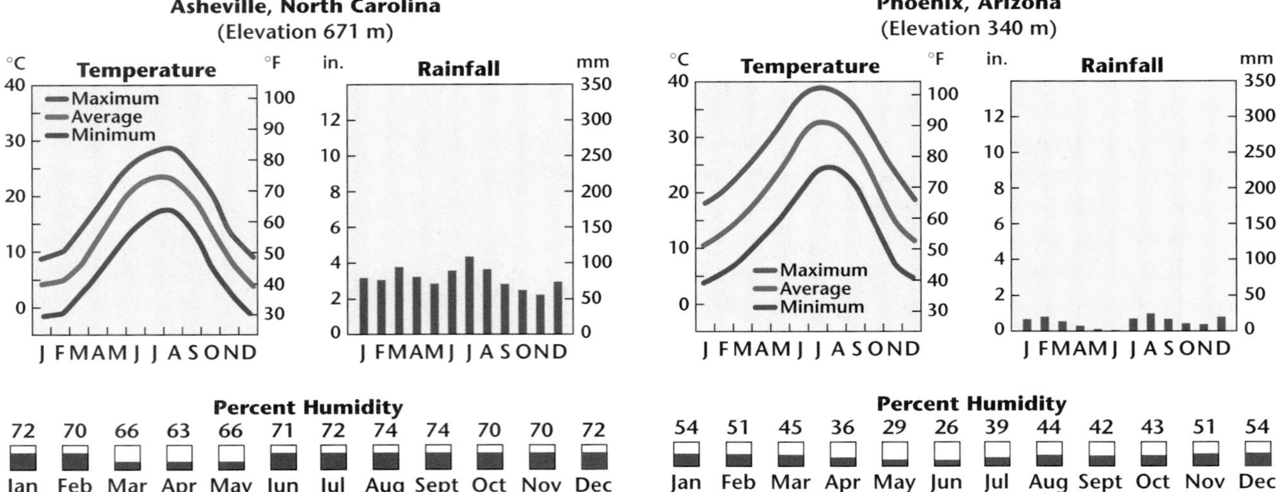

Figure 7-8 These graphs show a comparison of climatic conditions in Asheville and Phoenix.

shown in ***Figure 7-8.*** Because of these differences in their climates, rocks and minerals in Asheville experience a higher rate of mechanical and chemical weathering than those in Phoenix do.

Rock Type and Composition A wide variety of rocks and minerals cover Earth's surface. The characteristics of rocks, including how hard or resistant they are to being broken down, depend on their type and composition. In general, sedimentary rocks are more easily weathered than harder igneous and metamorphic rocks. The dramatic landscape in the Bisti Badlands of New Mexico exhibits rock layers with different degrees of resistance to weathering, as illustrated in ***Figure 7-9.***

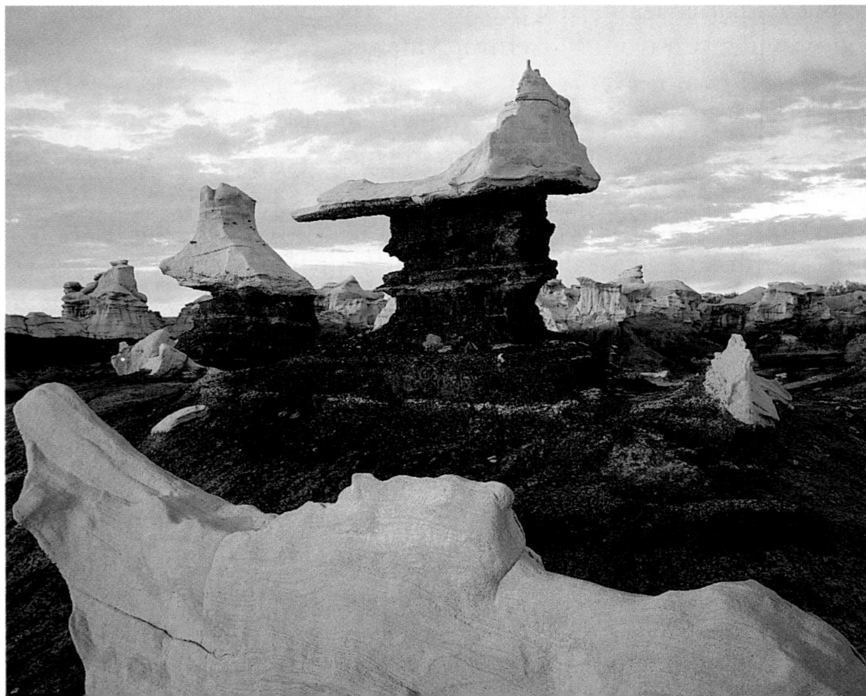

Figure 7-9 In the Bisti Badlands in New Mexico, these formations of resistant sandstone are situated on top of softer coal layers. ***What do you think caused the rocks to form in this way?***

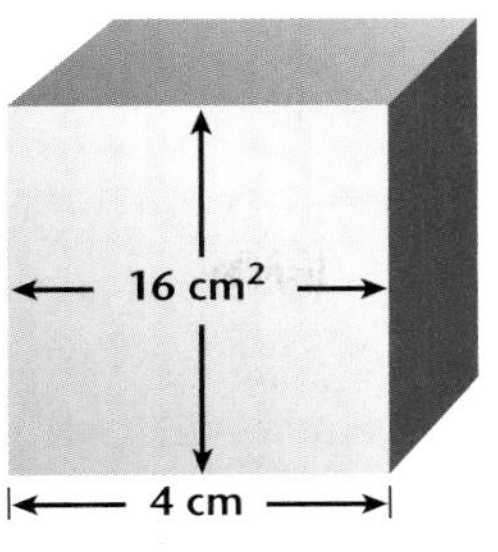

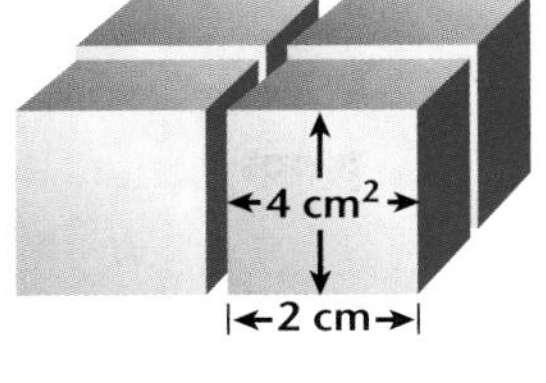

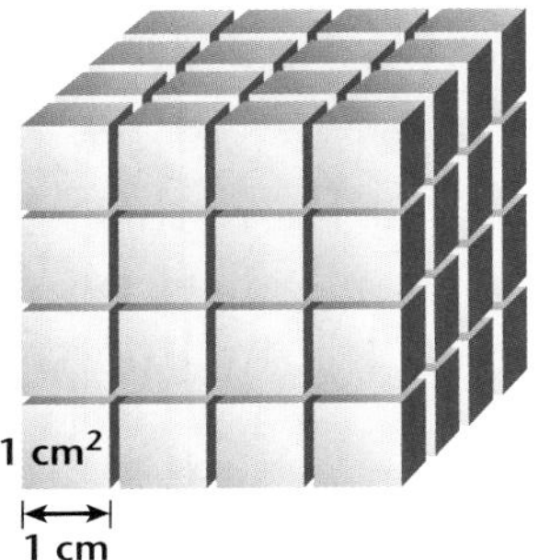

Figure 7-10 In this example, the original object has a surface area of 96 cm². When the same object is broken up into two pieces or more, the surface area increases.

Surface Area Mechanical weathering breaks up rocks into smaller pieces. As the pieces get smaller, their surface area increases, as illustrated in ***Figure 7-10.*** This means that more total surface area is available for chemical weathering. Thus, the greater the total surface area, the more weathering that occurs, as you learned in the *Discovery Lab.*

Topography and Other Variables Earth materials cover the surfaces of slopes and level areas. Materials on level areas are likely to remain in place as they undergo changes, whereas materials on slopes have a greater tendency to move as a result of gravity. As material moves down a slope, it exposes underlying rock surfaces and thus provides more opportunities for weathering to occur. As you learned earlier, organisms also affect the rate of weathering. Decaying organic matter and living plant roots release carbon dioxide, which combines with water to produce acid, which in turn increases the weathering rate.

Section Assessment

1. Distinguish between weathering and erosion.
2. List several variables that affect the rate of weathering.
3. What two climatic factors are most important in the weathering process?
4. **Thinking Critically** Describe how one variable may affect another variable in the weathering process.

Skill Review

5. **Making Graphs** Make a graph of the relationship between the rate of weathering and the surface area of a material. Plot the weathering rate on the *y*-axis and the surface area on the *x*-axis. For more help, refer to the *Skill Handbook*.

SECTION 7.2

Erosion and Deposition

Earth Sciences 9.b

OBJECTIVES

- **Analyze** *the impact of living and nonliving things on the processes of weathering and erosion.*
- **Describe** *the relationship of gravity to all agents of erosion.*

VOCABULARY

deposition
rill erosion
gully erosion

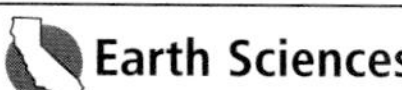

Earth Sciences

9.b Students know the principal natural hazards in different California regions and the geologic basis of those hazards.

As you have learned, erosion is the process that transports Earth materials from one place to another. A number of different agents transport weathered materials on Earth. Running water in streams and rivers, glaciers, wind, and ocean currents and waves all pick up and carry materials. Humans, plants, and animals also play a role in the erosional process. Erosion can result from the loss of plant cover, as shown in ***Figure 7-11.*** The land becomes barren as increasing amounts of soil are lost to wind and water erosion. At some point, the movement of transported materials will slow down. When this happens, the materials are dropped in another location in a process known as **deposition,** the final stage of the erosional process.

GRAVITY'S ROLE IN EROSION

Gravity is associated with many erosional agents, because the force of gravity tends to pull all materials downslope. Without gravity, glaciers would not move downslope and streams would not flow. Gravity is also an agent of mass movements such as landslides, mudflows, and avalanches, which you will learn about in Chapter 8.

EROSION BY RUNNING WATER

With the exception of the extremely strong winds associated with tornadoes and hurricanes, water has more power to move large particles of weathered material than wind does. As you might expect, stream erosion is greatest when a large volume of water is moving rapidly, such as during spring thaws and torrential downpours. Water flowing down steep slopes also has greater potential to erode Earth materials, because the steeper the slope, the faster the water flows. Not only does swiftly flowing water have greater erosional power than wind, but it can also carry more material along with it and over a greater distance.

Figure 7-11 Major erosion can occur on steep slopes as a result of the loss of plant cover due to the clearcutting of a forest.

Running water moves along Earth's surface from higher to lower elevations. Small streams at high elevations flow down to join larger streams at lower elevations. Such a network of streams drains an area called a watershed as the water works its way down toward the ocean.

The erosion by running water in small channels, on the side of a slope is called **rill erosion,** shown in ***Figure 7-12A.*** Rills commonly form on a slope. When a channel becomes deep and wide, it can evolve into **gully erosion,** as shown in ***Figure 7-12B.*** Gullies can be more than 3 m deep. They can be a major problem in farming and grazing areas.

Coastal Deposition and Erosion

Rocks exposed to their surrounding environment are slowly weathered away, as modeled in the *MiniLab* on this page. Each year, streams and rivers carry billions of metric tons of sediments and weathered materials to coastal areas. The Mississippi River alone carries 750 million metric tons of eroded material off the continent and into the Gulf of Mexico annually. When a

MiniLab

How do rocks weather?

Model how rocks are exposed to their surrounding environment and slowly weather away.

Procedure

1. Carve your name deeply into a bar of soap with a toothpick. Weigh the soap.
2. Measure and record the depth of the letters carved into the soap.
3. Place the bar of soap on its edge in a catch basin.
4. Sprinkle water over the bar of soap until a noticeable change occurs in the depth of the carved letters.
5. Measure and record the depth of the carved letters.

Analyze and Conclude

1. How did the depth of the letters carved into the bar of soap change?
2. Did the shape, size, or weight of the bar of soap change?
3. Where did the missing soap go?
4. What additional procedure could you follow to determine whether any soap wore away?

Figure 7-12 Rill erosion has occurred on these rocks in Badlands National Park **(A).** The removal of too much vegetation caused gully erosion in this farming area **(B).**

Figure 7-13 This photograph, taken aboard the space shuttle, shows the huge amount of sediment deposited at the Nile Delta.

river enters a large body of water, such as the ocean, the water slows down and deposits large amounts of sediments. The build-up of sediments forms deltas, such as the Nile Delta, shown in ***Figure 7-13.*** The volume of river flow and the action of tides determine the shapes of deltas, most of which contain fertile soil. Coastal areas also undergo erosion by ocean waves and wind. You will learn more about coastal erosion in the *Science & the Environment* feature at the end of this chapter.

In the ocean, weathering and erosional processes continue. The work of ocean currents, waves, and tides carves out cliffs, arches, and other features along the continents' edges. In addition, sand particles accumulate on shorelines and form dunes and beaches. Erosion of materials also occurs along the ocean floor and at continental and island shorelines. The constant movement of water and the availability of accumulated weathered material result in a continuous erosional process, especially along ocean shorelines. Sand along a shoreline is repeatedly picked up, moved, and deposited by ocean currents. In this way, sandbars form from offshore sand deposits. If the sandbars continue to be built up with sediments, they can become barrier islands. Many barrier islands, as shown in ***Figure 7-14,*** have formed along the Gulf and Atlantic Coasts of the United States.

Figure 7-14 A barrier island such as the Queens Atlantic Beach in New York was formed from the buildup of sandbars.

Figure 7-15 Groins at Cape May, New Jersey, are used to protect the beaches from wave erosion. Groins are vertical walls of rock placed perpendicular to the shore to trap sand from ocean currents.

Figure 7-16 This rock in Yosemite National Park, California, was polished by the activity of a glacier.

Erosion also occurs on islands, where the constant movement of water wears away at the shorelines. Changing tides and conditions associated with coastal storms can have a great impact on coastal erosion as well. Human development and population growth along shorelines have led to attempts to control the ocean's movements of sand. However, efforts to keep the sand on one beachfront disrupt the natural migration of sand along the shore, thereby depleting sand from another area. ***Figure 7-15*** shows one method used to help prevent beach erosion. You will learn more about ocean and shoreline features in Chapters 15 and 16.

Glacial Erosion

Although glaciers currently cover less than ten percent of Earth's surface, their erosional effects are large-scale and dramatic. Glaciers scrape and gouge out large sections of Earth's landscape. Because they are so dense, glaciers have the capacity to carry huge rocks and piles of debris over great distances. Glacial movements scratch and grind some surfaces, while they polish others, as shown in ***Figure 7-16.*** The landscape features left in the wake of glacial movements include valleys with majestic waterfalls, lakes, and variously shaped deposits of sediment. Such features are common in New England. The erosional effects of glaciers also include deposition. For example, soils in the northern sections of the United States are deposits of material once carried by glaciers. In these and other ways, glaciers continue to affect erosional processes on Earth, even though the time of the most recent ice age is long past. You will learn more about glaciers in the next chapter.

Figure 7-17 The plant's roots have protected the soil from the wind erosion that has eroded the surrounding area in Death Valley, California.

Wind Erosion

Wind is a major erosional agent in areas on Earth that experience both limited precipitation and high temperatures. Such areas typically have little vegetative cover to hold soil in place. Wind can easily pick up and move these fine, dry particles. When conditions become ideal for wind erosion, the effects can be dramatic and devastating. The abrasive action of wind-blown particles can damage both natural features and human-made structures. Wind erosion is common in Death Valley. ***Figure 7-17*** is in Death Valley. Shore areas also experience wind erosion. Even though winds can blow against the force

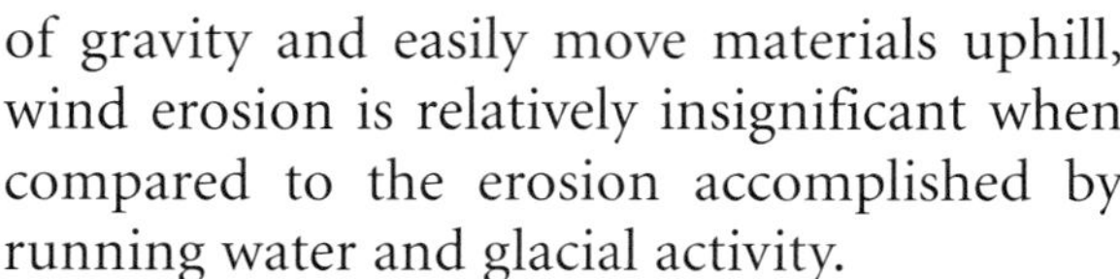

of gravity and easily move materials uphill, wind erosion is relatively insignificant when compared to the erosion accomplished by running water and glacial activity.

Wind Barriers One farming method that reduces the effects of wind erosion is the planting of wind barriers, also called windbreaks. Wind barriers are trees or other vegetation planted perpendicular to the direction of the wind. In many cases, a wind barrier may be simply a row of trees along the edge of a field. In addition to reducing soil erosion, wind barriers can trap blowing snow, conserve moisture, and protect crops from the effects of the wind.

Figure 7-18 The construction of a new highway in Ohio requires the removal of large amounts of soil.

ENVIRONMENTAL CONNECTION

Erosion by Plants, Animals, and Humans

Plants and animals living on the surface of Earth also play a role in erosion. As plants and animals carry on their life processes, they move Earth's surface materials from one place to another. For example, Earth materials are relocated as animals burrow into soil and shovel it to another place. Humans also excavate areas and move soil from one location to another. Planting a garden, developing a new athletic field, and building a highway, shown in ***Figure 7-18,*** are all examples of human activities that result in the moving of Earth materials from one place to another. The effects of erosion by the activities of plants, animals, and humans, however, are minimal in comparison to the erosional effects of water, wind, and glaciers.

Section Assessment

1. In the erosional process, what is gravity's role in relationship to the other agents of erosion?
2. Describe the agents of erosion and how they affect Earth's landforms.
3. What is the difference between rill erosion and gully erosion? Which is the most damaging?
4. **Thinking Critically** In what ways do the activities of humans affect the processes of erosion and weathering?

Skill Review

5. **Making and Using Tables** Make a data table that compares the various agents of erosion. In your table, rate each agent's overall ability to erode and list any conditions necessary for each type of erosion to occur. Note which erosional agent is most powerful and which is most dominant on Earth. For more help, refer to the *Skill Handbook.*

SECTION 7.3 Formation of Soil

I&E 1.c, 1.g, 1.m

Soil is an important natural resource because it is essential to life on Earth. It would be difficult to imagine a world without soil. Humans and other organisms are dependent on plants, which grow in soil, for food and other basic needs. If you were to make a list of all the things that humans obtain directly and indirectly from soil, you might be surprised by the number of items on your list. In addition to wood from trees, such things as oxygen from plants, food from plants, and meat from animals that are dependent on plants are all products of soil. Soil even helps to filter pollutants.

OBJECTIVES

- **Describe** *how soil forms.*
- **Explain** *the relationship between the organic and inorganic components of soil.*
- **Identify** *soil characteristics.*
- **Recognize** *soil horizons in a soil profile.*

VOCABULARY

soil
residual soil
transported soil
soil profile
soil horizon

DEVELOPMENT OF SOIL

Except for some steep mountain slopes and extremely cold regions, soil is found almost everywhere on Earth's surface. But what is soil? Weathered rock alone is not soil. **Soil** is the loose covering of broken rock particles and decaying organic matter, called humus, overlying the bedrock of Earth's surface. Soil is the result of chemical and mechanical weathering and biological activity over long periods of time. The soil-forming process begins when weathering breaks solid bedrock into smaller pieces. These pieces of rock continue to undergo weathering and break down into smaller and smaller pieces. Many organisms, such as bacteria, fungi, and insects, begin to live in these weathered materials. Over time, the organisms die, decay, and add nutrients to the weathered materials to form soil, which, in turn, supports a variety of life forms, as shown in ***Figure 7-19.***

The process of continual breakdown of organic materials is thus begun. Nutrients continue to be added to the soil, soil texture improves, and the soil's capacity to hold water increases. While all

Figure 7-19 Burrowing animals, insects, bacteria, and fungi help add organic matter to soil.

Figure 7-20 This freshly plowed field in southwestern Georgia has a residual soil that is red.

soils contain some organic matter in various states of decay, the amount of such matter varies widely among different types of soil. For example, forest soils contain a much higher percentage of organic matter than desert soils do.

Soil Composition

During the process of its development, soil forms in layers. The solid bedrock from which weathered pieces of rock first break off is known as the parent rock. As these pieces of weathered bedrock break off, they rest on top of the parent rock layer. The pieces of rock continue to weather, and the smaller pieces form a layer that rests on top of the larger pieces. Thus, the smallest pieces of weathered rock, along with living and dead organisms, remain in the very top layer. Rainwater seeps through this top layer of materials, dissolves soluble minerals, and carries them into the lower layers of the soil.

Soil located above its parent material is called **residual soil.** Kentucky's bluegrass soil is an example of residual soil, as are the red soils in Georgia, shown in ***Figure 7-20.*** In contrast, **transported soil** has been moved to a location away from its parent bedrock. Agents of erosion, such as running water, wind, and glaciers, may transport soil from its place of origin to new locations. For example, glaciers have transported sediments to form soil from other places to the northern regions of the United States. Streams and rivers, especially during times of flooding, also transport and deposit great amounts of soil on floodplains along their banks at downstream locations. Winds carry and deposit very fine material to new locations as well.

While the parent bedrock determines what kinds of minerals a soil contains, the proportion of minerals in a soil and in the parent bedrock may not be the same. Differences may occur as the result of chemical weathering. The length of time it takes for soil to form also depends on the type of parent rock, as well as the climatic conditions of an area. In general, however, the process of soil formation occurs over a very long period of time; it can take hundreds of years for only a centimeter of soil to form.

Soil Profiles

Digging a deep hole in the ground will expose a soil profile, as when heavy machinery digs out soil in the process of building roads or highways. A **soil profile** is the vertical sequence of soil layers, as illustrated in ***Figure 7-21A.*** Some soils have more distinct layers than

Figure 7-21 A soil profile is the vertical sequence of soil layers **(A).** A gardener is growing plants in rich, black soil **(B).**

others. For example, poorly developed soils show little distinction between layers. A distinct layer, or zone, within a soil profile is called a **soil horizon.** There are three major soil horizons: A, B, and C. High concentrations of organic matter and humus are found in A horizons. Soils rich in humus are usually dark colored; they range from gray to black. ***Figure 7-21B*** shows black soils. Horizons B and C, the layers under horizon A, are less-developed soil. Horizon B contains subsoils that are enriched with clay minerals. Many subsoils have a zone of accumulation consisting of soluble minerals that have been leached, or washed out, from the topsoil. Subsoils may be red or brown in color as a result of the presence of iron oxides. Accumulations of clay in the B horizon can cause the formation of a hard material, commonly called hardpan, which may be so dense that it allows little or no water to pass through it. Horizon C, below horizon B and directly above solid bedrock, contains weathered parent material. Horizons A, B, and C are distinct and well developed in mature soils. However, all horizons may not be present in a given soil.

Topography The topography of a region affects the thickness of developing soil. In sloped areas, where runoff readily occurs, the coarser particles of soil remain on the slopes, while the smaller particles move downslope. As a result, soils on slopes tend to be thin, coarse, and infertile, whereas soils formed in lower areas, such as in valleys, are thick and fertile. Because south-facing slopes receive the most direct sunlight, they have somewhat more vegetation and therefore thicker soils than slopes facing in other directions.

The development of mature soil with distinct horizons takes a very long time. Only over time can vegetation grow and mature in a soil and increase the rate of soil development. Vegetation contributes to the buildup of humus and supplies acids that further promote the weathering process.

Soil Types

A soil's appearance, rate of formation, and productivity are determined to a great extent by climate. Because soils form from different parent bedrock material and undergo different climatic conditions, soils vary greatly from one place to another. Other factors contribute to the development of soil, including the types of plants and animals living in the soil, the topography of the area, and the length of time that the soil has been forming. However, because climatic conditions are the main influence on soil development, soils are often classified based on the climates in which they form. The four major types of soil, are polar, temperate, desert, and tropical. ***Figure 7-22*** shows a map of major soil types.

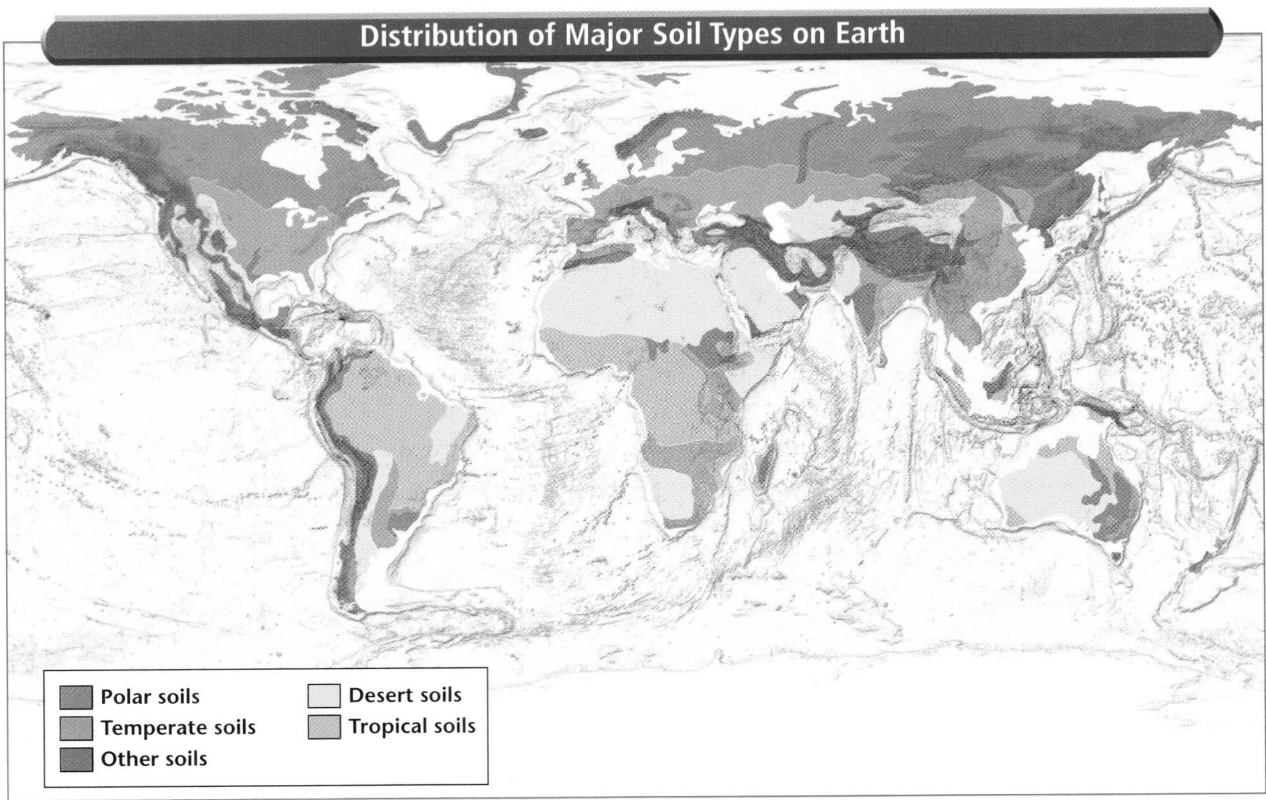

Figure 7-22 The major soil types include polar soils, temperate soils, desert soils, and tropical soils. Some climate regions have a variety of different soil types. They are identified on the map as "other." ***What soil types are found in the United States?***

Polar Soils Polar soils form at high latitudes and high elevations in places such as Greenland, Canada, and Antarctica. These soils have good drainage but no distinct horizons because they are very shallow, sometimes only a few centimeters deep. Permanently frozen ground, called permafrost, is often present under thin polar soils.

Temperate Soils Temperate soils vary greatly and are able to support such diverse environments as forests, grasslands, and prairies. While the temperate zone in general experiences annual rainfall greater than 50–60 cm, the specific amount of rainfall in an area determines the type of vegetation that will grow in temperate soils. Grasslands, which have an abundance of humus, are characterized by rich, fertile, soils, whereas forest soils are characterized by less deep and less fertile soils that contain aluminum-rich clays and iron oxides, such as those commonly found in the eastern portion of the United States. Soils in the drier, temperate prairies of the western United States support the growth of grasses and bushes. These areas experience annual rainfall of less than 50–60 cm.

Desert Soils Deserts receive low levels of precipitation—less than 25 cm per year. As a result, desert soils often have a high level of accumulated salts and can support only a limited amount of vegetation. Desert soils have little or no organic matter and a very thin A horizon. However, deserts often have abundant nutrients. During periods of precipitation deserts are able to support many plants that are adapted to survival during long periods of drought. Desert soils are also light-colored, coarse, and may contain salts and gypsum.

Figure 7-23 In the rain forests of Malaysia, the tropical soils are intensely weathered and contain very few nutrients.

Tropical Soils Tropical areas experience high temperatures and heavy rainfall. These conditions lead to the development of intensely weathered and often infertile soil, such as that shown in ***Figure 7-23.*** The intense weathering combined with a high degree of bacterial activity leave tropical soils with very little humus and very few nutrients. These soils experience much leaching of soluble materials, such as calcite and silica, but they have high concentrations of iron and aluminum. The characteristic red color of tropical soils is the result of the oxidation of iron. While these soils provide poor growth conditions, high-grade iron ore is mined from Brazilian, Australian, and Jamaican tropical soils.

Table 7-2 Soil Textures	
Soil Particles	**Size**
Very coarse sand	2–1 mm
Coarse sand	1–0.5 mm
Medium sand	0.5–0.25 mm
Fine sand	0.25–0.10 mm
Very fine sand	0.10–0.05 mm
Silt	0.05–0.002 mm
Clay	< 0.002 mm

Soil Textures

Particles of soil are classified according to size as being clay, silt, or sand, with clay being the smallest and sand being the largest, as shown in ***Table 7-2.*** The relative proportions of these particle sizes determine a soil's texture, as you will discover in the *Problem-Solving Lab* on this page. The proportions of different-sized particles present in a soil sample can be determined by first placing the sample along with water in a clear jar, shaking the jar, and allowing the particles to settle. With ample water, sediments will sort as they settle, and the percentage of settled clay, silt, and sand can then be estimated. This information, along with a soil textural triangle, shown in ***Figure 7-24,*** is used to determine a soil's texture. The texture of a soil affects its capacity to retain moisture and therefore its ability to support plant growth.

Soil Fertility

Soil fertility is the measure of how well a soil can support the growth of plants. Factors that affect soil fertility include the availability of minerals and nutrients, the number of microorganisms present, the amount of precipitation available, topography, and the level of acidity. Conditions necessary for growth vary with plant species. Farmers use

Problem-Solving Lab

Interpreting Data in a Table

Classify soils by texture Soils can be classified with the use of a soil textural triangle. Soil texture is determined by the relative proportions of particle sizes that make up the soil. The smallest particles are clay, and the largest are sand.

Soil Classification

Soil Sample	Percent Sand	Percent Silt	Percent Clay	Texture
1	50	40	10	Loam
2		20	30	
3	20		10	
4	20	20		

Analysis

1. Use the soil textural triangle shown in ***Figure 7-24*** to complete the data table. Record the percentages of particle sizes in the soil samples and the names of their textures.
2. Infer from the data table which soil sample has the greatest percentage of the smallest-sized particles.
3. Which soil sample has a sandy clay loam texture?

Thinking Critically

4. What can you conclude about the total of the percentages of sand, silt, and clay for each sample? Explain.
5. Name one characteristic of soil other than water-holding capacity that is determined by the soil's particle sizes.

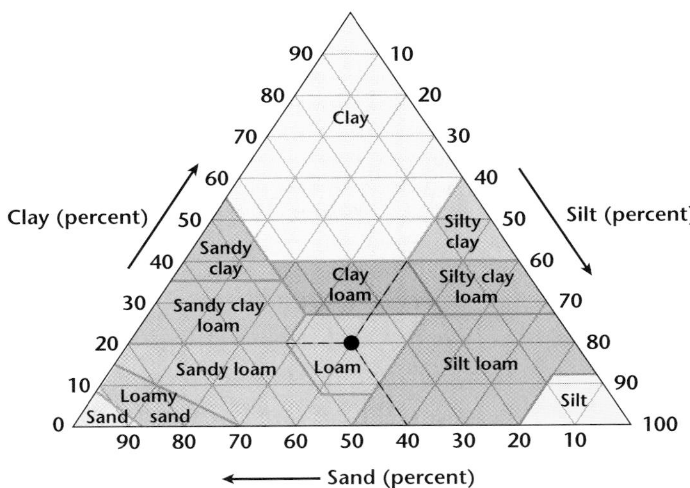

Figure 7-24 A soil textural triangle is used to determine a soil's texture. ***To determine the texture of a soil sample, find its percent for sand, silt and clay. Follow the percent lines for all three soils. The texture for the sample, will be where all three lines intersect. As an example, a soil sample of 40 percent silt, 40 percent sand, and 20 percent clay is the texture of loam.***

natural and commercially produced fertilizers to replace minerals and maintain soil fertility. Commercial fertilizers add nitrate, potassium, and phosphorus to soil. The planting of legumes, such as peas, beans, and clover, allows bacteria to grow on plant roots and replace nitrates in the soil. Pulverized limestone is often added to soil to reduce acidity and enhance crop growth. The addition of compost, organic mulch, and peat moss to soil also helps to maintain its fertility.

Soil Color A soil's composition and the climate in which it develops are the main factors that determine a soil's color. Topsoil is usually dark-colored because it is rich in humus. However, color alone is unreliable as an indicator of soil fertility. Red and yellow soils may be the result of oxidation of iron minerals. However, yellow soils are usually poorly drained and are often associated with environmental problems. Grayish or bluish soils are common in poorly drained regions where soils are constantly wet and lack oxygen.

SECTION ASSESSMENT

1. Explain the stages involved in the formation of soil.
2. Describe three characteristics of soil.
3. Explain the difference between temperate soils and tropical soils.
4. **Thinking Critically** How do the horizons in a typical soil profile differ from one another?

SKILL REVIEW

5. **Inferring** Infer what type of soil exists in your area and describe how you would determine whether your inference is correct. For more help, refer to the *Skill Handbook.*

earthgeu.com/self_check_quiz

Effects of Weathering

I&E

1.c Identify possible reasons for inconsistent results, such as sources of error or uncontrolled conditions.

Also covers: 1.g

Many factors affect the rate of weathering of Earth materials. Two major factors that affect the rate at which a rock weathers include the length of time it is exposed to a weathering agent and the composition of the rock.

Preparation

Problem

Investigate the relationship between time and the rate of weathering of halite chips.

Materials

plastic jar with lid
water (300 mL)
halite chips (100 g)
balance
timer
paper towels

Objectives

In this Geolab, you will:

- **Determine** the relationship between the length of time that rocks are exposed to running water and the degree of weathering of the rocks.
- **Describe** the appearance of weathered rocks.
- **Infer** what other factors may influence the rate of weathering.
- **Apply** your results to a real-world situation.

Safety Precautions

Wear splash-resistant safety goggles and an apron while you do this activity. Do not ingest the halite chips.

Weathering Data	
Average Shaking Time in Minutes	**Weight of Chips (g)**
3	________
6	________
9	________
12	________

Procedure

1. Soak 100 g of halite chips in water overnight.
2. As a class, decide on a uniform method of shaking the jars.
3. Pour off the water and place the halite chips in the plastic jar.
4. Add 300 mL of water to the jar.
5. Secure the lid on the jar.
6. Shake the jar for the assigned period of time.
7. Remove the water from the jar.
8. Use paper towels to dry the halite chips.
9. Use a balance to weigh the chips. Record your measurement in a data table similar to the one provided.

Analyze

1. Why did you need to soak the chips before conducting the investigation?
2. How did the mass of the rocks change with the length of time they were shaken?
3. How did the shape of the rocks change as a result of being shaken in a jar with water?
4. What factors could have affected a team's results?

Conclude & Apply

1. What real-world process did you model in this investigation?
2. How would acid precipitation affect this process in the real world?
3. How would the results of your investigation be affected if you used pieces of quartz instead of halite?

Science & the Environment

I&E 1.m

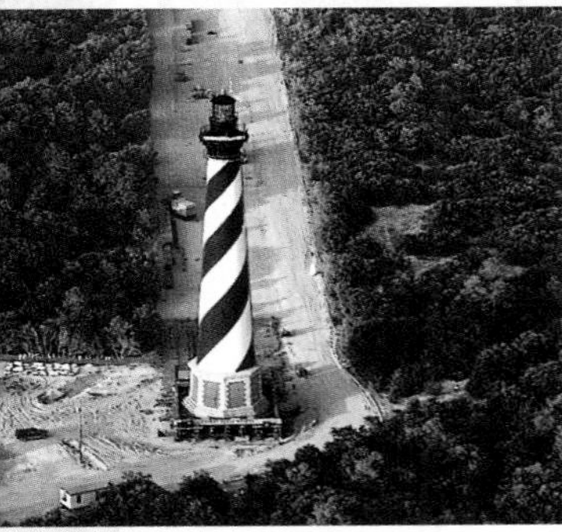

Shifting Sands

On June 17, 1999, thousands of people gathered on an island off North Carolina to witness an historic event. As the 4800 ton, 65-m tall Cape Hatteras lighthouse began its half-mile journey away from the sea to safety, people cheered. Engineers had scored a victory, however temporary, against the battering of the Atlantic Ocean on a beloved piece of American history.

Cape Hatteras lighthouse is the tallest brick lighthouse in the world. When the lighthouse was built in 1870, it was 500 m from the ocean. The strong beacon helped sailors navigate a coastline so dangerous that it was known as "The Graveyard of the Atlantic." By 1987, this famous light was only 50 m from the sea, and in danger of destruction.

Barrier Islands

The lighthouse is one of several found on barrier islands off the coast of North Carolina. These long, narrow islands of sand running parallel to the coast are relatively young, formed as rising global temperatures caused glaciers to begin melting 15 000 years ago. Sea level rise caused massive coastal flooding, separating dunes and beaches from the mainland and forming the barrier islands. These islands move constantly as wind, waves, and storms shift the unstable sand on which they are built.

Barrier islands are important. The islands absorb the first onslaught of waves and wind coming ashore from the Atlantic, sheltering the mainland from hurricanes and other storms.

Sea Level On the Rise

Earth has continued to warm and glaciers to melt since the last ice age ended, making sea level 100 m higher today than it was 15 000 years ago. The impact on barrier islands is enormous. Erosion on east-facing shorelines and sand accumulation on southwest-facing shorelines has resulted in southwest migration of North Carolina's barrier islands. Hatteras Island is moving, but it is not taking Cape Hatteras lighthouse with it.

Protecting structures built in coastal areas comes at great financial cost. The projected cost in the United States alone is at least $270 billion per 1 m rise in sea level. Various methods are used to protect property, including constructing bulkheads and levees, and pumping sand onto beaches to replace that lost to erosion. Moving Cape Hatteras lighthouse was a controversial method of saving the structure. The expenditure of nearly 12 million dollars on this project was questioned by many as a temporary fix for the permanent problem of barrier island erosion and migration.

Activity

A recent study projects a cost to U.S. taxpayers of between 270 and 450 billion dollars to protect coastal structures from destruction by a 1 meter sea level rise. Is this a wise expenditure of money? Debate this issue with other students in your class.

CHAPTER 7

Study Guide

Summary

SECTION 7.1

Weathering

Main Ideas

- The process of weathering breaks down Earth materials. Chemical weathering results in a change in the composition of a rock, whereas mechanical weathering results only in a change in a rock's size and shape.
- Temperature and pressure are major factors in the process of mechanical weathering. Changes in temperature can cause rocks to split.
- In chemical weathering, chemical reactions between rocks and water result in the formation of new minerals and the release of dissolved substances. The new minerals have different properties from those of the original rocks.

Vocabulary

chemical weathering (p. 155)
erosion (p. 153)
exfoliation (p. 155)
frost wedging (p. 154)
hydrolysis (p. 156)
mechanical weathering (p. 154)
oxidation (p. 156)
weathering (p. 153)

SECTION 7.2

Erosion and Deposition

Main Ideas

- Erosion is the process that moves weathered pieces of rock to new locations.
- Agents of erosion include moving water in streams and oceans, glaciers, wind, and gravity. Gravity is the driving force behind most agents of erosion.

Vocabulary

deposition (p. 162)
gully erosion (p. 163)
rill erosion (p. 163)

SECTION 7.3

Formation of Soil

Main Ideas

- Soil consists of weathered rock and humus, which is decayed organic matter in soil.
- Soil is residual or transported. Residual soil remains on top of its parent bedrock. Transported soil is moved to a location away from its parent bedrock by water, wind, or a glacier.
- A soil profile has horizons A, B, and C. Topsoil is located in horizon A, subsoil is in horizon B, and horizon C contains weathered rock from the bedrock.
- Characteristics of soil include texture, fertility, and color. Parent rock and environmental conditions determine a soil's composition.

Vocabulary

residual soil (p. 168)
soil (p .167)
soil horizon (p. 169)
soil profile (p. 168)
transported soil (p. 168)

CHAPTER 7

Assessment

Understanding Main Ideas

1. What erosional agent accounts for most of the erosion on Earth's surface?
a. water
b. wind
c. glaciers
d. living things

2. What is the underlying force of all agents of erosion?
a. magnetism
b. gravity
c. friction
d. light

3. The variables that most affect the weathering process are rock composition and what?
a. topography
b. surface area
c. living things
d. climate

4. Humus is found in which horizon?
a. A
b. B
c. C
d. D

5. What is the chemical reaction of oxygen with other substances called?
a. precipitation
b. hydrolysis
c. oxidation
d. humidity

6. What type of soil has the most humus?
a. polar
b. temperate
c. tropical
d. desert

7. On which side of a mountain slope is the greatest amount of vegetation likely to grow?
a. north
b. south
c. east
d. west

8. In which area is the topsoil most likely to be thickest?
a. on level land
b. on a hillside
c. on a mountain slope
d. in a river bed

9. Deep, rich soils are found in which regions?
a. temperate
b. desert
c. polar
d. tropical

10. What is the name of the soil type that is from a location that experiences high temperatures and high precipitation?
a. tropical
b. polar
c. desert
d. temperate

Applying Main Ideas

11. How does the size of an exposed rock affect its rate of weathering?

12. Describe how human activities can affect the rate of weathering.

13. What erosional process might convince a person not to purchase a home built on an ocean shore?

14. What do oxidation and hydrolysis have in common in relation to the weathering process?

15. What role does acid precipitation play in the weathering process?

16. How does the use of a wind barrier reduce erosion on a farm?

17. How do glaciers both remove and build up Earth's surface?

18. What unique feature does water exhibit in response to temperature changes?

Test-Taking Tip

PLAN YOUR WORK AND WORK YOUR PLAN Plan your workload so that you do a little each day rather than a lot all at once. The key to retaining information is to repeatedly review and practice it. Studying an hour a night for five days a week will help you remember more than cramming in a five-hour session on Saturday.

CHAPTER 7

Assessment

Thinking Critically

19. Name one reason why precipitation today is more acidic than precipitation in the 1800s.

20. If no water existed on Earth, how would erosional processes be affected?

21. Compare the rate of soil formation to the average human lifespan.

22. Describe how carbonic acid is formed.

23. In the blank circle below, divide and label the areas for each of the following components of a soil sample.

60% mineral matter
2% organic matter
30% air
8% water

24. Make a bar graph of the data in question 23.

25. Use the following terms to construct a concept map to organize the major ideas in Section 7.2, Erosion and Deposition. For more help, refer to the *Skill Handbook*.

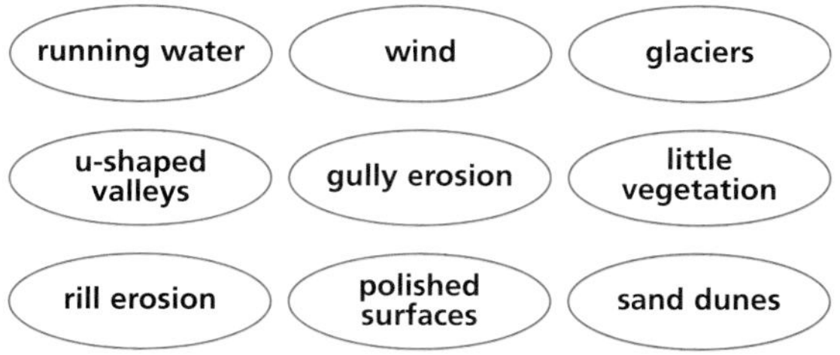

Standardized Test Practice

1. Which is NOT an agent of chemical weathering?
a. water
b. oxygen
c. carbon dioxide
d. wind

Interpreting Data Use the diagrams below to answer questions 2–4.

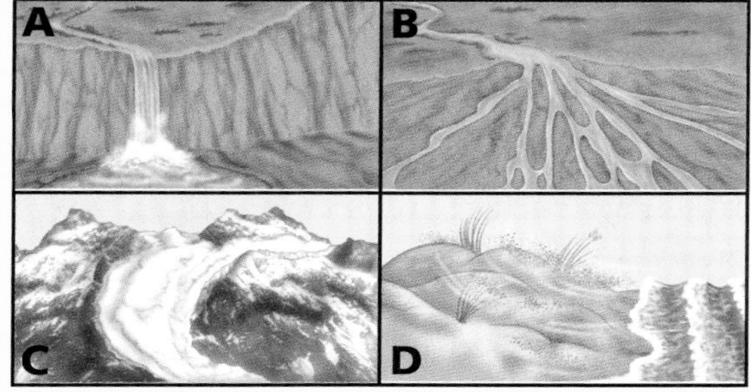

2. Which picture shows the erosional agent that was responsible for leaving behind U-shaped valleys, hanging valleys, lakes, and deposits of sediment in New England and New York State?
a. A **b.** B **c.** C **d.** D

3. Which picture shows the erosional agent responsible for dunes formed along the Gulf and Atlantic coasts of the U.S.?
a. A **b.** B **c.** C **d.** D

4. What common factor is responsible for three of the four erosional processes pictured?
a. wind
b. heat
c. human intervention
d. gravity

5. Which farming method is used to reduce wind erosion?
a. planting different crops
b. planting wind barriers
c. building earth mounds
d. building stone walls

Chapter 8

Mass Movements, Wind, and Glaciers

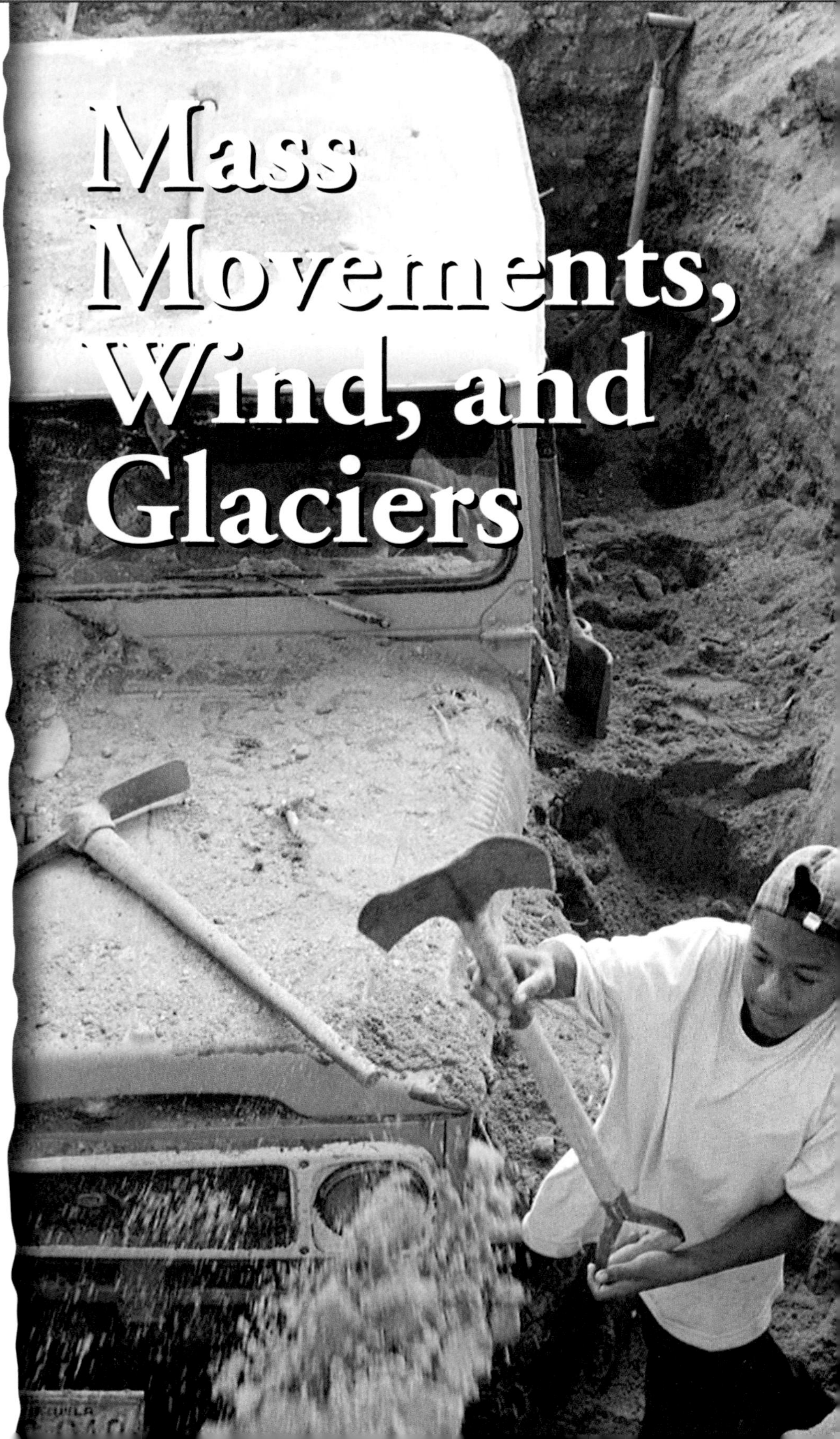

What You'll Learn

- How the processes of mass movements, wind, and glaciation change landscape features.
- What external features on Earth's surface are caused by mass movements, wind, and glaciers.

Why It's Important

Earth's external processes shape its surface. Some of the processes, such as landslides and avalanches, represent hazards. Mass movements, wind, and glaciers change the landscape and have an impact on human populations in many regions.

Earth Science Online

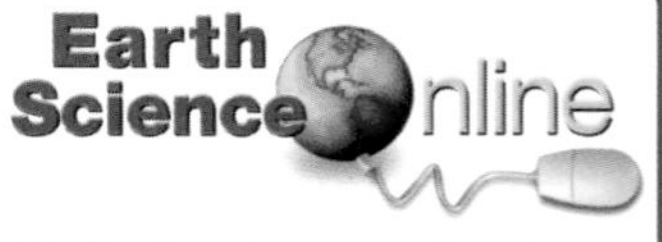

To find out more about mass movements, wind, and glaciers, visit the Earth Science Web Site at earthgeu.com

Discovery Lab

Model Sand-Slope Activity

Water affects sediment grains on slopes. If there is too little water, the sediments may not hold together, and as a result, they may move downslope. In this activity, you will demonstrate how the addition of water affects how sediments are held together.

1. Place 225 mL of sand in each of three separate containers, such as aluminum pie plates.
2. To the first container of sand, add 20 mL of water and mix well. To the second container of sand, add 100 mL of water and mix well. To the third container of sand, add 200 mL of water and mix well.
3. Empty the three mixtures of sand and water onto a tray or piece of cardboard. Keep each mixture separate.
4. Test each mixture for its ability to be molded and retain its shape. Compare your results for the three samples.

CAUTION: Always wear safety goggles and an apron in the lab.

Observe In your science journal, describe how the addition of water affected the sand's ability to be molded in the three samples.

SECTION 8.1 Mass Movements at Earth's Surface

Earth Sciences 9.b

OBJECTIVES

- **Identify** *factors that affect mass movements.*
- **Relate** *how mass movements affect people.*
- **Analyze** *the relationship between gravity and mass movements.*

VOCABULARY

mass movement
creep
mudflow
landslide
slump
avalanche

Every day, the landscape around us undergoes changes. If you compared old and new photographs of the landscape around your home, you might have to look very closely to notice some of the subtle differences. Other changes, such as landslides, occur quickly and have very noticeable, immediate effects.

MASS MOVEMENTS

How do landforms such as mountains, hills, and plateaus wear down and change? Landforms can change through processes involving wind, ice, and water, and sometimes through the force of gravity alone. The downslope movement of loose sediments and weathered rock resulting from the force of gravity is called **mass movement.** In the development of most of Earth's landforms, erosion is the step that follows weathering. After weathering processes weaken and break rock into smaller and smaller pieces, mass movements may occur and carry the debris downslope. Because climate has a major effect on the vegetation and the weathering activities that occur in a particular area, climatic conditions determine which materials and how much of each will be made available for mass movement.

All mass movements occur on slopes. Because very few places on Earth are completely flat, almost all of Earth's surface undergoes mass movement. Mass movements range from extremely slow motions to sudden slides, falls, and flows. The Earth materials that are moved range from fine mud to large boulders.

Earth Sciences

9.b Students know the principal natural hazards in different California regions and the geologic basis of those hazards.

VARIABLES THAT INFLUENCE MASS MOVEMENTS

Several variables influence the mass movements of Earth's material. One variable is the material's weight resulting from gravity, which works to pull the material down a slope. A second variable is the material's resistance to sliding or flowing. A third variable can be a trigger, such as an earthquake, that works to shake material loose from a slope, as shown in ***Figure 8-1A.*** Mass movement occurs when the forces working to pull material down a slope are stronger than the material's resistance to sliding, flowing, or falling. Some common types of mass movement are illustrated in ***Figure 8-1B.***

A

Figure 8-1 Sudden mass movements can be started by a trigger, such as an earthquake, which caused this landslide in Gallatin National Forest, Montana **(A).** There are three common types of mass movement. ***How are they alike? How are they different?*** **(B)**

B

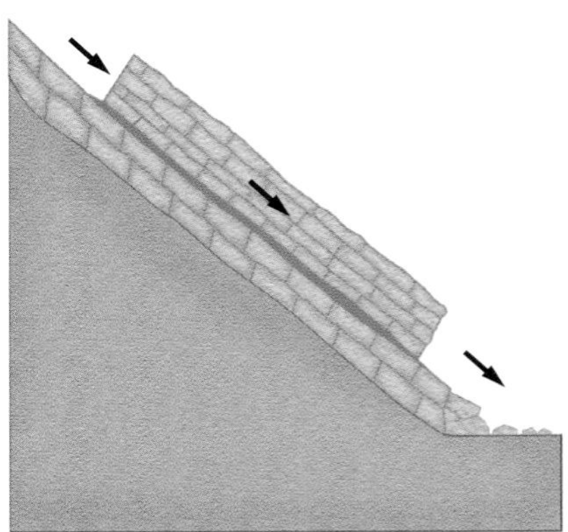

Slide
Moves as block of Earth material

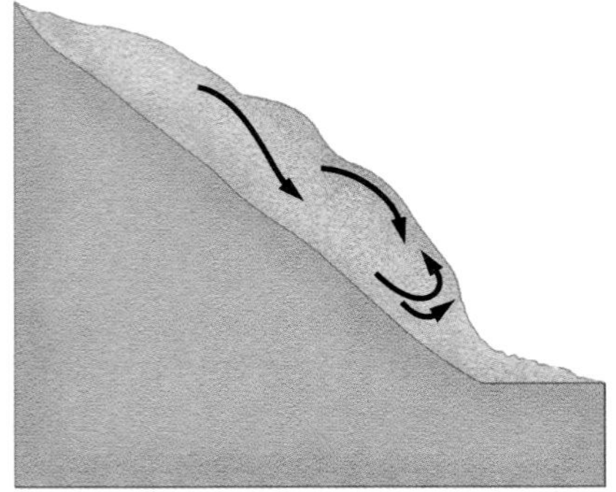

Flow
Movement involves mixing of particles within moving mass

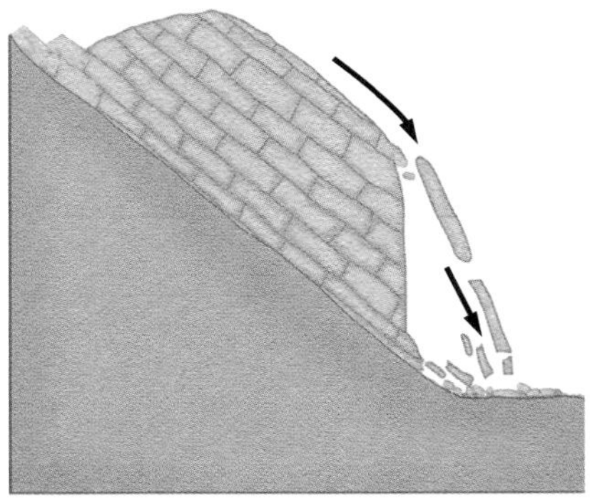

Fall
Free fall of Earth material

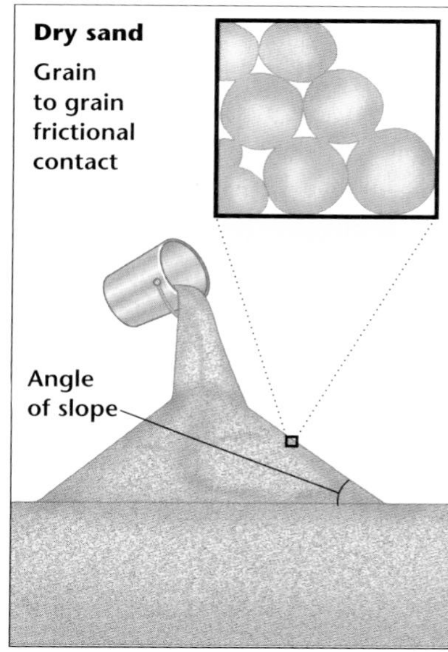

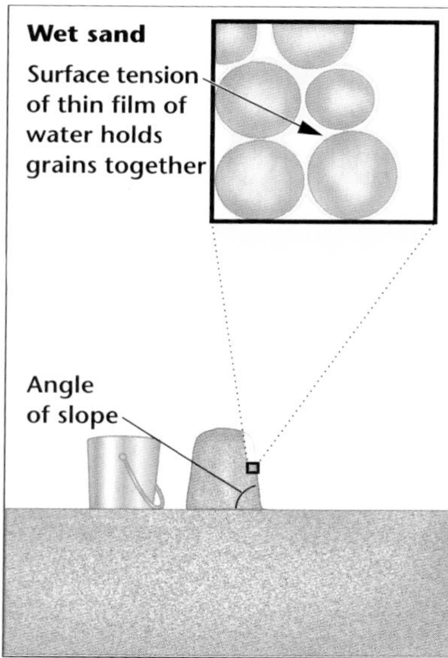

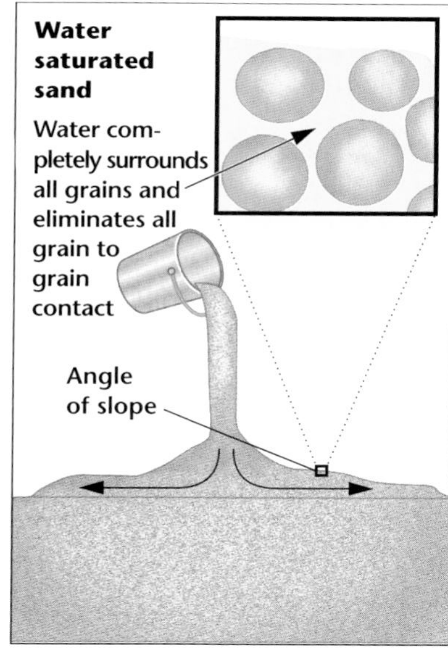

Figure 8-2 Water plays a key role in mass movements. Water acts as a lubricant between grains of sand to reduce friction between them.

Erosion and the undermining of soil at the foot of a slope increase the potential of Earth materials to move downhill, as does the increased weight from torrential rainfall. An important force that determines a material's resistance to downhill movement is friction between the material and the slope.

Water Water is a fourth variable that influences mass movements, as illustrated in ***Figure 8-2.*** On a slope, too little water may prevent sediment grains from holding together at all, thereby increasing the material's potential for movement. You could demonstrate this concept by trying to form dry, loose soil into a ball. It would be difficult to get the particles to hold together. However, if you added small amounts of water, the particles of soil would cling together and remain in the shape in which you formed them. Similarly, the addition of water to sediments on a slope helps to hold the grains together and makes the material more stable. On the other hand, too much water can make a slope unstable. Saturation by water greatly increases the weight of soils and sediments. In addition, as the water fills the tiny open spaces between grains, it may act as a lubricant between the grains to reduce the friction between them. Thus, the force of gravity is more likely to pull the saturated material downhill. While water is very important to the process of mass movement, it is important to note that water is not involved as a transport agent. In mass movements, water moves along with Earth materials. This is in contrast to stream transport, in which sediment moves along with the water.

Types of Mass Movements

Mass movements are classified as creep, flows, slides, and falls. Mass movements can move different types of materials in various ways. Let's investigate the different types of mass movement.

Creep The slow, steady, downhill flow of loose, weathered Earth materials, especially soils, is called **creep.** Because movement may be as little as a few centimeters per year, the effects of creep usually are noticeable only over long periods of time. One way to tell whether creep has occurred is to observe the positions of structures and objects. As illustrated in ***Figure 8-3,*** creep can cause the tilting of once-vertical utility poles, fences, and gravestones, the bending of trees, the cracking of walls, and the breaking of underground pipelines. Loose materials on almost all slopes, even the very gentlest, undergo creep. Soil creep moves huge amounts of surface material each year.

The slow, downhill movement of loose, water-logged materials that occurs in regions of permafrost is called solifluction. The material moved in solifluction is a mudlike liquid that is produced when water is released from melting permafrost during the warm season. The water saturates the surface layer of soil and is unable to move downward through the underlying permafrost. As a result, the surface layer can slide slowly down a slope. Solifluction may also occur in humid regions where the ground remains saturated year-round.

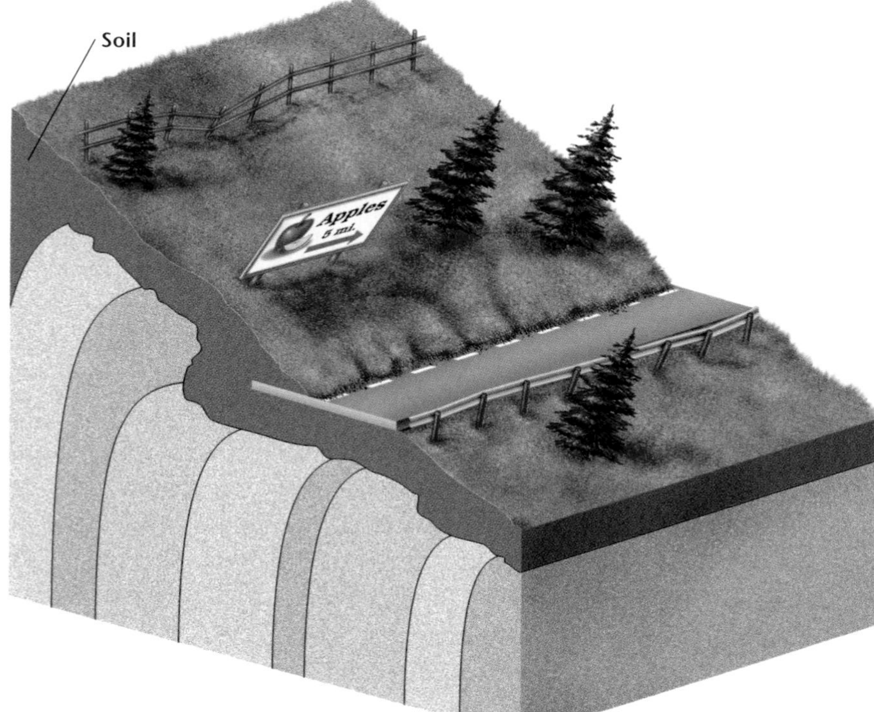

Figure 8-3 Creep can cause the slow tilting of various objects on slopes. ***Is there any evidence of creep in your area?***

Figure 8-4 Major mudflows occurred after the eruption of Mount St. Helens in 1980 **(A).** In 1998, a mudflow in Sarno, Italy, caused the deaths of 135 people and the destruction of many buildings **(B).**

Flows In some mass movements, Earth materials flow as if they were a thick liquid. The materials can move as slowly as a few centimeters per year or as rapidly as hundreds of kilometers per hour. Earth flows are moderately slow movements of soils, whereas **mudflows** are swiftly moving mixtures of mud and water. Mudflows can be triggered by earthquakes or similar vibrations and are common in volcanic regions where the heat from a volcano melts snow on nearby slopes that have fine sediment and little vegetation. The meltwater fills the spaces between the small particles of sediment and allows them to slide readily over one another and move down the slope. As shown in ***Figure 8-4A,*** when Mount St. Helens erupted in 1980, it triggered mudflows that traveled downhill at speeds of over 30 km/h. Although the area was sparsely populated, the mudflows damaged or destroyed more than 200 homes. In 1998, a mudflow in Italy caused many deaths and much destruction, as shown in ***Figure 8-4B.***

Mudflows are also common in sloped, semi-arid regions that experience intense, short-lived rainstorms. The Los Angeles Basin in southern California is an example of an area where mudflows are common. In such areas, periods of drought and forest fires can leave the slopes with little protective vegetation. When heavy rains eventually fall in these areas, they can cause massive, destructive mudflows. Mudflows are destructive in areas where urban development has spread to the bases of mountainous areas.

Figure 8-5 A landslide occurs when a sheet of loose Earth materials separates from the bedrock and moves quickly downslope **(A).** In March of 1998, a landslide caused about 120,000 m^3 of dirt and debris to flow into the Blackfoot River in Ovando, Montana. The river became well known after its portrayal in the movie *A River Runs Through It* **(B).**

Slides A rapid, downslope movement of Earth materials that occurs when a relatively thin block of loose soil, rock, and debris separates from the underlying bedrock is called a **landslide,** illustrated in ***Figure 8-5A.*** The material rapidly slides downslope as one block, with little internal mixing. Some landslides may reach speeds of 200 km/h. As shown in ***Figure 8-5B,*** a landslide mass eventually stops and becomes a pile of debris at the bottom of a slope, sometimes damming rivers and causing flooding. Landslides are common on steep slopes, especially when soils and weathered bedrock are fully saturated by water. This destructive form of mass movement causes almost $2 billion in damage and several deaths per year in the United States alone. You will explore the movement of a landslide in the *Mapping GeoLab* at the end of this chapter.

A rock slide is a type of landslide that occurs when a sheet of rock moves downhill on a sliding surface. During a rock slide, relatively thin blocks of rock are broken into smaller blocks as they move downslope. Often triggered by earthquakes, rock slides can move large amounts of material.

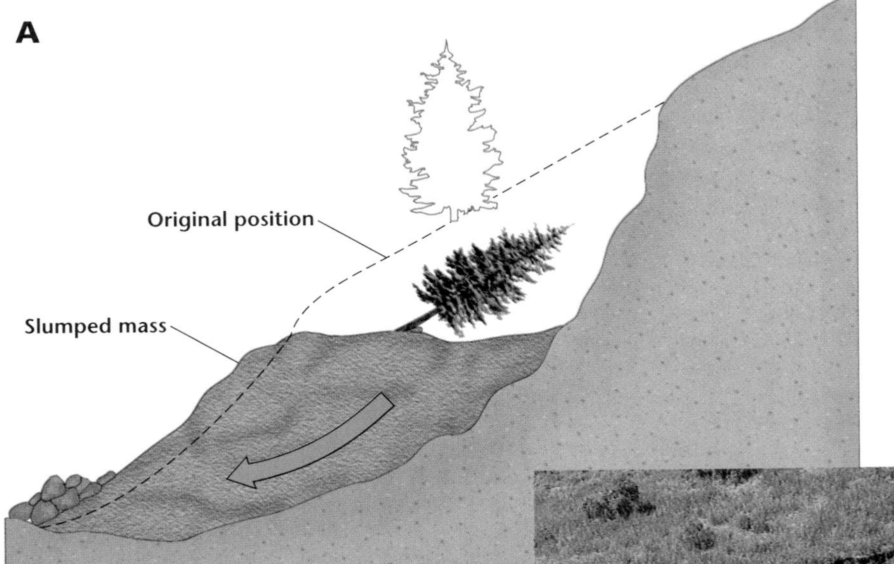

B

Figure 8-6 A slump occurs when Earth materials in a landslide rotate and slide along a curved surface **(A).** This slump, which occurred in Santa Barbara, California, was triggered by heavy rains **(B).**

Slumps As illustrated in ***Figure 8-6A,*** when the mass of material in a landslide rotates and slides along a curved surface, a **slump** results. Slumps, such as the one shown in ***Figure 8-6B,*** may occur in areas that have thick soils on moderate-to-steep slopes. Sometimes, slumps occur along highways where the slopes of soils are extremely steep. Slumps are common after rains, when water reduces the frictional contact between grains of soil and acts as a lubricant between surface materials and underlying layers. The weight of the additional water pulls material downhill. As with other types of mass movement, slumps can be triggered by earthquakes. Slumps leave crescent-shaped scars on slopes.

Avalanches Landslides that occur in mountainous areas with thick accumulations of snow are called **avalanches.** Avalanches usually occur on slopes of at least 35°. About 10 000 avalanches occur each year in the mountains of the western United States alone. Radiation from the Sun can melt surface snow, which then refreezes at night into an icy crust. Snow that falls on top of this crust can eventually build up, become heavy, slip off, and slide down a slope as an avalanche.

Another type of weak snow layer forms in early winter as the ground, which is still warm, melts the overlying snow. The snow then refreezes into a layer of jagged, slippery snow crystals. A vibrating trigger, even from a single skier, can send such an unstable layer sliding down a mountainside at speeds of up to 300 km/h. As shown in ***Figure 8-7,*** avalanches pose particular risks in places such as Switzerland, where more than 50 percent of the population lives in avalanche terrain. In the 1998–1999 season, for example, hundreds of major avalanches swept down the Swiss Alps.

Figure 8-7 In February of 1999, a deadly avalanche occurred in Switzerland. Several people were killed and many homes were destroyed. The damage was estimated at $100 million.

Rock Falls Rock falls commonly occur at high elevations, in steep road cuts, and on rocky shorelines. On high cliffs, rocks are loosened by physical weathering processes, such as freezing and thawing, and by plant growth. As rocks break up and fall directly downward, they may bounce and roll, ultimately producing a cone-shaped pile of coarse debris, called talus, at the base of the slope. On human-made rock walls, such as road cuts, rock falls are particularly common. "Falling Rock" warning signs can often be seen along highways that run through steep, rocky areas, as shown in ***Figure 8-8.*** Rock falls are less likely to occur in humid regions, where the rock is typically covered by a thick layer of soil, vegetation, and loose materials.

Figure 8-8 These rocks on a highway in Colorado were deposited by a rock fall. In areas where rock falls are common, signs are posted to warn motorists.

MASS MOVEMENTS AFFECT PEOPLE

While mass movements are natural processes, human activities often contribute to the factors that cause mass movements. Activities such as constructing heavy buildings, roads, and other structures can make slope materials unstable. In addition, poor maintenance of septic systems, which often leak, can trigger slides.

Dangerous Mudflows Human lives are in danger when people live on steep terrain or in the path of unstable slope materials. For example, in December of 1999, northern Venezuela experienced the heaviest rains that had fallen there in 100 years. Within several days, between 30 to 48 cm of unseasonal rainfall occurred in this area. The sudden saturation of sediments, combined with the area's steep topography and widespread deforestation, resulted in severe mudflows and landslides, as shown in ***Figure 8-9.*** Tens of thousands of people died, more than 114 000 people were left homeless, and 23 000 homes were destroyed. Entire villages were buried in mud, rock, and debris. The widespread loss of human life was primarily a result of the location of villages both high up in steep terrain and at the foot of unstable, saturated slopes.

REDUCING THE RISKS

Catastrophic mass movements are most common on slopes greater than 25° that experience annual rainfall of over 90 cm. The best way to minimize the destruction caused by mass movements is to avoid building structures on such steep and unstable slopes.

Figure 8-9 Deadly mudflows that occurred in Venezuela in 1999 caused extensive damage in Los Corales.

Figure 8-10 This steel net was installed along Route 101 near Quilcene, Washington, to protect motorists from rockslides **(A).** In Wolf Creek Pass, Colorado, a fence was built along this highway to protect motorists from rock slides **(B).**

Preventive Actions Although preventing mass-movement disasters is not an easy task, some actions can help to avoid the potential hazards. For example, a series of trenches can be dug to divert running water around a slope and control its drainage. Other approaches to controlling landslides include covering steep slopes with materials such as steel nets, and constructing protective fences along highways in areas where rock slides are common, as shown in ***Figure 8-10.*** Still other approaches involve the installation of retaining walls to support the bases of weakened slopes and prevent them from falling. Most of these efforts at slope stabilization and prevention of mass movements, however, are generally successful only in the short run. The best way to reduce the number of disasters related to mass movements is to educate people about the problems of building on steep slopes.

Section Assessment

1. Identify and describe one type of rapid mass movement and one type of slow mass movement.
2. Describe the underlying force behind all forms of mass movement and explain its role in the process of mass movement.
3. How does water affect the process of mass movement?
4. What precautions can humans take to avoid the dangers associated with mass movements?
5. **Thinking Critically** Explain how one particular human activity can increase the risk of mass movement and suggest a solution to the problem.

Skill Review

6. **Making Tables** Design a data table that shows the similarities and differences among the forms of mass movement discussed in this section. For more help, refer to the *Skill Handbook*.

SECTION 8.2 *Wind*

Moving air can pick up and transport Earth materials in the process of erosion. Unlike water, wind can transport sediments uphill as well as downhill. As an erosional agent, wind can modify and change landscapes in arid and coastal areas.

OBJECTIVES

- **Describe** *conditions that contribute to the likelihood that an area will experience wind erosion.*
- **Identify** *wind-formed landscape features.*
- **Describe** *how dunes form and migrate.*
- **Explain** *the effects of wind erosion on human activities.*

VOCABULARY

deflation
abrasion
ventifact
dune
loess

WIND EROSION AND TRANSPORT

A current of rapidly moving air can pick up and carry sediments in the same way that water does. However, except for the extreme winds of hurricanes, tornadoes, and other strong storms, winds generally cannot carry particles as large as those transported by moving water. Thus, the relative ability of wind to erode materials is less than that of other erosional agents, such as water and ice.

Winds transport materials by causing their particles to move in different ways. For example, wind can move sand on the ground in a rolling motion. A method of transport by which strong winds cause particles to stay airborne for long distances, as shown in ***Figure 8-11A,*** is called suspension. Another method of wind transport, called saltation, causes a bouncing motion of particles. Saltation accounts for most sand transport by wind. Both suspension and saltation are shown in ***Figure 8-11B.*** Most areas where wind transport and erosion occur experience limited amounts of precipitation, which helps to hold down sediments and allows plants to grow.

Figure 8-11 Dust storms, caused by suspension, are common in the Arizona desert **(A).** Strong winds can transport sediments by suspension and saltation **(B).**

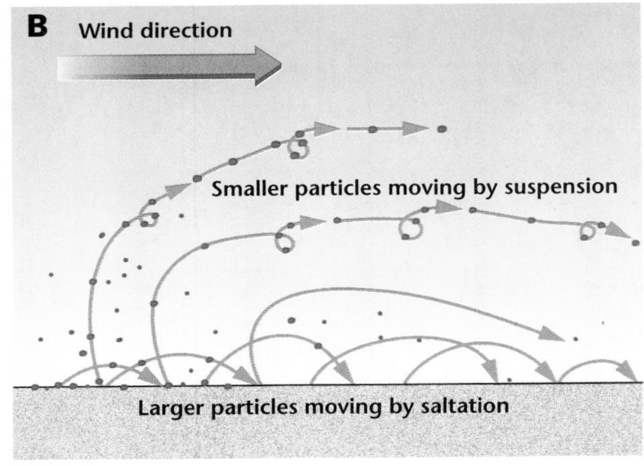

Figure 8-12 *What areas in the United States are subject to wind erosion? Why is wind erosion a problem in those areas?*

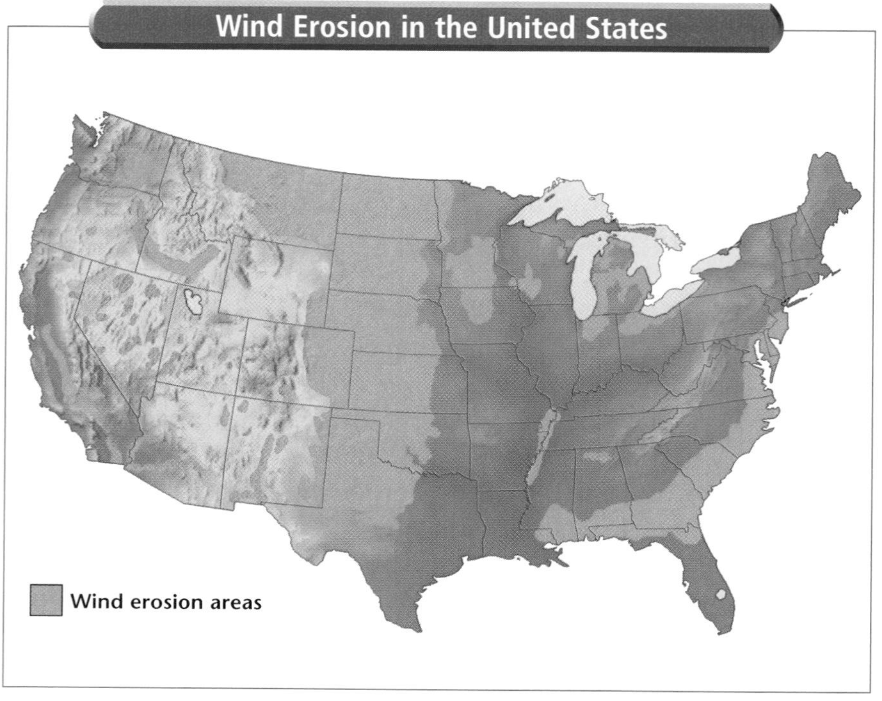

Figure 8-13 Desert pavement is the coarse sediment found on a desert floor **(A).** This desert pavement is located near the Gila River in New Mexico and Arizona **(B).**

A

B

Thus, wind transport and erosion primarily occur in areas with little vegetative cover, such as deserts, semi-arid areas, seashores, and some lakeshores. As shown in ***Figure 8-12,*** wind erosion is a problem in many parts of the United States. You will learn more about wind erosion in the *MiniLab* later in this section.

The lowering of the land surface that results from the wind's removal of surface particles is called **deflation.** The particles removed may be composed of any material. In areas of intense wind erosion, coarse gravel and pebbles are usually left behind as the finer surface material is removed by winds. The coarse surface left behind is called desert pavement, shown in ***Figures 8-13A and 8-13B.*** Deflation is a major problem in many agricultural areas of the world.

During the 1930s, portions of the Great Plains region, which stretches from Montana to Texas, experienced severe drought. The area was already suffering from the effects of poor agricultural practices, in which huge areas of natural vegetation were removed to clear the land for farming. Strong winds readily picked up the dry surface particles, which lacked any protective vegetation. Such severe dust storms resulted that daytime skies were often darkened and the region became known as the Dust Bowl. Today, the Great Plains are characterized by thousands of shallow depressions known as deflation blowouts. They are the result of the removal of surface sediment by wind erosion during the 1930s. The depressions range in size from a few meters to hundreds of meters in diameter.

Another process of erosion, called **abrasion,** occurs when particles such as sand rub against the surface of rocks or other materials. Abrasion occurs as part of the erosional activities of winds, streams, and glaciers. In wind abrasion, wind picks up materials such as sand particles and blows them against rocks and other objects. Because sand is often made of quartz, a very hard mineral, wind abrasion can be a very effective agent of erosion; windblown sand particles eventually wear away rocks, as shown in ***Figure 8-14.*** Structures such as telephone poles also may be worn away or undermined by wind abrasion, and paint and glass on homes and vehicles may be damaged by windblown sand. Materials that are exposed to wind abrasion show unique characteristics. For example, windblown sand causes rocks to become pitted and grooved. With continued abrasion, rocks become polished on the windward side and develop smooth surfaces with sharp edges, as shown in ***Figure 8-15A.*** In areas of shifting winds, abrasion patterns correspond to wind shifts, and different sides of rocks become polished and smooth. Rocks shaped by wind-blown sediments, such as those shown in ***Figure 8-15B,*** are called **ventifacts.**

Figure 8-14 The forces of abrasion carved this conglomerate rock called the Baja Mushroom, located at Puerto Ballandra, near La Paz, Baja, Mexico.

Figure 8-15 Rocks that are exposed to windblown sand become pitted and grooved **(A).** These ventifacts are located in Bull Pass, Antarctica **(B).**

A

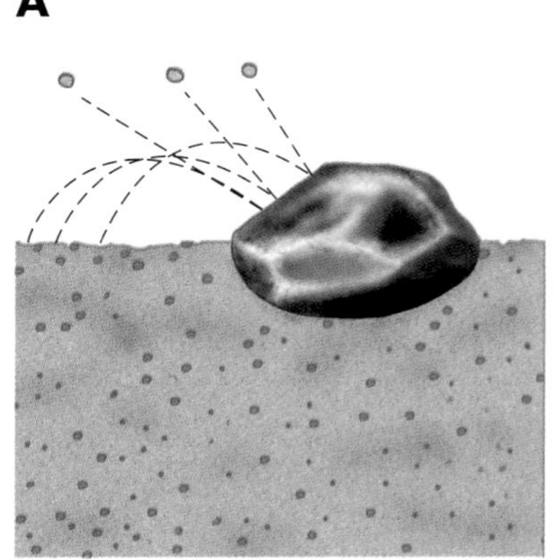

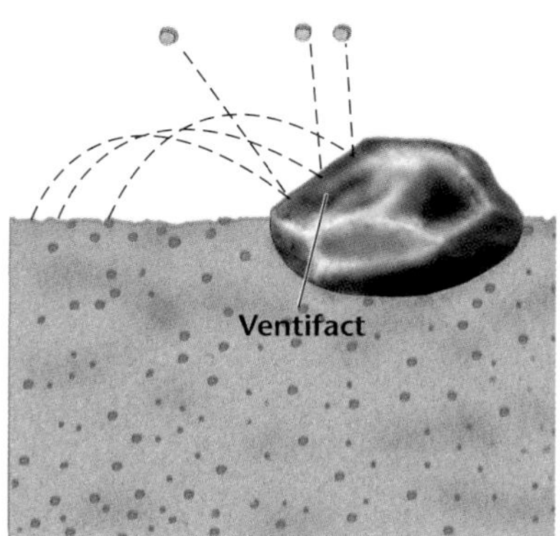

B

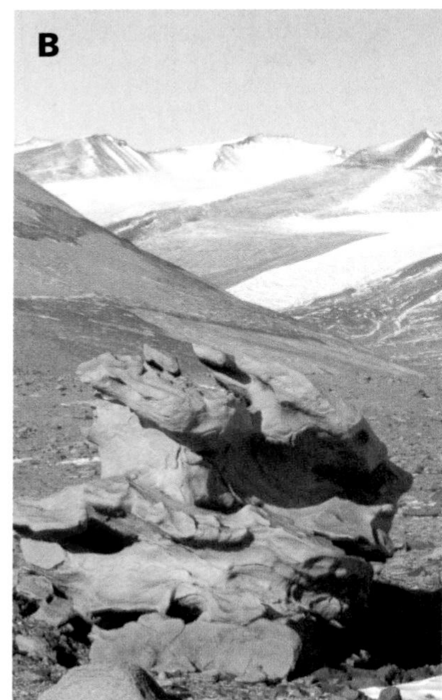

MiniLab

Where does wind erosion occur?

Interpret a wind erosion map to find out what parts of the United States are subject to wind erosion.

Procedure

Refer to the wind erosion map shown in *Figure 8-12* to answer the following questions.

Analyze and Conclude

1. Which areas of the United States experience wind erosion?
2. Where is the largest area of wind erosion? The second largest?
3. What coastal areas are subject to wind erosion?

Figure 8-16 The wind is lifting sand particles from the top of this sand dune in the Namibia Desert in Namibia, a country located on the west coast of Africa.

WIND DEPOSITION

Wind deposition occurs in areas where there are changes in wind velocity. As wind velocity decreases, some of the wind-blown sand and other materials can no longer stay airborne, and they drop out of the air stream to form a deposit on the ground.

Formation of Dunes In wind-blown environments, sand particles tend to accumulate where an object, such as a rock, landform, or piece of vegetation, blocks the particles' forward movement. Sand continues to be deposited as winds blow in one general direction. Over time, the pile of wind-blown sand develops into a **dune,** as shown in ***Figure 8-16.*** The conditions under which a dune forms determine its particular shape. These conditions include the availability of sand, wind velocity, wind direction, and the amount of vegetation present. All dunes have a characteristic profile. The gentler slope of a dune is located on the side from which the wind blows, the windward side. The steeper slope is on the side protected from the wind, called the leeward side.

The velocity of the wind above the ground surface determines the height of a dune. The heights of dunes are usually in the range of 12 to 25 m. The maximum height is variable, but the world's tallest dunes, in Saudi Arabia, measure more than 100 m in height.

Although quartz sand is the most common component of dunes, any dry, granular material can be formed into a dune, as long as winds continue to blow in a consistent direction and at a speed great enough to transport particles. Gypsum dunes are found at the White Sands National Monument in New Mexico, for example, and there are calcite dunes in Bermuda and areas in the Caribbean.

Table 8-1 Types of Sand Dunes

Type	Shape	Size	Area of formation
Barchan	Wind	Maximum size: 30 m high, 300 m point to point.	The most common dunes. Generally form in areas of constant wind direction. Migrate 8–15 m per year.
Transverse		Maximum height: 25 m.	Form in areas with strong winds and abundant sand.
Parabolic		Maximum height: 30 m.	Form in areas with moderate winds and some vegetation. Include extremely curved types called hairpin dunes. Common on seacoasts.
Longitudinal		Maximum height: 90 m. Can be 100 km long. Average dimensions are 8 m in height and 60 m in length.	Form in areas with high, somewhat variable winds and little sand.

Types of Dunes Dunes are classified according to their shapes, as shown in ***Table 8-1.*** Barchan dunes are solitary, crescent-shaped dunes that form in flat areas where there is little sand or vegetation. Transverse dunes are formed where there is plenty of sand, little or no vegetation, and strong, steady, prevailing winds. Transverse dunes form in a series of long ridges that are perpendicular to the direction of the wind. In humid areas, U-shaped dunes, called parabolic dunes, form between clumps of plants. Where there is limited sand available, strong prevailing winds shape longitudinal dunes, which are parallel to the wind direction.

On offshore islands and on lakeshores, dunes are formed by winds blowing off the water toward the shore. Coastal dunes protect against beach and coastal erosion by reducing the direct action of wind on beach sand. They also act as buffers against the action of waves and provide shelter for vegetation. Once dune vegetation is established, it helps to anchor coastal dunes.

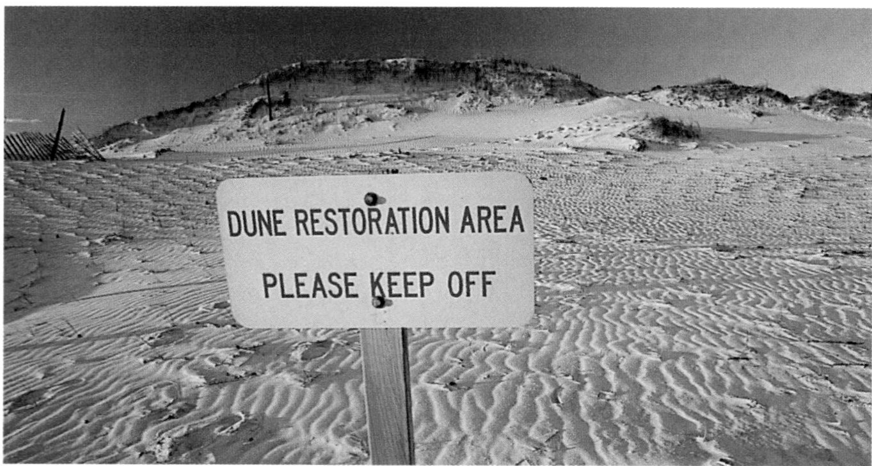

Figure 8-17 This sign was posted to make people aware that dunes are being restored in Okaloosa County, Florida.

Human activities, such as building in coastal-dune areas and removing dune vegetation, have disrupted dune growth and damaged dunes in many coastal areas of the United States. The destruction of dunes has led to increased beach erosion and nearshore flooding in these locations. Dune restoration areas, as shown in ***Figure 8-17,*** seek to restore and protect dunes in coastal areas.

As shown in ***Figure 8-18,*** dune migration is caused when prevailing winds continue to move sand from the windward side of a dune to its leeward side, thereby causing the dune to move slowly over time. As long as winds continue to blow, dunes continue to migrate and cover anything in their paths. Migrating dunes can block highways and cover farmland. Large dunes can even bury houses and other structures.

> **Earth Science Online**
>
> **Topic: Wind Erosion**
> To find out more about wind erosion, visit the Earth Science Web Site at earthgeu.com
>
> **Activity:** Research a dune restoration project. Write a newspaper article describing the project.

Loess Wind can carry fine, lightweight particles such as silt and clay in great quantities and for long distances. Many parts of Earth's surface are covered by thick layers of windblown silt, which are thought to have accumulated as a result of thousands of years of

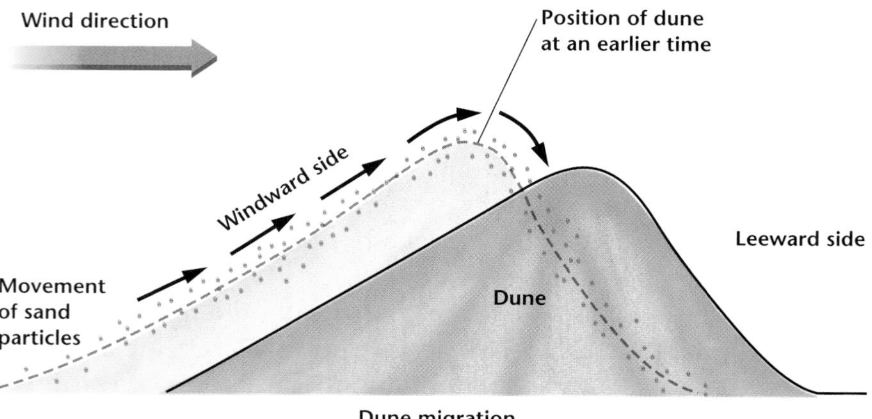

Figure 8-18 As dunes migrate, sand is moved from the windward side to the leeward side over a period of time.

Distribution of Loess Deposits in the United States

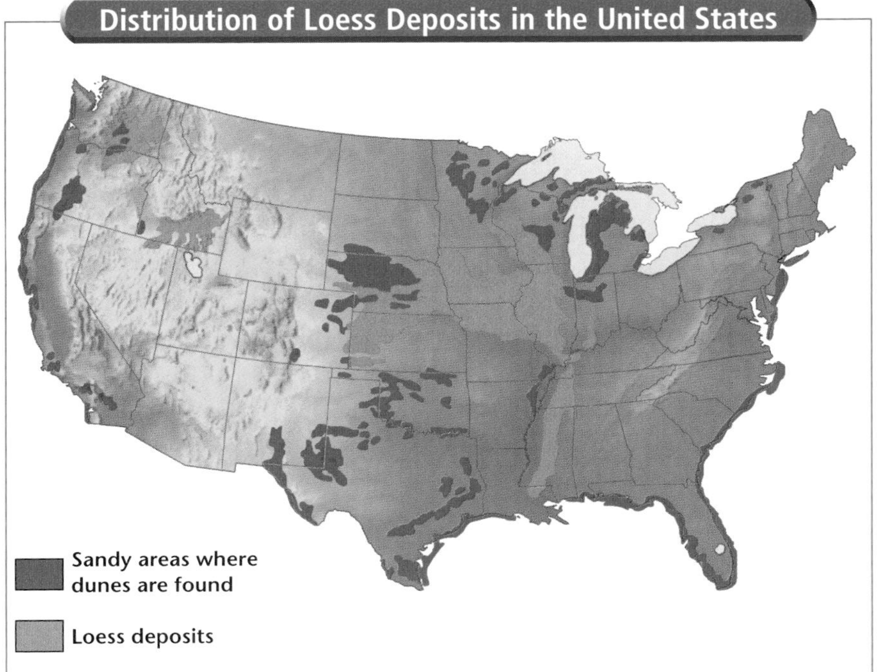

Figure 8-19 This map shows the distribution of loess soil in the United States. ***Where is the largest deposit of loess soil?***

dust storms. The source of these silt deposits may have been the fine sediments that were exposed when glaciers melted after the last ice age, more than 10 000 years ago. These thick, wind-blown silt deposits are known as **loess.** As shown in ***Figure 8-19,*** loess deposits are located in Illinois, Iowa, Missouri, South Dakota, Nebraska, Kansas, and Idaho. Where precipitation is adequate, loess soils are some of the most fertile soils on Earth because they contain abundant minerals and nutrients.

Section Assessment

1. What climatic conditions are most likely to produce wind erosion?
2. How does the vegetation growing in an area affect the wind's ability to modify the surface of Earth?
3. Draw a diagram showing how deflation occurs in wind erosion.
4. Why is wind abrasion such an effective agent of erosion?
5. How do dunes form?
6. **Thinking Critically** How can wind erosion directly affect human activities?

Skill Review

7. **Inferring** Describe the differences in appearance of wind-worn particles and water-worn particles. For more help, refer to the *Skill Handbook.*

SECTION 8.3 Glaciers

I&E 9.d

OBJECTIVES

- **Explain** *how glaciers form.*
- **Compare** *and* **contrast** *the conditions that produce valley glaciers and those that produce continental glaciers.*
- **Describe** *how glaciers modify the landscape.*
- **Recognize** *glacial landscape features.*

VOCABULARY

glacier
valley glacier
continental glacier
cirque
moraine
outwash plain
drumlin
esker

Glaciers formed much of the landscape that exists presently in the northern United States and elsewhere in the world. Glaciers shape the landscape by eroding, transporting, and depositing huge volumes of rocks and sediments. Today, scientists measure the movements of glaciers and changes in their sizes to track climatic changes. Air bubbles trapped deep in glacial ice can provide data about the composition of Earth's atmosphere at the time when these ancient ice layers were formed. Scientists can also study ice cores in glaciers to learn about Earth's environmental past, as shown in the *Problem-Solving Lab* in this section.

MOVING MASSES OF ICE

A large, moving mass of ice is called a **glacier.** Glaciers form near Earth's poles and in mountainous areas at high elevations. They currently cover only about 10 percent of Earth's surface, as shown in ***Figure 8-20.*** Even in the past, when glaciers were much more widespread than they are today, many areas of the world did not experience glacial activity. For example, during the last ice age, which began about 1.6 million years ago and ended over 10 000 years ago, ice probably covered only about 30 percent of Earth.

Areas of high latitude, such as Greenland and Antarctica, and areas of high elevation, such as the Alps, have cold temperatures year-round. Cold temperatures keep fallen snow from completely melting, and each year, the snow that has not melted accumulates in

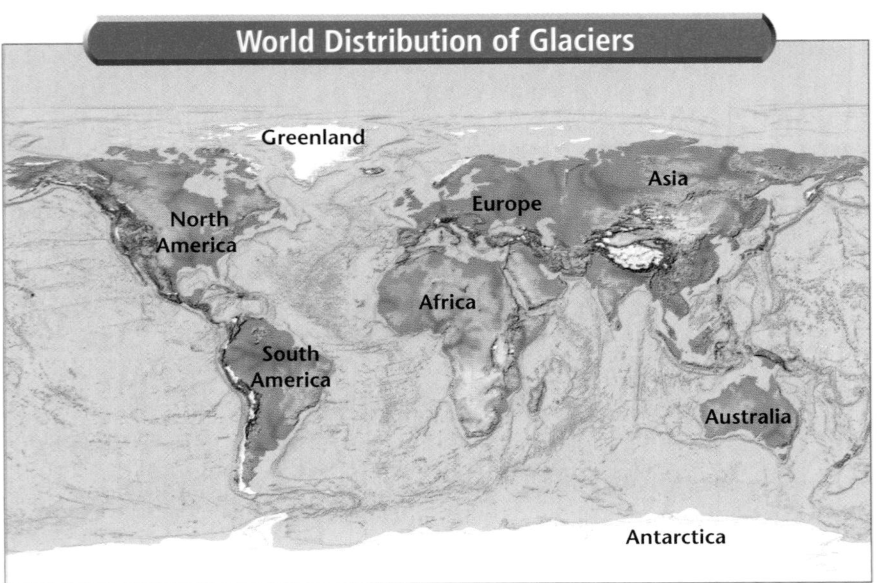

Figure 8-20 This map shows glacial distribution throughout the world. ***Locate the glacial areas in white. Are all glaciers located only in the polar regions?***

Figure 8-21 The Mendenhall Glacier is a valley glacier located in the Tongas National Forest near Juneau, Alaska.

an area called a snowfield. Thus, the total thickness of the snow layer increases as the years pass, and a glacier begins to form. The weight of the top layers of snow eventually exerts enough downward pressure to force the accumulated snow below to recrystallize into ice. This recrystallization is familiar to you if you have ever made a snowball: firmly compacting the snowball in your hands causes it to recrystallize, or partially melt to form ice. A glacier can develop in any location that provides the necessary conditions. For example, small glaciers form even in mountainous tropical areas along the equator, such as in Chile.

Valley Glaciers Glaciers can be classified as one of two types: valley glaciers or continental glaciers as shown in ***Figures 8-21*** and ***8-22.*** Glaciers that form in valleys in high, mountainous areas are called **valley glaciers.** A valley glacier is shown in ***Figure 8-21.*** The movement of a valley glacier occurs when the growing ice mass

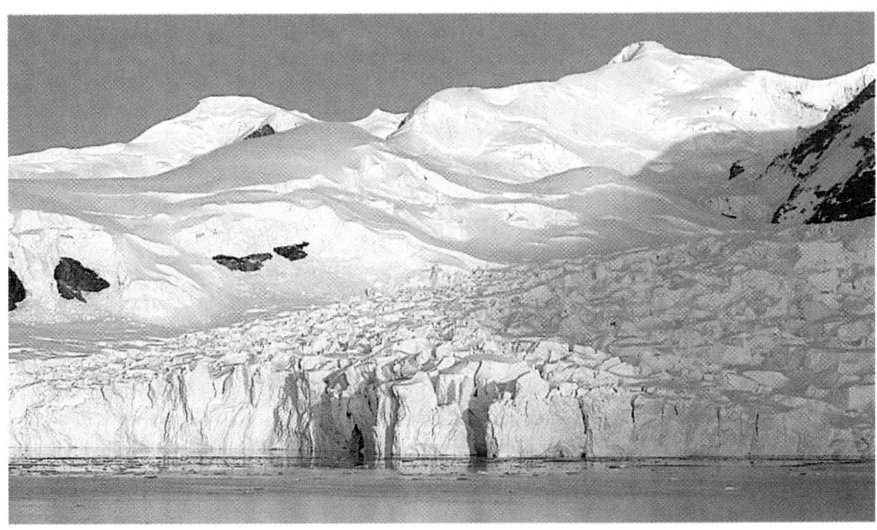

Figure 8-22 This continental glacier is located on Antarctic Peninsula.

becomes too heavy to maintain its rigid shape and begins to flow, much like a thick liquid. For most valley glaciers, flow begins when the accumulation of snow and ice exceeds 20 m in thickness. As a valley glacier moves, deep cracks in the surface of the ice, called crevasses, can form.

The speed of a valley glacier's movement is affected by the slope of the valley floor, the temperature and thickness of the ice, and the shape of the valley walls. The sides and bottom of a valley glacier move more slowly than the middle because friction slows down the sides and bottom, where the glacier comes in contact with the ground. Movement downslope is usually very slow, less than a few millimeters per day. You will learn more about the rates of glacial movement in the *Science & Math* feature at the end of this chapter. As valley glaciers flow downslope, their powerful carving action widens V-shaped stream valleys into U-shaped glacial valleys.

To learn more about glaciers, go to the **National Geographic Expedition** on page 874.

Continental Glaciers Glaciers that cover broad, continent-sized areas, such as the one shown in ***Figure 8-22,*** are called **continental glaciers.** They form under the same climatic conditions as valley glaciers, but they move in a different way. A continental glacier is

Problem-Solving Lab

Using Graphs

Observe how ice cores record history Scientists drill ice cores from glaciers and study them to learn about the past. Ice cores taken from the arctic region, for example, have been found to contain preserved radioactive fallout. Data collected from the study of these ice cores have been plotted on the graph. Use the graph to answer the following questions.

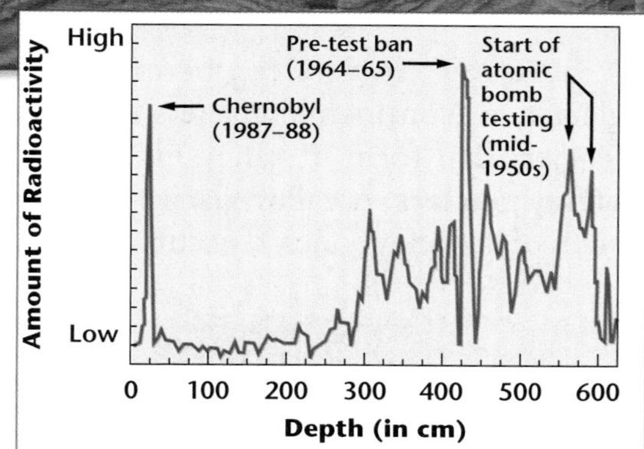

Analysis

1. At what depth in the ice cores was the highest amount of radioactivity found? At what depth was the lowest amount found?
2. Describe what happened to the amount of radioactivity in the ice cores between the pre-test ban and Chernobyl.
3. Infer what happened to the amount of radioactivity in the ice cores after Chernobyl.

Thinking Critically

4. What information or material other than radioactive fallout do you think ice cores might preserve within them?

thickest at its center. The weight of this thicker central region forces the rest of the glacier to flatten out in all directions. Continental glaciers, also called ice sheets, are much larger than valley glaciers. During periods in the past, when Earth experienced colder average temperatures than it does today, continental glaciers covered huge portions of Earth's surface. Today's continental glaciers cover much smaller areas, and they are confined to Greenland, northern Canada, and Antarctica.

Figure 8-23 Glacial striations are found on quartzite rocks located in Blue Mounds State Park in Minnesota.

Glacial Erosion

Of all the erosional agents, glaciers are the most powerful because of their great size, weight, and density. When a valley glacier moves, it breaks off pieces of rock through a process called plucking. Fallen rocks also accumulate along the edges of glaciers and give the sides a striped appearance. When glaciers with embedded rocks move over bedrock valley walls, they act like the grains on a piece of sandpaper, grinding out parallel scratches into the bedrock, as shown in ***Figure 8-23.*** Small scratches are called striations, and the larger ones are called grooves. Scratches and grooves provide evidence of a glacier's history and establish its direction of movement.

Glacial erosion can create certain features, as shown in ***Figure 8-24.*** In addition to carving U-shaped valleys, valley glaciers also scoop out deep depressions, called **cirques.** Where two cirques on opposite sides of a valley meet, they form a sharp, steep ridge called an arete. Where

Figure 8-24 Glacial features include hanging valleys, cirques, waterfalls, U-shaped valleys, horns, and aretes.

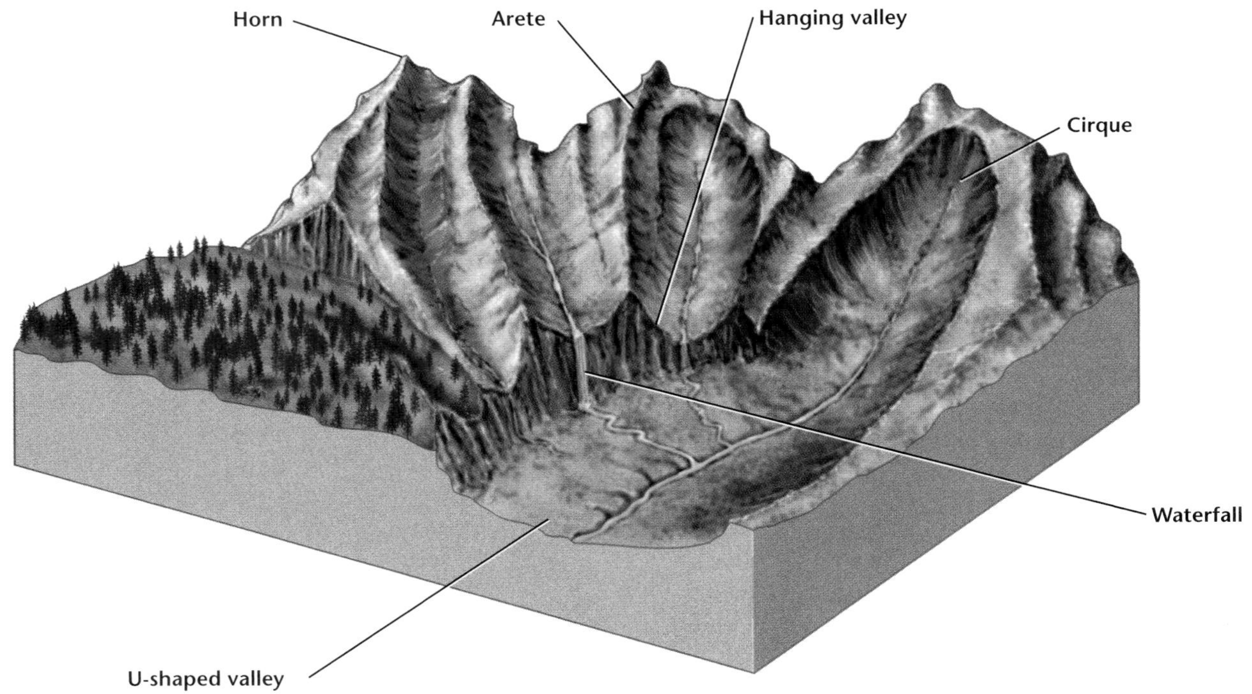

Figure 8-25 This terminal moraine was formed on Exit Glacier in Kenai Fjords National Park, in Alaska.

there are glaciers on three or more sides of a mountaintop, a steep, pyramid-shaped peak forms. This is known as a horn. The most famous example of this feature is Switzerland's Matterhorn. A tributary valley that enters a U-shaped valley from high up a mountain side is called a hanging valley.

GLACIAL DEPOSITION

Glacial till is the mixed debris that glaciers carry embedded in their ice and on their tops, sides, and front edges. When a glacier melts, glacial till is left behind. Ridges consisting of till deposited by glaciers are called **moraines.** Those at the foot of a large glacier are called terminal moraines, shown in ***Figure 8-25,*** and those at its sides are called lateral moraines. Where two glaciers join together, their lateral moraines combine to form a medial moraine.

Outwash When a glacier melts and begins to recede, meltwater floods the valley below. Meltwater contains gravel, sand, and fine silt formed from the grinding action of the glacier on underlying rock. When this sediment is deposited by meltwater, it is called outwash. The area at the leading edge of the glacier, where the meltwater streams flow and deposit outwash, is called an **outwash plain.**

Drumlins and Eskers Glaciers that move over older moraines form the material into elongated landforms called **drumlins.** A drumlin's steeper slope faces the direction from which the glacier came. Drumlin fields are found in Wisconsin, Massachusetts, and New York State. Long, winding ridges of layered sediments that are deposited by streams flowing under a melting glacier are called **eskers.**

Figure 8-26 This kettle lake, which was formed in glacial till, is located in Glacial Lakes State Park, Minnesota **(A).** This cirque lake is located in the western Cascade Mountains, in the state of Washington **(B).**

Glacial Lakes Sometimes, a large block of ice breaks off a glacier and is later covered by sediment. When the ice block melts, it leaves behind a depression called a kettle hole. After the ice block melts, the kettle hole fills with water from precipitation and runoff to form a kettle lake. Kettle lakes, such as the one shown in ***Figure 8-26A,*** are common in New England, New York State, and Wisconsin. As shown in ***Figure 8-26B,*** cirques also can fill with water, and they become cirque lakes. When a terminal moraine blocks off a valley, the valley fills with water to form a lake. Moraine-dammed lakes include the Great Lakes and the Finger Lakes of northern New York State which are long and narrow.

Mass movements, wind, and glaciers all contribute to the changing of Earth's surface. These processes constantly wear down landforms, and in many ways, they also impact human populations and activities.

SECTION ASSESSMENT

1. Explain how valley glaciers and continental glaciers form.
2. Draw a glacial landscape and label the glacial features.
3. How can a valley glacier modify the mountainous area where it forms? What glacial features form in mountainous areas?
4. How is a kettle lake formed?
5. **Thinking Critically** What evidence of past glaciers can be found on Earth today?

SKILL REVIEW

6. **Comparing and Contrasting** Make a data table that compares and contrasts the characteristics of valley glaciers and continental glaciers. For more help, refer to the *Skill Handbook.*

earthgeu.com/self_check_quiz

Internet GeoLab

Analyzing a Hazard Map

Earth Sciences

9.d Students know how to analyze published geologic hazard maps of California and how to use the map's information to identify evidence of geologic events of the past and predict geologic changes in the future.

It is just after noon, January 10, 2005, when mud and debris bury the small town of La Conchita, California. Houses are pushed from their foundations and into each other. Fifteen days of rain, steep mountain slopes, and the remnants of a 1995 landslide all contributed to this disaster.

Preparation

Problem

What can you do if you live near steep hills?

Hypothesis

Make a list of things you can do *before*, *during*, and *after* heavy rains to make sure you are safe.

Objectives

- **Analyze** and **interpret** a hazard map.
- **Recognize** the danger signals of a landslide.
- **Create** a plan for safety.

Data Sources

Go to the Earth Science Web site at earthgeu.com to find links to landslide safety data, or use information provided by your teacher.

Plan the Experiment

1. Find a source that explains the processes and dangers associated with landslides.
2. Check a hazard map to see if your area is in a danger zone, a moderate zone, or a small-risk zone.
3. Gather data about previous landslides in your area.
4. Make a list of safety practices for before, during, and after a storm.

Procedure

1. Visit links posted on earthgeu.com to find information about landslides and hazard safety.
2. Record your research in a data table.
3. Post your data at earthgeu.com.

Conclude & Apply

1. How can you tell from the map that a slope is steep?
2. How do patterns of storm-water drainage on slopes affect landslides?
3. What are some signs that might indicate potential landslides?
4. What are some signs of moving debris?
5. What should you do if a landslide is about to happen?
6. Why is it important to have laws about construction in areas susceptible to landslides?
7. The horseshoe-shaped outline is the site of a former landslide. What features indicate that this was a susceptible landslide area?

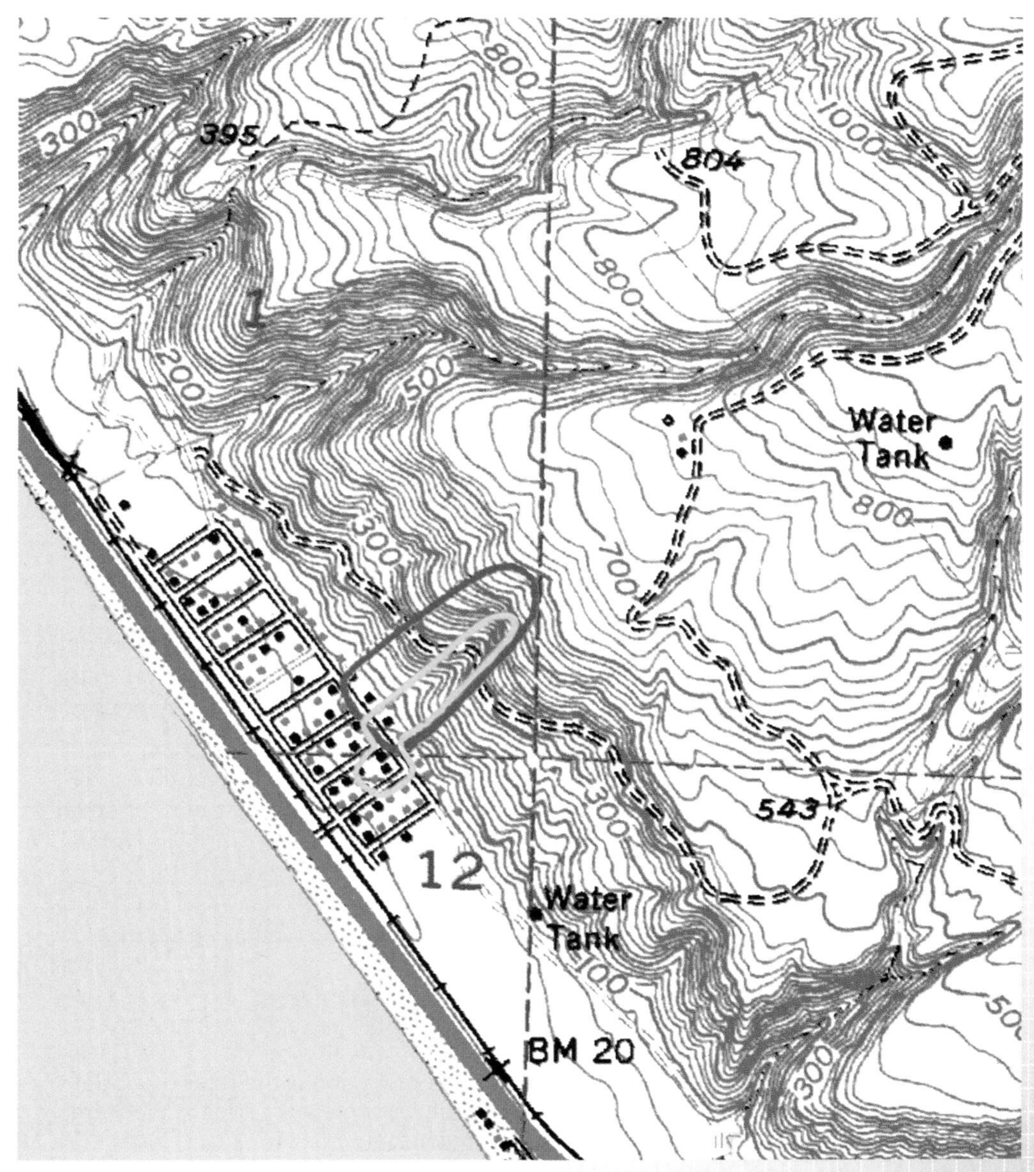

Science & Math

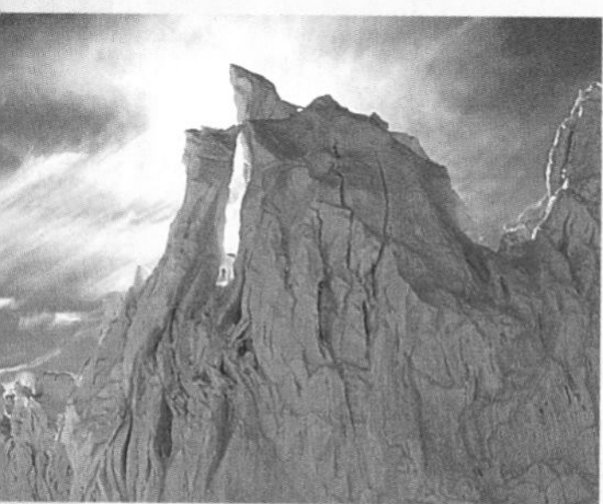

Rates of Glacial Movement

Rates are commonly used in everyday life. An example of a rate is the number of kilometers that a car travels per hour. A rate is a ratio of measurements, where one measurement is time. The most commonly used rate is distance per unit of time, also called speed.

Scientists who want to measure how fast a glacier moves typically refer to the overall speed of the glacier's movement. However, this is only an average, because a glacier moves at many different rates. A glacier is similar to a river in that the outside edges of a glacier move more slowly than the middle as a result of friction with the banks. A cross section of a glacier from top to bottom shows how different horizontal levels of the glacier move. The top surface of a glacier may move more quickly than the bottom surface.

Making the Measurements

To measure the rate of a glacier moving at various speeds, scientists calculate the glacier's average speed. The average speed of a glacier is the sum of the different speeds within the glacier divided by the total number of measurements.

Scientists can measure the difference in speed from the top surface of a glacier to the bottom by drilling into the glacier and placing a rod through the hole. After a period of time, the rod leans in the direction of the glacier's movement. The table at right shows data obtained from a rod placed in the Worthington Glacier in Alaska after 66 days. The distance the glacier moved at different depths is given in the table. The depths were measured from the surface of the glacier.

Procedure

1. Using the rod data, calculate the speed at which the glacier moved at each depth.

Rod Data from Worthington Glacier

Depth (m)	Distance (m)	Average Speed = Distance/Time
0	13.1	0.198 m/day
20	13.1	
60	12.8	
100	12.2	
140	11.2	
180	9.57	
	Average Speed of Glacier	

2. After you compute the speed calculations for all depths, average them together to find the average speed of the glacier.

Challenge

1. Note that the top surface of the glacier did not move as quickly as the portion of the glacier 20 m deep. What are some possible reasons for this?
2. Graph the average speed of the glacier at each depth. Do you notice a trend? If scientists drilled another 40 m, what would the speed of the glacier's movement be at that depth?

To find out more about the rates of glacial movement, visit the Earth Science Web Site at earthgeu.com

CHAPTER 8
Study Guide

Summary

SECTION 8.1

Mass Movements at Earth's Surface

Main Ideas

- Mass movement is the movement of Earth materials downslope as the result of the force of gravity. Almost all of Earth's surface undergoes mass movement.
- Mass movements may occur very slowly and become noticeable only over long periods of time. Creep is a form of slow mass movement. Rapid mass movements are noticeable.
- Variables involved in the mass movement of Earth materials include the material's weight, its resistance to sliding, and sometimes a trigger such as an earthquake. Water is important to the process of mass movement.
- Mass movements can cause great damage and loss of lives. Human activities may increase the potential for the occurrence of mass movements.

Vocabulary

avalanche (p. 187)
creep (p. 184)
landslide (p. 186)
mass movement (p. 181)
mudflow (p. 185)
slump (p. 187)

SECTION 8.2

Wind

Main Ideas

- Arid, semi-arid, and seashore environments are likely to experience wind erosion. Limited amounts of precipitation and protective vegetation commonly contribute to wind erosion in an area.
- Wind-carried sediments can cause abrasive action. Rocks exposed to continual wind abrasion often exhibit angular shapes with polished, smooth sides on the windward side. Features formed in wind-affected areas include deflation blowouts, dunes, and desert pavement. Dunes are classified by shape.
- The transport of Earth materials by wind can create problems for humans. Migrating dunes can block highways and cover structures.
- Loess soils deposited by wind are fertile soils because they contain minerals and nutrients.

Vocabulary

abrasion (p. 193)
deflation (p. 192)
dune (p. 194)
loess (p. 197)
ventifact (p. 193)

SECTION 8.3

Glaciers

Main Ideas

- Glaciers are large, moving masses of ice that form near Earth's poles and in mountainous areas at high elevations.
- Valley glaciers are formed in mountains, and continental glaciers are formed over broad regions of land. Valley glaciers move down mountainsides and form unique glacial features. Continental glaciers usually spread out from their centers.
- Features formed by glaciers include U-shaped valleys, hanging valleys and waterfalls in the mountains, moraines, drumlins, kettle holes along outwash plains, and several types of lakes.

Vocabulary

cirque (p. 201)
continental glacier (p. 200)
drumlin (p. 202)
esker (p. 202)
glacier (p. 198)
moraine (p. 202)
outwash plain (p. 202)
valley glacier (p. 199)

earthgeu.com/vocabulary_puzzlemaker

CHAPTER 8

Assessment

Understanding Main Ideas

1. What underlying force causes all forms of mass movement?
   a. friction
   b. gravity
   c. magnetism
   d. the Coriolis effect

2. Which of the following is an example of a slow mass movement?
   a. a mudflow
   b. a landslide
   c. creep
   d. an avalanche

3. Which of the following has the greatest erosional power?
   a. wind
   b. a landslide
   c. an avalanche
   d. a glacier

4. What is the movement of dunes called?
   a. ablation
   b. abrasion
   c. deflation
   d. migration

5. What percentage of Earth's surface is covered by glaciers?
   a. 5 percent
   b. 10 percent
   c. 15 percent
   d. 20 percent

6. Which feature is NOT formed by glaciers?
   a. moraines
   b. drumlins
   c. kettle holes
   d. dunes

7. Which state is most likely to experience wind erosion?
   a. Louisiana
   b. Kentucky
   c. Connecticut
   d. Utah

8. Which of the following has the fastest movement?
   a. solifluction
   b. creep
   c. mudflow
   d. earth flow

9. Which particles can wind move most easily?
   a. sand
   b. pebbles
   c. silt
   d. gravel

Applying Main Ideas

10. What human activity contributed to the dust storms in the Great Plains in the 1930s?
11. What mass movements are dependent on the addition of water?
12. Draw a simple diagram of the four major types of sand dunes. Then draw an arrow to show the direction of wind movement across each of the dunes.
13. Why is loess soil usually fertile?
14. How do particles eroded by wind differ from particles eroded by water?
15. What features on Earth's surface are characteristic of an outwash plain?
16. Give an example of one type of mass movement that could be caused by human activity.
17. Why do some glacial depositions have sorted sediments, while others do not?
18. How is it possible that glaciers exist at the equator?
19. How do valley glaciers form?
20. Why are some glacial lakes long and narrow, while others are round?
21. Why is wind abrasion such an effective agent of erosion?
22. What is one way to reduce the number of disasters related to mass movements?

Test-Taking Tip

USE THE BUDDY SYSTEM Study in groups. A small gathering of people works well because it allows you to draw from a broader base of skills and expertise. However, keep the group small and keep on target.

CHAPTER 8

Assessment

Thinking Critically

Use the following information and diagram to answer questions 23 and 24.

A person studying glaciers placed flags on tall rods across a valley glacier, as shown in diagram A. When the person returned to the site the following year, the flag rods were in the positions shown in diagram B.

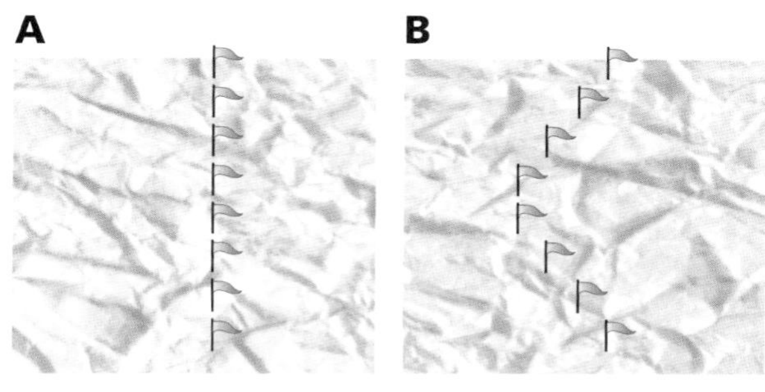

23. Based on the change in the positions of the flag rods, in which direction is the glacier moving?
a. north **b.** south **c.** east **d.** west

24. What can you tell about the rate of speed of different parts of the glacier?

25. **Concept Mapping** Use the following terms to construct a concept map of mass movements. For more help, refer to the *Skill Handbook.*

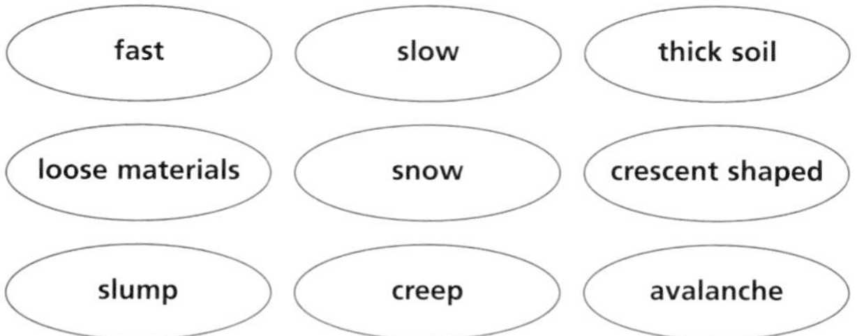

Standardized Test Practice

INTERPRETING DATA Use the table below to answer questions 1-2.

Region	Characteristics
A	semi-arid; experiences intense but brief rainstorms
B	permafrost; much loose, water-logged material
C	mountainous; thick accumulations of snow
D	thick soils on semi-steep and steep slopes; occasional earthquake activity
E	arid; high cliffs and rocky shorelines

1. Which mass movement is most likely to occur in Region A?
a. mudflow **c.** slump
b. avalanche **d.** rock fall

2. Which mass movement is most likely to occur in Region B?
a. solifluction **c.** avalanche
b. mudflow **d.** slump

3. What are dunes a result of?
a. wind erosion **c.** wind deposition
b. deflation **d.** abrasion

4. Which of the following is NOT a feature of valley glaciers?
a. cirques **c.** moraines
b. loess **d.** arete

5. What is it called when wind transports materials by causing a bouncing motion of particles?
a. suspension **c.** saltation
b. deflation **d.** abrasion

Chapter 9

Surface Water

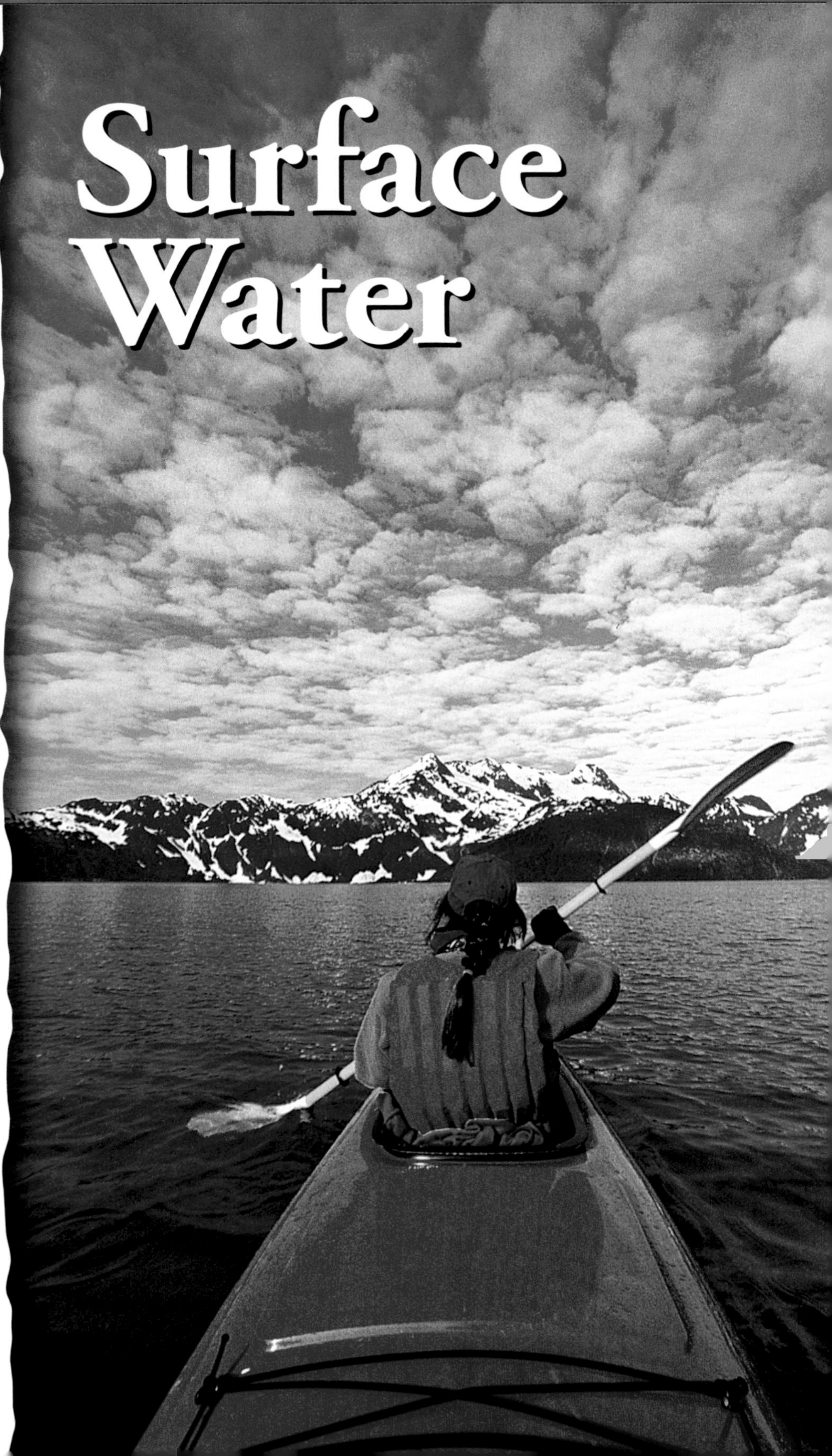

What You'll Learn

- What landscape features on Earth are formed and changed by surface water.
- How surface water moves materials and impacts humans.

Why It's Important

Landscape features formed by surface water are among the most numerous and visible features on Earth. Running water has the greatest impact on humans because we depend on streams for drinking-water supplies and irrigation. Humans also experience the negative effects of floods.

To find out more about surface water, visit the Earth Science Web Site at earthgeu.com

Discovery Lab

Modeling Water Movement

When water seeps into the ground, it moves at various rates through the different materials that make up Earth's surface. These Earth materials are comprised of different particle sizes. In this activity, you will investigate the movement of water as it seeps through two different kinds of Earth materials.

1. Place a small window screen on each of two clear plastic shoe boxes.
2. Place an 8 cm × 16 cm clump of grass or sod on one screen.
3. Place an 8 cm × 16 cm clump of barren soil on the other screen.
4. Lightly sprinkle 500 mL of water on each clump.

CAUTION: Always wear an apron in the lab.

Observe In your science journal, describe what happens to the water after five minutes. Measure how much water passes through each clump and collects in the plastic shoe box. Explain any differences in the amount of water collected in each plastic shoe box.

SECTION 9.1 *Surface Water Movement*

Earth Sciences 9.b

OBJECTIVES

- **Explain** *how surface water can move weathered materials.*
- **Explain** *how a stream carries its load.*
- **Describe** *how a floodplain develops.*

VOCABULARY

runoff
watershed
divide
solution
suspension
bed load
discharge
flood
floodplain

Earth's water supply is recycled in a continuous process called the water cycle. Water molecules move continuously through the water cycle following many pathways: they evaporate from a body of water or the surface of Earth, condense into cloud droplets, fall as precipitation back to Earth's surface, and soak into the ground. As part of a continuous cycle, the water eventually evaporates back into the atmosphere, again forms clouds, again falls as precipitation, and so on. Understanding the mechanics of the water cycle helps to explain the reasons for variations in the amount of water that is available throughout the world.

Often, a water molecule's pathway involves time spent within a living organism or as part of a snowfield, glacier, lake, or ocean. Although water molecules may follow a number of different pathways, the overall process is one of repeated evaporation and condensation powered by the Sun's energy. What happens once water reaches Earth's surface? Does all the water sink into the ground or evaporate?

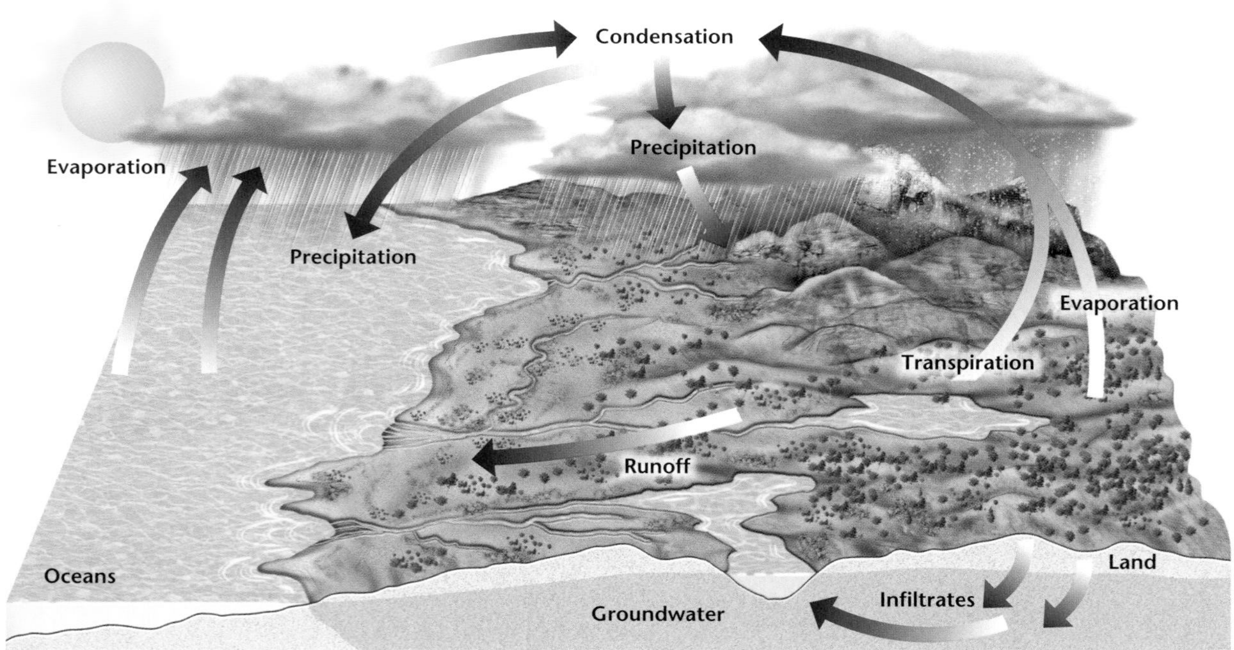

Figure 9-1 The water cycle, also referred to as the hydrologic cycle, is a never-ending, natural circulation of water through Earth's systems. The Sun provides the energy for the water cycle. Radiation from the Sun causes water to change to a gas called water vapor. The process of water vapor changing to a fluid is called condensation.

RUNOFF

As shown in ***Figure 9-1,*** precipitation falls to Earth's surface in the form of rain, snow, sleet, or hail. In most instances, solid forms of precipitation, such as snow, sleet, and hail, may eventually melt. Or they can also be incorporated into the mass of a glacier. Once water reaches Earth's surface, it can evaporate into the atmosphere, soak into the ground, or flow down slopes on Earth's surface. Water flowing downslope along Earth's surface is called **runoff.** Runoff may reach a stream, river, or lake, may evaporate, or it may accumulate as puddles in low-lying small depressions and eventually seep into the ground. During and after heavy rains, you can observe these processes occurring in your own yard or local park. Water that seeps into Earth's surface becomes groundwater.

A number of conditions determine whether water on Earth's surface will seep into the ground or become runoff. For water to enter the ground, there must be large enough pores or spaces in the ground's surface materials to accommodate the water's volume, as in the loose soil illustrated in ***Figure 9-2A.*** If the pores already contain water, the newly fallen precipitation will either remain standing on top of the ground or, if the area has a slope, run downhill. Water standing on the surface of Earth eventually evaporates or flows away.

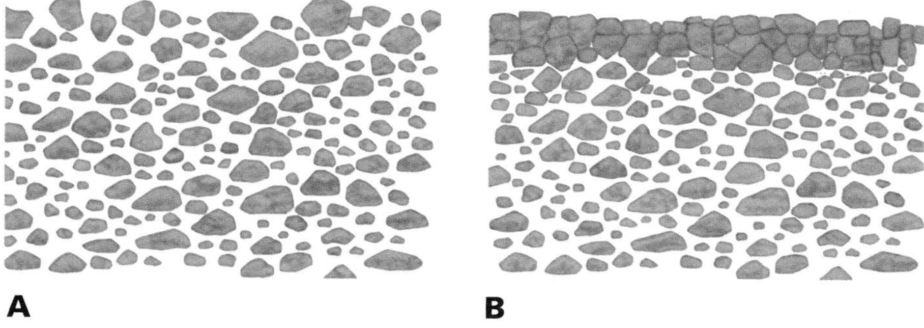

Figure 9-2 Soil that has open surface pores or spaces between particles allows water to infiltrate **(A).** Soil that has few or no pores or spaces can restrict water's ability to seep in **(B).**

Vegetation Soils that contain grasses or other vegetation allow more water to enter the ground than do soils with no vegetation. Precipitation falling on vegetation slowly flows down leaves and branches, and it eventually drops gently to the ground. In contrast, precipitation falls with far more force onto barren land. In such areas, soil particles clump together and form dense aggregates with few pores or spaces between them. The force of falling rain may then push the soil clumps together, thereby closing pores and allowing less water to enter, as illustrated in ***Figure 9-2B.*** This is why gardeners do not pack the soil around their plants. Compacting the soil reduces the spaces between the particles that are available for water to seep in, thus reducing the amount of water that is available to the plants' roots.

Rate of Precipitation Light, gentle precipitation infiltrates the dry ground. However, the rate of precipitation may temporarily exceed the rate of infiltration. For example, during heavy precipitation, water falls too quickly to soak into the ground and becomes runoff. Thus, a gentle, long-lasting rainfall is more beneficial to plants and causes less erosion by runoff than a torrential downpour. If you have a garden, remember that more water will enter the ground if you water your plants slowly and gently.

Soil Composition The physical and chemical composition of soil also affects its water-holding capacity. Soil consists of decayed organic matter, called humus, and minerals. Humus creates pores in the soil, thereby increasing a soil's ability to retain water. The minerals in soil have different particle sizes, which are classified as sand, silt, or clay. As you learned in Chapter 7, the percentages of particles of each size vary from soil to soil. Soil with a high percentage of coarse particles, such as sand, has relatively large pores between its particles that allow water to enter and pass through the soil quickly. In contrast, soil with a high percentage of fine particles, such as clay, clumps together and has few or no spaces between the particles.

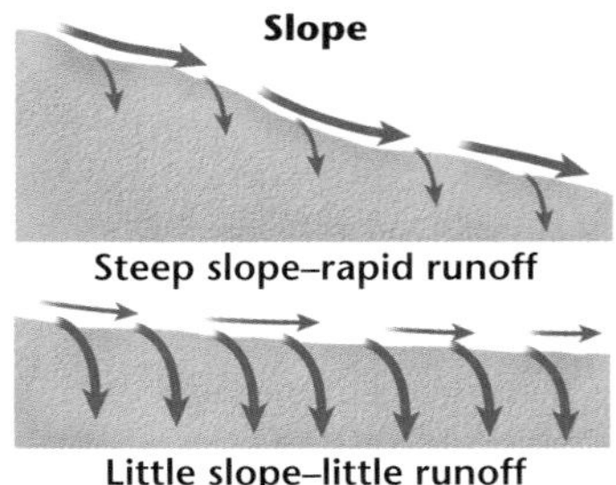

Figure 9-3 The angle of a slope is one variable that influences the movement of surface water on a slope.

Such small pores restrict both the amount of water that can enter the ground and the ease of movement of water through the soil.

Slope As you have learned, the slope of a land area plays a large role in determining the ability of water to enter the ground, as shown in ***Figure 9-3.*** Water from precipitation falling on slopes flows to areas of lower elevation. The steeper the slope, the faster the water flows. There is also greater potential for erosion on steep slopes. In areas with steep slopes, little water seeps into the ground before it runs off.

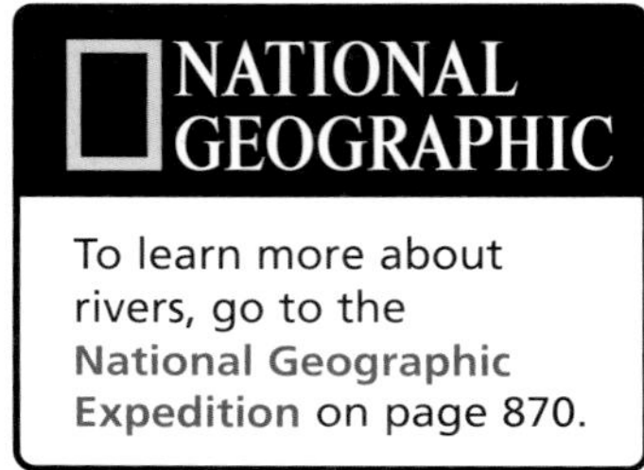

To learn more about rivers, go to the **National Geographic Expedition** on page 870.

Stream Systems

Precipitation that does not enter the ground usually runs off the surface quickly. Some surface water flows in thin sheets and eventually collects in small channels. As the amount of runoff increases, the channels widen, deepen, and become longer. Although it is common for these small channels to dry up shortly after precipitation stops, the channels again fill with water each time it rains and become larger and longer. If a sufficient supply of water develops, the water begins to flow more permanently in a channel and can become a stream.

All streams flow downslope in a watery path to lower elevations. However, the path of a stream can vary considerably, depending on the slope of the land and the type of material through which the stream flows. Some streams flow into lakes, while others flow directly into the ocean. Still others, called tributaries, flow into other streams, as shown in ***Figure 9-4.*** Each tributary increases the size of the stream it is joining and adds water to it. A large stream is called a river, and all its tributaries make up a stream, or river system. Small streams are called brooks and creeks. If there are any brooks or streams near your home, can you locate where they feed into other streams or lakes?

Figure 9-4 Numerous tributaries flow into several stream systems that drain into the Salton Sea in California.

Watersheds and Divides

All of the land area whose water drains into a stream system is called the system's **watershed,** or drainage basin. Watersheds can be relatively small or extremely large in area. A **divide** is a high land area that separates one watershed from another. Each tributary in a stream system has its own watershed and divides, but they are all part of the larger stream system to which the tributary belongs. The watershed of the Mississippi River, shown in ***Figure 9-5,*** is the largest in North America.

Stream Load

All the materials that the water in a stream carries is known as the stream's load. The living components of water include microscopic life-forms as well as larger plants and animals. The nonliving components of surface water include sediments, dissolved solids, and dissolved atmospheric gases, such as oxygen. There are three ways in which a stream carries its load.

Solution Material is carried in **solution** after it becomes dissolved in a stream's water. How much of a stream's load is carried in solution depends on the material through which the stream's water has passed. When water runs through or over rocks containing soluble

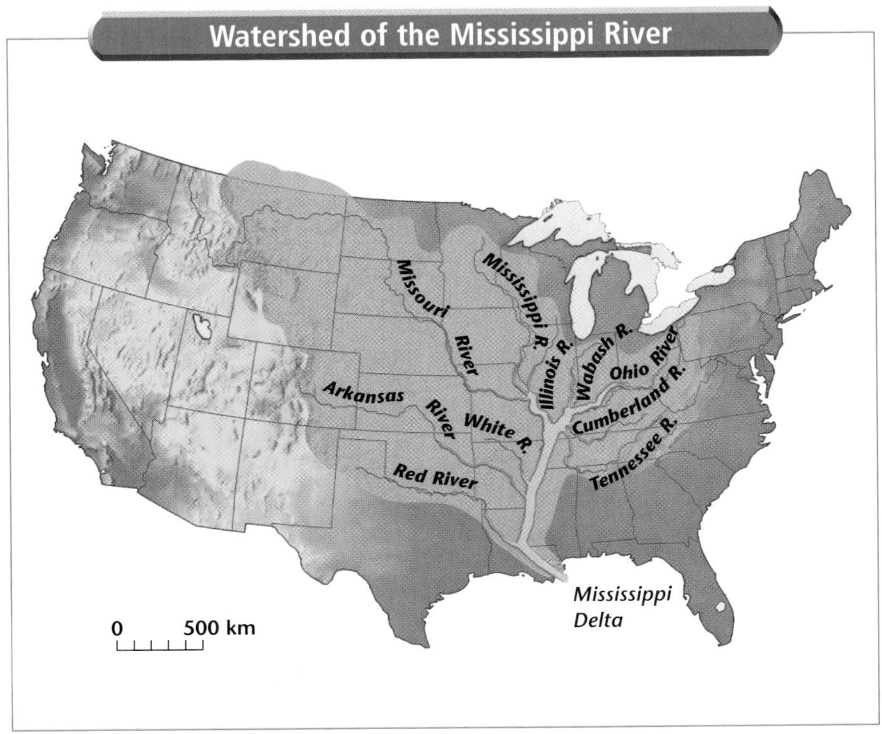

Figure 9-5 The watershed of the Mississippi River includes a large stream system. ***How many major rivers are part of the Mississippi watershed?***

minerals, it dissolves small amounts of the minerals and carries them away in solution. Water may readily dissolve calcium carbonate from limestone and marble, for example. Streams also commonly carry soluble magnesium compounds. Groundwater adds most of the dissolved load to stream water, while runoff adds only a very small amount.

The amount of dissolved material that water carries is often expressed in parts per million, or ppm, as shown in ***Table 9-1.*** For example, a measurement of 10 ppm means that there are 10 parts of dissolved material for every 1 million parts of water. The total concentration of materials in solution in streams averages 115–120 ppm, although some streams carry as little dissolved materials as 10 ppm. Values greater than 10 000 ppm have been observed for streams draining desert basins. Measuring the amount of material in solution helps scientists monitor water quality.

Suspension All particles small enough to be held up by the turbulence of a stream's moving water are carried in **suspension.** Particles, such as silt, clay, and sand, that are carried in suspension are part of a stream's suspended load. The amount of material in suspension varies with the volume and velocity of the stream water. Rapidly moving water can carry larger particles in suspension than slowly moving water can. As the velocity of water decreases, the heavier particles settle to the bottom, as you can see by doing the *Problem-Solving Lab* on the next page.

Table 9-1 Concentrations of Some Materials Dissolved in River Water and Seawater

Materials	Concentration (ppm)			
	Amazon River	Mississippi River	World Average (est.)	Average Seawater
Silica (SiO_2)	7.0	6.7	13.0	6.4
Calcium (Ca^{2+})	4.3	42.0	15.0	400.0
Sodium (Na^{+})	1.8	25.0	6.3	10 500.0
Potassium (K^{+})	—	2.9	2.3	380.0
Magnesium (Mg^{2+})	1.1	12.0	4.1	1350.0
Chloride (Cl^{-})	1.9	30.0	7.8	19 000.0
Fluoride (F^{-})	0.2	0.2	—	1.3
Sulfate (SO_4^{2-})	3.0	56.0	11.0	2700.0
Bicarbonate (HCO_3^{-})	19.0	132.0	58.0	142.0
Nitrate (NO_3^{-})	0.1	2.4	1.0	0.5

Source: J. D. Hem, *Study and Interpretation of the Chemical Characteristics of Natural Water,* U.S. Geological Survey Water-Supply Paper 1473, 1970, pp. 11, 12, and 50.

Bed Load Sediments that are too large or heavy to be held up by turbulent water are transported by streams in another manner. A stream's **bed load** consists of sand, pebbles, and cobbles that the stream's water can roll or push along the bed of the stream. The faster the water moves, the larger the particles it can carry both in suspension and as part of its bed load. As the particles move, they rub, scrape, and grind against one another or against the solid rock of the streambed in a process called abrasion, the wearing away of solid Earth material. This action contributes to the physical weathering of the stream's bottom and sides, and it provides an additional source of material to be eroded by the stream, either in solution or as part of the suspended load.

As gravity pulls stream water to lower elevations, the stream's load moves along with the water. The moving water continuously tosses and tumbles the weathered material, whose pieces become smooth and rounded over time, as shown in ***Figure 9-6.*** Most pebbles along the bottoms and sides of streams are round and polished as a result of this process.

Bed load sediments not only wear away one another, but they also abrade the surface of the streambed. Potholes may form on the bottoms of streams where pebbles have continued to swirl around in

Figure 9-6 The rounded shapes and smooth pebbles were caused by the stream erosion of the Snake River in Grand Teton National Park, Wyoming.

Problem-Solving Lab

Using Graphs

Predict how sediments move in a stream The velocity of water affects the transport of different-sized particles.

Analysis

1. Study the graph at right.
2. At what velocity would flowing water pick up a pebble?
3. Over what range of velocities would flowing water carry a pebble?

Thinking Critically

4. Infer which of the following objects would not fall into the same size range as a pebble: a large chicken egg, a baseball, a golf ball, a table-tennis ball, a volleyball, and a pea. How would you test your conclusions?

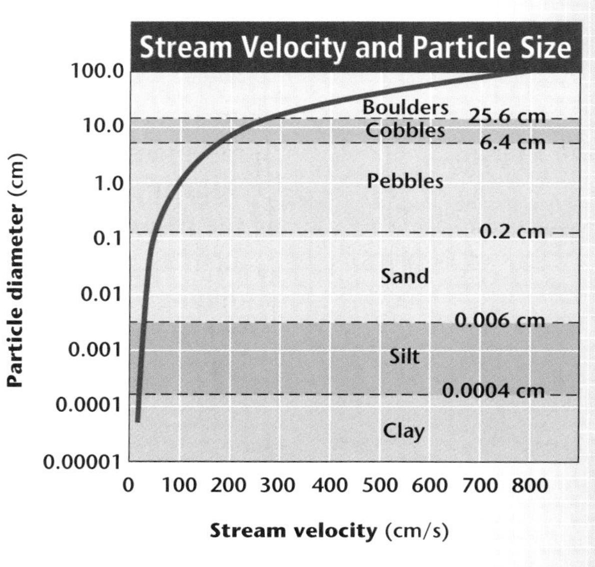

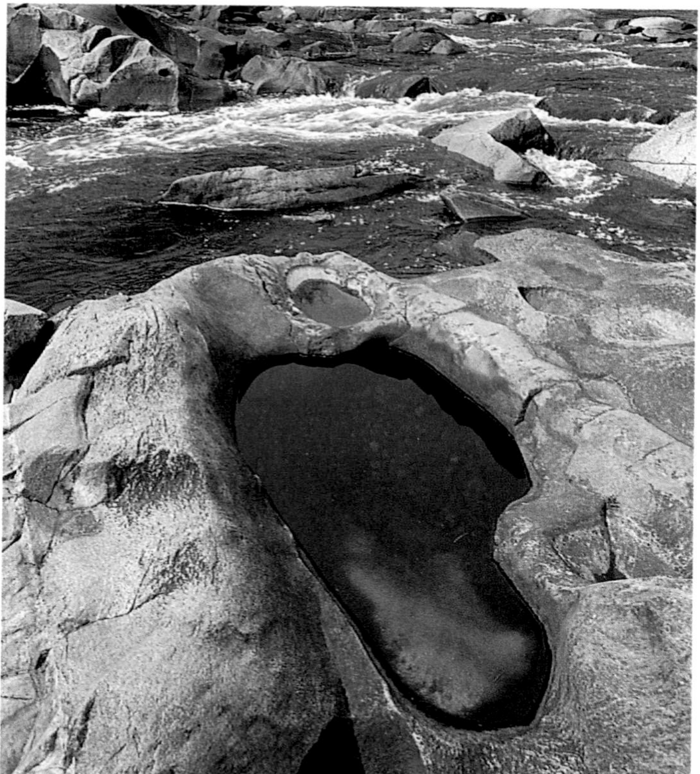

Figure 9-7 The potholes in this Wisconsin River streambed were scoured out by the abrasive action of the stream's bed load. The river is located at Grandfather Falls, Wisconsin.

one area and have slowly worn holes into solid rock. For example, potholes more than 3 m deep have formed near Little Falls, New York, in the Mohawk River Valley. Potholes can be found even in streambeds composed of very hard, exposed bedrock. The huge depressions in the streambed of the Wisconsin River, shown in ***Figure 9-7,*** were scoured out of granite. Large streambed potholes dramatically illustrate the powerful abrasive action caused by a stream's bed load.

STREAM VELOCITY AND CARRYING CAPACITY

The ability of a stream to transport material, referred to as its carrying capacity, depends on both the velocity and the amount of water moving in the stream. Study ***Figure 9-8*** as you read. The channel's slope, depth, and width all affect the speed and direction in which water moves within it. A stream's water moves more quickly where there is less friction; consequently, smooth-sided channels with great slope and depth allow water to move most rapidly. The total volume of moving water also affects a stream's carrying capacity. **Discharge** is the measure of the volume of stream water that flows over a particular location within a given period of time. Discharge is commonly expressed in cubic meters per second (m^3/s). The following formula is used to calculate the discharge of a stream:

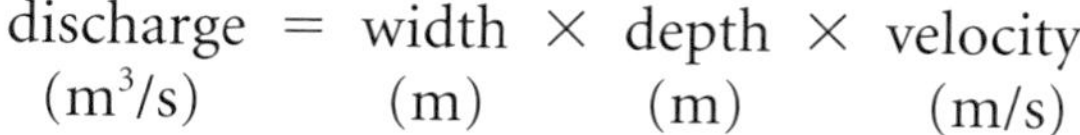

$$\underset{(m^3/s)}{\text{discharge}} = \underset{(m)}{\text{width}} \times \underset{(m)}{\text{depth}} \times \underset{(m/s)}{\text{velocity}}$$

The largest river in North America, the Mississippi, has a huge average discharge of 173 600 m^3/s. However, the Amazon River, the largest in the world, has an incredible discharge ten times that amount. The discharge from the Amazon River over a 24-hour period would supply New York City's water needs for nine years!

Earth Sciences

9.b Students know the principal natural hazards in different California regions and the geologic basis of those hazards.

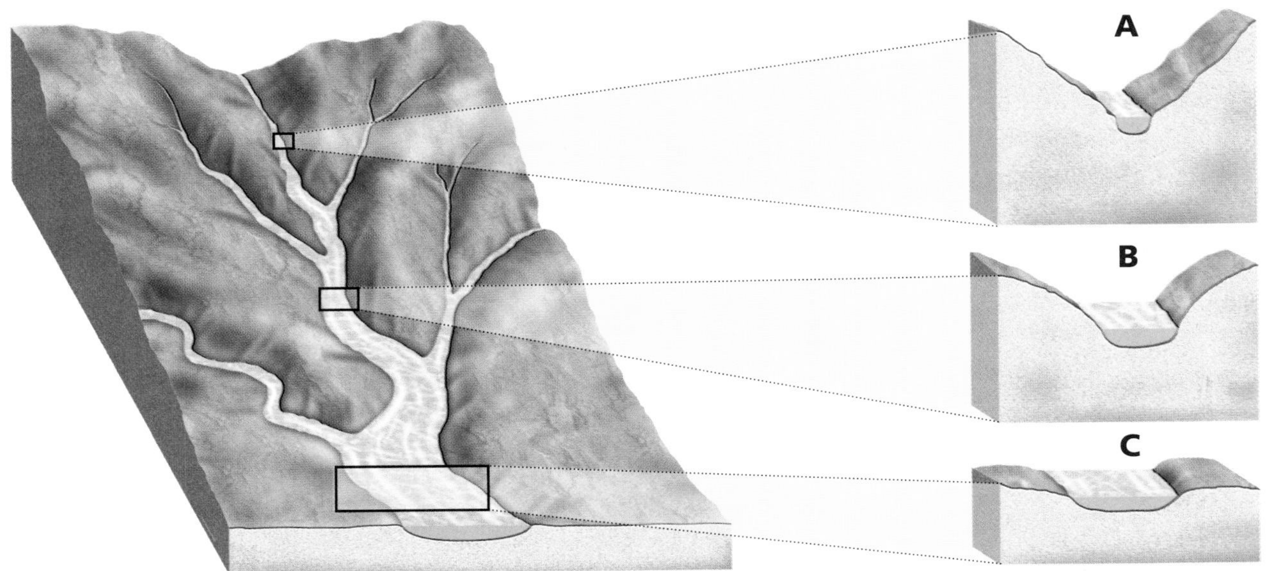

Figure 9-8 Describe the changes in the downstream direction of the stream's channel as the water flows from section A to section C.

As a stream's discharge increases, the stream's carrying capacity increases as well. The increased discharge results in a stream with greater carrying capacity as modeled in the *GeoLab* at the end of the chapter. Both water velocity and volume increase during times of heavy precipitation, rapid melting of snow, and flooding. In addition to increasing a stream's carrying capacity, these conditions heighten a stream's ability to erode the land over which it passes. As a result of an increase in erosional power, a streambed can widen and deepen, thereby increasing the stream slope and further adding to the stream's carrying capacity. As shown in ***Figure 9-9,*** the extraordinary power of water during such times can be especially hazardous for people who do not anticipate the dangers associated with flooding.

Figure 9-9 Several days of heavy rains in Buenos Aires, Argentina caused flood waters to strand traffic in the city.

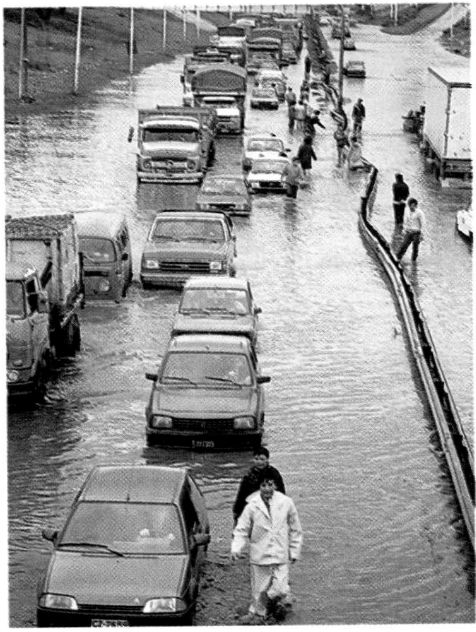

FLOODPLAINS

The amount of water being transported in a particular stream at any given time varies with weather conditions. Sometimes, more water pours into a stream than the banks of the stream channel can hold. A **flood** occurs when water spills over the sides of a stream's banks onto the adjacent land. The broad, flat area that extends out from a stream's bank and is covered by excess water during times of flooding is known as the stream's **floodplain.** Floodwater carries along with it a great amount of sediment eroded from Earth's surface and the sides of the stream channel. As floodwater recedes and its volume and speed decrease, the water drops its sediment load onto the stream's floodplain.

Figure 9-10 Floodplain deposits contain fertile soils.

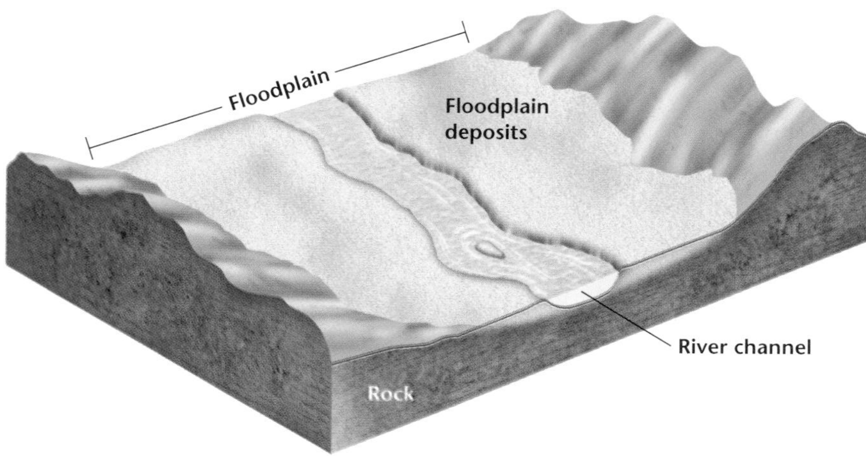

Topic: Floods
To find out more about floods and flood prevention, visit the Earth Science Web Site at earthgeu.com

Activity: Write a brief report to compare and contrast the flooding of the Yellow River in China, the Nile River in Egypt, and the Mississippi River in the United States.

Figure 9-10 illustrates a floodplain after a river overflows its channel. Floodplains develop highly fertile soils as more sediment is deposited with each subsequent flood. These fertile soils have historically enticed farmers to use the land for crop production, even at the risk of losing homes and crops to subsequent flooding.

FLOODS

Floods are a natural occurrence. When a stream reaches its flood stage, a flood can occur, as shown in ***Figure 9-11.*** Flood stage is the level at which a stream overflows its banks and the crest of the stream is the maximum height. Because it takes time for runoff to collect in streams, the water continues to rise and may reach its crest days after precipitation ends. The resulting flooding may occur over localized, small areas or across large regions. The flooding of a small area is known as an upstream flood. Sudden rainstorms that drop large amounts of rain within a short period of time cause upstream floods,

Figure 9-11 Davenport, Iowa was just one of many areas that was flooded by the Mississippi River in 1993.

as do dam failures. Although they are localized, upstream floods can do a great deal of damage within a very short period of time.

Heavy accumulations of excess water from large regional drainage systems result in downstream floods. Such floods occur during or after long-lasting, intense storms or spring thaws of large snowpacks. The tremendous volumes of water involved in a downstream flood can result in extensive damage. For example, the devastating floods in 1993 along the Mississippi River, which frequently causes downstream flooding, left landscape scars that are still visible today.

Figure 9-12 This USGS stream gauging station is located in the northwest section of Washington State.

Flood Monitoring and Warning Systems

In an attempt to provide warnings for people at risk, government agencies monitor potential flood conditions. The National Weather Service monitors changing weather conditions. Earth-orbiting weather satellites photograph Earth and collect and transmit information about weather conditions, storms, and streams. In addition, the U.S. Geological Survey (USGS) has established gauging stations, as shown in ***Figure 9-12,*** on approximately 4400 streams in the United States. The gauging stations provide a continuous record of the water level in each stream. Technological advances have made it possible for anyone with Internet access to obtain real-time data on streams through government-sponsored Web sites.

In areas that are prone to severe flooding, warning systems are the first step in implementing emergency management plans. Flood warnings and emergency plans often allow people to safely evacuate an area in advance of a flood.

Environmental Connection

Section Assessment

1. Describe ways in which moving water can carve a landscape.
2. Explain the three ways in which a stream carries its load.
3. What is the relationship between the carrying capacity of a stream and its discharge and velocity?
4. Explain why little water from runoff seeps into the ground in areas of steep slopes.
5. Discuss how a floodplain forms and why people live on floodplains.
6. **Thinking Critically** Under what conditions might a stream's volume increase, and under what conditions might it decrease? How would the size of the sediment particles in the stream's load differ in the two situations?

Skill Review

7. **Making Tables** Design a data table that compares how silt, clay, sand, and large pebbles settle to the bottom of a stream as the velocity of water decreases. For more help, refer to the *Skill Handbook.*

SECTION 9.2 Stream Development

OBJECTIVES

- **Describe** *some of the physical features of stream development.*
- **Explain** *the process of rejuvenation in stream development.*

VOCABULARY

stream channel
stream bank
meander
delta
rejuvenation

As a stream develops, it changes in shape, width, and size, as well as the landscapes over which it flows. Stream flow is part of a dynamic system that is greatly influenced by the varying environmental conditions of the stream's surroundings.

MOVING WATER CARVES A PATH

The first and foremost condition necessary for stream formation is an adequate supply of water. Precipitation provides the water for the beginnings of stream formation. In areas where precipitation falls infrequently, stream development and flow are also infrequent. For example, in some desert areas, where years pass between rainfalls, the streams that form are short-lived. However, most parts of the temperate and tropical regions on Earth experience precipitation on a regular basis.

The region where water first accumulates to supply a stream is called the headwaters. It is common for a stream's headwaters to be high in the mountains. Falling precipitation accumulates in small gullies at these higher elevations and forms briskly moving streams. As surface water first begins its flow, its path may not be well defined. In time, however, the moving water carves a narrow pathway into the sediment or rock called a **stream channel.** The channel widens and deepens as more and more water accumulates and cuts into Earth's surface. As shown in ***Figure 9-13,*** the moving water is held within the confines of the stream channel by the **stream banks,** the ground bordering the stream on each side. If you have ever fished in a stream, you might have sat on a stream bank to do so.

Figure 9-13 The river banks confine the water of the San Pedro River in the San Pedro Riparian National Conservation Area in Arizona.

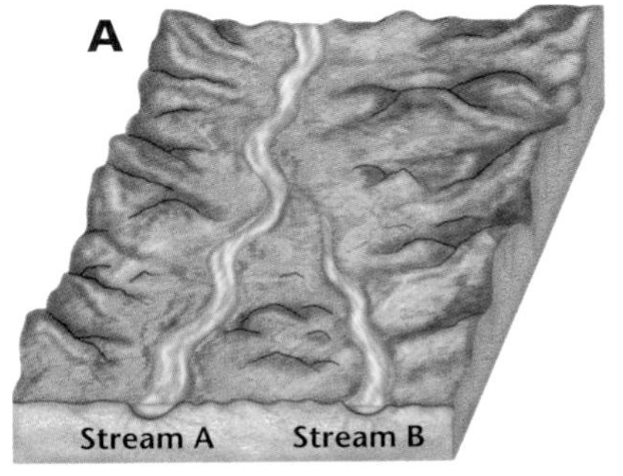

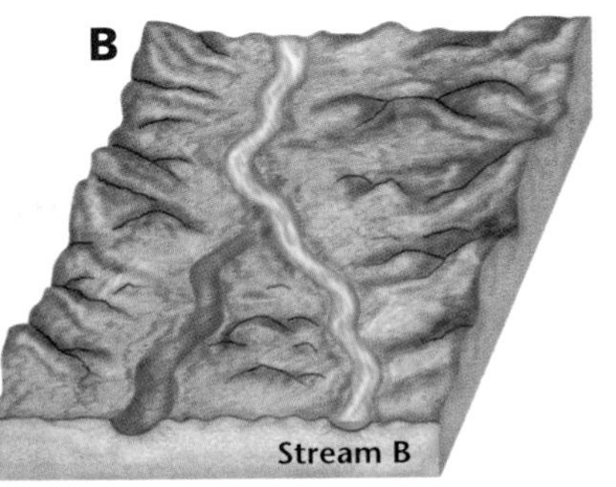

Figure 9-14 The headward erosion of stream B cuts into stream A and draws away its waters into one stream.

The process by which small streams erode away the rock or soil at the head of a stream is known as headward erosion. Headward erosion involves lengthening the stream. Streams at this point in their development are relatively small and narrow. These streams move swiftly over the rough terrain, and they often form waterfalls and rapids as they flow over steep inclines.

Sometimes, a stream erodes its way through the high area separating two drainage basins, joins another stream, and then draws away its water. This process is called stream capture, or stream piracy. As shown in ***Figure 9-14,*** the lower portion of the captured stream loses its water source, while the invading stream gains an additional source of water.

FORMATION OF STREAM VALLEYS

As a stream actively erodes its path through the sediment or rock, a V-shaped channel develops. V-shaped channels have steep sides and sometimes form canyons or gorges. The Grand Canyon is perhaps the best-known example of a V-shaped valley carved by a stream, the Colorado River. Ausable Chasm, in New York State, is another impressive, but very narrow, deep gorge carved by a stream. ***Figure 9-15*** shows the classic V-shaped valley created by the Yellowstone River.

Figure 9-15 A V-shaped valley was formed on the Lower Falls of the Yellowstone River in Wyoming.

A stream continues to erode until it reaches its base level, the elevation at which it enters another stream or body of water. The lowest base level possible for any stream is sea level, the point at which the stream enters the ocean. As a stream continues to erode its channel toward its base level, erosion will continue along the sides of the

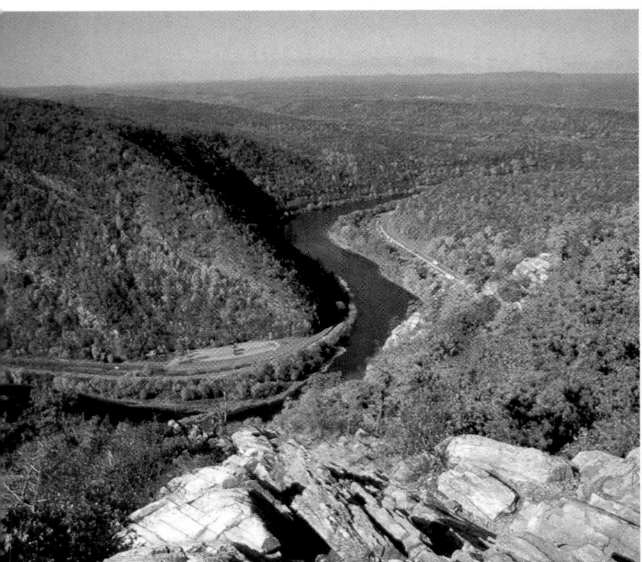

Figure 9-16 The Delaware Water Gap in Pennsylvania has been eroded into a wider, broader valley.

V-shaped channel. As shown in ***Figure 9-16,*** in time, a V-shaped valley will be eroded into a broader valley that has gentle slopes.

MEANDERING STREAMS

As stream channels develop into broader valleys, the volume of water and sediment that they are able to carry increases. In addition, a stream's slope, or gradient, decreases as it nears its base level, and as a result the channel gets wider. The decrease in gradient causes water to build up within the stream channel. Sometimes, the water begins to erode the sides of the channel in such a way that the overall path of the stream starts to bend or wind. As shown in ***Figure 9-17,*** a bend or curve in a stream channel caused by moving water is called a **meander.**

Water in the straight parts of a stream flows at different velocities, depending on the location of the water in the channel. In a straight length of a stream, water in the center of the channel is flowing at the maximum velocity. Water along the bottom and sides of the channel flows more slowly because it experiences friction as it moves against the land. In contrast, the water moving along the outside of a meander curve experiences the greatest rate of flow within the meander. The water that flows along this outside part of the curve continues to erode away the sides of the streambed, thus making the meander larger. Along the inside of the meander, the water moves more slowly and deposition is dominant. These differences in the rate of water flow within meanders cause the meanders to become more accentuated over time. ***Figure 9-18*** illustrates the processes of erosion and

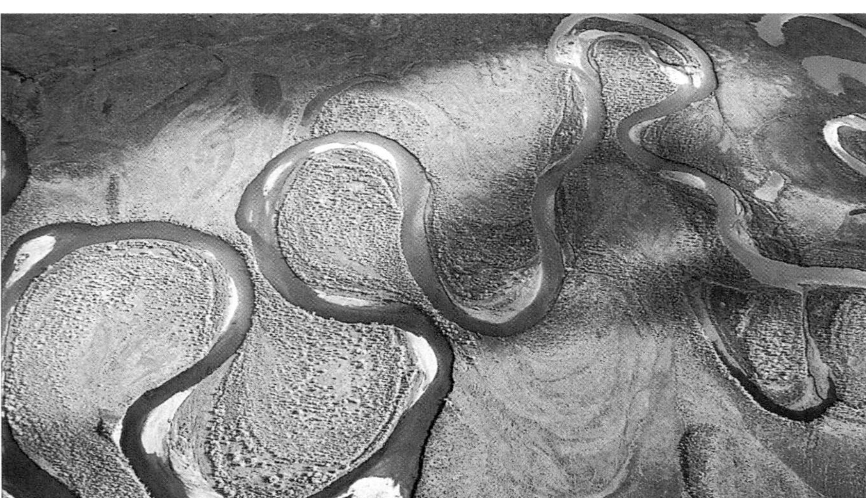

Figure 9-17 Several meanders are formed in the Tundra River in Yukon, Canada.

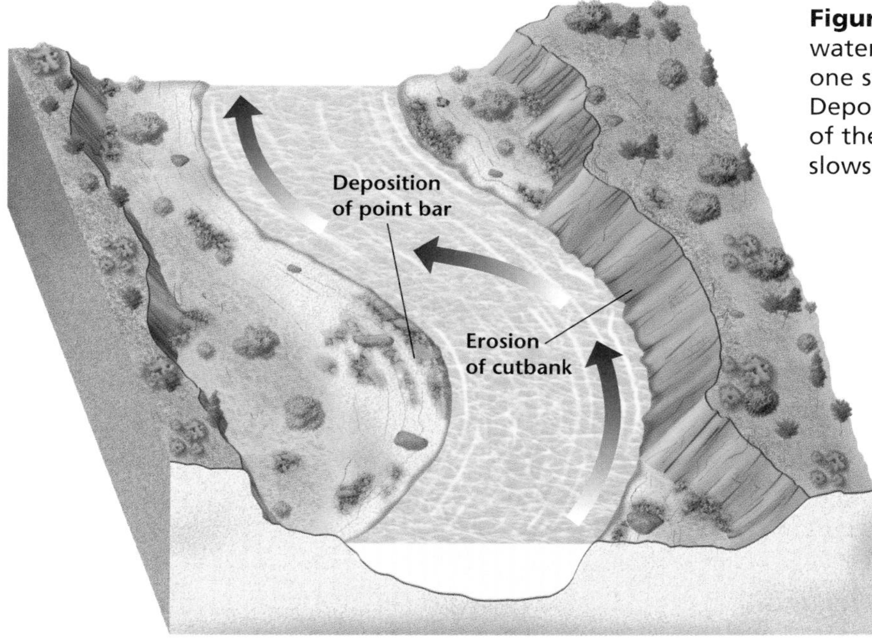

Figure 9-18 The high velocity of water in a meandering stream erodes one side of the stream's bank. Deposition occurs when the velocity of the water in a meandering stream slows down.

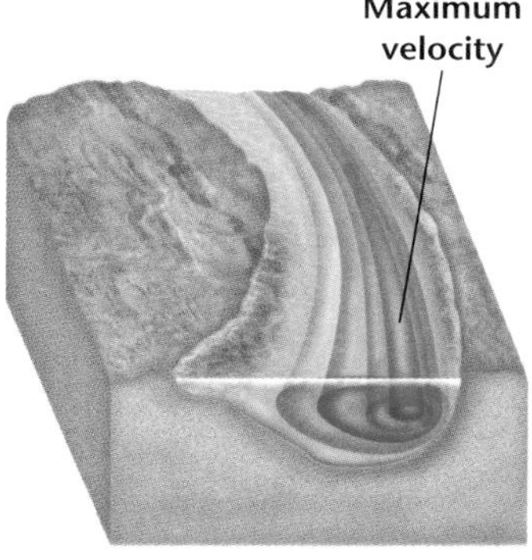

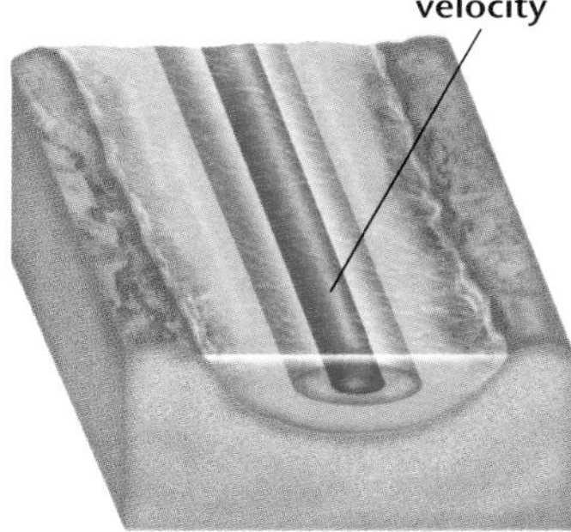

Figure 9-19 The maximum velocity of water in a stream will change its direction as the stream meanders.

deposition along a meander, and ***Figure 9-19*** shows the points of maximum water velocity within a meander and within a straight part of a stream. Stream meanders continue to develop and become larger and wider over time. After some degree of winding, however, it is common for a stream to cut off a meander and once again flow along a straighter path. The stream then deposits material along the adjoining meander and eventually blocks off its water supply, as shown in ***Figure 9-20.*** The blocked-off meander becomes an oxbow lake, which eventually dries up.

As a stream approaches a larger body of water or its ultimate end point, the ocean, the streambed's gradient flattens out and its channel becomes very wide. The area of the stream that leads into the ocean or another large body of water is called the mouth. The mouth of the Mississippi River is extremely wide.

Figure 9-20 The Devil's Elbow is an oxbow lake along the Congaree River in Congaree Swamp National Monument in South Carolina.

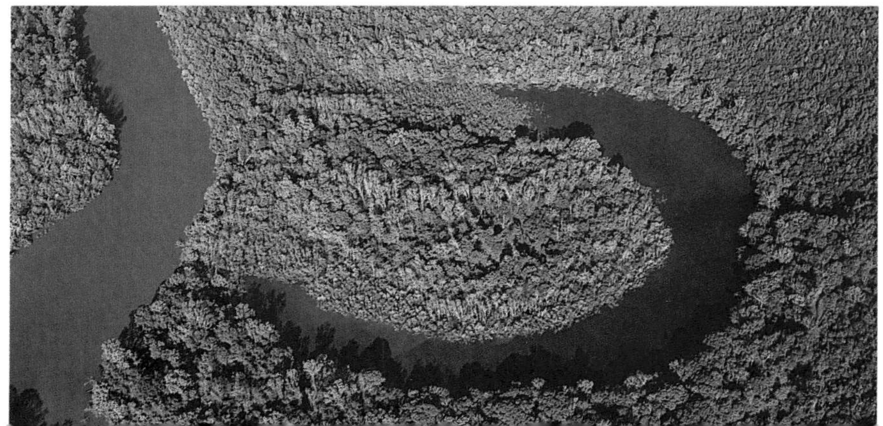

Figure 9-21 An alluvial fan was formed at Mormon Point beneath Black Mountain in Death Valley National Park in California.

Deposition of Sediment

Streams that lose velocity also lose their ability to carry sediment. A stream's velocity lessens and its sediment load drops when its gradient abruptly decreases. In dry regions, where mountain streams commonly flow down narrow valleys onto broad, flat, valley floors, a stream's gradient may suddenly decrease causing the stream to drop its sediment as a fan-shaped deposit called an *alluvial fan.* Alluvial fans are sloping depositional features formed at the bases of slopes and composed mostly of sand and gravel. They are found worldwide but are most common in dry, mountainous regions such as Death Valley, California, shown in ***Figure 9-21.***

Streams also lose velocity and the ability to carry sediment when they join larger bodies of quiet water. The triangular deposit that forms where a stream enters a large body of water is called a **delta,** named for the triangle-shaped Greek letter *delta* (Δ). Delta deposits usually consist of silt and clay particles. As a delta develops, sediments build up and slow the stream water, sometimes even blocking its movement. Smaller distributary streams then form to carry the stream water through the developing delta. The Mississippi River Delta, shown in ***Figure 9-22,*** began forming millions of years ago. Today, the city of New Orleans, Louisiana is located on that delta, an area that was under seawater only 5000 years ago.

Using Numbers If a stream's average velocity is 5 m/s, its width is 30 m, and its average depth is 10 m, what is the amount of the stream's discharge?

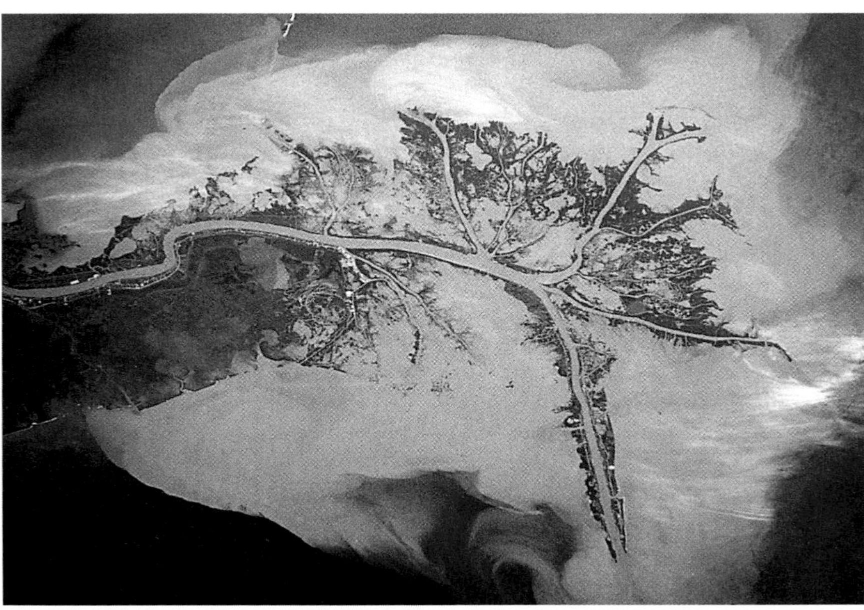

Figure 9-22 This photo shows a portion of the Mississippi River Delta. The delta consists of silt, sand, and clay deposits.

Figure 9-23 A steep-sided canyon occurs along a meander on the Escalante River in Glen Canyon National Recreation Area in Utah.

REJUVENATION

During the process of stream formation, downcutting, or the wearing away of the streambed, is a major erosional process until the stream reaches its base level, when downcutting stops. However, if the land over which the stream is flowing uplifts or if the base level lowers, the stream undergoes rejuvenation. *Rejuvenation* means "to make young again." During **rejuvenation,** the stream actively resumes the process of downcutting toward its base level. This causes an increase in the stream's rate of flow, and the stream's channel once again becomes V-shaped. If rejuvenation occurs in an area where there are meanders, deep sided canyons are formed. This effect is evident in Utah's Escalante River, shown in ***Figure 9-23.***

SECTION ASSESSMENT

1. Describe the formation of an oxbow lake.
2. Compare the rate of water flow on the inside of a meander curve with that on the outside of the curve.
3. Describe four changes that a stream undergoes as it works its way toward the ocean.
4. What are the differences between an alluvial fan and a delta?
5. **Thinking Critically** How does the type of bedrock over which a stream flows affect the time it takes for the stream to reach its base level?

SKILL REVIEW

6. **Making Graphs** Make a line graph that plots the direction of change in a hypothetical stream's rate of flow at the stream's headwaters, at midstream, and at its mouth. For more help, refer to the *Skill Handbook.*

earthgeu.com/self_check_quiz

SECTION 9.3

Lakes and Freshwater Wetlands

Earth Sciences 9.c **I&E** 1.a, 1.e, 1.j, 1.m

OBJECTIVES

- **Explain** *the formation of freshwater lakes and wetlands.*
- **Describe** *the process of eutrophication.*
- **Recognize** *the effects of human activity on lake development.*

VOCABULARY

lake
eutrophication
wetland

You have probably swum in, fished in, or gazed at the beauty of a lake. But, did you ever think about how lakes form? A **lake** is a depression in the surface materials of a landscape that collects and holds water. As shown in the *MiniLab* on the following page, surface materials determine where a lake can form. Lakes sometimes accumulate water from streams and runoff that flow into them. Lakes also receive water from local precipitation, springs, and other sources. Most lakes have outlets from which water flows to rivers and to the ocean. People sometimes build small lakes called ponds to serve as sources of water for livestock, to maintain fish supplies, to attract wildlife, or for their natural beauty. Reservoirs are lakes made for the primary purpose of storing water for a community's use.

ORIGINS OF LAKES

Natural lakes form in different ways in surface depressions and in low areas. As you have learned, oxbow lakes form when streams cut off meanders and leave isolated channels of water. Lakes can also form when stream flow becomes blocked by sediment from landslides. Other lakes, such as Utah's Great Salt Lake, shown in ***Figure 9-24,*** are remnants of prehistoric lakes that have receded to lower-lying areas.

Still other lakes have glacial origins, as you learned in Chapter 8. The basins of these lakes formed as glaciers gouged out the land during the ice ages. Most of the lakes in Europe and North America are in recently glaciated areas. Glacial moraines originally dammed some of these depressions and restricted the outward flow of water. The lakes that formed as a result are known as moraine-dammed lakes, shown in ***Figure 9-25.*** In another process, cirques carved high in the mountains by valley glaciers filled with water to form cirque lakes. Other lakes

Earth Sciences

9.c Students know the importance of water to society, the origins of California's fresh water, and the relationship between supply and need.

Figure 9-24 The Great Salt Lake in northern Utah has a much greater salinity than the oceans.

formed as blocks of ice left on the outwash plain ahead of melting glaciers eventually melted and left depressions called kettles. When these depressions filled with water, they formed kettle lakes.

Many lakes are found in areas where limestone is the dominant bedrock. As groundwater percolating through limestone bedrock slowly dissolves calcium carbonate, it leaves holes in the limestone and forms caverns. In some places, the ceilings of these caverns become so thin and weak that they collapse, which leaves depressions that may fill with water in time.

LAKES UNDERGO CHANGE

Water from precipitation, runoff, and underground sources can maintain a lake's water supply. Some lakes contain water only during times of heavy rain or excessive runoff from spring thaws. A depression that receives more water than it loses to evaporation or use by humans will exist as a lake for a long period of time. However, lakes are temporary water-holding areas; over hundreds of thousands of years, lakes usually fill in with sediment and become part of a new landscape.

Eutrophication Through the process of photosynthesis, plants add oxygen and waste products to lake water. Animals that live in a lake use the water's oxygen and add waste products to the water as they conduct their life processes. The decay process that occurs after plants and animals die also uses up dissolved oxygen supplies. The amount of dissolved oxygen helps determine the quality of lake water and its ability to support life.

MiniLab

Surface materials determine where a lake can form.

Model how different Earth materials may allow lakes to form. Lakes form when depressions or low areas fill with water.

Procedure ***CAUTION: Always wear safety goggles and an apron in the lab.***

1. Use three clear, plastic shoe boxes. Half fill each one with Earth materials: clay, sand, and gravel.
2. Slightly compress the material in each shoe box. Then make a shallow depression in each surface.
3. Slowly pour 500 mL of water into each of the depressions.

Analyze and Conclude

1. Describe what happened to the 500 mL of water that was added to each shoe box.
2. How is this activity similar to what actually happens on Earth's surface when a lake forms?
3. What can you infer about the Earth materials in which lakes most commonly form?

Figure 9-25 The moraine-dammed lakes in Banff National Park in Alberta, Canada were formed from glacial activity.

Figure 9-26 The aquatic species of this pond will change over the years because of the effects of eutrophication.

The process by which lakes become rich in nutrients from the surrounding watershed, thereby resulting in a change in the kinds of organisms in the lake, is called **eutrophication.** ***Figure 9-26*** shows a pond undergoing eutrophication. Although eutrophication is a natural process, it can be sped up with the addition of nutrients, such as fertilizers, that contain nitrogen and phosphorus. When this happens, the animal and plant communities in the lake can change rapidly. Algae growing in the water may suddenly multiply very quickly. The excessive algae growth in a lake or pond appears as green scum. Other organisms that eat the algae can multiply in numbers as well. The resulting overpopulation and decay of a large number of plants and animals depletes the water's oxygen supply. Fish and other sensitive organisms may die as a result of the lack of oxygen in the water.

Other major sources of nutrients that concentrate in lakes are animal wastes and phosphate detergents. Lakes can also suffer from the release of toxins from nearby industries and untreated sewage, as shown in the *Science & the Environment* feature at the end of the chapter.

Freshwater Wetlands A **wetland** is a land area that is covered with water for a large part of the year. Wetlands include environments commonly known as bogs, marshes, and swamps. They have certain soil types and support specific plant species.

A bog, shown in ***Figure 9-27,*** is an interesting wetland that deserves a closer look. Bogs are not stream-fed, but instead receive their water from precipitation. The waterlogged soil tends to be rich in *Sphagnum,* also called peat moss. The breakdown of peat moss produces acids, thereby contributing to the soil's acidity. The waterlogged, acidic soil supports unusual plant species, including insect-eating pitcher plants, sundew, and Venus' flytrap.

Figure 9-27 This bog in Norway has acid-rich soil that supports a variety of organisms.

Freshwater marshes frequently form along the mouths of streams and in areas with extensive deltas. The constant supply of water allows for the lush growth of marsh grasses. The shallow roots of the grasses anchor deposits of silt and mud on the delta, thereby slowing the water and expanding the marsh area. Grasses, reeds, sedges, and rushes, along with abundant wildlife, are common in marsh areas.

Swamps are low-lying areas often located near streams. Swamps may develop from marshes that have filled in sufficiently to support the growth of shrubs and trees. As these larger plants grow and begin to shade the marsh plants, the marsh plants die. Swamps that existed 250 million years ago developed into present-day coal reserves that are common in Pennsylvania and many other locations in the United States and around the world.

Figure 9-28 The wetlands in Bosque del Apache National Wildlife Refuge in New Mexico are home to migrating snow geese.

Wetlands play a valuable role in improving water quality. They serve as a filtering system that traps pollutants, sediments, and pathogenic bacteria contained in water sources. Wetlands also provide vital habitats for migratory waterbirds and homes for an abundance of other wildlife, as shown in ***Figure 9-28.*** Unfortunately, people's desire for land often conflicts with the need to preserve wetlands. In the past, it was common for wetland areas to be filled in to create more land on which to build. Government data reveal that from the late 1700s to the mid-1980s, the continental United States lost 50 percent of its wetlands. By 1985, it was estimated that 50 percent of the wetlands in Europe were drained. Now, however, the preservation of wetland areas has become a global concern.

Section Assessment

1. Describe the process of eutrophication.
2. What human activities affect the process of eutrophication?
3. What conditions are necessary for the formation of a natural lake?
4. **Thinking Critically** Describe a situation in which protection of wetlands may conflict with human plans for land use.

Skill Review

5. **Making Tables** Design a data table that compares the various types of lakes, their origins, and their characteristics. For more help, refer to the *Skill Handbook.*

Modeling Stream Velocity and Slope

Water in streams flows from areas of higher elevation to areas of lower elevation. The rate of stream flow varies from one stream to another and also in different areas of the same stream.

I&E
1.a Select and use appropriate tools and technology (such as computer-linked probes, spreadsheets, and graphing calculators) to perform tests, collect data, analyze relationships, and display data.

Also covers:
1.e, 1.j

Preparation

Problem

Determine how slope may affect stream-flow velocity.

Materials

1-m length of vinyl gutter pipe
ring stand and clamp
water source with long hose
protractor with plumb bob
sink or container to catch water
stopwatch
grease pencil
meterstick
paper
hole punch

Objectives

In this GeoLab, you will:

- **Measure** the time it takes for water to flow down a channel at different slopes and depths.
- **Organize** your data in a table.
- **Plot** the data on a graph to show how stream velocity is directly proportional to the stream channel's slope and depth.
- **Describe** the relationship between slope and rate of stream flow.

Safety Precautions

Always wear safety goggles in the lab.

Procedure

1. Use the hole punch to make 10 to 15 paper circles to be used as floating markers.
2. Use the illustration below as a guide to set up the protractor with the plumb bob.

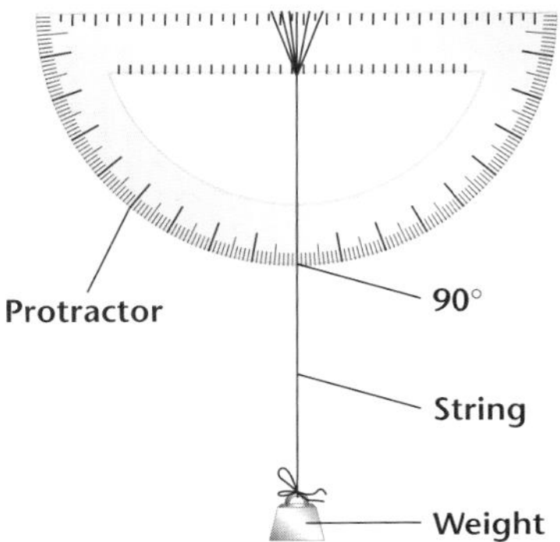

3. Use the grease pencil to mark two lines across the inside of the gutter pipe at a distance of 40 cm apart.
4. Use the ring stand and clamp to hold the gutter pipe at an angle of 10°. Place the end of the pipe in a sink or basin to collect the discharged flow of water.
5. Attach a long hose to a water faucet in the sink.
6. Keep the hose in the sink until you are ready to use it. Then turn on the water and adjust the flow until the water is moving quickly enough to provide a steady flow.
7. Bend the hose temporarily to block the water flow until the hose is positioned at least 5 cm above the top line marked on the pipe.
8. Keep the water moving at the same rate of flow for all slope angles being investigated.
9. Drop a floating marker approximately 4 cm above the top line on the pipe and into the flowing water. Measure the time it takes for the floating marker to move from the top line to the bottom line. Record the time in your science journal.
10. Repeat step 9 two more times.
11. Repeat steps 9 and 10 but change the slope to 20°, then 30°, and then 40°.
12. Make a line graph of the average stream-flow velocity.

Analyze

1. Why is it important to keep the water flow constant in this activity?
2. Which variables had to be controlled to avoid errors in your data?
3. Using your graph, predict the velocity of water flow for a 35° slope.

Conclude & Apply

1. What is the relationship between the rate of water flow and the angle of the slope?
2. Describe one reason why a stream's slope might change.
3. Where would you expect to find streams with the highest water-flow velocity?

Science & the Environment

The Jewel of Siberia

Pollution is threatening the ecosystems of Lake Baikal, the oldest, largest, and deepest freshwater lake on Earth. What is causing the problems? Are there any solutions for saving the lake?

A Natural Paradise

Known as the "Jewel of Siberia," Lake Baikal is the oldest and largest freshwater lake on Earth. Estimated to be 25 million years old, Lake Baikal contains 20% of Earth's unfrozen freshwater, that is, one-fifth of the world's fresh surface water. The lake contains approximately 80% of the former Soviet Union's freshwater supply and covers approximately 31 500 km^2. It reaches a maximum depth of approximately 1637 m making it the deepest lake in the world.

Fed by 330 tributaries, and surrounded by forests and mountain ranges, it is home to a wide variety of plant and animal species. The area is home to everything from microscopic organisms to large mammals including elk, moose, deer, and the brown bear.

One animal found only in this area is the Nerpa or Baikal seal. The Nerpa is the only known species of freshwater seal. It is believed that the seal may have migrated to the area in search of food while the lake was being formed thousands of years ago.

Threatened

Pollution has begun to slowly take its toll on animal and plant species in the region. Studies have reported that the fish population is dying out and thousands of the Baikal seals have died. High toxin levels from a nearby pulp and paper factory may be the cause. According to other studies, DDT and other pesticides have entered the waters via aerial spraying, and have been found in the lake's sediment.

Attempts to restrict the release of toxins into the freshwater lake have failed due to concern over the loss of industry and jobs. For example, if a lakeside pulp and paper factory were forced to close, many people would be left unemployed.

Many organizations have banded together in Russia to attempt to preserve the Lake Baikal area. There is even cooperation between groups in the United States and Russia who have been working together to come up with solutions to the problems facing the industry and the environment. Their activities range from efforts to save the Baikal seal to promoting tourism as a more attractive form of economic stability in the area.

Activity

Form small groups to research and discuss possible solutions to the problems threatening the ecosystems of Lake Baikal. How can the lake and its inhabitants be preserved without having to totally remove industry from the area? Visit the Earth Science Web Site at earthgeu.com to learn more about the struggle to save Lake Baikal and its many inhabitants.

Study Guide

Summary

SECTION 9.1

Surface Water Movement

Main Ideas

- Water on Earth may follow a variety of pathways as it is recycled through the processes of evaporation and condensation.
- Infiltration of water into the ground depends on the number of open pores or spaces in Earth materials and on the presence of unsaturated pores in the ground.
- All the land area that drains into a stream system is the system's watershed, or drainage basin. Elevated land areas called divides separate one watershed from another.
- A stream's load is all the material the stream carries, including material in solution, in suspension, and as bed load.
- A floodplain is a broad, flat area that extends out from a stream's bank during times of flooding.
- Flooding occurs in small, localized areas as upstream floods or in large, downstream floods. Damage from flooding can be devastating.

Vocabulary

bed load (p. 217)
discharge (p. 218)
divide (p. 215)
flood (p. 219)
floodplain (p. 219)
runoff (p. 212)
solution (p. 215)
suspension (p. 216)
watershed (p. 215)

SECTION 9.2

Stream Development

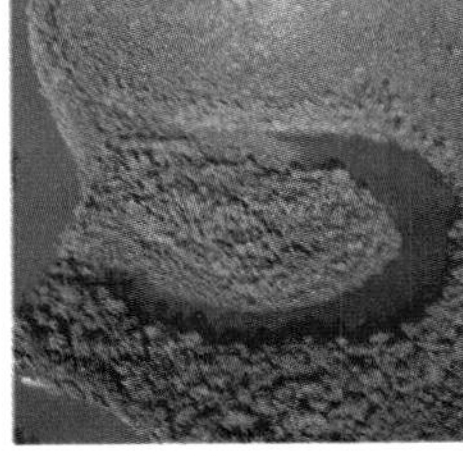

Main Ideas

- Water from precipitation gathers in gullies at a stream's source area, or headwaters. The stream's water flows in channels confined by the stream's banks.
- Alluvial fans and deltas form when stream velocity decreases and sediment is deposited. Alluvial fans are fan shaped, and they form where water flows down steep slopes onto flat plains. Deltas are triangular, and they form when streams enter large, relatively quiet bodies of water.

Vocabulary

delta (p. 226)
meander (p. 224)
rejuvenation (p. 227)
stream bank (p. 222)
stream channel (p. 222)

SECTION 9.3

Lakes and Freshwater Wetlands

Main Ideas

- Lakes form in a variety of ways when depressions on land fill with water. Lakes may be natural or human-made.
- Eutrophication is a natural nutrient enrichment process that may be sped up when nutrients from fertilizers, detergents, or sewage are added.
- Wetlands are low-lying areas that are periodically saturated with water and support specific plant species. Wetlands include bogs, marshes, and swamps.

Vocabulary

eutrophication (p. 230)
lake (p. 228)
wetland (p. 230)

earthgeu.com/vocabulary_puzzlemaker

CHAPTER 9
Assessment

Understanding Main Ideas

1. Which factor least affects the rate of runoff?
a. slope **c.** volume of runoff
b. vegetation **d.** nearness to water

2. What areas are most likely to contain fertile soils?
a. watersheds **c.** floodplains
b. dried-up streambeds **d.** mountainous areas

3. Which substance is most likely to be carried by a stream in solution?
a. quartz **c.** calcite
b. sand **d.** silt

4. What material plays a major role in the eutrophication of lakes?
a. iron **c.** ozone
b. phosphate **d.** salt

5. During the process of eutrophication, what happens to the oxygen present in a lake?
a. It increases. **c.** It stays the same.
b. It decreases. **d.** It evaporates.

6. What kind of streams form V-shaped valleys?
a. streams that are first forming
b. streams that carry much sediment
c. streams that move slowly
d. streams that have meanders

7. Where does water move most rapidly in the straight length of a stream?
a. along the bottom **c.** near the surface
b. along the sides **d.** in the center

8. If a stream is carrying sand, silt, clay, and small pebbles, which one is deposited last as the stream begins to slow down?
a. clay **c.** sand
b. silt **d.** small pebbles

9. Where do alluvial fans form?
a. on the outside of meanders
b. where streams enter the ocean
c. near lakes
d. along the bases of mountains

Applying Main Ideas

10. In what ways are a delta and an alluvial fan similar, and in what ways are they different?

11. Why is it important to preserve wetlands?

12. What means do governments use to try to prevent the loss of life and property in flood-prone areas?

Use the following aerial view of a stream to answer questions 13–15.

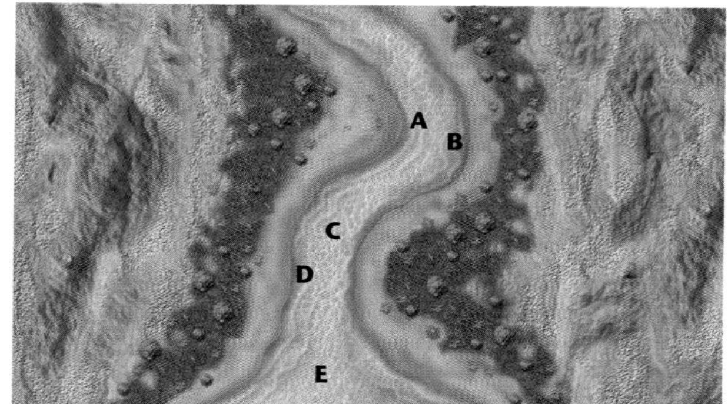

Test-Taking Tip

Beat the Clock—and Then Go Back
As you take a practice test, pace yourself to finish each section just a few minutes early so you can go back and check over your work. You will sometimes find a mistake or two.

earthgeu.com/chapter_test

CHAPTER 9
Assessment

13. At which location in the aerial view does the stream's water have the greatest velocity?
14. At which location is deposition most actively occurring?
15. At which location is erosion most actively occurring?
16. What is the discharge of a stream that has a velocity of 300 m/s and is 25-m wide and 3-m deep?

Thinking Critically

17. Why is a lake with a clay bottom able to hold more water than a lake with a sand bottom?
18. One morning, there was a torrential thunderstorm. In the afternoon, the skies cleared and a gardener decided to plant a tree. After digging in the ground only a short distance down, the gardener found that the ground was very dry. How could the ground be dry despite the heavy rains earlier in the day?
19. If floodplains are such hazardous areas to live in, why have so many people settled in these potential flood zones?
20. Use the following terms to construct a concept map to organize the major ideas in Section 9.1. For more help, refer to the *Skill Handbook.*

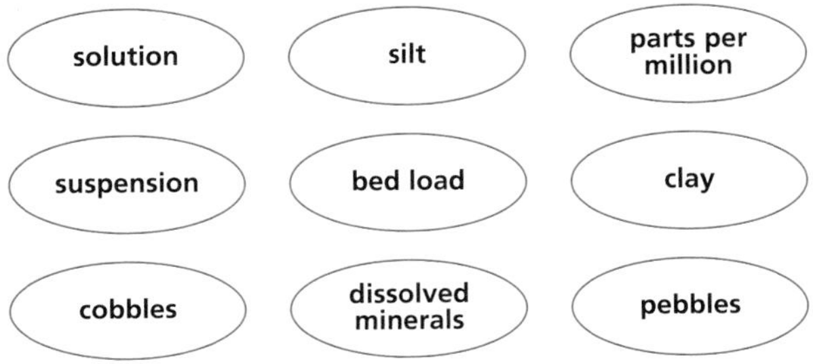

Standardized Test Practice

1. Which condition would create the most runoff?
   a. land covered with vegetation
   b. plants in densely packed soil
   c. light precipitation
   d. soil with a high percentage of sand

2. In which part of a meander does the water travel the fastest?
   a. the water that moves along the inside curve of the meander
   b. the water that moves along the bottom of the meander
   c. the water that moves along the outside curve of a meander
   d. all water flows at the same rate

3. Which of the following is NOT a value of wetlands?
   a. feeding lakes and deltas with nutrient- and oxygen-rich water
   b. filtering water by trapping pollutants, sediments, and pathogenic bacteria
   c. providing habitats for migratory birds and other wildlife
   d. preserving fossils due to the anaerobic and acidic conditions

4. As the velocity of a stream decreases, which transported particle size would settle to the stream's bottom first?
   a. clay
   b. silt
   c. pebble
   d. sand

5. Which condition helps determine the quality of lake water?
   a. the amount of nitrogen
   b. the amount of dissolved calcium carbonate
   c. the amount of potassium
   d. the amount of dissolved oxygen

Chapter 10

Groundwater

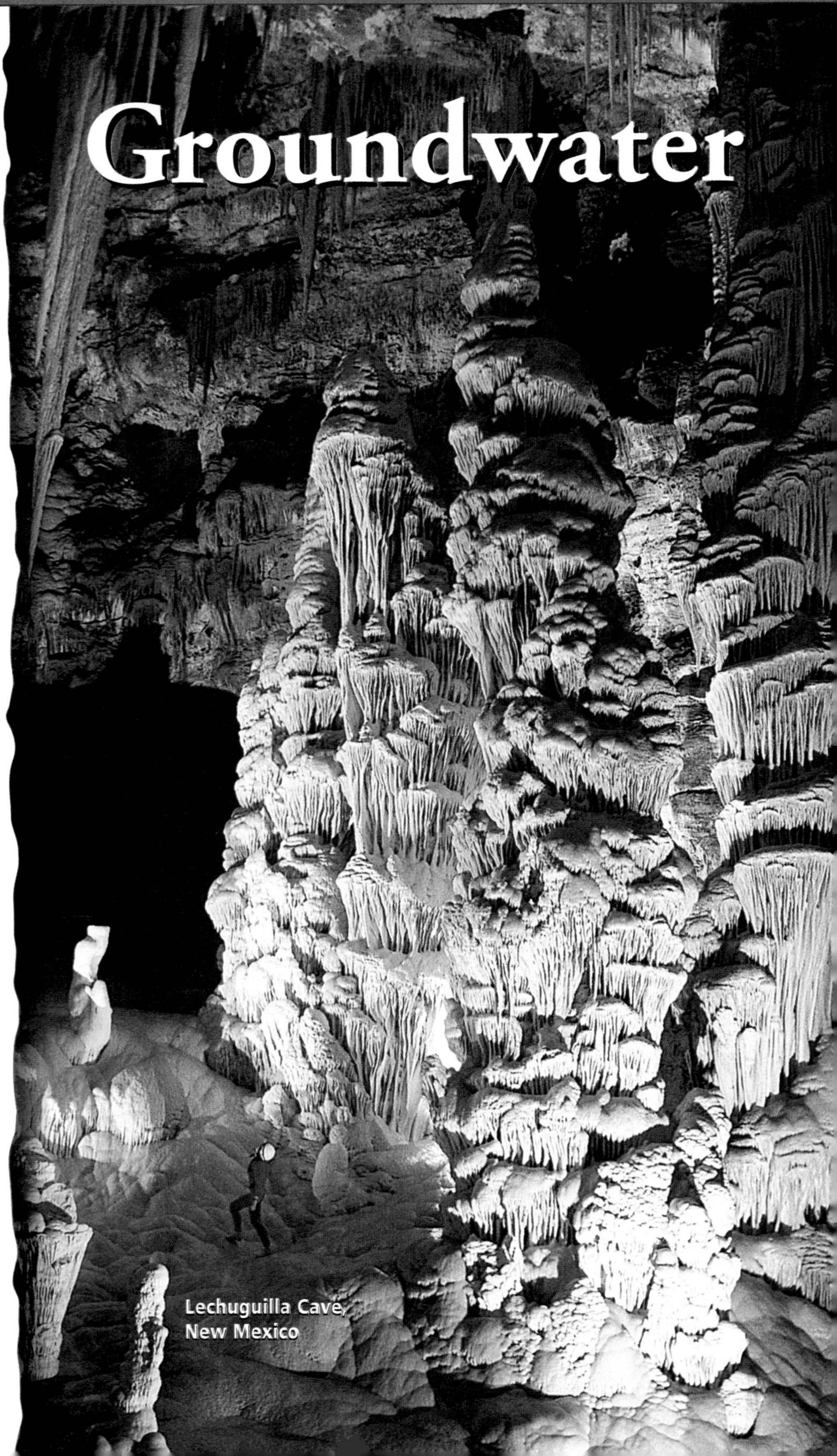

What You'll Learn

- How large amounts of water are stored underground.
- How groundwater dissolves limestone and forms caves and other natural features.
- How groundwater is removed from the ground by humans and what problems endanger our groundwater supply.

Why It's Important

Groundwater provides drinking water for half of the world's population and is a major source of the water used by agriculture and industry. However, groundwater supplies are threatened by overuse and pollution.

To find out more about groundwater, visit the Earth Science Web Site at earthgeu.com

Lechuguilla Cave, New Mexico

Discovery Lab

Model Underground Water Storage

Beneath your feet, there are vast amounts of water. This water fills in the pore spaces of sediments and rocks deep in the ground. In this activity, you will discover how much water can be stored in sand.

1. Fill a 250-mL graduated cylinder with dry sand.
2. Fill another 250-mL graduated cylinder with water.
3. Pour water from the second cylinder into the sand-filled cylinder until the water level is flush with the surface of the sand. Measure and record the volume of saturated sand in the cylinder.
4. Measure and record how much water is left in the second cylinder.

CAUTION: Always wear safety goggles and an apron in the lab.

Observe In your science journal, describe how much water is present in the saturated sand. Calculate the ratio of water volume to the volume of sand. Infer how many liters of water could be stored in a cubic meter of sand.

SECTION 10.1 *Movement and Storage of Groundwater*

Earth Sciences 9.c

OBJECTIVES

- **Describe** *how groundwater is stored and moves underground.*
- **Explain** *what an aquifer is.*

VOCABULARY

infiltration
porosity
zone of saturation
water table
permeability
aquifer

If you drill a deep enough hole anywhere on Earth, it will partially fill with groundwater, even in the desert! Groundwater is present everywhere beneath the surface of the land, but nevertheless is a small fraction of all the water on Earth.

THE HYDROSPHERE

The water on and in Earth's crust makes up the hydrosphere, named after *hydros,* the Greek word for "water." About 97 percent of the hydrosphere is contained in the oceans. The water contained by landmasses—nearly all of it freshwater—makes up only about 3 percent of the hydrosphere.

Freshwater is one of Earth's most abundant and important renewable resources. However, of all the freshwater, more than 90 percent is in the form of polar ice caps and glaciers. You may be surprised to

Table 10-1 World's Water Supply				
Location	**Surface Area (km^2)**	**Water Volume (km^3)**	**Percentage of Total Water**	**Estimated Average Residence Time of Water**
Oceans	361 000 000	1 230 000 000	97.2	Thousands of years
Atmosphere	510 000 000	12 700	0.001	Nine days
Rivers and streams	—	1200	0.0001	Two weeks
Groundwater: shallow, to a depth of 0.8 km	130 000 000	4 000 000	0.31	Hundreds to many thousands of years
Lakes (freshwater)	855 000	123 000	0.009	Tens of years
Ice caps and glaciers	28 200 000	28 600 000	2.15	Tens of thousands of years and longer

learn that most of the remaining freshwater is groundwater. All the rivers, streams, and lakes on Earth represent only a small fraction of Earth's liquid freshwater, as shown in ***Table 10-1.***

Precipitation and Groundwater

The ultimate source of all water on land is the oceans. Evaporation of seawater introduces water into the atmosphere in the form of invisible water vapor and visible clouds. Winds and weather systems move this atmospheric moisture all over Earth, much of it over the continents. Precipitation brings atmospheric moisture back to Earth's surface, mostly in the form of rain and snow. Some of this precipitation falls directly into the oceans, and some falls on land.

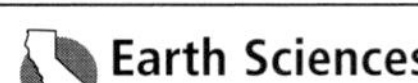

Earth Sciences

9.c Students know the importance of water to society, the origins of California's fresh water, and the relationship between supply and need.

Much of the precipitation that falls on land enters the ground through the process of **infiltration** and becomes groundwater. Only a small portion of precipitation becomes runoff and is returned directly to the oceans through streams and rivers. Solid precipitation, such as snow, may cover the ground for long periods of time before it melts and becomes runoff or infiltrates to become groundwater. Groundwater slowly moves through the ground, eventually returns to the surface through springs, and then flows back to the oceans.

Groundwater Storage

Puddles of water that are left after a rain quickly disappear, partly by evaporating and partly by percolating into the ground. On sandy soils, rain soaks into the ground almost immediately. Where does that water go? Subsurface Earth materials are not totally solid, but instead contain countless small openings, or pores, which make up a large portion of some of these materials, as you see in ***Figure 10-1.***

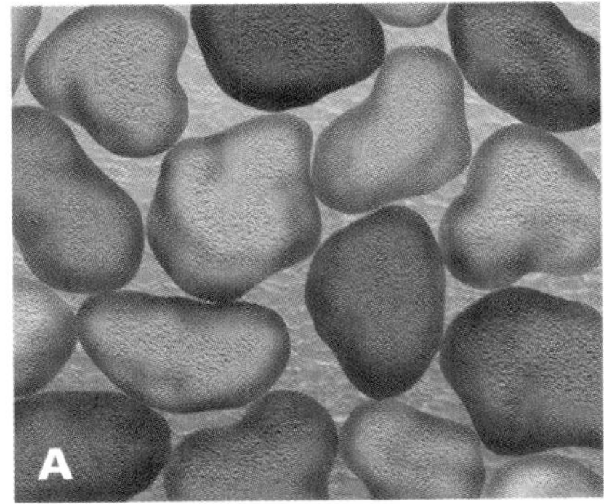

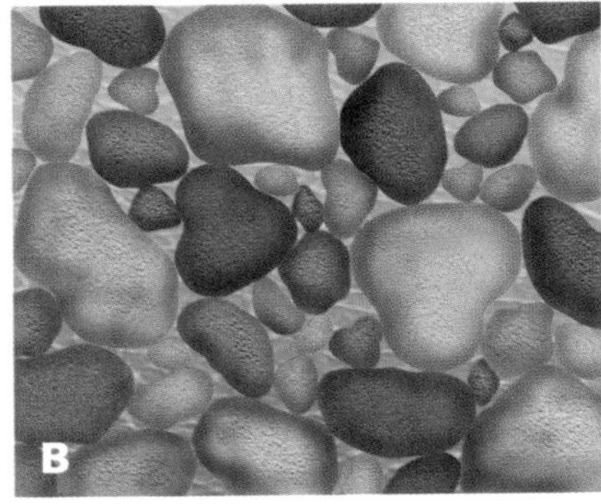

Figure 10-1 Pore spaces in sediments: the highest percentage of porosity is found in well-sorted sediments **(A)** while poorly sorted sediments **(B)** have a lower percentage.

The percentage of pore space in a material is called its **porosity.** Subsurface materials have porosities ranging from 2 or 3 percent to more than 50 percent. For example, the porosity of well-sorted sand is typically around 30 percent. In poorly sorted sediments, however, smaller particles of sediment occupy some of the pore spaces and reduce the overall porosity of these sediments. Similarly, the cement that binds the grains of sedimentary rocks together reduces the rocks' porosity. Nevertheless, enormous quantities of groundwater are stored in the pore spaces of rocks and sediments.

The Zone of Saturation

The depth below Earth's surface at which groundwater completely fills all the pores of a material is called the **zone of saturation.** The upper boundary of the zone of saturation is the **water table,** as shown in ***Figure 10-2.*** Strictly speaking, only the water in the zone of saturation is called groundwater. In the zone of aeration, which

Figure 10-2 Groundwater flows toward valleys where the water table is close to the surface. During dry periods the level of the water table falls.

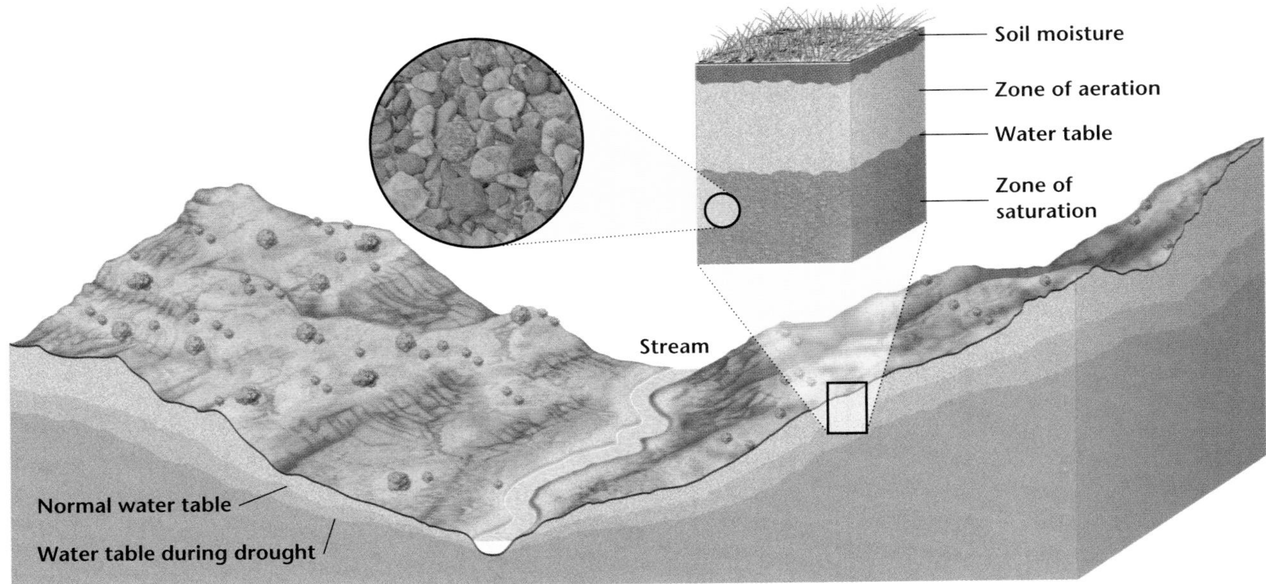

is above the water table, materials are moist, but the pores contain mostly air. Water in the zone of saturation can be classified as either gravitational water or capillary water. Gravitational water is water that trickles downward as a result of gravity. Capillary water is water that is drawn upward from the water table and is held in the pore spaces of rocks and sediments as a result of surface tension. Materials that are directly above the water table, especially fine-grained materials, are nearly saturated with capillary water. Capillary action is similar to the action of water that is drawn upward through the pore spaces of a paper towel when the end of it is dipped into water.

The Water Table The depth of the water table varies depending on local conditions. For example, in stream valleys, groundwater is close to Earth's surface, and thus the water table is a few meters deep at most. In swampy areas, the water table is almost at Earth's surface, whereas on hilltops or in arid regions, the water table can be tens to hundreds of meters or more beneath the surface. As shown in ***Figure 10-2,*** the topography of the water table follows the topography of the land above it. For example, the water table slopes toward valleys and forms hills under topographic hills. Water table topography forms in this way because water underground moves slowly and conforms to surface contours.

Because of its dependence on precipitation, the water table fluctuates with seasonal and other weather conditions. It rises during wet seasons, usually in spring, and drops during dry seasons, often in late summer.

Figure 10-3 In a saturated material, all grains are coated with a thin film of motionless water. In coarse-grained material like sand **(A)** this film occupies a relatively small fraction of the pore space, and moving water can pass freely through the pores. In fine-grained material like silt **(B)** this film occupies most of the pore space and blocks the movement of water. As a result, sand has a much higher permeability than silt.

A

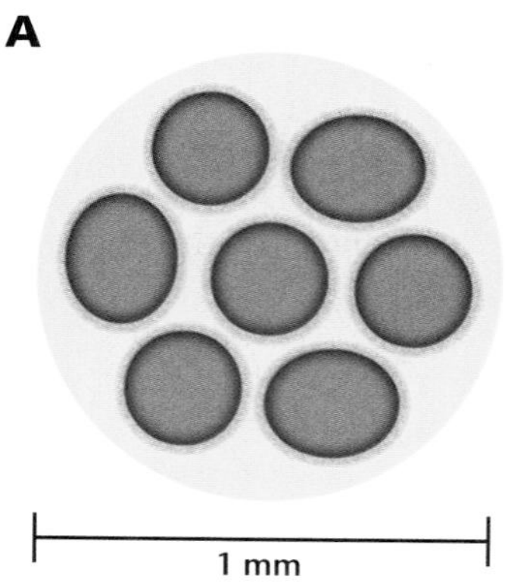

B

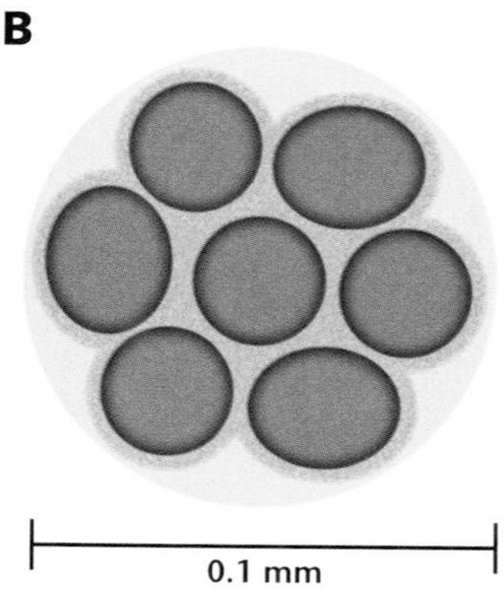

Groundwater Movement

Groundwater flows downhill in the direction of the slope of the water table. In most cases, this downhill movement is slow because the water has to squeeze through numerous tiny pores in the subsurface material. In fact, if the pores are small, not even individual water molecules can squeeze through. The ability of a material to let water pass through it is called **permeability.** Materials with large, connected pores, such as sand and gravel, as shown in ***Figure 10-3A,*** have high permeabilities and permit relatively high flow velocities, up to 1 m/h or more. Other permeable subsurface materials include sandstone, limestone, and all highly fractured bedrock.

Fine-grained materials typically have low permeabilities because their pores are so tiny, as shown in ***Figure 10-3B.*** These materials are said to be impermeable. Flow velocities in impermeable materials are so low that they are often measured in meters per year. Some examples of impermeable materials are silt, clay, and shale. Clay is so

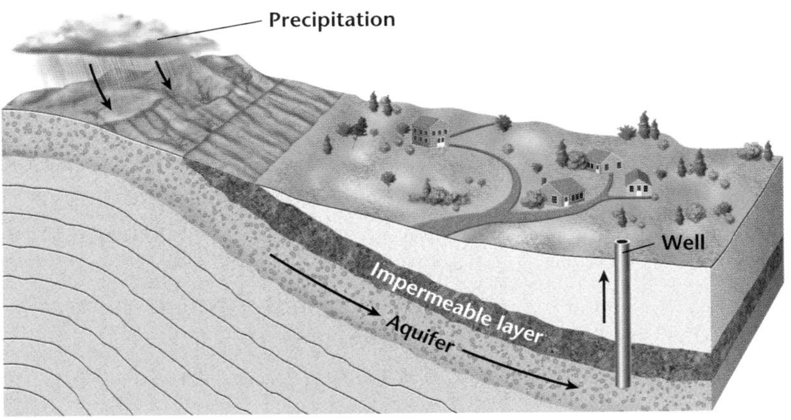

Figure 10-4 The aquifer is located in a permeable sandstone layer that is sealed beneath a capping layer of impermeable rock.

impermeable that a clay-lined depression will hold water. For this reason, clay is often used to line artificial ponds and landfills.

The flow velocity of groundwater primarily depends on the slope of the water table, because the force of gravity pulling the water downward is greater when the slope of the water table surface is steeper. You have experienced a similar effect if you have ever ridden a bicycle down a steep street and a gently sloping street. Although the flow velocity of groundwater is proportional to both the slope of the water table and the permeability of the material through which the water flows, permeability is the major factor. Thus, flow velocities through permeable materials are always higher than those through impermeable materials, regardless of the slope of the water table. Most groundwater flow takes place through permeable layers, called **aquifers,** such as the one shown in ***Figure 10-4.*** Impermeable layers, called aquicludes, are barriers to groundwater flow. In the next section, you'll discover what happens when groundwater moves slowly through materials.

Section Assessment

1. What is the greatest source of freshwater on Earth?
2. Where is the water table closest to Earth's surface: in the floodplain of a river, in a swamp, or on a hilltop?
3. What two factors determine the flow velocity of groundwater?
4. What is an aquifer?
5. **Thinking Critically** What is the difference between porosity and permeability in subsurface materials?

Skill Review

6. **Making and Using Tables** Design a data table that compares and contrasts the porosity and permeability of sand and a mixture of sand and gravel. Which material has the greater porosity? The greater permeability? For more help, refer to the *Skill Handbook*.

SECTION 10.2 Groundwater Erosion and Deposition

OBJECTIVES

- **Explain** *how groundwater dissolves and deposits rocks and minerals.*
- **Describe** *how caves form and how karst topography develops on Earth's surface.*

VOCABULARY

cave
sinkhole
karst topography
stalactite
stalagmite
travertine

In Chapter 3, you learned about the corrosive properties of acids. Acids are solutions that contain hydrogen ions. Most groundwater contains some acid, in most cases carbonic acid. Carbonic acid forms when carbon dioxide dissolves in water and combines with water molecules. This happens when rain falls through the atmosphere and interacts with carbon-dioxide gas or when groundwater percolates through carbon-rich, decaying organic material in soil. As a result of these processes, groundwater is usually slightly acidic and attacks carbonate rocks, especially limestone. Limestone consists mostly of calcium carbonate ($CaCO_3$), which dissolves readily in any kind of acid, the results of which are shown in ***Figure 10-5.***

DISSOLUTION BY GROUNDWATER

The process by which carbonic acid forms and dissolves calcium carbonate can be described by three simple chemical equations.

In the first process, carbon dioxide and water combine to form carbonic acid, as represented by the following equation.

$$CO_2 + H_2O \longrightarrow H_2CO_3$$

In the second process, the carbonic acid (H_2CO_3) molecules in the water split into hydrogen ions (H^+) and bicarbonate ions (HCO_3^-). This process is represented by the following equation.

$$H_2CO_3 \longrightarrow H^+ + HCO_3^-$$

In the third process, the hydrogen ions react with calcium carbonate and dissolve it, as represented by the following equation.

$$CaCO_3 + H^+ \longrightarrow Ca^{2+} + HCO_3^-$$

Figure 10-5 A viewing pagoda is standing among the massive limestone pillars of the stone forest in China. Carbonic acid is slowly dissolving the calcium carbonate in the limestone pillars.

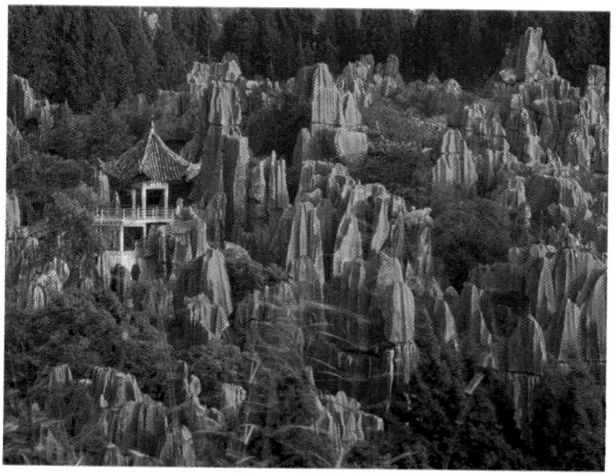

For every carbon dioxide molecule dissolved in groundwater, one hydrogen ion is produced and one calcium carbonate molecule is dissolved. The resulting calcium (Ca^{2+}) and bicarbonate (HCO_3^-) ions are then flushed away by the groundwater. Eventually, they precipitate out somewhere else. Precipitation of calcium carbonate occurs when the groundwater evaporates or when the gas carbon dioxide diffuses out of the water. Both the dissolution and formation of calcium carbonate play a major role in the formation of limestone caves.

Figure 10-6 In Mammoth Cave, Kentucky, there are a series of underground passages that extend to a length of more than 500 km.

Caves A natural underground opening with a connection to Earth's surface is called a **cave.** Some caves form three-dimensional mazes of passages, shafts, and great chambers that stretch for many kilometers. Many caves have structures that hang from the caves' ceilings. Some caves are dry, while others contain underground streams or lakes. Still others are totally flooded and can be explored only by cave divers. One of the most spectacular caves is the recently discovered Lechuguilla Cave of New Mexico, shown in the photograph at the beginning of this chapter. Another cave system in New Mexico, Carlsbad Caverns, includes a huge subterranean chamber over 1 km long and 100 m high. Mammoth Cave, in Kentucky, as shown in ***Figure 10-6,*** is composed of a series of connected underground passages.

Practically all caves of significant size are formed when groundwater dissolves limestone. Most caves develop in the zone of saturation just below the water table. As groundwater percolates through the cracks and joints of limestone formations, it gradually dissolves the adjacent rock and enlarges these passages to form an interconnected network of openings. Thus, the limestone formation becomes more permeable. The resulting increased downhill flow of groundwater gradually lowers the water table until much of the cave system is filled with air. New caves then form beneath the lowered water table. If the water table continues to drop, the thick limestone formations eventually become honeycombed with caves and caverns. This is a common occurrence in limestone regions that have been uplifted by tectonic forces.

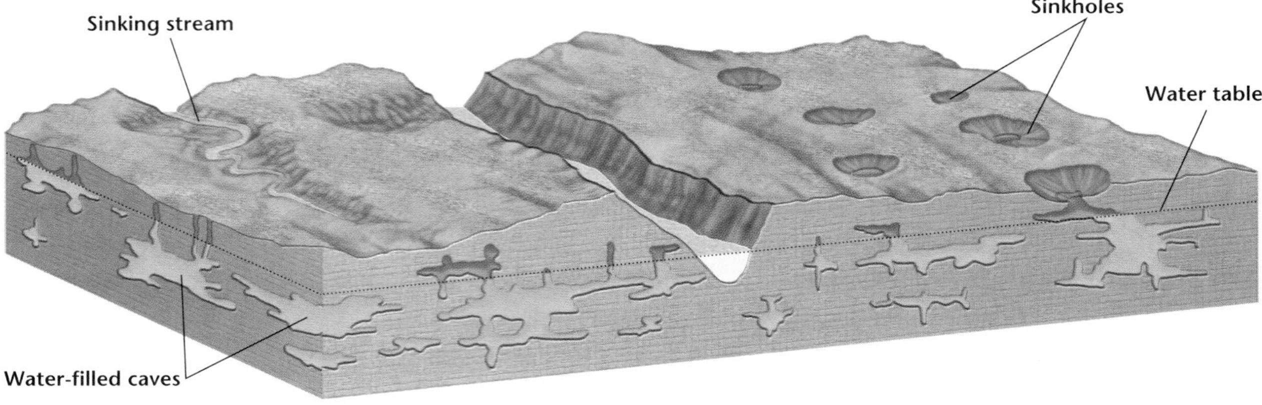

Figure 10-7 In the development of karst topography, caves form near or below the water table. Streams deepen their valleys. The water table drops and new caves form below the water table. Collapsing caves or dissolution of bedrock at the surface produce sinkholes.

Karst Topography ***Figure 10-7*** shows some of the characteristic surface features produced by the dissolution of limestone. The main feature is a **sinkhole,** as shown in ***Figure 10-8.*** A sinkhole is a depression in the ground caused by the collapse of a cave or by the direct dissolution of bedrock by acidic rain or moist soil. Another type of feature forms when a surface stream drains into a cave system, continues underground, and leaves a dry valley above. Such a stream, called a sinking stream, sometimes reemerges abruptly on Earth's surface as a karst spring.

Limestone regions that have sinkholes, sinks, and sinking streams are said to have **karst topography.** The word *karst* comes from the name of a limestone region in Croatia where these features are

Figure 10-8 Sinkholes developed near Roswell, New Mexico.

Figure 10-9 Dissolved minerals can build up thick deposits in plumbing pipes.

especially well developed. Prominent karst regions in the United States are located in Kentucky. The Mammoth Cave region in Kentucky has karst topography that contains tens of thousands of sinkholes. Most of the lakes in Central Florida are sinkholes.

Groundwater Deposits

You are probably aware that your tap water contains various dissolved materials. Some water contains sulfur compounds, and some contains dissolved iron compounds. Water that contains iron compounds typically leaves brownish or red stains on kitchen and bathroom fixtures.

Hard Water Water that contains high concentrations of calcium, magnesium, or iron is called hard water. Hard water is common in limestone areas where the groundwater is nearly saturated with calcium carbonate. Household use of hard water usually can cause a problem: deposits of calcium bicarbonate eventually clog water pipes, as shown in ***Figure 10-9.*** These problems can be controlled with a water softener, which removes dissolved ions from hard water. Water that contains few dissolved ions is called soft water.

Natural Deposits The most remarkable deposits produced by groundwater are the dripstone formations that decorate many caves above the water table. As their name indicates, these formations are built slowly as water drips through caves. Each drop of water hanging on the ceiling of a cave loses some of its carbon dioxide and deposits a tiny amount of calcium carbonate. Over many years, these deposits gradually form cone-shaped or cylindrical structures called

Using Numbers
Each drop of water deposits a 1-nm-thick layer of calcium carbonate at the tip of a stalactite growing in a cave. Given that 1 nm = 10^{-9} m, if a drop falls every 30 seconds, what length, in centimeters, will the stalactite be in 100 years? How many years will it take for the stalactite to reach a length of 5 m?

Figure 10-10 Stalactites, stalagmites, and dripstone columns are found in the Carlsbad Caverns of New Mexico.

stalactites that hang from the cave's ceiling like icicles. As the water drops splash to the floor of the cave, they gradually build mound-shaped dripstone deposits, called **stalagmites,** underneath the stalactites. In time, stalactites and stalagmites may grow together to form dripstone columns, such as the ones shown in ***Figure 10-10.*** These and other types of dripstone formations are composed of a type of limestone called **travertine.**

SECTION ASSESSMENT

1. What acid is most commonly present in groundwater?
2. How do caves form?
3. Compare the formation of stalactites and stalagmites.
4. What is karst topography?
5. **Thinking Critically** If you visited a region that consisted mostly of igneous rocks, would you expect to find karst topography? Explain.

SKILL REVIEW

6. **Concept Mapping** Use the following terms to construct a concept map to organize the major ideas in this section. For more help, refer to the *Skill Handbook.*

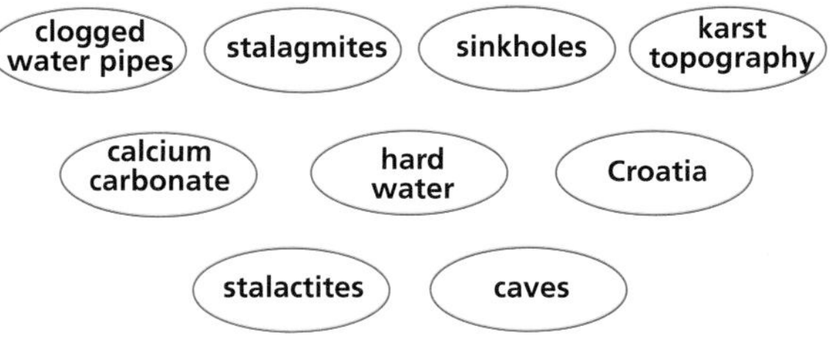

SECTION 10.3 Groundwater Systems

Earth Sciences 9.c I&E 1.h, 1.m

The average length of time that groundwater remains underground is several hundred years. Groundwater moves slowly but continuously through aquifers and eventually returns to Earth's surface. You may wonder how this can happen. Can groundwater flow upward against gravity? In some cases, it can, as you will learn in this section. In most cases, however, groundwater emerges wherever the water table intersects Earth's surface. Such intersections commonly occur in areas that have sloping surface topography. The exact places where groundwater emerges depend on the arrangement of aquifers and aquicludes in an area.

OBJECTIVES

- **Relate** *the different types of springs to common systems of aquifers.*
- **Explain** *how groundwater is withdrawn from aquifer systems by wells.*
- **Describe** *the major problems that threaten groundwater supplies.*

VOCABULARY

spring
hot spring
geyser
well
drawdown
recharge
artesian well

SPRINGS

You have learned that aquifers are permeable underground layers through which groundwater moves with relative ease, while aquicludes are impermeable layers. Aquifers are commonly composed of layers of sand and gravel, sandstone, and limestone. Because of its many solution cavities, limestone is usually highly permeable and permits high flow velocities. Underground streams in cavernous limestone formations may transport groundwater at a rate of several kilometers per day. In contrast, aquicludes, such as layers of clay or shale, block groundwater movement. As a result, groundwater tends to discharge at Earth's surface where an aquifer and an aquiclude come in contact, as shown in ***Figure 10-11.*** These natural discharges of groundwater are called **springs.**

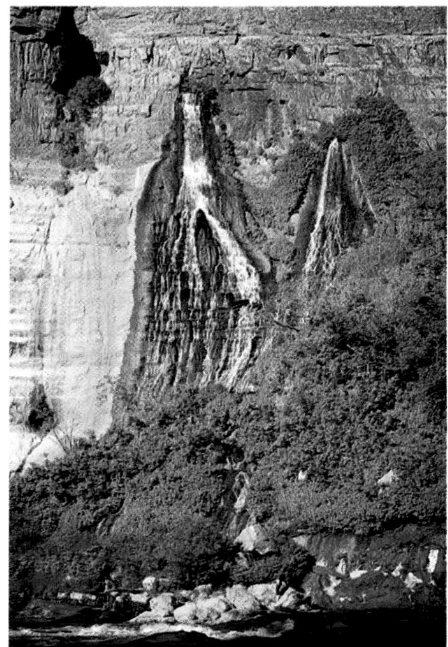

Figure 10-11 A spring occurs where an aquifer and an aquiclude come in contact at Earth's surface. At this point, the water flows out of the rock.

Earth Sciences

9.c Students know the importance of water to society, the origins of California's fresh water, and the relationship between supply and need.

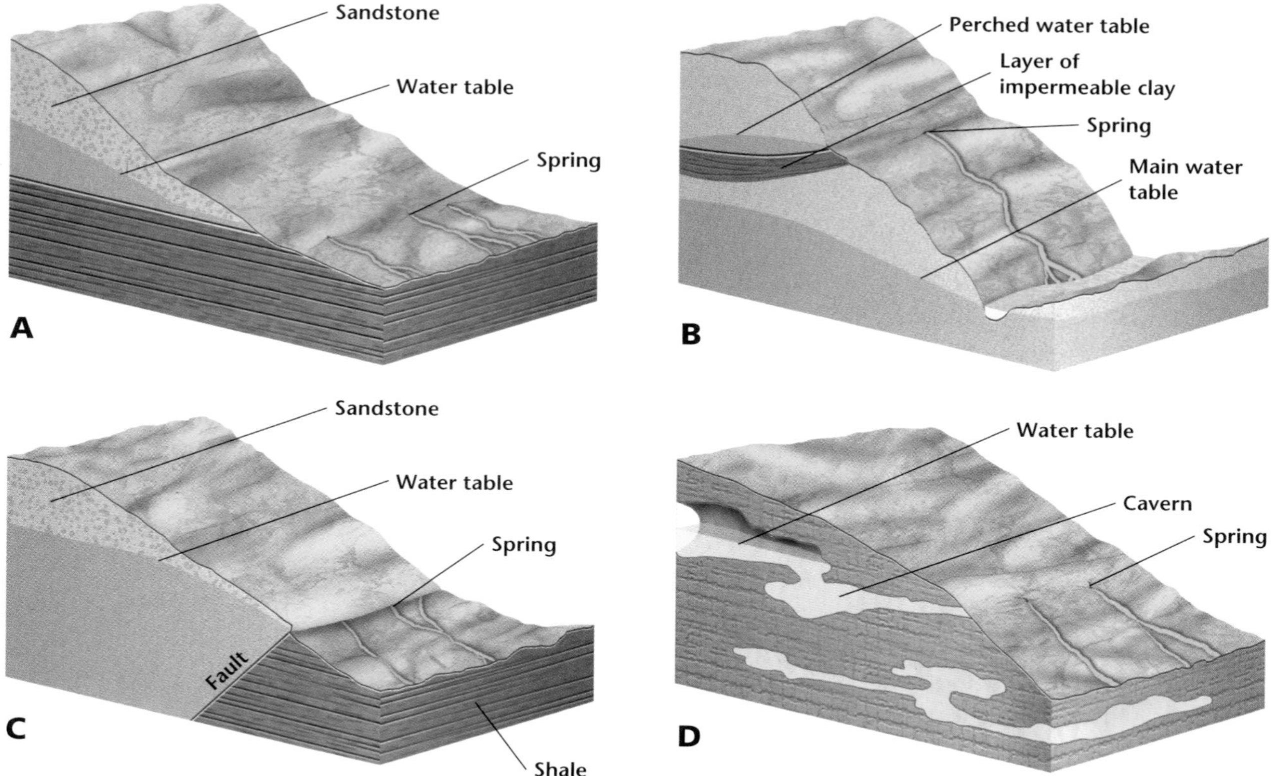

Figure 10-12 Springs occur in various places, such as at the sides of valleys and at the edges of perched water tables. The volume of water varies from the flow of a small spring to the rush of a large river.

Emergence of Springs The volume of water that is discharged by a spring may be a mere trickle, or, in karst regions, an entire river may emerge from the ground. Such a superspring is called a karst spring. Many of Florida's lakes are flooded sinkholes that are fed by karst springs whose discharge causes full-sized rivers to flow out of these lakes. In regions of near-horizontal sedimentary rocks, springs often emerge on the sides of valleys at about the same elevation, at the bases of aquifers, as shown in ***Figure 10-12A.*** Springs may also emerge at the edges of perched water tables. A perched water table, as shown in ***Figure 10-12B,*** is a zone of saturation that overlies an aquiclude that separates it from the main water table below. Other areas where springs tend to emerge are along faults, which are huge fractures that offset rock formations and sometimes block aquifers, as shown in ***Figure 10-12C.*** In limestone regions, springs discharge water from underground pathways, as shown in ***Figure 10-12D.***

Temperature of Springs Spring water is usually thought of as being cool and refreshing. Actually, the temperature of groundwater that is discharged through a spring is generally the average annual temperature of the region in which it is located. Thus, springs in New

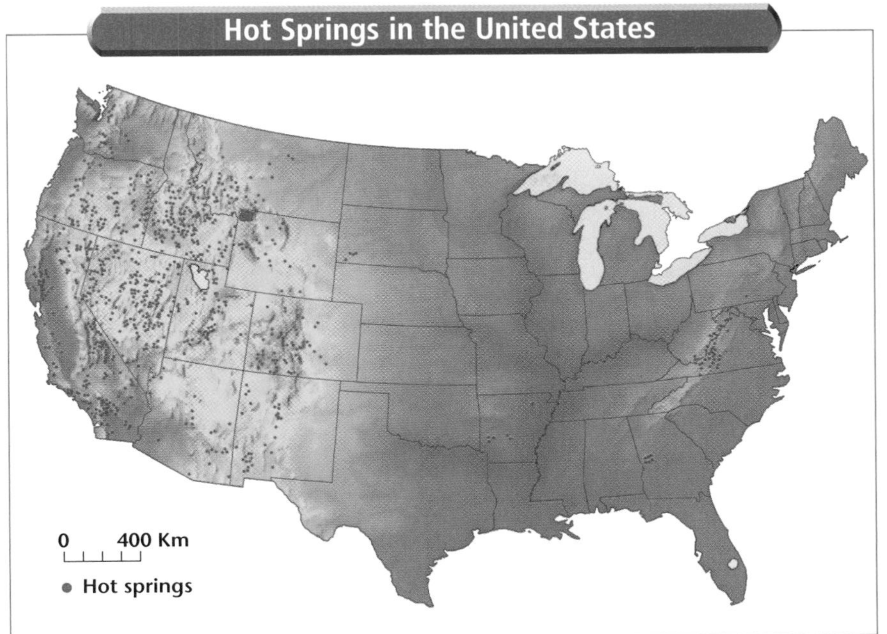

Figure 10-13 Hot springs occur in many areas of the United States.

England have year-round temperatures of about 10°C, while those in the Gulf states have temperatures of about 20°C.

Compared to air temperatures, groundwater is colder in the summer and warmer in the winter. However, in some regions of the United States, certain springs discharge water that is much warmer than the average annual temperature. These springs are called warm springs or hot springs, depending on their temperatures. **Hot springs** have temperatures higher than that of the human body. There are thousands of hot springs in the United States alone, as shown in ***Figure 10-13.*** Most of these are located in the western United States in areas where the subsurface is still quite hot from relatively recent igneous activity. A number of hot springs also occur in some eastern states. These eastern hot springs emerge from aquifers that descend to great depths in Earth's crust and allow deep, hot water to rise. The underground water is hot because temperatures in Earth's crust increase with depth by about 25°C for every kilometer. Among the most spectacular features produced by Earth's underground heat in volcanic regions are geysers, as shown in ***Figure 10-14.*** **Geysers** are explosive hot springs that erupt at regular intervals. One of the world's most famous geysers, Old Faithful, is located in Yellowstone National Park, Wyoming. Old Faithful erupts approximately every hour with a 40-m high column of boiling water and steam.

Figure 10-14 The Black Rock Desert geyser is a hot spring that erupts at regular intervals in Nevada.

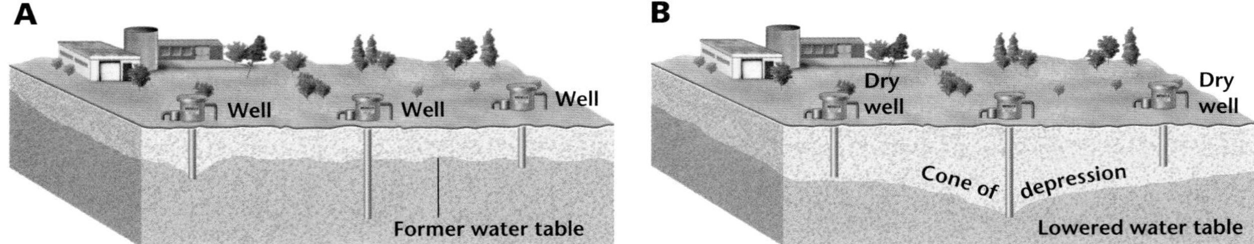

Figure 10-15 Wells must be drilled far enough below the water table so that they are not affected by seasonal water table fluctuations **(A)**. Overpumping of wells causes a lowering of the entire water table **(B)**.

WELLS

Wells are holes dug or drilled deep into the ground to reach a reservoir of groundwater. To produce water, a well must tap into an aquifer. The simplest wells are those that are dug or drilled below the water table, into the zone of saturation, and into what is called a water-table aquifer, as shown in ***Figure 10-15A.*** Initially, the water level in such a well is the same as the level of the water table. However, overpumping of the well lowers the water level in it and produces a cone of depression in the water table around the well, as shown in ***Figure 10-15B.*** The difference between the original water-table level and the water level in the pumped well is called the **drawdown.** If many wells withdraw water from a water-table aquifer, their cones of depression may overlap and cause an overall lowering of the water table, which can cause shallow wells to go dry. Water from precipitation and runoff is added back to the zone of saturation in the process of **recharge.** Groundwater recharge from precipitation and runoff sometimes replenishes the water withdrawn from wells. However, if recharge does not keep pace with groundwater withdrawal, the water table continues to drop until all wells in the area go dry.

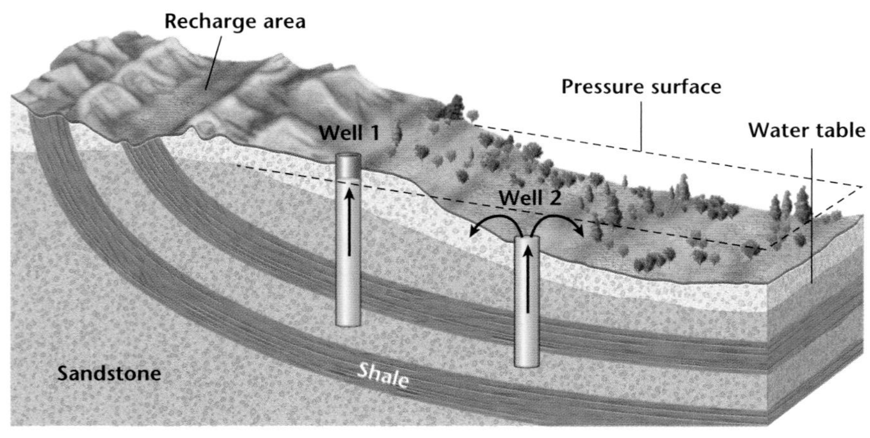

Figure 10-16 The level to which the water in an artesian well can rise is called the pressure surface.

Confined Aquifers

Water-table aquifers are unconfined and unprotected, and thus, they are easily polluted. Surface spills of pollutants often reach the water table and spread throughout aquifers. More reliable and less easily polluted water supplies can be found in deeper aquifers, called confined aquifers, which are generally sandwiched between aquicludes. The aquicludes form barriers that prevent pollutants from reaching such aquifers.

Artesian Wells Because the area of recharge is usually at a higher elevation than the rest of an aquifer, a confined aquifer contains water under pressure, as you can see by doing the *Problem-Solving Lab* on this page. The aquifer is called an artesian aquifer. When the rate of recharge is high enough, the pressurized water in a well drilled into a confined aquifer may spurt above the land surface in the form of a fountain called an **artesian well,** as shown in ***Figure 10-16.*** Similarly, a spring that discharges pressurized water is called an artesian spring.

Problem-Solving Lab

Using Tables and Making Graphs

Make inferences about the water levels of an artesian aquifer Artesian aquifers contain water under pressure. The table provides the following data about an artesian aquifer for three sites, spaced 100 m apart, along a survey line: elevations of the land surface, the water table, and the upper surface of the aquiclude on top of the artesian aquifer; and the artesian pressure surface, which is the level to which the artesian water can rise.

Site	Surface Elevation	Water Table Elevation	Aquiclude Elevation	Pressure Surface
1	396 m	392 m	388 m	394 m
2	394	390	386	393
3	390	388	381	392

Analysis

1. Plot the elevation data on a graph with the sites on the *x*-axis and the elevations on the *y*-axis. Make a cross section of the survey line from site 1 to site 3. Use a heavy line to indicate the land surface.
2. A well has been drilled at each site. The wells at sites 1 and 3 are 7 m deep. The well at site 2 was drilled into the artesian aquifer at a depth of 14 m. Sketch the wells at their proper depths on your cross section.

Thinking Critically

3. At what depth below the ground surface are the water levels in the three wells before they are pumped?
4. What would happen if a well was drilled into the confined aquifer at site 3?
5. At what sites could there be an artesian spring?

MiniLab

How does an artesian well work?

Model the changes that an artesian aquifer undergoes when a well is dug into it. What causes the water to rise above the ground surface?

Procedure ***CAUTION: Always wear safety goggles and an apron in the lab.***

1. Half-fill a plastic shoe box or other container with sand. Add enough water to saturate the sand. Cover the sand completely with a 1-or 2-cm layer of clay or a similar impermeable material.
2. Tilt the box at an angle of about 10°. Use a book for a prop.
3. Punch three holes through the clay, one each near the low end, the middle, and the high end of the box. Insert a clear straw through each hole into the sand below. Seal the holes around the straws.

Analyze and Conclude

1. Observe the water levels in the straws. Where is the water level the highest? The lowest?
2. Where is the water table in the box?
3. Where is the water under greatest pressure? Explain.
4. Predict what will happen to the water table and the surface if the water flows from one of the straws.

The name *artesian* is derived from the French province of Artois, where such wells were first drilled almost 900 years ago. To discover how an artesian well works, refer to the *MiniLab* on this page.

An important artesian aquifer in the United States is the Ogallala Aquifer, which is located in the Great Plains. This aquifer delivers water to a huge area stretching from South Dakota to Texas. The recharge areas of the Ogallala Aquifer are located in the Black Hills and the Rocky Mountains.

THREATS TO OUR WATER SUPPLY

Freshwater is Earth's most precious natural resource. Think about it! You can survive without gasoline, without electricity, and without most of the other materials that may seem to be essential, but you can't live without water. Human demands for freshwater are enormous. Not only is water used in households, but it is also used extensively in agriculture and industry, as shown in ***Figure 10-17.*** Groundwater supplies much of this water.

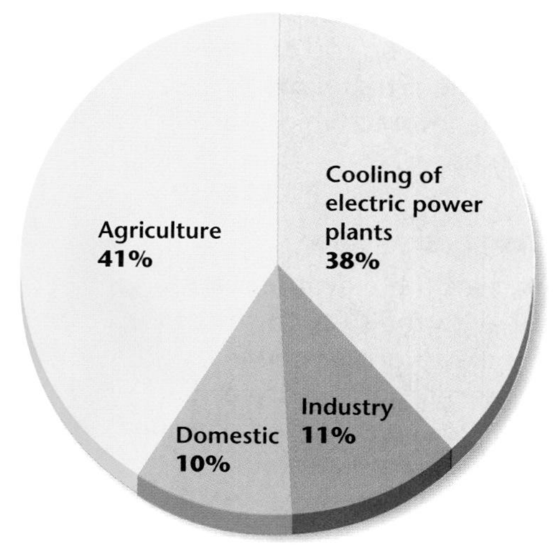

Figure 10-17 In the United States, the greatest amount of water usage is for agricultural activities, mostly for irrigation.

Overuse Groundwater supplies can be depleted. If groundwater is pumped out at a rate greater than the recharge rate, the groundwater supply will inevitably decrease, and the water table will drop. This is what is happening to the Ogallala Aquifer. Its water, used mostly for irrigation, is being withdrawn at a rate much higher than the recharge rate. You will learn more about the Ogallala Aquifer in the *Science & Environment* feature at the end of the chapter.

Subsidence Another problem caused by the excessive withdrawal of groundwater is ground subsidence, the sinking of land. The weight of the material overlying an aquifer is partly borne by water pressure. If that pressure is reduced, the weight of the overlying material is increasingly transferred to the aquifer's mineral grains, which then squeeze together more tightly. As a result, the land surface above the aquifer sinks.

Pollution in Groundwater In general, the most easily polluted groundwater reservoirs are water-table unconfined aquifers. Confined aquifers are affected less frequently by local pollution because they are protected by impermeable barriers. When the recharge areas of confined aquifers are polluted, however, those aquifers become contaminated as well.

The most common sources of groundwater pollution are sewage, industrial waste, landfills, and agricultural chemicals. These pollutants enter the ground above the water table, but they are eventually flushed downward by infiltrating precipitation and become mixed with the groundwater. Most sewage enters groundwater from sources such as faulty septic tanks. In highly permeable aquifers, all pollutants, including raw sewage, can spread quickly, as shown in ***Figure 10-18.***

Figure 10-18 Pollutants from sewage are traveling through a coarse-grained aquifer that will contaminate downslope wells.

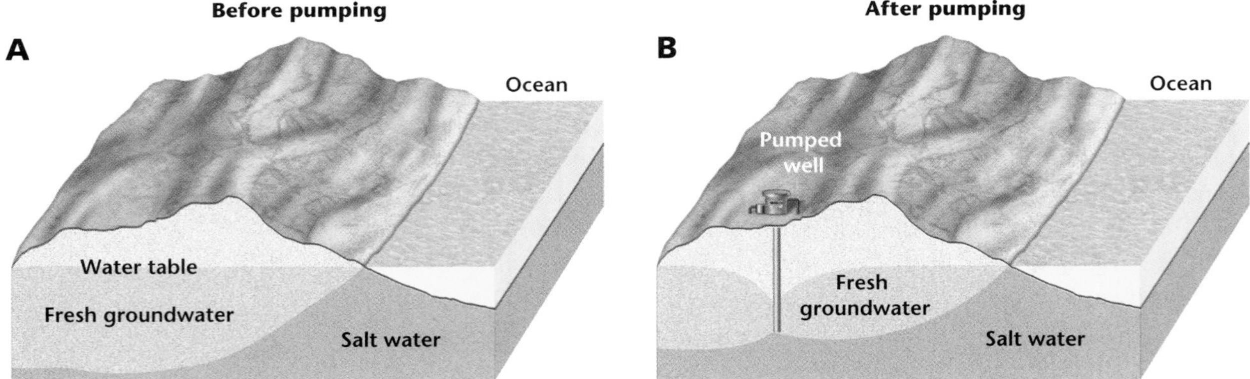

Figure 10-19 Saltwater incursion into groundwater can occur in coastal areas **(A).** Freshwater floats on top of denser seawater within the aquifer **(B).** Overpumping of wells draws the underlying salt-water into the wells and the freshwater zone.

Chemicals Chemicals dissolved or transported with groundwater are in the form of ions and molecules, so they cannot be filtered out in fine-grained sediments. For this reason, chemicals such as arsenic can contaminate any type of aquifer. They generally move downslope from a source in the form of a pollution plume, a mass of contaminants that spreads through the environment. Once chemical contaminants have entered groundwater, they cannot be easily removed.

Salt Not all pollutants are toxic or unhealthful in and of themselves. For example, ordinary table salt is widely used to season food. However, water is undrinkable when its salt content is too high. In fact, salt pollution is one of the major threats to groundwater supplies. In many coastal areas, the contamination of freshwater by salt water is the major problem. In such areas, the fresh groundwater near Earth's surface is underlain by denser, salty seawater as shown in ***Figure 10-19A.*** The overpumping of wells can cause the underlying salt water to rise into the wells and contaminate the freshwater aquifer, as shown in ***Figure 10-19B.***

Topic: Groundwater
To find out more about the ways to protect groundwater, visit the Earth Science Web Site at earthgeu.com

Activity: Write a TV ad describing measures being taken in your community to protect the groundwater.

Radon Another source of natural pollution is radioactive radon gas, which is one of the leading causes of cancer in the United States. This form of radon is generated by the radioactive decay of uranium in rocks and sediments, and it usually occurs in very low concentrations in all groundwater. However, some rocks, especially granite and shale, contain more uranium than others. The groundwater in areas where these rocks are present therefore contains more radioactive radon than normal. Some of this radon may seep into houses, and, because it is heavier than air, it can accumulate in poorly ventilated basements. The U. S. Environmental Protection Agency (EPA) advises homeowners in radon-prone regions to regularly have their homes tested for radon gas.

Table 10-2 Groundwater Pollution Sources
Accidental spills from vehicles
Leaks from storage tanks
Seepage from acid mine drainage
Seepage from faulty septic systems
Saltwater intrusion into aquifers near shorelines
Leaks from waste disposal sites

Protecting Our Water Supply

There are a number of ways in which groundwater resources can be protected and restored. First, all major pollution sources, which are listed in ***Table 10-2,*** need to be identified and eliminated. Pollution plumes that are already in the ground can be monitored with observation wells and other techniques. You will learn more about pollution plumes in the *Mapping GeoLab* at the end of this chapter. Most pollution plumes spread slowly. Thus, there is often time for alternate water supplies to be found. In some cases, pollution plumes can be stopped by the building of impermeable underground barriers. Polluted groundwater can be pumped out for chemical treatment on the surface.

While these measures can have limited success, they alone cannot save Earth's water supply. An important part of the solution is for humans to become more aware of how their activities impact the groundwater system.

Section Assessment

1. How are springs related to the water table?
2. What is the basic characteristic of an artesian well?
3. List four common sources of groundwater pollution.
4. Why are chemical contaminants a serious pollution problem in groundwater?
5. Artesian aquifers contain water under pressure. Explain why.
6. **Thinking Critically** What can you do to conserve and protect groundwater so that there will be safe and abundant water supplies in the future?

Skill Review

7. **Comparing and Contrasting** Compare and contrast different uses of water in the United States. For more help, refer to the *Skill Handbook.*

earthgeu.com/self_check_quiz

Mapping GeoLab

Mapping Pollution

***Y**ou can use a map to estimate the direction of groundwater flow and the movement of a pollution plume from its source, such as a leaking underground gasoline storage tank.*

I&E

1.h Read and interpret topographic and geologic maps.

Preparation

Problem

A major gasoline spill occurred at Jim's Gas Station near Riverside Acres, Florida. How can you determine the movement of the resulting pollution plume?

Materials

USGS topographic map of Forest City, Florida
transparent paper
ruler
graph paper
calculator

Procedure

1. Identify the lakes and swamps in the southeast corner of this map, and list their names or numbers and elevations in a data table. Note: The elevations are given or can be estimated from the contour lines.
2. Place the transparent paper over the southeast part of the map and trace the approximate outlines of these lakes or swamps, as well as the major roads. Enter lake or swamp elevations on your overlay, and indicate the location of Jim's gas station on Forest City Rd., about 1400 feet north of the Seminole County line (at the 96 foot elevation mark).
3. Add contour lines to your overlay using a contour interval of 10 feet.
4. Construct a cross section of the surface topography and the water table from Lake Lotus to Lake Lucien (through Jim's Gas Station).

Analyze

1. What is the slope of the water table at Jim's Gas Station?
2. What is the approximate direction of the water table slope at Jim's Gas Station?
3. In which direction will the pollution plume move?
4. Which settlements or houses are threatened by this pollution plume?

Conclude & Apply

1. How far below the surface is the water table in the highest area?
2. What is the relationship of the water table to the surface topography?

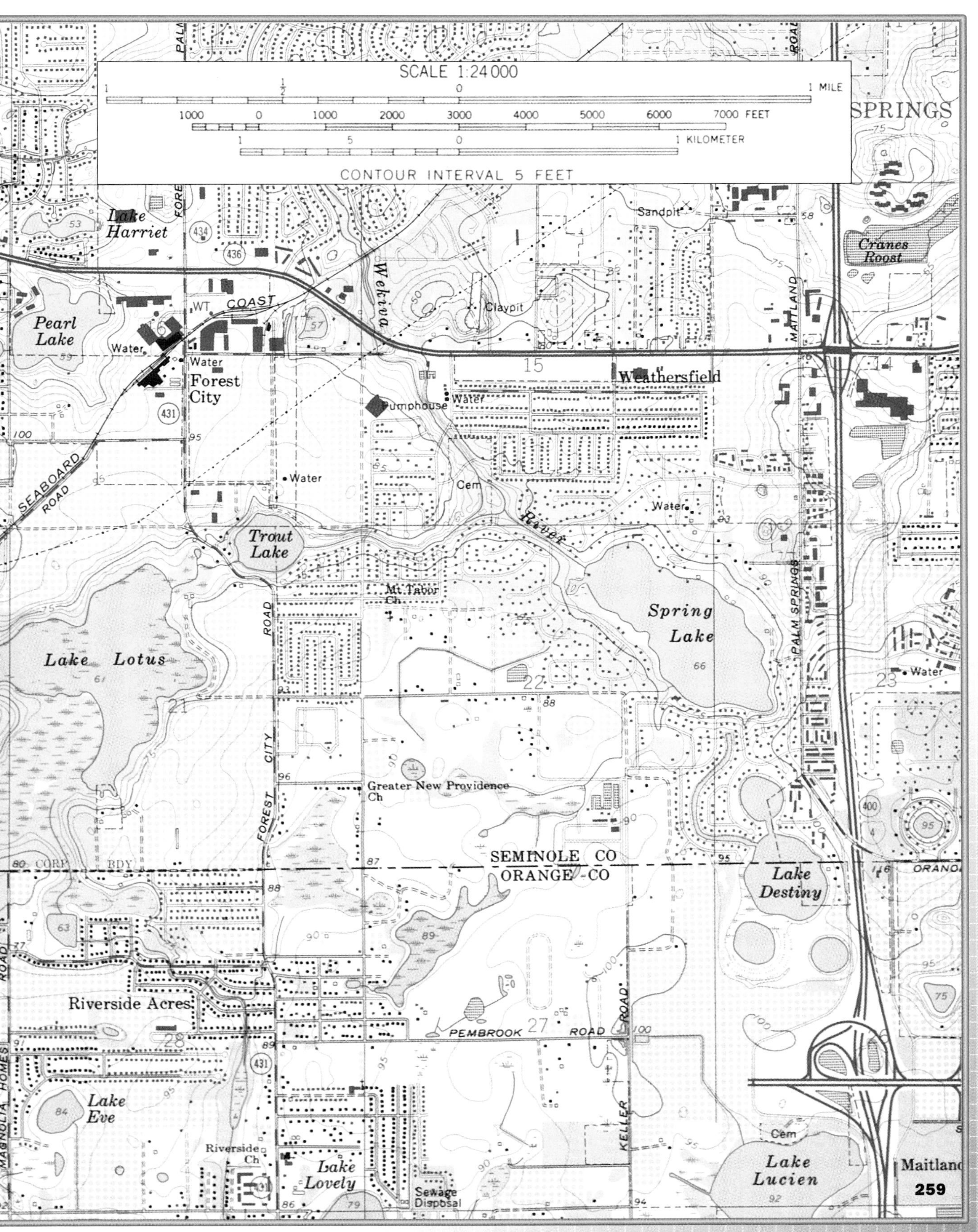
SCALE 1:24 000
1 MILE
1000 0 1000 2000 3000 4000 5000 6000 7000 FEET
1 KILOMETER
CONTOUR INTERVAL 5 FEET
SPRINGS
Lake Harriet
Cranes Roost
Sandpit
Pearl Lake
WT COAST
Claypit
Wekiva
Water
Forest City
Weathersfield
Pumphouse
SEABOARD ROAD
Cem
River
Trout Lake
Mt. Tabor Ch
Spring Lake
Lake Lotus
FOREST CITY ROAD
PALM SPRINGS
MAITLAND
Greater New Providence Ch
SEMINOLE CO
ORANGE CO
CORP BDY
Lake Destiny
Riverside Acres
PEMBROOK ROAD
KELLER ROAD
Lake Eve
MAGNOLIA HOMES ROAD
Riverside Ch
Lake Lovely
Sewage Disposal
Cem
Lake Lucien

Science & the Environment

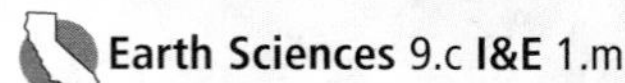

Earth Sciences 9.c **I&E** 1.m

The High Plains Aquifer

The High Plains Aquifer, also known as the Ogallala Aquifer, is located beneath parts of Colorado, Kansas, Nebraska, New Mexico, South Dakota, Texas, Oklahoma, and Wyoming. Environmentalists are concerned about the future of this source of fresh groundwater.

Long Term Effects of Water Use

Buried beneath 106 250 km^2 of sand and rock lies the largest groundwater system in the United States, the High Plains Aquifer, also known as the Ogallala Aquifer. The High Plains Aquifer, which contains the Ogallala Formation, is made of clay, gravel, sand, and silt. The result of deposition from an ancient snow melt in the Rocky Mountains region, the formation is divided into two sections. The upper section is known as the unsaturated zone. The lower section contains the Ogallala Aquifer, the water-bearing area. The Ogallala makes up approximately 80 percent of the High Plains Aquifer.

The aquifer's freshwater supply has been used for many years. Shortly after World War I, farmers began tapping into the groundwater reserves in Texas. By 1940, the aquifer had become a widely used source of irrigation in the Great Plains region as more and more farmers began to pump water from the aquifer.

By 1980, it was estimated that parts of the aquifer's water level were down as much as 30 meters. Because of concern about the decline in water levels, the U.S. Geological Survey began a monitoring program in 1988. The program revealed that from 1980 to 1994, water levels continued to decline in certain areas of the Texas High Plains, the Oklahoma Panhandle, and southwestern Kansas. In the eastern High Plains of Nebraska, higher than normal amounts of rainfall since 1980 have caused either a rise in the water level or a slower decline.

Demand placed upon the water supply over the years increased rapidly as more wells were drilled. In 1950, an estimated 8.6 trillion L of water were pumped from the aquifer. By 1980, the estimated amount had increased to 24 trillion L. The amount of water currently being pumped from the aquifer far exceeds the amount of water naturally replenished. Presently, 89 percent of the aquifer's water supply is still available despite the 170 000 wells drawing water. However, it has been predicted that by the year 2020, 25 percent of the water supply could be exhausted.

What Needs to be Done

While scientists cannot be sure that the aquifer will ever be completely depleted, there is reason to be concerned about the continuous lowering of water levels in the region. Wells in some areas overlying the aquifer have gone dry. Concerns about the depletion of the water supply have prompted several states to develop regulatory policies to protect this natural resource, which has often been taken for granted.

Activity

Research and construct a 3-dimensional model of the Ogallala Aquifer. Include the states that are part of the groundwater system. Label the aquifer's zone of saturation and water table.

CHAPTER 10

Study Guide

Summary

Section 10.1

Movement and Storage of Groundwater

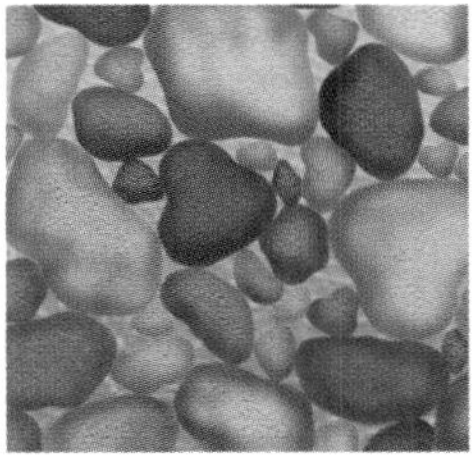

Main Ideas

- Some precipitation infiltrates the ground to become groundwater.
- Groundwater is stored below the water table in the pore spaces of rocks and moves through permeable layers called aquifers. Impermeable layers are called aquicludes.

Vocabulary

aquifer (p. 243)
infiltration (p. 240)
permeability (p. 242)
porosity (p. 241)
water table (p. 241)
zone of saturation (p. 241)

Section 10.2

Groundwater Erosion and Deposition

Main Ideas

- Groundwater dissolves limestone and forms underground caverns. Sinkholes form at Earth's surface when bedrock is dissolved or when caves collapse. Irregular topography caused by groundwater dissolution is called karst topography.
- The precipitation of dissolved calcium carbonate forms stalactites, stalagmites, and travertine deposits, including dripstone columns, in caves.

Vocabulary

cave (p. 245)
karst topography (p. 246)
sinkhole (p. 246)
stalactite (p.248)
stalagmite (p. 248)
travertine (p. 248)

Section 10.3

Groundwater Systems

Main Ideas

- The natural discharge of groundwater takes place through springs. Springs emerge where the water table intersects Earth's surface.
- Wells are drilled into the zone of saturation to provide water for human needs. The pumping of shallow wells produces cones of depression in the water table. Artesian wells tap deep, confined aquifers that contain water under pressure.
- In many regions, groundwater withdrawal exceeds groundwater recharge and causes considerable lowering of the water table as well as ground subsidence.
- The most common sources of groundwater pollution are sewage, industrial waste, landfills, and agricultural chemicals.

Vocabulary

artesian well (p. 253)
drawdown (p. 252)
geyser (p. 251)
hot spring (p. 251)
recharge (p. 252)
spring (p. 249)
well (p. 252)

earthgeu.com/vocabulary_puzzlemaker

CHAPTER 10

Assessment

Understanding Main Ideas

1. Where is most freshwater found on Earth?
   a. the oceans
   b. the atmosphere
   c. polar ice caps and glaciers
   d. lakes and rivers

2. What is a major source of freshwater in the United States?
   a. the Rocky Mountain snowpack
   b. the Mississippi River
   c. groundwater
   d. the Great Lakes

3. What happens to most of the precipitation that falls on land?
   a. It evaporates.
   b. It becomes runoff.
   c. It seeps into the ground.
   d. It becomes glacial ice.

4. What source usually replenishes groundwater?
   a. precipitation
   b. surface water
   c. underground streams
   d. municipal wastewater

5. Of the following materials, which is the most porous?
   a. a well-sorted sand
   b. a poorly sorted sand
   c. sandstone
   d. granite

6. Of the following materials, which is the most permeable?
   a. sandstone
   b. shale
   c. silt
   d. clay

7. What is the main characteristic of an aquifer?
   a. surface topography
   b. permeability
   c. subsidence
   d. dissolution

8. Which rock type is most easily dissolved by groundwater?
   a. sandstone
   b. granite
   c. limestone
   d. shale

9. What are the cone-shaped dripstone deposits that are found on the floor of caves?
   a. icicles
   b. rocksicles
   c. stalactites
   d. stalagmites

10. Which of the following are typical features of karst topography?
    a. moraines
    b. dunes
    c. sinkholes
    d. landslides

11. Where is groundwater closest to Earth's surface?
    a. in stream valleys
    b. on hilltops
    c. on mountaintops
    d. in arid regions

12. What does hard water usually contain?
    a. fluorine
    b. chloride
    c. carbonic acid
    d. calcium

13. What do artesian aquifers always contain?
    a. hot water
    b. water under pressure
    c. salt water
    d. steam

14. Which of the following is a common groundwater problem in coastal areas?
    a. saltwater contamination
    b. contamination by crude oil
    c. high sulfur content
    d. excessive recharge

Test-Taking Tip

BREATHE Oxygen helps to calm the anxiety associated with test-taking. When you start to feel your stomach lurch, take a deep breath and exhale slowly.

CHAPTER 10 Assessment

Applying Main Ideas

15. What is the difference between soft water and hard water?

16. What type of bedrock most likely exists in an area that has numerous sinkholes?

17. Subsurface temperatures increase with depth by about 25°C/km. What is the subsurface temperature 2 km below the surface in a region where the average annual surface temperature is 10°C?

18. A dripping water faucet in your home has produced brownish-red stains in the sink. What could have caused these stains?

19. Describe the processes involved in the formation of caves.

20. If the withdrawal of groundwater from an artesian aquifer exceeds the groundwater recharge, what consequence can be expected?

Thinking Critically

21. A well drilled into a water-table aquifer produces water only during springtime. Why?

Use the diagram below to answer question 22.

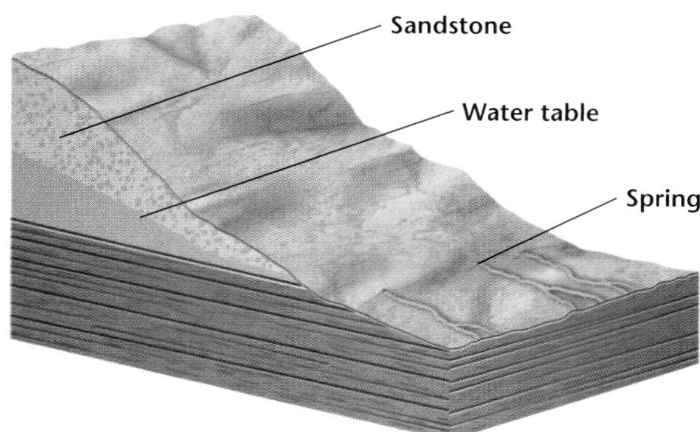

22. Make an inference on what can happen to the spring if the water table drops.

Standardized Test Practice

1. Which of the following materials would be best suited for lining a pond?
a. gravel
b. limestone
c. clay
d. sand

2. What are natural structures hanging from a cave's ceiling?
a. geyserites
b. travertines
c. stalagmites
d. stalactites

3. Which of the following usually describes the temperature of groundwater flowing through a natural spring?
a. hotter than the region's average temperature
b. cooler than the region's average temperature
c. the same temperature no matter where the spring is located
d. the same temperature as the region's average temperature

4. Which of the following water sources are the most easily polluted?
a. water-table aquifers
b. confined aquifers
c. artesian wells
d. hot springs

5. What is the composition of dripstone formations?
a. carbonic acid
b. carbon dioxide
c. iron oxide
d. calcium carbonate

UNIT 3

GeoDigest

For a **preview** of Earth's surface processes, study this GeoDigest before you read the chapters. After you have studied the chapters, you can use the GeoDigest to **review** the unit.

Surface Processes on Earth

Weathering, Erosion, and Soil

Chemical and mechanical weathering break down Earth materials. Chemical weathering causes a change in the composition of a rock. Agents of chemical weathering include hydrolysis, oxidation, acids from decaying organic matter, and acid precipitation. Each of these processes or substances combines with Earth materials, resulting in new combinations of minerals or in other substances. Mechanical weathering causes a change only in a rock's size and shape. Temperature and pressure are the major factors in mechanical weathering. Temperature changes can cause rocks to split. Pressure changes can cause rocks to crack or break apart.

Yosemite National Park, California

Erosion and Deposition Gravity is the driving force behind all agents of erosion, the process by which weathered pieces of rock are moved to new locations. Other agents of erosion include moving water, wind, and glaciers. Deposition occurs when the movement of transported materials slows down and they are dropped in a new location.

Formation of Soil Soils vary with climate and are classified as polar, temperate, desert, or tropical. A single centimeter of soil takes hundred of years to develop, but can erode away in just seconds. Soil is made of weathered rock and decayed organic matter called humus. Residual soil remains on top of its parent bedrock. Transported soil is moved away from its parent bedrock by weathering agents. A cross section of layers of soil is called a soil profile. The top layer, called horizon A, is topsoil. Horizons B and C are subsoil. Below horizon C is solid bedrock. Parent rock and environmental conditions determine a soil's composition. Soil texture is determined by the relative amounts of clay, sand, and silt the soil contains. Soil fertility is a soil's ability to grow crops. Farmers conserve soil through methods that include wind barriers.

Mass Movements, Wind, and Glaciers

The landscape is changed by mass movements, wind, and glaciation. Mass movement refers to the movement of Earth materials downslope as a result of gravity. Almost all of Earth's surface undergoes mass movements, which may be slow, as in creep, or rapid, as in landslides, mudflows, rock slides, rock falls, and avalanches. Mass movements are affected by the weight of the material involved, its level of saturation, its resistance to sliding, and sometimes, a trigger such as an earthquake. Mass movements can cause great damage and loss of life.

Wind Limited precipitation and scarce vegetation, conditions common to arid, semi-arid, and seashore environments, contribute to wind erosion. Wind-carried sediment causes abrasive action which wears down or polishes the sides of rocks that face the wind. Wind-formed Earth features include deflation blowouts, desert pavement, and sand dunes. Dunes are classified by shape as barchan, transverse, longitudinal, or parabolic. Wind-deposited soils called loess contain minerals and nutrients and are highly fertile.

Glaciers Large, moving masses of ice called glaciers form near Earth's poles and high in mountains, where cold temperatures keep fallen snow from completely melting. Over time, the weight of the snow exerts enough downward pressure to cause the accumulated snow to recrystallize into ice. Glacial features include U-shaped valleys, hanging valleys, and waterfalls in the mountains; moraines, drumlins, and kettles in outwash plains; and a variety of glacially formed lakes. Valley glaciers form in mountains and move downslope. Valley glaciers are much smaller than continental glaciers, which form over broad regions and spread out from their centers.

Tahquamenon River, Michigan

Sand overwhelms building, Namibia

Surface Water

Many landscape features on Earth are produced and changed by surface water. The amount of water in the ground depends on the number and sizes of pores in a particular Earth material and the amount of vegetation. A watershed or drainage basin is the land area drained by a stream system. Divides, which are raised areas of land, separate watersheds. All of the material carried by the stream, including material in solution, in suspension, and as bed load, is called the stream's load. Throughout history, humans built communities near water sources for survival and economic reasons. However, this practice has left humans vulnerable to dangerous floods. Weather and stream monitoring provides warnings of flooding.

Stream Development At a stream's source, or headwaters, water from precipitation begins its flow in channels confined by the stream's banks. Mountain streams have rapidly flowing water and often form waterfalls. When stream velocity

Earth Sciences 9.c

decreases, the stream's load is deposited in triangle-shaped alluvial fans or deltas. Alluvial fans form when streams flow out onto plains. Deltas form when streams enter large bodies of water. Uplifting of the land or lowering of the base level causes a stream to undergo rejuvenation and again begin to cut a V-shaped valley.

Lakes and Freshwater Wetlands Lakes form when depressions on land fill with water. Some lakes are human-made. When nutrients from fertilizers, detergents, or sewage enter a lake, eutrophication may be accelerated. The nutrients lead to an overabundance of some organisms and then a depletion of oxygen. Wetlands such as bogs, marshes, and swamps are low areas that regularly fill with water and support specific plants. Wetlands filter and clean water and are protected by law.

Groundwater

Groundwater is the largest source of freshwater available for human use. Groundwater is the portion of precipitation that infiltrates into the ground and is stored below the water table in the pore spaces of rocks. Groundwater moves through permeable layers called aquifers. Most groundwater contains carbonic acid which attacks carbonate rocks such as limestone which consists of calcium carbonate. The dissolution and precipitation of calcium carbonate plays a role in the formation of limestone caves. Caverns, sinkholes, karst topography, and travertine deposits are formed from groundwater action.

Groundwater Systems Springs, which are natural discharges of groundwater, emerge where the water table intersects Earth's surface. Wells drilled into the zone of saturation provide water for humans, but pumping of these wells may cause cones of depression in the water table. Artesian wells contain water under pressure from confined aquifers. When groundwater withdrawal exceeds groundwater recharge, it causes considerable lowering of the water table and ground subsidence. Pollution of aquifers may come from sewage, industrial waste, agricultural chemicals, and landfills.

Vital Statistics

Earth's Eight Longest Rivers

River	Length
Nile (Africa)	6650 km
Amazon (South America)	6400 km
Yangtze (Asia)	6300 km
Mississippi-Missouri (North America)	5971 km
Yenisey (Asia)	5540 km
Ob-Irtysh (Asia)	5410 km
Paraná (South America)	4880 km
Congo (Africa)	4700 km

FOCUS ON CAREERS

Landscaper

A landscaper makes a plan for outdoor scenery in an area and then follows the plan by planting the gardens and grounds. In addition to knowing about plants, a landscaper needs to know the soil characteristics and water drainage patterns in the area. Landscape architects have college degrees, but many others in the field start out with a high school degree and then gather the training they need on the job.

GeoDigest

ASSESSMENT

Understanding Main Ideas

1. What is the most powerful agent of erosion?
   a. water
   b. wind
   c. glaciers
   d. living things

2. What material makes up horizon B in a soil profile?
   a. topsoil
   b. loess
   c. subsoil
   d. bedrock

3. What is the underlying force of all agents of erosion?
   a. suspension
   b. friction
   c. magnetism
   d. gravity

4. Which of the following types of mass movement is not rapid?
   a. a landslide
   b. a mudflow
   c. an avalanche
   d. creep

5. What erosional agent causes deflation blowouts, desert pavement, and dunes?
   a. wind
   b. water
   c. earthquakes
   d. groundwater

6. Which of the following is a characteristic of valley glaciers?
   a. They form over broad regions.
   b. They move downslope.
   c. They are larger than continental glaciers.
   d. They completely melt each summer.

7. Which of the following occurs when a stream's velocity decreases?
   a. The stream load increases.
   b. The stream cuts a V-shaped valley.
   c. Deposition occurs.
   d. A drainage basin is formed.

8. Where do alluvial fans mostly occur?
   a. near lakes
   b. along the bases of mountains
   c. on the outside of meanders
   d. where streams enter the ocean

9. Of the following materials, which one is most permeable?
   a. mud
   b. clay
   c. silt
   d. gravel

10. What is the largest source of freshwater on Earth?
    a. lakes
    b. groundwater
    c. oceans
    d. glaciers

Thinking Critically

1. Describe the agents of erosion involved in chemical weathering.
2. Explain how groundwater works to form sedimentary rock.
3. How would you describe a soil's texture?
4. Discuss at least two factors that affect mass movement.
5. List at least two reasons why wetlands should be protected.

Himalayas

Unit 4

The Atmosphere and the Oceans

Off the Na Pali coast in Hawaii, clouds stretch toward the horizon. In this unit, you'll learn how the atmosphere and the oceans interact to produce clouds and crashing waves. You'll come away from your studies with a deeper understanding of the common characteristics shared by Earth's oceans and its atmosphere.

Unit Contents

Go to the **National Geographic Expedition** on page 880 to learn more about topics that are connected to this unit.

Kauai, Hawaii

Chapter 11

Atmosphere

What You'll Learn

- The composition, structure, and properties that make up Earth's atmosphere.
- How solar energy, which fuels weather and climate, is distributed throughout the atmosphere.
- How water continually moves between Earth's surface and the atmosphere in the water cycle.

Why It's Important

Understanding Earth's atmosphere and its interactions with solar energy is the key to understanding weather and climate, which control so many different aspects of our lives.

Earth Science Online

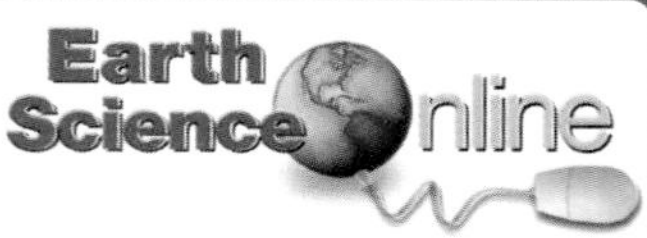

To find out more about the atmosphere, visit the Earth Science Web Site at earthgeu.com

Discovery Lab Dew Formation

Dew forms when moist air near the ground cools and the water vapor in the air changes into water droplets. In this activity, you will model the formation of dew.

1. Fill a glass about two-thirds full of water. Record the temperature of the room and the water.
2. Add ice cubes until the glass is full. Record the temperature of the water at 10-second intervals.
3. Observe the outside of the glass. Note the time and the temperature at which changes occurred on the outside of the glass.
4. Repeat the experiment outside. Record the temperature of the water and the air outside.

Observe In your science journal, describe what happened to the outside of the glass in step 3 and step 4. Relate your observations to the formation of dew. Graph the temperature of the water during both experiments. Did the results vary with location? Explain.

Atmospheric Basics

Earth Sciences 4.a, 4.b, 4.c, 5.a, 6.a, 8.a, 8.c

OBJECTIVES

- **Describe** *the composition of the atmosphere.*
- **Compare** *and* **contrast** *the various layers of the atmosphere.*
- **Identify** *three methods of transferring energy throughout the atmosphere.*

VOCABULARY

ozone
troposphere
stratosphere
mesosphere
thermosphere
exosphere
radiation
conduction
convection

Imagine living in the blazing heat of the Sahara desert, near the equator. Then imagine living in the frozen vastness above the arctic circle. Why are these places so different? The answer lies in how solar energy interacts with the atmosphere, and how the interactions combine to produce weather and climate.

ATMOSPHERIC COMPOSITION

The ancient Greeks thought that air was one of the fundamental elements that could not be broken down into anything else. Today, we know that air is a combination of many gases, each with its own unique characteristics. Together, these gases form Earth's atmosphere, which extends from Earth's surface to outer space.

About 99 percent of the atmosphere is composed of nitrogen and oxygen, with the remaining one percent consisting of small amounts of argon, hydrogen, carbon dioxide, water vapor, and other gases. The percentages of the main components, nitrogen and oxygen, are critical to life on Earth. If either were to change significantly, life as we know it could not exist. Among the lesser-percentage gases, however,

Figure 11-1 Nitrogen makes up 78 percent of the gases in Earth's atmosphere. Oxygen makes up 21 percent. The remaining one percent consists of small amounts of various other gases.

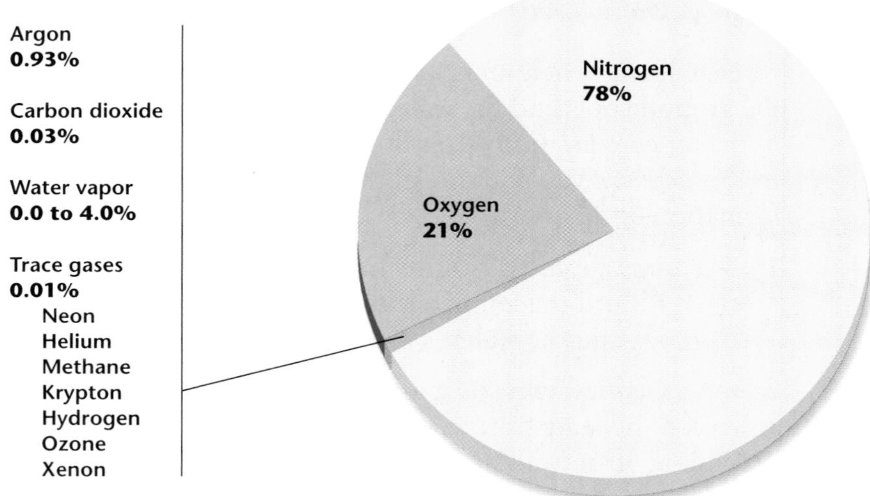

Earth Sciences

4.c Students know the different atmospheric gases that absorb the Earth's thermal radiation and the mechanism and significance of the greenhouse effect.

8.a Students know the thermal structure and chemical composition of the atmosphere.

8.c Students know the location of the ozone layer in the upper atmosphere, its role in absorbing ultraviolet radiation, and the way in which this layer varies both naturally and in response to human activities.

there is some variability, particularly in water vapor and carbon dioxide. ***Figure 11-1*** shows the composition of the atmosphere.

Key Atmospheric Gases The amount of water vapor in the atmosphere at any given time or place changes constantly. It can be as much as four percent of the atmosphere or as little as almost zero. The percentage varies with the seasons, with the altitude of a particular mass of air, and with the surface features beneath the air. Air over deserts, for instance, is drier than air over oceans. Carbon dioxide, another variable gas, makes up under one percent of the atmosphere. Why is it necessary to even mention these seemingly insignificant gases?

The level of both carbon dioxide and water vapor are critical because they play an important role in regulating the amount of energy the atmosphere absorbs. Water vapor, the gaseous form of water, is the source of clouds, rain, and snow. In addition, water is the only substance in the atmosphere that exists in three states: solid, liquid, and gas. This is important because when water changes from one state to another, heat is either absorbed or released, and this heat greatly affects the atmospheric motions that create weather and climate.

The atmosphere also contains solids in the form of tiny particles of dust and salt. Dust is carried into the atmosphere by wind. Salt is picked up from ocean spray. Dust and salt play a role in cloud formation, as you'll learn later. Ice is the third solid found in the atmosphere, usually in the form of hail and snow.

Ozone Another component of the atmosphere, **ozone** (O_3), is a gas formed by the addition of a third oxygen atom to an oxygen molecule (O_2). Ozone exists in small quantities mainly in a layer well above Earth's surface. It is important because it absorbs ultraviolet radiation from the Sun. If ozone did not control the amount of ultraviolet radiation reaching Earth's surface, our fragile skin could not tolerate exposure to the Sun for long. Evidence indicates that the ozone layer is thinning. You'll learn more about this issue in the *Science in the News* feature at the end of this chapter and in later chapters.

STRUCTURE OF THE ATMOSPHERE

The atmosphere is made up of several different layers, as shown in ***Figure 11-2.*** Each layer differs in composition and temperature.

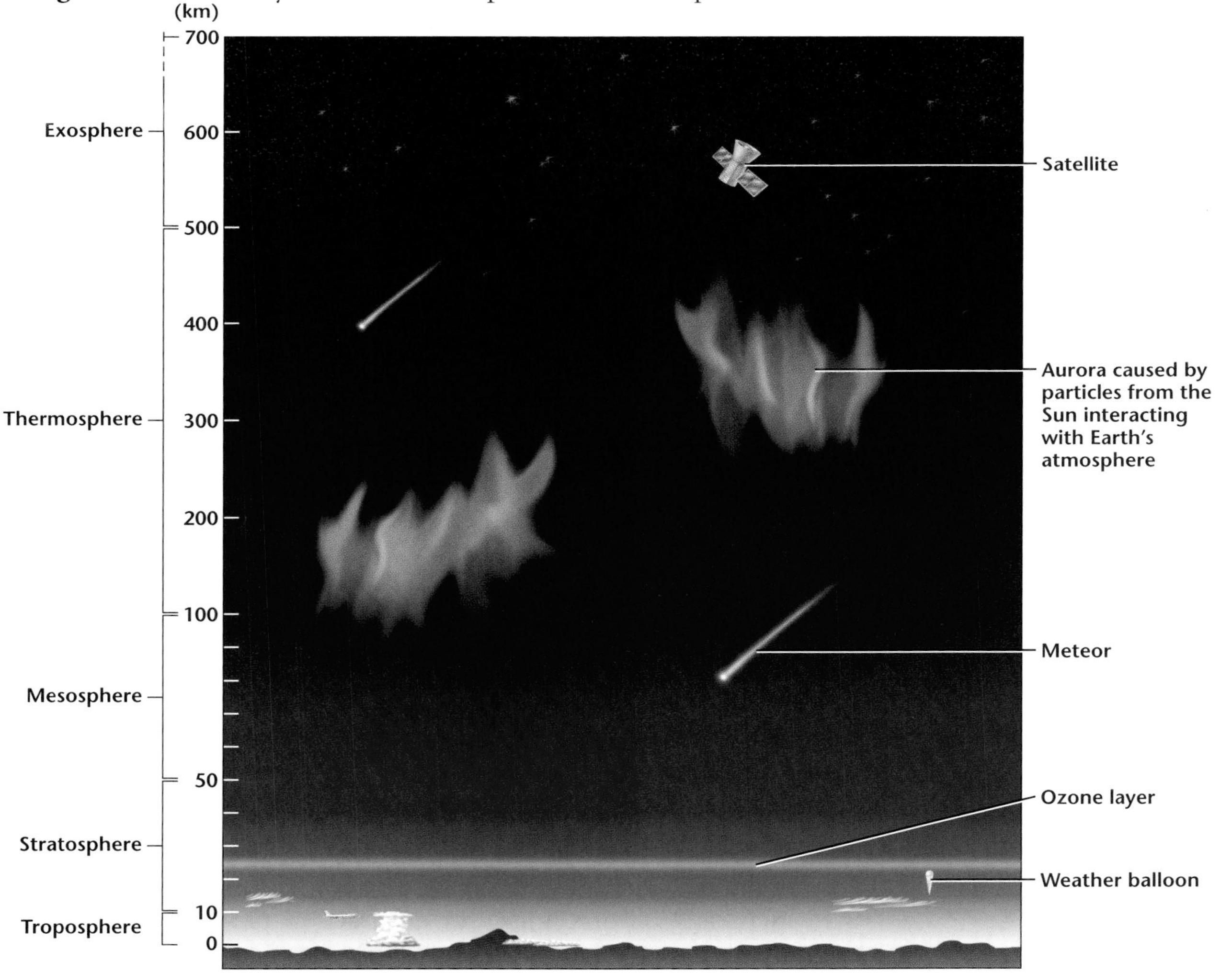

(km)
700
Exosphere 600
500
400
Thermosphere 300
200
100
Mesosphere
50
Stratosphere
10
Troposphere 0

Satellite
Aurora caused by particles from the Sun interacting with Earth's atmosphere
Meteor
Ozone layer
Weather balloon

Figure 11-2 The five main layers of the atmosphere vary in temperature and chemical composition.

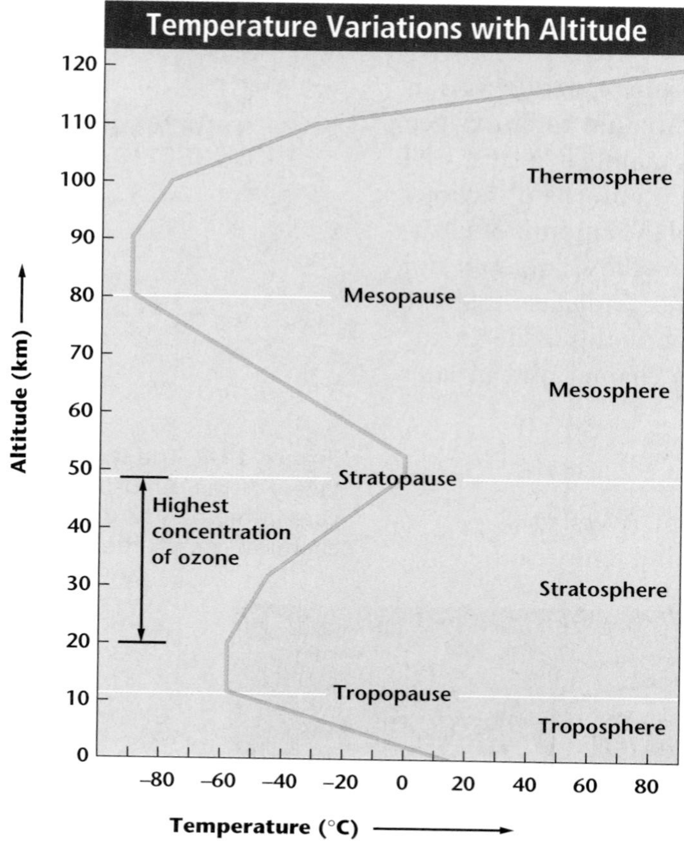

Figure 11-3 Differences in chemical composition cause air temperatures to vary throughout the atmosphere.

Lower Atmospheric Layers The layer closest to Earth's surface, the **troposphere,** contains most of the mass of the atmosphere, including water vapor. This is the layer in which most weather takes place and most air pollution collects. The troposphere is characterized by a general decrease in temperature from bottom to top. The upper limit of the troposphere, called the tropopause, varies in height. It's about 16 km above Earth's surface in the tropics and about 9 km or less at the poles. The tropopause is where the gradual decrease in temperature stops.

Above the tropopause is the **stratosphere,** a layer made up primarily of concentrated ozone. Ozone absorbs more ultraviolet radiation than does air in the troposphere. As a result, the stratosphere is heated, and air gradually increases in temperature to the top of the layer, called the stratopause, located about 50 km above Earth's surface.

Upper Atmospheric Layers Above the stratopause is the **mesosphere.** There is no concentrated ozone in the mesosphere, so the temperature decreases once again, as shown in ***Figure 11-3.*** The top of this layer, the mesopause, is the boundary between the mesosphere and the next layer, the thermosphere. The **thermosphere** contains only a minute portion of the atmosphere's mass. What air does exist in this layer increases in temperature once again, this time to more than 1000°C. In the thermosphere, however, the molecules that make up air are so sparse and widely spaced that, despite the high temperature, this layer would not seem warm to a human passing through it.

The ionosphere is part of the thermosphere. It is made up of electrically charged particles and layers of progressively lighter gases. The **exosphere** is the outermost layer of Earth's atmosphere. Light gases such as helium and hydrogen are found in this layer. Above the exosphere lies outer space. There is no clear boundary between the atmosphere and space. There are simply fewer and fewer molecules with increasing altitude until, for all practical purposes, you have entered outer space.

SOLAR FUNDAMENTALS

The Sun is the source of all energy in the atmosphere. This energy is transferred to Earth and throughout the atmosphere in three ways.

Radiation The Sun is shining on, and therefore warming, some portion of Earth's surface at all times. This method of energy transfer is called radiation. **Radiation** is the transfer of energy through space by visible light, ultraviolet radiation, and other forms of electromagnetic waves. All substances that have temperatures above absolute zero emit radiation. The higher the temperature of a substance, the shorter the wavelength it emits.

While Earth is absorbing solar radiation, it is also continuously sending energy back into space. As you can see from ***Figure 11-4,*** about 35 percent of incoming solar radiation is reflected into space by Earth's surface, the atmosphere, or clouds. Another 15 percent is absorbed by the atmosphere itself. This means that only about 50 percent of incoming solar radiation is absorbed directly or indirectly by Earth's surface. The rate of absorption for any particular area varies depending on the physical characteristics of the area and the amount of solar radiation it receives. Different areas absorb energy and heat up at different rates. For example, water heats up and cools down more slowly than land. And, as a general rule, darker objects absorb energy faster than lighter ones.

Earth Sciences

4.a Students know the relative amount of incoming solar energy compared with Earth's internal energy and the energy used by society.

4.b Students know the fate of incoming solar radiation in terms of reflection, absorption, and photosynthesis.

6.a Students know weather (in the short run) and climate (in the long run) involve the transfer of energy into and out of the atmosphere.

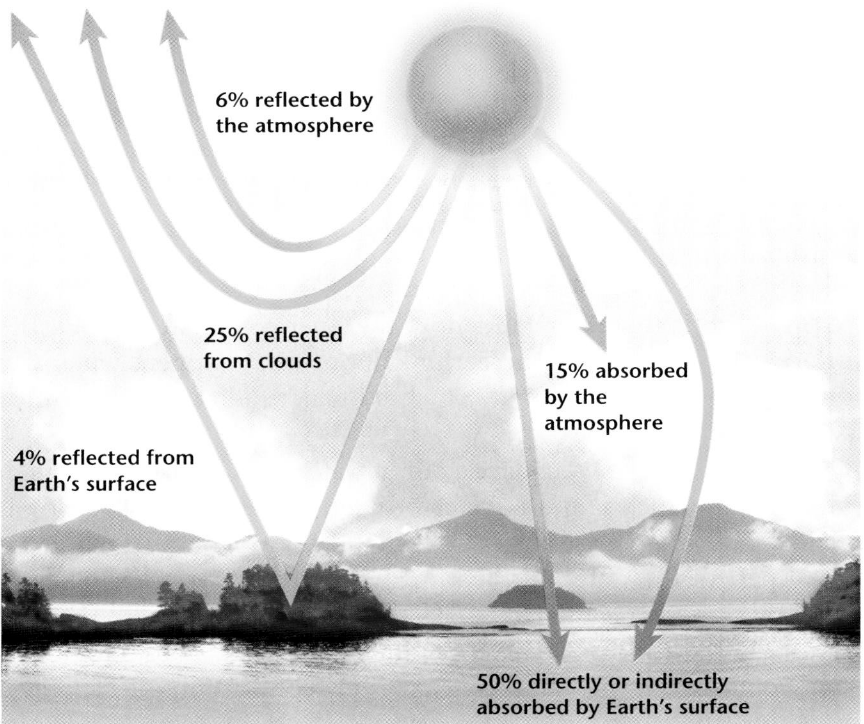

Figure 11-4 Over the course of a year, Earth sends back into space just about as much energy as it receives from the Sun. This is fortunate: if Earth sent back too much, it would gradually cool off, while if it sent back too little, it would warm up to potentially dangerous levels.

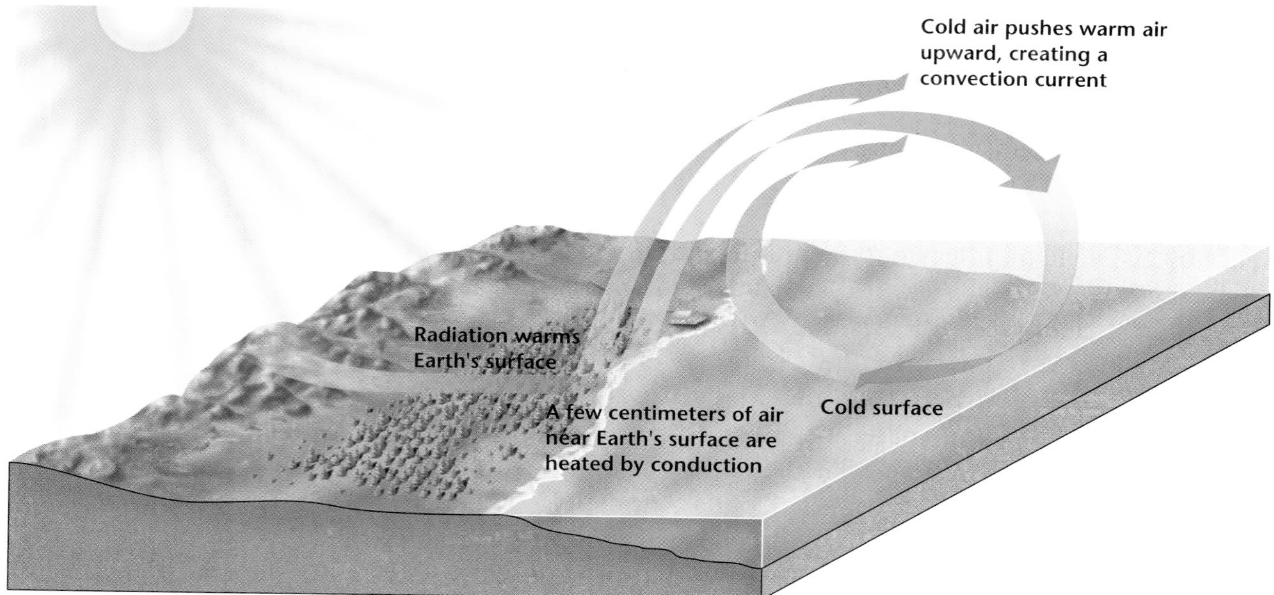

Figure 11-5 Energy is transferred throughout the atmosphere by the processes of conduction, convection, and radiation.

For the most part, solar radiation does not heat air directly. How, then, does air become warm? Most of the solar radiation that travels through the atmosphere does so at short wavelengths. The atmosphere does not easily absorb short wavelengths, so much of the solar radiation passes through the atmosphere and is absorbed by Earth's surface. The surface then radiates energy, but the radiation it gives off has a longer wavelength than the energy coming from the Sun. The energy radiated by Earth's surface does not pass back through the atmosphere. Rather, it is absorbed by the atmosphere and warms air through the processes of conduction and convection, which, along with radiation, make up the three methods of energy transfer illustrated in ***Figure 11-5.***

Earth Sciences

5.a Students know how differential heating of Earth results in circulation patterns in the atmosphere and oceans that globally distribute the heat.

Conduction To understand how the energy radiated by Earth's surface warms the atmosphere, think about what happens when you turn on a burner on the stove. The hot burner radiates energy much like Earth's surface does.

Now, imagine that you place a pot of water on the burner. Through **conduction,** which is the transfer of energy that occurs when molecules collide, energy is transferred from the bottom of the pot into the lowest part of the water. In the same way, energy is transferred from the particles of air near Earth's surface to the particles of air in the lowest layer of the atmosphere. For conduction to occur, substances must be in contact with one another. That's why conduction affects only a very thin atmospheric layer near Earth's surface.

Figure 11-6 Many different factors, including convection currents, cause the different types of weather shown here.

Convection Once the energy has made its way into the lower part of the atmosphere, can it ever move higher? Recall the pot of water. Energy has been transferred by conduction to the lowest layer of water molecules. This heated water expands, becomes less dense, and forms bubbles that rise. The rising bubbles bring the warm water to the top. The water at the top then cools, causing pockets of cool water to sink and become reheated when they come into contact with the bottom of the pot. This process is known as **convection,** the transfer of energy by the flow of a heated substance—in this case, the water. A similar process takes place in the atmosphere. Pockets of air near Earth's surface are heated, become less dense than the surrounding air, and rise. As the warm air rises, it expands and starts to cool. When it cools below the temperature of the surrounding air, it increases in density and sinks. As it sinks, it warms again and the process starts anew. Convection currents, as these movements of air are called, are among the main mechanisms responsible for the vertical motions of air, which in turn cause the different types of weather shown in ***Figure 11-6.***

Section Assessment

1. Describe the importance of water vapor in the atmosphere.
2. Why does temperature increase with height through the stratosphere?
3. Rank the main atmospheric gases in the troposphere in order from most abundant to least abundant. Do not include trace gases.
4. **Thinking Critically** Based on what you know about radiation and conduction, what conclusion might you make about summer temperatures in a large city compared with those in the surrounding countryside?

Skill Review

5. **Predicting** Of the three main processes of energy transfer throughout the atmosphere, which do you think plays the greatest role in warming the upper troposphere? Why? For more help, refer to the *Skill Handbook.*

State of the Atmosphere

Earth Sciences 5.a, 5.c, 8.a

OBJECTIVES

- **Describe** *the various properties of the atmosphere and how they interact.*
- **Explain** *why atmospheric properties change with changes in altitude.*

VOCABULARY

temperature
heat
dew point
condensation
lifted condensation level
temperature inversion
humidity
relative humidity

When people talk about the weather by saying that it's sunny or cloudy or cold, they're describing the current state of the atmosphere. Scientists describe the atmosphere, too, using words such as *temperature, air pressure, wind speed,* and *the amount of moisture in the air.* These are atmospheric properties that describe weather conditions. We'll examine each in turn, beginning with temperature.

TEMPERATURE VERSUS HEAT

Most of us tend to think of heat and temperature as being essentially the same thing. They are, in fact, two different concepts. **Temperature** is a measurement of how rapidly or slowly molecules move around. More molecules or faster-moving molecules in a given space generate a higher temperature. Fewer molecules or slower-moving molecules generate a lower temperature and cause a substance—air, for instance—to cool. **Heat,** on the other hand, is the transfer of energy that occurs because of a difference in temperature between substances. The direction of heat flow depends on temperature. Heat flows from an object of higher temperature to an object of lower temperature. How does this relate to the atmosphere? Heat is the transfer of energy that fuels atmospheric processes, while temperature is used to measure and interpret that energy.

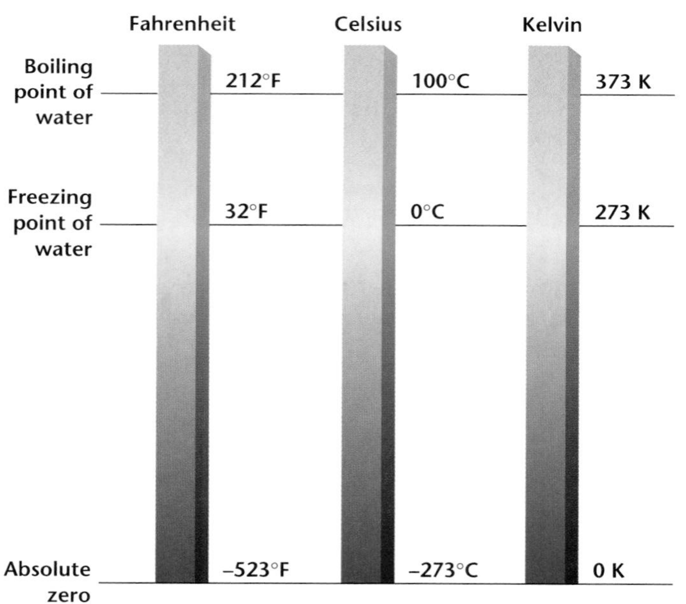

Figure 11-7 The Kelvin scale starts at 0 K, which corresponds to −273°C and −523°F.

Measuring Temperature

Temperature can be measured in degrees Fahrenheit (°F), in degrees Celsius (°C), or in kelvins (K). Fahrenheit is the scale most commonly used in the United States. Celsius, the scale used in this book, is convenient because the difference between its freezing and boiling points is exactly 100 degrees. The kelvin is the SI unit of temperature. The Kelvin scale measures the number of kelvins above absolute zero, which corresponds to approximately −273°C and −523°F. This scale is a more direct measure of molecular activity, because at absolute zero, molecular motion theoretically stops. Because nothing can be colder than absolute zero, there are no negative numbers on the Kelvin scale. ***Figure 11-7*** compares the different temperature scales.

Dew Point Another atmospheric measurement is the dew point. The **dew point** is the temperature to which air must be cooled at constant pressure to reach saturation. Saturation is the point at which the air holds as much water vapor as it possibly can. The dew point is important because until air is saturated, condensation cannot occur. **Condensation** occurs when matter changes state from a gas to a liquid. In this case, water vapor changes into liquid water and eventually falls as rain. Given its role in this process, the dew point is often called the condensation temperature.

VERTICAL TEMPERATURE CHANGES

The temperature on a mountaintop is cooler than at lower elevations because the temperature of the lower atmosphere decreases with increasing distance from its heat source—Earth's surface. Individual masses of air moving upward through the atmosphere experience a change in temperature, too. An air mass that does not exchange heat with its surroundings will cool off by about 10°C for every 1000-m increase in altitude. This is called the dry adiabatic lapse rate—the rate at which unsaturated air to which no heat is added or removed will cool. If the air is able to continue rising, eventually it will cool to its condensation temperature. The height at which condensation occurs is called the **lifted condensation level** (LCL). As shown in ***Figure 11-8,*** clouds form when water vapor condenses into water droplets, so the height of the LCL often corresponds to the base of clouds. Above the LCL, air becomes saturated and cools more slowly. The rate at which saturated air cools is called the moist adiabatic lapse rate. This rate ranges from about 4°C/1000 m in very warm air to almost 9°C/1000 m in very cold air.

8.a Students know the thermal structure and chemical composition of the atmosphere.

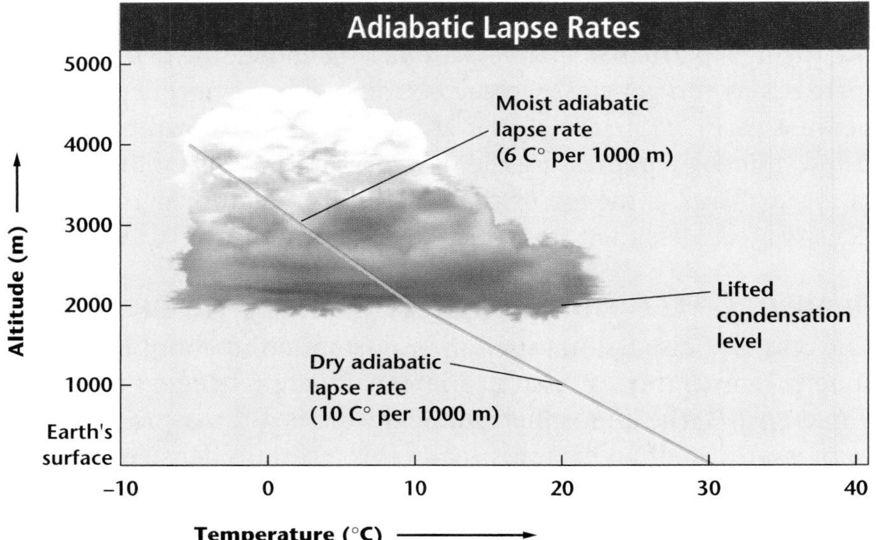

Figure 11-8 Condensation occurs at the lifted condensation level (LCL). Air above the LCL is saturated and thus cools more slowly than air below the LCL.

Table 11-1 Density Changes With Altitude

Altitude km	Density g/L	Altitude km	Density g/L
0	1.23	30	0.018
2	1.01	40	0.004
4	0.82	50	0.001
6	0.66	60	0.0003
8	0.53	70	0.00009
10	0.41	80	0.00002
15	0.19	90	0.000003
20	0.09	100	0.0000005

Air Pressure and Density

Just like water in the ocean, air has mass and constantly exerts pressure on our bodies. Why? The gravitational attraction between Earth and atmospheric gases causes particles of gas to be pulled toward the center of Earth. You don't notice this pressure because you have spent your whole life under it and are accustomed to it. A fish living deep in the ocean exists under pressure that would crush our bodies, but the fish survives because its body is adapted to such pressure. Just as water pressure increases with depth in the ocean, pressure increases as you near the bottom of the atmosphere because of the greater mass of the atmosphere above you. Conversely, atmospheric pressure decreases with height because there are fewer and fewer gas particles exerting pressure.

The density of air is proportional to the number of particles of air occupying a particular space. As ***Table 11-1*** shows, the density of air increases as you get closer to the bottom of the atmosphere. This is because gases at the top of the atmosphere press down on the air below, thereby compressing the particles and increasing the density of the air. Thus, at the top of a mountain, temperature, pressure, and density are all less than they are at lower elevations.

Table 11-2 Atmospheric Relationships

As T ↑, P ↑
As T ↓, P ↓
As T ↓, D ↑
As T ↑, D ↓

T = Temperature
P = Pressure
D = Density
↑ = Increases
↓ = Decreases

Pressure-Temperature-Density Relationship

The previous discussion raises an important point about the atmosphere: temperature, pressure, and density are related, as shown in ***Table 11-2.*** In the atmosphere, temperature is directly proportional to pressure. So, if an air mass maintains a certain density—that is, the number of gas particles in a fixed volume remains the same—as temperature increases or decreases, pressure does, too. By the same token, as pressure increases or decreases, temperature does, too. You will further explore this relationship in the *GeoLab* at the end of this chapter.

The relationship between temperature and density, on the other hand, is inversely proportional. So, if an air mass maintains a certain pressure, as temperature increases, density decreases, and as temperature decreases, density increases. This is why air rises when its temperature increases—it becomes less dense.

In most atmospheric interactions, however, neither density nor pressure remains unchanged, and this muddles the relationship among temperature, pressure, and density. Earlier, for example, we noted that both temperature and density decrease with increasing altitude in the troposphere. If density decreases with height, how can temperature decrease as well if it is inversely proportional to density? The answer lies in the fact that temperature varies with changes in both pressure and density. In this case, temperature is proportional to the ratio of pressure to density, which decreases with increasing altitude.

Earth Sciences

5.c Students know the origin and effects of temperature inversions.

Temperature Inversions In the atmosphere, the relationship between temperature and pressure is not always fixed. Although temperature and pressure in the overall troposphere decrease with height, there is an exception to this rule known as a temperature inversion. A **temperature inversion** is an increase in temperature with height in an atmospheric layer. It's called a temperature inversion because the temperature-altitude relationship is inverted, or turned upside down. This can happen in several ways. We'll consider one that involves the rapid cooling of land on a cold, clear, winter night when the wind is calm. Under these circumstances, the lower layers of the atmosphere are not receiving heat from Earth's surface—they're losing heat. As a result, the lower layers of air become cooler than the air above them, so that temperature increases with height and forms a temperature inversion. In some cities, such as the one shown in ***Figure 11-9,*** a temperature inversion can worsen air-pollution problems by acting like a lid to trap pollution under the inversion layer. In all cases, the presence or absence of inversions can have a profound effect on weather conditions, as you'll learn in the next chapter.

Figure 11-9 A temperature inversion in Long Beach, California, traps air pollution above the city.

Figure 11-10 When wind blows over these sand dunes in Namibia, it encounters more friction than when it blows over water.

Earth Sciences

5.a Students know how differential heating of Earth results in circulation patterns in the atmosphere and oceans that globally distribute the heat.

WIND

You may have entered a large, air-conditioned building on a hot summer day. As you opened the door, a sudden rush of cool air greeted you. This happened because the air conditioner created an imbalance between the warm, less-dense air outside the building and the cool, more-dense air inside. The cool air, being denser, had settled toward the bottom of the building. When the door opened, the cool, dense air rushed out to try to relieve the imbalance. The rush of air that you experienced is commonly known as wind.

In essence, the atmosphere works much like an air-conditioned building. Cool air, being more dense, sinks and forces warm, less-dense air upward. In the lower atmosphere, air generally moves from areas of high density to areas of low density. The air moves in response to density imbalances created by the unequal heating and cooling of Earth's surface. These imbalances, in turn, create areas of high and low pressure. In its simplest form, wind can be thought of as air moving from an area of high pressure to an area of low pressure. Wind is usually measured in miles per hour or kilometers per hour. Ships at sea usually measure wind in knots. One knot is equal to 1.85 km/h.

Like temperature and pressure, wind changes with height in the atmosphere. Why? Near Earth's surface, wind is constantly disrupted by the friction that results from its contact with trees, buildings, and hills—even the surface of water affects air motion. Farther up from Earth's surface, air encounters less friction, and wind speeds increase. Look at ***Figure 11-10.*** Would you expect the wind to blow more strongly over the ocean or across the dunes?

RELATIVE HUMIDITY

Just for fun, reach out and grab a handful of air. You may not know it, but you also grabbed some water vapor. Air in the lower portion of the atmosphere always contains at least some water vapor, even though that amount may be very small. The amount of water vapor in air is referred to as **humidity.**

Imagine now that you take your handful of air—and its water vapor—into a room full of dry air and let it go. Would that roomful of air have the same humidity as your handful? No, because the water vapor in that handful would be very small relative to how much water vapor that roomful of air could actually hold. The ratio of water vapor in a volume of air relative to how much water vapor that volume of air is capable of holding is called **relative humidity.** As the graph in the *Problem-Solving Lab* shows, relative humidity varies with temperature. Warm air is capable of holding more moisture than cool air. Thus, if the temperature of a room increased, the air in

Using Numbers At 20°C, a cubic meter of air can hold a total of 17 g of water vapor. What is the air's relative humidity if it holds only 6 g of water vapor?

Problem-Solving Lab

Interpreting Graphs

Determine relative humidity
Relative humidity is the ratio of water vapor in a given volume of air compared with how much water vapor that volume of air can actually hold. Use the graph at the right to answer the following questions.

Analysis

1. How much water vapor can a cubic meter of air hold at 25°C?
2. How much water vapor can the same volume of air hold at 15°C?

Thinking Critically

3. Why do the values in questions 1 and 2 differ?
4. If the relative humidity of the air in question 1 was 50 percent, how much water vapor would it hold?
5. If you wanted to decrease the relative humidity of a room, would you increase or decrease its temperature? Explain your answer.

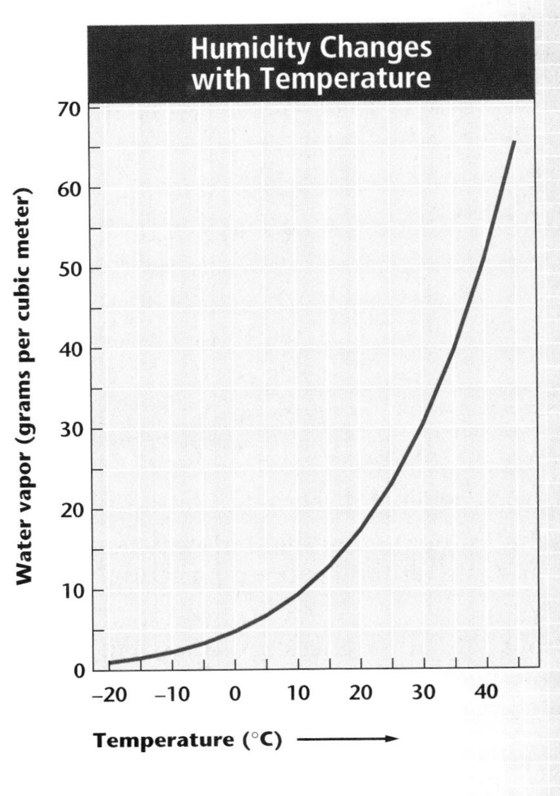

Figure 11-11 Clouds form when a mass of rising air becomes saturated and condenses its water vapor into large groups of water droplets.

the room would be capable of holding more moisture. If no additional water vapor was added to the air, its relative humidity would decrease. Conversely, if more water vapor was added to the air, its relative humidity would increase. Do the *Problem-Solving Lab* on the previous page to learn more about relative humidity.

Relative humidity is expressed as a percentage. If a certain volume of air is holding as much water vapor as it possibly can, then its relative humidity is 100 percent. If that same volume of air is holding half as much water vapor as it can, its relative humidity is 50 percent, and so on. Recall that air is saturated when it holds as much water vapor as it possibly can. As you'll see next, this has important implications for the development of precipitation and clouds such as those shown in ***Figure 11-11.***

SECTION ASSESSMENT

1. How is dew point related to saturation?
2. What is the relationship between temperature and altitude in a temperature inversion?
3. How does atmospheric pressure change with height in the atmosphere? Why does it change?
4. Compare and contrast humidity and relative humidity.
5. **Thinking Critically** Which would melt more ice—a pot of hot water or a tub of warm water? Explain your answer.

SKILL REVIEW

6. **Designing an Experiment** Design an experiment that shows how average wind speeds change over different types of surfaces. For more help, refer to the *Skill Handbook.*

SECTION 11.3

Moisture in the Atmosphere

Earth Sciences 7.c, 8.a, 8.c **I&E** 1.d, 1.m

Would you like to be able to predict the weather? To do so, you'll probably need to learn more about clouds. Certain types of clouds are associated with certain types of weather. Before learning about cloud types, however, you need to understand how clouds form.

OBJECTIVES

- **Explain** *how clouds are formed.*
- **Identify** *the basic characteristics of different cloud groups.*
- **Describe** *the water cycle.*

VOCABULARY

condensation nuclei
orographic lifting
stability
latent heat
coalescence
precipitation
water cycle
evaporation

CLOUD FORMATION

You know that air generally contains some amount of water vapor and that warm, less-dense air rises, while cool, more-dense air sinks. This tendency to rise or sink as a result of differences in density is called buoyancy. As you can see in ***Figure 11-12,*** clouds form when warm, moist air rises, expands, and cools in a convection current. As the air reaches its dew point, the water vapor in the air condenses around condensation nuclei. **Condensation nuclei** are small particles in the atmosphere around which cloud droplets can form. They come from a variety of sources, including sea salt and dust. When millions of these droplets collect, a cloud forms.

Clouds can also form when wind encounters a mountain and the air has no place to go but up. The effect is the same as with any rising air—it expands and cools. This method of cloud formation, shown in ***Figure 11-13A*** on the next page, is called **orographic lifting.** Another method of cloud formation involves the collision of air masses of different temperatures, as shown in ***Figure 11-13B*** on the next page. Recall that cold, more-dense air is heavier than warm, less-dense air, so it tends to collect near Earth's surface. As warmer air moves into the area, some of it will warm up the cold air, but the bulk of it will be forced to rise over the more-dense, cold air. As the warm air cools, the water vapor in it condenses and forms a cloud.

 **Earth Sciences**

8.a Students know the thermal structure and chemical composition of the atmosphere.

Figure 11-12 Clouds form when warm air is forced up in a convection current.

Stability Regardless of how a cloud forms, all rising air expands and cools. How rapidly any given mass of air cools determines its stability. **Stability** is the ability of an air mass to resist rising. Imagine an air mass that is warmer than the surface beneath it. Heat flows from the warmer air to the colder surface. The lower layer of the air mass thus loses heat and cools. The cooling air resists rising—it is stable. The rate at which an air mass cools depends in part on the temperature of the surface beneath the air. The temperature of surrounding air masses and the temperature of the air mass itself also play a role in determining the cooling rate.

Air can become unstable if it is cooler than the surface beneath it. In this case, heat flows from the warmer surface to the cooler air. The air warms and becomes less dense than the surrounding air. The less-dense air mass rises. If temperature conditions are right and the air mass rises rapidly, it can produce the type of clouds associated with thunderstorms.

Latent Heat As water vapor in the air condenses, heat is released. Where does this heat come from? It takes energy to change liquid water into a gaseous state. The energy that is transferred to the gas doesn't just go away; it is stored in the water vapor and will not be released into the air until condensation occurs. The stored energy is called **latent heat.** Until condensation occurs, latent heat is not available to warm the atmosphere.

When condensation takes place, latent heat is released and warms the air. At any given time, the amount of water vapor present in the atmosphere is a significant source of energy because of the latent heat it contains. When condensation occurs, this latent heat can provide energy to a weather system, thereby increasing its intensity.

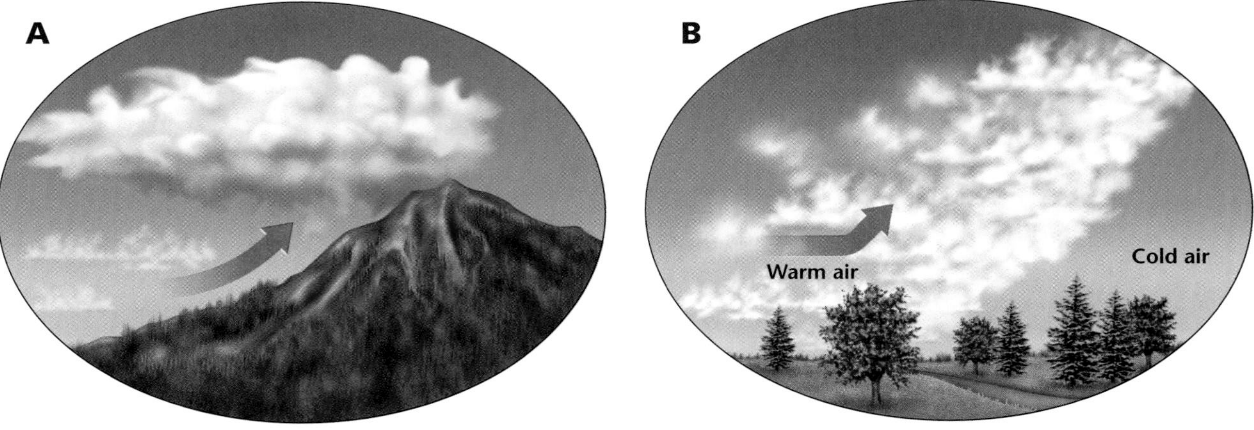

Figure 11-13 Clouds form when warm moist air is forced to rise over a mountain **(A)** and when two air masses of different temperatures meet **(B)**.

TYPES OF CLOUDS

When a mass of rising air reaches its lifted condensation level or LCL, water vapor condenses into droplets of liquid water or ice, depending on the temperature. If the density of these droplets is great enough, they become visible in the form of a cloud. While this is the basic principle behind the formation of all clouds, this process can take place at many different altitudes—sometimes even in contact with Earth's surface, in which case it is known as fog. In addition to forming at different heights, clouds form in different shapes, depending on the factors involved in their formation.

Clouds are generally classified according to a system originally developed by English naturalist Luke Howard in 1803. As shown in ***Table 11-3,*** the modern system groups clouds by the altitude at which they form and by their shape. Low clouds typically form below 2000 m. Middle clouds form mainly between 2000 m to 6000 m. High clouds composed of ice crystals form above 6000 m. The final group of clouds includes those that spread throughout all altitudes—at the same time, no less. These are vertical development clouds.

Topic: Clouds
To find out more about clouds, visit the Earth Science Web Site at earthgeu.com

Activity: Make a poster or media presentation showing the types of clouds you observed during a one-week period.

Table 11-3 Cloud Classification

Height		Shape	
Prefix		Prefix	
Cirro	Describes high clouds with bases starting above 6000 m.	**Cirrus**	Latin meaning: "hair." Describes wispy, stringy clouds.
Alto	Describes middle clouds with bases between 2000 m to 6000 m.	**Cumulus**	Latin meaning: "pile or heap." Describes puffy, lumpy-looking clouds.
Strato	Refers to low clouds below 2000 m.	**Stratus**	Latin meaning: "layer." Describes featureless sheets of clouds.
		Nimbus	Latin meaning: "cloud." Describes low, gray rain clouds.

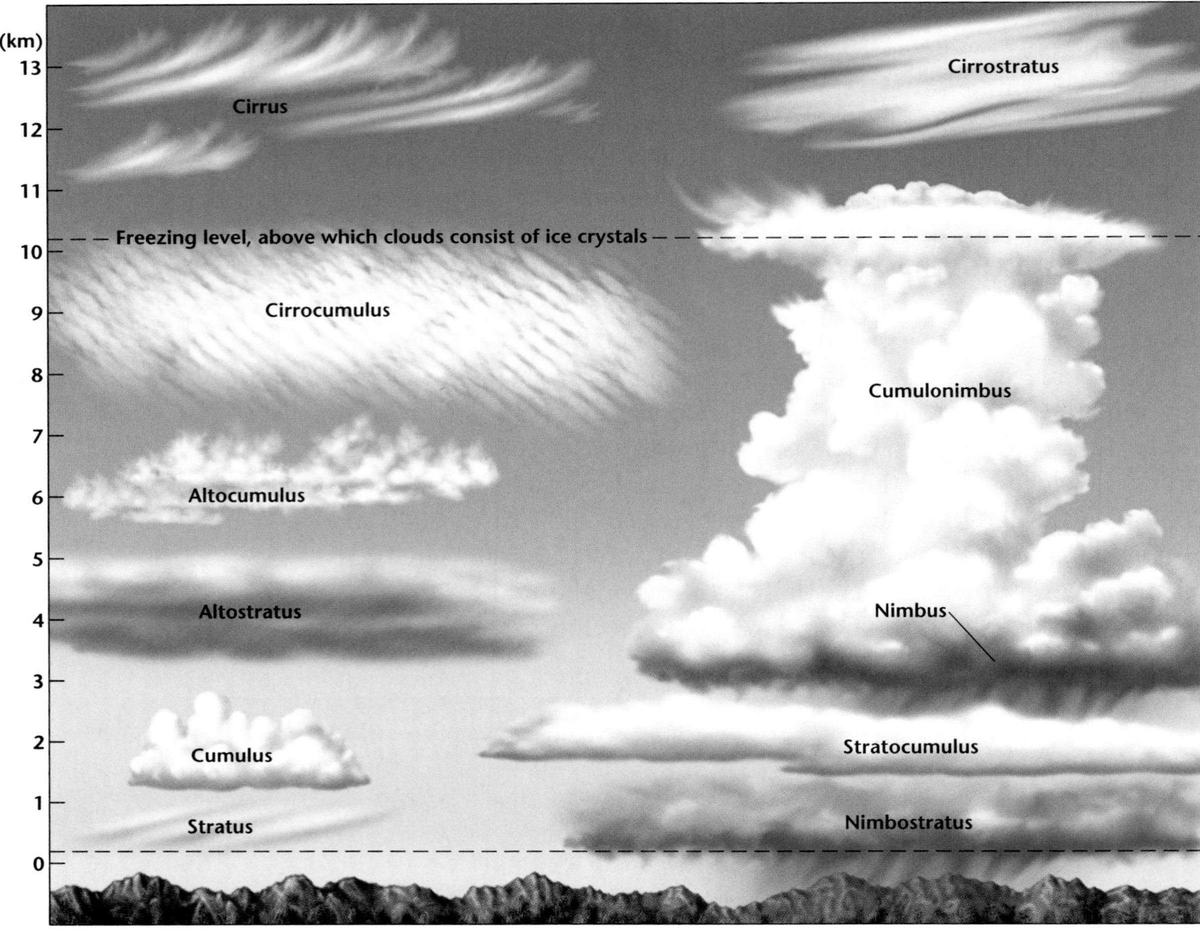

Figure 11-14 Clouds form at different heights and in different shapes. ***Compare and contrast cirrus and stratus clouds.***

Low Clouds Imagine a warm, summer afternoon. The Sun is beating down, heating Earth's surface. In areas where the heating is particularly intense, such as fields with dark soil, conduction causes air above the surface to become warmer than the surrounding air. As the temperature rises, the air expands. Its density becomes lower than that of surrounding air and it begins to rise and cool by further expansion. When it reaches its LCL, it becomes saturated, and the water vapor it contains condenses into water droplets. These droplets eventually become numerous enough to form a visible cloud. If the air stays warmer than the surrounding air, the cloud will continue to grow. If the air does not stay warmer than the surrounding air, the cloud will flatten out and winds will spread it horizontally into stratocumulus or layered cumulus clouds. Another type of low cloud that forms at heights below 2000 m is a stratus, a layered cloud that covers much or all of the sky in a given area. Stratus clouds often form when fog lifts away from Earth's surface. ***Figure 11-14*** shows these and other types of clouds.

Middle Clouds Altocumulus and altostratus clouds, which form at heights between 2000 m and 6000 m, can be either all liquid or a mixture of liquid and ice crystals. This is due to the cooler temperatures generally present at the heights at which these clouds form. Middle clouds are usually layered. Altocumulus clouds often resemble white fish scales. Altostratus clouds are dark but thin veils of clouds that sometimes produce mild precipitation.

High Clouds Because they form above heights of 6000 m, where temperatures are below freezing, high clouds are made up of ice crystals. Thus, some, such as cirrus clouds, often have a wispy, indistinct appearance. Another type of cirriform cloud, called a cirrostratus, forms as a continuous layer that sometimes covers the sky. Cirrostratus clouds can vary in thickness from being almost transparent to being dense enough to block out the Sun or Moon.

Clouds of Vertical Development If the air that makes up a cumulus cloud is unstable enough, the cloud will be warmer than the surface or surrounding air and will continue to grow. As it rises, water vapor condenses, and the air receives additional warmth from the release of latent heat. The cloud can grow through middle altitudes as a towering cumulus; if conditions are right, it can reach nearly 18 000 m. Its top is then composed of ice crystals. Strong winds can spread it into a familiar anvil shape. A puffy, white cumulus cloud can thus develop into a full-fledged cumulonimbus, as shown in ***Figure 11-15.*** What began as a small mass of moist air is now an atmospheric giant, capable of producing the torrential rains and strong winds that are characteristic of thunderstorms.

Figure 11-15 Cumulonimbus clouds, such as this one, in Arizona, are associated with thunderstorms.

PRECIPITATION

When cloud droplets collide, they join together to form a larger droplet in a process called **coalescence.** As the process continues, the droplet eventually becomes too heavy to be held aloft. At this point, gravity takes over and the droplet falls to Earth as precipitation. **Precipitation** includes all forms of water, both liquid and solid, that fall from clouds. Rain, snow, sleet, and hail are the four main types of precipitation. Coalescence is the primary process responsible for the formation of precipitation from warm clouds. Precipitation from cold clouds generally involves the interaction of ice and

MiniLab

What affects the formation of clouds and precipitation?

Model the water cycle.

Procedure

1. Pour about 125 mL of warm water into a clear, plastic bowl.
2. Loosely cover the top of the bowl with plastic wrap. Overlap the edges by about 5 cm.
3. Fill a self-sealing plastic bag with ice cubes, seal it, and place it in the center of the plastic wrap on top of the bowl. Push the bag of ice down so that the plastic wrap sags in the center, but doesn't touch the surface of the water.
4. Use tape to seal the plastic wrap around the bowl.
5. Observe the surface of the plastic wrap directly under the ice cubes every 10 minutes for one-half hour, or until the ice melts.

Analyze and Conclude

1. What formed on the underside of the wrap? Infer why this happened.
2. Relate your observations to processes in the atmosphere.
3. Predict what would happen if you repeated this activity with hotter water.

Figure 11-16 Note the distinctive layers in the cross-section of the hailstone. ***Infer how the layers formed.***

water molecules in the clouds. Do the *MiniLab* on this page to model the formation of clouds and precipitation.

Why are there so many variations in precipitation? When precipitation forms at cold temperatures, it takes the form of ice crystals or snow. Sometimes, convective currents carry the droplets up and down through freezing and nonfreezing air, thereby forming ice pellets or sleet. If that up-and-down motion is especially strong and takes place over large stretches of the atmosphere, it can form very large ice pellets known as hail. ***Figure 11-16*** shows a sample of hail.

THE WATER CYCLE

The total amount of water on Earth is constant, and probably has been for millions of years. More than 97 percent of Earth's water is salt water found in oceans. Only three percent is freshwater, and two-thirds of this is frozen in ice caps at the poles. At any one time, only a small percentage of water is present in the atmosphere. Still, this water is vitally important because as it continually moves between the atmosphere and Earth's surface, it nourishes living things. The constant movement of water between the atmosphere and Earth's surface is known as the **water cycle.**

The water cycle, shown in ***Figure 11-17,*** receives its energy from the Sun. Radiation from the Sun causes liquid water to change into a gas. The process of water changing from a liquid to a gas is called **evaporation.** This is the first step in the water cycle. Water evaporates from lakes, streams, and oceans

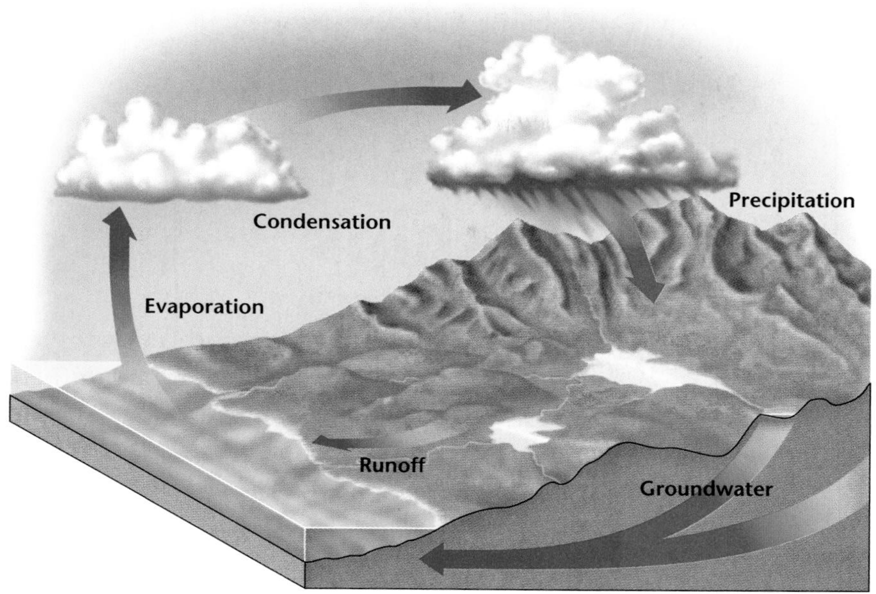

Figure 11-17 Water moves from Earth to the atmosphere and back to Earth in the water cycle.

 **I&E**

1.d Formulate explanations by using logic and evidence.

and rises into Earth's atmosphere. As water vapor rises, it cools and changes back into a liquid. This process, as you have learned, is called condensation, the second step of the water cycle. When water vapor condenses, it forms clouds.

In the third step of the water cycle, water droplets combine to form larger drops that fall to Earth as precipitation. This water soaks into the ground and enters lakes, streams, and oceans, or it falls directly into these bodies of water and eventually evaporates, and the water cycle continues.

SECTION ASSESSMENT

1. Explain why a cumulonimbus cloud is not considered to be a low, middle, or high cloud.
2. Describe the process that causes a water droplet to fall to Earth as precipitation.
3. What determines whether precipitation will fall as rain or snow?
4. **Thinking Critically** Based on what you have learned about latent heat, explain why the lapse rate of moist air is less than that of dry air.

SKILL REVIEW

5. **Concept Mapping** Use the following terms to construct a concept map that describes the processes of the water cycle. For more help, refer to the *Skill Handbook.*

water cycle | evaporation | condensation | precipitation | water changes from liquid to gas | water changes from gas to liquid | water falls as rain, snow, sleet, or hail

earthgeu.com/self_check_quiz

Interpreting Pressure-Temperature Relationships

As you go up a mountain, both temperature and air pressure decrease. These effects are easily explained. Temperature decreases as you get farther away from the atmosphere's heat source, Earth's surface. Pressure decreases as you ascend the mountain because there are fewer and fewer particles of air above you. Pressure and temperature, however, are also related through the expansion and compression of air, regardless of height. In this activity, you will demonstrate that relationship.

Preparation

Problem

Demonstrate the relationship between temperature and pressure.

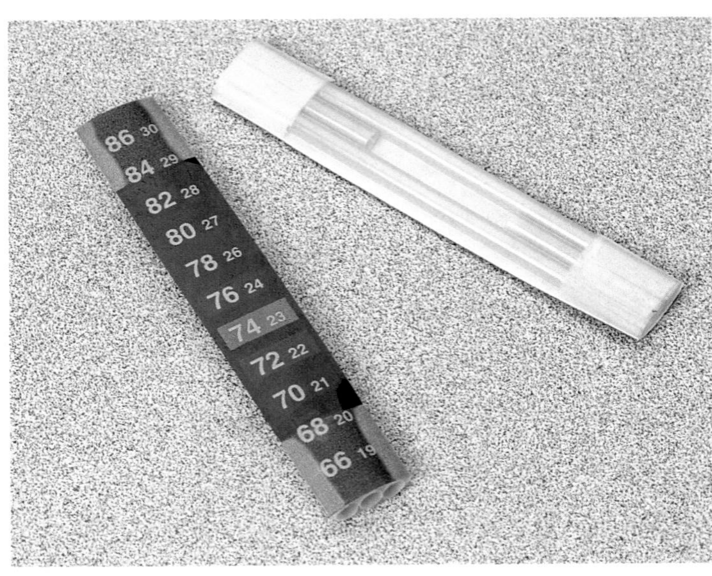

Objectives

In this GeoLab, you will:

- **Model** the temperature and pressure changes that take place as a result of the expansion and compression of air.
- **Relate** the changes to processes in the atmosphere.

Materials

clean, clear, plastic 2-L bottle with cap
plastic straws
scissors
thin, liquid-crystal temperature strip
tape
watch or timer

Safety Precautions

Always wear safety goggles and an apron in the lab.

Procedure

1. Cut two pieces of straw, each the length of the temperature strip. Then cut two 2-cm pieces of straw. Lay the two long pieces on a table. Place the two shorter pieces within the space created by the longer pieces so that the four pieces form a supportive structure for the temperature strip, as shown in the photograph on the previous page.
2. Tape the four pieces of straw together. Place the temperature strip lengthwise upon the straws. Tape the strip to the straws.
3. Slide the temperature strip-straw assembly into the clean, dry bottle. Screw the cap on tightly.
4. Place the sealed bottle on the table so that the temperature strip faces you and is easy to read. Do not handle the bottle any more than is necessary so that the temperature inside will not be affected by the warmth of your hands.
5. Record the temperature of the air inside the bottle as indicated by the temperature strip.
6. Next, position the bottle so that about half its length extends beyond the edge of the table. Placing one hand on each end of the bottle, push down on both ends so that the bottle bends in the middle. Hold the bottle this way for two minutes. During this time, your partner should record the temperature every 15 seconds.
7. Release the pressure on the bottle. Observe and record the temperature every 15 seconds for the next two minutes.

Analyze

1. What was the average temperature of the air inside the bottle as you applied pressure to the bottle? How did this differ from the average temperature of the bottled air when you released the pressure on the bottle?
2. Make a graph of the temperature changes you recorded throughout the experiment.
3. Explain how these temperature changes are related to changes in pressure.

Conclude & Apply

1. Predict how the experiment would change if you took the cap off the bottle.
2. Given your observations and what you know about the behavior of warm air, would you expect the air over an equatorial desert at midday to be characterized by high or low pressure?

Science in the News

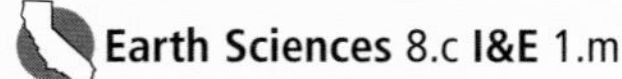

Earth Sciences 8.c I&E 1.m

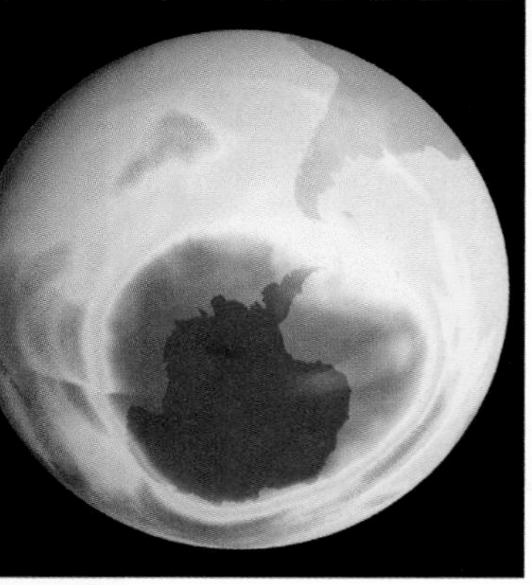

The Montreal Protocol and the Ozone Layer

In 2002, scientists reported that the amount of ozone-destroying chemicals in the atmosphere, particularly chlorine, was decreasing. This decrease is largely due to the efforts of the countries that have signed the Montreal Protocol, an international treaty which set restrictions on the global production and use of chlorofluorocarbons (CFCs) and other ozone-destroying chemicals that can reach the atmosphere.

The ozone layer, located in the stratosphere, absorbs up to 99 percent of incoming ultraviolet radiation. Overexposure to ultraviolet radiation can lead to the development of skin cancer and cataracts.

Chemical Reactions

Once they enter the atmosphere, CFCs, chemicals that were used in refrigerators and air conditioners, and halons, used in fire extinguishers, are broken down by ultraviolet light. The products of these breakdowns include highly reactive chlorine and bromine. The chlorine and bromine atoms destroy ozone molecules during chemical reactions. A single chlorine or bromine atom can destroy hundreds of ozone molecules before it reacts with another gas, thus ending the destruction. Because of all this, a small amount of reactive chlorine or bromine can have a large impact on the ozone layer.

International Effort

After the discovery of the "hole" in the ozone layer, an international effort began to stop the destruction of the ozone layer. At its inception in 1987, 24 countries signed the Montreal Protocol. The Montreal Protocol called for a phase-out in the production and use of most ozone-destroying chemicals by developed countries by the year 2005. Developing countries around the world are working to achieve the same goal by 2015. By 2003, 186 countries had signed and were abiding by the restrictions of the Montreal Protocol.

Future Outlooks

Recent studies indicate that actions taken as a result of the Montreal Protocol restrictions have an effect on levels of chlorine in the atmosphere, which are decreasing each year. Scientists warn that even complete compliance with the Montreal Protocol will still leave the ozone layer vulnerable for the next decade. However, based on current trends in data, a return to pre-1980 ozone amounts over Antarctica is expected by the middle of this century.

Activity

Research the latest information about the Montreal Protocol. What amendments were added in 1999? What has occurred since 1999? Write a short report explaining the latest goals of and changes to the Montreal Protocol.

CHAPTER 11
Study Guide

Summary

SECTION 11.1

Atmospheric Basics

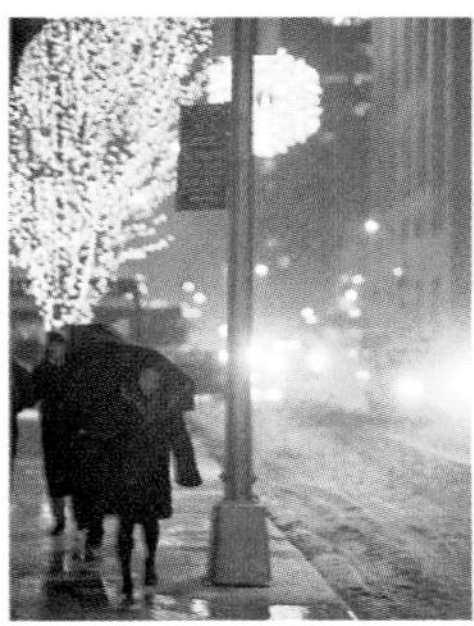

Main Ideas

- Earth's atmosphere is made of a combination of several gases, primarily nitrogen and oxygen. It also contains small amounts of water vapor, carbon dioxide, ozone, and dust, which play key roles in the production of weather and climate.
- The atmosphere consists of several layers characterized by differences in temperature. The most important for weather is the lowest layer, the troposphere, where most of the mass of the atmosphere is found.
- The Sun is the source of energy in Earth's atmosphere. Solar energy absorbed by Earth's surface is transferred throughout the atmosphere by the processes of radiation, conduction, and convection.

Vocabulary

conduction (p. 276)
convection (p. 277)
exosphere (p. 274)
mesosphere (p. 274)
ozone (p. 273)
radiation (p. 275)
stratosphere (p. 274)
thermosphere (p. 274)
troposphere (p. 274)

SECTION 11.2

State of the Atmosphere

Main Ideas

- Heat is the transfer of energy that occurs because of a difference in temperature between substances. Temperature is the measure of how rapidly or slowly molecules move around. Atmospheric temperature generally decreases with altitude.
- Air has mass and exerts a force called atmospheric pressure. Because there are fewer molecules of gas in the upper atmosphere, atmospheric pressure decreases with increasing altitude.
- Wind is the movement of air that results from differences in pressure. Wind speed is affected by friction; mountains, forests, and buildings slow wind down.

Vocabulary

condensation (p. 279)
dew point (p. 279)
heat (p. 278)
humidity (p. 283)
lifted condensation level (p. 279)
relative humidity (p. 283)
temperature (p. 278)
temperature inversion (p. 281)

SECTION 11.3

Moisture in the Atmosphere

Main Ideas

- Clouds are formed as warm, moist air is forced upward, expands, and cools. Orographic lifting is a method of cloud formation that involves air moving up the side of a mountain. Clouds may also form when air masses of different temperatures collide.
- Clouds are generally classified according to the altitudes at which they form and their shapes.
- As cloud droplets collide, they coalesce into larger droplets, which may fall to Earth as precipitation. The four main types of precipitation are rain, snow, sleet, and hail.
- In the water cycle, water continually moves between Earth's surface and the atmosphere through the processes of evaporation, condensation, and precipitation.

Vocabulary

coalescence (p. 289)
condensation nuclei (p. 285)
evaporation (p. 290)
latent heat (p. 286)
orographic lifting (p. 285)
precipitation (p. 289)
stability (p. 286)
water cycle (p. 290)

earthgeu.com/vocabulary_puzzlemaker

CHAPTER 11

Assessment

Understanding Main Ideas

1. What process describes the change of state of water from a liquid to a gas?
   a. condensation
   b. evaporation
   c. melting
   d. drying

2. Condensation nuclei are involved in the formation of which of the following?
   a. cloud droplets
   b. ozone
   c. dry ice
   d. carbon dioxide

3. Which atmospheric layer contains most of the mass of Earth's atmosphere?
   a. tropopause
   b. troposphere
   c. stratosphere
   d. mesosphere

4. Which object would heat up most rapidly?
   a. water
   b. asphalt
   c. grass
   d. cement

5. What percentage of incoming solar radiation does Earth's atmosphere absorb?
   a. 100 percent
   b. 15 percent
   c. 50 percent
   d. 35 percent

Use this diagram to answer questions 6–8.

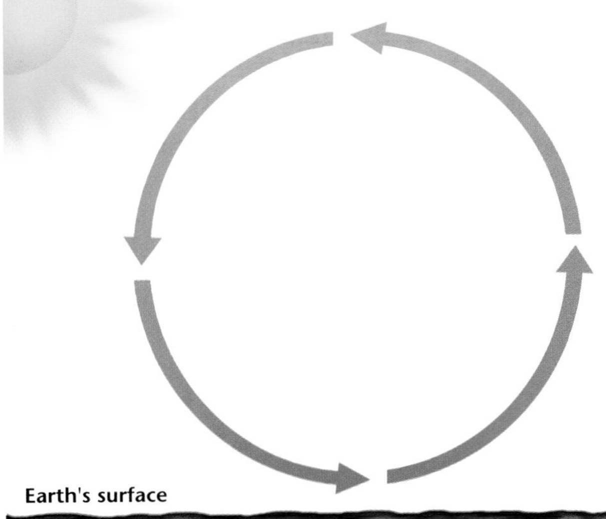

6. What type of energy transfer is shown in the diagram on this page?
   a. radiation
   b. convection
   c. conduction
   d. vaporization

7. In the diagram, what type of air is rising?
   a. warm
   b. cool
   c. dry
   d. cold

8. Which statement describes the rising air in the diagram in relation to the surrounding air?
   a. The rising air is more dense.
   b. The rising air is less dense.
   c. The rising air is cooler.
   d. The rising air is thicker.

9. What do we call the temperature at which air becomes saturated?
   a. humidity
   b. the lapse rate
   c. the LCL
   d. the dew point

10. What type of cloud is a stratus cloud?
    a. low
    b. vertical development
    c. high
    d. middle

11. What causes wind?
12. Describe a temperature inversion.
13. What is latent heat?
14. How do clouds form? How are they classified?
15. Explain the three main processes involved in the water cycle.

Test-Taking Tip

Quiet Zone It's best to study in a similar environment to the one in which you'll be tested. Thus, try to study in a quiet, disturbance-free, well-lit place. Avoid blaring stereos, video games, chatty friends, and television screens.

CHAPTER 11
Assessment

Applying Main Ideas

16. If clouds absorb only a small amount of solar radiation, how is Earth's atmosphere heated?

17. Which two gases make up most of Earth's atmosphere? List other important atmospheric gases.

18. Use the following terms to construct a concept map that shows the three methods of energy transfer in the atmosphere.

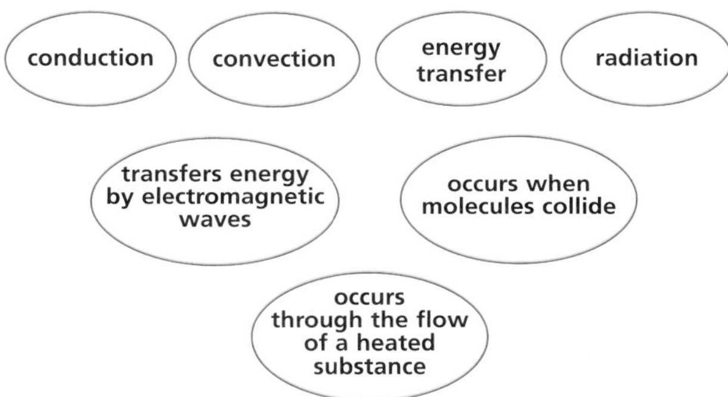

19. What type of cloud might produce the most intense precipitation? Explain your answer.

Thinking Critically

20. A summer rain could begin as a snowstorm in the clouds overhead. How is this possible?

21. Given that air cannot rise through a stable inversion layer, predict the effects of a temperature inversion on a heavily populated, highly industrial city located beneath the inversion.

22. Given the varying depths of the troposphere, why do holes in the ozone layer tend to appear mainly in the polar regions?

23. A spoon that sits in a bowl of hot soup feels hot when touched. How was energy transferred to the spoon?

Standardized Test Practice

INTERPRETING DATA Use the diagram below to answer questions 1 and 2.

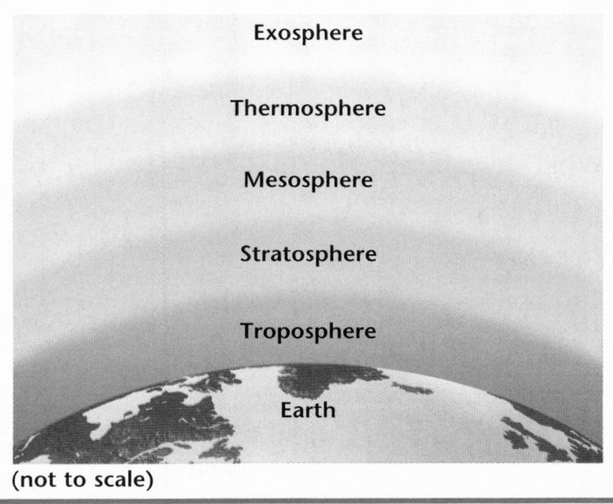

(not to scale)

1. In which layer of Earth's atmosphere is air most likely warmed by conduction?
- **a.** troposphere
- **b.** stratosphere
- **c.** thermosphere
- **d.** exosphere

2. Which of the following is NOT true of ozone?
- **a.** It absorbs ultraviolet radiation.
- **b.** Its concentration is decreasing or thinning.
- **c.** It is concentrated in the atmospheric layer called the mesosphere.
- **d.** It is a gas formed by the addition of one oxygen atom to an oxygen molecule.

3. Which of the following is most likely to cause orographic lifting?
- **a.** a sandy beach
- **b.** a flowing river
- **c.** a rocky mountain
- **d.** a sand dune

4. Which clouds are most likely to form when fog lifts away from Earth's surface?
- **a.** cumulus
- **b.** cirrostratus
- **c.** stratus
- **d.** altocumulus

earthgeu.com/standardized_test

Chapter 12

Meteorology

What You'll Learn

- What determines global weather patterns.
- How air masses move and change.
- How the strengths and weaknesses of weather forecasts differ.
- How to create a weather chart.

Why It's Important

Few aspects of the environment have as much impact on our everyday decisions as weather does. A basic knowledge of weather processes can make those decisions easier and sometimes far safer.

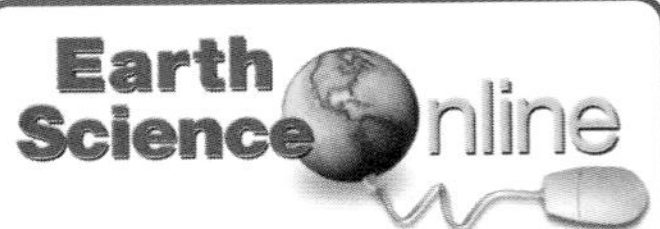

To find out more about meteorology, visit the Earth Science Web Site at earthgeu.com

Discovery Lab Model a Cold Air Mass

An air mass is a large body of air that takes on the characteristics of the area over which it forms. You can demonstrate the formation of a cold air mass using simple materials.

1. Place a tray full of ice cubes on a table with a pencil underneath each end of the tray so that the tray is slightly elevated.
2. Slide a liquid-crystal temperature strip underneath the ice-cube tray.
3. Rest another temperature strip on two pencils on top of the tray.
4. Record the temperature of each strip at one-minute intervals for about five minutes.

CAUTION: Always wear protective clothing in the lab.

Observe In your science journal, make a graph showing the temperature changes for each temperature strip. What happened to the temperature of the air beneath the tray and the air above the tray? Explain how this model represents a cold air mass.

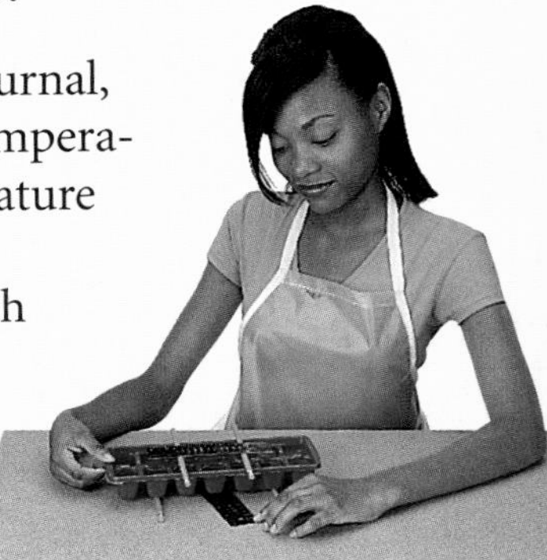

SECTION 12.1 The Causes of Weather

Earth Sciences 5.a, 6.a

OBJECTIVES

- **Compare** *and* **contrast** *weather and climate.*
- **Analyze** *how imbalances in the heating of Earth's surface create weather.*
- **Describe** *how and where air masses form.*

VOCABULARY

meteorology
weather
climate
air mass
air mass modification

Have you ever sat back and watched the sky on a lazy summer afternoon? You might have noticed clouds of different shapes, or felt the warmth of the sun against your face or an occasional puff of wind as it cooled your skin. All of these phenomena are part of a highly organized sequence of events with specific causes. Those events and the factors that cause them are all part of meteorology. **Meteorology** is the study of atmospheric phenomena. The root of the word *meteorology—meteor*—is the name given in modern times to a flaming rock falling through space. But the ancient Greek meaning of *meteor* was "high in the air," and it is this meaning that pertains to meteorology.

Clouds, raindrops, snowflakes, fog, dust, and rainbows are all types of atmospheric "meteors." The primary types are cloud droplets and forms of precipitation that contain water in any phase; they are known as hydrometeors. Smoke, haze, dust, and other condensation nuclei are called lithometeors. Thunder and lightning are examples of electrometeors, which are visible or audible manifestations of atmospheric electricity. These various phenomena are the objects and events that meteorologists study.

Weather and Climate

Atmospheric phenomena, shown in ***Figure 12-1,*** interact to affect the environment and life on Earth. This is basically what we call weather. **Weather** is the current state of the atmosphere. When we speak of weather, we are referring mainly to short-term variations in the atmosphere. These variations can take place over minutes, hours, days, weeks, or months. Long-term variations in weather for a particular area make up the **climate** of that area. Climate is usually averaged over the course of 30 years or more. You'll learn more about climate in Chapter 14. For now, simply recognize that meteorology, weather, and climate are related. Meteorology is the study of the atmosphere; weather is the current state of the atmosphere, including short-term variations that affect our lives; and climate describes the average weather over a long period of time.

A Question of Balance

As you've learned, the Sun heats the surface of Earth, and Earth radiates back to space about as much energy as it receives over the course of a year. In meteorology, a crucial question is how that radiation is distributed around the planet. You know that the Sun feels hotter during the afternoon, when its rays strike Earth more directly, than it does in the early morning or evening, when its rays strike Earth at a low angle. The Sun's rays are more spread out when they strike Earth at a low angle, as you'll see in the *MiniLab* later in this chapter. The same amount of energy is spread over a larger area. As shown in ***Figure 12-2,*** the solar radiation reaching Earth's surface at the poles is therefore less intense. This explains, in part, why the tropics are warmer than the poles. But why don't the tropics become steadily warmer if the Sun is always directly overhead? How do regions manage to maintain fairly constant average temperatures?

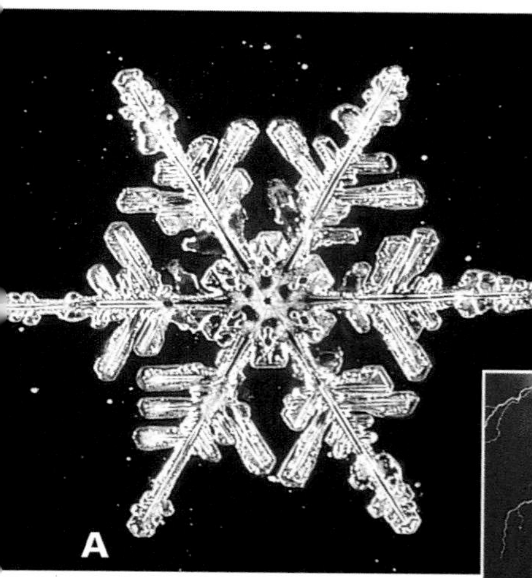

Figure 12-1 Snowflakes **(A),** lightning **(B),** and fog **(C)** are types of atmospheric phenomena.

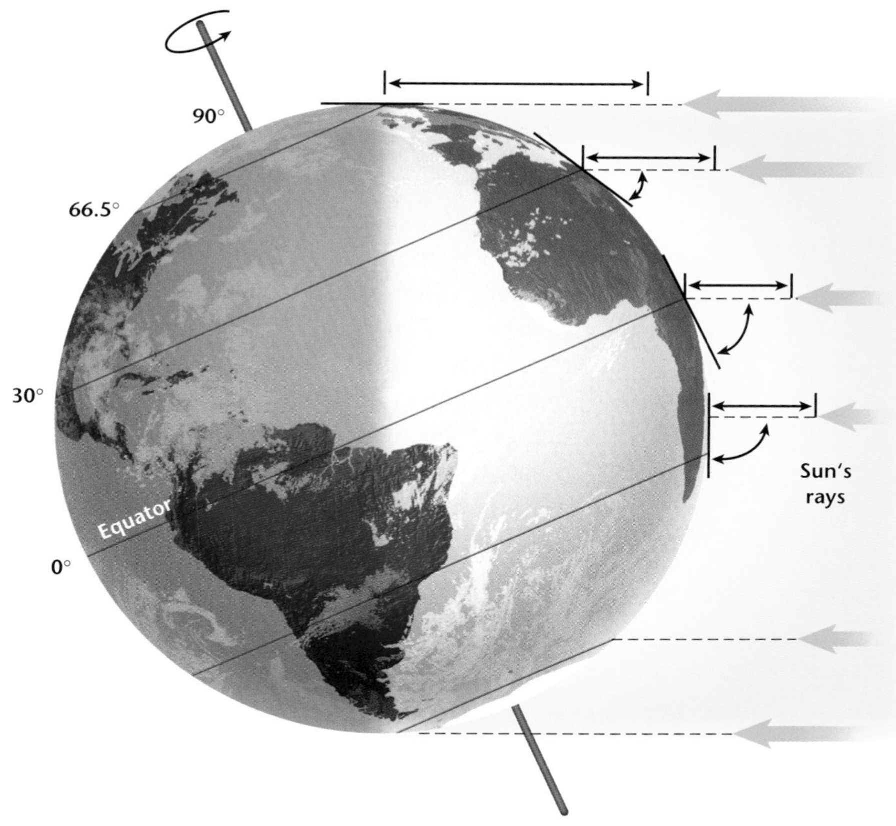

Figure 12-2 The Sun's rays strike Earth more directly at the tropics than they do at the poles. At the poles, the same amount of solar radiation is spread over a larger area than at the equator. Therefore, polar regions are never very warm.

Earth Sciences

5.a Students know how differential heating of Earth results in circulation patterns in the atmosphere and oceans that globally distribute the heat.

6.a Students know weather (in the short run) and climate (in the long run) involve the transfer of energy into and out of the atmosphere.

Balancing the Budget The tropics and other places maintain fairly constant average temperatures because heat energy is redistributed around the world. The continual motion of air and water reallocates heat energy among Earth's surface, oceans, and atmosphere and brings it into balance. Virtually everything that we consider to be weather—every atmospheric motion from the tiniest convection current to thunderstorms to large-scale weather systems—is part of this constant redistribution of Earth's heat energy.

Air Masses

In Chapter 11, you learned that when air over a warm surface, such as a parking lot, becomes warmer than the surrounding air because of conduction, the warm air rises. Now, imagine this same process taking place over thousands of square kilometers. Imagine that the warm air remains over this same area for days or weeks. The result is the formation of an air mass. An **air mass** is a large body of air that takes on the characteristics of the area over which it forms. Meteorologists call the region over which an air mass forms the source region. Air masses form over land or water. Those that form

MiniLab

How does the angle of the Sun's rays differ?

Model the angle at which sunlight reaches Earth's surface. This angle greatly affects the intensity of solar energy received in any one place.

Procedure

1. Hold a flashlight several centimeters above a piece of paper and point the flashlight straight down.
2. Use a pencil to trace the outline of the light on the paper. The outline models how the Sun's rays strike the equator.
3. Keeping the flashlight at the same distance above the paper, tilt the top of the flashlight to roughly a 30° angle.
4. Trace the new outline of the light. This is similar to how the Sun's rays are received at the poles.

Analyze and Conclude

1. Describe how the outline of the light differed between step 1 and step 3. Explain why it differed.
2. How do you think the change in area covered by the light affects the intensity of light received at any one place?
3. The flashlight models solar radiation striking the surface of Earth. Knowing this, compare how much heat energy is absorbed near the equator and near the poles.

over land have less exposure to large amounts of moisture, so they are drier than those that form over water. Air masses take on the temperature of the source region, too.

Classifying Air Masses Air masses are classified exactly as we have already described them: according to their source regions. The main types of air masses, shown in ***Figure 12-3,*** are warm and dry continental tropical (cT), warm and humid maritime tropical (mT), cold and dry continental polar (cP), cold and humid maritime polar (mP), and arctic (A). An arctic air mass is basically the same as a continental polar air mass, but much colder. It's the type you may have heard most about because it brings the most frigid outbreaks of winter. Extremely cold arctic air masses are usually associated with very high pressure as a result of the massive sinking of cold air over a large area.

Source Regions All five main types of air masses can be found in North America because of the continent's proximity to the source regions associated with each air mass. Maritime polar air forms over the cold waters of the North Atlantic and North Pacific. It primarily affects the West Coast, bringing occasionally heavy rains in winter. Continental polar air forms over the interior of Canada and Alaska and can be quite frigid in the winter, when nights are long. In the summer, however, cP air can bring pleasant relief from heat and humidity because of its cool and relatively dry composition.

The origins of maritime tropical air are tropical and subtropical oceans, such as the Caribbean Sea and the Gulf of Mexico. In the summer, mT air brings hot, oppressively humid weather to the eastern two-thirds of the United States and Canada. The desert Southwest and Mexico are the source regions

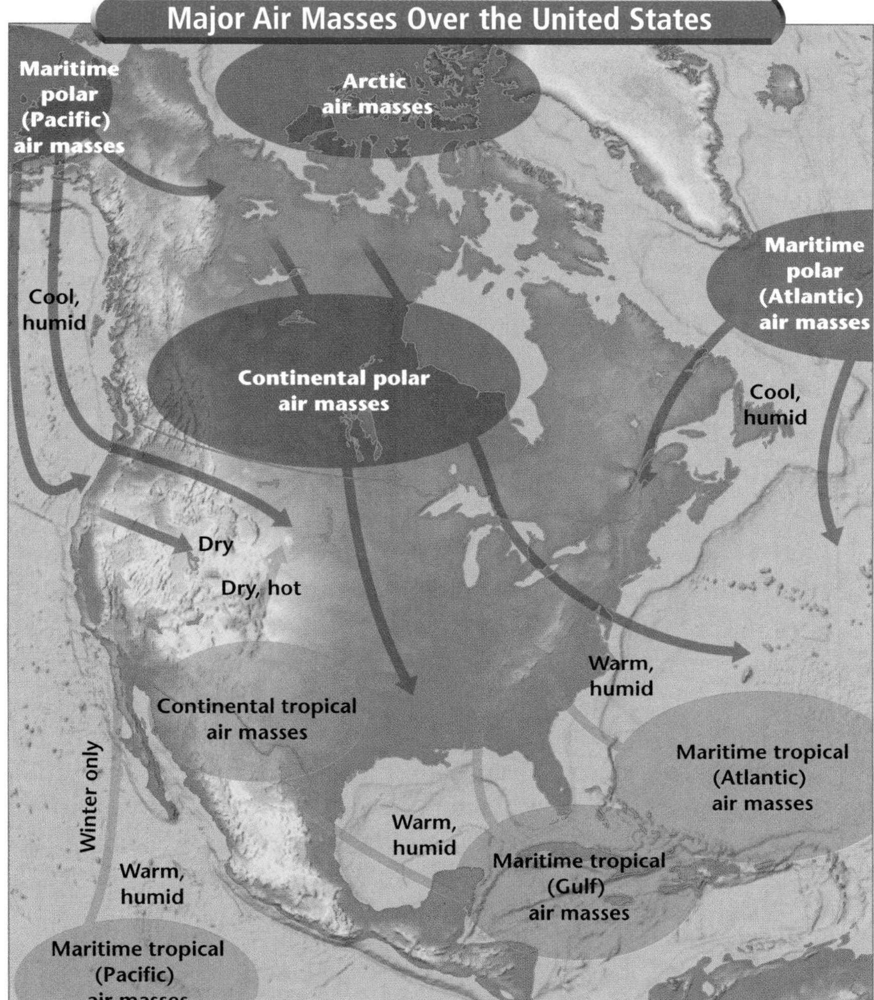

Figure 12-3 Each of the major air masses that affects weather in the United States has a similar temperature and moisture content as the area over which it formed.

of continental tropical air, which is hot and dry, especially in summer. Arctic air develops over latitudes above 60°N in the ice- and snow-covered regions of Siberia and the Arctic Basin. During the winter, this area receives almost no solar radiation but continues to radiate heat out to space, so it can get very cold indeed. In addition to temperature and humidity, another important characteristic of an air mass is its stability. The stability of air is an important factor in its ability to produce clouds and precipitation.

Air Mass Modification Like the warm air over a large city during a summer afternoon, air masses do not stay in one place indefinitely. Eventually, they move, transferring heat from one area to another and thus establishing the heat balance discussed earlier. As an air mass moves, it may travel over land or water that has different

Table 12-1 Air Mass Characteristics				
	Source Region Stability		Characteristics	
Air Mass Type	Winter	Summer	Winter	Summer
A	Stable		Bitter cold, dry	
cP	Stable	Stable	Very cold, dry	Cool, dry
cT	Unstable	Unstable	Warm, dry	Hot, dry
mP (Pacific)	Unstable	Unstable	Mild, humid	Mild, humid
mP (Atlantic)	Unstable	Stable	Cold, humid	Cool, humid
mT (Pacific)	Stable	Stable	Warm, humid	Warm, humid
mT (Atlantic)	Unstable	Unstable	Warm, humid	Warm, humid

characteristics from those of its source region. The air mass then starts to acquire some of the characteristics of the new surface beneath it. When this happens, it is said to undergo **air mass modification,** which is the exchange of heat or moisture with the surface over which an air mass travels. ***Table 12-1*** summarizes the characteristics of the main types of air masses before modification.

All air masses become modified to some extent as they move away from their source regions. Eventually, an air mass becomes modified to such a degree that its characteristics are almost the same as the new surface over which it is traveling. At this point, the air mass has lost its original identity and is now simply part of the air over the new source region it has encountered.

SECTION ASSESSMENT

1. What is the difference between weather and climate? How do both relate to the science of meteorology?
2. What must happen to keep the poles from steadily cooling off and the tropics from heating up over time?
3. What method of heat transfer plays the primary role in the formation of an air mass?
4. Describe how the moisture in a maritime polar (mP) air mass that formed over the North Pacific Ocean would modify as it moved inland over the western coast of North America.
5. **Thinking Critically** Explain why an arctic air mass is usually more stable than a maritime tropical air mass. In other words, why does the arctic air resist rising more than the tropical air does?

SKILL REVIEW

6. **Predicting** Which type of air mass would you expect to become modified more quickly: an arctic air mass moving over the Gulf of Mexico in winter or a maritime tropical air mass moving into the southeastern United States in summer? For more help, refer to the *Skill Handbook.*

earthgeu.com/self_check_quiz

SECTION 12.2 Weather Systems

Earth Sciences 5.a, 5.b, 6.a

OBJECTIVES

- **Describe** *how the rotation of Earth affects the movement of air.*
- **Compare** *and* **contrast** *wind systems.*
- **Identify** *the various types of fronts.*

VOCABULARY

Coriolis effect
trade winds
prevailing westerlies
polar easterlies
jet stream
front

If Earth were either all land or all water and did not rotate on its axis, a large convection cell would form in each hemisphere with the colder and denser air at the poles sinking to the surface and flowing toward the tropics. There, it would force the warm air already at the equator to rise, and then it would cool and flow back toward the poles. The problem with this proposal is that Earth does rotate from west to east. This rotation causes the **Coriolis effect,** wherein moving particles such as air are deflected to the right in the northern hemisphere and to the left in the southern hemisphere. The Coriolis effect, illustrated in ***Figure 12-4A,*** combines with the heat imbalance found on Earth to create distinct global wind systems that transport colder air to warmer areas and warmer air to colder areas. The end result is the balancing of heat energy on Earth.

GLOBAL WIND SYSTEMS

There are three basic zones, or wind systems, in each hemisphere, as shown in ***Figure 12-4B.*** The first, known as the **trade winds,** occurs at 30° north and south latitude. There, air sinks, warms, and moves toward the equator in a westerly direction. When the air reaches the equator, it rises again and moves back toward latitude 30°, where it sinks and the process starts anew. The circulation pattern for the trade

Figure 12-4 The rotation of Earth causes the Coriolis effect **(A),** which, along with the heat imbalance on Earth, creates the three major global wind systems **(B).** The convection cells show the movement of air in each zone.

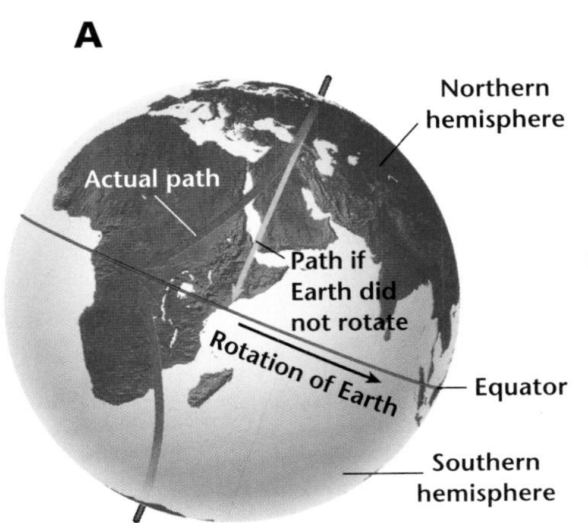

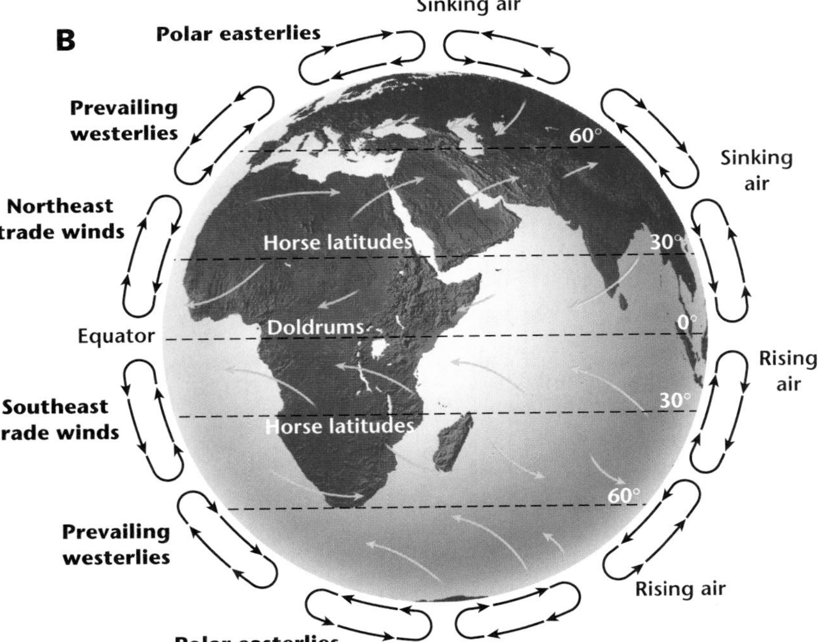

winds closely fits a model proposed by English scientist George Hadley in 1735; thus, this zone is sometimes known as the Hadley cell.

Around 30° latitude, the sinking air associated with the trade winds creates a belt of high pressure that in turn causes generally weak surface winds. Spanish sailors called this belt the horse latitudes because their ships became stranded in these waters as a result of the near-calm winds. According to legend, they could no longer feed or water their horses and were forced to throw them overboard.

Near the equator, the trade winds from both hemispheres move together from two different directions, as shown in ***Figure 12-4.*** The air converges, is forced upward, and creates an area of low pressure. This process, called convergence, can occur on a small or large scale. Near the equator, it occurs over a large area called the intertropical convergence zone (ITCZ). As ***Figure 12-5*** shows, the ITCZ migrates south and north of the equator as the seasons change. In essence, it follows the path of the Sun's rays, which are directly over the equator in September and March. Because the ITCZ is a region of rising air, it is characterized by a band of cloudiness and occasional showers that help provide the moisture for many of the world's tropical rain forests. Note that the ITCZ is also called the doldrums. As in the horse latitudes, sailing ships were often stranded in this belt of light winds.

Earth Sciences

5.a Students know how differential heating of Earth results in circulation patterns in the atmosphere and oceans that globally distribute the heat.

5.b Students know the relationship between the rotation of Earth and the circular motions of ocean currents and air in pressure centers.

6.a Students know weather (in the short run) and climate (in the long run) involve the transfer of energy into and out of the atmosphere.

Other Wind Zones The second wind system, the **prevailing westerlies,** flows between 30° and 60° north and south latitude in a circulation pattern opposite that of the trade winds. In this zone, surface winds move toward the poles in a generally easterly direction, as shown in ***Figure 12-4.*** Note that wind is named for the direction

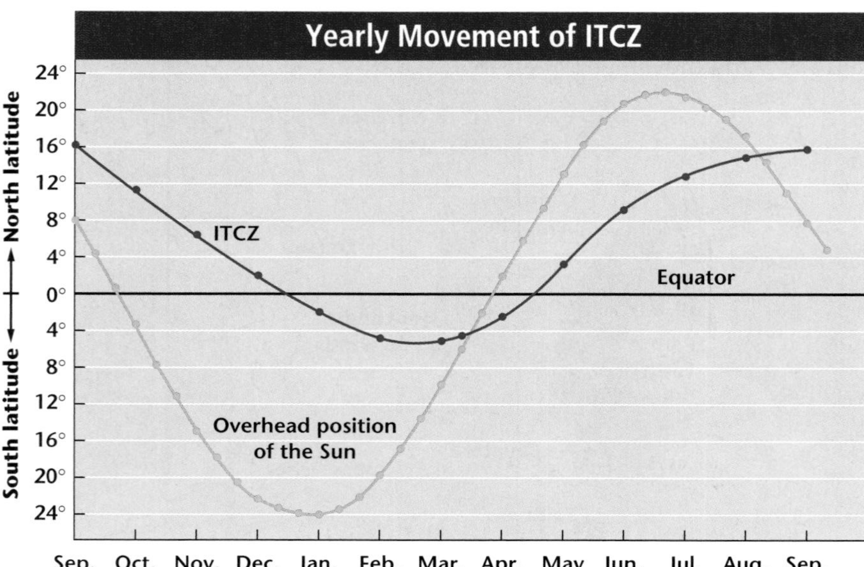

Figure 12-5 The mean position of the ITCZ basically follows the path of the Sun's rays throughout the year. The two paths don't match exactly because the ITCZ responds gradually to changes in the Sun's position, and thus lags behind the Sun.

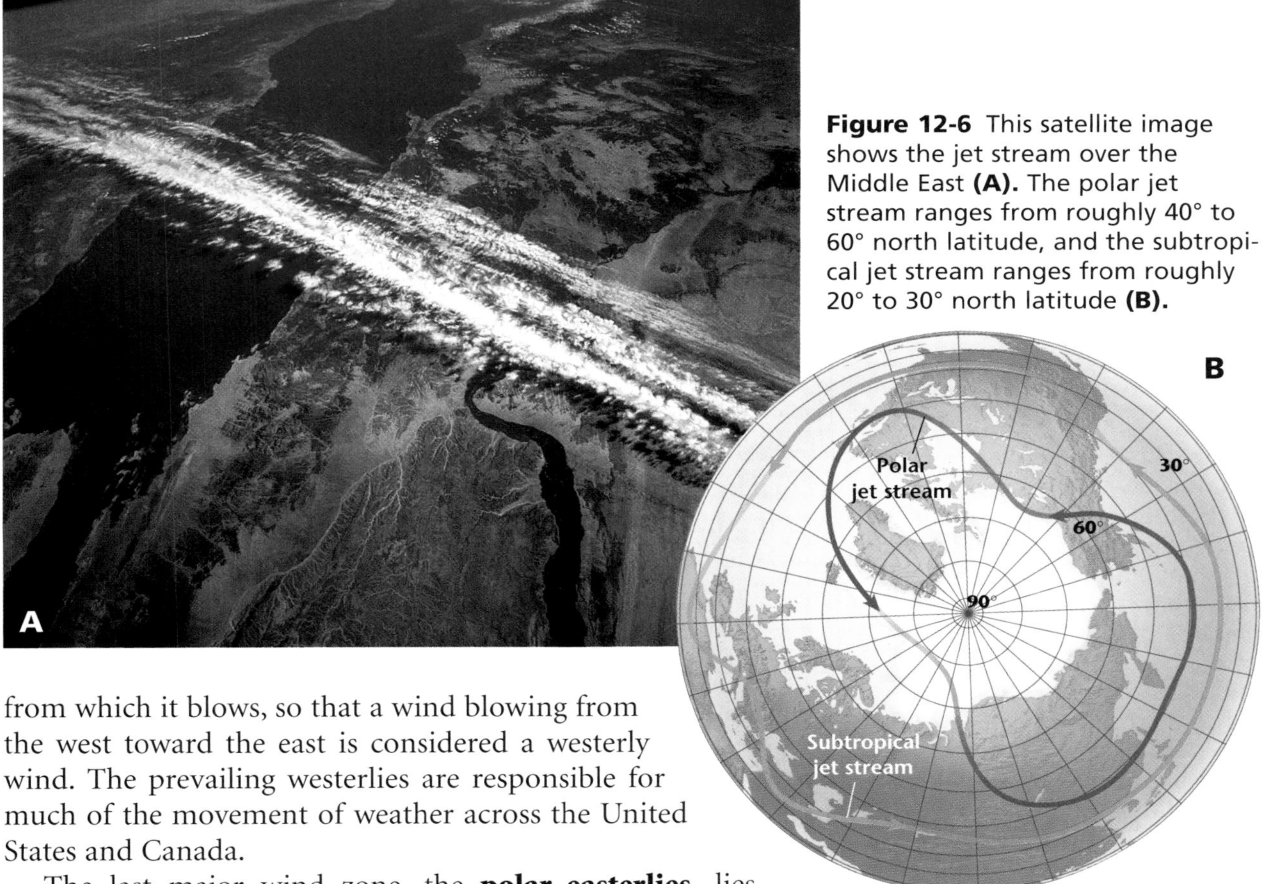

Figure 12-6 This satellite image shows the jet stream over the Middle East **(A).** The polar jet stream ranges from roughly 40° to 60° north latitude, and the subtropical jet stream ranges from roughly 20° to 30° north latitude **(B).**

from which it blows, so that a wind blowing from the west toward the east is considered a westerly wind. The prevailing westerlies are responsible for much of the movement of weather across the United States and Canada.

The last major wind zone, the **polar easterlies,** lies between 60° latitude and the poles. Similar to the trade winds, the polar easterlies flow from the northeast to the southwest in the northern hemisphere. Note that global wind direction reverses in the southern hemisphere. South of the equator, for instance, the polar easterlies flow from the southeast to the northwest. In both hemispheres, the polar easterlies are characterized by cold air.

JET STREAMS

Earth's weather is strongly influenced by atmospheric conditions and events that occur at the boundaries between wind zones. On either side of these imaginary boundaries, both surface and upper-level air differs greatly in temperature and pressure. Remember that wind, temperature, and pressure are related. Differences in temperature and pressure cause wind. Therefore, a large temperature gradient in upper-level air should result in strong westerly winds, and indeed, this is what happens. Narrow bands of fast, high-altitude, westerly winds called **jet streams** flow at speeds up to 185 km/h at elevations of 10.7 km to 12.2 km. Jet streams, shown in ***Figure 12-6,*** are so named because they resemble jets of water. The most significant one, the polar jet stream, separates the polar easterlies from the prevailing westerlies. A second version, the subtropical jet stream, is located where the trade winds meet the prevailing westerlies.

Large-Scale Weather Systems Disturbances form along jet streams and give rise to large-scale weather systems that transport surface cold air toward the tropics and surface warm air toward the poles. Keep in mind that the position of the jet stream varies. It can dive almost directly south or north, instead of following its normal westerly direction. It can also split into different branches and later reform into a single stream. Whatever form or position it takes, the jet stream represents the strongest core of westerly winds. Together, these winds form a sort of atmospheric railroad track, with large-scale weather systems serving as the atmospheric trains. Weather systems generally follow the path of these winds. The jet stream also affects the intensity of weather systems by moving air of different temperatures from one region to another. Thus, despite its altitude, it has a significant impact on weather.

FRONTS

The different temperatures and pressures of air masses have other consequences apart from the jet stream. In the middle latitudes, air masses with different characteristics sometimes collide, forming a front. A **front** is the narrow region separating two air masses of different densities. The density differences are caused by differences in temperature, pressure, and humidity. Fronts can stretch over thousands of kilometers across Earth's surface. The interaction between the colliding air masses can bring dramatic changes in weather. As shown in ***Figure 12-7,*** there are four main types of fronts: cold fronts, warm fronts, stationary fronts, and occluded fronts.

Figure 12-7 These diagrams show the structures of the four main types of fronts. Symbols below each diagram indicate how the fronts are represented on a weather map. A cold front often moves quickly. Thunderstorms may form along the front **(A).** In a warm front, precipitation often occurs over a wide band. High-altitude cirrus clouds may form as water vapor condenses **(B).** Light wind and precipitation are sometimes associated with a stationary front **(C).** Strong winds and heavy precipitation may occur along an occluded front **(D).**

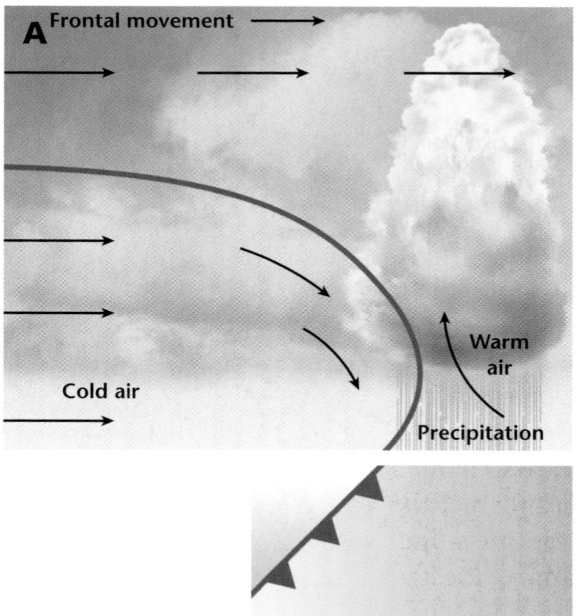

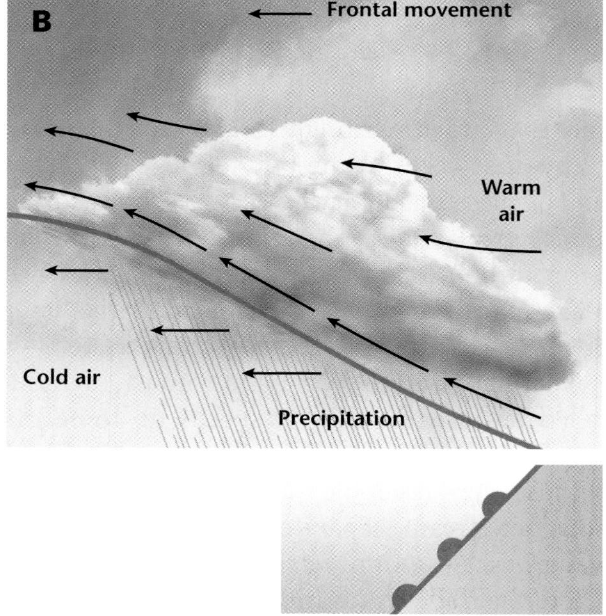

Cold Fronts In a cold front, shown in ***Figure 12-7A,*** cold, dense air displaces warm air and forces the warm air up along a steep front. As the warm air rises, it cools and condenses. Clouds, showers, and sometimes thunderstorms are associated with cold fronts. A cold front is represented on a weather map as a solid blue line with blue triangles that point in the direction of the front's motion.

Warm Fronts In a warm front, advancing warm air displaces cold air, as shown in ***Figure 12-7B.*** Because the air ahead of a warm front moves more slowly than does an advancing cold air mass, the warm air encounters less friction with the ground and thus develops a gradual frontal slope rather than a steep boundary. A warm front is characterized by extensive cloudiness and precipitation. On a weather chart, a warm front appears as a solid red line with regularly spaced, solid red semicircles pointing in the direction of the front's motion.

Topic: Weather
To find out more about weather fronts, visit the Earth Science Web Site at earthgeu.com

Activity: Obtain a weather map showing today's weather. Label the weather fronts.

Stationary Fronts Sometimes, two air masses meet and neither advances into the other's territory. In this case, the boundary between the air masses stalls. This type of front, called a stationary front, frequently occurs when two air masses have become so modified in their travels that the temperature and pressure gradients between them are small. Stationary fronts seldom have extensive cloud and heavy precipitation patterns; any patterns that do occur are somewhat similar to those of a warm front. A stationary front is represented on a weather map by a combination of short segments of cold- and warm-front symbols as shown in ***Figure 12-7C.***

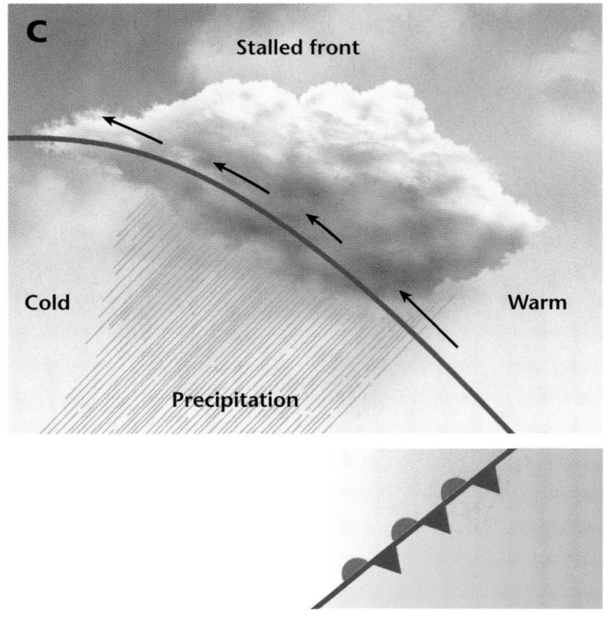

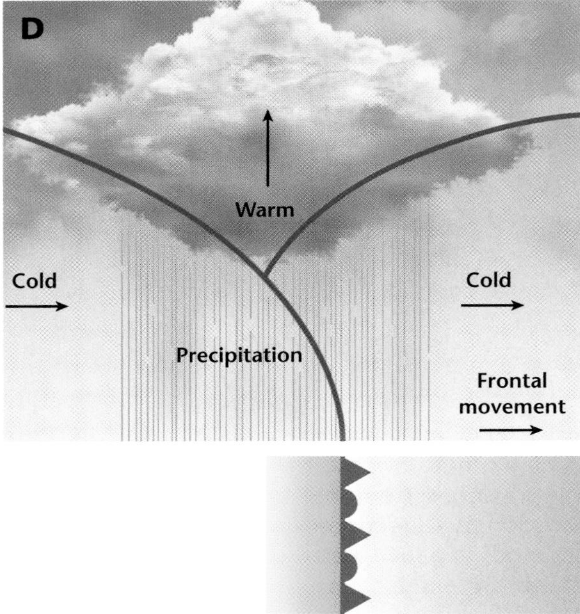

Using Numbers
Most weather occurs in the troposphere between the surface of Earth and an altitude of 11 km. Suppose the temperature in an area drops about 7°C for each kilometer of increase in altitude. If the surface temperature is 15°C, what is the temperature at an altitude of 3 km?

Occluded Fronts Sometimes, a cold air mass moves so rapidly that it overtakes a warm front. The advancing cold air wedges the warm air upward, as shown in ***Figure 12-7D*** on page 309. Recall that a warm front involves warm air gliding over a cold air mass. When the warm air is lifted, this cold air mass collides with the advancing cold front. The warm air is thus squeezed upward between the two cold air masses. This is called an occluded front and is represented on a weather map by a line with alternating purple triangles and semicircles that point toward the direction of motion. Precipitation is common on both sides of an occluded front.

Pressure Systems

You have learned that at Earth's surface, rising air is associated with low pressure and sinking air is associated with high pressure. Rising or sinking air, combined with the Coriolis effect, results in the formation of rotating low- and high-pressure systems in the atmosphere. Air in these systems moves in a general circular motion around either a high- or low-pressure center.

High-Pressure Systems In a surface high-pressure system, air sinks, so that when it reaches Earth's surface, it spreads away from the center. The deflection of air to the right caused by the Coriolis effect makes the overall circulation around a high-pressure center move in a clockwise direction in the northern hemisphere, as shown in ***Figure 12-8A.*** Keep in mind that the Coriolis effect is reversed in the southern hemisphere; there, high-pressure systems rotate in a counterclockwise direction. Some high-pressure systems are associated with cold air masses that move and modify; others, such as subtropical high-pressure systems, are more stationary.

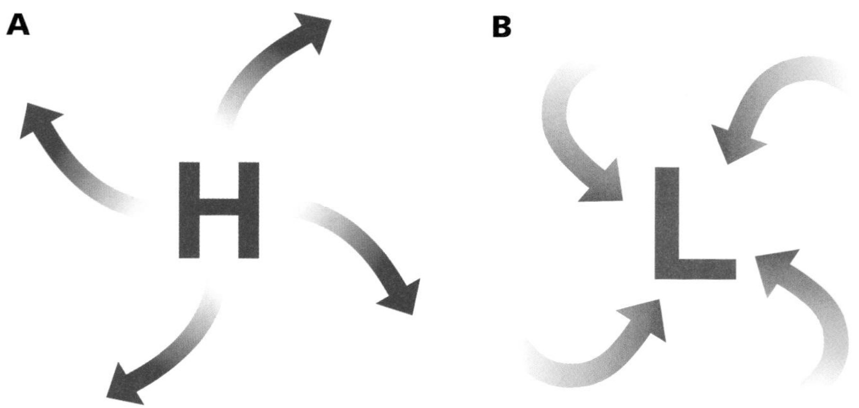

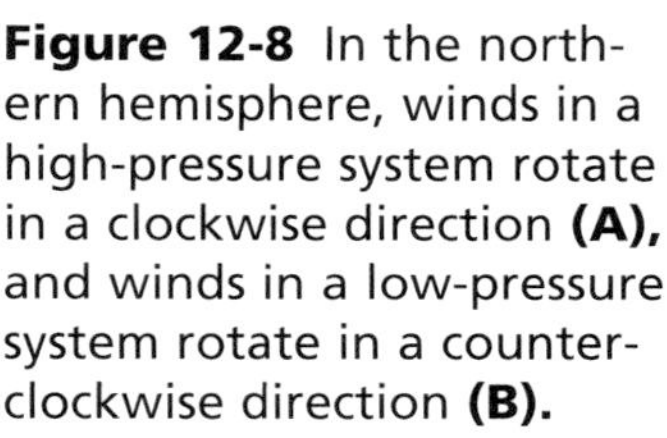
Figure 12-8 In the northern hemisphere, winds in a high-pressure system rotate in a clockwise direction **(A),** and winds in a low-pressure system rotate in a counterclockwise direction **(B).**

Low-Pressure Systems In surface low-pressure systems, air rises. The rising air must be replaced by air from outside the system, so the net flow is inward toward the center and then upward. In contrast to air in a high-pressure system, air in a low-pressure system in the northern hemisphere moves in a counterclockwise direction, as shown in ***Figure 12-8B.*** This movement is reversed in the southern hemisphere.

Recall from Chapter 11 that it's difficult for clouds to form when air is sinking, as it does in high-pressure systems. Thus, high-pressure systems are usually associated with fair weather, while low-pressure systems are associated with clouds and precipitation. In fact, most of Earth's subtropical oceans are dominated by large high-pressure systems with generally pleasant conditions. One of the main producers of inclement weather in the middle latitudes, meanwhile, is a specific type of low-pressure system called a wave cyclone. A wave cyclone usually begins along a stationary front. Some imbalance in temperature, pressure, or density causes part of the front to move south as a cold front and another part of the front to move north as a warm front. This sets up a counterclockwise or cyclonic circulation, as shown in ***Figure 12-9.*** Eventually, if upper-level conditions are favorable, a fully developed low-pressure system forms. Pushed by the prevailing westerlies, this system may travel thousands of kilometers affecting large areas in the middle latitudes.

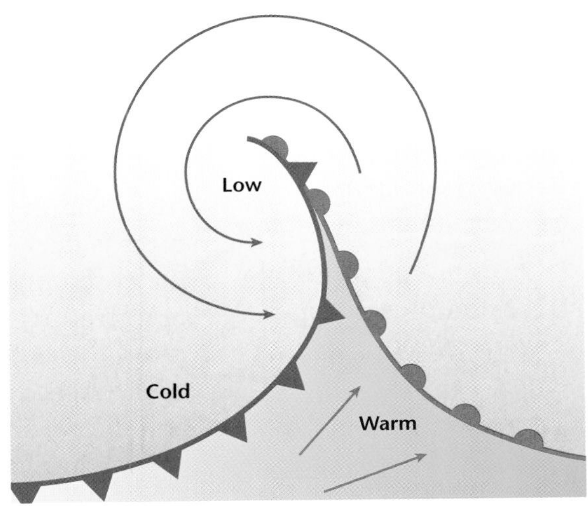

Figure 12-9 The counterclockwise circulation is characteristic of a wave cyclone.

Section Assessment

1. Describe the main global wind systems. Give characteristics of each.
2. Explain why most tropical rain forests are located near the equator.
3. How does the jet stream affect the movement of air masses?
4. What is the Coriolis effect? How does it affect air in the northern and southern hemispheres?
5. Compare and contrast a low-pressure system and a high-pressure system.
6. **Thinking Critically** Based on what you know about the three major zones of global air circulation, form a hypothesis about why most of the world's deserts are located between 10° and 30° north and south latitudes.

Skill Review

7. **Interpreting Scientific Illustrations** Refer to the illustrations of the four types of fronts shown in ***Figure 12-7.*** Sketch the fronts in your science journal or use a computer graphics program to make a model of them. Label the warm and cold air masses. Indicate the direction of their movement and describe the type of weather associated with each front. For more help, refer to the *Skill Handbook.*

SECTION 12.3

Gathering Weather Data

OBJECTIVES

- **Recognize** *the importance of accurate weather data.*
- **Describe** *the technology used to collect weather data.*
- **Analyze** *the strengths and weaknesses of weather observation systems.*

VOCABULARY

thermometer
barometer
anemometer
hygrometer
ceilometer
radiosonde
Doppler effect

When you visit a doctor, she or he first measures your temperature and blood pressure to make an accurate diagnosis and prescribe treatment. If the doctor's data are incomplete or inaccurate, the diagnosis is likely to be inaccurate as well. The same principle applies to meteorology. Meteorologists measure the atmospheric variables of temperature, air pressure, wind, and relative humidity to make accurate weather forecasts. The quality of the data is critical. In fact, two of the most important factors in weather forecasting are the accuracy and the density of the data—density in this case refers to the amount of data available. Just as a doctor uses several different instruments to assess your health, meteorologists use several types of technology to gather information about the atmosphere.

SURFACE DATA

One of the most common weather instruments is a **thermometer,** a device used to measure temperature. Usually, thermometers contain liquids such as mercury or alcohol, which expand when heated. The height of the liquid column indicates temperature. Another common weather instrument, the barometer, also uses mercury to obtain weather data. **Barometers** measure air pressure. In a mercury barometer, changes in air pressure are indicated by changes in the height of a column of mercury. An aneroid barometer contains a vacuum

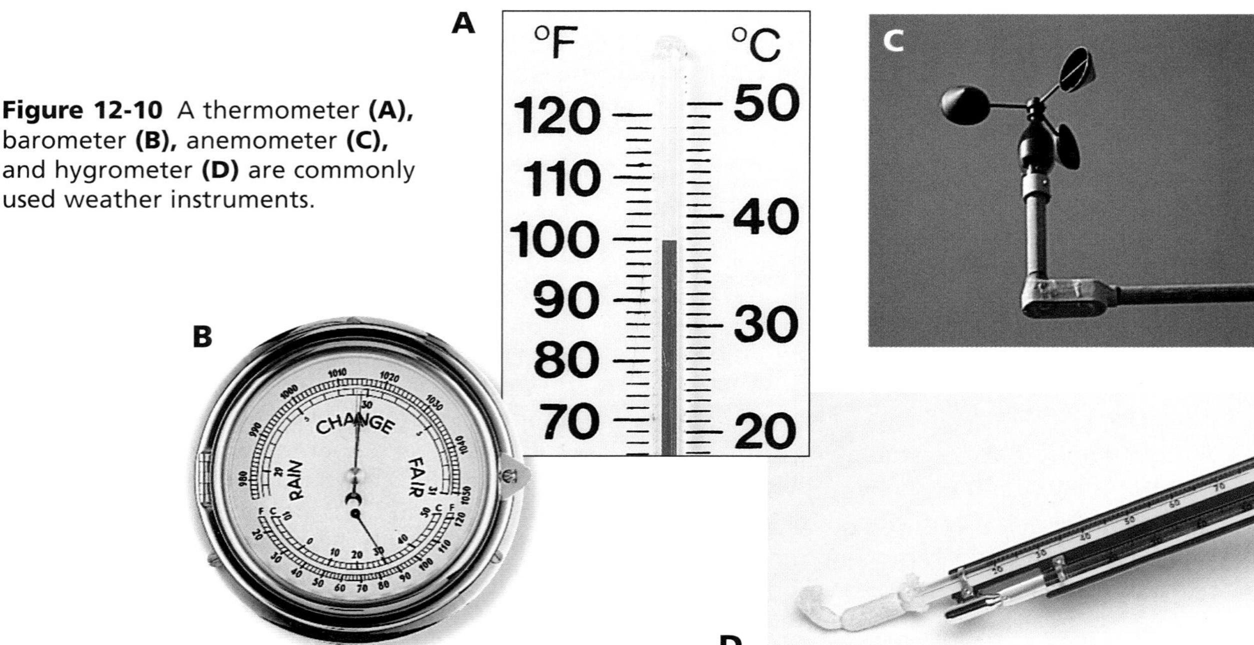

Figure 12-10 A thermometer **(A)**, barometer **(B)**, anemometer **(C)**, and hygrometer **(D)** are commonly used weather instruments.

inside a metal chamber. The chamber contracts or expands with changes in air pressure. A thermometer and a mercury barometer are shown in ***Figures 12-10A*** and ***12-10B.***

Other Surface Instruments An **anemometer,** shown in ***Figure 12-10C,*** is used to measure wind speed. The simplest type of anemometer has cupped arms that rotate as the wind blows. A **hygrometer** measures relative humidity. One type of hygrometer, shown in ***Figure 12-10D,*** uses wet- and dry-bulb thermometers. As water evaporates from the wet bulb, the bulb cools, creating a temperature difference between the wet bulb and the dry bulb. This temperature difference is used in conjunction with a relative humidity chart to determine relative humidity. See *Appendix F* for an example of a relative humidity chart.

Automated Surface Observing System To make weather forecasts, meteorologists analyze and interpret data gathered from weather instruments. In this regard, timing is crucial. Data must be gathered at the same time at many different locations. Why? It would do no good to analyze how temperature and air pressure are interacting in the atmosphere if the two variables were measured at different times. Meteorologists need an accurate "snapshot" of the atmosphere at a particular moment in time to develop a reliable forecast. Thus, the National Weather Service in the United States has established a surface observation network across the country. Made up of some 1700 official sites, the network gathers data in a consistent manner at regular intervals—usually a minimum of once an hour. Most of these data are collected by the Automated Surface Observing System (ASOS), shown in ***Figure 12-11.*** In addition to the weather instruments already discussed, the ASOS uses a rain gauge for measuring rainfall, as well as a **ceilometer,** which measures the height of cloud layers and estimates the amount of sky covered by clouds. You'll learn more about the ASOS in this chapter's *Science & Technology* feature.

Figure 12-11 This automated weather system measures surface data.

UPPER-LEVEL DATA

While surface weather data are important, the weather that we experience is largely the result of changes that take place high in the troposphere. To make accurate forecasts, meteorologists must gather atmospheric data at heights of up to 30 000 m. This is a more formidable task than gathering surface data, and therefore it requires more sophisticated technology.

Figure 12-12 Radiosondes are used to gather upper-level weather data.

At present, the instrument of choice for gathering upper-level data is a balloon-borne package of sensors called a **radiosonde,** shown in ***Figure 12-12.*** The sensors on a radiosonde measure temperature, air pressure, and humidity. These readings are constantly sent back by radio signal to a ground station that tracks the movements of the radiosonde. Tracking is a crucial component of upper-level observations—meteorologists can determine wind speed and direction by tracking how fast and in what direction the radiosonde is moving. The various data are plotted on a chart, giving meteorologists a profile of the temperature, pressure, humidity, wind speed, and wind direction of a particular part of the troposphere. Such charts are used to forecast atmospheric changes that affect surface weather. While radiosondes provide accurate snapshots of atmospheric conditions, they are quite expensive. It is hoped that in the future, data from satellites will replace or greatly supplement radiosonde observations.

WEATHER RADAR

There are thousands of surface observation sites and 100 upper-level observation sites across the United States. Yet the data from these sites cannot pinpoint where rain is falling at any given moment. For that purpose, a weather radar system is needed. The term *radar* stands for "radio detecting and ranging." A radar system is made of several parts. A transmitter generates electromagnetic waves, which leave the transmitter through antennae. In weather radar systems, the waves are programmed to ignore small cloud droplets and to bounce off large raindrops. The large raindrops scatter some of the radio waves. These scattered waves, or echoes as they are often called, are received by other antennae. An amplifier increases the wave signals of the scattered waves. A computer then processes the signals and displays them on a screen. From this, meteorologists can compute the distance to the raindrops and the location of the rain relative to the receiving antennae. The radar system rotates in a circle, allowing meteorologists to gauge where rain is falling within the radar's range—usually an area with a diameter of about 400 km.

Doppler Radar Many advanced weather radar systems take advantage of a phenomenon called the Doppler effect. The **Doppler effect** is the change in wave frequency that occurs in energy, such as sound or light, as that energy moves toward or away from an observer. You've probably noticed that sounds produced by a horn from an approaching train change once the train has passed. Look at ***Figure 12-13.*** As the train approaches, the frequency and pitch of the sound coming from the horn are high. As the train passes, the frequency and pitch lower. This is the Doppler effect in action. Meteorologists use Doppler radar, which is based on the Doppler effect, to plot the speed at which raindrops move toward or away from a radar station. Because the motion of the moving raindrops is caused by wind, Doppler radar provides a good estimation of the wind speeds associated with precipitation areas, including those that are experiencing severe weather such as thunderstorms and tornados. The ability to measure wind speeds gives Doppler radar a distinct advantage over conventional weather radar systems.

Weather Satellites

In addition to communications, one of the main uses of satellites in orbit around Earth is to observe weather. Cameras mounted aboard a weather satellite take photos of Earth at regular intervals. These photos are beamed back to ground stations and their data are plotted on maps. Unlike weather radar, which tracks precipitation but not clouds, satellites track clouds but not necessarily precipitation. By combining data from the two types of technology, meteorologists can determine where both clouds and precipitation are occurring.

Figure 12-13 As the train approaches, the sound waves ahead of it are compressed. These short waves have a high frequency, so the horn sounds high. Behind the train, the sound waves are stretched out. These longer waves have a lower frequency, so the horn sounds lower.

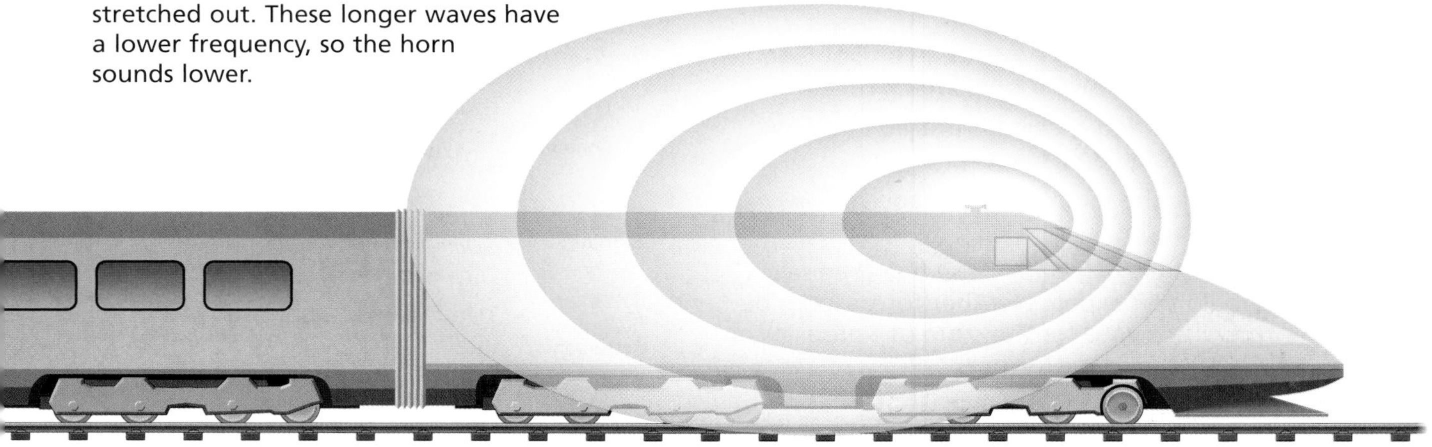

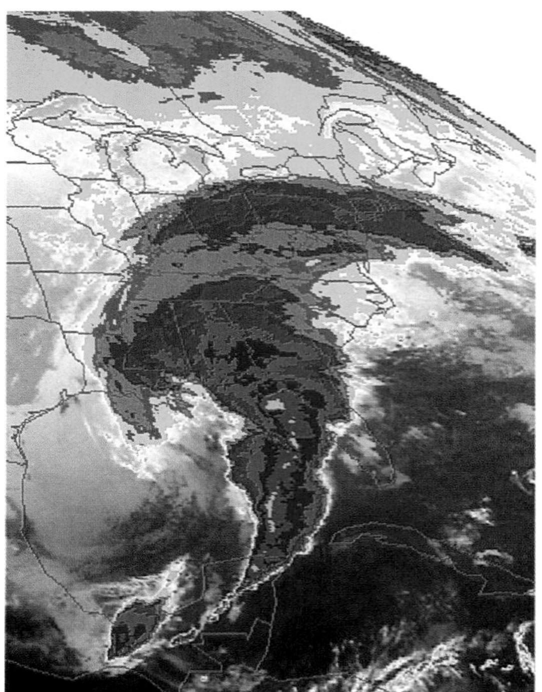

Figure 12-14 This infrared image shows a huge storm system over the eastern United States.

Infrared Imagery Weather satellites use both visible light and invisible radiation to observe the atmosphere. The satellites discussed thus far use cameras that need visible light to take photos. When such a satellite is observing a portion of Earth that is in darkness, however, its cameras are useless. Thus, some satellites are designed to use infrared imagery. Infrared imagery detects differences in thermal energy, which are used to map either cloud cover or surface temperatures. In an infrared image, such as the one shown in ***Figure 12-14,*** objects that radiate warmth at slightly different frequencies show up as different colors. As you learned in Chapter 11, different types of clouds form at different levels of the atmosphere, which are characterized by different temperatures. Infrared images allow meteorologists to determine the temperature of a cloud. From this, they can infer what type it is and estimate its height. Infrared imagery is especially useful in detecting strong thunderstorms that extend to great heights in the atmosphere and consequently show up as very cold areas on an infrared image. Because the strength of a thunderstorm is related to its height, infrared imagery can be used to establish a storm's potential to produce severe weather.

Section Assessment

1. If your goal was to vastly improve the density of weather data in the United States, would you focus on gathering more surface data or more upper-level data? Explain.
2. What is the main advantage of Doppler radar over conventional weather radar?
3. Compare and contrast infrared imagery and visible-light imagery.
4. What is the main disadvantage of radiosondes?
5. **Thinking Critically** All else being equal, would you expect weather forecasts to be more accurate for the state of Kansas or a Caribbean island? Why?

Skill Review

6. **Concept Mapping** Use the following terms to construct a concept map about instruments that gather surface weather data. For more help, refer to the *Skill Handbook.*

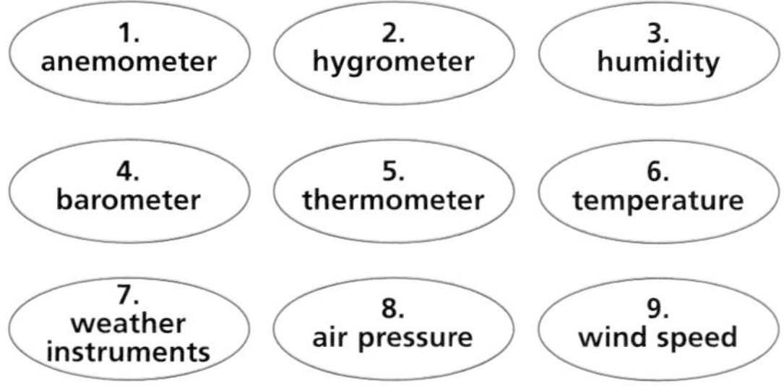

SECTION 12.4 Weather Analysis

I&E 1.d, 1.m

After weather observations are gathered, meteorologists plot the data on a map, using station models for individual cities or towns. A **station model** is a record of weather data for a particular site at a particular time. Meteorological symbols, such as the ones shown in ***Figure 12-15,*** are used to represent weather data in a station model. (For a more complete list of the symbols that meteorologists use to represent weather data, see *Appendix E.*) A station model allows meteorologists to fit a large amount of data into a small space. It also gives meteorologists a uniform way of communicating weather data. You'll use station models and weather maps to forecast the weather in the *Mapping GeoLab* at the end of this chapter.

OBJECTIVES

- **Analyze** *a basic surface weather chart.*
- **Distinguish** *between analog and digital forecasting.*
- **Describe** *problems with long-term forecasts.*

VOCABULARY

station model
isopleth
digital forecast
analog forecast

SURFACE ANALYSIS

Station models provide information for individual sites. To plot data nationwide or globally, meteorologists use **isopleths,** which are lines that connect points of equal or constant values. The values represent different weather variables, such as pressure or temperature. Lines of equal pressure, for example, are called isobars; lines of equal temperature are called isotherms. The lines themselves are similar to the contour lines—lines of equal elevation—that you studied in Chapter 2. Just as you can make inferences about elevation by studying contour intervals on a map, you can also make inferences about weather by studying isobars or isotherms on a map. For instance, you can tell how fast wind is blowing in an area by noting how closely isobars are spaced. Isobars that are close together indicate a large

I&E

1.d Formulate explanations by using logic and evidence.

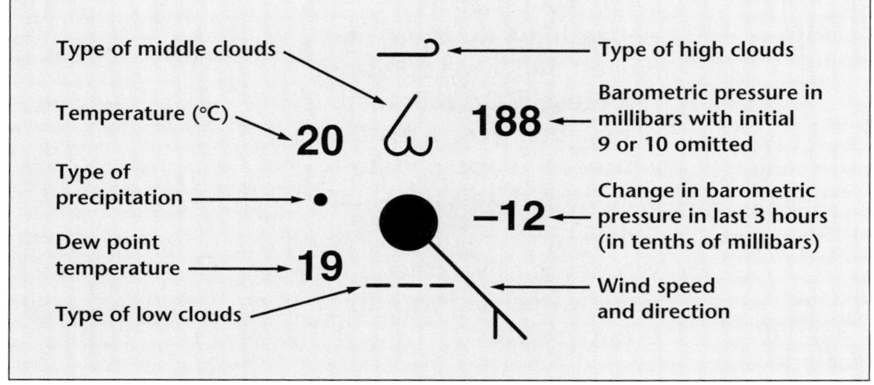

Figure 12-15 A station model shows weather data for a particular area at a particular time.

pressure difference over a small area. A large pressure difference causes strong winds. Conversely, isobars that are spread far apart indicate a small difference in pressure. Winds in these areas would be light. As shown in ***Figure 12-16,*** isobars also indicate the locations of high- and low-pressure systems. This information is especially useful when combined with isotherms, which identify temperature gradients and, consequently, frontal systems. Using isobars, isotherms, and station-model data, meteorologists can analyze current weather conditions for a particular time and place. This is crucial information—meteorologists must understand current weather conditions before they can move on to forecasting the weather. You'll learn more about isobars in the *Problem-Solving Lab* on this page.

SHORT-TERM FORECASTS

In the early days of weather forecasting, meteorologists simply observed current weather conditions, compared these conditions to those that had occurred a day or two before, and then extrapolated the changes a day or two into the future. The resulting positions of

Problem-Solving Lab

Interpreting Scientific Illustrations

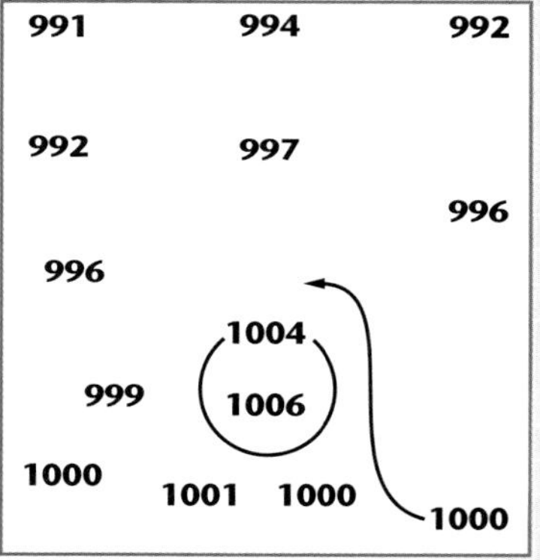

Create and analyze isobars on a weather map Areas of high and low pressure can be indicated on a weather map by lines of approximately equal pressure called isobars.

Analysis

1. On a blank piece of paper, trace the diagram shown here, along with the pressure values at various locations, which are given in millibars (mb).
2. A 1004-mb isobar that encircles one location on this map has been drawn and labeled. Complete and label the 1000-mb isobar that has been started. Finally, draw a 996-mb isobar and a 992-mb isobar. The isobars may not completely encircle a location in a map of this scale.

Thinking Critically

3. What is the contour interval of the isobars on this map?
4. A blue letter *H* and a red letter *L* in the centers of closed isobars mark the areas of highest and lowest pressure, respectively. On your map, place a blue *H* or a red *L*—whichever is appropriate—inside the closed 1004-mb isobar.
5. What type of weather is commonly associated with this pressure system?

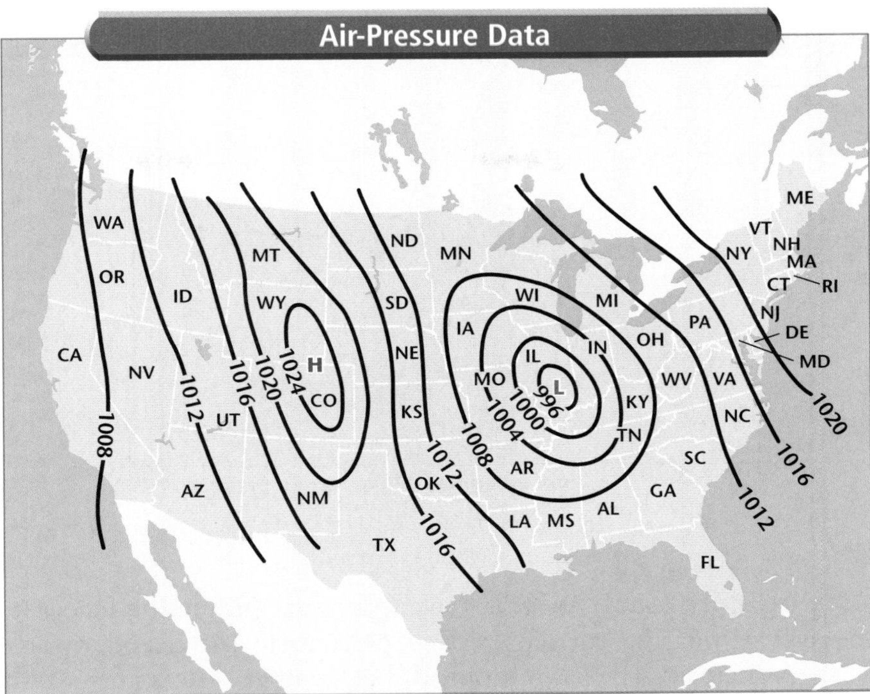

Figure 12-16 This map shows air-pressure data for the United States. ***Where would you expect light winds?***

the weather systems served as the basis for their forecasts. Weather forecasting, however, is too complicated to rely on extrapolating the past movements of weather systems. Weather systems change directions, speed, and intensity with time. These changes take place in response to changes in the upper atmosphere, so a reliable forecast must analyze data from different levels in the atmosphere.

Digital Forecasts The key to unlocking the forecast puzzle lies in the fact that the atmosphere behaves much like a fluid. Thus, we can apply many of the same principles to the atmosphere and its variables, such as temperature, pressure, density, and so on, that we can apply to a fluid. Furthermore, these principles can be expressed in mathematical equations to determine how atmospheric variables change with time. For meteorologists to solve these equations on a global or national level would take an impossibly large amount of time. Fortunately, high-speed computers can do the job. A forecast such as this that relies on numerical data is known as a **digital forecast.** Digital forecasting is the main method used by modern meteorologists, such as the one shown in ***Figure 12-17.*** It is highly dependent on the density of the data available—basically, the more data, the more accurate the forecast.

Figure 12-17 This meteorologist is preparing a weather forecast.

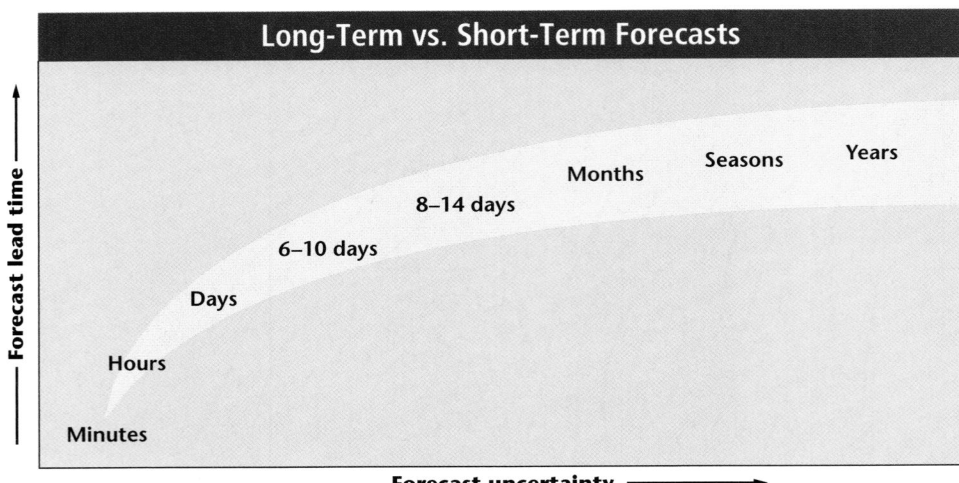

Figure 12-18 This graph shows that forecast uncertainty increases with time.

Analog Forecasts Another type of forecast, an **analog forecast,** involves comparing current weather patterns to patterns that took place in the past. The assumption is that weather systems behave in a similar fashion. Analog forecasting is so called because meteorologists look for a pattern from the past that is analogous, or similar to, a current pattern. To ensure an accurate analog forecast, meteorologists must find a past event that is similar to a current event through all levels of the atmosphere, and also over a large area. This is the main drawback of analog forecasting. Still, analog forecasting is useful for conducting monthly or seasonal forecasts, which are based mainly on the past behavior of cyclic weather patterns. Let's explore the strengths and weaknesses of long-term forecasts.

Long-Term Forecasts

Regardless of the forecasting method used, all forecasts become less reliable when they attempt to predict long-term changes in the weather. Why? Even high-tech computers cannot model every factor that affects the weather. Recall that mountains, valleys, rivers, lakes, cities, and countless other features on Earth's surface affect the amount of heat absorbed in a particular location. This, in turn, affects the pressure and therefore the wind of that area, which in turn affects cloud formation and virtually all other aspects of the weather. Over time, all these factors interact to create progressively more complicated scenarios.

The most accurate and detailed forecasts are short-term in nature, as shown in ***Figure 12-18.*** For hourly forecasts, extrapolation is a reliable forecasting method because the current weather is dominated by small-scale weather features that are readily observable by radar and satellite. Forecasts in the one- to three-day range, however, are no longer based on the movement of observed clouds and precipitation, which change by the hour. These forecasts are dependent on the behavior of larger surface and upper-level features, such as low-pressure systems. A one- to three-day forecast can somewhat accurately predict whether the day will be rainy or dry, and, if rainy, when the precipitation will occur. At this range, however, the forecast will not be able to pinpoint an exact temperature or sky condition at a specific time.

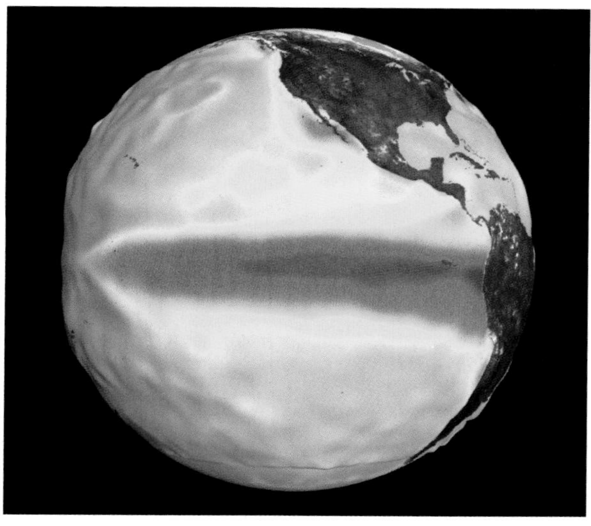

Figure 12-19 Changes in ocean-surface temperatures can trigger changes in weather patterns. This satellite image shows a cyclic event known as El Niño wherein the Pacific Ocean warms along the equator and triggers short-term climatic changes.

Accuracy Declines with Time At the four- to seven-day range, forecasts must attempt to predict changes in surface weather systems based on circulation patterns throughout the troposphere and lower stratosphere. Meteorologists can estimate each day's weather but can offer little detail as to when or what exact weather conditions will occur. At the one- to two-week range, forecasts are based on changes in large-scale circulation patterns. Thus, these forecasts are vague and based mainly on analogous conditions.

Long-term forecasts involving months and seasons are based largely on patterns or cycles. Several of these cycles, such as the one shown in ***Figure 12-19,*** involve changes in the atmosphere, ocean currents, and solar activity, all of which might be occurring at the same time. The key to future improvement in weather forecasts lies in identifying the many influences involved, understanding these influences and how they interact, and finally, determining their ultimate effect on weather over progressively longer periods of time.

SECTION ASSESSMENT

1. Find an example of a station model in your local newspaper and describe the symbols on the model.
2. Compare and contrast analog and digital forecasting.
3. Explain why long-term forecasts aren't always accurate.
4. **Thinking Critically** Based on what you have learned about digital forecasting, what single improvement do you think would be necessary to increase the reliability of this type of forecasting?

SKILL REVIEW

5. **Forming a Hypothesis** For a time period of three days or less, hypothesize which would be more accurate: digital forecasting or analog forecasting. Explain your hypothesis. For more help, refer to the *Skill Handbook.*

Mapping GeoLab

Interpreting a Weather Map

It's time to put your knowledge of meteorology into action. The surface weather map on the following page shows actual weather data for the United States. In this activity, you will use the station models, isobars, and pressure systems on the map to forecast the weather.

Preparation

Problem

How can you use a surface weather map to interpret information about current weather and to forecast future weather?

Materials

pencil
ruler

Procedure

1. The map scale is given in nautical miles. Refer to the scale when calculating distances.
2. The unit for isobars is millibars (mb). In station models, pressure readings are abbreviated. For example, 1021.9 mb is plotted on a station model as 219 but read as 1021.9.
3. Wind shafts point in the direction from which the wind is blowing.

Analyze

1. What is the contour interval of the isobars?
2. What are the highest and lowest isobars? Where are they located?
3. In which direction are the winds blowing across Texas and Louisiana?
4. What and where are the coldest and warmest temperatures that you can find in the continental United States?

Conclude & Apply

1. Would you expect the weather in Georgia and Florida to be clear or rainy? Why?
2. Both of the low-pressure systems in eastern Canada and off the Oregon coast are moving toward the east at about 15 mph. What kind of weather would you predict for Oregon and for northern New York for the next few hours? Explain.

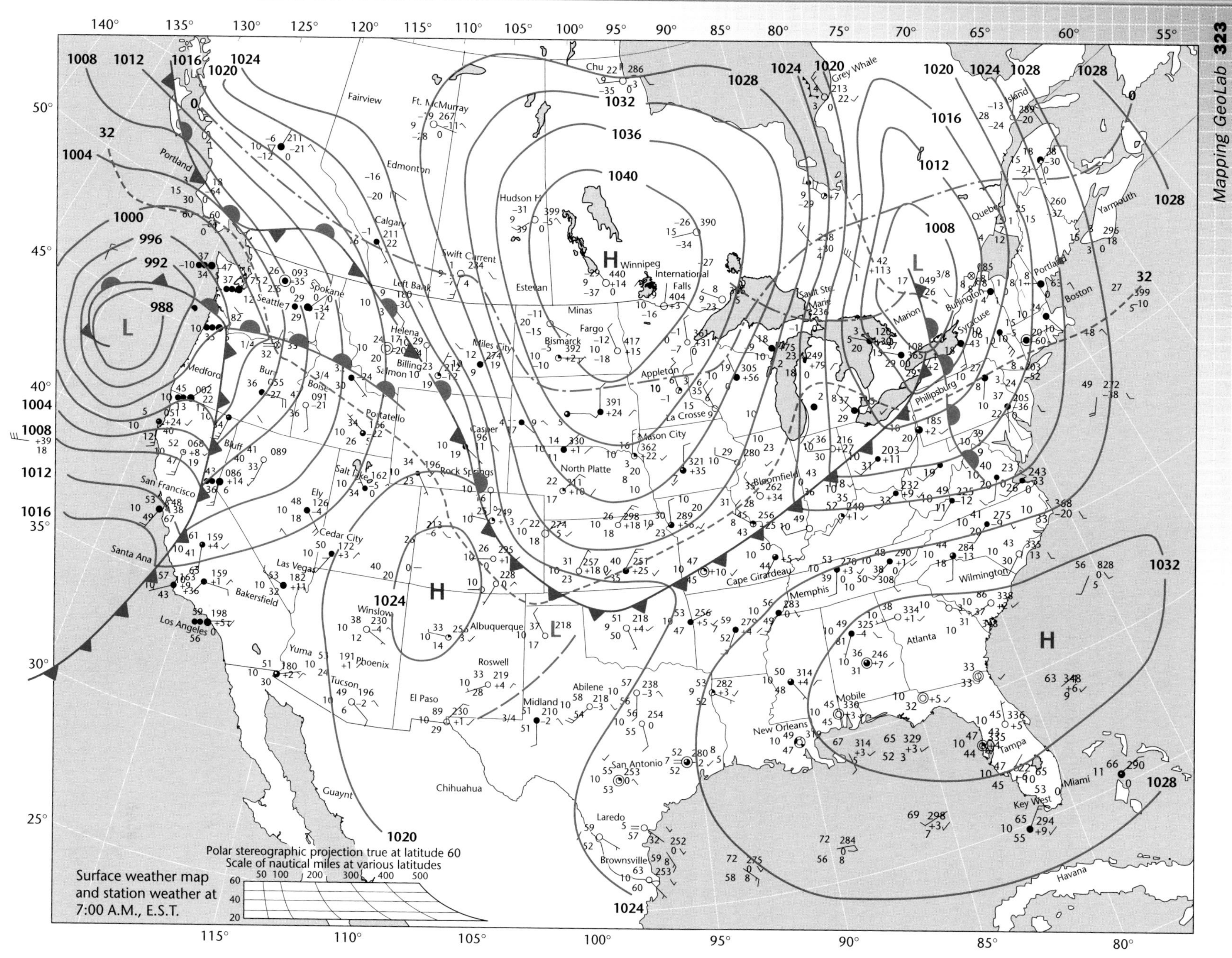
Surface weather map
and station weather at
7:00 A.M., E.S.T.
Polar stereographic projection true at latitude 60
Scale of nautical miles at various latitudes

Science & Technology

I&E 1.m

Tracking Atmospheric Change

On July 16, 1999, John F. Kennedy Jr. piloted a small private plane bound for Martha's Vineyard. The plane never reached its destination. Searchers determined that the plane crashed into the Atlantic Ocean, killing all three people on board. While no one will know exactly what happened, the dense fog may have been a factor in the crash.

Safety is a priority in the aviation industry. Pilots must be aware of weather conditions to avoid crashing when landing or taking off. Prior to 1990, National Weather Service personnel were responsible for gathering and communicating weather data to pilots. These professionals collected data on air pressure, wind speed, temperature, cloud cover, and precipitation.

ASOS

Concern about possible human errors prompted scientists to develop a more efficient system for transmitting weather data to pilots. The resulting computerized system, called the Automated Surface Observing System (ASOS), is now the cornerstone of weather forecasting and communication in this country.

The Pros and Cons

Today, more than 1000 ASOS units are in operation at major airports, continuously recording air pressure, temperature, wind speed and direction, runway visibility, cloud ceiling, and precipitation intensity. Data are automatically updated every minute. While human observations are based on what can be seen from a given vantage point, computerized observations are not affected by varying light and terrain conditions. For this reason, many aviation professionals believe that ASOS data are more consistent than manually collected data.

Some aviation professionals, however, disagree. The National Air Traffic Controllers Association believes that human observers were replaced by machines primarily to save money, and that the loss of trained weather observers is detrimental to aviation safety. On the evening that Kennedy's plane went down, the ASOS indicated that visibility was 13 to 16 km. In actuality, visibility was reported to be so poor by other pilots that the lights of Martha's Vineyard could not be seen from the air. The National Weather Service admits that the ASOS needs to be refined and continues to implement upgrades. Still, this group insists that widespread use of the ASOS will reduce weather-related aviation accidents.

Activity

Does your local airport use the ASOS? Go to the Earth Science Web Site at earthgeu.com to research which airports use the ASOS, or contact your local airport. Write to airport officials for statistics on aviation accidents both prior to and after the installation of the ASOS. Has there been a change in the number of aviation accidents since the airport implemented the ASOS? Use this information to write a short opinion piece on the continued use of the ASOS.

CHAPTER 12

Study Guide

Summary

SECTION 12.1

The Causes of Weather

Main Ideas

- Meteorology is the study of the atmosphere. Weather is the current state of the atmosphere, and climate is the average weather over a long period of time.
- An air mass is a large body of air that takes on the characteristics of the area over which it forms.

Vocabulary

air mass (p. 301)
air mass modification (p. 304)
climate (p. 300)
meteorology (p. 299)
weather (p. 300)

SECTION 12.2

Weather Systems

Main Ideas

- The Coriolis effect deflects air to the right in the northern hemisphere and to the left in the southern hemisphere. The Coriolis effect combines with the heat imbalance found on Earth to form the trade winds, prevailing westerlies, and polar easterlies.
- Weather in the middle latitudes is strongly influenced by fast-moving, high-altitude jet streams.
- A front is the boundary between two air masses of different densities. The four types of fronts are cold fronts, warm fronts, occluded fronts, and stationary fronts.

Vocabulary

Coriolis effect (p. 305)
front (p. 308)
jet stream (p. 307)
polar easterlies (p. 307)
prevailing westerlies (p. 306)
trade winds (p. 305)

SECTION 12.3

Gathering Weather Data

Main Ideas

- Two of the most important factors in weather forecasting are the accuracy and the density of the data. Surface data are easier to gather than upper-level data.
- The most common instrument for collecting upper-level data is a balloon-borne radiosonde. Radiosondes measure temperature, pressure, humidity, wind speed, and wind direction.
- Weather radar pinpoints exactly where precipitation occurs. Weather satellites use both visible-light imagery and infrared imagery to observe weather conditions on Earth.

Vocabulary

anemometer (p. 313)
barometer (p. 312)
ceilometer (p. 313)
Doppler effect (p. 315)
hygrometer (p. 313)
radiosonde (p. 314)
thermometer (p. 312)

SECTION 12.4

Weather Analysis

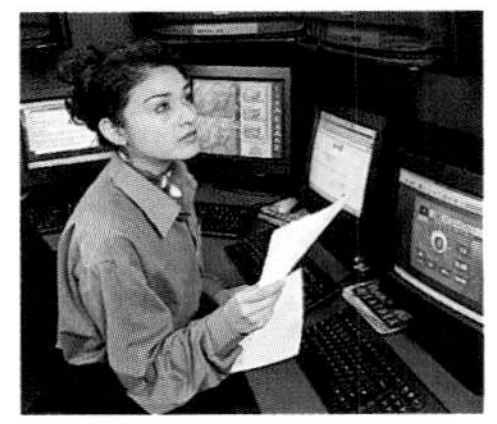

Main Ideas

- A station model is a record of weather data for a particular site at a particular time. On a weather map, lines of equal pressure are called isobars and lines of equal temperature are called isotherms.
- Digital forecasting uses numerical data. Analog forecasting compares current weather patterns to patterns that took place in the past. All forecasts become less reliable when they attempt to predict long-term changes in the weather.

Vocabulary

analog forecast (p. 320)
digital forecast (p. 319)
isopleth (p. 317)
station model (p. 317)

earthgeu.com/vocabulary_puzzlemaker

CHAPTER 12

Assessment

Understanding Main Ideas

1. Which term best describes a snowflake?
   a. hydrosphere
   b. hydrometeor
   c. lithometeor
   d. electrometeor

2. What winds blow between 30° and 60° north and south latitude?
   a. trade winds
   b. prevailing westerlies
   c. polar easterlies
   d. jet streams

3. What would be the most likely classification for an air mass originating over Alaska and Canada?
   a. mT
   b. cP
   c. cT
   d. mP

4. What would be the most likely dominant air mass over the eastern United States in summer?
   a. cT
   b. cP
   c. mT
   d. mP

5. What instrument is used to measure the heights of the bases of clouds?
   a. radiosonde
   b. ceilometer
   c. hygrometer
   d. barometer

6. Which term describes changes in air motion resulting from Earth's rotation?
   a. jet stream
   b. convergence
   c. Coriolis effect
   d. Hadley cell

7. What forecast method is best for researching past weather events?
   a. digital
   b. analog
   c. extrapolation
   d. numerical

8. Which of the following is NOT characteristic of a high-pressure system?
   a. sinking air
   b. dense air
   c. fair weather
   d. thunderstorms

9. Which of the following would NOT be included in a station model?
   a. humidity
   b. wind
   c. pressure
   d. temperature

10. What does an anemometer measure?
    a. humidity
    b. air pressure
    c. wind speed
    d. wind direction

11. Describe the relationship among meteorology, weather, and climate.

Use the surface weather chart below to answer questions 12 and 13.

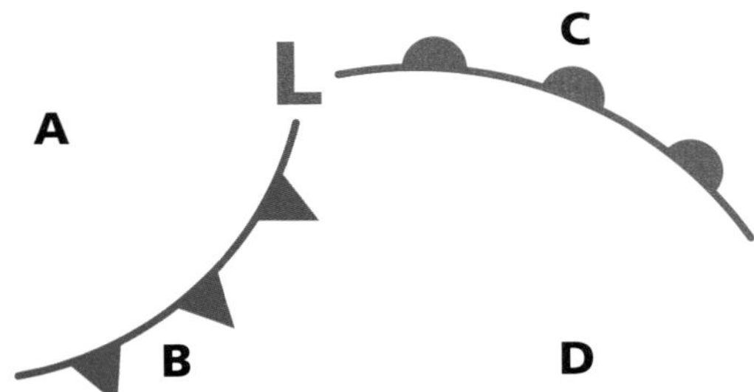

12. What location will soon experience a warm front?

13. At what location would you forecast the greatest probability of thunderstorms? At what location would you forecast the weather to turn colder over the next day?

Test-Taking Tip

Focus While you take a test, pay absolutely no attention to anyone other than the teacher or proctor. If someone tries to talk with you during a test, don't answer. You'll become distracted. Also, the proctor may think that you were cheating. Don't take the chance. Focus on the test, and nothing else.

Assessment

Applying Main Ideas

14. Is meteorological data easier to gather on land or on water? Explain your answer.

15. What is Doppler radar?

16. What happens to an air mass as it moves away from its source region?

17. Why are rain showers common near the ITCZ?

18. In your own words, describe the global wind systems.

19. Construct a station model using the following data: temperature: –5°F; dew point: –12°F; wind: north, 20 knots; barometric pressure: 1038.5 mb; sky: clear.

Thinking Critically

20. Forecast the weather for the next 24 hours for an area experiencing the same conditions as those included in the station model you constructed in question 19.

21. Like the north pole, the south pole receives little solar radiation during the winter. Unlike the north pole, however, the south pole does not send outbreaks of extremely frigid air as far as the subtropics. Why? (Hint: You may want to study a world map to answer this question.)

22. You hear on a news report that an area has received nearly twice its normal snowfall during the winter. What can you infer about the position of the jet stream from this report?

23. Review ***Figure 12-4*** on page 305, which shows global wind systems, then note the relative positions of North America and Europe on a world map. Hypothesize why the winds that blow between 30° north and south latitude and the equator are called the trade winds.

Standardized Test Practice

1. Which of the following types of air masses are most likely to form over land near the equator?

a. mP **c.** cP
b. mT **d.** cT

2. Which wind system flows between 30° and 60° latitude north and south of the equator in an easterly direction toward the poles?

a. trade winds **c.** prevailing westerlies
b. Coriolis effect **d.** polar easterlies

INTERPRETING DATA Use the photo to answer question 3.

3. Meteorologists use many different instruments to gather atmospheric information. What is the instrument shown here called?

a. a hygrometer **c.** a radiosonde
b. a ceilometer **d.** radar

4. What does Doppler radar monitor?

a. the motion of moving raindrops
b. atmospheric pressure
c. temperature, air pressure, and humidity
d. the height of cloud layers

5. The data gathered by Doppler radar can be used to make a type of forecast that relies on numerical data. What is this type of forecast called?

a. an analog forecast **c.** an isopleth
b. a digital forecast **d.** ASOS

Chapter 13

The Nature of Storms

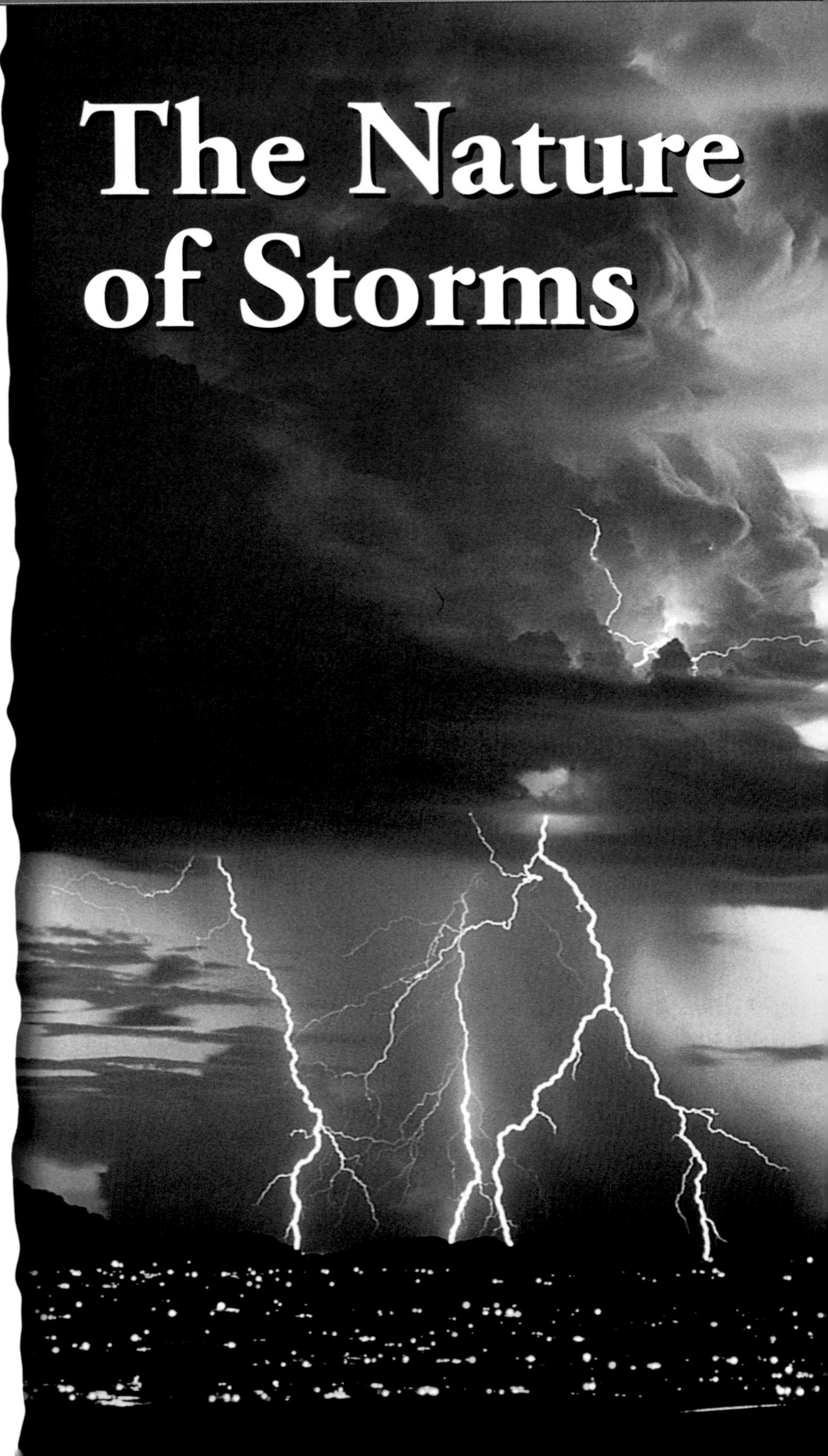

What You'll Learn

- How thunderstorms, tornadoes, and hurricanes form.
- What the effects of severe weather are.
- How repetitive weather patterns can cause droughts, floods, and other hazards.

Why It's Important

Severe weather can result in extensive property damage and loss of life. To implement safety measures and make other preparations for severe weather, it's necessary to understand when and where severe weather is likely to occur.

Earth Science Online

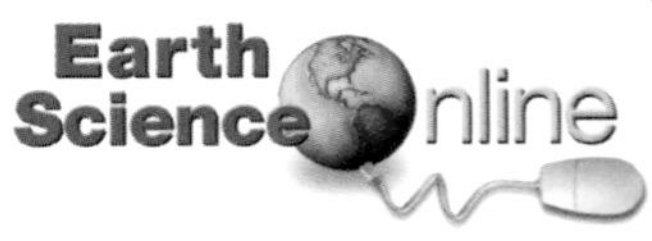

To find out more about severe weather, visit the Earth Science Web site at earthgeu.com

Discovery Lab Model Thunder

Did you know that lightning causes thunder? During a thunderstorm, lightning can reach temperatures of 30 000°C. This extreme heat causes the air around the lightning to expand rapidly, then quickly cool and contract. The rapid expansion of air generates sound waves heard as thunder. You can model thunder using a paper bag.

1. Blow into a brown paper lunch bag until it is full of air.
2. Hold the top of the bag firmly in one hand and twist it so that the air inside is trapped. Take care not to tear the bag.
3. Strike the bag sharply with your other hand so that the bag breaks.

CAUTION: Always wear safety goggles in the lab.

Observe What did you hear when the bag broke? How is this similar to the thunder produced by a lightning bolt? Light moves much faster than sound. Knowing this, what can you infer about the movement of a thunderstorm if the amount of time between when you see the lightning and hear the thunder increases between each lightning flash?

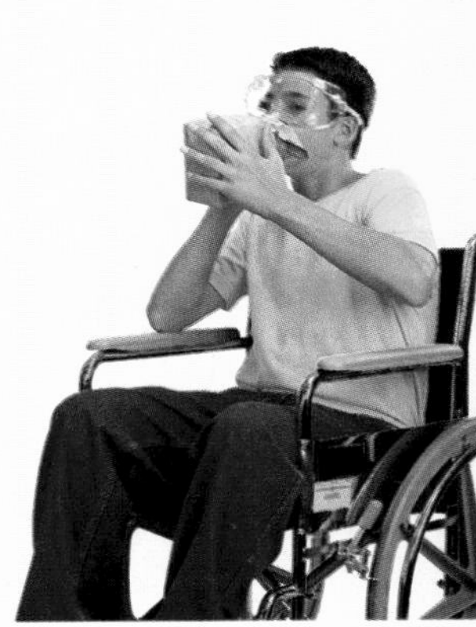

Section 13.1 *Thunderstorms*

I&E 1.i

OBJECTIVES

- **Identify** *the processes that form thunderstorms.*
- **Compare** *and* **contrast** *different types of thunderstorms.*
- **Describe** *the life cycle of a thunderstorm.*

VOCABULARY

air-mass thunderstorm
sea-breeze thunderstorm
frontal thunderstorm

At any given moment, nearly 2000 thunderstorms are occurring around the world. Most do little more than provide welcome relief on a muggy summer afternoon. Some, however, grow into atmospheric monsters capable of producing hail the size of baseballs, swirling tornadoes, and surface winds of more than 160 km/h. These severe thunderstorms can also provide the energy for nature's most destructive storms: hurricanes. All thunderstorms, regardless of intensity, have certain characteristics in common.

How Thunderstorms Form

In Chapter 11, you learned that under the right conditions, convection can cause a cumulus cloud to grow into a cumulonimbus cloud. You also learned that cumulonimbus clouds produce thunderstorms. What conditions are necessary for this to happen? For a thunderstorm to form, three conditions must exist. First, there must be an abundant source of moisture in the lower levels of the atmosphere.

As this moisture condenses, it releases latent heat. The release of latent heat keeps the cloud warmer than the air around it, which is crucial in maintaining the upward motion of the cloud. Second, some mechanism must lift the air so that the moisture can condense and release latent heat. You'll read about these mechanisms on the next page. Last, the portion of the atmosphere through which the cloud grows must be unstable. In other words, the air must continue to cool with increasing altitude for the growing cloud to stay warmer than the surrounding air. Recall that air can rise only if it's warmer than the air around it. If an air mass is stable, even the release of latent heat will not keep that air warmer than the air around it. The upward motion and growth of the cloud will stop.

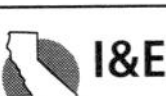

I&E

1.i Analyze the locations, sequences, or time intervals that are characteristic of natural phenomena (e.g., relative ages of rocks, locations of planets over time, and succession of species in an ecosystem).

Limits to Growth If the three conditions just described are met, the air will keep rising, causing more moisture to condense and creating more latent heat. This process will continue until the rising air meets a layer of stable air that it cannot overcome, or until the rate of condensation, which diminishes with height, is insufficient to generate enough latent heat to keep the cloud warmer than the surrounding air. This second factor limits most cumulonimbus clouds to a height of around 18 000 m. Because of factors which you'll learn about later in this section, typical thunderstorms last only about 30 minutes, and individual storms are only about 24 km in diameter. ***Figure 13-1*** shows which areas of the United States experience the most thunderstorms annually.

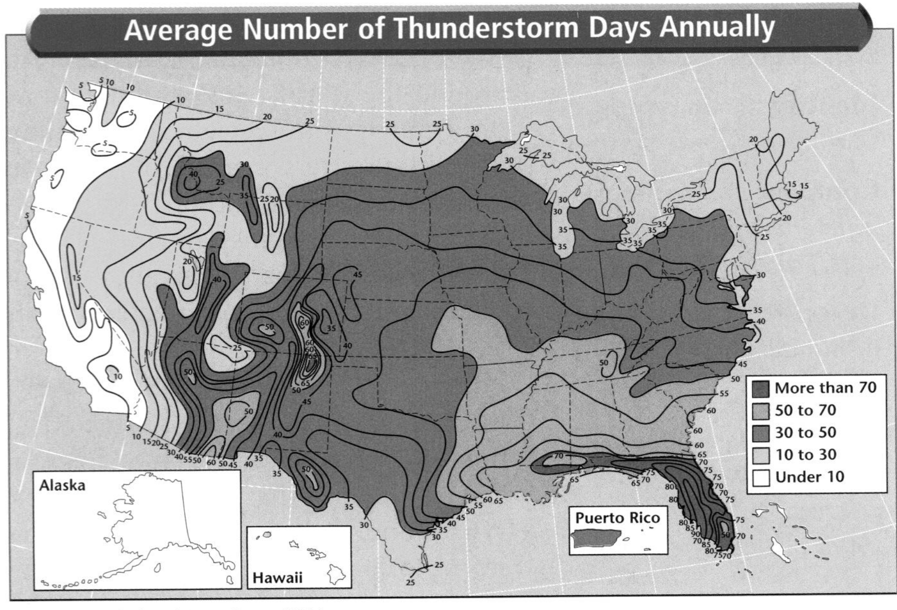

Source: National Climatic Data Center, NOAA

Figure 13-1 Geography and the movement of air masses both play roles in making thunderstorms most common in the southeastern United States.

AIR-MASS THUNDERSTORMS

Earlier you learned that some mechanism must lift air through a growing cloud so that its moisture can condense and release latent heat. Thunderstorms are often classified according to the mechanism that caused the air to rise. If the air rose because of unequal heating of Earth's surface within one air mass, the thunderstorm is called an **air-mass thunderstorm.** The unequal heating of Earth's surface reaches its maximum during mid-afternoon. Thus, air-mass thunderstorms are most common then.

There are two common types of air-mass thunderstorms. Mountain thunderstorms occur when an air mass rises as a result of orographic lifting, which, as you learned in Chapter 11, involves air moving up the side of a mountain. Sea-breeze thunderstorms are common along coastal areas during the summer, especially in the tropics and subtropics. **Sea-breeze thunderstorms** are local air-mass thunderstorms caused in part by extreme temperature differences between the air over land and the air over water, as shown in ***Figure 13-2.***

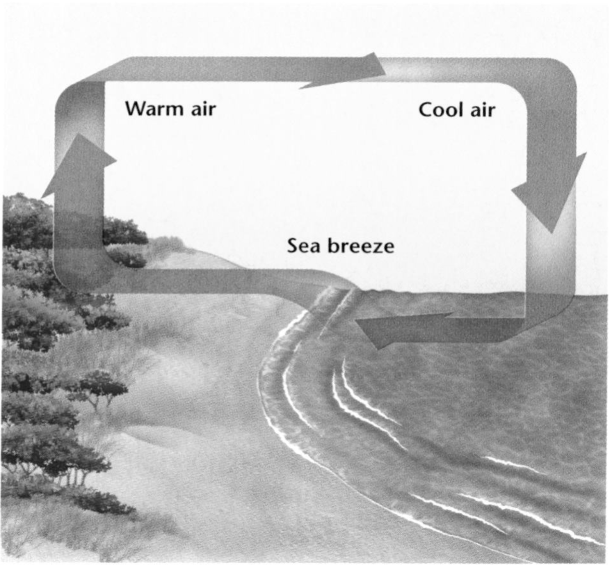

Figure 13-2 During the day, cool air over the ocean moves inland and creates a sea breeze. The cool air forces warm air over the land to rise. The rising air cools and sinks, creating a convection cell. These conditions can produce strong updrafts that result in thunderstorms.

FRONTAL THUNDERSTORMS

The second main classification of thunderstorms is **frontal thunderstorms,** which are produced by advancing cold fronts and, more rarely, warm fronts. In a cold front, cold air pushes warm air rapidly up the steep cold-front boundary. This rapid upward motion can produce a line of thunderstorms, sometimes hundreds of kilometers long, along the leading edge of the cold front. Cold-front thunderstorms get their initial lift from the push of the cold air. Because they are not dependent on daytime heating for their initial lift, cold-front thunderstorms can persist long into the night.

Less frequently, thunderstorms can develop along the advancing edge of a warm front. In a warm front, a warm air mass slides up and over a cold air mass. The boundary between the two air masses is not steep; thus, the air rises gradually. However, if the warm air behind the warm front is unstable and moisture levels are sufficiently high, a relatively mild thunderstorm can develop.

STAGES OF DEVELOPMENT

A thunderstorm usually has three stages: the cumulus stage, the mature stage, and the dissipation stage. The stages are classified according to the direction in which the air is moving.

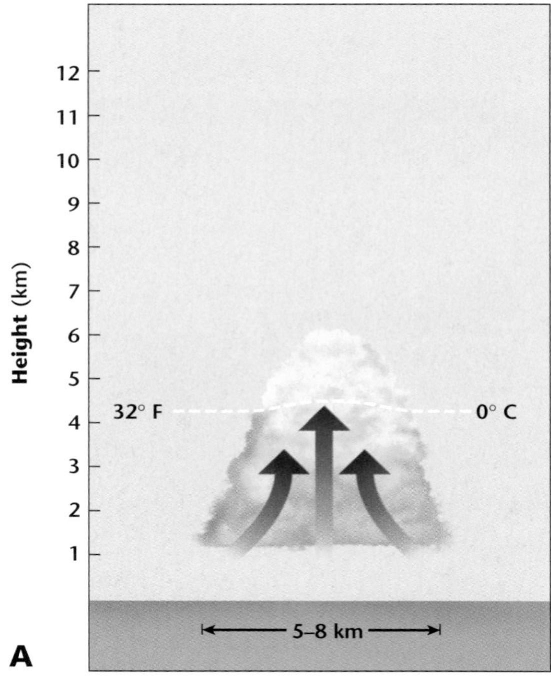

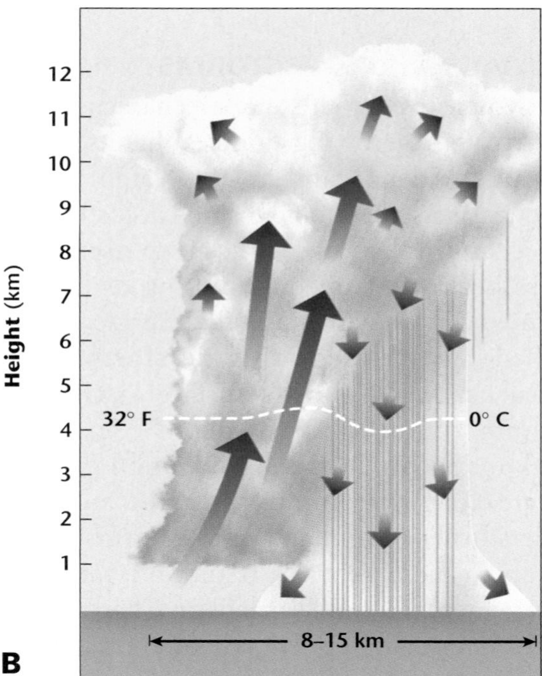

Figure 13-3 The cumulus stage of a thunderstorm is characterized mainly by updrafts **(A).** The mature stage is characterized by strong updrafts and downdrafts **(B).** The storm loses energy in the dissipation stage **(C).**

Cumulus Stage In the cumulus stage, air starts to rise nearly vertically upward, as shown in ***Figure 13-3A.*** This creates updrafts, which transport moisture to the upper reaches of the cloud. The moisture condenses into visible cloud droplets and releases latent heat. As the cloud droplets coalesce, they form larger and larger droplets, which eventually fall to Earth as precipitation. This begins the mature stage of a thunderstorm.

Mature Stage Precipitation in a thunderstorm is composed of water droplets that formed at high, cool levels of the atmosphere. As the precipitation falls, it cools the air around it. The newly cooled air is more dense than the surrounding air, so it sinks rapidly to the ground along with the precipitation. This creates downdrafts. As ***Figure 13-3B*** shows, the updrafts and downdrafts form a convection cell that produces the gusty surface winds associated with thunderstorms. In the mature stage, nearly equal amounts of updrafts and downdrafts exist side by side in the cumulonimbus cloud.

Dissipation Stage The production of downdrafts is ultimately the thunderstorm's undoing. The convection cell can exist only if there is a steady supply of warm, moist air at Earth's surface. Once that supply runs out, the updrafts slow and eventually stop. In a thunderstorm, shown in the photo on the next page, the supply of warm, moist air runs out because the cool downdrafts spread in all directions when they reach Earth's surface. This cools the areas from which the storm

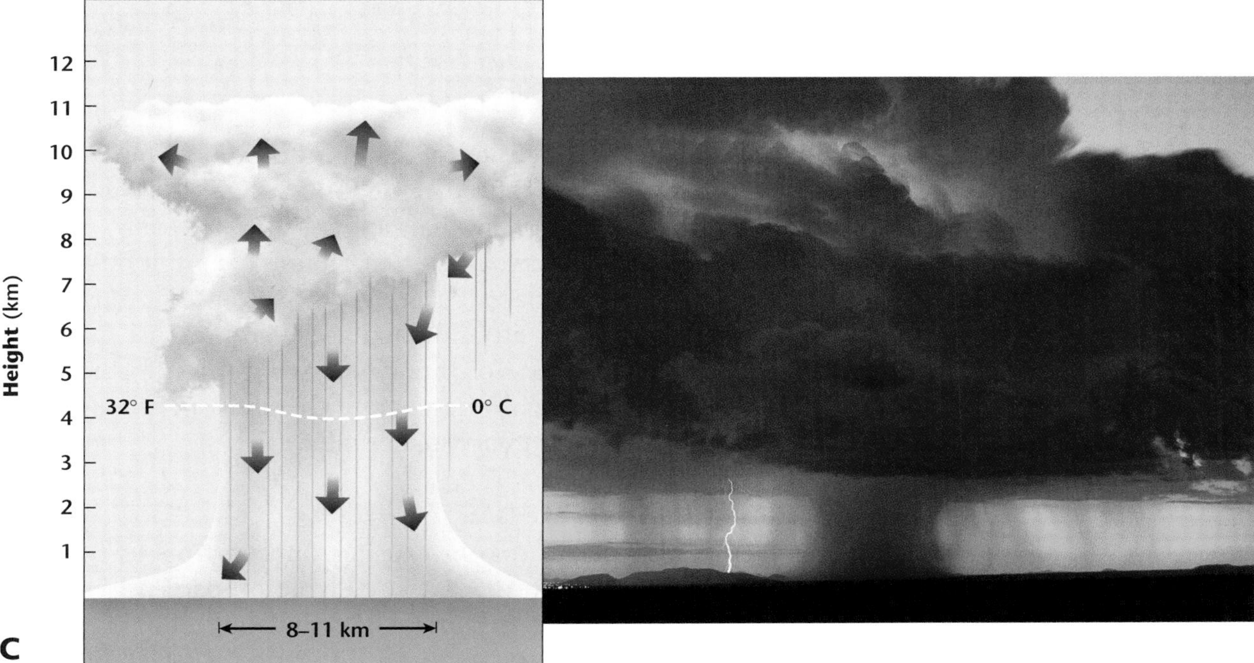

draws its energy. Without the warm air, the updrafts cease and precipitation can no longer form. The storm is then in the dissipation stage, as shown in ***Figure 13-3C.*** This stage, which is characterized primarily by lingering downdrafts, will last until the cloud runs out of previously formed raindrops. Next, you'll explore the destructive forces that can be unleashed when a severe thunderstorm strikes.

Section Assessment

1. It's 2:00 A.M. in the northeastern United States. A thunderstorm rumbles on the horizon. What type is it most likely to be? Why?
2. What conditions must be present for a thunderstorm to form?
3. Explain why a cold-front thunderstorm is usually more severe than a warm-front thunderstorm.
4. **Thinking Critically** In the tropics, where the tropopause is higher than in other areas, cumulonimbus clouds commonly reach towering heights of 15 000 m. Why is the height of the tropopause a factor in how tall a cumulonimbus cloud can grow?

Skill Review

5. **Concept Mapping** Use the following phrases to complete an events-chain concept map about the life cycle of a thunderstorm. For more help, refer to the *Skill Handbook.*

1. water droplets coalesce
2. updrafts stop
3. surface air is warmed by conduction
4. precipitation ends
5. water vapor condenses
6. warm air rises, creating updrafts
7. surface air is cooled by downdrafts
8. latent heat is released
9. precipitation and downdrafts begin

SECTION 13.2 Severe Weather

I&E 1.i

OBJECTIVES

- **Explain** *why some thunderstorms are more severe than others.*
- **Recognize** *the dangers of severe thunderstorms, including lightning, hail, high winds, and floods.*
- **Describe** *how tornadoes form.*

VOCABULARY

supercell
downburst
tornado
Fujita tornado intensity scale

All thunderstorms are not created equal. Some die out within minutes, while others flash and thunder throughout the night. What makes one thunderstorm more severe than another? Occasionally, weather events come together in such a way that there is a continuous supply of surface moisture. This happens along a cold front that moves into warmer territory and can lift and condense a continuous supply of warm air. In this case, a line of thunderstorms can last for hours or even days as they continually regenerate themselves with the new, warm air that is introduced into the updrafts.

SEVERE THUNDERSTORMS

Other factors also play a role in causing some storms to be more severe than others. Cold fronts are usually accompanied by upper-level, low-pressure systems that are marked by pools of cold air. This cold, high air increases the temperature difference between the upper and lower parts of the storm, which causes the air to become more unstable. As the instability of the air increases, the strength of the storm's updrafts and downdrafts intensifies. The storm is then considered to be severe. Severe thunderstorms can produce some of the most violent weather conditions on Earth. They may develop into self-sustaining, extremely powerful storms called **supercells,** which are characterized by intense, rotating updrafts. ***Figure 13-4B*** shows an illustration of a supercell. These furious storms can last for several hours and can have updrafts as strong as 240 km/h.

Figure 13-4 An anvil-shaped cumulonimbus cloud is characteristic of many severe thunderstorms **(A).** The most severe thunderstorms are supercells **(B).**

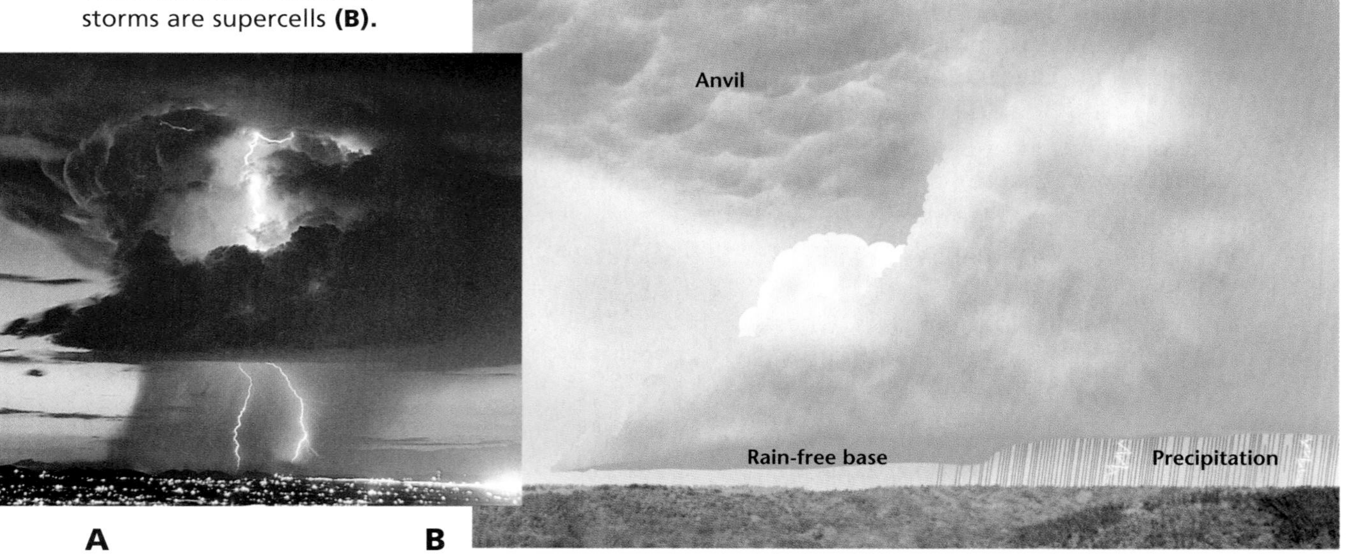

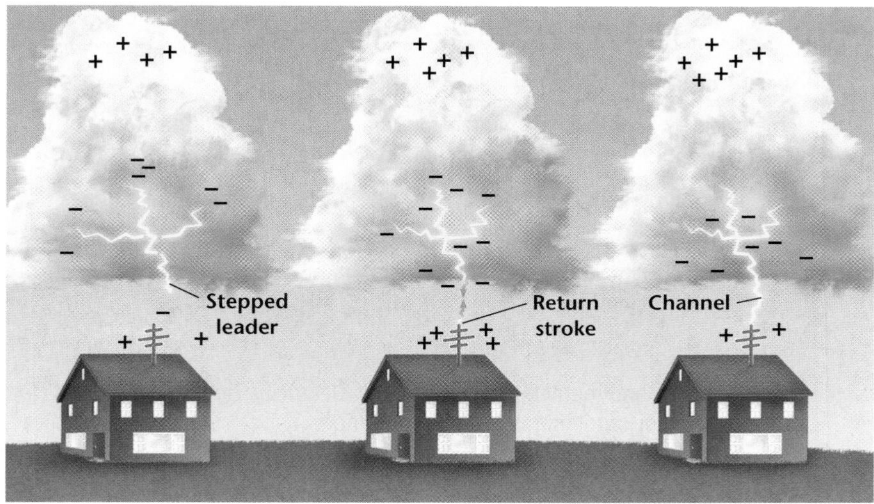

Source: NOAA

Figure 13-5 When a stepped leader nears an object on the ground, a powerful surge of electricity from the ground moves upward to the cloud, and lightning is produced.

Of the estimated 100 000 thunderstorms that occur each year in the United States, only about ten percent are considered to be severe, and fewer still reach classic supercell proportions. But when certain atmospheric conditions come together in the right way, the results can be spectacular and sometimes deadly, as you'll learn next.

LIGHTNING

Have you ever touched a metal object on a dry winter day and been zapped by a spark of electricity? If so, you were, in a sense, playing with lightning. Lightning is electricity caused by the rapid rush of air in a cumulonimbus cloud. A lightning bolt forms when friction between the updrafts and downdrafts within a cumulonimbus cloud separates electrons from some of their atoms either in the cloud or near the ground. The atoms that lose electrons become positively charged ions. Other atoms receive the extra electrons and become negatively charged ions. As ***Figure 13-5*** shows, this creates regions of air with opposite charges. To relieve the electrical imbalance, an invisible channel of negatively charged air, called a stepped leader, moves from the cloud toward the ground. When the stepped leader nears the ground, a channel of positively charged ions, called the return stroke, rushes upward to meet it. The return stroke surges from the ground to the cloud, illuminating the channel with about 100 million V of electricity. That illumination is lightning. You'll learn more about lightning in the *Science & Technology* feature at the end of this chapter.

The Power of Lightning A lightning bolt heats the surrounding air to about 30 000°C. That's about five times hotter than the surface of the Sun! The thunder you hear is the sound produced as this superheated air rapidly expands and contracts. Because sound waves travel

Table 13-1 Thunderstorm and Lightning Safety

When Thunderstorms Approach . . .
• Remember: If you can hear thunder, you are close enough to the storm to be struck by lightning. Go to a safe shelter immediately.
• Move to a sturdy building or car. Do not take shelter in small sheds, under isolated trees, or in convertible automobiles.
• If lightning is occurring and a sturdy shelter is not available, get inside a hard-topped automobile and keep the windows up.
• Get out of boats and away from water.
• Telephone lines and metal pipes can conduct electricity. Unplug appliances not necessary for obtaining weather information. Avoid using any electrical appliances. Use phones ONLY in an emergency.
If You Are Caught Outdoors and No Shelter Is Nearby . . .
• Find a low spot away from trees, fences, and poles. Make sure the place you choose is not subject to flooding.
• If you are in the woods, take shelter under the shorter trees.
• If you feel your skin tingle or your hair stand on end, squat low to the ground on the balls of your feet. Place your hands on your knees with your head between them. Make yourself the smallest target possible, and minimize your contact with the ground.

Source: NOAA

Topic: Thunderstorms
To find out more about thunderstorms, visit the Earth Science Web site at earthgeu.com

Activity: Design a bookmark with 10 facts on thunderstorms. Include safety tips.

more slowly than light waves, you may see lightning well before you hear thunder, even though they are generated at the same time.

Each year in the United States, lightning accounts for about 7500 forest fires, which result in the loss of millions of acres of forest. In addition, lightning strikes in the United States cause a yearly average of 300 injuries and 93 deaths to humans. ***Table 13-1*** lists safety tips to follow to avoid property damage and loss of life from lightning strikes.

THE FURY OF THE WIND

Recall that rain-cooled downdrafts descend to Earth's surface during a thunderstorm and spread out as they reach the ground. Sometimes, however, instead of dispersing that downward energy over a large area underneath the storm, the energy becomes concentrated in a local area. The resulting winds are exceptionally strong, with speeds of more than 160 km/h. Violent downdrafts that are concentrated in a local area are called **downbursts.**

Based on the size of the area they affect, downbursts are further classified as either macrobursts or microbursts. Macrobursts can cause a path of destruction up to 5 km wide. They have wind speeds of more than 200 km/h and can last up to 30 minutes. Smaller in size, though

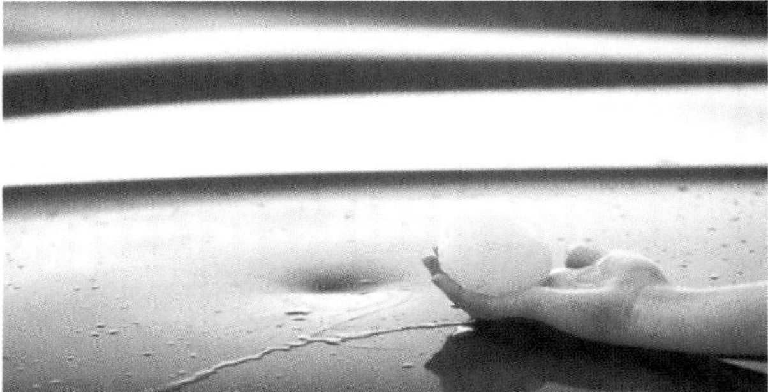

Figure 13-6 This car was damaged by large hailstones similar to the one shown here.

deadlier in force, microbursts affect areas of less than 3 km but can have winds exceeding 250 km/h. Despite lasting less than 10 minutes on average, a microburst is especially deadly because its smaller size makes it extremely difficult to detect and thus prepare for.

HAIL

Each year in the United States, almost $1 billion in damage is caused by another danger associated with thunderstorms: hail. Hail is precipitation in the form of balls or lumps of ice. It can do tremendous damage to crops, particularly in the Central United States, where hail occurs most frequently. Hail is most common during the spring growing season. ***Figure 13-6*** shows the damage that hail can cause.

Hail forms because of two characteristics common to thunderstorms. First, water droplets exist in the liquid state in the parts of a cumulonimbus cloud where the temperature is actually below freezing. When these supercooled water droplets encounter ice pellets, the water droplets freeze on contact and cause the ice pellets to grow larger. The second characteristic that allows hail to form is an abundance of strong updrafts and downdrafts existing side by side within a cloud. The growing ice pellets are caught alternately in the updrafts and downdrafts, so that they are constantly encountering more supercooled water droplets. The ice pellets keep growing until they are too heavy for even the strongest updrafts to keep aloft, and they finally fall to Earth as hail.

FLOODS

Sometimes, the wind currents in the upper atmosphere that cause weather systems to move are weak, and the weather systems and resulting storms move slowly. When this happens, a storm may dump its rain over a limited location, rather than spreading it over a large area. Floods such as the one in ***Figure 13-7*** can occur. The situation can worsen if there is abundant moisture available not just at Earth's surface, but also throughout the atmosphere. This makes the whole process of condensation, coalescence, and precipitation much more

Figure 13-7 This rural community was devastated by a flood in Arizona.

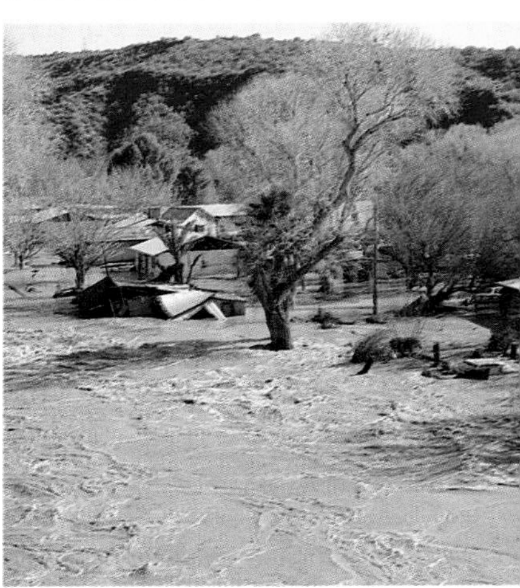

efficient and thus produces more rainfall. If the rain falls faster than the ground can absorb it, or faster than streams and rivers can transport it out of the area, flooding can occur. Floods are the main cause of thunderstorm-related deaths in the United States each year.

TORNADOES

Of all the dangers associated with thunderstorms, the most impressive by far is a tornado. A **tornado** is a violent, whirling column of air in contact with the ground. Before a tornado reaches the ground, it is called a funnel cloud. Tornadoes are often associated with supercells, the most severe thunderstorms. The air in a tornado is made visible by dust and debris drawn into the swirling column, or by the condensation of water vapor into a visible cloud. Over the area it covers, few storms on Earth can match a tornado's violence.

I&E

1.i Analyze the locations, sequences, or time intervals that are characteristic of natural phenomena (e.g., relative ages of rocks, locations of planets over time, and succession of species in an ecosystem).

A tornado forms when wind speed and direction change suddenly with height, a phenomenon known as wind shear. Under the right conditions, this can produce a horizontal rotation near Earth's surface, as shown in ***Figure 13-8.*** If this rotation takes place close enough to the thunderstorm's updrafts, the twisting column of wind can be tilted from a horizontal to a vertical position. As updrafts accelerate the rotation, air is removed from the center of the column, which in turn lowers air pressure in the center. The extreme pressure gradient between the center and the outer portion of the tornado produces the violent winds associated with tornadoes. Although tornadoes rarely exceed 200 m in diameter and usually last only a few minutes, they can be extremely destructive. In fact, they are classified according to their destructive force.

Tornado Classification Tornadoes can vary greatly in size and intensity. They are classified according to the **Fujita tornado intensity scale,** which ranks tornadoes according to their path of destruction, wind speed, and duration. The Fujita scale was named

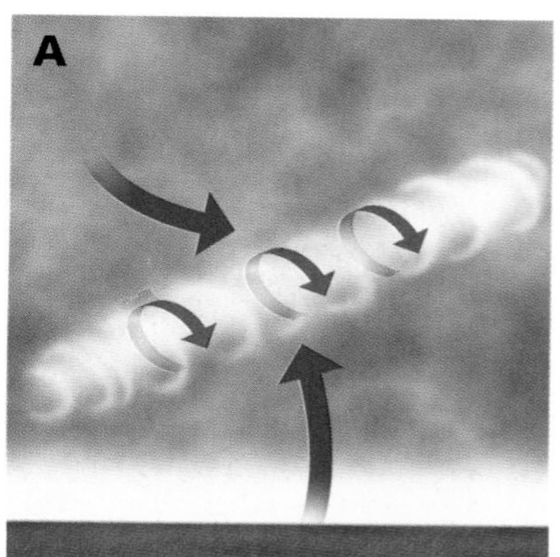

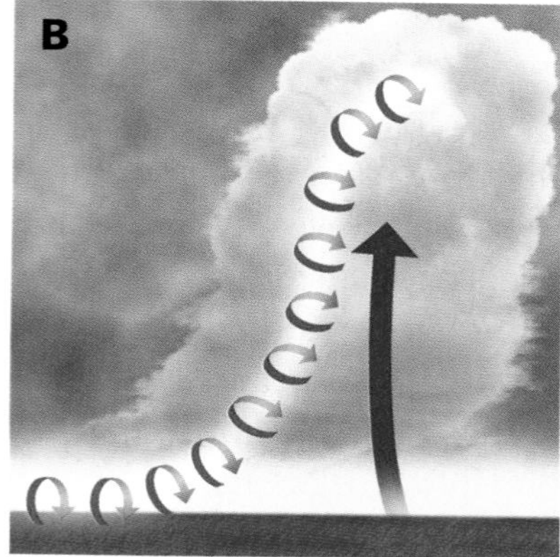

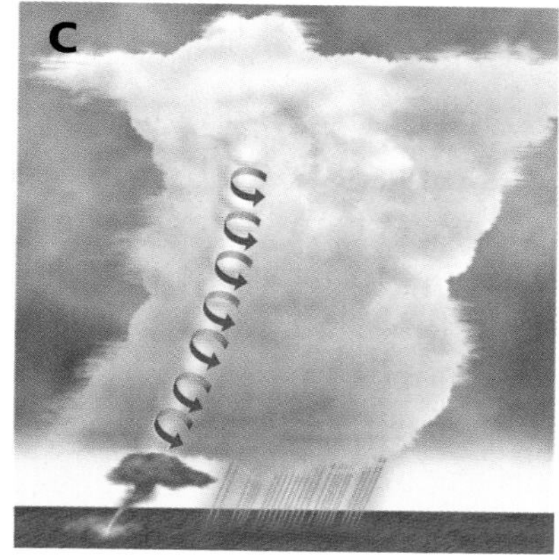

Figure 13-8 A change in wind direction and speed creates a horizontal rotation in the lower atmosphere **(A).** Strong updrafts tilt the rotating air from a horizontal to a vertical position **(B).** A tornado forms within the rotating winds **(C).**

for Japanese tornado researcher Dr. Theodore Fujita. The scale ranges from F0, which is characterized by winds of up to 118 km/h, to the incredibly violent F5, which can pack winds of more than 500 km/h. Most tornadoes do not exceed the F1 category. In fact, only about one percent ever reach the violent categories of F4 and F5. Those that do, however, can lift entire buildings from their foundations and toss automobiles and trucks around like toys. The Fujita scale is shown in ***Table 13-2.***

Tornado Distribution While tornadoes can occur at any time and at any place, there are some times and locations that are more conducive to their formation. Most tornadoes—especially the violent ones—form in the spring during the late afternoon and evening, when the temperature contrasts between polar air, which still has winter characteristics, and tropical air, which is steadily becoming warmer, are the greatest. These large temperature contrasts often spark the development of supercells, which are each capable of producing several strong tornadoes. Large temperature contrasts occur most frequently in the Central United States, where cold continental polar air collides with maritime tropical air moving northward from the Gulf of Mexico. More than 700 tornadoes touch down each year in the United States. Many of these occur in a region called "Tornado Alley," which extends from northern Texas through Oklahoma, Kansas, and Missouri.

Tornado Safety In the United States, an average of 80 deaths and 1500 injuries result from tornadoes each year. In an ongoing effort to reduce tornado-related fatalities, the National Weather Service issues tornado watches and warnings before a tornado actually strikes. These advisories are broadcast on local radio stations when tornadoes are

Table 13-2 Fujita Scale

Weak Tornadoes (F0 and F1)

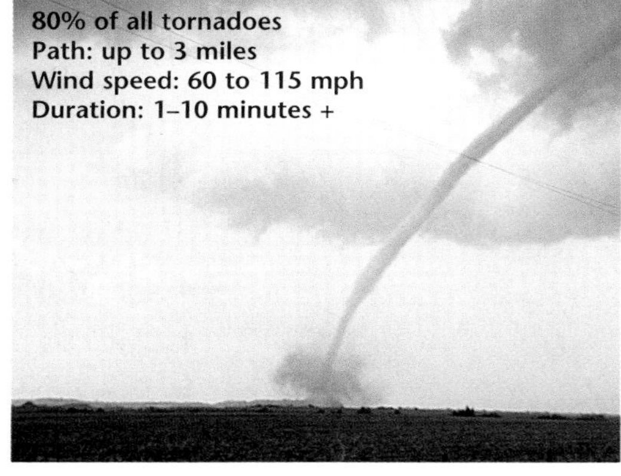

Strong Tornadoes (F2 and F3)

Violent Tornadoes (F4 and F5)

Table 13-3 Tornado Safety
If a Warning Is Issued or If Threatening Weather Approaches . . .
• If you are in a home or building, move to a predesignated shelter, such as a basement.
• If an underground shelter is not available, move to an interior room or hallway on the lowest floor and get under a sturdy piece of furniture.
• Stay away from windows.
• Get out of automobiles.
• Do not try to outdistance a tornado in a car; instead, leave the car immediately.
• If you are caught outside or in a vehicle, lie flat in a nearby ditch or depression.
• Mobile homes, even when tied down, offer little protection from tornadoes and should be abandoned.

Source: NOAA

indicated on weather radar or spotted in the region. During a severe thunderstorm, the presence of dark, greenish skies, a towering wall of clouds, large hailstones, and a loud, roaring noise similar to that of a freight train are signs of an approaching or developing tornado. ***Table 13-3*** lists safety measures recommended by the National Weather Service in the event of a tornado. The agency stresses that despite advanced tracking systems, some tornadoes develop exceedingly quickly. In these cases, advance warnings may not be possible. However, the threat of tornado-related injury can be substantially decreased when people seek shelter at the first sign of threatening skies. In the next section, you'll learn about another type of severe weather: tropical storms.

Section Assessment

1. Describe two characteristics of thunderstorms that lead to hail formation.
2. Compare and contrast a macroburst and a microburst.
3. What type of front would you expect to be associated with flooding? Why?
4. Why are some thunderstorms more severe than others?
5. If the time between when you see lightning and hear thunder is increasing, a storm is moving away from you. Why is this true?
6. **Thinking Critically** Based on what you know about stepped leaders and return strokes, why are tall objects more likely to be struck by lightning than shorter ones?

Skill Review

7. **Recognizing Cause and Effect** In the United States, most thunderstorms occur in Florida, yet the central states experience the strongest tornadoes. Why doesn't Florida have more violent tornadoes? For more help, refer to the *Skill Handbook.*

earthgeu.com/self_check_quiz

SECTION 13.3 *Tropical Storms*

Earth Sciences 5.b I&E 1.a

If you wanted to search for the origin of the most violent type of storm on Earth, the last place you'd probably look would be the calm, sunny tropics. However, during summer and fall, the sunny tropics are the birthing grounds of large, rotating, low-pressure storms called **tropical cyclones.** The strongest of these cyclonic storms are known in the United States and other parts of the Atlantic Ocean as hurricanes. ***Figure 13-9*** illustrates the rotating nature of a typical hurricane.

OBJECTIVES

- **Identify** *where tropical cyclones originate.*
- **Describe** *the life cycle of a tropical cyclone.*
- **Recognize** *the dangers of hurricanes.*

VOCABULARY

tropical cyclone
eye
eyewall
Saffir-Simpson hurricane scale
storm surge

TROPICAL CYCLONES

Unlike midlatitude storms that derive their energy from the contrast between warm and cold air masses, tropical cyclones thrive on the tremendous amount of energy in warm, tropical oceans. As water evaporates from the ocean surface, latent heat is stored. This latent heat is later released when the air begins to rise and water vapor condenses into clouds and rain. The air usually rises because of some sort of existing weather disturbance moving across the tropics. Many such disturbances originate along the Intertropical Convergence Zone (ITCZ), which you learned about in Chapter 12. As these disturbances produce more precipitation, more energy is released. In addition, the rising air creates an area of low pressure at the ocean surface. As more warm air moves toward the low-pressure center to replace the air that has risen, the Coriolis effect causes the moving air to turn counterclockwise in the northern hemisphere. This produces the cyclonic rotation of a tropical cyclone.

As the moving air approaches the center of the growing storm, it rises, rotates faster and faster, and increases in speed as more energy is released through condensation. In the process, air pressure in the center of the system continues to decrease, while surface wind speeds

Earth Sciences

5.b Students know the relationship between the rotation of Earth and the circular motions of ocean currents and air in pressure centers.

Figure 13-9 The characteristic rotating nature of cyclonic storms is evident in this hurricane that formed over the Atlantic Ocean.

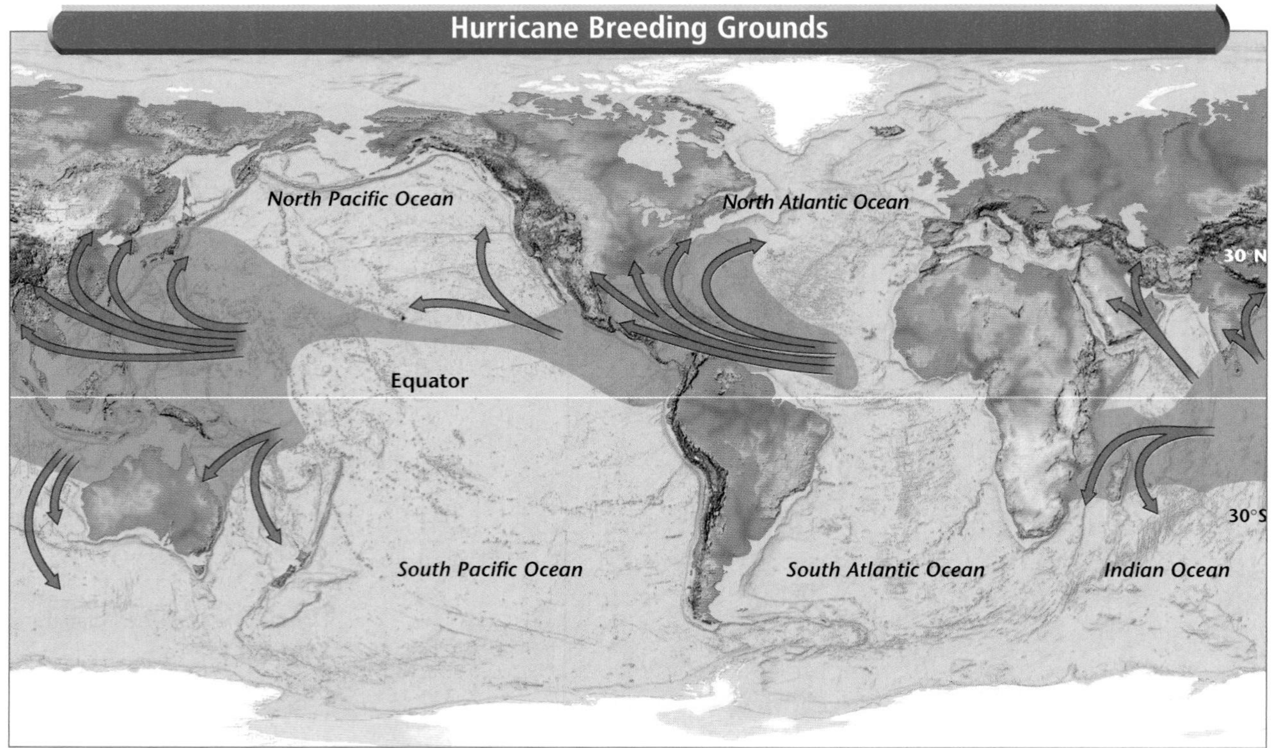

Figure 13-10 Hurricanes form in all of Earth's tropical oceans except in the relatively cool waters of the South Pacific and South Atlantic Oceans.

increase—sometimes in excess of 240 km/h. This process will continue as long as atmospheric conditions allow warm air to be fed into the system at the surface and to be removed from the system in the upper atmosphere.

Formation of Tropical Cyclones Tropical cyclones require two basic conditions to form: an abundant supply of very warm ocean water and some sort of disturbance to lift warm air and keep it rising. These conditions exist in all tropical oceans except the South Atlantic Ocean and the Pacific Ocean west of the South American Coast. Ocean waters in these areas are somewhat cooler. In addition, the ITCZ is positioned farther north. As a consequence, tropical cyclones do not occur in these areas. They do occur in the large expanse of warm waters in the western Pacific Ocean, where they are known as typhoons. To people living near the Indian Ocean, they are known as cyclones. Near the Atlantic Ocean, the Caribbean Sea, the Gulf of Mexico, and along the western coast of Mexico, they are called hurricanes. The map in ***Figure 13-10*** shows where hurricanes generally form. They occur most frequently in the late summer and early fall, when Earth's oceans contain their greatest amount of stored heat energy.

Movement of Tropical Cyclones Like all large-scale storms, tropical cyclones move according to the wind currents that steer them. Recall that many of the world's oceans are home to subtropical high-pressure systems that are present to some extent throughout the year. In the deep tropics, tropical cyclones are often caught up in the circulation of these high-pressure systems. They move steadily toward the west, then eventually turn poleward when they reach the far edges of the high-pressure systems. There, they are guided by prevailing westerlies and begin to interact with midlatitude systems. At this point, the interaction of the various wind and weather systems makes the movement of the storms unpredictable.

Stages of Tropical Cyclones A traveling tropical disturbance, which can cause air in a developing tropical cyclone to rise, is the first stage of a tropical cyclone. Disturbances can originate either from the ITCZ or as weak, low-pressure systems called tropical waves. These disturbances are common during the summer and early fall. Sometimes, midlatitude weather disturbances can move into the tropics, become stranded there, and gradually acquire tropical characteristics. Whatever their origin, only a small percentage of tropical disturbances ever develop into full-fledged hurricanes. This is because conditions throughout the atmosphere must be such that rising air can be dispersed into the upper atmosphere. ***Figure 13-11*** shows a cross section of a hurricane.

Figure 13-11 In this hurricane cross section, the rising, moist air—indicated by small red arrows—forms clouds in bands around the eye. The photo shows the eye of a hurricane that formed over the Pacific Ocean in 1991.

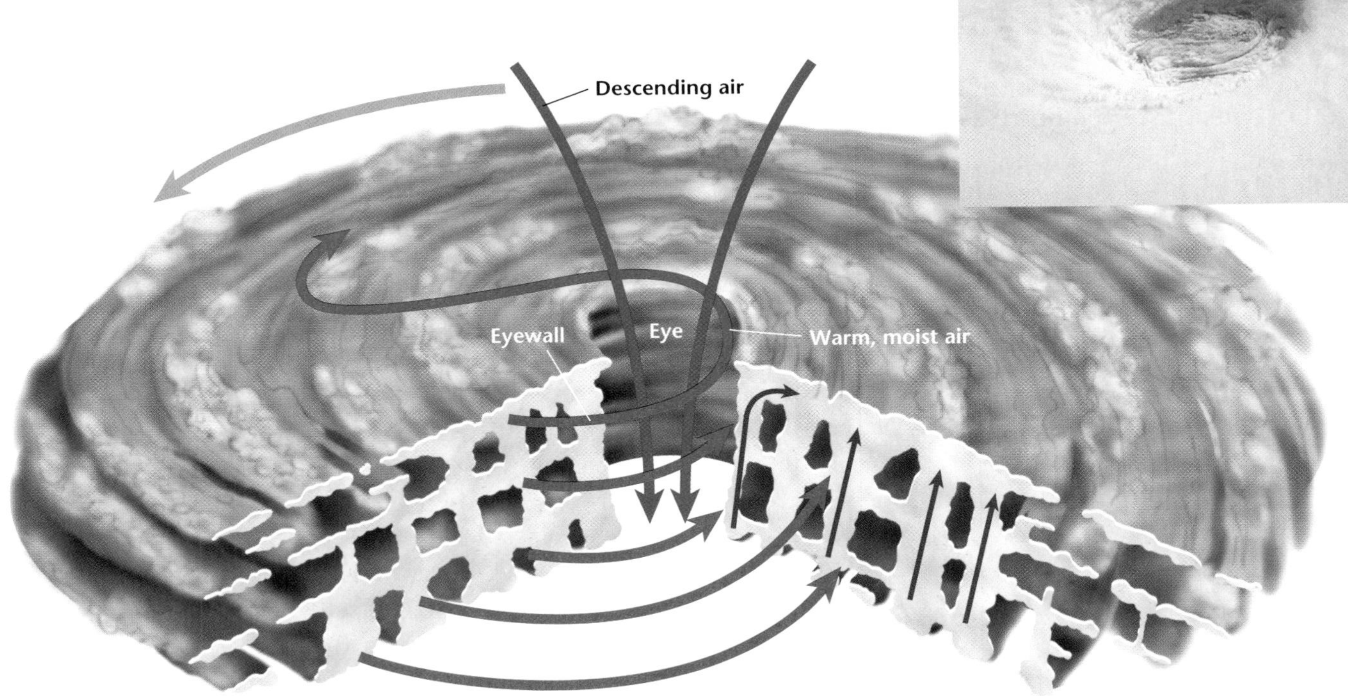

Table 13-4 Saffir-Simpson Hurricane Scale

Scale Number (Category)	Sustained Winds (mph)	Damage	Examples of Hurricanes and the States Affected
1	74–95	Minimal	Florence, 1988 (LA) Charley, 1988 (NC)
2	96–110	Moderate	Frances, 2004 (FL) Floyd, 1999 (NE U.S.)
3	111–130	Extensive	Fran, 1996 (SC) Ivan, 2004 (AL)
4	131–155	Extreme	Charley, 2004 (FL) Hugo, 1989 (SC)
5	> 155	Catastrophic	Camille, 1969 (LA/MS) Andrew, 1992 (S. FL)

Source: National Weather Service

Using Numbers Suppose that a hurricane has been spotted at 25°N, 50°W, which is roughly 2900 km from Miami, Florida. The hurricane is moving west at 25 km/h. How long will it take the hurricane to reach Miami?

When a disturbance over a tropical ocean acquires a cyclonic circulation around a center of low pressure, it has reached the next developmental stage, which is known as a tropical depression. When wind speeds around the low-pressure center of a tropical depression exceed 65 km/h, the system is called a tropical storm. If air pressure continues to fall and winds around the center reach at least 120 km/h, the storm is officially classified as a hurricane. Once winds reach these speeds, another phenomenon takes place—the development of a calm center of the storm called an **eye.** The strongest winds in a hurricane are usually concentrated in a band immediately surrounding the eye called the **eyewall.**

Classifying Hurricanes

The **Saffir-Simpson hurricane scale** classifies hurricanes according to wind speed, air pressure in the center, and potential for property damage. As shown in ***Table 13-4,*** the Saffir-Simpson hurricane scale ranges from Category 1 hurricanes, which have minimum wind speeds of 74 mph (120 km/h), to the monstrous Category 5 storms, which can have winds in excess of 155 mph (250 km/h). Once a hurricane reaches Category 3 status, it is considered to be a major hurricane, with good reason. Most of the deadliest hurricanes that strike the United States were classified as major hurricanes.

Running Out of Energy A hurricane will last until it can no longer produce enough energy to sustain itself. This usually happens when the storm moves over land and no longer has access to the warm ocean surface from which it draws its energy, or when the storm moves over colder water. During its life cycle, a hurricane can undergo several fluctuations in intensity as it interacts with other atmospheric systems.

Hurricane Hazards

Hurricanes can cause a lot of damage, particularly along coastal areas where human populations have increased. Much of this damage is associated with violent winds. The strongest winds in a hurricane are usually confined to the eyewall, the band about 40 to 80 km wide that surrounds the calm eye. Outside of the eyewall, winds taper off with distance from the center, although winds of more than 60 km/h can extend as far as 400 km from the center of a hurricane.

Storm Surges Strong winds moving onshore in coastal areas are partly responsible for another major hurricane threat: storm surges. A **storm surge** occurs when hurricane-force winds drive a mound of ocean water toward coastal areas, where it washes over the land. Storm surges can sometimes reach 6 m above normal sea level, as shown in ***Figure 13-12.*** When this occurs during high tide, the surge can cause enormous damage. In the northern hemisphere, a storm surge occurs primarily on the right side of a storm relative to its eye, where the strongest onshore winds occur.

The heat released through the condensation of vast amounts of water vapor fuels hurricanes. This condensation also produces great amounts of rain. Thus, floods are an additional hurricane hazard, particularly if the storm moves over mountainous areas, where orographic lifting enhances the upward motion of air.

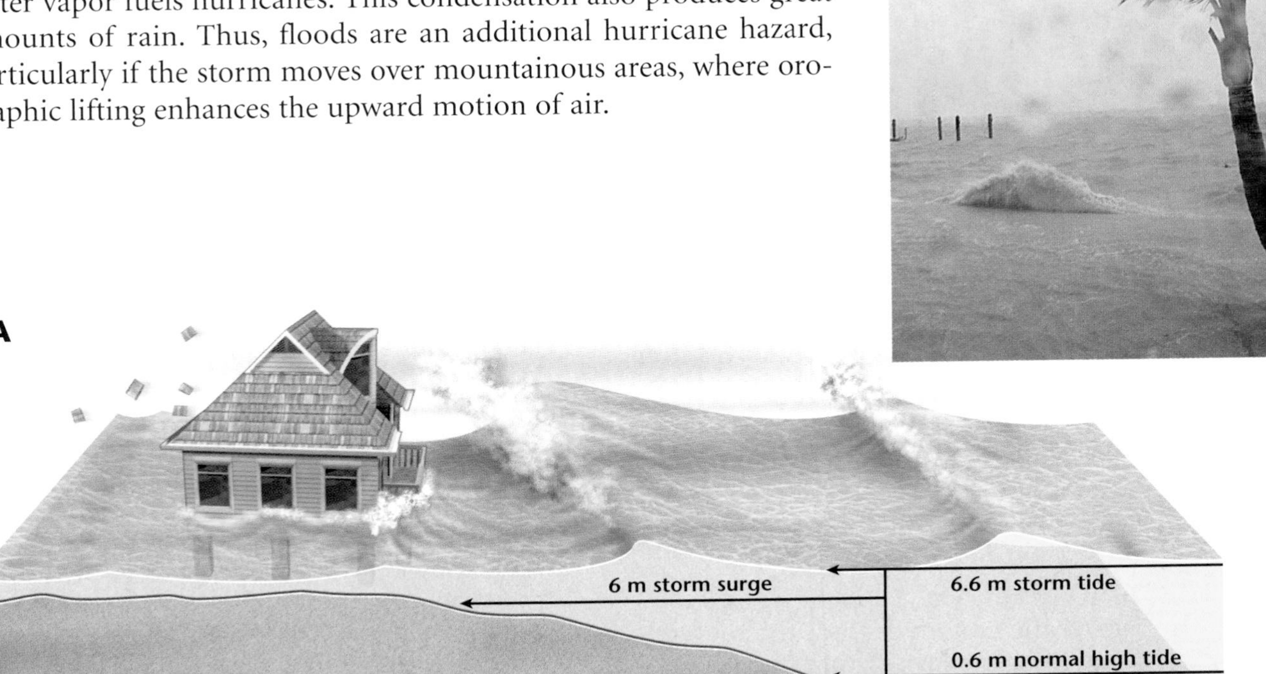

Figure 13-12 Storm surges can sometimes reach 6 m above normal sea level and cause enormous damage **(A).** This storm surge in the Florida Keys was caused by Hurricane Irene **(B).**

Table 13-5 Hurricane Safety

- Turn the refrigerator to the maximum cold setting and open it only when necessary.
- Turn off utilities if told to do so by authorities.
- Unplug small appliances.
- Turn off propane tanks.
- Fill bathtubs and large containers with water for sanitary purposes.

If Winds Become Strong . . .

- Stay away from windows and doors even if they are covered. Take refuge in a small interior room, closet, or hallway.
- Close all interior doors. Secure and brace external doors.
- If you are in a two-story house, go to an interior first-floor room, such as a bathroom or closet.
- If you are in a multiple-story building and away from water, go to the first or second floor and take refuge in a hall or other interior room away from windows.
- Lie on the floor under a table or other sturdy object.

Source: NOAA

Hurricane Advisories The National Hurricane Center, which is responsible for tracking and forecasting the intensity and motion of tropical cyclones in the western hemisphere, issues a hurricane warning at least 24 hours before a hurricane strikes. The center also issues regular advisories that indicate a storm's position, strength, and movement. Using this information, people can then track a storm on a hurricane-tracking chart, such as the one you'll use in the *Internet GeoLab* at the end of this chapter. This type of awareness, combined with proper safety precautions such as those listed in ***Table 13-5,*** has greatly reduced death tolls associated with hurricanes in recent years.

SECTION ASSESSMENT

1. Identify the four main stages of a tropical cyclone.
2. Describe the changing wind systems that guide a tropical cyclone as it moves from the tropics to the midlatitudes.
3. Why don't tropical cyclones form in the South Atlantic Ocean or off the western coast of South America?
4. What two conditions must exist for a tropical cyclone to form?
5. **Thinking Critically** Suppose that you live on the eastern coast of the United States and are advised that the center of a hurricane is moving inland 70 km north of your location. Would you predict that a storm surge will be a major problem in your area? Why or why not?

SKILL REVIEW

6. **Making and Using Tables** Research at least ten hurricanes that have occurred throughout the world since 1980. Based on the Saffir-Simpson scale, make a data table showing wind speed, air pressure in the center, and property damage associated with each hurricane. For more help, refer to the *Skill Handbook.*

SECTION 13.4 Recurring Weather

On a hot, summer day, a sudden thunderstorm is a welcome event. Such rains are not so welcome, however, when they continue for hours or even days. Persistent or repetitive weather can negatively affect agriculture, transportation, and recreation.

OBJECTIVES

- **Describe** *recurring weather patterns and the problems they create.*
- **Identify** *atmospheric events that cause recurring weather patterns.*

VOCABULARY

drought
heat wave
cold wave
wind-chill factor

FLOODS AND DROUGHTS

An individual thunderstorm can unleash enough rain to produce floods, and hurricanes are notorious for their torrential downpours. Floods can also occur, however, when weather patterns cause even mild storms to persist over the same area. For example, a storm with a rainfall rate of 1.5 cm/h is not much of a problem—providing that it lasts only an hour or two. If this same storm were to remain over one spot for 18 hours, however, total rainfall would be 27 cm, which is more than enough to create flooding in most areas. You will learn more about floods in the *MiniLab* on the following page.

On the other hand, too much dry weather can create nearly as much havoc as too much rainfall. **Droughts** are extended periods of well-below-normal rainfall. One of the most extreme droughts occurred during the 1930s in the Central United States. ***Figure 13-13*** shows a dust storm that occurred in the Dust Bowl, which was the name given to the affected states. This extended drought put countless farmers out of business, as rainfall for several seasons was inadequate to grow crops.

Figure 13-13 The dust blowing over this highway was caused by a severe drought in the Central United States during the 1930s.

MiniLab

How can mild rains cause floods?

Model the effects of repeated, slow-moving storms that drop rain over the same area for a long period of time.

Procedure

1. Place an ice-cube tray on the bottom of a large sink or tub.
2. Pour water into a clean, plastic dishwashing-detergent bottle until it is two-thirds full. Replace the cap on the bottle.
3. Hold the bottle upside down with the cap open about 8 cm above one end of the ice-cube tray. Gently squeeze the bottle to maintain a constant flow of water into the tray. Slowly move the bottle from one end of the tray to the other over the course of 30 seconds. Try to put approximately equal amounts of water in each ice-cube compartment.
4. Measure the depth of water in each compartment. Calculate the average depth.
5. Repeat steps 1–4, but move the bottle across the ice-cube tray in 15 seconds.

Analyze and Conclude

1. How did the average depth of the water differ in steps 4 and 5? How might you account for the difference?
2. Based on these results, infer how the speed of a moving storm affects the amount of rain received in any one area.
3. How could you alter the experiment to simulate different rates of rainfall?

Droughts are usually the result of shifts in global wind patterns that allow large high-pressure systems to persist for weeks or months over continental areas. Under a dome of high pressure, air sinks on a large scale. Because the sinking air will resist any attempt to lift moisture through it, condensation cannot occur, and drought will set in until global patterns shift enough to move the high-pressure system out of the way.

Heat Waves An unpleasant side effect of droughts often comes in the form of **heat waves,** which are extended periods of above-normal temperatures. Heat waves can be formed by the same high-pressure systems that cause droughts. As the air under a large high-pressure system sinks, it warms by compression and causes above-normal temperatures. The high-pressure system also blocks cooler air masses from moving into the area, so there is little relief from the heat. Because it is difficult for condensation to occur under the sinking air of the high-pressure system, there are few, if any, clouds to block the blazing sunshine. To make matters worse, the jet stream, or "atmospheric railway," that weather systems follow is farther north and weaker during the summer. Thus, the upper-air currents that might guide the high-pressure system are so weak that the system scarcely moves.

Even increasing humidity does not ease the discomfort of a heat wave. Human bodies cool by evaporating moisture from the surface of the skin. In the process, heat is removed from the body. If air is humid, the rate of evaporation is reduced, which diminishes the body's ability to regulate internal temperature. In heat waves, this can lead to serious health problems such as heatstroke, sunstroke, and even death.

Table 13-6 The Heat Index

Relative Humidity (%)	Air Temperature (°F)										
	70	75	80	85	90	95	100	105	110	115	120
	Apparent Temperature (°F)										
0	64	69	73	78	83	87	91	95	99	103	107
10	65	70	75	80	85	90	95	100	105	111	116
20	66	72	77	82	87	93	99	105	112	120	130
30	67	73	78	84	90	96	104	113	123	135	148
40	68	74	79	86	93	101	110	123	137	151	
50	69	75	81	88	96	107	120	135	150		
60	70	76	82	90	100	114	132	149			
70	70	77	85	93	106	124	144				
80	71	78	86	97	113	136					
90	71	79	88	102	122						
100	72	80	91	108							

Source: National Weather Service, NOAA

Because of the extreme dangers posed by the lethal combination of heat and humidity, the National Weather Service routinely reports the heat index, shown in ***Table 13-6.*** Note that the National Weather Service uses the Fahrenheit scale in the heat index because most U.S. citizens are most familiar with this scale. The heat index assesses the effect of the body's increasing difficulty in regulating its internal temperature as relative humidity rises. For example, an air temperature of 85°F (29°C) combined with relative humidity of 80 percent would require the body to cool itself at the same rate as if the air temperature were 97°F (36°C). Do the *Problem-Solving Lab* on the following page to learn more about heat waves.

Cold Waves

The flip side of a heat wave is a **cold wave,** which is an extended period of below-normal temperatures. Interestingly, cold waves are also brought on by large, high-pressure systems. However, cold waves are caused by systems of continental polar or arctic origin. During the arctic winter, little sunlight is available to provide warmth. At the same time, the snow-covered surface is constantly radiating its limited heat back to space. The combined effect of these two factors is the development of large pools of extremely cold air over polar continental areas. Because cold air sinks, the pressure near the surface increases, creating a strong high-pressure system.

Because of the location and the time of year in which they occur, winter high-pressure systems are much more influenced by the jet stream than are summer high-pressure systems. Moved along by the jet stream, these high-pressure systems rarely linger in any area. However, the winter location of the jet stream may remain essentially unchanged for days or even weeks. This means that several polar high-pressure systems can follow the same path and subject the same areas to bout after bout of numbing cold. ***Figure 13-14*** shows some effects of prolonged periods of cold weather.

Because wind transports heat away from the body, the effects of cold air are worsened by wind. This phenomenon is known as the **wind-chill factor.** The wind-chill factor is measured by the wind-chill index, which estimates the heat loss from human skin caused by the combination of cold air and wind. This index estimates how cold the air actually feels to the human body. As with the heat index, the National Weather Service records the wind-chill index in U.S. units

Problem-Solving Lab

Making and Using Graphs

Charting a heat wave The following data represent the daily maximum and minimum temperatures for ten consecutive summer days in a major city.

Daily Temperatures

Day	Maximum	Minimum	Average
1	92	76	
2	91	75	
3	94	78	
4	95	75	
5	93	77	
6	96	76	
7	94	80	
8	96	72	
9	92	74	
10	94	68	

Analysis

1. Copy the data table in your science journal. Calculate the average temperature for each day, then include those temperatures in your data table.
2. Plot the daily maximum temperatures on a graph with the days on the *x*-axis and the maximum temperatures on the *y*-axis. Connect the data points to show how the maximum temperature changed over the ten-day period.
3. Repeat step 2 for the minimum and average temperatures.

Thinking Critically

4. A heat wave is defined as two or more consecutive days with an average temperature of 85°F or higher. On what day did the city begin its heat wave? How long did the heat wave last?
5. Calculate the average temperature for the days of the heat wave only. Compare this to the average temperature of the remaining days.
6. What safety measures could residents of the city take to minimize the effects of a heat wave?

Figure 13-14 An ice storm in 1990 damaged utility lines in Watertown, New York **(A).** Nearly 2.3 m of snow fell near Lake Ontario in 1996 **(B).**

for the sake of convenience. While the wind-chill index is helpful, it does not account for individual variations in sensitivity to cold, the effects of physical activity, or humidity. Some scientists, noting that this system has been in place since the 1940s, are calling for the development of new methods that more accurately estimate the effects of cold weather on the human body.

Section Assessment

1. Why are droughts usually associated with high-pressure systems?
2. Describe a situation wherein a relatively light rain could cause flooding.
3. Compare and contrast a cold wave and a heat wave.
4. What is the wind-chill factor? What does the wind-chill index measure?
5. Using *Table 13-6,* estimate the heat index for air with a temperature of 80°F and relative humidity of 90 percent.
6. Extreme floods occur more often in summer than any other time of the year. Use your knowledge of the jet stream to explain why this is true.
7. **Thinking Critically** Air in a summer high-pressure system warms by compression. Based on what you know about molecular motion, explain why air in a winter high-pressure system doesn't warm by compression, too.

Skill Review

8. **Forming a Hypothesis** A key requirement for the formation of snow is cold air. Yet some parts of the United States have more annual snowfall than Canada, which is farther north and should therefore be colder. Form a hypothesis to explain the apparent discrepancy. For more help, refer to the *Skill Handbook.*

earthgeu.com/self_check_quiz

Internet GeoLab

Tracking a Hurricane

Hurricanes are violent storms. That's why it's important to have plenty of advance warning before they hit land. By tracking the changing position of a storm on a chart and connecting these positions with a line, you can determine a hurricane's path.

I&E

1.a Select and use appropriate tools and technology (such as computer-linked probes, spreadsheets, and graphing calculators) to perform tests, collect data, analyze relationships, and display data.

Preparation

Problem

What information can you obtain by studying the path of a hurricane?

Hypothesis

Gather information about the path of a hurricane. **Form a hypothesis** about how the hurricane's path can be used to predict the strength of the storm and where most damage might be inflicted.

Objectives

- **Gather** and **communicate** data about hurricanes.
- **Plot** data on a hurricane-tracking chart.
- **Predict** where storm-inflicted damage might occur.

Data Sources

Go to the Earth Science Web site at earthgeu.com to find links to hurricane data, or use information provided by your teacher. Make copies of the hurricane-tracking chart in this lab or download a chart from the Web site.

Plan the Experiment

1. Find a resource that lists major hurricanes that have occurred within the past five years. The Earth Science Web site provides a list of sites that have information about hurricanes.
2. Choose a hurricane to research. Some recent major hurricanes include Hurricane Claudette, Hurricane Isabel, and Hurricane Floyd.
3. Gather data about the hurricane from the links on the Earth Science Web site or the library.

Procedure

1. Incorporate your research into a data table. Add any additional information that you think is important.
2. Go to the Earth Science Web site at earthgeu.com to post your data.
3. Visit sites listed on the Earth Science Web site for information on other major hurricanes.

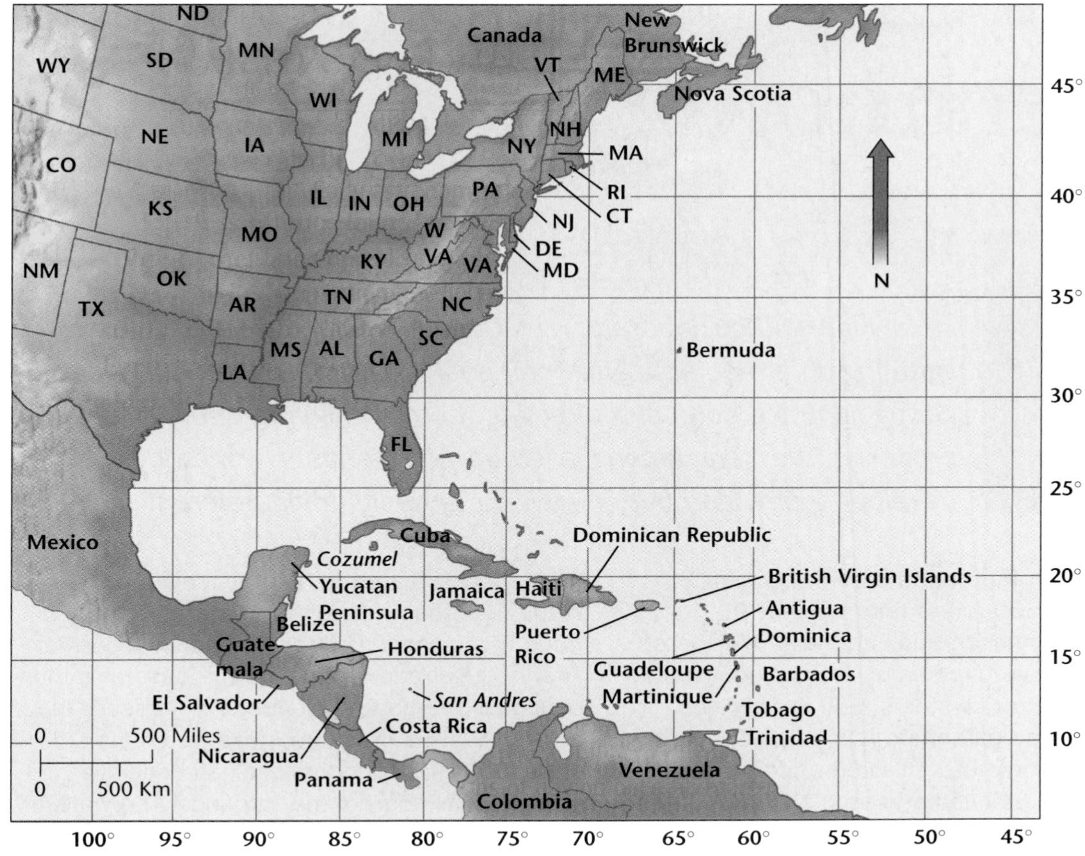

Conclude & Apply

Sharing Your Data Find this *Internet GeoLab* on the Earth Science Web site at earthgeu.com. Post your data in the table provided for this activity. Use the additional data from other students to complete your chart and answer the Conclude & Apply questions.

1. Plot the position, air pressure, wind speed, and stage of the hurricane at six-hour intervals throughout its existence.
2. Plot the changing position of the hurricane on your hurricane-tracking chart.
3. What was the maximum wind speed in knots that the hurricane reached?
4. Multiply the value from question 3 by 1.15 to find the wind speed in miles per hour. Based on this value, how would the hurricane be classified on the Saffir-Simpson scale?
5. Using your completed hurricane-tracking chart, list the landmasses over which the hurricane passed.
6. Where would you expect the storm surge to have been greatest? Explain. Compare your answer to the information you gathered on the damage inflicted by the storm. Was your answer correct?
7. How was the hurricane's strength affected when its center passed over land?

Science & Technology

Taming Lightning

In a high-voltage laboratory in Canada, scientists experiment with lightning on a regular basis. Their goal is to one day work outside, triggering lightning bolts and directing them safely away from people and property. They are among a group of scientists worldwide who are developing new and better ways to tame lightning before it strikes.

Each year in the United States, lightning accounts for 7500 forest fires, roughly 93 deaths, several hundred injuries, and millions of dollars in damage to communications equipment, buildings, electrical systems, and power lines. To guard against loss of life and property, most homes and commercial structures are equipped with lightning-protection systems. These systems use lightning rods—slender metal rods placed upon rooftops—to gather positive charges from the ground. The positive charges attract the negative charges in the base of a thundercloud and neutralize them split seconds before they coalesce into a lightning strike. Aluminum or copper cables act as conducters, connecting the lightning rods to ground terminators, which are metal rods buried beneath the soil. The function of these rods is to guide the electrical current harmlessly into the ground.

Lightning-protection systems such as these have been around since Benjamin Franklin flew his kite in a storm some 200 years ago, demonstrating conclusively that lightning is indeed electricity. Today, one of the most promising areas of lightning-protection research involves something that Benjamin Franklin never had at his disposal—laser beams.

Home-Grown Lightning

Let's return to the high-voltage lab in Canada. There, scientists use two huge, circular electrodes to re-create the natural conditions that result in a shocking bolt of lightning. One electrode is suspended about 6 m above the other. The top electrode represents the base of a thundercloud. The bottom electrode represents the ground. Just as in nature, negative charges flow down from the "thundercloud" and positive charges flow up from the "ground." The wave of current that surges upward to meet the downward discharge results in the bright, jagged flash known as lightning.

The goal of these scientists, however, is not simply to model lightning. They want to learn how to harness lightning—to trigger controlled strikes and guide them to safe locations. To do this, the scientists aim a laser beam through a hole in the center of the bottom electrode. The beam displaces electrons from charged particles in the air, simultaneously provoking a lightning strike and providing a guided path for the discharge to follow. Instead of a jagged bolt of lightning, the controlled strike is as straight as a laser beam.

Activity

Where's the safest place to be during a thunderstorm? If you're caught outside in the open, what should you do? Research lightning safety tips, then develop a safety brochure for distribution at your school.

CHAPTER 13
Study Guide

Summary

SECTION 13.1

Thunderstorms

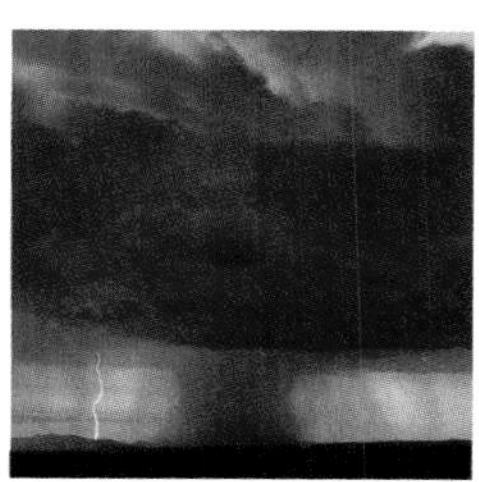

Main Ideas

- For a thunderstorm to occur, there must be abundant moisture in the lower levels of the atmosphere and a mechanism to lift the moisture so it can condense. In addition, the air must be unstable so that the growing cloud will continue to rise.
- Thunderstorms are classified according to the mechanism that caused the air to rise. In an air-mass thunderstorm, the cloud rises because of unequal heating of Earth's surface within one air mass. In a frontal thunderstorm, the air rises because it is pushed up by an advancing air mass.

Vocabulary

air-mass thunderstorm (p. 331)
frontal thunderstorm (p. 331)
sea-breeze thunderstorm (p. 331)

SECTION 13.2

Severe Weather

Main Ideas

- Lightning is produced when an advancing stepped leader unites with an upward-moving return stroke. Thunder is the sound made by the rapid expansion of air around the lightning bolt as a result of extreme heating of the lightning channel.
- Thunderstorms can damage property and cause loss of life. The hazards of thunderstorms include lightning, violent winds, hail, floods, and tornadoes.
- The Fujita tornado intensity scale classifies tornadoes according to wind speed, path of destruction, and duration.

Vocabulary

downburst (p. 336)
Fujita tornado intensity scale (p. 338)
supercell (p. 334)
tornado (p. 338)

SECTION 13.3

Tropical Storms

Main Ideas

- Tropical cyclones derive their energy from the evaporation of warm ocean water and the release of heat.
- The Saffir-Simpson hurricane scale classifies hurricanes according to intensity.
- Hurricane hazards include violent winds, floods, and storm surges. The National Hurricane Center tracks hurricanes and issues advance warnings to help reduce loss of life.

Vocabulary

eye (p. 344)
eyewall (p. 344)
Saffir-Simpson hurricane scale (p. 344)
storm surge (p. 345)
tropical cyclone (p. 341)

SECTION 13.4

Recurring Weather

Main Ideas

- Examples of persistent weather events include floods, droughts, cold waves, and heat waves.
- The heat index assesses the impact of humidity combined with excessive heat on the human body. The wind-chill index estimates the heat loss from human skin caused by a combination of cold air and wind.

Vocabulary

cold wave (p. 349)
drought (p. 347)
heat wave (p. 348)
wind-chill factor (p. 350)

earthgeu.com/vocabulary_puzzlemaker

CHAPTER 13 Assessment

Understanding Main Ideas

1. Which of the following would work against the development of a thunderstorm?
a. rising air
b. stable air
c. moisture
d. unstable air

2. Which of the following does NOT describe a type of damaging thunderstorm wind?
a. downburst
b. microburst
c. land breeze
d. macroburst

3. Flooding is most likely to take place because of rains associated with what type of front?
a. stationary front
b. occluded front
c. cold front
d. warm front

4. During what stage of a tropical cyclone does an eyewall develop?
a. tropical depression
b. tropical storm
c. hurricane
d. tropical wave

5. What is the first stage of a lightning bolt?
a. return stroke
b. stepped leader
c. positive charge
d. downdraft

6. Which of the following does NOT play a key role in the development of hail?
a. supercooled water
b. strong downdrafts
c. warm ocean water
d. strong updrafts

7. Heat waves involve high-pressure systems that cause air to sink and warm by which of the following processes?
a. compression
b. conduction
c. evaporation
d. condensation

8. Which of the following weather hazards involves lack of moisture?
a. hail
b. drought
c. storm surge
d. flood

9. What percentage of tornadoes are classified as F4 or F5 on the Fujita tornado intensity scale?
a. one percent
b. ten percent
c. 50 percent
d. 75 percent

10. Which of the following factors, if increased, would make a thunderstorm severe?
a. temperature
b. surface moisture
c. duration
d. conduction

11. Which way do hurricanes rotate in the southern hemisphere?
a. south
b. clockwise
c. counterclockwise
d. north

12. In which ocean would you NOT expect to experience a tropical cyclone?
a. West Pacific
b. Indian
c. North Atlantic
d. South Atlantic

13. What weather events are cold waves most often associated with?
a. floods
b. polar high-pressure systems
c. tropical high-pressure systems
d. droughts

14. Compare and contrast tornadoes and hurricanes.

15. Why are cold fronts more likely to produce severe thunderstorms than warm fronts?

Test-Taking Tip

Carefully Observe Scientific Illustrations If a test question requires you to interpret a scientific illustration, look very closely at the details of the illustration. Your answer may depend on a small detail.

CHAPTER 13
Assessment

Applying Main Ideas

16. Using ***Table 13-6,*** determine the heat index if the temperature is 90°F and relative humidity is 60 percent.

17. Using ***Table 13-4,*** classify a hurricane with a maximum wind speed of 120 mph.

Use the illustration of a hurricane in the northern hemisphere to answer question 18.

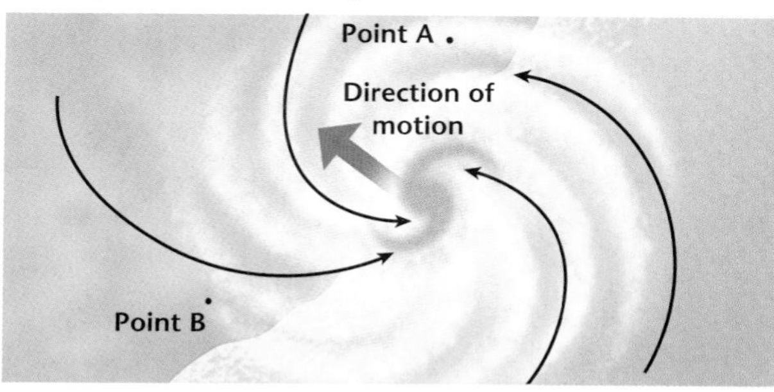

18. Would a storm surge be more likely to occur at point A or point B? Why?

19. How might you prepare for a tornado? What safety measures would you recommend?

20. In which oceans would you NOT expect to experience a tropical cyclone? Why?

Thinking Critically

21. Extreme cold waves are more common in the northern hemisphere than in the southern hemisphere. Why?

22. Tropical cyclones are never observed within about 5° north and south latitudes. What do you think might account for this?

23. Supercells that produce tornadoes often produce large hailstones as well. Explain.

24. Why are boats on lakes or on the ocean especially vulnerable to lightning strikes?

Standardized Test Practice

INTERPRETING SCIENTIFIC ILLUSTRATIONS
Use the illustration below to answer questions 1 and 2.

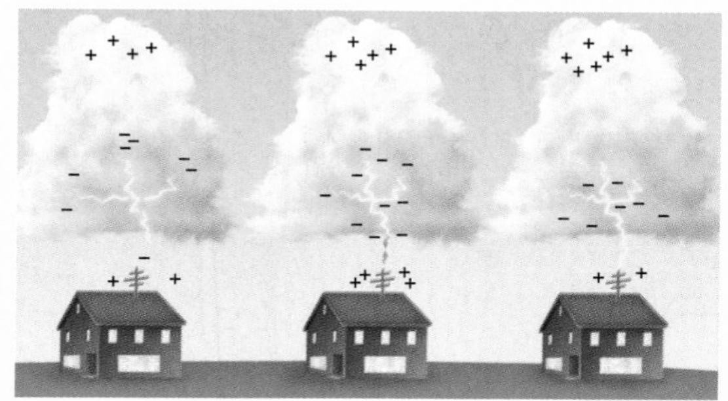

1. Which type of cloud is lightning associated with?
a. altocumulus **c.** cirrus
b. stratocumulus **d.** cumulonimbus

2. Lightning is the illumination that occurs when an invisible channel of negatively charged air descends to the ground and a channel of positively charged ions rushes upward to meet it. What is the channel of positively charged ions called?
a. return stroke **c.** ground stroke
b. stepped leader **d.** electronic leader

3. What occurs when winds of at least 120 km/h drive a mound of ocean water toward coastal areas?
a. downburst **c.** storm surge
b. cold wave **d.** tornado

4. Which factor is NOT associated with a heat wave?
a. a high-pressure system
b. a weakened jet stream
c. above-normal temperatures
d. increased cloud cover

Chapter 14

Climate

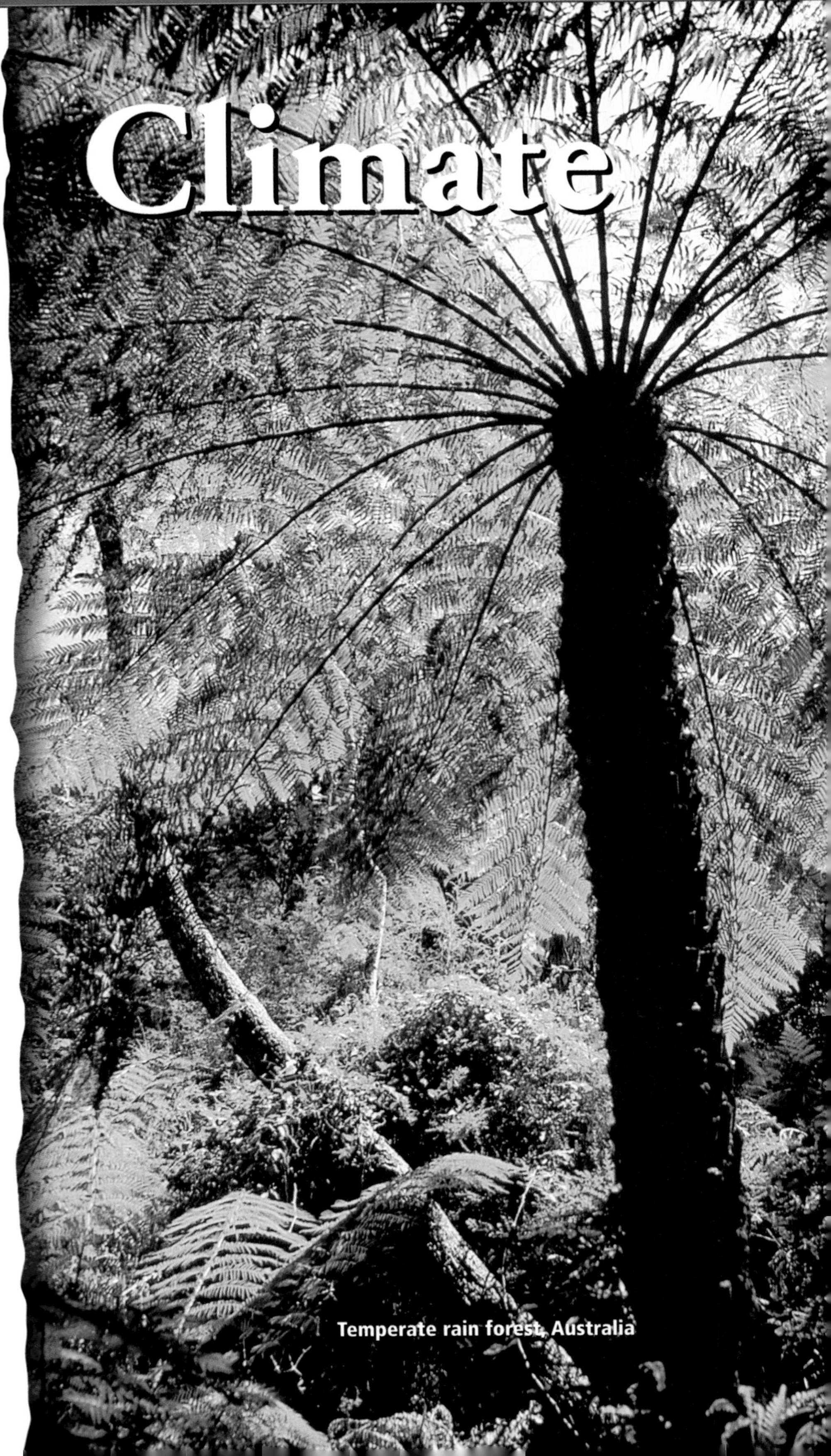

What You'll Learn

- What causes different climates.
- How climates are classified.
- How climates change as a result of natural events and human activities.

Why It's Important

This rain forest is one example of the wide variety of plants found in different climates. Climate affects where we live, what we wear, and what we eat. Changes in climate can have far-reaching effects on agriculture, industry, transportation, and recreation.

To find out more about climate, visit the Earth Science Web Site at earthgeu.com

Temperate rain forest, Australia

Discovery Lab Model Cloud Cover

Some areas are generally more cloudy than others. This affects both the temperature and the amount of precipitation that these areas receive. In this activity, which should be done only when the weather forecast calls for clear, calm skies overnight, you'll model the effect of cloud cover on local temperatures.

1. On a calm, clear afternoon, lay two sheets of dark construction paper on the grass in an open area. Place a rock on each sheet of paper to prevent them from blowing away.
2. Open an umbrella and prop it on the ground over one of the sheets of paper.
3. The next morning, observe what has happened to the sheets of paper.

Observe In your science journal, describe any differences in dew formation that you observed. How is the umbrella in this activity similar to clouds in the atmosphere? Based on your observations, infer how temperatures during the night might differ between climates with extensive cloud cover and climates with fewer clouds.

Section 14.1 What is climate?

Earth Sciences 5.f, 6.a, 6.b

OBJECTIVES

- **Describe** *different types of climate data.*
- **Recognize** *limits associated with the use of normals.*
- **Explain** *why climates vary.*

VOCABULARY

climatology
climate
normal
tropics
temperate zone
polar zone

Fifty thousand years ago, the United States had much different weather patterns than those that exist today. The average temperature was several degrees cooler, and the jet stream was probably farther south. Understanding and predicting such climatic changes are the basic goals of climatology. **Climatology** is the study of Earth's climate and the factors that affect past, present, and future climatic changes.

Climate: More Than Just Average Weather

Climate, as you'll recall from Chapter 12, describes the long-term weather patterns of an area. These patterns include much more than average weather conditions. Climate also describes annual variations of temperature, precipitation, wind, and other weather variables. Studies of climate show extreme fluctuations of these variables over time. For example, climatic data can indicate the warmest and coldest temperatures ever recorded for a location. This type of information, combined with comparisons between recent conditions and long-term averages, can be used by companies to decide where to locate new facilities and by people who have medical conditions that require them to live in certain climates.

NORMALS

The data used to describe an area's climate are compiled from meteorological records, which are continuously gathered at thousands of locations around the world. These data include daily high and low temperatures, amounts of rainfall, wind speed and direction, humidity, and air pressure. Once the data are gathered, they are averaged on a monthly or annual basis for a period of at least 30 years to determine the **normals,** or standard values, for a location. The *Problem-Solving Lab* below lists some normals for Jacksonville, Florida.

Problem-Solving Lab

Making and Using Tables

Infer climatic conditions from normals Normals offer a comprehensive look at local weather conditions over relatively long periods of time. Use the data provided in the table to answer the following questions about the climate of Jacksonville, Florida.

Analysis

1. According to normal daily maximum temperatures, during what months would you expect the temperature to reach at least 90°F?
2. What were the highest and lowest temperatures ever recorded in this city, and in what month and year?

Thinking Critically

3. Use graph paper to plot the monthly values for normal daily maximum temperatures which cover the 30-year time period from 1966 through 1996. Next, use the monthly values to calculate the average daily maximum temperature for the 30-year period.
4. Which months were warmer than the average daily maximum temperature of the 30-year period? Which months were colder?

Normals for Jacksonville, Florida

Temperature F°	Time Period (Years)	Jan	Feb	Mar	Apr	May	Jun	Jul	Aug	Sep	Oct	Nov	Dec
Normal Daily Maximum	30	64.2	67.0	73.0	79.1	84.7	89.3	91.4	90.7	87.2	80.2	73.6	66.8
Highest Daily Maximum	55	85	88	91	95	100	103	105	102	100	96	88	84
Year of Occurrence		1947	1962	1974	1968	1967	1954	1942	1954	1944	1951	1986	1994
Normal Daily Minimum	30	40.5	43.3	49.2	54.9	62.1	69.1	71.9	71.8	69.0	59.3	50.2	43.4
Lowest Daily Minimum	55	7	19	23	34	45	47	61	63	48	36	21	11
Year of Occurrence		1985	1996	1980	1987	1992	1984	1972	1984	1981	1989	1970	1983

While normals offer valuable information, they must be used with caution. Weather conditions on any given day might differ widely from normals. For instance, the average high temperature in January for a city might be 0°C. However, it's possible that no one day in January had a high of exactly 0°C. Normals are not intended to describe usual weather conditions. They are simply the average values over a long period of time.

Another issue complicates the use of normals. While climate describes the average weather conditions for a region, normals apply only to the specific place where the meteorological data were collected. Most meteorological data are gathered at airports, which cannot operate without up-to-date, accurate weather information. Do you know anyone who lives at an airport? Probably not. In fact, many airports are located well outside city limits because of the noise and traffic that they generate. When climatic normals are based on airport data, they may differ quite a lot from actual weather conditions in nearby cities. Why? Changes in elevation and other factors such as proximity to large bodies of water can cause climates to vary, as you'll learn next.

 **Earth Sciences**

5.f Students know the interaction of wind patterns, ocean currents, and mountain ranges results in the global pattern of latitudinal bands of rain forests and deserts.

6.a Students know weather (in the short run) and climate (in the long run) involve the transfer of energy into and out of the atmosphere.

6.b Students know the effects on climate of altitude, elevation, topography, and proximity to large bodies of water and cold or warm ocean currents.

What Causes Climates?

One glance at the map shown in ***Figure 14-1*** shows that climates around the country vary greatly. For example, average daily temperatures are much warmer in Dallas, Texas, than in Minneapolis,

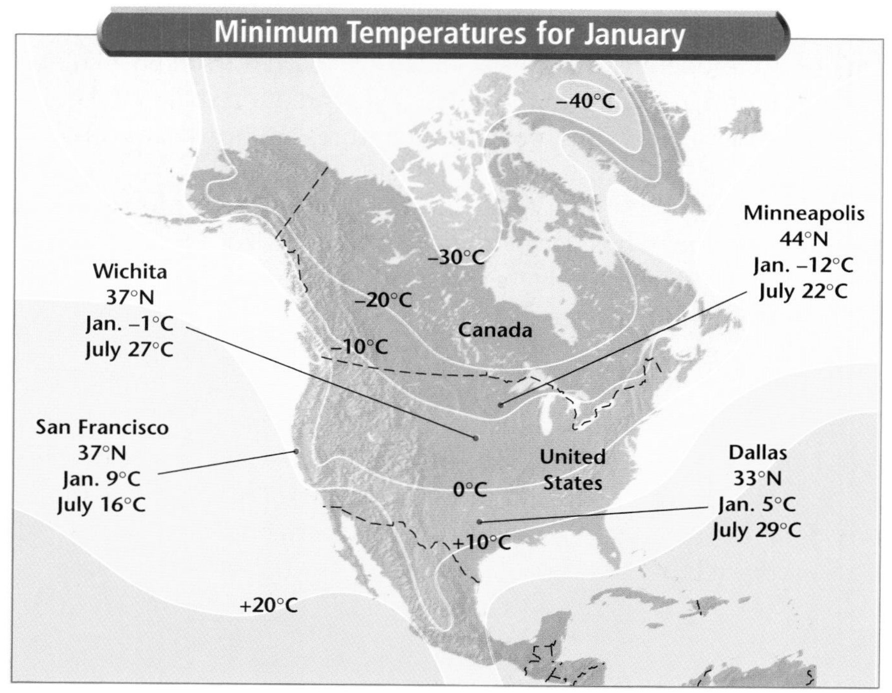

Figure 14-1 This map shows daily minimum temperature in January across the United States. The latitudes of the cities are shown because, as you'll learn on the next page, latitude greatly affects climate.

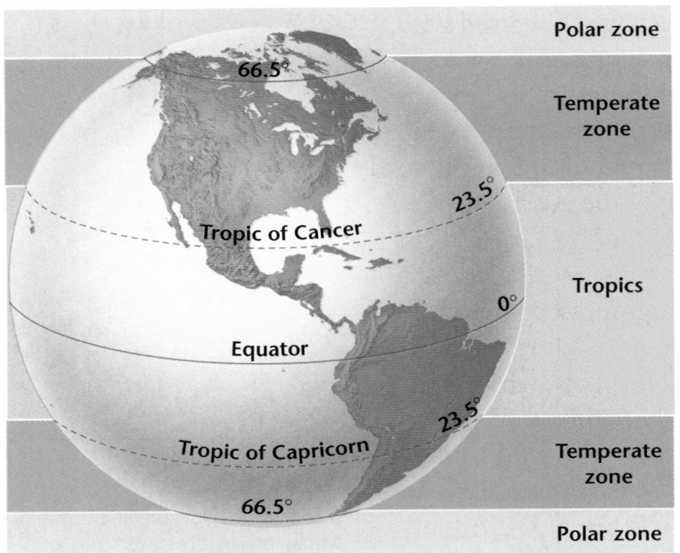

Figure 14-2 Latitude has a great effect on climate. The amount of solar radiation received on Earth decreases in intensity from the equator to the poles.

Minnesota. There are several reasons for such climatic variations, including latitude, topography, closeness of lakes and oceans, availability of moisture, global wind patterns, ocean currents, and air masses.

Latitude Recall that different parts of Earth receive different amounts of solar radiation. The amount of solar radiation received by any one place varies because Earth is tilted on its axis, and this affects how the Sun's rays strike Earth's surface. As ***Figure 14-2*** shows, the area between 23.5° south of the equator and 23.5° north of the equator, known as the **tropics,** receives the most solar radiation because the Sun's rays strike that area from almost directly overhead. As you might expect, temperatures in the tropics are generally warm year-round. The **temperate zones** lie between 23.5° and 66.5° north and south of the equator. As their name implies, temperatures in these regions are moderate. The **polar zones** are located from 66.5° north and south of the equator to the poles. Solar radiation strikes the polar zones at a low angle. Thus, polar temperatures are nearly always cold.

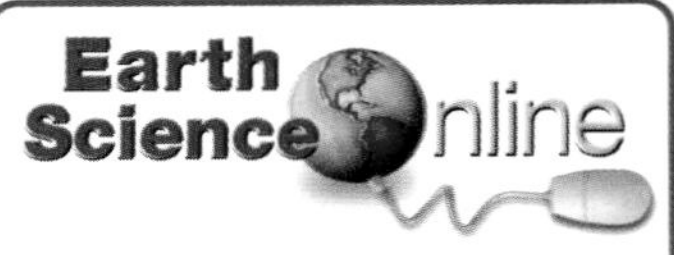

Topic: Tropics
To find out more about the tropics, visit the Earth Science Web Site at earthgeu.com

Activity: Design a Venn diagram to compare and contrast the three major types of tropical climates.

Topographic Effects Water heats up and cools down more slowly than land. Thus, large bodies of water affect the climates of coastal areas. Many coastal regions are warmer in the winter and cooler in the summer than inland areas of similar latitude.

Also, temperatures in the lower atmosphere generally decrease with altitude. Thus, mountain climates are usually cooler than those at sea level. In addition, climates often differ on either side of a mountain. Air rises up one side of a mountain as a result of orographic lifting. The rising air cools, condenses, and drops its moisture. The climate on this side of the mountain—the windward side—is usually wet and cool. On the opposite side of the mountain—the leeward side—the air is dry, and it warms as it descends. For this reason, deserts are common on the leeward sides of mountains, as shown in ***Figure 14-3.***

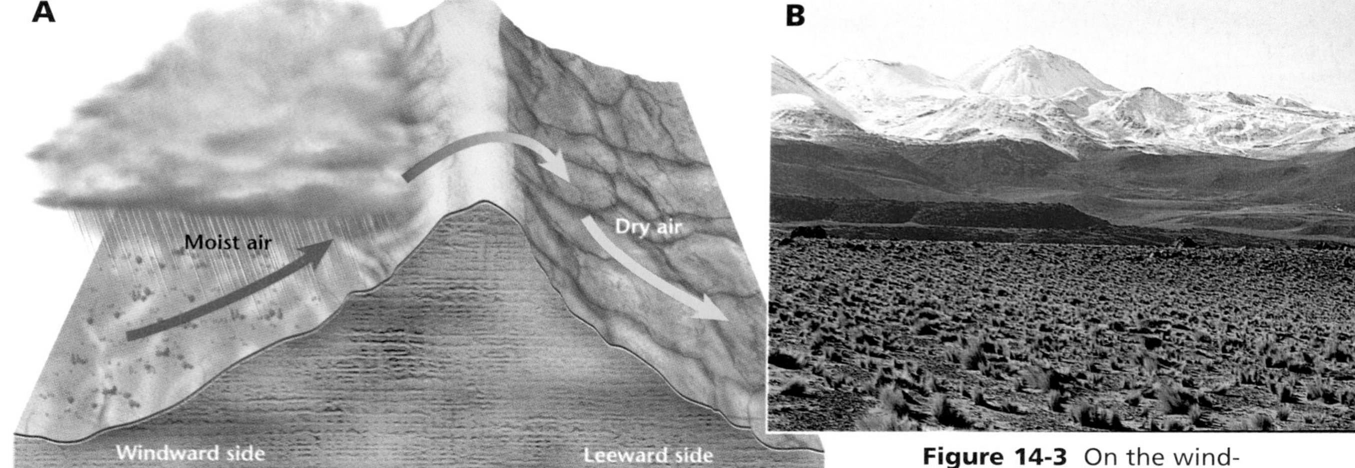

Figure 14-3 On the windward side of a mountain, moist air is forced upward, cools, condenses, and drops its moisture **(A).** The air and the climate on the leeward side of the mountain are dry. Deserts such as the Atacama in Chile are common on leeward sides of mountains **(B).**

Air Masses Two of the main causes of weather are the movement and interaction of air masses. Air masses affect climate, too. They have distinct regions of origin, caused primarily by differences in the amount of solar radiation. The properties of air masses are also dependent on whether they formed over land or water.

Average weather conditions in and near regions of air-mass formation are fairly similar to those exhibited by the air masses themselves. For example, consider an island in the tropical Atlantic Ocean. Because this island is located in an area where maritime tropical (mT) air masses dominate the weather, the island's average weather conditions, or climate, have maritime tropical characteristics.

Section Assessment

1. Compare and contrast temperatures in the tropics, temperate zones, and polar zones.
2. Infer how climate data can be used by farmers.
3. What are some limits associated with the use of normals?
4. Describe two topographic features that cause variations in climate.
5. **Thinking Critically** Average daily temperatures for one city, located at 15° south latitude, are 5°C cooler than average daily temperatures for a second city, located at 30° south latitude. What might account for the cooler temperatures in the first city, which lies so near the equator?

Skill Review

6. **Forming a Hypothesis** Suppose that meteorological data for an area are normally gathered at an airport located 10 km from a large lake. Hypothesize how normals for the area might change if the data were gathered from the edge of the lake. For more help, refer to the *Skill Handbook.*

SECTION 14.2 Climate Classification

Earth Sciences 5.e, 6.b

OBJECTIVES

- **Describe** *the criteria used to classify climates.*
- **Compare** *and* **contrast** *different climates.*

VOCABULARY

Koeppen classification system
microclimate
heat island

Picture a parched desert with wind-blown dunes stretching toward the horizon. Now, imagine a glistening iceberg floating amid a polar sea. These images represent vastly different climates. What criteria would you use to classify them? Temperature is an obvious choice, as is amount of precipitation. The **Koeppen classification system,** a widely used classification system for climates, uses both of these criteria. Developed by Russian-born German climatologist Wladimir Koeppen (1846–1940), the system is based on the average monthly values of temperature and precipitation. It also takes into account the distinct vegetation found in different climates.

Figure 14-4 Koeppen's classification system, shown here in a modified version, is made up of six main divisions.

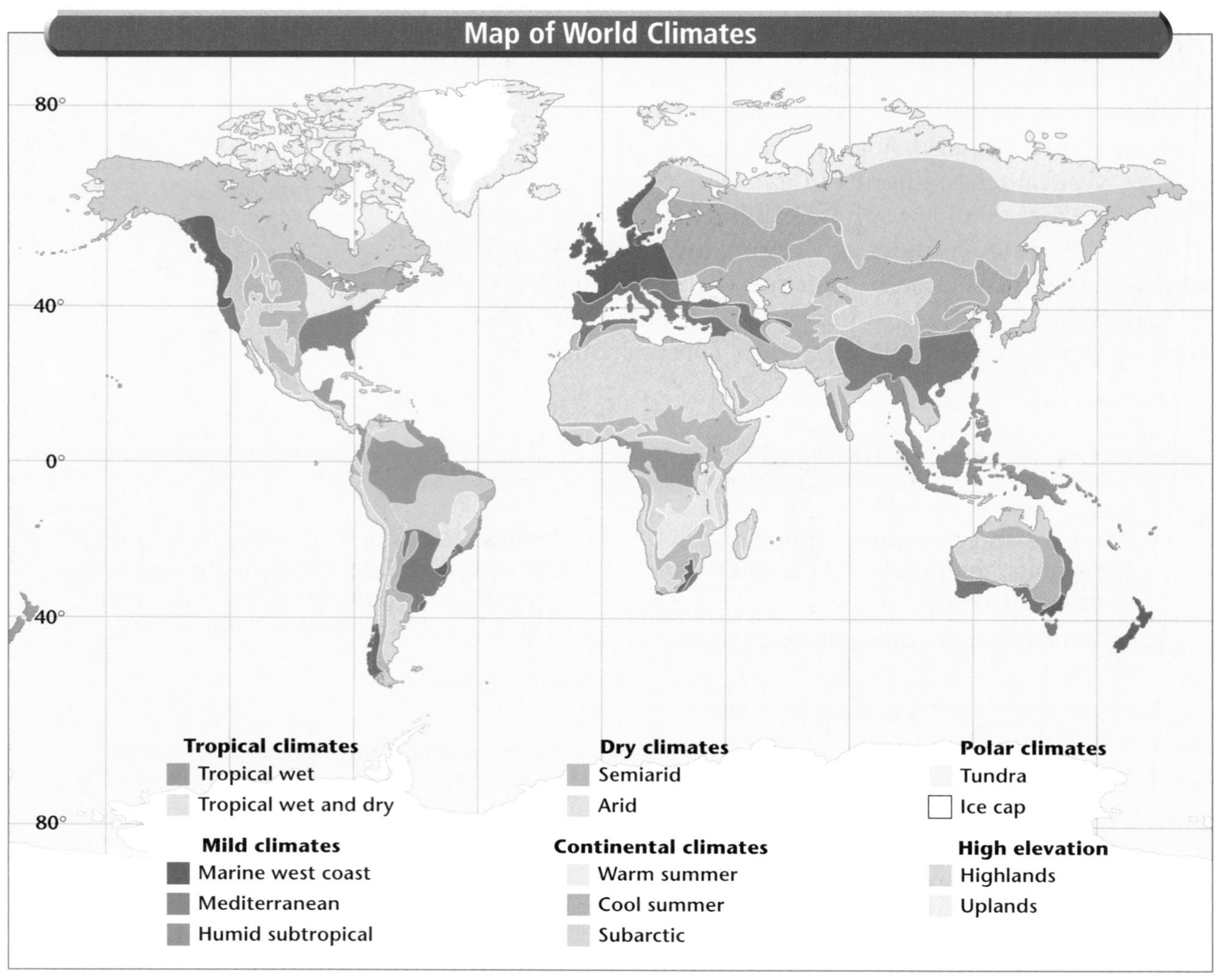

KOEPPEN CLASSIFICATION SYSTEM

Koeppen decided that a good way to distinguish among different climatic zones was by natural vegetation. Palm trees, for instance, are not located in polar regions; they are largely limited to tropical and subtropical regions. Koeppen later realized that quantitative values would make his system more objective and therefore more scientific. Thus, he revised his system to include the numerical values of temperature and precipitation. A map of global climates according to a modified version of Koeppen's classification system is shown in ***Figure 14-4.***

Estimating Use the map in ***Figure 14-4*** to determine the approximate percentage of land covered by tropical wet climates.

Tropical Climates Constant high temperatures characterize tropical climates. In some tropical areas, the heat is accompanied by up to 600 cm of rain each year. The combination of heat and rain produces tropical rain forests, which contain some of the most dramatic vegetation on Earth. You saw an example of a tropical rain forest in the photograph at the beginning of this chapter. Tropical regions are almost constantly under the influence of maritime tropical air. The transition zones that border the rainy tropics north and south of the equator, known as tropical wet and dry zones, have distinct dry winter seasons as a result of the occasional influx of dry continental air masses. Tropical wet and dry zones include savannas, as shown in ***Figure 14-5.*** These tropical grasslands are found in Africa, among other places.

 **Earth Sciences**

5.e Students know rain forests and deserts on Earth are distributed in bands at specific latitudes.

6.b Students know the effects on climate of altitude, elevation, topography, and proximity to large bodies of water and cold or warm ocean currents.

Dry Climates Dry climates, which cover about 30 percent of Earth's land area, make up the largest climatic zone. Most of the world's deserts, such as the Sahara, the Gobi, and the Australian, are classified as dry climates. In these climates, continental tropical (cT) air dominates, precipitation is low, and vegetation is scarce. Many of these areas are located near the tropics. Thus, intense amounts of solar radiation result in high rates of evaporation and few clouds. Overall, evaporation rates exceed precipitation rates. The resulting moisture deficit gives this zone its name. Within this classification, there are two subtypes: arid regions or deserts, and semi-arid regions or steppes. Steppes are more humid than deserts; they generally separate arid regions from bordering wet climates.

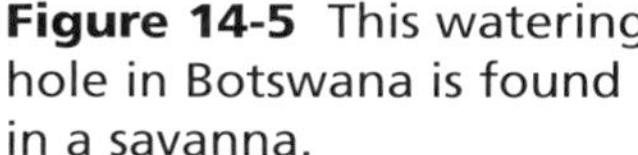
Figure 14-5 This watering hole in Botswana is found in a savanna.

Figure 14-6 The Golden Gate Bridge in San Francisco, California, is nearly hidden beneath a dense layer of fog. Fog is characteristic of marine west coast climates.

Mild Climates Mild climates can be classified into three subtypes: humid subtropical climates, marine west coast climates, and mediterranean climates. Humid subtropical climates are influenced by the subtropical high-pressure systems that are normally found over oceans in the summer. The southeastern United States has this type of climate. There, warm, muggy weather prevails during the warmer months and dry, cool conditions predominate during the winter. The marine west coast climates are dominated by the constant inland flow of air off the ocean, which creates mild winters and cool summers, with abundant precipitation throughout the year. An example of this type of climate is shown in ***Figure 14-6.*** Mediterranean climates, found in Italy and parts of Spain, among other places, are influenced by the Mediterranean Sea. Summers in these climates are generally warm because the lack of cool ocean currents in the Mediterranean Sea results in relatively warm water temperatures.

Continental Climates Continental climates are also classified into three subtypes: warm summer climates, cool summer climates, and subarctic climates. Located in the zone dominated by the polar front, continental climates are battlegrounds for clashing tropical and polar air masses. Thus, these zones experience rapid and sometimes violent changes in weather. Both summer and winter temperatures can be extreme because the influence of polar air masses is strong in winter, while warm tropical air dominates in summer. The presence of warm, moist air causes summers to be generally wetter than winters, especially in latitudes that are relatively close to the tropics.

Polar Climates To the north of continental climates lie the polar climates, the coldest regions on Earth. Just as the tropics are known for their year-round warmth, polar climates are known for their constant cold—the mean temperature of the warmest month is less than 10°C. Precipitation is generally low because cold air holds less moisture than warm air. Also, the amount of heat radiated by Earth's surface is too low to produce the strong convection currents needed to release heavy precipitation. ***Figure 14-7A*** shows an ice-cap polar climate.

A variation of the polar climate is found at high elevations. This type of climate includes parts of the Andes Mountains of South America, shown in ***Figure 14-7B,*** which lie near the equator. The intense solar radiation found near such equatorial regions is offset by the decrease in temperature that occurs with altitude.

Figure 14-7 Penguins are one of the few species that can survive in Antarctica's ice-cap polar climate **(A).** Llamas are common in the high-elevation climates of the Andes Mountains **(B).**

MICROCLIMATES

Sometimes, the climate of a small area can be much different from that of the larger area surrounding it. A localized climate that differs from the main regional climate is called a **microclimate.** If you climb to the top of a mountain, you can experience a type of microclimate; the climate becomes cooler with increasing elevation. You'll learn more about microclimates in the *Design Your Own GeoLab* at the end of this chapter. ***Figure 14-8*** shows a microclimate in a city.

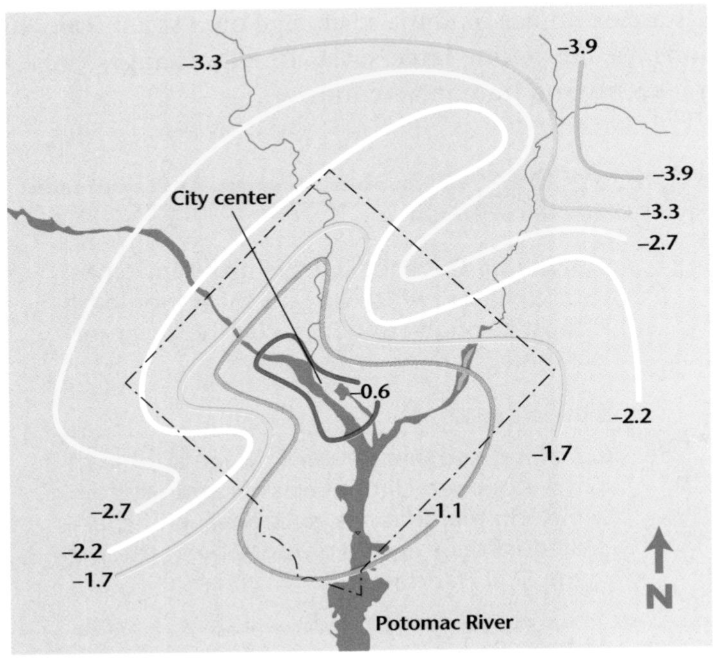

Figure 14-8 This diagram shows winter temperatures in Washington, D.C. The buildings and paved surfaces of the city create a microclimate. The temperature in the center of the city is –0.6°C, nearly 3°C warmer than temperatures in some parts of the surrounding area.

Figure 14-9 These images show differences in daytime temperatures between an urban area **(A)** and a suburban area **(B).** The coolest temperatures are represented by blue; the warmest temperatures are represented by red.

Heat Islands The mere presence of a building can create microclimates in the area immediately surrounding it. How? The building casts shadows that lower air temperature. The presence of many concrete buildings and large expanses of asphalt can create **heat islands,** wherein the climate is warmer than in surrounding rural areas. This effect was recognized as long ago as the early nineteenth century, when Londoners noticed that the temperature in their city was noticeably warmer than in the surrounding countryside.

The heat-island effect occurs because large areas of asphalt and concrete radiate far more heat into the air than do grasslands, wooded areas, and bodies of water. This causes mean temperatures in large cities to be significantly warmer than in surrounding areas, as shown in ***Figure 14-9.*** The heat-island effect also causes greater changes in temperature with altitude, which sparks strong convection currents. This in turn produces increased cloudiness and up to 15 percent more total precipitation in cities.

Heat islands are examples of climatic change on a small scale. In the next sections, we'll examine large-scale climatic changes caused by both natural events and human activities.

Section Assessment

1. Compare and contrast the five main climate types.
2. What criteria is the Koeppen climate classification system based on?
3. What are microclimates? What climatic effects do heat islands have on large cities?
4. Describe the climate of your area. Which zone do you live in? What type of air masses generally affect your climate?
5. **Thinking Critically** Of the different types of climates, which do you think would be most strongly influenced by the polar jet stream? Why?

Skill Review

6. **Making and Using Tables** Make a table of the Koeppen climate classification system. Include major zones, subzones, and characteristics of each. For more help, refer to the *Skill Handbook*.

earthgeu.com/self_check_quiz

SECTION 14.3

Climatic Changes

Earth Sciences 5.g, 6.c

Some years may be warmer, cooler, wetter, or drier than others, but during the average human lifetime, climates do not appear to change significantly. However, a study of Earth's history over hundreds of thousands of years shows that climates always have been, and currently are, in a constant state of change. These changes usually take place over extremely long time periods. Geologic records show that in the past, Earth was sometimes much colder or warmer than it is today.

OBJECTIVES

- **Distinguish** *among different types of climatic changes.*
- **Recognize** *why climatic changes occur.*

VOCABULARY

ice age
season
El Niño
Maunder minimum

ICE AGES

A good example of climatic change involves glaciers, which have alternatively advanced and retreated over the past 2 million years. At times, much of Earth's surface was covered by vast sheets of ice. During these periods of extensive glacial coverage, called **ice ages,** average global temperatures decreased by an estimated 5°C. Although this may not seem like a large decrease, global climates became generally colder and snowfall increased, which sparked the advance of existing ice sheets. Ice ages alternate with warm periods called interglacial intervals—we are currently experiencing such an interval. The most recent ice age ended only about 10 000 years ago. In North America, glaciers spread from the east coast to the west coast and as far south as Indiana, as shown in ***Figure 14-10.*** The results of this glacial period are apparent in the Great Lakes and the Finger Lakes of central New York, which were scoured out as the glaciers retreated.

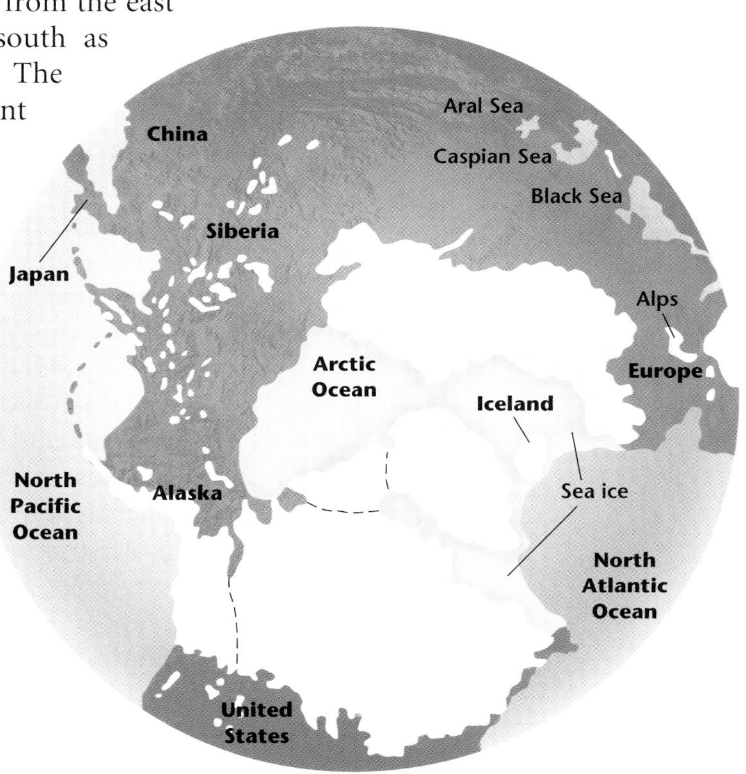

Figure 14-10 The last ice age covered large portions of North America, Europe, and Asia. Average global temperatures were roughly 5°C lower than normal.

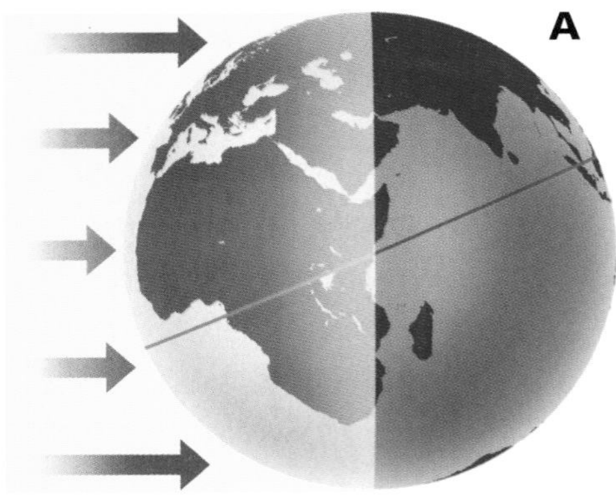

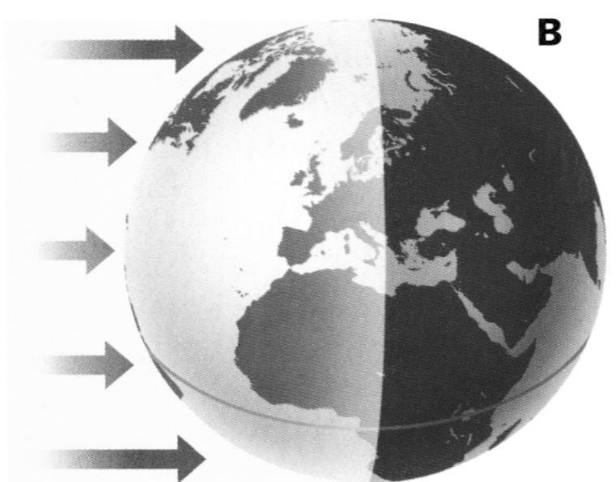

Figure 14-11 When the north pole is pointed toward the Sun, the northern hemisphere experiences summer and the southern hemisphere experiences winter **(A).** During spring and fall, neither pole points toward the Sun **(B).**

SHORT-TERM CLIMATIC CHANGES

While ice ages take place over many thousands of years, other climatic changes take place in much shorter time periods. The most obvious of these are **seasons,** which are short-term periods of climatic change caused by regular variations in daylight, temperature, and weather patterns. These variations are the result of changes in the amount of solar radiation an area receives. As ***Figure 14-11*** shows, the tilt of Earth on its axis as it revolves around the Sun causes different areas of Earth to receive different amounts of solar radiation. During summer in the northern hemisphere, the north pole is tilted toward the Sun, and this hemisphere experiences long hours of daylight and warm temperatures. At the same time, it is winter in the southern hemisphere. The south pole is tilted away from the Sun, and the southern hemisphere experiences long hours of darkness and cold temperatures. Throughout the year, the seasons are reversed in the north and south hemispheres.

El Niño Other short-term climatic changes are caused by **El Niño,** a warm ocean current that occasionally develops off the western coast of South America. In the Southeast Pacific Ocean, atmospheric and ocean currents along the coast of South America normally move north, transporting cold water from the Antarctic region. Meanwhile, the trade winds and ocean currents move westward across the tropics, keeping warm water in the western Pacific. This circulation, driven by a semipermanent high-pressure system, creates a cool, dry climate along much of the northwestern coast of South America.

Occasionally, however, for reasons that are not fully understood, this high-pressure system and its attendant trade winds weaken drastically, which allows the warm water from the western Pacific to surge eastward toward the South American coast. The sudden presence of this warm water heats the air near the surface of the water. Convection currents strengthen, and the normally cool and dry northwestern coast of South America becomes much warmer and wetter. The increased precipitation pumps large amounts of heat and moisture into the upper atmosphere, where upper-level winds transport the hot, moist air eastward across the tropics. This hot, moist air in the upper atmosphere is responsible for dramatic climatic changes. Sharp temperature differences in the upper air allow the jet stream to shift farther south. This causes weather systems to take a more southerly track, bringing violent storms to California and the Gulf Coast, which are usually south of the storm tracks.

Figure 14-12 During El Niño, some areas of the world experience extreme droughts while other areas are ravaged by heavy floods.

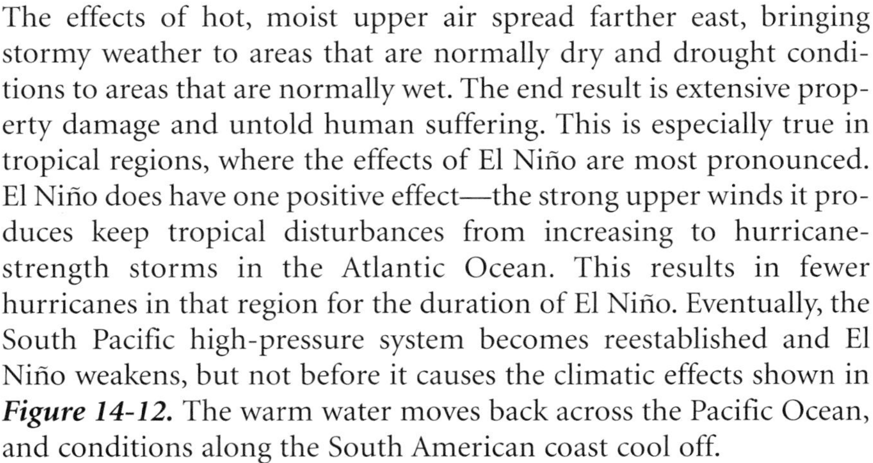

The effects of hot, moist upper air spread farther east, bringing stormy weather to areas that are normally dry and drought conditions to areas that are normally wet. The end result is extensive property damage and untold human suffering. This is especially true in tropical regions, where the effects of El Niño are most pronounced. El Niño does have one positive effect—the strong upper winds it produces keep tropical disturbances from increasing to hurricane-strength storms in the Atlantic Ocean. This results in fewer hurricanes in that region for the duration of El Niño. Eventually, the South Pacific high-pressure system becomes reestablished and El Niño weakens, but not before it causes the climatic effects shown in ***Figure 14-12.*** The warm water moves back across the Pacific Ocean, and conditions along the South American coast cool off.

 **Earth Sciences**

5.g Students know features of ENSO (El Niño southern oscillation) cycle in terms of sea-surface and air temperature variations across the Pacific and some climatic results of this cycle.

6.c Students know how Earth's climate has changed over time, corresponding to changes in Earth's geography, atmospheric composition, and other factors, such as solar radiation and plate movement.

Change Can Be Natural

Much discussion has taken place in recent years about whether Earth's climate is changing as a result of human activities. We'll discuss this in the next section. For now, it's important to note that climatic changes occurred long before humans came on the scene. Studies of tree rings, ice-core samples, fossils, and radiocarbon samples provide evidence of past climatic changes. These changes in Earth's climate were caused by natural events such as variations in solar activity, changes in Earth's tilt and orbit, and volcanic eruptions.

Figure 14-13 Very few sunspots occurred during the Maunder minimum, and temperatures were lower than normal. Thus, scientists theorize solar activity may be linked to climatic changes.

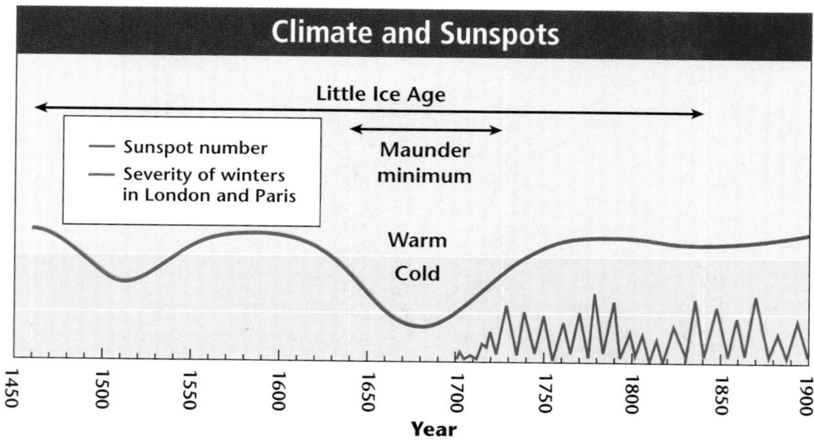

Solar Activity Evidence of a possible link between solar activity and Earth's climate was provided by English astronomer E. W. Maunder in 1893. The existence of sunspot cycles lasting approximately 11 years had been recognized since the days of Galileo. However, Maunder found that from 1645 to 1716, sunspot activity was scarce to nonexistent. This period of very low sunspot activity, called the **Maunder minimum,** closely corresponds to an unusually cold climatic episode called the "Little Ice Age." During this time, much of Europe experienced bitterly cold winters and below-normal temperatures year-round. Residents of London are said to have ice-skated on the Thames River in June. The relationship between climate and periods of low sunspot activity is illustrated in ***Figure 14-13.*** Studies indicate that increased solar activity coincides with warmer-than-normal climates, while periods of low solar activity, such as the Maunder minimum, coincide with cold climatic conditions.

Figure 14-14 Scientists theorize that a more elliptical orbit around the Sun could produce significant changes in Earth's climate.

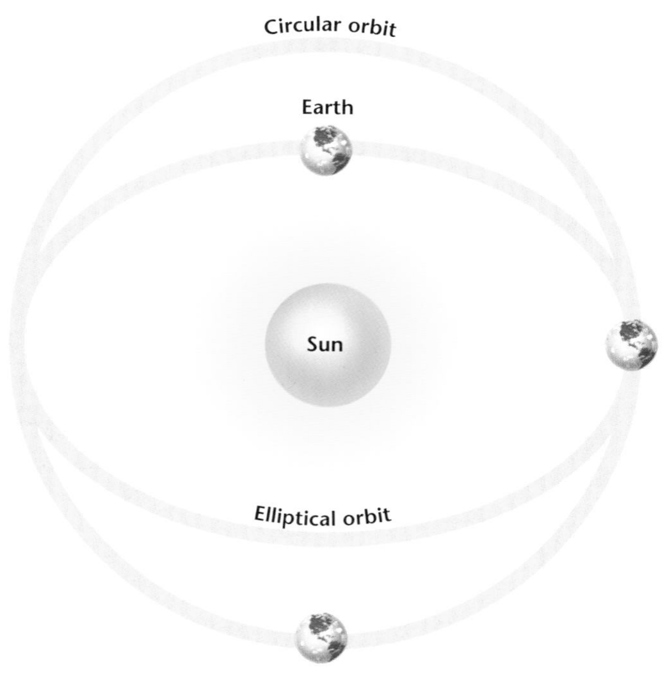

Earth's Orbit Climatic changes may also be triggered by changes in Earth's axis and orbit. The shape of Earth's elliptical orbit appears to change, becoming more elliptical, then more circular, over the course of a 100 000-year cycle. As ***Figure 14-14*** shows, when the orbit elongates, Earth passes closer to the Sun, and temperatures become warmer than normal. When the orbit is more circular, Earth is farther from the Sun and temperatures dip below average. The amount of radiation

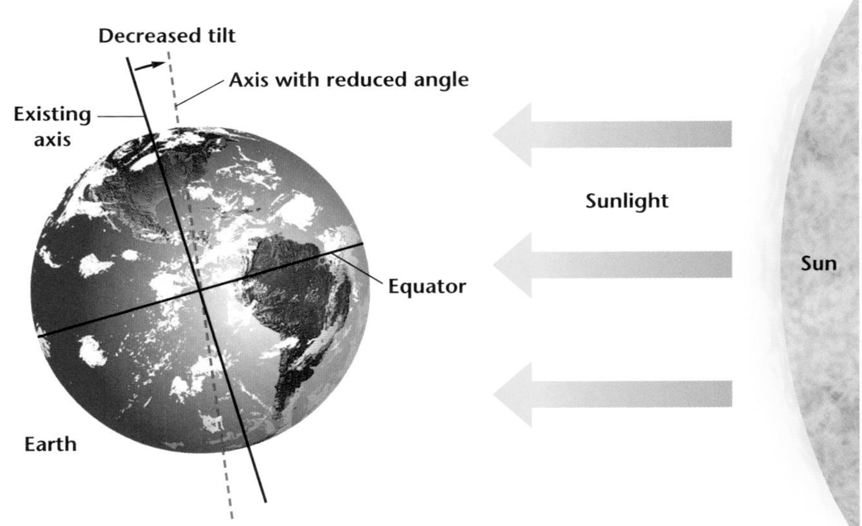

Figure 14-15 If the angle of tilt of Earth's axis decreased, there would be less temperature contrast between summer and winter.

Earth receives when its orbit elongates is much higher than when its orbit is more circular.

As you know, seasons are caused by the tilt of Earth's axis. At present, the angle of the tilt is 23.5°. However, the angle of tilt varies from a minimum of 22.1° to a maximum of 24.5° every 41 000 years. Scientists theorize that these changes in angle cause seasons to become more severe. For example, a decrease in the angle of the tilted axis, as shown in ***Figure 14-15,*** might cause a decrease in the temperature difference between winter and summer. Winters would be warmer and wetter, and summers would be cooler. The additional snow in latitudes near the poles would not melt in summer because temperatures would be cooler than average. This could result in expanded glacial coverage. In fact, some scientists hypothesize that changes in the angle of Earth's tilted axis cause ice ages.

Earth's Wobble Another movement of Earth may be responsible for climatic changes. Over a period of about 26 000 years, Earth wobbles as it spins on its axis. Currently, the axis points toward the North Star, Polaris, as shown in ***Figure 14-16.*** Because of Earth's wobbling, however, the axis will tilt toward another star, Vega, by about the year 14 000. Currently, winter occurs in the northern hemisphere when Earth is closest to the Sun, and summer occurs when Earth is farthest from the Sun. When the axis tilts toward Vega, however, winter will occur in the northern hemisphere when Earth is farthest from the Sun, and summer will occur when Earth is closest to the Sun. This will cause warmer summers and colder winters than those that we now experience.

Figure 14-16 By about the year 14 000, Earth's axis will point toward the star, Vega. Winter will then occur in the northern hemisphere when Earth is farthest from the Sun, causing winters to be colder.

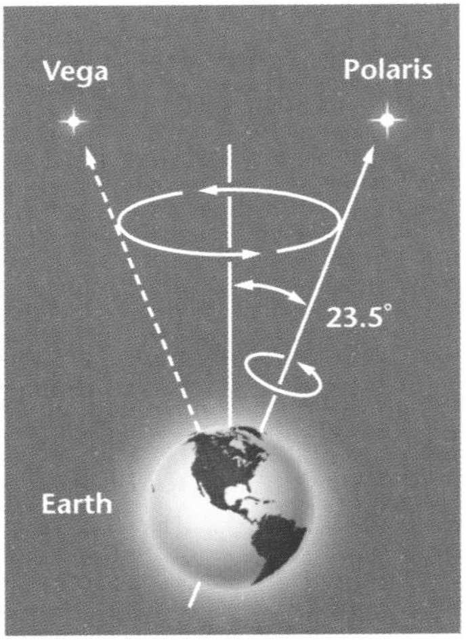

Figure 14-17 The dust and gases released by this volcanic eruption in New Guinea blocked incoming solar radiation and affected global climates.

Volcanic Activity Climatic changes can also be triggered by the immense quantities of dust released into the atmosphere during major volcanic eruptions, shown in ***Figure 14-17.*** Volcanic dust can remain suspended in the atmosphere for several years, blocking incoming solar radiation and thus lowering global temperatures. Some scientists theorize that periods of high volcanic activity cause cool climatic periods. This theory is supported by records over the past century because several large eruptions have been followed by below-normal global temperatures. For instance, the ash released during the 1991 eruption of Mt. Pinatubo in the Philippines resulted in slightly cooler temperatures around the world the following year. Generally, volcanic eruptions appear to have only short-term effects on climate. These effects, as well as the others you've read about thus far, are a result of natural causes. In the next section, you'll learn about climatic changes caused by human activities.

Section Assessment

1. What three changes in Earth's movement in space might result in long-term climatic changes?
2. What are seasons? What causes them?
3. Explain how El Niño might affect weather in California and along the Gulf Coast.
4. Why are the greatest effects of El Niño experienced mainly in the tropics?
5. How does volcanic activity affect climate? Are these effects examples of short-term or long-term climatic change?
6. **Thinking Critically** What might be the effect on seasons if Earth's orbit became more elliptical and, at the same time, the angle of the tilt of Earth's axis increased?

Skill Review

7. **Concept Mapping** Use the following phrases to complete a concept map of the effects of El Niño. For more help, refer to the *Skill Handbook.*

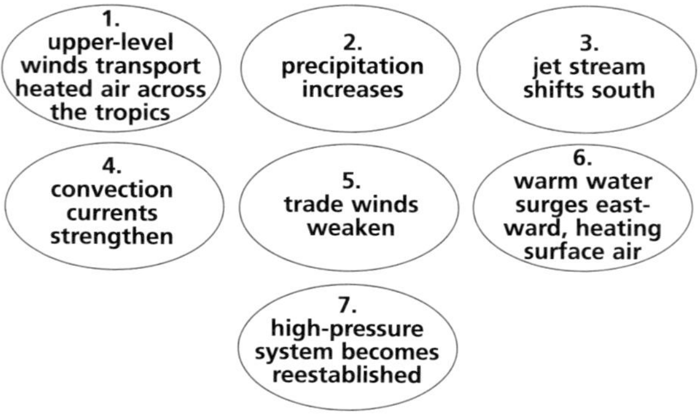

SECTION 14.4 The Human Factor

Earth Sciences 4.c, 4.d, 5.g, 6.c, 7.a, 7.b, 8.b **I&E** 1.a, 1.d, 1.j, 1.l

One of the most significant influences on Earth's climate is the atmosphere. As you learned in Chapter 11, solar radiation that is not reflected by clouds passes freely through the atmosphere. It's then absorbed by Earth's surface and released as long-wavelength radiation. This radiation is absorbed by atmospheric gases such as water vapor, methane, and carbon dioxide. The atmospheric gases then reradiate the stored energy, so that Earth receives energy from two sources: the Sun and the atmosphere.

OBJECTIVES

- **Compare** *and* **contrast** *the greenhouse effect and global warming.*
- **Identify** *how humans impact climate.*

VOCABULARY

greenhouse effect
global warming

THE GREENHOUSE EFFECT

The retention of heat by the atmosphere results in the **greenhouse effect,** which is the natural heating of Earth's surface caused by certain atmospheric gases called greenhouse gases. Without the greenhouse effect, which is illustrated in ***Figure 14-18,*** life as we know it could not exist on Earth. Our planet would be cold, like Mars, which has an extremely thin atmosphere and surface temperatures that dip to –90°C. On the other hand, a marked increase in the greenhouse effect might cause our planet to be hot, like Venus, which, because of

Figure 14-18 Solar radiation reaches Earth's surface and is reradiated as long-wavelength radiation. This radiation cannot escape through the atmosphere, and is absorbed and re-released by atmospheric gases. This process is called the greenhouse effect because it is similar to the way that heat is trapped and released in a greenhouse.

MiniLab

How does the atmosphere affect the transfer of energy?

Model the greenhouse effect.

Procedure

1. On a clear day, place a cardboard box outside in a shaded area. Prop two thermometers against the box. Make sure the thermometers are not in direct sun.
2. Cover one thermometer with a clean glass jar.
3. Observe and record the temperature changes of each thermometer every two minutes over a 30-minute period.

Analyze and Conclude

1. Make a graph showing how the temperatures of the two thermometers changed over time.
2. Based on your graph, which thermometer experienced the greatest increase in temperature? Why?
3. Relate your observations to the greenhouse effect in the atmosphere.

Earth Sciences

4.c Students know the different atmospheric gases that absorb the Earth's thermal radiation and the mechanism and significance of the greenhouse effect.

Also covers 4.d, 5.g, 6.c, 7.a, 7.b, 8.b **I&E** 1.d

its thick atmosphere, has surface temperatures of 470°C. You'll model the greenhouse effect in the *MiniLab* on this page.

Scientists theorize that it is possible to increase or decrease the greenhouse effect by changing the amount of atmospheric greenhouse gases, particularly carbon dioxide (CO_2). Any increase in the amount of these gases would theoretically result in the increased absorption of radiation. Levels of atmospheric carbon dioxide are increasing. This in turn can lead to a rise in global temperatures, known as **global warming.**

Global Warming

Temperatures worldwide have indeed shown an upward trend over the past 200 years, with several of the warmest years on record having occurred within the last two decades. If the trend continues, polar ice caps might melt, sea level might rise and flood coastal cities, deserts could spread into fertile regions, and the frequency and severity of storms could increase.

Based on available evidence, many scientists agree that global warming is occurring. They disagree, however, about what is causing this warming. As you've learned, natural cycles of Earth and the Sun can affect climate. Some scientists hypothesize that these natural changes adequately explain the increased temperatures. Mounting evidence indicates, however, that the warming trend is a result of increases in atmospheric carbon dioxide. Global warming remains a controversial issue. Neither viewpoint can be proven or disproven conclusively; it might very well be that there are several factors involved. However, if increased carbon dioxide is responsible, two logical questions follow: What is causing the increase? Can anything be done to stop it?

Impact of Human Activities

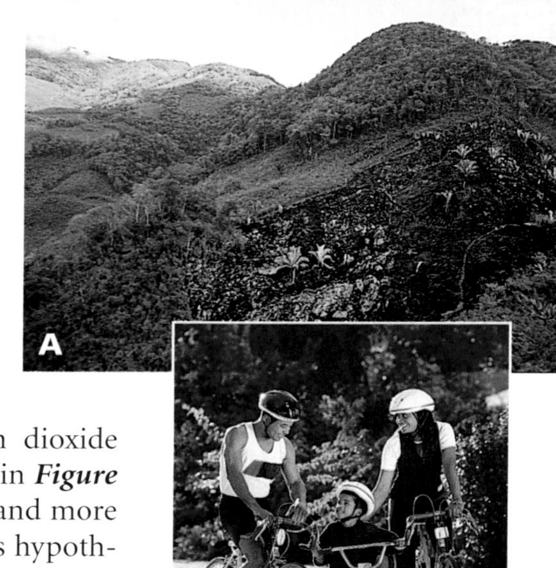

To find a possible cause for rising levels of atmospheric carbon dioxide, we need look no further than our driveways. Automobile exhaust is a prime source of atmospheric CO_2, as are industrial emissions. In fact, almost any process that involves the burning of fossil fuels results in the release of carbon dioxide and other gases into the atmosphere. In addition, deforestation, the mass removal of trees, plays a role in increasing levels of atmospheric CO_2. During photosynthesis, vegetation removes carbon dioxide from the atmosphere. When trees such as the ones shown in ***Figure 14-19A*** are cut down, rates of photosynthesis are reduced and more carbon dioxide remains in the atmosphere. Many scientists hypothesize that deforestation intensifies global warming trends.

Figure 14-19 Deforestation in places such as Guatemala may increase global warming **(A).** Automobile exhaust may also increase global warming; thus, riding bikes is one way to help the environment **(B).**

Environmental Efforts

Because global warming appears to be linked to human activities that cause pollution or widespread deforestation, we must closely examine those activities and work to reduce their environmental impact. Individuals can combat global warming by conserving energy, which in turn reduces the consumption of fossil fuels. Some easy ways to conserve energy include turning off appliances and lights when a room is not in use, turning down thermostats in the winter, and recycling. An additional option is shown in ***Figure 14-19B.*** You'll learn more about global warming in the *Science & Math* feature at the end of this chapter.

Section Assessment

1. Why do some scientists theorize that global warming might not be the result of increases in atmospheric carbon dioxide?
2. What are some possible consequences of global warming?
3. Describe some human activities that may have an impact on Earth's climate. List several actions you can take to reduce this impact.
4. Compare and contrast global warming and the greenhouse effect.
5. **Thinking Critically** Based on what you have learned in this section, explain why tropical rain forests are often called the "lungs" of our planet.

Skill Review

6. **Designing an Experiment** When green plants photosynthesize, they take carbon dioxide out of the atmosphere. Design an experiment in which you could use green plants and glass jars to study the effects of carbon dioxide on global warming. For more help, refer to the *Skill Handbook*.

earthgeu.com/self_check_quiz

DESIGN YOUR OWN GeoLab

Microclimates

Microclimates can be caused by tall buildings, large bodies of water, and mountains, among other things. In this activity, you'll observe different microclimates and then attempt to determine which factors strengthen microclimates and how these factors change with distance from Earth's surface.

I&E

1.a Select and use appropriate tools and technology (such as computer-linked probes, spreadsheets, and graphing calculators) to perform tests, collect data, analyze relationships, and display data.

Also covers:
1.d, 1.j

Preparation

Problem

Which type of surface creates the most pronounced microclimate?

Possible Materials

thermometer
psychrometer
paper strip or wind sock
meterstick
relative humidity chart *(Appendix F)*

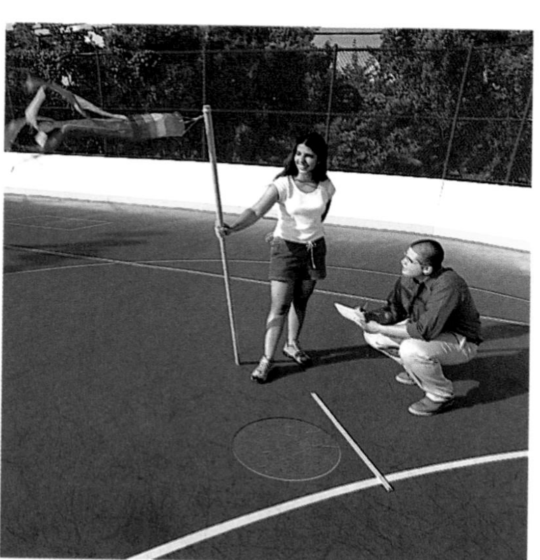

Hypothesis

Hypothesize how different areas of Earth's surface, such as grassy lawns, asphalt parking lots, and bodies of water, affect local climates. Consider also whether distance from the ground might affect temperature, relative humidity, and wind speed.

Objectives

In this GeoLab, you will:

- **Design** and **carry out** an experiment to study microclimates both at Earth's surface and above its surface.
- **Observe** and **record** temperature, relative humidity, and wind speed.
- **Infer** how different surfaces and changes in height above these surfaces affect microclimates.

Safety Precautions

Be careful when you handle glass thermometers, especially those that contain mercury. If the thermometer breaks, do not touch it. Have your teacher properly dispose of the glass and the mercury.

Plan the Experiment

1. As a group, agree upon and write out your hypothesis. List the steps needed to test your hypothesis. Include in your plan how you will use your equipment to measure temperature, relative humidity, and wind speed at different surfaces and at various heights above these surfaces.
2. Select your sites. Possible sites include a grassy playground area, a paved parking lot, and a swimming pool.
3. Be sure to control variables. For instance, different members of your group should make observations at each site at the same time each day. You'll also need to record weather variables at several different distances above each surface. Possible distances include 5 cm, 1 m, and 2 m.
4. Make a map of your test sites. Design and construct data tables for recording your observations. You'll need separate data tables for each site.
5. Read over your entire plan to make certain all steps are in logical order. Identify constants and variables in your plan.
6. Make sure your teacher approves your plan before you proceed with your experiment.
7. Carry out your plan.

Analyze

1. **Comparing and Contrasting** Map your data. Color code the areas on your map to show which surfaces had the highest and lowest temperatures, the highest and lowest relative humidity, and the greatest and least wind speed. On your map, include data for surface areas only.
2. **Making and Using Graphs** Graph your data for each site, showing differences in temperature with height. Plot temperature on the x-axis and height on the y-axis. Repeat this step for relative humidity and wind speed.
3. **Interpreting Scientific Illustrations** Analyze your maps, graphs, and data to find patterns. Which surfaces had the most pronounced microclimates? Did height above the surface affect your data? Infer why or why not.
4. **Thinking Critically** Analyze your hypothesis and the results of your experiment. Was your hypothesis supported? Explain.

Conclude & Apply

1. Why did some areas have more pronounced microclimates than others? Which factors seemed to contribute most to the development of microclimates?
2. Which variable changed most with height: temperature, relative humidity, or wind speed? Which variable changed least? Infer why some variables changed more than others with height.

Science & Math

Earth Sciences 4.b, 4.c, 6.c, 7.d **I&E** 1.l

Global Warming

According to the National Academy of Scientists, Earth's surface temperature has risen about one degree Fahrenheit in the past 100 years. This increase in temperature can be correlated to an increase in the concentration of carbon dioxide and other greenhouse gases in the atmosphere. How might this increase in temperature affect Earth's climate?

Carbon dioxide is one of the greenhouse gases that helps keep temperatures on Earth warm enough to support life. However, a build-up of carbon dioxide and other greenhouse gases such as methane and nitrous oxide can lead to global warming, an increase in Earth's average surface temperature. Since the industrial revolution in the 1800s, atmospheric concentrations of carbon dioxide have increased by almost 30 percent, methane concentrations have more than doubled, and nitrous oxide concentrations have increased approximately 15 percent. Scientists attribute these increases to the burning of fossil fuels for automobiles, industry, and electricity, as well as deforestation, increased agriculture, landfills, and mining.

Impact on Climate

Based on current data, scientists estimate that Earth's average surface temperature could rise between 1 and 4.5 degrees Fahrenheit in the next 50 years. As the temperature increases, evaporation rates increase, which can lead to increased precipitation. With increased evaporation rates, soil may become drier and intense rainstorms more frequent. Some scientists estimate that sea levels may rise as much as two feet along most of the U.S. coast.

Procedure

1. Create a line graph which displays the information shown in the data table below.

Challenge

1. During which time period did Earth's atmosphere experience a slow, steady increase in carbon dioxide? Describe the carbon dioxide increase in the last 50 years.
2. Assuming that carbon dioxide levels continue to increase at the current rate, predict the global atmospheric CO_2 concentration in parts per million in the year 2015.

Global Levels of Atmospheric CO_2

Year	Global Atmospheric CO_2 Concentration (parts per million)	Year	Global Atmospheric CO_2 Concentration (parts per million)
1745	279	1949	311
1791	279.7	1958	312
1816	283.8	1965	318
1843	287.4	1974	330.1
1854	288.2	1984	344.3
1874	289.5	1994	358.8
1894	297	1995	361
1909	299.2	1996	363
1921	301.6	1997	364
1935	306.6	1998	367

CHAPTER 14

Study Guide

Summary

SECTION 14.1

What is climate?

Main Ideas

- Climate describes the long-term weather patterns of a region. Climatological data include annual variations of temperature, precipitation, wind, and other weather variables, as well as extreme fluctuations in these variables.
- The factors that influence climate include latitude, topography, closeness of lakes and oceans, availability of moisture, global wind patterns, ocean currents, and air masses.

Vocabulary

climate (p. 359)
climatology (p. 359)
normal (p. 360)
polar zone (p. 362)
temperate zone (p. 362)
tropics (p. 362)

SECTION 14.2

Climate Classification

Main Ideas

- The Koeppen classification system divides climates into five basic types according to temperature, rainfall, and vegetation.
- A microclimate is a localized climate that differs from the surrounding regional climate. In cities, the numerous concrete buildings and large expanses of asphalt can create heat islands, wherein the climate is warmer than in surrounding rural areas.

Vocabulary

heat island (p. 368)
Koeppen classification system (p. 364)
microclimate (p. 367)

SECTION 14.3

Climatic Changes

Main Ideas

- Earth's climate is in a constant state of change. These changes usually take place over extremely long time periods. Fossils, ice cores, and other geologic records show that Earth was sometimes much colder or warmer than it is today.
- Periods of extensive glacial coverage, called ice ages, are examples of long-term climatic changes. Examples of short-term climatic changes include the seasons and the effects of El Niño.
- Some changes in Earth's climate may be caused by a combination of numerous natural cycles involving solar activity, changes in the tilt of Earth's axis and its orbit, and volcanic eruptions.

Vocabulary

El Niño (p. 370)
ice age (p. 369)
Maunder minimum (p. 372)
season (p. 370)

SECTION 14.4

The Human Factor

Main Ideas

- The greenhouse effect is the retention of heat by atmospheric gases that helps to keep Earth warm enough to sustain life. An increase in greenhouse gases may lead to global warming.
- Some scientists theorize that human activities such as the burning of fossil fuels and deforestation cause global warming.

Vocabulary

global warming (p. 376)
greenhouse effect (p. 375)

earthgeu.com/vocabulary_puzzlemaker

CHAPTER 14

Assessment

Understanding Main Ideas

1. Which of the following does NOT influence climate?
   a. latitude
   b. satellites
   c. mountains
   d. large bodies of water

2. A heat island is an example of what type of climate?
   a. tropical climate
   b. microclimate
   c. dry climate
   d. polar climate

3. Which of the following is NOT a factor used to classify climates in the Koeppen classification system?
   a. temperature
   b. moisture
   c. wind
   d. vegetation

4. Which of the following is an example of a long-term climatic change?
   a. fall
   b. ice ages
   c. summer
   d. El Niño

5. What is El Niño?
   a. a warm wind
   b. a cool ocean current
   c. a cool wind
   d. a warm ocean current

6. Which of the following would normals best describe?
   a. a specific location
   b. a land mass
   c. an ocean
   d. a mountain

7. What is the Maunder minimum?
   a. a period of low sunspot activity
   b. an ice age
   c. a volcanic eruption
   d. a cycle of Earth's orbit

8. Which of the following would NOT be likely to produce a microclimate?
   a. an ocean shoreline
   b. a valley
   c. a flat prairie
   d. a large city

9. Which greenhouse gas is most associated with global warming?
   a. methane
   b. ozone
   c. carbon dioxide
   d. carbon monoxide

10. If it's winter in the northern hemisphere, what season is it in the southern hemisphere?
    a. summer
    b. fall
    c. winter
    d. spring

11. El Niño develops because of a weakening of what?
    a. the polar front
    b. the trade winds
    c. the prevailing westerlies
    d. the jet stream

12. Which of the following is not a natural cause of long-term climatic change?
    a. solar activity
    b. industrialization
    c. changes in the tilt of Earth's axis
    d. changes in Earth's rotation

13. What is an ice age? When did the last ice age end?
14. What is deforestation? How is it linked to global warming?
15. On which side of a mountain would a desert most likely form? Why?
16. Compare and contrast a steppe and a savanna.
17. When Earth's orbit elongates, are temperatures warmer or cooler than normal? Explain.

Test-Taking Tip

PLAN AHEAD Find out where and when your test will take place. If the location is unfamiliar, go there ahead of time to be sure you can find your way. Find out if you will need identification or other items on the day of the test.

CHAPTER 14

Assessment

Applying Main Ideas

18. The graph on this page shows sunspot activity from 1860 to 1995. Predict when the next maximum will occur.

Use this graph to answer question 18.

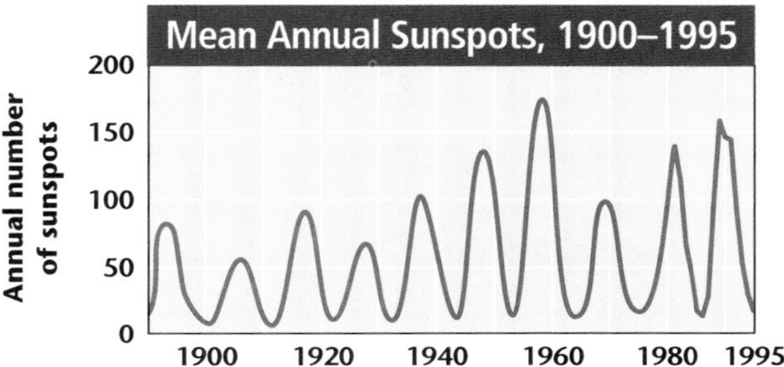

19. Based on the data shown in the *Problem-Solving Lab* on page 360, what months are good for planting crops that cannot survive frosts?

20. Use a world map and ***Figure 14-4*** to determine the climate classifications of the following cities: Paris, France; Athens, Greece; London, England; and Sydney, Australia.

Thinking Critically

21. Which would El Niño affect more: Anchorage, Alaska, or Los Angeles, California? Why?

22. How might a mountain valley produce a microclimate?

23. Based on the results of the *Discovery Lab* that you performed at the beginning of this chapter, would you expect temperatures during the night to drop more sharply in marine climates or continental climates? Why?

24. A large, three-story shopping mall is built near a town. Would you expect temperatures in the immediate region to increase, decrease, or remain the same? Explain.

Standardized Test Practice

INTERPRETING DATA Use the table below to answer questions 1 and 2.

Location	Climate Description
New Caledonia, South Pacific	Constant high temperatures, plenty of rain
Southern Israel	Humid in summer, dry in winter
Gobi Desert, Mongolia	Continental tropical air, low precipitation, scarce vegetation
Bogotá, Colombia	Mild winters, cool summers, abundant precipitation
Yukon, Canada	Year-round cold, low precipitation

1. According to the modified Koeppen classification system, Southern Israel has what kind of climate?
a. tropical
b. dry
c. humid subtropical
d. continental

2. Where is a steppe most likely to be found?
a. New Caledonia
b. Gobi Desert
c. Bogotá
d. Yukon

3. Where is a heat island most likely to be found?
a. a farm
b. a beach
c. a mountaintop
d. an inner city

4. Why is deforestation often linked to global warming?
a. It increases the amount of dry land on Earth's surface.
b. It releases toxic gases and pollutants into the atmosphere.
c. It increases the amount of CO_2 released into the atmosphere.
d. It decreases the amount of CO_2 released into the atmosphere.

earthgeu.com/standardized_test

Chapter 15

Physical Oceanography

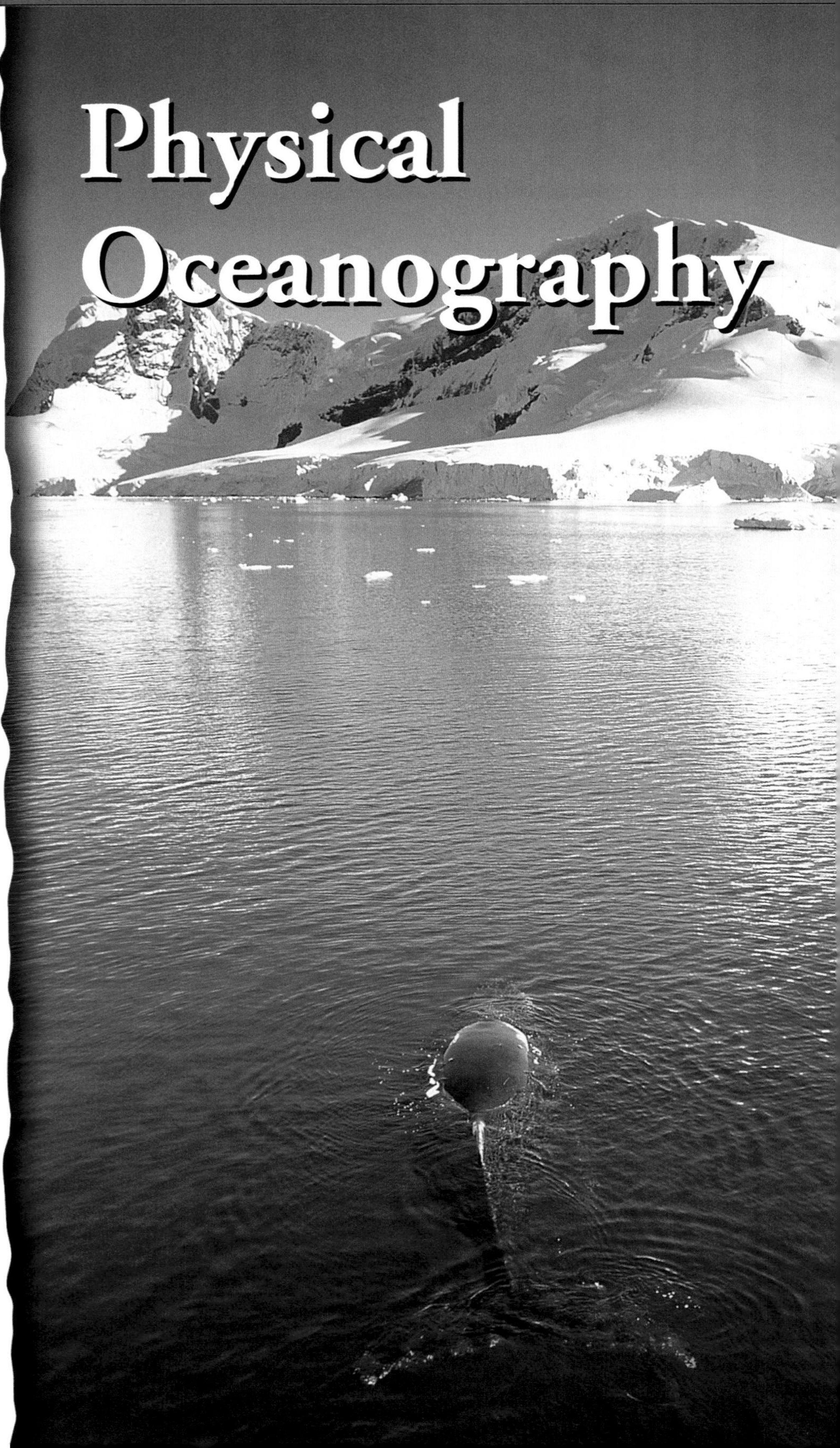

What You'll Learn

- How scientists study oceans, how the oceans formed, and where they are located.
- How the physical and chemical properties of seawater differ from those of freshwater.
- What causes tides, waves, and ocean currents.

Why It's Important

More than 71 percent of Earth's surface is covered by oceans. These vast bodies of water affect weather, climate, food supplies, recreation, global trade, and marine life, such as this humpback whale in Paradise Bay, Antarctica.

To learn more about oceanography, visit the Earth Science Web Site at earthgeu.com

Discovery Lab

Measure Earth's Land and Water

Earth is often referred to as the "blue planet" because so much of its surface is made up of water. If you study a globe or a photograph of Earth taken from space, you can clearly see that oceans cover much more of Earth than landmasses do. This activity will help you to quantify the amount of water on Earth's surface.

1. Stretch a piece of string about 1 m in length around the equator of a globe.
2. Use a blue marker to color the sections of the string that cross the oceans.
3. Measure the length of the globe's equator, then measure the length of each blue section on the string. Add the lengths of the blue sections.
4. Divide the total length of the blue sections by the length of the globe's equator.

Observe What percentage of the globe's equator is made up of oceans? What percentage of the globe's equator is made up of land? Study the globe again. Are the oceans separate bodies of water, or do they interconnect? Describe your observations in your science journal.

SECTION 15.1 *The Oceans*

Earth Sciences 1.c

OBJECTIVES

- **Identify** *methods used by scientists to study Earth's oceans.*
- **Discuss** *the origin and composition of the oceans.*
- **Describe** *the distribution of oceans and major seas.*

VOCABULARY

oceanography
side-scan sonar
sea level

Since prehistoric times, people have used Earth's oceans for travel and recreation and to obtain food. Early Polynesians and Phoenicians were accomplished sailors who discovered new lands and sea routes for commerce. These seafarers acquired considerable knowledge of the oceans, but they lacked the technology to explore the ocean depths. Such exploration had to wait until the late 1800s, when the British *Challenger* expedition became the first research ship to use relatively sophisticated measuring devices to study the oceans. *Challenger* also was the first expedition devoted exclusively to the scientific study of Earth's oceans, known as **oceanography.** The discipline of oceanography is usually considered to have started with the *Challenger.*

MODERN OCEANOGRAPHY

The *Challenger* expedition investigated ocean currents, water temperature and chemical composition, seafloor sediments and topography, and marine life. The expedition used nets, bottom dredges, and other tools to gather enough research to fill 50 thick volumes. Later expeditions, such as that of the German research ship *Meteor* in the 1920s, used sonar for the first time to map the seafloor features of the South Atlantic Ocean, including the Mid-Atlantic Ridge. *Sonar* stands for *so*und *na*vigation and *r*anging. It uses the return time of an echo and the known velocity of sound in water to determine water depth. The velocity of sound in water is 1500 m/s. To determine ocean depth in a particular place, scientists send a sonar signal to the ocean floor and time its return, or echo. They multiply the time by 1500 m/s, then divide by 2 to calculate the distance to the ocean floor. ***Figure 15-1*** has more information about sonar.

Advanced Technology Recent technological advances such as the one shown in ***Figure 15-2*** have tremendously expanded scientific knowledge of the oceans. Satellites such as the *Topex/Poseidon* continually monitor the ocean's surface temperatures, currents, and wave conditions. Submersibles, or underwater vessels, investigate the deepest ocean trenches. Large portions of the seafloor have been mapped using **side-scan sonar,** a technique that directs sound waves to the seafloor at an angle, so that the sides of underwater hills and other topographic features can be mapped. You'll

Figure 15-1 A ship equipped with sonar can gather data about deep-lake and ocean-floor features **(A).** Computers use the resulting data to make maps of underwater features such as this color-enhanced map of Lake Tahoe in Nevada **(B).**

Figure 15-2 The *Topex/Poseidon* satellite orbits 1331 km above Earth, gathering data about the oceans.

Earth Sciences

1.c Students know the evidence from geological studies of Earth and other planets suggests that the early Earth was very different from Earth today.

learn more about seafloor topography in the next chapter. For now, let's examine what all the various oceanography studies have taught us about the ocean's origin and composition.

ORIGIN OF THE OCEANS

Have you ever wondered whether Earth has always had oceans? Several geological clues indicate that oceans have existed almost since the beginning of geologic history. Studies of radioactive isotopes indicate that Earth is about 4.6 billion years old. Scientists have found rocks nearly as old that formed from sediments deposited in water. Ancient lava flows are another clue—some of these lava flows have glassy crusts that form only when molten lava is chilled rapidly under water. Radioactive studies and lava flows offer evidence that there has been abundant water throughout Earth's geologic history.

Where did the water come from? Scientists hypothesize that Earth's water could have originated from two sources. Comets, such as the one shown in ***Figure 15-3,*** travel throughout the solar system and occasionally collide with Earth. These impacts release water, possibly enough to have filled the ocean basins over geologic time. In addition, studies of meteorites, which are composed of the same material that may have formed the early planets, indicate that meteorites contain up to 0.5 percent water. If the early Earth contained the same percentage of water, it would have been more than sufficient to form the early oceans. However, some mechanism must have existed to allow the water to rise from deep in Earth's interior to its surface. Scientists theorize that that mechanism was volcanism.

Figure 15-3 Comets such as Hyakutake are composed of dust and rock particles mixed with frozen water and gases. Early in Earth's geologic history, collisions with comets may have released water that, over an extremely long period of time, formed the first oceans.

Figure 15-4 In addition to comets, water for Earth's early oceans may have come from volcanic eruptions. An intense period of volcanism occurred shortly after the planet formed. This volcanism released large quantities of water vapor and other gases into the atmosphere. The water vapor eventually condensed into oceans.

Volcanism During volcanic eruptions, significant quantities of gases are emitted. These volcanic gases consist mostly of water vapor and carbon dioxide. Shortly after the formation of Earth, when the young planet was much hotter than it is today, an episode of massive, violent volcanism took place over the course of perhaps several hundred million years. As shown in ***Figure 15-4,*** this volcanism released huge amounts of water vapor, carbon dioxide, and other gases, which combined to form Earth's early atmosphere. As Earth's crust cooled, the water vapor gradually condensed into oceans. By the time the oldest known crustal rock formed some 4 billion years ago, Earth's oceans may have been close to their present size. Water is still being added to the hydrosphere by volcanism, but some water molecules in the atmosphere are continually being destroyed by ultraviolet radiation from the Sun. These two processes balance each other. What do you think would happen over geologic time if they didn't?

DISTRIBUTION OF EARTH'S WATER

The oceans contain 97 percent of the water found on Earth. Another 3 percent is freshwater located in the frozen ice caps of Greenland and Antarctica and in rivers, lakes, and underground sources. The percentage of ice on Earth has varied over geologic time from near zero to perhaps as much as 10 percent of the hydrosphere. Thus, global **sea level,** which is the level of the oceans' surfaces, has risen and fallen by hundreds of meters in response to melting ice during warm periods and expanding glaciers during ice ages. Other processes that affect sea level are tectonic forces that lift or lower portions of the seafloor. A rising seafloor causes a rise in sea level, while

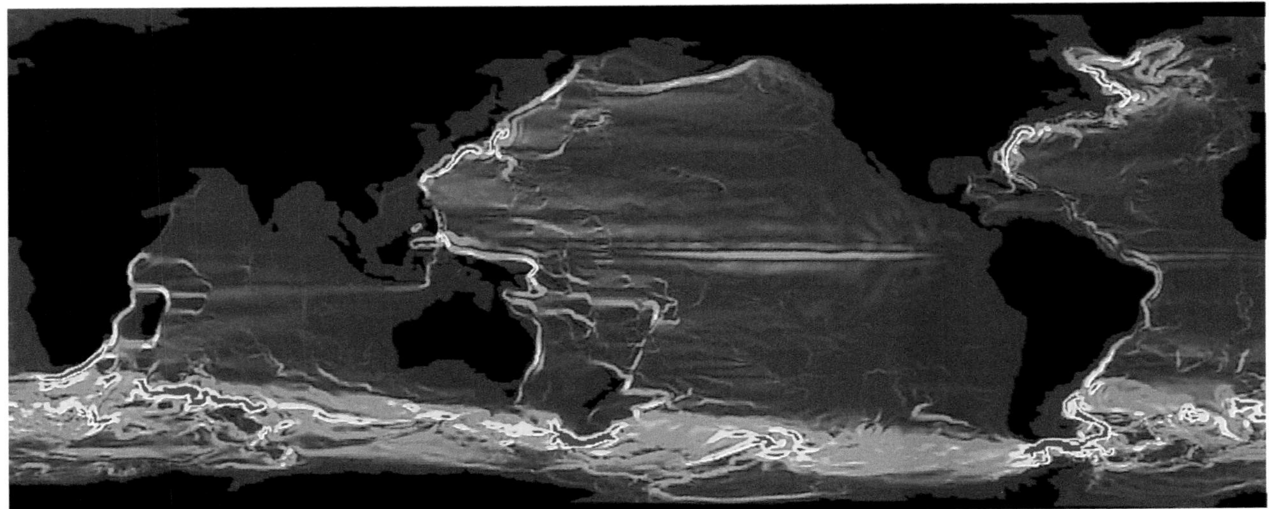

Figure 15-5 The vast interconnected oceans of Earth, the "blue planet," can clearly be seen in this computer image. The colors represent the speeds of ocean currents. Red is the fastest, blue is the slowest.

a sinking seafloor causes sea level to drop. At present, average global sea level is slowly rising at a rate of 1 to 2 mm per year in response to melting glaciers.

The Blue Planet As shown in the image of Earth in ***Figure 15-5,*** Earth is known as the "blue planet" for good reason—approximately 71 percent of its surface is covered by oceans. The average depth of these oceans is 3800 m. Earth's landmasses are like huge islands, almost entirely surrounded by water. Because most landmasses are in the northern hemisphere, oceans cover only 61 percent of the surface there. However, 81 percent of the southern hemisphere is covered by water. ***Figure 15-6*** shows the distribution of water in the northern and southern hemispheres. Note that all the oceans are really one vast, interconnected body of water. They have been divided into specific oceans and seas largely because of historic and geographic considerations.

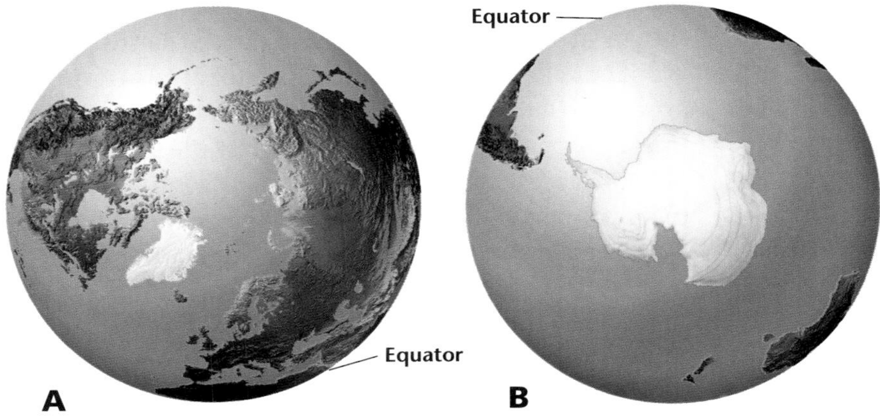

Figure 15-6 The northern hemisphere is covered by slightly more water than land **(A).** The southern hemisphere, however, is covered mainly by water **(B).**

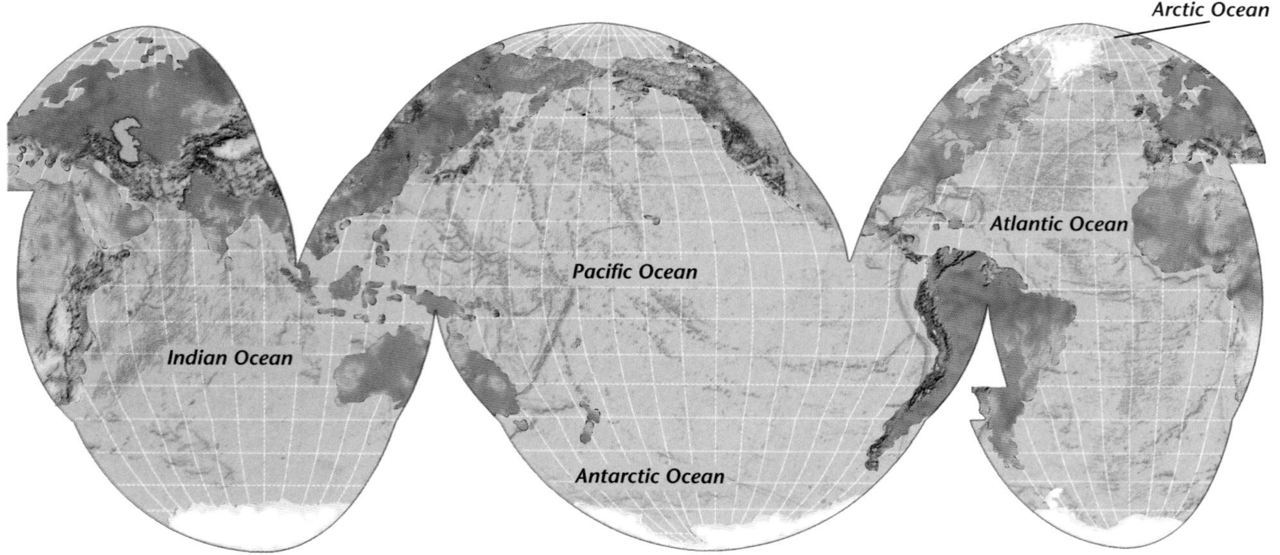

Figure 15-7 The Pacific, Atlantic, and Indian Oceans stretch from Antarctica to the north. The smaller Arctic and Antarctic Oceans are located near the north and south poles, respectively.

Major Oceans As ***Figure 15-7*** shows, there are three major oceans: the Pacific, the Atlantic, and the Indian. The Pacific Ocean is the largest. Containing roughly half of Earth's seawater, it is larger than all of Earth's landmasses combined. The second-largest ocean, the Atlantic, extends for more than 20 000 km from Antarctica to the arctic circle. North of the arctic circle, the Atlantic Ocean is often referred to as the Arctic Ocean. The third-largest ocean, the Indian, is located mainly in the southern hemisphere. The storm-lashed region surrounding Antarctica, south of 50° south latitude, is known as the Antarctic Ocean.

Figure 15-8 Ice is present in this polar sea year round.

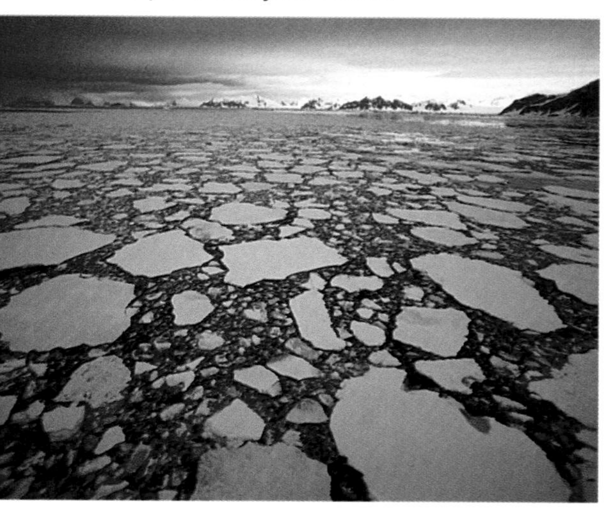

Sea Ice The Arctic and Antarctic Oceans are covered by vast expanses of sea ice, particularly during the winter. In summer, the ice breaks up somewhat, as shown in ***Figure 15-8.*** Ice is less dense than water, so it floats. When sea-ice crystals first form, a sort of ice-crystal slush develops at the surface of the water. The thickening ice eventually solidifies into individual round pieces called pancake ice. Eventually, these pieces of pancake ice thicken and freeze into a continuous ice cover called pack ice. In the coldest parts of the Arctic and Antarctic Oceans, there is no summer thaw, and the pack ice is generally several meters thick. In the winter, the pack-ice cover may be more than 1000 km wide.

Figure 15-9 The Aral Sea, located in Asia, is an example of a land-locked body of water.

Seas Seas are smaller than oceans and are partly or mostly land-locked. A prominent example, the Aral Sea, is shown in ***Figure 15-9.*** Another example, the Mediterranean Sea, is located between Africa and Europe. It was the first sea to be explored and mapped by ancient peoples such as the Egyptians, Phoenicians, Greeks, and Romans. Notable seas in the northern hemisphere include the Gulf of Mexico, the Caribbean Sea, and the Bering Sea, which is located between Alaska and Siberia. Keep in mind that all seas and oceans belong to one global ocean whose waters are thoroughly mixed. As a result, ocean water everywhere contains nearly identical proportions of dissolved salts, as you'll learn in the next section.

Section Assessment

1. What is oceanography? What was learned from the *Challenger* expedition?
2. What is sonar? Which research vessel first used sonar to map the Mid-Atlantic Ridge?
3. What evidence indicates that oceans formed early in Earth's geologic history?
4. Where did the water in Earth's early oceans come from?
5. **Thinking Critically** The Great Lakes contain as much water as some seas. Why aren't they considered to be seas?

Skill Review

6. **Measuring in SI** Calculate the distance to the ocean floor if a sonar signal takes six seconds to return to a ship's receiver. For more help, refer to the *Skill Handbook*.

earthgeu.com/self_check_quiz

SECTION 15.2 Seawater

Earth Sciences 5.d I&E 1.i

OBJECTIVES

- **Compare** *and* **contrast** *the physical and chemical properties of seawater.*
- **Explain** *ocean layering.*
- **Describe** *the formation of deep-water masses.*

VOCABULARY

salinity
temperature profile
thermocline

Earth Sciences

5.d Students know properties of ocean water, such as temperature and salinity, can be used to explain the layered structure of the oceans, the generation of horizontal and vertical ocean currents, and the geographic distribution of marine organisms.

Have you ever accidentally swallowed a gulp of seawater? If so, you've noticed its salty taste. Seawater is a solution of about 96.5 percent water and 3.5 percent dissolved salts. The most abundant salt in seawater is sodium chloride (NaCl). Other salts present in seawater are chlorides and sulfates of magnesium, potassium, and calcium. In fact, most elements on Earth are present in seawater. Because these substances are dissolved, they are in the form of ions. ***Table 15-1*** shows the concentrations of the most important ions in the oceans.

CHEMICAL PROPERTIES OF SEAWATER

Salinity is a measure of the amount of dissolved salts in seawater. Oceanographers express salinity as grams of salt per kilogram of water, or parts per thousand (ppt). The total salt content of seawater is, on average, 35 ppt, or 3.5 percent. In addition to salt ions, seawater contains dissolved gases and nutrients. The dissolved gases are mostly oxygen, nitrogen, and carbon dioxide, and the dissolved nutrients are commonly nitrates, phosphates, and silicates. As you might guess, the nutrients and dissolved gases in seawater greatly affect life in the oceans.

Sodium chloride crystals

Table 15-1 Major Ions in Seawater

Ion	Chemical Symbol	ppt in seawater
Chloride	Cl^-	19.35
Sodium	Na^+	10.76
Sulfate	SO_4^{2-}	2.71
Magnesium	Mg^{2+}	1.29
Calcium	Ca^{2+}	0.41
Potassium	K^+	0.39
Bicarbonate	HCO_3^-	0.14
Bromide	Br^-	0.067
Strontium	Sr^{2+}	0.008
Boron	B^{3+}	0.004
Fluoride	F^-	0.001
Total		~35.00

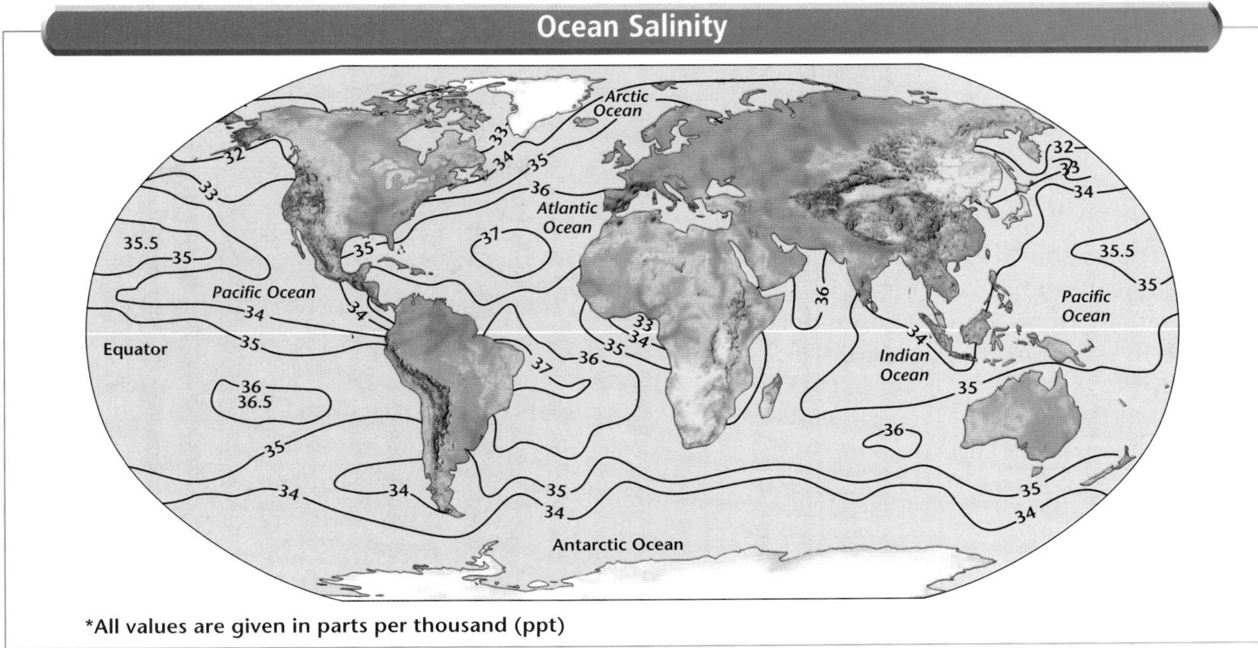

Figure 15-10 Ocean salinities vary from place to place. High salinities are common in areas with high rates of evaporation. Low salinities often occur where rivers empty into oceans.

Variations in Salinity Although the average salinity of the oceans is 35 ppt, actual salinities vary from place to place, as shown in ***Figure 15-10.*** In subtropical regions where rates of evaporation exceed those of precipitation, salt ions left behind by the evaporation of water molecules accumulate in the surface layers of the ocean. There, salinities may be as high as 37 ppt. In equatorial regions where precipitation is abundant, salinities are lower. Even lower salinities of 32 or 33 ppt occur in polar regions where seawater is diluted by melting sea ice. The lowest salinities often occur where large rivers empty into the oceans. Even though salinities vary, the relative proportion of major sea salts is always constant because all ocean water continually intermingles and is thoroughly mixed. Do the *MiniLab* on the following page to further analyze the salinity of seawater.

Sources of Sea Salt Geological evidence indicates that the salinity of ancient seas was not much different from that of today's oceans. One line of evidence is based on the proportion of magnesium in the calcium-carbonate shells of some marine organisms. That proportion depends on the overall salinity of the water in which the shells form. Present-day shells, such as the one shown in ***Figure 15-11,*** contain about the same proportion of magnesium as similar shells throughout geologic time.

Just as the proportion of sea salts has remained the same over time, so too have the sources of sea salts. In addition to water vapor, volcanic gases contain chlorine and sulfur dioxide. These gases

Figure 15-11 By comparing the proportion of magnesium in present-day shells, such as this nautilus, with those of ancient shells, scientists have determined that the salinity of Earth's oceans has remained nearly the same over geologic time.

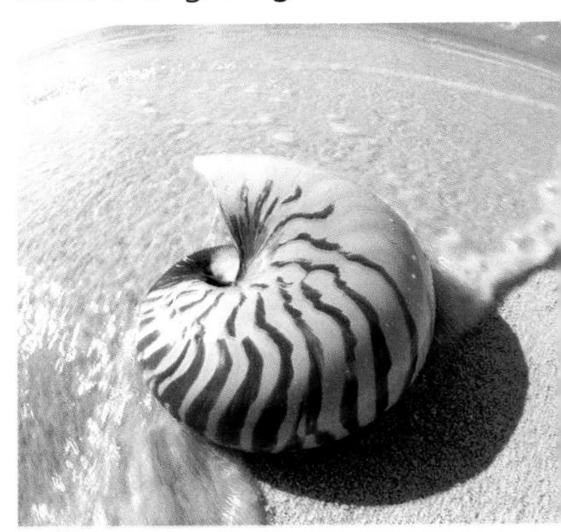

MiniLab

What is the chemical composition of seawater?

Determine the chemical composition of seawater using the following ingredients. The salinity of seawater is commonly measured in parts per thousand (ppt).

sodium chloride (NaCl)	23.48 g
magnesium chloride ($MgCl_2$)	4.98 g
sodium sulfate (Na_2SO_4)	3.92 g
calcium chloride ($CaCl_2$)	1.10 g
potassium chloride (KCl)	0.66 g
sodium bicarbonate ($NaHCO_3$)	0.19 g
potassium bromide (KBr)	0.10 g

Procedure

1. Carefully measure the ingredients and put them all in a large beaker.
2. Add 965.57 g of distilled water and mix.

Analyze and Conclude

1. How many grams of solution do you have? What percentage of this solution is made up of salts?
2. Given that 1 percent is equal to 10 ppt, what is the salinity of your solution in parts per thousand?
3. Identify the ions in your solution.
4. Infer how your solution differs from actual seawater.

dissolve in water and form the chlorine and sulfate ions of seawater. The weathering of crustal rocks generates most of the other abundant ions in seawater. Sodium, calcium, and potassium come from the weathering of feldspars. Iron and magnesium come from the weathering of minerals and rocks rich in these elements. These ions are then flushed into rivers and transported to oceans.

Removal of Sea Salts Although salt ions are continuously added to seawater, the salinity of seawater does not increase. Why? Because salts are removed from the ocean at the same rate as they are added. The removal of sea salts involves several processes. Some precipitate from seawater near arid, coastal regions such as the one shown in ***Figure 15-12.*** This process removes immense quantities of sodium chloride, calcium sulfate, and other sea salts. In addition, small salty spray droplets from breaking waves are picked up by winds and deposited inland. Marine organisms also remove ions from seawater to build their shells, bones, and teeth. As these organisms die, their solid parts accumulate on the seafloor and become incorporated into the bottom sediments. All these

Figure 15-12 Salts that have precipitated from seawater form deposits along the coast of Baja, Mexico.

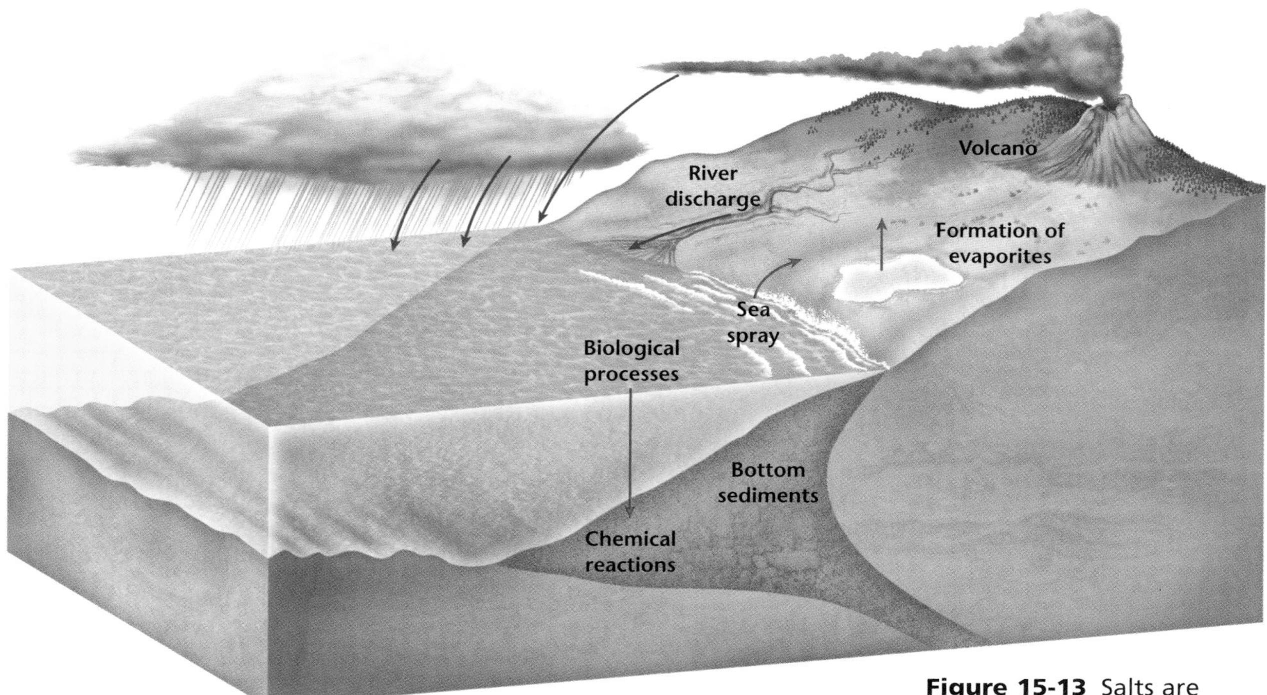

Figure 15-13 Salts are added to seawater by volcanic eruptions and by the weathering and erosion of rocks. Salts are removed from seawater by the formation of evaporites and biological processes. Salty droplets also are deposited inland by winds.

processes remove immense quantities of salt ions from the ocean. Thus, the existing salinity of seawater represents a balance between the processes that remove salts and those that add them, as shown in ***Figure 15-13.***

PHYSICAL PROPERTIES OF SEAWATER

The presence of various salts causes the physical properties of seawater to be quite different from those of freshwater. Freshwater has a maximum density of 1.00 g/cm^3. Because salt ions are heavier than water molecules, they increase the density of water. Seawater is therefore denser than freshwater, and its density varies, depending on its salinity. Temperature also affects density—cold water is denser than warm water. Because of salinity and temperature variations, the density of seawater ranges from about 1.02 g/cm^3 to 1.03 g/cm^3. These variations may seem small, but they are significant. They affect many oceanic processes, which you'll learn about in the next chapter. Variations in salinity also cause the freezing point of seawater to be somewhat lower than that of freshwater. Freshwater freezes at 0°C. Because salt ions interfere with the formation of hydrogen bonds, the freezing point of seawater is –2°C.

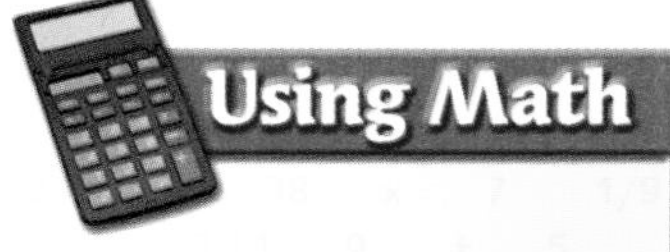

Using Numbers If the density of a sample of seawater is 1.02716 g/mL, calculate the mass of 4.00 mL of the sample.

Absorption of Light If you've ever swum in a lake, you may have noticed that the intensity of light decreases with depth. The water may be clear, but if the lake is deep, the bottom waters will be dark.

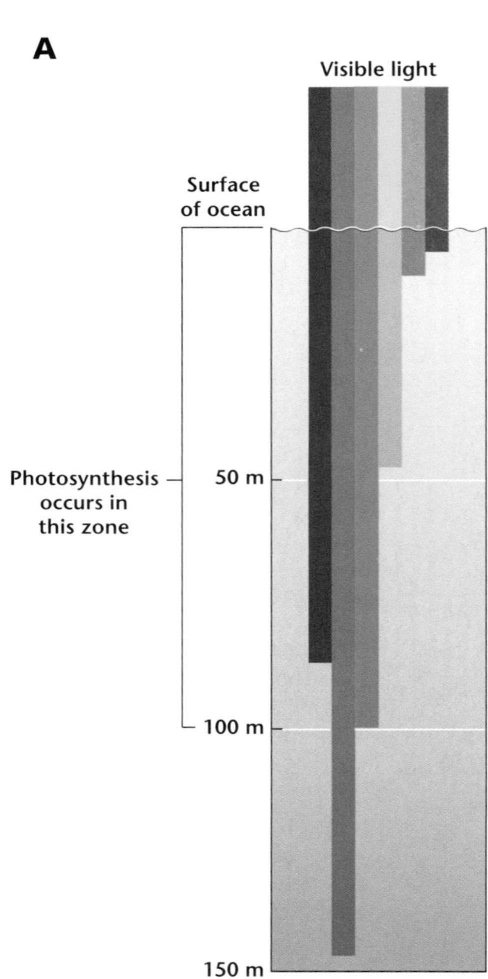

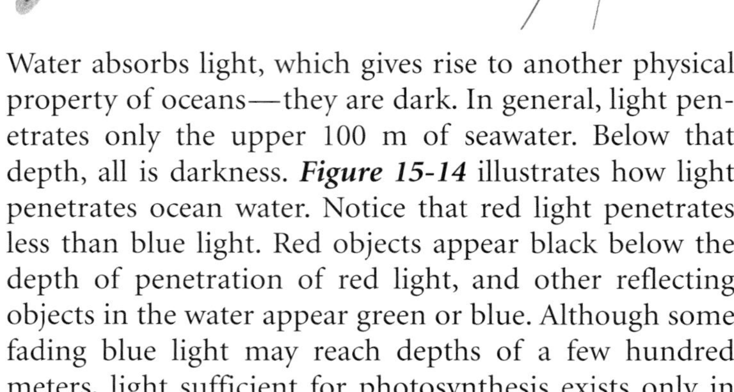

Figure 15-14 Red light does not penetrate as far as blue light in the ocean **(A).** Thus, red marine organisms such as this deep-sea shrimp **(B)** appear black below a depth of 10 m. This helps them escape predators.

I&E

1.i Analyze the locations, sequences, or time intervals that are characteristic of natural phenomena (e.g., relative ages of rocks, locations of planets over time, and succession of species in an ecosystem.)

Water absorbs light, which gives rise to another physical property of oceans—they are dark. In general, light penetrates only the upper 100 m of seawater. Below that depth, all is darkness. ***Figure 15-14*** illustrates how light penetrates ocean water. Notice that red light penetrates less than blue light. Red objects appear black below the depth of penetration of red light, and other reflecting objects in the water appear green or blue. Although some fading blue light may reach depths of a few hundred meters, light sufficient for photosynthesis exists only in the top 100 m of the ocean.

Ocean Layering

Ocean surface temperatures range from –2°C in polar waters to 30°C in equatorial regions, with the average surface temperature being 15°C. Ocean water temperatures, however, decrease significantly with depth. Thus, deep ocean water is always cold, even in tropical oceans. ***Figure 15-15*** shows a typical ocean **temperature profile,** which plots changing water temperatures with depth. Such profiles vary, depending on location and season. In the temperature profile shown here, beneath roughly 100 m, temperatures decrease continuously with depth to around 4°C at 1 km. The dark waters below 1 km have fairly uniform temperatures of less than 4°C. Based on temperature variations, the ocean can be divided into three layers, as shown in ***Figure 15-16.*** The first is a relatively warm, sunlit, surface layer some 100 m thick. Under this is a transitional layer known as the **thermocline,** which is characterized by rapidly decreasing temperatures with depth. The bottom layer is cold and dark with temperatures near freezing. Both the thermocline and the warm surface layer are absent in polar seas, where water temperatures are cold from top to bottom. In general, ocean layering is caused by density differences. Because cold water is more dense than warm water, cold water sinks to the bottom, while less-dense, warm water is found near the ocean's surface.

WATER MASSES

The temperature of the bottom layer of ocean water is near freezing even in tropical oceans, where surface temperatures are warm. Where does all this cold water come from? The source is Earth's polar seas. Recall that high salinity and cold temperatures cause seawater to become more dense. When seawater freezes during the arctic or antarctic winter, sea ice forms. However, salt ions aren't incorporated into the growing ice crystals and accumulate beneath the ice. Consequently, the cold water beneath the ice becomes saltier and denser than the surrounding seawater, and this saltier water sinks. This salty water then migrates toward the equator as a cold, deep water mass along the ocean floor. Other cold, deep water masses form when surface currents in the ocean bring relatively salty midlatitude or subtropical waters into polar regions. In winter, these waters become colder and denser than the surrounding polar surface waters, and thus, they sink.

Three water masses account for most of the deep water in the Atlantic Ocean. Antarctic Bottom Water forms when antarctic seas freeze during the winter. With temperatures below 0°C, this water

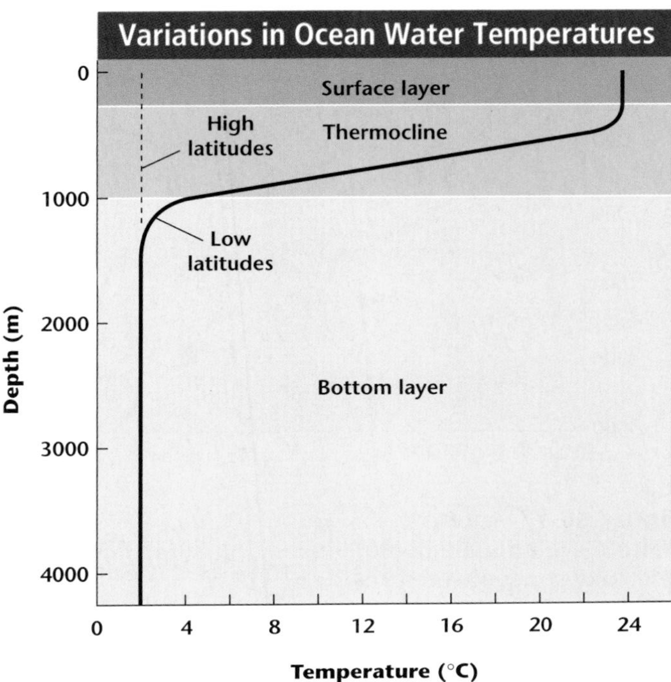

Figure 15-15 Ocean water temperatures decrease with depth. Areas near the equator have warmer ocean surface temperatures than do midlatitudes or areas near the poles.

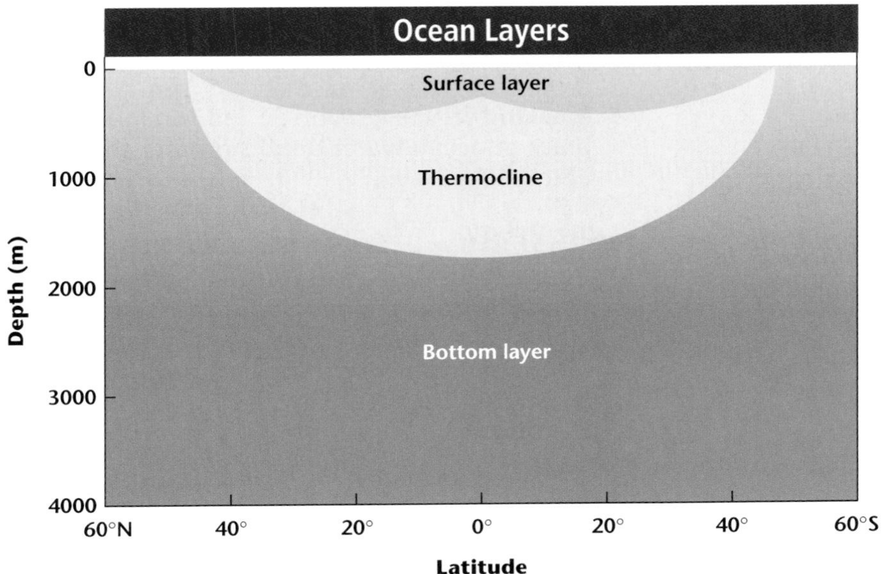

Figure 15-16 Based on water temperatures, the ocean can be divided into three layers: the relatively warm surface layer, the transitional thermocline, and the cold bottom layer.

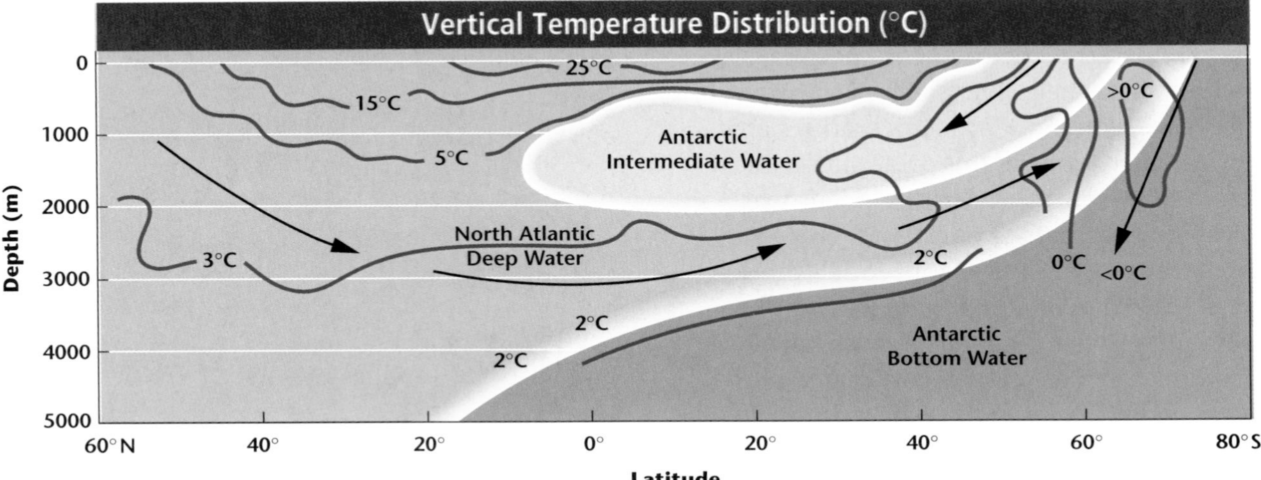

Figure 15-17 Antarctic Bottom Water is the densest and coldest deep water mass. It is overridden by the slightly warmer and less dense North Atlantic Deep Water. Antarctic Intermediate Water is still warmer and less dense, and thus it overrides the other two deep water masses.

mass is the coldest and densest in all the oceans, as shown in ***Figure 15-17.*** North Atlantic Deep Water forms in a similar manner offshore from Greenland. It is warmer and less dense than Antarctic Bottom Water and thus overrides it. Antarctic Intermediate Water forms when the relatively salty waters of the Antarctic Ocean decrease in temperature during winter and sink. Being slightly warmer and less dense than North Atlantic Deep Water, Antarctic Intermediate Water overrides the other two water masses. While the Atlantic Ocean contains all three major deep-water masses, the Indian and Pacific Oceans contain only the two deep antarctic water masses. You'll model water masses in the *GeoLab* at the end of this chapter. In the next section, you'll learn about other water movements in the ocean.

Section Assessment

1. What is the most abundant salt in seawater? How do salts enter the ocean?
2. How does the salinity of seawater affect its density?
3. The salinity of seawater is higher in subtropical regions than at the equator. Why?
4. Explain why North Atlantic Deep Water and Antarctic Intermediate Water override Antarctic Bottom Water.
5. Which is more dense, cold freshwater or warm seawater? Explain.
6. **Thinking Critically** Why do red fish look black at ocean water depths greater than about 10 m?

Skill Review

7. **Recognizing Cause and Effect** Based on what you have learned about the freezing point of seawater, explain why salt is often used to de-ice roads in the winter. For more help, refer to the *Skill Handbook.*

SECTION 15.3

Ocean Movements

Earth Sciences 5.a, 5.b, 5.d, 6.b **I&E** 1.g

OBJECTIVES

- **Describe** *the physical properties of waves.*
- **Explain** *how tides form.*
- **Compare** *and* **contrast** *various ocean currents.*

VOCABULARY

wave
crest
trough
breaker
tide
density current
surface current
upwelling

Oceans are never completely motionless. Their most obvious movement is the constant motion of the waves. A **wave** is a rhythmic movement that carries energy through space or matter—in this case, ocean water. Ocean waves are generated mainly by wind flowing over the water's surface. As an ocean wave passes, the water moves up and down in a circular pattern and returns to its original position, as shown in ***Figure 15-18A.*** Only the energy moves steadily forward. The water itself moves in circles until the energy passes, but it does not move forward.

WAVE CHARACTERISTICS

In the open ocean, a typical wave has the characteristics shown in ***Figure 15-18B.*** The highest point of a wave is the **crest,** and the lowest point is the **trough.** The vertical distance between crest and trough is the wave height; the horizontal crest-to-crest distance is the wavelength. The wavelength determines the depth to which the wave disturbs the water. That depth, called the wave base, is equal to half the wavelength. The wavelength also determines the speed with which waves move through deep water. Wave speed increases with wavelength.

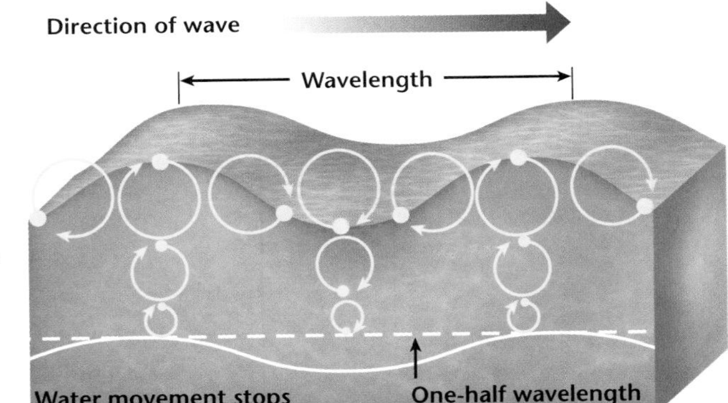

Figure 15-18 In a wave, water moves in circles that decrease in size with depth. At a depth equal to half the wavelength, water movement stops **(A).** Wave characteristics include wave height, wavelength, crest, and trough **(B).**

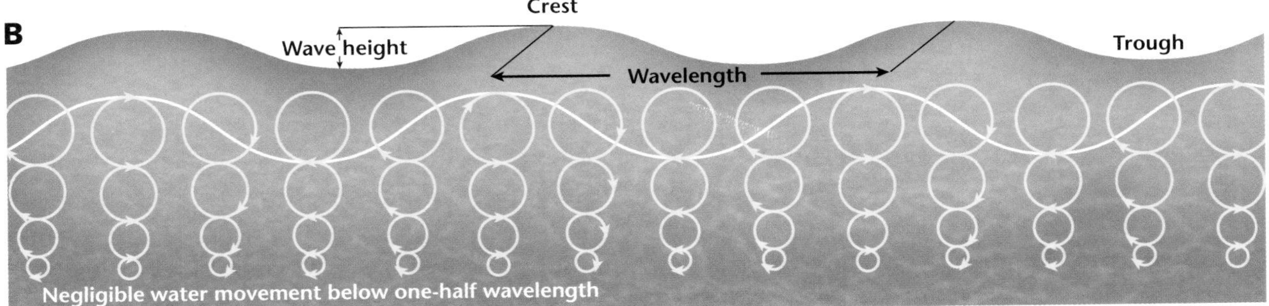

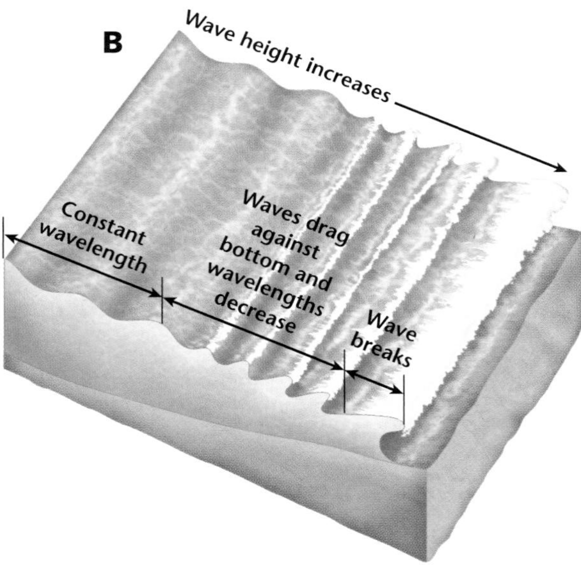

Figure 15-19 A breaker **(A)** forms when wavelength decreases and wave height increases as the wave nears the shore **(B).**

Wave Height Wave heights depend upon three factors: wind speed, wind duration, and fetch. Fetch refers to the expanse of water that the wind blows across. Large storm waves can be much higher than average. For instance, hurricanes can generate waves more than 10 m high. The greatest wave height ever recorded was more than 30 m. This monstrous wave occurred in the North Pacific.

Breaking Waves As ocean waves reach the shallow water near shorelines, they begin to lose energy because of friction with the ocean bottom. This causes the waves to slow down. As the water becomes shallower, incoming wave crests gradually catch up with the slower wave crests ahead. As a result, the crest-to-crest wavelength decreases. The incoming waves become higher, steeper, and unstable, and their crests collapse forward. Collapsing waves, such as those shown in ***Figure 15-19,*** are called **breakers.** The formation of breakers is also influenced by the motion of wave crests, which are less affected by friction than wave troughs and thus overrun the troughs. The collapsing crests of breakers moving at high speeds toward shore play a major role in shaping shorelines. You'll learn more about breakers and shoreline processes in the next chapter.

Topic: High Tides
To learn more about tides, visit the Earth Science Web Site at earthgeu.com

Activity: Research the best places on Earth to observe high tides. Note these places on a map of the world.

TIDES

Tides are the periodic rise and fall of sea level. The highest level to which water rises is known as high tide, and the lowest level is called low tide. Because of differences in topography and latitude, the tidal range—the difference between high tide and low tide—varies from place to place. In the Gulf of Mexico, the tidal range is less than 1 m. In New England, it can be as high as 6 m. The greatest tidal range

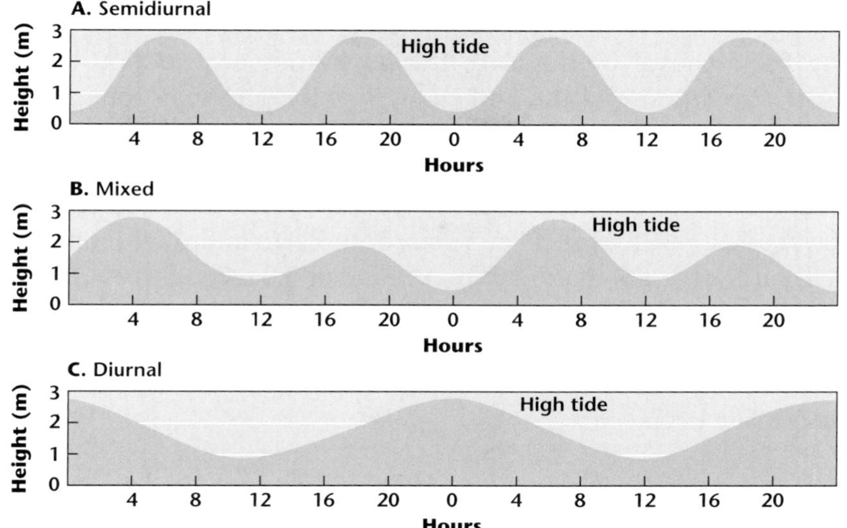

Figure 15-20 Differences in topography and latitude cause three different daily tide cycles. Areas with semidiurnal cycles experience two high tides per day. Areas with mixed cycles have one pronounced and one smaller high tide each day. Areas with diurnal cycles have one high tide per day.

occurs in the Bay of Fundy between New Brunswick and Nova Scotia, Canada, where it is as high as 15 m. Generally, a daily cycle of high and low tides takes 24 hours and 50 minutes. As shown in ***Figure 15-20,*** the daily cycle can follow three distinct patterns. You'll learn about tides in the *Problem-Solving Lab* on this page.

Problem-Solving Lab

Making and Using Graphs

Analyze a tidal record The water levels shown in the data table were measured over a 24-hour period.

Tidal Record

Time (h)	Water Level (m)	Time (h)	Water Level (m)
00:00	3.08	13:00	2.78
01:00	3.35	14:00	2.81
02:00	3.33	15:00	2.59
03:00	3.03	16:00	2.16
04:00	2.53	17:00	1.64
05:00	1.95	18:00	1.15
06:00	1.44	19:00	0.84
07:00	1.10	20:00	0.78
08:00	1.02	21:00	1.00
09:00	1.21	22:00	1.46
10:00	1.59	23:00	2.07
11:00	2.07	24:00	2.67
12:00	2.51		

Analysis

1. Plot these values on graph paper with time on the *x*-axis and water level on the *y*-axis.
2. Estimate the approximate times and water levels of high tides and low tides.

Thinking Critically

3. Refer to ***Figure 15-20*** to determine the tidal pattern shown in your graph.
4. What is the tidal range for this area?
5. Predict the water level at the next high tide. Estimate when that high tide will occur.

Causes of Tides

The basic causes of tides are the gravitational attraction among Earth, the Moon and the Sun, as well as the fact that gravitational attraction decreases with distance. Consider the Earth-Moon system. As shown in ***Figure 15-21,*** the Moon does not actually orbit Earth. Rather, both Earth and the Moon orbit around a common center of gravity. As a result of their motions, both Earth and the Moon experience differing gravitational forces. The unbalanced forces generate tidal bulges on opposite sides of Earth. The gravitational effect on Earth's oceans is similar to what happens to the liquid in a coffee cup inside a car as the car goes around a curve. The liquid sloshes toward the outside of the curve.

The Sun's Influence The gravitational attraction of the Sun and Earth's orbital motion around the Sun also generate tides. However, even though the Moon is much smaller than the Sun, lunar tides are more than twice as high as those caused by the Sun because the Moon is much closer to Earth. Consequently, Earth's tidal bulges are always aligned with the Moon. Although the Sun's tidal effect is smaller than that of the Moon, it is still significant because of the Sun's great mass. Depending on the phases of the Moon, solar tides can either enhance or diminish lunar tides, as illustrated in ***Figure 15-22.*** Notice that large tidal ranges, called spring tides, occur when the Moon is either full or new. These phases of the Moon occur when the Sun, the Moon, and Earth are aligned. During spring tides, high tides are higher than normal and low tides are lower than normal. Small tidal ranges, called neap tides, occur when there is a first- or third-quarter Moon. During these times, the Sun, the Moon, and

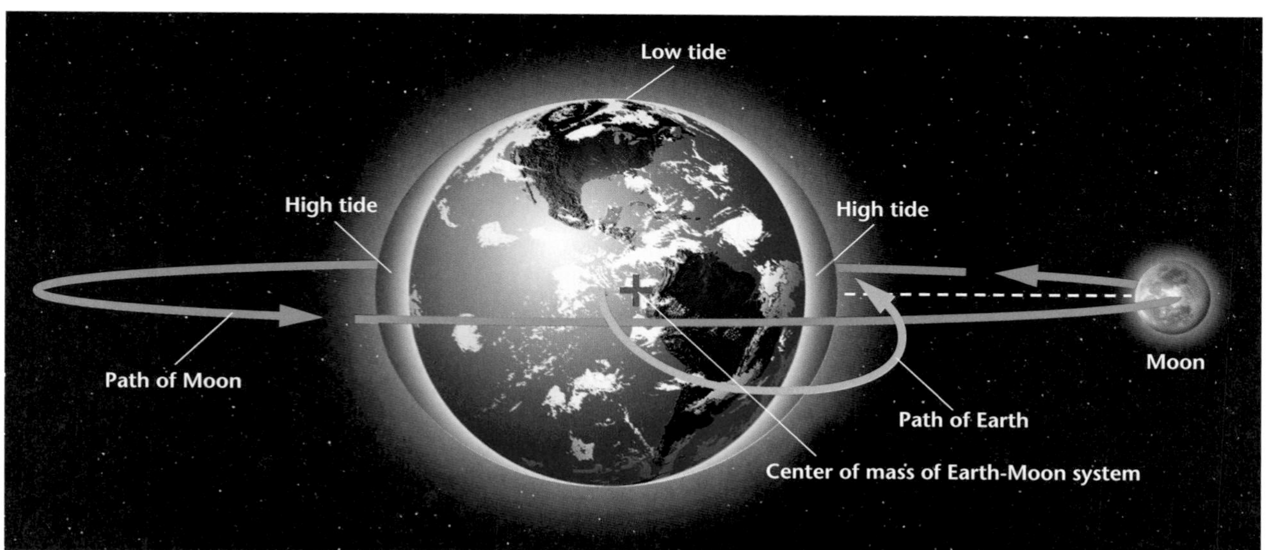

Figure 15-21 The Moon and Earth revolve around a common center of gravity and experience unbalanced gravitational and centrifugal forces. These forces cause tidal bulges on opposite sides of Earth.

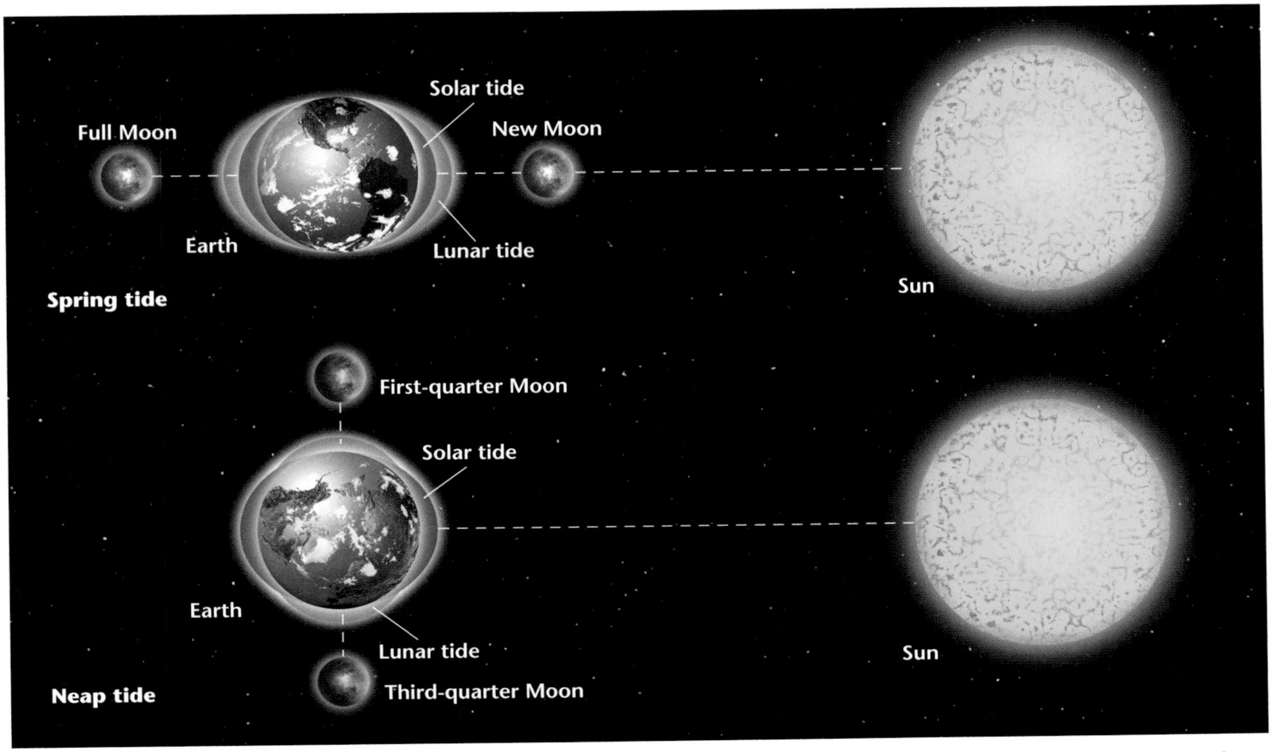

Solar tide
Full Moon
New Moon
Earth
Lunar tide
Sun
Spring tide
First-quarter Moon
Solar tide
Earth
Lunar tide
Sun
Third-quarter Moon
Neap tide

Figure 15-22 Spring tides occur when the Sun, the Moon, and Earth are aligned **(A).** Neap tides occur when the Sun, the Moon, and Earth form a right angle **(B).**

Earth form a right angle. During neap tides, high tides are lower and low tides are higher than normal. Spring and neap tides alternate every two weeks. On average, spring tides are three times higher than neap tides.

Ocean Currents

Recall the discussion of Antarctic Bottom Water in the previous section. The movement of Antarctic Bottom Water is an example of an ocean current. In this case, the current is called a **density current** because it is caused by differences in the temperature and salinity of ocean water, which in turn affect density. Density currents move slowly in deep ocean waters.

More noticeable than underwater density currents are wind-driven surface currents. **Surface currents** affect mainly the upper few hundred meters of the ocean, and they can move as fast as 100 km per day. Driven by Earth's global wind systems, surface currents follow predictable patterns. In the northern hemisphere, tropical trade winds blow from east to west. The resulting tropical ocean currents also flow from east to west. In northern midlatitudes, the prevailing westerlies and resulting ocean currents move from west to east. In northern polar regions, polar easterly winds push surface waters from east to west.

Earth Sciences

5.a Students know how differential heating of Earth results in circulation patterns in the atmosphere and oceans that globally distribute the heat.

Earth Sciences

5.b Students know the relationship between the rotation of Earth and the circular motions of ocean currents and air in pressure centers.

Gyres If Earth had no landmasses, the global ocean would have simple belts of easterly and westerly surface currents. But the continents deflect ocean currents to the north and south so that closed circular current systems, called gyres, develop. As shown in ***Figure 15-23,*** there are five major gyres: the North Pacific, the North Atlantic, the South Pacific, the South Atlantic, and the Indian Ocean. Because of the Coriolis effect, which you learned about in Chapter 12, the gyres of the northern hemisphere circulate in a clockwise direction, and those of the southern hemisphere circulate in a counterclockwise direction. The parts of all gyres closest to the equator move towards the west as equatorial currents. When these currents encounter a landmass, they are deflected toward the poles. These poleward-flowing waters carry warm, tropical water into higher, colder latitudes. A well-known example of a warm current is the Kuroshio, or Japan Current in the western North Pacific.

After these warm waters enter polar regions, they gradually cool and, deflected by landmasses, move back toward the equator. The resulting currents then bring cold water from higher latitudes into tropical regions. An example of such cold ocean currents is the California Current in the eastern North Pacific. You'll learn more about currents in the *Science in the News* feature at the end of this chapter.

Figure 15-23 The Coriolis effect deflects water and other free-moving objects to the right north of the equator and to the left south of the equator. Thus, gyres in the northern hemisphere circulate in a clockwise direction. The motion is reversed in the southern hemisphere.

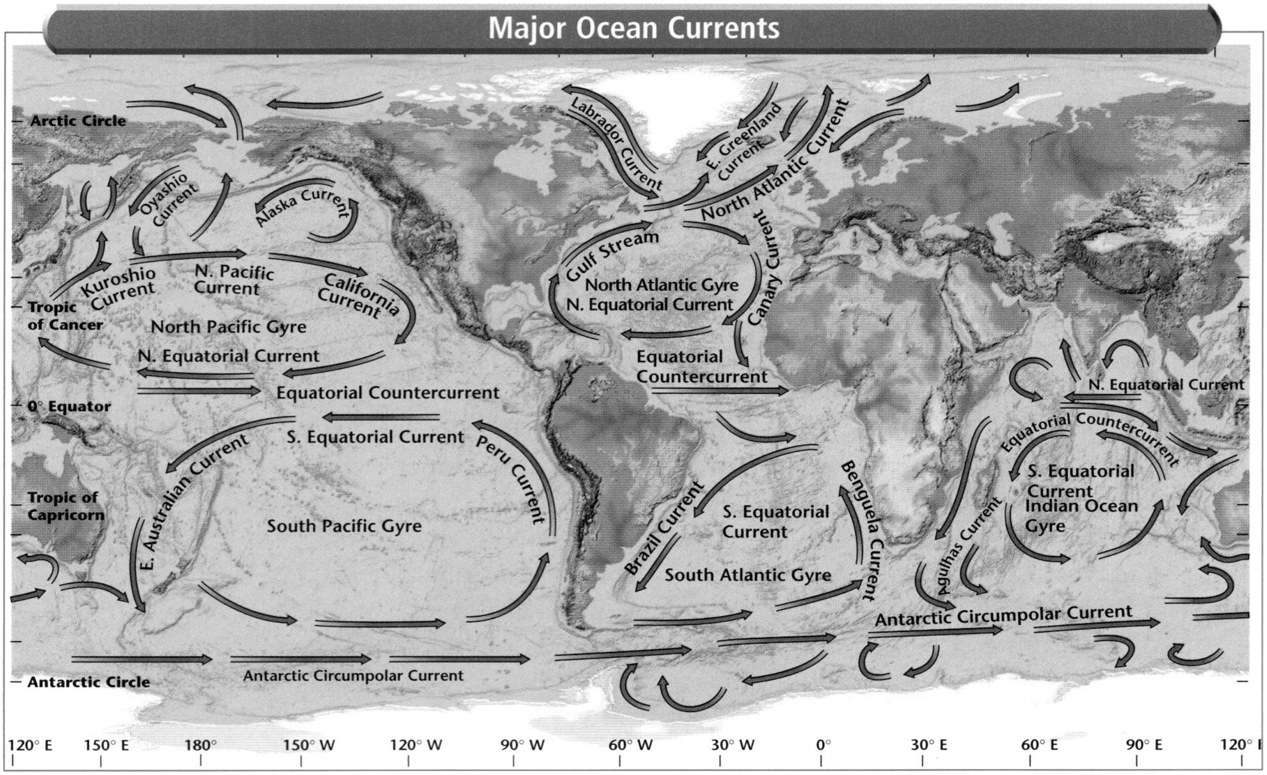

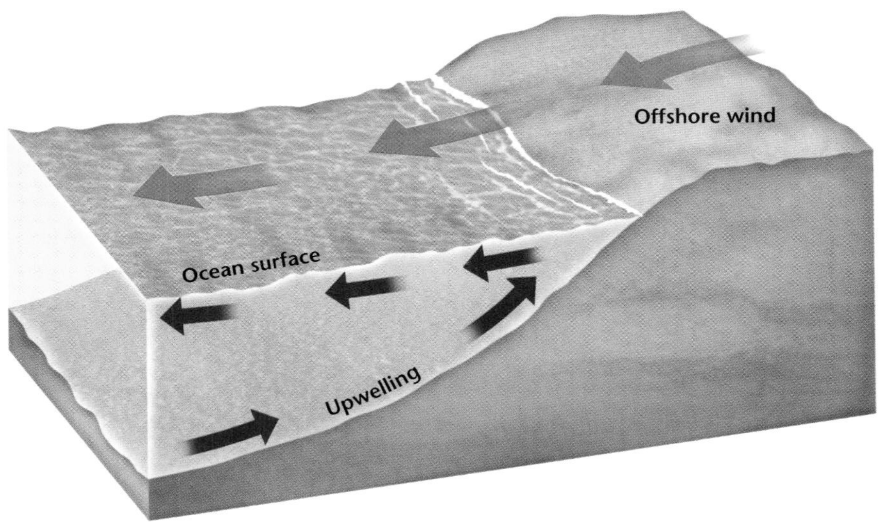

Figure 15-24 Upwelling occurs when the trade winds blow surface water offshore, and deep, colder water rises to the surface.

Earth Sciences

5.d Students know properties of ocean water, such as temperature and salinity, can be used to explain the layered structure of the oceans, the generation of horizontal and vertical ocean currents, and the geographic distribution of marine organisms.

UPWELLING

In addition to moving horizontally, ocean water moves vertically. The upward motion of ocean water is called **upwelling.** Upwelling waters originate from the bottom of the ocean and thus are cold. Areas of upwelling exist mainly off the western coasts of continents in the trade-wind belts. As ***Figure 15-24*** shows, the trade winds blow surface water offshore, and the surface water is replaced by upwelling deep water. Upwelling waters are rich in nutrients, which support abundant populations of marine life. Consequently, some of the world's richest fishing grounds are found off the coasts of Peru and California.

SECTION ASSESSMENT

1. Describe how water moves as a wave passes.
2. What three factors determine the height of a wave?
3. What causes tides? Compare and contrast a spring tide and a neap tide.
4. Why are upwelling waters always cold?
5. **Thinking Critically** Upwelling currents are rich in nutrients. Predict the effects on marine ecosystems if these currents stopped.

SKILL REVIEW

6. **Concept Mapping** Use the following phrases to complete a concept map of wave characteristics. For more help, refer to the *Skill Handbook.*

lowest point of wave | wave characteristics | crest
wavelength | wave height | trough
horizontal crest-to-crest distance | highest point of wave | vertical distance between crest and trough

earthgeu.com/self_check_quiz

Modeling Water Masses

I&E

1.g Recognize the usefulness and limitations of models and theories as scientific representations of reality.

The water in the oceans is layered because water masses with higher densities sink below those with lower densities. The density of seawater depends on its temperature and salinity. In this activity, you'll model different types of water masses to observe the effects of density firsthand.

Preparation

Problem

Determine how changes in salinity and temperature affect water density.

Materials

scale
graduated 500-mL cylinder
100-mL glass beakers (4)
water
red, yellow, and blue food coloring
salt
thermometer
eyedropper
graph paper
pencil
ruler
calculator

Objectives

In this GeoLab you will:

- **Compare** and **contrast** the movement of different water samples.
- **Determine** the relative densities of the water samples.
- **Predict** the arrangement of layers in a body of water.
- **Construct** and **interpret** a temperature profile.

Safety Precautions

Always wear safety goggles and an apron in the lab. Wash your hands after completing the lab.

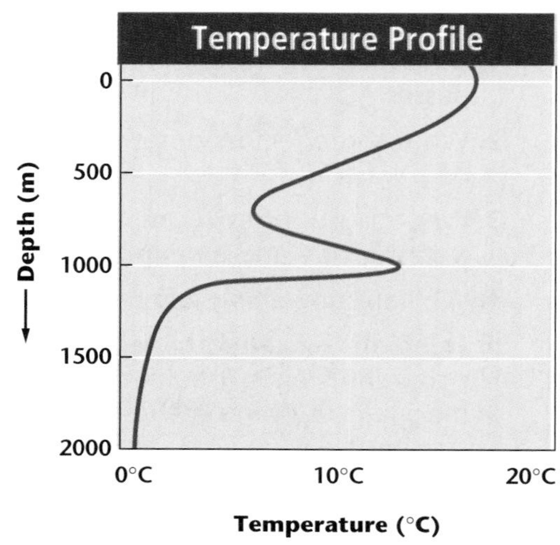

Procedure

1. Mix 200 mL of water and 7.5 g of salt in the graduated cylinder. Pour equal amounts of the salt solution into two beakers. Fill each of the two other beakers with 100 mL of freshwater.
2. Put a few drops of red food coloring in one of the salt solutions. Put a few drops of yellow food coloring in the other salt solution. Put a few drops of blue food coloring in one of the beakers of freshwater. Do not add food coloring to the other beaker of freshwater.
3. Place the beakers with the red salt solution and the blue freshwater in the refrigerator. Refrigerate them for 30 minutes.
4. Measure and record the temperature of the water in all four beakers.
5. Put several drops of the cold, red saltwater into the beaker with the warm, yellow saltwater and observe what happens. Record your observations.
6. Put several drops of the cold, blue freshwater into the beaker with the warm, clear freshwater and observe what happens. Record your observations.
7. Put several drops of the cold, blue freshwater into the beaker with the warm, yellow saltwater and observe what happens. Record your observations.

Analyze

1. In your science journal, describe the movement of the cold, red saltwater in step 5. Compare this to the movement of the cold, blue freshwater in step 7. What accounts for the differences you observed?
2. Based on your observations, list the water samples by color in order of increasing density.
3. If you poured the four water samples into the graduated cylinder, how would they arrange themselves into layers by color, from top to bottom?

Conclude & Apply

1. Assume that four water masses in a large body of water have the same characteristics as the water in the four beakers. The warm water layers are 100 m thick, and the cold layers are 1000 m thick. Graph the temperature profile of the large body of water.
2. What is the salinity in parts per thousand of the combined saline solutions? (Hint: ppt equals grams of salt per kilogram of solution. Assume that 200 mL of water has a mass of 200 g. Be sure to include the mass of the salt in the total mass of the solution.)
3. The temperature profile on the opposite page was constructed from measurements taken in the Atlantic Ocean off the coast of Spain. Study the profile, then infer why a high-temperature layer exists beneath the thermocline. Is this layer denser than the colder water above? Explain.

Science in the News

Caught in the Current

"The gale was still blowing from the northeast on January 21, drifting snow from the continental ice shelf...Held fast in the ice, the Endurance *was being carried with the rest of the pack by the Weddell Sea's current; soon she would be moving away from land."—from* The Endurance: Shackleton's Legendary Antarctic Expedition, *by Caroline Alexander (Knopf, 1999).*

Alexander's book tells the story of the *Endurance* and the 28 men who sailed in the ship to Antarctica in 1914. The explorers on board planned the first overland crossing of the Antarctic continent. The ship left South Georgia Island in the South Atlantic in December 1914. During the next two years, the expedition suffered both horrible misfortune and incredible luck.

Frozen Solid

In mid-December, the *Endurance* neared the coast of Antarctica. The pack ice that year was especially dense, and by January 1915, the ship had frozen solidly into the ice just 126 km from its destination—126 km of solid ice. The ship and its crew could only drift aimlessly with the ice.

In this part of the sea a clockwise current churns the ice pack in endless circles. The ice crunches against the Antarctic Peninsula, generating massive waves of pressure. The *Endurance*, still frozen in the ice, was carried by the current farther and farther from land. In November 1915, the pressure of the ice pack crushed the ship to pieces. The crew escaped but was forced to camp on the drifting ice in subzero temperatures with thin tents as their only shelter. They managed to salvage some supplies and three small boats.

As the Current Turns

The crew hoped that the ice would carry them toward land. However, the ice carried them into the open waters of the South Atlantic. Desperate, they launched their boats into the ocean. They stayed awake for days, battling rough waves and towering icebergs that threatened to crush the small boats to pieces. In April 1916, they finally reached a small barren island. The largest of the boats continued on a perilous journey to South Georgia Island for help. The boat was crewed by Shackleton and five others. These men faced an almost impossible task: to travel in a tiny boat more than a thousand kilometers across the roughest expanse of sea in the world.

Despite the odds, Shackleton and his crew did reach South Georgia Island. By this time, winter was approaching and it was several months before they could rescue the shipwrecked members of the *Endurance's* crew. All 28 men survived the adventure. They failed in their mission to cross Antarctica, but their survival remains a tale of great glory.

Activity

Plot a course for a journey by boat from the coast nearest you to Antarctica. Use the map of ocean currents in ***Figure 15-23*** to identify the currents you would use and those you would avoid.

CHAPTER 15

Study Guide

Summary

SECTION 15.1

The Oceans

Main Ideas

- Oceanography is the scientific study of Earth's oceans. Oceanographers use sonar, satellites, and submersibles, among other tools, to explore the ocean.
- Earth's first oceans likely formed more than 4 billion years ago. Some water may have come from impacting comets or from deep within Earth's interior. Scientists theorize that water from within Earth's interior was released by volcanism.
- Approximately 71 percent of Earth's surface is covered by oceans. The major oceans are the Pacific, Atlantic, Indian, Arctic, and Antarctic.

Vocabulary

oceanography (p. 385)
sea level (p. 388)
side-scan sonar (p. 386)

SECTION 15.2

Seawater

Main Ideas

- Seawater contains 96.5 percent water and 3.5 percent dissolved salts. The average salinity of seawater is 35 ppt. The salinity of the ocean remains constant because salts are removed from the ocean at the same rate as they are added.
- Ocean surface temperatures range from –2°C in polar waters to 30°C in equatorial waters. Seawater density changes with changes in salinity and temperature.
- Ocean water temperatures decrease with depth. The ocean can be divided into three layers: the surface layer, the transitional thermocline, and the bottom layer.

Vocabulary

salinity (p. 392)
temperature profile (p. 396)
thermocline (p. 396)

SECTION 15.3

Ocean Movements

Main Ideas

- Ocean waves are generated by wind. Water in a wave moves in a circular motion but does not move forward. When waves reach shallow water, friction with the ocean bottom slows them, and they become breakers.
- Tides are caused by the gravitational attraction among Earth, the Moon, and the Sun. Lunar tides are twice as high as solar tides.
- Density currents are deep currents generated by salinity and temperature differences. Wind-driven surface currents affect the upper few hundred meters of the ocean. Upwelling occurs when winds push surface water aside and the surface water is replaced by cold, deep water.

Vocabulary

breaker (p. 400)
crest (p. 399)
density current (p. 403)
surface current (p. 403)
tide (p. 400)
trough (p. 399)
upwelling (p. 405)
wave (p. 399)

earthgeu.com/vocabulary_puzzlemaker

CHAPTER 15
Assessment

Understanding Main Ideas

1. Which of the following is used to measure ocean depth?
a. bottom dredges
b. nets
c. sonar
d. tidal patterns

2. Which of the following are the most common gases emitted by volcanoes?
a. hydrogen and helium
b. oxygen and nitrogen
c. water vapor and carbon dioxide
d. chlorine and hydrogen

3. What is the average depth of the oceans?
a. 380 m
b. 38 m
c. 3800 m
d. 3 km

4. What is the average salinity of seawater?
a. 100 ppt
b. 50 ppt
c. 35 ppt
d. 3.5 ppt

5. What is the average temperature of deep water below the thermocline?
a. 15°C
b. more than 4°C
c. less than 4°C
d. 0°C

6. What basic motion does water follow during the passage of a wave?
a. forward
b. backward
c. up and down
d. circular

7. Which of the following does not affect wave height in deep water?
a. wavelength
b. wind duration
c. wind speed
d. fetch

8. Which type of seawater has the greatest density?
a. warm, with low salinity
b. warm, with high salinity
c. cold, with low salinity
d. cold, with high salinity

9. To what average depth does light penetrate in the ocean?
a. 1 m
b. 10 m
c. 100 m
d. 1000 m

10. What type of high tides occur during a full Moon?
a. spring tides
b. neap tides
c. tidal ranges
d. tidal cycles

11. What is the the densest water mass in the Atlantic Ocean?
a. North Atlantic Deep Water
b. surface water
c. Antarctic Bottom Water
d. Antarctic Intermediate Water

12. The Arctic Ocean is the northern part of which body of water?
a. Atlantic Ocean
b. Pacific Ocean
c. Bering Sea
d. Indian Ocean

13. Explain why the Moon exerts a greater tidal influence than the Sun.

14. What distinguishes a sea from an ocean?

15. Where in the oceans are the highest values of salinity found? Explain.

16. What would be the wave base for a wave that is 200 m long?

17. Which gyre would have clockwise circulation: the North Pacific, the South Pacific, the South Atlantic, or the Indian Ocean? Explain.

Test-Taking Tip

Maximize Your Score If possible, find out how your standardized test will be scored. In order to do your best, you need to know if there is a penalty for guessing, and if so, how much of one. If there is no random-guessing penalty, you should always fill in an answer.

earthgeu.com/chapter_test

Assessment

Applying Main Ideas

18. Why does a wave break?
19. Copy the illustration on this page. Then use the following terms to label the characteristics of an ocean wave: *crest, trough, wave height,* and *wavelength.*

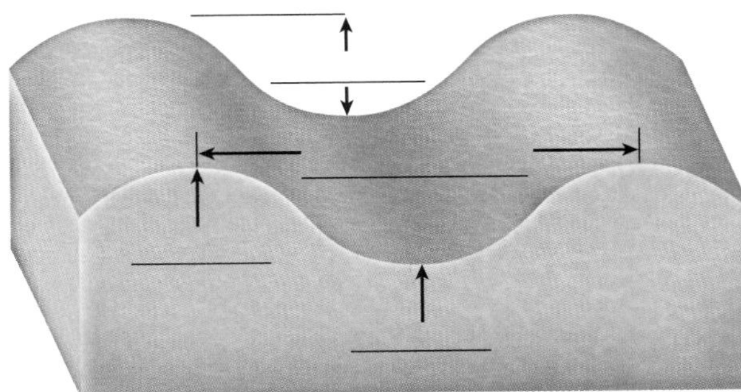

20. Cold water masses are generally denser than warm water masses, yet warm water from the Mediterranean Sea sinks to a depth of more than 1000 m when it flows into the Atlantic Ocean. Why?

Thinking Critically

21. One of the effects of El Niño, which you learned about in the previous chapter, is that the trade winds reverse direction. Predict how this might affect upwelling off the coast of Peru.
22. Based on what you have learned about water density, describe the movement of freshwater from a river as it flows into the sea.
23. Surface currents can affect coastal climates. Would the Gulf Stream and the Peru Current, both of which are surface currents, have the same effect on coastal climate? Explain.
24. Use your knowledge of global warming to hypothesize why sea level is rising.

Standardized Test Practice

1. Which sea was the first to be mapped?
   a. the Bering Sea
   b. the Caribbean Sea
   c. the Gulf of Mexico
   d. the Mediterranean Sea

2. Which region's seawater is most likely to have the highest concentration of dissolved salts?
   a. an equatorial region
   b. a subtropical region
   c. a polar region
   d. a delta where rivers empty into oceans

INTERPRETING SCIENTIFIC ILLUSTRATIONS
Use the illustration below to answer questions 3 and 4.

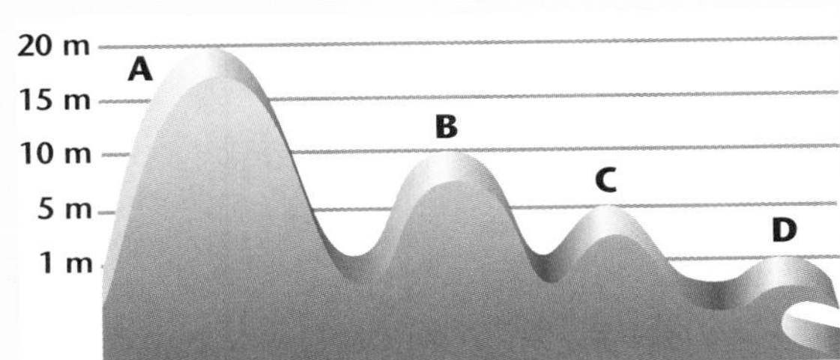

3. Which wave is most likely caused by a strong hurricane?
   a. A
   b. B
   c. C
   d. D

4. Why is Wave D most likely collapsing?
   a. friction from the ocean floor
   b. storm activity
   c. increased crest-to-crest wavelength
   d. opposing tidal movement

5. Which ocean movement is slow-moving and occurs in deep waters?
   a. surface currents
   b. upwelling
   c. density currents
   d. gyres

Chapter 16

The Marine Environment

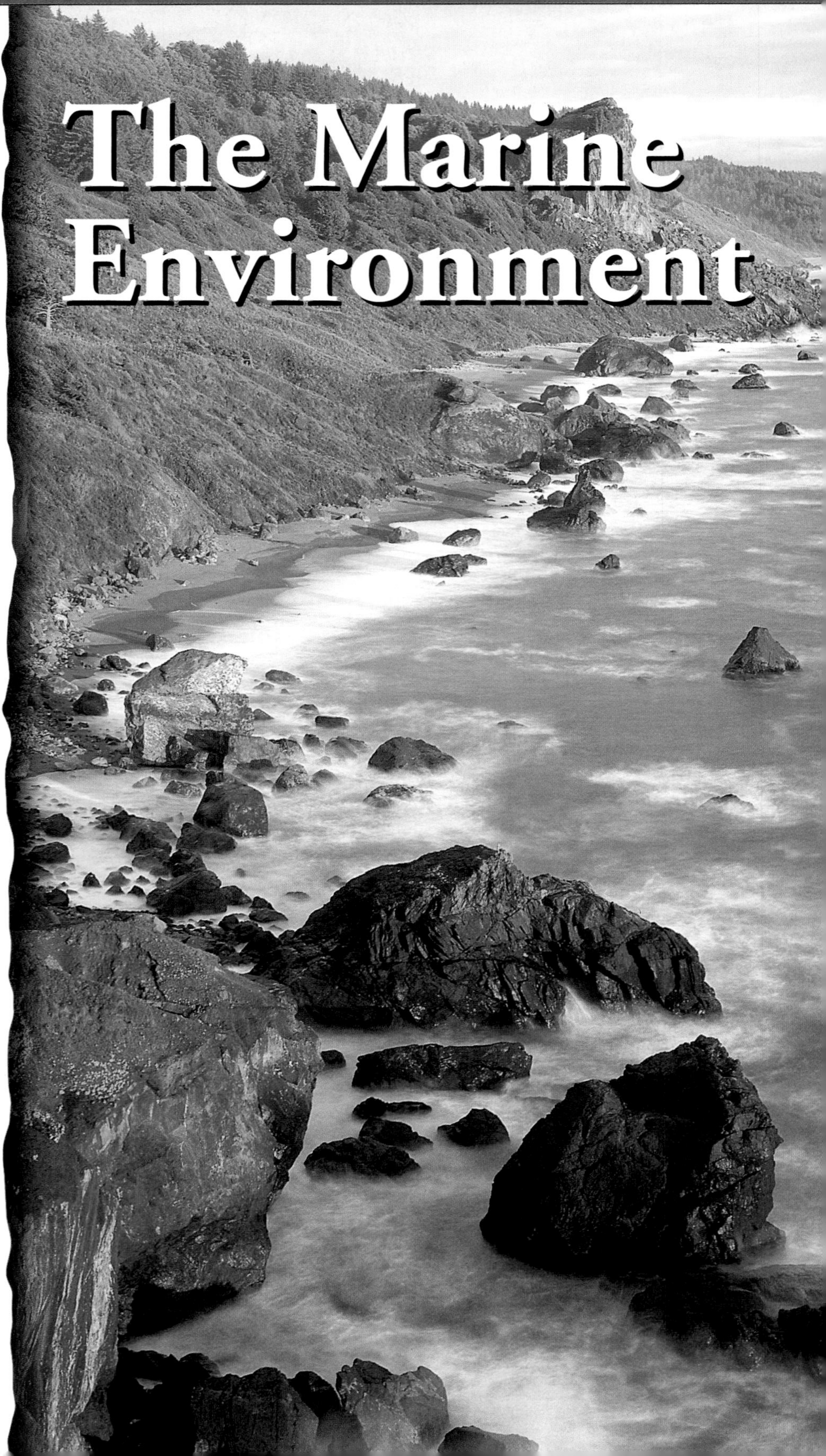

What You'll Learn

- How wave action affects shorelines and produces erosional and depositional coastal features.
- What major features and sediments are found on the ocean floor.

Why It's Important

The oceans cover 71 percent of Earth's surface and have a major impact on weather and climate. Shoreline features provide recreational opportunities as well as protection from major storms along coastlines.

To learn more about the marine environment, visit the Earth Science Web Site at earthgeu.com

Discovery Lab — Composition of Chalk

Although you may not live anywhere near a coast, parts of your environment were shaped by the ocean. For example, you may be just a few meters away from former seafloor deposits that are now part of the bedrock underground. One such seafloor deposit is chalk. How can you tell that chalk formed on the seafloor?

1. Grind up a small piece of natural chalk into a powder. Make a slide of the powdered chalk.
2. Observe the chalk powder through a microscope. Describe the powder. Are the grains irregular in shape or size? Do some of the grains have patterns?
3. Analyze the powder and hypothesize the origin of the chalk.

CAUTION: ***Always wear safety goggles and an apron in the lab. Use caution to prevent chalk dust from becoming airborne.***

Observe In your science journal, describe the composition of the powdered chalk. What is the origin of the chalk? On what evidence do you base your conclusion?

SECTION 16.1 *Shoreline Features*

OBJECTIVES

- **Explain** *how shoreline features are formed and modified by marine processes.*
- **Describe** *the major erosional and depositional shoreline features.*

VOCABULARY

wave refraction
beach
estuary
longshore bar
longshore current
barrier island

Some of the most interesting places on our planet are the seashores, the places where the land meets the sea. They are places of continuous, often dramatic geologic activity, places where you can see geological changes occurring almost daily. Shorelines are shaped by the action of waves, tides, and currents. As waves erode some coastlines, they create some of the most impressive rock formations on Earth. In other areas, waves deposit loose material and build wide, sandy beaches. To understand how waves act in different areas of the coast, let's reexamine the behavior of breakers.

Erosional Landforms

You learned in Chapter 15 that waves increase in height and become breakers as they approach a shoreline. Large breakers can hurl thousands of metric tons of water, together with suspended rock fragments, against a shore with such destructive force that they are capable of eroding even solid rock. This destructive action of breakers is most evident at rocky headlands, which are points of land sticking out into the ocean.

Figure 16-1 Wave crests advance toward the shoreline and slow down when they encounter shallow water. This causes the wave crests to bend toward the headlands and move in the direction of the arrows. Wave rays, drawn perpendicular to the wave crests, show the direction of wave travel and bending of wave crests.

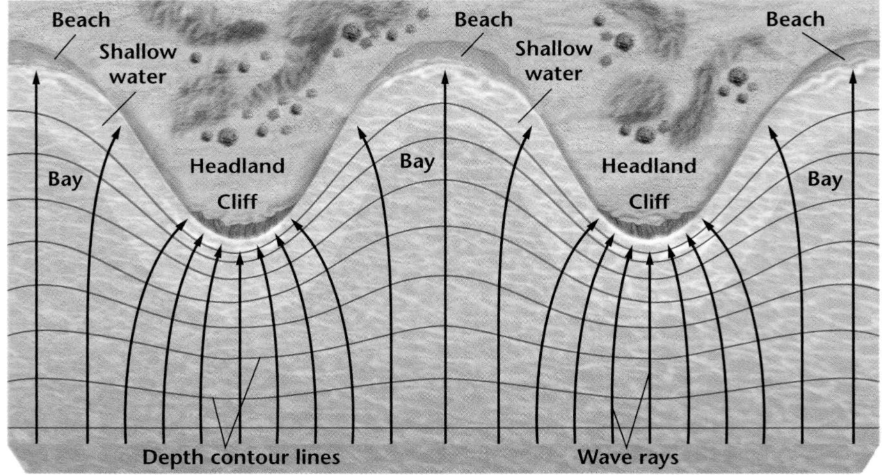

Waves move faster in deep water than in shallow water. This difference in wave speed causes initially straight wave crests to bend when part of the crest moves into shallow water, a process known as **wave refraction,** illustrated in ***Figure 16-1.*** Along an irregular coast with headlands and bays, the wave crests bend towards the headlands. As a result, most of the breaker energy is concentrated along the relatively short section of the shore around the tips of the headlands, while the remaining wave energy is spread out along the much longer shoreline of the bays. The headlands thus undergo severe erosion. The material eroded from the headlands is swept into the bays, where it is deposited in the form of crescent-shaped beaches. Can you guess what the long-term effect of this process is? The headlands are worn back and the bays are filled in until the shoreline straightens. Given enough time, irregular shorelines are straightened by wave action.

Landforms of Rocky Headlands Many headlands have spectacular rock formations. Generally, as a headland is gradually worn away, a flat erosional surface called a wave-cut platform is formed. The wave-cut platform terminates against a steep wave-cut cliff, as illustrated in ***Figure 16-2.*** Differential erosion, the removal of weaker rocks or rocks near sea level, produces many of the other characteristic landforms of rocky headlands. As shown in ***Figure 16-3,*** sea stacks are isolated rock towers or similar erosional remnants left on wave-cut platforms, and sea arches are formed as stronger rocks are undercut by wave erosion. Sea caves are tubelike passages

Figure 16-2 A headland can be modified by wave erosion. The dotted lines indicate the original shape of the headland.

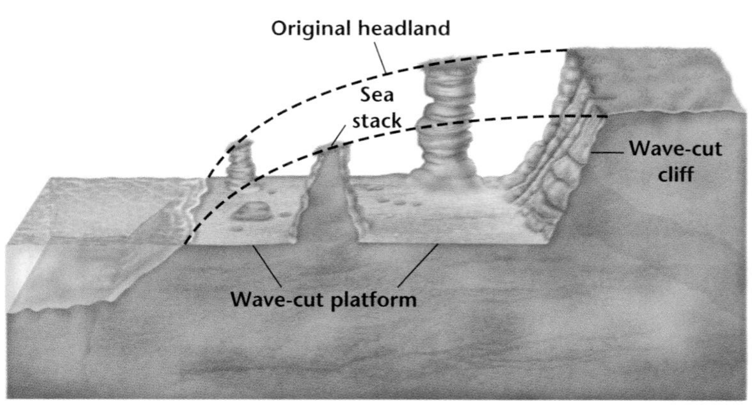

blasted into the headland at sea level by the never-ending assault of the breakers.

BEACHES

Have you ever visited any of the beaches along the coastline of the United States? Long stretches of our coastlines are lined with wide, sandy beaches. A **beach** is a sloping band of sand, pebbles, gravel, or mud at the edge of the sea. Beaches are composed of loose sediments deposited and moved about by waves along the shoreline. The size of sediment particles depends on the energy of the waves striking the coast and on the source of the sediment. Beaches pounded by large waves or formed on rocky coasts usually consist of coarse materials such as pebbles and cobbles.

The composition of beach material also depends upon the source of the material. Some Hawaiian beaches consist of black sand, tiny grains of minerals derived from the volcanic rocks that make up most of the Hawaiian Islands. The white and pink sand that form the beaches of southern Florida and the Bahamas has the consistency of cornmeal; these beaches are composed mostly of small fragments of local corals and seashells. Beaches near the mouths of large rivers are composed of the sandy sediments that are washed in by river water and made up of small grains of quartz and feldspar.

ESTUARIES

If you look at the map of the eastern coast of the United States shown in ***Figure 16-4,*** you will see rivers and streams entering the ocean. The area where the lower end of a freshwater river or stream enters the ocean is an **estuary.**

Figure 16-3 These sea stacks and sea arches along the rugged coastline of Washington State were formed by wave refraction at a rocky headland.

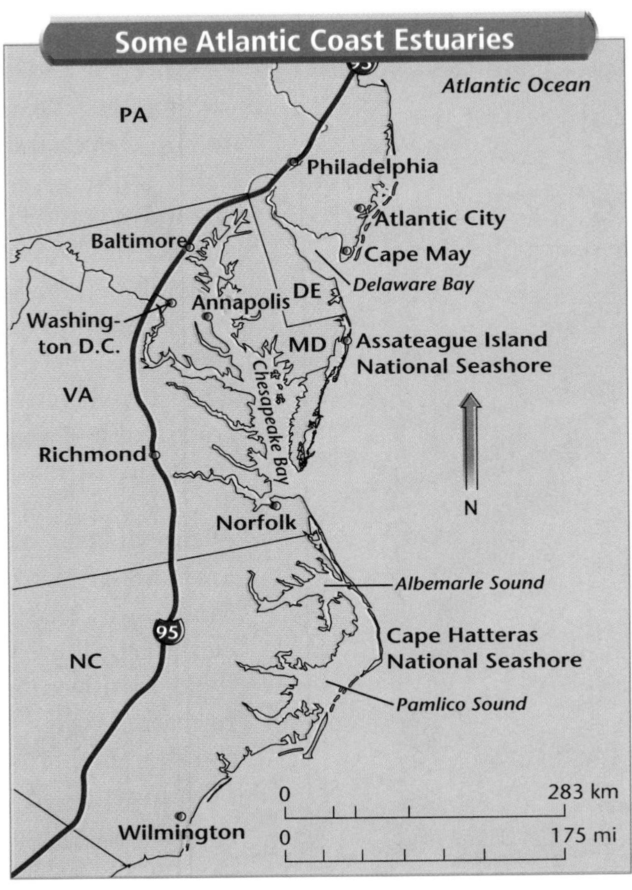

Figure 16-4 Estuaries provide an abundant supply of food and shelter to the young of commercially-important marine organisms. Sea grasses in estuaries also trap sediment and help filter out some water pollutants.

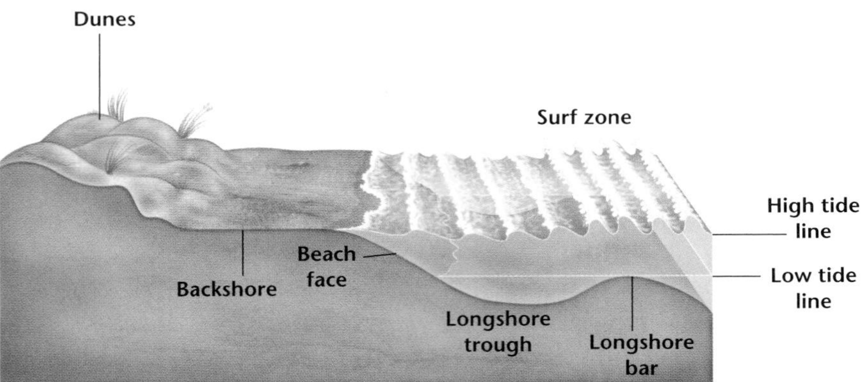

Figure 16-5 The sloping beach face is shaped by daily wave action, while the dunes behind the beach are affected only by large storm waves. Coastal dunes form from beach sand blown inland. The submerged longshore bar is located in the surf zone.

The water in estuaries is brackish, a mixture of freshwater and salt water. Estuaries are nurseries to the young of many different species, including ocean fishes. ***Figure 16-4*** reveals many large estuaries, such as Chesapeake Bay and Pamlico Sound.

LONGSHORE CURRENTS

Suppose you stood on a beach at the edge of the water and began to walk out into the ocean. As you walked, the water would get deeper for a while, but then it would become shallow again. The shallow water offshore lies above a sand bar, called the **longshore bar,** that forms in front of most beaches, as illustrated in ***Figure 16-5.*** Waves break on the longshore bar in the area known as the surf zone. The deeper water closer to shore than the longshore bar is called the longshore trough. The waves striking the beach are almost parallel to the shoreline, although the waves seaward of the longshore bar are generally not parallel to the shore. This is another case of wave refraction. The slowing of the waves in shallow water causes the wave crests to bend towards the shore. As water from incoming breakers spills over the longshore bar, a current flowing parallel to the shore, called the **longshore current,** is produced. This current varies in strength and direction from day to day. Over the course of a year, because of prevailing winds and wave patterns, one direction usually dominates.

Movement of Sediments Longshore currents move large amounts of sediments along the shore. Fine-grained material such as sand is suspended in the turbulent, moving water, and larger particles are pushed along the bottom by the current. Additional sediment is moved back and forth on the beach face by incoming and retreating waves. Incoming waves also move sediment at an angle to the shoreline in the direction of wave motion. Overall, the transport of sediment is in the direction of the longshore current. On both the Atlantic and Pacific Coasts of the United States, longshore transport is generally to the south.

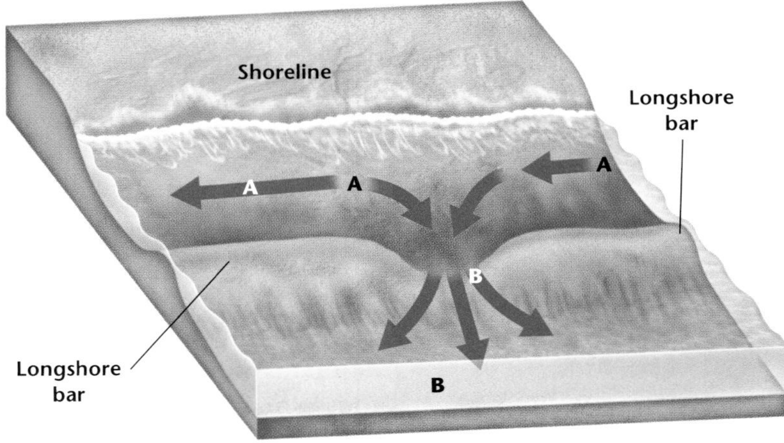

Figure 16-6 Longshore currents **(A)** are driven by incoming waves. Rip currents **(B)** return water through gaps in the longshore bar out to sea. Rip currents spread out and weaken beyond the longshore bar.

Rip Currents Wave action also produces rip currents, which flow out to sea through gaps in the longshore bar. Rip currents return the water spilled into the longshore trough to the open ocean. These dangerous currents can reach speeds of several kilometers per hour. If you are ever caught in a rip current, you should not try to swim against it, but rather swim parallel to the shore to get out of it. ***Figure 16-6*** illustrates both longshore and rip currents.

Depositional Features of Seashores

As a result of wave erosion, longshore transport, and sediment deposition, most seashores are in a constant state of change. Sediments are eroded by large storm waves and deposited wherever waves and currents slow down. Sediments moved and deposited by longshore currents build various characteristic coastal landforms, such as spits and barrier islands, illustrated in ***Figure 16-7.*** A narrow bank of sand that projects into the water from a bend in the coastline is called a spit. A spit, which forms where a shoreline changes direction, is protected from wave action. When a growing spit crosses a bay, a baymouth bar forms.

Figure 16-7 Depositional features of coastlines include spits, baymouth bars, lagoons, and barrier islands.

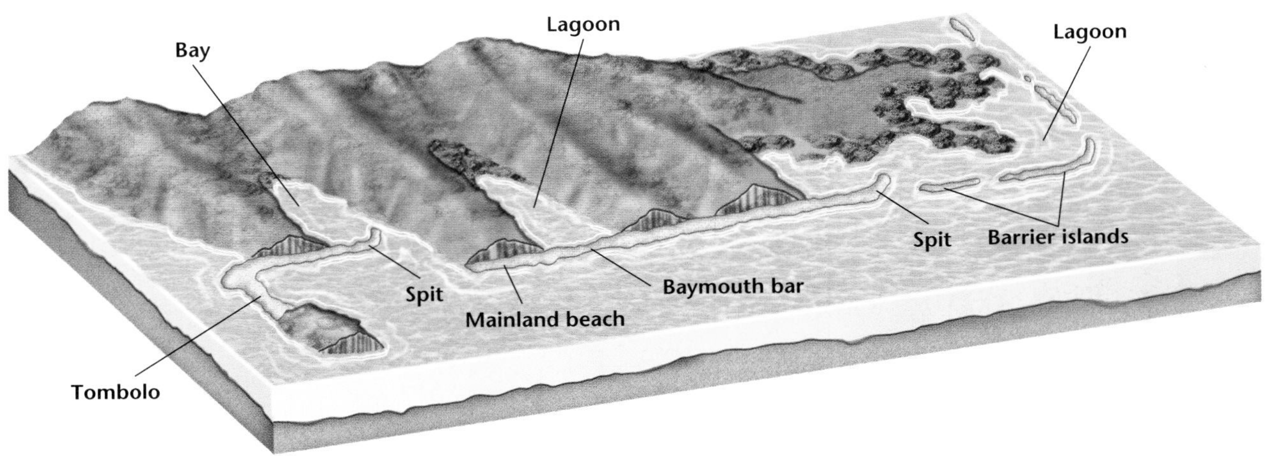

Barrier islands are long ridges of sand or other sediment, deposited or shaped by the longshore current, that are separated from the mainland. Barrier islands can be several kilometers wide and tens of kilometers long. Most of the Gulf Coast and the eastern coast south of New England are lined with an almost continuous chain of barrier islands. The shallow, protected bodies of water behind baymouth bars and barrier islands are called *lagoons,* which essentially are saltwater coastal lakes that are connected to the open sea by shallow, restricted outlets. Another, somewhat peculiar coastal landform is a tombolo, a ridge of sand that forms between the mainland and an island and connects the island to the mainland. When this happens, the island is no longer an island, but the tip of a peninsula.

All of these depositional coastal landforms, including large barrier islands, are unstable and temporary. Occasionally, major storms sweep away entire sections of barrier islands and redeposit the material elsewhere. Even in the absence of storms, however, changing wave conditions can slowly erode beaches and rearrange entire shorelines. For example, the shoreline of Cape Cod, Massachusetts, is retreating by as much as 1 m per year. ***Figure 16-8*** shows some results of retreating shorelines.

You may wonder how longshore transport can build coastal features that rise well above sea level. Several factors play a role in this. At high tide, a longshore current can deposit sediment on a beach so that it extends in the direction of the longshore current. In addition, storm waves can pile up submerged sediments to heights well above the level of the highest tides. Wherever sediments are exposed at low tide, winds pick up dry sand and build sand dunes.

Figure 16-8 Shore erosion during a storm undermined the cliffs on which these houses were built, along the coast of Pacifica, California **(A).** The ocean has eroded the beach near Galveston, Texas, to the extent that the bases of telephone poles along a beach road now are under water **(B).**

Protective Structures

In many coastal areas, protective structures such as seawalls, groins, jetties, and breakwaters are built in an attempt to prevent beach erosion and destruction of oceanfront properties. ***Figure 16-9*** illustrates the effects of building structures in areas of longshore transport. These artificial structures interfere with natural shoreline processes and can have unexpected negative effects. For example, seawalls built along the shore to protect beachfront properties from powerful storm waves reflect the energy of such waves back towards the beach, where they worsen beach erosion. Eventually, seawalls are undercut and have to be rebuilt larger and stronger than before. Groins are wall-like structures built into the water perpendicular to the shoreline for the purpose of trapping beach sand. Groins interrupt natural longshore transport and deprive beaches down the coast of sand. The result is aggravated beach erosion down the coast from groins. Similar effects are caused by jetties, which are walls of concrete built to protect a harbor entrance from drifting sand. Jetties trap sand upshore from a harbor and prevent sand from reaching the beaches downshore. Eventually, sand drifts around the jetty and closes the harbor entrance anyway, unless it is removed periodically by dredging. Breakwaters are built in the water parallel to straight shorelines to provide anchorages for small boats.

Breakwaters affect the longshore current in much the same way as offshore islands do. The current slows down behind the breakwater and is no longer able to move its load of sediment, which is then deposited behind the breakwater. If the accumulating sediment is left alone, it will eventually fill the anchorage. To prevent this, all such anchorages have to be dredged regularly at great expense. In general, protective structures cause an overall loss of the sediments that maintain beaches.

Figure 16-9 The entrance to Channel Island Harbor in Oxnard, California, is protected by jetties and a breakwater **(A).** Jetties deprive downshore beaches of sand **(B).** Breakwaters cause beach sand to accumulate and eventually close the anchorage **(C).**

B

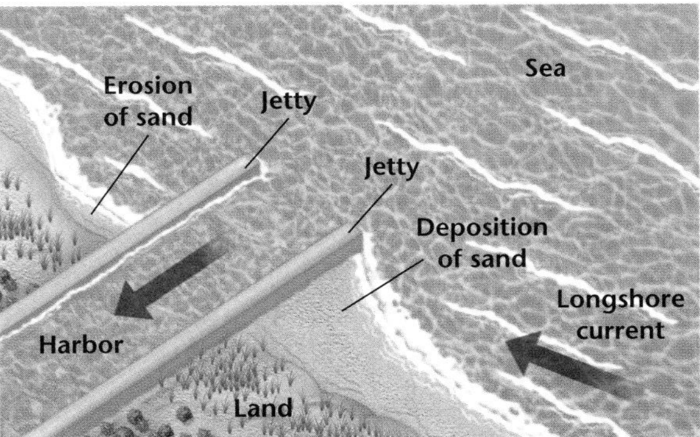

C

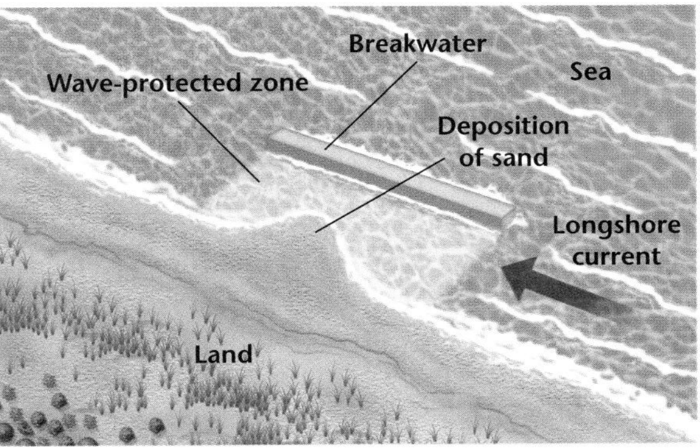

Using Numbers The highest point in the state of Florida is Walton County, at an elevation of just 105 m above sea level. If the sea level continues to rise at the highest estimated rate of 3.9 mm/y, in how many years will all of Florida be under water?

CHANGES IN SEA LEVEL

At the height of the last ice age, approximately 10 000 years ago, the global sea level was about 130 m lower than it is at present. Since that time, the melting of most of the ice-age glaciers has raised the ocean to its present level. In the last 100 years, the global sea level has risen 10 to 15 cm. It continues to rise slowly; estimates suggest a rise in sea level of 1.5 to 3.9 mm/year. Many scientists contend that this continuing rise in sea level is the result of global warming. Over the last century, Earth's average surface temperature has increased by approximately 0.5°C. As Earth's surface temperature rises, seawater warms up and as it warms, it also expands, which adds to the total volume of the seas. In addition, higher temperatures on Earth's surface cause glaciers to melt, and the meltwater flowing into the oceans increases their volume. Scientists predict that global sea levels could rise another 30 cm in the next 70 years.

Effects of Sea Level Changes If Earth's remaining polar ice sheets, in Greenland and Antarctica, melted completely, their meltwaters would raise sea level by another 70 m. This rise would totally flood some countries, such as the Netherlands, along with some coastal cities in the United States, such as New York City, and low-lying states such as Florida and Louisiana. Fortunately, this isn't likely to happen anytime soon. However, if Earth's surface temperature continues to rise, an unstable part of the Antarctic ice sheet eventually could melt and cause a rise in sea level of about 6 m. Many of the barrier islands of the Atlantic and Gulf Coasts may be former coastal dunes that were drowned by rising sea levels. Other features produced by rising sea levels are the fjords of Norway, shown in ***Figure 16-10.*** Fjords are deep coastal valleys that were scooped out by glaciers during the ice age and later flooded when these glaciers melted.

Figure 16-10 Fjords are flooded U-shaped valleys carved by glaciers. Fjords may be up to 1200 m deep.

Figure 16-11 This series of elevated marine terraces can be found on San Clemente Island in California.

Effects of Tectonic Forces Other processes that affect local sea levels are tectonic uplift and sinking. If a coastline sinks, there is a relative rise in sea level along that coast. A rising coastline, on the other hand, produces a relative drop in sea level. As a result of tectonic forces in the western United States, much of the West Coast is being pushed up much more quickly than the sea level is rising. Because much of the West Coast was formerly under water, it is called an emergent coast. Emergent coasts tend to be relatively straight because the exposed seafloor topography is much smoother than typical land surfaces with hills and valleys. Other signs of an emergent coast are former shoreline features such as sandy beach ridges located far inland. Among the most interesting of these features are elevated marine terraces, former wave-cut platforms that are now high and dry, well above current sea level. ***Figure 16-11*** shows striking examples of such platforms. Some old wave-cut platforms in southern California are hundreds of meters above current sea level. You will identify an emergent coast in the *Mapping GeoLab* at the end of this chapter.

Section Assessment

1. Irregular shorelines have headlands and bays. Which of these experiences the most severe erosion by breakers? Why?
2. What are sea stacks, and how are they formed?
3. What effect does a seawall have on a beach?
4. If a coast has elevated marine terraces, is it rising or sinking? Explain.
5. **Thinking Critically** Resort communities such as Ocean City, Maryland, are built on barrier islands. These communities spend thousands of dollars each year to add sand to the beaches along the shoreline. Explain why this is necessary.

Skill Review

6. **Predicting** Are rip currents most dangerous on calm days, on stormy days with winds blowing from the land, or on stormy days with winds blowing from the ocean? Explain. For more help, refer to the *Skill Handbook*.

SECTION 16.2 The Seafloor

OBJECTIVES

- **Explain** *the reason for the existence of continents and ocean basins.*
- **Compare** *the major geologic features of continental margins and ocean basins.*
- **Describe** *the different types of marine sediments and their origin.*

VOCABULARY

continental margin
continental shelf
continental slope
turbidity current
continental rise
abyssal plain
deep-sea trench
mid-ocean ridge
seamount

If you were asked to draw a map of the seafloor, what kind of topographic features would you include? Until recently, most people had little knowledge of the features of the ocean floor. However, modern oceanographic techniques, including satellite imagery, reveal that the topography of the ocean bottom is as varied as that of the continents. *Appendix C* on pages 912–913 shows the major features of the seafloor that have been revealed by modern scientific methods.

OCEANIC AND CONTINENTAL CRUST

The topography of the seafloor is surprisingly rough and irregular, with numerous high mountains and deep depressions. The deepest place on the seafloor, the Marianas Trench in the Pacific Ocean, is just over 11 km deep. This is deeper than the height of Mount Everest, the tallest mountain on Earth.

Recall that Earth has two types of crust: continental crust, with an average thickness of 40 km; and thin oceanic crust, with an average thickness of 6 or 7 km. Crustal elevation depends on crustal thickness, and thus the thick continental crust is always associated with higher elevations on land, and the thin oceanic crust is always associated with the deep ocean basins. You will find out more about surface elevations on Earth in the *Problem-Solving Lab* on the next page. Note that part of the continental section is actually below sea level and that the ocean covers parts of the continents. These submerged parts of continents are called **continental margins.** They represent the shallowest parts of the ocean. As shown in ***Figure 16-12,*** a continental margin includes

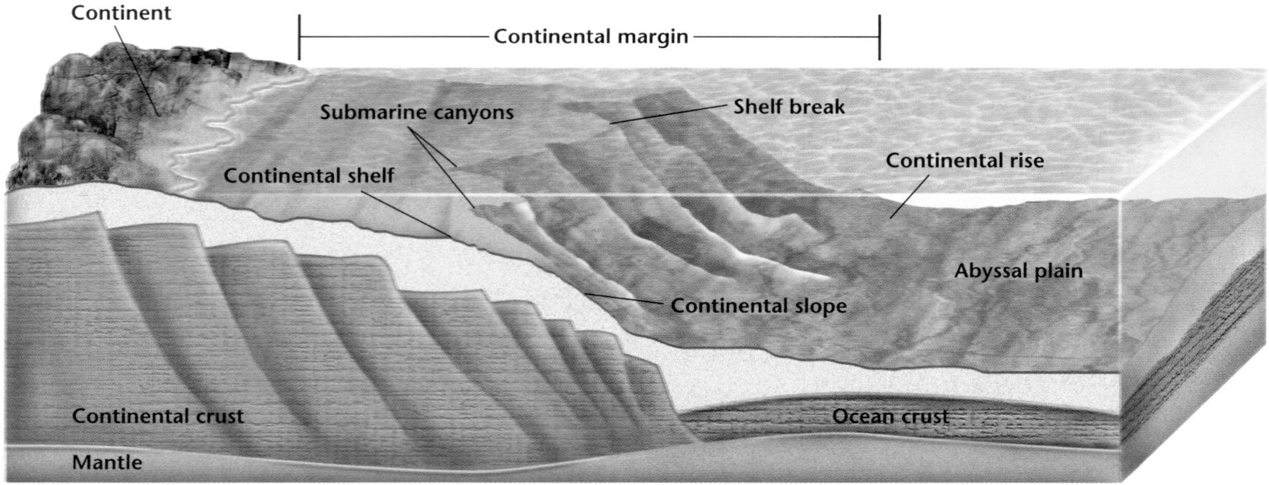

Figure 16-12 The major features of the continental margin are illustrated in this diagram.

the continental shelf, the continental slope, and the continental rise. Study ***Figure 16-12*** as you read about the features in this section.

CONTINENTAL SHELVES

The continental margins are the areas where the edges of continents meet the ocean. The shallowest part of a continental margin extending seaward from the shore is the **continental shelf.** Although continental shelves vary greatly in width, the average width is 60 km. On the Pacific Coast of the United States, the continental shelf is only

Problem-Solving Lab

Interpreting Graphs

Compare surface elevations A useful comparison of the heights of the continents to the depths of the oceans is given by the curve in the graph below. Note that the curve has two relatively flat sections, one near sea level, and another at a depth of about 5 km. The flat section near sea level represents the continents; the lower flat section represents the ocean basins.

Analysis

1. How tall is the highest mountain on Earth's surface in km approximately?
2. At about what depth would you begin to find trenches on the ocean floor?
3. What percentage of Earth's surface is above current sea level?
4. What percentage of Earth's surface is represented by the continental margin?

Thinking Critically

5. The oceanic crust is that part of the crust that is at a depth of 2 km or more below sea level. What percentage of Earth's surface lies above the oceanic crust?
6. What is the total average difference in surface elevations on Earth?

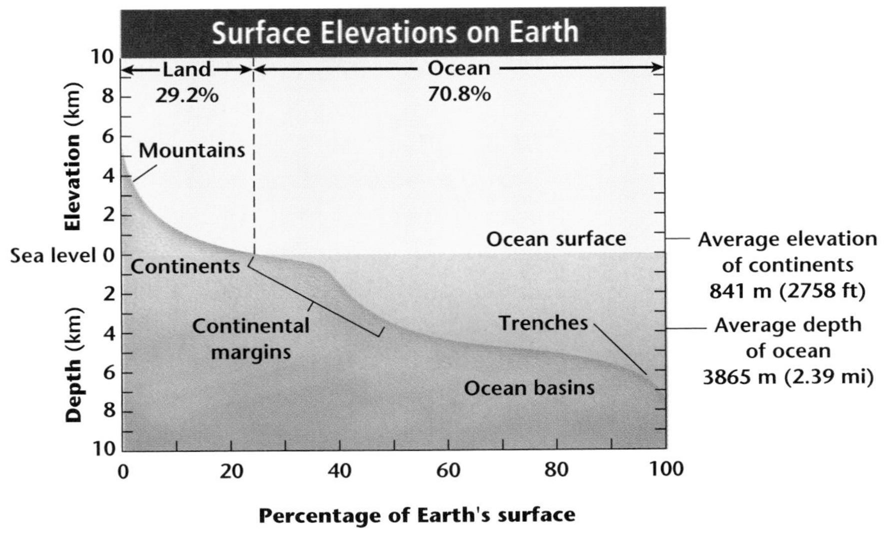

a few kilometers wide, whereas the continental shelf of the Atlantic Coast is hundreds of kilometers across. The average depth of the water above continental shelves is about 130 m. Recall that sea level during the last ice age was some 130 m lower than at present; therefore, most of the world's continental shelves must have been above sea level at that time. As a result, present day coastlines are radically different from the way they were during the last ice age. At that time, Siberia was attached to North America by the Bering land bridge, Great Britain was attached to Europe, and a large land mass existed where today there are only the widely scattered islands of the Bahamas. When Earth's surface began to warm after the last ice age, and the continental ice sheets began to melt, the sea gradually covered up the continental shelves. Beaches, river valleys, and other coastal landforms from that time are now submerged and located far beyond the present shoreline. Large numbers of commercially valuable fishes now inhabit the shallow, nutrient-rich waters of the continental shelves. In addition, the thick sedimentary deposits on the shelves are significant sources of oil and natural gas.

Continental Slopes

Beyond the continental shelves, the seafloor drops away quickly to depths of several kilometers, with slopes averaging nearly 100 m/km. These sloping regions are the **continental slopes.** To geologists, the continental slope is the true edge of a continent because it generally marks the edge of the continental crust. In many places, this slope is cut by deep submarine canyons, similar to canyons on land, some of which are comparable in size to the Grand Canyon of Arizona. How do you think these canyons formed? On land, canyons like these are cut by rivers. But the sea level never dropped below the edge of the continental shelves, and the water from freshwater rivers flowing into the ocean is less dense than seawater, which means that it floats at the ocean's surface and thus cannot erode the seafloor. These submarine canyons were cut by **turbidity currents,** which are rapidly flowing water currents along the bottom of the sea that carry heavy loads of sediments, similar to mudflows on land. Turbidity currents, illustrated in ***Figure 16-13,*** may originate as underwater landslides on the continental slope that are triggered by earthquakes, or they may originate from sediment stirred up by large storm waves on the continental shelf. Once formed, a turbidity current can reach speeds exceeding 30 km/h and effectively erode bottom sediments and bedrock. The sediments carried down the

Figure 16-13 Turbidity currents flow along the seafloor because the seawater-sediment mixture of the current is denser than seawater alone.

continental slope by these currents eventually come to rest at the bottom of the slope and beyond. The gently sloping accumulation of deposits from turbidity currents that forms at the base of the continental slope is called a **continental rise.** A continental rise may be several kilometers thick. The rise gradually gets thinner and eventually merges with the sediments of the seafloor beyond the continental margin. In some places, especially around the Pacific Ocean, the continental slope ends in deeper depressions, known as deep-sea trenches, in the seafloor. In such places, there is no continental rise at the foot of the continental margin.

Ocean Basins

Beyond the continental margin are ocean basins, which are deeper parts of the seafloor that lie above the thin, basaltic, oceanic crust. Ocean basins represent about 60 percent of Earth's surface and contain some of Earth's most interesting topography. ***Figure 16-14*** shows the topography of the ocean basin beneath the Atlantic Ocean.

Figure 16-14 Features of the ocean basin can be identified in this physiographic map of the Atlantic Ocean.

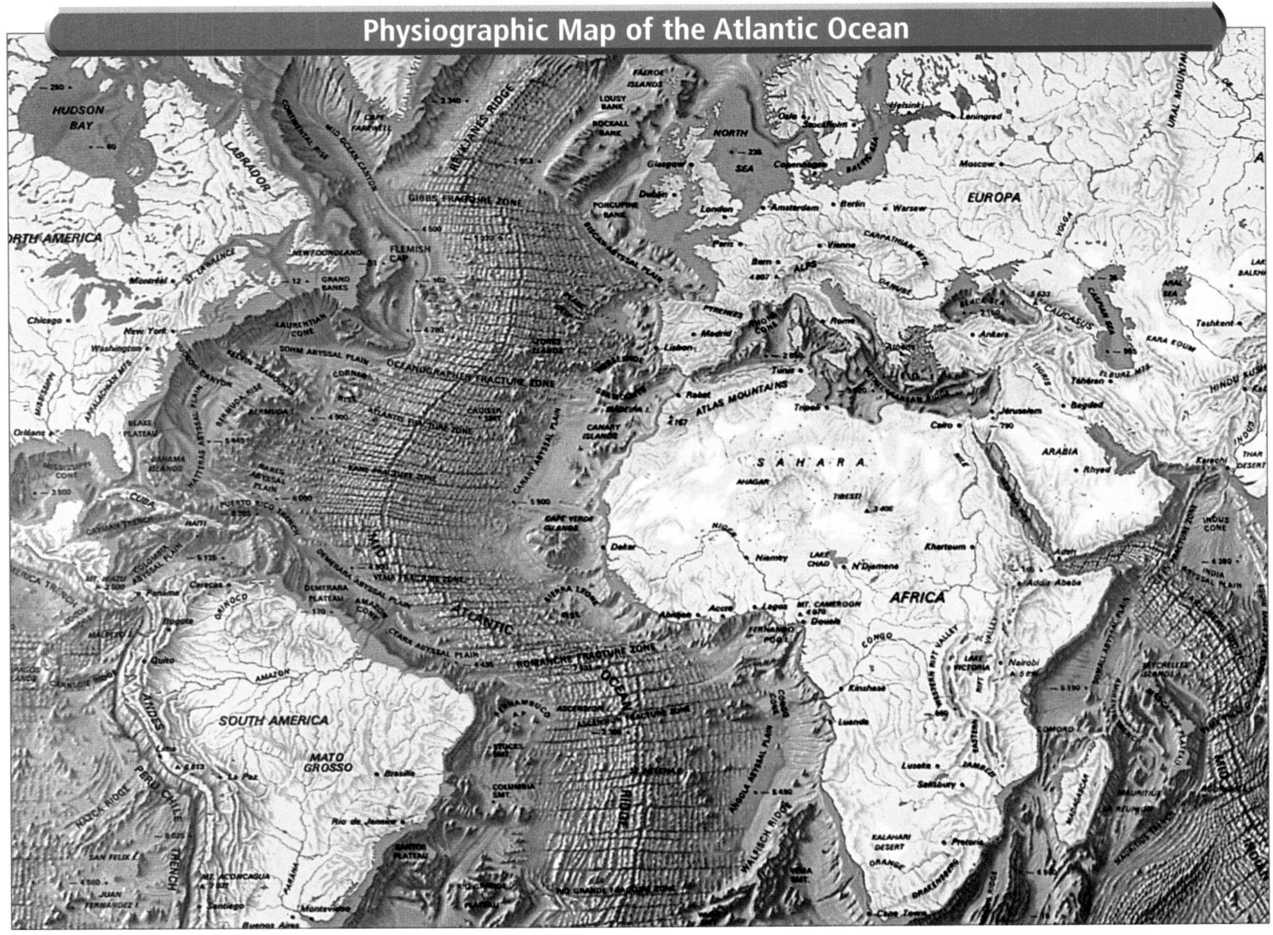

Abyssal Plains The smooth parts of the ocean floor 5 or 6 km below sea level are called abyssal plains. **Abyssal plains** are plains covered with hundreds of meters of fine-grained muddy sediments and sedimentary rocks that were deposited on top of basaltic volcanic rocks. These plains, extending seaward from the continental margins, are probably the flattest surfaces on Earth, and cover large areas of Earth's surface.

Deep-Sea Trenches The deepest parts of the ocean basins are the **deep-sea trenches,** which are elongated, sometimes arc-shaped depressions in the seafloor several kilometers deeper than the adjacent abyssal plains. Many deep-sea trenches lie next to chains of volcanic islands, such as the Aleutian Islands of Alaska, and most of them are located around the margins of the Pacific Ocean, as you can see in ***Figure 16-15.*** Deep-sea trenches are relatively narrow, about 100 km across, but they may extend for thousands of kilometers. Their significance will be discussed in more detail in Chapter 17.

Mid-Ocean Ridges The most prominent features of the ocean basins are the **mid-ocean ridges,** which run through all the ocean basins and have a total length of over 65 000 km, more than Earth's

Figure 16-15 The seafloor of the Pacific is characterized by many deep-sea trenches.

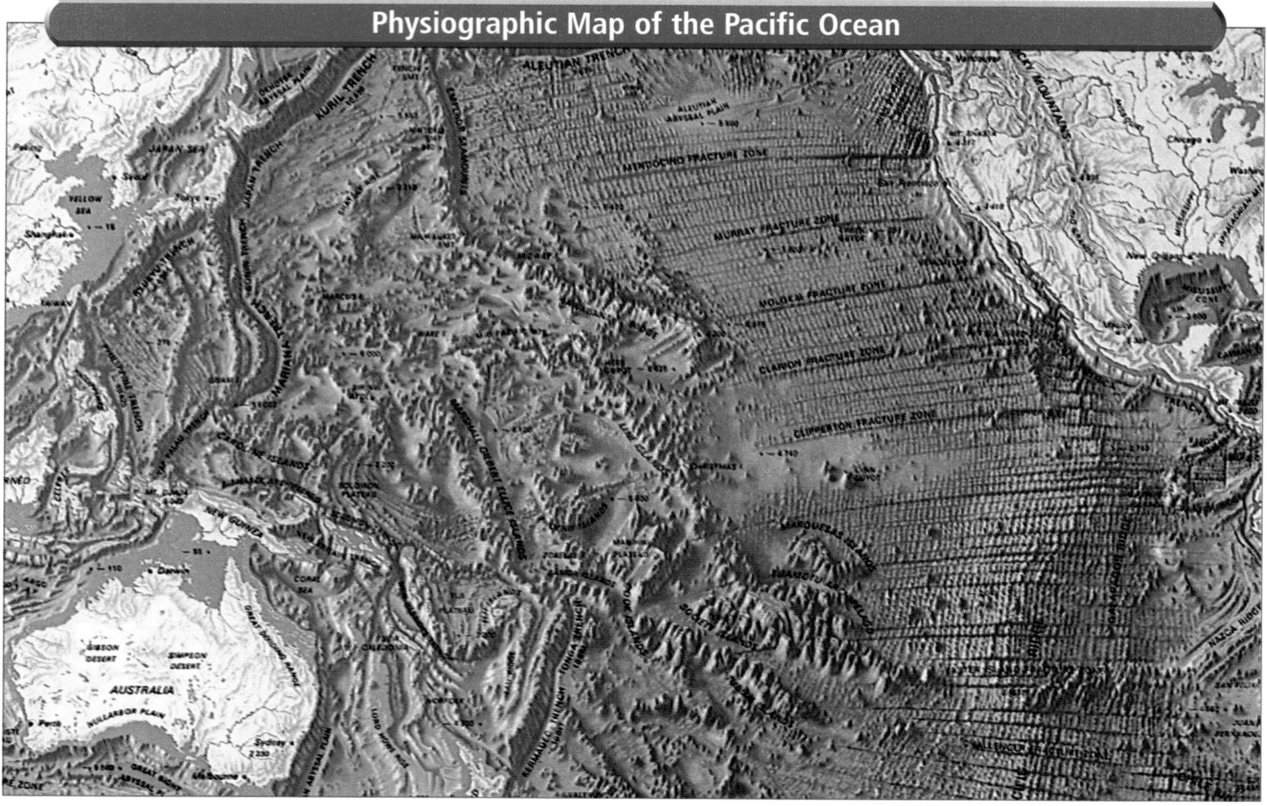

circumference. Mid-ocean ridges have an average height of 1500 m, but they may be thousands of kilometers wide. The highest peaks in mid-ocean ridges are over 6 km tall and emerge from the ocean as volcanic islands. Mid-ocean ridges are sites of frequent volcanic eruptions and earthquake activity. The crests of these ridges often have valleys called rifts running through their centers. Rifts may be up to 2 km deep.

Mid-ocean ridges do not form continuous lines. The mid-ocean ridges break into a series of shorter, stepped sections, which run at right angles across each mid-ocean ridge. The areas where these breaks occurs are called fracture zones. Fracture zones are about 60 km wide, and they curve gently across the seafloor, sometimes for thousands of kilometers. Volcanic and earthquake activity occurs frequently in fracture zones. Some volcanic islands, such as the Azores in the Atlantic Ocean and the Galápagos Islands in the Pacific Ocean, lie along fracture zones.

Figure 16-16 Unique communities of organisms can be found around black smokers.

Hydrothermal Vents Have you ever heard of hydrothermal vents on the seafloor? A hydrothermal vent is a hole in the seafloor through which fluid heated by magma erupts. Most hydrothermal vents are located along the bottom of the rifts in mid-ocean ridges. When the heated fluid that erupts from these vents contains metal oxides and sulfides, they immediately precipitate out of the fluid and produce thick, black, smokelike plumes. This type of hydrothermal vent, known as a black smoker, ejects superheated water with temperatures of up to 350°C. ***Figure 16-16*** illustrates the black smokers found in the rift valley of a mid-ocean ridge. A second type of vent, known as a white smoker, ejects warm water. Smokers are caused by seawater circulating through the hot crustal rocks in the centers of mid-ocean ridges. The fundamental cause of mid-ocean ridges and the volcanic activity associated with them is plate tectonics, which will be discussed in Chapter 17.

Topic: Organisms
To learn more about hydrothermal vent communities, visit the Earth Science Web Site at earthgeu.com

Activity: Research the types of organisms found near hydrothermal vents. List four types of organisms found in these communities.

SEAFLOOR VOLCANOES

Satellite radar imagery has revealed that the ocean floor is dotted with tens of thousands of solitary mountains. These mountains are not located near areas of active volcanism. How, then, did they form? You have learned that the ocean basins are volcanically active at mid-ocean ridges and fracture zones. The almost total absence of earthquakes in most other areas of the seafloor suggests that volcanism in those areas must have ceased a long time ago. Thus, most of the mountains on the

MiniLab

How fast do sediment grains sink?

Investigate how grain size affects settling speed.

Procedure ***CAUTION: Always wear safety goggles and an apron in the lab.***

1. Obtain five round pebbles and sand grains with approximate diameters of 0.5 mm, 1 mm, 2 mm, 5 mm, and 10 mm.
2. Draw a data table in your science journal with these headings: Type of Particle, Diameter (mm), Distance (cm), Time (sec), Settling Speed (cm/s).
3. Measure the diameters of each specimen using a set of sieves. Record these measurements in your data table.
4. Fill a 250-mL graduated cylinder with cooking oil.
5. Drop the largest specimen into the oil. Measure the time it takes for the specimen to sink to the bottom of the cylinder. If the specimen doesn't fall quickly, measure the time it takes to fall a given distance. Record this time in your data table.
6. Repeat step 5 for the remaining specimens.
7. Calculate the settling speed for each specimen and fill in your data table.
8. Plot the settling speed (cm/s) against particle diameter (mm) on a graph.

Analyze and Conclude

1. How do settling speeds change as particle sizes decrease?
2. How much faster does a 10-mm particle sink compared to a 1-mm particle?
3. How long would it take a 1-mm sand grain and a 0.001-mm clay particle to settle to the bottom of the ocean at a depth of 5 km?

seafloor probably are extinct volcanoes. Investigations of individual volcanoes on the seafloor have revealed that there are two types: seamounts and guyots. **Seamounts** are submerged basaltic volcanoes more than 1 km high. Many linear chains of seamounts are stretched out across the Pacific Ocean Basin in roughly the same direction. Guyots, also called tablemounts, are large, extinct, basaltic volcanoes with flat, submerged tops.

While extinct volcanoes on land erode within a few million years, this doesn't happen on the deep seafloor, because currents are generally too weak to erode solid rock and no other mechanisms of erosion exist. Once they are formed, seafloor structures persist practically forever. The only process that modifies them after they are formed is sedimentation; the oldest seamounts are covered with thick marine sediments.

Marine Sediments

The sediments that cover the ocean floor come from a variety of sources, but most come from the continents. Land-derived sediments include mud and sand washed into the oceans by rivers, as well as dust and volcanic ash blown over the ocean by winds. Much of the coarser material supplied by rivers settles out near shorelines or on beaches, but fine-grained material such as silt and clay settles so slowly through water that some tiny particles take centuries to reach the bottom. You will examine how quickly sediment settles in the *MiniLab* on this page.

Ocean currents disperse fine silt, clay, and volcanic ash throughout the ocean basins, and thus the dominant type of sediment on the deep ocean floor is fine-grained, deep-sea mud. Deep-sea mud usually has a reddish color because the iron present in some of the sediment grains becomes oxidized during their journey to the ocean bottom. Closer to land, the sediments become mixed with coarser materials such as sand, but some sandy sediments occasionally reach the abyssal plains in particularly strong turbidity currents.

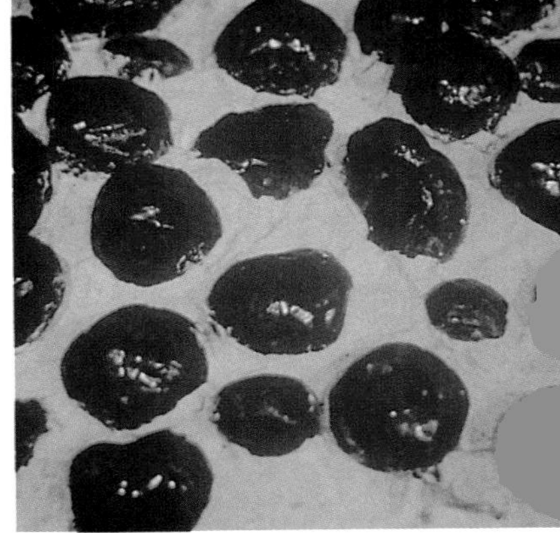

Figure 16-17 Scientists estimate that manganese nodules such as these cover 20 to 50 percent of the Pacific seafloor.

Ooze Another major source of deep-sea sediments is the shells and hard parts of marine organisms. You can find out more about one marine organism in the *Science & the Environment* feature at the end of this chapter. When these organisms die, their shells rain down on the ocean floor and accumulate there. Sediments containing a large percentage of particles derived from once-living organisms are called oozes. Most of these particles are small and consist of either calcium carbonate or silica.

The oozes and deep-sea muds of the deep ocean typically accumulate at a rate of only a few millimeters per thousand years. Although the wreck of the *Titanic* has been resting on the ocean bottom since 1912, it has acquired a fine dusting of sediments barely a fraction of a millimeter thick.

To learn more about the *Titanic,* go to the National Geographic Expedition on page 880.

Manganese Nodules Another type of sediment, manganese nodules consist of oxides of manganese, iron, copper, and other valuable metals that precipitated directly from seawater. Their growth rates are incredibly slow, and thus they are measured in millimeters per million years. Manganese nodules usually resemble potatoes of variable sizes, as shown in ***Figure 16-17,*** and cover huge areas of the seafloor.

Section Assessment

1. What is the relationship between crustal thickness and surface elevation?
2. How are submarine canyons formed?
3. Which sediment grains sink faster, pebbles or sand grains?
4. What is the difference in origin between deep-sea muds and oozes?
5. **Thinking Critically** If there is little volcanic activity on abyssal plains, yet they are dotted with thousands of seamounts, where did these extinct volcanoes come from?

Skill Review

6. **Concept Mapping** Use the following terms to complete the concept map below: the continental shelf, the continental rise, turbidity currents, submarine canyons. For more help, refer to the *Skill Handbook.*

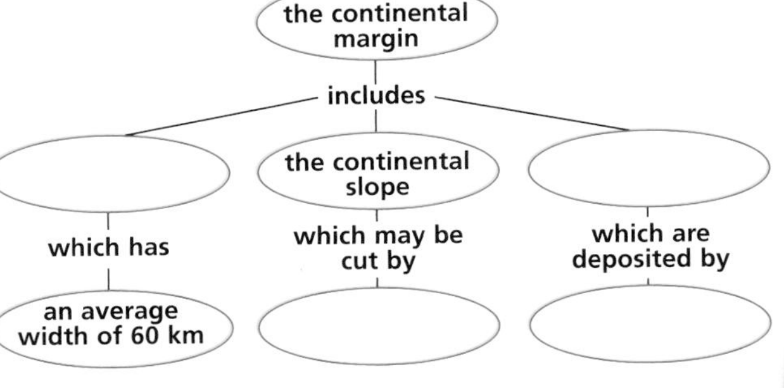

Mapping GeoLab

Identifying Coastal Landforms

Topographic maps of coastal areas show a two-dimensional representation of coastal landforms. You can identify an emergent coast by the landforms along the coastline as well as landforms found inland.

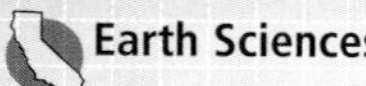

Earth Sciences

9.d Students know how to analyze published geologic hazard maps of California and know how to use the map's information to identify evidence of geologic events of the past and predict geologic changes in the future.

I&E 1.h

Preparation

Problem

How can you identify and describe the coastal landforms of an emergent coast on a topographic map?

Materials

metric ruler
drafting compass
pencil
graph paper
calculator

Procedure

1. Determine the map scale and the contour interval.
2. On the inset map, plot a west-east cross section of the coast just north of Islay Creek from the 60 ft depth contour to a point 5000 feet inland. Use a scale of 1:24 000 and a vertical exaggeration of 4.
3. Use both maps to answer the following questions.

Analyze

1. What kind of coastal landform is the Morro Rock Peninsula?
2. What kind of feature is Pillar Rock, and how was it formed?
3. On what coastal feature is Morro Bay State Park located? How was the feature formed?
4. What are the irregular sand hills in Morro Bay State Park?
5. What is the direction of the longshore transport along Morro Bay? Explain.
6. Your west-east cross section shows an elevated flat area next to the shoreline. What kind of coastal landform is this? How was it formed?
7. If sea level dropped 10 m, how would the shoreline change? How far would it move seaward? Would it become more regular or irregular? What would happen to Morro Bay?
8. If sea level rose 6 m, how would the coastal region change? Name three major changes.

Conclude & Apply

1. Is this portion of the California coast emergent or submergent? What features of this coastline provide evidence for your answer?

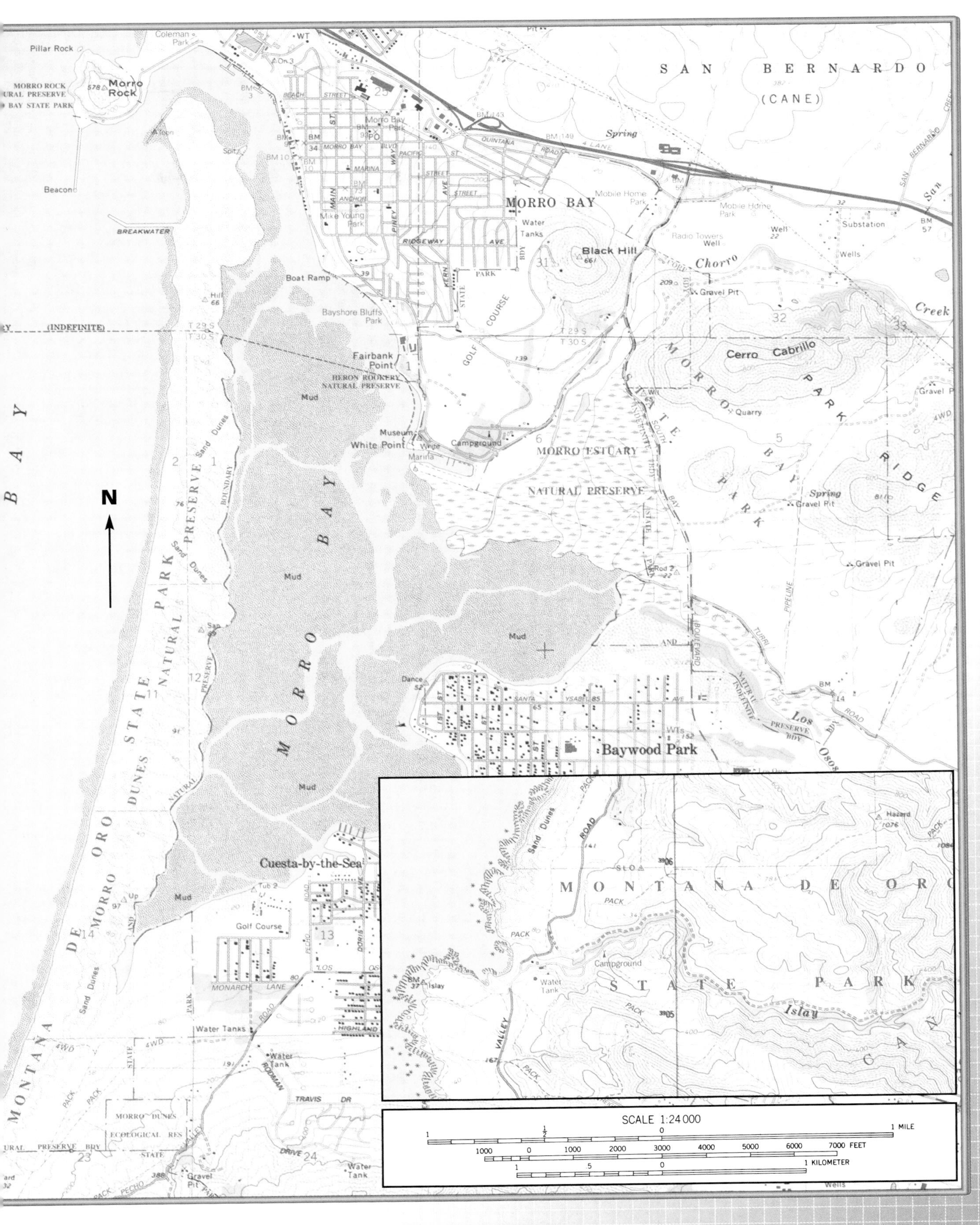

Morro Rock
MORRO BAY
Black Hill
Cerro Cabrillo
MORRO BAY STATE PARK
MORRO ESTUARY NATURAL PRESERVE
Baywood Park
Cuesta-by-the-Sea
MORRO DUNES NATURAL PRESERVE
MONTANA DE ORO STATE PARK
SAN BERNARDO (CANE)
Chorro Creek
Los Osos
N
SCALE 1:24 000
1 MILE
7000 FEET
1 KILOMETER

Science & the Environment

Deep Sea Dangler

Imagine the darkest darkness, the coldest cold, the heaviest pressure. Off in the distance, there is a light, a tiny pinprick of white. It acts as a beacon, drawing you forward. You approach the light with curiosity, when—suddenly—you are pulled forward into the belly of a giant beast. That is the experience of a tiny fish being eaten by a bigger fish deep beneath the ocean's surface.

The animal luring its prey out of the dark with its own light is an anglerfish. Light from the Sun only penetrates to a depth of about 200 m. Below that depth, organisms such as the anglerfish often create their own light.

A Fishing Fish

Like a human angler with a fishing rod, an anglerfish dangles bait in the water from a fin on its back that sticks out in front of the fish. The tip of the fin glows—which is an exciting sight in the darkness. Curious fishes are attracted to the light. Slowly, the anglerfish pulls the bait—and its dinner—closer and closer to its mouth. When it opens its huge jaws, water rushes into its mouth. The prey animal is caught up in the current and ends up in the anglerfish's stomach.

Food is scarce in the deep ocean, so fishes can't afford to let any food go by, even food that seems too big. Deep-sea fishes may have mouths lined with teeth that slant backwards so dinner can't get out once it is caught. Many of these fish also have expandable stomachs. Some deep-sea fishes scavenge for food near the surface at night, but spend the day in the ocean depths. These fishes are eaten by predators like the anglerfish, which stay below 900 m in depth all the time.

Anglerfish Life Cycle

Anglerfish lay eggs in the deep ocean. The eggs float to the ocean surface, where they hatch. The young drift downward until, by maturity, they are submerged in darkness. Male anglerfish are tiny; they attach to the side of the female. In the darkness, this is a way for these fish to make sure that they find a mate. The female is the only one that glows; thus, she does all the fishing and eating for both of them. The male gets food from her bloodstream.

Glowing From Within

Below 1800 m, every swimming animal glows in some way. Living things that glow are bioluminescent. Bioluminescence is a cold light produced by living things. Many bioluminescent fishes get their light from glowing bacteria that live in the fishes. In this symbiotic association, the bacteria get food from the fish, and the bacteria lure food for the fish.

Activity

Use library resources or go to earthgeu.com to research food chains that end with a deep-sea predator such as the anglerfish. Make a diagram of one of these food chains.

CHAPTER 16
Study Guide

Summary

SECTION 16.1

Shoreline Features

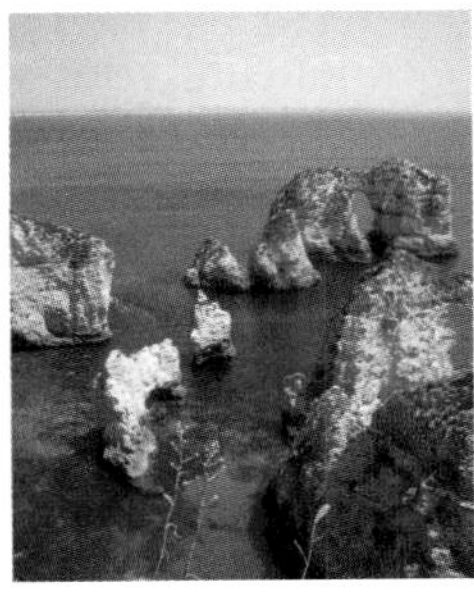

Main Ideas

- Wave erosion of headlands produces wave-cut platforms and cliffs, sea stacks, sea arches, and sea caves. Wave refraction concentrates breaker action on headlands.
- Beaches consist of loose sediment deposited along the shoreline. Wave action and longshore currents move sediment along the shore and build barrier islands and other depositional features. Artificial protective structures interfere with longshore transport.
- Sea levels in the past were 130 m lower than at present. When the land is rising, coasts are emergent and relatively straight.

Vocabulary

barrier island (p. 418)
beach (p. 415)
estuary (p. 415)
longshore bar (p. 416)
longshore current (p. 416)
wave refraction (p. 414)

SECTION 16.2

The Seafloor

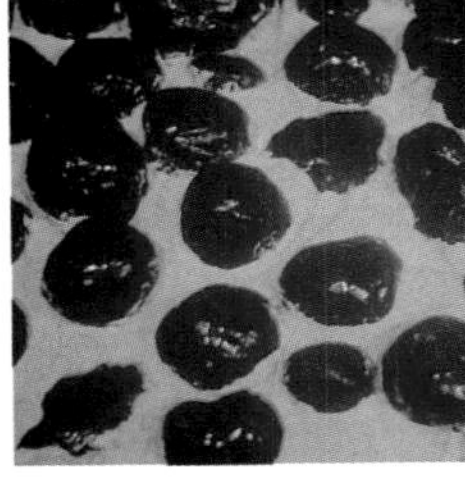

Main Ideas

- The oceans cover the thin oceanic crust and the lower parts of the thicker continental crust. The submerged part of a continent is the continental margin, the shallowest part of the ocean.
- A continental margin consists of the continental shelf, the continental slope, and the continental rise. Turbidity currents cut submarine canyons in the continental slopes and deposit their sediments in the form of continental rises.
- The flat part of the seafloor is the abyssal plain. Most deep-sea trenches are in the Pacific Ocean. Mid-ocean ridges extend through all ocean basins. Countless active and extinct volcanoes are on the mid-ocean ridges and deep seafloor.
- Most deep-sea sediments are fine-grained and accumulate slowly. Sediments may be derived from land or living organisms, or they may precipitate from seawater. Oozes are rich in sediment derived from organisms. Deep-sea muds are mostly derived from the land. Manganese nodules are precipitated from seawater.

Vocabulary

abyssal plain (p. 426)
continental margin (p. 422)
continental rise (p. 425)
continental shelf (p. 423)
continental slope (p. 424)
deep-sea trench (p. 426)
mid-ocean ridge (p. 426)
seamount (p. 428)
turbidity current (p. 424)

CHAPTER 16 Assessment

Understanding Main Ideas

1. Which coastal features are usually found in the bays along irregular coasts with headlands?
   a. sea stacks **c.** wave-cut platforms
   b. wave-cut cliffs **d.** beaches

2. Which of the following coastal landforms is NOT produced by longshore transport?
   a. a barrier island **c.** a baymouth bar
   b. a sand spit **d.** an estuary

3. What percentage of Earth's surface is below sea level?
   a. 10 percent **c.** 50 percent
   b. 30 percent **d.** 70 percent

4. What do the sediments of the abyssal plains mostly consist of?
   a. sand and gravel **c.** seashells
   b. oozes **d.** mud

5. Where are most deep-sea trenches located?
   a. in the Atlantic Ocean **c.** in the Pacific Ocean
   b. in the Indian Ocean **d.** in the Arctic Ocean

6. Which is the longest mountain system on Earth?
   a. the Hawaii-Emperor seamount chain
   b. the mid-ocean ridge system
   c. the Himalayas
   d. the Rocky Mountains

Use the map below to answer questions 7 and 8.

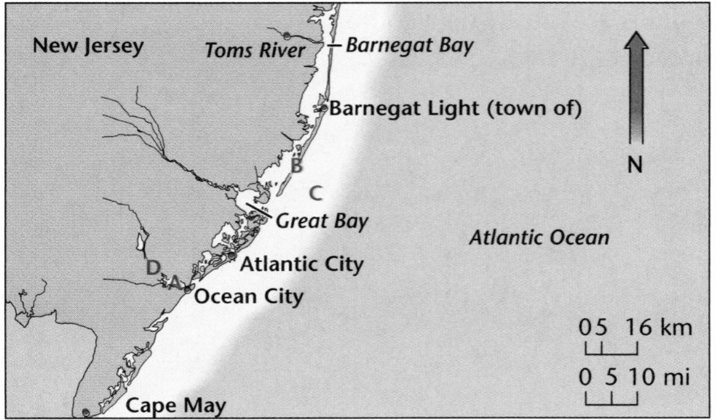

7. Which letter indicates the continental shelf?
   a. A **b.** B **c.** C **d.** D

8. Which feature is indicated by the letter A?
   a. a guyot **c.** a continental rise
   b. a continental slope **d.** an estuary

9. Under which circumstances do waves move faster?
   a. in shallow water
   b. over the longshore bar
   c. in deep water
   d. behind breakwaters

10. Which features are not caused by differential erosion?
    a. wave-cut platforms **c.** sea stacks
    b. barrier islands **d.** sea arches

11. Which marks the true edge of a continent?
    a. submarine canyon **c.** continental shelf
    b. continental slope **d.** abyssal plain

12. Which seafloor feature can be found along rifts in the mid-ocean ridges?
    a. hydrothermal vents **c.** deep sea trenches
    b. manganese nodules **d.** seamounts

Test-Taking Tip

PROCESS OF ELIMINATION On any multiple-choice test, there are two ways to find the correct answer to each question. You can either choose the right answer immediately or you can eliminate the answers that you know are wrong. It may be easier to find wrong answers than right ones. Find the ones you know are wrong and cross them out. You may be surprised at how few choices are left!

CHAPTER 16

Assessment

13. Which represents the flattest part of Earth's surface?
    a. deep-sea trenches
    b. continental margins
    c. abyssal plains
    d. mid-ocean ridges

14. Which features of the seafloor are cut by turbidity currents?
    a. longshore bars
    b. submarine canyons
    c. abyssal plains
    d. baymouth bars

15. Which is not associated with mid-ocean ridges?
    a. black smokers
    b. guyots
    c. fracture zones
    d. hydrothermal vents

Applying Main Ideas

16. Describe the effect that wave refraction has on incoming wave crests that approach the shore of a straight coast at an angle.
17. Explain how incoming waves along a shoreline create the longshore current.
18. Is global sea level currently rising, falling, or staying the same? During the last ice age, was the sea level higher, lower, or the same as at present? Explain.
19. Explain the relationship between oozes and the sedimentary rock known as chalk.
20. Why are the continental shelves considered part of the continents when they are presently covered by the oceans?

Thinking Critically

21. Is it possible to have submergent coasts when global sea level is falling? Explain your reasoning.
22. Why are submergent coasts more irregular than emergent coasts?
23. Why do geologists think that seamounts are extinct volcanoes?

Standardized Test Practice

Interpreting Scientific Illustrations

Use the illustration below to answer questions 1–5.

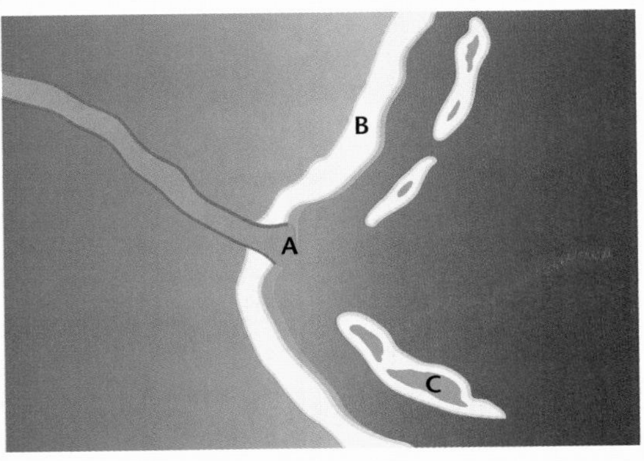

1. What shoreline feature is indicated by letter A?
   a. a barrier island
   b. a spit
   c. an estuary
   d. a bay

2. What type of water typically would be found in the area indicated by letter A?
   a. freshwater
   b. salt water
   c. brackish water
   d. very cold water

3. What shoreline feature is indicated by letter B?
   a. a tombolo
   b. a spit
   c. a lagoon
   d. a beach

4. What shoreline feature is indicated by letter C?
   a. a barrier island
   b. the longshore bar
   c. an estuary
   d. a beach

5. Which of these forms the area indicated by letter C?
   a. large storm waves
   b. gaps in the longshore bar
   c. the longshore current
   d. a rip current

For a **preview** of Earth's atmosphere and oceans, study this GeoDigest before you read the chapters. After you have studied these chapters, you can use the GeoDigest to **review** the unit.

The Atmosphere and the Oceans

Atmosphere

Atmospheric Basics Earth's atmosphere is made up primarily of nitrogen and oxygen. The atmosphere consists of several layers of different temperatures. The lowest layer, the troposphere, is the most important for weather. In this layer, temperatures generally decrease with altitude. Solar energy absorbed by Earth's surface is transferred through the atmosphere by radiation, conduction, and convection. The weight of air exerts a force called atmospheric pressure. In the troposphere, atmospheric pressure generally decreases with increasing altitude. Wind is the movement of air resulting from differences in pressure. Wind speed is affected by friction. Clouds form when warm, moist air is forced upward, expands, and cools. Clouds are classified according to the altitudes at which they form and according to their shapes. Precipitation such as rain, snow, sleet, and hail occurs when cloud droplets collide to form larger droplets. The water cycle is the continual movement of water between Earth's surface and the atmosphere through the processes of evaporation, condensation, and precipitation.

Lightning

Meteorology

Weather Basics Meteorology is the study of the atmosphere. Weather is the current state of the atmosphere. Climate is the average of weather patterns of an area over a long period of time. Differing amounts of solar radiation received by different parts of Earth lead to uneven heating; the motion of water and air balances this heat. The Coriolis effect deflects air and, with the heat imbalance, creates global wind systems. An air mass is a body of air that takes on the same characteristics as the land over which it formed. A front is the boundary between two air masses. High-pressure systems cause fair weather. Low-pressure systems cause precipitation and clouds.

Weather Analysis Accuracy and density of data are important in forecasting weather. Balloon-borne instruments called radiosondes collect data about temperature, pressure, humidity, wind speed, and wind direction from high in the atmosphere. Weather radar locates precipitation, while Doppler radar measures wind speed. Weather satellites use visible light and infrared imagery to record weather. Symbols are used with station models to record weather for a given place and time. Isobars and isotherms are used on weather maps to connect lines of equal pressure and equal temperature, respectively.

GeoDigest

Most modern forecasts use digital forecasting methods. Analog forecasting compares current and past weather patterns.

The Nature of Storms

Severe Weather Abundant moisture in the lower atmosphere, a mechanism to lift the moisture, and unstable air are necessary for a thunderstorm to form. The way in which the air rises—whether by the unequal heating of Earth's surface or by the push of an advancing front—determines the type of storm that develops. The three stages of a thunderstorm are the cumulus stage, the mature stage, and the dissipation stage. Thunderstorm hazards include lightning, violent winds, hail, floods, and tornadoes. The Fujita tornado intensity scale classifies tornadoes. The Saffir-Simpson hurricane scale rates hurricane intensity.

Climate

Climatic Basics Climate is the long-term weather pattern of a region and includes annual variations of temperature, precipitation, and wind. Data covering at least 30 years are averaged on a monthly or yearly basis to determine a region's normals. Latitude, topography, bodies of water, moisture, wind patterns, ocean currents, and air masses are factors that influence climate. The tropical latitudes are always warm. The temperate zones have moderate climates. Polar zones are always cold.

Time lapse photo of a hurricane

Climatic Changes Climatic changes are indicated by fossils, ice cores, and other evidence. Ice ages, or periods of extensive glacial coverage, are examples of long-term climatic changes. Seasons are examples of short-term climatic changes. Climatic changes may be caused by changes in solar activity, the tilt of Earth's axis, Earth's orbit, and volcanic eruptions. Some human activities may cause climatic change. Global warming may be caused by a rise in atmospheric carbon dioxide. The burning of fossil fuels and deforestation may contribute to global warming, but scientists are still investigating its causes.

Vital Statistics

Deadliest Hurricanes of the Twentieth Century (Western Hemisphere)

Location	Year	Lives Lost
Hurricane Mitch, Central America	1998	11 000
Galveston, Texas	1900	8000
Hurricane Fifi, Honduras	1974	8000
Dominican Republic	1930	8000
Hurricane Flora, Haiti and Cuba	1963	7200

GeoDigest

British Virgin Islands

Physical Oceanography

The Oceans The first oceans probably formed more than 4 billion years ago. The water may have come from impacting comets or from volcanic eruptions which released water from Earth's interior. About 71 percent of Earth's surface is covered by oceans. Seawater is 96.5 percent water and 3.5 percent dissolved salts, which are added and removed from the ocean at the same rate. Seawater density changes with temperature and salinity. Water temperature decreases with depth, forming three layers. Ocean waves are generated by wind. Water in a wave moves in a circular motion but does not move forward. When waves reach shallow water they become breakers. Tides are caused by the gravitational attraction of the Sun and the Moon. Density currents are deep currents caused by differences in temperature and salinity. Upwelling occurs when winds push surface water aside and it is replaced by cold, deep water.

The Marine Environment

Shoreline and Seafloor Wave erosion and refraction shape shorelines. Beaches are made of loose sediment deposited along shorelines. Longshore currents move sediment along the shore, building barrier islands and other depositional features. Sea level has changed over time. The submerged part of a continent is the continental margin. The oceans cover thinner, oceanic crust. The flat part of the seafloor is the abyssal plains. Deep-sea trenches are found mainly in the Pacific Ocean. Mid-ocean ridges extend through all ocean basins. Active and inactive volcanoes are found on mid-ocean ridges and the seafloor. Deep-sea sediments are fine-grained and accumulate slowly. They may be derived from land or living organisms or may precipitate out of seawater.

FOCUS ON CAREERS

Mariculturist

Mariculturists farm marine organisms such as salmon, oysters, mussels, clams, scallops, shrimp, crayfish, prawns, and seaweed. Marine organisms may be sold as food or they may be raised for research purposes. Many mariculturists have a degree in fisheries. Some mariculturists receive advanced training in field techniques and biology of the species that they plan to raise.

GeoDigest

ASSESSMENT

Understanding Main Ideas

1. What forms when warm, moist air is forced upward, expands, and cools?
   a. clouds
   b. density currents
   c. atmospheric pressure
   d. altitude

2. What happens to atmospheric pressure in the troposphere with increasing altitude?
   a. It gets thicker. **c.** It increases.
   b. It gets warmer. **d.** It decreases.

3. What do we call the average weather patterns for an area over a long period of time?
   a. the Coriolis effect **c.** climate
   b. meteorology **d.** pressure systems

4. What kind of imagery do satellites use to record weather from space?
   a. analog **c.** symbolic
   b. infrared **d.** directional

5. What helps shape the shoreline?
   a. waves
   b. reflection
   c. atmospheric pressure
   d. solar energy

6. Which scale is used to classify tornadoes?
   a. the Fujita intensity scale
   b. the Saffir-Simpson scale
   c. the Richter scale
   d. the heat index

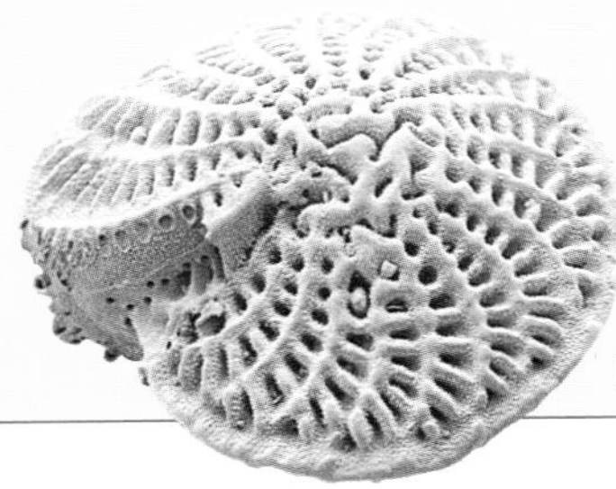

Foraminifera

7. How many years of weather are usually averaged together to determine a region's normals?
   a. 3 **c.** 300
   b. 30 **d.** 3000

8. Which of the following provides evidence of climatic change?
   a. deforestation **c.** tilt of Earth's orbit
   b. fossil fuels **d.** ice cores

9. What percentage of Earth is covered by water?
   a. 3.5 percent **c.** 35 percent
   b. 7.1 percent **d.** 71 percent

10. What is the flat part of the seafloor called?
    a. abyssal plain
    b. deep-sea trench
    c. continental margin
    d. mid-ocean ridge

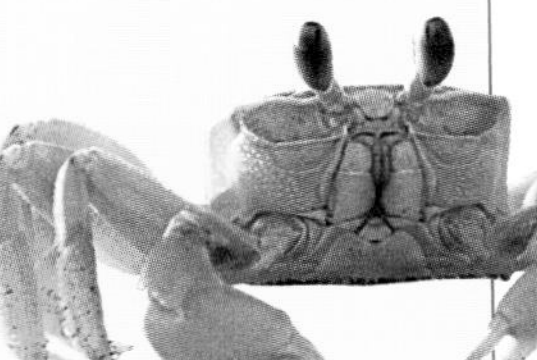

Fiddler crab

Thinking Critically

1. If a meteorologist wanted to know how fast a storm was moving, would he or she use traditional weather radar or Doppler radar? Why?
2. You watch a wave approach the shore. Is the water that breaks onshore the same water that you observed offshore? Explain your answer.
3. Suppose you collected a sample of deep-sea sediments. How could you determine the derivation of these sediments?

Unit 5

The Dynamic Earth

The mountain shown in the photograph is Mount Fuji, an inactive volcano in Japan that towers above the surrounding landscape. Its snow-covered peak rises more than 3700 m above sea level to overlook five, interconnected lakes. How did this magnificent structure form? Earth's outermost layers are broken into enormous slabs that shift slowly across the planet's surface. The movements, which occur at rates of only a few centimeters per year, cause mountains to form, volcanoes to erupt, and earthquakes to shake our planet. In this unit, you will explore how and why these slabs of rock move to change Earth.

Unit Contents

Go to the **National Geographic Expedition** on page 886 to learn more about topics that are connected to this unit.

Mount Fuji, Japan

Chapter 17

Plate Tectonics

What You'll Learn

- Why continental drift was not accepted when it was first proposed and what kinds of evidence led to its acceptance.
- How Earth's tectonic plates interact.
- What causes tectonic plates to move.

Why It's Important

Understanding the theory of plate tectonics is important because interactions between the enormous slabs of Earth's crust and rigid upper mantle result in the formation of many of Earth's surface features, including volcanoes and some mountain ranges. Many earthquakes are also caused by plate movements.

To learn more about plate tectonics, visit the Earth Science Web Site at earthgeu.com

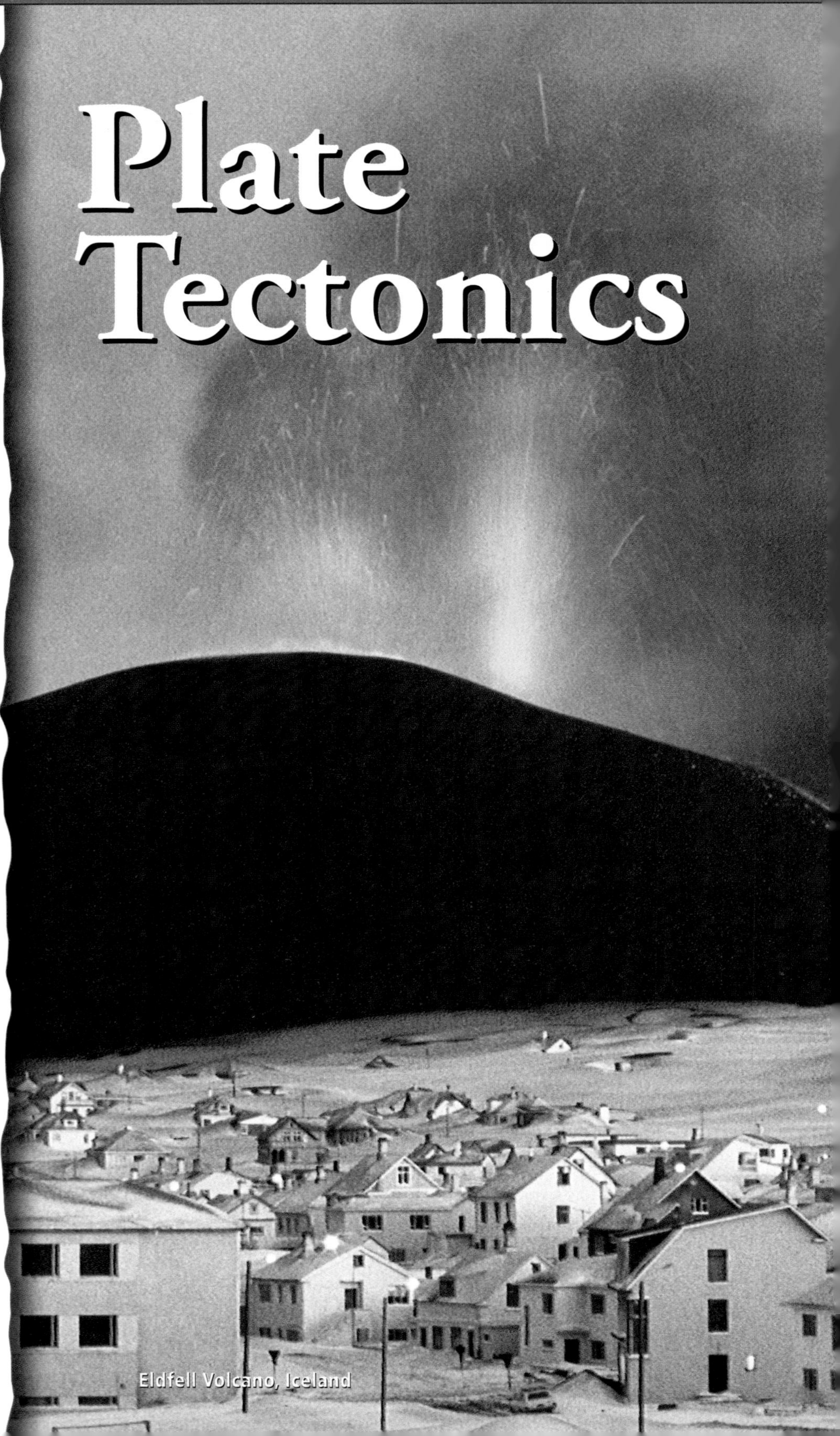

Eldfell Volcano, Iceland

Discovery Lab

Coming Apart at the Seams?

Southwestern California is separated from the rest of the state by a system of cracks along which movement takes place. These cracks are called faults. One of these cracks, as you may already know, is the San Andreas Fault. Movement along this fault is sending southwestern California to the northwest in relation to the rest of North America at a rate of about 5 cm/y.

1. Use a metric ruler and the map scale to determine the actual distance between San Francisco and Los Angeles.
2. At the current rate of movement, when will these two cities be next-door neighbors?

Infer Infer what might be causing these large pieces of land to move. Describe your inference in your science journal.

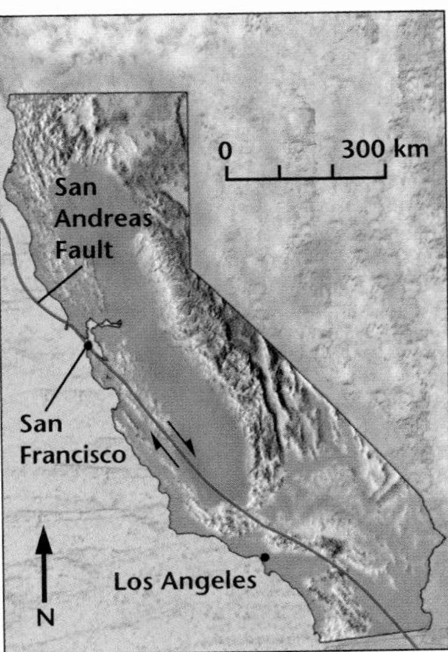

SECTION 17.1 *Drifting Continents*

I&E 1.k, 1.n

OBJECTIVES

- **Describe** *one piece of early evidence that led people to suggest that Earth's continents may have once been joined.*
- **Discuss** *evidence of continental drift.*
- **Explain** *why continental drift was not accepted when it was first proposed.*

VOCABULARY

continental drift
Pangaea

With the exception of events such as earthquakes, volcanic eruptions, and landslides, Earth's surface appears to remain relatively unchanged during the course of an average human lifetime. On the geologic time scale, however, Earth's surface is changing at rates almost too great to imagine! South America is moving away from Africa at a rate of 2 to 3 cm/y. The volcanic islands that make up Hawaii are migrating toward the northwest at a rate of 8 to 9 cm/y. And Mt. Everest, Earth's highest point, is slowly rising. What could be causing such enormous pieces of land to move?

EARLY OBSERVATIONS

Some of the first people to consider the idea of moving landmasses were early mapmakers. In the late 1500s, Abraham Ortelius, a Dutch mapmaker, noticed the apparent fit of continents on either side of the Atlantic Ocean. He (incorrectly) proposed that North and South America had been separated from Europe and Africa by earthquakes and floods. Over the next 300 years, many people noticed and commented on the matching coastlines. In the late 1800s, Eduard Suess, an Austrian geologist, hypothesized that the present southern continents had once been joined as a single landmass that he named Gondwanaland. The first time that the idea of moving continents was proposed as a serious scientific hypothesis, however, occurred in the early 1900s. In 1912, a German scientist named Alfred Wegener presented his ideas about continental movement to the scientific community.

Continental Drift

Wegener called his hypothesis **continental drift,** which proposed that Earth's continents had once been joined as a single landmass. He called this supercontinent **Pangaea,** a Greek word that means "all the earth." Wegener proposed that Pangaea began to break apart about 200 million years ago. Since that time, he reasoned, the continents had continued to slowly move to their present positions, as shown in ***Figure 17-1.***

Wegener was one of the first supporters of the concept of drifting continents to base his hypothesis on more than just the puzzlelike fit of continental coastlines on either side of the Atlantic Ocean. For Wegener, these gigantic puzzle pieces were just the beginning. He also collected and organized rock, fossil, and climatic data to support his hypothesis.

Evidence from Rock Formations Wegener reasoned that when Pangaea began to break apart, large geologic structures, such as mountain ranges, would have fractured as the continents separated. Using this reasoning, Wegener hypothesized that there should be areas of similar rock types on opposite sides of the Atlantic Ocean. He observed that some of the rocks of the Appalachian Mountains in the United States shared similar features with rocks in Greenland and Europe. These similar groups of rocks, all older than 200 million years, supported Wegener's idea that the continents had once been joined. What other groups of rocks shown in ***Figure 17-2*** suggest the existence of a single landmass in the past?

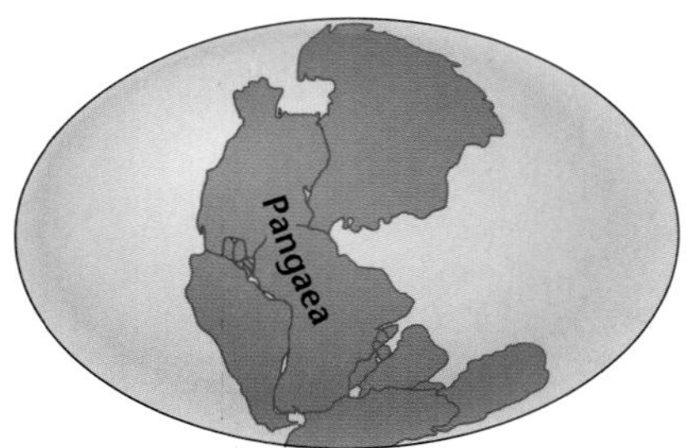

200 million years ago

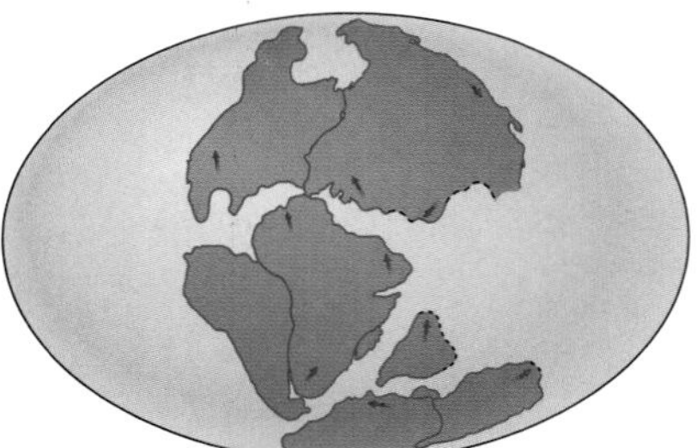

180 million years ago

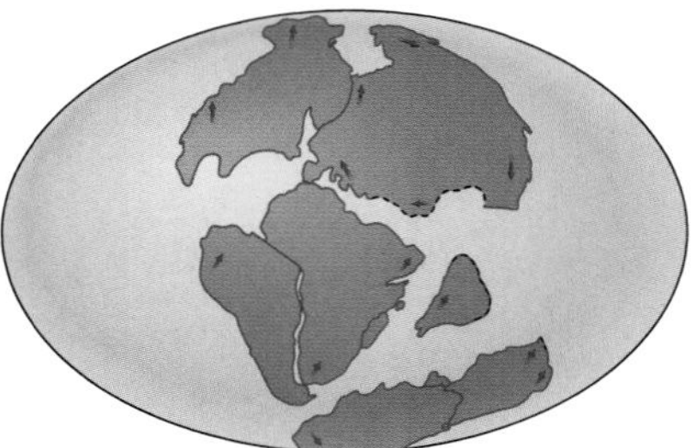

135 million years ago

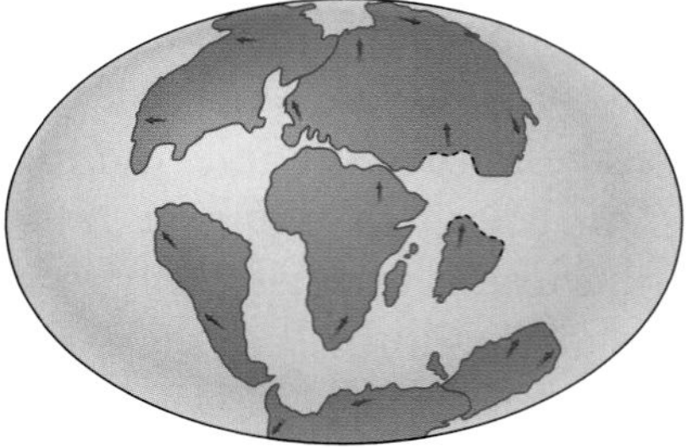

65 million years ago

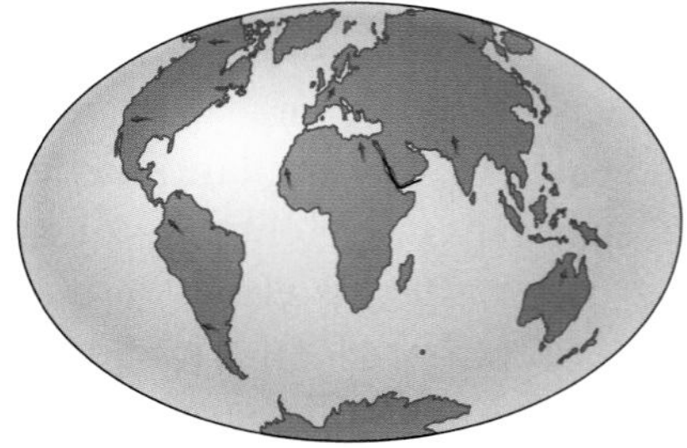

Present

Figure 17-1 Pangaea was an ancient landmass that was made up of Earth's continents. Pangaea began to break apart about 200 million years ago.

Evidence from Fossils Wegener also gathered evidence of the existence of Pangaea from fossils. Similar fossils of several different animals and plants that once lived on land had been found on widely separated continents, as shown in ***Figure 17-2.*** Wegener reasoned that the land-dwelling animals, such as *Kannemeyerid* and *Labyrinthodont,* could not have swum the great distances that now exist between continents. Wegener also argued that because fossils of *Mesosaurus,* an aquatic reptile, had been found only in freshwater rocks, it was unlikely that this species could have crossed the oceans. The ages of these different fossils also predated Wegener's time frame for the breakup of Pangaea, which gave him the confidence he needed to move forward with his hypothesis.

Another fossil that Wegener used to support his hypothesis of continental drift was *Glossopteris,* a seed fern that resembled low shrubs. Fossils of this plant had been found on many continents, which today have very different climates. Because he had a strong background in meteorology, Wegener was able to reason that the area separating these fossil finds was too large to have had a single climate. This led him to conclude that the rocks containing these fossil ferns had once been joined. Wegener also argued that because *Glossopteris* grew in temperate climates, the places where these fossils had been found were once closer to the equator.

 **I&E**

1.k Recognize the cumulative nature of scientific evidence.

1.n Know that when an observation does not agree with an accepted scientific theory, the observation is sometimes mistaken or fraudulent (e.g., the Piltdown Man fossil or unidentified flying objects) and that the theory is sometimes wrong (e.g., the Ptolemaic model of the movement of the Sun, Moon, and planets).

Figure 17-2 Alfred Wegener used fossils and the similarity of rock formations on opposite sides of the Atlantic Ocean as evidence that Earth's continents were once joined.

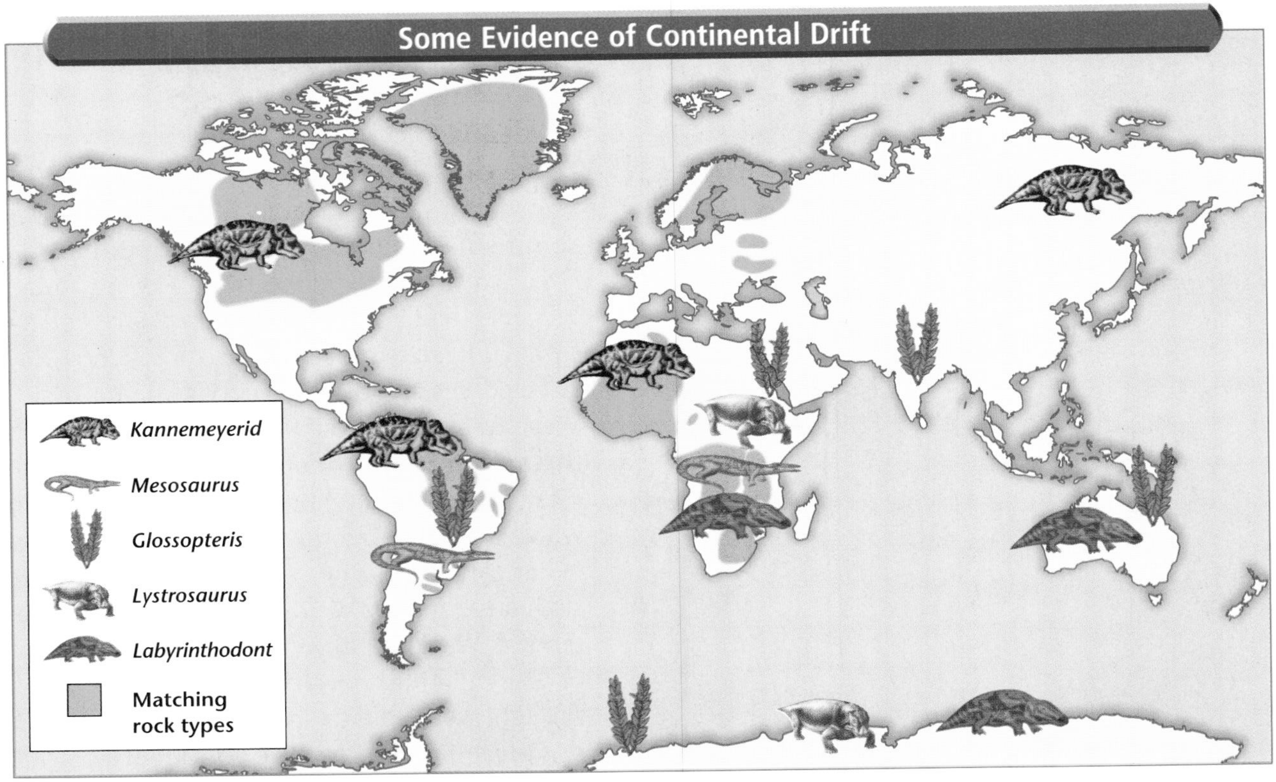

Some Evidence of Continental Drift

Kannemeyerid
Mesosaurus
Glossopteris
Lystrosaurus
Labyrinthodont
Matching rock types

Ancient Climatic Evidence Recall from Chapter 6 that sedimentary rocks provide clues to past environments and climates. By studying various sedimentary rocks, Wegener found evidence of vast climatic changes on some continents. Coal deposits, for example, had been found in Antarctica. Coal forms from dead swamp plants. Swamps are areas of wet, spongy land often covered by water. The existence of coal beds in Antarctica, then, indicated that this frozen land once had a temperate, rainy climate. Wegener used this evidence to conclude that Antarctica must have been closer to the equator sometime in the geologic past.

Another piece of climatic evidence came from glacial deposits found in Africa, India, Australia, and South America. The presence of these 290-million-year-old deposits, some of which are shown in ***Figure 17-3,*** suggested to Wegener that these areas had once been covered by thick ice caps. Because continental glaciers do not presently exist on these continents, Wegener proposed that they once were located near the south pole, as shown in ***Figure 17-4,*** before Pangaea began to fracture. Furthermore, because the south pole is not presently located near Africa or India, Wegener suggested two possibilities to explain the glacial deposits. The first was that the pole had shifted its position, and the second was that these landmasses had

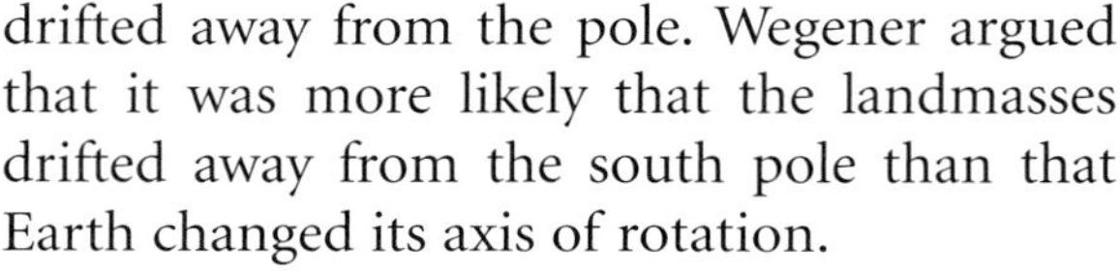

drifted away from the pole. Wegener argued that it was more likely that the landmasses drifted away from the south pole than that Earth changed its axis of rotation.

Figure 17-3 These glacial deposits in Africa helped support Wegener's hypothesis of continental drift.

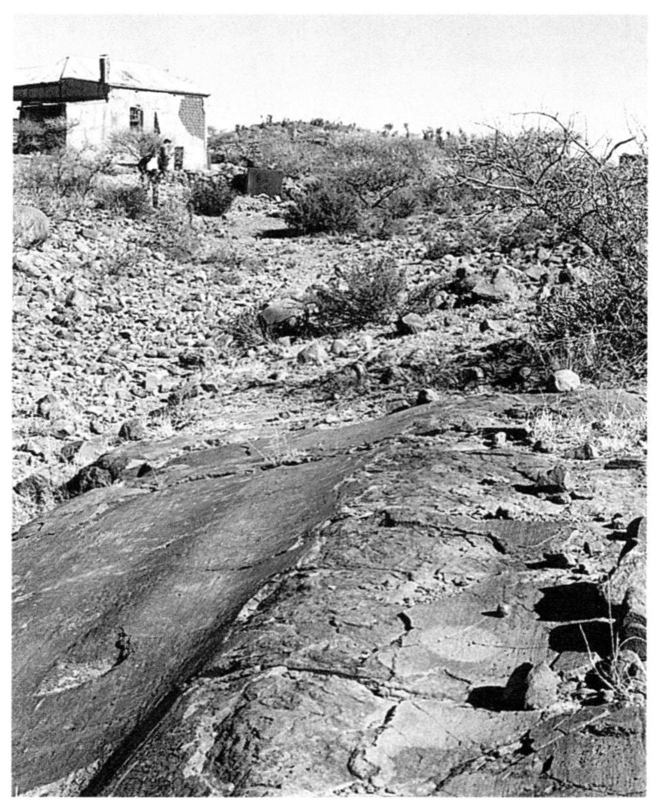

A Rejected Hypothesis

In the early 1900s, many people in the scientific community strongly believed that the continents and ocean basins were permanent, fixed features of Earth's surface. Even though Wegener had compiled an impressive collection of data, most scientists of his day rejected his hypothesis of continental drift. Wegener's hypothesis had two major flaws that prevented it from being widely accepted. First, Wegener could not satisfactorily explain what was causing the continents to move. What force could be strong enough to move such large masses of rock over such great distances? Wegener suggested that the rotation of Earth could be responsible. Physicists, though, were able to show that this force was not great enough to move continents.

Scientists also had questions about how the continents were moving. Wegener had proposed that the continents were plowing through a stationary ocean floor. His peers, however, argued that continents could not push through the ocean floor without fracturing, because crustal rock is too brittle. Also, because no evidence for such fracturing had been found, geologists argued that the continents could not be moving as Wegener had proposed. These two unanswered questions—what forces could move continents and how continents could move without shattering—were the main reasons that the hypothesis of continental drift was rejected when it was first proposed.

Alfred Wegener did not give up when his hypothesis was not accepted by most other scientists of his time. He continued to search for evidence to support his hypothesis of continental drift. Wegener died in 1930 on an expedition to Greenland, but the controversy over his hypothesis remained alive for several decades after his death. It wasn't until the early 1960s that new evidence revealed a process that could explain why and how the continents move. This evidence, which was found on the seafloor, will be discussed in the next section.

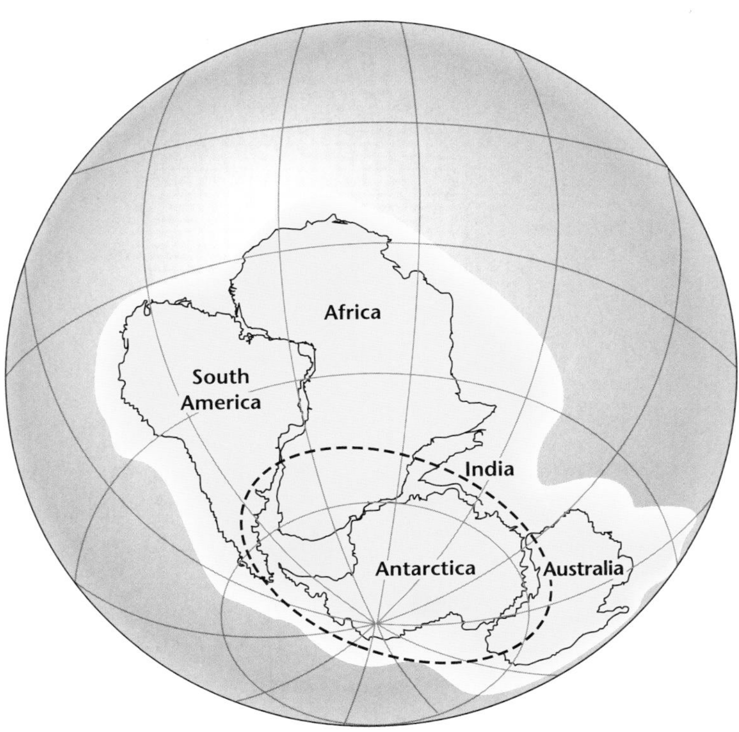

Figure 17-4 Glacial deposits nearly 300 million years old on several continents led Wegener to propose that these landmasses may have once been joined and covered with ice. The extent of the ice is shown by the dashed line.

SECTION ASSESSMENT

1. What early evidence suggested that Earth's continents might be moving?
2. How do ancient glacial deposits in Africa, India, Australia, and South America support the idea of continental drift?
3. How did Alfred Wegener use rock and fossil evidence to support his hypothesis?
4. Why was Wegener's hypothesis rejected by most scientists of the early 1900s?
5. **Thinking Critically** Oil deposits approximately 200 million years old have been discovered in Brazil. Where might geologists find oil deposits of a similar age? Explain.

SKILL REVIEW

6. **Comparing and Contrasting** Compare and contrast the different types of evidence used by Wegener to support his idea of continental drift. Which type do you think provides the strongest support for his hypothesis? Why? For more help, refer to the *Skill Handbook*.

earthgeu.com/self_check_quiz

SECTION 17.2 Seafloor Spreading

Earth Sciences 3.a I&E 1.k

OBJECTIVES

- **Summarize** *the evidence that led to the discovery of seafloor spreading.*
- **Explain** *the significance of magnetic patterns on the seafloor.*
- **Explain** *the process of seafloor spreading.*

VOCABULARY

magnetometer
paleomagnetism
magnetic reversal
isochron
seafloor spreading

Until the mid-1900s, most people, including many scientists, thought that the ocean floor, unlike the continents, was essentially flat. Many people also had the misconceptions that oceanic crust was unchanging and was much older than continental crust. Advances in technology during the 1940s and 1950s, however, proved all of these widely accepted ideas to be wrong.

HELP FROM TECHNOLOGY

One advance that allowed scientists to study the ocean floor in great detail was the development of echo-sounding methods. One type of echo sounding is sonar, which uses sound waves to measure water depth. As shown in ***Figure 17-5,*** regular pulses of sound are sent out from a device aboard a ship. The sound waves travel through the water and are reflected from the ocean floor. The time it takes for these waves to travel from the device and back to a receiver can be used to calculate the distance from the sonar device to the ocean floor. Look again at ***Figure 17-5.*** How would the travel times of waves differ at various points along this hypothetical ocean bottom?

Another technological advance that was used to study the ocean floor was the magnetometer. A **magnetometer** is a device that can detect small changes in magnetic fields. Magnetometers towed by ships record the magnetic field strength in the rocks that make up the ocean floor. The measurements are then used to construct magnetic maps of the seafloor. You'll learn more about magnetism and how it supports the hypothesis of continental drift later in this section.

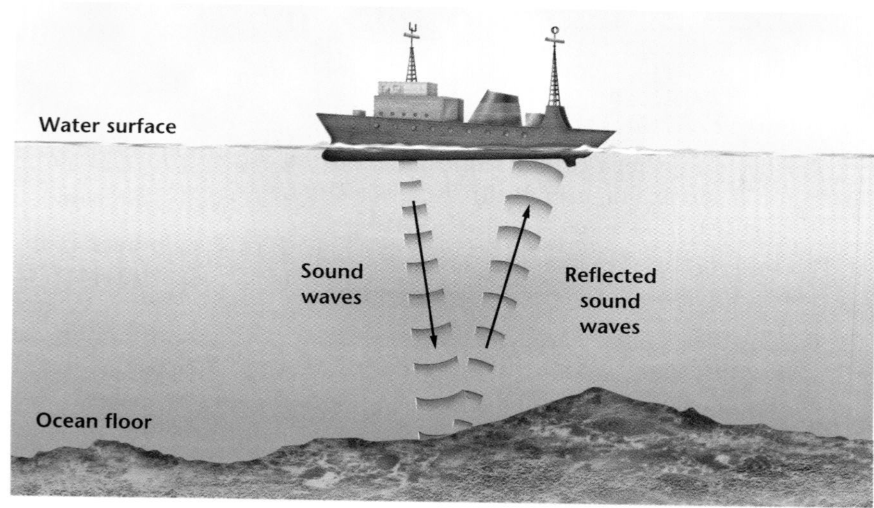

Figure 17-5 Sonar is an echo-sounding technique that has been used to map ocean-floor topography. The travel times of the sound waves can be used to calculate the distance to the ocean floor.

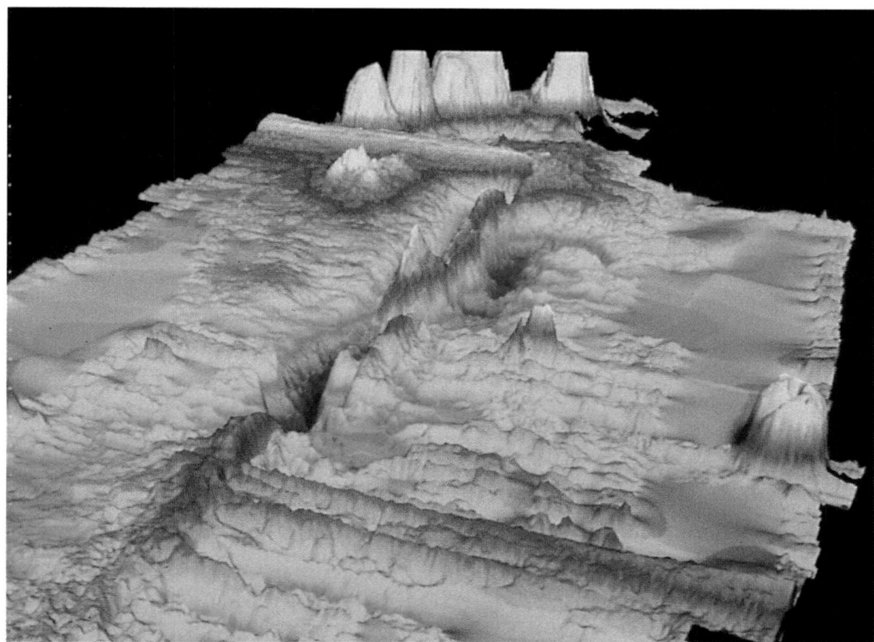

Figure 17-6 This is a false-color photo of a part of the East Pacific Rise. The red areas are areas of volcanism. Earthquakes are common in the areas colored blue.

Ocean Floor Topography

The maps made from the data collected by sonar and magnetometers surprised many scientists. Vast, underwater mountain chains called ocean ridges were discovered. These features of the ocean floor form the longest continuous mountain range on Earth. When they were first discovered, these ridges generated much discussion simply because of their size. Later, scientists discovered that earthquakes and volcanism are common along the ridges. A false-color photo of part of the East Pacific Rise, which is a ridge in the Pacific Ocean, is shown in ***Figure 17-6.***

Maps generated with sonar data also revealed that these underwater mountain chains had counterparts called deep-sea trenches. Recall from Chapter 16 that a deep-sea trench is a narrow, elongated depression in the seafloor with very steep sides. Trenches can be thousand of kilometers long and are extremely deep. The deepest trench is in the Pacific Ocean. This chasm, called the Mariana Trench, is just over 11 km deep!

These two topographic features of the ocean floor—deep-sea trenches and ocean ridges—puzzled geologists for over a decade after their discovery. What could have formed an underwater mountain range that extended around Earth? What is the source of the volcanism that is associated with these mountains? What kinds of forces could depress Earth's crust far enough to create trenches nearly six times as deep as the Grand Canyon? You'll find out the answers to these questions later in this chapter.

 **Earth Sciences**

3.a Students know features of the ocean floor (magnetic patterns, age, and sea-floor topography) provide evidence of plate tectonics.

I&E 1.k Recognize the cumulative nature of scientific evidence.

Ocean Rocks and Sediments

In addition to making maps, scientists collected samples of deep-sea sediments and the underlying crust to try to better understand Earth's ocean floors. Analysis of the rocks and sediments produced two important discoveries. First, the ages of the rocks that make up the seafloor vary in different places, and these variations change in a predictable way. Rock samples taken from areas near ocean ridges were younger than samples taken from areas near deep-sea trenches. Detailed analysis showed that the age of oceanic crust consistently increases with distance from a ridge, shown in ***Figure 17-7.*** Scientists also discovered from the rock samples that the oldest part of the seafloor is geologically young at about 180 million years old. Why are ocean-floor rocks so young compared to continental rocks, some of which are 3.8 billion years old? Geologists knew that oceans had existed for more than 180 million years. Where, then, they wondered, is the ocean crust from those earlier oceans?

The second discovery involved the sediments that are deposited on the ocean floor. Measurements showed that the thickness of ocean-floor sediment is, in general, much less than expected. Ocean-floor sediments are typically a few hundred meters thick. Large areas of continents, on the other hand, are blanketed with sedimentary rocks from a few kilometers to about 20 kilometers thick. Scientists knew that erosion and deposition were at work in Earth's oceans. Why, then, they asked, aren't seafloor sediments as thick as their continental counterparts? Could the relatively thin blanket of ocean sediments be a result of the age of the ocean crust? Careful observations of ocean-floor sediments also revealed that the thickness of the sediments increases with distance from an ocean ridge, as shown in ***Figure 17-7.***

Topic: Ocean Floor Depths
To learn more about Earth's ocean floors, visit the Earth Science Web Site at earthgeu.com

Activity: Research the deepest places in the ocean. Draw a graph comparing the five deepest places on the ocean floor.

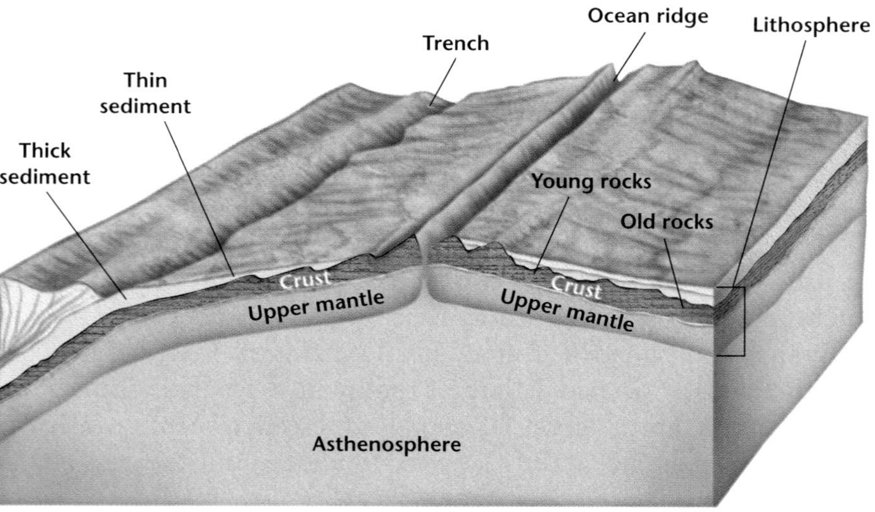

Figure 17-7 The ages of ocean crust and the thicknesses of ocean-floor sediments increase with distance from an ocean ridge.

MAGNETISM

Prior to their studies of the seafloor, scientists knew that rocks containing iron-bearing minerals provided a record of Earth's magnetic field. The study of this magnetic record is called **paleomagnetism.** Basalt, because it is rich in iron-bearing minerals, provides an accurate record of ancient magnetism. As basaltic lava cools, the iron-bearing minerals become oriented parallel to Earth's magnetic field. These minerals are, in effect, compass needles. When the lava hardens, the magnetic orientation is locked in place and provides a record of Earth's magnetic field at that time.

The Geomagnetic Time Scale Studies of continental basalt flows in the early 1960s revealed a pattern of magnetic reversals over geologic time. A **magnetic reversal** is a change in Earth's magnetic field. A magnetic field that has the same orientation as Earth's present field is said to have normal polarity. A magnetic field that is opposite to the present field has reversed polarity. The data gathered from continental basalt flows allowed scientists to construct the geomagnetic time scale, shown in ***Figure 17-8.***

To find out how the continental basalt-flow data compared with the basalts that make up the ocean floor, scientists proposed that magnetometers be towed behind ships to measure the magnetic field of the ocean floor. A very interesting pattern emerged. In some places, the magnetic field strength was greater than normal. In other places, the field strength was lower than normal. In places where the magnetic readings of the ocean floor matched Earth's present field, the two fields combined. This produced a stronger-than-normal reading (+). In places where the magnetic data were reversed in relation to Earth's present magnetic field, the two fields partially cancelled one another. In these cases, a lower-than-normal reading (−) was recorded, as shown in ***Figure 17-9.***

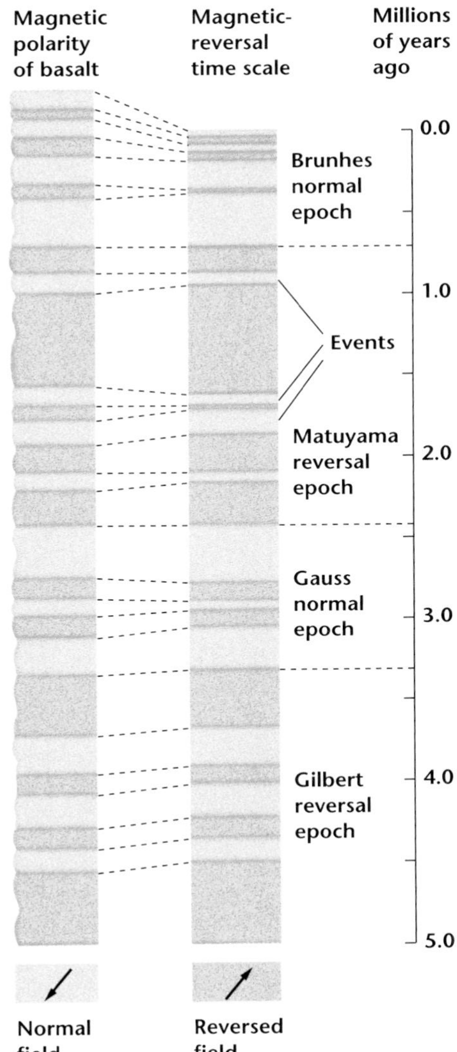

Figure 17-8 Continental basalt flows were used to construct the geomagnetic time scale. Long term changes in Earth's magnetic field are called epochs. Short term changes are called events.

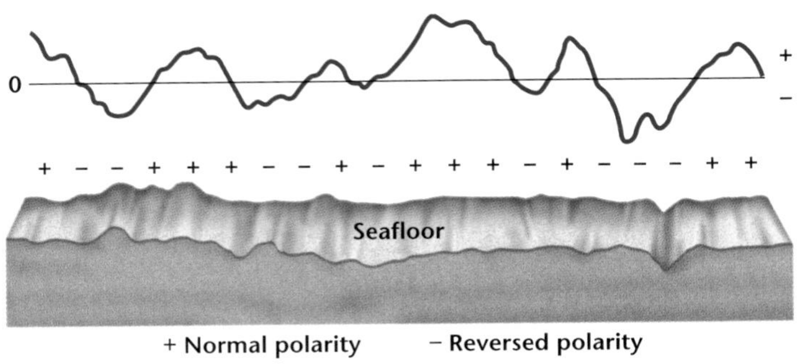

Figure 17-9 A stronger-than-normal magnetic reading (+) indicates a normal field. A weaker-than-usual reading (−) indicates a field with reversed polarity.

Figure 17-10 Reversals in the polarity of Earth's magnetic field are recorded in the rocks that make up the ocean floor. ***What is the polarity of the rocks closest to the ridge?***

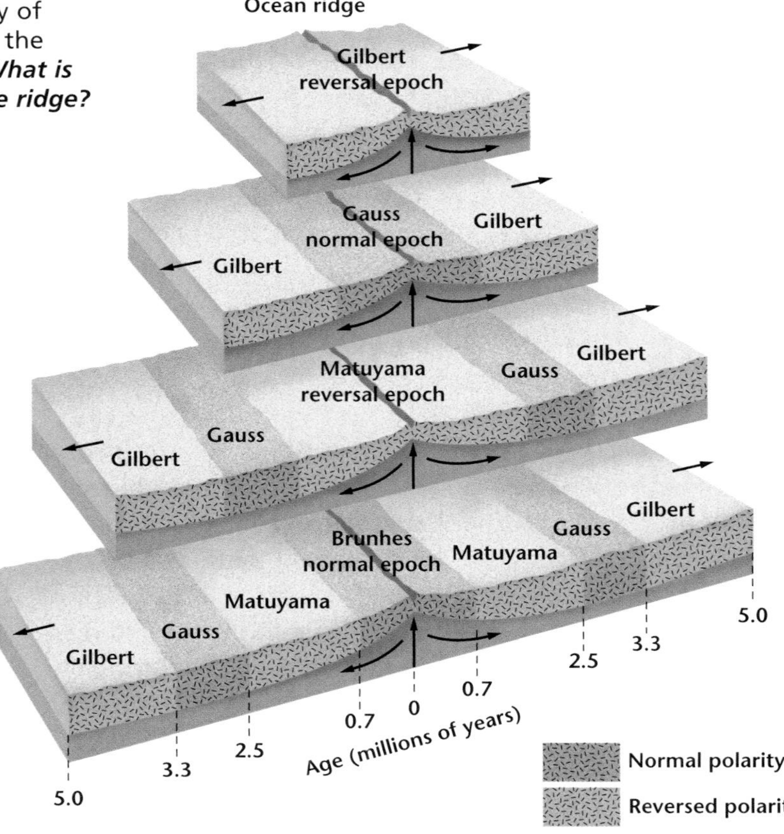

Using Numbers
Refer to ***Figure 17-8.*** What is the average time interval between reversals? Over the last 4.5 million years, how many years has the field experienced reversed polarity? Normal polarity? What is the percentage of the total time spent in each orientation?

Magnetic Symmetry As more data were collected, scientists noticed that the positive and negative areas formed a series of stripes that were parallel to ocean ridges. Compare the magnetic pattern on opposite sides of the ocean ridge shown in ***Figure 17-10.*** How would you describe this pattern? The magnetic pattern on one side of the ridge is a mirror image of the pattern on the other side of the ridge.

As scientists sought to explain these parallel stripes, they made another important discovery. The magnetic data collected from the ocean floor, shown in ***Figure 17-10,*** matched the pattern of magnetic reversals that had been found in basalt flows on land, which is shown in ***Figure 17-8.*** By matching the patterns on the seafloor with the known pattern of reversals on land, scientists were able to determine the age of the ocean floor from a magnetic recording. This method enabled scientists to quickly create isochron maps of the ocean floor. An **isochron** is a line on a map that connects points that have the same age. In the isochron map shown in ***Figure 17-11,*** note that relatively young ocean-floor crust is near ocean ridges, while older ocean crust is found along deep-sea trenches. You will generate and analyze an isochron map of a part of the seafloor in the *Mapping GeoLab* at the end of this chapter.

Seafloor Spreading

Once scientists had compiled all the topographic, sedimentary, age, and magnetic data from the seafloor, an American scientist named Harry Hess proposed a theory that could explain their observations. This theory, called **seafloor spreading,** states that new ocean crust is formed at ocean ridges and destroyed at deep-sea trenches. What actually happens during seafloor spreading? Magma, because it is hotter and less dense than surrounding mantle material, is forced toward the crust along an ocean ridge and fills the gap that is created as shown in ***Figure 17-12A,*** on page 454. When the magma hardens, a small amount of new ocean floor is added to Earth's surface, as illustrated in ***Figure 17-12B.*** As spreading along a ridge continues, more magma is forced upward and hardens. Each cycle of spreading and the intrusion of magma results in the formation of another small section of ocean floor, which slowly moves away from the ridge, as shown in ***Figure 17-12C.***

Figure 17-11 Each colored band on this isochron map of the ocean floor represents the age of that strip of the crust. ***What pattern do you observe?***

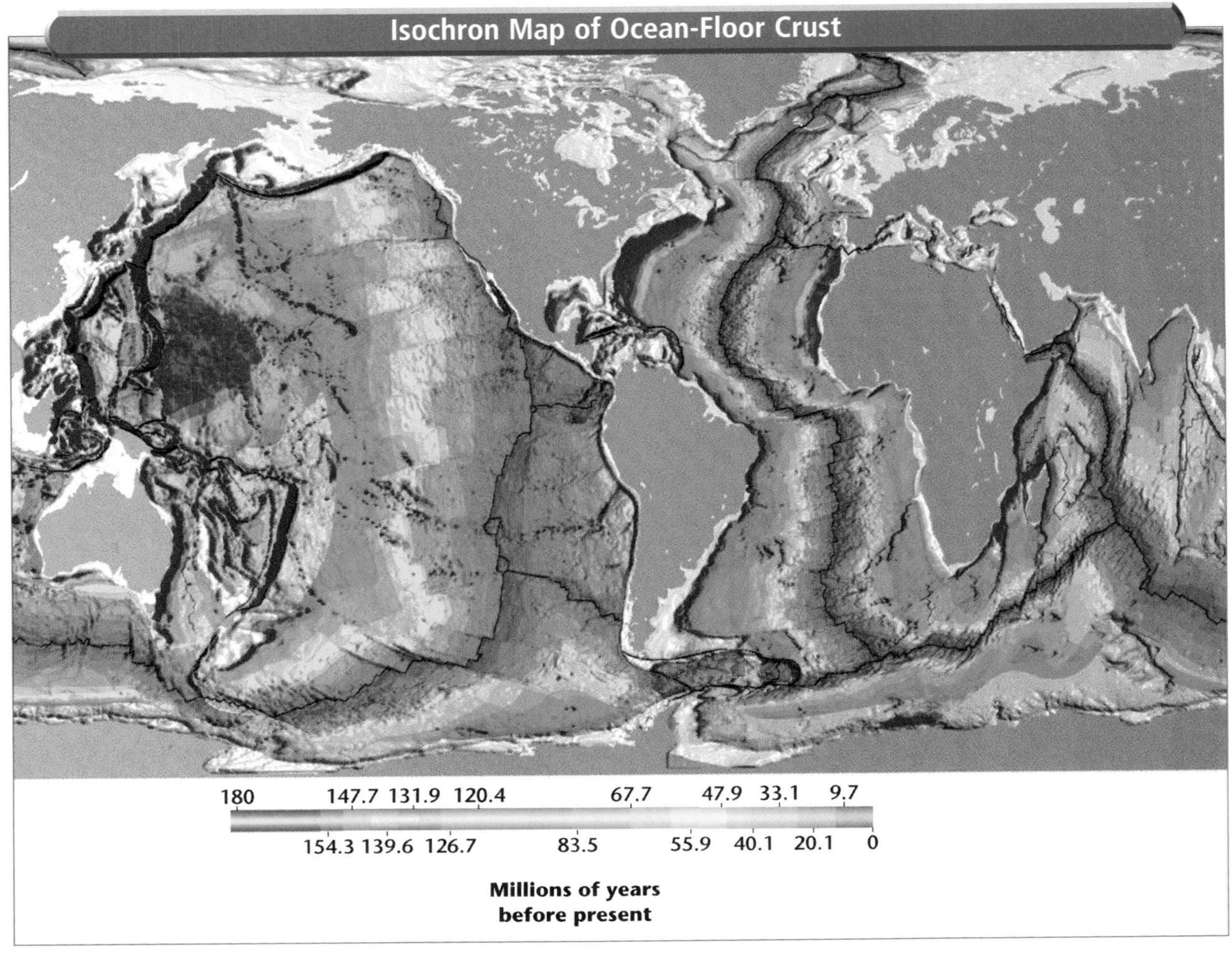

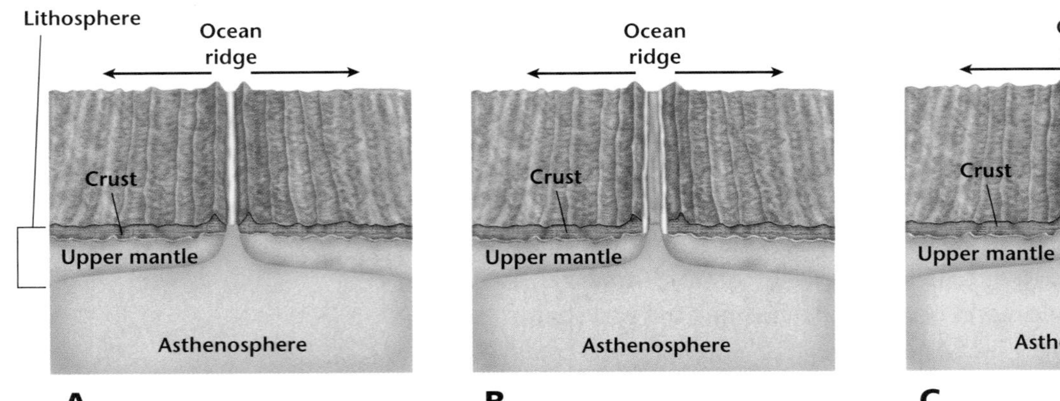

Figure 17-12
Magma that forms in the asthenosphere intrudes into the ocean floor along a ridge and fills the gap that is created **(A).** When the molten material solidifies, new ocean crust is formed **(B).** This continuous intrusion and consequent spreading results in the addition of more crust to the ocean floor **(C).**

The Missing Link Seafloor spreading was the missing link needed by Wegener to complete his model of continental drift. Recall that while Wegener compiled many data to support his idea that the continents are drifting across Earth's surface, he could not explain what caused the landmasses to move or how they moved. Seafloor spreading provides the answer to the *how* question. Continents are not pushing through ocean crust, as Wegener proposed; they are merely passengers that ride with ocean crust as it slowly moves away from ocean ridges. In the next section, you will learn what causes slabs of Earth's crust and rigid upper mantle to move.

SECTION ASSESSMENT

1. How do ocean ridges and deep-sea trenches support the theory of seafloor spreading?
2. Explain how ocean-floor rocks and sediments are evidence of seafloor spreading.
3. Compare and contrast normal magnetic polarity and reversed magnetic polarity.
4. Explain how an isochron map of the ocean floor supports the theory of seafloor spreading.
5. **Thinking Critically** Refer to ***Figure 17-11.*** Why are the magnetic bands in the eastern Pacific Ocean so far apart compared to the magnetic bands along the Mid-Atlantic Ridge?

SKILL REVIEW

6. **Concept Mapping** Make an events-chain concept map to show the sequence of steps involved in seafloor spreading. For more help, refer to the *Skill Handbook.*

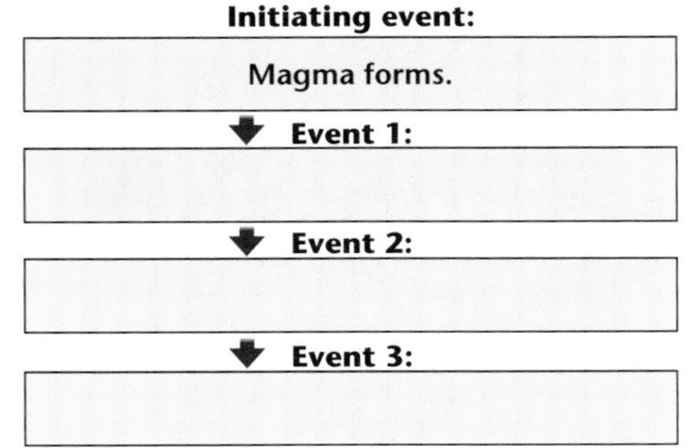

SECTION 17.3 Theory of Plate Tectonics

Earth Sciences 3.a, 3.b, 3.d, 3.f **I&E** 1.k

Have you ever wondered why some regions of Earth are dotted with many active volcanoes while other regions have none? Or why earthquakes occur more frequently in certain areas than in others? The answers to these and many other questions are explained by the **theory of plate tectonics,** which states that Earth's crust and rigid upper mantle are broken into enormous slabs called plates. There are a dozen or so major plates and several smaller ones, as shown in ***Figure 17-13.*** Refer to this figure as you read about tectonic plates. Tectonic plates move in different directions and at different rates over Earth's surface. Rates of movement have been measured with the use of a sophisticated system of receivers and satellites that you'll learn about in the *Science & Technology* feature at the end of this chapter.

OBJECTIVES

- **Explain** *the theory of plate tectonics.*
- **Compare and contrast** *the three types of plate boundaries and the features associated with each.*

VOCABULARY

theory of plate tectonics
divergent boundary
rift valley
convergent boundary
subduction
transform boundary

PLATE BOUNDARIES

Tectonic plates interact at places called plate boundaries. At some boundaries, plates come together, or converge. At others, plates move away from one another, or diverge. At the third type of boundary, plates move horizontally past one another. Each type of boundary has certain geologic characteristics and processes associated with it.

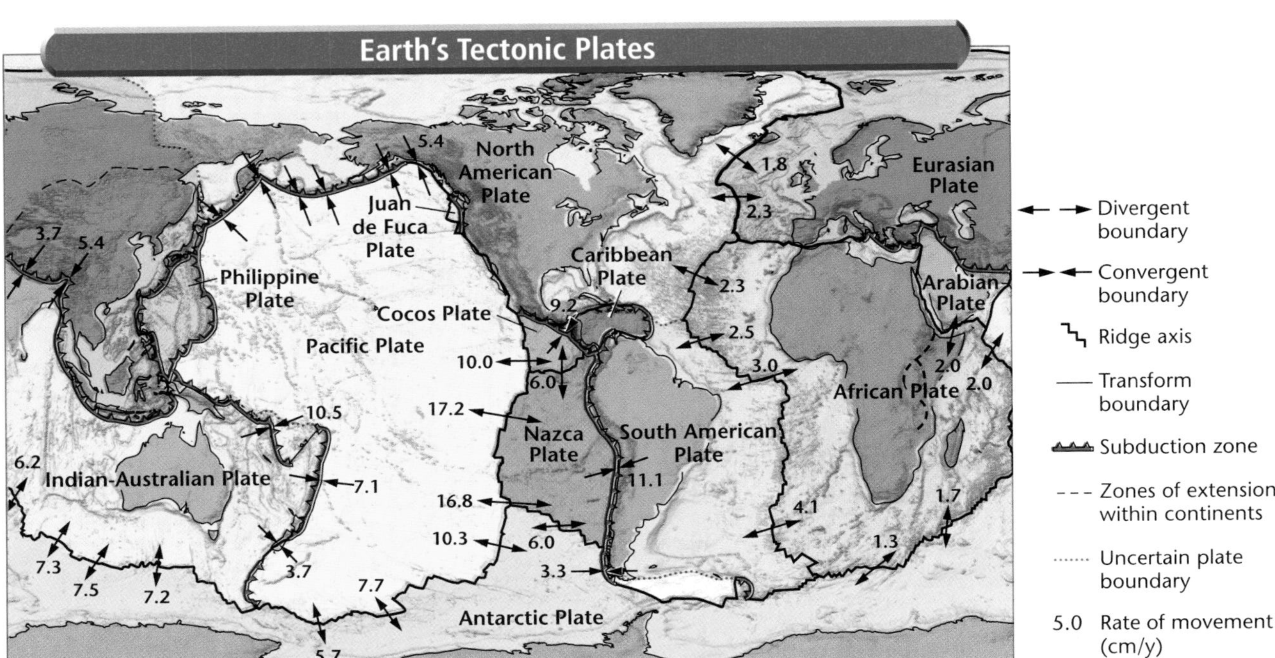

Figure 17-13 Earth's crust and rigid upper mantle are broken into enormous slabs called tectonic plates that interact at places called boundaries.

MiniLab

Model ocean-basin formation

Model the formation of the South Atlantic Ocean.

Procedure

1. Use a world map to create paper templates of South America and Africa.
2. Place the two continental templates in the center of a piece of 11" × 17" paper and fit them together along their Atlantic coastlines.
3. Carefully trace around the templates with a pencil. Remove the templates and label the diagram "150 million years ago."
4. Use an average spreading rate of 4 cm/y and a map scale of 1 cm = 500 km to create a series of maps that accurately show the development of the Atlantic Ocean at 30-million-year intervals, beginning 150 million years ago.

Analyze and Conclude

1. Compare your last map with a world map. Is the actual width of the South Atlantic Ocean the same on both maps?
2. What might have caused any difference between the width in your model and the actual width of the present South Atlantic Ocean?

Divergent Boundaries Places where two tectonic plates are moving apart are called **divergent boundaries.** Most divergent boundaries are found on the seafloor, where they form ocean ridges, as shown in ***Figure 17-14.*** The actual plate boundary is located in a rift, or fault-bounded valley, which forms along the axis of a ridge. It is in this central rift that the process of seafloor spreading begins. The formation of new ocean crust at most divergent boundaries accounts for the high heat flow, volcanism, and earthquakes associated with these boundaries.

Over millions of years, the process of seafloor spreading along a divergent boundary may cause an ocean basin to grow wider. Scientists have measured the rate of growth of the Atlantic Ocean at about 2 to 3 cm/y. How does this compare to the rate used in the *MiniLab* on this page? Can you explain the difference between the actual size of the Atlantic Ocean Basin and the size of the basin in your model?

Although most divergent boundaries form ridges on the ocean floor, some divergent boundaries form on continents. When continental crust begins to separate, the stretched crust forms a long, narrow depression called a **rift valley.** A rift valley is currently forming in East Africa. The rifting might eventually lead to the formation of a new ocean basin.

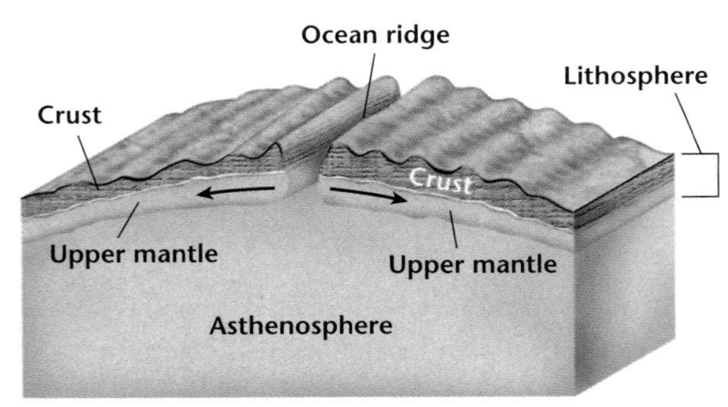

Figure 17-14 Divergent boundaries are places where plates separate. An ocean ridge is a divergent boundary on the ocean floor.

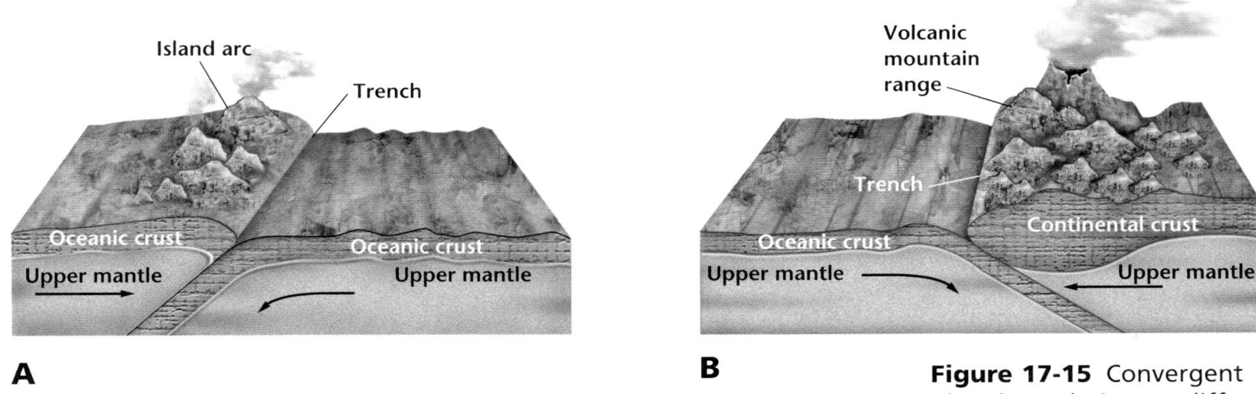

Figure 17-15 Convergent plate boundaries are differentiated according to the type of crust involved. There are three types of convergent boundaries: oceanic-oceanic **(A),** oceanic-continental **(B),** and continental-continental **(C).**

Convergent Boundaries Places where two tectonic plates are moving toward each other are **convergent boundaries.** There are three types of convergent boundaries, which are classified according to the type of crust involved. Recall from Chapter 1 that oceanic crust is made mostly of basalt. Continental crust is composed mostly of granite and sedimentary rocks, both of which are less dense than basalt. The three types of convergent boundaries are oceanic crust converging with oceanic crust, oceanic crust converging with continental crust, and continental crust converging and colliding with continental crust. These three types of tectonic boundaries are shown in ***Figure 17-15.***

What is happening at the oceanic-oceanic boundary shown in ***Figure 17-15A?*** One of the two plates is descending beneath the other in a process called **subduction.** A subduction zone forms when one oceanic plate, which has become denser as a result of cooling, descends below another plate. The process of subduction creates a deep-sea trench. The subducted plate descends into the mantle and melts, thereby recycling the oceanic crust formed at the ridge. Some of the magma that forms is forced back to the surface, erupts, and forms an arc of volcanic islands that parallel the trench.

Earth Sciences

3.a Students know features of the ocean floor (magnetic patterns, age, and sea-floor topography) provide evidence of plate tectonics.

3.b Students know the principal structures that form at the three different kinds of plate boundaries.

3.d Students know why and how earthquakes occur and the scales used to measure their intensity and magnitude.

3.f Students know the explanation for the location and properties of volcanoes that are due to hot spots and the explanation for those that are due to subduction.

I&E 1.k Recognize the cumulative nature of scientific evidence.

Figure 17-16 Mt. Cleveland is a volcano in Alaska that formed as the result of convergence along an oceanic-oceanic boundary.

Some examples of trenches and island arcs are the Mariana Trench and Mariana Islands in the West Pacific Ocean and the Aleutian Trench and Aleutian Islands in the North Pacific Ocean. A volcanic peak in the Aleutian Islands is shown in ***Figure 17-16.***

Subduction also occurs when an oceanic plate converges with a continental plate, as you can see in ***Figure 17-15B.*** Note that the denser oceanic plate is subducted. Oceanic-continental convergence also produces a trench and volcanic arc. However, instead of an arc of volcanic islands forming, as in oceanic-oceanic plate convergence, a series of volcanoes erupt along the edge of the continental plate. The result of this type of subduction is a mountain range with many volcanoes. The Peru-Chile Trench and the Andes Mountains, which are located along the western coast of South America, formed in this way.

The third type of convergent boundary forms when two continental plates collide. This convergent boundary forms when an ocean basin between converging oceanic and continental plates is

Problem-Solving Lab

Interpreting Scientific Illustrations

Determine how plate motions change along a transform boundary The figure at right shows the Gibbs Fracture Zone, which is a segment of the Mid-Atlantic Ridge located south of Iceland and west of the British Isles. Copy this figure.

Analysis

1. In what direction is the seafloor moving at each location? Draw arrows on your copy to indicate these directions at each location.
2. Compare the direction of motion for the following pairs of locations: A and D, B and E, and C and F.

Thinking Critically

3. Which locations are on the North American Plate?
4. Which portion of the fracture zone is the boundary between North America and Europe?
5. At which location is the oldest crust located?

entirely subducted. One continent is pulled into the subduction zone, but it can't be subducted because continental rocks are too buoyant to be forced into the mantle. As a result, the colliding edges of the continents are crumpled and uplifted to form a mountain range, as shown in ***Figure 17-15C.*** The Himalayas formed in this way.

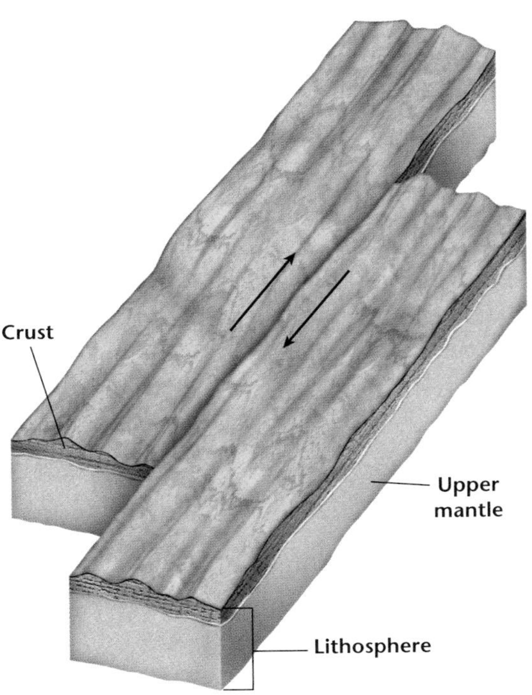

Figure 17-17 Plates move horizontally past each other along a transform plate boundary.

Transform Boundaries A place where two plates slide horizontally past each other, as shown in ***Figure 17-17,*** is a **transform boundary.** How does this type of boundary compare to divergent and convergent boundaries? Along divergent boundaries, new crust is formed at ridges, and at convergent boundaries, old crust is destroyed by subduction. At transform boundaries, however, crust is only deformed or fractured. Transform boundaries are characterized by long faults, sometimes hundreds of kilometers in length, and by shallow earthquakes. Most transform boundaries offset sections of ocean ridges, as you observed in the *Problem-Solving Lab* on the previous page. Rarely do transform boundaries occur on continents. The San Andreas Fault is probably the best-known exception. Recall from the *Discovery Lab* at the beginning of this chapter that the San Andreas Fault system separates southwestern California from the rest of the state. Movements along this transform fault system are responsible for most of the earthquakes that strike California every year.

SECTION ASSESSMENT

1. Explain the theory of plate tectonics.
2. Explain what happens at the three types of plate boundaries.
3. What geologic features are associated with each type of convergent boundary?
4. Where do transform boundaries most commonly occur?
5. **Thinking Critically** Transform boundaries change, or transform, the relative movement of the plates involved. Describe how newly formed crust moves between parts of a ridge that are offset by a transform boundary.

SKILL REVIEW

6. **Making and Using Tables** Make a table that summarizes the features associated with each type of plate boundary. For more help, refer to the *Skill Handbook.*

SECTION 17.4

Causes of Plate Motions

Earth Sciences 3.a, 3.b

OBJECTIVES

- **Explain** *the process of convection.*
- **Summarize** *how convection in the mantle is related to the movements of tectonic plates.*
- **Compare and contrast** *the processes of ridge push and slab pull.*

VOCABULARY

ridge push
slab pull

Today, plate tectonics is no longer in question. The directions and rates of plate movements have been measured. One of the main questions about the theory of plate tectonics, however, has remained unanswered ever since Alfred Wegener first proposed continental drift. What force or forces cause tectonic plates to move? Even today, what actually causes the plates to move is not well understood. One of the leading hypotheses proposes that large-scale motion in the mantle is the mechanism that drives the movement of tectonic plates.

MANTLE CONVECTION

Recall from Chapter 11 that the transfer of thermal energy by the movement of heated matter is called convection. The heating of matter causes it to expand and to decrease in density. The warmed matter then rises as a result of buoyancy. The cooler part of the matter sinks as a result of gravity. This up-and-down flow produces a pattern of motion called a convection current. Convection currents aid in the transfer of thermal energy from the warmer regions of matter to cooler regions. A convection current can be observed in the series of photographs shown in ***Figure 17-18.***

Convection currents in the mantle are thought to be the driving mechanism of plate movements. Recall that even though the mantle is a solid part of it, the asthenosphere can flow like a soft, pliable plastic. Convection currents in this part of the mantle are set in motion by the transfer of energy between Earth's hot interior and its cooler exterior. Hot mantle material is less dense than cooler mantle

Figure 17-18 Convection currents transfer thermal energy in unequally heated matter.

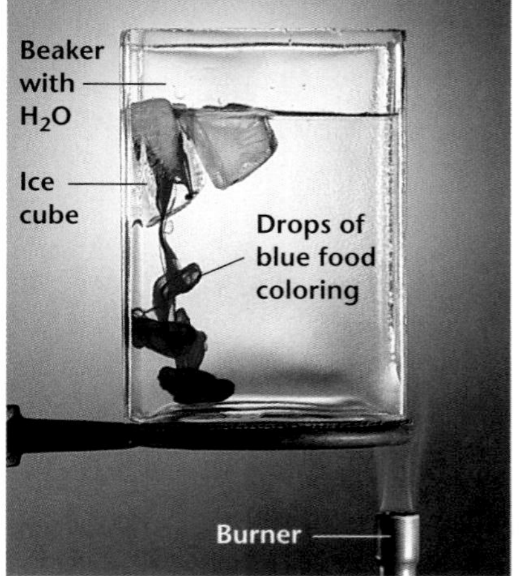

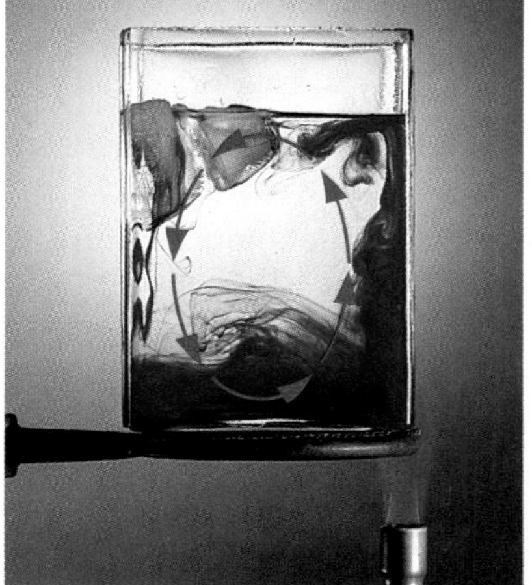

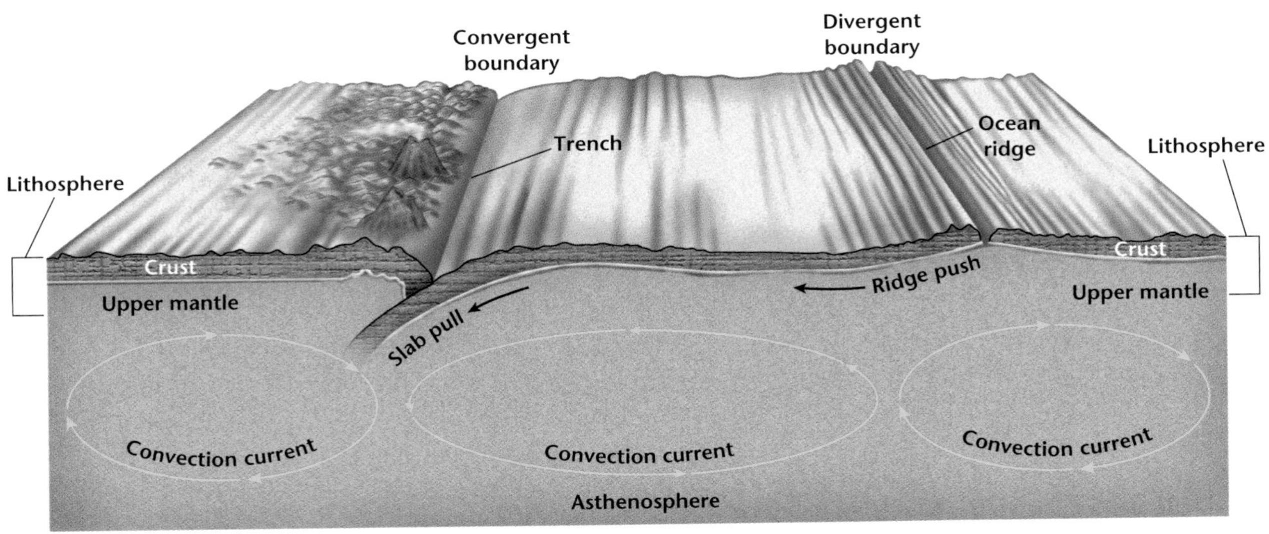

Figure 17-19 Ridge push and slab pull are tectonic processes associated with convection currents in the mantle.

material and, thus, the hot material is slowly forced toward the crust. Cooler parts of this material sink back toward the core. The convection currents that result can be thousands of kilometers across but flow at rates of only a few centimeters per year. Scientists hypothesize that these convection currents are probably set in motion by subducting slabs, thus causing plates to move.

How are convergent and divergent movements related to the flow direction of mantle convection? The rising part of a convection current spreads out as it reaches the upper mantle and causes both upward and lateral forces. These forces lift and split the lithosphere at divergent plate boundaries. As the plates separate, material rising from the mantle supplies the magma that hardens to form new ocean crust. The downward part of a convection current occurs where a sinking force pulls tectonic plates downward at convergent boundaries.

 **Earth Sciences**

3.b Students know the principal structures that form at the three different kinds of plate boundaries.

Push and Pull Exactly how is mantle convection related to the movements of tectonic plates? Scientists hypothesize that there are several processes involved. Study ***Figure 17-19.*** During the formation of an ocean ridge, forces in the mantle cause the asthenosphere to rise. The weight of the uplifted ridge is thought to push an oceanic plate toward the trench formed at the subduction zone in a process called **ridge push** as shown in ***Figure 17-19.*** In addition to ridge push, the horizontal flow at the top of a convection current could create drag on the lithosphere and thereby contribute to plate motion.

A combination of mechanisms may be involved in plate motions at subduction zones as well. A sinking region of a mantle convection current could suck an oceanic plate downward into a subduction zone. The weight of a subducting plate helps pull the trailing lithosphere into the subduction zone in a process called **slab pull,** as shown in ***Figure 17-19.***

Unanswered Questions While most scientists agree that convection currents in the mantle are related to the movement of tectonic plates, there are still unanswered questions about how these currents originate and what their actual sizes are. Are mantle convection currents permanent features? Do they shift their positions through geologic time? How does a convection current start? What causes the movement of a convection current to stop? Is convection taking place only in the upper mantle? Or, do these enormous convection currents include the lower mantle? ***Figure 17-20*** shows an image of the mantle 2700 km below Earth's surface that was generated from earthquake data. This image suggests that the convection currents associated with plate movements might include the lower mantle. Some geologists working in this branch of Earth science have even suggested that subducted slabs, over time, might eventually reach Earth's outer core.

Other remaining questions concern relationships between convection currents and the overlying plates. For example, does downward convection cause a subduction zone, or does the process of subduction create the convection current? Most studies show that the process of slab pull is the most important force driving tectonic plate motions.

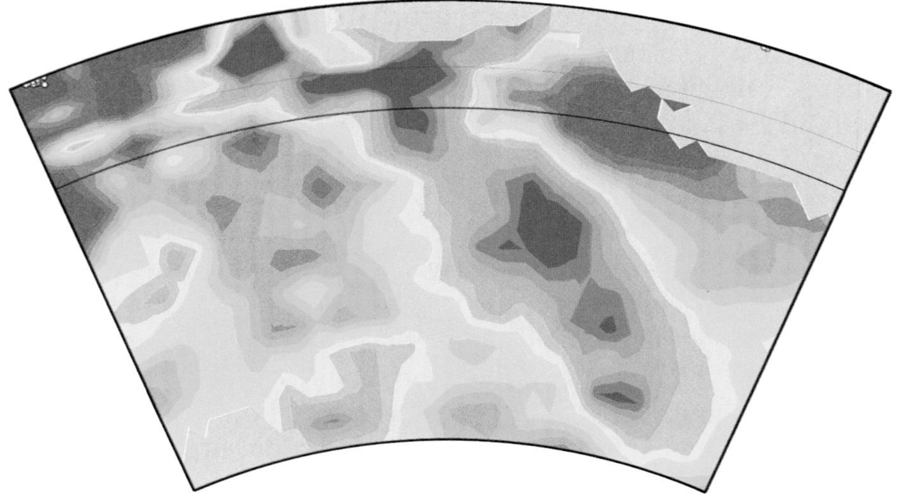

Figure 17-20 This is an image of Earth's mantle under the southern United States. The image was generated using the travel times of earthquake waves. The elongated blue feature is an area of the mantle where subduction is thought to have occurred in the past.

Figure 17-21 Spreading in the Great Rift Valley of East Africa, shown here, is related to convection currents in Earth's mantle.

A similar set of questions surround the formation of divergent plate boundaries. The Great Rift Valley, in Africa, part of which is shown in ***Figure 17-21,*** is evidence that a divergent boundary can develop on a continent. But what causes an upward convection current to form beneath a continent? One hypothesis is that large continental masses ultimately cause their own breakup by acting as insulating blankets. The underlying mantle then becomes warmer and causes the upward leg of a convection current to develop, which eventually causes the continent to split.

The questions presented here are just some of those that geologists hope to answer in the future. What questions do you have about what causes Earth's crust and rigid upper mantle to slowly but continuously change over geologic time?

Section Assessment

1. Make and label a diagram that explains the process of convection.
2. What are the relationships among mantle convection, ocean ridges, and subduction zones?
3. Hypothesize what might happen to mantle convection currents beneath a continental-continental convergent boundary.
4. How have earthquake data been used to explore the size of convection currents in the mantle?
5. **Thinking Critically** If a subducting slab is pulling the lithosphere into the mantle and a continent is pulled into the subduction zone, what do you think might happen?

Skill Review

6. **Making Diagrams** Make a diagram that contrasts the tectonic processes of ridge push and slab pull. Include captions in your diagram. For more help, refer to the *Skill Handbook*.

earthgeu.com/self_check_quiz

Mapping GeoLab

Making a Paleomagnetic Map

Iron-bearing minerals in rocks record the orientation of Earth's magnetic field at the time of their formation. These preserved magnetic alignments in ocean-floor rocks can be used to date different parts of the seafloor and to determine rates of spreading along divergent plate margins.

Earth Sciences

3.a Students know features of the ocean floor (magnetic patterns, age, and sea-floor topography) provide evidence of plate tectonics.

Preparation

Problem

How can you use paleomagnetic data to interpret information about ocean-floor rocks?

Materials

tracing paper
metric ruler
No. 2 pencil
colored pencils
Figure 17-8

Procedure

1. Use the tracing paper and the No. 2 pencil to carefully copy the base map on the following page. Be sure to include the magnetic survey lines, the map scale, and the location indicated by the letter *X* on your tracing.
2. Transfer the magnetic survey data from PM1 to your tracing by placing the survey line on your map over the PM1 data curve on the following page. Be sure to align the mid-ocean ridge with the dashed line before you begin transferring the line. Label the line.
3. Repeat step 2 for the other four lines.
4. Next, use your ruler to draw a series of parallel lines between PM1 and PM2 to connect the corresponding positive and negative magnetic reversals. Draw the lines past PM2, but stop the lines at the transform boundary.
5. Draw another series of lines between survey lines PM3 and PM4 and between PM4 and PM5. Again, end these lines at the transform boundary.
6. The positive magnetic reading along the mid-ocean ridge represents the Brunhes Magnetic Epoch. The first negative anomaly on either side of the ridge marks the beginning of the Matuyama Magnetic Epoch. Use ***Figure 17-8*** to identify the magnetic reversals.
7. Assign a color for each magnetic reversal. Then, color the corresponding stripe on each side of the ridge.

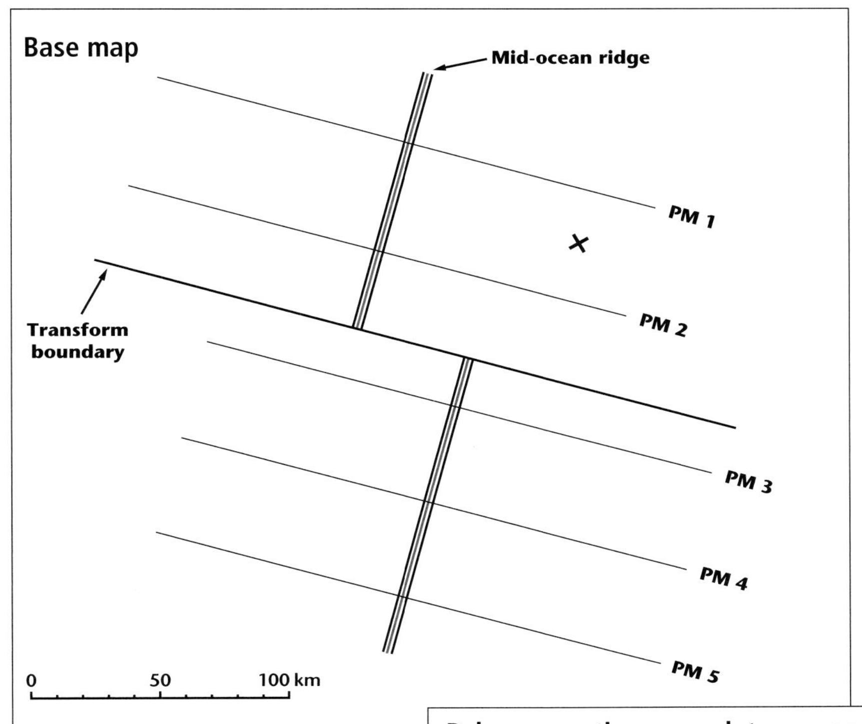

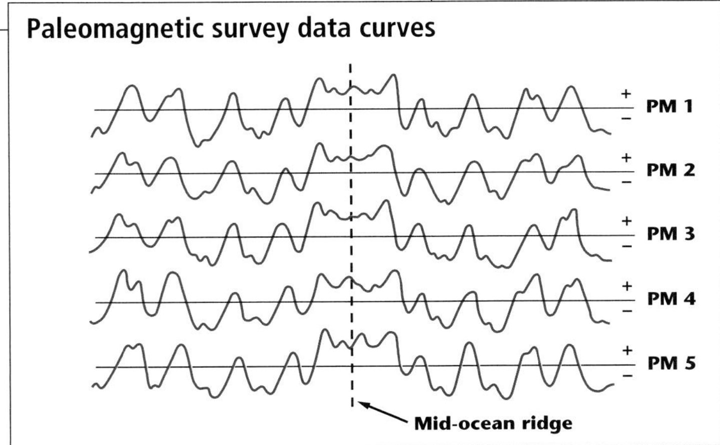

Analyze

1. Why was it necessary to color corresponding stripes on each side of the ridge in step 7?
2. Use ***Figure 17-8*** to determine the age of the seafloor at location *X*.

Conclude & Apply

1. What might cause an individual magnetic stripe of ocean-floor crust to vary in width?
2. What is the average spreading rate along this section of the mid-ocean ridge?

Science & Technology

The Global Positioning System

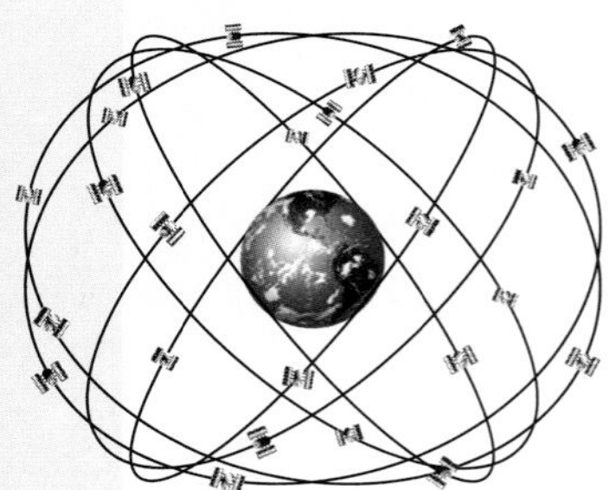

An ambulance driver speeds toward the location of someone having a heart attack. A hiker in unfamiliar territory successfully returns to her campsite. The driver of an armored vehicle transports hundreds of thousands of dollars from businesses to a local bank. What do these people have in common? Each of them is likely to be navigating using a system of satellites currently in orbit around Earth, a system that most people aren't even aware exists.

The Global Positioning System, or GPS, was developed by the United States Department of Defense to track tanks, planes, and ships. Twenty-four satellites, roughly 20 000 km above Earth, are part of this system, each circling on a different orbital path about once every 12 hours. The number of satellites, their distances from Earth, and their orbital periods make the GPS incredibly accurate and essentially give every square meter on our planet a unique address.

How does this system work? A receiver on Earth picks up signals coming from at least three or four satellites in space. The time it takes for the signals to travel from the satellites to the receiver can be used to calculate the latitude, longitude, and elevation of the location of the receiver.

Everyday Uses

The development of the GPS forever changed navigation and mapping. While this system is still operated by the military, there are thousands of civilian users all over the world. Fire, police, and rescue departments, public transportation systems, long-haul trucking companies, taxi services, and fishing fleets all use the GPS to keep drivers and passengers safe, to navigate in unfamiliar territory, and to protect expensive cargo from theft or loss. Hand-held GPS receivers are also used by hikers, hunters, drivers, and boaters.

Scientific Uses

In addition to its everyday uses, the GPS is a valuable tool for collecting information about Earth's surface. Scientists use the GPS to map volcanic regions, where temperature extremes and steep, rocky slopes make it very difficult to use traditional mapping systems. The GPS is also useful in determining sites for nuclear power plants and radioactive waste disposal sites, facilities that must be constructed in areas that are relatively inactive geologically.

The pinpoint accuracy of the GPS has also allowed geophysicists to measure how tectonic plates move in relation to each other over time. For example, the expansion of Iceland, which sits astride the Mid-Atlantic Ridge, is taking place at about 12 mm/year. This rate of movement would have been impossible to measure without the use of the GPS.

Internet

Go to the Earth Science Web Site at earthgeu.com to find out how the GPS has been used to help predict earthquakes that occur along the transform boundary called the San Andreas Fault.

CHAPTER 17

Study Guide

Summary

SECTION 17.1

Drifting Continents

Main Ideas

- The matching coastlines of continents on opposite sides of the Atlantic Ocean suggest that the continents were once joined.
- Continental drift states that Earth's continents were joined as a single landmass that broke apart and sent the continents adrift.
- Wegener supported his hypothesis of continental drift with rock types, fossils, and ancient climatic data. His hypothesis was not accepted at first because he couldn't explain how the continents moved or what caused their motion.

Vocabulary

continental drift (p. 444)
Pangaea (p. 444)

SECTION 17.2

Seafloor Spreading

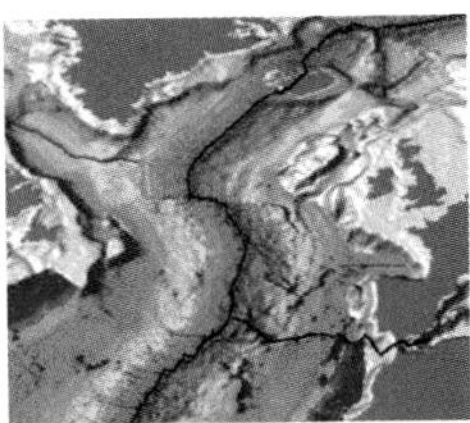

Main Ideas

- Sonar and magnetic studies of ocean rocks and sediments led to the proposal of the theory of seafloor spreading.
- Magnetic patterns on the seafloor are symmetric in relation to ocean ridges, indicating that ocean crust on either side of the ridge is moving away from the ridge at essentially the same rate.
- During seafloor spreading, magma rises and hardens to form new crust, which becomes part of the ocean floor. Each cycle of spreading and intrusion results in the formation of another small section of ocean floor, which slowly moves away from the ridge.

Vocabulary

isochron (p. 452)
magnetic reversal (p. 451)
magnetometer (p. 448)
paleomagnetism (p. 451)
seafloor spreading (p. 453)

SECTION 17.3

Theory of Plate Tectonics

Main Ideas

- Plate tectonics states that Earth's crust and rigid upper mantle are broken into large slabs of rock called plates, which move in different directions and at different rates over Earth's surface.
- At divergent plate boundaries, plates move apart. At convergent boundaries, plates come together. At transform boundaries, plates slide horizontally past each other.
- High heat flow, volcanism, and earthquakes are associated with divergent boundaries; trenches, island arcs, and folded mountains with convergent boundaries; and faults and earthquakes with transform boundaries.

Vocabulary

convergent boundary (p. 457)
divergent boundary (p. 456)
rift valley (p. 456)
subduction (p. 457)
theory of plate tectonics (p. 455)
transform boundary (p. 459)

SECTION 17.4

Causes of Plate Motions

Main Ideas

- Convection is the transfer of energy via the movement of heated matter. Convection currents in the mantle are the result of energy transfer between Earth's hot interior and cooler exterior.
- Ridge push occurs when the elevation of a ridge pushes a plate toward a subduction zone. Slab pull occurs as the weight of the subducting plate pulls a plate into a subduction zone.

Vocabulary

ridge push (p. 461)
slab pull (p. 462)

CHAPTER 17
Assessment

Understanding Main Ideas

1. What was Wegener's hypothesis called?
   a. seafloor spreading
   b. plate tectonics
   c. continental drift
   d. slab pull

2. The fit of the coastlines of which of the following continents led people to suggest that the continents had drifted over time?
   a. North and South America
   b. North America and Africa
   c. South America and Africa
   d. Europe and North America

3. Which of the following evidence was used by Wegener to support his hypothesis that the continents had once been joined?
   a. rock types and ages
   b. plate tectonics
   c. slab pull and ridge push
   d. fossils of ocean plants

4. Which of the following was NOT used by Wegener to support his hypothesis of continental drift?
   a. fossils of land-dwelling animals
   b. paleomagnetic data
   c. coal beds in Antarctica
   d. glacial deposits

5. Why was the hypothesis of continental drift rejected when it was proposed by Wegener?
   a. Wegener thought that the south pole had changed location.
   b. Wegener thought that Earth's rotation was the driving force.
   c. Wegener couldn't explain how or why continents moved.
   d. Wegener died in Greenland in 1930.

6. How did the use of sonar change scientists' ideas about ocean-floor topography?

7. Explain how sediments on the ocean floor vary in relation to an ocean ridge.

8. Differentiate among the following terms: *magnetic reversal, reversed field,* and *normal field.*

9. Explain the theory of plate tectonics.

10. Compare and contrast the three types of convergent plate boundaries.

Applying Main Ideas

Use the diagram below to answer questions 11, 12, and 13.

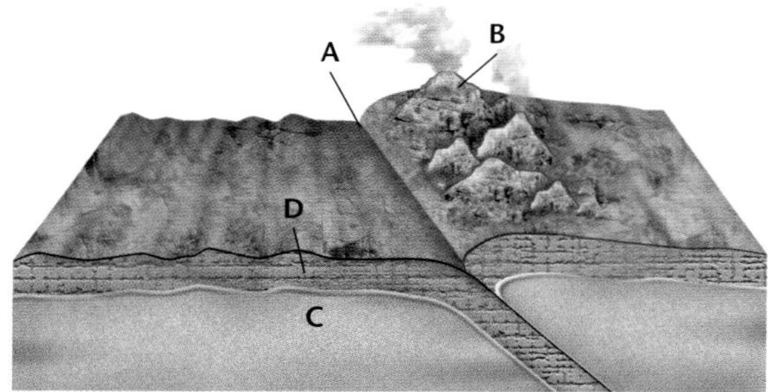

11. What type of boundary is shown?

12. Copy the figure and identify the features labeled A, B, C, and D.

13. Use arrows to label the directions of plate movements.

14. The Andes Mountains are tectonically active. What type of plate boundary is involved in their continued formation?

Test-Taking Tip

READ QUESTIONS CAREFULLY Watch the little words. Underline words like *least, not, except,* and *but* when you see them in test questions. They change the meaning of the question!

earthgeu.com/chapter_test

CHAPTER 17
Assessment

15. Explain the relationship among an ocean ridge, a rift, and a divergent plate boundary.

16. Explain how movements at divergent and convergent plate boundaries are related to flow direction of mantle convection currents.

Use *Figure 17-8* on page 451 to answer questions 17, 18, and 19.

17. Which is longer, an event or an epoch?

18. How long did the Gauss Epoch last?

19. How many reversals occurred in the Gilbert Epoch? How many normal events occurred during this epoch?

20. Why did the idea of continental drift become more accepted after seafloor spreading was proposed?

21. Explain what causes the paleomagnetic patterns on the seafloor.

22. Refer to ***Figure 17-13*** on page 455. Find and identify a place where the Mid-Atlantic Ridge surfaces.

23. Study the rates of plate motions shown in ***Figure 17-13*** on page 455. How far has the island of Hawaii moved since you were born?

Thinking Critically

24. Locate Africa in ***Figure 17-13*** on page 455. Identify the major plate boundaries affecting this continent. In which direction is Africa moving in relation to these boundaries?

25. What might cause corresponding magnetic stripes of ocean-floor crust to have different widths?

26. Why do you think earthquakes are common along plate boundaries?

27. Most earthquakes associated with convergent boundaries originate much deeper than earthquakes associated with divergent boundaries. Explain.

Standardized Test Practice

1. Which piece of evidence did NOT advance the hypothesis of continental drift?
   a. similar rock types on different continents
   b. crustal rock on the ocean floor
   c. similar fossils on different continents
   d. coal beds in Antarctica

Interpreting Scientific Illustrations

Use the diagram to answer questions 2–3.

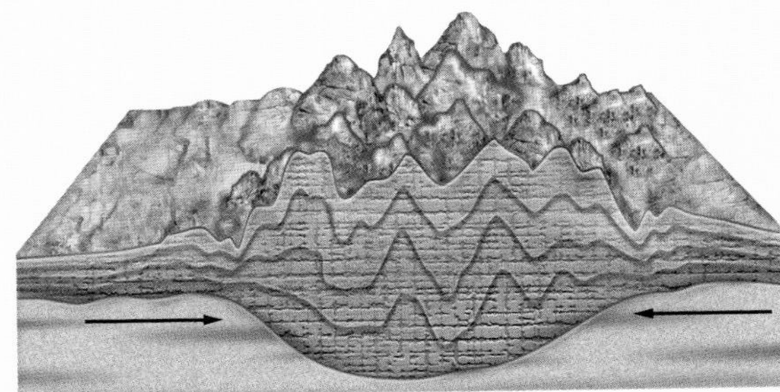

2. What type of plate boundary is shown?
   a. an ocean ridge
   b. a continental-continental boundary
   c. a transform boundary
   d. an oceanic-continental boundary

3. Which of the following features forms along this type of boundary?
   a. subduction zones **c.** island arcs
   b. oceanic trenches **d.** folded mountains

4. The weight of a subducting plate helps pull the trailing lithosphere into a subduction zone in which of the following processes?
   a. slab pull **c.** slab push
   b. ridge pull **d.** ridge push

Chapter 18

Volcanic Activity

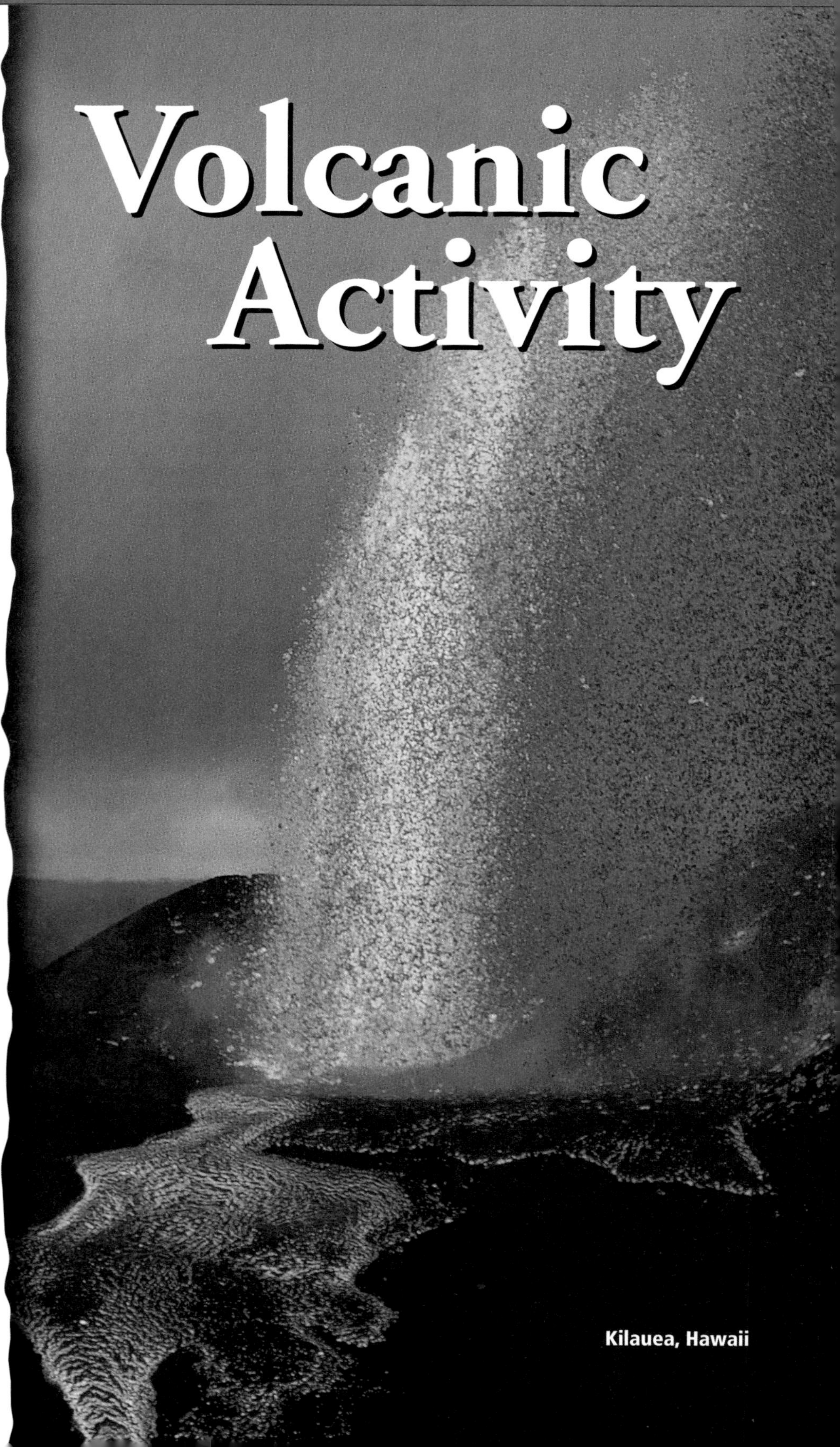

What You'll Learn

- How magma forms.
- What kinds of features form as the result of igneous activity within Earth.
- How volcanoes form and how they can be classified.

Why It's Important

Many of Earth's internal processes help to shape our planet's surface. Igneous activity deep within Earth and at its surface produce many of the mountains and rock formations on Earth.

To learn more about volcanic activity, visit the Earth Science Web Site at earthgeu.com

Kilauea, Hawaii

Discovery Lab Model Magma Movement

Magma is molten rock beneath Earth's surface that rises because it is less dense than the surrounding rock. In this activity, you will model how magma moves within Earth.

1. Fill a 250-mL beaker with 175 mL of ice-cold water.
2. Carefully fill a 100-mL beaker with very hot tap water. Add 2–3 drops of food coloring to the water and stir well.
3. Carefully fill a dropper with the hot, colored water.
4. Slowly insert the full dropper into the 250-mL beaker until the tip of the dropper is 1 cm from the bottom of the beaker. Squeeze the dropper and keep the bulb depressed as you slowly pull the dropper back out of the cold water.

CAUTION: Always wear safety goggles and an apron in the lab.

Observe In your science journal, describe what happened to the colored water when it entered the beaker. How might this be similar to what happens to magma beneath Earth's surface? Infer what would have happened if you had released the hot water at the surface of the cold water.

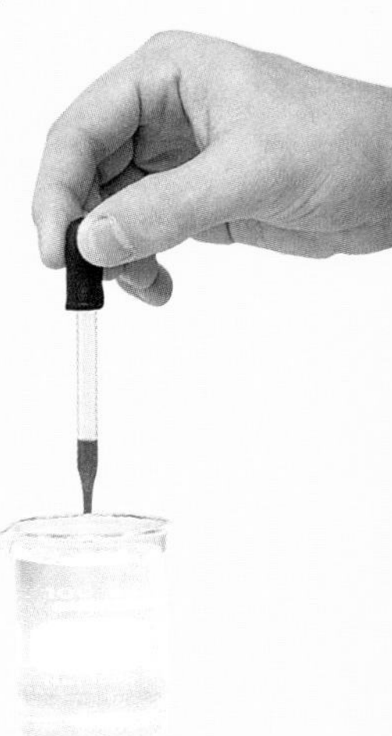

Section 18.1 Magma

Earth Sciences 3.e

OBJECTIVES

- **Describe** *factors that affect the formation of magma.*
- **Compare** *and* **contrast** *the different types of magma.*

VOCABULARY

viscosity

Volcanic eruptions are spectacular events. The ash that spews from some volcanoes can form billowy clouds that travel around the world before raining back down to Earth. The red-hot lava that erupts from other volcanoes, such as the Hawaiian volcano Kilauea shown on the facing page, can destroy everything in their paths. In the last 10 000 years, more than 1500 different volcanoes have erupted—providing evidence that Earth is indeed geologically active. Where do ash, lava, and other types of volcanic debris come from?

How Magma Forms

All volcanoes are fueled by magma deep beneath Earth's surface. Recall from Chapter 5 that magma is a mixture of molten rock, suspended mineral grains, and dissolved gases deep beneath Earth's surface. Magma forms when temperatures are high enough to melt the rocks involved. Depending on their composition, most rocks begin to melt at temperatures between 800°C and 1200°C. Such

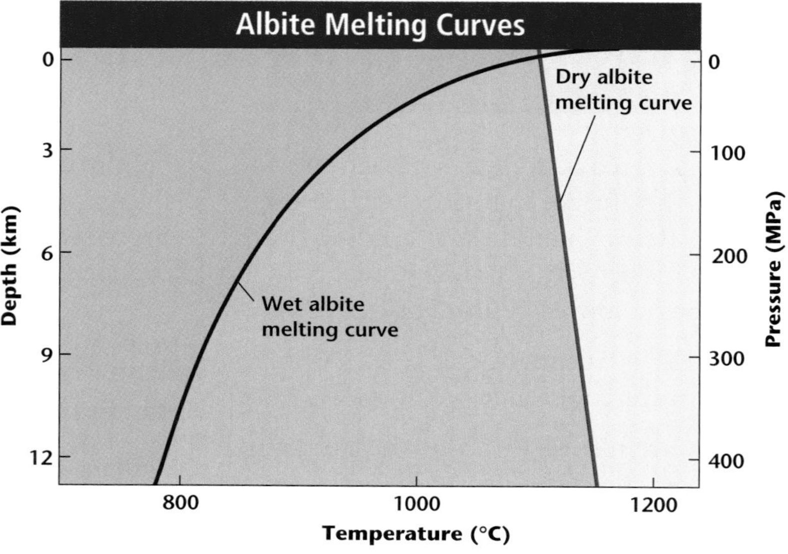

Figure 18-1 Both pressure and the presence of water affect the melting temperature of minerals and thus, rocks.

temperatures exist at the base of the lithosphere and in the asthenosphere, the plasticlike portion of the mantle directly beneath the lithosphere. Recall that temperature increases with depth beneath Earth's surface. If rocks melt at temperatures found in the asthenosphere, and temperature increases with depth, then why isn't the entire mantle liquid? What other factors, besides temperature, affect the formation of magma?

Earth Sciences

3.e Students know there are two kinds of volcanoes: one kind with violent eruptions producing steep slopes and the other kind with voluminous lava flows producing gentle slopes.

Pressure Pressure is one factor that determines whether rocks will melt to form magma. Like temperature, pressure increases with depth because of the weight of overlying rocks. Laboratory experiments have shown that as pressure increases, the temperature at which a substance melts also increases. ***Figure 18-1*** shows two melting curves for a variety of feldspar called albite. Find the line that represents the dry melting curve. Note that at Earth's surface, dry albite melts at about 1100°C, but at a depth of about 12 km, the melting point of dry albite is about 1150°C. At a depth of about 100 km, the melting point of dry albite increases to 1440°C. The effect of pressure explains why most of the rocks in Earth's lower crust and upper mantle do not melt to form magma, even though the temperatures are high enough.

Water The presence of water also influences whether a rock will melt. Recall that water can be found in the pore spaces of some rocks and can be bound into the crystal structure of some minerals. Even a small amount of water can have a significant effect on a mineral's, and thus a rock's, melting point. At any given pressure, a wet mineral or rock will melt at a lower temperature than the same mineral or rock under dry conditions. Locate the melting curve of wet albite in

Figure 18-1. How does the melting point of wet albite compare to that of dry albite at a depth of 3 km? At a depth of 12 km?

TYPES OF MAGMA

Recall from Chapter 5 that the three major igneous rock types are basalt, andesite, and granite. These rocks form from three major types of magma: basaltic magma, andesitic magma, and rhyolitic magma. The term *rhyolitic* is used to describe the magma that solidifies to form granite because magmas are named after extrusive rocks.

Basaltic magma has the same composition as basalt. Locate the Hawaiian Islands in ***Figure 18-2,*** which shows some of Earth's active volcanoes. The volcanoes that make up the Hawaiian Islands, which include Kilauea and Mauna Loa, are made of basalt. Surtsey, which formed south of Iceland in 1963, is another basaltic volcano.

Andesitic magma has the same composition as andesite. Mount St. Helens in Washington State and Tambora in Indonesia are two andesitic volcanoes. You will find out more about Tambora in the *Science & the Environment* feature at the end of this chapter. Rhyolitic magma has the same composition as granite. The dormant volcanoes in Yellowstone National Park in the western United States were fueled by rhyolitic magma.

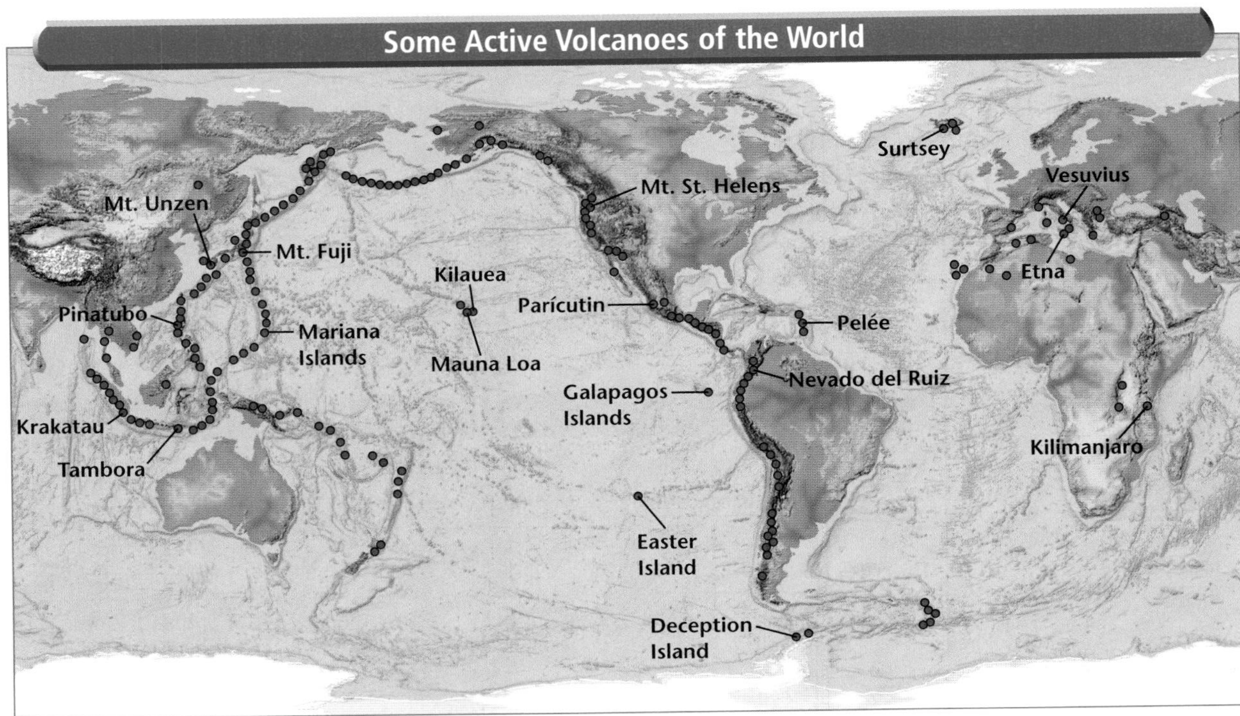

Figure 18-2 Compare this map of some of Earth's active volcanoes to the map shown in **Figure 17-13** on page 455. ***Where are most active volcanoes located?***

MiniLab

How does silica affect lava flow?

Model the changes in lava viscosity with the addition of silica.

CAUTION: Always wear safety goggles and an apron in the lab.

Procedure

1. Pour 120 mL of dishwashing liquid into a 250-mL beaker.
2. Stir the liquid with a stirring rod. Describe the viscosity.
3. Add 30 g of NaCl (table salt) to the liquid. Stir well. Describe what happens.
4. Repeat step 3 three more times.

Analyze and Conclude

1. What do the liquid and NaCl represent?
2. How does an increase in silica affect lava viscosity?
3. Basaltic eruptions are called flows because of the way they move across Earth's surface. What can you infer about the silica content of a basaltic flow?

Magma Composition What accounts for the different types of magma? A number of factors determine the composition of magma, as shown in ***Table 18-1.*** One of these factors is **viscosity,** the internal resistance to flow. Substances such as honey, liquid soap, and motor oil have a higher viscosity than water, vinegar, and gasoline. Refer to ***Table 18-1.*** What kind of magma has a viscosity similar to that of honey? You can model the effect of silica content on viscosity in the *MiniLab* on this page.

Basaltic Magma Basaltic magma typically forms when rocks in the upper mantle melt. Most basaltic magma rises relatively rapidly to Earth's surface and reacts very little with crustal rocks because of its low viscosity. Because basaltic magma contains small amounts of dissolved gases and silica, the volcanoes it fuels erupt relatively quietly.

Andesitic Magma Andesitic magma is found along continental margins, where oceanic crust is subducted into Earth's mantle. The source material for this magma can be either oceanic crust or oceanic sediments. As shown in ***Table 18-1,*** andesitic magma contains about 60 percent silica. This high silica content results in its having an intermediate viscosity. Thus, the volcanoes it fuels are said to have intermediate eruptions.

Table 18-1 Magma Composition and Characteristics

Composition	Source Material	Viscosity	Gas Content	Silica Content	Explosiveness	Location of Magma
Basaltic magma	Upper mantle	Low	1–2%	about 50%	Least	Both oceanic and continental crust
Andesitic magma	Oceanic crust and oceanic sediments	Intermediate	3–4%	about 60%	Intermediate	Continental margins associated with subduction zones
Rhyolitic magma	Continental crust	High	4–6%	about 70%	Greatest	Continental crust

Rhyolitic Magma Rhyolitic magma forms when molten material rises and mixes with the overlying silica- and water-rich continental crust. The high viscosity of rhyolitic magma inhibits its movement. This resistance to flow, along with the large volume of gas trapped within this magma, makes the volcanoes fueled by rhyolitic magma very explosive.

Figure 18-3 Basaltic lava has a low viscosity, and thus, flows relatively quickly from a volcano. The basaltic volcano shown here is Mauna Loa, one of the many volcanoes that make up the Hawaiian Islands.

VISCOSITY

The viscosity of magma and of its surface counterpart, lava, depends on both temperature and composition. The hotter the magma or lava, the lower the viscosity. The temperatures of basaltic lavas are generally between 1000°C and 1250°C. Rhyolitic lava temperatures are usually between 700°C and 900°C. Which type of lava, basaltic or rhyolitic, has a greater viscosity as a result of its temperature? What do you think happens to viscosity as magma or lava cools?

The amount of silica in magma or lava increases the viscosity, as you discovered in the *MiniLab* on the previous page. Thus, magmas and lavas high in silica have higher viscosities than magmas and lavas low in silica. As shown in ***Table 18-1,*** rhyolitic magmas have the highest silica content, basaltic magmas the lowest, and andesitic magmas have silica contents between these two extremes. Based on composition, which type of lava, basaltic lava or andesitic lava, has a lower viscosity? Basaltic lavas, because of their low silica content, have a lower viscosity than andesitic lavas. The basaltic lava flows that often erupt from Mauna Loa in Hawaii, which is shown in ***Figure 18-3,*** have been clocked at 16 km/h!

SECTION ASSESSMENT

1. Describe three factors that affect the formation of magma.
2. How does the presence of water affect the melting temperature of a rock?
3. Compare and contrast the properties of the three types of magma.
4. Refer to ***Table 18-1.*** Where does andesitic magma form? What is the source material of this type of magma?
5. Explain the relationship between the viscosity of a magma and its temperature.
6. **Thinking Critically** A volcano violently erupted in Indonesia in 1883. What can you infer about the composition of the magma that fueled the volcano? If people witnessed the eruption, what do you think they were able to observe about the lava flow?

SKILL REVIEW

7. **Concept Mapping** Use the following terms to construct a concept map to organize the major ideas in this section. For more help, refer to the *Skill Handbook.*

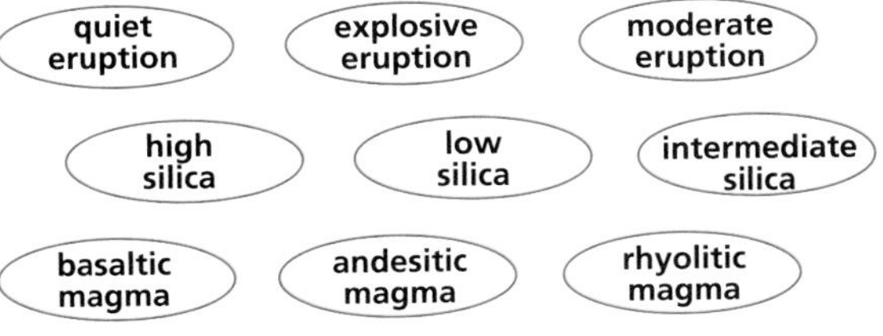

SECTION 18.2 Intrusive Activity

Earth Sciences 3.b

OBJECTIVES

- **Explain** *how magma affects overlying crustal rocks.*
- **Compare** *and* **contrast** *intrusive igneous rock bodies.*

VOCABULARY

pluton
batholith
stock
laccolith
sill
dike

Magma, because it is molten, is less dense than surrounding rocks. This density difference, which you modeled in the *Discovery Lab*, forces magma to move upward and eventually come into contact with, or intrude, the overlying crust. Intruding magma can affect the crust in several ways, as shown in ***Figure 18-4.*** Magma can force the overlying rock apart and enter the newly formed fissures. Magma can also cause blocks of rock to break off and sink into the magma, where the rocks may eventually melt. Finally, magma can melt the rock into which it intrudes. But what happens deep in the magma chamber as the magma slowly cools?

PLUTONS

Recall from Chapter 5 that when magma cools, minerals form. Over a very long period of time, these minerals will combine to form intrusive igneous rock bodies. Some of these rock bodies are thin, ribbonlike features only a few centimeters thick and several hundred meters long. Others are very large, ranging in size from about 1 km^3 to hundreds of cubic kilometers. These intrusive igneous rock bodies, called **plutons,** can be exposed at Earth's surface as a result of uplift and erosion and are classified based on their size, shape, and relationship to surrounding rocks.

Batholiths and Stocks The largest plutons are called **batholiths.** These irregularly shaped masses of coarse-grained igneous rocks cover at least 100 km^2 and take millions of years to form. Batholiths are common in the interiors of major mountain chains. Many batholiths in North America are composed primarily of granite, the

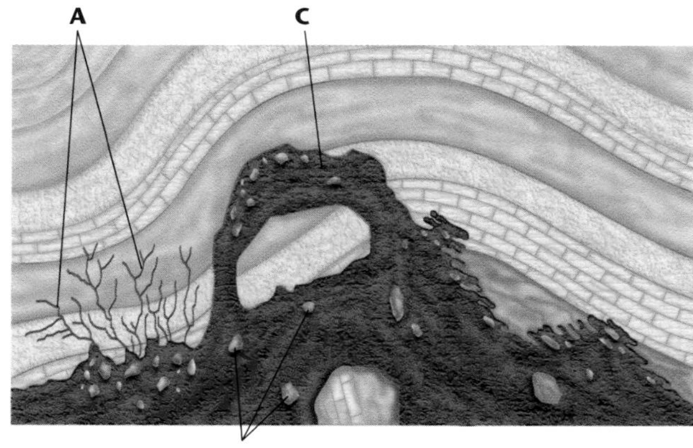

Figure 18-4 Magma can enter fissures in rocks **(A).** Magma can also cause blocks of rock to break off the overlying rock into which the magma intrudes. These blocks of rock become part of the magma body **(B).** Magma can cause the rocks with which it comes in contact to melt **(C).**

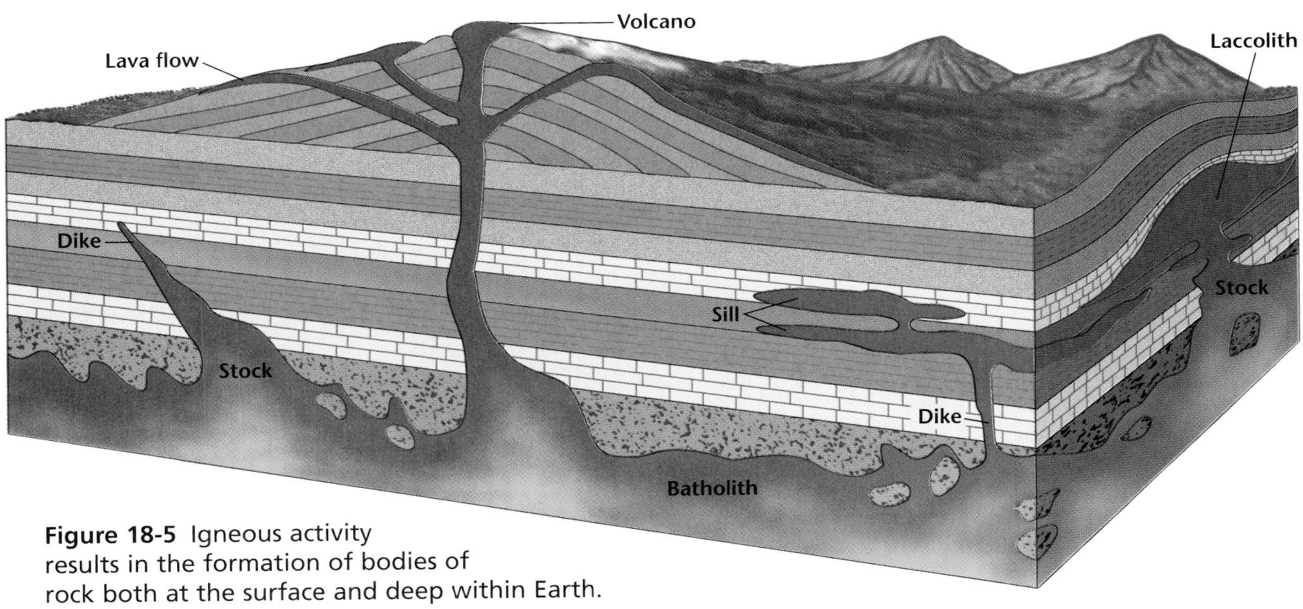

Figure 18-5 Igneous activity results in the formation of bodies of rock both at the surface and deep within Earth.

most common rock type found in plutons. However, gabbro and diorite, the intrusive equivalents of basalt and andesite, are also found in batholiths. The largest batholith in North America, the Coast Range Batholith in British Columbia, is more than 1500 km long. Irregularly shaped plutons that are similar to batholiths but smaller in size are called **stocks.** Both batholiths and stocks, as shown in ***Figure 18-5,*** cut across older rocks and generally form 10–30 km beneath Earth's surface.

Laccoliths Sometimes, when magma intrudes into parallel rock layers close to Earth's surface, some of the rocks bow upward as a result of the intense heat and pressure of the magma body. When the magma solidifies, a laccolith forms. As shown in ***Figure 18-5,*** a **laccolith** is a mushroom-shaped pluton with a round top and flat bottom. Compared to batholiths and stocks, laccoliths are relatively small; they are, at most, up to 16 kilometers wide. Laccoliths exist in the Black Hills of South Dakota, the Henry Mountains of Utah, and the Judith Mountains of Montana, among other places.

Sills and Dikes A **sill** is a pluton that forms when magma intrudes parallel to layers of rock, as shown in ***Figure 18-5.*** A sill can range from only a few centimeters to hundreds of meters in thickness. The Palisades Sill, which is exposed in the cliffs above the Hudson River near New York City, is about 300 m thick. What effect do you think this sill, shown on the next page, had on the sedimentary rocks into which it intruded?

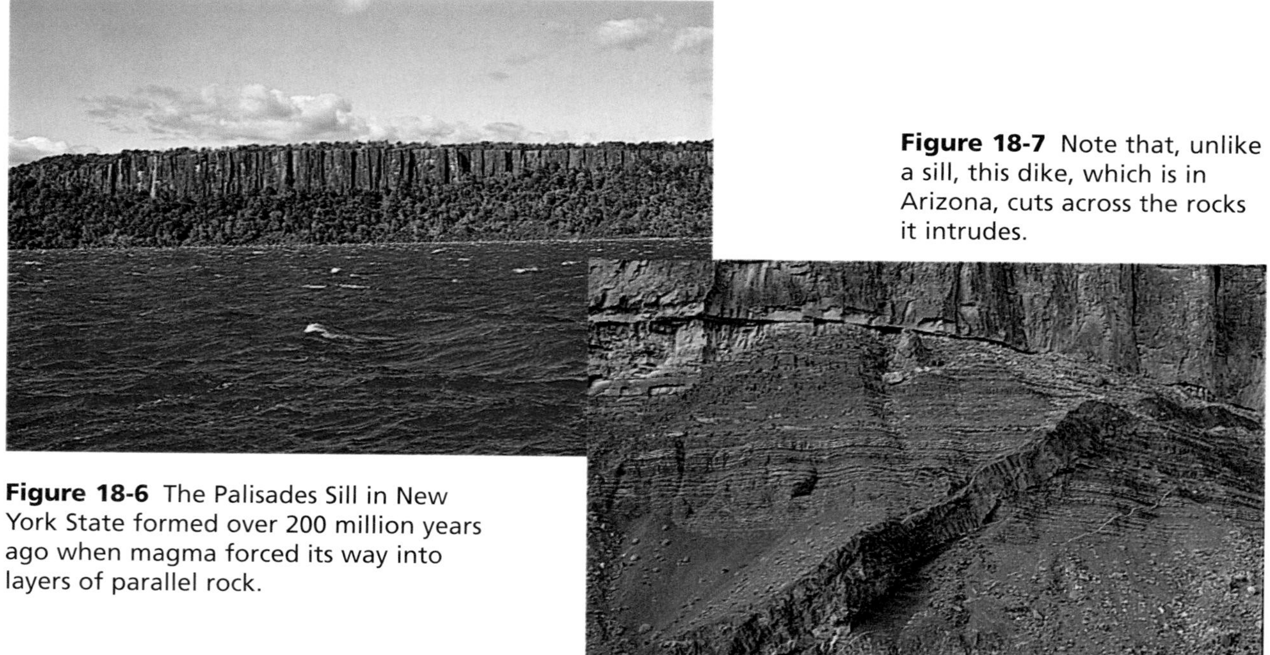

Figure 18-6 The Palisades Sill in New York State formed over 200 million years ago when magma forced its way into layers of parallel rock.

Figure 18-7 Note that, unlike a sill, this dike, which is in Arizona, cuts across the rocks it intrudes.

Unlike the sill shown in ***Figure 18-6,*** which is parallel to the rocks it intrudes, a **dike** is a pluton that cuts across preexisting rocks, as shown in ***Figure 18-7.*** Dikes often form when magma invades cracks in surrounding rock bodies. Most dikes are a few centimeters to several meters wide and up to tens of kilometers long. The Great Dike in Zimbabwe, Africa, however, is an exception: it is about 8 km wide and 500 km long.

While the textures of sills and dikes vary, many of these plutons are coarse grained. Recall from Chapter 5 that grain size is related to the rate of cooling. Coarse-grained sills and dikes are thought to have formed deep in Earth's crust, where the magma cooled relatively slowly to yield large mineral grains.

PLUTONS AND TECTONICS

Many plutons are formed as the result of mountain-building processes. In fact, batholiths are found at the cores of many of Earth's mountain ranges. Where did the enormous volume of magma that cooled to form these igneous bodies come from? Recall from Chapter 17 that many major mountain chains formed along continental-continental convergent plate boundaries. Scientists hypothesize that these collisions might have forced continental crust down into the upper mantle, where it melted, intruded into the overlying rocks, and eventually cooled to form batholiths.

Earth Sciences

3.b Students know the principal structures that form at the three different kinds of plate boundaries.

Figure 18-8 The granite cliffs that tower over Yosemite National Park in California are parts of the Sierra Nevada batholith that have been exposed at Earth's surface.

Batholiths are also thought to have formed as a result of oceanic-oceanic convergence. Again, recall from Chapter 17 that when two oceanic plates converge, one plate is subducted into the mantle. Parts of this subducted plate melt to form magma. The Sierra Nevada batholith, which has been exposed at Earth's surface as a result of uplift and erosion, formed from at least five episodes of igneous activity beneath what is now California. The famous granite cliffs found in Yosemite National Park, some of which are shown in ***Figure 18-8,*** are relatively small parts of this extensive batholith.

The plutons that form deep beneath Earth's surface represent the majority of igneous activity on our planet. Nevertheless, most people think of volcanoes when they hear the words *igneous activity.* These often-spectacular examples of igneous activity at Earth's surface are discussed in the next section.

SECTION ASSESSMENT

1. Discuss three ways in which magma affects the crust into which it intrudes.
2. What are plutons, and how are they classified?
3. How are sills and dikes similar? How do they differ? Give an example of each.
4. What is a laccolith?
5. **Thinking Critically** Sometimes, the texture in the same sill varies: finer grains are found along the margins and coarser grains are found toward the middle of the pluton. What might cause this difference in texture?

SKILL REVIEW

6. **Making a Table** Make a table in which you compare and contrast the different types of intrusive igneous bodies. For more help, refer to the *Skill Handbook.*

SECTION 18.3 Volcanoes

Earth Sciences 3.b, 3.e, 3.f

OBJECTIVES

- **Describe** *the major parts of a volcano.*
- **Compare** *and* **contrast** *shield, cinder-cone, and composite volcanoes.*
- **Contrast** *the volcanism that occurs at plate boundaries.*
- **Explain** *the relationship between volcanism and hot spots.*

VOCABULARY

vent
crater
caldera
shield volcano
cinder-cone volcano
composite volcano
tephra
pyroclastic flow
hot spot

What comes to mind when you hear the word *volcano?* Do you picture clouds of ash and jagged rocks being thrown violently into the air? Or do you envision rivers of reddish-orange lava flowing down the slopes of a steep volcanic peak? Both of these represent volcanic activity on Earth's surface. Volcanism produces various features that alter Earth's landscape. In this section, you will examine some of these features, beginning with the one created at the point where magma reaches the surface: the vent.

ANATOMY OF A VOLCANO

At the beginning of this chapter, you learned that magma chambers deep within Earth fuel the volcanoes that erupt at the planet's surface. Also recall that when magma reaches Earth's surface, it is called lava. Lava erupts through an opening in the crust called a **vent.** As lava flows out onto the surface, it cools and solidifies around the vent. Over time, the lava can accumulate to form a mountain known as a volcano. At the top of a volcano, around the vent, is a bowl-shaped depression called a **crater.** The crater is connected to the magma chamber by the vent. Locate the crater of the volcano shown in ***Figure 18-9.***

Figure 18-9 A crater is the bowl-shaped depression that surrounds the central vent at a volcano's summit. The volcano shown below is one of many that dot the northern Arizona landscape near Flagstaff.

Figure 18-10 The caldera now known as Crater Lake is located in Oregon. This caldera formed about 6600 years ago as a result of numerous volcanic eruptions of Mount Mazama.

Figure 18-11 Crater Lake formed as the result of many eruptions.

Mount Mazama erupted many times.

Mount Mazama

Magma chamber

The top of partially empty magma chamber collapsed.

Pyroclastic flow

Pyroclastic flow

Steam caused explosions.

Caldera eventually filled with water to form lake.

Wizard Island

Crater Lake

Volcanic craters are usually less than 1 km in diameter. Larger depressions called **calderas,** which can be up to 50 km in diameter, however, can form when the summit or the side of a volcano collapses into the magma chamber that once fueled the volcano. The caldera now known as Crater Lake formed in this way, as shown in ***Figure 18-11.*** The caldera walls, which are visible in the photograph in ***Figure 18-10,*** form cliffs that tower nearly 600 m above the water's surface. Wizard Island, which is located in the center of the lake, is actually a small volcanic cone that formed after the caldera collapsed.

TYPES OF VOLCANOES

The appearance of a volcano depends on two factors: the type of material that forms the volcano and the type of eruptions that occur. Based on these two criteria, three major types of volcanoes have been identified: shield volcanoes, cinder-cone volcanoes, and composite volcanoes. Each differs in size, shape, and composition.

Shield Volcanoes A **shield volcano** is a mountain with broad, gently sloping sides and a nearly circular base. Shield volcanoes form when layer upon layer of basaltic lava accumulates during

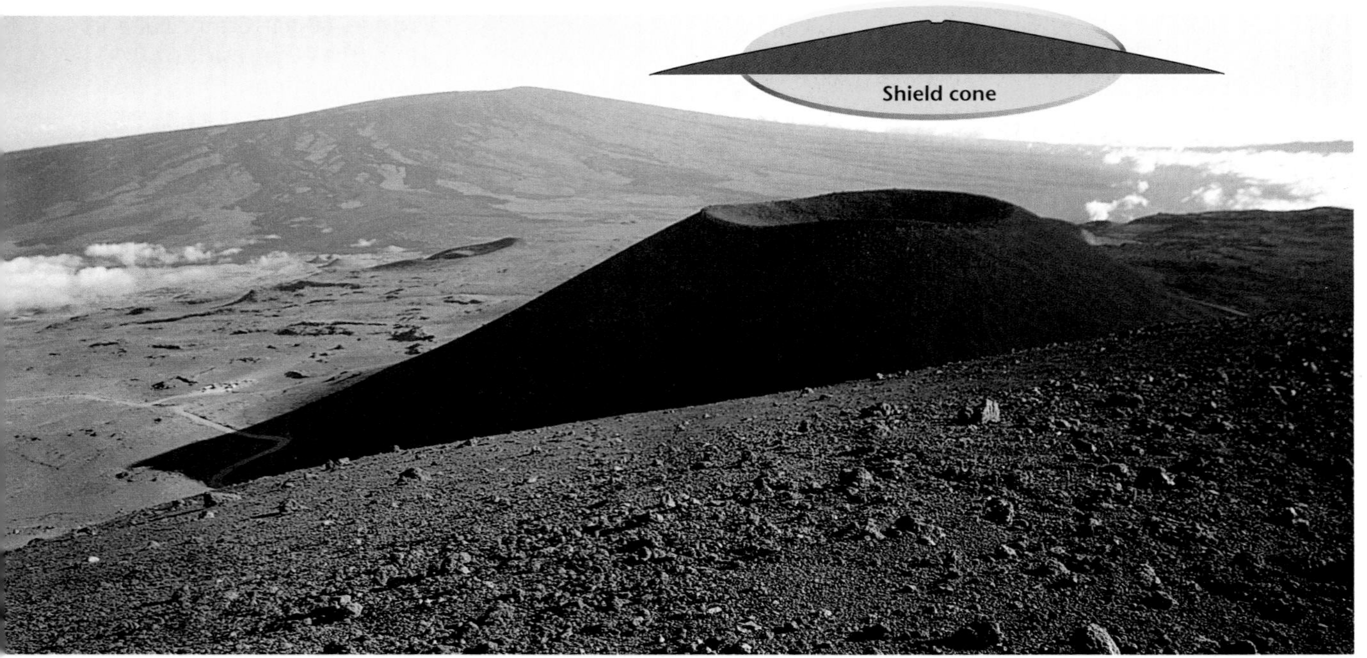

Figure 18-12 Mauna Loa, shown in the distance, is a shield volcano in Hawaii. A small cinder-cone volcano on the flank of Mauna Kea is visible in the foreground.

nonexplosive eruptions. Recall that eruptions involving basaltic lava are less explosive than other eruptions. This is because basaltic lava has a low viscosity as a result of the relatively small amounts of gases and silica it contains. The shield volcanoes that make up the Hawaiian Islands are made of basalt. Mauna Loa, which is shown in ***Figure 18-12,*** is one such volcano.

Cinder-Cone Volcanoes A **cinder-cone volcano** forms when material ejected high into the air falls back to Earth and piles up around the vent. Cinder-cone volcanoes have steep sides, as shown in ***Figure 18-13,*** and are generally small; most are less than 500 m high. The magma that fuels cinder-cone volcanoes contains more water and silica than the magma that fuels shield volcanoes. This more viscous magma also contains large volumes of gases, which make cinder-cone volcanoes more explosive in nature than shield volcanoes.

To learn more about volcanoes, go to the **National Geographic Expedition** on page 886.

Composite Volcanoes **Composite volcanoes** form when layers of volcanic fragments alternate with lava. As with cinder-cone volcanoes, the magma that forms composite volcanoes commonly contains large amounts of silica, water, and gases. Composite volcanoes are much larger than cinder-cone volcanoes, and, because of their violently explosive nature, they are potentially dangerous to humans and the environment. Two composite volcanoes of the Cascade Range in the western United States, Mount St. Helens and Mount Rainier, are shown in ***Figure 18-14.***

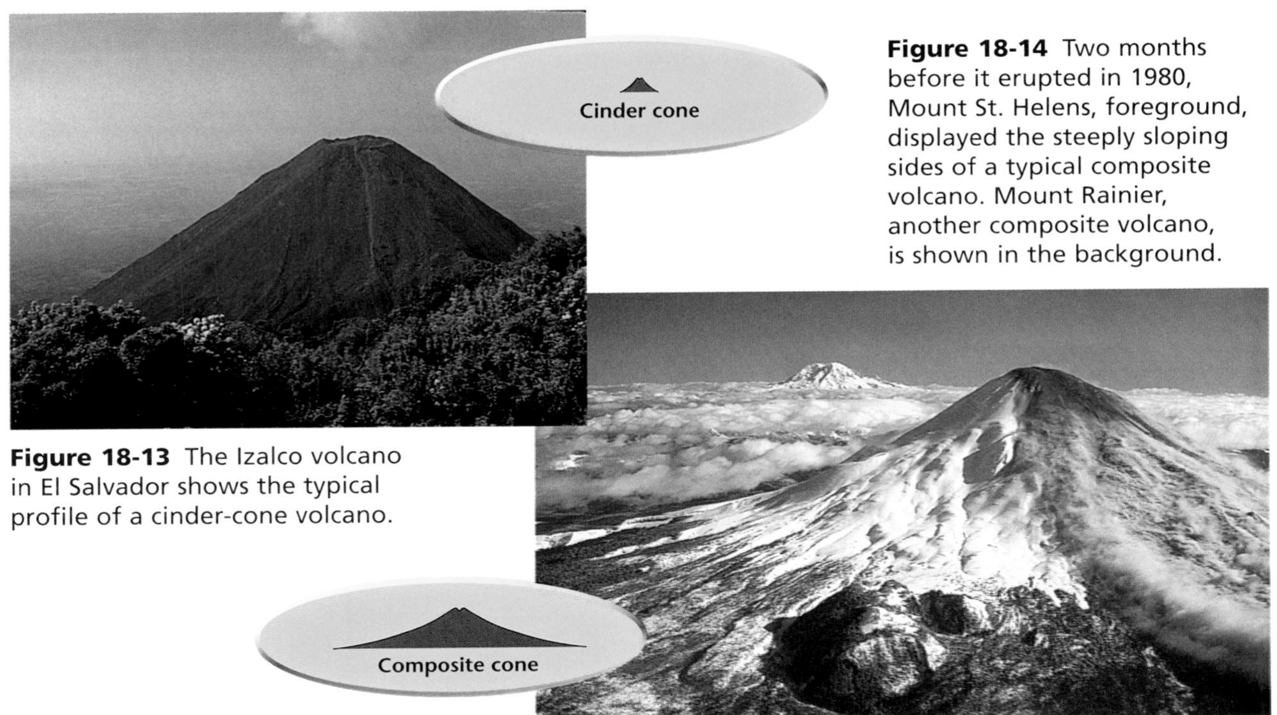

Figure 18-14 Two months before it erupted in 1980, Mount St. Helens, foreground, displayed the steeply sloping sides of a typical composite volcano. Mount Rainier, another composite volcano, is shown in the background.

Figure 18-13 The Izalco volcano in El Salvador shows the typical profile of a cinder-cone volcano.

Size and Slope Look at the small sketches that show the relative sizes of the three types of volcanoes in ***Figures 18-12*** through ***18-14.*** These diagrams are drawn to scale. As you can see, shield volcanoes are by far the largest. The smallest volcanoes are cinder-cone volcanoes, which often form on or very near larger volcanoes. Notice, too, that cinder-cone volcanoes have the steepest slopes, while shield volcanoes have the gentlest slopes. The slopes of cinder-cone and composite volcanoes are concave, and the slopes of shield volcanoes are straight. These differences in both size and slope are the result of many factors, including the different kinds of materials that make up each volcano, the vegetation that grows on the volcano's slopes, local climate, and the eruptive history of the volcano.

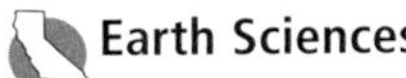

Earth Sciences

3.b Students know the principal structures that form at the three different kinds of plate boundaries.

3.e Students know there are two kinds of volcanoes: one kind with violent eruptions producing steep slopes and the other kind with voluminous lava flows producing gentle slopes.

Volcanic Material

Rock fragments thrown into the air during a volcanic eruption are called **tephra.** Tephra can be newly cooled and hardened lava, mineral grains that started to crystallize prior to the eruption, or pieces of the volcanic cone. Tephra are classified by size. The smallest fragments, called dust, are less than 0.25 mm in diameter. Ash, another kind of tephra, is larger than dust but less than 2 mm in diameter. Somewhat larger fragments of tephra are called lapilli, an Italian word that means "little stones." Lapilli are larger than 2 mm but less than 64 mm in diameter. The largest tephra thrown from a volcano

Figure 18-15 More than 29 000 people died as a result of the pyroclastic flow that accompanied the 1902 eruption of Mount Pelée on the island of Martinique. Note that much of the city of St. Pierre was destroyed.

can be the size of a car or a small building. When these large volcanic fragments are angular, they are called volcanic blocks. Volcanic blocks as large as houses have been ejected more than 10 km into the air during some eruptions. When blobs of lava are forcefully ejected from a volcano, they may cool to form rounded or streamlined tephra called volcanic bombs. Volcanic bombs may harden in the air or they may flatten and solidify after they hit the ground.

Topic: Mt. St. Helens
To learn more about damage caused by volcanoes, visit the Earth Science Web Site at earthgeu.com

Activity: Research the eruption of Mt. St. Helens in 1980. Prepare a brief report or media presentation describing the change in the landscape that occurred.

Pyroclastic Flows Some tephra cause tremendous damage and kill thousands of people. Violent volcanic eruptions can send clouds of gas, ash, and other tephra down a slope at incredible speeds. This rapidly moving volcanic material, which is called a **pyroclastic flow,** can travel at speeds of nearly 200 km/h and may contain hot, poisonous gases. The temperature at the center of a pyroclastic flow can exceed 700°C. One of the most widely known and deadly pyroclastic flows occurred in 1902 on Mount Pelée, on the island of Martinique in the Caribbean Sea. More than 29 000 people suffocated or were burned to death. What little was left of the town of St. Pierre after the eruption is shown in ***Figure 18-15.***

WHERE DO VOLCANOES OCCUR?

The distribution of volcanoes on Earth's surface is not random. Most volcanoes form at plate boundaries. In fact, about 80 percent of all volcanoes are found along convergent boundaries, and about 15 percent are found along divergent boundaries. Only about 5 percent of extrusive igneous activity occurs far from plate boundaries.

Convergent Volcanism Recall from Chapter 17 that plates come together along convergent boundaries. Also recall that convergence involving oceanic plates creates subduction zones, places where slabs of oceanic crust descend into the mantle and eventually melt. The magma generated is forced upward through the overlying plate and forms volcanoes when it reaches the surface.

The volcanoes associated with convergent plate boundaries form two major belts, as shown in ***Figure 18-16.*** The larger belt, the Circum-Pacific Belt, is also called the Pacific Ring of Fire. It stretches along the western coasts of North and South America, across the Aleutian Islands, and down the eastern coast of Asia. Volcanoes in the Cascade Range of the western United States, and Mount Pinatubo in the Philippines are some of the volcanoes in the Circum-Pacific Belt. The smaller belt, which is called the Mediterranean Belt, includes Mount Etna and Mount Vesuvius, two composite volcanoes in Italy.

Divergent Volcanism Volcanic activity is also common along divergent plate boundaries, where two plates are moving apart. Magma is forced upward into the fractures and faults that form as the plates separate. These areas of major faults and fractures are called rift zones. Most of the world's rift volcanism occurs under water along ocean ridges. Recall from Chapter 17 that this type of volcanism results in the formation of new ocean floor during the

Update For an online update of recent volcanic eruptions, visit the Earth Science Web Site at earthgeu.com

Earth Sciences

3.f Students know the explanation for the location and properties of volcanoes that are due to hot spots and the explanation for those that are due to subduction.

Figure 18-16 Most of Earth's volcanoes form two distinct volcanic belts: the larger Circum-Pacific Belt and the much smaller Mediterranean Belt.

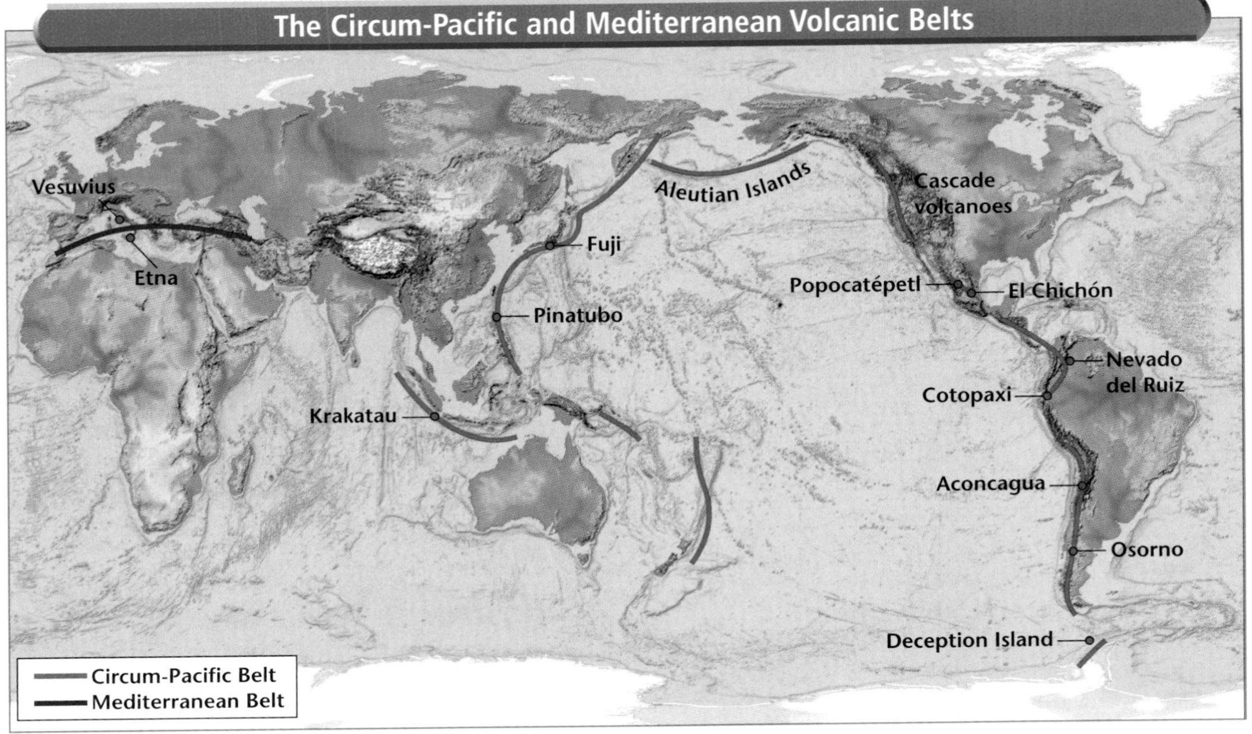

Using Numbers Look at ***Figure 18-17.*** Note that the distance from Daikakuji Seamount to Hawaii is about 3500 km. Daikakuji is 43 million years old. What is the average speed of the Pacific Plate?

process of seafloor spreading. One of the few places where rift volcanism can be observed above sea level today is in Iceland. This island is a part of the Mid-Atlantic Ridge, and consequently, several active volcanoes dot the landscape.

Hot Spots Some volcanoes are located far from plate boundaries. These volcanoes form as the result of **hot spots,** which are unusually hot regions of Earth's mantle where high-temperature plumes of mantle material rise toward the surface. Plumes originate deep within the mantle, or perhaps even near the core-mantle boundary. The intense heat of the plumes melts rock, which is then forced upward toward the crust as magma. The magma, in turn, melts through the crust to form volcanoes. While a plume does move vertically, it does not move laterally. As a result, a trail of progressively older volcanoes forms as a plate moves over a hot spot.

Some of Earth's best known volcanoes formed as a result of hot spots under the Pacific Ocean. The Hawaiian Islands, for example, continue to rise above the ocean floor as the Pacific Plate moves slowly over a hot spot. The volcanoes on the oldest island, Kauai, are inactive because the island no longer sits above the hot spot. The world's most active volcano, Kilauea, is on the big island of Hawaii and is currently located over the hot spot. Another volcano, Loihi, is forming on the seafloor east of the big island of Hawaii and may eventually break the ocean surface to form a new island.

Problem-Solving Lab

Making and Using Graphs

Calculate and graph how fast lava flows On June 8, 1783, the Laki fissure zone in Iceland began to erupt in what would become the largest flood basalt in recent history. A flood basalt forms when lava flows from fissures to create a vast plain or plateau. The Laki eruption resulted in a total volume of 14.73 km^3 of basalt, which covered 565 km^2. The lava erupted from fissures located 45 km from the coast, and flowed at speeds averaging 0.4 km/h.

Analysis

1. Design a data table to show the distance traveled by the lava over a five-day period. Calculate the distance every 12 hours.
2. Plot the data on a graph: put time on the *x*-axis and distance on the *y*-axis.

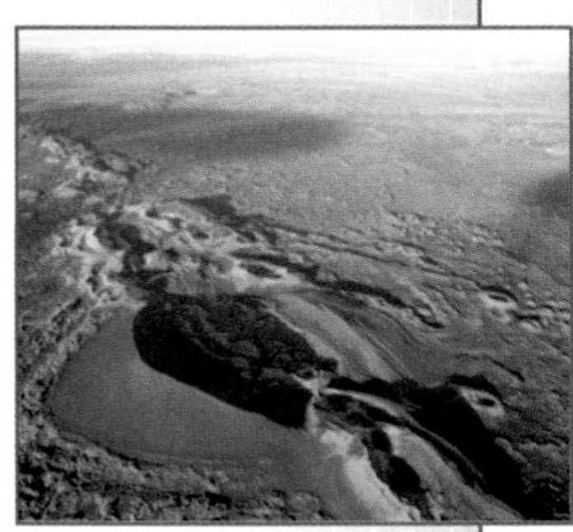

Thinking Critically

3. How long did it take the lava to reach the coast?
4. How many kilometers did the lava travel in three days?

The chains of volcanoes that form over hot spots provide important information about plate motions. The rate and direction of motion can be calculated from the positions of these volcanoes. Even changes in plate motion that occurred in the distant past can be determined. Look at ***Figure 18-17.*** Note that the Hawaiian Islands are at one end of the 5800-km Hawaiian-Emperor volcanic chain. The oldest seamount, Meiji, is at the other end of the chain and is about 75–80 million years old, which indicates that this hot spot has existed for at least that many years. The bend in the chain at Daikakuji Seamount records a change in the direction of the Pacific Plate that occurred about 43 million years ago.

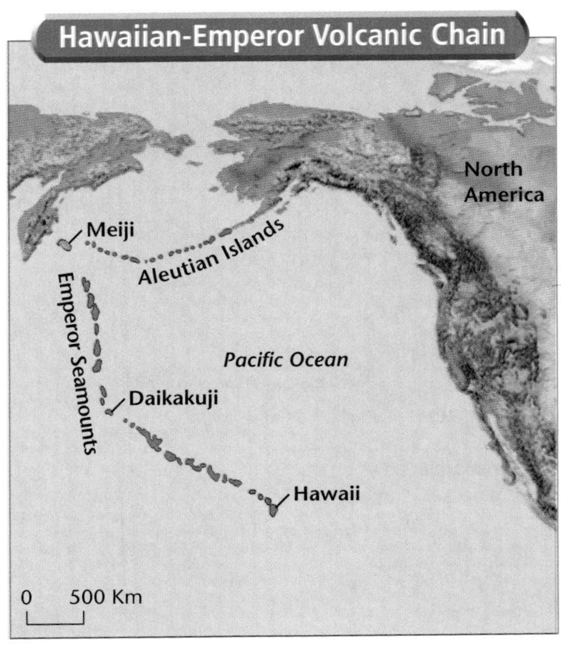

Figure 18-17 The Emperor Seamounts and the Hawaiian Islands continue to form as the Pacific Plate moves over a stationary hot spot in the mantle.

In addition to seamount chains, hot spots can result in the formation of flood basalts. Flood basalts erupt from fissures rather than a central vent and form flat plains or plateaus rather than volcanic mountains. The volume of basalt in these eruptions can be tremendous. The Columbia River Basalts in the northwestern United States, for example, contain 170 000 km^3 of basalt. The volume of basalt in the Deccan Traps in India is estimated to be 512 000 km^3. The volume of basalt in the Laki eruption in Iceland, which you can analyze in the *Problem-Solving Lab* on page 486, is small by comparison at 14.73 km^3.

Volcanic activity is proof that Earth is a dynamic planet. And, while many volcanic eruptions can be spectacular events, these geologic phenomena can pose risks to humans and their environment. In the *Internet GeoLab* that follows, you will research and rank some of Earth's potentially dangerous volcanoes.

SECTION ASSESSMENT

1. What is a volcanic crater, and how does it differ from a caldera?
2. Describe the different kinds of tephra.
3. Explain why volcanic blocks would be uncommon on shield volcanoes.
4. What is a pyroclastic flow? What are the characteristics of a pyroclastic flow that make them so dangerous?
5. Where are Earth's major volcanic belts located?
6. What are hot spots?
7. **Thinking Critically** The slopes of composite volcanoes are notoriously unstable and prone to landslides. Why?

SKILL REVIEW

8. **Comparing and Contrasting** Compare and contrast the characteristics of the three major types of volcanoes. For more help, refer to the *Skill Handbook*.

earthgeu.com/self_check_quiz

Internet GeoLab

Ranking Hazardous Volcanoes

Earth Sciences

3.e Students know there are two kinds of volcanoes: one kind with violent eruptions producing steep slopes and the other kind with voluminous lava flows producing gentle slopes. **Also covers:** 1.a

Some volcanoes can be explosively dangerous. Along with clouds of ash and other volcanic debris that can linger in the air for years after an eruption, pyroclastic flows, landslides, and mudflows are common volcanic hazards. An explosive volcano may not be a hazard to human life and property, however, if it is located in a remote area or erupts infrequently. A number of factors must be taken into account to determine if a particular volcano poses a risk.

Preparation

Problem

Which volcanoes on our planet pose the greatest risk to human life and property?

Mount St. Helens May 18, 1980

Hypothesis

Form a hypothesis about where you think the most hazardous volcanoes are located on Earth. Think about the potential risk to people and property near the volcano when formulating your hypothesis.

Objectives

- **Gather** and **communicate** data about three volcanoes in different parts of the world.
- **Form conclusions** about the hazards posed by the volcanoes based on their location, size, lava type, and eruptive history.

Data Sources

Go to the Earth Science Web Site at earthgeu.com to find links to volcano data on the Internet. You can also use current reference books and scientific magazines to aid in the collection of data.

Ranking of Some Hazardous Volcanoes			
Volcano name			
Country			
Location of volcano (latitude and longitude)			
Type of volcano			
Composition of lava/ Explosiveness			
Date of last eruption			
Eruption interval (number of eruptions over a period of time)			
Height of volcano			
Distance to nearest population center			
Approximate number of people living near the volcano			
Type(s) of potential hazards			
Human hazard ranking (high, medium, low)			

Procedure

1. Select a country and find out if there are any volcanoes in that country. If there are no volcanoes, choose another country. If there are a lot of volcanoes in that country, narrow your search.
2. Repeat step 1 for at least two other volcanoes. Copy and complete the data table with the information about each of the volcanoes you selected.

Conclude & Apply

Sharing Your Data Find this Internet GeoLab on the Earth Science Web Site at earthgeu.com. Post your data in the table provided for this activity. Use the additional data from other students to complete your chart and answer the Conclude & Apply questions.

1. Which of the volcanoes you researched threatens the greatest number of people? Where is this volcano located?
2. Analyze the data posted by others at the earthgeu.com site. Which country has the greatest number of potentially dangerous volcanoes? Why?
3. Which country has the greatest total population threatened by volcanoes?

Science & the Environment

Earth Sciences 3.e

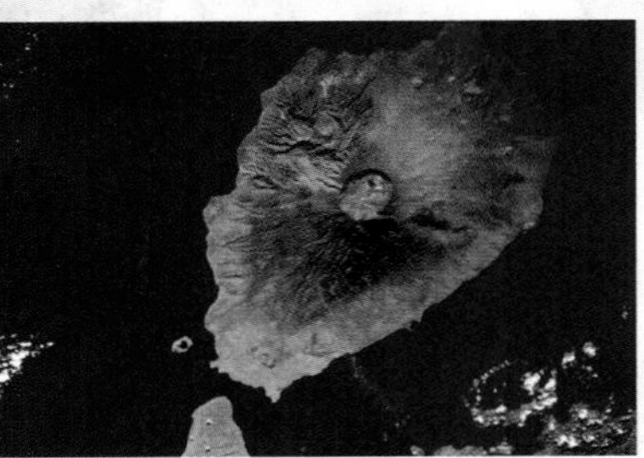

Tambora

The Year Without a Summer

In 1816, Chauncey Jerome, a Connecticut clockmaker, reported that the clothes his wife had hung out to dry the day before had frozen overnight. This would not have been significant, except for the date—it was June 10.

What Jerome and other New Englanders experienced during the cold summer of 1816 was directly linked to an event that had occurred one year earlier and thousands of kilometers away. On April 5, 1815, Mount Tambora, a dormant volcano in Indonesia, came alive in a series of explosive eruptions.

Tambora's Direct Impact

Historic reports attest to Tambora's explosive power. The eruption was heard on Jakarta, which is more than 1200 km away. Over the course of that April week, the volcano ejected an estimated 150 km^3 of tephra into the ocean and onto surrounding islands. By contrast, the volume of debris erupted during the 1980 explosion of Mount St. Helens was only 1 km^3.

Volcanic ash from Tambora hung thickly in the sky and caused a three-day period of darkness within 600 km of the island. By the time the eruptions stopped, more than a third of the 3900-m mountain had been blown off. Some 12 000 people were killed directly by volcanic fallout, and more than 80 000 died soon after from famine and disease. The disaster, however, was not over. Its global effects would be felt around the world the following year—the year without a summer.

Indirect Consequences

Tambora spewed an immense amount of volcanic dust and gases such as sulfur dioxide into the atmosphere. These particles prevented sunlight from reaching Earth's surface. In effect, the short wavelengths of incoming sunlight, which are similar in size to particles of dust and gas, collided with the particles and were reflected back into space. The problem was worsened when heat radiated from Earth's surface, which takes the form of longer wavelengths, escaped into space.

The net result was wildly fluctuating weather on a global scale. A snowfall in southern Italy, unusual in itself, caused widespread alarm because the snow was tinted red from the volcanic ash. In New England, summer temperatures dipped and soared from about 2°C to over 31°C within a matter of days. Crops were devastated.

At the time, the cause of the climatic changes was not understood; no one linked the changes to the eruption of Tambora. Today, however, we know that volcanic gases can linger in the atmosphere for years after an eruption and wreak havoc on the weather.

Internet

Not all volcanic eruptions have negative effects. Go to earthgeu.com for links to information on the eruptions that occurred in northern Arizona some 1000 years ago. How did these eruptions affect the Sinagua? What positive impacts did the eruptions have?

CHAPTER 18

Study Guide

Summary

Section 18.1

Magma

Main Ideas

- Temperature, pressure, and the presence of water are factors that affect the formation of magma.
- As pressure increases, the temperature at which a substance melts also increases. At any given pressure, the presence of water will cause a substance to melt at a lower temperature than the same substance under dry conditions.
- There are three major types of magma: basaltic magma, andesitic magma, and rhyolitic magma. These magmas differ in the source rock from which they form, viscosity, silica content, gas content, and explosiveness. Basaltic magma is the least explosive magma; rhyolitic magma is the most explosive.

Vocabulary

viscosity (p. 474)

Section 18.2

Intrusive Activity

Main Ideas

- Magmatic intrusions affect the crust in several ways. Magma can force overlying rock apart and enter the newly formed fissures. Magma can also cause blocks of rock to break off and sink into the magma chamber. Magma can melt the rock into which it intrudes.
- Batholiths, stocks, sills, dikes, and laccoliths are plutons that are classified according to their size, shape, and relationship to surrounding rocks. Batholiths are the largest plutons and often form the cores of many of Earth's major mountain chains.

Vocabulary

batholith (p. 476)
dike (p. 478)
laccolith (p. 477)
pluton (p. 476)
sill (p. 477)
stock (p. 477)

Section 18.3

Volcanoes

Main Ideas

- Lava flows onto Earth's surface through a vent. Over time, multiple lava flows may accumulate to form a volcano. A crater is a depression that forms around the vent at the summit of a volcano. A caldera is a large crater that forms when a volcano collapses during or after an eruption.
- There are three types of volcanoes: shield volcanoes, cinder-cone volcanoes, and composite volcanoes.
- Rock fragments ejected during eruptions are called tephra.
- Most volcanoes form along convergent and divergent plate boundaries. Volcanoes also form over hot spots, which are unusually hot areas in the mantle that are stationary for long periods of time.
- Flood basalts form when lava flows from fissures to form flat plains or plateaus.

Vocabulary

caldera (p. 481)
cinder-cone volcano (p. 482)
composite volcano (p. 482)
crater (p. 480)
hot spot (p. 486)
pyroclastic flow (p. 484)
shield volcano (p. 481)
tephra (p. 483)
vent (p. 480)

earthgeu.com/vocabulary_puzzlemaker

CHAPTER 18
Assessment

Understanding Main Ideas

1. Which of the following does NOT play a role in magma formation?
   a. temperature
   b. pressure
   c. presence of water
   d. tephra type

2. Which of the following is true?
   a. An increase in pressure results in a higher melting temperature of a dry substance.
   b. A decrease in pressure increases the temperature at which a dry substance melts.
   c. The addition of water increases the melting temperature of a substance.
   d. An increase in pressure decreases the melting temperature of a dry substance.

3. Which of the following melts to form rhyolitic magma?
   a. continental crust
   b. oceanic crust
   c. oceanic sediment
   d. the upper mantle

4. Which type of pluton is completely parallel to the rock layers into which it intrudes?
   a. dike
   b. sill
   c. laccolith
   d. stock

5. The Hawaiian volcanoes formed as a result of which of the following?
   a. divergence
   b. a hot spot
   c. subduction
   d. subsidence

6. Which of the following is NOT true?
   a. An increase in silica increases the viscosity of a magma.
   b. Andesitic magma has both an intermediate gas content and explosiveness.
   c. An increase in temperature increases a magma's viscosity.
   d. Basaltic magma has a low viscosity and contains little gas.

7. What is the largest type of tephra?
   a. ash
   b. volcanic blocks
   c. dust
   d. lapilli

8. Which of the following has broad, gently sloping sides and a circular base?
   a. hot spot
   b. cinder-cone volcano
   c. composite cone
   d. shield volcano

9. What is the Ring of Fire, and why does it exist?

10. Explain the relationship between hot spots and volcanism.

Use the table to answer questions 11–15.

Economic Losses (millions of dollars) from the 1980 Eruption of Mount St. Helens

Sector	Federal	Private	State	Local
Forestry	168.0	218.1	63.7	——
Clean-up	307.9	9.7	5.0	41.3
Property	43.6	44.8	2.5	16.0
Agriculture	——	39.1	——	——
Income	——	8.9	——	——
Transportation	——	——	——	2.1

11. What was the total economic cost of cleaning up after the eruption?

12. What was the total economic loss from this eruption?

13. What percent of the total loss was caused by property damage?

14. Which sector suffered the smallest loss?

15. Which sector suffered the greatest economic loss? What percent of the total was this?

Test-Taking Tip

STANDARDIZED TEST FORMS Fill in one answer bubble as you answer each question. If you need to skip a question, make sure you skip the corresponding bubble on the answer sheet also.

Assessment

Applying Main Ideas

16. As rhyolitic magma rises to Earth's surface, pressure decreases and water escapes from the magma. What effect does this have on the melting temperature? How might this cause the magma to solidify before reaching the surface?
17. How does magma affect the rocks into which it intrudes?
18. Hawaiian lava flows can travel great distances through underground passageways called lava tubes. Why would lava flow faster through a lava tube than it would above ground?
19. Describe batholiths and explain where and how they form.
20. What is a laccolith and how does it form?
21. Explain the relationship among a vent, a crater, and a caldera.

Thinking Critically

22. Soils that form from volcanic debris are very productive. What are some reasons for the high fertility of volcanic soils?
23. Pumice, a volcanic glass that contains such a large percentage of holes that it floats in water, is almost never basaltic in composition. Why?
24. Which type of volcano would you expect to produce the largest volume of tephra? Explain.
25. Why do shield volcanoes have gentle slopes and large bases?
26. Geothermal energy associated with magma chambers close to Earth's surface can be used to produce electricity. Name several places in the United States where the use of this energy might be possible.

Standardized Test Practice

INTERPRETING DIAGRAMS Use the diagram below to answer the following questions.

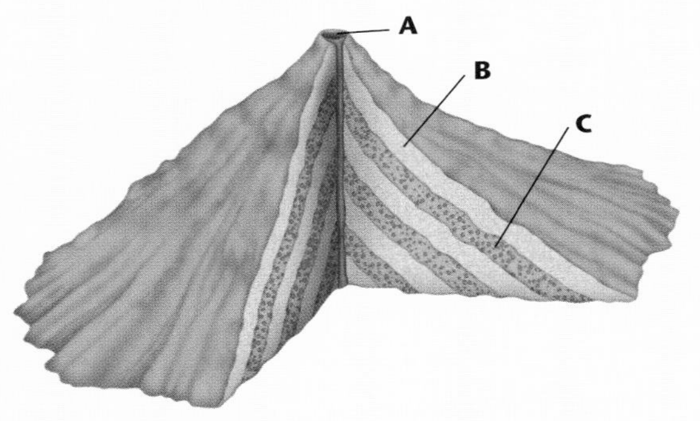

1. What kind of volcano is shown in the diagram?
   a. cinder-cone volcano
   b. composite volcano
   c. shield volcano
   d. hot-spot volcano

2. What kind of volcanic feature is designated by the letter A?
   a. the vent
   b. the magma chamber
   c. the crater
   d. the sill

3. What type of material makes up the layer designated by the letter B?
   a. lava
   b. flood basalts
   c. tephra
   d. volcanic gases

4. What type of material makes up the layer designated by the letter C?
   a. lava
   b. flood basalts
   c. tephra
   d. volcanic gases

5. Which of the following is NOT true of this type of volcano?
   a. It erupts violently.
   b. The magma that fuels it is rich in silica.
   c. It forms over a hot spot in Earth's mantle.
   d. It has concave slopes.

Chapter 19

Earthquakes

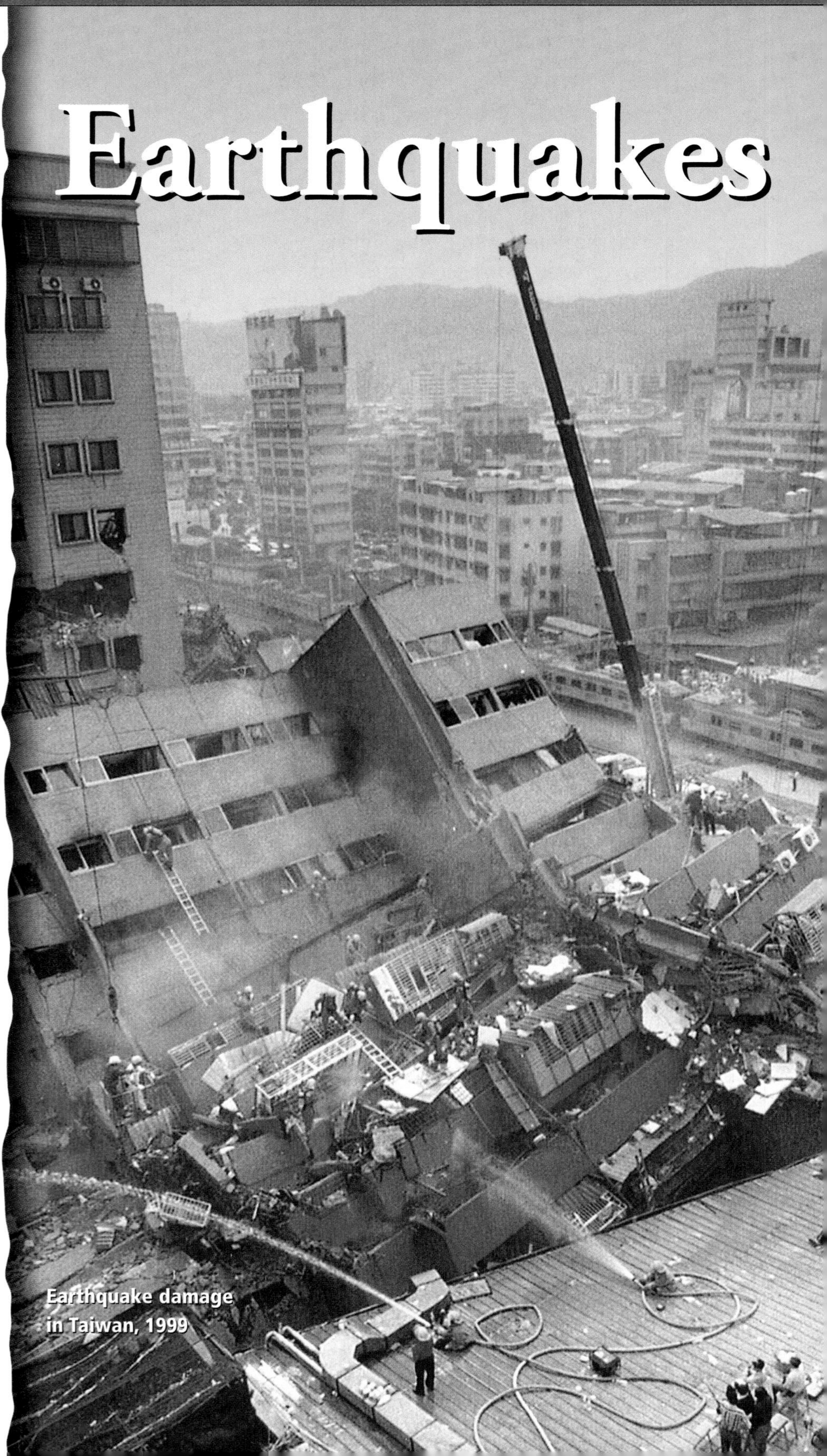

What You'll Learn

- What causes earthquakes and how they affect Earth's surface.
- How earthquakes and the destruction they cause are measured.
- What factors determine seismic risk.

Why It's Important

Earthquakes are natural phenomena that can cause vast amounts of damage as well as many deaths. Understanding what causes earthquakes is essential to our being prepared for these natural disasters.

To learn more about earthquakes, visit the Earth Science Web Site at earthgeu.com

Earthquake damage in Taiwan, 1999

Discovery Lab Model An Earthquake

Earthquakes are natural vibrations of the ground. Most quakes are caused by movement along enormous fractures in Earth's crust. In this activity, you will model how movements along these fractures can cause earthquakes.

1. Slide the largest surfaces of two smooth wooden blocks against each other. Describe the movement.
2. Cut two pieces of coarse-grained sandpaper so that they are about 1 cm larger than the largest surface of each block.
3. Place the sandpaper, coarse side up, against the largest surface of each block. Wrap the paper over the edges of the blocks and secure it with thumbtacks.
4. Slide the sandpaper-covered sides of the blocks against each other. What happens?

Observe and Infer In your science journal, compare the two movements. Infer which of the two scenarios models what happens during an earthquake.

SECTION 19.1 Forces Within Earth

Earth Sciences 3.d, 9.b

OBJECTIVES

- **Define** *stress and strain as they apply to rocks.*
- **Distinguish** *among the three types of faults.*
- **Contrast** *three types of seismic waves.*

VOCABULARY

stress
strain
fault
primary wave
secondary wave
surface wave
focus
epicenter

Earthquakes are natural vibrations of the ground caused by movement along gigantic fractures in Earth's crust, or sometimes, by volcanic eruptions. If you've experienced an earthquake or heard about quakes in the news, you know that they can be extremely destructive. There are some instances in which a single earthquake has killed more than 100 000 people. Earthquakes have even destroyed entire cities. Anyone living in an area prone to earthquakes should be aware of the potential danger posed by these events and how to minimize the damage that they cause.

STRESS AND STRAIN

Most earthquakes occur when rocks fracture, or break, deep within Earth. Fractures form when **stress,** the forces per unit area acting on a material, exceeds the strength of the rocks involved. There are three kinds of stress that act on Earth's rocks: compression, tension, and shear. Compression is stress that decreases the volume of a material, tension is stress that pulls a material apart, and shear is stress that causes a material to twist. The deformation of materials in response to stress is called **strain.** ***Figure 19-1*** illustrates the strain caused by compression, tension, and shear.

Laboratory experiments on rock samples show a distinct relationship between stress and strain. When the stress applied to a rock is plotted against strain, a stress-strain curve, like the one shown in ***Figure 19-2,*** is produced. A stress-strain curve usually has two segments: a straight segment and a curved segment. Low stresses produce the straight segment, which represents the elastic strain of a material. Elastic strain causes a material to bend and stretch, and can be demonstrated by gently applying tension to a rubber band. When this tensional stress is released, the rubber band returns to its original size and shape. In ***Figure 19-2,*** note that elastic strain is proportional to stress, and thus, if the stress is reduced to zero, the strain, or deformation, disappears.

Ductile Deformation When stress exceeds a certain value, however, a material undergoes ductile deformation, shown by the curved segment of the graph in ***Figure 19-2.*** Unlike elastic strain, this type of strain produces permanent deformation, which means that the material stays deformed even if the stress is reduced to zero. A rubber band undergoes ductile deformation when it is stretched beyond its elastic limit. This permanent deformation results in an increase in size and produces slight tears or holes in the band. When stress exceeds the strength of a material, the material breaks, or fails, as designated by the *X* on the graph. From experience, you probably know that exerting too much tension on a rubber band will cause it to snap.

Most materials exhibit both elastic and ductile behavior. Brittle materials, such as glass, certain plastics, and dry wood, fail before much ductile deformation occurs. Ductile materials such as rubber,

Undeformed material

A Compressional strain

B Tensional strain

C Shear strain

Figure 19-1 Compression causes a material to shorten **(A).** Tension causes a material to lengthen **(B).** Shear causes distortion of a material **(C).**

Earth Sciences

3.d Students know why and how earthquakes occur and the scales used to measure their intensity and magnitude.

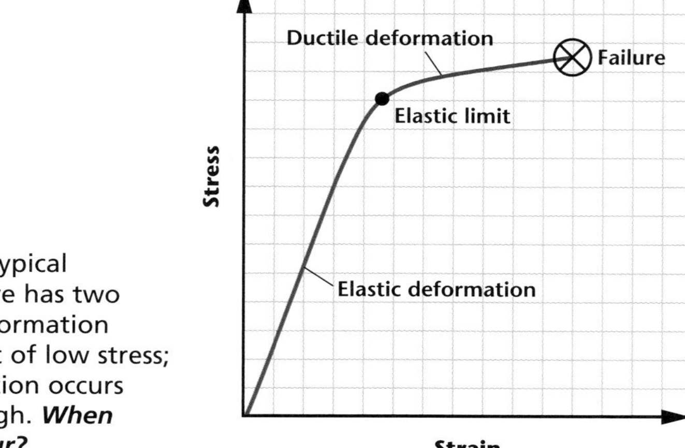

Figure 19-2 A typical stress–strain curve has two parts. Elastic deformation occurs as a result of low stress; ductile deformation occurs when stress is high. ***When does failure occur?***

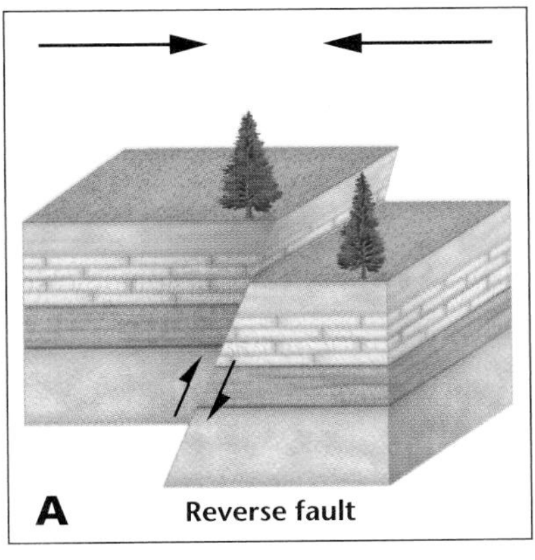

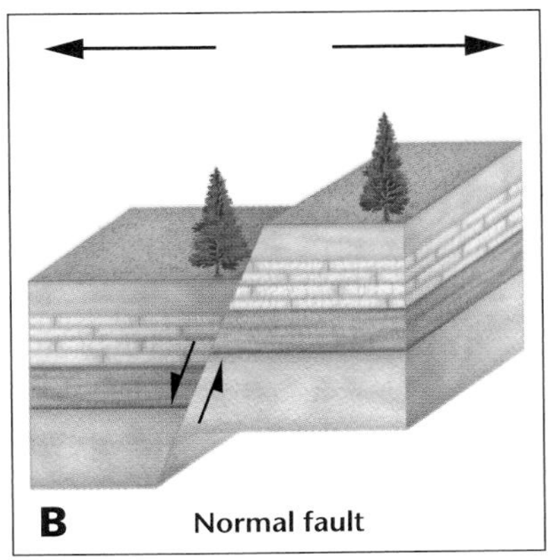

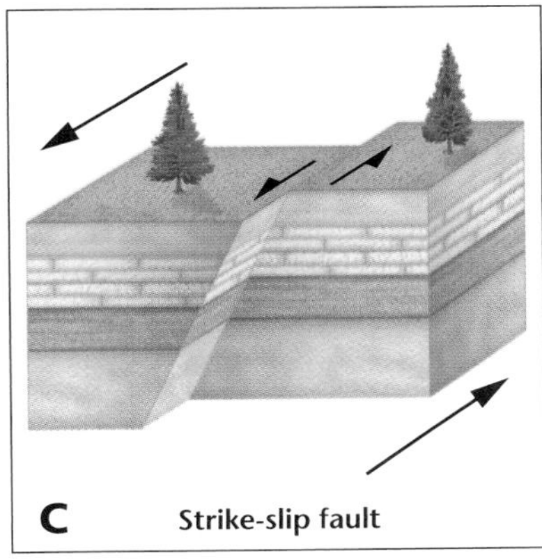

Figure 19-3 Reverse faults form when horizontal stress is exerted on a rock body from opposite sides **(A).** Normal faults form when bodies of rock are pulled from opposite sides **(B).** Strike-slip faults are caused by horizontal shear stress **(C).**

silicon putty, and metals, on the other hand, can undergo a great deal of ductile deformation before failure occurs, or, they may not fail at all. Most rocks are brittle under the relatively low temperatures that exist in Earth's crust but become ductile at the higher temperatures present at greater depths.

Faults

Many kinds of rocks that make up Earth's crust fail when stress is applied too quickly, or when stress is great. The resulting fracture or system of fractures, along which movement occurs, is called a **fault.** The surface along which the movement takes places is called the fault plane. The orientation of the fault plane can vary from nearly horizontal to almost vertical. In diagrams, small arrows along the fault plane indicate the direction of movement of the rocks involved.

Types of Faults There are three basic types of faults, as shown in ***Figure 19-3.*** Reverse faults are fractures that form as a result of horizontal compression. Note that the compressional force results in a horizontal shortening of the crust involved. What evidence in ***Figure 19-3A*** indicates this shortening? Normal faults are fractures caused by horizontal tension. Movement along a normal fault is partly horizontal and partly vertical. The horizontal movement along a normal fault occurs in such a way as to extend the crust. Note in ***Figure 19-3B*** that the two trees separated by the normal fault are farther apart than they were before the faulting.

Strike-slip faults are fractures caused by horizontal shear. The movement along a strike-slip fault is mainly horizontal, as shown in ***Figure 19-3C.*** The San Andreas Fault, which runs through California, is a strike-slip fault. This fault is one of thousands of faults responsible for many of the state's earthquakes. Motion along this fault has offset features that were originally continuous across the fault, as shown in ***Figure 19-4.***

Figure 19-4 The orange trees in the background have moved to the right relative to those in the foreground as the result of the 1940 Imperial Valley earthquake along the San Andreas Fault.

EARTHQUAKE WAVES

Most earthquakes are caused by movements along faults. Recall from the *Discovery Lab* that some slippage along faults is relatively smooth. Other movements, modeled by the sandpaper-covered blocks, show that irregular surfaces in rocks can snag and lock. As stress continues to build in these rocks, they reach their elastic limit, break, and produce an earthquake.

> **Earth Sciences**
>
> **9.b** Students know the principal natural hazards in different California regions and the geologic basis of those hazards.

Types of Seismic Waves The vibrations of the ground during an earthquake are called seismic waves. Every earthquake generates three types of seismic waves. **Primary waves,** or P-waves, squeeze and pull rocks in the same direction along which the waves are traveling, as shown in ***Figure 19-5A.*** Note how a volume of rock, which is represented by the small red square, changes shape as a P-wave passes through it. **Secondary waves,** or S-waves, cause rocks to move at right angles in relation to the direction of the waves, as shown in ***Figure 19-5B.*** **Surface waves** are a third type of seismic wave that move in two directions as they pass through rock. An up-and-down movement similar to that of an ocean wave occurs as a surface wave travels through a rock. A surface wave also causes rocks to move from side to side as it passes, as shown in ***Figure 19-5C.***

As you might guess from the name, surface waves travel along Earth's surface. P-waves and S-waves, on the other hand, pass through Earth's interior. For this reason, P-waves and S-waves are also called body waves. The first body waves generated by a quake

Figure 19-5 A P-wave causes rock particles to move back and forth as it passes **(A).** An S-wave causes rock particles to move at right angles to the direction of the wave **(B).** A surface wave causes rock particles to move both up and down and from side to side **(C).**

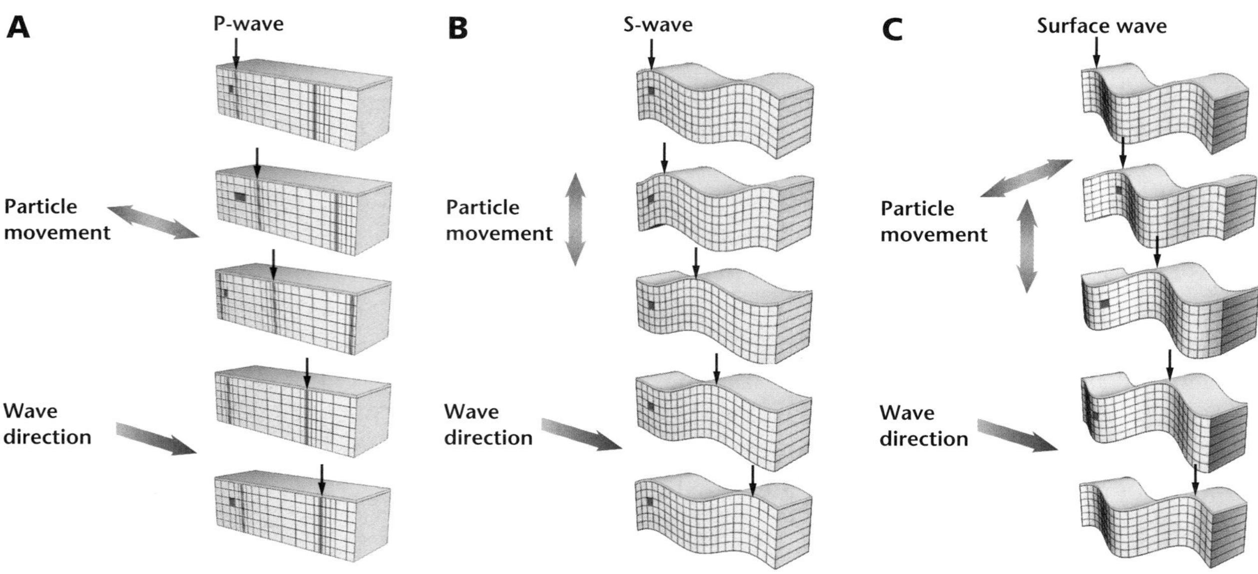

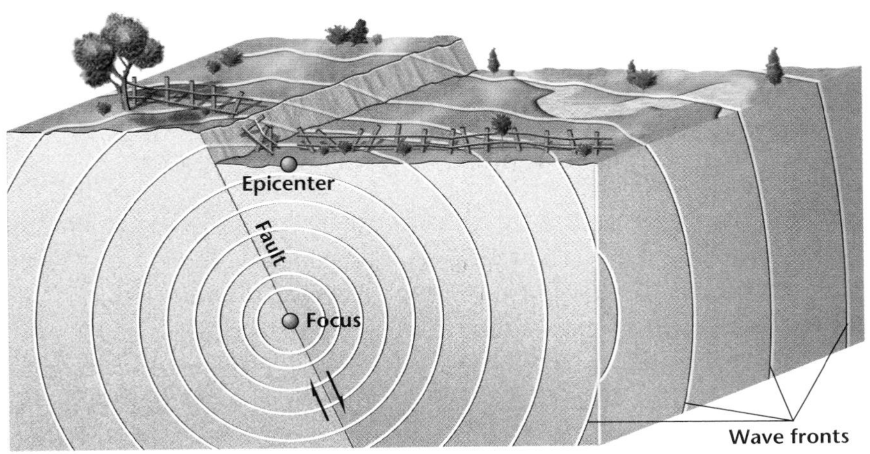

Figure 19-6, The focus of an earthquake is the point of initial fault rupture. The surface point directly above the focus is the epicenter.

spread out from the point of failure of rocks at depth. This point, where an earthquake originates, is the **focus** of the earthquake. The focus is usually at least several kilometers below Earth's surface. The point on Earth's surface directly above the focus is the earthquake's **epicenter.** Locate the focus and the epicenter in the diagram shown in ***Figure 19-6.***

You've just learned that Earth's rocks deform when stress is exerted on them. If stress exceeds a certain limit, the rocks fracture to form faults. Movement along faults causes most earthquakes. How are these earth-shattering events measured and what do they tell us about Earth's interior? You'll find out the answer to this question in the next section.

SECTION ASSESSMENT

1. What are stress and strain?
2. Describe a typical stress-strain curve and the relationship between the segments of the curve and stress and strain.
3. Compare and contrast the three types of faults.
4. What causes most earthquakes?
5. **Thinking Critically** Most earthquakes are shallow. Use the concepts of elastic and ductile deformation to explain this fact.

SKILL REVIEW

6. **Concept Mapping** Use the following terms to complete the concept map to organize some of the major ideas in this section. For more help, refer to the *Skill Handbook.*

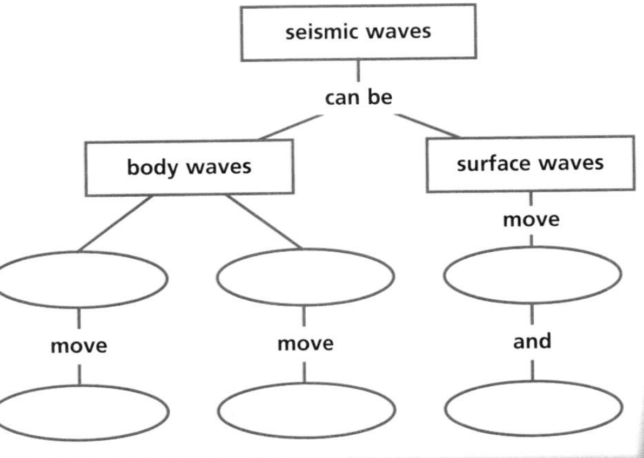

earthgeu.com/self_check_quiz

SECTION 19.2

Seismic Waves and Earth's Interior

Earth Sciences 3.d

OBJECTIVES

- **Describe** *how a seismometer works.*
- **Explain** *how seismic waves have been used to determine the structure and composition of Earth's interior.*

VOCABULARY

seismometer
seismogram

Earth Sciences

3.d Students know why and how earthquakes occur and the scales used to measure their intensity and magnitude.

The study of earthquake waves is called seismology. In view of the potential for major disaster, it should come as no surprise to you that many scientists study earthquakes. Some seismologists, however, study earthquakes for another reason. The seismic waves that shake the ground during a quake also penetrate Earth's interior. This has provided information that has enabled Earth scientists to construct models of Earth's internal structure. Thus, even though seismic waves can wreak havoc on the surface, they are invaluable for their contribution to our understanding of Earth's interior.

SEISMOMETERS AND SEISMOGRAMS

Vibrations sent out by earthquakes shake the entire globe. Although most of the vibrations can't be felt great distances from a quake's epicenter, they can be detected and recorded by sensitive instruments called seismographs, or **seismometers**. Some seismometers consist of a rotating drum covered with a sheet of paper, a pen or other such recording tool, and a mass. Seismometers vary in design, as shown in ***Figure 19-7,*** but all include a frame that is anchored to the ground and a mass that is suspended from a spring or wire. Because of inertia, the mass tends to stay at rest as the ground and, consequently, the frame, vibrate during an earthquake. The relative motion of the mass in relation to the frame is then registered on the paper with the recording tool, or is directly recorded onto a computer disk. The record produced by a seismometer is called a **seismogram,** a portion of which is shown in ***Figure 19-8.***

Figure 19-7 ***Which of these seismometers records horizontal motion during an earthquake? Which detects vertical motion?***

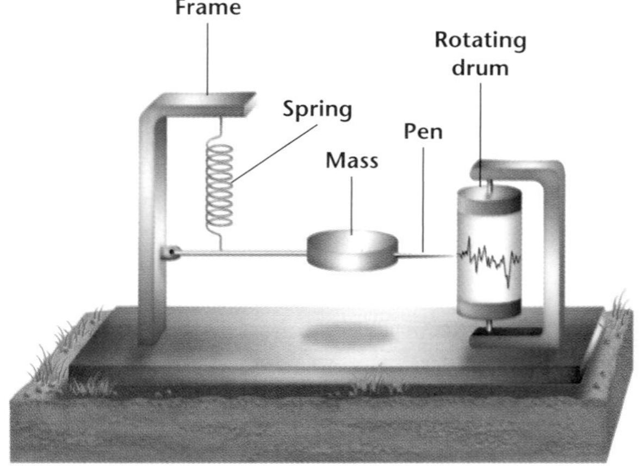

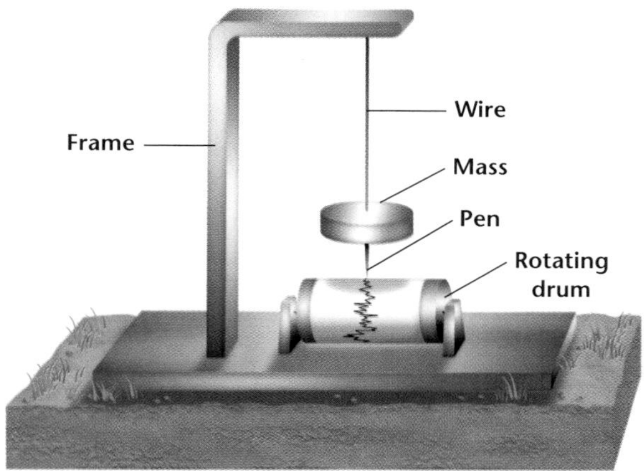

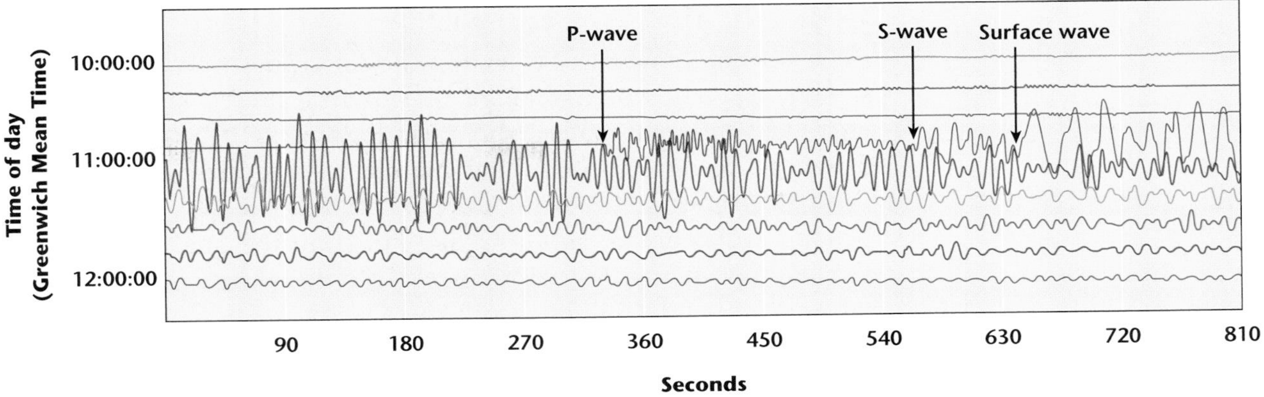

Figure 19-8 *Use this section of a seismogram, which contains colored lines to make it easier to read, to determine which of the three types of seismic waves is the fastest. Which type is the slowest?*

Travel-Time Curves Seismic waves that travel from the epicenter of an earthquake are recorded by seismometers housed in distant facilities. Over many years, the arrival times of seismic waves from countless earthquakes at seismic facilities all over the globe have been collected. Using these data, seismologists have been able to construct global travel-time curves for the initial P-waves and S-waves of an earthquake, as shown in ***Figure 19-9.*** These general curves have provided the average travel times of all seismic waves for different distances, no matter where on Earth an earthquake occurs. You can make and use a travel-time curve by doing the *Problem-Solving Lab* on the next page.

Look at ***Figure 19-9.*** Note that for any distance from the epicenter, the P-waves always arrive first at a seismic facility. Note, too, that with increasing travel distance, the time separation between the curves for the P-waves and S-waves increases. This means that waves recorded on seismograms from more distant facilities are farther apart than waves recorded on seismograms at stations closer to the epicenter. This separation of seismic waves on seismograms can be used to determine the distance from the epicenter of a quake to the seismic facility that recorded the seismogram. Can you guess how? This method of precisely locating an earthquake's epicenter will be discussed later in this chapter.

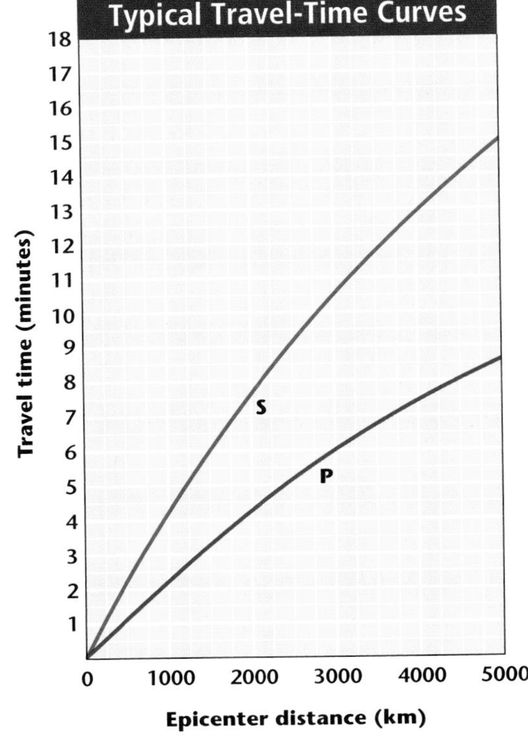

Figure 19-9 Travel-time curves show the time it takes for P-waves and S-waves to travel to seismic stations located at different distances from an earthquake's epicenter.

Clues to Earth's Interior

Most of the knowledge of Earth's interior comes from the study of seismic waves, which change speed and direction when they encounter different materials. Note in ***Figure 19-10*** that P-waves and S-waves traveling through the mantle follow fairly direct paths. P-waves that strike the core, however, are refracted, or bent, so that beyond a distance of about 11 000 km from the quake's epicenter, they disappear. The P-waves refracted into the core reemerge at a distance of about 16 000 km from the epicenter. The region between these two distances doesn't receive direct P-waves and is known as the P-wave shadow zone.

What happens to the S-waves generated by an earthquake? S-waves do not enter Earth's core because they cannot travel through liquids. Thus, like P-waves, they also do not reappear beyond the

Problem-Solving Lab

Making and Using Graphs

Model seismic-wave travel times

Seismic waves can be compared to runners. Suppose that Pam and Sam are running the 100-m dash. Pam runs the 100 m in 12 s, and Sam runs it in 14.5 s.

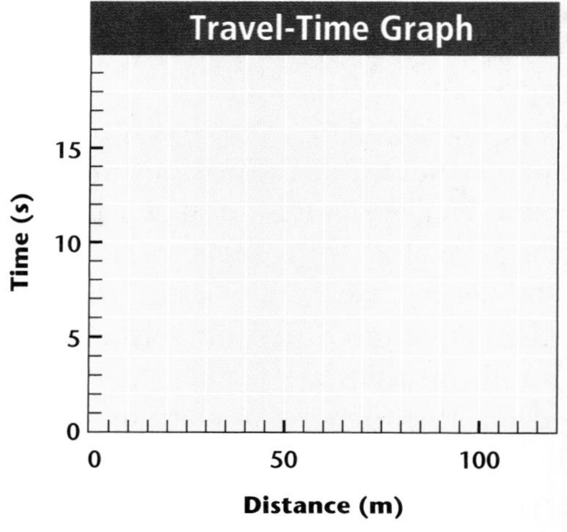

Procedure

1. Copy the graph shown. Plot two points to show the running times for Pam and Sam. Draw a straight line from each point to the origin. These are Pam's and Sam's travel-time curves.
2. Measure the separation of the two lines at a distance of 50 m. Is it more or less than 1 s?
3. Slide a ruler, parallel to the vertical axis, along the curves until you find the distance at which the separation is 1 s. What is that distance?

Analysis

4. Is there any other distance with the same separation?
5. What is the average speed, in m/s, of each runner?

Thinking Critically

6. Double-check your answer by dividing the distance in step 3 by each runner's speed to get each runner's time to that point. Do these times differ by 1 s?
7. What do Pam and Sam represent in terms of seismic waves? Discuss some ways in which a seismic travel-time curve differs from yours.

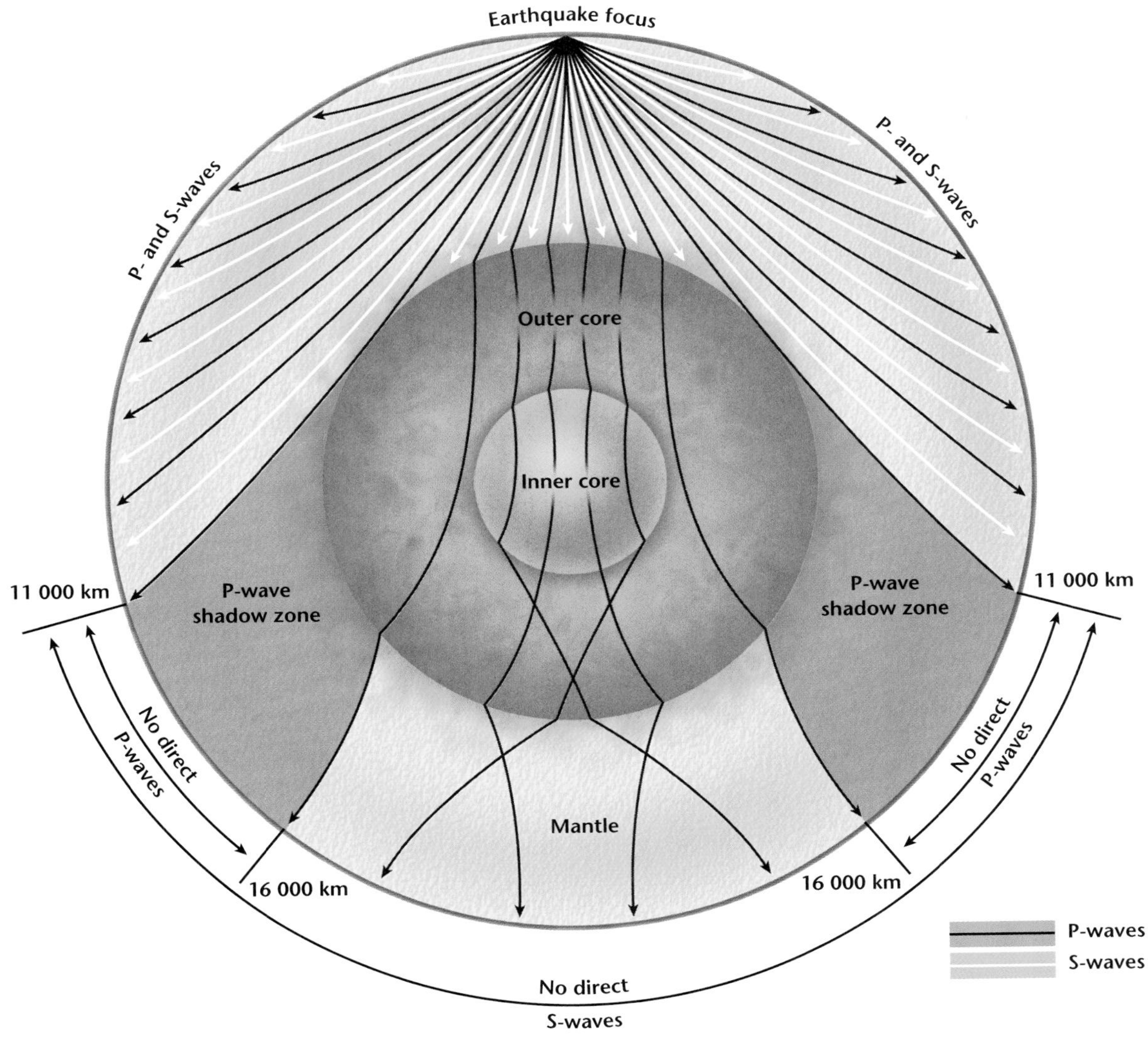

Figure 19–10 Refraction of P-waves into the outer core generates a P-wave shadow zone on Earth's surface where no direct P-waves appear on seismograms. Other P-waves are reflected and refracted by the inner core. S-waves cannot travel through liquids, and thus, don't reappear beyond the P-wave shadow zone.

11 000-km distance. This disappearance of S-waves has allowed seismologists to reason that Earth's outer core must be liquid. Detailed studies of how other seismic waves reflect deep within Earth show that Earth's inner core is solid.

Earth's Internal Structure The travel times and behavior of seismic waves provide a detailed picture of Earth's internal structure. These waves also provide clues about the composition of the various parts of Earth. By studying seismic-wave travel times, scientists have

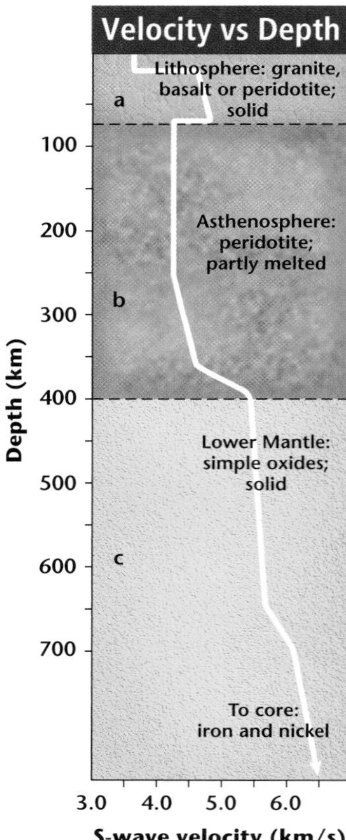

Figure 19-11 Observations of seismic wave velocities have enabled scientists to subdivide Earth's interior into various layers.

determined that the lithosphere, which includes the crust and the top of the upper mantle, is made up primarily of the igneous rocks granite, basalt, and peridotite as shown in ***Figure 19-11.*** The crustal part of the lithosphere is composed of either granite or basalt. The mantle section of the lithosphere is made of a dense, coarse-grained, intrusive rock called peridotite, which is made mostly of the mineral olivine. Much of the partially melted mantle, or asthenosphere, is also thought to be peridotite. Earth's lower mantle is solid and is probably composed of simple oxides containing iron, silicon, and magnesium. Seismic waves traveling through the core, together with gravity studies, indicate that this inner part of the planet is very dense and is probably made of a mixture of iron and nickel.

Earth's Composition The composition data obtained from seismic waves is supported by studies of meteorites, which are solid, interplanetary bodies that enter Earth's atmosphere. You might be thinking, but how can objects from space help to determine the composition of Earth's interior? Meteorites are pieces of asteroids, which are rocky bodies that orbit the Sun. Asteroids are thought to have formed in much the same way and at the same time as the planets in our solar system. Meteorites consist mostly of iron, nickel, and chunks of rock similar to peridotite in roughly the same proportions as the rocks thought to make up Earth's core and mantle.

Meteorites, together with the data provided by the travel times of seismic waves, have enabled scientists to indirectly probe Earth's interior. What other kinds of information can these waves provide? As you'll find out in the next section, seismic waves are used to determine the strength of an earthquake as well as the precise location of its epicenter.

SECTION ASSESSMENT

1. Explain how a seismometer works.
2. What is a seismogram?
3. What is a seismic travel-time curve, and how is it used to study earthquakes?
4. What is the P-wave shadow zone, and what causes it?
5. How have scientists determined the composition of Earth's mantle and core?
6. **Thinking Critically** As shown in ***Figure 19-8,*** on page 501, surface waves are the last to arrive at a seismic station. Why then do they cause so much damage?

SKILL REVIEW

7. **Recognizing Cause and Effect** Is there any way for P-waves to appear in the shadow zone? Explain. For more help, refer to the *Skill Handbook.*

earthgeu.com/self_check_quiz

SECTION 19.3 *Measuring and Locating Earthquakes*

Earth Sciences 3.d

How many earthquakes do you suppose occur each year? If you were to rely only on news accounts of these events, you might guess a dozen or so, at most. It may surprise you to learn that more than one million earthquakes occur each year! However, more than 90 percent of these are not felt and cause little, if any, damage. The earthquakes that make the news are major seismic events that cause much damage, such as the one that occurred in Bam, Iran in December 2003.

OBJECTIVES

- **Compare** *and* **contrast** *earthquake magnitude and intensity and the scales used to measure each.*
- **Explain** *why data from at least three seismic stations are needed to locate an earthquake's epicenter.*
- **Describe** *Earth's seismic belts.*

VOCABULARY

magnitude
Richter scale
moment magnitude scale
modified Mercalli scale

EARTHQUAKE MAGNITUDE AND INTENSITY

The amount of energy released during an earthquake is measured by its **magnitude.** Many news accounts describe the magnitude of an earthquake on a numerical scale called the Richter scale, which was devised by an American seismologist named Charles Richter. An earthquake's rating on the **Richter scale** is based on the size of the largest seismic waves generated by the quake. Each successive number in the scale represents an increase in seismic-wave size, or amplitude, of a factor of 10. For example, the seismic waves of a magnitude-8 earthquake on the Richter scale are ten times larger than those of a magnitude-7 earthquake, and 100 times larger than those of a magnitude-6 earthquake. The differences in the amounts of energy released by earthquakes are even greater than the differences between the amplitudes of their waves. Each increase in magnitude corresponds to about a 32-fold increase in seismic energy. Thus, an earthquake of magnitude-8 releases about 32 times the energy of a magnitude-7 earthquake, and over 1000 times the energy of a magnitude-6 earthquake. Some of the damage caused by a quake measuring 8.6 on the Richter scale is shown in ***Figure 19-12.***

 **Earth Sciences**

3.d Students know why and how earthquakes occur and the scales used to measure their intensity and magnitude.

Figure 19-12 The damage shown here was caused by an 8.6-magnitude earthquake that struck Anchorage, Alaska in 1964.

Moment Magnitude Scale While the Richter scale can still be used to describe the magnitude of an earthquake, most seismologists today use a scale called the moment magnitude scale to measure earthquake magnitude. The **moment magnitude scale** takes into account the size of the fault rupture, the amount of movement along the fault, and the rocks' stiffness. Unlike Richter-scale values, which are based on the largest seismic waves generated by a quake, moment magnitude values are estimated from the size of several types of seismic waves produced by an earthquake.

Modified Mercalli Scale Another way to assess earthquakes is to measure the amount of damage done to the structures involved. This measure, called the intensity of an earthquake, is determined using the **modified Mercalli scale,** which rates the types of damage and other effects of an earthquake as noted by observers during and after its occurrence. This scale uses the Roman numerals I to XII to designate the degree of intensity. Specific effects or damage correspond to specific numerals; the higher the numeral, the worse the damage. A simplified version of the modified Mercalli intensity scale is shown in ***Table 19-1.*** Use the information given in this scale to try to rate the intensity of the earthquake that caused the damage shown in ***Figure 19-13.***

Figure 19-13 *Use the modified Mercalli scale to determine the intensity of the earthquake that caused the damage to this grocery store in Washington State.*

Table 19-1 Modified Mercalli Intensity Scale	
I.	Not felt except under unusual conditions.
II.	Felt only by a few persons. Suspended objects may swing.
III.	Quite noticeable indoors. Vibrations are like the passing of a truck.
IV.	Felt indoors by many, outdoors by few. Dishes and windows rattle. Standing cars rock noticeably.
V.	Felt by nearly everyone. Some dishes and windows break, and some plaster cracks.
VI.	Felt by all. Furniture moves. Some plaster falls and some chimneys are damaged.
VII.	Everybody runs outdoors. Some chimneys break. Damage is slight in well-built structures but considerable in weak structures.
VIII.	Chimneys, smokestacks, and walls fall. Heavy furniture is overturned. Partial collapse of ordinary buildings occurs.
IX.	Great general damage occurs. Buildings shift off foundations. Ground cracks. Underground pipes break.
X.	Most ordinary structures are destroyed. Rails are bent. Landslides are common.
XI.	Few structures remain standing. Bridges are destroyed. Railroad ties are greatly bent. Broad fissures form in the ground.
XII.	Damage is total. Objects are thrown upward into the air.

Earthquake intensity depends primarily on the amplitude of the surface waves generated. Because surface waves, like body waves, gradually decrease in size with increasing distance from the focus of an earthquake, the intensity also decreases as the distance from a quake's epicenter increases. Maximum intensity values are observed in the region near the epicenter; Mercalli values decrease to I at distances very far from the epicenter.

Modified Mercalli scale intensity values of places affected by an earthquake can be compiled to make a seismic-intensity map. Contour lines join points that experienced the same intensity. The maximum intensity is usually, but not always, found at the quake's epicenter. You will generate a seismic-intensity map when you do the *MiniLab* on the next page.

Topic: Bam, Iran
To learn more about damage caused by earthquakes, visit the Earth Science Web Site at earthgeu.com

Activity: Research the December 2003 earthquake in Bam, Iran. Describe what an eyewitness may have seen.

Depth of Focus Earthquake intensity is related to earthquake magnitude. Both measurements reflect the size of the seismic waves generated by the quake. Another factor that determines the intensity of an earthquake is the depth of the quake's focus. An earthquake can be classified as shallow, intermediate, or deep depending on the location of the quake's focus. Because a deep-focus earthquake produces smaller vibrations at the epicenter than a shallow-focus quake, a shallow-focus, moderate quake of magnitude-6, for example, may generate a greater maximum intensity than a deep-focus quake of

MiniLab

How is a seismic–intensity map made?

Make a Map using seismic–intensity data.

Procedure

1. Trace the map onto paper. Mark the locations indicated by the letters on the map.
2. Plot these Mercalli intensity values on the map next to the correct letter: A, I; B, III; C, II; D, III; E, IV; F, IV; G, IV; H, V; I, V; J, V; K, VI; L, VIII; M, VII; N, VIII; O, III.
3. Draw contours on the map to separate the intensity values.

Analyze and Conclude

1. What is the maximum intensity value?
2. Where is the maximum intensity value located?
3. Where is the earthquake's epicenter?

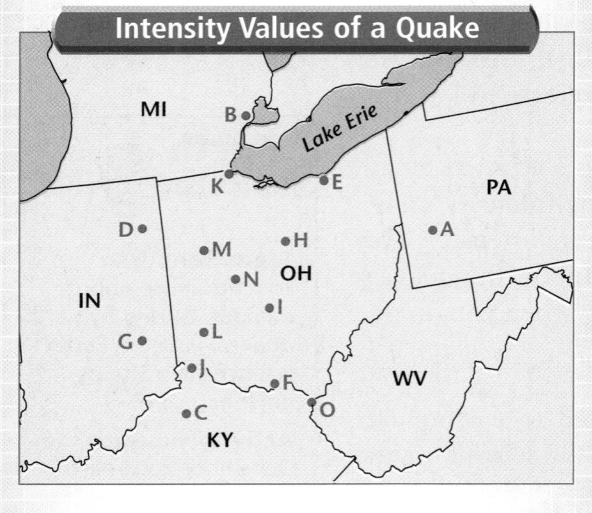

magnitude-8. Catastrophic quakes with high intensity values are almost always shallow-focus events.

Locating an Earthquake

The exact location of an earthquake's epicenter and the time of the quake's occurrence are initially unknown. All epicenter locations, as well as times of occurrence, however, can be easily determined using seismograms and travel-time curves.

Distance to an Earthquake Look again at ***Figure 19-9*** on page 501. Suppose the separation time for the P-waves and S-waves is six minutes. Based on known travel times of seismic waves, the distance between the earthquake's epicenter and the seismic station that recorded the waves can only be 4500 km—no more, no less. This is because the known travel time over that distance is eight minutes for P-waves and 14 minutes for S-waves. At greater distances from the epicenter, the travel times for both types of waves increase. This results in a larger P-S separation because S-waves lag behind the faster P-waves. The distance to a quake's epicenter, then, is determined by the P-S separation. By measuring this separation on any seismogram as well as the distance on a travel-time graph at which the P-curve and S-curve have the same separation, the distance to a quake's epicenter can be determined. This distance is called the epicentral distance.

The distance between an earthquake's epicenter and a single seismic station does not provide sufficient information to determine the location of that epicenter—it could be located in any direction from the seismic station. The only thing that is certain is that the epicenter is located somewhere on a circle centered on the seismic station. The radius of this circle is equal to

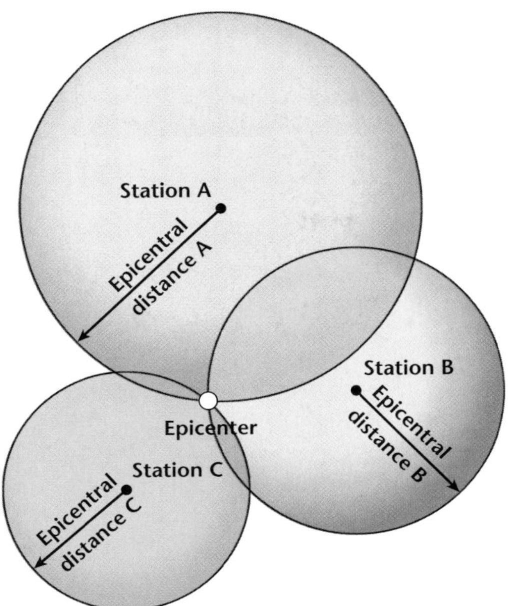

Figure 19-14 To determine the location of an earthquake's epicenter, the locations of three seismic stations are plotted on a map. A circle whose radius is equal to the corresponding epicentral distance is plotted around each station. The point of intersection of these circles is the earthquake's epicenter.

the epicentral distance. If the distances between three or more seismic stations and an earthquake's epicenter are known, the exact location of the epicenter can be determined, as shown in ***Figure 19-14.*** You will locate an actual epicenter in the *GeoLab* at the end of this chapter.

Time of an Earthquake The time of occurrence of an earthquake can be easily calculated, again by using the travel-time graph shown in ***Figure 19-9.*** The exact arrival times of the P-waves and S-waves at a seismic station can be read from the seismogram. The travel time of either wave at the epicentral distance of that station can be read from the travel-time graph. The time of occurrence of the earthquake is then determined by subtracting the appropriate travel time from the known arrival time of the wave.

Update For an online update of recent earthquakes, visit earthgeu.com and select the appropriate chapter.

Seismic Belts

Over the years, seismologists have collected and plotted the locations of numerous earthquake epicenters. The global distribution of these epicenters, shown in ***Figure 19-15*** on the next page, reveals an interesting pattern. Earthquake locations are not randomly distributed. The majority of the world's earthquakes occur in relatively narrow seismic belts that separate large regions with little or no seismic activity. What causes this pattern of seismic activity?

Look back at ***Figure 17-13*** on page 455, which shows Earth's tectonic plates and the boundaries between them. How does this map compare with the map shown in ***Figure 19-15?*** By comparing the two maps, you can see that most earthquakes are associated with tectonic plate boundaries. Almost 80 percent of all earthquakes occur in the Circum-Pacific Belt. Another zone of significant seismic activity stretches across southern Europe and Asia. About 15 percent of the

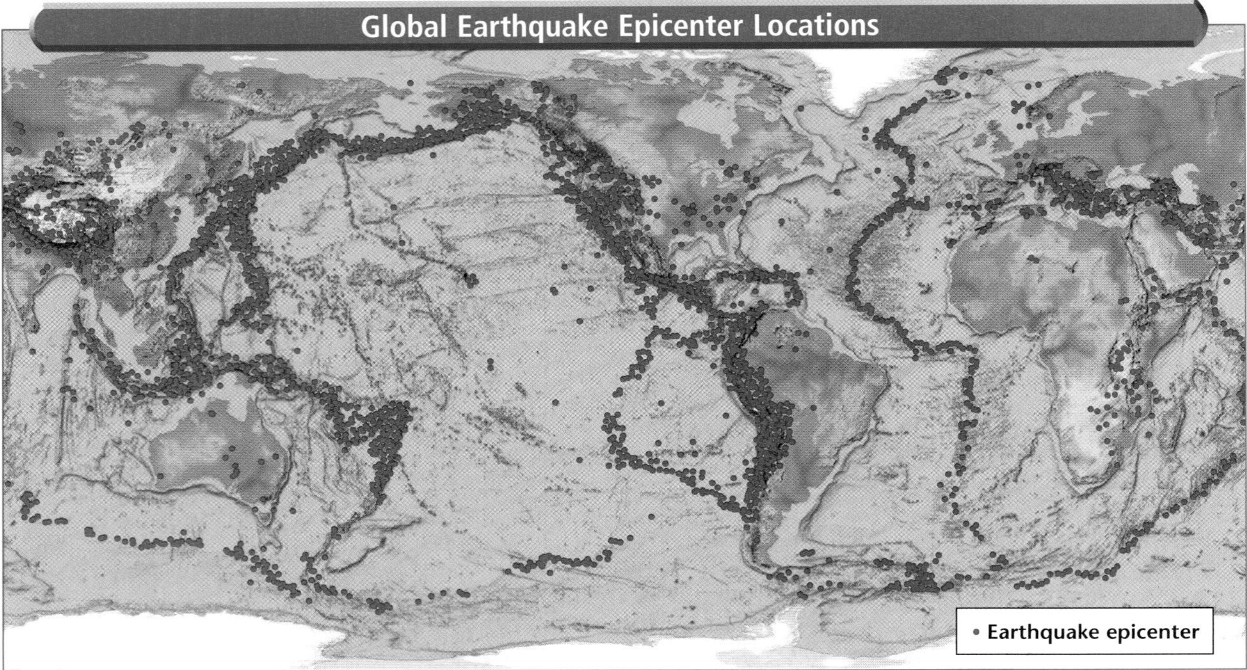

Figure 19-15 This map shows the locations of earthquake epicenters.

world's earthquakes take place in this region, which is sometimes called the Mediterranean-Asian Belt. Most of the remaining earthquakes occur in narrow bands that run along the crests of ocean ridges. A very small percentage of earthquakes happen far from tectonic plate boundaries and are distributed more or less at random.

Look again at ***Figure 19-15.*** Do you live in an area prone to earthquakes? If so, how can you minimize the damage done by these seismic events?

Section Assessment

1. What is earthquake magnitude and how is it measured?
2. What is earthquake intensity and how is it measured?
3. Explain why earthquake data from at least three seismic stations are needed to locate an earthquake's epicenter.
4. **Thinking Critically** The separation of P-waves and S-waves on a seismogram recorded 4500 km from the epicenter of an earthquake is six minutes. On another seismogram that separation is seven minutes. Is the second station closer to or more distant from the epicenter? Explain.

Skill Review

5. **Interpreting Scientific Diagrams** Use ***Figure 19-15*** above to determine which of the following countries has the most earthquakes: Ireland, Pakistan, South Africa, or Australia. Why does this country have more earthquakes than the other three countries? For more help, refer to the *Skill Handbook.*

earthgeu.com/self_check_quiz

SECTION 19.4 Earthquakes and Society

Earth Sciences 9.b

Most earthquake damage results from the prolonged shaking of the ground by surface waves. During major quakes, this shaking can last longer than a minute. Many structures cannot withstand such violent motion. Collapsing buildings are responsible for many earthquake-related deaths. What other types of damage are caused by earthquakes? What kinds of factors affect the damage done during a quake? Is it possible to predict earthquakes?

OBJECTIVES

- **Discuss** *factors that affect the amount of damage done by an earthquake.*
- **Explain** *some of the factors considered in earthquake probability studies.*
- **Define** *seismic gaps.*

VOCABULARY

tsunami
seismic gap

SOME EARTHQUAKE HAZARDS

The damage produced by an earthquake is directly related to the strength or quality of the structures involved. The most severe damage occurs to unreinforced buildings made of stone, concrete, or other brittle building materials. Wooden structures, on the other hand, are remarkably resilient and generally sustain significantly less damage. Many high-rise, steel-frame buildings also sustain little damage during an earthquake because they are reinforced to make them earthquake resistant. Some buildings in earthquake-prone areas, such as California, even rest on large rubber structures that absorb most of the vibrations generated during a quake.

Structural Failure In many earthquake-prone areas, buildings are destroyed as the ground beneath them shakes. In some cases, the supporting walls of the ground floor fail and cause the upper floors, which initially remain intact, to fall and collapse as they hit the ground or lower floors. The resulting debris resembles a stack of pancakes, and thus, the process has been called "pancaking". This type of structural failure is shown in the photograph on page 494, and was a common result of the quake that rocked Turkey in 1999. You'll learn more about this quake and the damage it caused in the *Science in the News* feature at the end of this chapter.

Another type of structural failure is related to the height of a building. During the 1985 Mexico City earthquake, for example, most buildings between 5 and 15 stories tall collapsed or were otherwise completely destroyed as shown in ***Figure 19-16.*** Similar structures that were either shorter or taller, however, sustained only minor

Figure 19-16 The buildings damaged or destroyed during the 1985 Mexico City quake vibrated with the same period as the seismic waves.

damage. Can you guess why? The shaking caused by the quake had the same period of vibration as the natural sway of the intermediate buildings, which caused them to sway violently during the quake. The ground vibrations, however, were too rapid to affect taller buildings, whose periods of vibration were longer than the earthquake waves, and too slow to affect shorter buildings, whose periods of vibration were shorter.

 **Earth Sciences**

9.b Students know the principal natural hazards in different California regions and the geologic basis of those hazards.

Land and Soil Failure In addition to their effects on structures made by humans, earthquakes can wreak havoc on Earth itself. In sloping areas, earthquakes may trigger massive landslides. Most of the estimated 30 000 deaths caused by the 7.8-magnitude earthquake that struck in Peru in 1970 resulted from a landslide that buried several towns. In areas with fluid-saturated sand, seismic vibrations may cause subsurface materials to liquefy and behave like quicksand, generating landslides even in areas of low relief. Soil liquefaction can also cause trees and houses to fall over or to sink into the ground and can cause underground pipes and tanks to rise to the surface.

In addition to causing soil liquefaction, earthquake waves can be amplified as they travel through a soil. Because soft materials have little resistance to deformation, seismic waves are amplified in such materials but are muted in more-resistant materials. Consequently, wave size and earthquake intensity are greatest in soft, unconsolidated sediments and relatively small in hard, resistant rocks such as granite. The severe damage to structures in Mexico City during the 1985 earthquake is attributed to the fact that Mexico City is built on soft sediments. The thickness of the sediments caused them to resonate with the same frequency as that of the surface waves generated by the quake. This produced reverberations that greatly enhanced the ground motion and the resulting damage.

Figure 19-17 This fault scarp was produced by faulting associated with an earthquake that struck Mount Borah, Idaho, in 1983.

Fault Scarps Fault movements associated with earthquakes can produce areas of great vertical offset where the fault intersects the ground surface. These offsets are called fault scarps. As shown in ***Figure 19-17,*** a distinct fault scarp formed as the result of an earthquake that struck central Idaho in 1983. The magnitude-7.6 quake that rocked Taiwan in

Figure 19-18 Vertical offset along a fault caused this waterfall to form during the 1999 Taiwan earthquake.

1999 produced vertical offsets of up to 10 m, the greatest fault movement observed in recent history. An 8-m-high waterfall that formed where the fault crosses a river is shown in ***Figure 19-18.***

Tsunami Another type of earthquake hazard is a **tsunami,** a large ocean wave generated by vertical motions of the seafloor during an earthquake. These motions displace the entire column of water overlying the fault, creating bulges and depressions in the water. The disturbance then spreads out from the epicenter in the form of extremely long waves. While these waves are in the open ocean, their height is generally less than 1 m. When the waves enter shallow water, however, they may form huge breakers with heights occasionally exceeding 30 m! These enormous wave heights, together with open-ocean speeds between 500 and 800 km/h, make tsunamis dangerous threats to coastal areas both near and far from the quake's epicenter. The Indian Ocean tsunami of December 26, 2004 originated with a magnitude 9.0 earthquake in the ocean about 160 km west of Sumatra. The 30-m-tall tsunami radiated across the Indian Ocean and struck the coasts of Indonesia, Sri Lanka, India, Thailand, Somalia, and several other nations. The death toll from the tsunami exceeded 225,000, making it one of the most devastating natural disasters in modern history.

Using Numbers The speed of tsunamis in the open ocean can reach 800 km/h. At this speed, how long will it take a tsunami to travel from Japan to California, a distance of about 9000 km? If the waves are 160 km long, how far apart in time are the wave crests?

Seismic Risk

Recall that most earthquakes occur in areas called seismic belts. The probability of future quakes is much greater in these belts than elsewhere around the globe. The past seismic activity in any region is also a reliable indicator of future earthquakes and can be used to generate seismic-risk maps. A seismic-risk map of the United States is shown in ***Figure 19-19*** on the next page. Can you locate the areas of highest seismic risk on the map? In addition to Alaska, Hawaii, and some western states, there are several regions of relatively high

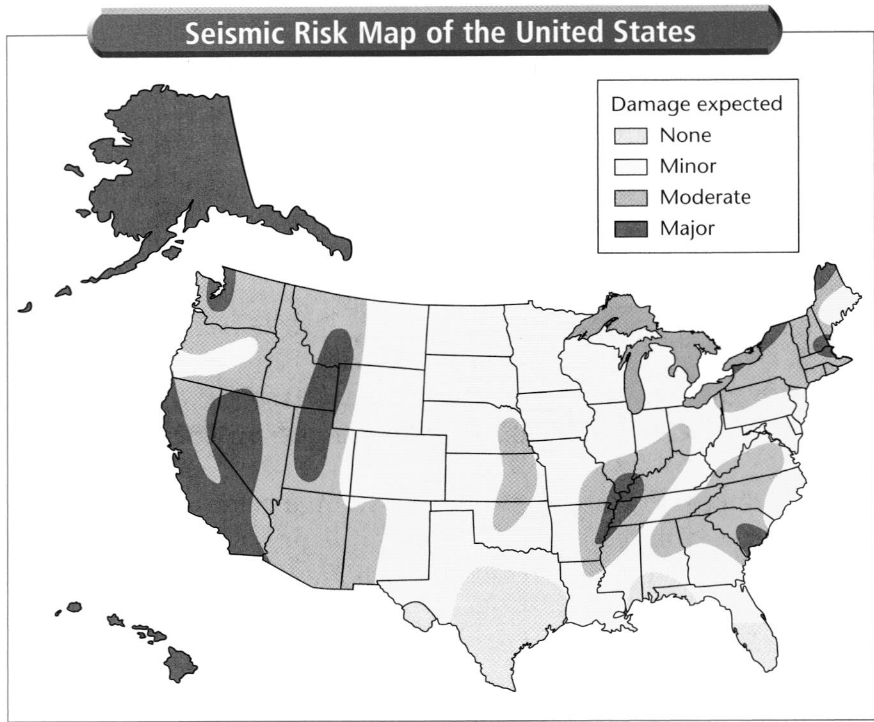

Figure 19-19 Areas of high seismic risk in the United States include Alaska, Hawaii, and some of the western states.

seismic risk in the central and eastern United States. These regions have suffered disastrous earthquakes in the past and probably will experience significant seismic activity in the future. Locate your state in ***Figure 19-19.*** What is the seismic risk of your area?

EARTHQUAKE PREDICTION

To minimize the damage and deaths caused by quakes, seismologists are searching for ways to predict these events. Earthquake prediction research is largely based on probability studies. The probability of an earthquake's occurring is based on two factors: the history of earthquakes in an area and the rate at which strain builds up in the rocks.

Earthquake History Earthquake recurrence rates can indicate that the fault involved ruptures repeatedly at regular intervals to generate similar quakes. The earthquake-recurrence rate at Parkfield, California, for example, shows that a sequence of quakes of approximately magnitude-6 shook the area about every 22 years from 1857 until 1966. This record indicates a 90-percent probability that a major quake will rock the area within the next few decades. Several kinds of instruments, including the lasers shown in ***Figure 19-20,*** are in place around Parkfield in an attempt to predict future quakes.

Probability forecasts are also based on the location of seismic gaps. **Seismic gaps** are sections of active faults that haven't experienced significant earthquakes for a long period of time. One seismic gap in the San Andreas Fault cuts through San Francisco. This section of the fault hasn't ruptured since the devastating earthquake that struck the city in 1906. Because of this inactivity, seismologists currently predict that there is a 67-percent probability that the San Francisco area will experience a magnitude-7 or higher quake within the next 30 years.

Figure 19-20 Lasers are just one technique being used to study earthquake probability in the area around Parkfield, California. Lasers can record very small movements along a fault.

Strain Accumulation The rate at which strain builds up in rocks is another factor used to determine the earthquake probability along a section of a fault. To predict when a quake might occur, scientists make several measurements. The strain accumulated in a particular part of the fault, together with how much strain was released during the last quake along that section of the fault, are two important factors in earthquake probability studies. Another factor is how much time has passed since an earthquake has struck that section of the fault.

Earthquake prediction is still a relatively new branch of geology. Being able to predict these destructive events can prevent damage to property, possibly reduce the number of injuries as a result of the quake, and, most importantly, save many lives.

Section Assessment

1. Describe structural damage caused by earthquakes.
2. What are some of the effects caused by soil liquefaction?
3. How are tsunamis generated?
4. What are some of the factors considered in earthquake probability studies?
5. **Thinking Critically** Which structure is less likely to suffer severe damage during an earthquake: a high-rise, steel-frame hotel built on sediments, or a wood-frame house built on bedrock? Explain.

Skill Review

6. **Comparing and Contrasting** Compare and contrast the seismic risk of your state to at least three neighboring states. For more help, refer to the *Skill Handbook*.

earthgeu.com/self_check_quiz

Locating an Epicenter

The separation of P- and S-waves on a seismogram allows you to estimate the distance between the seismic station that recorded the data and the epicenter of that earthquake. If the epicentral distance from three or more seismic stations is known, then the exact location of the quake's epicenter can be determined.

Preparation

Problem

Determine the epicenter location and the time of occurrence of an actual earthquake, using the travel times of P- and S-waves recorded at three seismic stations.

Materials

Figures 17-13, 19-8, and *19-9*
map on facing page
drafting compass
tracing paper
calculator
metric ruler

Objectives

In this GeoLab, you will:

- **Determine** the arrival times of P- and S-waves from a seismogram.
- **Interpret** travel-time curves.
- **Plot** an epicenter location on a map.
- **Relate** seismic data to plate tectonics.

Safety Precaution

Procedure

1. The seismogram in ***Figure 19-8*** shows the arrival time of the first P-wave at 10 h, 50 min, 32 s GMT, Greenwich mean time. Estimate the arrival time of the first S-wave to the nearest tenth of a minute.
2. Subtract this S-wave time from the initial P-wave time. What is the P-S separation on the seismogram, in minutes and tenth of minutes? Enter this value in the data table for the Berkeley seismic station.

GeoLab Data Table

Seismic Station	P-S Separation *(min)*	Epicenter Distance *(km)*	Map Distance *(cm)*
Berkeley, CA			
Boulder, CO	3.6		
Knoxville, TN	4.9		

3. The P-S separation observed on two other seismograms, which are not shown, are also listed in the table. Use the travel-time curves in ***Figure 19-9*** to determine the distances at which the P- and S-curves are separated by the time intervals listed in the table. Enter these distances in the table under the Epicenter Distance.
4. Carefully trace the map on this page. Accurately mark the three seismic station locations.
5. Determine the epicentral distances on your tracing, using a scale of about 0.9 cm = 500 km. Enter your values in the table under the Map Distance.
6. Use the compass to draw circles around each station on the map with the radius of each circle equal to the map distance, in cm, for that station.
7. Mark the point of intersection. This is the epicenter of the earthquake.
8. Determine the time of occurrence of this earthquake by reading the P-wave travel time from ***Figure 19-9*** for the epicentral distance for Berkeley. Subtract this from the initial P-wave arrival time at Berkeley, which was 10 h, 50.5 min. Express this time in terms of hours, minutes, and seconds.

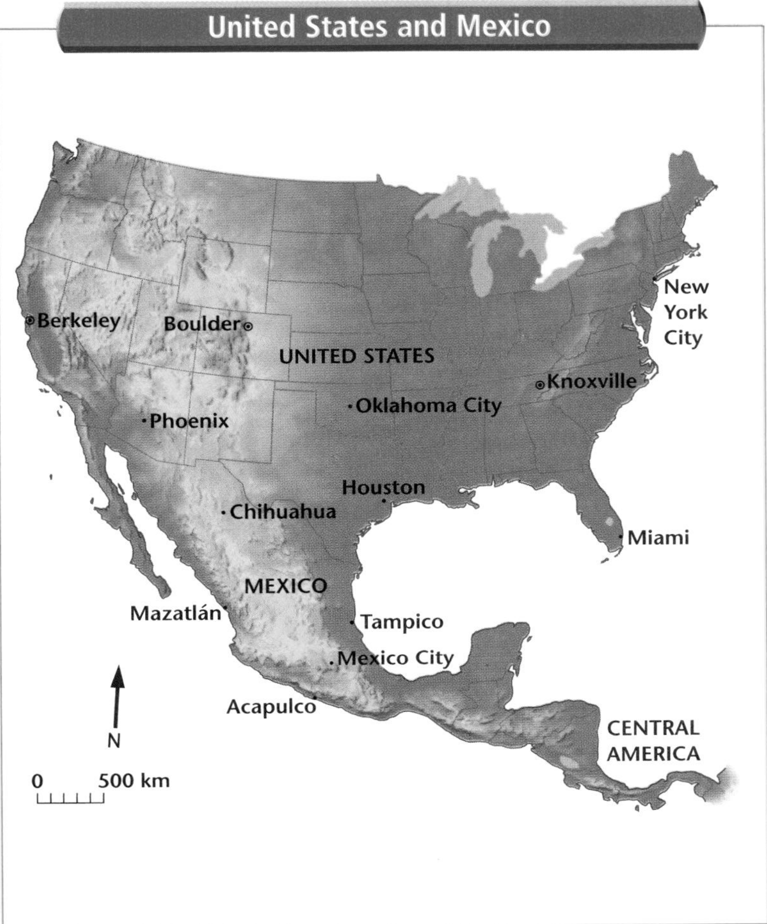

Analyze

1. Where is this epicenter located?
2. In which major seismic belt did this earthquake occur?
3. Use ***Figure 17-13*** to determine which plates form the boundary associated with this earthquake.

Conclude & Apply

1. What type of plate boundary is this?
2. Briefly describe the relative motion of the plates involved.
3. Describe the tectonic motions that caused the earthquake.

Science in the News

Earthquake in Iran

At 5:26 am local time on December 26, 2003, an earthquake rocked the city of Bam, located in southeastern Iran. Some residents were already sleeping outside due to a foreshock that occurred the previous evening. The earthquake had a magnitude of 6.6 and caused extensive damage in terms of human life and property, including leveling an ancient fortress that had been a World Heritage Site.

It is estimated that as many as 40,000 people were killed in the quake, up to 30,000 more were injured, and more than 75,000 people were left without homes. Eighty-five percent of the buildings in the city were damaged or destroyed, and the water supply system was damaged. A 2,000-year-old citadel, Arg-e Bam, thought to be the world's greatest mud fortress, was reduced to rubble as a result of the quake. The historic fortress was the city's most popular tourist attraction.

Tectonic Factors

The southeastern portion of Iran is a seismically active area. Although no earthquakes have previously been reported in Bam, major earthquakes have occurred in the region northwest of Bam. Between 1981 and 1998, four earthquakes with magnitudes greater than 5.6 occurred in this area. The origin of the December 26 quake is thought to be the Bam fault, a fault line that runs north-south through the region. According to the U.S. Geological Survey, the quake occurred as a result of stresses created by the motion of the Arabian plate northward against the Eurasian plate at a rate of approximately 3 cm/yr. The horizontal and vertical movement along a strike-slip fault resulted in the destruction of homes and buildings throughout Bam.

The Aftermath

Immediately following the quake, rescue efforts and humanitarian aid were launched by international organizations such as the United Nations and the Red Cross. Food, water, shelter, and medical services were provided to victims through relief efforts and donations. Search teams found one man alive beneath rubble 13 days after the quake struck. Initial estimates for recovery and rebuilding of the city of Bam were between $70 million and $1 billion dollars. The World Health Organization estimated that 30 million dollars would be needed to reestablish Bam's health services. Two hospitals were destroyed and half of Bam's health workers were killed in the earthquake. A minimum estimate of two years was given by the United Nations Office of Coordination of Human Affairs for rebuilding the fallen city.

Activity

Dealing with disasters of the magnitude of the Bam earthquake can seem overwhelming. Contact emergency preparedness agencies in your area to find out what steps have been taken to locally prepare for natural disasters. What kinds of disasters do these agencies primarily prepare for? Make a pamphlet to summarize your findings.

CHAPTER 19
Study Guide

Summary

SECTION 19.1

Forces Within Earth

Main Ideas

- Stress is a force per unit area that acts on a material. The deformation of materials in response to stress is called strain.
- Reverse faults form as a result of horizontal compression; normal faults, horizontal tension; strike-slip faults, horizontal shear.
- P-waves squeeze and pull rocks in the same direction along which the waves travel. S-waves cause rocks to move at right angles to the direction of the waves. Surface waves cause both an up-and-down and a side-to-side motion as they pass through rocks.

Vocabulary

epicenter (p. 499)
fault (p. 497)
focus (p. 499)
primary wave (p. 498)
secondary wave (p. 498)
strain (p. 496)
stress (p. 495)
surface wave (p. 498)

SECTION 19.2

Seismic Waves and Earth's Interior

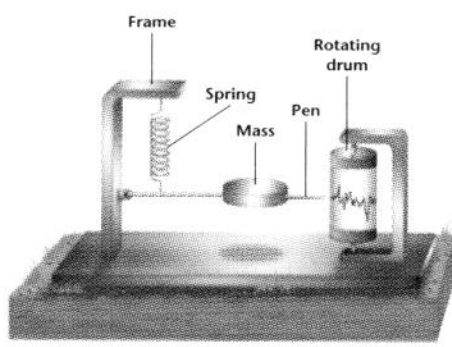

Main Ideas

- A seismometer has a frame that is anchored to the ground and a suspended mass. Because of inertia, the mass tends to stay at rest as the ground and, thus, the frame vibrate during a quake. The motion of the mass in relation to the frame is registered and recorded.
- Seismic waves are reflected and refracted as they strike different materials. Analysis of these waves has enabled scientists to determine the structure and composition of Earth's interior.

Vocabulary

seismogram (p. 500)
seismometer (p. 500)

SECTION 19.3

Measuring and Locating Earthquakes

Main Ideas

- Earthquake magnitude is a measure of the energy released during a quake and can be measured on the Richter scale. Intensity is a measure of the damage caused by a quake and is measured with the modified Mercalli scale.
- Data from at least three seismic stations are needed to locate an earthquake's epicenter.
- Most earthquakes occur in areas associated with plate boundaries called seismic belts.

Vocabulary

magnitude (p. 505)
modified Mercalli scale (p. 506)
moment magnitude scale (p. 506)
Richter scale (p. 505)

SECTION 19.4

Earthquakes and Society

Main Ideas

- Earthquakes cause structural collapse, landslides, soil liquefaction, fissures, fault scarps, uplift or subsidence, and tsunamis. Factors that affect the extent of damage done by a quake include the type of subsurface as well as the quality, height, and structure of buildings and other structures involved.
- The probability of an earthquake is based on the history of quakes in an area and the rate at which strain builds in the rocks.
- Seismic gaps are places along an active fault that haven't experienced significant earthquakes for a long period of time.

Vocabulary

seismic gap (p. 515)
tsunami (p. 513)

earthgeu.com/vocabulary_puzzlemaker

CHAPTER 19
Assessment

Understanding Main Ideas

1. What is stress?
   a. movement of waves parallel to rock particles
   b. deformation of a material caused by applied forces
   c. forces per unit area acting on a material
   d. unit of measure on the Richter scale

2. What is strain?
   a. forces per unit area acting on a material
   b. deformation of a material caused by applied forces
   c. unit of measure on the Mercalli scale
   d. travel time of seismic waves

3. Which type of seismic wave causes rock particles to move in the same direction as the wave movement?
   a. P-wave
   b. S-wave
   c. tension wave
   d. shear wave

4. What part of Earth doesn't receive direct P-waves from a quake?
   a. epicenter
   b. focus
   c. shadow zone
   d. mantle

5. Which is used to measure magnitude?
   a. Richter scale
   b. Mercalli scale
   c. shadow zone
   d. seismic gap

6. What is earthquake intensity?
   a. a measure of the energy released
   b. a measure of seismic risk
   c. a measure of damage done
   d. a measure of the quake's focus

7. What is a seismic gap?
   a. a large fault scarp
   b. a part of an active fault that hasn't recently experienced seismic activity
   c. the time separation between P- and S-waves
   d. the liquefaction of soil during a quake

8. Compare and contrast the three types of faults.
9. Draw three diagrams to show how each type of seismic wave moves through rocks.
10. Explain how a seismometer works.
11. How have seismic waves been used to determine Earth's structure and composition?
12. Why are data from at least three seismic stations needed to locate an epicenter?

Use this figure to answer questions 13–15.

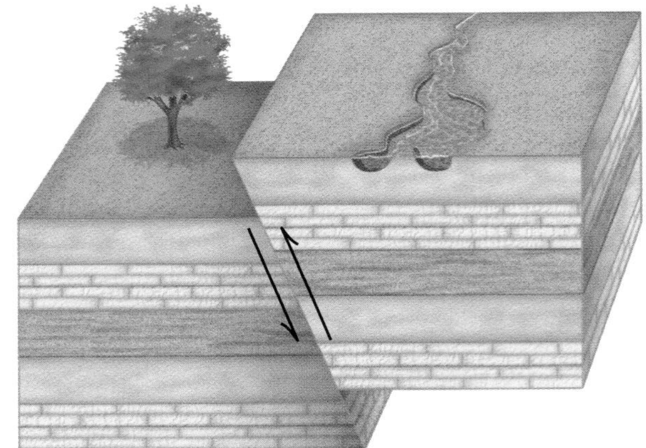

13. What type of fault is shown?
14. What type of force caused this fault to form?
15. Where is the fault plane?
16. Explain the relationship between worldwide earthquake distribution and tectonic boundaries.
17. What factors affect the damage done by an earthquake?

Test-Taking Tip

MORE THAN ONE GRAPHIC If a test question refers to more than one table, graph, diagram, or drawing, use them all. If you answer based on just one graphic, you'll probably miss an important piece of information.

earthgeu.com/chapter_test

CHAPTER 19 Assessment

18. What factors are studied in the field of earthquake probability?

Applying Main Ideas

19. Explain why a stress-strain curve usually has two segments.
20. Why do surface waves cause so much destruction?
21. Why are two types of seismometers generally used to record the same earthquake?
22. How were P-waves and S-waves used to determine the physical state of Earth's core?
23. Compare and contrast the Richter scale and the moment magnitude scale.
24. Explain how shear stress is different from tension and compression.
25. Compare and contrast the composition of Earth's mantle and core with the composition of meteorites.

Thinking Critically

26. How do you think a thin, plastic, ruler would react to a small amount of stress? What would happen if the stress applied exceeded the elastic limit of the ruler?
27. Describe several reasons why an earthquake of magnitude-3 can cause more damage than a quake of magnitude-6.
28. Why are tsunamis so destructive?
29. If rocks below the lithosphere are too hot to undergo brittle fracture, how is it possible to have deep-focus earthquakes beneath island arcs?
30. Refer to ***Figure 19-19*** on page 514. Explain why some areas in the eastern part of the United States are prone to major earthquake damage even though these places are far from present tectonic plate boundaries.

Standardized Test Practice

1. What happens to a material when it undergoes stress that exceeds its strength?
   a. The material undergoes ductile formation.
   b. The material is deformed permanently.
   c. The material returns to its original state.
   d. The material breaks or fails.

2. What is the order in which seismic waves are recorded by a seismometer?
   a. S-wave, P-wave, surface wave
   b. surface wave, P-wave, S-wave
   c. P-wave, S-wave, surface wave
   d. S-wave, surface wave, P-wave

INTERPRETING DATA Use the table below to answer questions 3 and 4.

Some Earthquakes In Recent History

Location	Year	Richter Magnitude
Chile	1960	8.5
California	1906	7.9
Alaska	1964	8.6
Columbia	1994	6.8
Taiwan	1999	7.6

3. Approximately how much more energy was released by the Chilean quake than the Taiwan earthquake?
   a. twice as much
   b. ten times as much
   c. thirty two times as much
   d. one thousand times as much

4. Approximately how much larger was the amplitude of the waves generated by the Alaskan quake than the Taiwan quake?
   a. about twice as large
   b. about ten times as large
   c. about one hundred times as large
   d. about one thousand times as large

Chapter 20

Mountain Building

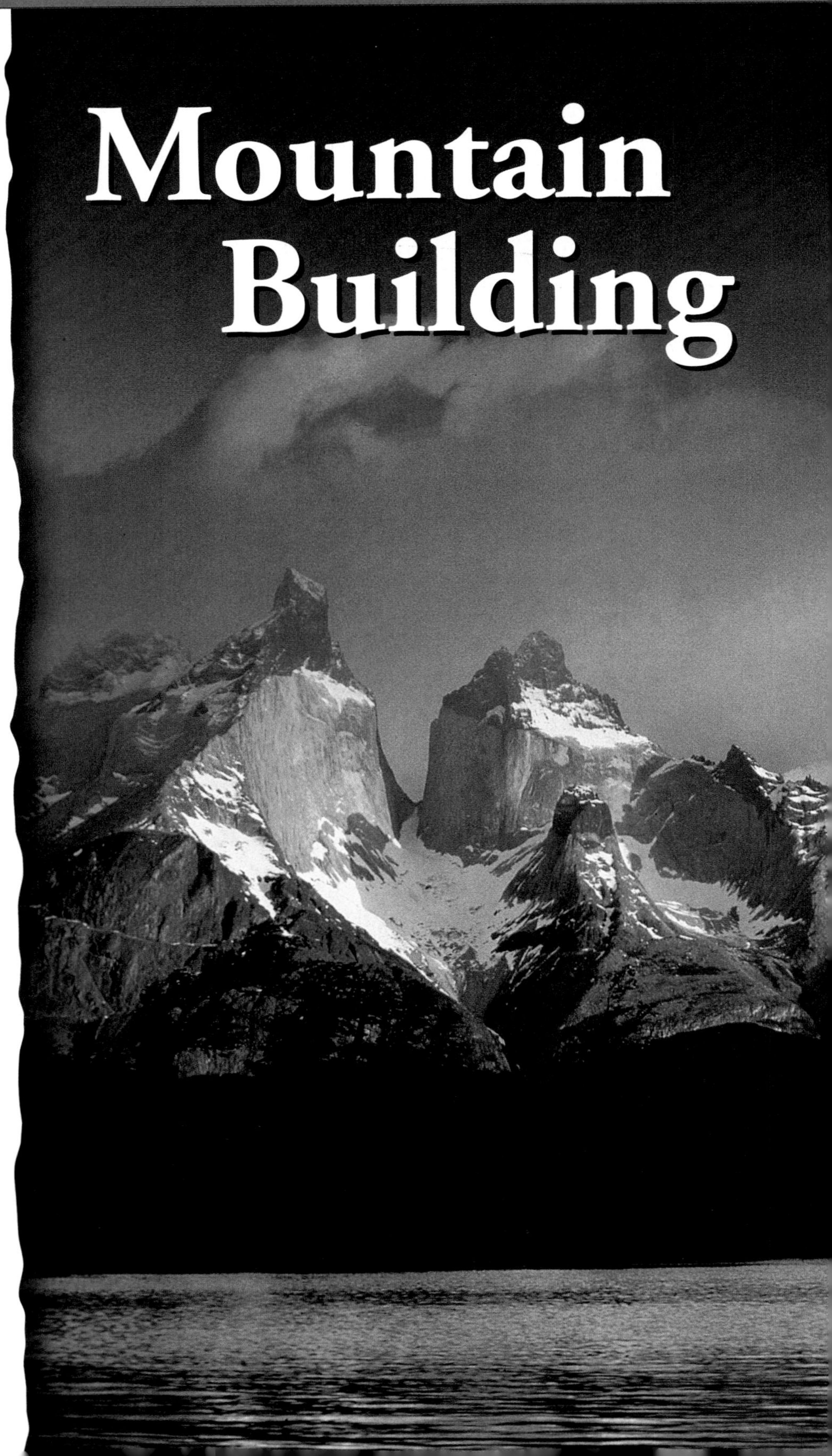

What You'll Learn

- Why Earth's crust displaces the mantle on which it rests.
- How different processes create mountains that rise above Earth's surface.

Why It's Important

All mountains rise above the surrounding land, yet each of these awesome structures is unique. Understanding the various processes involved in mountain building is critical to our understanding of the dynamic planet on which we live.

To learn more about mountain building, visit the Earth Science Web Site at earthgeu.com

Discovery Lab — Model Crustal Differences

Continental and oceanic crust have different densities. In this activity, you will model how both kinds of crust displace the mantle.

1. Obtain three wood blocks from your teacher. Determine the mass and volume of each. Calculate the density of each block. Record all of these values in a data table.
2. Half-fill a clear, plastic container with water. Place both of the 2-cm-thick blocks in the container.
3. Use a ruler to measure how much of each block is above the water surface. Record the measurements.
4. Replace the hardwood block with the 4-cm-thick softwood block.
5. Measure and record how much of each block is above the water surface.

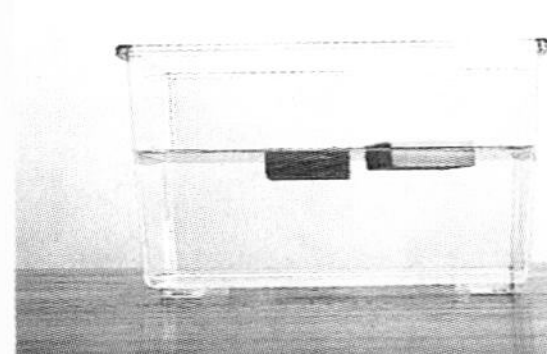

CAUTION: Always wear safety goggles and an apron in the lab.

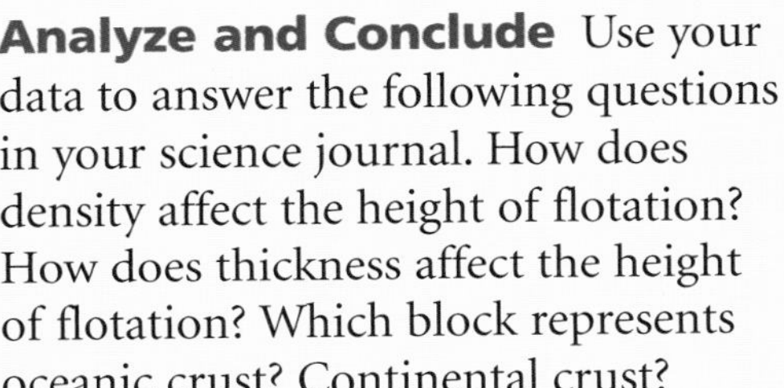

Analyze and Conclude Use your data to answer the following questions in your science journal. How does density affect the height of flotation? How does thickness affect the height of flotation? Which block represents oceanic crust? Continental crust?

SECTION 20.1 Crust-Mantle Relationships

OBJECTIVES

- **Describe** *the elevation distribution of Earth's surface.*
- **Explain** *isostasy and how it pertains to Earth's mountains.*
- **Describe** *how Earth's crust responds to the addition and removal of mass.*

VOCABULARY

isostasy
isostatic rebound

Mountains are spectacular features of Earth's crust that rise high above their surroundings. Mountains can occur as individual peaks such as the Towers of Paine in southern South America, which are shown on the facing page, or as immense ranges that snake for many kilometers along the landscape. Why do these geologic wonders rise high above Earth's surface, and how are such vast masses of rock supported? The answers to these questions lie in the relationships between Earth's crust and the underlying mantle.

EARTH'S TOPOGRAPHY

When you look at a globe or a map of Earth's surface, you immediately notice the oceans and continents. From these models of Earth, you can estimate that about 70 percent of Earth's surface is below sea level, and that the remaining 30 percent lies above the ocean's surface. What isn't obvious from most maps and globes, however, is the change in elevation, or topography, of the crust. Look at ***Figure 20-1*** on the next page, which is a map of the general topography of Earth's crust. Where are

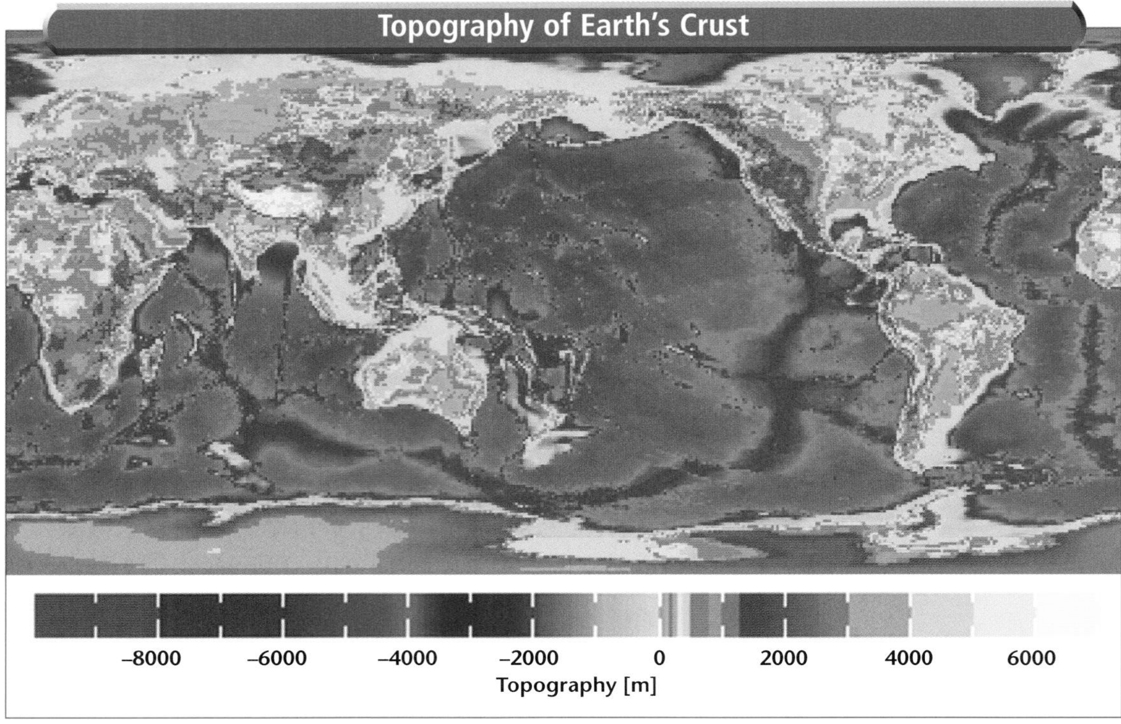

Figure 20-1 The highest point on Earth is Mt.Everest in Asia. The lowest point on Earth is the Mariana Trench, which is in the Pacific Ocean.

Earth's highest elevations? Where are Earth's lowest elevations?

When Earth's topography is plotted on a graph, a pattern in the distribution of elevations emerges, as shown in ***Figure 20-2.*** Note that there are two main elevation modes. Most of Earth's elevations cluster around these two modes: 0 to 1 km above sea level and 4 to 5 km below sea level. These two modes dominate Earth's topography and reflect the basic differences in density and thickness between continental and oceanic crust.

You observed in the *Discovery Lab* at the beginning of this chapter that blocks of wood with different densities displaced different amounts of water, and thus floated at various heights above the surface of the water. The block with the greatest density displaced the most water and floated lower in the water than the less-dense blocks. The results of this simple experiment are similar to the relationship that exists between Earth's crust and mantle. Recall from Chapter 1 that oceanic crust is composed mainly of basalt and that continental crust is composed primarily of granite. The average density of basalt is about 2.9 g/cm^3, while the average density of granite is about 2.8 g/cm^3. The slightly higher density of oceanic crust causes it to displace more of the mantle—which has a density of about 3.3 g/cm^3—than the same thickness of continental crust does.

Using Numbers
Suppose a mountain is being uplifted at a rate of 1 m every 1000 years. It is also being eroded at a rate of 1 cm/y. Is this mountain rising faster than it is being eroded? Explain.

Differences in elevation, however, are not caused by density differences alone. Recall from the *Discovery Lab* what happened when the thicker wood block was placed in the water. It displaced more water than the other two blocks, but, because of its density, it floated higher in the water than the other two blocks. Continental crust, which is thicker and less dense than oceanic crust, behaves similarly. It extends deeper into the mantle because of its thickness, and it rises higher above Earth's surface than oceanic crust because of its lower density, as shown in ***Figure 20-3.***

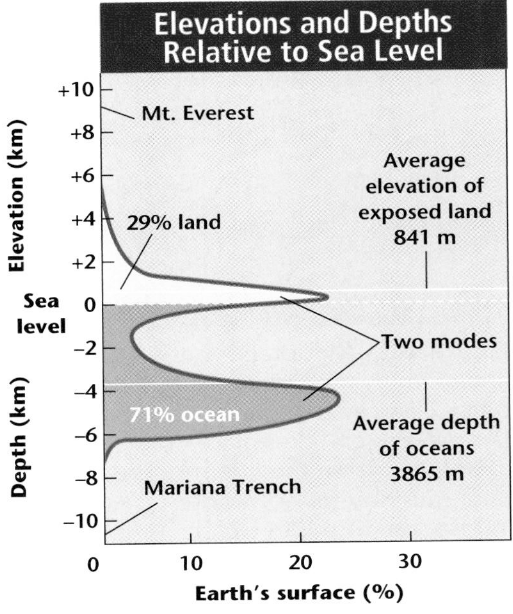

Figure 20-2 About 29% of Earth is land; 71% of Earth is water. Two elevations dominate Earth's surface: 0 to 1 km above sea level and 4 to 5 km below sea level. The average elevation above sea level is 841 m. The average depth of Earth's oceans is 3865 m.

ISOSTASY

The displacement of the mantle by Earth's continental and oceanic crust is a condition of equilibrium called **isostasy.** The crust and mantle are in equilibrium when the force of gravity on the mass of crust involved is balanced by the upward force of buoyancy. This balance is familiar to you if you have ever watched people get in and out of a small boat. As the people boarded the boat, it sank deeper into the water. Conversely, as the people got out of the boat, it displaced less water and floated higher in the water.

A similar sinking and rising that result from the addition and removal of mass occurs with the crust that makes up Earth's mountains. Gravitational and seismic studies have detected thick roots of continental material that extend into the mantle below Earth's mountain ranges. According to the principle of isostasy, parts of the crust will rise or subside until these parts are buoyantly supported by their roots. In other words, a mountain range requires large roots to counter the enormous mass of the range above Earth's surface. Continents and mountains are said to float on the mantle because they are less dense than the underlying mantle and therefore project into the mantle to

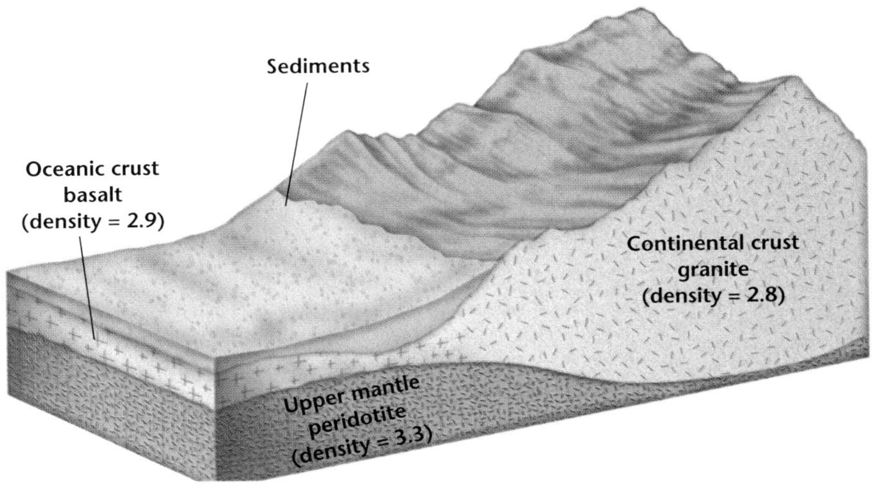

Figure 20-3 Continental crust is less dense and thicker than oceanic crust, and it thus extends higher above Earth's surface and deeper into the mantle than oceanic crust.

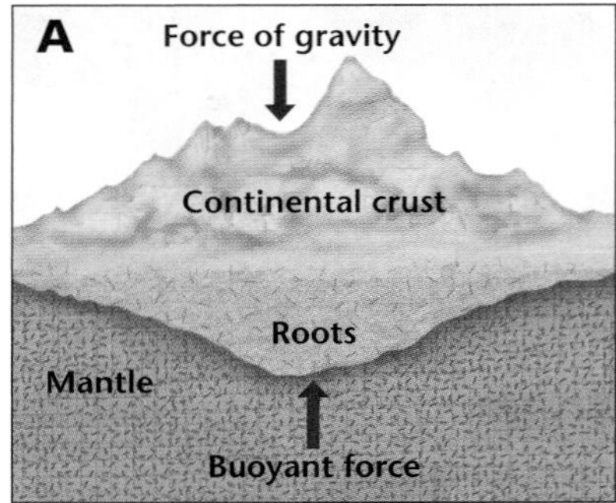

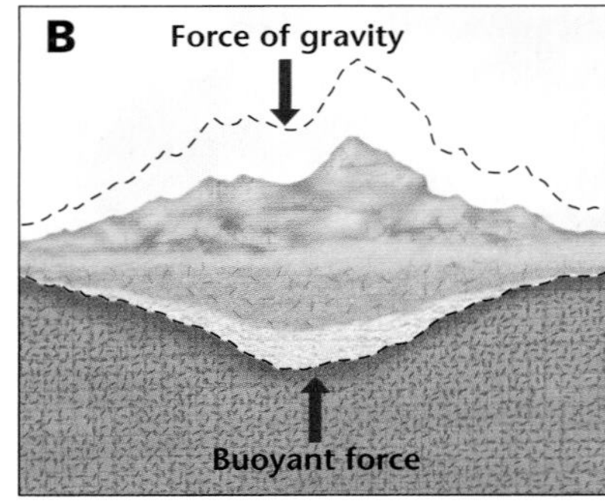

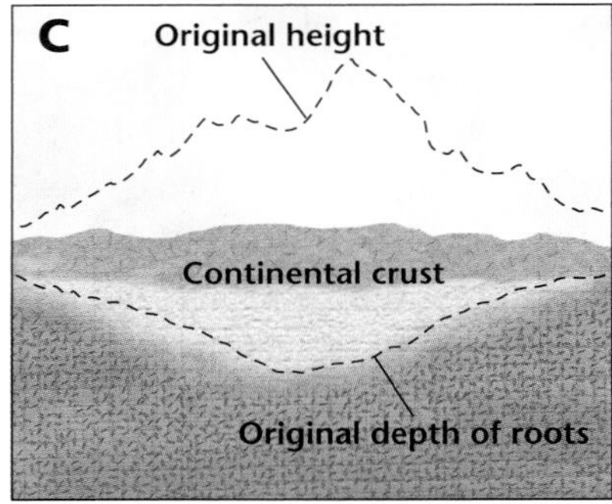

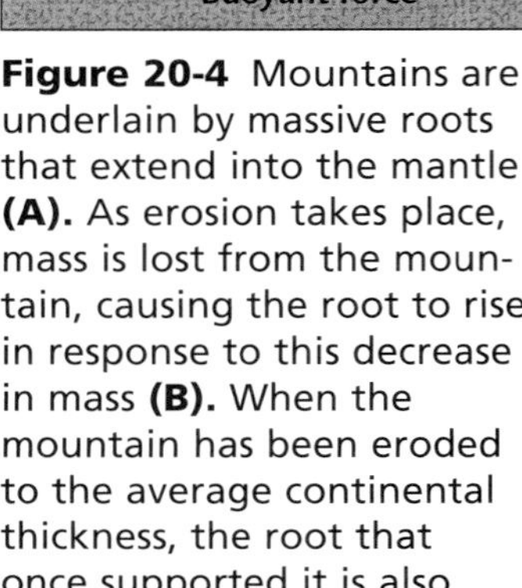

Figure 20-4 Mountains are underlain by massive roots that extend into the mantle **(A).** As erosion takes place, mass is lost from the mountain, causing the root to rise in response to this decrease in mass **(B).** When the mountain has been eroded to the average continental thickness, the root that once supported it is also gone **(C).**

provide the necessary buoyant support. What do you think happens when mass is removed from a mountain or mountain range?

Isostasy and Erosion You might recall from Chapter 17 that the Appalachian Mountains formed millions of years ago when the North American continent collided with the European continent. Rates of erosion on land are such that these mountains should have been completely eroded long ago. Why, then, do these mountains still exist? As mountains rise above Earth's surface, deep roots form until isostatic equilibrium is achieved and the mountains are buoyantly supported. As peaks are eroded, mass decreases, and the roots become smaller, as shown in ***Figure 20-4.*** A balance between erosion and the decrease in the size of the root will continue for hundreds of millions of years until both the mountains and their roots disappear. This slow process of the

Problem-Solving Lab

Making and Interpreting Graphs

Graph isostatic rebound The rate of isostatic rebound changes over time. An initially rapid rate often declines to a very slow rate. Use the data in the table to generate a graph of isostatic rebound with time.

Isostatic Rebound Data

Years Before Present	Total Amount of Rebound (m)
8000	54
6000	80
4000	93
2000	100
0	104

Analysis

1. How much of the total rebound occurred during the first 2000 years?

Thinking Critically

2. Predict how much rebound will still occur. Approximately how long will this take?
3. Study your graph. Describe how the rate of isostatic rebound decreases with time.

crust's rising as the result of the removal of overlying material is called **isostatic rebound.** You can explore how the rate of isostatic rebound changes with time in the *Problem-Solving Lab* on page 526.

Crustal movements resulting from isostasy are not restricted to Earth's continents. Individual volcanic mountains called seamounts can form on the ocean floor as a result of a plate's moving over a hot spot in Earth's mantle. On the geologic time scale, these mountains form very quickly. What do you think happens to the seafloor after these seamounts form? The seamounts are added mass. As a result of isostasy, the oceanic crust around these peaks displaces the underlying mantle until equilibrium is achieved.

You've just learned that the elevation of Earth's crust depends upon the thickness of the crust as well as its density. You also learned that a mountain peak is countered by a root. Mountain roots can be many times as deep as a mountain is high. Mt. Everest, shown in ***Figure 20-5,*** towers nearly 9 km above sea level and is the tallest peak in the Himalayan Mountains. Some parts of the Himalayas are underlain by crustal roots nearly 80 km thick! You'll learn more about Mt. Everest in the *Science & Math* feature at the end of this chapter. Where do the immense forces required to produce such crustal thickening originate? You'll find out in the next section.

Figure 20-5 Mt. Everest, a peak in Asia, is currently the highest mountain on Earth. It is underlain by a very deep crustal root that supports its mass.

SECTION ASSESSMENT

1. If continental crust were thinner than its average thickness of 40 km, would it depress the mantle more or less than it does now? Explain.
2. What is isostasy?
3. Describe the distribution of Earth's elevations and explain what causes this distribution.
4. Why is the crust thicker beneath continental mountain ranges than it is under flat-lying stretches of landscape?
5. **Thinking Critically** The area around the Great Lakes was once covered by thick sheets of ice. Use the principle of isostasy to explain how the melting of these ice sheets has affected the land around the lakes.

SKILL REVIEW

6. **Recognizing Cause and Effect** Explain what happens in terms of isostasy to the land surrounding a mountain range as sediments are eroded from the mountains onto the nearby land. For more help, refer to the *Skill Handbook.*

SECTION 20.2 Convergent-Boundary Mountains

Earth Sciences 3.b

OBJECTIVES

- **Compare** *and* **contrast** *the different types of mountains that form along convergent plate boundaries.*
- **Explain** *how the Appalachian Mountains formed.*

VOCABULARY

orogeny

A quick glance at a world map will show that Earth's landscape is dotted with numerous mountain peaks and ranges. The Cascades and the Appalachians, for example, run north-south on either side of the United States. The Andes form the western border of South America, and the majestic Himalayas separate Nepal from Tibet. Mt. Kilimanjaro is a volcano that rises high above the African continent. Mauna Loa, another volcanic peak, is located in Hawaii. Most of these ranges and peaks, like most earthquakes and volcanoes, have formed as a result of tectonic interactions.

OROGENY

The processes that form all mountain ranges are called **orogeny.** Orogeny results in broad, linear regions of deformation known as orogenic belts. Most orogenic belts, as shown in ***Figure 20-6,*** are associated with plate boundaries. The greatest variety and the tallest of these belts are found at convergent boundaries. The compressive forces at these

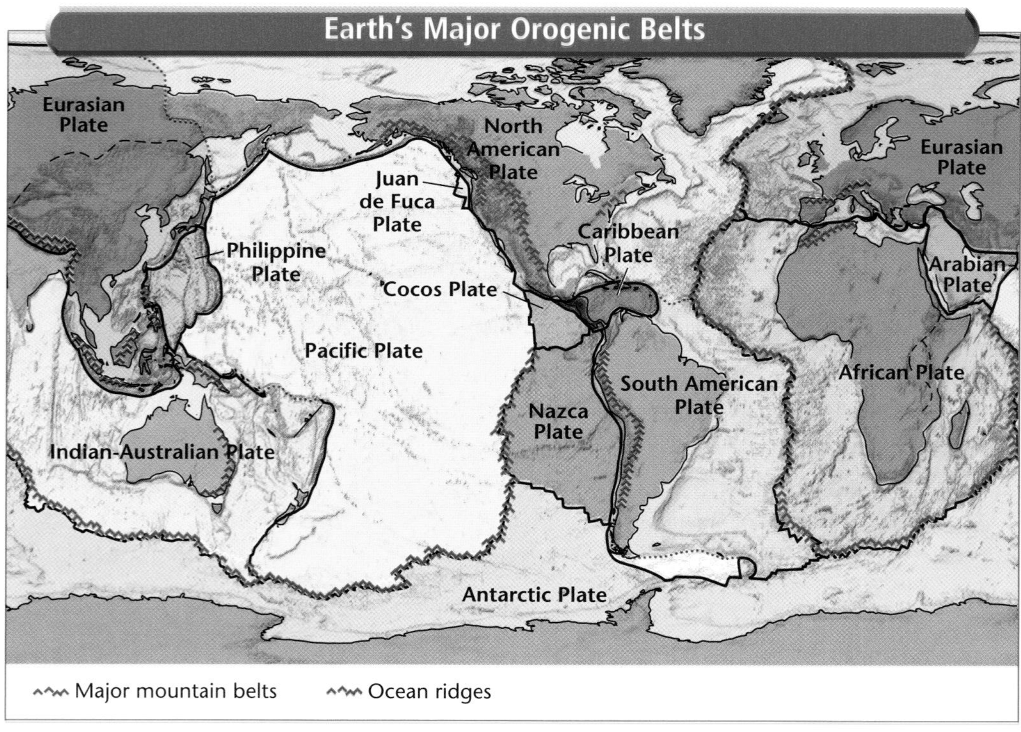

Figure 20-6 Most of Earth's mountain ranges have formed along plate boundaries.

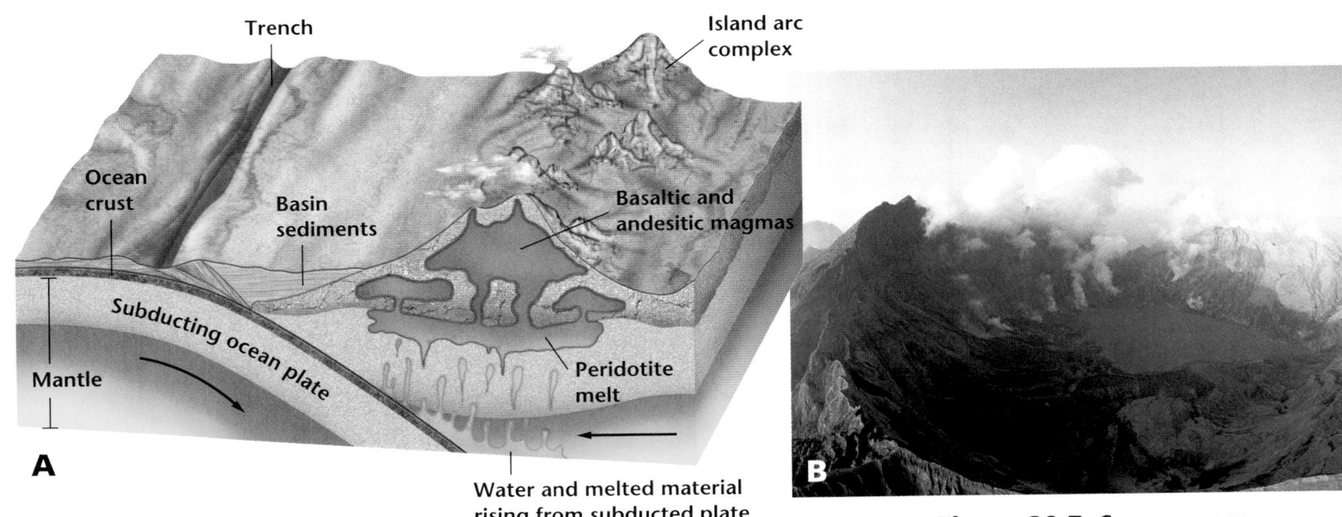

Figure 20-7 Convergence between two oceanic plates results in the formation of individual volcanic peaks that make up an island arc complex **(A).** Mt. Pinatubo, shown here **(B),** is one of several volcanic peaks that make up the island arc complex known as the Philippine Islands.

boundaries may cause the intense deformation—folding, faulting, metamorphism, and igneous intrusions—that is characteristic of orogenic belts. Interactions at each type of convergent boundary create different types of mountain ranges.

Oceanic-Oceanic Convergence When an oceanic plate converges with another oceanic plate, one plate descends into the mantle to create a subduction zone. As parts of the subducted plate melt, magma is forced upward to form a series of volcanic peaks called an island arc complex. The tectonic relationships and processes associated with oceanic-oceanic convergence are detailed in ***Figure 20-7A.*** Note that the crust along the island arc thickens to form a root. According to the principle of isostasy, it is the displacement of the mantle by this root that provides the necessary buoyancy for a mountain peak.

What kinds of rocks make up island arc complexes? Recall from Chapter 19 that seismic studies indicate that much of the mantle is made of an igneous rock called peridotite. The water released from a subducted plate and the sediments it carries causes the peridotite to melt. The melted rocks, along with bits of the subducted plate, are forced upward toward the surface, where more melting occurs. As the melted material comes into contact with the crust, magmas with different compositions form. Eventually, basaltic and andesitic magmas rise to the surface and erupt to form the island arc complex.

In addition to the volcanic rocks that make up an island arc complex, some large complexes contain sedimentary rocks. Between an island arc and a trench is a depression, or basin, which fills with sediments eroded from the island arc. If subduction continues for a long enough period of time, some of these sediments can be uplifted,

Earth Sciences

3.b Students know the principal structures that form at the three different kinds of plate boundaries.

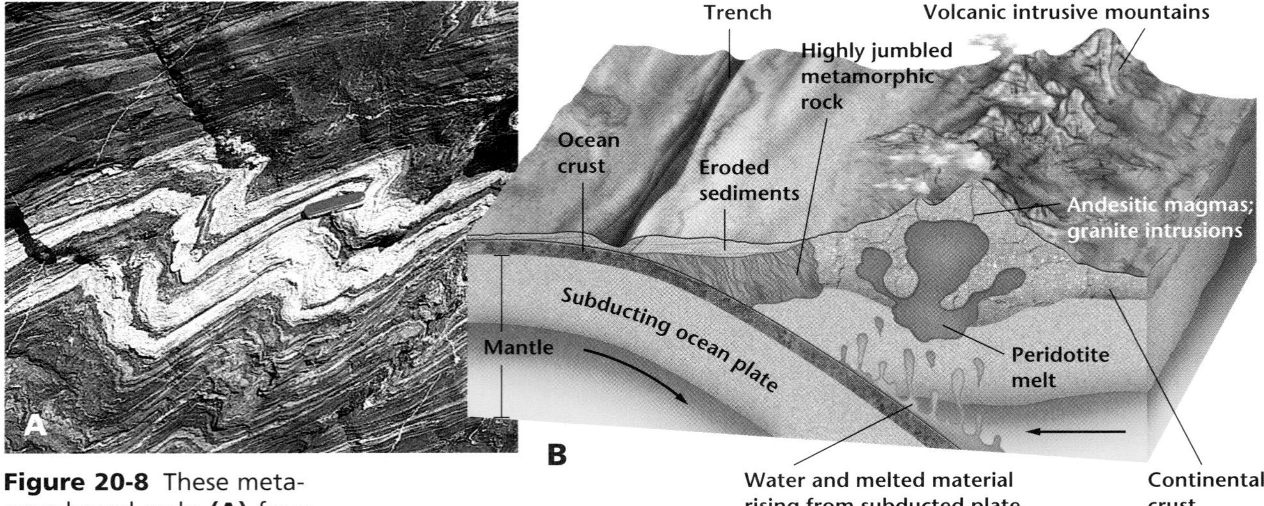

Figure 20-8 These metamorphosed rocks **(A)** from Catalina Island, California, formed as the result of convergence of an oceanic plate with a continental plate. At an oceanic-continental boundary **(B)** compression causes continental crust to fold and thicken. Igneous activity and metamorphism are also common along such boundaries.

folded, faulted, and thrust against the island arc to form a complex mass of sedimentary and island-arc volcanic rocks.

Oceanic-Continental Convergence Oceanic-continental boundaries are very similar to oceanic-oceanic boundaries in that convergence along both creates subduction zones and trenches. The similarity ends there, however, because convergence between oceanic and continental plates can produce major mountain belts. When an oceanic plate converges with a continental plate, the descending oceanic plate forces the edge of the continental plate upward. This uplift marks the beginning of orogeny, as detailed in ***Figure 20-8B.*** In addition to uplift, compressive forces may cause the continental crust to fold and thicken. As the crust thickens, higher and higher mountains form. Deep roots develop to support these enormous masses of rocks.

Another important orogenic process that occurs along an oceanic-continental boundary is the formation of magma, as illustrated in ***Figure 20-8B.*** As the subducting plate sinks into the mantle, parts of the plate begin to melt. As the magma moves upward through the continental crust, the magma becomes rich in silica and gives rise to granitic intrusions and volcanoes fueled by andesitic magma.

Sediments eroded from volcanic intrusive mountains may fill the low areas between the trench and the coast. These sediments, along with ocean sediments and material scraped off the descending plate, are shoved against the edge of the continent to form a jumble of highly folded, faulted, and metamorphosed rocks. The metamorphosed rocks shown in ***Figure 20-8A*** formed when the Pacific Plate subducted beneath the North American Plate millions of years ago.

Topic: Mt. Everest
To take a virtual tour of Mt. Everest, visit the Earth Science Web Site at earthgeu.com

Activity: High Altitudes
Research the effects of high altitudes on climbers. How does the density of air at 29 000 feet affect the human body? How do climbers cope with this effect?

Continental-Continental Convergence Earth's tallest mountain ranges, including the Himalayas, are formed at continental-continental plate boundaries. Because of its relatively low density, continental crust cannot be subducted into the mantle when two plates converge. Instead, the energy associated with the collision is transferred to the crust involved, which becomes highly folded and faulted, as shown in ***Figure 20-9A.*** Compressional forces break the crust into thick slabs that are thrust onto each other along low-angle faults. This process can double the thickness of the deformed crust. Deformation can also extend laterally for hundreds of kilometers into the continents involved. The magma that forms as a result of continental-continental mountain building hardens beneath Earth's surface to form granite batholiths.

Another common characteristic of mountains that form when two continents collide is the presence of marine sedimentary rock near the mountains' summits. Where do you think this rock comes from? Such rock forms from the sediments deposited in the ocean basin that existed between the continents before their collision. Mount Godwin Austen, also known as K2 in the western Himalayas, for example, is composed of thousands of meters of marine limestone that sits upon a granite base. The limestone represents the northern portions of the old continental margin of India that were pushed up and over the rest of the continent when India began to collide with Asia about 50 million years ago.

To learn more about Mt. Everest, the tallest peak in the Himalayas, go to the **National Geographic Expedition** on page 864.

Figure 20-9 Intense folding and faulting along continental-continental boundaries produce some of the highest mountain ranges on Earth **(A).** K2, shown here **(B),** is the second-highest peak in the Himalayas; only Mt. Everest is taller.

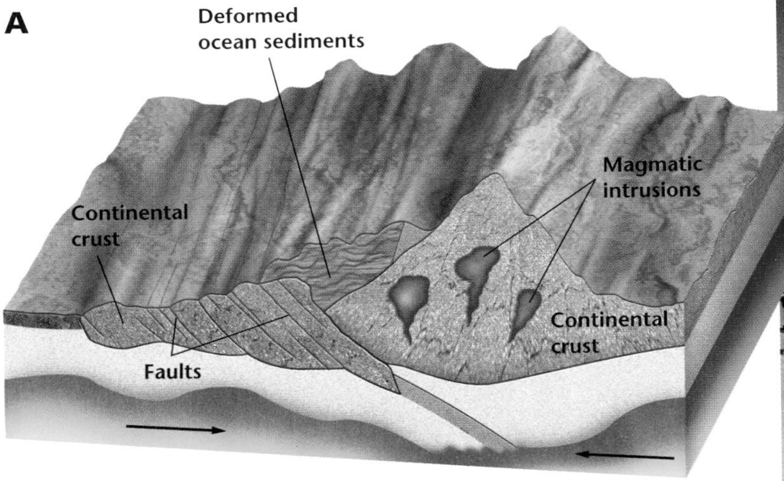

The Appalachian Mountains–A Case Study

Recall from Chapter 17 that Alfred Wegener used the matching rocks and geologic structures in the Appalachians and mountains in Greenland and northern Europe to support his hypothesis of continental drift. In addition to Wegener, many other scientists have studied the Appalachians. In fact, the geology of this mountain range, which is located in the eastern United States, has been the subject of many studies for more than a hundred years. Based on these studies, geologists have divided the Appalachian Mountain Belt into several distinct regions, including the Valley and Ridge, the Blue Ridge, and the Piedmont Provinces. Each region is characterized by rocks that show different degrees of deformation. Rocks of the Valley and Ridge Province, for example, some of which are shown in ***Figure 20-10,*** are highly folded; most of the rocks that make up the Piedmont Province are not. Why are these regions so different? What kinds of processes led to their formation?

The Early Appalachians The tectonic history of the Appalachian Mountains began about 700 to 800 million years ago when ancestral North America separated from ancestral Africa along two divergent boundaries to form two oceans. The ancestral Atlantic Ocean was located off the western coast of ancestral Africa. A shallow, marginal sea formed along the eastern coast of ancestral North America. A continental fragment was located between the two divergent boundaries.

About 700 to 600 million years ago, the directions of plate motions reversed. The ancestral Atlantic Ocean began to close as the plates converged. This convergence resulted in the formation of a subduction zone and a volcanic island arc east of ancestral North America, as shown in ***Figure 20-11A.***

Figure 20-10 These folded rocks in West Virginia are part of the Valley and Ridge Province of the Appalachian Mountains.

Figure 20-11 The Appalachians formed millions of years ago as a result of convergence.

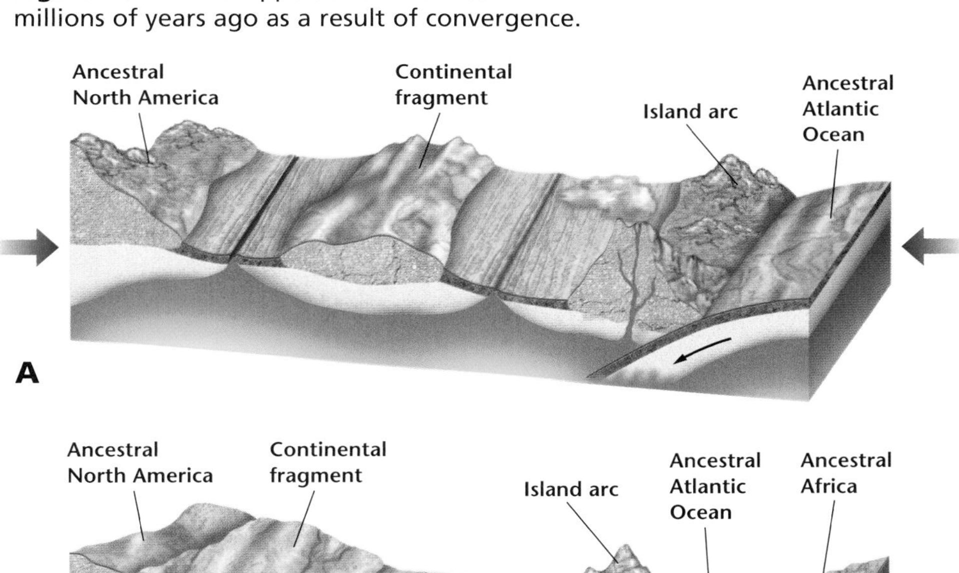

700-600 M.Y.B.P.* Convergence causes the ancestral Atlantic Ocean to begin to close. An island arc develops east of ancestral North America.

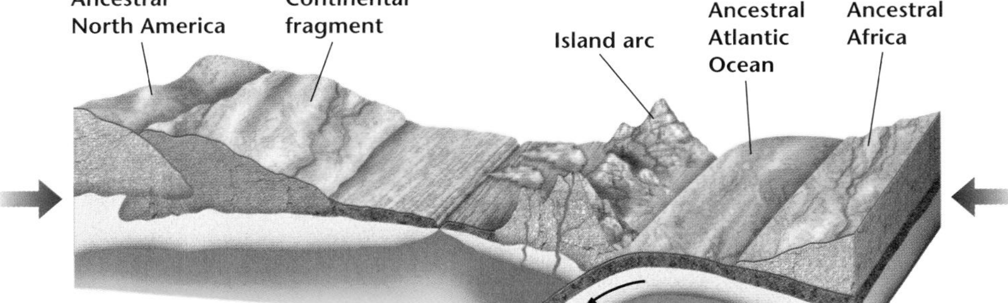

500-400 M.Y.B.P. The continental fragment, which eventually becomes the Blue Ridge Province, becomes attached to ancestral North America.

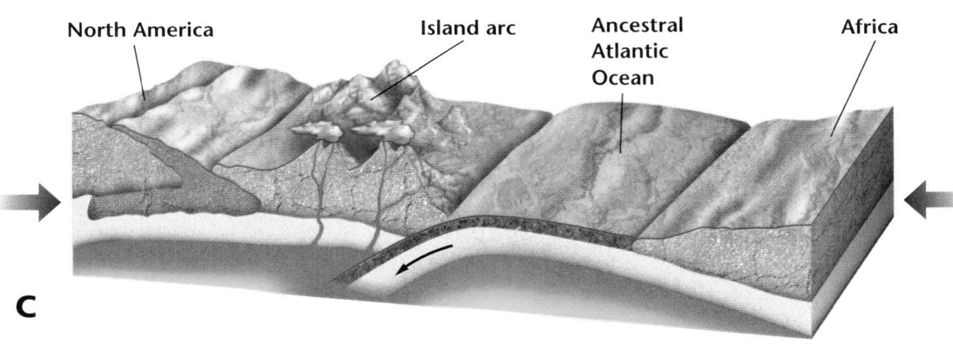

400-300 M.Y.B.P. The island arc becomes attached to ancestral North America and the continental fragment is thrust farther onto ancestral North America. The arc becomes the Piedmont Province.

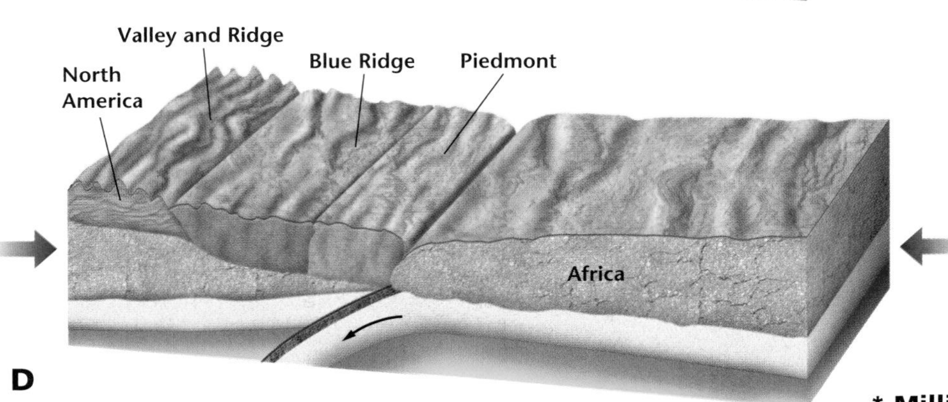

300-260 M.Y.B.P. Ancestral Africa collides with ancestral North America to close the ancestral Atlantic Ocean. Compression forces the Blue Ridge and Piedmont rocks farther west and the folded Valley and Ridge Province forms.

*** Million Years Before Present**

Figure 20-12 The metamorphosed rocks in the foreground are parts of the Blue Ridge Province of North Carolina.

About 200 million years passed before the continental fragment became attached to ancestral North America, as shown in ***Figure 20-11B*** on page 533. These highly metamorphosed rocks, some of which are shown in ***Figure 20-12,*** were thrust over younger rocks to become the Blue Ridge Province.

The Final Stages of Formation Between about 400 and 300 million years ago, the island arc became attached to North America, as shown in ***Figure 20-11C*** on page 533. Evidence of this event is preserved in the Piedmont Province as a group of metamorphic and igneous rocks. These rocks were also faulted over the continent, pushing the Blue Ridge rocks farther west.

Between about 300 and 260 million years ago, the ancestral Atlantic Ocean closed as ancestral Africa, Europe, and South America collided with ancestral North America to form Pangaea. This collision resulted in extensive folding and faulting, as illustrated in ***Figure 20-11D*** on page 533, to form the Valley and Ridge Province. When rifting caused Pangaea to break apart about 200 million years ago, the modern Atlantic Ocean formed.

The Appalachian Mountains are only one example of the many mountain ranges that have formed along convergent boundaries. In the next section, you'll find out about the orogeny that takes place along divergent plate boundaries, as well as some of the types of mountains that form far from plate margins.

SECTION ASSESSMENT

1. Describe how mountains form along a continental-continental plate boundary.
2. How do the mountains that form at oceanic-oceanic plate boundaries differ from the mountains that form at oceanic-continental plate boundaries?
3. **Thinking Critically** Locate the Aleutian Islands on the map shown in ***Figure 20-6.*** How do you think these mountain peaks formed?

SKILL REVIEW

4. **Sequencing** Sequence the events that resulted in the formation of the Appalachian Mountains. For more help, refer to the *Skill Handbook.*

SECTION 20.3 Other Types of Mountains

Earth Sciences 3.b

When ocean ridges were first discovered, they caused quite a stir in the scientific community simply because of their size. These mountains form a continuous chain that snakes along Earth's ocean floor for over 65 000 km! In addition to their being much longer and taller than most of their continental counterparts, these mountains formed as a result of different orogenic processes.

OBJECTIVES

- **Describe** *the mountain ranges that form along ocean ridges.*
- **Compare** *and* **contrast** *uplifted and fault-block mountains.*
- **Describe** *the mountains that form as a result of hot spots in Earth's mantle.*

VOCABULARY

pillow basalt
uplifted mountain
fault-block mountain

Earth Sciences

3.b Students know the principal structures that form at the three different kinds of plate boundaries.

DIVERGENT-BOUNDARY MOUNTAINS

Ocean ridges are regions of very broad uplift that seems to be related to the rising convection cells that form deep in the mantle beneath these ridges. As matter is heated, it expands, which results in a decrease in density. Magma is less dense than surrounding mantle material, and thus it is forced upward, where it warms the overlying lithosphere. As a result of this increase in temperature, the lithosphere along a divergent boundary bulges upward and stands higher than the surrounding ocean crust to form a gently sloping mountain range, as shown in ***Figure 20-13.*** As newly formed lithosphere moves away from the central rift, it cools, contracts, and becomes more dense.

Ocean ridge mountain ranges can be thousands of kilometers wide. In the *MiniLab* on page 536, you will compare the size of one ocean ridge, the Mid-Atlantic Ridge, with the size of the continental United States.

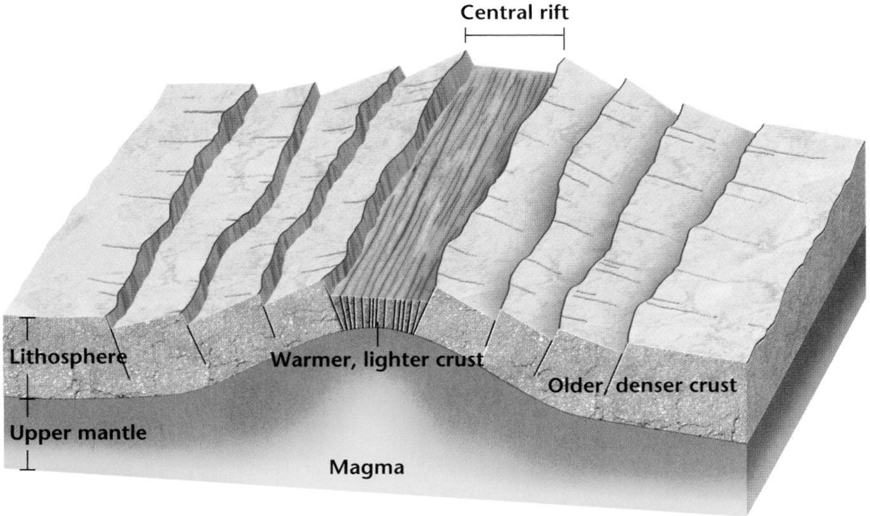

Figure 20-13 An ocean ridge is a broad, topographic high that forms as lithosphere bulges upward due to an increase in temperature along a divergent boundary.

MiniLab

How large is an ocean ridge?

Compare the width of part of an ocean ridge with the size of the United States.

Procedure

1. Obtain physiographic maps of the Atlantic Ocean floor and North America.
2. Use tracing paper to copy the general outline of a section of the Mid-Atlantic Ridge. The length of the section should be long enough to stretch from San Francisco to New York City. Mark the ridge axis on your tracing.
3. Place the same tracing paper on the map of North America with the ridge axis running east-west. Trace the general outline of the United States onto the paper.

Analyze and Conclude

1. How wide is the Mid-Atlantic Ridge?
2. Are there any parts of the United States that are not covered by your tracing?
3. If a mountain range the size of the Mid-Atlantic Ridge were located in the United States as you have drawn it, how would it affect the major river drainage patterns and climates in various parts of North America?

Ocean–Ridge Rocks Ocean ridges are composed mainly of igneous rocks. Recall from Chapter 17 that as tectonic plates separate along an ocean ridge, hot mantle material is forced upward. The partial melting of this material results in a mixture that accumulates in a magma chamber beneath the ridge. From the chamber, the mixture intrudes into the overlying rock to form a series of vertical dikes that resemble a stack of index cards standing on edge, as shown in *Figure 20-14B.* Some of the magma also pushes through the dikes and erupts onto the seafloor to form igneous rocks called **pillow basalts,** which, as you can see in *Figure 20-14A,* resemble a pile of sandbags.

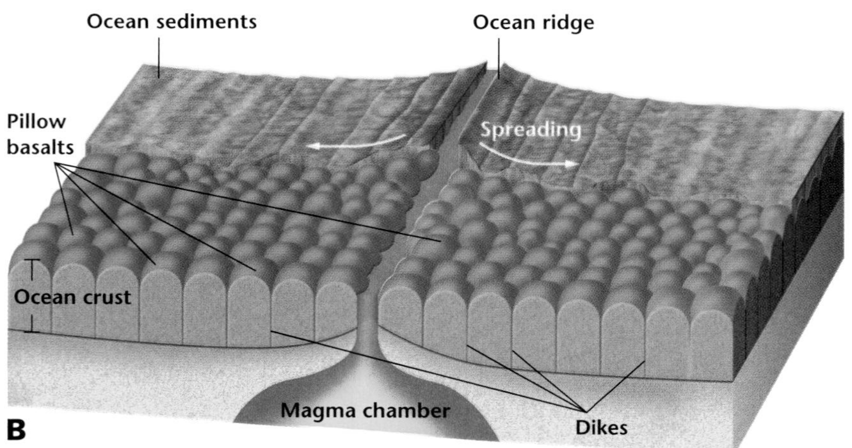

Figure 20-14 Vertical dikes overlain by pillow basalts, which are shown in the photo **(A),** are characteristic of ocean-ridge rocks **(B).**

Nonboundary Mountains

You've just learned that island arc complexes, intrusive volcanic mountain ranges, highly folded continental mountains, and ocean ridges are all associated with plate boundaries. While these types of mountains make up the majority of ranges and peaks on Earth, some mountains and peaks form in places far removed from tectonic boundaries. Three nonboundary types of mountains are uplifted mountains, fault-block mountains, and some volcanoes.

Uplifted Mountains As shown in ***Figure 20-15B,*** some mountains form when large regions of Earth have been slowly forced upward as a unit; these mountains are called **uplifted mountains.** The Adirondack Mountains in New York State, shown in ***Figure 20-15A,*** are uplifted mountains. Generally, the rocks that make up uplifted mountains undergo less deformation than rocks associated with plate boundary orogeny, which are highly folded, faulted, and metamorphosed.

The cause of large-scale regional uplift is not well understood. It is possible that warmer regions of the mantle heat these portions of the lithosphere. The heat causes the density of the crust to decrease, resulting in slow uplift as that section rebounds in response to isostasy. Another possible cause is upward movement in the mantle, which lifts regions of the crust without causing much deformation. Regional uplift can form broad plateaus, such as the Colorado Plateau, which extends through Colorado, Utah, Arizona, and New Mexico. Erosional forces eventually carve uplifted areas to form mountains, valleys, and canyons.

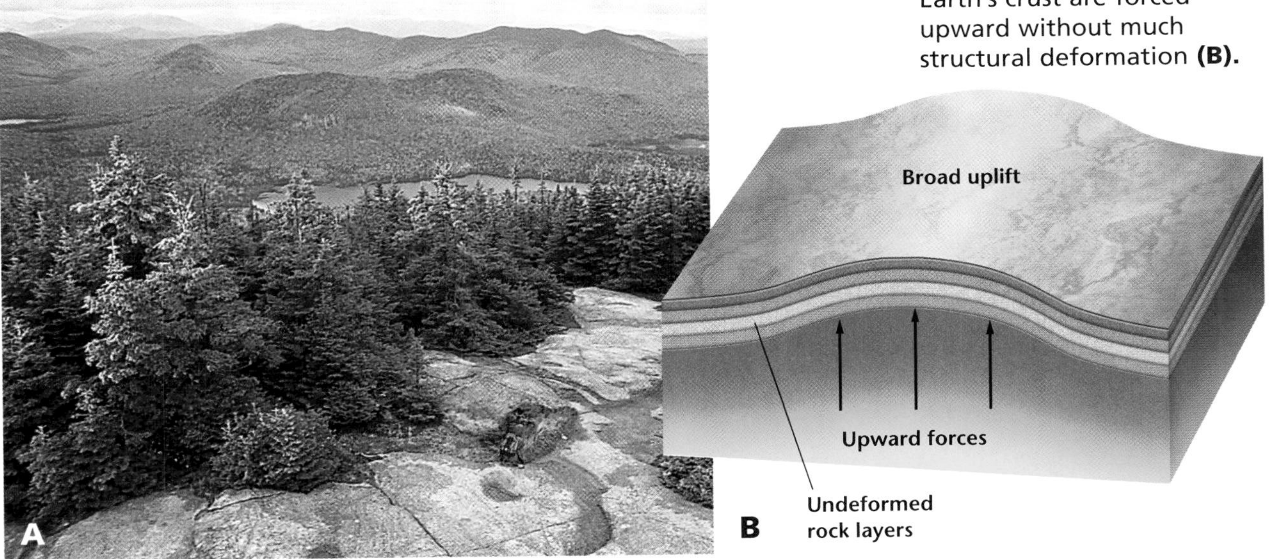

Figure 20-15 The Adirondack Mountains of New York State, shown in the photo **(A),** are uplifted mountains. Uplifted mountains form when large sections of Earth's crust are forced upward without much structural deformation **(B).**

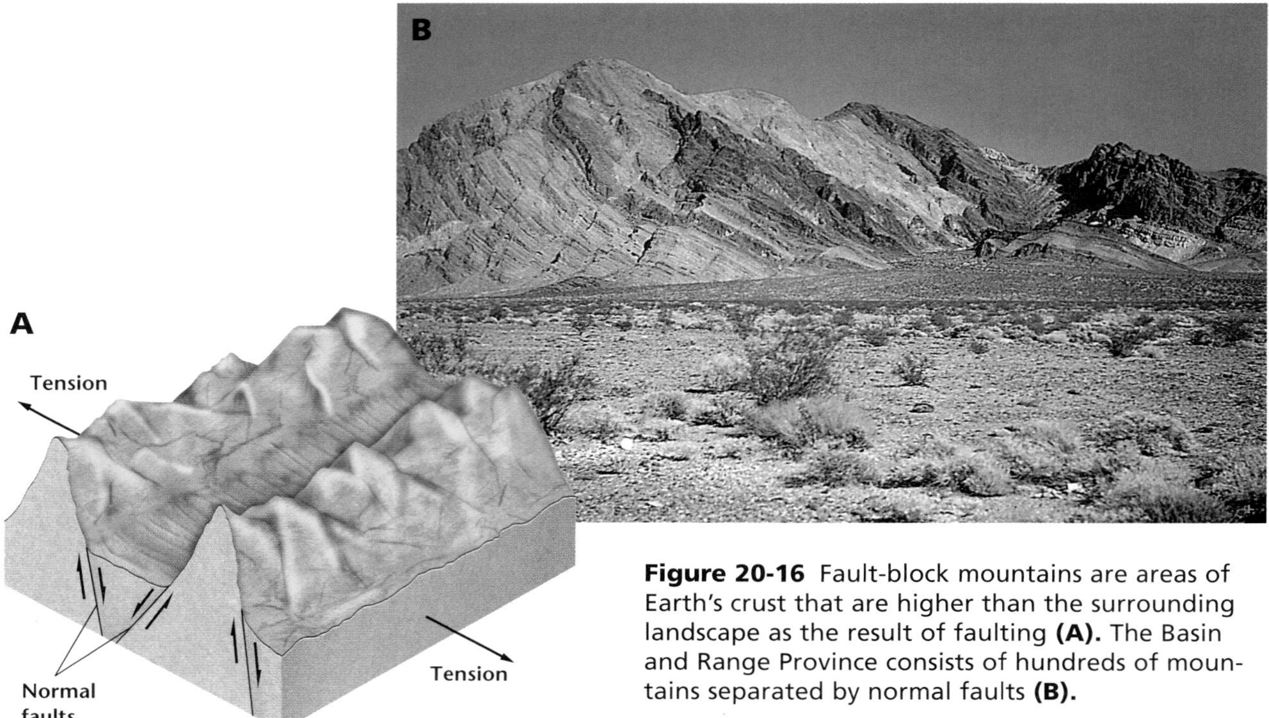

Figure 20-16 Fault-block mountains are areas of Earth's crust that are higher than the surrounding landscape as the result of faulting **(A).** The Basin and Range Province consists of hundreds of mountains separated by normal faults **(B).**

Fault-Block Mountains Another type of mountain that is not necessarily associated with plate boundaries is a fault-block mountain. **Fault-block mountains** form when large pieces of crust are tilted, uplifted, or dropped downward between large faults, as shown in ***Figure 20-16A.*** The Basin and Range Province of the southwestern United States and northern Mexico, a part of which is shown in ***Figure 20-16B,*** consists of hundreds of nearly parallel mountains separated by normal faults. The Grand Tetons in Wyoming are also fault-block mountains. You'll explore the topography of this range in the *Mapping Geolab* at the end of this chapter.

Volcanic Peaks Volcanoes that form along oceanic-continental convergent margins are usually parts of large mountain ranges. Volcanoes that form over hot spots, however, are generally solitary peaks that form far from tectonic plate boundaries. Recall from Chapter 18 that a hot spot is a region in Earth's mantle that is much hotter than the surrounding area. As a tectonic plate moves over a hot spot, plumes of mantle material are forced through the crust to form a volcanic peak. As the plate continues to move over the hot spot, a chain of volcanoes forms. The shield volcanoes that make up the state of Hawaii are volcanic peaks that formed as the Pacific Plate moved over a hot spot in the mantle. Mauna Kea, which is shown in ***Figure 20-17,*** is one of these volcanic peaks.

Figure 20-17 Mauna Kea, and the small volcanoes that dot its flanks are some of the many volcanic peaks that make up the Hawaiian Islands, which formed as the Pacific Plate moved over a hot spot in Earth's mantle.

While all mountains are similar in that they tower high above the surrounding land, individual peaks and chains are unique, as you have discovered in this chapter. Some peaks and chains form along tectonic plate boundaries, while others form far from these boundaries. Some mountains are produced by faulting and folding; others form as the result of igneous activity and crustal uplift. No matter how they form, all mountains are evidence that Earth, unlike some of its neighbors, is truly a dynamic planet.

Section Assessment

1. What kinds of rocks are associated with ocean ridges?
2. Explain why an ocean ridge is higher than the surrounding crust.
3. How do volcanoes that form when a plate moves over a hot spot in the mantle differ from volcanoes that form along convergent plate boundaries?
4. Compare and contrast the formation of uplifted and fault-block mountains.
5. **Thinking Critically** Would you expect a volcano that forms on a continent to depress the crust as much as a volcano that forms on the ocean floor? Explain your reasoning.

Skill Review

6. **Concept Mapping** Use the following terms to construct a concept map that contrasts the mountain types discussed in this section. For more help, refer to the *Skill Handbook.*

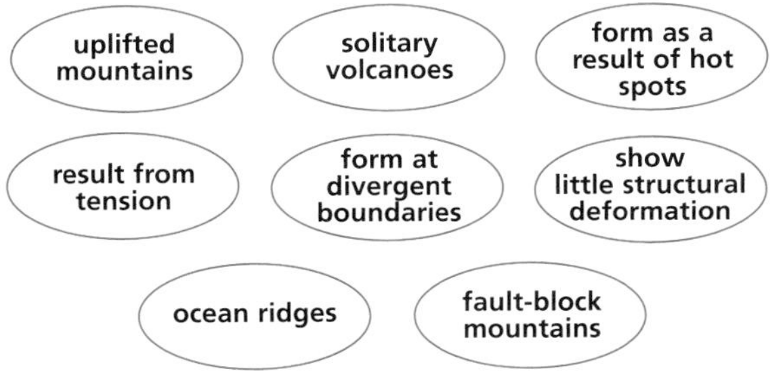

Mapping GeoLab

Making a Map Profile

A map profile, which is also called a cross section, is a side view of a geographic or geologic feature constructed from a topographic map. You will construct and analyze a profile of the Grand Tetons, a mountain range in Wyoming that formed when enormous blocks of rocks were faulted along their eastern flanks, causing the blocks to tilt to the west.

 I&E

1.h Read and interpret topographic and geologic maps.

Preparation

Problem

How do you construct a map profile?

Materials

metric ruler
graph paper
sharp pencil

Procedure

1. On the graph paper, make a grid like the one shown on the facing page.
2. Place the edge of a paper strip along the profile line AA' and mark where each major contour line intersects the strip.
3. Label each intersection point with the correct elevation.
4. Transfer the points from the paper strip to the profile grid.
5. Connect the points with a smooth line to construct a profile of the mountain range along line AA'.
6. Label the major geographic features on your profile.

Analyze

1. Describe how the map profile changes with distance from point A.
2. What is the elevation of the highest point on the map profile? The lowest point?
3. What is the average elevation shown in the profile?
4. Calculate the total relief shown in the profile.

Conclude & Apply

1. Is your map profile a scale model of the topography along line AA'? Explain.
2. What determined the scale of this map profile?
3. Why are map profiles made from topographic maps often exaggerated vertically?

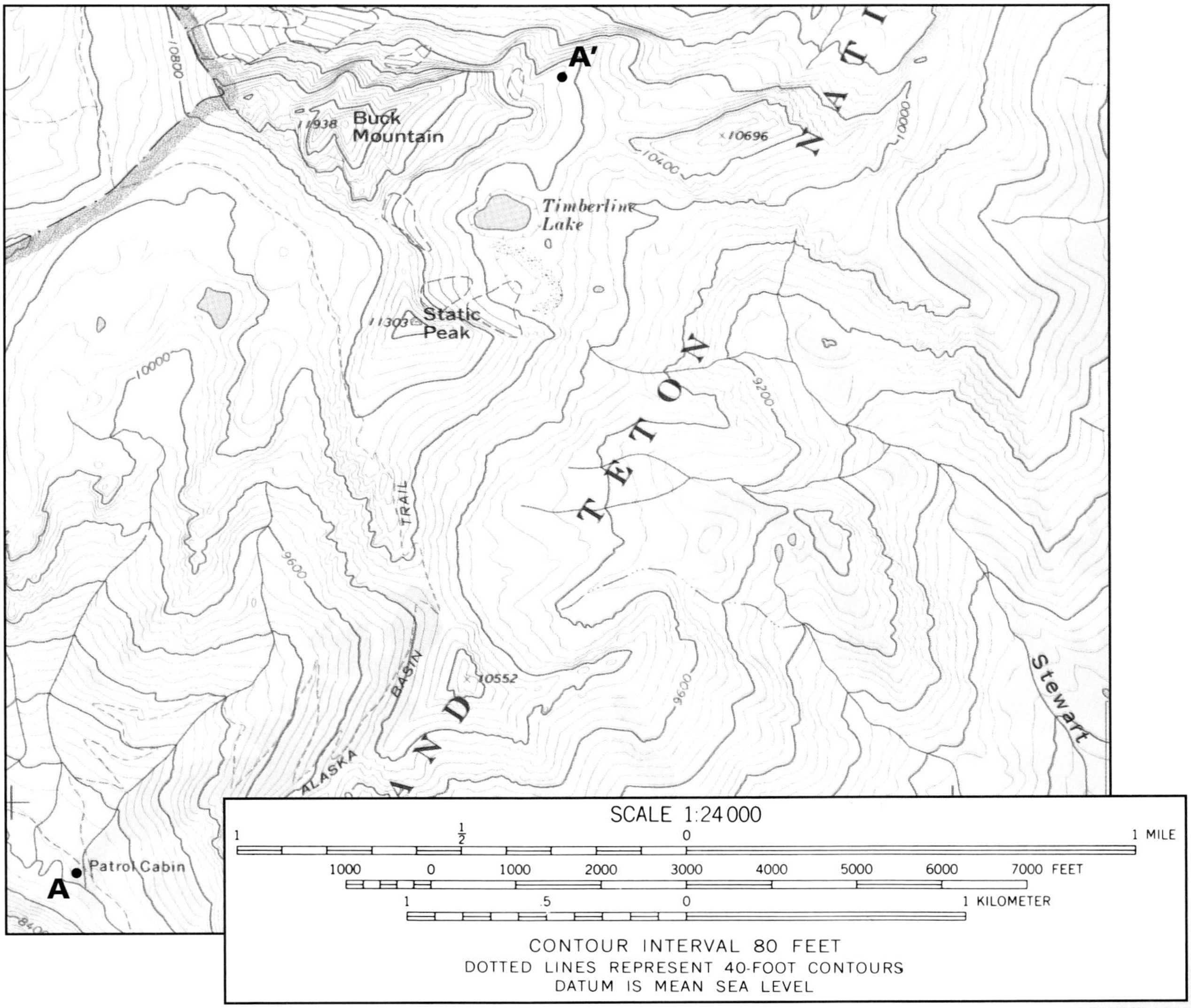

A'

Buck Mountain

11938

10800

×10696

10400

10000

Timberline Lake

Static Peak

11303

10000

TETON

NATI

9200

TRAIL

9600

BASIN

×10552

9600

Stewart

ALASKA

AND

Patrol Cabin

A

8400

SCALE 1:24 000

1 ½ 0 1 MILE

1000 0 1000 2000 3000 4000 5000 6000 7000 FEET

1 .5 0 1 KILOMETER

CONTOUR INTERVAL 80 FEET

DOTTED LINES REPRESENT 40-FOOT CONTOURS

DATUM IS MEAN SEA LEVEL

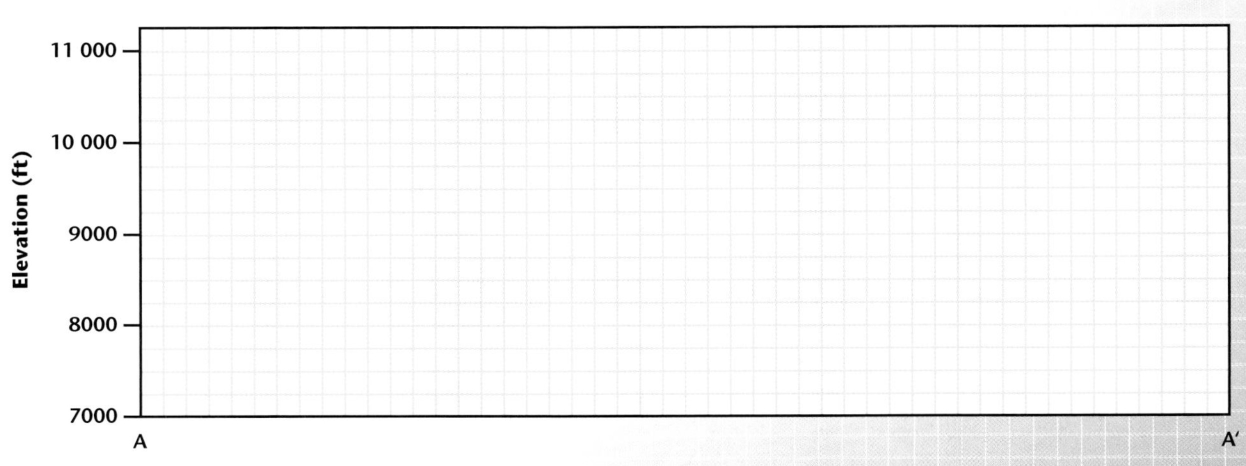

Science & Math

Tom Whittaker on Everest, 1998

The Roof of the World

Since 1953, more than 600 people have reached the summit of Mt. Everest to stand nearly 9 km above sea level on Earth's highest point. Brutal cold, oxygen deprivation, and treacherous conditions have claimed the lives of hundreds who attempted the climb. Preserved from decay by the dry, cold conditions and high altitude, most of their bodies remain on Everest, silent witnesses to the awesome forces that continue to build this mountain.

Since British surveyor Sir George Everest first measured the height of this Himalayan peak in 1852, many explorers have dreamed of reaching Mt. Everest's summit. One hundred years would pass from the initial measurement to the first successful attempt to reach the summit by Sir Edmund Hillary and Tenzing Norgay. An explorer named George Mallory led the first attempts to scale Everest during the 1920s. He did not return from his last attempt. The discovery of his body by climbers in 1999 reminded many of Mallory's reply when asked why he was trying to reach Mt. Everest's peak. His famous answer, "Because it's there," echoes the sentiments of many who have followed in his footsteps.

Measuring a Mountain

Some of those who have climbed Mt. Everest in the past 50 years have had another reason to scale this peak: to measure the elevation of Earth's tallest point. In 1954, an elevation of nearly 8848 m was determined by averaging altitude measurements taken from 12 different points around the mountain. Climbers with the Millennium Expedition, which took place from 1998 to 2000, utilized the highly accurate Global Positioning System to calculate an elevation of 8850 m for Earth's highest point.

Is Mt. Everest getting taller, or is the difference in elevation a result of the different instruments used to measure this mountain? The answer could be both. The collision between two tectonic plates is forcing the Indian subcontinent beneath Asia, causing Everest to rise at a rate of about 5 to 8 mm/y. Readings from GPS instruments on the mountain also suggest that Everest and other peaks in the range are moving toward China at about 6 cm/y.

Technology—Then and Now

The elevation determined in 1954 by the Survey of India was calculated by picking the unweighted mean of altitudes determined from the 12 survey stations around the mountain. These measurements varied by about 5 m. The data gathered by the GPS to calculate the elevation have a margin of error of just over 2 m.

Activity

Is Mt. Everest actually 2 m taller than it was in 1954? Use an average rate of uplift of 6.5 mm/y to determine how much Mt. Everest has risen since it was first measured. How does this compare with the newly calculated elevation?

CHAPTER 20
Study Guide

Summary

Section 20.1
Crust-Mantle Relationships

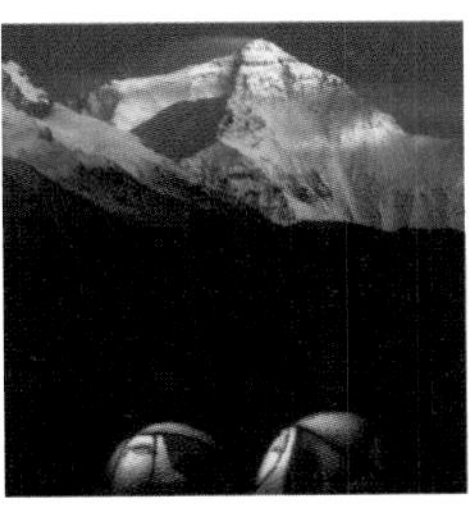

Main Ideas

- Earth's elevations cluster around two intervals: 0 to 1 km above sea level and 4 to 5 km below sea level. These modes reflect the differences in density and thickness of the crust.
- Isostasy is a condition of equilibrium. According to this principle, the mass of a mountain above Earth's surface is supported by a root that projects into the mantle. The root provides buoyancy for the massive mountain.
- The addition of mass to Earth's crust depresses the crust; the removal of mass from the crust causes the crust to rebound in a process called isostatic rebound.

Vocabulary

isostasy (p. 525)
isostatic rebound (p. 527)

Section 20.2
Convergent-Boundary Mountains

Main Ideas

- Orogeny is the cycle of processes that form mountain belts. Most mountain belts are associated with plate boundaries.
- Island arc complexes are volcanic mountains that form as a result of the convergence of two oceanic plates.
- Highly deformed mountains with deep roots may form as a result of the convergence of an oceanic plate and a continental plate.
- Earth's tallest mountains form along continental-continental plate boundaries, where the energy of the collision causes extensive deformation of the rocks involved.
- The Appalachian Mountains, which are located in the eastern United States, formed millions of years ago mainly as the result of convergence between two tectonic plates.

Vocabulary

orogeny (p. 528)

Section 20.3
Other Types of Mountains

Main Ideas

- At a divergent boundary, newly formed lithosphere moves away from the central rift, cools, contracts, and becomes more dense to create a broad, gently sloping mountain range called an ocean ridge. Rocks that make up ocean ridges include dikes and pillow basalts.
- Regional uplift can result in the formation of uplifted mountains that are made of nearly horizontal, undeformed layers of rock.
- Fault-block mountains form when large pieces of the crust are tilted, uplifted, or dropped downward between normal faults.
- Most solitary volcanic peaks form as a tectonic plate moves over a hot spot in Earth's mantle.

Vocabulary

fault-block mountain (p. 538)
pillow basalt (p. 536)
uplifted mountain (p. 537)

earthgeu.com/vocabulary_puzzlemaker

CHAPTER 20 Assessment

Understanding Main Ideas

1. What causes differences in elevation on Earth?
   a. density and thickness of the crust
   b. vertical dikes and pillow basalts
   c. seamounts and hot spots
   d. uplifted and faulted mountains

2. Which of the following is *not* associated with orogeny at convergent boundaries?
   a. island arcs
   b. highly folded and faulted ranges
   c. ocean ridges
   d. deformed sedimentary rocks

3. What type of mountains are generally made up of undeformed rocks?
   a. fault-block mountains
   b. uplifted mountains
   c. convergent-boundary mountains
   d. continental mountains

4. What type of mountain would you expect to find at a convergent boundary involving two oceanic plates?
   a. fault-block mountain
   c. uplifted mountain
   b. volcanic mountain
   d. an ocean ridge

5. What is isostasy?
   a. a convergent-boundary mountain
   b. a condition of equilibrium
   c. a fault-block mountain
   d. a difference in crustal densities

6. Adding mass to the crust causes
   a. the crust to rebound.
   b. the mantle to rebound.
   c. the crust to become depressed.
   d. the mantle to displace the crust.

7. Explain why continental crust can displace more of the mantle than oceanic crust can.

8. What happens to a mountain's root as the mountain is eroded?

9. What type of plate boundary is most often associated with orogenic belts? Why?

10. Describe three mechanisms of crustal thickening that occur at convergent boundaries.

11. Explain why ocean ridges rise high above the surrounding ocean floor.

12. What processes might be responsible for regional uplift of continental crust?

13. Discuss the processes involved in the formation of the Appalachian Mountains.

Use the figure below to answer questions 14–16.

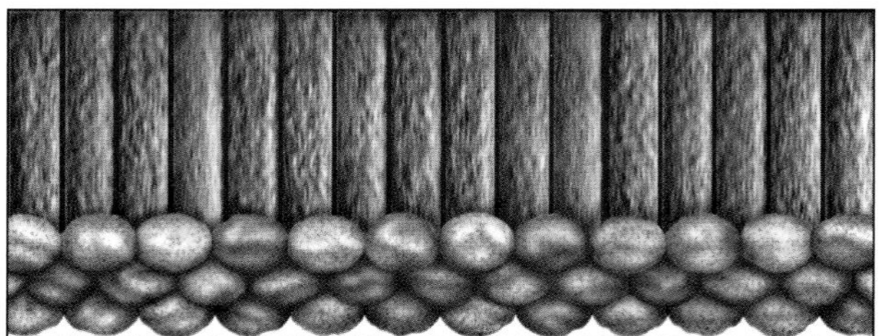

14. A geologist observed these two igneous rock beds in a coastal mountain range. Based on what you've learned in this chapter, where did these rocks form?

15. Explain how the two rock beds formed.

16. Are the rock beds in their original positions? Explain.

Test-Taking Tip

TEST YOURSELF Have a classmate make a practice test for you that covers the material discussed in this chapter. Take the test under test-like conditions. Show your practice test to one of your teachers for an objective assessment of your performance.

earthgeu.com/chapter_test

Assessment

Applying Main Ideas

17. If Earth's mantle were denser than it is now, would continental crust displace more or less of it? Explain.
18. Explain what caused an island arc to develop east of the North American continent early in the tectonic history of the Appalachian Mountains.
19. Refer to ***Figure 20-11.*** Describe how the deformation due to orogeny changes among the provinces of the Appalachian Mountains.
20. What type of faulting occurs at a continental-continental convergent boundary? Explain your answer.
21. How are ocean ridges and fault-block mountains similar? How do they differ?

Thinking Critically

22. Thick ice sheets once covered the Great Lakes region. The area north of the Great Lakes has rebounded more than areas south of the lakes. What conclusion can you make about the thickness of the ice in these two areas? Explain your reasoning.
23. Suppose a mountain range is 4 km high. Explain why more than 4 km of material would have to be eroded before the mountain would be completely level.
24. When does the isostatic rebound of an area stop?
25. Why are dikes rather than sills common along ocean ridges?
26. The Appalachians mark the western side of the convergent boundary between North America and Africa. What kinds of structures do you think would be found on the eastern side of this boundary? Where would you look for these structures?

Standardized Test Practice

1. Why would a certain thickness of continental crust displace less of the mantle than the same thickness of oceanic crust?
   a. Continental crust is more dense.
   b. Continental crust is less dense.
   c. Continental crust is mainly basalt.
   d. Continental crust is closer to the mantle.

2. Which of the following can NOT form as the result of oceanic-oceanic convergence?
   a. rift zones
   b. trenches
   c. subduction zones
   d. island arc complexes

3. Which type of mountains form as the result of uplift far from plate boundaries?
   a. ocean ridges
   b. uplifted mountains
   c. faulted mountains
   d. volcanic ranges

INTERPRETING DIAGRAMS Use the diagram and ***Figure 20-11*** to answer questions 4 and

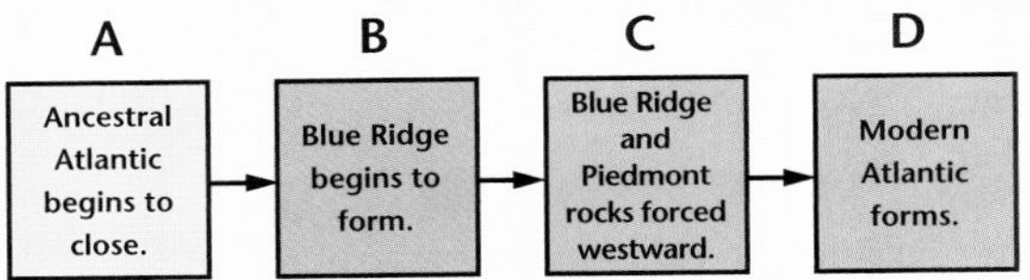

4. Which of the following occurred between events B and C?
   a. The island arc attached to North America.
   b. Plate motions reversed.
   c. Africa collided with North America.
   d. The island arc developed.

5. Approximately when did event C occur?
   a. 800 to 700 M.Y.B.P.
   b. 700 to 600 M.Y.B.P.
   c. 500 to 400 M.Y.B.P.
   d. 300 to 200 M.Y.B.P.

UNIT 5

GeoDigest

For a **preview** of Unit 5, study this GeoDigest before you read the chapters in the unit. After you have studied these chapters, you can use the GeoDigest to **review** the unit.

Earth Sciences 3.a, 3.e

The Dynamic Earth

Plate Tectonics

Drifting Continents and Seafloor Spreading Continental drift, first hypothesized by Alfred Wegener, states that Earth's continents were once joined as a single landmass that broke up and drifted apart. Wegener used matching coastlines of Earth's continents, similar rocks and fossils, and ancient climatic data to support his hypothesis. Wegener, however, couldn't explain how or why the continents moved. The answer, seafloor spreading, was found by studying ocean-floor rocks and sediments. Magnetic patterns of ocean-floor rocks are symmetric in relation to ocean ridges, which indicates that oceanic crust on either side of the ridge is moving away from the ridge. Seafloor spreading occurs as magma rises toward the crust, cools and hardens to fill the gap that forms, and becomes a new section of oceanic crust.

The Theory of Plate Tectonics Earth's crust and the top of the upper mantle are broken into large slabs of rock called plates, which move at different rates over Earth's surface. Interactions occur at plate boundaries. At a divergent boundary, plates move apart; high heat flow, volcanism, and earthquakes are associated with divergent boundaries. At a convergent boundary, plates come together; deep-sea trenches, island arcs, and folded mountain ranges are associated with this type of boundary. At a transform boundary, plates slide past one another horizontally; faults and shallow earthquakes are associated with transform boundaries. Convection currents in Earth's mantle are related to plate movements. Convection currents transfer energy through the movement of matter. Heat causes matter to expand and decrease in density. Warm matter is thus forced upward and cool matter is pulled downward. Ridge push is a tectonic process that occurs when the weight of an ocean ridge pushes a plate toward a subduction zone. Slab pull is a tectonic process that occurs when the weight of the subducting plate pulls a plate into a subduction zone.

Soufrière, a composite cone in the West Indies

Volcanic Activity

Magma The composition of magma is affected by temperature, pressure, and the presence of water. There are three main types of magma—basaltic, andesitic, and rhyolitic— that differ in the source rock from which they form, and also in composition, viscosity, silica content, amount of dissolved gases, and explosiveness.

Intrusive Activity Magma can affect the crust in several ways. Magma can force overlying rock apart and enter the fissures formed, can cause blocks of rock to break off and sink into the magma chamber, and can melt the rock into which it intrudes. Plutons, which include batholiths, stocks, sills, dikes, and laccoliths, are classified according to their size, shape, and relationship to surrounding rocks.

Volcanoes A volcano forms when lava repeatedly flows out through a vent and accumulates. A crater is a depression that forms around the vent at the summit of a volcano; a caldera is a large crater that forms when a volcano collapses during or after an eruption. There are three types of volcanoes—shield volcanoes, cinder-cone volcanoes, and composite volcanoes. Shield volcanoes are large, gently sloping volcanoes composed of basalt. Cinder-cone volcanoes have steep sides and are made of volcanic fragments. Composite volcanoes have relatively steep slopes and are made of layers of volcanic fragments that alternate with lava. Most volcanoes form along subduction zones and rifts, but they may also form over hot spots, which are especially hot areas in Earth's mantle.

Hot Spring, Africa

Earthquakes

Forces Within Earth Stresses exist within Earth. Stress is the forces per unit area that act on a material. Strain is the deformation of a material in response to stress. Stress is released at breaks in Earth's crust called faults. Reverse faults form as a result of horizontal compression; normal faults as a result of horizontal tension; and strike-slip faults as a result of horizontal shear. Movements along many faults cause earthquakes. Earthquakes generate waves that pass through Earth in specific ways. P-waves squeeze and pull rocks in the same direction in which the waves travel. S-waves cause rocks to move at right angles to the direction of the waves. Surface waves cause both up-and-down and side-to-side motions. Seismic waves are reflected and refracted as they strike different materials. Analysis of these different waves have enabled scientists to infer the structure of Earth's interior.

Measuring and Locating Earthquakes
Most earthquakes occur along plate boundaries in areas called seismic belts. Earthquake epicenters are located with data from at least three seismic stations. Magnitude is a measurement of the energy released during an earthquake and is measured on the Richter scale or the moment-magnitude scale. Intensity is the measure of the damage caused by an earthquake as measured on the modified Mercalli scale.

Vital Statistics

Highest Peak per Continent

Continent	Peak	Height
Africa	Kilimanjaro	5895 m
Antarctica	Vinson Massif	4897 m
Asia	Everest	8850 m
Australia	Kosciusko	2228 m
Europe	Elbrus	5642 m
North America	McKinley	6194 m
South America	Aconcagua	6960 m

GeoDigest

Earthquake damage, Guatemala

Earthquakes and Society Earthquakes can cause buildings and other structures to collapse, soil to behave like quicksand, fissures and fault scarps to form, uplift and subsidence of the crust, and tsunamis to form. The type of subsurface, the height and structure of buildings, the distance to the epicenter, and the depth of the earthquake's focus all affect the extent of damage done by an earthquake. The probability of an earthquake is determined by the history of earthquakes in an area and the rate at which strain builds up in the rocks.

Mountain Building

Isostasy According to the principle of isostasy, the mass of a mountain above Earth's surface is supported by a root that projects into the mantle. The root provides buoyancy for the mountain. Adding mass to Earth's crust results in a depression of the crust. Removing mass causes the crust to rebound. Earth's elevations cluster around two intervals: 0 to 1 km above sea level and 4 to 5 km below sea level. These modes reflect the differences in density and thickness of Earth's two kinds of crust.

Types of Mountains The cycle of processes that forms mountain belts is called orogeny. Island-arc complexes are volcanic mountains that form as the result of convergence of two oceanic plates. When an oceanic plate converges with a continental plate, highly deformed mountains with deep roots form. When two continental plates converge and collide, very tall mountains result, along with extensive deformation of the rocks involved. Ocean ridges are mountains that form on ocean floors along divergent boundaries; they form topographic highs as the result of density differences—crust along the ridges is warmer and less dense than older, cooler crust farther from the ridges. Some mountains form far from plate boundaries. Uplifted mountains are made of nearly horizontal, undeformed layers of rock and form as the result of regional uplift. Fault-block mountains form when large pieces of the crust are tilted, uplifted, or dropped downward between large faults. Solitary volcanic peaks form as a result of a plate's moving over a hot spot in Earth's mantle.

FOCUS ON CAREERS

Volcanologist

A volcanologist is a scientist who studies volcanoes. A volcanologist must have at least a bachelor's degree, but often pursues a master's or doctoral degree in order to specialize in a specific aspect of volcanology. Some volcanologists work close to active volcanoes; others study data in labs, perhaps searching for relationships that will allow the prediction of eruptions.

ASSESSMENT

Understanding Main Ideas

1. The study of which of the following led to the proposal of seafloor spreading?
   a. ocean rocks and sediments
   b. soil liquefaction
   c. seismic wave travel times
   d. P-wave motions

2. Which of the following is the probable cause of plate movements?
   a. ocean currents
   b. topographic highs
   c. convection currents in the mantle
   d. strain

3. In addition to temperature and pressure, what other factor affects the composition of magma?
   a. seismic gap
   b. the presence of water
   c. explosiveness
   d. seismic waves

4. What are batholiths, stocks, sills, dikes, and laccoliths?
   a. types of plutons
   b. kinds of magma chambers
   c. seismic wave types
   d. topographic lows

5. Which of the following forms when a volcano collapses during an eruption?
   a. a vent
   b. a crater
   c. a lava pipe
   d. a caldera

6. Which type of fault forms as a result of horizontal tension?
   a. reverse
   b. normal
   c. strike-slip
   d. shear

7. Which of the following is not caused by earthquakes?
   a. collapse of buildings
   b. flood basalt
   c. soil liquefaction
   d. tsunamis

8. How is earthquake magnitude measured?
   a. on the modified Mercalli scale
   b. by seismic gaps
   c. on the Richter scale
   d. by the earthquake history of an area

9. Which type of mountain forms when large pieces of Earth's crust are tilted, uplifted, or dropped down between large faults?
   a. uplifted mountains
   b. island arcs
   c. fault-block mountains
   d. hot spot volcanoes

10. What is stress?
    a. the deformation of a material
    b. the force that acts on a material
    c. a series of P-waves
    d. a series of S-waves

Thinking Critically

1. Describe the process of seafloor spreading.
2. How do scientists determine the probability of an earthquake?

Sierra Nevada, California

Unit 6

Geologic Time

Volcanic eruptions, glaciations, tectonic collisions and mountain building have all occurred many times in various places on Earth. What information do we have today that tells us of these events? The rocks themselves! It has been said that a picture tells a thousand words. The photograph at right is worth many more than that. Recorded in the rock layers is evidence of geologic events that have helped to shape Earth's history. Along with these dramatic events, countless species of plants and animals have appeared and disappeared. Fossils of these organisms are evidence of these occurrences. By studying the characteristics of layered rocks and the changes in life through time, geologists have been able to unravel Earth's history.

Unit Contents

Go to the National Geographic Expedition on page 892 to learn more about topics that are connected to this unit.

Capital Reef National Park, Utah

Chapter 21

Fossils and the Rock Record

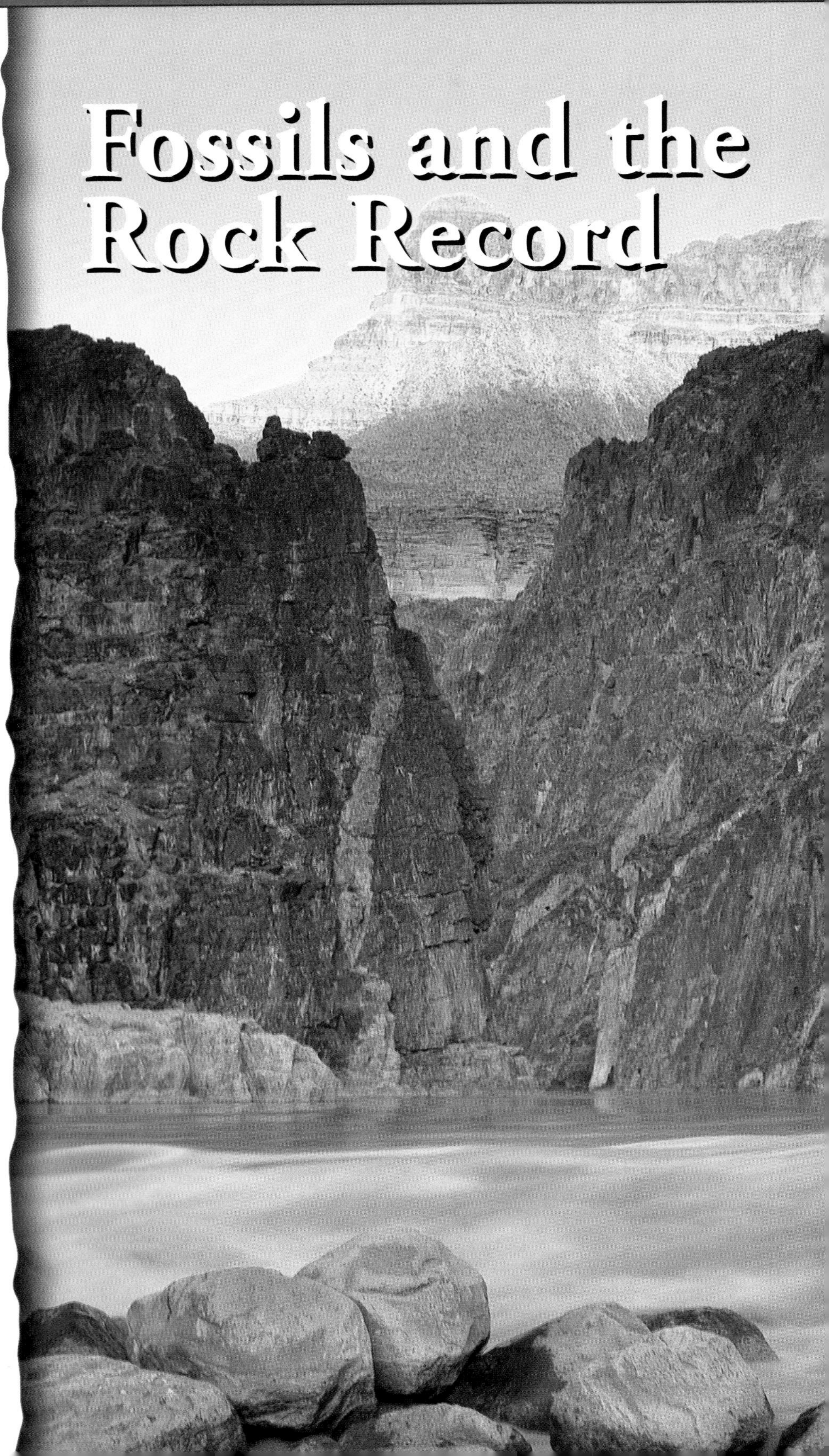

What You'll Learn

- How geologists divide Earth's long history.
- How certain geologic principles can be used to interpret age relations in layered rocks.
- How different techniques to determine the ages of rocks are used.
- What fossils are, how they form, and how they are used to interpret Earth's history.

Why It's Important

Fossils and rocks contain a record of Earth's history and can be used to make predictions about Earth's future. Some fossils can help identify potential sites of energy resources.

To learn more about fossils and the rock record, visit the Earth Science Web Site at earthgeu.com

Discovery Lab Fossil Hunt Activity

Have you ever found shells at a beach, along a river, or by a pond? If so, did you wonder where they came from or what type of animal might have lived in them? The shape, size, and composition of shells provide clues about the environment in which individual animals once lived. In this activity, you will make inferences about shells that you examine.

1. Obtain a mixture of sand and microfossils from your teacher.
2. Place the mixture on a petri dish or a small, shallow tray.
3. Use tweezers or a small, dry paintbrush to separate the fossils from the sandy sediment.
4. Categorize the fossils by shape, size, and composition.

CAUTION: Always wear safety goggles and an apron in the lab. Wash your hands when lab is completed.

Observe In your science journal, explain how fossils can help determine the age of sediment or a rock. Does categorizing the fossils provide any further clues about the environment in which the fossiliferous sediment formed? Explain.

SECTION 21.1 The Geologic Time Scale

Earth Sciences 1.c

OBJECTIVES

- **Describe** *the geologic time scale.*
- **Distinguish** *among the following geologic time scale divisions: eon, era, period, and epoch.*

VOCABULARY

geologic time scale
eon
era
period
epoch

A hike down the Kaibab Trail in the Grand Canyon reveals the multicolored layers of rock that make up the canyon walls. These layers, or strata, are made of different types of sedimentary rock. Some of the rock layers have fossils in them. At the bottom of the Grand Canyon, is the Colorado River, which has been cutting downward through the rocks of the canyon for millions of years. Also at the bottom are rocks that date back 400 million years or more. These rocks record the many advances and retreats of oceans and the development of plants and animals. By studying the characteristics of rocks and the fossils within them, geologists can interpret the environments the rocks were deposited in, reconstruct Earth's history, and possibly predict events or conditions in the future.

THE ROCK RECORD

To help in the analysis of Earth's rocks, geologists have divided the history of Earth into time units based upon the fossils contained

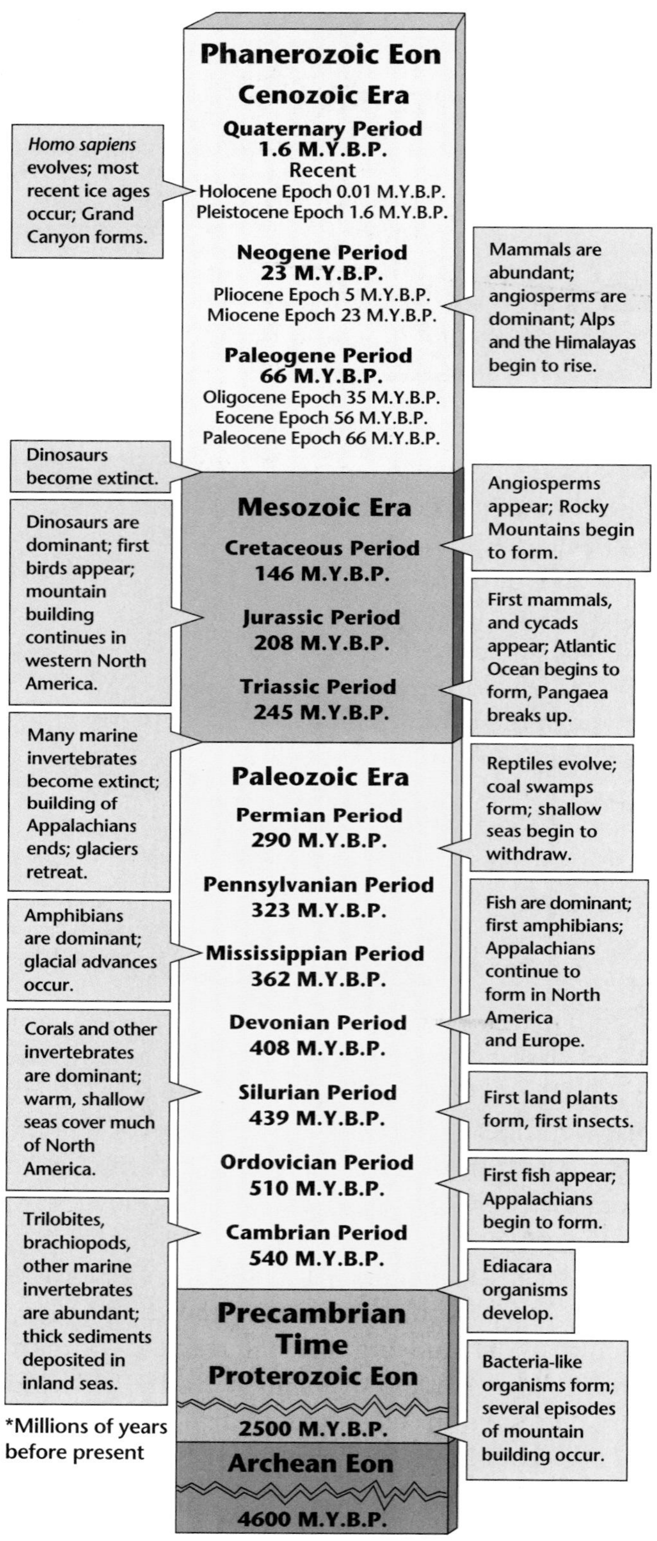

Figure 21-1 Earth's long history is divided into specific units of time in the geologic time scale.

within the rocks. These time units are part of the **geologic time scale,** a record of Earth's history from its origin 4.6 billion years ago to the present. Since the naming of the first geologic time period, the Jurassic, in 1797, development of the time scale has continued to the present. The names of the periods do not change, but the years marking the beginning and end of each unit of time are continually being refined. The geologic time scale is shown in ***Figure 21-1.*** This scale enables scientists from around the world to correlate the geologic events, environmental changes, and the development of life-forms that are preserved in the rock record.

GEOLOGIC TIME

The oldest division of time is at the bottom of the geologic time scale. Moving upward on the scale, each division is younger, just as the rock layers in the rock record grow generally younger as you move upward. The time scale is divided into units called eons, eras, periods, and epochs. An **eon** is the longest time unit and is measured in billions of years. The Archean, the Proterozoic, and the Phanerozoic are eons. An **era** is the next-longest span of time; it is measured in hundreds of millions to billions of years. Eras are defined by the differences in life-forms found in rocks; the names of eras are based on the relative ages of these life-forms. For example, in Greek, *paleo* means "old," *meso* means "middle," and *ceno* means "recent." *Zoic* means "of life" in Greek, and thus *Mesozoic* means "middle life." Precambrian Time, which makes up approximately 90 percent of geologic time, is divided into the Archean and Proterozoic Eons. The Proterozoic is the more recent of the two, and the end of it is marked by the first appearance of organisms with hard parts. All life-forms up until then had soft bodies and no shells or skeletons. Some of these resembled organisms that exist today, such as sponges, snails, and worms, while others cannot be accurately assigned to any known animal or plant group.

To learn more about dinosaurs and their evolution, go to the **National Geographic Expedition** on page 892.

Plants and Animals Evolve During the Paleozoic Era, the oceans became full of a wide diversity of plants and animals. Trilobites dominated the oceans in the Cambrian Period; land plants appeared and were followed by land animals; and swamps provided the plant material that became the coal deposits of the Pennsylvanian. The end of the Paleozoic Era is marked by the largest extinction event in Earth's history. As many as 90 percent of all marine invertebrate species became extinct. The era following the Paleozoic Era, the Mesozoic Era, is known for the emergence of dinosaurs, but other important developments occurred then as well. Reef-building corals and large predatory reptiles developed in the oceans. Amphibians began living on land as well as in water. Dinosaur populations began a slow decline in numbers throughout the Cretaceous Period as mammals evolved and grew in number. Flowering plants and trees evolved during the Cretaceous.

The end of the Mesozoic is also marked by a large extinction event. In addition to the remaining dinosaurs, many other groups of organisms, became extinct. Mammals increased both in number and diversity in the Cenozoic. Human ancestors developed at this time. Grasses and flowering plants expanded on land while ocean life remained relatively unchanged throughout this era.

Topic: Time Scales
To find out more about the geologic time scale, visit the Earth Science Web Site at earthgeu.com

Activity: Compare the span of time represented by the geologic time scale to the span of time represented by a 24-hour day. Apply the geologic time span to the 24-hour time span. How much time in a 24-hour day is represented by each era? Each period?

Periods of Geologic Time **Periods** are defined by the life-forms that were abundant or became extinct during the time in which specific rocks were deposited. Periods are usually measured in terms of tens of millions of years to hundreds of millions of years. Some were named for the geographic region in which the rocks of that age were first observed, studied, and described. For example, the Mississippian Period was named for the distinctive limestone bluffs along the Mississippi River, as shown in ***Figure 21-2.*** The Jurassic Period was named for the rocks that were described in the Jura Mountains in Europe.

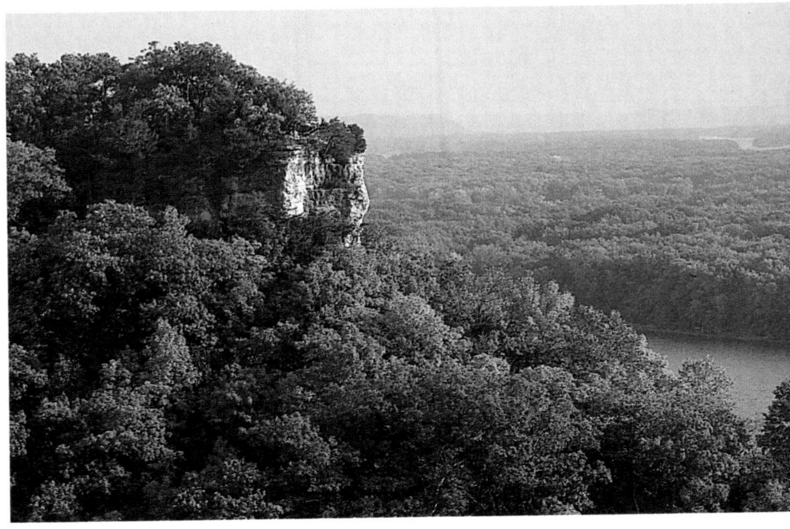

Figure 21-2 These Mississippian-aged limestone bluffs border the Mississippi River in Iowa.

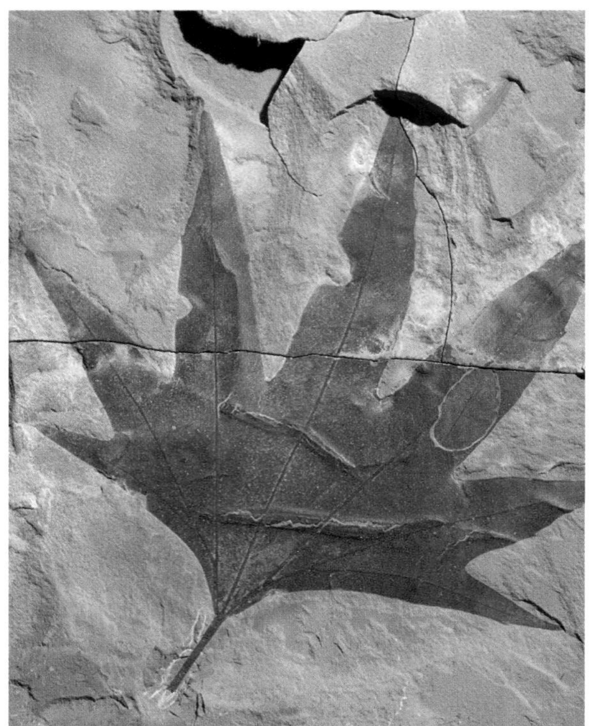

Figure 21-3 This fossil sycamore leaf is preserved in the Eocene-aged, Green River Formation in Wyoming.

Historically, the Cenozoic Era was divided into two periods, the Tertiary and the Quaternary. Currently, however, the Cenozoic is divided into three periods: the Paleogene, Neogene, and Quaternary. In contrast to the boundaries between the Paleozoic and the Mesozoic Eras, the boundaries between the periods of the Cenozoic are not marked by extinction events.

Epochs of Geologic Time **Epochs** are even smaller divisions of geologic time and are usually measured in millions of years to tens of millions of years. The fossil record of the Cenozoic Era is relatively complete because there has been less time for weathering and erosion to remove evidence of this part of Earth's history. Thus, rocks and fossils from this era are easily accessed and studied. Accordingly, the Cenozoic Periods have been further divided into epochs, such as the Paleocene and the Oligocene. Different groups of organisms have been used to distinguish the various epochs. For example, marine fossils were used to mark the Oligocene Epoch, and terrestrial plant fossils, such as those shown in ***Figure 21-3,*** were used to mark the Eocene Epoch.

Regardless of how a geologic period was defined, each unit contains specific characteristics that set it apart from the rest of geologic history. In the *Design Your Own GeoLab* at the end of this chapter, you will find out what makes each time unit unique.

Earth Sciences
1.c Students know the evidence from geologic studies of Earth and other planets suggests that early Earth was very different from Earth today.

Section Assessment

1. How did geologists determine the divisions of the geologic time scale?
2. What does the geologic time scale indicate about the change in life-forms over time?
3. What do the names of the three eras of the Phanerozoic mean?
4. What major change occurred in life-forms at the end of the Proterozoic?
5. How were the geologic time periods named? On what basis are they defined?
6. **Thinking Critically** Explain why the use of living faunas is acceptable for defining the periods and epochs of the Cenozoic Era.

Skill Review

7. **Graphing** Make a bar graph that shows the relative percentages of time that each period of the geologic time scale spans. For more help, refer to the *Skill Handbook.*

SECTION 21.2 Relative-Age Dating of Rocks

I&E 1.i

As late as the turn of the nineteenth century, the majority of the world believed that Earth was only about 6000 years old. This age had been determined by Archbishop James Ussher of Ireland, who used a chronology of human and Earth history to calculate Earth's age. As early as 1770, James Hutton, a Scottish physician and geologist, had begun to observe and to attempt to explain Earth's landscapes. Hutton's observations in Great Britain helped him to develop the principle of uniformitarianism, which attempts to explain the forces that continually change the surface features of Earth. Such processes include mountain building, erosion, earthquakes, and sea-level changes. The principle of **uniformitarianism** states that the processes occurring today have been occurring since Earth formed. Only the rate, intensity, and scale with which they occur have changed. For example, if you stand on the shore of an ocean watching the waves come in, you are observing a process that has not changed since the oceans were formed. The waves crashing on a Cambrian shore, a Jurassic shore, and a modern shore all share the same process. The resulting sediments and rocks all record a beach environment, where the sediments become finer with distance from shore and the fossils within the rocks preserve evidence of the life-forms that lived during the time of deposition.

OBJECTIVES

- **Apply** *the principles for determining relative age to interpret rock sequences.*
- **Describe** *an unconformity and how it is formed within the rock record.*

VOCABULARY

uniformitarianism
original horizontality
superposition
cross-cutting relationships
unconformity
correlation

I&E

1.i Analyze the locations, sequences, or time intervals that are characteristic of natural phenomena (e.g., relative ages of rocks, locations of planets over time, and succession of species in an ecosystem).

PRINCIPLES FOR DETERMINING RELATIVE AGE

The concept of relative-age dating places the ages of rocks and the events that formed them in order, but without exact dates. This is done by comparing one event or rock layer to another.

Figure 21-4 The Colorado River, in Grand Canyon National Park, has cut through rock layers that span the Triassic through the Precambrian.

MiniLab

How is relative age determined?

Demonstrate how the principles of superposition, original horizontality, and cross-cutting relationships are used to determine the relative ages of rock layers.

Procedure

1. Draw a diagram of an outcrop with four horizontal layers. Label the layers 1 through 4.
2. Draw a vertical intrusion from layer 1 to layer 3.
3. Label the bottom-left corner of the diagram *X* and the top-right corner *Y*.
4. Cut the paper diagonally from *X* to *Y*. Move the left-hand piece 1.5 cm along the cut.

Analyze and Conclude

1. How can you determine the relative ages of the strata in your diagram?
2. How does the principle of cross-cutting relationships explain the age of the vertical intrusion?
3. What does line XY represent? Is line XY older or younger than the vertical intrusion and surrounding strata? Explain.

Geologic Principles Many different horizontal or nearly horizontal layers of rocks make up the walls of the Grand Canyon, shown in ***Figure 21-4.*** Most of the rocks are sedimentary and were originally deposited millions of years ago by water or wind. The principle of **original horizontality** states that sedimentary rocks are deposited in horizontal or nearly horizontal layers. While we may not know the actual ages of the rock layers, we can assume that the oldest rocks are at the bottom and that each successive layer going toward the top is younger. Thus, we can infer that the Moenkopi Formation, which rims the top of the Grand Canyon, is much younger than the Vishnu Group found at the bottom of the gorge as shown in ***Figure 21-5.*** This is an application of the principle of **superposition,** which states that in an undisturbed rock sequence, the oldest rocks are at the bottom and each successive layer is younger than the layer beneath.

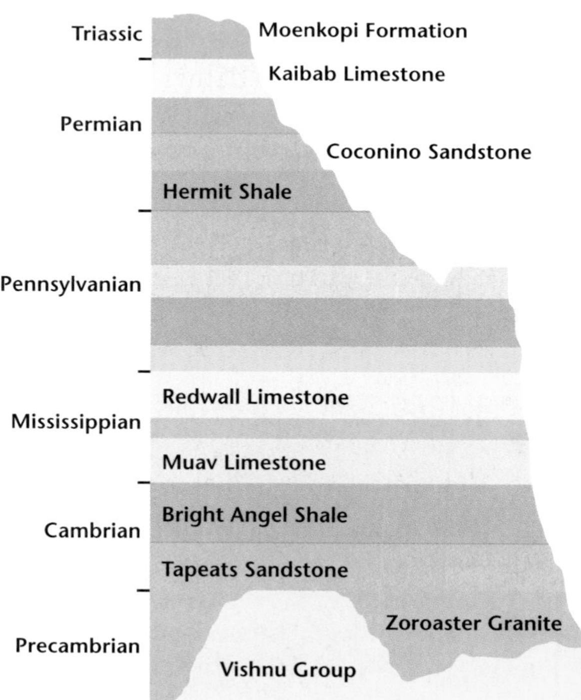

Figure 21-5 Using the principle of superposition, geologists have determined the relative ages of these rock layers.

Rocks exposed in the deepest part of the Grand Canyon are some of the oldest in North America. These are mostly igneous and metamorphic rocks. Within the Vishnu Group at the bottom of the Grand Canyon sequence are dikes of granite. The principle of **cross-cutting relationships** states that an intrusion or a fault is younger than the rock it cuts across. Therefore, the granite is younger than the schist, because the granite cuts across the schist. In earthquake-prone areas, such as California, and in ancient, mountainous regions, such as the Adirondacks of New York, there are many faults. As you learned in Chapter 20, a fault is a fracture in Earth along which movement takes place. A fault is younger than the strata and surrounding geologic features because it cuts across them. You used geologic principles to determine relative ages of rocks in the *MiniLab* on the previous page.

Inclusions Relative age also can be determined where an overlying rock layer contains particles of rock material from the layer beneath it. The bottom layer was eroded, and the loose material on the surface became incorporated in the newly deposited top layer. These particles, called inclusions, indicate that the rocks in the lower layer are older than those on top. As you learned in Chapter 6, once a rock has been eroded, the resulting sediment may be transported and redeposited and recemented many kilometers away. In this case, although this newly formed rock may be Jurassic in age, the grains that make up the rock may be Cambrian in age. Another example of the use of eroded sediments to determine the relative ages of rocks is a cooled lava flow that has bedrock particles trapped within it. An inclusion layer that is formed during a lava flow is illustrated in ***Figure 21-6.***

OTHER MEANS OF DETERMINING RELATIVE AGE

The fact that Earth is constantly changing as a result of processes such as weathering, erosion, earthquakes, and volcanism makes it difficult to find an undisturbed sequence of rock layers. For example, if rocks that record a volcanic eruption or the last occurrence of a particular fossil are eroded away, then the record of that particular

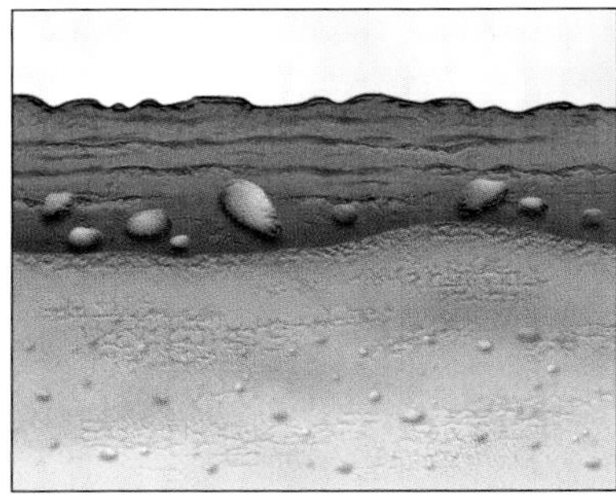

Figure 21-6 This pahoehoe lava flow in Hawaii most likely contains pieces, or inclusions, of the aa lava flow beneath it **(A).** When lava or sediments are deposited on top of an eroded surface that contains loose fragments, the fragments become incorporated as inclusions in the top layer **(B).**

event has been lost. Further changes may occur if the area is covered by a river during a flood or by the sea. Additionally, an erosional surface might become buried by the deposition of younger rocks. This buried erosional surface results in a gap in the rock record and is called an **unconformity.** When horizontal sedimentary rocks overlie horizontal sedimentary rocks, the unconformity is called a disconformity. A different type of unconformity exists when sedimentary rocks overlie nonsedimentary rocks such as granite or marble. Such an unconformity suggests a possible uplifting of the marble or granite and exposure at the surface by weathering and erosion. The contact point between the nonsedimentary and sedimentary rock is called a nonconformity. The formation of unconformities are illustrated in ***Figure 21-7*** on the following page.

When horizontal sedimentary rocks are uplifted and tilted, they are exposed to the processes of weathering and erosion. When deposition resumes, horizontal layers of sedimentary rocks are laid down on top of the erosional surface. The layers beneath the eroded surface of the folded layers remain intact, but they are at an angle to the eroded surface. This type of unconformity is called an angular unconformity. You will use several geologic principles to interpret the geologic history of an area in the *Problem-Solving Lab* on the this page.

Problem-Solving Lab

Interpreting Diagrams

Interpret the relative ages of rock layers Use the diagram to answer the following questions.

Analysis

1. Which is the oldest rock unit in the diagram?
2. An unconformity exists between which two layers of rock? Explain.
3. What happened to the rock that came in contact with the molten material of the intruded dike?
4. Explain why the rock layers and features on the left side of the diagram do not match the rock layers and features on the right side.

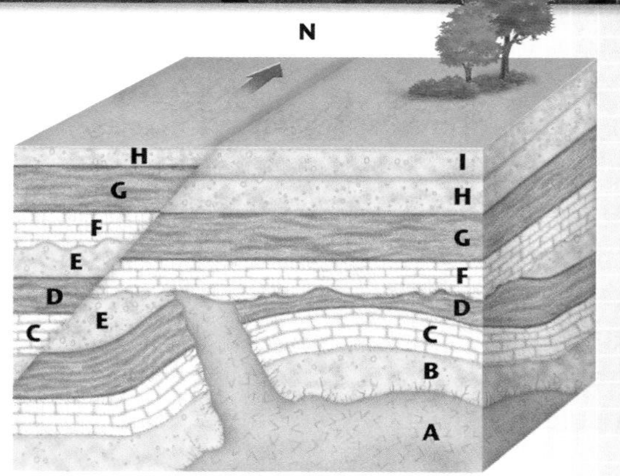

Thinking Critically

5. Which is the younger feature in the outcrop, the dike or the folded strata? Explain.
6. List the order of geologic events represented by this diagram. Which geologic principles did you use?

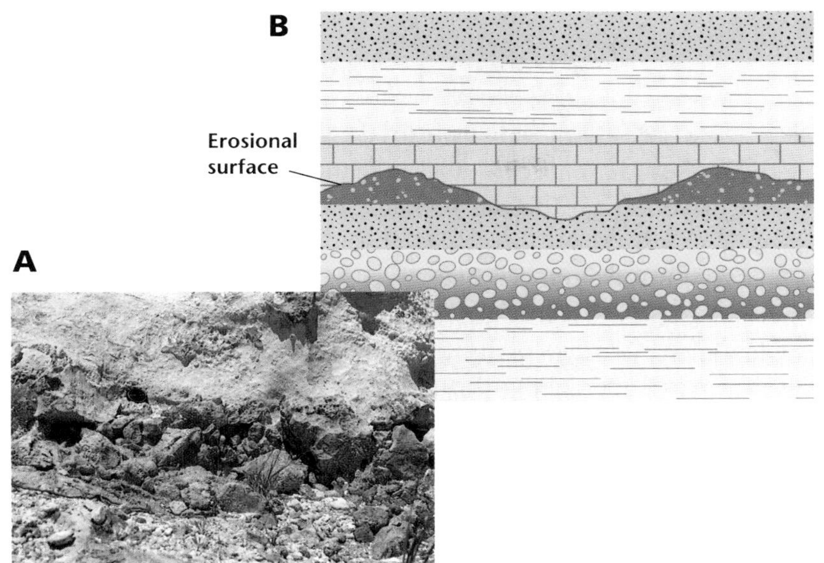

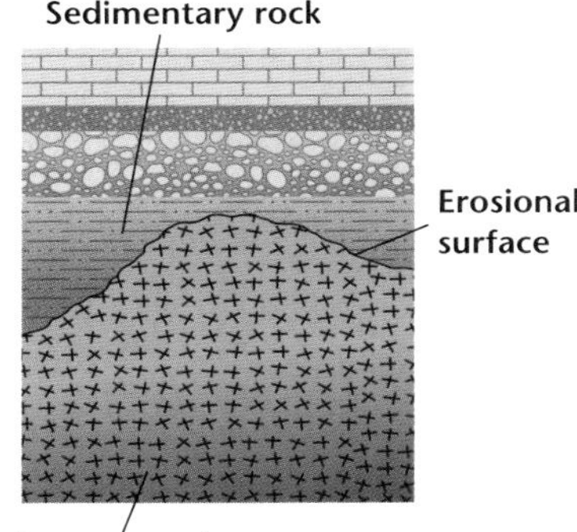

Figure 21-7 This disconformity on San Salvador Island, Bahamas, was formed by a soil that developed on top of a fossilized coral reef **(A).** A disconformity forms when a sedimentary rock layer is deposited on top of an eroded sedimentary rock layer **(B).** A nonconformity forms when a sedimentary rock layer is deposited on top of an eroded metamorphic or igneous rock layer **(C).**

Correlation of Rock Strata The Permian Kaibab Formation rims the top of the Grand Canyon, but is also found about 300 km away at the bottom of a 200-m gorge in Capitol Reef National Park in Utah. How do geologists match rock layers such as these, which are far apart from each other? One method is by correlation. **Correlation** is the matching of outcrops of one geographic region to another. Geologists examine rocks for distinctive fossils and unique rock or mineral features to help correlate the rock layers. This information can be used to help in the exploration for oil or valuable minerals. For example, if a sandstone layer in one area contains oil, it is possible that the same layer in a different area also contains oil. Correlation allows geologists to accurately locate that same sandstone layer in another location.

Section Assessment

1. How would a geologist use the principle of superposition to determine the relative ages of the rocks in the Grand Canyon?
2. What is an unconformity?
3. Explain how inclusions at the base of a lava flow can help determine the relative age of the layers.
4. A fault or a dike cuts across a sequence of rocks. What does this suggest about the relative ages of the rocks?
5. **Thinking Critically** Explain how the principle of uniformitarianism is used to interpret Earth's past.

Skill Review

6. **Interpreting Data** Discuss how a sequence of strata can be correlated from one side of a canyon to another. For more help, refer to the *Skill Handbook*.

SECTION 21.3 Absolute-Age Dating of Rocks

Earth Sciences 1.c I&E 1.i, 1.k, 1.l

OBJECTIVES

- **Explain** *the several different methods used by scientists to determine absolute age.*
- **Describe** *how objects are dated by the use of certain radioactive elements.*
- **Explain** *how annual tree rings and glacial varves are used to date geologic events.*

VOCABULARY

radioactive decay
radiometric dating
half-life
dendrochronology
varve
key bed

As you have learned, relative-age dating is a method of comparing past geologic events based on the observed order of strata in the rock record. In contrast, absolute-age dating enables scientists to determine the actual age of a rock, fossil, or other object. Scientists have devised a method for dating very old objects using the decay rate of radioactive isotopes. These isotopes are found in igneous and metamorphic rocks, some fossils, and organic remains. Radioactive substances emit nuclear particles at a constant rate. As the numbers of protons and neutrons change with each nuclear emission, the element is converted to a different element. The original radioactive element is referred to as the "parent," and the new element is referred to as the "daughter." For example, a radioactive isotope of uranium, U-238, will decay into an isotope of lead, Pb-206, over a specific span of time, as illustrated in ***Figure 21-8.*** The emission of radioactive particles and the resulting change into other elements over time is called **radioactive decay.** Once the emission of these atomic particles begins, the rate remains constant regardless of environment, pressure, temperature, or any other physical changes. Thus, these atomic particles become accurate indicators of the absolute age of an object.

USE OF RADIOACTIVE ISOTOPES

In a process called **radiometric dating,** scientists attempt to determine the ratio of parent nuclei to daughter nuclei within a given sample of a rock or fossil. This ratio is then used to determine the absolute age of the rock or fossil. As the number of parent atoms decreases, the number of daughter atoms increases by the same amount and indicates the increasing age of an object. Because it often takes a long time for the entire amount of an isotope to decay, geologists use the length of time it takes for one-half of the original amount to decay. This

Figure 21-8 The decay of U-238 to Pb-206 follows a specific and never-changing path.

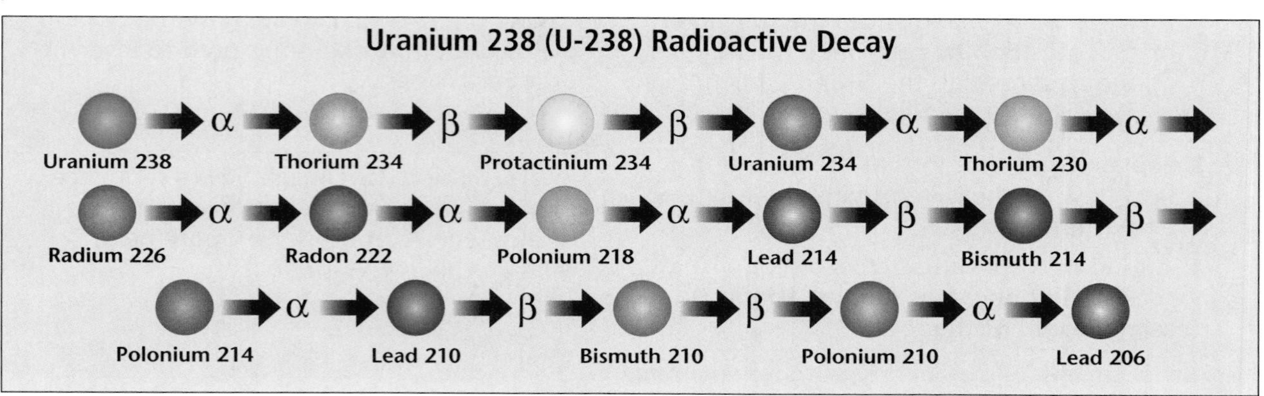

Table 21-1 Half-Lives of Selected Radioactive Isotopes		
Radioactive Isotope	**Approximate Half-Life**	**Decay Product**
Rubidium-87	48.6 billion years	Strontium-87
Thorium-232	14.0 billion years	Lead-208
Potassium-40	1.3 billion years	Argon-40
Uranium-238	4.5 billion years	Lead-206
Uranium-235	0.7 billion years	Lead-207
Carbon-14	5730 years	Nitrogen-14

period of time is called the **half-life.** ***Table 21-1*** lists some common radioactive isotopes and their half-lives.

Carbon-14 Another radioactive isotope that is commonly used to determine the absolute age of an object is carbon-14 (C-14). This isotope is especially useful for finding the age of materials that are of organic origin, such as amber, humanoid bones, papyrus, and charcoal fragments. This is because all organic materials contain carbon. C-14 decays into the stable, nonradioactive element nitrogen-14 (N-14). The half-life of C-14 is 5730 years, as shown in ***Table 21-2.*** When state-of-the-art technology is used, C-14 dating is accurate for objects up to 75 000 years old.

If U-238 is used for an object that is only a few hundred thousand years old, the ratio of parent to daughter atoms will be too large to be useful; therefore, a radioactive isotope with a shorter half-life than U-238, such as U-235, which has a half-life of 700 000 000 years, must be used. Conversely, for the dating of a particularly old rock sample, a radioactive isotope with a longer half-life must be used. Otherwise, there may come a point when the ratio of parent-to-daughter atoms is too small to measure but the age of the rock has not yet been determined. In essence, the isotope used for dating depends on the estimated general age of the rock or object being dated.

Using Numbers A granite sample from Canada was dated using uranium-235, which has a half-life of 700 000 000 years. The rock was calculated to be 2.8 billion years old. How many half-lives have elapsed since the rock formed?

Table 21-2 Radioactive Decay of Carbon-14 to Nitrogen-14				
	Percent Parent Element	**Percent Daughter Element**	**Elapsed Years**	**Number of Half-Lives**
Time 1	100	0	0	0
Time 2	50	50	5730	1
Time 3	25	75	11 560	2
Time 4	12.5	87.5	17 090	3

Update For an online update on the oldest rocks found to date, visit the Earth Science Web Site at earthgeu.com

 I&E

1.i Analyze the locations, sequences, or time intervals that are characteristic of natural phenomena (e.g., relative ages of rocks, locations of planets over time, and succession of species in an ecosystem).

OTHER WAYS TO DETERMINE AGE

Determining the relative or absolute age of an object or event is not limited to the use of rocks or chemical elements. Naturally occurring materials, such as trees, lake-bottom sediment, and volcanic ash can also be used to help geologists determine the age of an object or event, such as a forest fire, a drought, a flood, or a volcanic eruption.

Tree Rings With the use of a technique from the science of forestry, the age of a tree can be determined by counting the number of annual tree rings in a cross section of the tree. During the spring months, a tree experiences its greatest growth, while in the winter, its growth is less. Thus, the widths of tree rings are directly related to the climatic conditions during growth periods. A pair of spring and winter growth rings represents an annual tree ring. **Dendrochronology** is the science of comparing annual growth rings in trees to date events and changes in past environments. For example, in Mesa Verde National Park in Colorado, the age of the wooden rafters used to build the pueblos of the Anasazi have been determined with the use of dendrochronology. ***Figure 21-9*** shows a pueblo from Mesa Verde National Park. The Anasazi were a group of Native Americans that lived in the southwestern United States. It has been calculated by other methods, that the pueblos were built between A.D. 1150 and A.D. 1200. It also has been determined that pueblos in the southwestern United States were abandoned by the Anasazi around A.D. 1300, most likely because of a severe drought that lasted from A.D. 1276 to A.D. 1299.

Seasonal Climatic Changes About 11 000 years ago, continental glaciers covered the northern part of the United States. During the summer months, the ice would partially melt. Large volumes of water containing fine glacial sediment were carried downstream and deposited in large lakes. Summer deposits are generally light-colored and relatively thick compared to the thinner, organically enriched,

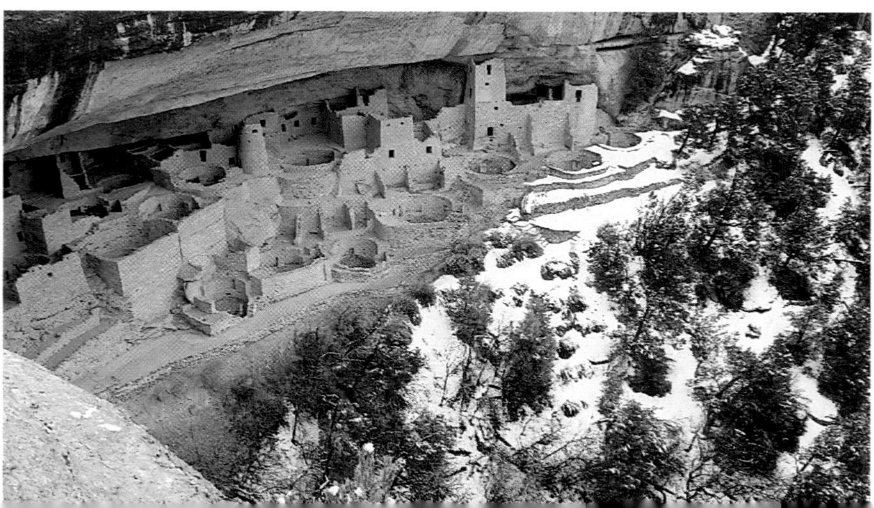

Figure 21-9 The Cliff Palace, in Mesa Verde National Park, was built by an ancient group of Native Americans known as the Anasazi.

and dark-colored sediments of winter. These bands of alternating light- and dark-colored sediments of sand, clay, and silt are called **varves.** Varves are similar to tree rings in that they show evidence of cyclic events—in this case, a cycle of summer to winter. Varves from different lakes can be compared to determine the ages of glacial lake sediments from about 15 000 to 12 000 years ago.

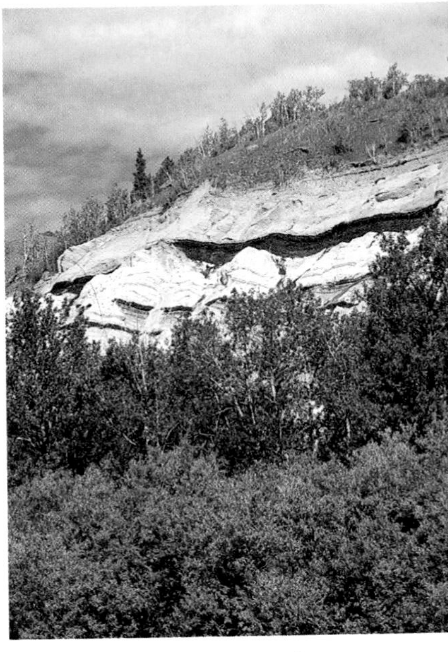

Figure 21-10 This coal seam near Healy, Alaska can be used to correlate rock layers across a large area because it is exposed in many outcrops throughout the area.

Distinctive Sediment Layers Scientists hypothesize that about 66 million years ago, an asteroid 10 km in diameter struck Earth in the region known today as the Yucatan Peninsula in Mexico. The blast threw out large amounts of crushed rock into Earth's atmosphere. Much later, when the asteroid's impact debris settled onto the surface of Earth, it formed a sediment layer that can be found in many parts of the world today. This layer lies between rocks deposited at the end of the Cretaceous Period of the Mesozoic Era and the beginning of the Paleogene Period of the Cenozoic Era. These geologic time units are shown in ***Figure 21-1.*** When such a layer is formed by an instantaneous or short-lived event, geologists may be able to determine the time of the event through radiometric dating. The layer then becomes a time marker called a **key bed,** which can be used to correlate rock layers across large areas. Key beds, such as the black coal bed shown in ***Figure 21-10,*** contain material that is distinctive and easy to recognize in the rock record.

Volcanic eruptions also create key beds. For example, when Mount St. Helens erupted in 1980, vast amounts of volcanic ash were distributed over many states. The ash will eventually become a thin, clay layer and will mark the date of the eruption. Clay layers such as this one occur throughout the rock record and attest to Earth's volcanic history. You will learn more about this history in Chapter 22.

SECTION ASSESSMENT

1. What is the difference between relative-age dating and absolute-age dating?
2. A scientist finds the charred remains of a tree in a layer of volcanic ash thought to be from the eruption of Mt. Mazama some 6600 years ago. Which radioactive isotope, U-238 or C-14, would you use to verify the actual age of the charred wood? Explain.
3. What is a key bed?
4. **Thinking Critically** Potassium-40 decays to the noble gas argon-40. What problems might arise when these radioactive isotopes are used for age dating?

SKILL REVIEW

5. **Comparing and Contrasting** Compare and contrast the uses of tree rings and varves in relative-age dating. For more help, refer to the *Skill Handbook.*

SECTION 21.4 Remains of Organisms in the Rock Record

OBJECTIVES

- **Define** *fossil.*
- **Explain** *several methods by which fossils can be preserved.*
- **Describe** *the characteristics of an index fossil.*
- **Discuss** *how fossils can be used to interpret Earth's past physical and environmental history.*

VOCABULARY

fossil
evolution
original preservation
altered hard part
permineralization
index fossil
mold
cast

Fossils are the evidence or remains of once-living plants or animals. They provide evidence of the past existence of a wide variety of life-forms, most of which have become extinct. The fossil record also provides evidence that populations have undergone change through time in response to changes in their environments. This change in populations as a result of environmental change is **evolution.** Fossils preserved in the rock record also provide information about past environmental conditions. They can even be used to correlate rock layers from one area to another.

TYPES OF FOSSILS

Fossils with **original preservation** are the soft and hard parts of plant and animal remains that have not undergone any kind of change since the organisms' deaths. Such fossils are uncommon because their preservation requires extraordinary circumstances such as freezing, drying out, or oxygen-free environments. In Alaska, original woody parts of plants are imbedded in the permafrost from 10 000-year-old bogs. Soft parts of mammoths and saber-toothed cats are preserved in the sticky ooze of the La Brea Tar Pits in California. Tree sap from prehistoric trees that hardened into amber sometimes has fossil insects imbedded in it. You will read about the recent discovery of a mammoth carcass that is at least 20 000 years old in the *Science in the News* feature at the end of this chapter. Soft parts are also preserved when plants or animals have been dried out and their remains have been mummified. Most mummified remains are found in dry caves or are buried in desert sands. For example, in 1935, the mummified remains of a Native American were found in Mammoth Cave National Park in Kentucky. ***Figure 21-11*** shows another fossil with original preservation.

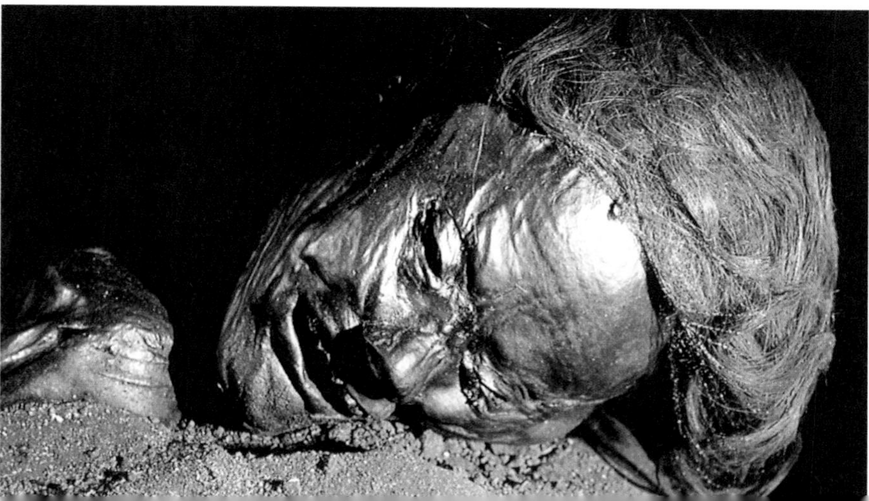

Figure 21-11 The Graubelle man was found in 1952 in a bog in Jutland, Denmark. It dates from between A.D. 80 to 170 B.C. Fossils like these are called bog bodies.

Figure 21-12 One year after the 1980 eruption of Mount St. Helens, in Washington State, these trees are slowly being converted to petrified wood **(A).** This petrified tree stump in Theodore Roosevelt National Park, North Dakota, stands as testimony to an ancient volcanic eruption **(B).**

Altered Hard Parts When all the organic material has been removed and the hard parts of a plant or animal have been changed either by mineral replacement or by recrystallization, their fossils are referred to as **altered hard parts.** The process by which pore spaces are filled in with mineral substances is called **permineralization.** For example, when volcanic ash spreads out and settles over a large geographic area, entire forests may become buried. Over a long period of time, quartz and other minerals in the ash combine with groundwater and slowly fill in the spaces between the cellular walls of the trees and silicify them. A common name for this type of fossilized tree is petrified wood. Modern and fossil examples are shown in ***Figure 21-12.*** Permineralization also occurs when groundwater combines with minerals in sediments. The mineral-rich water then reaches a buried organism and the replacement process begins, as illustrated in ***Figure 21-13.***

Changes in temperature and pressure may also result in changes in shell or bone material. Shells, such as those of clams, or exoskeletons, such as those of corals, may be affected by a process called recrystallization. The exterior of the shell remains the same, but the shell microstructures are destroyed during this process.

Figure 21-13 The shells of spectacularly preserved Brachiopods, such as this *Echinauris,* from Permian-aged rocks in the Glass Mountains of Texas, were replaced by silica during permineralization **(A).** This process is illustrated in **(B).**

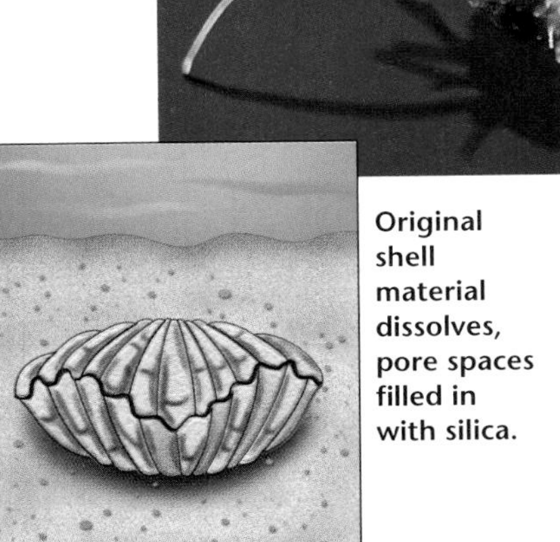

B

Burial

Organic material decays

Mineral-rich ground water seeps in

Original shell material dissolves, pore spaces filled in with silica.

Figure 21-14 The characteristic shape of this mollusk, *Ecphora,* makes it easy to recognize and identify.

Index Fossils Some fossils are more useful than others in relative age dating. **Index fossils** are remains of plants or animals that can be used by geologists to correlate rock layers over large geographic areas or to date a particular rock layer. An index fossil is easily recognized, abundant, and widely distributed geographically. It must also have lived during a short period of time. For example, the mollusk *Ecphora,* shown in ***Figure 21-14,*** is an excellent index fossil for the Mesozoic because of its distinctive shape and its abundance.

Molds and Casts Some fossils do not contain any shell or bone material. They may be molds and casts of shelled organisms, such as clams. A **mold** is formed when the original shell parts of an organism within a sedimentary rock are weathered and eroded. A hollowed-out impression, or mold, of the shells is left in their place. This cavity might later become filled with minerals or sediment to create a **cast** of the organism. A mold and cast are illustrated in ***Figure 21-15.***

Indirect Evidence of Past Life Indirect evidence of plant and animal life are called trace fossils. Some examples are worm trails, burrows, and footprints. Trace fossils such as those shown in ***Figure 21-16*** can provide information about how an organism lived, how it moved, or how it obtained food. Dinosaur trackways in Texas and Connecticut provide scientists with clues about the size and walking characteristics of dinosaurs. Other trace fossils include gastroliths, smooth and rounded rocks that dinosaurs had in their stomachs to help them digest and grind their food, and coprolites, the remains of solid waste materials of animals. By analyzing the content of coprolites, scientists can learn about the eating habits of ancient animals.

Figure 21-15 This mold and cast of the trilobite *Elrathia kingii,* from the Cambrian of Utah **(A),** shows the results of the process illustrated in **B.**

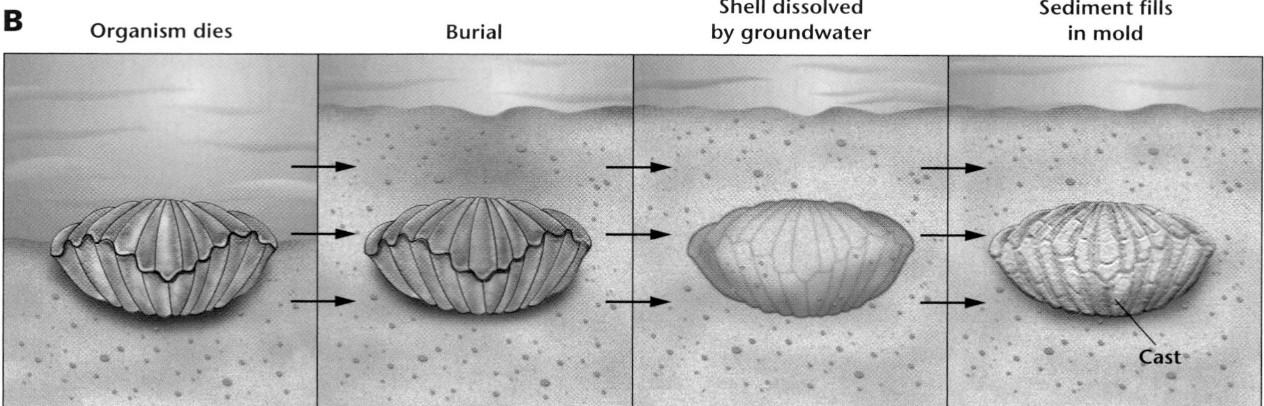

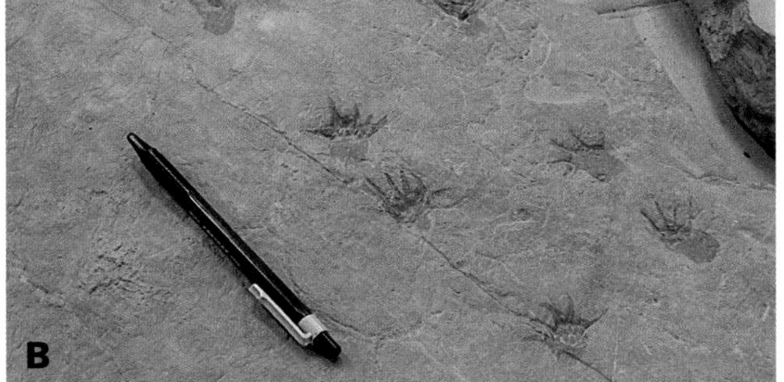

Why Study Fossils?

The study of fossils allows scientists to interpret and describe Earth's history. Fossils from different geologic time periods describe how organisms have changed through time. Fossils also show evidence of ancient environmental conditions. They also may help scientists find patterns and cycles that can be used to predict future phenomena, such as climatic changes. The study of fossils further allows geologists to locate energy resources. For example, petroleum geologists use certain index microfossils to determine whether oil might be present at a particular site. These fossils provide information about the ages of rocks and, in some cases, information that indicates whether the temperature and pressure conditions needed to form oil or gas were present in those layers.

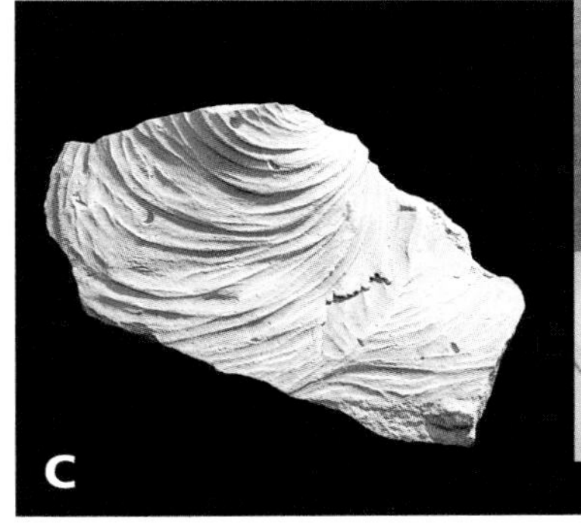

D

Figure 21-16 The tracks made by this horseshoe crab **(A)** may someday become preserved as did the tracks of this small Permian reptile **(B).** Zoophycus **(C)** is interpreted to be the feeding trace of a wormlike, burrowing organism **(D).** The sweeping lines are distinctive trace fossils.

Section Assessment

1. What is a fossil?
2. What is the difference between a fossil with original preservation and an altered hard part?
3. What are the characteristics of an index fossil?
4. Explain why the eruption of Mount St. Helens in 1980 resulted in the formation of a key bed.
5. **Thinking Critically** How might a mold or cast of a fossil help scientists to interpret the type of environment it lived in?

Skill Review

6. **Concept Mapping** Rearrange the following terms to construct a sequencing concept map to illustrate the formation of a mold and a cast. For more help, refer to the *Skill Handbook*.

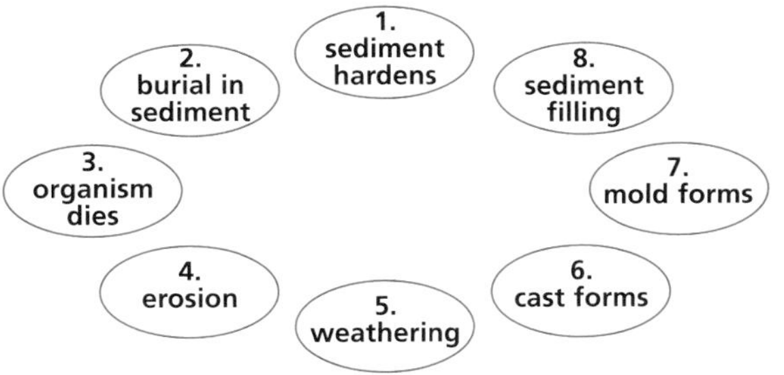

DESIGN YOUR OWN GeoLab

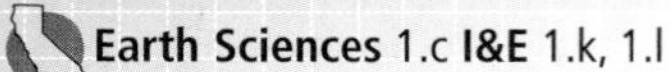

Interpreting History-Shaping Events

What do volcanic eruptions, mountain building, flooding, and drought have in common? They are all events that in some way affect life and the surface of Earth. How strong an impact does each event have on the future of Earth? How different would things be if certain events in Earth's history had not happened?

Preparation

Problem

What are the most important events in Earth's history? Where do they fit in the long history of Earth's development? Why are these events important? Do some geologic time periods contain more history-shaping events than others?

Possible Materials

paper	colored pencils
pencil	geologic time scale
posterboard	calculator
meterstick	encyclopedia
tape measure	reference books

Hypothesis

Brainstorm about Earth's history and the changes that Earth has experienced over time. Hypothesize which events had the most impact on the direction that Earth's development has taken. Determine where additional data might be available and collect resources to use as your references. Describe the best way to list and illustrate your choices.

Objectives

In this GeoLab, you will:

- **Hypothesize** about important events in Earth's development.
- **Explain** why such events had a significant impact on Earth's history.
- **Communicate** your results and interpretations.

Plan the Experiment

1. Review the list of events in the table on the facing page.
2. As a group, decide on and make a list of events that you think can help support your hypothesis.
3. Choose two other resources and use them to find at least ten more events to add to your list.
4. Design and construct a way to exhibit and explain your results.
5. Check your plan. Make sure your teacher has approved your plan before you proceed.
6. Carry out your plan.

Analyze

1. **Interpreting Observations** Did more history-shaping events seem to have occurred early in Earth's history or later on? Explain.
2. **Comparing and Contrasting** Plot your list of events on a copy of the geologic time scale. Compare and constrast the number of events in each era. Does any geologic time period contain more history-shaping events than others? Explain.
3. **Observing and Inferring** Choose one event in the Mesozoic and infer how Earth's history might have progressed had the event not happened.

Conclude & Apply

1. How do extinction events influence the development of life on Earth?
2. How do mountain-building events and glaciations affect the development of life on Earth?
3. If another planet experienced the same events that you chose, would that planet be identical to Earth? What would be similar or different?

Earth History-Shaping Events

Origin of the solar system
Earth forms, 4.6 B.Y.B.P.
Oceans form, 4.0 B.Y.B.P
Primitive algae evolves, 3.3 B.Y.B.P
First fossil evidence of multicellular organisms, 1.2 B.Y.B.P.
Trilobites are abundant, algal reefs form, 600 M.Y.B.P.
First corals evolve, invertebrates dominate oceans, 500 M.Y.B.P.
First land plants evolve, insects appear, 440 M.Y.B.P.
Fishes are abundant, early amphibians evolve, 400 M.Y.B.P.
Forests that become coal swamps are present, 300 M.Y.B.P.
Earliest reptiles evolve, 300 M.Y.B.P.
Alleghenian Orogeny occurs, 270 M.Y.B.P.
Trilobites become extinct, 270 M.Y.B.P.
Earliest dinosaurs appear, 225 M.Y.B.P.
Pangaea breaks up, 225 M.Y.B.P.
Earliest mammals evolve, 200 M.Y.B.P.
Dinosaurs are abundant, 180 M.Y.B.P.
The first birds evolve, 180 M.Y.B.P.
Asteroid impact, 66 M.Y.B.P.
Modern groups of mammals appear, 60 M.Y.B.P.
Earliest horses evolve, 60 M.Y.B.P.
Large mammals evolve, 40 M.Y.B.P.
Carnivores abundant, 11 M.Y.B.P.
Adirondack mountains uplifted, 11 M.Y.B.P.
First humanoids evolve *(Australopithecus africanus),* 3 M.Y.B.P.
Ice Age of the Pleistocene begins, 1 M.Y.B.P.
Mammoths and mastodons are abundant, 1 M.Y.B.P.
Large ice sheets retreat ~ 10 000 years ago
Mount Vesuvius erupts and destroys Pompeii, A.D. 79
New Madrid Earthquake, 1811–1812
Chicago Fire, 1871
Krakatoa eruption in Java, 1883
Mt. St. Helens eruption in Washington, U.S., 1980

Science in the News

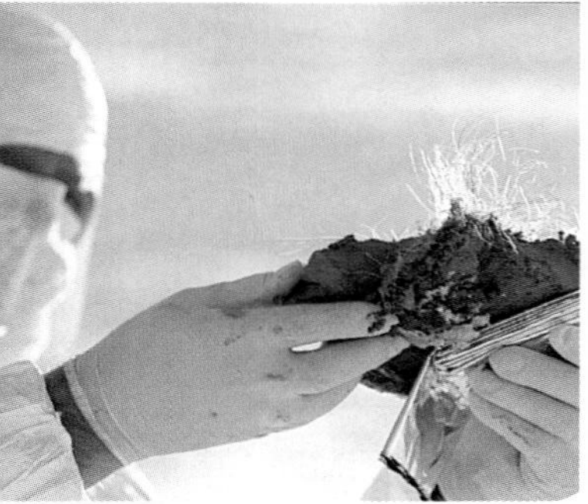

Frozen Mammoth

The sound of jackhammers echoed across the tundra. As a helicopter hovered nearby, the last few cuts were made into the permanently frozen ground. Many hours later, a cube of frozen earth weighing more than 26 metric tons went soaring across the Siberian steppe as it dangled from the bottom of the helicopter.

A chunk of tundra may not seem to be important, but this chunk found in 1997, contained a valuable prize: a complete specimen of a woolly mammoth that had died about 20 000 years ago and became buried in the permafrost. Many fossils represent only the partial remains of once-living organisms. Thus, a complete mammoth with bones, skin, hair, and internal organs intact represents a unique opportunity for scientists to investigate the lifestyle of this animal and the environment in which it lived.

The mammoth was a member of one of the great herds of the species *Mammuthus primigenius.* These herds roamed the vast grasslands of Siberia during the Pleistocene Epoch. The mammoths, cousins of today's elephants, were the largest land mammals ever. They ranged from 2.5 to 4.2 m high at the shoulder, had trunks and tusks like today's elephants, and were covered with thick hair for protection from the cold. This mammoth was a tusked male, about 3.5 m tall, and was about 47 years old when it died.

Solving the Mystery

The mammoth was flown to an ice cave near the town of Khatanga in Central Siberia, where scientists used hair dryers to thaw it out one tiny piece at a time. Every square centimeter of the mammoth and the soil surrounding it will be examined. One thing scientists hope to find out is why mammoths became extinct. They may have been unable to adapt to climatic changes, they may have been over-hunted by early humans, or they may have died of disease. Tissue samples from the mammoth may help to solve this mystery.

"Pleistocene Park"

The amount of knowledge to be gained from the mammoth is tremendous. Some scientists even hope to find cells in the mammoth that could be cloned, and from them, to grow a living woolly mammoth. Other scientists hope to find sperm cells that they can use to cross-breed it with a living female elephant. The hybrid offspring would also be fertilized with mammoth sperm, and a nearly pure-bred mammoth would result after several generations. In a scene right out of "Pleistocene Park," the woolly mammoth might be returned to the steppe as an ecological tourist attraction.

Activity

The woolly mammoth is one of several species in the family Proboscidea. Visit the Earth Science Web Site at earthgeu.com to research the other members of this family. Make a poster telling when, where, and how they lived.

Study Guide

Summary

	Main Ideas	Vocabulary
SECTION 21.1 **The Geologic Time Scale** 	• Geologists have separated Earth's history into divisions based upon the fossil record. • The divisions of the geologic time scale, in descending order and decreasing length of time spans, are eons, eras, periods, and epochs.	eon (p. 554) epoch (p. 556) era (p. 554) geologic time scale (p. 554) period (p. 555)
SECTION 21.2 **Relative Age-Dating of Rocks** 	• The principles of uniformitarianism, original horizontality, superposition, and cross-cutting relationships are used to interpret Earth's rock record and, thus, to describe the planet's history. • Unconformities caused by weathering and erosion or by periods of nondeposition mark missing layers in the rock record.	correlation (p. 561) cross-cutting relationships (p. 559) original horizontality (p. 558) superposition (p. 558) unconformity (p. 560) uniformitarianism (p. 557)
SECTION 21.3 **Absolute Age-Dating of Rocks** 	• Absolute-age dating measures the actual age of an object such as a mineral, rock, or fossil. • Radioactive decay is the emission of particles from a radioactive atom. The decay rate can be used to determine the age of a rock or fossil. The time it takes a radioactive element to decay to 50 percent of its original mass is known as its half-life. • Tree rings and varves can also determine the dates of events and changes in the environment. Volcanic ash and meteorite-impact debris create key beds that mark the time of the event.	dendrochronology (p. 564) half-life (p. 563) key bed (p. 565) radioactive decay (p. 562) radiometric dating (p. 562) varve (p. 565)
SECTION 21.4 **Remains of Organisms in the Rock Record** 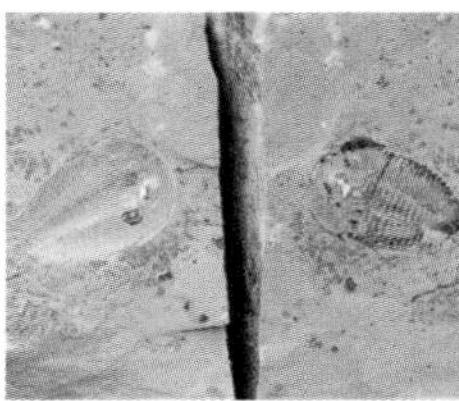	• The remains and evidence of plants and animals that once lived on Earth are called fossils. • Fossils preserved in the rock record provide information about past environmental conditions, evolutionary changes in life-forms, and help geologists to correlate rock layers from one area to another.	altered hard part (p. 567) cast (p. 568) evolution (p. 566) fossil (p. 566) index fossil (p. 568) mold (p. 568) original preservation (p. 566) permineralization (p. 567)

earthgeu.com/vocabulary_puzzlemaker

CHAPTER 21

Assessment

Understanding Main Ideas

1. Which geologic principle is used when a geologist observes an outcrop of rocks and determines that the bottom layer is the oldest?
a. uniformitarianism
b. original horizontality
c. superposition
d. inclusion

2. What is a magma intrusion that cuts across pre-existing rock layers called?
a. sill
b. lava flow
c. fault
d. dike

3. Which term does NOT describe a gap in geologic time?
a. unconformity
b. nonconformity
c. disconformity
d. key bed

4. Which of the following is NOT a characteristic of an index fossil?
a. was commonplace while alive
b. existed for a long period of time
c. is geographically widespread
d. is easily recognizable

5. How old is a mammoth's tusk if there is only 25 percent C-14 remaining in the sample?
a. 5700 years
b. 11 400 years
c. 17 100 years
d. 22 800 years

6. Trees that have been buried by volcanic ash are preserved in what manner?
a. original preservation
b. permineralization
c. mummification
d. recrystallization

7. What feature is formed when a sedimentary rock layer overlies a nonsedimentary rock layer?
a. unconformity
b. nonconformity
c. disconformity
d. contact metamorphism

8. Based on radioactive elements, what is the calculated age of Earth?
a. 4.6 billion years
b. 5 million years
c. 15 billion years
d. 1 million years

9. Which type of fossil forms when an organism's hard parts dissolve and leave a cavity that later fills with sediment?
a. mold
b. coprolite
c. cast
d. gastrolith

10. What are glacial sediments that show cyclic deposition called?
a. annual rings
b. varves
c. tillites
d. unconformities

Applying Main Ideas

11. Why is it difficult to interpret the rock record of Precambrian Time?

12. What evidence is preserved to indicate that a large meteorite impact occurred at the end of the Mesozoic?

13. How would you explain a horizontal sedimentary rock layer containing Cambrian fossils lying on top of a rock layer containing Ordovician fossils?

14. Does an outcrop containing Jurassic dinosaur fossils overlying Pennsylvanian coal deposits represent uninterrupted deposition? Explain.

Test-Taking Tip

WRITE IT DOWN! Most tests ask you a large number of questions in a small amount of time. Write down your work wherever possible. Do math on paper, not in your head. Underline and reread important facts in passages and diagrams—don't try to memorize them.

earthgeu.com/chapter_test

Assessment

15. What radioactive isotope would be best for determining the age of prehistoric human remains?
16. How is an index fossil like a key bed?

Thinking Critically

Use the diagram below to answer questions 17–23.

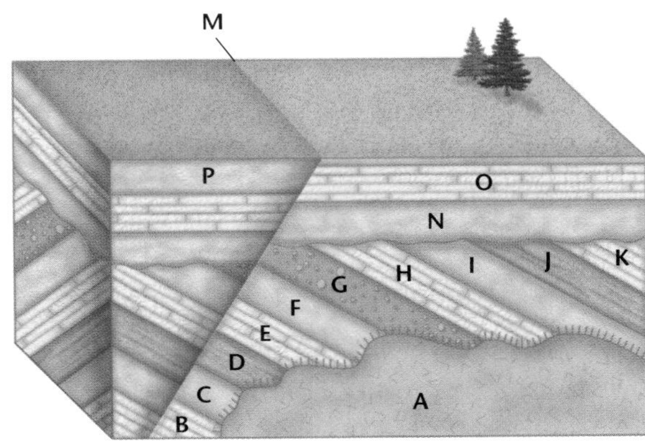

17. Which is the oldest rock unit in the diagram?
18. What has happened to the sedimentary rocks since they were first deposited?
19. Which is the youngest feature in the diagram, the fault or Layer P? Explain.
20. An unconformity exists between which two layers of rock? Explain your answer.
21. Why is rock unit P thicker on the left side of the diagram than on the right side?
22. Explain why the rock layers and features on the left side of the diagram do not match the rock layers and features on the right side.
23. List the order of geologic events represented in the diagram. List the principles you used to place the structures in order from oldest to youngest.

Standardized Test Practice

1. Which of the following constitutes the fewest number of years?
   a. eon
   b. era
   c. period
   d. epoch

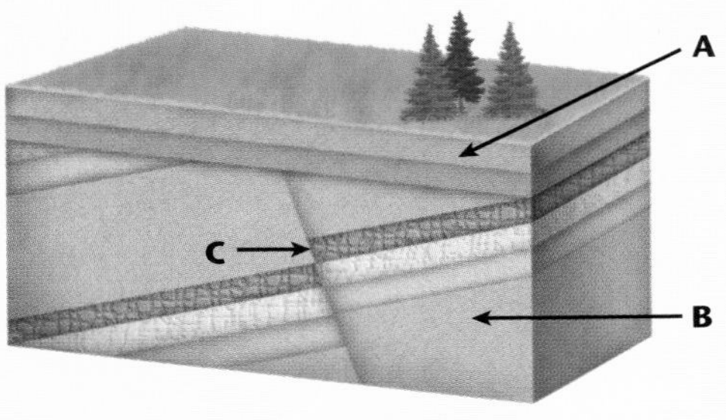

Interpreting Scientific Illustrations

Use the diagram of the rock region above to answer questions 2 and 3.

2. Which principle for determining relative age is relevant to point A of this diagram of a rock region?
   a. The principle of original horizontality.
   b. The principle of superposition.
   c. The principle of cross-cutting relationships.
   d. The principle of uniformitarianism.
3. Which principle is relevant to point C of the diagram?
   a. The principle of original horizontality.
   b. The principle of superposition.
   c. The principle of cross-cutting relationships.
   d. The principle of uniformitarianism.
4. In which of the following do original structures of an organism remain?
   a. mold fossil
   b. permineralized fossil
   c. cast fossil
   d. all of the above

Chapter 22

The Precambrian Earth

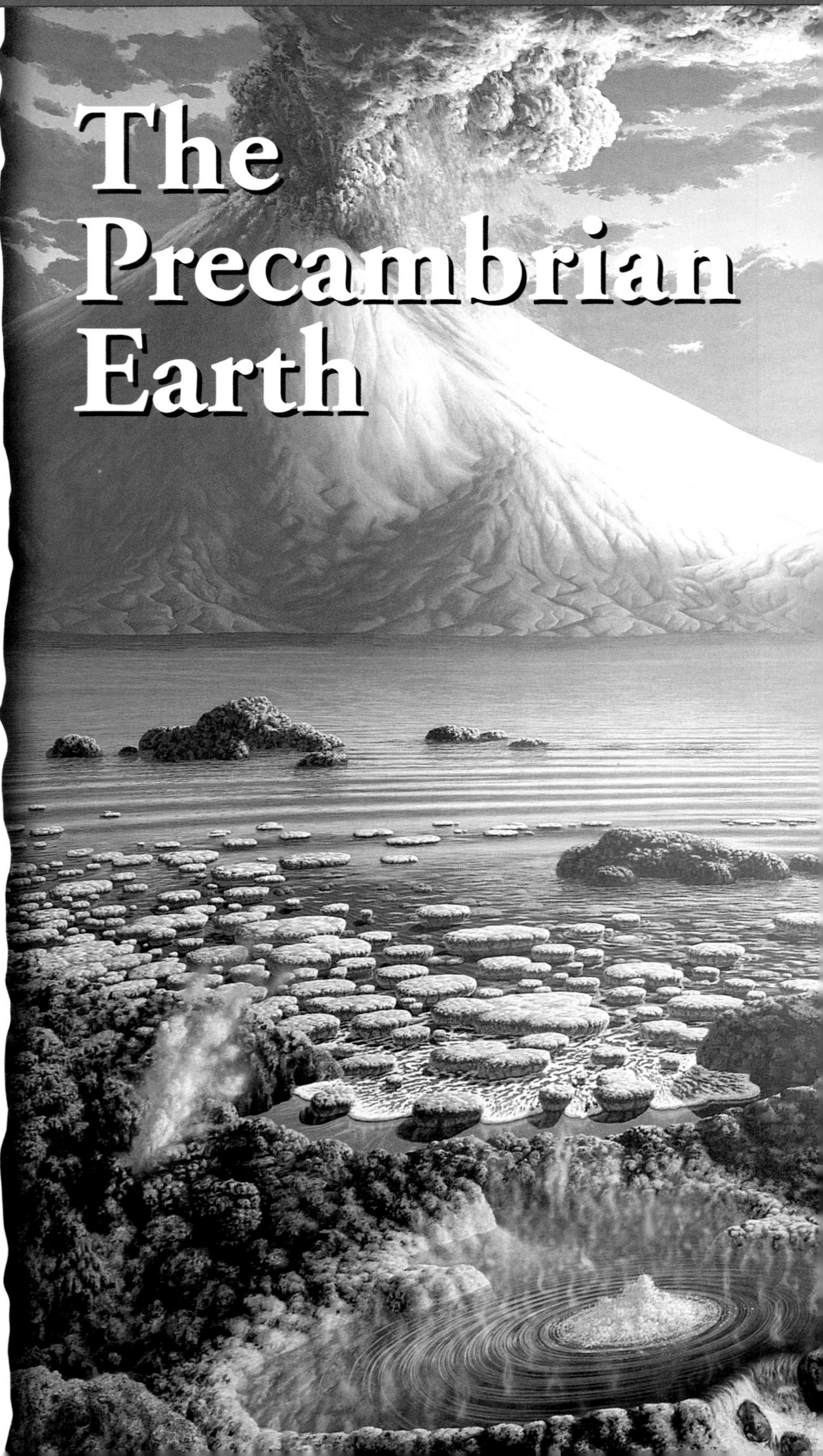

What You'll Learn

- How the age of Earth is determined.
- How the continents, atmosphere, and oceans formed.
- When life first appeared on Earth.
- What kinds of organisms populated the Precambrian Earth.

Why It's important

Most of Earth's history occurred during the Precambrian. During this time, the crust, atmosphere and oceans formed and life first appeared. Early life-forms produced oxygen through photosynthesis, and, thus, changed the atmosphere and the history of life on Earth.

To find out more about Precambrian Earth, visit the Earth Science Web Site at earthgeu.com

Discovery Lab Density Separation

Earth's core, mantle, and crust have different average densities. The core is the densest, and the crust is the least dense. Scientists hypothesize that when Earth formed, temperatures were hot enough for the materials that make up Earth to act, in part, like a liquid and flow. In this activity, you will model how liquids of different densities react when they are mixed together.

1. Fill a 250-mL beaker with 50 mL of tap water.
2. Add 2–3 drops of dark food coloring to the water.
3. Poor 175 mL of vegetable oil into the beaker and stir the contents.

CAUTION: ***Always wear safety goggles and an apron in the lab.***

Observe In your science journal, describe what happened to the colored water and vegetable oil in the beaker. Explain how this is similar to what happened to the core and mantle when Earth formed.

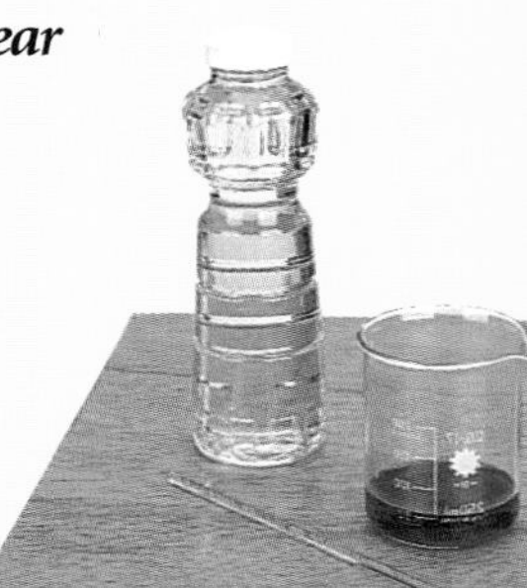

SECTION 22.1 *The Early Earth*

Earth Sciences 1.c, 4.a

OBJECTIVES

- **Describe** *the evidence used to determine the age of Earth.*
- **Understand** *why scientists theorize that the early Earth was hot.*

VOCABULARY

zircon
asteroid
meteorite

For most of Earth's history, there was nothing like the plants and animals that exist today. In Chapter 21, you learned about the geologic time scale and how Earth's history is divided into time periods. In this chapter, you will learn about the earliest part of the geologic time scale, the Precambrian.

EARTH'S "BIRTH"

For about the first 4 billion years of Earth's 4.6-billion-year existence, most of the life-forms that inhabited Earth were unicellular organisms. What evidence did these organisms leave of their existence? What was Earth like when these organisms lived? How did Earth change during the Precambrian? How did these changes set the stage for the animal life that exists today? Answers to these questions not only help us to understand the history of Earth, but they also serve as a model for the search for life on other planets. In 1996, the announcement that a meteorite from Mars might contain microscopic fossils of bacteria rekindled scientific interest in the search for life elsewhere in the universe. You can read more about this meteorite in this chapter's *Science in the News* feature. It may be possible to

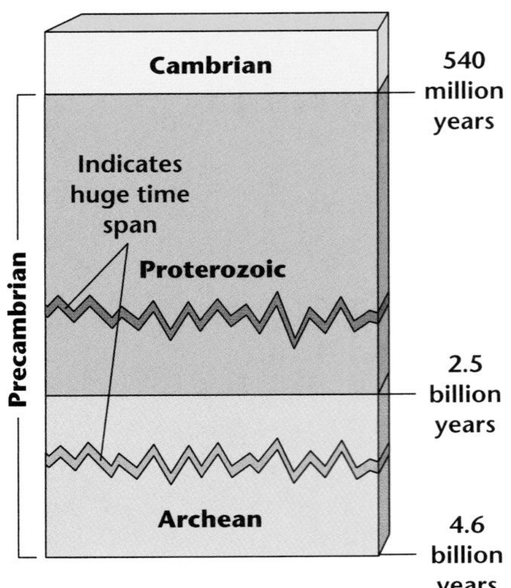

Figure 22-1 Most of Earth's history is contained within the 4 billion years that make up the Precambrian.

identify clues to the possible existence of life on other planets, even if rocks from those planets are the only evidence we have. After all, rocks are all that we have left of the Precambrian Earth, and as you will learn, there is evidence of life's humble beginnings on Earth in Precambrian rocks. The Precambrian portion of the geologic time scale is shown in ***Figure 22-1.***

HOW OLD IS EARTH?

We know that Earth must be at least as old as the oldest rocks in the crust. Radiometric dating has determined that the age of the oldest rocks on Earth is between 3.96 to 3.8 billion years. But the rocks that form Earth's crust have been eroded over time. Evidence of 4.1- to 4.2-billion-year-old crust exists in the mineral zircon that is contained in metamorphosed sedimentary rocks in Australia. **Zircon** is a very stable mineral that commonly occurs in small amounts in granite. Radiometric dating has determined that the zircon grains in these sedimentary rocks are between 4.1 and 4.2 billion years old. The zircon existed before it became cemented into the sedimentary rocks, and scientists theorize that the zircon is the eroded residue left behind from 4.1- to 4.2-billion-year-old granitic crustal rocks. Based on this evidence, Earth must be at least 4.2 billion years old.

Meteorites, such as the one shown in ***Figure 22-2,*** have been radiometrically dated at between 4.5 and 4.7 billion years old. Most astronomers agree that the solar system, including Earth, formed at the same time, and therefore, Earth and meteorites should be about the same age. In addition, the oldest rock samples from the Moon, collected during the Apollo missions, are approximately 4.6 billion years old. Thus, taking all of the evidence into consideration, scientists commonly agree that the age of Earth is 4.6 billion years.

> **Earth Sciences**
>
> **1.c** Students know the evidence from geologic studies of Earth and other planets suggests that early Earth was very different from Earth today.
>
> **4.a** Students know the relative amount of incoming solar energy compared with Earth's internal energy and the energy used by society.

EARTH'S HEAT SOURCES

Earth was most likely extremely hot shortly after it formed, and there were three likely sources of this heat. The first source was radioactivity. Radioactive isotopes were more abundant during the past because, over time, radioactive decay has reduced the original amount of Earth's radioactive isotopes. One product of radioactive decay is energy, which generates heat. Much of Earth's current internal heat is attributed to the energy released by radioactivity. Because there were more radioactive isotopes in Earth's distant past, scientists infer that more heat was being generated then and that Earth was hotter.

The second source of Earth's heat was the impact of asteroids and meteorites. **Asteroids** are metallic or silica-rich objects that are

1 km to 950 km in diameter. Today, most asteroids orbit the Sun between the orbit of Mars and Jupiter. Meteoroids are small asteroids or fragments of asteroids. When meteoroids fall to Earth, we call them **meteorites.** Evidence from the surfaces of the Moon and other planets suggests that there were many more meteoroids and asteroids distributed throughout the early solar system than there are today, and therefore collisions were much more common. For this reason, scientists infer that for the first 500 to 700 million years of Earth's history, bombardment by meteorites and asteroids was common. These impacts generated a tremendous amount of thermal energy.

Figure 22-2 This 10 000-year-old, 16-ton meteorite was found in Oregon on the tribal lands of the Willamette Native Americans.

The third source of Earth's heat was gravitational contraction. As a result of meteor bombardment and the subsequent accumulation of meteorite material on Earth, the size of Earth increased. The weight of the material caused gravitational contraction of the underlying zones. The energy of the contraction was converted to thermal energy. The new material also caused a blanketing effect, which prevented the newly generated heat from escaping.

The combined effects of radioactive decay, meteorite and asteroid bombardment, and gravitational contraction made a hot and rather inhospitable beginning for Earth. However, cooling and subsequent crystallization laid the foundation for Earth's crust to form and prepared Earth for the next phase in its development.

Section Assessment

1. What is the age of Earth?
2. Describe the evidence used to determine the age of Earth.
3. How is zircon used to date igneous rocks?
4. Which of Earth's early sources of heat are not major contributors to Earth's present-day internal heat?
5. **Thinking Critically** If most astronomers hypothesize that the solar system formed all at once, why is it important that we use the age of the oldest rocks on Earth to determine the age of Earth rather than using only the age of meteorites?

Skill Review

6. **Comparing and Contrasting** Compare and contrast Earth's three early sources of heat. For more help, refer to the *Skill Handbook*.

SECTION 22.2

Formation of the Crust and Continents

Earth Sciences 1.c

OBJECTIVES

- **Explain** *the origin of Earth's crust.*
- **Describe** *the formation of the Archean and Proterozoic continents.*

VOCABULARY

differentiation
Precambrian shield
Canadian Shield
microcontinent
Laurentia

Were continents always present on Earth's surface? Early in the formation of Earth, the planet was molten, and numerous elements and minerals were mixed throughout the magma. Over time, the minerals became concentrated in specific zones and Earth became layered. As the magma reached the surface and cooled, landmasses began to form.

FORMATION OF THE CRUST

When Earth formed, iron and nickel, which are dense elements, concentrated in its core. Minerals with low densities tend to crystallize from magma at cooler temperatures than denser minerals do. Therefore, near the surface of Earth, where it is cooler, the rocks are generally composed of a high proportion of the less-dense minerals. For example, granite is common at Earth's surface. Granite is mainly composed of feldspar, quartz, and mica, which, as you learned in Chapter 4, are minerals with low densities. Lava flowing from the hot and partly molten interior of Earth concentrated the less-dense minerals near the surface of Earth over time. In contrast, the denser minerals, which crystallize at higher temperatures, concentrated deeper within Earth and formed the rocks that make up Earth's mantle. The process by which a planet becomes internally zoned when heavy materials sink toward its center and lighter materials accumulate near its surface is called **differentiation.** The differentiated zones of Earth are illustrated in ***Figure 22-3.***

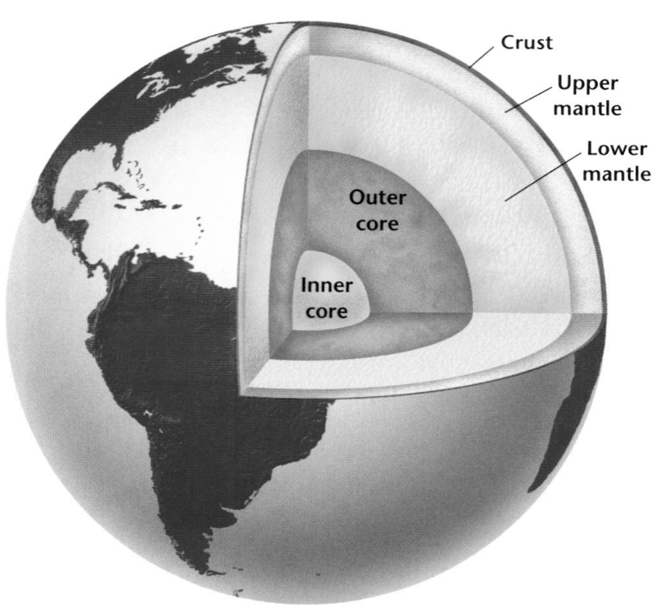

Figure 22-3 Earth's layers formed as a result of differentiation. The density of the minerals found within each layer decreases toward the crust.

Geologists hypothesize that Earth's earliest crust formed as a result of the cooling of the uppermost mantle. Thus, the crust likely consisted of iron and magnesium-rich minerals similar to those found in basalt. As these minerals weathered, they formed sediments that covered the early crust. Geologists also hypothesize that as sediment-covered slabs of the crust were recycled into the mantle at subduction zones, the slabs partly melted and generated magmas with different mineral compositions. These magmas crystallized to form the first granitic continental crust, which was rich in feldspar, quartz, and mica. The formation of

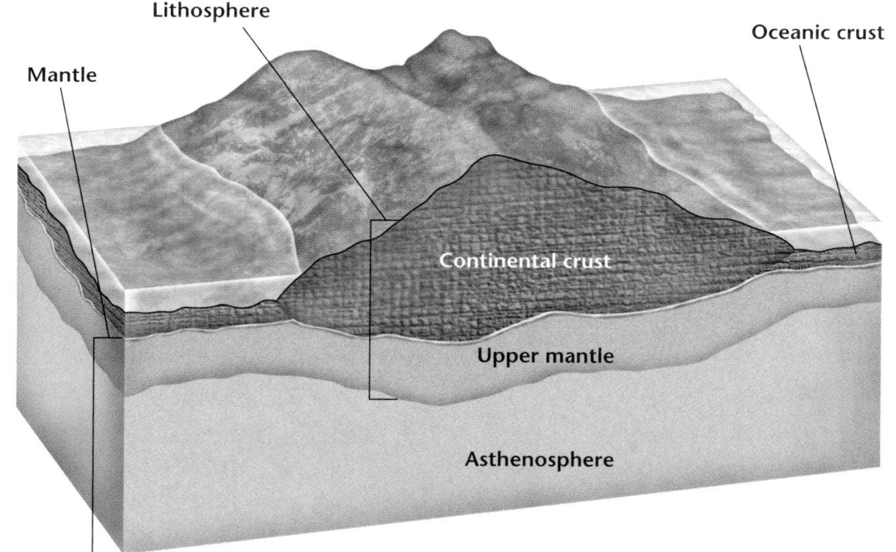

Figure 22-4 The difference in density between the heavier ocean crust and the lighter continental crust allows the continental crust to float higher on the mantle, even when its thickness is greater.

granitic crust was a slow process. Most geologists hypothesize that the formation of the majority of crustal rocks that make up the low-density, granitic cores of continents was completed by about 2.5 billion years ago. The rocks of the earliest crust no longer exist because they were recycled in subduction zones long ago.

 **Earth Sciences**

1.c Students know the evidence from geologic studies of Earth and other planets suggests that early Earth was very different from Earth today.

Less-dense material has a tendency to float on more-dense material. In the *Discovery Lab* at the beginning of this chapter, you observed that oil floats on top of water. This happens because oil is less dense than water. For this same reason, continental crust "floats" on top of the mantle below it. In addition, basaltic crust is more dense than granitic crust, and therefore, it does not float as high on the mantle as granitic, continental crust does. This difference in density is what causes the basaltic ocean floor to be lower in elevation than the less-dense granitic continental crust, as illustrated in ***Figure 22-4.***

THE CORES OF THE CONTINENTS

Today's continents each contain a core of Archean and Proterozoic rock called a **Precambrian shield.** In some areas, the Precambrian shields are exposed at the surface, whereas in other areas, younger sedimentary rocks bury them. The buried and exposed parts of a shield together compose the craton, which is the stable part of a continent. In North America, the Precambrian shield is called the **Canadian Shield** because much of it is exposed in Canada. As shown in ***Figure 22-5*** on page 582, the Canadian Shield is also exposed in the northern parts of Minnesota, Wisconsin, and Michigan; in the Adirondack Mountains of New York; and over a large part of Greenland.

Figure 22-5 The oldest rocks in North America are found in the Precambrian shield rocks in Canada. They were the first-formed rocks of the North American continent.

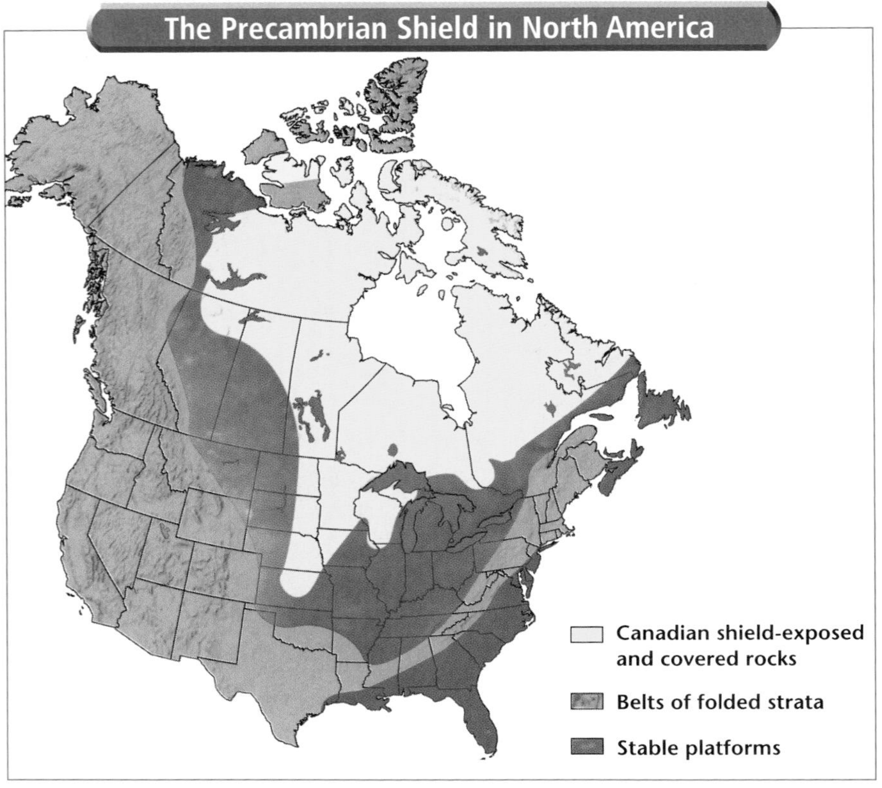

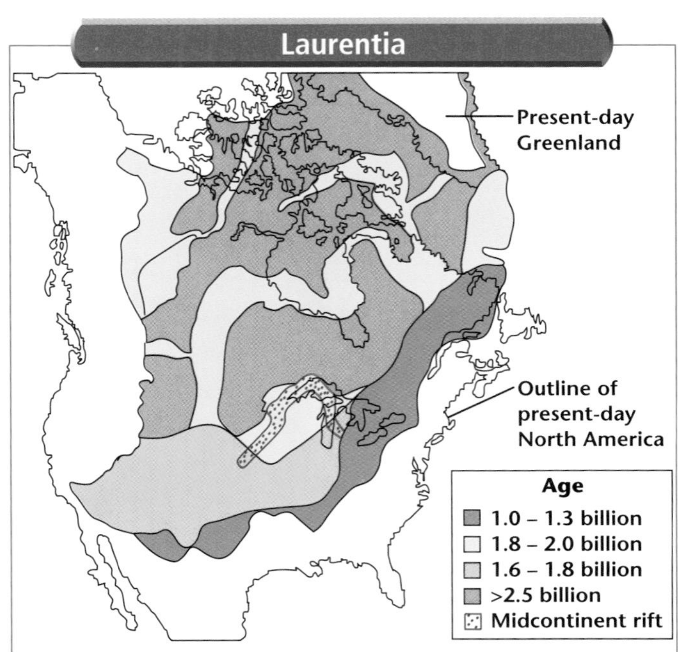

Figure 22-6 From the beginning of its formation, the continent of Laurentia resembled the familiar shape of present-day North America.

Growth of Continents

Early during the Proterozoic, small pieces of continental crust, called **microcontinents,** that formed during the Archean began to collide as a result of plate tectonics. The impact of the collisions jammed the microcontinents together, and they became larger continents. At each of these collision sites, the Archean microcontinents were sutured or fused together at orogens. These orogens are belts of rocks that were deformed by the immense energy of the colliding continents. The resulting mountain ranges have been deeply eroded since that time. By about 1.8 billion years ago, the core of modern-day North America had been assembled; it formed the ancient continent known as **Laurentia,** as shown in ***Figure 22-6.*** You will explore the technique of interpreting continental growth in the *Mapping GeoLab* at the end of this chapter.

Near the end of the Early Proterozoic, between 1.8 and 1.6 billion years ago, volcanic island arcs collided with the southern margin of Laurentia. This added more than 1000 km of continental crust to southern Laurentia. The final phase of Proterozoic growth of Laurentia is called the Grenville Orogeny. Recall that an orogeny is a mountain-building event. The Grenville Orogeny occurred between 1.2 billion and 900 million years ago and added a considerable amount of continental crust to the southern and eastern margins of Laurentia. Also by the end of the Proterozoic, nearly 75 percent of present-day North America had formed. The remaining 25 percent, as you will learn in Chapter 24, was added to the eastern and western margins of the North American craton during the Phanerozoic.

By the end of the Proterozoic, all of the major masses of continental lithosphere had formed on Earth. The lithospheric plates were moving around, periodically colliding with each other and suturing together. By the end of the Proterozoic, so many of these collisions had occurred that Rodinia, the first supercontinent, had formed. It was positioned so that the equator ran through Laurentia, as shown in ***Figure 22-7.*** Rodinia began to break apart at the end of the Proterozoic and continued to do so during the Early Phanerozoic. During this time, Earth also acquired an atmosphere and oceans. You will learn how they formed in the next section.

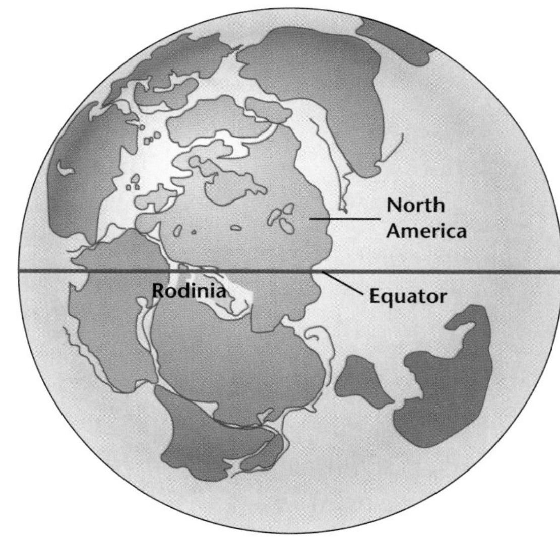

750 million years ago

Figure 22-7 Rodinia was the first supercontinent to form on Earth's surface. Similar rock types in eastern North America and western Africa are evidence of its existence.

SECTION ASSESSMENT

1. Describe the origin of Earth's crust.
2. How did the Archean and Proterozoic continents form?
3. How does a planet become internally zoned?
4. **Thinking Critically** The oceans are underlain by basaltic crust, and the continents are underlain by granitic crust. What would Earth be like if all of its crust were made of the same material?

SKILL REVIEW

5. **Sequencing** Suppose that you are a reporter about to witness the formation of a large continent. In your science journal, write your step-by-step eyewitness account of the event. For more help, refer to the *Skill Handbook.*

SECTION 22.3 Formation of the Atmosphere and Oceans

Earth Sciences 1.c, 8.b I&E 1.m

OBJECTIVES

- **Describe** *the formation of Earth's atmosphere and oceans.*
- **Identify** *the origin of oxygen in the atmosphere.*
- **Explain** *the evidence that oxygen existed in the atmosphere during the Proterozoic.*

VOCABULARY

cyanobacteria
stromatolite
banded iron formation
red bed

Earth Sciences

1.c Students know the evidence from geologic studies of Earth and other planets suggests that early Earth was very different from Earth today.

If you could travel back in time to the Early Precambrian, what would you take with you? Probably the most important thing that you could take for your survival would be a supply of oxygen! This is because Earth's early atmosphere was nothing like what it is today. The oxygen that early forms of algae produced through the process of photosynthesis affected the development of life on Earth in two very important ways. First, it changed the composition of the atmosphere and thus made life possible for oxygen-breathing animals. Second, it produced the ozone layer that filters ultraviolet (UV) radiation. Scientists refer to these types of processes, which modify a system, as feedback.

THE PRECAMBRIAN ATMOSPHERE

Hydrogen and helium probably dominated Earth's earliest atmosphere. However, because of their small masses, these gases could not remain near Earth for long. Earth's gravity is not strong enough to keep hydrogen and helium from escaping to space. However, gases that have greater masses, such as carbon dioxide and nitrogen, cannot escape Earth's gravity. This is why Earth's atmosphere is rich in carbon dioxide and nitrogen today.

There was considerable volcanic activity during the Early Precambrian. In Chapter 18, you learned that lava is not the only substance that erupts from volcanoes. Tremendous amounts of gases are also vented during volcanic eruptions in a process called outgassing, shown in ***Figure 22-8.*** The most abundant gases vented from volcanoes are water vapor (H_2O), carbon dioxide (CO_2), nitrogen (N_2), and carbon monoxide (CO). Many geologists hypothesize that outgassing formed Earth's early atmosphere. Thus, the early atmosphere must have contained large concentrations of water vapor, carbon dioxide, and nitrogen. In addition, the early atmosphere most likely contained methane (CH_4) and ammonia (NH_3), both of which may have formed as a result of chemical reactions among the volcanic gases. The argon (Ar) that is present in today's atmosphere began to accumulate during the Early Precambrian. It forms when the radioactive isotope potassium-40 (K-40) decays to argon-40 (Ar-40). Through this reaction, the amount of argon has increased in the atmosphere at about the same rate that K-40 decays.

Figure 22-8 Steam and gas from Poas Volcano in Poas National Park, Costa Rica, rise high above the volcano's summit.

Oxygen in the Atmosphere

No wonder life as we know it today did not exist during the Precambrian—there was no oxygen in the atmosphere to breathe! Volcanoes do not commonly give off oxygen; so, scientists do not think that the oxygen in Earth's atmosphere came from volcanoes. Where did the oxygen come from? The oldest known fossils which can help answer this question, are preserved in rocks that are about 3.5 billion years old. These fossils are the remains of tiny, threadlike chlorophyl bearing filaments of **cyanobacteria.** Such fossils are contained in 3.46-billion-year-old rocks called the Warrawoona Group, from western Australia. ***Figure 22-9*** compares fossilized cyanobacteria and modern cyanobacteria. Like their present-day counterparts, ancient cyanobacteria used photosynthesis to produce the nutrients they needed to survive. In the process of photosynthesis, solar energy is used to convert carbon dioxide and water into sugar. Oxygen is given off as a waste product.

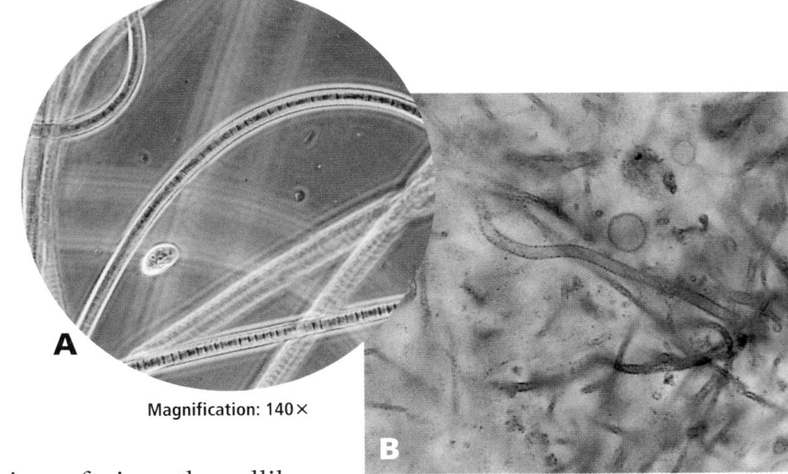

Figure 22-9 The cells of modern cyanobacteria are often identical in shape and size to some fossil cyanobacteria that are billions of years old. This micrograph is of the filamentous cyanobacterium, *Oscillatoria* **(A).** This fossil cyanobacterium is from the Gunflint Chert, Ontario, Canada **(B).**

Oxygen Producers Could microscopic cyanobacteria have produced enough oxygen to change the composition of Earth's early atmosphere? The abundance of cyanobacteria increased throughout the Archean until they became truly abundant during the Proterozoic. Large mats and mounds of billions of cyanobacteria, called **stromatolites,** dominated the shallow oceans of the Proterozoic. Modern stromatolites, such as those found in Australia today and shown in ***Figure 22-10,*** do not differ much from their ancient counterparts. The oldest stromatolites are preserved in 3.4-billion-year-old rocks from the Swaziland Supergroup of South Africa.

Environmental Connection

Evidence in the Rocks One way to test the hypothesis that there was oxygen in Earth's early atmosphere is to look for oxidized iron, or iron oxides, in Archean and Proterozoic rocks. The iron in rocks

Figure 22-10 Colonies of cyanobacteria form modern stromatolites in Hamelin Pool, Australia. Millions of individual cyanobacteria cells make up each colony **(A).** Fossil stromatolites are nearly identical to modern stromatolites. They provide evidence for the existence of a shallow sea in the areas where they are found **(B).**

Figure 22-11 The beautiful rocks of the Banded Iron Formation give testimony to the oxygenated atmosphere of the Precambrian. This sample is from the 2.1 billion year old Negaunee Formation, Michigan **(A).** The Empire Iron Mine in Ishpeming, Michigan is well known for its iron production **(B).**

Earth Sciences

8.b Students know how the composition of Earth's atmosphere has evolved over time and know the effect of outgassing, the variations of carbon dioxide concentration, and the origin of atmospheric oxygen.

reacts with free oxygen in the atmosphere to form iron oxides. Iron oxides are identified by their red color and provide undeniable evidence of free oxygen in the atmosphere. Some metamorphosed Archean sedimentary rocks contain mineral grains that would have been oxidized if there had been oxygen in the atmosphere. However, these minerals were not oxidized, which indicates that there was little or no free oxygen in the atmosphere throughout most of the Archean. However, there is evidence that near the end of the Archean and by the beginning of the Proterozoic, photosynthesizing stromatolites in shallow marine water increased oxygen levels in localized areas. These locally high concentrations of oxygen in otherwise oxygen-poor, shallow, ocean water allowed unique deposits to form. These deposits, which consist of alternating bands of chert and iron oxides are called **banded iron formations.** Today, these formations are mined as a source of iron, as shown in ***Figure 22-11.*** The *Problem-Solving Lab* should give you an example

Problem-Solving Lab

Profits from the Precambrian

Calculate mining costs Precambrian rocks contain many important mineral deposits, such as uranium, which is used in nuclear reactors. In a uranium oxide (U_3O_8) ore deposit in southern Ontario, the ore-containing rocks are an average of 3 m thick over an area that is 750 m long and 1500 m wide. Geochemical analysis of the deposit indicates that there are, on average, 0.9 kg of uranium oxide per metric ton of rock. Additionally, 0.3 cubic meters of the uranium-bearing rock weighs one ton (2000 lbs).

Analysis

1. How many pounds of ore does this new deposit contain?
2. It will cost $45 per cubic yard and 10 years to mine and extract the ore from the deposit. How much will this cost?

Thinking Critically

3. The current market price is $9.25 per pound of uranium oxide. Based on your answer to question 2, can the ore be mined for a profit?

of mining for a profit. Many sedimentary rocks that are younger than 1.8 billion years are rusty red in color and are called **red beds.** The presence of red beds in rocks that are Proterozoic and younger is strong evidence that the atmosphere by this time contained free oxygen. You will observe oxidation when you complete the *MiniLab* on this page.

Importance of Oxygen

Oxygen is important not only because most animals require it for respiration, but also because it provides protection against UV radiation from the Sun. If you have ever read a label on sunscreen, you know that UV radiation can be harmful. Today, only a small fraction of the UV radiation that the Sun radiates toward Earth reaches its surface. This is because Earth is naturally protected from this radiation by ozone that is present in the lower part of Earth's upper atmosphere. An ozone molecule (O_3) consists of three oxygen atoms bonded together. Ozone forms when high-energy UV radiation splits oxygen gas molecules (O_2) and the single oxygen atoms combine with other oxygen molecules. This ozone layer filters out much of the Sun's UV radiation. Oxygen in Earth's atmosphere that was produced mainly through photosynthesis also contributes to the ozone layer. Early life, mainly the cyanobacteria that made up stromatolites, modified the atmosphere by generating large amounts of oxygen. Some of this oxygen also formed ozone, which, in turn, filtered out UV radiation so that other forms of life could survive on Earth's surface. It appears that nearly all the oxygen that we breathe today, and the oxygen that all animals have breathed in the geologic past, was released into the atmosphere by photosynthesis.

MiniLab

Why are red beds red?

Model the formation of red beds with iron, oxygen, and water.

Procedure *CAUTION: Steel wool can be sharp. Wear gloves in the lab.*

1. Place 40 mL of white sand in a 150-mL beaker.
2. Add water so that the total volume is 120 mL.
3. Add 15 mL of bleach.
4. Place a piece of steel wool about the size of your thumbnail, in the beaker. Cover the beaker with a petri dish and allow it to stand in a quiet place for one day.
5. Remove the steel wool and stir the contents of the beaker. Allow the mixture to settle for five minutes after stirring.
6. Slowly pour off the water so that the iron-oxide sediment is left behind.
7. Stir the mixture again, then spoon some of the sand onto a watch glass and allow it to dry.

Analyze and Conclude

1. In your science journal, describe how the color of the sediment changed.
2. Where does the iron in the experiment come from?
3. Where in nature does the red in rocks come from?
4. What do you think is the function of the bleach?

FORMATION OF THE OCEANS

Oceans are thought to have originated largely from the same process of outgassing that formed the atmosphere. A major component of the gas was water vapor. As the early atmosphere and the surface of Earth cooled, the water vapor condensed to form liquid water. You have probably observed the result of condensation on the sides of a cold glass of water. During the Archean, the entire atmosphere was rich with water vapor. When it began to cool, the result was a tremendous amount of rain, which slowly filled the low-lying, basalt-floored basins, thus forming the oceans. Rainwater dissolved the soluble minerals exposed at Earth's surface and just as they do today, rivers, runoff and groundwater transported these minerals to the oceans. These dissolved minerals made the oceans of the Precambrian salty just as they make the oceans salty today.

Topic: Ozone Levels
To find out more about ozone in Earth's atmosphere, visit the Earth Science Web Site at earthgeu.com

Activity: Research the monthly concentrations of atmospheric ozone for a given year in the Antarctic or Arctic. Graph the ozone concentrations for each month in the year.

Another source of water may have played an important role in adding water vapor to Earth's atmosphere. A recent but controversial hypothesis suggests that some of Earth's water may have come from outer space! Earth is constantly bombarded with very small comets made of frozen gas and water. Based on the current rate of micro-comet bombardment, some scientists calculate that a significant portion of Earth's surface waters might be extraterrestrial in origin.

Oxygen Causes Change The Precambrian began with an oxygen-free atmosphere and simple life-forms. Cyanobacteria then evolved, and their oxygen contribution caused the atmosphere to become filled with oxygen. This oxygen not only enabled new life-forms to evolve, but it also protected Earth's surface from the Sun's UV rays. Oceans formed from abundant water vapor in the atmosphere and possibly from outer space. Earth was then a hospitable place for new life-forms to inhabit.

SECTION ASSESSMENT

1. What are banded iron formations?
2. Describe the origin of the oxygen in the atmosphere.
3. Explain the relationship between red beds and oxygen in the atmosphere.
4. **Thinking Critically** If cyanobacteria had not produced as much oxygen as they did, how might life on Earth be different from how it is today?

SKILL REVIEW

5. **Comparing and Contrasting** Compare and contrast the formation of the atmosphere and the formation of the oceans. For more help, refer to the *Skill Handbook*.

SECTION 22.4 *Early Life on Earth*

Earth Sciences 1.c

Of all the questions that humans have ever asked, none fascinates us more than those about the origin of life. "Where did life come from?" is a question that has been explored from many different perspectives. Today, we know that life comes from other life through reproduction. But where did the first life come from? What does science tell us about the origin of life?

OBJECTIVES

- **Describe** *the experimental evidence of how life developed on Earth.*
- **Distinguish** *between prokaryotes and eukaryotes.*
- **Identify** *when the first multicellular animals appeared in geologic time.*

VOCABULARY

amino acids
hydrothermal vent
prokaryote
eukaryote
Varangian Glaciation
Ediacara fauna

ORIGIN OF LIFE ON EARTH

You have learned that Earth is about 4.6 billion years old and that fossil evidence indicates that life existed on Earth about 3.5 billion years ago. Thus, life must have begun during Earth's first 1.1 billion years. Earth probably could not have supported life until about 3.9 billion years ago because meteorites were constantly striking its surface. If life did begin during this time of meteorite bombardment, it is unlikely that it could have survived for long. This places the origin of life somewhere between 3.9 and 3.5 billion years ago.

Experimental Evidence During the first half of the twentieth century, scientists hypothesized that the early atmosphere contained carbon dioxide, nitrogen, water vapor, methane, and ammonia, but no free oxygen. They also theorized that numerous storms produced lightning and that the surface of Earth was relatively warm. Molecular biologists in the 1920s also suggested that an atmosphere containing abundant ammonia and methane but lacking free oxygen would be an ideal setting for the "primordial soup" in which life may have begun. A young graduate student named Stanley Miller, who was working with his graduate advisor, Nobel prize-winning chemist Harold Urey in 1953, was aware of these hypotheses.

Miller and Urey decided to create their own primordial soup. They set up an apparatus, like that shown in ***Figure 22-12,*** that contained a chamber filled with hydrogen, methane, and ammonia to simulate the early atmosphere. This atmospheric chamber was connected to a lower chamber that was designed to catch any particles that condensed in the atmospheric chamber.

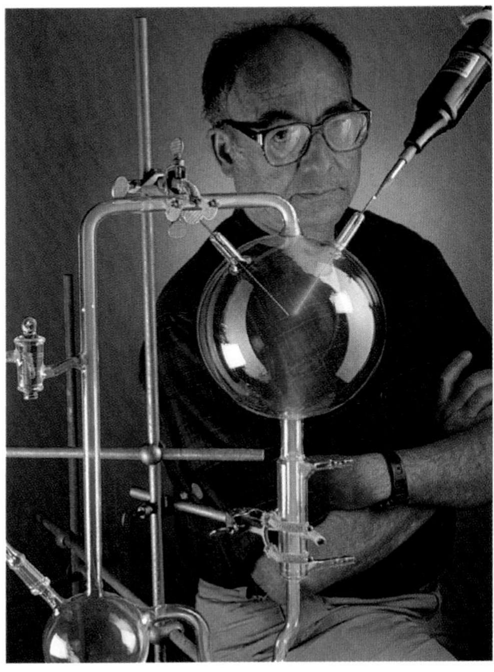

Figure 22-12 Dr. Stanley Miller is shown with a replica of the apparatus used to model Earth's early atmosphere in the Miller-Urey experiment.

Using Numbers
Some scientists argue that the influx of small comets into the atmosphere could add about one inch of water to Earth's surface every 20 000 years. The average depth of Earth's oceans is 3795 meters. If this influx has been constant for the past 4.6 billion years, what should be the minimum average depth of the oceans? How can you account for the difference in water volume?

Earth Sciences

1.c Students know the evidence from geologic studies of Earth and other planets suggests that early Earth was very different from Earth today.

Miller and Urey added sparks from tungsten electrodes to simulate lightning in the atmosphere. Only one week after the start of the experiment, the lower chamber contained a murky, brown liquid—the primordial soup! The "soup" that formed in this experiment contained organic molecules such as formaldehyde (H_2CO), and four different amino acids. **Amino acids** are the building blocks of proteins, and proteins are the basic substances from which life is built.

Continued experiments showed that 13 of the 20 amino acids known to occur in living things could be formed using experimental set-ups similar to the Miller-Urey method. Further experiments demonstrated that heat, cyanide, and certain clay minerals could cause amino acids to join together in chains like proteins. Proteins provide structure for tissues and organs, and are important agents in cell metabolism. Thus, the discovery that amino acids could be formed in this way was amazing. What Miller and Urey demonstrated, is that however life first formed, the basic building blocks of life were most likely present on Earth during the Archean.

The Role of RNA Not much is known about the next step required for the development of life. It is one thing for the proteins that are required for life to exist on the Archean Earth, but quite another for organic life to actually exist. One essential characteristic of life is the ability to reproduce. The nucleic acids RNA and DNA, shown in ***Figure 22-13,*** are the basic requirements for reproduction.

In modern organisms, DNA carries the instructions necessary for cells in all living things to function. Both RNA and DNA need enzymes to replicate and at least one of them is necessary for the synthesis of enzymes. Recent experiments have shown that types of RNA molecules, called ribozymes, can act as an enzyme. These RNA ribozymes can, therefore, replicate without

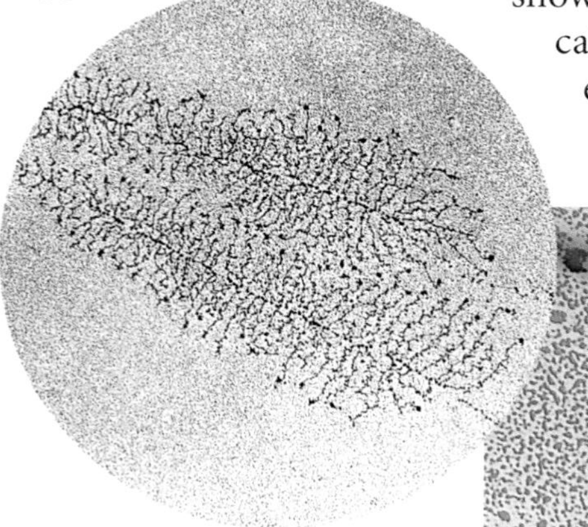

Figure 22-13 These are ribonucleic acid (RNA) polymerase molecules **(A).** This is a single strand of deoxyribonucleic acid (DNA) **(B).**

the aid of enzymes. This suggests that RNA molecules may have been the first replicating molecules on Earth. An RNA-based world may have been intermediate between an inorganic world and the DNA-based organic world that followed. The remaining mystery is figuring out how the first RNA molecule formed because RNA cannot be easily synthesized under the conditions that likely existed at the surface of the Archean Earth.

Figure 22-14 These sulfur encrusted, underwater geothermal vents bubble as volcanic gasses escape. They are located off Dobu Island, Papua, New Guinea.

Hydrothermal Vents and the Beginnings of Life

The Urey-Miller model places the origin of life in shallow surface waters. But, life on Earth may have originated deep in the ocean, near active volcanic seafloor rifts. Ocean water seeps through the cracks in the ocean floor and is heated by the magma at the rifts. This heated water rises and is expelled from the ocean floor at hot-water vents called **hydrothermal vents,** as shown in ***Figure 22-14.*** All of the energy and nutrients necessary for the origin of life are present at these deep-sea hydrothermal vents. In fact, amino acids have been found there. This has led some scientists to hypothesize that during the Archaean, near hydrothermal vents, amino acids joined together on the surfaces of clay minerals to form proteins. Other scientists contend that this is not possible in such an environment. It is important to note that life is not being synthesized at these vents today, only because amino acids are quickly devoured by organisms that live near the vents. But in the Archaean, no organisms existed to eat the amino acids being produced.

Update For an on-line update on recent discoveries of life-forms at hydrothermal vents, visit the Earth Science Web Site at earthgeu.com

PROTEROZOIC LIFE

At the beginning of the Proterozoic, life-forms were still quite simple. The only evidence of life-forms that existed before the Proterozoic is the fossilized remains of unicellular organisms called prokaryotes. A **prokaryote** is an organism that is composed of a single cell, which does not contain a nucleus and is the simplest kind of cell. All prokaryotes, including the cyanobacteria that make up stromatolites, belong to Kingdom Monera.

A **eukaryote** is an organism that is composed of cells that contain a nucleus. One way to determine whether an organism is a prokaryote or a eukaryote is by its size. As a general rule, eukaryotes are larger than prokaryotes. This general observation is useful in determining whether a fossil was a prokaryote or a eukaryote because it is rare for a fossil to be preserved in enough detail to determine whether its cells had nuclei. The oldest known fossil eukaryotes occur in a 2.1-billion-year-old banded iron formation in northern Michigan.

There is growing evidence that a widespread glaciation, which occurred between 800 and 700 million years ago, played a critical role in the extinction of many members of a group of possible eukaryotes, the acritarchs. This glaciation event, called the **Varangian Glaciation,** was so widespread that some geologists liken Earth at that time to a giant snowball. Evidence from ancient glacial deposits suggests that glacial ice advanced nearly to the equator. Shortly after the ice retreated toward the poles, 700 million years ago, multicellular organisms first appeared in the fossil record.

EDIACARA FOSSILS

In 1947, the impressions of soft-bodied organisms were discovered in Late Proterozoic rocks in the Ediacara Hills of southern Australia. These fossils are collectively referred to as the **Ediacara fauna.** ***Figure 22-15*** shows an interpretation of the Ediacaran world. There has been much debate in the scientific community about the precise nature of these remarkable fossils. It is generally agreed that these fossils represent animals that were composed of different types of eukaryotic cells. Scientists are unsure, however, whether the Ediacara fauna are relatives of modern animal groups or whether they were completely different types of organisms.

The discovery of the Ediacara fauna at first seemed to solve one of the great mysteries in geology—why were there no fossils of the ancestors of the complex and diverse organisms that existed during the Cambrian Period. The Ediacara fauna seem to provide fossil

Figure 22-15 The Ediacara fauna contained a wide variety of organisms. It included floating organisms as well as those which were attached to the sea floor and possibly some organisms that were actively mobile.

evidence of an ancestral stock of complex Proterozoic animals. Indeed, many of the Ediacara fossils look quite similar in overall body shape to jellyfish, sea pens, segmented worms, arthropods, and echinoderms—just the kind of ancestral stock that geologists had been hoping to find.

Some scientists, however, hypothesize that the Ediacara fauna does not represent an ancestral stock of any modern group. These scientists consider the similarity in shape to animals in other phyla coincidental and that the Ediacara fauna represents a virtual dead end. None of the Ediacara fossils shows any evidence of a mouth, anus, or gut, and there is little evidence these animals could move. Arthropods, for example, leave tracks and trails when they move across the seafloor, but there is no evidence of such trace fossils associated with the Ediacara fossils. This has led some geologists to hypothesize that the Ediacara organisms were relatively immobile and that they fed by passively absorbing nutrients from seawater. These geologists point out that in the absence of any animal predators, there would have been no disadvantage to being a defenseless creature basking in the warm seawater and absorbing nutrients.

In recent years, geologists have found Ediacara fossils in all parts of the world. This suggests that these organisms were widely distributed throughout the shallow oceans of the Late Proterozoic. They seemed to have flourished between 670 and 570 million years ago. Then, in an apparent mass extinction, they disappeared before organisms that are likely related to modern phyla took over the oceans of the world.

Topic: Ediacara Fossils
To find out more about the Ediacara fauna, visit the Earth Science Web Site at earthgeu.com

Activity: Research the locations of Ediacara fossils. List some locations in your state where they may be found.

Section Assessment

1. Explain why the Miller-Urey experiment was important.
2. What kind of organisms do the earliest fossils represent?
3. What is the significance of the Varangian Glaciation?
4. Discuss the differences between prokaryotes and eukaryotes.
5. **Thinking Critically** Describe how early life might have changed if some of the Ediacaran fauna had been able to move and if predators had been present in their environment.

Skill Review

6. **Concept Mapping** Rearrange the following events into an events chain that describes the results of the Miller-Urey experiment. For more help, refer to the *Skill Handbook.*

chamber with cyanide, formaldehyde, and amino acids	heat and clay minerals added
chains of amino acids	simulated atmospheric lightning
chamber with hydrogen, methane, and ammonia	

Mapping GeoLab

Mapping Continental Growth

Plotting the distribution of the ages of rocks onto a map helps geologists to reconstruct the history of continental accretion. During the Precambrian, microcontinents and island arcs collided to form what would become the modern continents.

Preparation

Problem

How can the distribution of the ages of rocks plotted on a map be used to interpret the growth of a continent?

Materials

paper
pencil
colored pencils
metric ruler

Procedure

1. Your teacher will set up locations with a rock sample at each location.
2. Make an outline map of your classroom similar to the map on the next page, using the scale 1 cm = 100 km.
3. Visit each location where a rock sample has been set out. Plot each location and record the age of each rock on the map.
4. After you have recorded all the locations, use a pencil to draw lines on the map that separate rocks of different ages. Be careful not to simply connect the dots.
5. Use a different colored pencil to shade in the areas on the map that contain rocks of the same age. These are your geologic age provinces.
6. Make a key for your map by drawing a small rectangle for each different geologic age province. Name the oldest province "Province A," the next oldest "Province B," and so on for all provinces.

Analyze

1. Use the ruler to measure the east-to-west width of Province A. Convert the map scale to ground distance by using the scale 1 cm = 1 km.
2. Why do some of your classmates have different answers? Who is right?
3. Where is the oldest province located relative to all the other provinces?

Conclude & Apply

1. Based on the distribution of the geologic age provinces, describe the sequence of collisional events that formed the craton represented by your map.
2. According to your map, where would you find metamorphic rocks? What type of metamorphism would have occurred?
3. If your map represents an area composed of Precambrian-aged rocks, would the mountains that formed from collisions still be high and rugged? Explain.
4. Compare the distribution of age provinces on your map with ***Figure 22-6.*** What are the similarities?
5. Based on what you learned in this activity, describe the formation of the North American Craton.

Locality Data

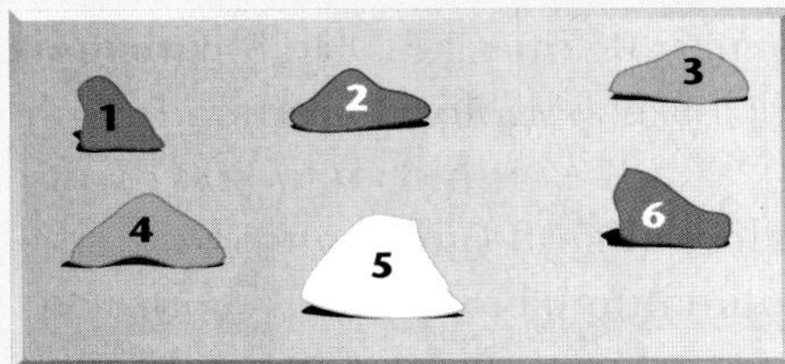

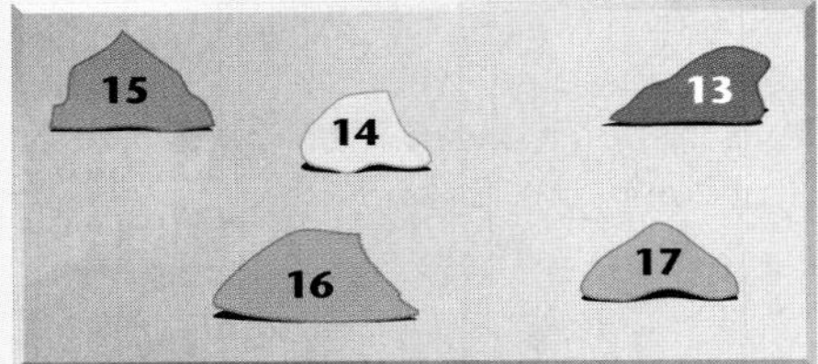

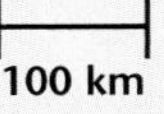

Science in the News

I&E 1.m

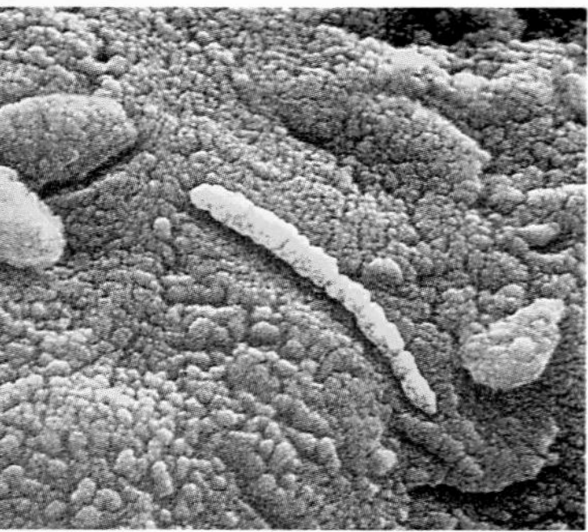

Martians or Meteorites?

Some of the rarest meteorites found on Earth come from Mars. One of these, ALH 84001, caused quite a stir in the scientific community when some scientists claimed to have found fossils in it. Many fossils, such as dinosaur bones and impressions of fern fronds, are obviously the remains of once-living things. Less-obvious evidence of life are the microscopic spheres and rods found in ALH 84001.

ALH 84001 is a potato-shaped rock that weighs 1.9 kg. It was found in the Allan Hills Ice Field, Antarctica, in 1984. The meteorite formed about 4.5 billion years ago, when Mars did. The "fossils," shown above, are found in cracks in the rock and are more than one million times smaller than a typical bacterial cell and are 1/100th the diameter of human hair.

Mystery on Earth

A similar situation took place in the 1860s, when features that were assumed to be fossils were found in limestone in Canada. The features were millimeter-thick bands of dark minerals separated into blobs and layers by light minerals. This "fossil" was named *Eozoon canadense,* the "dawn animal of Canada." This discovery created quite a stir, because *Eozoon canadense* came from the Precambrian, in some of the oldest rocks in North America. Scientists hailed it as "the greatest discovery in geology for half a century."

Not everyone was so taken with *Eozoon canadense,* however. Within months, the first doubts that it was truly a fossil were published.

Gradually, the weight of evidence turned against the possibility that *Eozoon canadense* had an organic origin. There were three main lines of evidence. First, it was found that the original layering of the rock cut across the "fossil," rather than being parallel to it as it should if the animal had actually lived on the bottom of an ancient sea. Second, mapping showed that *Eozoon canadense* occurred around igneous intrusions. The heat from the cooling magma, it seemed, had created the appearance of fossils. Finally, nearly identical features were found in limestone blocks ejected by a volcano, again showing that heat, not life, had formed these features.

What are they?

What about the so-called fossils in the Martian meteorite? Many scientists now do not think that they are fossils. One reason is their size. Their volume is 2000 times smaller than the smallest known living things—parasitic organisms that live in other cells. Many scientists think that something so tiny probably could not contain enough genetic information to direct life processes. The search goes on for extraterrestrial life.

Activity

What other kinds of geologic features can be mistaken for fossils? Visit the Earth Science Web Site at earthgeu.com or a library to find more examples of pseudofossils. Choose one and, in your science journal, describe how it forms.

CHAPTER 22
Study Guide

Summary

SECTION 22.1
The Early Earth

Main Ideas

- Geologists have used radiometric dating to show that Earth must be at least 4.2 billion years old.
- Because the solar system formed all at the same time, Moon rocks and meteorites that are approximately 4.6 billion years old suggest that Earth is 4.6 billion years old too.
- The early Earth was a very hot place because of abundant radioactive isotopes, bombardment by meteorites, and gravitational contraction.

Vocabulary

asteroid (p. 578)
meteorite (p. 579)
zircon (p. 578)

SECTION 22.2
Formation of the Crust and Continents

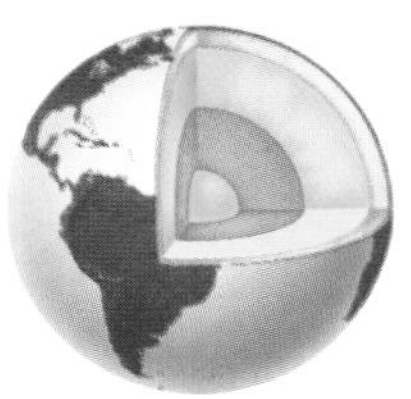

Main Ideas

- Earth's early crust formed by the cooling of the uppermost mantle. This early crust weathered and formed sediments.
- Sediment-covered slabs of this early crust were subducted and generated magmas that contained granitic minerals.
- During the Archean, microcontinents collided with one another throughout the Proterozoic and formed the cores of the continents. By the end of the Proterozoic, the first supercontinent, Rodinia, had formed.

Vocabulary

Canadian Shield (p. 581)
differentiation (p. 580)
Laurentia (p. 582)
microcontinents (p. 582)
Precambrian shield (p. 581)

SECTION 22.3
Formation of the Atmosphere and Oceans

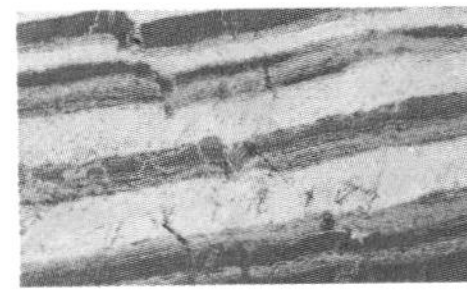

Main Ideas

- Earth's early atmosphere and the oceans formed mainly by the process of outgassing.
- Nearly all of the oxygen in the atmosphere is a result of photosynthesis.
- Certain minerals oxidize, or rust, in the presence of free oxygen. Proterozoic red beds are sedimentary rock deposits that contain oxidized iron. They are the evidence that there was free oxygen in the atmosphere during the Proterozoic.

Vocabulary

banded iron formation (p. 586)
cyanobacteria (p. 585)
red bed (p. 587)
stromatolite (p. 585)

SECTION 22.4
Early Life on Earth

Main Ideas

- All the ingredients were present on the early Earth to form proteins, the building blocks of life. Amino acids, the molecules that make up proteins, were present on the surface of the early Earth.
- Prokaryotic cells are generallly small and contain no nuclei. Eukaryotic cells contain nuclei and are generally larger and more complex than prokaryotic cells.
- The first evidence of multicellular animals are fossils of 2.1 billion year old eukaryotic algae.

Vocabulary

amino acids (p. 590)
Ediacara fauna (p. 592)
eukaryote (p. 591)
hydrothermal vent (p. 591)
prokaryote (p. 591)
Varangian Glaciation (p. 592)

earthgeu.com/vocabulary_puzzlemaker

CHAPTER 22

Assessment

Understanding Main Ideas

1. What is the commonly accepted age of Earth?
   a. 4.6 million years
   b. 46 million years
   c. 4.6 billion years
   d. 46 billion years

2. Which of the following was not a source of heat for the early Earth?
   a. meteor bombardment
   b. gravitational contraction
   c. radioactivity
   d. hydrothermal energy

3. What are small asteroids called?
   a. comets
   b. meteoroids
   c. cratons
   d. microcontinents

4. What is the process by which a planet becomes internally zoned when heavy materials sink toward its center and lighter materials accumulate near its surface?
   a. photosynthesis
   b. dewatering
   c. accretion
   d. differentiation

5. Where is most of the North American Precambrian shield exposed at the surface?
   a. Canada
   b. Minnesota
   c. Wisconsin
   d. Michigan

6. What mineral can be used to radiometrically date Earth's age?
   a. zircon
   b. quartz
   c. hematite
   d. feldspar

7. Refer to ***Figure 22-6.*** What name is given to the core of the modern-day North American continent that formed in the Proterozoic?
   a. Baltica
   b. Yavapai
   c. Grenville
   d. Laurentia

8. What is the name of the first supercontinent, which formed near the end of the Proterozoic?
   a. Laurentia
   b. Grenville
   c. Rodinia
   d. Pangaea

9. What volcanic process most likely formed Earth's atmosphere?
   a. differentiation
   b. outgassing
   c. crystallization
   d. photosynthesis

10. Why is ozone a necessary component of Earth's atmosphere?

11. Why is Earth's atmosphere rich in nitrogen (N) and carbon dioxide (CO_2) today?

Applying Main Ideas

12. Rearrange the following phrases to create a cycle map that describes the formation of Earth's early crust.

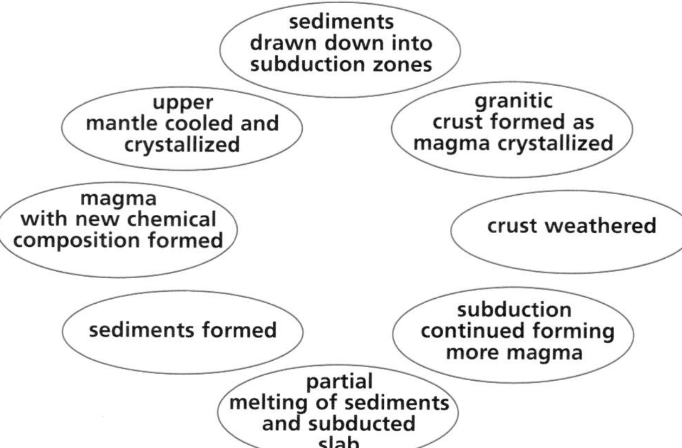

Test-Taking Tip

Keep a Clear Mind When you take a test, each new question should be a clean slate. Once you have read a question, considered the answers, and chosen one, put that question behind you. Don't let one or two troublesome questions distract you while you're working on other questions.

Assessment

13. Explain how geologists have determined the age of Earth.
14. Discuss the relationships among the formation of the continents, the atmosphere, and the oceans.
15. What is the geologic significance of banded iron formations?
16. What geologic evidence suggests that free oxygen was accumulating in Earth's atmosphere during the Proterozoic?
17. What is the difference between prokaryotes and eukaryotes? Which appeared first in the fossil record?
18. What characteristics of continental crust allow it to "float" higher on the mantle than oceanic crust?
19. Why are orogens deformed?
20. What is the significance of the Ediacara fauna?
21. Discuss the evidence that suggests that most members of the Ediacara fauna were immobile.

Thinking Critically

22. Explain how the production of oxygen through photosynthesis by cyanobacteria affected the composition of the atmosphere and the development of other organisms.
23. A rock sample from Mars is reported to contain fossil evidence of life. What kind of fossil would you expect it to be? Explain your answer.
24. Where in North America would you look if you wanted to find evidence of Archean life? Explain your answer.
25. When making a map of geologic age provinces, as you did in the *Mapping GeoLab* in this chapter, why did you draw the lines between the data points instead of connecting them?
26. How might Earth's surface be different if water vapor had not been a product of outgassing?

Standardized Test Practice

1. Which of the following is NOT a likely source of the Precambrian Earth's heat?
   - **a.** radioactivity
   - **b.** asteroid impact
   - **c.** increased solar activity
   - **d.** gravitational contraction

2. What does orogeny refer to?
   - **a.** the drifting of microcontinents
   - **b.** the building of mountain ranges
   - **c.** the formation of volcanic islands
   - **d.** the breaking apart of the supercontinents

3. Which of the following was NOT a source of information about the early presence of oxygen on Earth?
   - **a.** red beds
   - **b.** banded iron formations
   - **c.** stromatolites
   - **d.** meteorites

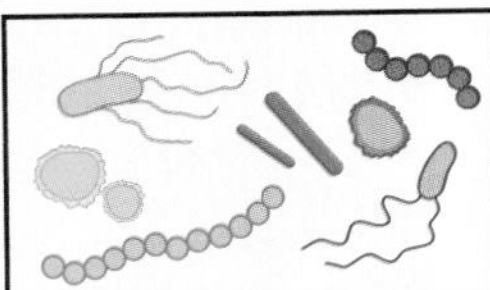

Group A

Group B

Interpreting Scientific Illustrations
Use the diagrams to answer questions 4 and 5.

4. How do members of Group A differ from members of Group B?
   - **a.** They belong to the Kingdom Plantae.
   - **b.** They can be found in Proterozoic fossils.
   - **c.** They contain no nuclei.
   - **d.** They are all unicellular.

5. Where did members of Group B probably originate?
   - **a.** glaciers
   - **b.** hydrothermal vents
   - **c.** Australian fauna
   - **d.** oil deposits

Chapter 23

The Paleozoic Era

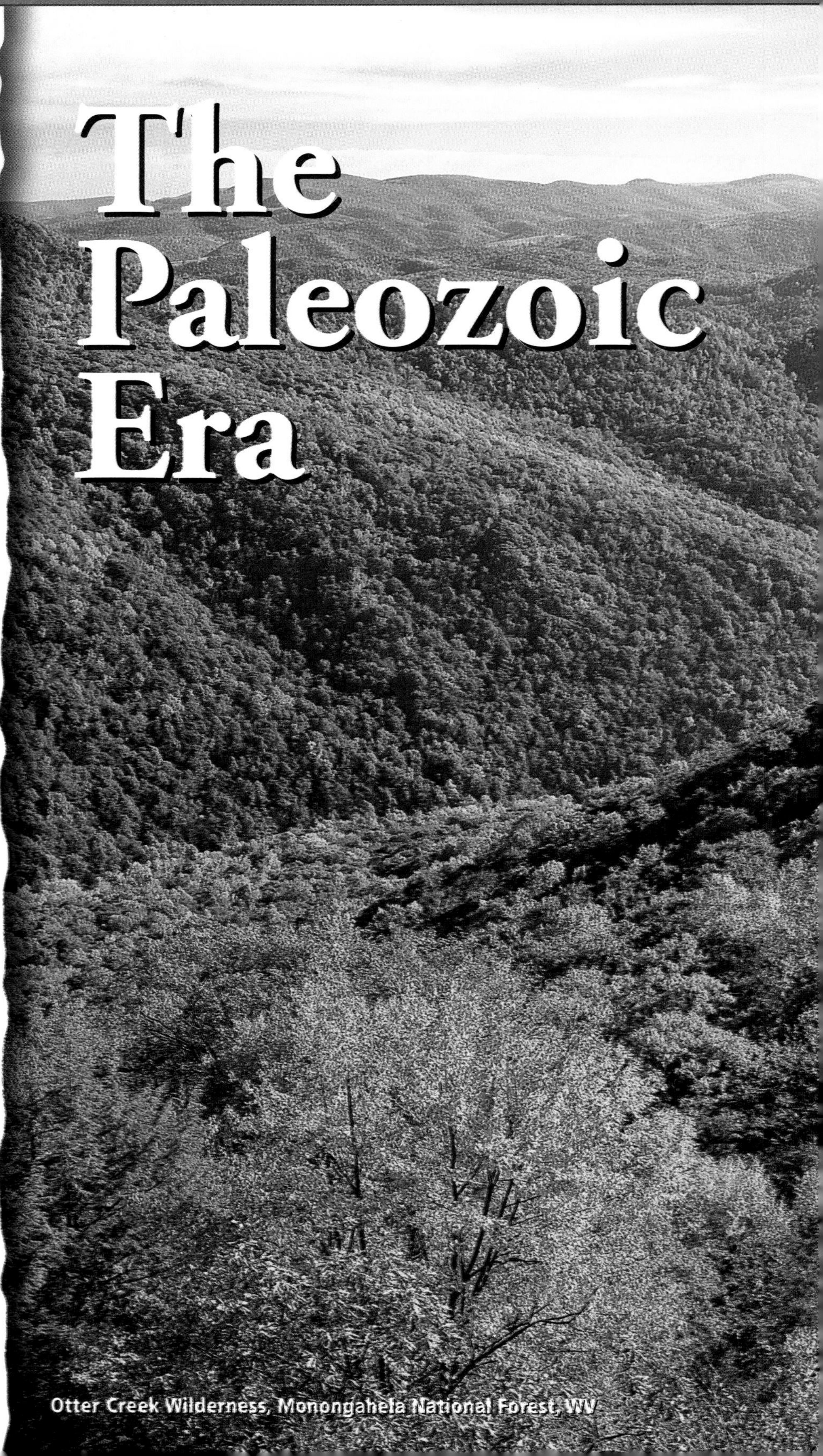

Otter Creek Wilderness, Monongahela National Forest, WV

What You'll Learn

- How the Appalachian and Ouachita Mountains formed.
- When and on what scale three mass extinctions occurred.
- Why coal is common in Pennsylvanian-aged rocks.
- How the development of seeds and eggs affected the evolution of life.

Why It's Important

The tectonic setting of eastern North America provided a hospitable environment not only for a wide variety of animals and plants, but also for the vast swamps that ultimately formed the rich coal deposits of eastern North America.

Earth Science Online

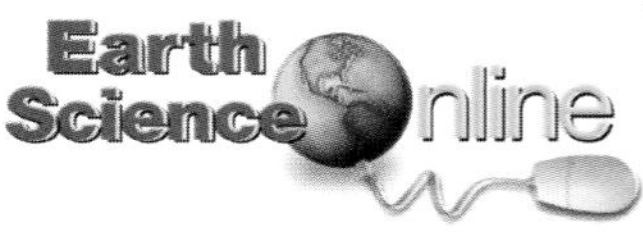

To find out more about the Paleozoic Era, visit the Earth Science Web Site at earthgeu.com

Discovery Lab Where is oil found?

Many sedimentary rocks contain oil and water in the pore spaces between their grains. For example, as you will learn later in this chapter, there is abundant oil in the Paleozoic-aged, sedimentary rocks of West Texas. In this activity, you will observe how oil or water can be stored in solid rock.

1. Place an unglazed brick or sandstone sample on your table.
2. Using a dropper, slowly squeeze three to five drops per minute, of water or oil, for ten minutes.

Infer In your science journal, sketch a cross-section of the rock or brick. Include both before and after you added the water to it. Include in your sketches what the inside of the brick might look like. Infer what happened to the water.

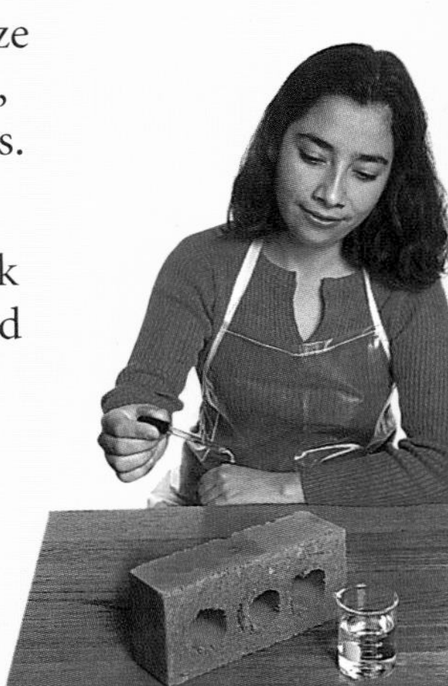

SECTION 23.1 The Early Paleozoic

Earth Sciences 3.c

OBJECTIVES

- **Describe** *the Cambrian paleogeography of Laurentia.*
- **Discuss** *the concept of a passive margin.*
- **Describe** *the Cambrian fauna.*

VOCABULARY

paleogeography
passive margin
transgression
regression
Burgess Shale

Clues from the Paleozoic Era help us to understand how the diversity of life developed. Today, the Appalachian Mountains of the eastern United States border a tectonically calm coastline. But this area was much different in the past. The Paleozoic story explains how early life-forms moved from water to land, why the Appalachians exist, and why they contain the vast coal deposits that fueled the industrial revolution. The Paleozoic portion of the geologic time scale is shown in ***Figure 23-1*** on page 602.

Continental Setting

Geologists refer to the ancient geographic setting of an area as its **paleogeography.** The supercontinent Rodinia, was present at the end of the Proterozoic. By the Cambrian, the ancient North American continent of Laurentia had split off from Rodinia, was located near the equator, and was surrounded by ocean. In addition, it was almost completely covered by a shallow, tropical sea.

Throughout the Cambrian Period, there was no plate tectonic activity on Laurentia. There were no collisional tectonic events, so there were no mountain ranges actively forming at that time. The edge of a continent is called a margin; when there is no tectonic activity

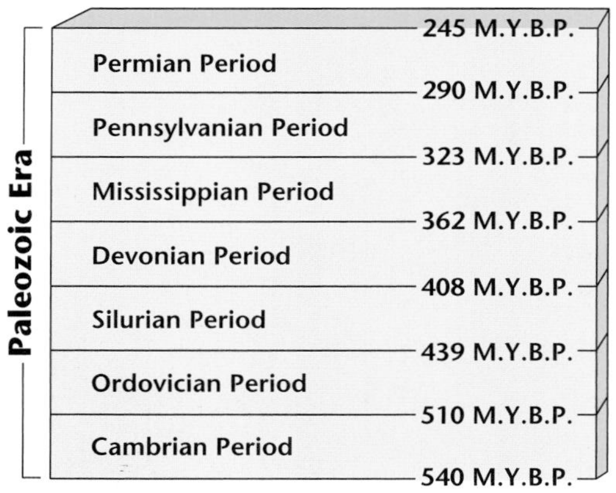

Figure 23-1 The Paleozoic Era spans 290 million years and is divided into seven periods. The rocks of the Paleozoic Era, tell of the great changes in both the tectonic setting and the life-forms that developed during that era.

along a margin, it is called a **passive margin.** Laurentia was completely surrounded by passive margins throughout the Cambrian Period.

Characteristic Sediments The paleogeographic setting of Laurentia, as shown in ***Figure 23-2,*** set the stage for a characteristic pattern of sandstone-shale-limestone deposits that represent increasing water depth from shore. The exposed land surface was flat, and the floor of the shallow sea that covered Laurentia dipped gently away from the shoreline. On land, the exposed Precambrian rocks were deeply eroded. Large, sandy beaches formed as fragments of quartz were weathered from the rocks of the Precambrian Shield and transported to Laurentia's shoreline. Clay-sized sediments were carried by strong tides and deposited in slightly deeper water. Carbonate sediment accumulated in even deeper water as organisms whose skeletons were composed of calcium carbonate died and fell to the seafloor. Over time, the sand that had been deposited on the beaches became sandstone, the clay-sized sediments compacted to form shale, and the carbonate sediment became limestone. The most famous location where the sandstone-shale-limestone deposits of the Cambrian are exposed is the Grand Canyon. This sequence is shown in ***Figure 23-3.***

Earth Sciences

3.c Students know how to explain the properties of rocks based on the physical and chemical conditions in which they formed, including plate tectonic processes.

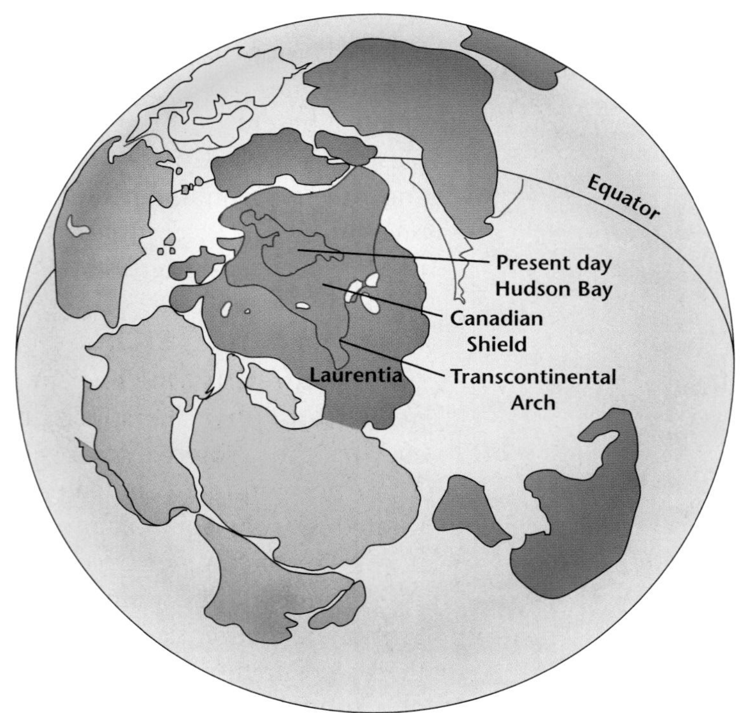

Figure 23-2 Laurentia was positioned at the equator during the Early Paleozoic. Strong, tropical storms, much like today's hurricanes, contributed to erosion and formation of sandy beaches. Corals thrived in the warm ocean waters and ultimately contributed to the formation of limestone.

Changes in Sea Level

The sandstone-shale-limestone sequence deposited during the Cambrian was a side-by-side, or lateral, sequence from shallow to deeper water. Why, then, are the layers of sandstone, shale, and limestone stacked one on top of the other in the Grand Canyon? The sediments themselves reflect the energy of the water and often, the water depth. Thus, any changes in sediments may indicate changes in sea level.

A **transgression** occurs when sea level rises and the shoreline moves further inland. Think of a sandy beach. In water slightly deeper than that at the beach, there is clay-rich sediment, and beyond this, in even deeper water, is carbonate-rich sediment. As sea level rises, the water floods inland and the shoreline moves inland. The beach is now located where dry land had been. The area of slightly deeper water where clay-sized sediments are deposited also moves shoreward on top of the old beach. The result of the transgression is deeper-water deposits overlying shallower-water deposits, as shown in ***Figure 23-4.***

A **regression** occurs when sea level falls and causes the shoreline to move seaward. This results in shallow-water deposits overlying deeper-water deposits. A stacked sequence of limestone-shale-sandstone is evidence of a regression. As sea level rises or falls, sediments that are lateral to each other become stacked one on top of another.

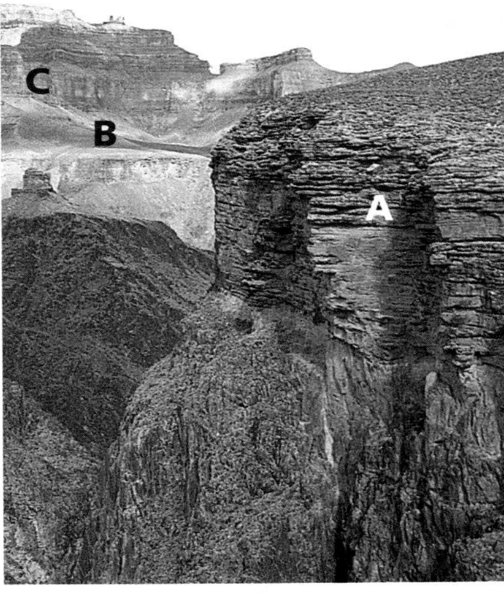

Figure 23-3 This sequence of rocks is evidence of a rise in sea level during the Cambrian. The beach environment that existed during the Cambrian is preserved in the Tapeats Sandstone **(A).** Slightly deeper water is represented by the Bright Angel Shale **(B).** The fossiliferous Muav Limestone represents the deepest, offshore environment **(C).**

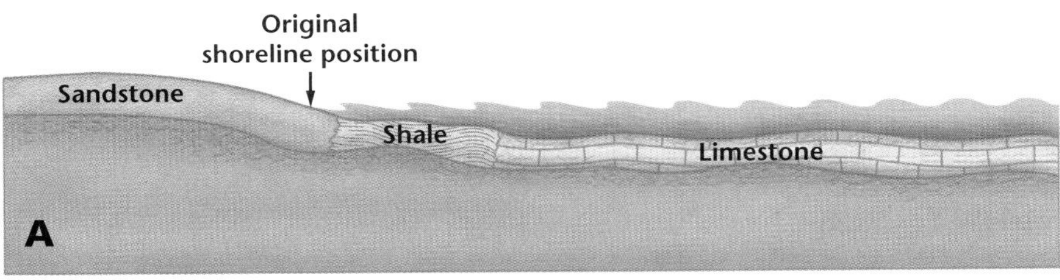

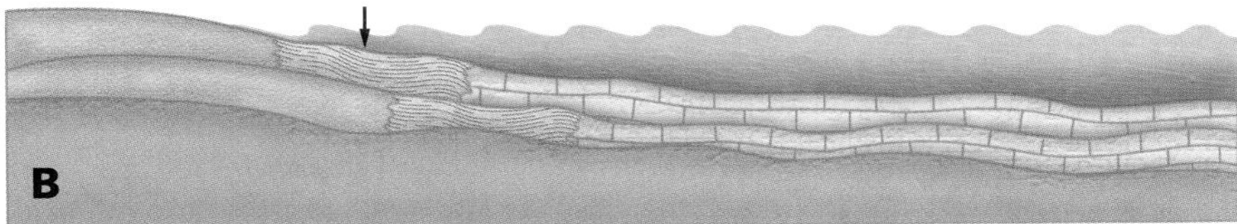

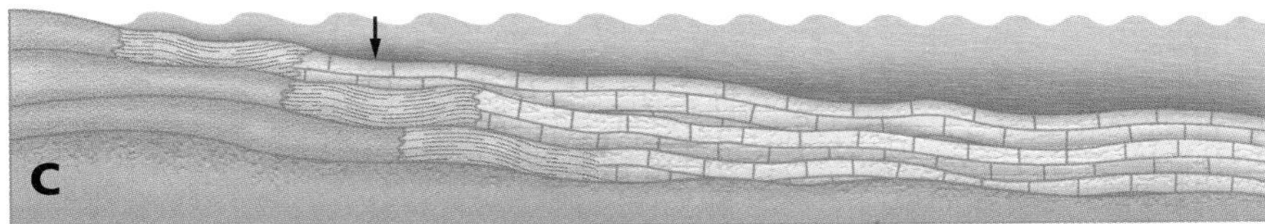

Figure 23-4 Time 1—sand is deposited at the shoreline and shale and limestone are deposited farther out **(A).** Time 2—sea level has risen, the shoreline has moved inland and shale is deposited at the original shoreline position **(B).** Time 3—sea level has risen again and limestone is deposited at the original shoreline position. The result is a sedimentary sequence of sandstone–shale–limestone **(C).**

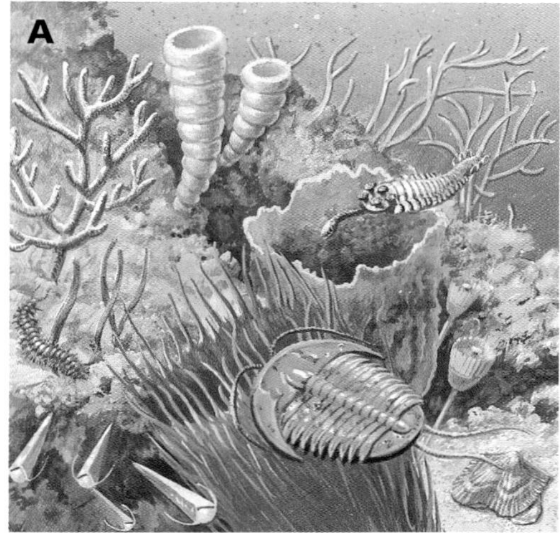

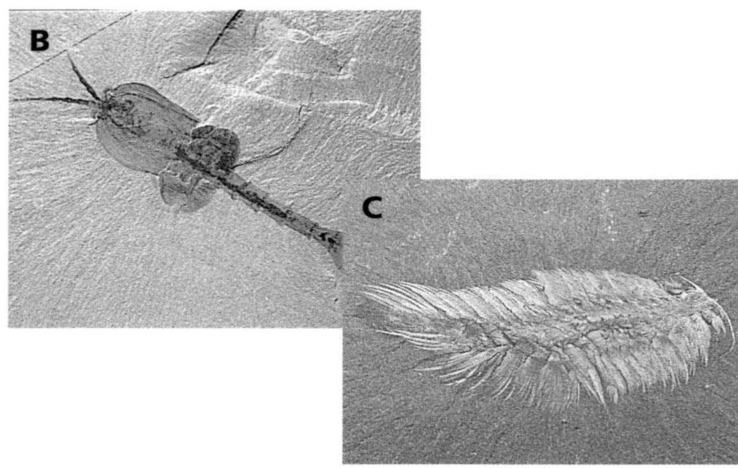

Figure 23-5 The Burgess Shale fauna were found in an isolated quarry near the town of Field, British Columbia. The unusual organisms lived a quiet life in the absence of predators **(A).** *Waptia fieldensis* was an arthropod **(B).** *Canadia spinosa* was a polychaete or bristle worm **(C).**

You have learned that these distinct sequences of rock are the result of changes in sea level. While sea level was changing throughout the Paleozoic, so were the types of plants and animals that lived in the seas.

EARLY PALEOZOIC LIFE

There was such an increase in diversity and abundance of life-forms near the beginning of the Cambrian Period that some call this the Cambrian "explosion." Organisms representing all but one of the major marine groups appeared at this time. The development of mineralized skeletons or hard parts also mark the Cambrian explosion.

Some of the best fossilized Cambrian organisms come from the **Burgess Shale,** in the Canadian Rocky Mountains. A spectacular array of soft-bodied organisms, as well as organisms with hard parts, are preserved. This deposit includes wormlike animals of unknown affinity and many other animals that do not fit into any living phylum, as shown in ***Figure 23-5.***

SECTION ASSESSMENT

1. What was the paleogeography of Laurentia during the Cambrian?
2. What is a passive margin?
3. Briefly describe the significance of the Cambrian fauna.
4. Describe how sedimentary sequences change when sea level lowers.
5. **Thinking Critically** If there were a major ice age today, explain the effect it would have on the sediments that are currently being deposited in the oceans and at sea level.

SKILL REVIEW

6. **Communicating** In your science journal, explain the relationship between the meanings of the words Cambrian *explosion* and the *explosion* of an object. For more help, refer to the Skill Handbook.

SECTION 23.2

The Middle Paleozoic

Earth Sciences 3.c

OBJECTIVES

- **Describe** *the Middle Paleozoic paleogeography.*
- **Explain** *the concept of an active margin and the formation of a clastic wedge.*
- **Describe** *the Middle Paleozoic fauna.*
- **Define** *the concept of mass extinction.*

VOCABULARY

Taconic Orogeny
Caledonian Orogeny
Acadian Orogeny
Antler Orogeny
Paleozoic fauna
vascular plant
mass extinction

The passive margin that existed around Laurentia continued into the Early Ordovician Period. The paleogeography of Laurentia, as shown in ***Figure 23-6A,*** was still equatorial, with the paleo-equator running from approximately modern-day New Mexico through Minnesota.

SEA LEVEL CHANGES AGAIN

Sea level rose during the Early Ordovician and, once again, a beach environment covered much of Laurentia's margins. The base of the rock layers that were deposited is marked throughout much of central North America by the pure quartz sand of the St. Peter Sandstone, shown in ***Figure 23-7*** on page 606. The sandstone is overlain by minor amounts of shale and by extensive limestone deposits. These limestone deposits contain a tremendously diverse array of organisms, including the first corals that built organic reefs.

Organic reefs are structures composed of carbonate skeletons made by living organisms. Corals require warm, clear, shallow water of normal marine salinity in order to thrive. For this reason, they are confined to latitudes between 30° north and south of the equator. During the Middle Ordovician, corals and a group of sponges called stromatoporoids became common and began to build reefs in the shallow sea that covered Laurentia.

Reefs affect the environments in which they grow. They commonly form in long, linear mounds parallel to islands or continents, where they absorb the energy of waves that crash against them on their oceanward side. This protects the environments behind reefs from the waves' energy. The calm areas behind reefs, are called lagoons.

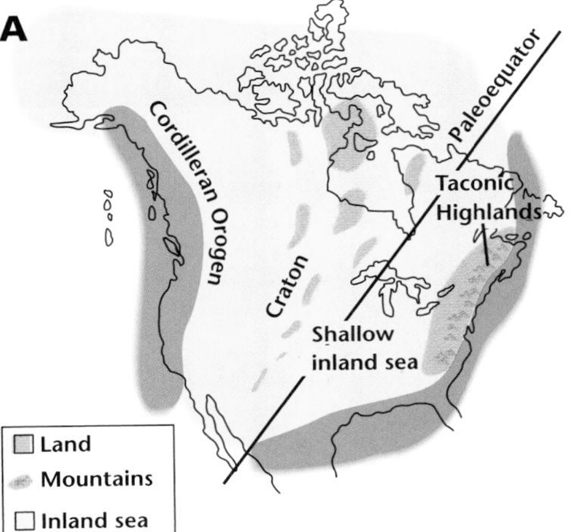

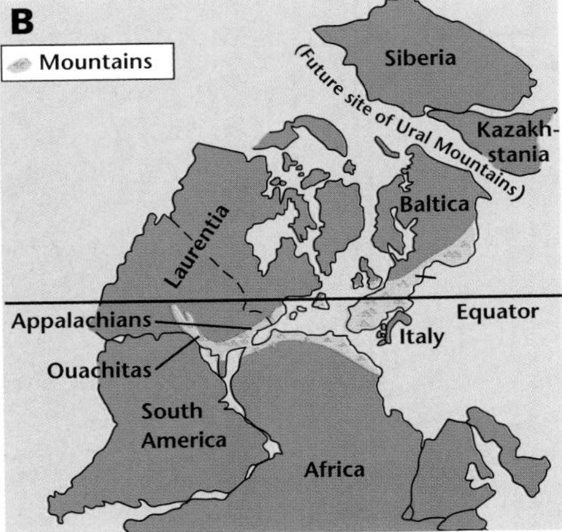

Figure 23-6 Laurentia was positioned across the equator during the Early Ordovician. The Taconic Orogeny formed the Taconic Highlands along Laurentia's eastern margin **(A)**. Orogenic activity continued during the Pennsylvanian. Mountains formed along the line of collision. All of the continents were assembled as one supercontinent, Pangaea **(B)**.

Figure 23-7 Glass is made from silica sand such as the very pure and clear St. Peter sandstone.

Here fragile organisms can thrive as shown in ***Figure 23-8.*** Reefs also can restrict water flow from the lagoon to the ocean. The shallow, quiet water in lagoons warms in the tropical sunlight and water evaporates at a high rate. When this happens, lagoon waters may become oversaturated with calcium and sodium. These elements combine with other elements or compounds in the water and precipitate out of solution as evaporite minerals such as gypsum and halite.

The Great Lakes area of North America contains huge deposits of Silurian-aged evaporite minerals that are mined commercially. Surrounding many of these evaporite deposits are reefs that almost certainly played a role in restricting the flow of water, thus making conditions right for the deposition of evaporite minerals. In the Michigan Basin, periodic restriction and deposition of evaporites resulted in deposits of halite and anhydrite more than 700 m thick! The Ohio Basin and the Appalachian Basin also were sites of evaporite deposition during the Late Silurian. ***Figure 23-9*** shows the locations of these basins. Today, these evaporite deposits are important sources of gypsum, which is used to make plaster and drywall, and halite, which is mainly used as road salt.

MIDDLE PALEOZOIC TECTONICS

During the Middle Ordovician, an ocean-continent collisional boundary, an active margin, developed in what is now eastern North America. As shown in ***Figure 23-6A,*** this resulted in a mountain building event called the **Taconic Orogeny,** named for the Taconic

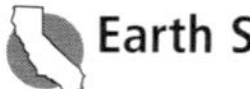

Earth Sciences

3.c Students know how to explain the properties of rocks based on the physical and chemical conditions in which they formed, including plate tectonic processes.

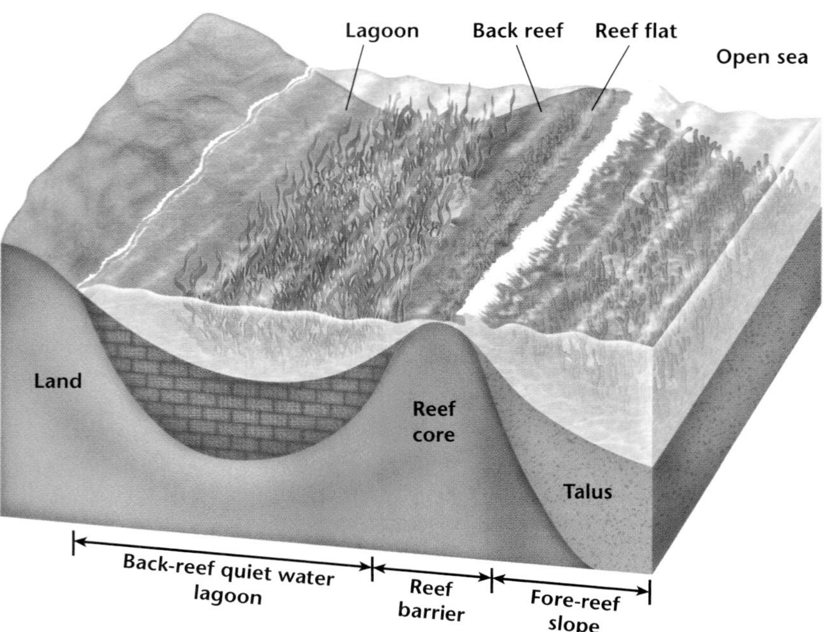

Figure 23-8 The sheltered, quiet water behind a coral reef is an ideal place for fragile organisms to thrive.

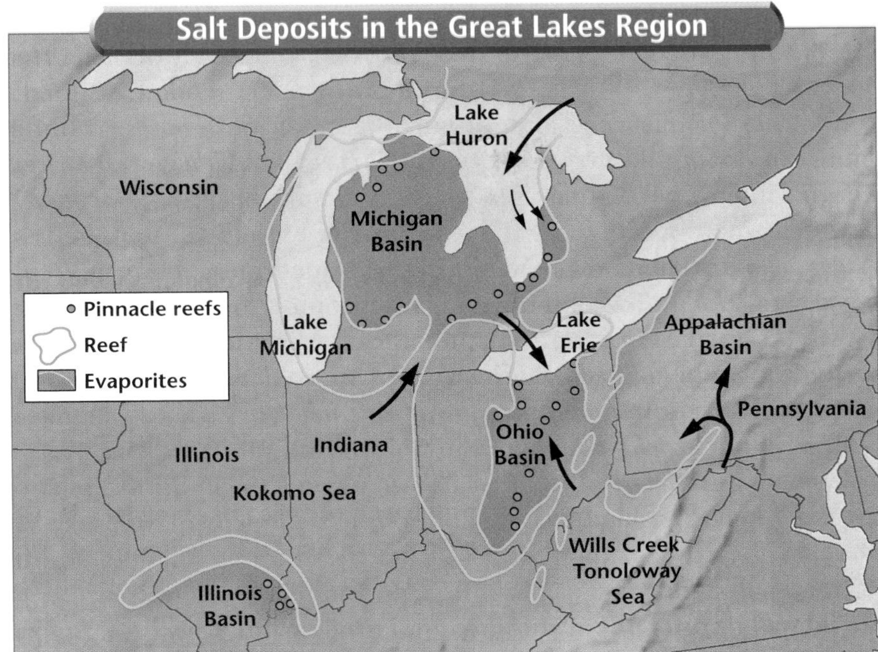

Figure 23-9 The barrier reefs that grew within these basins most likely restricted the flow of water from time to time. This, coupled with increased evaporation, resulted in the formation of halite and other evaporite minerals.

Mountains of eastern New York State. A series of volcanic islands also formed as a result of this mountain building event.

How do we know that this Taconic Orogeny occurred? Geologists look for clues and evidence in the rocks like detectives at a crime scene. One form of evidence is angular unconformities. Rocks that are older than the Middle Ordovician of eastern New York are tilted at a different angle than those of the younger Silurian rocks. This means that something, such as a tectonic collision, tilted the rocks after the Middle Ordovician but before the Silurian. Volcanic activity also provides evidence of the Taconic Orogeny. Lava flows and volcanic ash deposits that are Middle-to-Late Ordovician in age exist in present-day eastern North America. There are also igneous intrusions and regional metamorphic features that have been radiometrically dated as being between 480 and 440 million years old. The final piece of evidence of the Taconic Orogeny is the wedge of sediment, some of which is shown in ***Figure 23-10,*** that formed as the mountains from the Taconic Orogeny eroded. In a cross section, such a deposit has a triangular wedge shape, and is called a clastic wedge.

Orogenies and Deformation Tectonism continued during the Late Silurian and into the Devonian when Laurentia collided with Baltica, an ancient continent that consisted of what are now northern Europe and Russia west of the Ural Mountains. This collision joined Laurentia and Baltica into a larger continent known as Laurasia, pictured in ***Figure 23-6B*** on page 605.

Figure 23-10 Erosion and transport of material from rising mountains to the east resulted in the formation of these coarse-grained, layered deposits. They are part of a large clastic wedge called the Queenston Delta and are exposed in Genesee Gorge in Letchworth State Park, NY.

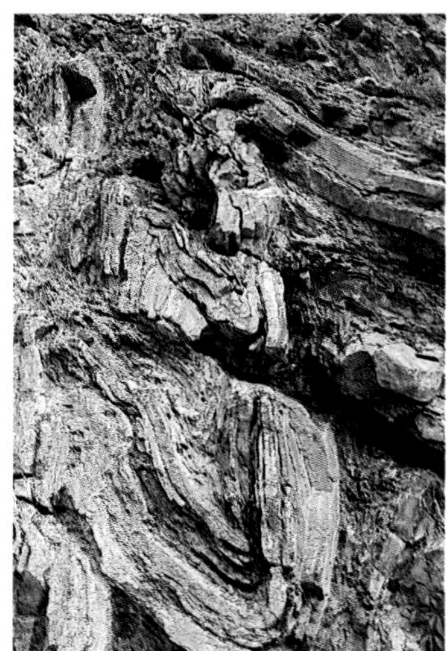

Figure 23-11 These rocks were deformed and faulted during the collisional events of the Taconic and Acadian Orogenies.

This collisional tectonic event, known as the **Caledonian Orogeny,** closed the ocean that had separated these two continents. Shortly after Baltica collided with Laurentia, a microcontinent called Avalonia, which is now Newfoundland, collided with the southeastern margin of Laurasia. This tectonic event, known as the **Acadian Orogeny,** affected the same general area as the Ordovician Taconic Orogeny. The deformation caused by the Acadian Orogeny added folds, faults, and igneous intrusions to the already deformed rocks of the Taconic Orogeny. ***Figure 23-11*** shows some of the complex folds that resulted. During the Late Devonian and into the early Mississippian, the passive western margin of Laurentia was affected by a collisional tectonic event called the **Antler Orogeny.** It was most likely caused by a microcontinent or island arc that collided with what is now western North America, as shown in ***Figure 23-12.*** As you will learn in the next chapter, one of the problems with interpreting the Paleozoic geology of western North America is that it has been complicated by numerous younger orogenic events.

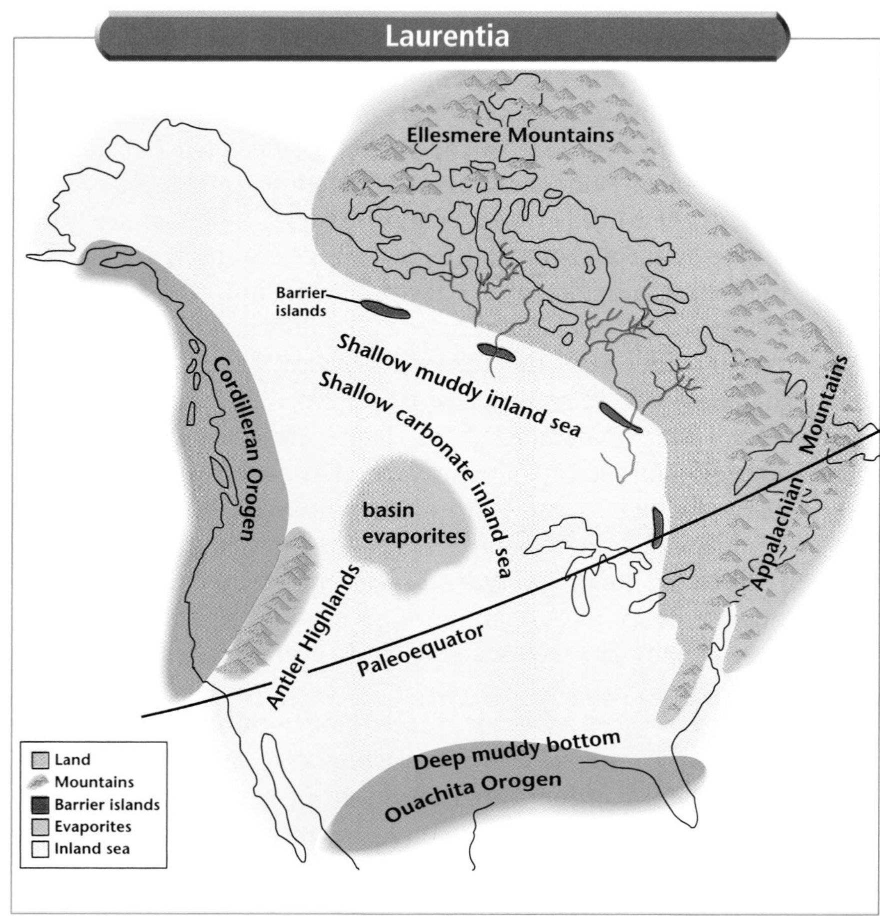

Figure 23-12 The mountains of the Antler Orogeny bordered an inland sea to the east. The collision of a microcontinent with Laurentia is thought to be the cause of the orogeny.

Middle Paleozoic Life

The Middle Paleozoic seas were dominated by animals that are collectively called the **Paleozoic fauna.** The animals that dominated the Cambrian seas were replaced during the Early Ordovician by a variety of new organisms. Representatives of the Paleozoic fauna are shown in ***Figure 23-13.***

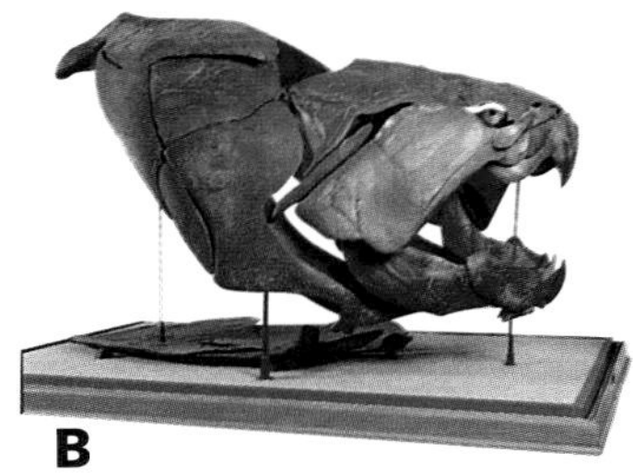

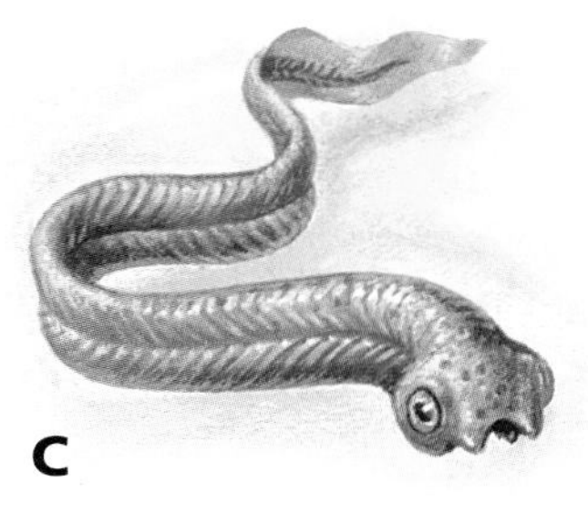

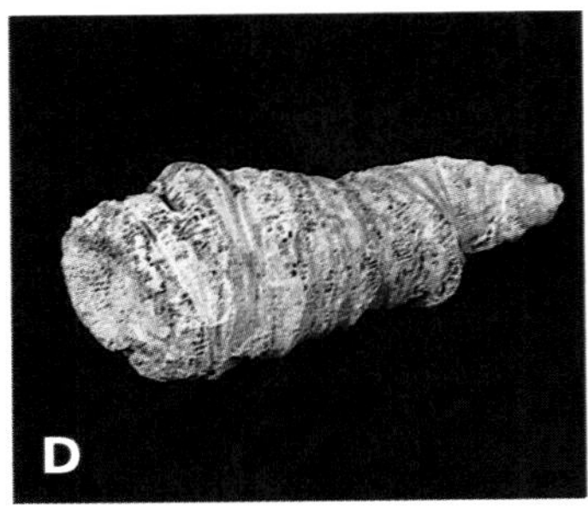

Figure 23-13 Graptolites lived in shallow, marine waters **(A).** The Early Paleozoic fishes had internal skeletons of cartilage. This *Dunkleosteus,* from Ohio was a top predator in the Devonian Seas **(B).** Conodonts were eel-like swimmers whose only hard-parts were tooth-like structures. They are excellent index fossils because they are widespread, existed during a short period of time, and have distinct characteristics **(C).** The growth lines on this Devonian horn coral, *Heliophyllum,* help determine the length of a Devonian year **(D).**

Biologists have observed that modern corals deposit thin layers of carbonate, called growth lines, to their skeletons each day. Seasonal variations in length of day cause yearly changes in the spacing of these growth lines. The Devonian horn coral, *Heliophyllum,* pictured in ***Figure 23-13D,*** displays these types of growth lines. Based upon the spacing and number of growth lines in a cycle, paleontologists have deduced that, during the Devonian, the days were about 22 hours long and there were about 400 days in a year. The corals thus indicate that Earth was rotating more rapidly during the Paleozoic and, like a top, has been slowing ever since. You will calculate the lengths of days of other periods in the *Problem-Solving Lab* on page 610.

Life Moves to Land

The seas were not the only places where life was developing during the Middle Paleozoic. For the first time in the history of life on Earth, land was colonized. Fossilized plant parts and spores suggest that some form of land plant existed during the Late Ordovician. **Vascular plants** have tissue for circulating water and nutrients through their stems and leaves. It was the development of this tissue that allowed plants to spread out on land. Fossils that undisputedly represent vascular plants are found in Early Silurian rocks. These plants were small, leafless, grew along the ground, and reproduced by spores.

New Plants Emerge By the Late Devonian, three important groups of vascular spore-bearing plants were living on land. One group, the ferns, still thrive today. The second group, the sphenopsids, were joint-stemmed plants. A modern species of sphenopsid is the horsetail or scouring rush, shown in ***Figure 23-14,*** whose common name is derived from its use by early plains settlers to scour their dishes. The last group is the lycopods. The leaves of the lycopods grew directly from their stems, leaving diamond-shaped scars on the fossilized stems.

Figure 23-14 This modern horsetail **(A)** and this fossil horsetail from Italy **(B)** exhibit strikingly similar features such as the individual joints and clusters of leaves along the stem. This fossil fern is from Piesberg, Germany **(C)**.

During the Late Devonian, the seed ferns developed. The most important aspect of these plants was the development of seeds. Seeds contain their own moisture and food source. They are covered by a hard coating that prevents drying and protects them. Seeds allowed plants to spread out and colonize dry land. Until the development of the seed, plants required water to complete their reproductive cycles. Thus, with the development of seeds, the surface of the continents was changed forever.

MASS EXTINCTIONS

Two of the greatest extinction events in Earth's history occurred during the Middle Paleozoic. When an unusually large number of organisms become extinct over a relatively short period of geologic time, it

Problem-Solving Lab

Interpreting Graphs

Determine the length of a Paleozoic day Geologists study the shapes and compositions of fossil organisms to interpret how and in what types of environments they lived. Fossils can even be used to interpret climatic changes. For example, rugose corals exhibit lines on their shells that represent yearly growth cycles. Ultimately, paleontological data such as this can be used to interpret the past and to predict future events. Use the figure at the right to answer the following questions.

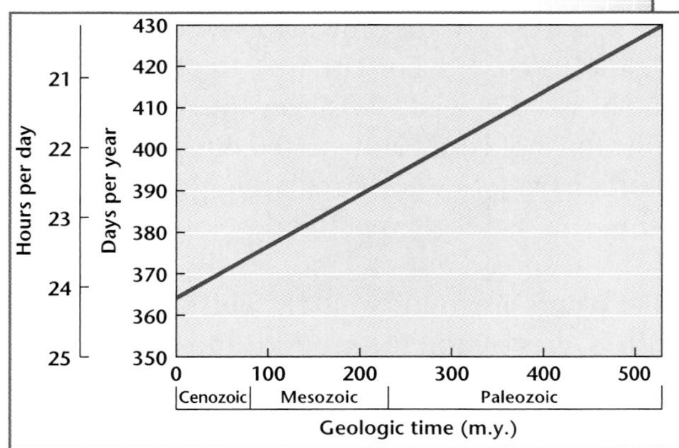

Analysis

1. How many hours were in a day 400 million years ago?
2. How many hours were in a day 200 million years ago?

Thinking Critically

3. When will there be 24.5 hours in a day?

is called a **mass extinction.** The first mass extinction occurred at the end of the Ordovician Period. Approximately 57 percent of all marine genera became extinct. Those that survived suffered large but not complete losses in their numbers of species and genera.

What caused the Late Ordovician mass extinction? One clue is that there are Late Ordovician glacial deposits preserved in what is now northern Africa. Another clue is that species that were adapted to warm environments were the hardest hit. Geologists hypothesize that an episode of global cooling, plus the rapid lowering of sea level that accompanied the glaciations, caused the extinction of many species that were adapted to warm environments.

Update To find out more about present-day extinctions, visit the Earth Science Web Site at earthgeu.com

Global Cooling and Overturning A second mass extinction occurred during the late Devonian when approximately 50 percent of the marine genera were wiped out, including many species of corals and stromatoporoids. Again, the polar communities were not severely affected. Was global cooling again the cause? There is evidence that glaciation occurred on some continents at this time. The cooling of Earth's atmosphere may have caused a tremendous disturbance in the ocean system. This created overturning, a process in which oxygen-poor, deep ocean water rises up, creating surface waters that contain little or no oxygen.

Evidence of overturning exists in the form of black shale. Black sediment indicates a lack of oxygen. Black shale deposits from the Late Devonian are common worldwide, and therefore, suggests that the seawater at that time contained reduced amounts of oxygen. The lack of oxygen in the water stressed the organisms to the point at which many species became extinct.

Section Assessment

1. Where was Laurentia positioned on Earth during the Middle Paleozoic?
2. Explain how evaporite deposits formed during the Middle Paleozoic.
3. What kinds of evidence do geologists use to determine whether an area was subjected to an orogeny?
4. What were the common causes of the Late Ordovician and Late Devonian mass extinctions?
5. **Thinking Critically** In Chapter 19, you learned about different kinds of faults. What kinds of faults would you expect to have been associated with the Acadian Orogeny? Why?

Skill Review

6. **Recognizing Cause and Effect** Explain how the appearance of land plants led to the appearance of land animals.

earthgeu.com/self_check_quiz

SECTION 23.3 The Late Paleozoic

Earth Sciences 3.c I&E 1.d, 1.l

OBJECTIVES

- **Describe** *the formation of Pangaea.*
- **Explain** *how cyclothems formed.*
- **Identify** *the importance of amniote eggs.*
- **Discuss** *the causes of the Late Permian mass extinction.*

VOCABULARY

Gondwana
cyclothem
Ouachita Orogeny
Ancestral Rockies
Alleghenian Orogeny
amniote egg

During the Late Paleozoic, the supercontinent Pangaea formed. In the previous section, you learned that Laurentia, Baltica, and the microcontinent Avalonia were joined together by the end of the Devonian. ***Figure 23-15*** shows how, as a result of similar collisional events, South America, Africa, India, and Antarctica joined to form the large continent in the southern hemisphere called **Gondwana.**

SEA LEVEL AND DEPOSITION

The Late Paleozoic began with Laurasia still covered by a shallow tropical sea. This setting provided the perfect environment for the deposition of carbonate sediment. Thus, Mississippian rocks throughout North America are predominantly limestone. The end of the Mississippian was marked by a major regression of the sea.

The Pennsylvanian Period began with a slow transgression. The sediments that were deposited have quite different characteristics from those of the underlying Mississippian carbonates. The Pennsylvanian and Permian rocks in central and eastern North America are predominantly river and delta deposits.

The Pennsylvanian Period is known for the coal deposits that accumulated in heavily vegetated lowland swamps. As you have learned, coal forms through the compaction of plant material. Why

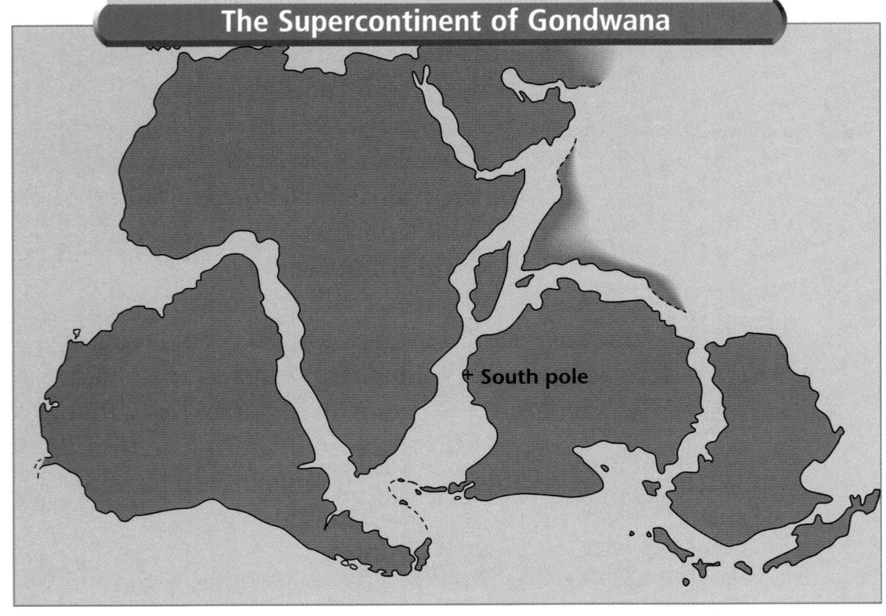

Figure 23-15 The mountain building and subsequent erosion that occurred as Gondwana formed resulted in the formation of clastic wedges in front of the mountains. The mountain building events are sometimes referred to as the Gondwanan Orogeny.

are coal deposits not common in older rocks? The answer lies in the development of plant life on land. Not enough plant material had accumulated on land to form significant coal deposits until the Mississippian. Because a shallow sea covered Laurasia during the Mississippian, and because coal deposits generally form on land, there are few Mississippian-aged coal deposits in North America.

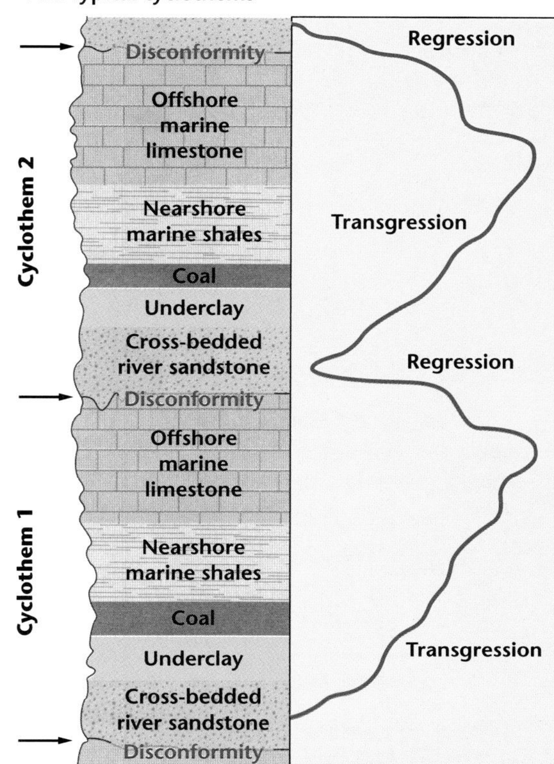

Figure 23-16 The general rock sequence of a cyclothem from base to top, is sandstone-clay-coal-shale-limestone.

Cyclothems Pennsylvanian rocks in North America display a repeating cyclic pattern of sediments stacked one on top of another. Such a sequence as pictured in ***Figure 23-16*** is called a **cyclothem.** Cyclothems record two changes in sea level. A regression is represented by layers of marine limestone, and sandstone. A transgression is represented by sandstone, clay, coal, shale and marine limestone. In some parts of Laurentia, as many as 40 to 50 of these sequences were stacked one on top of the other. What could have caused the seas to transgress and regress 50 times? Most geologists hypothesize that glaciation was responsible. Glaciers grow by the accumulation of snow. Most of the water that freezes into snow comes from the evaporation of seawater. Thus, when glaciers are growing, sea level is falling. Gondwana was located in part over the south pole during the Pennsylvanian and large, glacial ice sheets developed on land. It is likely that glaciers grew during cooler periods during the Pennsylvanian and caused sea level to drop, which, in turn, caused a regression. The glaciers partially melted during warmer periods and caused sea level to rise, which resulted in a transgression. In this way, numerous cyclothems were deposited one on top of the other.

Reefs and Evaporites

During the Permian, much of what is now eastern North America was dry land and a shallow sea covered most of what is now western North America. A spectacular fossilized barrier reef complex was formed during the Permian in what is now western Texas, southeastern New Mexico, and northern Mexico. Sponges and algae built this reef complex, known as the Great Permian Reef Complex, the remains of which are shown in ***Figure 23-17*** on page 614. Because these barrier reefs restricted the flow of water, large evaporite deposits formed behind the reefs during the Middle and Late Permian.

Earth Sciences

3.c Students know how to explain the properties of rocks based on the physical and chemical conditions in which they formed, including plate tectonic processes.

Figure 23-17 The best exposure of the Great Permian Reef Complex is at Guadalupe Mountains National Park in West Texas.

ENVIRONMENTAL CONNECTION

These evaporites and the Great Permian Reef Complex are very important deposits. As you learned in the *Discovery Lab* at the beginning of this chapter, pore spaces in rocks and sediment can be filled with oil or water. The pore spaces in Permian reefs and in the surrounding coarse-grained rocks are filled with oil making these reefs important oil reservoirs. The evaporites associated with these reefs are important for a different reason. The thick salt deposits have very low permeability and, thus, virtually no liquids move through them. This makes them an excellent environment for long-term storage of nuclear waste. In the thick salt deposits in southeastern New Mexico, a large facility has been built for the disposal of various kinds of nuclear waste.

Figure 23-18 These rocks are part of the Fountain Formation, a unit whose sediments were derived from erosion of Pennsylvanian-aged arkoses. As a result of continued faulting, the layers are standing literally on end.

Continental Collisions and Mountain Building

The Late Paleozoic was a time of active mountain building. Gondwana collided with the southeastern margin of Laurasia during the **Ouachita Orogeny.** The Ouachita Orogeny formed the Ouachita Mountains of Arkansas and Oklahoma and the mountains in the Marathon area of West Texas.

The collision between Gondwana and Laurasia was so intense that it caused the crust to uplift inland as far as present-day Colorado. Geologists call the mountain range that formed the **Ancestral Rockies.** Large vertical faults lifted the Paleozoic rocks more than 2 km. The uplifted rocks eventually eroded and were deposited as red sandstone beds. The most famous of these deposits are the beautiful exposures at Garden of the Gods in Colorado Springs, shown in ***Figure 23-18.***

As the collision between the two supercontinents continued, it closed the ocean that separated Gondwana from Laurasia. The part of Gondwana that is now Africa began to collide with Laurasia during the Late Pennsylvanian, causing the **Alleghenian Orogeny.**

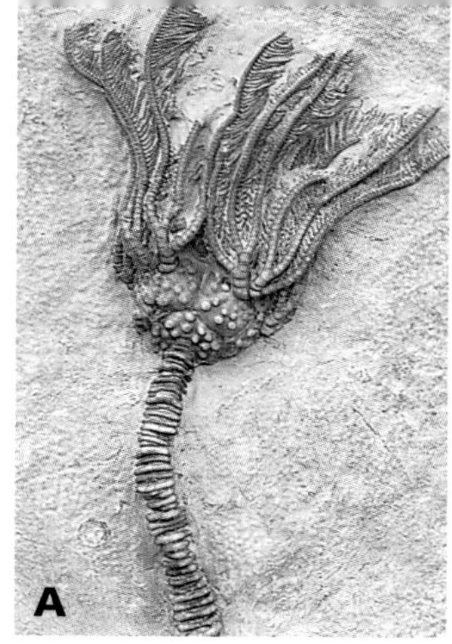

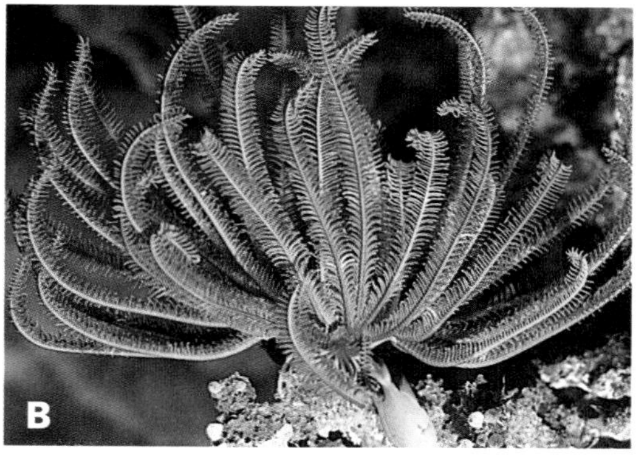

Figure 23-19 This fossil crinoid is from the Silurian **(A).** Living crinoids feed by spreading out their arms to catch and filter food particles from the water and pass them down to their mouths through grooves in their arms **(B).**

It continued through the Permian along the southeastern margin of Laurasia. The Alleghenian Orogeny was the last of the three major mountain-building events to affect what is now eastern North America and resulted in the formation of the Appalachian Mountains. Pangaea had formed. This is summarized in ***Figure 23-6B*** on page 605.

LATE PALEOZOIC LIFE

The invertebrate marine organisms that flourished during the Middle Paleozoic continued to dominate the marine environment. One group, the crinoids, shown in ***Figures 23-19A*** and ***B,*** became dominant during the Mississippian. Crinoids were so abundant that some Mississippian limestones are made almost entirely of fossilized crinoid stem fragments.

Changes in the Ocean Following the extinction of many fishes during the Late Devonian, sharks and other ray-finned fishes, some of which still exist today, became more abundant. Lobe-finned fishes appeared during the Late Devonian, but became more abundant during the Late Paleozoic. Their club-shaped fins were supported by thick bones that made it possible for some of these fishes to walk on land. Of the three groups of lobe-finned fishes, two of them, the lungfishes and the coelacanths, are still living today. The third group of lobe-finned fishes gave rise to the amphibians.

On land, the true ferns, sphenopsids, lycopods, and seed ferns that appeared during the Late Devonian had thoroughly

Figure 23-20 A reconstruction of the life-forms on land during the Late Paleozoic.

MiniLab

Collisions and Shelves

Model the difference in continental shelf area between individual continents and one supercontinent.

Procedure

1. Using 250 g of modeling clay, make a sphere and flatten it into a disk that is 1/2 cm thick. This represents a craton.
2. Divide another 250 g of clay into 2 equal spheres and flatten them as above.
3. Roll 250 g of modeling clay into 3 long cylinders with 1/2 cm diameters. Wrap the cylinders around the edges of the clay disks. These represent continental shelves.
4. Using the formula area = πr^2, calculate the area of the large craton and the area of the large craton plus the continental shelf. Subtract the craton area from the total area. This equals the area of the continental shelf.
5. Repeat step 4 for each of the small models.

Analyze and Conclude

1. Which has more shelf area, two small continents or one large continent? Why?
2. Tropical oceans contain the greatest diversity of animals. If there is only one supercontinent how does this further limit the amount of habitat space?
3. Explain how reduced habitat space, Pangaea, and the mass extinction at the end of the Permian are related.

invaded the low, swampy land of the Pennsylvanian as illustrated in ***Figure 23-20.*** The warm, wet lowlands that developed at the base of the newly formed Appalachian Mountains was a perfect environment for plants to grow in and ultimately gave rise to the great coal swamps of the Pennsylvanian. They were a breeding ground for insects as well as plants. In fact, the largest insects that ever lived were preserved in coal swamps, including cockroaches that reached more than 10 cm in length and centipedes that were longer than 30 cm. The largest of all was a dragonfly with a 74-cm wingspan.

The coal swamps also housed amphibians. Although they were most likely sluggish and ate only fish, insects, vegetation, and smaller amphibians, they were the top carnivores on the land. Amphibians have one major limitation: they must lay their eggs in water because their eggs have no protective coating to prevent them from drying out on land. Thus, although early amphibians could colonize the land, they had to remain close to a source of water. Sometime during the Late Mississippian, this dependence on water was overcome. Reptiles that evolved from the early amphibians developed a new type of egg, the **amniote egg.** It has a shell that protects the embryo, which is surrounded by a liquid-filled sac that contains a food sac and a waste sac. Early reptiles were small, agile, and had more advanced teeth and jaws than amphibians. During the Permian, the more rapidly moving and stronger reptiles became abundant on land and displaced the amphibians as the top land carnivores.

The Permian Mass Extinction

The largest mass extinction in the history of life on Earth defines the end of the Paleozoic Era. It is called the Permo-Triassic Extinction Event and it marks the end of nearly 95 percent of all species including trilobites. You will learn more about this event in the *Science & the Environment* feature at the end of this chapter.

In contrast to the mass extinctions of the Middle Paleozoic, this extinction did not affect only the marine animals. Many of the terrestrial organisms that thrived in the swamps of the Late Paleozoic became extinct as well. More than 65 percent of the amphibians and reptiles did not survive, nor did almost one third of all insects. What could have caused such a widespread catastrophe?

Calculating Percentages Is Earth currently in the middle of a mass extinction? If ten million species exist today, and 5.5 species extinctions occur every day, calculate how many years it would take for 96 percent of the species living today to become extinct.

A Major Marine Regression One possible cause of the Permo-Triassic Extinction Event was a major marine regression. Now, as then, most marine animals live in the relatively shallow water called the continental shelf. When sea level is high, the entire continental shelf is flooded, and there is quite a large area where marine animals can live. During regression, however, the continental shelf can become very narrow, leaving little space on the continental shelf for marine animals to live. This situation would have been particularly critical when there was only one continent. You learned about the consequences of reduced continental shelf space in the *MiniLab* on the previous page.

The end of the Paleozoic was a stressful time for life on Earth. Marine habitats shrank as a result of the major regression that occurred around Pangaea. Whether additional stresses were acting on life at that time is not known. What is known, however, is that regardless of the ultimate cause, life on Earth was irrevocably changed at the end of the Paleozoic.

Section Assessment

1. Explain the geologic events that resulted in the formation of Pangaea.
2. What is the relationship between cyclothems and glaciation?
3. What do scientists hypothesize were the main causes of the Late Permian mass extinction?
4. **Thinking Critically** In the previous section, you learned about the development of the seed. In this section, you learned about the development of the amniote egg. Explain the similarities between these two important developments in the history of life on Earth.

Skill Review

5. **Concept Mapping** Reorganize the events below into an events chain that describes the formation of Pangaea. Ouachita Orogeny occurs; Ancestral Rockies uplifted; North America joined to Gondwana; Pangaea formed; Part of Gondwana collides with eastern Laurasia; Alleghenian Orogeny occurs; Part of Gondwana collides with SE Laurasia.

Symmetry, Shape and Shells

 I&E

1.d Formulate explanations by using logic and evidence.

Brachiopods and bivalves have been present in Earth's oceans since the Cambrian. Both have two shells and live in marine environments. But the similarity ends there. How can you tell the two apart? Oysters are bivalves that are known for the pearls they secrete inside their shells. Can you distinguish an oyster from a brachiopod? If you were searching for pearls, you would want to know how!

Preparation

Problem

Distinguish between brachiopods and bivalves and interpret the environment where a brachiopod lived based on its shell.

Materials

fossil brachiopods (4), each belonging to a different species
fossil bivalves (4), each belonging to a different species
paper
pencil

Objectives

In this GeoLab you will:

- **Determine** if a fossil is a brachiopod or a bivalve.
- **Describe** the symmetry of fossil brachiopods and bivalves.
- **Infer** the environment in which different fossil brachiopods lived.

Safety Precautions

Always wear safety goggles and an apron in the lab.

Fossil Data		
Specimen	**Brachiopod or Bivalve**	**Deeper water, low energy environment or shallow water, high-energy environment**
1		
2		
3		

Procedure

1. Design a data table like the one above. You may choose to add more columns to record additional data.
2. Examine the fossils provided by your teacher.
3. Determine where the plane of symmetry is for each specimen. An organism that can be divided into two nearly identical halves has bilateral symmetry.
4. Identify the specimens as brachiopods or bivalves based on their symmetry and record this in your data table. Brachiopod symmetry runs across both shells. Bivalve symmetry runs between the shells.
5. Divide the brachiopods into two groups based on whether you think they lived in a deeper water, low energy environment, or in a shallow water, high-energy environment. Record this in your data table.

Analyze

1. **Interpreting Observations** Explain how symmetry is useful in determining whether a fossil is a brachiopod or a bivalve.
2. **Applying and Interpreting** If you only had one shell, how could you determine if it was the shell of a brachiopod or the shell of a bivalve?
3. **Comparing and Contrasting** Explain the similarities and differences between a streamlined auto and a smooth brachiopod, in terms of their place in wind or water.

Conclude & Apply

1. What principle did you use to determine the environment in which the fossil brachiopods lived? Explain.
2. Hypothesize about the reasons for the different shell types for brachiopods that live in different environments.
3. All living brachiopods pump water through their shells and filter organic particles out of that water to feed. Some brachiopod shells close along a straight line, whereas others close along a zig-zag line. What is the benefit of having a zig-zag opening for a filter feeding brachiopod?

Science & the Environment

I&E 1.l

Trilobite

Mass Extinctions

Most organisms that have ever lived on Earth are now extinct. For shallow-water, ocean-dwelling species, the average lifetime of a species is about 4 million years. That works out to an average of two or three species becoming extinct every year. The fossil record also indicates that, at certain times, many species became extinct over a short time.

The extinction of the dinosaurs and many other forms of life at the end of the Cretaceous Period has received much attention. For sheer numbers, though, these extinctions can't compare with the extinctions that ended the Paleozoic Era.

The Big Extinction Event

At the end of the Mesozoic, 65 to 75 percent of all species on Earth became extinct. When the Paleozoic ended, as many as 95 percent of all species became extinct. This event is called the Permo-Triassic Extinction Event.

Some marine organisms that became extinct during this Extinction Event were fusilinid foraminifera, rugose corals, and trilobites. On land, therapsids or mammal-like reptiles, became extinct, and insects suffered their only major extinction in their 390 million year history.

Paleontologists once thought that the Permo-Triassic extinctions took place over millions of years. However, recent research indicates that the extinctions were much more rapid. The causes are not well understood, but the extinctions appear to be the result of a fatal combination of changes in sea level, reduced oxygen levels, massive volcanic eruptions and global warming.

Opportunities Lost and Gained

Although the biosphere took millions of years to recover from each mass extinction, such extinctions are not entirely negative. The Permo-Triassic Extinction Event cleared the way for new kinds of life, including the dinosaurs. The extinction at the end of the Mesozoic cleared the way for mammals to dominate. Without those extinctions, humans might not be here.

Many scientists believe that we are currently in the midst of another mass extinction. Over a thousand species of animals, and probably even more species of plants, are in danger of becoming extinct within our lifetime. This extinction is not caused by a regression or climatic changes, but by the impact of human activities.

Drugs to fight everything from AIDS to cancer have been found in organisms ranging from sponges to trees. Plants might be discovered that can be used to feed the world's hungry people. If a species becomes extinct, we lose the possibility of discovering any potential uses it might have had.

Activity

Go to earthgeu.com to find links to more information about the use of drugs derived from plants or animals to fight disease. Summarize where the drug is found, whether the plant or animal is in danger of becoming extinct, and how it is used to fight disease in humans.

CHAPTER 23
Study Guide

Summary

SECTION 23.1

The Early Paleozoic

Main Ideas

- The ancient North American continent of Laurentia was located near the equator and surrounded by ocean during the Cambrian Period. A shallow sea covered most of Laurentia.
- Laurentia was completely surrounded by passive margins throughout the entire Cambrian Period.
- Many new organisms developed during the Cambrian explosion. Fossils of trilobites and articulate brachiopods are particularly common in Cambrian rocks.
- When environments change position laterally due to changes in sea level, adjacent depositional facies overlie each other in vertical succession.

Vocabulary

Burgess Shale (p. 604)
paleogeography (p. 601)
passive margin (p. 602)
regression (p. 603)
transgression (p. 603)

SECTION 23.2

The Middle Paleozoic

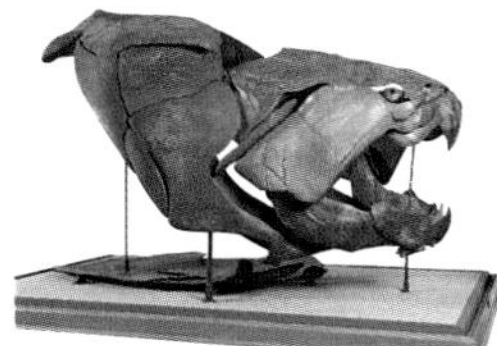

Main Ideas

- High evaporation rates in lagoon settings cause the water to become oversaturated with calcium and sodium. These elements combine with other elements or compounds in the water and precipitate out of solution as the evaporite minerals.
- Clastic wedges provide evidence for orogenic events.
- The seas were dominated by articulate brachiopods, corals, mollusks, bryozoans, crinoids, graptolites and conodonts. Fishes were the top predators of the seas during the Devonian. Ferns, sphenopsids, and lycopods covered the landscape by the Late Devonian.
- Two mass extinctions occurred. A mass extinction occurs when an unusually large number of organisms become extinct over a relatively short period of geologic time.

Vocabulary

Acadian Orogeny (p. 608)
Antler Orogeny (p. 608)
Caledonian Orogeny (p. 608)
mass extinction (p. 611)
Paleozoic fauna (p. 609)
Taconic Orogeny (p. 606)
vascular plant (p. 609)

SECTION 23.3

The Late Paleozoic

Main Ideas

- Pangaea formed as Laurasia (North America + Europe) collided with Gondwana (South America, Africa, India, Australia and Antarctica).
- Cyclothems consist of transgressive and regressive rock sequences stacked one on top of another. They represent cycles of glacial-interglacial periods.
- Seeds evolved and allowed plants to colonize dry land. The amniote egg evolved and allowed reptiles to colonize dry land.
- One possible cause of the Permo-Triassic Extinction Event was regression. Marine habitats around Pangaea shrank. Feedback from this event caused global warming, which affected organisms on land as well as in the sea.

Vocabulary

Alleghenian Orogeny (p. 614)
amniote egg (p. 616)
Ancestral Rockies (p. 614)
cyclothem (p. 613)
Gondwana (p. 612)
Ouachita Orogeny (p. 614)

earthgeu.com/vocabulary_puzzlemaker

CHAPTER 23

Assessment

Understanding Main Ideas

1. Where was Laurentia located during the Paleozoic?
   a. over the south pole
   b. over the north pole
   c. over the equator
   d. where it is today

2. During what period was Laurentia completely surrounded by passive margins?
   a. the Cambrian
   b. the Ordovician
   c. the Devonian
   d. the Permian

3. What term describes a drop in sea level and the resulting seaward movement of the shoreline?
   a. regression
   b. excursion
   c. explosion
   d. transgression

4. What are the most common fossils of the Cambrian fauna?
   a. sharks
   b. trilobites
   c. crinoids
   d. rugose corals

5. What is the calm area behind a reef called?
   a. a clastic wedge
   b. a passive margin
   c. a lagoon
   d. a continental shelf

6. Which of the following is not an evaporite mineral?
   a. anhydrite
   b. quartz
   c. halite
   d. gypsum

7. What evaporite mineral is used to make plaster and drywall?
   a. gypsum
   b. aragonite
   c. halite
   d. quartz

8. What is a triangular-shaped sedimentary rock deposit that was deposited adjacent to an uplifted area called?
   a. a passive margin
   b. an active margin
   c. a clastic wedge
   d. an orogeny

9. What landmass collided with Laurentia and caused the Acadian Orogeny?
   a. Gondwana
   b. Avalonia
   c. Baltica
   d. Siberia

10. What group of fishes were able to walk on land?
    a. ray-finned fishes
    b. sharks
    c. lobe-finned fishes
    d. Dunkleosteus

11. What group of fishes gave rise to the amphibians?
    a. acanthodians
    b. sharks
    c. lobe-finned fishes
    d. Dunkleosteus

Applying Main Ideas

12. What is special about the fossils of the Burgess Shale?

Use the figure below to answer question 13.

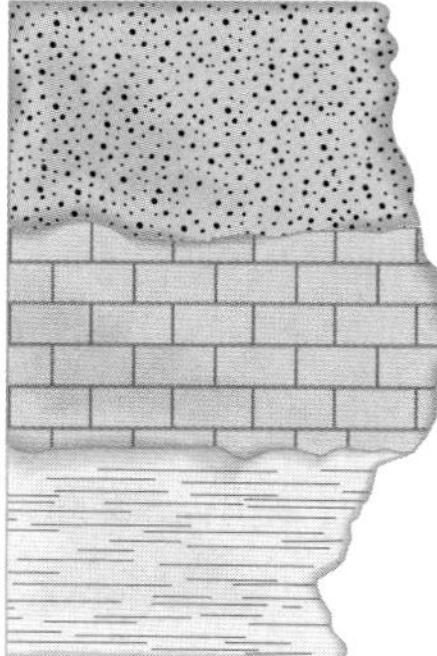

13. What change(s) in sea level is/are represented by the sequence of rocks?

Test-Taking Tip

DRAWINGS AND DIAGRAMS WITH TEXT
If a written description is included with a drawing or diagram, read it carefully. Sometimes the written description can clarify or change the meaning of the graphic.

earthgeu.com/chapter_test

Assessment

14. Describe the differences between the Cambrian and Paleozoic fauna.
15. What kind of clues do geologists use to determine whether an area was the site of an orogeny in the geologic past?
16. How did paleontologists use rugose corals to determine that Earth rotated more rapidly in the geologic past than it does now?
17. How did the Pennsylvanian cyclothems form?

Thinking Critically

18. Explain why many paleontologists consider the development of the seed and the amniote egg to be two of the most important events in Earth's history.
19. How do the many coral reef deposits in the Great Lakes area of North America support the hypothesis that Laurentia was positioned across the equator in the Late Paleozoic?
20. Explain how identical clastic wedges formed on either side of the mountains that were formed by the Acadian Orogeny.
21. Distribution of plants occurs with the help of animals that eat them. Explain how the evolution of the seed ensured the success of this type of plant distribution.

Standardized Test Practice

1. What is it called when the sea level rises and shorelines move inland?
   - **a.** regression
   - **b.** passive Margin
   - **c.** transgression
   - **d.** laurentia

2. Which tectonic event was probably caused by a small microcontinent or island arc that collided with present-day western North America?
   - **a.** Taconic Orogeny
   - **b.** Caledonian Orogeny
   - **c.** Acadian Orogeny
   - **d.** Antler Orogeny

INTREPRETING SCIENTIFIC ILLUSTRATIONS
Use the figure to answer question 3.

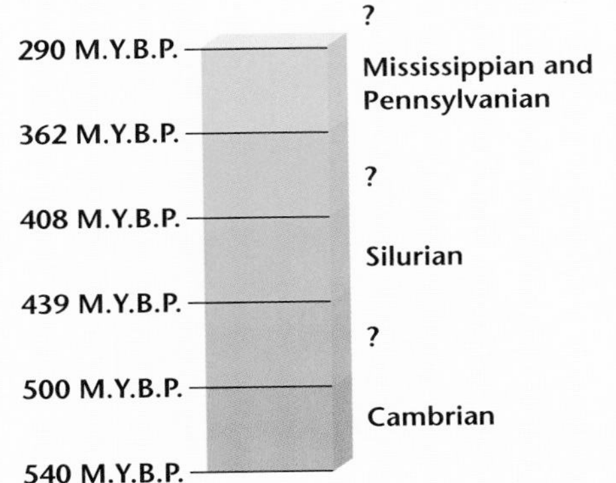

3. During the last period of the Paleozoic Era, much of what is now western North America was covered with dry land. What was this period called?
   - **a.** Pennsylvanian
   - **b.** Devonian
   - **c.** Permian
   - **d.** Ordovician

4. What was formed in North America when Gondwana and Laurasia collided?
   - **a.** Ancestral Rocky Mountains
   - **b.** Appalachian Mountains
   - **c.** Ouachita Mountains
   - **d.** Great Permian Reef

Chapter 24

The Mesozoic and Cenozoic Eras

What You'll Learn

- How plate tectonics shaped the landscape of western North America.
- What the characteristics of a dinosaur are.
- How dinosaurs and many other organisms became extinct at the end of the Mesozoic Era.

Why It's Important

As the physical geology of Earth changed, so did the biosphere. Reptiles ruled the land during the Mesozoic, but their reign ended abruptly with the dawn of the Cenozoic.

To find out more about the Mesozoic and Cenozoic Eras, visit the Earth Science Web Site at earthgeu.com

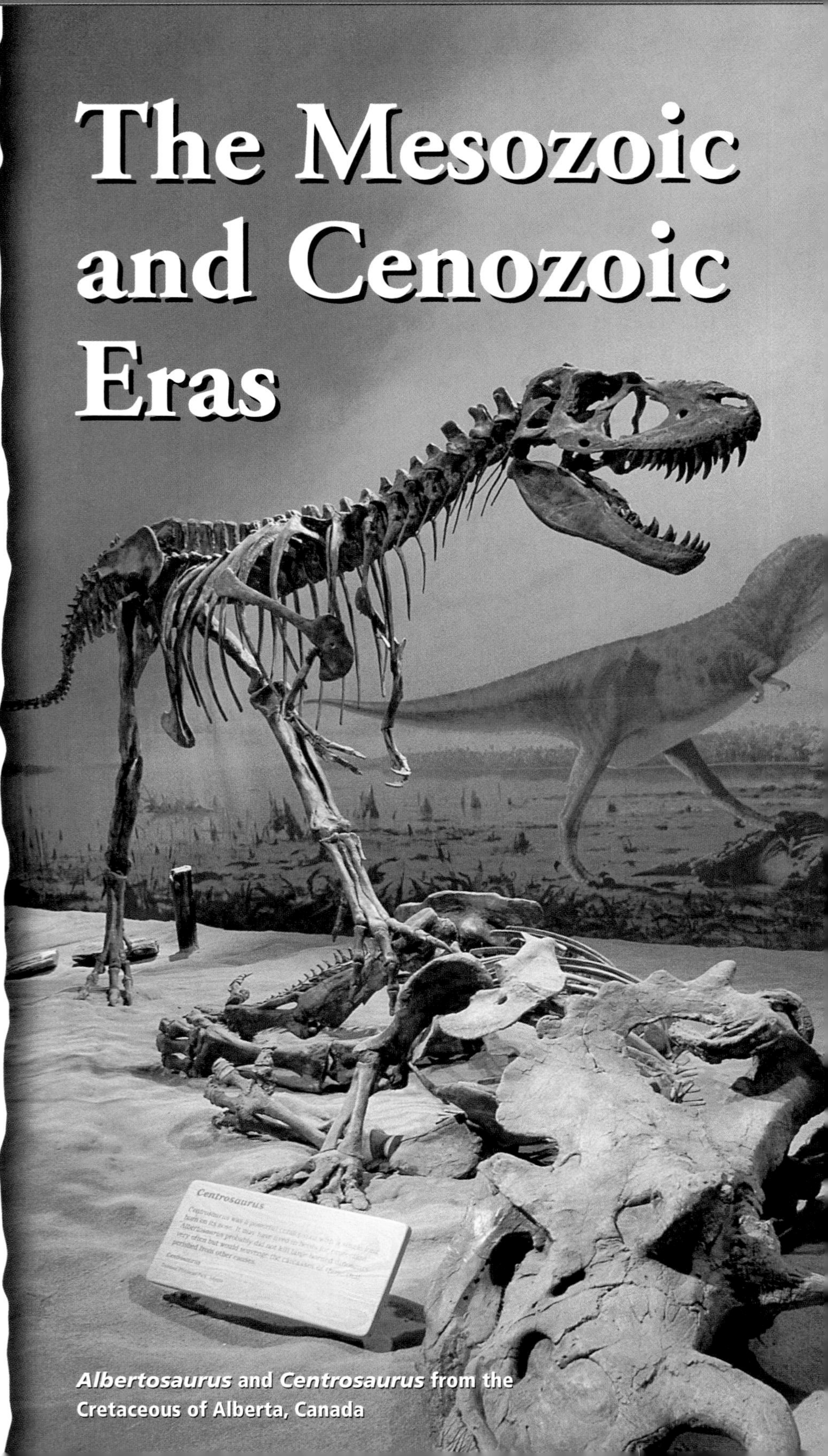

Albertosaurus and *Centrosaurus* from the Cretaceous of Alberta, Canada

Discovery Lab

Determining the Effect of Shape on Buoyancy

How can the shape of a microfossil help paleontologists determine whether it floated or whether it lived on the seafloor? In this lab, you will learn how the shape of a fossil shell affected the animal's buoyancy, and how this information helps paleontologists determine how the animal lived.

1. Fill a 1000-mL beaker with water.
2. Using plasticene, make two solid spheres that are each 1.5 cm diameter.
3. Deform one of the spheres by pinching the plasticene into 5 "spines" that extend around the sphere.
4. Drop both spheres into the beaker at the same time and observe what happens.

Observe In your science journal, record your observations. What effect does the shape of the spheres have on their buoyancy? The photos on this page show marine microfossils. Based on your observations in this lab, which one do you think floated in the water? Which one lived on the seafloor? Explain your reasoning.

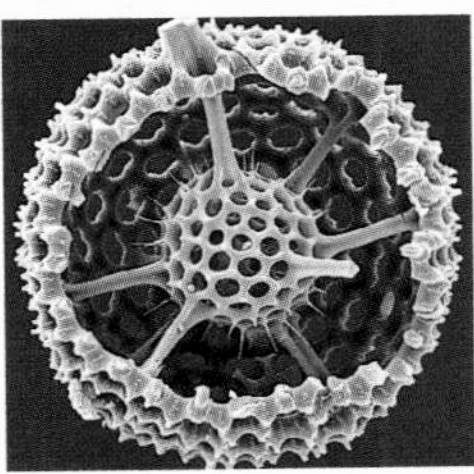

SECTION 24.1 *Mesozoic Paleogeography*

Earth Sciences 3.c

OBJECTIVES

- **Explain** *the breakup of Pangaea.*
- **Distinguish** *between the different tectonic characteristics of the Mesozoic Orogenies.*

VOCABULARY

Cordillera

The Mesozoic Era consisted of the Triassic, Jurassic, and Cretaceous periods, as shown in ***Figure 24-1*** on page 626. Movies such as *Jurassic Park* and *The Lost World* have popularized the Mesozoic as the age of the dinosaurs. What was the world like in which dinosaurs such as those pictured on page 624 lived? Why did this chapter in Earth's history end? These and other questions make the Mesozoic one of the most fascinating times of study in Earth's history.

THE BREAKUP OF PANGAEA

An important event that occurred during the Mesozoic Era was the breakup of Pangaea. Because heat causes solid objects to expand, the heat coming from within Earth beneath Pangaea caused the continent to expand. By the Late Triassic, the brittle lithosphere of Pangaea had cracked and broken apart. As some of the large cracks, or rifts, widened and as the landmasses spread apart, the ocean flooded the rift valleys. This resulted in the formation of new oceans that divided the newly separated continents.

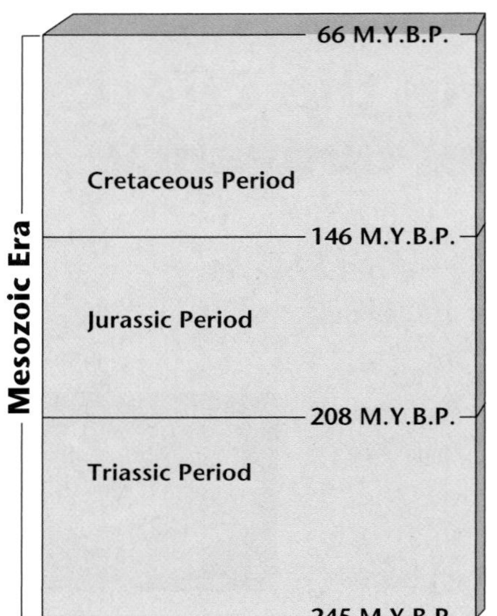

Figure 24-1 Each Mesozoic period is distinct. Pangaea broke apart and new seaways formed during the Triassic. Dinosaurs dominated the land during the Mesozoic. Flowering plants evolved during the Cretaceous. You'll learn more about these events as you study this chapter.

The breakup of Pangaea resulted in the formation of the Atlantic Ocean. As North America rifted away from Europe and Africa, some of the spreading areas joined together to form a long, continuous rift system called the Mid-Atlantic Ridge. The Mid-Atlantic Ridge is still active today.

ACTIVE TECTONISM IN WESTERN NORTH AMERICA

In contrast to the passive margin that existed along eastern North America during the Mesozoic, active subduction along the western coast continued through the Middle Triassic. Geologists refer to the mountain ranges that formed in western North America during this time as the **Cordillera,** which means "mountain range" in Spanish.

Deformation along the western margin of North America increased substantially when Pangaea broke apart. Three major episodes of orogenies—mountain building—occurred along the western margin of North America during the Mesozoic. Different types of deformation occurred during each of these orogenies.

The oldest orogeny was characterized by a tremendous number of igneous intrusions. Large bodies of granite called batholiths exist throughout the cordillera. The spectacular exposure of Half-Dome at Yosemite National Park was intruded during this orogeny.

The next orogeny was characterized by low-angle thrust faulting and folding. This was caused by collisional tectonism along the western margin of North America. This type of deformation began in the Late Jurassic and continued through the Late Cretaceous. The thrust faults run north-south and place older rocks on top of younger rocks in Utah, Idaho, Wyoming, western Canada, and Montana, shown in ***Figure 24-2.***

Earth Sciences

3.c Students know how to explain the properties of rocks based on the physical and chemical conditions in which they formed, including plate tectonic processes.

Figure 24-2 These Paleozoic limestones in the Sawtooth Mountains, MT were faulted and thrust eastward during the second Mesozoic orogeny.

The third Mesozoic orogenic event was characterized by vertical uplifts. This orogeny mainly affected the area east of the folds and faults caused by the second orogeny. Deformation caused by the third orogeny began during the Late Cretaceous and continued into the Cenozoic.

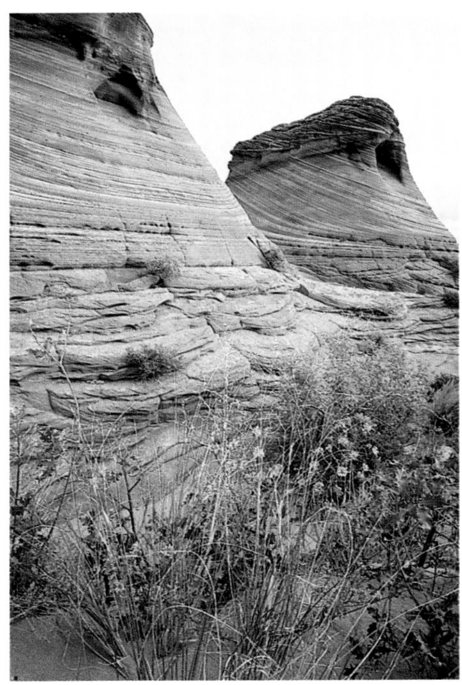

Figure 24-3 The effects of the wind is preserved in these sand dunes in the Vermillion Cliffs Wilderness, AZ.

SEAWAYS AND SAND DUNES

Throughout the Early and Middle Triassic, the supercontinent Pangaea and a single global ocean defined Earth's paleogeography. As Pangaea began to split apart, numerous rift basins formed in eastern North America, and large blocks of crust collapsed to form deep valleys. The Triassic ended with a rapid drop in sea level that caused sedimentation in the western United States to change dramatically during the Late Triassic and Early Jurassic. Western North America became much more arid, and it was covered by a thick blanket of sand. Strong winds shaped the sand into dunes. Evidence of this ancient desert is preserved in large-scale, cross-bedded sandstone deposits some of which are shown in ***Figure 24-3.***

Sea level rose again in the Jurassic, and a shallow sea covered central North America. The Appalachian Mountains still rose high in the east, and the newly formed mountains of the Cordillera rose high in the west. As the mountains continued to uplift in the west, large river systems transported sediments from the mountains into the sea. The deposits of the Late Jurassic river systems are preserved today as multicolored sandstones, siltstones, and mudstones. They are well known for large numbers of dinosaur fossils.

The ocean continued to rise onto North America during the Cretaceous Period, and the Gulf of Mexico flooded the entire southeastern margin of North America. As a result, a sea covered the interior of North America from Texas to Alaska.

SECTION ASSESSMENT

1. Explain why Pangaea rifted apart.
2. Describe the difference between the three major episodes of mountain building that occurred during the Mesozoic.
3. **Thinking Critically** What might happen if a continent stops rifting apart?

SKILL REVIEW

4. **Recognizing Cause and Effect** Explain how heat that is generated in Earth's interior can cause continents to rift apart, as in the case of Pangaea. For more help, refer to the *Skill Handbook.*

SECTION 24.2 Mesozoic Life

Earth Sciences 1.f

OBJECTIVES

- **Discuss** *why many paleontologists theorize that birds are descended from dinosaurs.*
- **Describe** *how paleontologists distinguish among reptile, dinosaur, and mammal fossils.*
- **Explain** *the evidence indicating that a meteorite impact caused the Cretaceous-Paleogene mass extinction event.*

VOCABULARY

modern fauna
angiosperm
dinosaur
Ornithischia
Saurischia
ectotherm
endotherm
iridium

The Mesozoic is commonly referred to as the Age of Reptiles. However, the Mesozoic also was a time of some other very important biological firsts, such as the first mammals, the first birds, and the first flowering plants. The mass extinction at the end of the Paleozoic Era had left Earth's biosphere quite barren, and therefore ripe for the appearance of new organisms.

As Pangaea began to split apart during the Triassic, much of the habitat on the continental shelves that was lost during the formation of Pangaea became available again. New marine organisms, which are collectively called the **modern fauna,** evolved which filled this habitat. The modern fauna includes crabs, lobsters, shrimps, sponges, sea urchins, modern corals, snails, and clams. The major marine vertebrate groups include bony fishes, sharks, aquatic reptiles, and aquatic mammals.

LIFE IN THE OCEANS

Much attention is given to the large animals of the Mesozoic. It is important, however, to remember that then, as now, the base of the food chain that supported all the large animals consisted of tiny, ocean-dwelling organisms called phytoplankton, shown in ***Figure 24-4.*** Phytoplankton float near the surface of oceans and lakes and make their own food through the process of photosynthesis. You examined the relationship between shape and floating ability in the *Discovery Lab* on page 625.

ENVIRONMENTAL CONNECTION

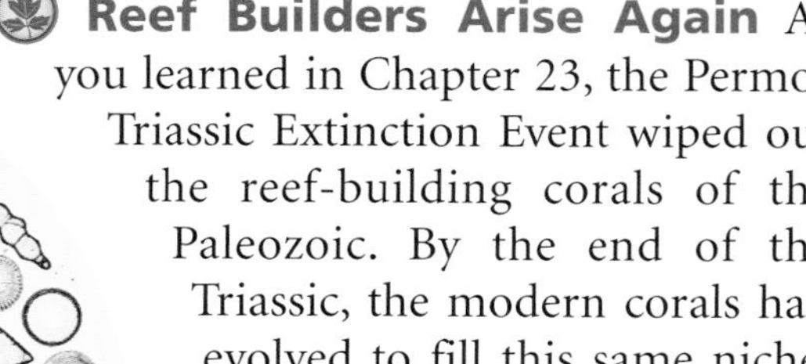

Reef Builders Arise Again As you learned in Chapter 23, the Permo-Triassic Extinction Event wiped out the reef-building corals of the Paleozoic. By the end of the Triassic, the modern corals had evolved to fill this same niche. In addition to corals, a new group of clams called rudists developed the ability to build reefs during the Cretaceous. Rudists are important because the reefs that they built were very porous and today contain some of the largest Cretaceous oil deposits in areas such as West Texas.

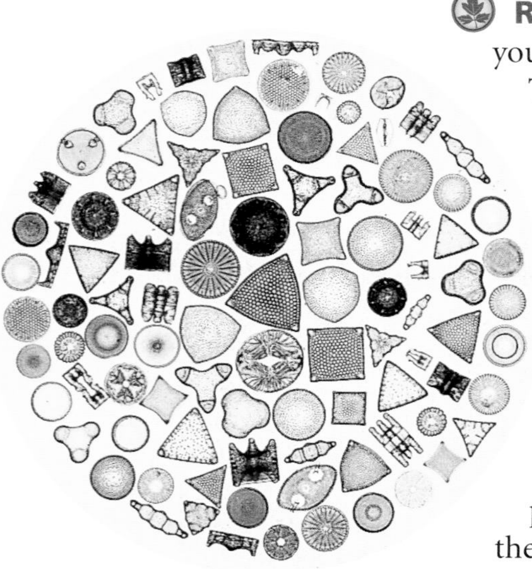

Magnification 48×

Figure 24-4 These phytoplankton are diatoms; tiny freshwater or marine plants. Upon death, their silica shells contribute to the ocean floor sediment.

Important Index Fossils Ammonites are related to modern nautiluses, octopuses, and squids. Ammonites were abundant and diverse throughout the Late Paleozoic and Mesozoic, and their abundance indicates that they were very successful predators. Several ammonites are shown in ***Figure 24-5.*** Ammonite fossils are widespread and abundant. Because ammonite species were also short-lived, they are excellent index fossils.

Figure 24-5 These ammonites are from the Lower Jurassic of Europe. Their distinct shell shape and ornamentation adds to their usefulness as index fossils. The use of ammonites to determine the age of Mesozoic rocks is usually much more accurate than radiometric dating.

While the ammonites were successful Mesozoic marine predators, they were by no means the top predators of the Mesozoic. In fact, swimming reptiles ruled the Mesozoic oceans. Ichthyosaurs resembled modern dolphins while plesiosaurs were more like walruses and seals. Predatory mosasaurs lived only during the Cretaceous, but for that time in the history of life, the sea was theirs to rule. These Mesozoic reptiles are shown in ***Figure 24-6.***

Life on the Land

Life on land changed dramatically as the cool climate that characterized the end of the Paleozoic over many parts of Earth came to an end during the Mesozoic. The large, temperate coal swamps dried up, and the climate gradually warmed. Fossils of insects that dominated the Paleozoic coal swamps, such as flies, mosquitoes, wasps, and bees, were also present but were not as common in Mesozoic rocks.

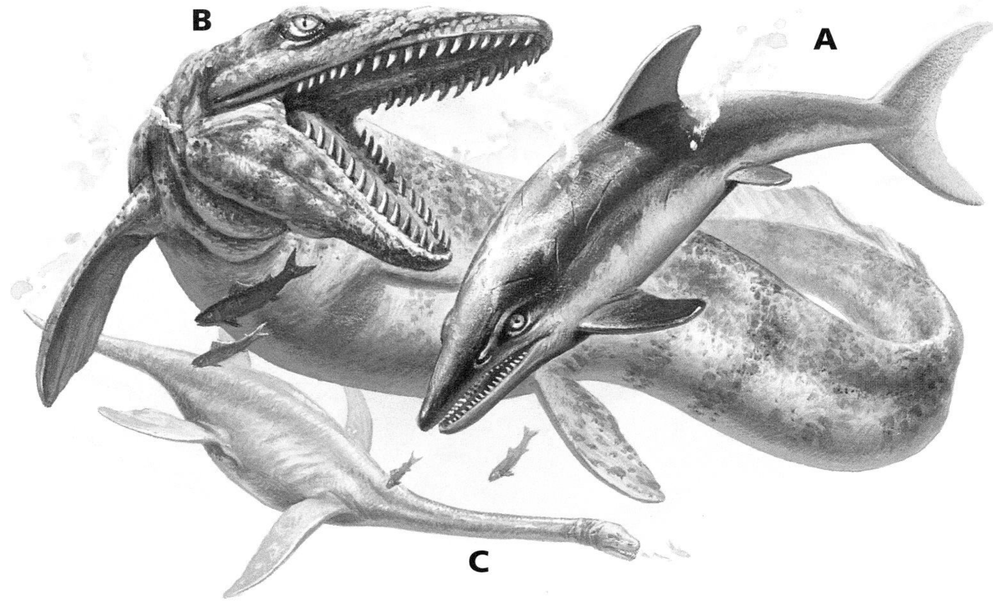

Figure 24-6 The presence in the oceans of aggressive marine predators, such as ichthyosaurs **(A)**, mosasaurs **(B)**, and plesiosaurs **(C)**, no doubt kept their intended prey on alert.

Figure 24-7 Cycads have a limited distribution today. They grow only in the southern hemisphere, in Central America and in Florida. Only one species of ginkgo survives today, *Ginkgo biloba,* and it is quite common.

The land flora also changed substantially during the Mesozoic. Paleobotanists rightly call the Mesozoic the Age of Cycads. Cycads are seed plants that do not have true flowers. During the Jurassic, tall cycad trees along with ginkgoes and conifers dominated the landscape, as shown in ***Figure 24-7.***

Angiosperms Evolve A new kind of plant evolved during the Cretaceous that would come to dominate the terrestrial landscape. This important group of plants was the **angiosperms,** which are seed-bearing plants that have flowers. Before the Cretaceous, there were no flowers. By the Middle Cretaceous, however, flowering plants were common. By the end of the Cretaceous, the land was covered with flowering trees, shrubs, and vines.

Early Mammals Small, primitive mammals evolved during the Late Triassic. Mammals are easily recognizable today because they are warm-blooded, have hair or fur, and mammary glands. How do scientists identify fossils of mammals? One method is to examine the structure of the lower jaw, middle ear, and teeth. A reptile has one ear bone and multiple lower jawbones, whereas a mammal has one lower jawbone and three ear bones. Early mammals arose from mammal-like reptiles. The evolution of the mammal jaw is illustrated in ***Figure 24-8.*** Mammals have two sets of teeth during their lives, and their teeth are differentiated as incisors, canines, and molars. In contrast, reptiles generally have only one kind of tooth, and their teeth continually grow and replace older teeth throughout their lives.

Figure 24-8 Fossils clearly show the progressive enlargement of one of the lower jawbones in mammal-like reptiles and its evolution into a single jawbone, which is characteristic of mammals. At the same time, the other jawbones decreased in size and became ear bones.

Primitive mammal-like reptile

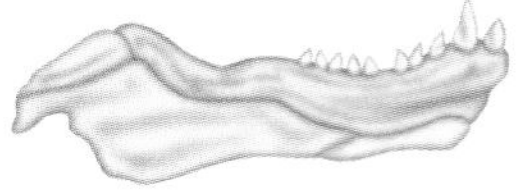

Primitive therapsid

Advanced therapsid

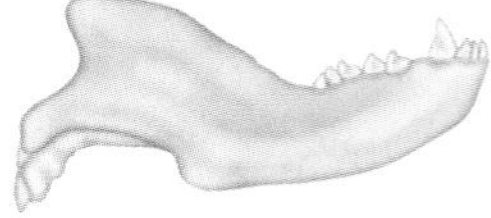

Mammal

Flying Reptiles Pterosaurs, as shown in ***Figure 24-9,*** were flying reptiles that dominated the air during the Mesozoic. The earliest pterosaurs were small and had long tails and wingspans of less than 60 cm. Their descendents had no tails and were generally much larger. One species had a wingspan of more than 12 m. Pterosaurs had light, hollow bones, like modern birds. The modification that allowed them to fly was the growth of a membrane from a greatly lengthened fourth finger.

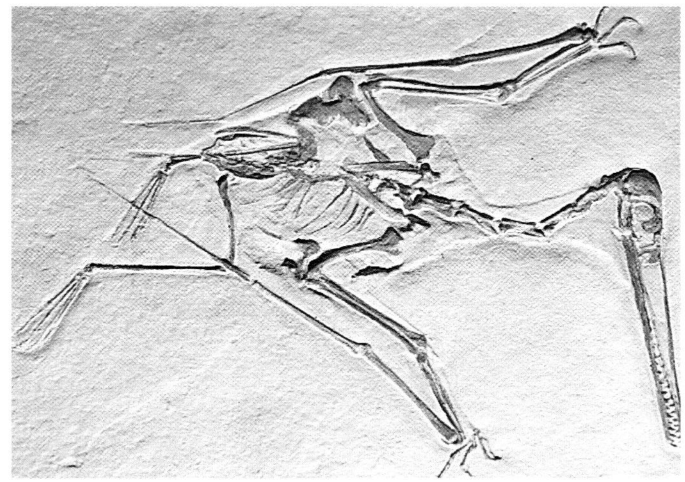

Figure 24-9 This pterosaur fossil was found in Jurassic-aged rocks in Germany. The membrane that was attached to the fourth finger was actually a flap of tough skin that stretched from the finger to the sides of the pterosaur's body.

Two groups of reptiles that arose during the Mesozoic and still exist are the crocodiles and turtles. Why crocodiles and turtles survived the great extinction at the end of the Mesozoic, while the mosasaurs, plesiosaurs, icthyosaurs, and dinosaurs did not, puzzles paleontologists to this day.

Dinosaurs Rule the Land

No other group of animals has captured the human imagination as much as the dinosaurs have. **Dinosaurs** were a group of reptiles that developed an upright posture about 228 million years ago. Even though our understanding of these rulers of the Mesozoic land has changed, our fascination with dinosaurs has remained. Dinosaurs came in all sizes, from the very small to the extraordinarily large, and all were terrestrial. You can learn how to estimate a dinosaur's weight in the *Science and Math* feature on page 644. Although the largest dinosaurs were most likely slow and plodding animals, many of them were quick and agile. All reptiles other than dinosaurs have a sprawling posture, that is, their legs are not set directly underneath their bodies. In contrast, as shown in ***Figure 24-10,*** dinosaurs' legs were set directly underneath their bodies because their hips and ankles were different from those of other reptiles.

Figure 24-10 The characteristic sprawling posture of reptiles is shown here. Dinosaurs, with their straight shoulders, ankles, and hips, stood apart from the reptiles.

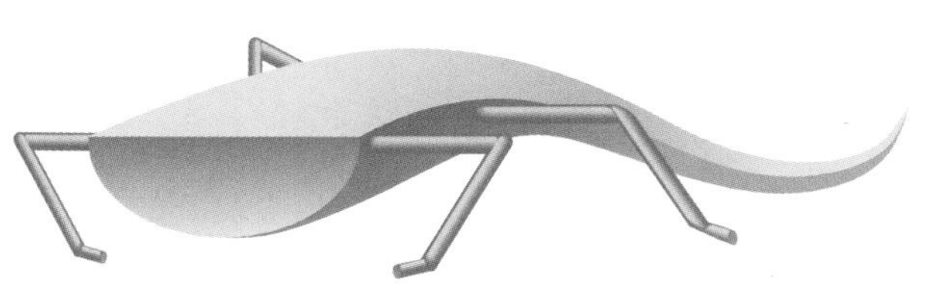

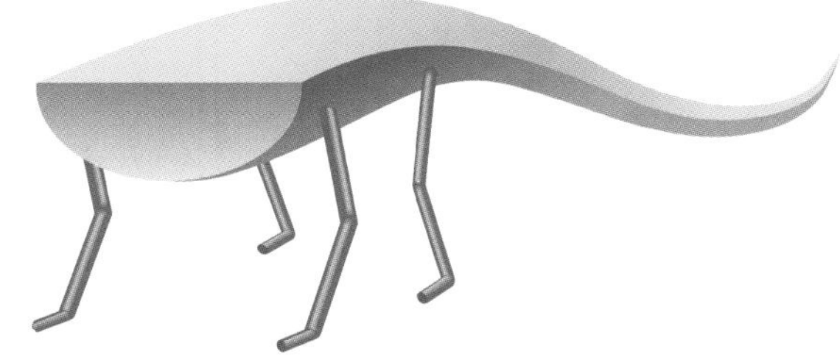

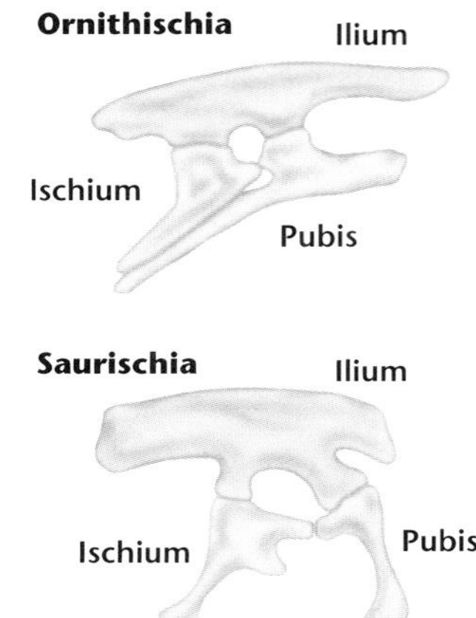

Figure 24-11 The arrangement of hip bones in dinosaurs serves as a basis for their classification.

Dinosaur Hips Two major groups of dinosaurs are recognized based on their hip structure: Ornithischia and Saurischia. Three bones comprised the hip, as shown in ***Figure 24-11.*** The ischium and pubis were parallel to one another in ornithischian dinosaurs. This is similar to the orientation of these bones in modern birds, which is why scientists named this group of dinosaurs **Ornithischia,** meaning "bird-hipped". This name is unfortunate, however, because birds likely did not evolve from Ornithischia. The ischium and pubis were at an angle to one another in Saurischia, similar to the orientation observed in modern lizards. **Saurischia** means "lizard hipped." Scientists hypothesize that birds are actually descended from the Saurischia.

There were five different groups of ornithischian dinosaurs: stegosaurs, ankylosaurs, pachycephalosaurs, ceratopsians, and ornithopods. All ornithischian dinosaurs were plant eaters, also called herbivores. There were two different groups of saurischian dinosaurs: sauropods and theropods. Although they shared a similar hip structure, these two groups of dinosaurs were quite different. The sauropods were all quadrupedal—walked on four legs—plant eaters, and some grew to enormous sizes. They were the largest land animals to have ever lived. In contrast to the sauropods, all theropods were bipedal—walked on two legs—carnivores. Your research results from the *GeoLab* on page 642 will help you interpret what dinosaurs ate. ***Figure 24-12*** shows some representative Mesozoic dinosaurs.

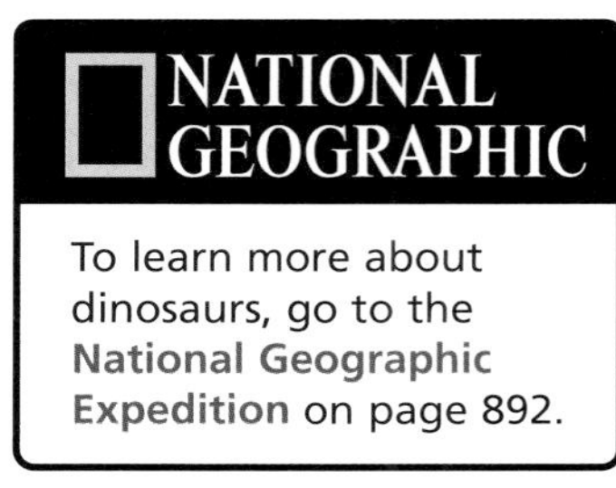

To learn more about dinosaurs, go to the **National Geographic Expedition** on page 892.

Figure 24-12 Herbivores such as *Parasaurolophus* (right) and *Triceratops* (center) had to constantly be on the lookout for carnivores such as *Tyrannosaurus.*

Dinosaurs to Birds? The idea that birds are related to dinosaurs stems from the amazing similarities between theropods and the oldest known bird, *Archaeopteryx.* Fossils of feather impressions and a wishbone provide clear evidence that *Archaeopteryx* was definitely a bird, even though it did have teeth and a theropod-like skeleton.

Ectotherm or Endotherm? All living reptiles are **ectotherms,** meaning that their body temperatures vary in response to outside temperatures. All living mammals and birds are **endotherms,** meaning that they maintain relatively constant body temperatures, regardless of temperatures outside. Some paleontologists hypothesize that at least some groups of dinosaurs were endotherms. One reason is that bones of endotherms typically have more passageways for blood than bones of ectotherms do, and dinosaur bones have numerous passageways. Critics of the hypothesis that some dinosaurs were endotherms correctly point out that the bones of crocodiles and turtles also have numerous passageways, and yet they are ectotherms.

A recent discovery of a possible fossilized heart in a dinosaur supports the hypothesis that some dinosaurs were endothermic. The heart has four chambers and one aortal arch, which is a condition that exists only in endothermic animals. Whether dinosaurs were ectotherms or endotherms is still controversial, but for now, there seems to be evidence that perhaps some dinosaurs were endotherms.

Using Numbers The end of the Permian is marked by an extinction event in which 80 percent of all marine genera became extinct, and the end of the Cretaceous is marked by an extinction event in which 50 percent of all marine genera became extinct. How many times greater was the Permian extinction event than the Cretaceous extinction event?

Mass Extinctions

A major mass extinction event ended the Mesozoic. Most major groups of organisms were devastated and all known species of dinosaurs, pterosaurs, ammonites, mosasaurs, and plesiosaurs

Earth Sciences

1.f Students know the evidence for the dramatic effects that asteroid impacts have had in shaping the surface of planets and their moons and in mass extinctions of life on Earth.

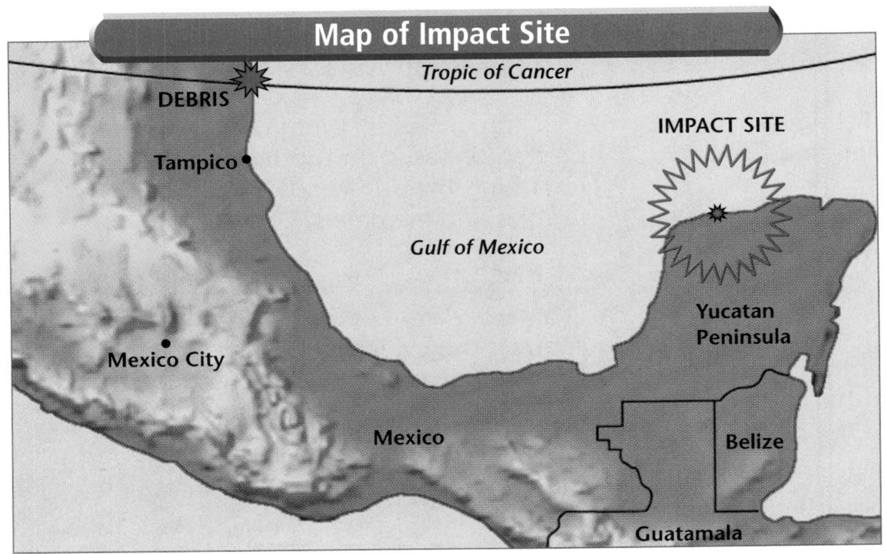

Figure 24-13 The circular shape and underlying layer of melted iridium-rich rock provides evidence of a meteorite impact. It was named the Chicxulub crater after a nearby village.

Topic: Meteor Impacts and Extinctions
To find out more about mass extinction, visit the Earth Science Web Site at earthgeu.com

Activity: Prepare a map of the world showing three large meteor impacts besides the Chicxulub impact. Indicate the approximate dates on your map. Do any of these dates correspond to other mass extinctions?

became extinct. Geologists theorize that a large meteorite, at least 10 km in diameter, slammed into the Yucatan Peninsula shown in ***Figure 24-13*** on page 633, at a speed of up to 240 000 kph. Such an impact would have blown 25 trillion metric tons of rock and sediment high into the atmosphere. Organisms lucky enough to have survived the impact would have faced a millennium of greenhouse warming and increased levels of UV radiation from the Sun.

The chemistry of a clay layer that separates Cretaceous from Paleogene rocks in Italy provides evidence of such a meteorite impact. An unusually high amount of **iridium,** a metal that is rare in rocks at Earth's surface but is relatively common in meteorites and asteroids, is found not only in Italy but in Cretaceous-Paleogene boundary sites worldwide. Soot and charcoal, which are evidence of widespread fires, are also common in the sediments.

A buried crater in the Gulf of Mexico contains iridium, which has a radiometrically dated age of approximately 65 million years. This age is close to the time of the Late Cretaceous mass extinctions. It is important to note, however, that elevated amounts of iridium are also present in Earth's interior, and thus, volcanic deposits can be enriched with iridium. During the Late Cretaceous, massive volcanic eruptions occurred in India.

Today, most scientists agree that both a large meteorite impact and massive volcanism occurred at the end of the Cretaceous. In the midst of a stressful time of climatic cooling, changing plant populations and a gradual decline in dinosaur diversity and abundance, a large meteorite struck Earth. The extraordinary stress that the impact added to an already stressed ecosystem likely caused the climax of the Cretaceous-Paleogene mass extinction.

Section Assessment

1. How do paleontologists distinguish between fossils of mammals and fossils of reptiles?
2. Discuss how *Archaeopteryx* supports the hypothesis that birds are descended from dinosaurs.
3. Explain why a cycad is not an angiosperm.
4. What is the main characteristic that separated dinosaurs from all other reptiles?
5. Describe the differences between Saurischia and Ornithischia.
6. **Thinking Critically** What conclusions can be stated, based on tooth characteristics, regarding the differences in diet between reptiles and mammals?

Skill Review

7. **Comparing and Contrasting** Compare and contrast the evidence for meteorite impact and volcanism at the end of the Cretaceous. For more help, refer to the *Skill Handbook.*

SECTION 24.3 Cenozoic Paleogeography

Earth Sciences 3.f I&E 1.a, 1.e, 1.l

You are living during the Cenozoic, which means "recent life." The Cenozoic Era encompasses approximately the last 66 million years of Earth's history. The Cenozoic portion of the geologic time scale is shown in ***Figure 24-14.*** The final breakup of Pangaea occurred during the Cenozoic. Earth's life-forms and surface features continued to change, evolving into their present form. The Cenozoic has been a time of major climate changes, which have been in part, caused by the positions of the continents.

OBJECTIVES

- **Describe** *the type of tectonism that characterized the Cenozoic orogeny.*
- **Understand** *the extent of glaciation that occurred in North America.*

VOCABULARY

Basin and Range Province
Tethys Sea

The Ice Ages As Australia split apart from Antarctica during the Middle-to-Late Eocene, the warm climate began to deteriorate. A change in ocean circulation is thought to be the cause. When Antarctica and Australia were connected as shown in ***Figure 23-15*** on page 612, a current of warm water from the north moderated the temperature of Antarctica. When Antarctica split apart from Australia during the Oligocene, it was isolated over the south pole. A cold current began to flow around Antarctica, and glaciers began to form.

The climate began to warm again during the Early Miocene. The glaciers on Antarctica began to melt, and the sea rose onto the margin of North America. In the *Problem-Solving Lab* on page 637, you will consider glacial melting and the length of time involved. Glaciers returned to Antarctica during the Middle and Late Miocene. During the Pliocene, the water of the Arctic Ocean began to freeze to form an arctic ice cap, which set the stage for the ice ages of the Late Pliocene and the Pleistocene.

During the Late Pliocene through the Pleistocene, the northern hemisphere experienced extensive glaciation, or an ice age. Glaciers from the arctic advanced and retreated in at least four stages over North America. The paths of the Ohio River and the Missouri River roughly mark the southernmost point to which glaciers advanced in North America. During the peak of Pleistocene glaciation, glaciers up to 3 km-thick covered some areas north of these rivers. You will model the effect glaciers have on sediment deposition when you complete the *MiniLab* on page 636.

Figure 24-14 Although the last Ice Age began about 2 million years ago during the Pleistocene, ice age conditions really began about 40 million years ago during the Oligocene Epoch.

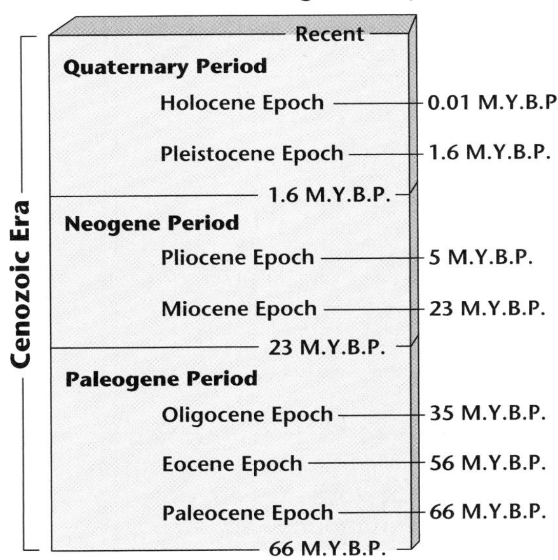

TECTONIC EVENTS

Western North America had been tectonically active throughout the Cenozoic. The orogenic events that occurred at the end of the Mesozoic uplifted massive

MiniLab

Glaciers and Deposition

Model the deposition of sediment by melting glaciers.

Procedure

1. Pour water into a large, wide-mouthed jar until it is approximately 3/4 full.
2. Add a mixture of clay, silt, sand, and pebbles to the jar. Put the lid tightly on the jar.
3. Shake the jar for 30 seconds and allow the contents to settle.
4. Finely crush enough ice to fill approximately three-fourths of a second large, wide-mouthed jar. Stir a mixture of clay, silt, sand, and pebbles into the ice and pour the mixture into the jar.
5. Allow the ice to melt and the particles to settle overnight.

Analyze and Conclude

1. Describe the differences in the way the sediments settled in the two jars.
2. Compare and contrast the sorting of the grain sizes in the two jars.
3. How could geologists use this information to determine whether sediment had been deposited by a glacier or by running water?

blocks of crust to form the Rocky Mountains. Large basins that formed adjacent to the Rocky Mountains were filled with as much as 3000 m of sediment from the uplifted and eroded mountains. The sediment that filled these basins contains beautifully preserved fish, insect, frog, plant, and bird fossils. A fossil fish from the most famous of these deposits, the Green River Formation in Wyoming, is shown in ***Figure 24-15.*** The basins in Wyoming that filled with huge, swampy, river deposits provided an ideal environment for the accumulation of vast amounts of coal. These coal seams are close to Earth's surface, and some are more than 50 m thick. Wyoming is one of the largest coal mining regions in the world. The coal there is especially valuable because it has a very low sulfur content, and thus, it burns cleanly.

Subduction in the West Volcanism returned to the western coast of North America at the end of the Eocene. The Cascade Mountains in the Pacific Northwest are the result of the subduction of an oceanic plate beneath the western coast of North America. During the Miocene, the North American Plate was forced over the East Pacific Rise resulting in the creation of the San Andreas Fault. Because there is currently

Earth Sciences

3.f Students know the explanation for the location and properties of volcanoes that are due to hot spots and the explanation for those that are due to subduction.

Figure 24-15 Fossil fish, like this 50 million year old *Phareodus sp.* from the Eocene-aged, Green River formation in Wyoming, are preserved in spectacular detail.

no subduction beneath southern and central California, the volcanoes in most of California are geologically inactive.

The subduction of the East Pacific Rise beneath the North American plate coincides with pull-apart, or extensional, tectonism in the southwestern United States and north-central Mexico. A series of mountains that trend north-to-slightly-northeast are separated by long, linear valleys and extend from Nevada and western Utah to north-central Mexico. This area, shown in ***Figure 24-16,*** is called the **Basin and Range Province.** As extensional tectonism has pulled the crust apart, large blocks of the crust have dropped down along normal faults to form the basins, leaving other blocks at higher elevations to form the mountain ranges. This extension is still occurring today.

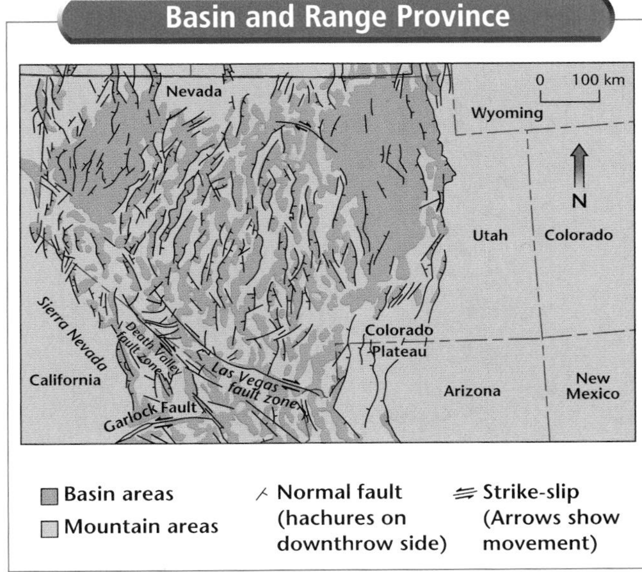

Figure 24-16 The extensive development of basins and mountain ranges in this area is a clear example of the results of extensional tectonism.

Hot Spots in the West In Chapter 18, you learned that hot spots are mantle plumes that rise to Earth's surface. Some of the hot spots that occur in the western United States are related to the continuing subduction along the western coast of North America. Yellowstone National Park is famous for its beautiful geysers and hot springs. The land that makes up Yellowstone National Park is situated on a hot spot that has been active since the Early Cenozoic. The rocks at Yellowstone

Problem-Solving Lab

Using Math

New Continental Shorelines? If all the ice on Earth were to melt, sea level would rise to approximately 50 m above its current level. Sea level is rising today at an average rate of 2 mm per year. Assuming that sea level will continue to rise at this rate, how long will it take for all the ice on Earth to melt and sea level to rise to its maximum? Use the relationship distance = rate × time. Remember that 1000 mm = 1 m.

Analysis

1. What assumption is this calculation based on?
2. How realistic are the average rates?
3. What are some possible shortcomings of this calculation?
4. What major cities would this rise in sea level endanger?

Thinking Critically

5. Where, if at all, would new coral reefs grow? Why?

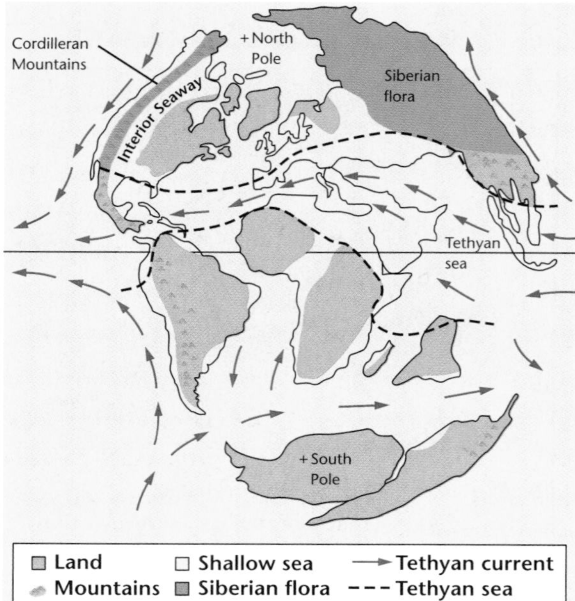

Figure 24-17 The Tethys Sea had a strong, westward-flowing current that transported organisms across large distances. Cretaceous-aged fossils that resemble fossils from the Tethys Sea area are found as far west as the Hawaiian Islands.

indicate that, in the past, this hot spot generated tremendous volcanic activity. Volcanic rocks that are Pleistocene in age record episodes of explosive volcanism. In fact, the yellow color of the extrusive volcanic rock rhyolite is responsible for the name of the river and the national park.

Continental Collisions While the final breakup of Pangaea occurred during the Cenozoic, plate tectonics also brought continents together during this time. The spectacular result of one such collision is the Himalayan Mountains. India traveled north and collided with the southern margin of Asia. The force of this tremendous continent-to-continent collision resulted in the formation of the Himalayan Mountains, which contain the point of highest elevation on Earth, Mt. Everest. The rocks at the top of Mt. Everest are Ordovician marine limestones. Tectonic forces have pushed what was the seafloor during the Ordovician to the top of the world!

Africa also drifted north after the breakup of Pangaea and collided with the connected landmass of Europe and Asia, or Eurasia. This continent-to-continent collision formed the Alps. Before Africa collided with Eurasia, a narrow sea called the **Tethys Sea** separated the two continents, as shown in ***Figure 24-17.*** The collision between Africa and Eurasia continues today, and it is almost certain that in the geologic future, the last remnant of the Tethys Sea will dry up.

Section Assessment

1. What kind of tectonic deformation characterized the Cenozoic?
2. What was the southern boundary of Pleistocene glaciation in the Central United States?
3. What is the relationship between the East Pacific Rise, the North American Plate and the San Andreas Fault?
4. **Thinking Critically** The positioning of Antarctica over the south pole helped trigger the Cenozoic ice ages. There would be a major rise of sea level if all the ice on Antarctica melted. If Antarctica were to move north of the south pole, as India did, it is likely that all the ice on the continent would melt. Explain why the movement of Antarctica should or should not be considered in predicting how a sea level change would affect the climate today.

Skill Review

5. **Recognizing Cause and Effect** Discuss the tectonic activity on the western coast of North America and the effect it had on the Basin and Range Province. For more help, refer to the *Skill Handbook.*

earthgeu.com/self_check_quiz

SECTION 24.4 Cenozoic Life

The modern marine fauna, including clams, snails, sea urchins, crustaceans, bony fishes, and sharks, survived the Cretaceous mass extinction to populate the modern oceans. Whales and dolphins evolved during the Cenozoic as completely aquatic mammals. Walruses and sea lions returned to the oceans. These mammals are still partly terrestrial today, but, like plesiosaurs were, they are much more at home in the water than on land.

OBJECTIVES

- **Describe** *the landscape of the Oligocene in Central North America.*
- **Discuss** *the changes in animals in North America during the Cenozoic.*
- **Identify** *the characteristics of primates.*
- **Explain** *what separates hominids from the other hominoids.*

VOCABULARY

primate
hominoid
hominid
Homo sapiens

LIFE ON LAND

Most of the currently living groups of mammals had evolved by the Eocene. Forests dominated North America during the Paleocene and Eocene. As the climate cooled during the Late Eocene, however, the forests gave way to open land. Grasses, which were important to many large mammals, appeared during the Eocene. Grasses spread out over the plains, and by the Late Oligocene, grassy savannas, like those in East Africa today, were common from Texas to South Dakota. The grasslands supported a large diversity of mammals, most of which are members of groups living today. These included dogs, cats, rodents, rabbits, camels, horses, pronghorn antelope, and mastodons. The rocks in Badlands National Park in South Dakota, shown in ***Figure 24-18,*** contain a treasure trove of fossils from the Oligocene. The rocks are made of clay, silt, and sand that were deposited in marshes and lakes and by slowly moving streams.

Pleistocene Mammals As the Pliocene ice age began, the great savannas were replaced by more arid land. The change in climate caused many of the savanna mammals to become extinct. A new group of animals evolved to populate the land as the Late Pliocene-Pleistocene ice age sent a chill across North America. Several of these animals are shown in ***Figure 24-19*** on page 640.

Figure 24-18 The Badlands of South Dakota may resemble a moon-scape but they were far from a lifeless plain. In fact, the abundant and diverse mammal fossils found here have inspired some paleontologists to call the Oligocene the Golden Age of Mammals.

Figure 24-19 The woolly mammoth, dire wolf, and the sabre-toothed cat are a few of the Pleistocene animals that evolved abilities to withstand the cold.

PRIMATES AND HUMANS

One of the most difficult problems to overcome in studies of the origin of humans and our relationship to other primates is the scarcity of fossils. This makes this area of study dynamic, partly because the discovery of a single new fossil can dramatically change our understanding.

Primates **Primates** are distinguished from other mammals by a grasping hand with an opposable thumb and two eyes directed forward that result in stereoscopic vision. Such vision allows primates to judge distance quite accurately. Other primate characteristics include smaller, fewer, and less-specialized teeth than other mammals and a relatively large brain.

Our species, *Homo sapiens*, belongs to a group of primates called hominids. In turn, hominids are part of a larger group called **hominoids** that includes the great apes. The fossil record of hominids extends back approximately 4.4 million years. What is it that separates a hominid from the other hominoids? The most recognizable feature is that all **hominids** are bipedal. That is, they have an upright posture resulting from a modification of the hipbone and they walk on two legs. Hominids also have larger brains, smaller canine teeth, and smaller faces than other hominoids, and they use sophisticated tools and have greater manual dexterity.

Update To find out more about hominid fossil finds, visit the Earth Science Web Site at earthgeu.com

The Rise of *Homo Sapiens* Tracing the ancestry of **Homo sapiens,** the species to which humans belong, to earlier hominids began in the 1850s, when a fossilized skull was discovered in Neander Tal near Dusseldorf, Germany. *Tal* is the German word for "valley". Since that time, paleoanthropologists have scoured Earth looking for fossils of hominids in an attempt to piece together the ancestry of *Homo sapiens*.

What about the fossils from Neander Tal? These hominids are commonly called Neanderthals. Most fossil evidence indicates that Neanderthals were most likely a side branch of *H. sapiens* and not direct ancestors of modern humans. The Neanderthals were hunters that inhabited Europe and the Near East approximately 200 000 to 30 000 years ago. Neanderthals differed from modern humans in a number of ways; differences in their skulls are shown in ***Figure 24-20.*** Neanderthals had heavy brows, mouths that projected forward, and receding chins. They had short, thick limbs and more massive, muscular bodies. Perhaps most surprising is that they had slightly larger brains than modern humans do. Neanderthals did live in caves and used a variety of tools. There is evidence that they buried their dead and placed items such as tools in their graves.

It may seem strange that scientists do not have a more complete understanding of the relationships among the hominids. After all, the fossil skeleton of one of the earliest known hominid species, *Australopithecus afarensis,* provides evidence that bipedal, upright-walking hominids existed at least 3.5 million years ago. However, compared to the rest of the fossil record, there are relatively few hominid fossils. As a result, our understanding of the evolution of many other groups of animals is much better than our understanding of the evolution of hominids. Subsequently, each new find yields information that could dramatically change previous ideas.

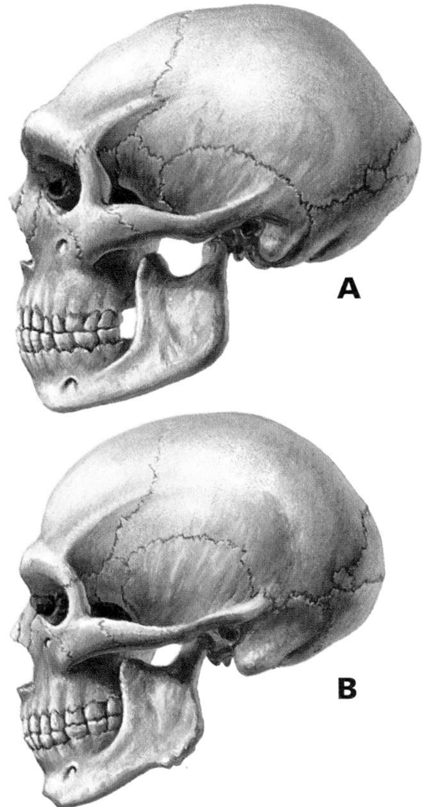

Figure 24-20 A characteristic of Neanderthals is a gap that occurs between the rear teeth and the jaw bone **(A).** Modern humans do not possess this **(B).**

SECTION ASSESSMENT

1. What kind of plants dominated the landscape of Central North America during the Oligocene?
2. How did the flora and fauna change in response to climatic changes during the Cenozoic?
3. What characteristics distinguish primates from other mammals?
4. What separates hominids from the other hominoids?
5. Discuss why it is difficult to reconstruct the evolutionary history of hominids.
6. **Thinking Critically** Explain why the evolution of grasses was a significant event for the mammals of central North America.

SKILL REVIEW

7. **Communicating** Suppose you are a journalist. In your science journal, write a press-release describing the recent discovery of a group of Neanderthal hominids. For more help, refer to the *Skill Handbook.*

Internet GeoLab

Huge Appetites

I&E

1.a Select and use appropriate tools and technology (such as computer-linked probes, spreadsheets, and graphing calculators) to perform tests, collect data, analyze relationships, and display data.

Also covers: 1.l

The study of how an organism interacts with its environment is called ecology. Ecology includes how an organism obtains energy from its environment. Animals do this by eating. Determining the diet of modern animals is relatively easy to do. We can observe them in their habitat and watch what they eat, or we can examine their feces. Paleoecology is the ecology of ancient organisms. Part of dinosaur paleoecology includes determining what and how dinosaurs ate. Imagine how much food some dinosaurs must have eaten!

Preparation

Problem

How do paleontologists tell what types of food different dinosaurs ate?

Hypothesis

What kind of evidence might you use to determine what type of diets dinosaurs ate? What are the diets of different animals today? Think about the characteristics of these different animals. Do most meat eaters share certain characteristics? What about plant eaters? **Form a hypothesis** about the skeletal characteristics of plant eaters and meat eaters.

Objectives

- **Gather** data and **communicate** interpretations about the characteristics of meat eaters and plant eaters.
- **Form conclusions** about the characteristics of plant eating and meat eating dinosaurs.
- **Discover** how sauropods might have shared food resources.

Data Sources

Go to the Earth Science Web Site at earthgeu.com to find links to fossil data on the Internet. You can also visit your library or local natural history museum to gather information about dinosaur diets.

This Late Cretaceous *Tyrannosaurus* is from Alberta, Canada.

Plan the Experiment

1. Find a resource that describes skeletal characteristics of meat-eating and plant-eating dinosaurs. The Earth Science Web Site lists sites with information about dinosaurs.
2. Gather information from the links on the Earth Science Web Site or the library about the environments that these two types of dinosaurs lived in and which dinosaurs lived in the same environments.
3. Design a data table to record your research results. Include categories such as Dinosaur Name, Meat or Plant Eater, Food Preference, Skeletal Characteristics, Jaw and Teeth Characteristics, and so on.

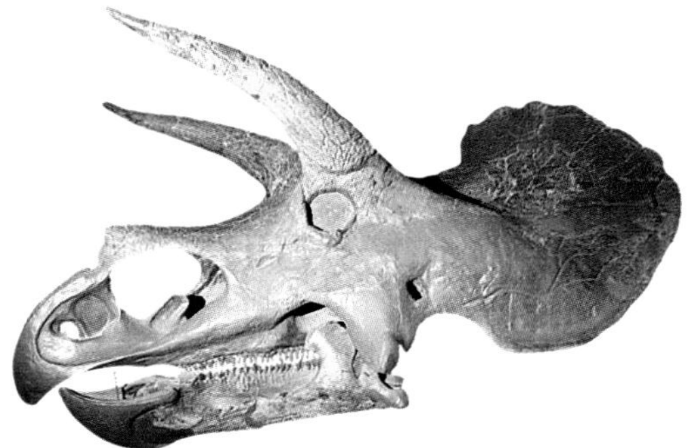

Triceratops was an herbivore.

Procedure

1. Complete your data table, including all information that you think is important.
2. Go to the Earth Science Web Site at earthgeu.com to post your data.
3. Visit sites listed on the Earth Science Web Site for more information on the diets of dinosaurs.

Conclude & Apply

Sharing Your Data Find this Internet GeoLab on the Earth Science Web Site at earthgeu.com. Post your data in the table provided for this activity. Use the additional data from other students to complete your chart and answer the Conclude and Apply questions.

1. What part of a dinosaur skeleton is most important in determining its diet? Why? What is the likelihood that this part of a skeleton will be preserved?
2. What are some other characteristics associated with dinosaur skeletons that help paleontologists determine what their diets were like?
3. Which were more abundant, meat-eating dinosaurs or plant-eating dinosaurs? Why?
4. How did sauropods share food resources? Describe the evidence used by paleontologists to determine how sauropods shared food resources.
5. How could the same evidence that is used to determine the diets of dinosaurs be used for other animals?

Science & Math

Model of *Parasaurolophus,* actual dinosaur was 10 m tall.

I&E 1.e

Weighing a Dinosaur

***H**ow do you weigh a dinosaur? Since the last one died 66 million years ago, you can't exactly ask one to step on a scale.*

Weight is a much better indicator of overall size than length or height. A python may be longer than an elephant, and a giraffe may be taller, but most people would agree that the heavier elephant was the bigger animal.

How to Weigh a Dinosaur

Two methods have been employed to estimate the weights of dinosaurs. The first uses a formula derived from the cross-sectional area of the leg bones. Since the legs supported the animal, their strength is an approximation of that animal's weight. Heavier animals have thicker bones.

The second method estimates the volume of a dinosaur, then multiplies the volume by the specific gravity. Most living animals have a specific gravity of around 0.9 kilograms per liter, but this can vary from 0.8 to just over 1. But how do you obtain the volume of a dinosaur? Calculating the volumes of simple shapes like cubes and spheres is easy, but dinosaur shapes were not simple.

One way is to measure the volume of an accurate scale model, then multiply that volume by the scale cubed (since the real dinosaur increased in length, width, and height by the scaled amount). The easiest way to obtain the volume is to immerse the model in water and measure how much water is displaced.

Procedure

1. Obtain several waterproof models of dinosaurs. You will also need a ruler, a calculator, and a graduated cylinder large enough to hold your biggest model.
2. Calculate the scale of each model. Measure the length of the model. The length of the actual dinosaur divided by the length of the model is the scale. Example: a model *Tyrannosaurus* is 30 cm long. A real *Tyrannosaurus* skeleton is 12 meters (1200 cm) long, so the scale is 1/40.
3. Fill the empty graduated cylinder with enough water to cover the model. Record that amount. Completely submerge the model and record the new level of the water. The difference between the first reading and second is the volume of the model.
4. Multiply the volume of the model by the cube of the scale. The result is the volume of the living dinosaur.
5. Multiply the volume of the living dinosaur by the density of living tissue (about 0.9 g/cc). The result is the weight of the living dinosaur, in grams. Convert that number into kilograms.

To find out more about dinosaur sizes and weights, visit the Earth Science Web Site at earthgeu.com

CHAPTER 24 Study Guide

Summary

Section 24.1

Mesozoic Paleogeography

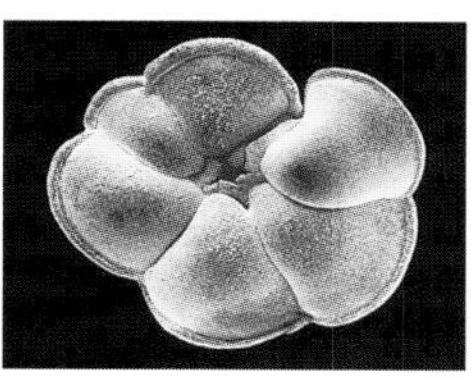

Main Ideas

- Geologists hypothesize that Pangaea broke apart as heat built up beneath it. Expansion occurred and ultimately resulted in the rifting apart of Pangaea.
- The first orogeny is characterized by igneous intrusions, whereas the second orogeny is characterized by thrust faulting and folding.

Vocabulary

Cordillera (p. 626)

Section 24.2

Mesozoic Life

Main Ideas

- The modern marine fauna include crabs, lobsters, shrimps, sponges, sea urchins, modern corals, snails, and clams. The major vertebrate groups of the modern fauna include bony fishes, sharks, aquatic reptiles, and aquatic mammals.
- The oceans contained vast numbers of ammonites that are now index fossils. The most common land plants were cycads, and the dominant land animals were dinosaurs.
- An upright posture distinguishes dinosaurs from other reptiles.

Vocabulary

angiosperm (p. 630)
dinosaur (p. 631)
ectotherm (p.633)
endotherm (p. 633)
iridium (p. 634)
modern fauna (p. 628)
Ornithischia (p. 632)
Saurischia (p. 632)

Section 24.3

Cenozoic Paleogeography

Main Ideas

- The Cenozoic tectonism is characterized by vertical normal faulting.
- During the Pleistocene, glaciers extended as far south as the courses of the Ohio and Missouri Rivers.

Vocabulary

Basin and Range Province (p. 637)
Tethys Sea (p. 638)

Section 24.4

Cenozoic Life

Main Ideas

- Large mammals evolved to feed on the abundant grasslands that developed during the Cenozoic. As the ice ages approached, many of the mammals that lived on these grasslands became extinct and were replaced by large mammals that were adapted to the cold and arid land south of the glaciers.
- Primates are mammals that developed specialized traits, including a grasping hand with an opposable thumb; stereoscopic vision; smaller, fewer, and less-specialized teeth, and a relatively large brain.

Vocabulary

hominid (p. 640)
hominoid (p. 640)
Homo sapiens (p. 640)
primate (p. 640)

CHAPTER 24

Assessment

Understanding Main Ideas

1. Which term describes down-dropped blocks of crust bounded by steeply dipping normal faults?
   a. volcanic arcs
   b. basins
   c. red beds
   d. batholiths

2. What are tiny organisms that float in the oceans and make their own food through the process of photosynthesis called?
   a. echinoids
   b. gastropods
   c. phytoplankton
   d. teleosts

3. What makes a dinosaur different from other reptiles?
   a. Dinosaurs were large.
   b. Dinosaurs laid eggs.
   c. Dinosaurs were carnivores.
   d. Dinosaurs had upright postures.

4. The Saurischia were one main group of dinosaurs. What was the other one?
   a. Ornithischia
   b. Sugoschia
   c. *Australopithecus*
   d. *Smilodon*

5. Which of the following best describes all theropod dinosaurs?
   a. herbivores
   b. carnivores
   c. omnivores
   d. quadrupedal

6. Which of the following best describes all sauropod dinosaurs?
   a. herbivores
   b. carnivores
   c. omnivores
   d. bipedal

7. What seed-bearing, flowering plants first appeared during the Cretaceous?
   a. angiosperms
   b. phytoplankton
   c. prosimians
   d. cycads

8. What flying reptiles became extinct at the end of the Cretaceous?
   a. plesiosaurs
   b. mosasaurs
   c. icthyosaurs
   d. pterosaurs

9. What region in the southwestern United States and North-Central Mexico is characterized by northeast-trending mountains and long, linear valleys?
   a. Basin and Range
   b. Colorado Plateau
   c. Sierra Nevada
   d. Mississippi Embayment

10. Use the following terms to complete the concept map below: pterosaurs, ichthyosaurs, primitive mammals, reptiles, rudists & corals, cycads, ammonites, angiosperms, dinosaurs, phytoplankton, turtles, mosasaurs, predatory reptiles.

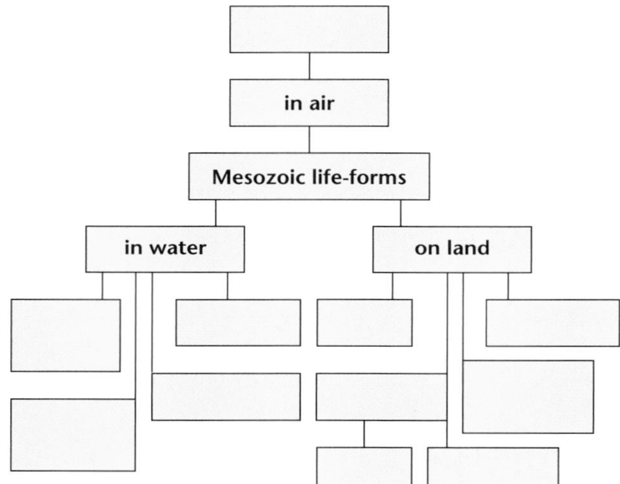

Test-Taking Tip

IF IT LOOKS TOO GOOD TO BE TRUE... Beware of answer choices in multiple-choice questions that seem ready-made and obvious. Remember that only one answer choice for each question is correct. The rest are made up by the test-makers to distract you. This means that sometimes they look very appealing. Check each answer choice carefully before finally selecting it.

Assessment

11. What mountain range formed as a result of the collision of India with Asia?
- **a.** Rocky Mountains
- **b.** Andes Mountains
- **c.** Himalayan Mountains
- **d.** Alps

12. What genus do Neanderthals belong to?
- **a.** *Australopithecus*
- **b.** *Ardipithicus*
- **c.** *Smilodon*
- **d.** *Homo*

Applying Main Ideas

13. Discuss how and when the Mid-Atlantic Ridge formed.

14. What evidence indicates that during the Late Triassic and Early Jurassic, western North America was arid?

15. What evidence suggests that dinosaurs were endothermic?

16. How did the paleogeography of Antarctica and Australia affect the climate during the Oligocene?

17. What characteristics separate primates from other mammals?

Thinking Critically

18. Compare the body shapes of ichthyosaurs and plesiosaurs to those of dolphins and whales. What can you conclude about these body shapes?

19. If a meteorite struck Earth at the end of the Cretaceous, does this mean that the impact caused the extinction of the dinosaurs? Explain why or why not.

20. Cite several reasons why the evolutionary history of ammonites is better supported than theories about hominid evolution.

21. Do the characteristics of today's coral reefs suggest that they might become oil reservoirs as the rudist reefs did? Explain.

Standardized Test Practice

1. What characterizes the second orogeny that affected western North America in the Mesozoic Era?
- **a.** low-angle faults
- **b.** block faults
- **c.** igneous intrusions
- **d.** clastic wedges

INTERPRETING DATA Use the table below to answer questions 2 and 3.

Mass Extinction Theories	
Evidence for meteor impact	Unusually high levels of iridium in Cretaceous-Paleogene boundary sediments; Discovery of Chicxulub crater
Evidence for massive volcanic activity	Volcanic eruptions during the Late Cretaceous in India; Unusually high levels of iridium, soot, and charcoal in Cretaceous-Paleogene boundary sediments

2. How does the presence of iridium at the Cretaceous-Paleogene boundary support the theory of massive volcanic activity?
- **a.** Iridium is deposited after a large fire is extinguished.
- **b.** Iridium is a common byproduct of combustion reactions.
- **c.** Iridium is found in abundance in Earth's core.
- **d.** All of the above.

3. Underneath the Chicxulub crater is a large layer of melted rock and a layer of jumbled rocks. Why are these rocks jumbled?
- **a.** They have been broken up by water erosion.
- **b.** They contain a higher level of iridium.
- **c.** They are pieces of the meteorite that broke off upon impact.
- **d.** They fell into the crater after the impact.

earthgeu.com/standardized_test

UNIT 6

GeoDigest

For a **preview** of geologic time, study this GeoDigest before you read the chapters. After you have studied geologic time, use the GeoDigest to **review** the unit.

Geologic Time

Fossils and the Rock Record

Geologists have separated Earth's history into divisions based upon the fossil record and organized these divisions into the geologic time scale. The divisions found on the geologic time scale—in descending order of length—are eons, eras, periods, and epochs. The principles of uniformitarianism, original horizontality, superposition, and cross-cutting relationships are used to interpret Earth's rock record—and thus describe our planet's history. Unconformities caused by weathering and erosion or by periods of nondeposition indicate missing layers in the rock record. Fossils are the remains and evidence of plants and animals that once lived on Earth. Fossils preserved in the rock record provide information about past environmental conditions and evolutionary changes in life-forms. They thus help to correlate rock layers from one area to another.

Absolute Age of Rocks Absolute age dating measures the actual age of an object such as a mineral, rock, or fossil. Radioactive dating uses the decay rates of various types of radioactive minerals to determine the actual age of a rock or fossil. Radioactive decay is the random emission of particles from the nucleus of a radioactive atom at a constant rate. The time it takes a radioactive isotope to decay fifty percent of its original mass to a nonradioactive element is known as the mineral's half-life. Tree rings and varves are other methods of determining the dates of events and changes in the environment. Key beds in the rock record that mark the time of their occurrence include volcanic ash and meteorite-impact debris that spread out over large areas of Earth in a small amount of time.

The Precambrian

Geologists have used radiometric dating to show that Earth must be at least 4.2 billion years old. Because all of the objects in the solar system formed at the same time, and Moon rocks and meteorites are dated at 4.6 billion years old, Earth is assumed to be 4.6 billion years old also. Early Earth was hot because of abundant radioactive isotopes, bombardment by meteorites, and gravitational contraction. Earth's crust formed when the uppermost portion of the mantle cooled. The early crust weathered and formed sediments. Sediment-covered slabs of early crust were subducted and generated magmas with granitic compositions. During the Archaean Eon, granitic crust formed microcontinents. The microcontinents collided with one another throughout the Proterozoic Eon to form the cores of today's continents. Earth's early

Fossil snake

atmosphere and the oceans formed through the process of outgassing. Nearly all of the oxygen in the atmosphere is a result of photosynthesis. Certain minerals oxidize, or rust, in the presence of free oxygen. Proterozoic red beds are sedimentary rock deposits that contain oxidized iron. They are evidence that there was free oxygen in the atmosphere during the Proterozoic Eon.

Early Life on Earth All the ingredients were present on the early Earth to form proteins, which are the building blocks of life. Experiments have demonstrated that amino acids, the molecules that make up proteins, were likely abundant on the surface of early Earth. The first life was likely prokaryotic cells, which are small and contain no nuclei. Eukaryotic cells, which contain nuclei and are generally larger and more complex than prokaryotic cells, emerged later. The first evidence of multicellular organisms are fossils of eukaryotic algae from a 2.1 billion year old banded iron formation in northern Michigan. By about 670 million years ago, the multicellular Ediacaran organisms began to flourish throughout the world.

The Paleozoic Era

Early in the Paleozoic, a shallow sea covered the ancient North American continent of Laurentia, which was located near the equator. Laurentia's continental margin was passive—no tectonic activity was occurring. The fauna included many new organisms that evolved during the Cambrian explosion. Fossils of trilobites and articulate brachiopods are common. Sea level changes cause depositional environments to change position laterally. This results in adjacent depositional sequences overlying each other in vertical succession.

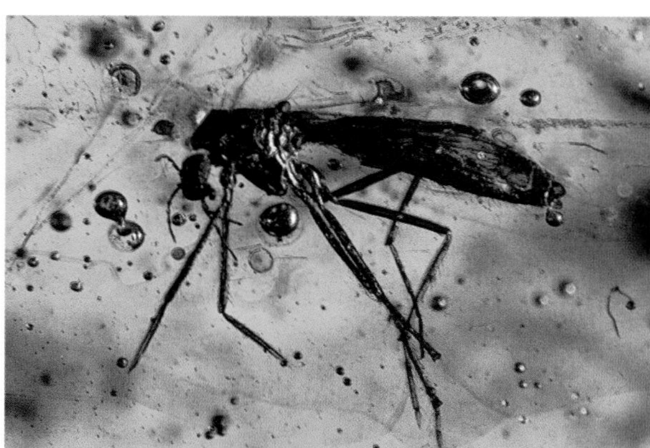

Insect in amber

Middle Paleozoic Deposits of evaporite minerals such as gypsum, and halite formed in the mid-Paleozoic as areas of ocean water were isolated and evaporated. Collisional tectonism occurred, causing mountain building along active continental margins. Clastic wedges formed as a result of mountain building as deposits of sedimentary rocks formed adjacent to uplifted areas in wedge-shaped formations. Articulate brachiopods, corals, and many other organisms dominated the seas. Fishes evolved as top ocean predators and plants moved onto land. At the end of the Ordovician and in the Late Devonian, large numbers of organisms became extinct in relatively short periods of time. This phenomenon is called a mass extinction.

Late Paleozoic In the Late Paleozoic, continents collided to form Pangaea. Cycles of glacial/interglacial periods during the Pennsylvanian are represented by cyclothems—stacked deposits of

Vital Statistics

Half-Lives of Radioactive Isotopes

Isotope	Half-life
Carbon 14	5730 (+/- 30) years
Uranium 235	700 million years
Rubidium 87	4.9 billion years
Uranium 238	4.5 billion years

alternating transgressive and regressive rock sequences. The ocean shrank during glacial periods and expanded during interglacial periods. Seeds and the amniote egg developed during the Late Paleozoic, developments that allowed plants and reptiles to move onto dry land. Regression of the ocean and climate change led to a mass extinction at the end of the Permian Period.

The Mesozoic and Cenozoic Eras

Mesozoic Era In the Mesozoic, Pangaea broke up. Geologists hypothesize that the size of the supercontinent led to a heat buildup and expansion of the landmass, resulting in rifting that broke it up. Two orogenies occurred. The first orogeny is characterized by igneous intrusions whereas the second orogeny is characterized by thrust faulting and folding. Ammonites were prevalent in the Mesozoic ocean; the most common land plants were cycads; and the dominant land animals were dinosaurs. High levels of iridium, and the Chicxulub structure in the rocks at the Cretaceous-Paleogene boundary indicate that a meteorite hit Earth at the end of the Cretaceous Period. This meteorite may have led to the extinction of the dinosaurs. Dinosaurs are distinguished from reptiles by their upright posture. Birds may be descended from dinosaurs. The fossil *Archaeopteryx*, which contains features of both dinosaurs and birds, is cited as evidence for the relationship. The fossils of reptiles are distinguished from the fossils of mammals by tooth shape, and the number of lower jawbones.

Cenozoic Era The major orogeny of the Cenozoic is characterized by vertical normal faulting. During the Pleistocene ice ages, glaciers extended as far south as the courses of the Ohio and Missouri rivers. By the Oligocene, grassy savannas covered much of Central North America. Large mammals evolved to feed in these abundant grasslands. As the ice ages approached, many grazers became extinct and were replaced by large mammals that were adapted to cold and arid climates. Primates emerged during the Cenozoic. Primates are mammals that developed specialized traits such as a grasping hand with an opposable thumb; smaller, fewer, and less specialized teeth; stereoscopic vision; and a relatively large brain.

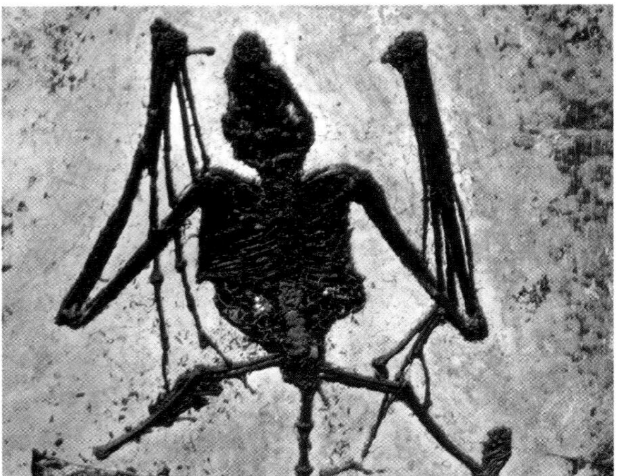

Fossil bat

FOCUS ON CAREERS

Paleoecologist

Paleoecologists study the ecology of ancient animals and plants. Where did a fossil specimen live? What did it eat? What ate it? What were the limiting factors in its environment? Paleontology, biology, and ecology are put together to complete pictures of ancient environments. Curiosity and patience are essential to a paleoecologist, but advanced degrees are also required.

GeoDigest

ASSESSMENT

Understanding Main Ideas

1. What feature, caused by weathering and erosion or by periods of nondeposition, indicates missing layers in the rock record?
   a. uniformitarianism
   b. unconformity
   c. horizontality
   d. cross-cutting

2. What is the random emission of particles from the nucleus of a radioactive atom at a constant rate called?
   a. radioactive minerals
   b. half-life
   c. radioactive decay
   d. uranium

3. What formed when the uppermost portion of Earth's mantle cooled?
   a. meteorites
   b. radioactive isotopes
   c. sediment
   d. Earth's crust

4. Earth's early atmosphere and the oceans formed through what process?
   a. photosynthesis
   b. free oxygen
   c. outgassing
   d. Proterozoic red beds

5. The first life on Earth was a small cell with no nucleus called what?
   a. a prokaryotic cell
   b. a eukaryotic cell
   c. algae
   d. Ediacaran organisms

6. What forms when areas of ocean water are isolated and evaporate?
   a. collisional tectonism
   b. clastic wedges
   c. evaporite minerals
   d. brachiopods

Fossil conifer cone

7. What sequence represents alternating glacial/interglacial periods during the Pennsylvanian?
   a. cyclothems
   b. amniote egg
   c. clastic wedges
   d. microcontinent

8. What were the most common land plants of the Mesozoic?
   a. ammonites
   b. conifers
   c. cycads
   d. embayments

9. Which of these land features was prominent in Central North America during the Oligocene?
   a. grasslands
   b. forests
   c. primates
   d. glaciation

10. Which of the following is a trait of primates?
    a. more-specialized teeth
    b. smaller brain
    c. hooves
    d. stereoscopic vision

Thinking Critically

1. Put the following geologic time divisions in order from shortest to longest: period, eon, epoch, era.
2. Describe why early Earth was hot.
3. Discuss how paleontologists distinguish reptiles from mammals in the fossil record.

Unit 7

Resources and the Environment

Herds of scimitar-horned oryx meander across the rolling hills and meadows of The Wilds, a unique, 4000 ha wildlife preserve and environmental education center in rural Ohio. The Wilds was developed on land donated by a local power company after the area was strip-mined for coal. When resources such as coal are removed by surface mining, reclamation not only can restore the area to its former contours, but also can provide new educational and recreational activities for local residents.

Unit Contents

Go to the **National Geographic Expedition** on page 898 to learn more about topics that are connected to this unit.

Scimitar-horned Oryx,
The Wilds, Cumberland, Ohio

Chapter 25

Earth Resources

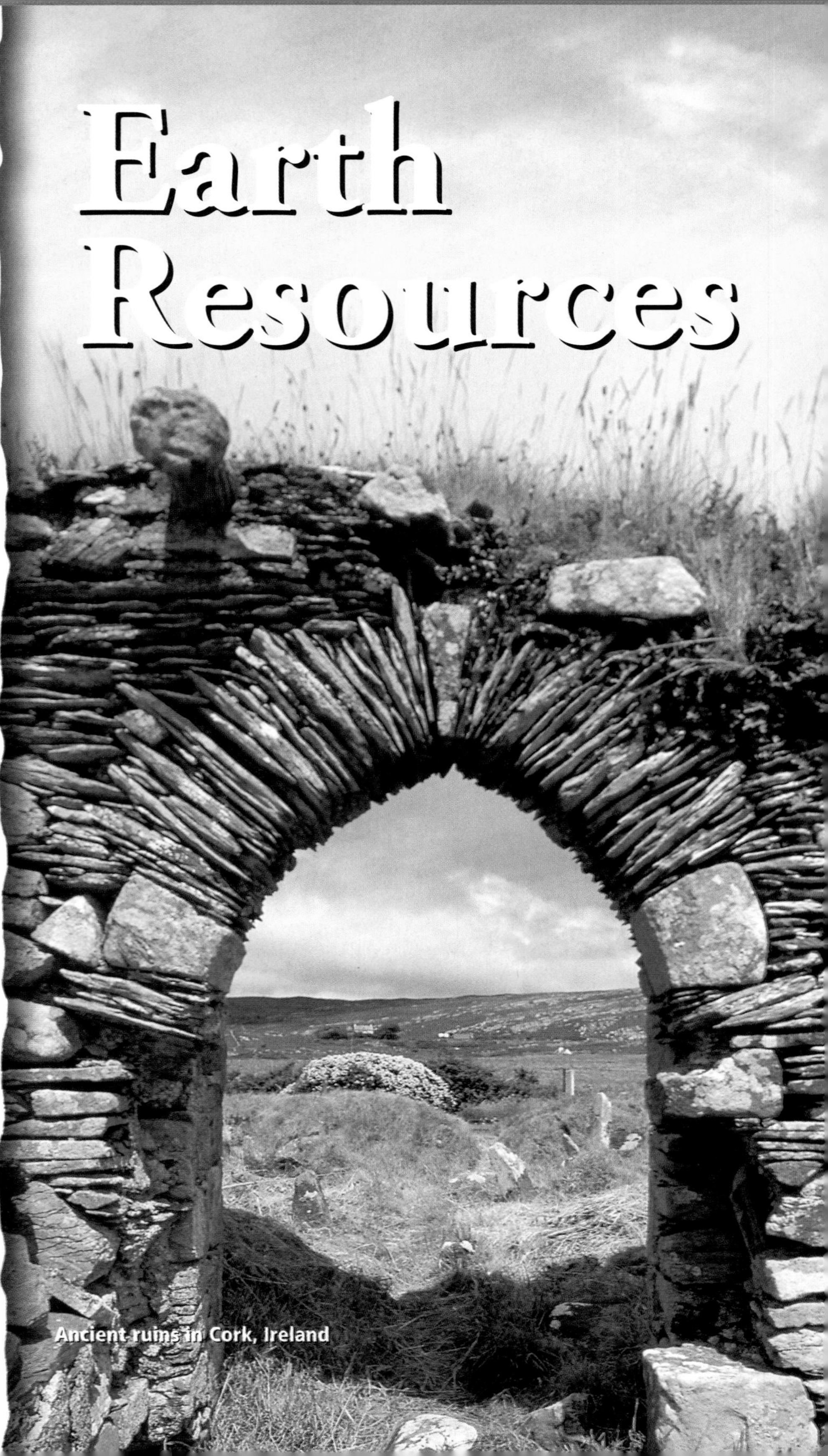

What You'll Learn

- What materials are considered to be Earth resources.
- Which Earth resources are renewable and which are nonrenewable.
- How Earth resources are used.

Why It's Important

Earth resources can be derived from either living or nonliving things. Many Earth resources are essential for life. Once used, some resources cannot be replaced, whereas others can be replaced in relatively short periods of time. The use of Earth resources must be balanced for life on Earth to continue.

To learn more about earth resources, visit the Earth Science Web Site at earthgeu.com

Ancient ruins in Cork, Ireland

Discovery Lab Origins of Resources

All the material goods that you use every day are matter. One way in which matter can be classified is whether it comes from living things or from nonliving things. In this activity, you will identify the origins of some common materials.

1. In your science journal, make a data table with the headings "Item," "Living," "Nonliving," "Easily Replaced," and "Not Replaceable."
2. Look around your classroom and list common items that you see in the first column of your table.
3. Classify the matter in each item as coming from either living or nonliving things, and as being either easily replaced or not replaceable, by placing check marks in the appropriate columns.

Classify Compare your data table with those of several other students. Which of the items were identified as coming from living things? Which came from nonliving things? Which are easily replaced? Which are not replaceable? What criteria did you use to classify each item? Were you unable to classify any item using just these categories? Explain.

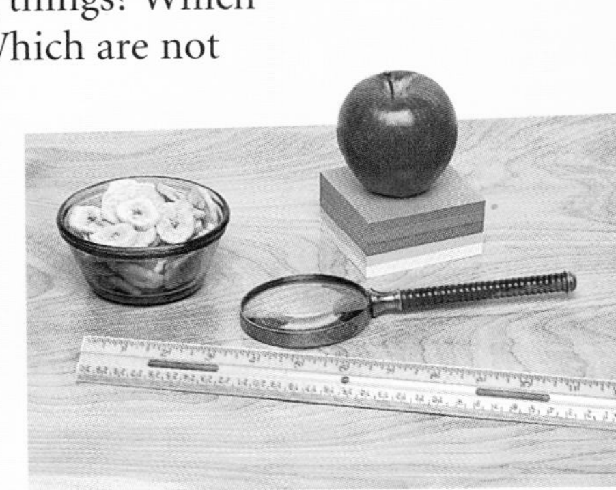

SECTION 25.1 What are resources?

OBJECTIVES

- **Distinguish** *between renewable and nonrenewable resources.*
- **Identify** *renewable and nonrenewable resources.*

VOCABULARY

natural resource
renewable resource
sustainable yield
nonrenewable resource

Did you eat an apple or a banana for breakfast this morning? Every day, you eat food and drink water because these resources are necessary for you to live. You and every other living thing on Earth must have certain resources to grow, develop, maintain life processes, and reproduce. In addition to food and water, most animals also need shelter. Think about the resources used to provide shelter for you and your family. Maybe your home is made of brick or stone, or perhaps it has wood shingles or aluminum siding. All of these materials come from Earth.

NATURAL RESOURCES

The resources that Earth provides are known as **natural resources.** Natural resources include Earth's air, water, and land; all living things; and nutrients, rocks, and minerals in the soil and deep in Earth's crust. Recall from Chapter 3 that neither matter nor energy can be

Figure 25-1 Trees are renewable resources because they are living things that reproduce within a relatively short period of time.

created or destroyed, but both can be changed from one form to another. You will find out how some natural resources are transformed in the *Science in the News* feature at the end of this chapter. One way in which natural resources are changed is through cycling, as in the carbon, nitrogen, and water cycles that you learned about in previous chapters.

Have you ever recycled an aluminum can? If you recycle, you probably are already aware that some resources cannot be replaced in a reasonable amount of time. If you have ever mowed a lawn or planted a garden, you know that some other natural resources can be used and replaced through natural processes in a short period of time. Both types of natural resources are necessary for life on Earth.

RENEWABLE RESOURCES

Do you live in an area that has an autumn season? During the autumn, as the amount of sunlight declines, deciduous trees stop producing chlorophyll and become dormant so that they can survive the cold winter season. When the leaves of trees do not contain the green pigment chlorophyll, many other colors can be seen, as shown in ***Figure 25-1.*** Eventually, all of these colorful leaves fall to the ground. However, in the spring, new leaves appear and the trees continue their life cycles. If you cut down a tree, you can replace that tree by planting a seedling in its place. Trees are examples of **renewable resources,** which are natural resources that it is possible to use indefinitely without causing a reduction in the available supply. Renewable resources include fresh air; fresh surface water in lakes, rivers, and streams; most groundwater; fertile soil; elements that cycle through Earth's systems, such as nitrogen, carbon, and phosphorus; and all living things. Resources that exist in an inexhaustible supply, such as solar energy, also are renewable resources. Renewable resources are replaced through natural processes at a rate that is equal to, or greater than, the rate at which they are being used.

Living Things Organisms in the biosphere are important renewable resources. Plants and animals reproduce, and therefore, as long as some mature individuals of a species survive, they can be replaced. Crops can be planted every spring and harvested every fall from the same land as long as the Sun shines, the rain falls, and the required nutrients are provided by organic matter or fertilizers. Animals that are raised for food, such as chickens and cattle, also can be replaced in short periods of time. Forests that are cut down for the production of paper products can be replanted and ready for harvest again in 10 to 20 years. Trees that are cut down for timber also can be replaced after a period of up to 60 years. Humans who use natural resources responsibly are practicing management techniques to replace resources as they are used, as shown in ***Figure 25-2.*** The replacement of renewable resources at the same rate at which they are consumed results in a **sustainable yield.**

Sunlight Some of Earth's renewable resources are not provided by Earth. The Sun provides an inexhaustible source of energy for all processes on Earth. Sunlight is considered to be a renewable resource because it will continue to be available for at least the next 5 billion years.

Figure 25-2 Paper companies may cut down entire forests to provide wood for pulp and paper, but they also manage the forests and replant on a regular schedule so that the forest resources are always available.

NONRENEWABLE RESOURCES

Suppose you visit a fine jewelry store. You notice that diamonds, such as those shown in ***Figure 25-3,*** are very expensive. Why are diamonds so expensive? After all, they form through geologic processes, just like quartz and feldspar do. Diamonds are expensive because the supply of diamonds is limited. When all the diamond mines that currently exist have been exhausted, no more natural diamonds will become available. Diamonds are an example of a **nonrenewable resource,** a resource that exists in a fixed amount in various places in Earth's crust and can be replaced only by geological, physical, and chemical processes that take hundreds of millions of years. Resources such as fossil fuels, diamonds and other gemstones, and elements such as gold, copper, and silver are therefore considered to be nonrenewable. Nonrenewable resources are exhaustible because they are being extracted and used at a much faster rate than the rate at which they were formed.

Figure 25-3 Diamonds form through geologic processes. ***Why is a diamond ring so expensive?***

DISTRIBUTION OF RESOURCES

Do you live in an area that has coal mines, oil wells, or deposits of bauxite, the ore that contains aluminum? Perhaps you live near a scenic river or a hot spring. Wherever you live, you probably have noticed that natural resources are not distributed evenly on Earth. The availability of natural resources helps determine the wealth and the power of countries around the world. Countries with many natural resources, such as the United States, are able to support higher living standards for their citizens than countries with fewer resources. However, smaller countries may have an abundance of

Figure 25-4 This graph shows the percentage of crude oil that is consumed by the United States and worldwide. Note that the United States consumes 27 percent of the total crude oil used each day.

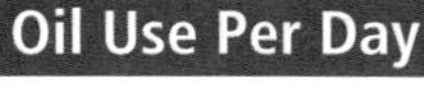

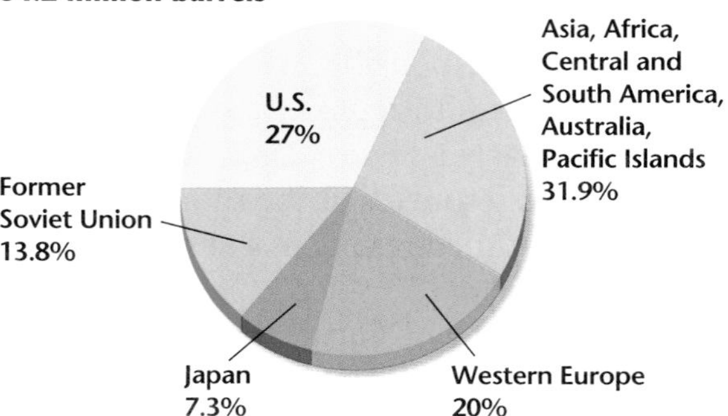

one natural resource that is needed by many other countries. Surinam and Guyana, in South America, for example, have some of the richest reserves of bauxite in the world. Saudi Arabia and Kuwait have some of the richest petroleum reserves.

The United States has a high standard of living and it consumes approximately 30 percent of Earth's mineral and energy resources each year, even though it has only 6 percent of the world's population. Some countries with larger populations have lower standards of living than the United States, and thus, these countries do not consume as many resources. One nonrenewable resource that developed countries consume in ever-increasing amounts is crude oil. ***Figure 25-4*** shows the percentage of total worldwide consumption of crude oil versus the amount used in the United States daily.

Section Assessment

1. What is a natural resource?
2. Explain the difference between a renewable resource and a nonrenewable resource.
3. Name three renewable resources and three nonrenewable resources.
4. Explain why it is important to know whether a resource that you are using is renewable or nonrenewable.
5. **Thinking Critically** Fossil fuels are derived from the remains of once-living organisms, and living things are renewable resources. Why, then, are fossil fuels considered to be nonrenewable resources?

Skill Review

6. **Concept Mapping** Make a concept map of the major ideas in this section using the section headings and the vocabulary terms. For more help, refer to the *Skill Handbook*.

earthgeu.com/self_check_quiz

Section 25.2 Land Resources

Earth Sciences 9.a

In the springtime, many people visit garden centers and buy sand, mulch, peat moss, topsoil, and different kinds of rocks for landscaping purposes. These materials all are derived from land, a valuable natural resource. Land provides places for humans and other organisms to live and interact. Land also provides spaces for the growth of crops, forests, grasslands, and for wilderness areas.

OBJECTIVES

- **Describe** *why land is considered to be a natural resource.*
- **Recognize** *the need to protect Earth's land as a resource.*
- **Explain** *how humans adjust to the uneven distribution of land resources.*

VOCABULARY

desertification
bedrock
aggregate
ore
gangue

Protected Land

Of all the land in the United States, 42 percent is certified as public land, which consists of forests, parks, and wildlife refuges. Of this public land, 73 percent is located in Alaska, and 22 percent is located in the western states. These land areas are federally administered to protect timber, grazing areas, minerals, and energy resources. Some public land, such as national forests, is managed for sustainable yield and includes multiple-use areas where resources are used for many purposes, including recreation. Public land includes grasslands, prairies, deserts, scrub forests, and other open spaces. Some of these more remote areas eventually may become wilderness areas, places that are maintained in their natural state and protected from development.

The national park system preserves scenic and unique natural landscapes, preserves and interprets the country's historic and cultural heritage, protects wildlife habitats and wilderness areas, and provides areas for various types of recreation. About 49 percent of the land in the national park system is designated as wilderness.

National wildlife refuges provide protection of habitats and breeding areas for wildlife, and some provide protection for endangered species, as shown in ***Figure 25-5.*** Other uses of the land in wildlife refuges, such as fishing, trapping, farming, and logging, are permitted as long as they are compatible with the purpose of each individual refuge.

Soil

Do you know what is in the soil under your feet? It can take up to 1000 years to form just a few centimeters of topsoil, yet it can be lost in a matter of minutes as a result of erosion by wind or water. Plowing and leaving bare ground without plant cover can increase topsoil loss. The loss of topsoil makes soil less fertile and less able to hold water. The result is poorer crops. Today, topsoil is eroding faster than it forms on about

Figure 25-5 The Aransas National Wildlife Refuge was established in 1937 to protect the vanishing wildlife of coastal Texas. The whooping crane, an endangered migratory bird species in the United States, makes its winter home in this refuge.

Figure 25-6 When animals are allowed to graze on marginal land in arid or semi-arid climates, the topsoil can be eroded easily, and desertification can result.

one-third of Earth's croplands. Each decade, Earth loses about seven percent of its topsoil, yet the eroded croplands must feed an ever-increasing human population. In arid and semi-arid areas of the world, the loss of topsoil leads to **desertification,** which is the process whereby productive land becomes desert. Desertification can occur when too many grazing animals are kept on arid lands, as shown in ***Figure 25-6,*** or when soil is compacted by large herds of heavy animals or heavy farm equipment.

Desertification is a growing problem in cattle-producing areas in North Africa south of the Sahara, in the Middle East, in the western half of the United States, and in Australia. Desertification also occurs when trees and shrubs are cut down for use as fuel in areas with few energy resources. When these plants are not replaced, erosion can lead to the loss of topsoil. Desertification can be prevented by reducing overgrazing and by planting trees and shrubs to anchor soils and retain water.

BEDROCK

Underneath the topsoil is a layer of soil consisting of inorganic matter, including broken-down rock, sand, silt, clay, and gravel. This deeper soil layer lies on a base of unweathered parent rock called **bedrock.** Bedrock is solid rock, and it may consist of limestone, granite, marble, or other rocks that can be mined in quarries, as shown in ***Figure 25-7.*** Slabs of bedrock are often cut from quarry faces. Such large pieces of bedrock are used in the construction of buildings, monuments, flooring, and fireplaces.

Figure 25-7 The finest-grained marble in the world is obtained from the bedrock at Carrara, Italy.

AGGREGATES

Have you ever watched a highway being built? You may have seen construction workers place layers of crushed or broken stone, pebbles, or sand on the ground before they began to build the highway surface. The materials used for this first layer come from an **aggregate,** which is a mixture of gravel, sand, and crushed stone that naturally accumulates on or close to Earth's surface. Some aggregates are found on floodplains in river valleys and in alluvial fans in mountainous areas. Other aggregrates were deposited by glacial activity in moraines, eskers, kames, and outwash plains. Some aggregates contain erratics, shown in ***Figure 25-8,*** which are rocks or rock fragments deposited by glaciers far from their origins.

Aggregates used in construction are mixed with cement, lime, gypsum, or other materials to form concrete or mortar. An aggregate provides volume and stability to the resulting mixture, and it also makes the finished surface more resistant to erosion and weathering. The most commonly used natural aggregates are sand, crushed or broken rocks, and gravel.

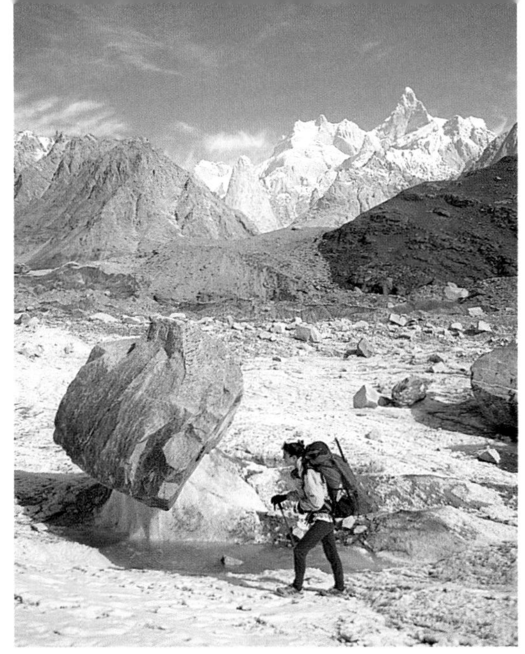

Figure 25-8 Glaciers often deposit erratics in areas with rocks that are vastly different. Some erratics are deposited on softer rocks, which erode more quickly, and thus, the erratics are left perched on top.

ORES

An **ore** is a natural resource that can be mined at a profit; that is, it can be mined as long as its value on the market is greater than the cost of its extraction. For example, the mineral hematite (Fe_2O_3) is an iron ore because it contains 70 percent iron by weight. Other minerals, such as limonite, also contain iron, but they are not considered to be ores because the percentage of iron contained in them is too low to make extraction profitable. Ores can be classified by the manner in which they formed. Some ores are associated with igneous rocks, whereas others are formed by processes at Earth's surface.

Settling of Crystals Ores associated with igneous rocks may contain iron, chromium, and platinum. Chromium and platinum ores can form when minerals crystallize and settle to the bottom of a cooling body of magma. Chromite ore deposits are often found near the bases of sills and other igneous intrusions. One of the largest deposits of chromium and platinum in the world is the Bushveldt Complex in South Africa.

Earth Sciences

9.a Students know the resources of major economic importance in California and their relation to California's geology.

Hydrothermal Fluids The most important sources of metallic ore deposits are hydrothermal fluids. Hot water and other fluids may be part of the magma that is injected into surrounding rock during the last stages of magma crystallization. Because atoms of metals such as copper and gold do not fit into the crystals of feldspar and other minerals during the cooling process, they become concentrated in the remaining magma. Eventually, a solution rich in metals and silica moves into the surrounding rocks to create ore deposits, known as hydrothermal veins. Hydrothermal veins, such as the one shown in ***Figure 25-9,*** commonly form along faults and joints.

Figure 25-9 This gold nugget contains a hydrothermal vein that is composed of quartz.

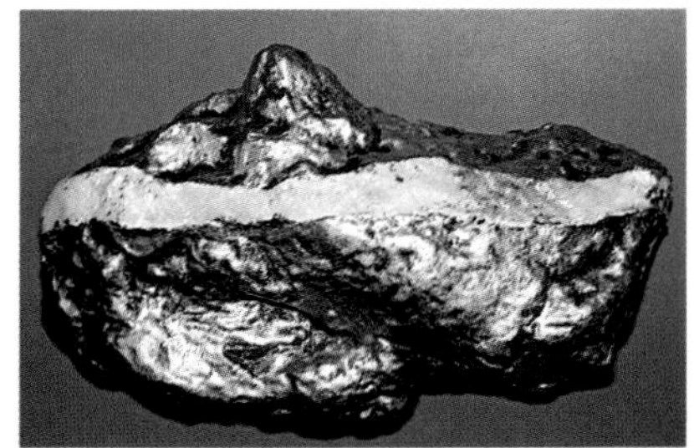

Chemical Precipitation Ores of manganese and iron most commonly originate from chemical precipitation in layers. Iron ores in sedimentary rocks are often found in bands made up of alternating layers of iron–bearing minerals and chert. The origin of these ores, called banded iron formations, is not fully understood. Banded iron formations may have resulted from volcanic activity or weathering and then may have been deposited in layers in shallow, water-filled basins during the Precambrian to form sedimentary rocks. Banded iron deposits are discussed in more detail in Chapter 22.

Placer Deposits Some sediments, such as grains of gold and silver, are heavy. When stream velocity decreases, as, for example, when a stream flows around a bend, heavy sediments may be dropped by the water and deposited in bars of sand and gravel. Sand and gravel bars that contain heavier sediments such as gold nuggets, gold dust, diamonds, platinum, and gemstones, as well as rounded pebbles of tin and titanium oxides, are known as placer deposits. Some of the gold found during the Gold Rush in California during the late 1840s was located in placer deposits.

Concentration by Weathering Some ores form when the minerals in rocks are concentrated by weathering. For example, aluminum forms in bauxite through weathering in tropical climates. Other metals that become more concentrated as rocks weather include nickel, copper, silver, lead, tin, mercury, uranium, and manganese.

OTHER LAND RESOURCES

Are there many brick buildings where you live? Bricks are made from clay, another resource found on land. Clay is a sediment, and a group of minerals, made up of tiny particles with diameters of less than 0.004 mm. In addition to bricks, clay is used to make china, ceramics, tiles, and pottery. It is also used in the paper-making process.

Figure 25-10 The Wieliczka Salt Mine in Poland has been mined continuously since the thirteenth century. Today, it is well known for its many rock-salt structures, including the underground Queen Kinga's Chapel. The statues, chandeliers, railings, altars, and candlesticks are carved from rock salt.

Salt, or sodium chloride, also occurs in deposits both on Earth's surface and underground. One of the most famous salt mines in the world, the Wieliczka Salt Mine in Poland, is shown in ***Figure 25-10.*** Other mineral resources found on land include gypsum, which is used to make plaster; talc, which is used in cosmetics; and graphite, which is used as a lubricant. Both salt and gypsum deposits can form when seawater evaporates.

USING LAND RESOURCES

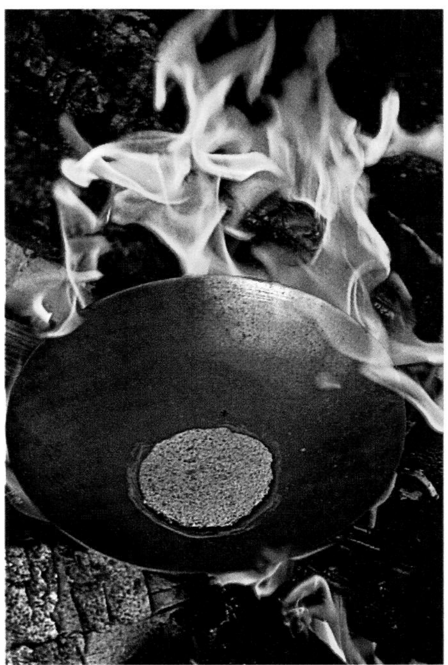

Figure 25-11 Small-scale miners in the Amazon River Basin use mercury to extract gold.

Although many of the resources that you have learned about in this section can be extracted with little impact on the surrounding environment, the extraction of others can have negative impacts. Mines that are used to remove materials from the ground surface destroy the original ground contours. Open-pit mines leave behind waste rock that can weather and release pollutants into the air and water. The extraction of mineral ores often involves grinding parent rock to separate the ore. The material left after the ore is extracted, called **gangue,** may release harmful chemicals into groundwater or surface water. Sometimes, chemicals that harm the environment are used to separate ores. Mercury is used to extract gold from alluvial deposits of sand along rivers in the Amazon River Basin. Liquid mercury dissolves the gold particles in these deposits, forming a solution. When the solution is heated, as shown in ***Figure 25-11,*** the mercury evaporates, leaving the gold behind. Miners who have inhaled mercury vapors and people who live downstream from the mining operations have been poisoned by mercury. Mining sometimes exposes other materials, such as pyrite, that form acids as they weather and pollute groundwater. In addition to causing environmental problems, mining itself is a dangerous activity. In fact, the National Safety Council has identified mining as the most dangerous occupation in the United States: of all occupations, it has the highest yearly death rate.

SECTION ASSESSMENT

1. Describe two resources found on land.
2. Why is the loss of topsoil through erosion considered to be a worldwide problem?
3. Name five ways in which ores can form, and give an example of a mineral that forms in each way.
4. What is the difference between an aggregate and an ore?
5. **Thinking Critically** What options would humans have if a land resource became depleted?

SKILL REVIEW

6. **Predicting** Many developing countries would like to have the same standard of living as that enjoyed by citizens of the United States. As these countries become industrialized, what may happen to the demand for land resources? How can this demand be met? For more help, refer to the *Skill Handbook.*

earthgeu.com/self_check_quiz

SECTION 25.3 Air Resources

Earth Sciences 7.a

OBJECTIVES

- **Recognize** *that the atmosphere is an Earth resource.*
- **Describe** *the importance of clean air.*

VOCABULARY

pollutant
air pollution

Have you ever gone outside after a rainstorm and noticed how clean and fresh the air smelled? Most of the time, people don't think about air. However, air contains substances that all organisms need to survive, including nitrogen, oxygen, carbon dioxide, hydrogen, methane, and ozone. Water vapor can make up as much as five percent of air by volume. For humans and all other animals, the most important component of air is oxygen. Oxygen makes up 21 percent of air.

ORIGIN OF OXYGEN

Most organisms on Earth require oxygen or carbon dioxide to maintain their life processes. Oxygen has not always been a part of Earth's atmosphere. Scientists hypothesize that 4.6 to 4.5 billion years ago, Earth's atmosphere was similar to the mixture of gases released by erupting volcanoes. These gases include carbon dioxide, nitrogen, and water vapor. As Earth cooled and became more solid, rains washed most of the carbon dioxide out of the atmosphere and into the oceans. Early life-forms in the seas used carbon dioxide during photosynthesis and released oxygen and water vapor. Over time, oxygen in the atmosphere built up to levels that allowed the evolution of organisms that required oxygen for life processes.

DISRUPTING EARTH'S CYCLES

The geochemical cycles of Earth's atmosphere are in a delicate balance. Volcanic eruptions release various gases and dust particles into the atmosphere. Photosynthetic organisms in the oceans and on land take in and use carbon dioxide and release oxygen. Other organisms take in this oxygen and release carbon dioxide. Life on Earth continues to survive as a result of this balanced gas exchange.

However, human activities are disrupting these cycles. For example, humans burn fossil fuels to produce electricity and burn forests to clear land. These two activities release carbon dioxide into the atmosphere. Increased amounts of carbon dioxide are thought to play a role in global warming, which is the gradual rising of Earth's average surface temperature. The human alteration of the carbon cycle has the potential to change global climate and therefore the environments of food-producing regions. Rainfall patterns in the tropical rain forests of the Amazon River Basin have already changed as a result of the loss of forest cover. One of these rain forests is shown in ***Figure 25-12.*** You will find out how deforestation is affecting tropical rain forests worldwide in the *Problem-Solving Lab* on the next page.

Figure 25-12 The forests of the Amazon River Basin are sometimes called the lungs of Earth. They help to balance gas exchange for the entire planet.

Humans also disrupt other geochemical cycles. By burning fossil fuels and using fertilizers that contain nitrogen, humans release about three times as much nitrogen oxide and ammonia gas into the atmosphere as do the natural processes of the nitrogen cycle. In the atmosphere, nitrogen oxides are converted to nitric acid, which returns to Earth in acid precipitation and damages surface water, plants, and soil. Human activities also release sulfur into the atmosphere when coal and oil are burned to produce electricity. Sulfur in the atmosphere is converted to sulfuric acid, which also returns to Earth in the form of acid precipitation. Both the excess nitrogen oxides and sulfur are **pollutants,** which are substances that can adversely affect the survival, health, or activities of organisms. These are only a few of the chemicals that human activities release into the atmosphere. Small amounts of toxic metals, such as lead, cadmium, and arsenic, also are released. When pollutants in air occur in quantities that become harmful to human health and the health of the environment, **air pollution** results.

Earth Sciences

7.a Students know the carbon cycle of photosynthesis and respiration and the nitrogen cycle.

Problem-Solving Lab

Interpreting Graphs

Calculate the rate of deforestation Many experts are concerned about the loss of the forest cover in tropical rain forests worldwide. In the Amazon River Basin, for example, scientists estimate that 1 hectare (ha, about 2.47 acres) of forest is cut down each hour. Nearly 20 million ha of rain forest is destroyed each year worldwide. If this rate continues, there will be no tropical rain forests left in just 40 years. The graph indicates the fate of the world's tropical rain forests if the current rate of deforestation continues.

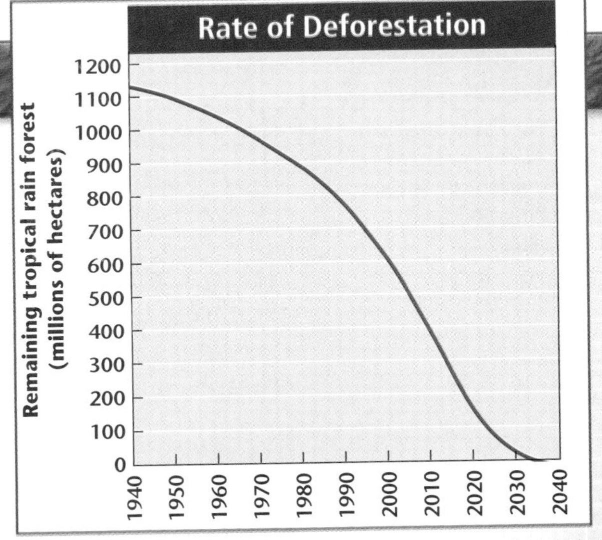

Analysis

1. How much tropical rain forest has been depleted since the year you were born?
2. According to the graph, when will all the rain forests be depleted?

Thinking Critically

3. What are the rates of deforestation for the 1950–1985 and 1985–2000 time periods?
4. How do the rates of deforestation between 1950–1985 and 1985–2000 compare?
5. What might be the reason for the change in the rate of deforestation between these two time periods?

Figure 25-13 Air pollution can be caused by natural phenomena, such as volcanic eruptions **(A)** and forest fires **(B).**

Sources of Air Pollution

Air pollution has both natural and human origins. Two natural sources of air pollution are shown in ***Figure 25-13.*** Human sources of air pollution include gases, smoke, and dust. One of the biggest sources of air pollution is the burning of fossil fuels. Power plants that generate electricity burn coal and oil, which produce many types of air pollution. However, the single largest source of air pollution in the United States is the exhaust from motor vehicles that burn fossil fuels in the form of gasoline. In the United States, motor vehicles cause 90 percent of the carbon monoxide pollution in cities. In cities such as Los Angeles, Rome, and Mexico City, shown in ***Figure 25-14,*** motor vehicles are responsible for 80 to 88 percent of the air pollution.

Air pollution can make humans ill. When humans inhale harmful gases, the gases can be absorbed by the bloodstream and interfere with various body systems. Carbon monoxide, a colorless and odorless gas, interferes with the body's ability to absorb oxygen and causes headaches, chest pains, dry throat, and nausea. Pollution can also cause burning eyes, irritated throats, and breathing difficulties. Some chemical air pollutants can cause cancer, birth defects, brain damage, long-term injury to lungs, and even death.

As clean air in the troposphere moves across Earth's surface, it collects both naturally occurring and human-made pollutants. What happens to these pollutants? They may be transported, diluted, transformed, or removed from the atmosphere.

Transport and Dilution Some pollutants may be carried downwind from their origin. Transport depends upon wind direction and speed, topographical features, and the altitude of the pollutants. For example, hills, valleys, and buildings interrupt the flow of winds and

Figure 25-14 Air pollution from vehicles in Mexico City is so concentrated that people who have breathing difficulties often listen to the radio to find out whether it is safe for them to go outdoors.

thus influence the transport of pollutants. Many of the pollutants in the acid precipitation that falls in the Adirondack Mountains of New York State were transported from coal-burning power plants in the midwestern states. If air movement in the troposphere is turbulent, some pollutants are diluted and spread out, which reduces their concentration.

Transformation and Removal Other pollutants undergo chemical changes, called photochemical changes, that are triggered by reactions with ultraviolet (UV) radiation. Photochemical smog, for example, forms when a mixture of nitrogen oxides and volatile organic compounds interact under the influence of sunlight.

Some other air pollutants undergo physical changes. For example, dry particles may clump together and become heavy enough to fall back to Earth's surface. These and other air pollutants are removed from the atmosphere in precipitation, which includes snow, mist, and fog as well as rain.

Topic: Indoor Air Pollution
To find out more about air pollution, visit the Earth Science Web Site at earthgeu.com

Activity: Research some ways to reduce indoor air pollution. Design a brochure that informs readers about reducing indoor air pollution.

INDOOR AIR POLLUTION

Have you ever shopped at a fabric store? Some people cannot even enter such a store because they are sensitive to the chemical formaldehyde, which is used in fabrics to prevent damage from insects. About 90 percent of the furniture sold in the United States also contains formaldehyde. Formaldehyde is just one of the many air pollutants that occurs indoors, as shown in ***Figure 25-15.***

Figure 25-15 Major indoor air pollutants are shown in this diagram of a new house.

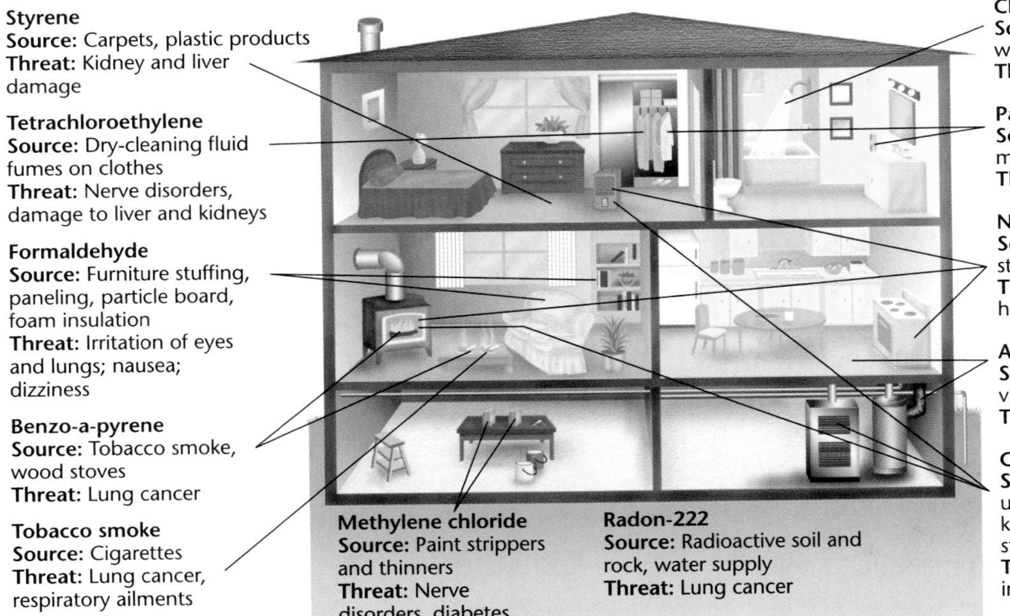

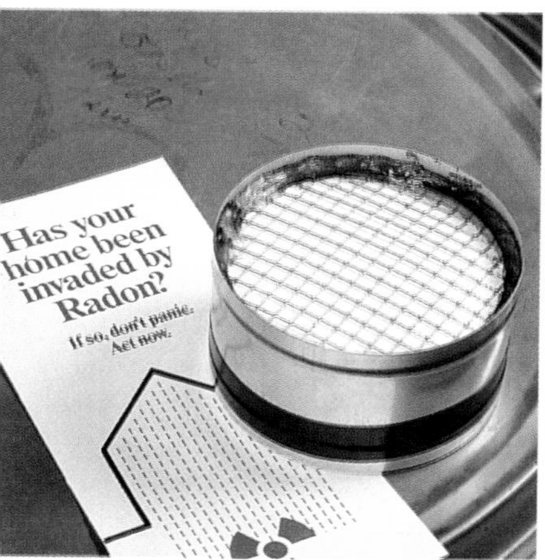

Figure 25-16 A radon test kit is a simple device that measures the amount of radon gas infiltrating a building over a week's time.

"Sick" Buildings Studies conducted by the United States Environmental Protection Agency (EPA) and by scientists in European countries have linked indoor air pollutants to headaches, coughing, sneezing, burning eyes, nausea, chronic fatigue, and flu-like symptoms. When these symptoms are experienced by 20 percent of the occupants of a building, the building is said to be "sick." Often, these symptoms disappear when the affected people go outside. New buildings are more likely to be "sick" than older buildings. This is because newer buildings tend to be airtight so that heating and cooling costs can be kept to a minimum, and because new furniture and carpeting release many indoor air pollutants, including styrene and formaldehyde.

Radon Gas The gas known as radon-222 is colorless, odorless, tasteless, and naturally occurring. Radon-222 is produced by the radioactive decay of uranium-238. Small amounts of uranium-238 are found in most soils and rocks, and in underground deposits. Usually, radon gas from such deposits seeps upward through the soil and is released into the atmosphere, where it is diluted to harmless levels. However, when buildings are constructed with hollow concrete blocks, or when they have cracks in their foundations, radon gas can enter and build up to high levels indoors. Once indoors, radon gas decays into radioactive elements that can be inhaled. Scientists have traced approximately 13 000 lung-cancer deaths in the United States each year to high levels of radon gas in homes.

Because it is impossible to see or smell a buildup of radon gas in a building, the EPA suggests that people test the radon levels in their homes and offices. Radon test kits, such as the one shown in ***Figure 25-16,*** measure the levels of radon in buildings.

Section Assessment

1. Why is air considered to be an Earth resource?
2. How did oxygen originate on Earth?
3. Explain how the oxygen and carbon-dioxide cycles on Earth are related.
4. Describe how air can be polluted by both natural processes and human activities.
5. **Thinking Critically** Explain why photochemical smog is a major problem in large cities that have little public transportation.

Skill Review

6. **Comparing and Contrasting** Compare and contrast the components of indoor air pollution and the components of atmospheric air pollution. Is one type of air pollution more damaging than the other? Explain. For more help, refer to the *Skill Handbook.*

earthgeu.com/self_check_quiz

SECTION 25.4 Water Resources

Earth Sciences 9.a, 9.c **I&E** 1.a, 1.l

When astronauts first took photographs of Earth from space, many people were surprised to see how much of Earth's surface is covered by water. One such photograph is shown in ***Figure 25-17.*** The oceans contain 97 percent of the planet's water, which means that only 3 percent of Earth's water is freshwater. Of this freshwater, about 2.997 percent is either locked up in ice caps and glaciers or stored as groundwater that is too deep to extract. This leaves only 0.003 percent of Earth's total volume of water available to humans for domestic, agricultural, and industrial purposes. This fraction of freshwater is in the form of surface water, water vapor, and obtainable groundwater.

OBJECTIVES

- **Explain** *the importance of clean freshwater.*
- **Analyze** *how water is distributed and used on Earth.*
- **Identify** *ways in which humans can reduce the need for increasing production of freshwater resources.*

VOCABULARY

desalination

The Importance of Water

About 71 percent of Earth's surface is covered by water. However, this is not the only reason that Earth is sometimes called the "water planet." The world's oceans help regulate climate, provide habitats for marine organisms, dilute and degrade many pollutants, and even have a role in shaping Earth's surface. Freshwater is an important resource for agriculture, transportation, recreation, and numerous other human activities. In addition, the organisms that live on Earth are made up mostly of water. Most animals are about 50 to 65 percent water by weight, and even trees may be composed of up to 60 percent water. Without water, life as we know it could not exist on Earth.

Liquid Water Why is water such an important resource? Water is a unique substance with many desirable qualities. Water can exist as a liquid over a wide range of temperatures because of the hydrogen bonds between water molecules. Recall from Chapter 3 that water molecules are polar molecules with positive and negative ends. Hydrogen bonds form when the positive ends of some water molecules are attracted to the negative ends of other water molecules.

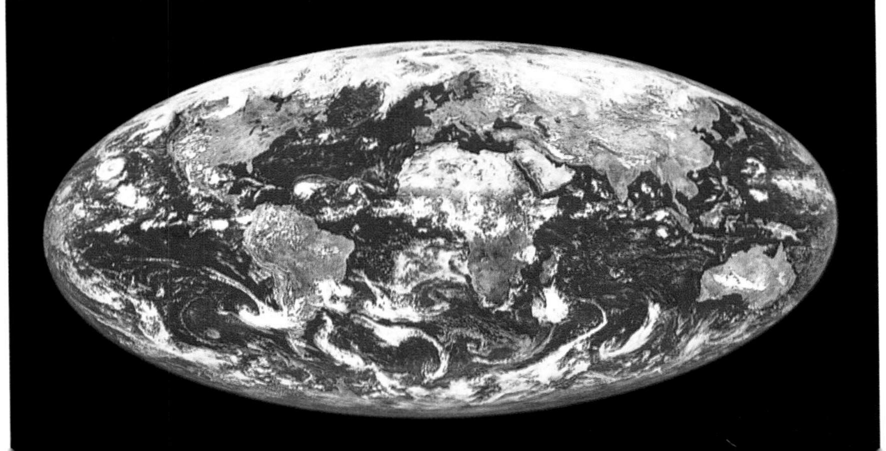

Figure 25-17 This composite photograph was created by a satellite as it orbited Earth. It depicts the entire surface of Earth, showing the relative sizes of land masses and oceans and their true colors. From this photograph, it is easy to see why Earth is sometimes called the "water planet."

Figure 25-18 Perspiration is the body's way of helping you cool off. As the water in perspiration evaporates, it removes heat from your skin.

The hydrogen bonds cause water's surface to contract and allow water to adhere to and coat a solid. These properties enable water to rise from the roots of a plant through its stem to its leaves. Water also has a high boiling point, 100°C, and a low freezing point, 0°C. As a result, water remains a liquid in most of the environments on Earth.

Heat-Storage Capacity Liquid water can store a large amount of heat without a correspondingly high increase in temperature. This property protects organisms that live in water from abrupt temperature changes, and it is also responsible for water's ability to regulate Earth's climate. Because of this same property, water is used as a coolant for automobile engines, power plants, and other heat-generating processes. Have you ever perspired heavily while participating in an outdoor activity on a hot day? Evaporation of perspiration from your skin, shown in ***Figure 25-18,*** helps you cool off because water absorbs large quantities of heat as it changes into water vapor.

Water as a Solvent Liquid water also can dissolve a wide variety of compounds. This ability enables water to carry nutrients into, and waste products out of, the tissues of living things. The diffusion of water across cell membranes enables all cells to regulate their internal pressure. Water also dilutes water-soluble waste products of humans and thus serves as an all-purpose cleanser.

Solid Water Expands Unlike most liquids, water expands when it freezes. Because ice has a lower density than liquid water, it floats on top of water, as shown in ***Figure 25-19.*** As a result, bodies of water freeze from the top down. If water did not have this property, ponds and streams would freeze solid, and aquatic organisms would die each winter. The expansion of water as it freezes also can fracture rocks when ice crystals form in preexisting cracks and force the cracks to widen. Thus, ice forming in cracks becomes part of the weathering process.

Figure 25-19 Ice is less dense than liquid water, and therefore it floats on top of water.

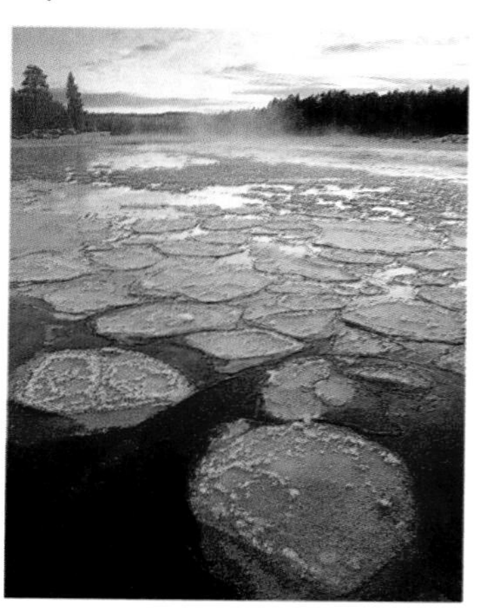

Location of Freshwater Resources

Freshwater resources are not distributed evenly across Earth's landmasses. Although the United States has plenty of freshwater, much of it is concentrated in certain areas or has been contaminated by agricultural or industrial processes. The eastern states receive ample precipitation, and most freshwater in these states is used for cooling, energy production, and manufacturing. By contrast, western states often have too little precipitation. Thus, in the West, the largest use of freshwater is for irrigation. Water tables in the West are dropping as farmers and cities continue to sink wells into aquifers and use the groundwater faster than it can be recharged.

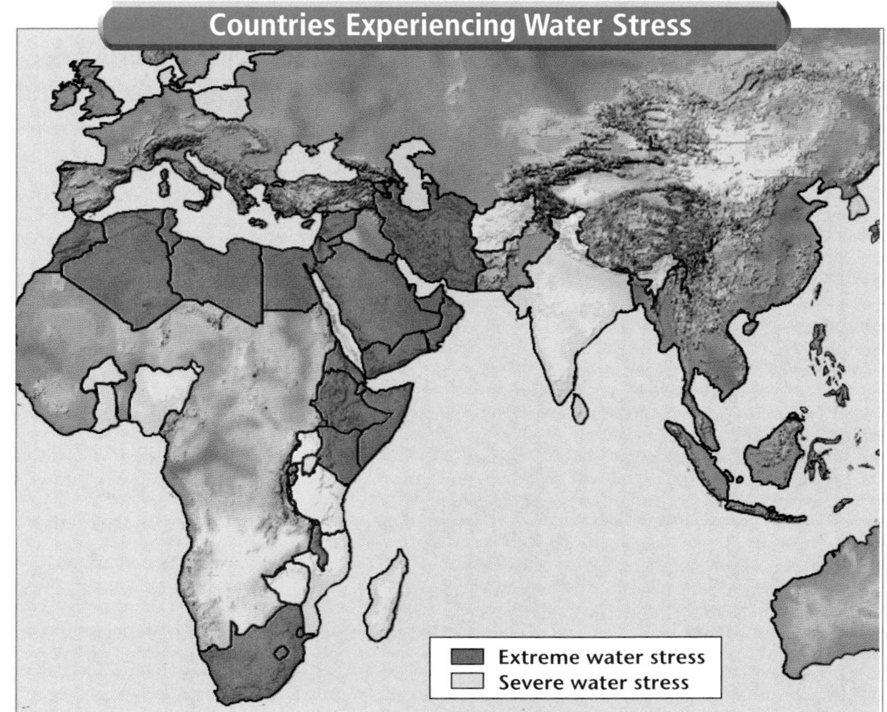

Figure 25-20 Countries experience severe water stress when there is less than 1000 m^3 of freshwater per person. Extreme water stress occurs when there is less than 500 m^3 of freshwater per person. Increased populations will impose even greater demands on the freshwater available in these countries.

Earth Sciences

9.c Students know the importance of water to society, the origins of California's fresh water, and the relationship between supply and need.

Worldwide, water distribution is a continuing problem, even though most continents have plenty of water. Since the 1970s, scarcity of water has caused the deaths of more than 24 000 people worldwide each year and created huge numbers of environmental refugees. In areas where water is scarce, women and children often walk long distances each day to collect a meager supply of water for domestic uses. Millions of people also try to survive on land that is prone to drought. About 25 countries, primarily in Africa, experience chronic water shortages, as shown in ***Figure 25-20.*** That number is expected to rise to 90 countries by the year 2025.

Use of Freshwater Resources

As you learned in Chapters 9 and 10, freshwater on Earth is held either in surface waters, such as lakes, rivers, and streams, or in the ground as groundwater. Recall that the upper surface of groundwater is called the water table, and that the water-saturated layers of sand, gravel, or bedrock through which groundwater flows are called aquifers. Aquifers are refilled naturally as rain percolates downward through soil and rock in the process known as natural recharge. Many humans worldwide rely on wells drilled into the ground that tap aquifers for freshwater supplies.

The current rate of withdrawal of freshwater from both surface and groundwater sources worldwide is five times greater than it

To learn more about water, go to the **National Geographic Expedition** on page 898.

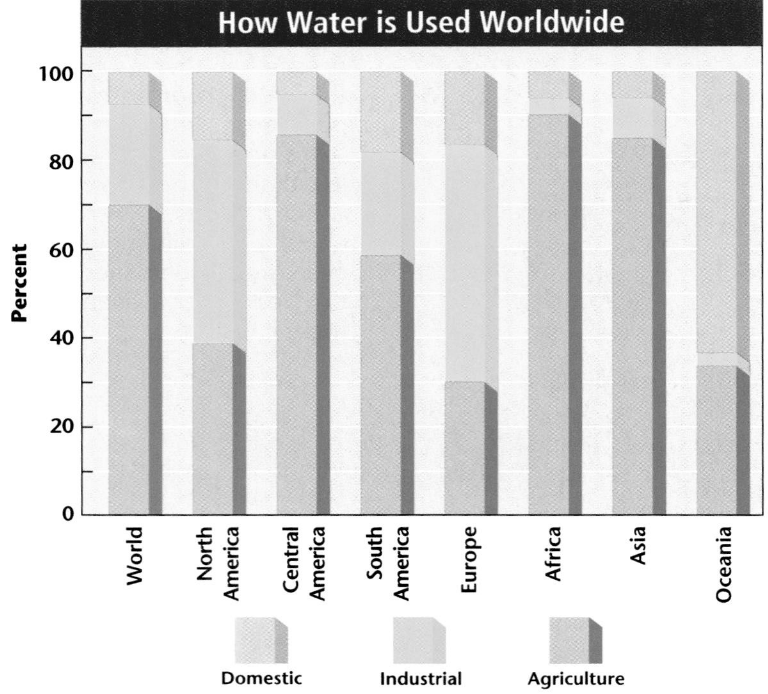

Figure 25-21 This bar graph identifies the uses of freshwater worldwide. ***Where in the world is the most water used for industrial purposes?***

was just 50 years ago. This increase has occurred primarily to meet the drinking-water and agricultural needs of an increasing human population. Withdrawal rates for freshwater resources are expected to double again within the next 20 years.

Uses of freshwater vary worldwide, but about 70 percent of the water withdrawn each year is used to irrigate 18 percent of the world's croplands. However, much of the water used for irrigation is not used by the plants; nearly 80 percent of this water evaporates or seeps into the ground before it can be used by crops. About 23 percent of freshwater is used for cooling purposes in power plants, for oil and gas production, and in industrial processing. Domestic and municipal uses account for only seven percent of the freshwater withdrawn from surface and groundwater resources. The uses of freshwater worldwide are shown in ***Figure 25-21.***

MANAGING FRESHWATER RESOURCES

The dam shown in ***Figure 25-22*** is being built to hold back the floodwaters of the Yangtze River in China. Called the Three Gorges Project, the construction of this dam will provide freshwater and supply power to 150 million people. However, the dam will also flood large areas of farmland and displace about 1 million people who live nearby. Some critics think that the Three Gorges Project will ruin the water quality of the Yangtze River and create more severe flooding as the dam fills up with sediment and eventually overflows. However, most countries manage their supplies of freshwater by building dams, by transporting surface water, or by tapping groundwater. Some countries also have had success removing the salts from seawater to provide needed freshwater supplies.

Figure 25-22 The purpose of the Three Gorges Dam is to reduce China's reliance on coal for energy production and to regulate the annual flooding of areas along the Yangtze River.

Dams and Reservoirs Building dams is one of the primary ways that countries try to manage their freshwater resources. Large dams are built across river valleys, usually to control flooding downstream, and the reservoirs behind dams capture the rivers' flow as well as rain and melting snow. The water captured in these reservoirs can be released as necessary to provide water for irrigation and municipal

Figure 25-23 Roman aqueducts, such as this one still standing in France, were considered to be architectural wonders when they were first built **(A).** This present-day aqueduct transports water from water-rich northern California to semiarid and arid areas in southern California **(B).**

uses, such as in homes and businesses, or to produce hydroelectric power. Reservoirs also provide opportunities for recreational activities, such as fishing and boating. Dams and reservoirs currently control between 25 and 50 percent of the total runoff on every continent.

Transporting Surface Water If you were to visit Europe, you would likely see many ancient aqueducts like the one shown in ***Figure 25-23A.*** The Romans built aqueducts to bring water from other locations into their cities 2000 years ago. Today, many countries use aqueducts, tunnels, and underground pipes to bring water from areas where it is plentiful to areas in need of freshwater supplies.

The California Water Project is one example of the benefits, as well as the costs, of transporting surface water. In California, about 75 percent of the precipitation occurs north of the city of Sacramento, yet 75 percent of the state's population lives south of that city. The California Water Project uses a system of dams, pumps, and aqueducts to transport water from northern California to southern California. Most of this water, 82 percent, is used for agriculture. The residents of Los Angeles and San Diego are withdrawing groundwater faster than it is being replenished. As a result, there is a demand for even more water to be diverted to the south. However, the residents of northern California object, because the diversion of more water would harm the Sacramento River and threaten fisheries. As this example illustrates, conflicts over the transport of surface water will probably increase as human populations increase and create higher demands for water.

Earth Sciences

9.a Students know the resources of major economic importance in California and their relation to California's geology.

Update For an online update of the progress of the Three Gorges Project in China, visit earthgeu.com and select the appropriate chapter.

Tapping Groundwater Most people in the United States obtain drinking water by turning on a faucet in their kitchens or bathrooms. But do you know where this water comes from? In this country, about 23 percent of all freshwater used is groundwater pumped

MiniLab

Hard Water

Determine the hardness of water samples by observing how easily soap suds can be produced.

Procedure

1. Obtain six clean baby-food jars. Label them A through F.
2. Measure 20 mL of one water sample. Pour the water into the jar marked A.
3. Repeat step 2 four more times, using a different water sample for jars B through E.
4. Measure 20 mL of distilled water. Pour this water into jar F.
5. Make a data table in your science journal. In the first column, write the letters A–F.
6. Place one drop of liquid soap in sample jars A through E. Do not place any soap in jar F. Tighten the lids. Then shake each jar vigorously for five seconds.
7. Using the following rating scale, record in your data table the amount of suds in each jar: 1—no suds, 2—few suds, 3—moderate amount of suds, 4—lots of suds.

Analyze and Conclude

1. List the water samples in order from hardest to softest.
2. What is the difference between hard and soft water?
3. What are some disadvantages of hard water?

from aquifers. In some states, such as Florida, Hawaii, and Nebraska, more than 90 percent of the population depends upon groundwater from aquifers for drinking water. Sometimes, groundwater contains substances that make it difficult to use for domestic purposes. For example, water that contains calcium and magnesium ions, known as hard water, does not form suds when soap is added. You will determine the hardness of water samples in the *MiniLab* on this page.

Groundwater normally moves from points of high elevation and pressure to points of lower elevation and pressure. This movement of water is relatively slow; water moves through the ground at a rate of only about 1 m/year. If the withdrawal rate of an aquifer exceeds its natural recharge rate, the water table around the withdrawal point is lowered. This lowering of the water table is known as drawdown. If too many wells are drilled into the same aquifer in a limited area, the drawdown can lower the water table below the bottoms of the wells, and as a result, the wells will run dry. Because groundwater is the source of many streams in the United States, groundwater depletion also affects stream flow. In coastal areas, drawdown of groundwater can also result in the intrusion of salt water into shallow aquifers.

Desalination With all the water available in the oceans, some countries have explored the possibility of removing salt from ocean water to provide freshwater in a process called **desalination.** Desalination occurs when salt water is distilled. The water is first heated until it evaporates, then it is condensed and collected. This evaporation process leaves the salts behind. Most countries that use desalination to produce freshwater use solar energy to evaporate sea water. Although the evaporation of seawater by solar energy is a slow process, it is an inexpensive way to provide needed freshwater. Some desalination plants use fuel to distill seawater, but because this process is expensive, it is used primarily to provide drinking water. You will find out how a simple desalinator works in the *Design Your Own GeoLab* at the end of this chapter.

Figure 25-24 Trickle irrigation uses a system of perforated pipes to release trickles of water next to plant roots. This minimizes evaporation and seepage and brings 80 to 90 percent of the irrigation water directly to the crops.

Reducing Freshwater Use The increasing need for freshwater supplies has led to some extremely creative solutions. You may have heard about a plan to tow an Antarctic iceberg to the Middle East to provide needed freshwater to arid countries in that region. However, most experts agree that the best way to meet the need for freshwater is to use available supplies more efficiently. For example, irrigation of field crops loses vast amounts of freshwater to evaporation. Farmers can prevent evaporation of irrigation water by changing their irrigation methods. Trickle irrigation, shown in ***Figure 25-24,*** provides water directly to plant roots, and thus considerably reduces evaporation rates. Some farmers also monitor the soil and provide irrigation only when necessary. Water can be used more efficiently by industries when they use recycled water instead of clean, freshwater for manufacturing processes.

Domestic uses of water, such as flushing toilets, bathing, and washing dishes and clothing, account for about 78 percent of the water used in a typical home in the United States. In the summer, watering lawns and gardens may account for 80 percent of a home's daily water usage. Many of these domestic uses can be reduced by installing low-flow toilets, using plants that are drought-resistant for landscaping, and fixing leaky pipes and faucets.

SECTION ASSESSMENT

1. Why is clean water important to life on Earth?
2. How does the distribution of freshwater resources affect humans?
3. Describe three ways in which humans provide for their freshwater needs.
4. How can the amount of water used for irrigation be reduced?
5. **Thinking Critically** Many people from the northeastern part of the United States have moved to the sunny southwestern states of Arizona and New Mexico. These new residents increase local populations. How does this relocation of people affect the demand for freshwater resources?

SKILL REVIEW

6. **Predicting** Some aquifers are found deep underground and receive very little recharge. These aquifers are sometimes referred to as fossil aquifers. Once water is taken from a fossil aquifer, it rarely fills up again. Is water from fossil aquifers a renewable resource? Explain. For more help, refer to the *Skill Handbook.*

DESIGN YOUR OWN GeoLab

Designing a Solar Desalinator

 I&E

1.a Select and use appropriate tools and technology (such as computer-linked probes, spreadsheets, and graphing calculators) to perform tests, collect data, analyze relationships, and display data.

Most of Earth's surface is covered with salty ocean water. Ocean water can be used for drinking water and other purposes if the salts are first removed. Solar energy can be used to evaporate water from seawater, leaving the salts behind. The evaporated water can then be condensed into freshwater.

Preparation

Problem

How can you build a small-scale, working solar desalinator?

Possible Materials

clear plastic or Plexiglas
large pans to hold water
salt water
collecting containers
lamp
glass pan or beaker
hot plate

Hypothesis

The Sun's energy can be collected to desalinate salt water.

Objectives

In this GeoLab, you will:

- **Design** a model of a working solar desalinator.
- **Assemble** the model from design plans.
- **Test** the effectiveness of the design model.
- **Analyze** the model to suggest possible improvements.

Safety Precautions

Always wear safety goggles and an apron in the lab. Be careful when handling hot materials.

Plan the Experiment

1. Use the library and go to earthgeu.com to identify designs of solar desalinators.
2. Draw a design for your model desalinator. (Hint: Solar energy must be collected in some way that allows sunlight to enter and causes an increase in temperature inside the container so that water in saturated air can condense and be collected.)
3. Make a list of the materials you will need, and then collect them.
4. Construct the desalinator you designed.
5. Test the desalinator by recording how long it takes to collect the purified water and how much water was collected.
6. Test the water to see if it has been purified by boiling the water away. If any salt remains in the container after the water has evaporated, your desalinator did not remove all of the salts from the salt water.

Analyze

1. **Interpreting Scientific Illustrations** Draw the desalinator that you constructed.
2. **Interpreting Observations** How well did your desalinator work? On what criteria did you base the effectiveness of your desalinator?
3. **Observing and Inferring** What problems did you encounter in this investigation?
4. **Comparing and Contrasting** Compare and contrast your desalinator with one of your classmates'. What were the advantages or disadvantages of your design?

Conclude & Apply

1. What factors affected the efficiency of the desalinator?
2. How did your solar desalinator's efficiency compare with the efficiencies of other students' models?
3. How could you improve your desalinator?
4. What conclusions could be drawn from your investigation regarding the viability and use of solar-powered desalinators?

Science in the News

I&E 1.l

Glass from the Past

In the autumn of 1739, Caspar Wistar opened a glass factory in southern New Jersey—the first successful glassworks in North America. He chose a location that had the chief resources he needed to make and ship glass: sand, wood, and water.

From glass fragments, old documents, and even old advertisements, historians have pieced together the history of Wistar's glassworks. It was likely a wooden building constructed around a large, beehive-shaped furnace. The furnace was built of clay bricks and was divided into three levels. The bottom level held the fire; it had an opening at one end where wood could be added and the condition of the coals could be checked. The middle level of the furnace held specially-made pots into which the ingredients for the glass itself were placed. The pots had to withstand the extreme heat of molten glass—a broken pot meant that the furnace shut down, and production ceased. Completed pieces were placed in the top level of the furnace, the annealing chamber, where they cooled slowly so that they would retain their strength.

Ingredients for Success

The ingredients for making glass have remained the same since the earliest days of glassmaking, about 4000 years ago. The primary ingredient in all glass is sand. In the eighteenth century, glass was made from sand dug from the ground. The dry ingredients, sand, potash—which is potassium carbonate found in wood ashes—lime, and other minerals used for coloring, were mixed in large troughs. The dry mixture was then placed in the hot pots in the middle level of the furnace. The ingredients melted together to form molten glass.

In addition to sand, glass makers needed a continuous supply of wood because the fire in the furnace could never be allowed to go out. Wistar built his factory near both wood and sand; thus, he never ran out of either resource.

Importance of Water

Being close to water was important in the location of Wistar's factory for two reasons. First, having water nearby meant that fires could be put out quickly. Many early glassworks were destroyed by fire. Second, being close to water made the shipping of finished pieces much easier. The fewer times that glass had to be handled on its way to the marketplace, the less breakage there was. Given the heaviness of glass, shipping it by water was the transportation method of choice.

Activity

A variety of minerals and compounds from Earth's crust are used to make different colors of glass. Use the library or go to earthgeu.com to research the different materials used to produce different colors of glass. Record your findings in a data table. Are any of the materials you found surprising? Why?

CHAPTER 25

Study Guide

Summary

SECTION 25.1

What are resources?

Main Ideas

- Natural resources are the resources that Earth provides, including air, water, land, organisms, rocks, minerals, and nutrients.
- Renewable resources are replaced by natural processes at a rate that is equal to or greater than the rate at which they are being used.
- Nonrenewable resources exist in a fixed amount and can be replaced only by geological, physical, and chemical processes that take hundreds of millions of years.

Vocabulary

natural resource (p. 655)
nonrenewable resource (p. 657)
renewable resource (p. 656)
sustainable yield (p. 657)

SECTION 25.2

Land Resources

Main Ideas

- Land resources include topsoil, rocks, and minerals. Land also provides space for agriculture, housing, roadways, and protected areas such as national forests, wildlife refuges, and national parks.
- Topsoil is a complex mixture of decaying organic matter, eroded rock, minerals, nutrients, oxygen, and water. In arid areas, loss of topsoil can lead to desertification.
- Bedrock is unweathered parent rock.
- Aggregates, including sand, gravel, and crushed stone, are found in glacial deposits.
- An ore is a natural resource that can be mined at a profit. Ores may be associated with igneous rocks or formed by processes at Earth's surface.

Vocabulary

aggregate (p. 660)
bedrock (p. 660)
desertification (p. 660)
gangue (p. 663)
ore (p. 661)

SECTION 25.3

Air Resources

Main Ideas

- The atmosphere contains mostly nitrogen and oxygen, as well as various other gases in smaller amounts. Early Earth had no oxygen; this was supplied by photosynthetic organisms.
- The geochemical cycles of Earth's atmosphere are delicately balanced. Human activities disrupt this balance, and air pollution results.
- Clean air is necessary to most organisms. Both outdoor and indoor air pollution are harmful to living things.

Vocabulary

air pollution (p. 665)
pollutant (p. 665)

SECTION 25.4

Water Resources

Main Ideas

- Freshwater is necessary to all life and to many Earth processes. Water is recycled continually through the water cycle.
- Water has unique properties that allow life to exist on Earth.
- Water is not evenly distributed on Earth's surface.
- Water-management methods distribute freshwater resources more evenly through the use of dams, aqueducts, and wells.

Vocabulary

desalination (p. 674)

earthgeu.com/vocabulary_puzzlemaker

CHAPTER 25

Assessment

Understanding Main Ideas

1. Which of the following is a renewable resource?
   a. oil
   b. natural gas
   c. trees
   d. coal

2. What portion of Earth's atmosphere consists of oxygen?
   a. 21 percent
   b. 78 percent
   c. 3 percent
   d. trace amounts

3. What is the origin of oxygen in Earth's atmosphere?
   a. photosynthetic organisms
   b. volcanic eruptions
   c. meteorites
   d. burning fossil fuels

4. Marble is what type of land resource?
   a. an aggregate
   b. an ore
   c. soil
   d. bedrock

5. Which of the following is a nonrenewable resource?
   a. bauxite
   b. carbon
   c. water
   d. nitrogen

6. Of Earth's surface waters, what percent is freshwater available for human use?
   a. 97 percent
   b. 3 percent
   c. less than 1 percent
   d. 21 percent

7. Which of these is NOT a way to manage water resources?
   a. building dams
   b. using aqueducts
   c. settling crystals
   d. desalination

8. What is the process by which productive land becomes desert?
   a. deforestation
   b. desertification
   c. desalination
   d. respiration

Use the following table to answer questions 9 and 10.

Efficiencies of Irrigation Methods in Texas

Irrigation Method	Percent Efficiency	Water Needed to Add 100 mm to Root Zone
Conventional furrow	60%	167 mm
Furrow with surge valve	80%	125 mm
Low-pressure sprinkler	80%	125 mm
Trickle irrigation	95%	105 mm

9. Which of the irrigation methods requires the most water to add 100 mm to the root zone of plants?
   a. low-pressure sprinkler
   b. conventional furrow
   c. trickle irrigation
   d. furrow with surge valve

10. Which of the irrigation methods appears to be the most efficient use of water?
    a. low-pressure sprinkler
    b. conventional furrow
    c. trickle irrigation
    d. furrow with surge valve

11. Which air pollutant is more harmful to human health inside a building than outside?
    a. smog
    b. radon gas
    c. carbon dioxide
    d. nitrogen oxide

Test-Taking Tip

TABLES If a test question involves a table, skim the table before reading the question. Read the title, column heads, and row heads. Then read the question and interpret the information in the table.

12. Which of these is NOT a manner in which ores form?
    a. crystal settling
    b. weathering
    c. chemical precipitation
    d. photochemical changes

13. Which of the following is NOT a property of water?
    a. It has a high heat-storage capacity.
    b. It exists mostly as a gas.
    c. It expands when it freezes.
    d. It dissolves many compounds.

Applying Main Ideas

14. What is the atmosphere's role in the exchange of gases on Earth?
15. Why is the loss of the forest cover in the Amazon River Basin a worldwide concern?
16. What is the best way to prevent loss of topsoil?
17. How would planting forests help control desertification in arid areas?

Thinking Critically

18. If Earth processes recycle water resources, why is water pollution a problem?
19. Most of the water resources on Earth are salt water. Why can't all of the human population's freshwater needs be supplied by the desalination of salt water?
20. Suppose that you recently moved into a new house. Shortly thereafter, you began to feel ill. How can you determine whether the house is "sick"?
21. Volcanic eruptions and other natural events result in air pollution. Why is the air pollution produced by human activities a concern?

Standardized Test Practice

INTERPRETING SCIENTIFIC ILLUSTRATIONS

Use the illustrations below to answer questions 1 and 2.

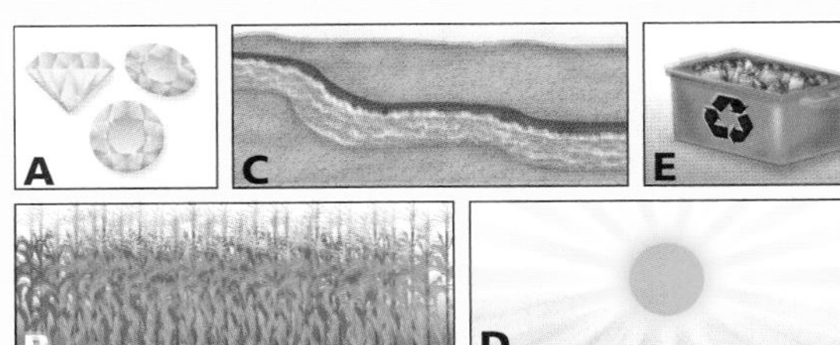

1. Which shows a nonrenewable resource?
    a. A
    b. B
    c. C
    d. D

2. Which resource is replaced through natural processes more quickly than it is used?
    a. B
    b. C
    c. D
    d. E

3. Which of the following is NOT an example of a material that is mined at a profit?
    a. hematite
    b. diamonds
    c. limonite
    d. chromium

4. Which is the greatest source of air pollution in the U.S.?
    a. power plants
    b. industrial smokestacks
    c. agricultural processes
    d. motor vehicles

5. Which is the most expensive and least commonly used method of providing water to areas that need it?
    a. tapping groundwater
    b. aqueducts
    c. desalination
    d. dams

earthgeu.com/standardized_test

Chapter 26

Energy Resources

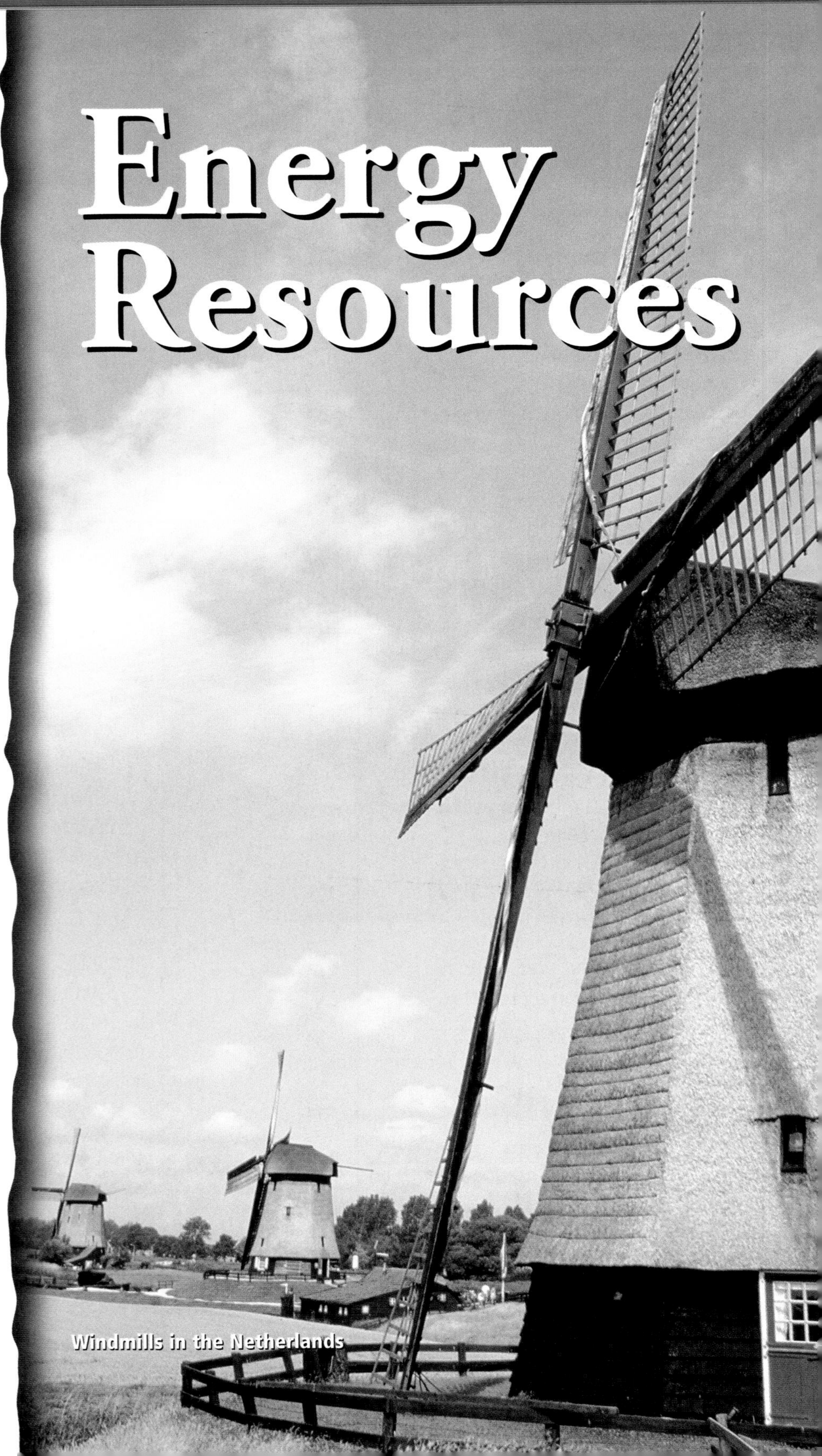

What You'll Learn

- What energy resources are found on Earth.
- What alternatives to traditional energy resources exist.
- How conservation can extend both traditional and alternative energy resources.

Why It's Important

Life on Earth could not exist without energy resources. Many commonly used energy resources are nonrenewable; thus, energy conservation and the development of alternative energy resources are necessary to ensure a continuous energy supply.

To find out more about energy resources, visit the Earth Science Web Site at earthgeu.com

Windmills in the Netherlands

Discovery Lab Sources of Energy

Energy cannot be created or destroyed, but it can change form and be transferred. Thus, the same energy can be used over and over again. In this activity, you will observe a type of energy transfer that occurs every day.

1. Add 200 mL of water to a 250-mL glass beaker.
2. Place the beaker on a hot plate.
3. Turn on the hot plate. Observe what happens to the water as it heats up and begins to boil.

CAUTION: Always wear safety goggles and an apron in the lab. Allow the beaker to cool before moving it at the end of the activity.

Observe In your science journal, trace the energy source used to bring the water to a boil back to its origin. Describe what happened to the energy as it was used to heat and boil the water. In your description, include an explanation of the source of most energy on Earth. Infer where the energy went when the water began to boil.

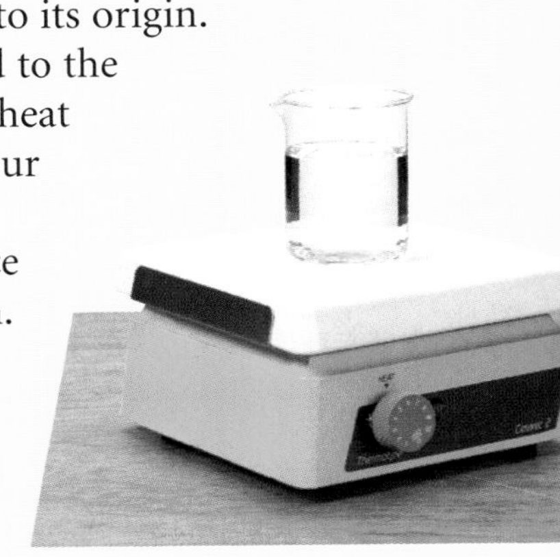

SECTION 26.1 Conventional Energy Resources

Earth Sciences 4.a, 4.b, 7.b

OBJECTIVES

- **Recognize** *the Sun as the ultimate source of most energy on Earth.*
- **Describe** *how energy changes from one form to another.*
- **Identify** *materials that are used as fuels.*
- **Explain** *how fossil fuels form.*

VOCABULARY

fuel
peat
fossil fuel

What kinds of activities do you engage in each morning? Do you turn on lights or run water for a shower? In the kitchen, you might toast bread or use a microwave oven to heat up your breakfast. You may ride a bus to school or drive a car. All of these activities depend upon energy. Where does most of the energy that you use each morning come from? The energy that humans and all other organisms use comes mostly from the Sun.

TRANSFER OF SOLAR ENERGY

How is solar energy used by organisms? Green plants, protists such as algae, and cyanobacteria are producers that capture the Sun's energy in the process of photosynthesis. In these photosynthetic organisms, solar energy is used for maintenance, growth, and reproduction. Whatever energy is not used right away is stored by the organisms. When consumers eat producers, they use that stored

Figure 26-1 Wheat plants in a field trap the Sun's energy during photosynthesis **(A).** When you eat a breakfast cereal made from wheat **(B),** you are consuming solar energy in another form.

energy for their own life processes. For example, when you eat a breakfast cereal made from grain such as oats or wheat, as illustrated in ***Figure 26-1,*** you are consuming the energy stored by those green plants. In this way, trapped solar energy is transferred through the food chains found in most ecosystems. A food chain is a model that shows how solar energy flows from the Sun to producers and then to consumers in an ecosystem.

Earth Sciences

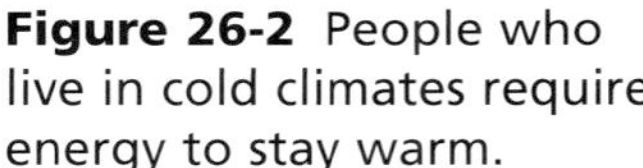

4.b Students know the fate of incoming solar radiation in terms of reflection, absorption, and photosynthesis.

Humans also need energy to keep them warm in cold climates, to cook food, to pump water, to grind grain, and to provide light. The energy for all of these purposes also comes primarily from the Sun. Traditional sources of energy, such as wood and peat, are derived from producers such as plants. Even gasoline and kerosene are derived from decayed organisms that first obtained energy from the Sun. When organic materials such as these are burned, the energy stored in them is released.

Figure 26-2 People who live in cold climates require energy to stay warm.

Traditional Sources of Energy

Do you live in an area that has four seasons each year? As you can see in ***Figure 26-2,*** some people live in climates that are very cold for part of the year. Humans have been able to survive in such cold climates primarily because of their ability to alter the environment to meet their needs. Living in cold areas requires humans to use energy to provide heat. Most humans also use energy to provide light and to cook food. The energy for all of these activities is provided by **fuels,** which are materials that are burned to produce heat or power. Probably the earliest use of fuels occurred when humans found pieces of wood that had been struck by lightning and were still burning, and then used them to start fires back at their homesteads. Archaeologists have discovered fire pits in caves that provide evidence that humans burned wood to cook their food many thousands of years ago. Traditional fuels include renewable resources such as wood, dried field crops, and dried fecal material from animals such as cows and bison. In fact, any material that is in good supply and also burns can be used as fuel.

The total amount of living things in an ecosystem is its biomass. Thus, fuels derived from living things are called biomass fuels. In many developing countries, biomass fuels are used to provide energy for cooking and heating. By far, the most commonly used traditional biomass fuel is wood. Today, wood is the primary source of energy for more than half of the world's population.

Earth Sciences

4.a Students know the relative amount of incoming solar energy compared with Earth's internal energy and the energy used by society.

Wood Humans have been using wood as an energy source for thousands of years. While wood is currently the primary source of energy for only about four percent of households in the United States, roughly 1.5 billion people throughout the world use wood as their primary source of fuel for heating and cooking. Many of these people live in developing countries, which use half of the world's wood supply. Unfortunately, the need to use wood as a fuel has led to deforestation in many areas of the world. As the forests near villages are cut down for fuel, people travel farther and farther away to gather the wood they need. In some parts of the world, this demand for wood has led to the complete removal of forests, which, in turn, has resulted in erosion and the loss of topsoil. In industrialized countries such as the United States and Canada, trees are cut down for lumber and paper production rather than fuel. However, these uses of forest resources can have the same negative impact on the environment.

Field Crops When wood is scarce, humans use other materials, including field crops, as fuel. The simplest way to use field crops, such as corn, hay, and straw, is to burn them directly. Crop residues left after harvest, including the stalks, hulls, pits, and shells from corn, oats, rice, wheat, and nuts, are other sources of energy. All of these can be burned to provide heat. Crops and their residues are most commonly burned for fuel on farms and in homes.

Fecal Material Feces are the solid wastes of animals. In many cases, dried feces contain undigested pieces of grass that help the material to burn. Fecal material from cows often meets the energy needs of people in developing countries that have limited forest resources. People who live in villages in India, Pakistan, and Afghanistan collect animal dung for fuel and dry it on the outside walls of their stables or compounds, as illustrated in ***Figure 26-3.***

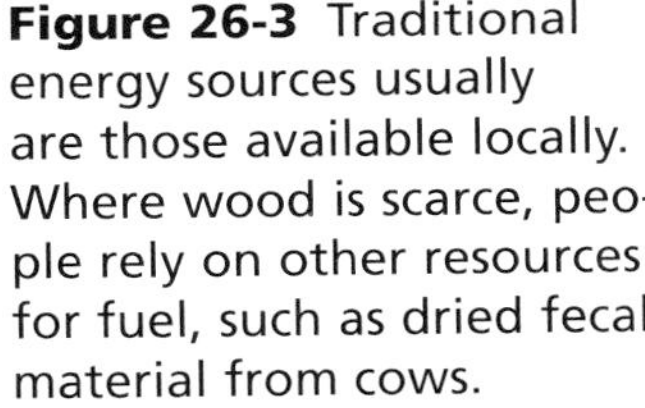

Figure 26-3 Traditional energy sources usually are those available locally. Where wood is scarce, people rely on other resources for fuel, such as dried fecal material from cows.

Figure 26-4 Peat is cut into blocks, dried in the Sun, and then burned in stoves and furnaces to provide heat for homes. When it burns, peat has an earthy smell that many people enjoy.

Peat Bogs are poorly drained areas with spongy, wet ground that is composed mainly of dead and decaying plant matter. Plants in bogs include *Sphagnum* moss, which forms large mats on top of the water. When plants in a bog die, they fall into the water. Bog water is acidic and has low levels of oxygen; these conditions slow down or stop the growth of the bacteria that decompose dead organic matter, including plants. As a result, dead and partially decayed plant material builds up on the bottom of the bog. Over time, as the plant material is compressed by the weight of water and by other sediments that accumulate above, it becomes a light, spongy material called **peat,** shown in ***Figure 26-4.*** Most of the peat used as fuel today is several thousands of years old.

Peat has been used as a low-cost fuel for centuries because it can easily be cut out of a bog, dried in the sun, and then burned directly in a stove or furnace to produce heat. Highly decomposed peat burns with greater fuel efficiency than wood. Today, peat is still used to heat homes in Ireland, England, parts of northern Europe, and the United States.

Fossil Fuels

Peat is one of the **fossil fuels,** which are energy sources that formed over geologic time as a result of the compression and partial decomposition of plants and other organic matter. Although peat and all fossil fuels originally formed from once-living things, these energy sources are considered to be nonrenewable because their formation occurred over thousands or even millions of years. The formation of peat is the first step in the development of coal.

Fossil fuels also include coal, natural gas, and petroleum. The high concentration of carbon and hydrogen in fossil fuels makes them very efficient energy sources. Most industrialized countries of the world today, including the United States, depend primarily on coal, natural gas, and petroleum to fuel power plants that provide electricity and to fuel vehicles. You can find out how one oil company preserves the environment while prospecting for fossil fuels in the *Science & the Environment* feature at the end of this chapter. Although fossil fuels are diverse, all of them originated from organic matter trapped in sedimentary rock.

Earth Sciences

7.b Students know the global carbon cycle: the different physical and chemical forms of carbon in the atmosphere, oceans, biomass, fossil fuels, and the movement of carbon among these reservoirs.

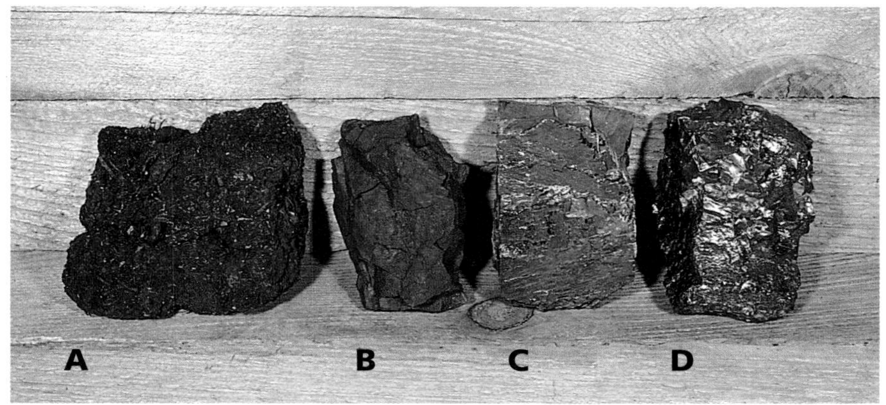

Figure 26-5 Peat **(A)** is light and spongy. Lignite **(B)** is a soft, brown coal. Bituminous coal **(C)** and anthracite **(D)** differ mainly in hardness, color, and carbon content.

Coal During periods of coal formation, tectonic plate movements caused some landmasses to move near Earth's equator. As a result, these areas experienced humid, tropical conditions that supported abundant plant growth. Generations of swamp plants, such as ferns and sedges, grew in the warm, tropical swamps. As each generation died, the organic material settled to the bottom of the swamp and became covered with subsequent generations of dead plants. The limited supply of oxygen was used up quickly, which resulted in a slow rate of decay. Over time, oxygen and hydrogen were lost from the organic matter, and the concentration of remaining carbon increased. Eventually, this compressed organic matter became coal.

Coal can be classified according to the amount of pressure under which it formed and the amount of time involved. ***Figure 26-5*** shows types of coal. When peat continues to be compressed, it becomes a type of coal called lignite, a soft, brown, low-grade coal. Over time, and under increasing pressure, lignite develops into higher grades of coal as it changes from soft bituminous coal to hard anthracite, the highest grade of coal. Carbon concentrations in lignite are generally around 40 percent. In bituminous coal, carbon concentrations can be as high as 85 percent, and in anthracite, these concentrations reach 90 to 95 percent. The higher the carbon concentration, the hotter and cleaner the coal burns.

Anthracite is the most efficient and most cleanly burning coal. However, less than one percent of the coal reserves in the United States are anthracite. Most coal reserves in the United States are bituminous coal; thus, many of the electric power plants in the United States burn this type of coal. When bituminous coal burns, it releases carbon and sulfur and nitrogen oxides into the air, causing air pollution. Although lignite has a low sulfur content—less than 1 percent—and is less expensive than bituminous coal, lignite is a less-efficient fuel; more of it must be burned than other types of coal to provide the same amount of energy.

Topic: Fossil Fuels
To find out more about fossil fuels, visit the Earth Science Web Site at earthgeu.com

Activity: World Reserves
Research the amount of petroleum that geologists think is left in the world (reserves). How many barrels are left? How many years will it last?

MiniLab

Oil Migration

Model the migration of oil and natural gas upward through layers of porous rocks.

Procedure

1. Pour 20-mL of cooking oil into a 100-mL graduated cylinder.
2. Carefully pour sand into the graduated cylinder until the sand-oil mixture reaches the 40-mL mark.
3. Now add a layer of colored aquarium gravel above the sand until the gravel reaches the 70-mL mark.
4. Pour tap water into the graduated cylinder until the water reaches the 100-mL mark.
5. Let stand and observe for 5 minutes.

Analyze and Conclude

1. What does the cooking oil represent? What do the sand and aquarium gravel represent?
2. What happens when water is added to the mixture in the graduated cylinder? Why does adding water cause this change?
3. Predict what might occur in the graduated cylinder if a carbonated soft drink was added to the mixture instead of water. What would the bubbles represent?

Petroleum and Natural Gas The word *petroleum* comes from the Greek word *petra,* meaning "rock," and the Latin word *oleum,* meaning "oil"; thus, petroleum was originally known as rock oil. Today, the term *petroleum* is used to refer to the natural crude oil found underground and on Earth's surface in natural seeps, which are areas on Earth's surface where shallow deposits of crude oil ooze upward into pits or creeks, or along beaches. One such seep is illustrated in ***Figure 26-6.*** Crude oil is a mixture of compounds of hydrogen and carbon called hydrocarbons, which can be burned to release energy. Crude oil that is collected on Earth's surface or pumped out of the ground is refined into a wide variety of petroleum products, such as gasoline and kerosene.

Most geologists hypothesize that oil originated organically, in a manner similar to the formation of coal. Millions of years ago, much of Earth's land surface was covered by shallow seas. Rivers carrying mud and silt, along with other sediments, emptied into these seas. Organisms that died in or near the water became part of the sediment load and fell to the bottom of the seas. As layers of sediment accumulated, they were pressed down by the weight of overlaying layers and eventually became sedimentary rocks.

Figure 26-6 La Brea Tar Pits in Hancock Park, Los Angeles, California, are fossil-bearing seeps that ooze crude oil.

Most scientists hypothesize that crude oil and natural gas originated with once-living organisms partly because sedimentary rocks associated with oil deposits, such as sandstone and shale, contain fossils of ancient organisms. Also, because little oxygen could reach the layers of organic matter at the bottom of the seas, bacteria that do not require oxygen partially decomposed the accumulated organisms, and released a waste product called methane, which is one of the components of natural gas.

Migration Crude oil and natural gas migrate sideways and upward from their place of formation. As they migrate, they move through the pores of permeable sedimentary rocks such as limestone and sandstone. These pores in permeable rocks are the reservoirs in which crude oil and gas accumulate. As the oil and gas rise upward, they displace some, but not all, of the water that originally filled the pores. You can find out how oil migrates in the *MiniLab* on the previous page. Oil and gas continue to rise until they reach a barrier of impermeable rock, such as slate or shale, that prevents their continued upward movement. This barrier effectively seals the reservoir and creates a trap for the petroleum. In some petroleum traps, the natural gas forms a gas cap above the oil, but at high pressures, the gas may form a layer below the crude oil. Geologic formations such as faults and anticlines can trap petroleum deposits, as shown in ***Figure 26-7.*** Because most geologists accept the hypothesis that oil and natural gas originated with the sedimentation of once-living organisms, the search for crude oil and natural gas often begins in areas with thick beds of sedimentary rocks. Today, geologists search for oil deposits using remote sensors, magnetometers, and seismographic equipment that create subsurface maps.

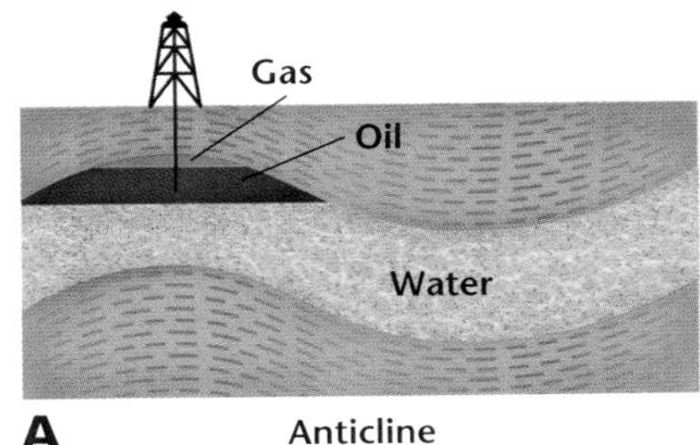

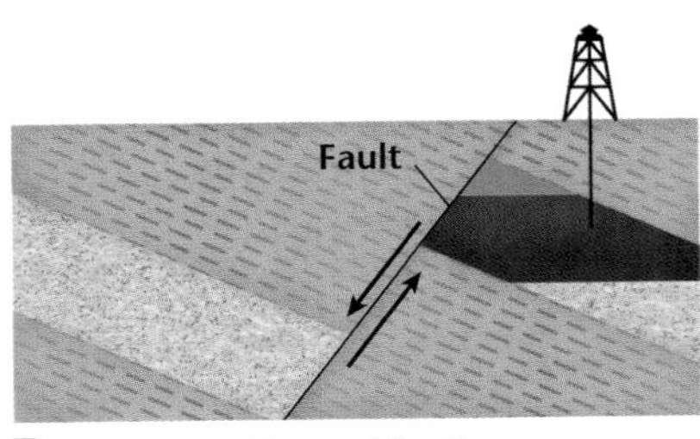

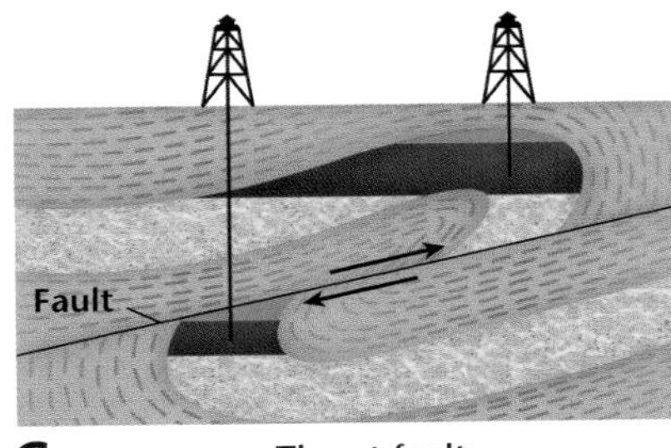

Figure 26-7 These diagrams show typical structural traps for oil and gas deposits.

SECTION ASSESSMENT

1. What is the primary source of energy on Earth?
2. How does coal form?
3. How does petroleum form?
4. **Thinking Critically** Explain how the energy released by a burning candle originated from the Sun.

SKILL REVIEW

5. **Comparing and Contrasting** Compare and contrast the formation of peat and the formation of crude oil. How are these two energy sources alike? How are they different? For more help, refer to the *Skill Handbook.*

SECTION 26.2

Alternative Energy Resources

Earth Sciences 4.a, 7.b, 9.a

OBJECTIVES

- **Identify** *alternative energy resources.*
- **Compare** *the advantages and disadvantages of the various alternative energy resources.*

VOCABULARY

photovoltaic cell
geothermal energy
biogas
gasohol

As you have learned, many of the fuels used today are renewable resources, including wood. Most people, however, rely on nonrenewable fossil fuels for their energy needs. Recently, it has become clear that humans are using up nonrenewable fuels at an alarming rate. Even though there are known reserves of fossil fuels around the world, development of such reserves may be too dangerous, too expensive, or too damaging to the environment to be practical. Some experts estimate that petroleum resources may be used up within the next 60 years. Scientists, private companies, and government agencies are all studying renewable alternatives to traditional energy resources. These alternative energy resources include solar energy, wind, water, geothermal energy, nuclear energy, and biomass.

SOLAR ENERGY

Have you ever used a calculator like the one shown in ***Figure 26-8?*** This calculator has batteries, but it also has a solar collector that uses the Sun's energy to provide power. As you have learned, the Sun is the ultimate source of most energy on Earth. The main advantages of solar energy are that it is free and it doesn't cause any kind of pollution.

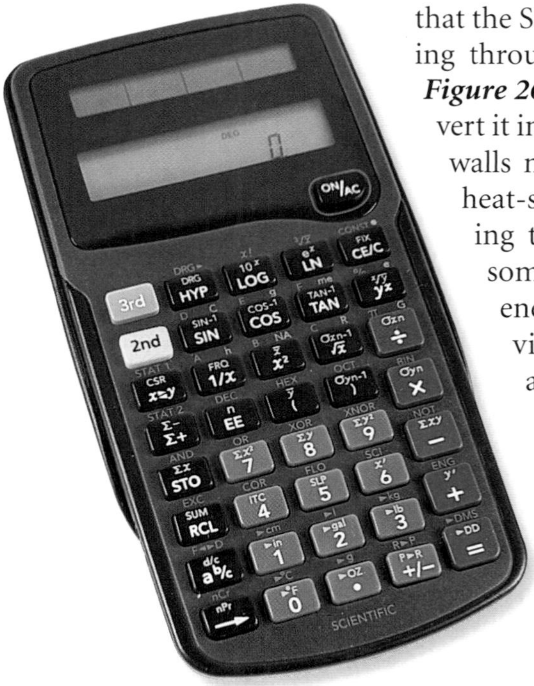

Figure 26-8 This hand-held calculator uses solar energy for power.

Passive Solar Heating Have you ever sat on the vinyl seat of a car that had been in direct sunlight for a few hours? If so, you know that the Sun can heat up the inside of a car or a building just by shining through the windows. The sunporch of the house shown in ***Figure 26-9*** uses this principle to capture sunlight directly and convert it into heat. The Sun's energy also can be captured in floors and walls made of concrete, adobe, brick, stone, or tile, which have heat-storing capacities. These materials collect solar energy during the daytime and slowly release it during the evening. In some warm climates, these materials alone provide enough energy to keep a house warm. Passive solar designs can provide up to 70 percent of the energy needed to heat a house, as well as up to 60 percent of the energy needed to cool it. Although a passive solar house can be slightly more expensive to build than a traditional home, the cost of operating such a house is 30 to 40 percent lower.

Active Solar Heating Even in areas that do not receive consistent sunlight, the Sun's energy can still be used for heating. Active solar-heating systems include collectors such as solar panels that absorb solar energy and fans or pumps that distribute that energy throughout the house. Solar panels mounted on the roof, as shown in the house in ***Figure 26-9,*** have unobstructed exposure to the Sun. Heat collected by these solar panels can be used to heat a house directly, or it can be stored for later use in insulated tanks that contain rocks, water, or a heat-absorbing chemical. Solar panels mounted on a roof can heat water up to 65°C (149°F), which is hot enough to wash dishes and clothing.

Figure 26-9 This house incorporates both passive and active solar heating in its design. Deciduous trees help block the Sun in the summer to keep the house cool. In the winter trees lose their leaves, allowing the Sun to warm the house directly.

Solar Cookers Have you ever heard a weather forecaster say that temperatures will be hot enough to cook eggs on a sidewalk? The Sun's energy can cook food when it is focused correctly. Solar cookers can be used effectively where fuels are scarce or expensive, as in countries that have cut down most of their forests. A solar cooker can be as simple as an enclosed box with reflectors to direct the Sun's rays inside the box. More-sophisticated types of solar cookers, such as the parabolic cooker shown in ***Figure 26-10,*** can provide enough heat to boil water by focusing sunlight on one point. When the Sun's rays are focused in this way, however, they can damage eyesight, and therefore dark glasses must be worn when solar cookers are used.

Photovoltaic Cells All of the uses of solar energy described so far rely on direct sunlight. Using direct sunlight is relatively easy, but energy is also needed during hours of darkness and on cloudy days. On overcast days and in areas that don't get much direct sunlight, solar energy cannot be used directly. In addition, solar energy is difficult to store. An economical and practical method of storing large amounts of solar energy for long periods of time has not yet been developed. If such a method were to be developed, there might be no need for any other energy resources.

Until such a method is developed, solar energy is converted into electrical energy by **photovoltaic cells,** which are thin, transparent wafers made up of layers of boron- and phosphorus-enriched silicon. When sunlight falls on a photovoltaic cell, it releases a flow of electrons that creates an electrical current. Although a photovoltaic cell produces only a small amount of electricity, many such cells can be wired together in a panel that provides 30 to 100 W of power. In the same way, several panels wired together increase the amount of power produced. The electricity produced by photovoltaic cells can be stored in batteries.

Figure 26-10 This parabolic solar cooker focuses sunlight on the spot where the cooking pot is placed.

Figure 26-11 A power tower is surrounded by banks of solar panels that reflect and concentrate sunlight onto the tower, where the sunlight is collected and stored in batteries.

Photovoltaic cells are reliable, quiet, and typically should last more than 30 years. They can be installed quickly and can be moved easily. Large-scale groups of cell panels can be set up in deserts and in other land areas that are not useful for other purposes. Today, more than 20 public utility companies in the United States use photovoltaic cells in their operations. Power towers are being used to collect solar energy and produce electricity, as shown in ***Figure 26-11.*** Some scientists estimate that power towers may someday supply 30 percent of the electric power used worldwide.

Energy from Water

Are you familiar with the waterfall pictured in ***Figure 26-12?*** This is Niagara Falls, a waterfall in the Niagara River that straddles the border between the United States and Canada. This waterfall produces electricity for both countries. Water from the falls is diverted into massive turbines. As water falls over the turbines, they turn, producing mechanical energy that drives a generator and produces electrical energy. Energy produced in this way is called hydroelectric power.

The power of falling water also can be harnessed to produce electricity when a dam is built across a large river to create a reservoir. The water stored in the reservoir flows through huge pipes at controlled rates and causes turbines to spin to produce electricity. Today, hydroelectric power provides about 20 percent of the world's electricity and 6 percent of its total energy. Approximately 10 percent of the electricity used in the United States is generated by water, while Canada obtains more than 70 percent of its electricity from this source. Many of the hydroelectric power resources of North America and Europe have already been developed, but in Africa, Latin America, and Asia, many potential sites for hydroelectric power plants have not yet been explored.

One advantage of hydroelectric power is that it is nonpolluting. Dams built to harness hydroelectric energy provide additional

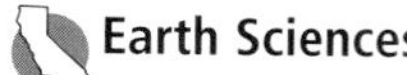

Earth Sciences

4.a Students know the relative amount of incoming solar energy compared with Earth's internal energy and the energy used by society.

Figure 26-12 The water diverted from Niagara Falls powers huge turbines. Hydroelectric power presently provides 26 percent of the electricity needs of Upstate New York.

benefits in the form of recreational opportunities, drinking water, flood control, and water for irrigation. Dams also have negative impacts, however. When the reservoirs behind dams fill, they flood large areas and force people to move, destroy wildlife habitats, interrupt migration routes for fish, and change the natural pattern of water flow. This causes sediments to accumulate in the reservoir, streambeds downstream to erode, and water quality to degrade.

Earth Sciences

9.a Students know the resources of major importance in California and their relation to California's geology.

Energy from the Oceans Ocean water is another potential source of energy. The kinetic energy in waves, which is created primarily by wind, can be used to generate electricity. Barriers built across estuaries or inlets can capture the energy associated with the ebb and flow of tides for use in tidal power plants. One such plant exists at the mouth of La Rance River in France. While power from moving ocean water is renewable and nonpolluting, barriers in the ocean can change the water level and may disrupt coastal and marine ecosystems.

To learn more about water, go to the **National Geographic Expedition** on page 898.

Geothermal Energy

Most of the energy sources you have studied in this chapter so far came from the Sun. However, one energy source used today originates from Earth's own internal heat. Some of the hot springs at Yellowstone National Park, in the western United States, regularly shoot out geysers, tall fountains of steam mixed with hot water. Old Faithful is one of the best-known geysers in the world. What causes geysers? Water trapped underground in fractures or in porous rock is heated by Earth's internal heat. Some of the water becomes steam. When the heated water and steam escape through cracks in Earth's crust, they explode upwards in spectacular displays. Energy produced by naturally occurring steam and hot water is called **geothermal energy.** While some geothermal energy escapes from Earth in such small amounts that it is barely noticeable, large amounts of geothermal energy are released at other surface locations. In these areas, which usually coincide with plate boundaries, geothermal energy can be harnessed to heat homes and businesses, used in power plants to produce electricity, and even used to provide recreational opportunities, as illustrated in ***Figure 26-13.***

The U.S. Department of Energy estimates that if the geothermal reservoirs in the United States were developed, they could provide up to 30 times as much energy as the country currently uses.

Figure 26-13 Geothermal reservoirs are most common in areas of high volcanic and seismic activity. In Reykjavik, Iceland, almost 80 percent of the buildings are heated by power plants that draw hot water directly from geothermal wells underneath the city.

Advantages and Disadvantages of Geothermal Energy

One advantage of geothermal energy is that it is abundant and reliable at the sites where it occurs. However, as the water heated by geothermal energy is tapped, cooler water replaces it. To provide continuous power, geothermal energy reservoirs must be managed carefully. For example, in Rotorua, New Zealand, homes are heated with geothermal energy, but the availability of water is decreasing and restrictions are now being placed on its use. Geothermal steam is generally pollution-free, but water heated by geothermal energy frequently contains large amounts of minerals that can clog pipes and pollute surface water. These problems can be eliminated with systems that hold hot water and steam from geothermal reservoirs in closed containers. A greater disadvantage of geothermal energy is that its development can disrupt ecosystems and can cause local air and water pollution. Also, geothermal energy is useful only near sites where it exists, because transporting it is not practical.

Topic: Alternative Energy
To find out more about alternative energy sources, visit the Earth Science Web Site at earthgeu.com

Activity: Fuel Cells
Research fuel cells as alternatives to petroleum to power automobiles. What issues must be resolved to make fuel cells practical?

Wind Energy

Have you ever seen a windmill? Windmills in the Netherlands have been capturing wind power for human use for more than 2000 years. Today, wind farms, such as the one shown in ***Figure 26-14,*** are replacing the more traditional windmills that farmers once used to pump water from underground wells. The windmills on a wind farm are more properly called wind turbines, because they convert the energy of the wind to mechanical energy, which is then used to produce electrical energy. Wind energy increases with the cube of the wind speed. For example, when wind velocity doubles, the wind's capacity to generate power increases 8 times.

Most of the wind farms in the United States are in California, yet nearly all of the energy needs of the country could be met if wind farms were built in just three states that experience consistent, steady winds: North Dakota, South Dakota, and Texas. Wind turbines currently provide three percent of the electricity used in Denmark. Experts suggest that wind power could supply more than 10 percent of the world's electricity by the year 2050.

Figure 26-14 Wind farms such as this one in California produce one percent of the state's electricity.

Advantages and Disadvantages of Wind Energy

Wind is a virtually unlimited energy resource at favorable sites worldwide. Locations at high altitudes generally produce the strongest, most consistent winds. Another advantage of using wind energy is that wind farms can be built quickly and expanded as needed. They are nonpolluting and do not require water for cooling purposes, and the land underneath wind turbines can be used for cattle grazing or other farming activities. As a result, wind energy is one of the least expensive ways to produce electricity.

Why isn't wind power used to provide more of the world's electricity? Wind power is economical only in areas with steady winds. When the wind dies down, people have to rely on backup systems for power, including traditional fossil fuel-burning power plants. Other disadvantages of wind farms are that they are not very attractive and they have been shown to interfere with and even kill migrating birds, as well as birds of prey. Windmills also can be noisy and interfere with radio and television reception.

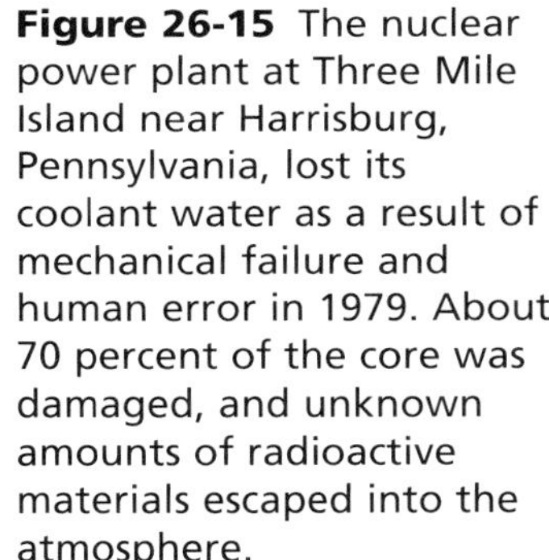

Using Math

Calculating Wind Speed Wind energy increases with the cube of the wind speed. What increase in wind energy would occur if the wind speed quadrupled?

Nuclear Energy

As you learned in Chapter 3, atoms lose particles in the process of radioactive decay. One process by which atomic particles are given off is called nuclear fission. Nuclear fission is the process in which a heavy nucleus (mass number greater than 200) divides to form smaller nuclei and one or two neutrons. This process releases a large amount of energy. Radioactive elements consist of atoms that have a natural tendency to undergo nuclear fission. Uranium is one such radioactive element that is commonly used in the production of nuclear energy. Nuclear energy is one other energy source that does not come directly from the Sun.

In the late 1950s, power companies in the United States began developing nuclear power plants because scientists suggested that nuclear power could produce electricity at a much lower cost than coal and other types of fossil fuels. Another advantage was that nuclear power plants do not produce carbon dioxide or any other greenhouse gases. After 50 years of development, however, 424 nuclear reactors in 25 countries currently are producing only 17 percent of the world's electricity. Construction of new nuclear power plants in Europe has come to a halt, and no new nuclear plants have been built in the United States since 1978.

What happened to the promise of nuclear power? Poor management, high operating costs, poor reactor designs, and public concerns about safety and disposal of radioactive wastes contributed to the decline of nuclear power. In addition, nuclear accidents such as those at Three Mile Island, shown in ***Figure 26-15,*** and at

Figure 26-15 The nuclear power plant at Three Mile Island near Harrisburg, Pennsylvania, lost its coolant water as a result of mechanical failure and human error in 1979. About 70 percent of the core was damaged, and unknown amounts of radioactive materials escaped into the atmosphere.

Figure 26-16 Bales of bagasse are burned to produce the energy that powers this steam locomotive in Java, Indonesia.

Chernobyl, Ukraine, in 1986, alerted people worldwide about the hazards of nuclear power plants. Because of its hazards, nuclear power is no longer considered to be the solution to providing for the world's energy needs, although nuclear power plants continue to provide energy in many countries.

BIOMASS

Biomass is a renewable energy resource as long as the organisms that provide the biomass are replaced. Biomass fuels include wood, dried field crops, and dried fecal materials from animals. One way to produce biomass fuel is to plant large numbers of rapidly growing plants, such as cottonwood trees, in biomass plantations. After harvest, these plants can be burned directly, converted into gas, or fermented into alcohol fuel.

Bagasse, which is the residue of sugar cane after the juice has been extracted, is another source of biomass that is burned to produce power as illustrated in ***Figure 26-16***. The burning of bagasse produces approximately 10 percent of Hawaii's electricity supply, and thus, it eliminates the need for approximately 2.7 million barrels of oil each year. Other types of biomass fuels are produced when bacteria and chemical processes are used to convert solid biomass into gaseous and liquid biofuels, such as biogas, liquid ethanol, and liquid methanol. A disadvantage of biomass fuels is that when they are burned, they release carbon dioxide and particulate matter into the atmosphere. Biomass is the main source of energy for more than half of the world's population.

Earth Sciences

7.b Students know the global carbon cycle: the different physical and chemical forms of carbon in the atmosphere, oceans, biomass, fossil fuels, and the movement of carbon among these reservoirs.

Biogas **Biogas** is a mixture of gases that includes 50 to 70 percent methane gas and 30 to 48 percent carbon-dioxide gas. Plant and animal wastes can be converted into methane gas in simple containers, called digesters, by the action of bacteria. In China, more than 8 million biogas digesters are in use in individual households. In a biogas digester the gas is separated from the solid wastes and piped into homes for use as a cooking fuel. The leftover solid wastes then can be used as fertilizer on food crops, because the high temperatures inside the digester destroy harmful bacteria.

Ethanol and Methanol Liquid ethanol is another name for grain alcohol. Ethanol can be made from sugar and grain crops, including sugar cane, sugar beets, sorghum, and corn. Currently, ethanol produced from corn is used in gasoline mixtures around the world. Gasoline mixed with ethanol makes **gasohol,** which can be burned in conventional gasoline engines. The use of gasohol can extend gasoline supplies and reduce dependency on foreign petroleum reserves. Ethanol fuels burn more cleanly than pure gasoline. Liquid methanol, which is wood alcohol, is made mostly from natural gas, but it can also be made from wood, wood wastes, agricultural wastes, sewage sludge, garbage, or coal.

Figure 26-17 Shale oil is extracted from oil shale, a fine-grained rock. Some oil shale can actually ignite and burn on its own, as this photo shows.

Energy from Oil Shale and Tar Sand

You have learned that crude oil and natural gas can be found in porous sedimentary rocks. Sometimes, other hydrocarbon mixtures become trapped in different types of rocks. For example, oil shale, shown in ***Figure 26-17,*** is a fine-grained rock that contains a solid, waxy mixture of hydrocarbon compounds called kerogen. Oil shale can be mined, then crushed and heated until the kerogen vaporizes. The kerogen vapor can then be condensed to form a heavy, slow-flowing, dark-brown oil known as shale oil. Shale oil is processed to remove nitrogen, sulfur, and other impurities before it can be sent through pipelines to a refinery. At present, the cost of processing oil shale is higher than the cost of crude oil sold by countries that have abundant oil supplies.

Tar sand is a mixture of clay, sand, water, and bitumen, which is a heavy, black, high-sulfur oil. Tar sand also can be mined, then heated until the bitumen fluid softens and floats to the top. Bitumen can be purified and upgraded into a type of crude oil. However, the processing of oil shale and tar sand requires large amounts of energy and produces air and water pollution.

Section Assessment

1. Identify one alternative energy resource that is associated with each of Earth's systems: the atmosphere, hydrosphere, biosphere, and lithosphere.
2. Compare passive and active solar energy.
3. What is gasohol?
4. What alternative energy source would be the least damaging to the environment if the required technology could be developed to harness and use it? Explain.
5. **Thinking Critically** Although solar energy could supply all of the world's energy needs, why isn't it used to do so?

Skill Review

6. **Making Tables** Prepare a table that compares the advantages and disadvantages of alternative energy resources, including solar energy, hydroelectric energy, geothermal energy, wind, nuclear energy, and biomass. In your table, include the following headings: Location, Limits to Use, Health Hazards, Affordability, Major Advantage, and Major Disadvantage. For more help, refer to the *Skill Handbook.*

earthgeu.com/self_check_quiz

SECTION 26.3

Conservation of Energy Resources

Earth Sciences 4.a I&E 1.a, 1.j, 1.m

OBJECTIVES

- **Recognize** *the need for the conservation of energy resources.*
- **Identify** *ways to conserve energy resources.*

VOCABULARY

energy efficiency
cogeneration
sustainable energy

As you have learned, traditional energy resources such as fossil fuels are nonrenewable and in limited supply. Yet industrialized countries continue to consume these resources at ever increasing rates. ***Figure 26-18*** compares the energy resources used in industrialized countries to those used in developing countries. The graphs in ***Figure 26-18*** show that renewable resources account for 41 percent of the energy used in developing countries, in comparison to industrialized countries where renewable resources account for only 10 percent of the energy used. Experts have concluded that the best way to meet energy needs is a combination of improved energy efficiency and increased use of locally available, renewable energy resources. This means that it is better to use a variety of energy resources at all times than to depend upon a single, nonrenewable energy resource such as oil, coal, or natural gas. For example, a community that has hydroelectric energy resources may also use solar energy to generate electricity in months when water levels are low.

Energy Efficiency

Energy efficiency is the use of energy resources in the ways that are most productive. This means using the same amount of a resource but getting more from it. To find ways to use resources more efficiently, scientists study exactly how energy resources are used and where improvements are needed. Using resources more efficiently is a type of conservation.

Figure 26-18 This graph shows the relative amounts of energy used by industrialized and developing countries worldwide.

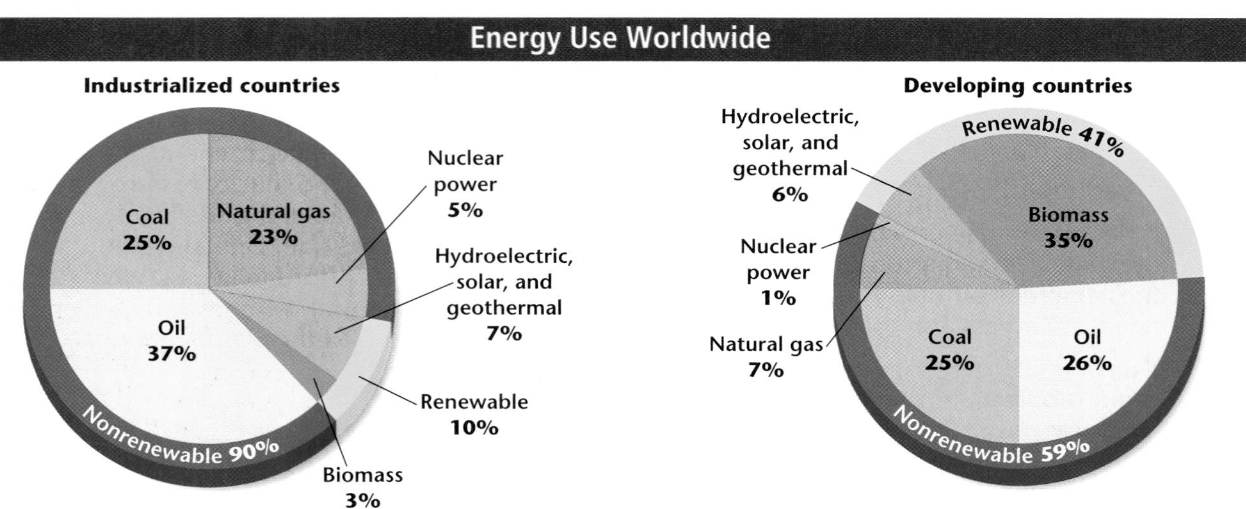

How can energy efficiency be improved? Energy analysts have suggested several ways of doing so. People can recycle old appliances and vehicles, and purchase newer, more energy-efficient models. They also can improve the energy efficiency of older homes by adding insulation, installing solar panels, or by installing new windows, or they may purchase newer, energy-efficient homes. Local power companies can use energy from alternative resources in areas where they are available to decrease their dependence on petroleum. Governments also can help by offering tax savings to people who buy more-efficient vehicles and appliances, and by funding research and development projects related to energy efficiency.

Earth Sciences

4.a Students know the relative amount of incoming solar energy compared with Earth's internal energy and the energy used by society.

Conservation of Energy Resources Do you wet your toothbrush, then turn off the water while you brush your teeth? When you leave an empty room, do you turn off the lights? You can probably think of many other ways that you could conserve energy at home, at school, and in the workplace. Conserving energy is, in the long run, less expensive than finding new energy sources. You can find out how energy use has changed in the *Problem-Solving Lab* on this page.

Problem-Solving Lab

Changes in Energy Resource Use

Analyze how the use of energy resources has changed Many types of energy resources are used throughout the world. Over time, fluctuations occur in the amount of each resource used. The data in the table show the changes that have occurred in world energy use between 1900 and 1997. Are these changes good or bad?

Analysis

1. Plot the data in the table on a graph. Use a different color for each year.

Thinking Critically

2. Of all the energy used between 1900 and 1997, what percentage was nonrenewable?
3. What trend in the use of renewable energy sources is evident?
4. What concerns are reflected by the data? How can these concerns be addressed?

World Energy Use

Energy Source	Percent of Total Energy Provided	
	1900	1997
Coal	55	22
Oil	2	30
Natural gas	1	23
Nuclear power	0	6
Alternatives (biomass, hydroelectric energy, wind, solar energy, geothermal energy)	42	19

Getting More for Less

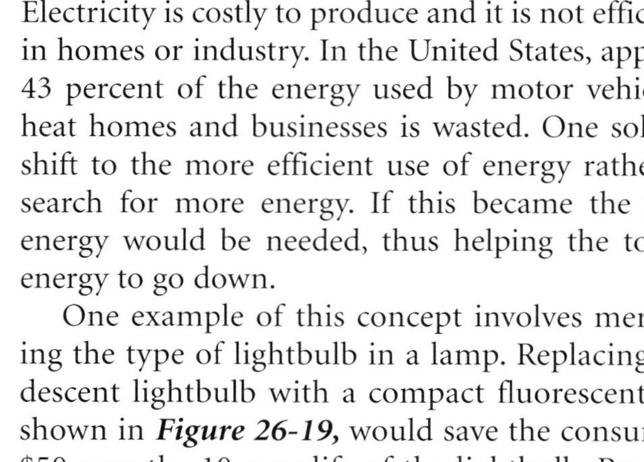

The usual approach to energy use in industrialized countries has been to spend more to get more. Higher demand requires a greater supply and results in higher costs. The price that people in these countries pay for energy is high. This is especially true of electrical energy. Electricity is costly to produce and it is not efficiently used in homes or industry. In the United States, approximately 43 percent of the energy used by motor vehicles and to heat homes and businesses is wasted. One solution is to shift to the more efficient use of energy rather than the search for more energy. If this became the norm, less energy would be needed, thus helping the total cost of energy to go down.

Figure 26-19 Fluorescent lightbulbs like this one can be used in most lamps to save energy.

One example of this concept involves merely changing the type of lightbulb in a lamp. Replacing an incandescent lightbulb with a compact fluorescent lightbulb, shown in ***Figure 26-19,*** would save the consumer \$35 to \$50 over the 10-year life of the lightbulb. Replacing just 25 incandescent lightbulbs in a house with fluorescent ones could save between \$87 to \$125 each year in electricity costs. Use of fluorescent lightbulbs or other energy-efficient lighting could save businesses in the United States alone billions of dollars per year in the cost of electricity. In addition, less energy would be used. This would help to reduce the amount of coal or other fossil fuels needed for generation of electricity, which could in turn decrease the amount of carbon dioxide and sulfur dioxide emitted into the atmosphere. The net effect would be a reduction in air pollution.

Cogeneration When power plants generate electricity, waste heat is given off during the process. However, it is possible to recover this waste heat and use it to produce another form of energy. The production of two usable forms of energy, such as steam and electricity, at the same time from the same process is called **cogeneration.** Cogeneration can produce income and reduce the need for additional energy resources. One secondary use of the heat given off by the generation of electricity is the warming of buildings or water. Another is the operation of electrical devices in the power plant, such as scrubbers, which remove sulfur from the air emitted from smokestacks. Cogeneration has enabled Central Florida to operate the nation's cleanest coal-powered electric facility. Sweden has achieved an 85-percent energy efficiency rating while releasing only a fraction of the nitrogen-oxide and sulfur-oxide emissions that are permitted for coal-powered facilities in the United States.

Figure 26-20 Electric cars generally are smaller than gasoline-powered vehicles, and they have a limited range. However, for trips close to home at moderate speeds, these vehicles are extremely efficient and non-polluting.

Improving Efficiency in Transportation Transportation is necessary to move food and other goods from one place to another, and to move people from their homes to workplaces, schools, stores, and other places. Although transportation requires the use of fuel, conservation practices can help reduce dependency on the fuel resources used for transportation.

The use of fuel-efficient vehicles is one way to reduce the amount of petroleum resources consumed. Automobile manufacturers now have the ability to build vehicles that achieve high rates of fuel efficiency without sacrificing performance. Laws that lower speed limits help improve fuel efficiency, because engines burn fuel more completely at lower speeds. The future of this industry looks especially promising as hybrid and electric cars, such as the one shown in ***Figure 26-20,*** begin to reach the consumer market.

Figure 26-21 Special lanes for car pools encourage people to leave their cars at home and travel with a friend or two to work each day, thus reducing total vehicle emissions.

People who live in metropolitan areas can improve energy efficiency by using public transportation. When it is necessary to drive private automobiles, carpooling can reduce the number of vehicles on the highways and reduce gasoline consumption. Carpooling also eases congestion on major highways in and around large cities. Some metropolitan areas, such as Washington, DC, encourage carpooling by providing express lanes for cars with multiple passengers, as illustrated in ***Figure 26-21.*** In Europe, mass transportation includes long-distance rail systems, as well as electric trams and trolleys in the major cities.

People who live in rural areas are often dependent on automobiles. In many rural areas, modes of transportation other than the automobile are limited or nonexistent. However, with the increasing importance of computers and access to the Internet, more jobs can be performed from home. The use of bicycles for short distances is another option in some places. In China, a country of 1.3 billion people, approximately 300 million bicycles are in use.

Improving Efficiency in Industry While industries use one-third of all energy produced in the United States, cogeneration has allowed some industries to increase production while leveling off their energy use. This has been accomplished in part by the use of more efficient machinery. Industries can further improve their energy efficiency by making greater efforts to reduce their use of both materials and the energy used to produce those materials. For example, packaging can be reduced overall, and unnecessary packaging can be eliminated. These efforts would cut down on resource use, lower costs, and also reduce the amount of solid waste.

Increasing Efficiency at Home People can do many things in their own homes to conserve energy. For example, fluorescent lights last longer than incandescent bulbs and need to be replaced less frequently. The use of energy-efficient appliances can also make a significant difference in energy consumption. This is especially true of appliances that consume large amounts of energy, such as refrigerators, water heaters, and ovens.

The use of more-efficient insulation on existing homes can result in dramatic savings on heating costs, especially in climates with cold winters. As warm air rises, heat escapes through windows, chimneys, and roofs. Weather-stripping around doorways and caulking around older windows can help keep cold air outside and warm air inside. Insulating pipes and water heaters also reduces energy consumption.

Figure 26-22 Insulation is rated by its R-value, which is the resistance to heat flow. An R-value of 18 indicates that this insulation is more efficient at retaining energy than insulation with an R-value of 3.

Building materials and windows are rated according to their insulation abilities. Construction materials are labeled with these ratings, known as R-values, as shown in ***Figure 26-22.*** The use of materials with high insulation values can significantly reduce energy consumption. Replacing older windows can save so much money in reduced energy costs that the windows pay for themselves in just a few years.

When new structures are built, the use of energy-efficient materials and windows can have a major impact on future energy needs. Designs for new buildings that incorporate passive and active solar heating also can reduce the need for the consumption of traditional energy resources. Find out more about solar heating in the *Design Your Own GeoLab* at the end of this chapter. Some window manufacturers now triple-glaze their windows, as shown in ***Figure 26-23,*** or place an inert gas between the panes to reduce energy loss. Superinsulation and air barriers in new homes built in Minnesota recently resulted in heating savings of 68 percent. Newly constructed buildings that are designed to save energy cost more initially, but they can save money and resources in the long run.

Figure 26-23 A triple-glazed window is one that has either three panes of glass or two panes of glass with a middle layer of plastic film. Some triple-glazed windows also have an inert gas, such as argon or krypton, between the layers to improve the insulating ability of the windows.

Sustainable Energy

All humans have needs for energy, but these needs vary. Energy resources on Earth are interrelated, and they affect one another. **Sustainable energy** involves the global management of Earth's natural resources to meet current and future energy needs without causing environmental damage. A good management plan incorporates both conservation and energy efficiency. The development of new technology to extend current resources and provide additional energy resources is a vital part of such a plan. Global cooperation can help ensure the necessary balance between protection of the environment and economic growth. The achievement of these goals will depend on the commitment made by all to ensure that future generations have access to the energy resources required to maintain a high quality of life on Earth.

Section Assessment

1. Why should you be concerned about energy efficiency?
2. Describe three ways in which you could conserve electrical energy in your home.
3. How does cogeneration save energy resources?
4. Why is it important to conserve resources instead of seeking new sources of energy?
5. **Thinking Critically** Why is there such a difference in energy consumption among different countries, such as the United States and India?

Skill Review

6. **Concept Mapping** Use the following terms to construct a concept map of the major concepts in this section. For more help, refer to the *Skill Handbook.*

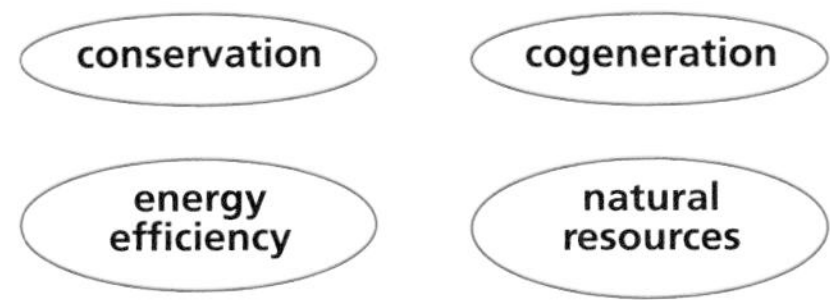

DESIGN YOUR OWN GeoLab

Designing an Energy-Efficient Building

I&E 1.a, 1.j

Buildings can be designed to conserve heat energy. Some considerations involved in the design of a building that conserves heat include the materials that will be used in construction, the materials that will store heat, and the overall layout of the building.

Preparation

Problem

How can a building be designed to conserve energy? What building materials will work best, and what other factors need to be considered?

Possible Materials

glass or clear plastic
sturdy cardboard boxes
scissors
tape
glue
thermometers
paints of various colors
materials to cover the building (paper, aluminum foil, foamboard, and so on)
interior materials (stones, mirrors, fabric, and so on)
light source

Hypothesis

Brainstorm a list of design features that might contribute to the energy efficiency of a building. Hypothesize how you could incorporate some of these features into an energy-efficient building. Find out what materials are used in heat-efficient homes and research local sources of materials for your design.

Decide how you will determine the heat efficiency of the building you construct. Be sure to plan for a control building for comparison.

Objectives

In this Geolab, you will:

- **Research** what materials are used in the construction of energy-efficient buildings.
- **Design** a building that is energy efficient.
- **Construct** the building that you design.
- **Determine** the heat efficiency of the building by comparing it to a control building.
- **Interpret** the data that you collect to determine your success in developing an energy-efficient building.

Safety Precautions

Be careful when you are using scissors. Make sure to handle the light source carefully when it is hot. Always wear safety goggles and an apron in the lab.

Plan the Experiment

1. Review the data that you collected about building energy-efficient buildings. Also review your list of possible design features.
2. Design your building. Make a list of the heat-conserving issues that you addressed.
3. Decide on the materials that you will use to build your house. Collect those materials.
4. Construct the building and a control building for comparison.
5. Devise a way to test the heat-holding ability of each building.
6. Proceed with the test on each building. To test the buildings' heat energy efficiency, it may be necessary to heat the buildings and determine how long heat is conserved within each one. ***CAUTION:*** *Make sure the heat source is far enough away from the building materials so that they do not burn or melt.*
7. Record your data in a table. Then, make a graph of your data.
8. Make modifications to the design to improve the building's efficiency.

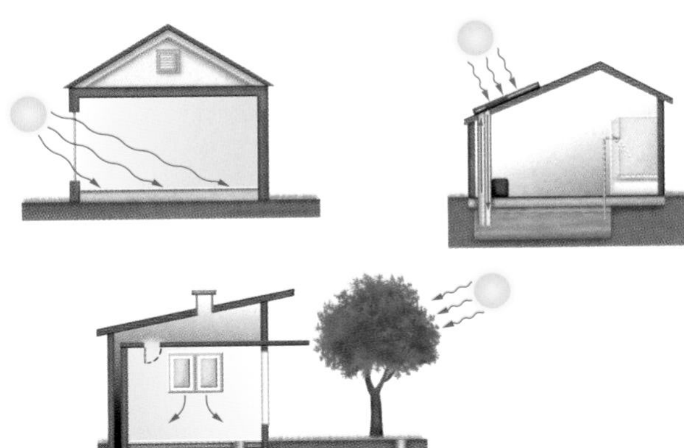

Analyze

1. **Checking Your Hypothesis** Was the building that you designed more energy-efficient than the control building? Why did you construct a control building?
2. **Interpreting Observations** What problems did you encounter, and how did you solve them?
3. **Observing and Inferring** How did your observations affect decisions that you might make if you were to repeat this lab? Why do you think your design worked or did not work?
4. **Comparing and Contrasting** Compare and contrast the building you designed and the control building. Compare and contrast your design and the designs of your classmates.
5. **Thinking Critically** Suppose you could use only naturally occurring materials. Would that limit your design? Explain your answer.

Conclude & Apply

1. How could you incorporate some of your design elements in your own home?
2. How could your design be improved?
3. How could using different energy sources affect your results?

Science & the Environment

 I&E 1.m

Preserving the Rain Forest

Papua New Guinea's rain forest is home to a diverse population of living things. Huge butterflies glide through the air along with more than 700 species of tropical birds. This paradise also contains extinct volcanoes and beautiful waterfalls.

Imagine the distress the people of Papua New Guinea must have felt when a multinational petroleum company announced that it had plans to search for oil in their rain forest home. Plans to begin drilling for oil in the Kikori area of Papua New Guinea met with local opposition immediately. The oil company responded by teaming up with the government of Papua New Guinea to develop a comprehensive environmental plan. This plan called for the study of archaeological, cultural, and socioeconomic impacts that the extraction of oil would have on the rain forest. In addition, the company enlisted the help of an international wildlife organization to study the environmental impacts of oil extraction. The Kikori Integrated Conservation and Development Plan is the result of that study. It includes a major biodiversity survey of the area, experimental projects in ecotourism and ecoforestry, and training of personnel in conservation management. The project has now become a model for development in Papua New Guinea.

Protecting the Ecosystem

Since the beginning of oil drilling in the Kikori area, most of the rain forest has been left intact. The clearings for oil drilling equipment are small; only as much room as is needed has been cleared of vegetation. When crews are finished working in an area, it is reseeded with native plants. Only essential roads have been built; most roads are narrow and hard to see from above. Supplies are brought in by boats or seaplanes.

Birds Tell the Tale

Once oil drilling operations in tropical rain forests begin, wildlife often leave. However, in the Kikori area, tropical birds such as the bird of paradise remain. Bird watchers are amazed to see that many endangered species of birds still call the rain forest home. These include the double-wattled cassowary, shown in the photograph above, a flightless bird that is related to emus.

Other Benefits

The oil company has built schools and trained local residents in health and sanitation methods. In addition, the company has donated money to a fund that protects tropical birds. In these ways, oil companies can continue to search for and extract oil while preserving the environment.

Activity

Research the endangered bird species found in Papua New Guinea. Choose one bird species and report on how it nests, what it eats, what part of the rain forest it lives in, and so on. Find a photograph of the bird in a book or go to earthgeu.com and make a drawing of the bird to include in your report.

CHAPTER 26
Study Guide

Summary

Section 26.1

Conventional Energy Resources

Main Ideas

- The Sun is the ultimate source of most energy on Earth. The Sun's energy is transferred from photosynthetic organisms to all other living things.
- Materials derived from living things, known as biomass, have been used as renewable fuels by humans for thousands of years.
- Wood continues to serve as a fuel for over half of the world's population.
- Fossil fuels, such as natural gas, coal, and petroleum, formed from organisms that lived millions of years ago. The burning of these fossil fuels releases sulfur into the atmosphere, and thus contributes to air pollution.

Vocabulary

fossil fuel (p. 686)
fuel (p. 684)
peat (p. 686)

Section 26.2

Alternative Energy Resources

Main Ideas

- Alternative energy resources, such as solar energy, water, geothermal energy, wind, nuclear energy, and biomass, can supplement dwindling conventional energy resources.
- Solar energy is unlimited, but technological advances are needed to find practical solutions to collect and store it.
- Hydroelectric power is derived from the energy of moving water and is commonly used in the production of electricity. Geothermal energy is a product of Earth's internal heat. Its usefulness is limited to areas where it is found near Earth's surface. Wind is a source of energy in areas that have consistently strong winds.
- Nuclear energy results when atoms of radioactive elements emit particles in the process known as fission.
- Oil shale and tar sand contain secondary oil resources that are expensive to extract.

Vocabulary

biogas (p. 696)
gasohol (p. 696)
geothermal energy (p. 693)
photovoltaic cell (p. 691)

Section 26.3

Conservation of Energy Resources

Main Ideas

- Energy resources will last longer if conservation and energy efficiency measures are developed and used. Energy efficiency results in the use of fewer resources to provide more usable energy.
- Cogeneration, in which two usable forms of energy are produced at the same time from the same process, saves resources in the long run.
- The achievement of sustainable energy use will ensure that current and future energy needs are met while maintaining standards of living and at the same time protecting the environment.

Vocabulary

cogeneration (p. 700)
energy efficiency (p. 698)
sustainable energy (p. 703)

CHAPTER 26
Assessment

Understanding Main Ideas

1. What is the ultimate source of most energy on Earth?
   a. tides
   b. radioactivity
   c. the Sun
   d. the mantle

2. Which product can be made from crude oil?
   a. kerosene
   b. peat
   c. cornmeal
   d. biogas

3. Which is NOT derived from living things?
   a. petroleum
   b. coal
   c. peat
   d. photovoltaic cells

4. Which is NOT a biomass energy resource?
   a. wood
   b. sugar cane
   c. fecal material
   d. wind

5. In which process is the Sun's energy captured and used for food production in living things?
   a. photosynthesis
   b. respiration
   c. radioactivity
   d. combustion

6. What organic fuel is derived from moss and other bog plants?
   a. bagasse
   b. peat
   c. biogas
   d. oil shale

7. Which is NOT a fossil fuel?
   a. crude oil
   b. bituminous coal
   c. lignite
   d. biogas

8. What percentage of the world's electricity is provided by falling water?
   a. 50 percent
   b. 20 percent
   c. 30 percent
   d. 60 percent

9. How many nuclear reactors are producing electricity in the world today?
   a. 25
   b. 50
   c. 17
   d. 424

10. Which is NOT a type of fuel?
    a. wood
    b. kerosene
    c. the Sun
    d. coal

Use the diagram to answer questions 11, 12, and 13.

11. The diagram represents a house in New York with a glass-enclosed porch. Which direction should the porch be facing to take advantage of passive solar heating?
    a. north
    b. south
    c. east
    d. west

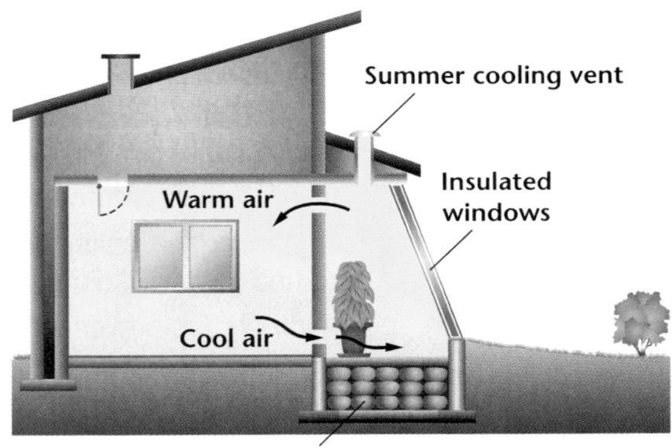

12. What material should be used as flooring in the porch to reduce the need for a furnace to heat the room?
    a. wall-to-wall carpeting
    b. slate
    c. oak
    d. vinyl tile

Test-Taking Tip

DIAGRAMS If a test question requires you to understand a diagram, check the labels carefully. Then test yourself by mentally explaining the diagram.

earthgeu.com/chapter_test

13. To make full use of the energy-conservation abilities of this house, what landscape plants should be planted in front of the porch?
    a. tall deciduous trees
    b. short evergreen bushes
    c. short evergreen trees
    d. ornamental grasses

Applying Main Ideas

14. Why isn't wind energy used to provide electricity in most parts of the northeastern United States?

15. What are two problems associated with the use of solar energy?

16. How is the production of oil from oil shale similar to coal mining?

17. Describe five ways in which you could improve energy efficiency in your home.

18. Explain why nuclear energy is no longer considered to be a solution to providing for the world's energy needs.

Thinking Critically

19. How can a household that uses only electricity be responsible for depleting fossil fuel reserves?

20. Why is the deforestation of tropical rain forests a global concern?

21. What might be some negative consequences of a nation being dependent on foreign energy resources?

22. Explain how using closed containers in geothermal reservoirs is similar to saving energy in cogeneration.

Standardized Test Practice

INTERPRETING SCIENTIFIC ILLUSTRATIONS Use the illustration below to answer questions 1 and 2.

1. How could this kitchen be made more energy efficient?
   a. by maintaining older appliances instead of replacing them with newer ones
   b. by replacing the fluorescent light bulb with an incandescent one
   c. by washing the dishes in the dishwasher instead of the sink
   d. by replacing the old windows with newer ones

2. If this kitchen was located in a house in China, which alternative energy source would most likely be used in it?
   a. bagasse
   b. biogas
   c. gasohol
   d. oil shale

3. Which type of coal is the most efficient and burns most cleanly?
   a. peat
   b. lignite
   c. bituminous coal
   d. anthracite

4. Which is NOT a good way to conserve transportation energy?
   a. drive at a lower speed
   b. make frequent stops
   c. work from home
   d. use a hybrid or electric car

Chapter 27

Human Impact on Earth Resources

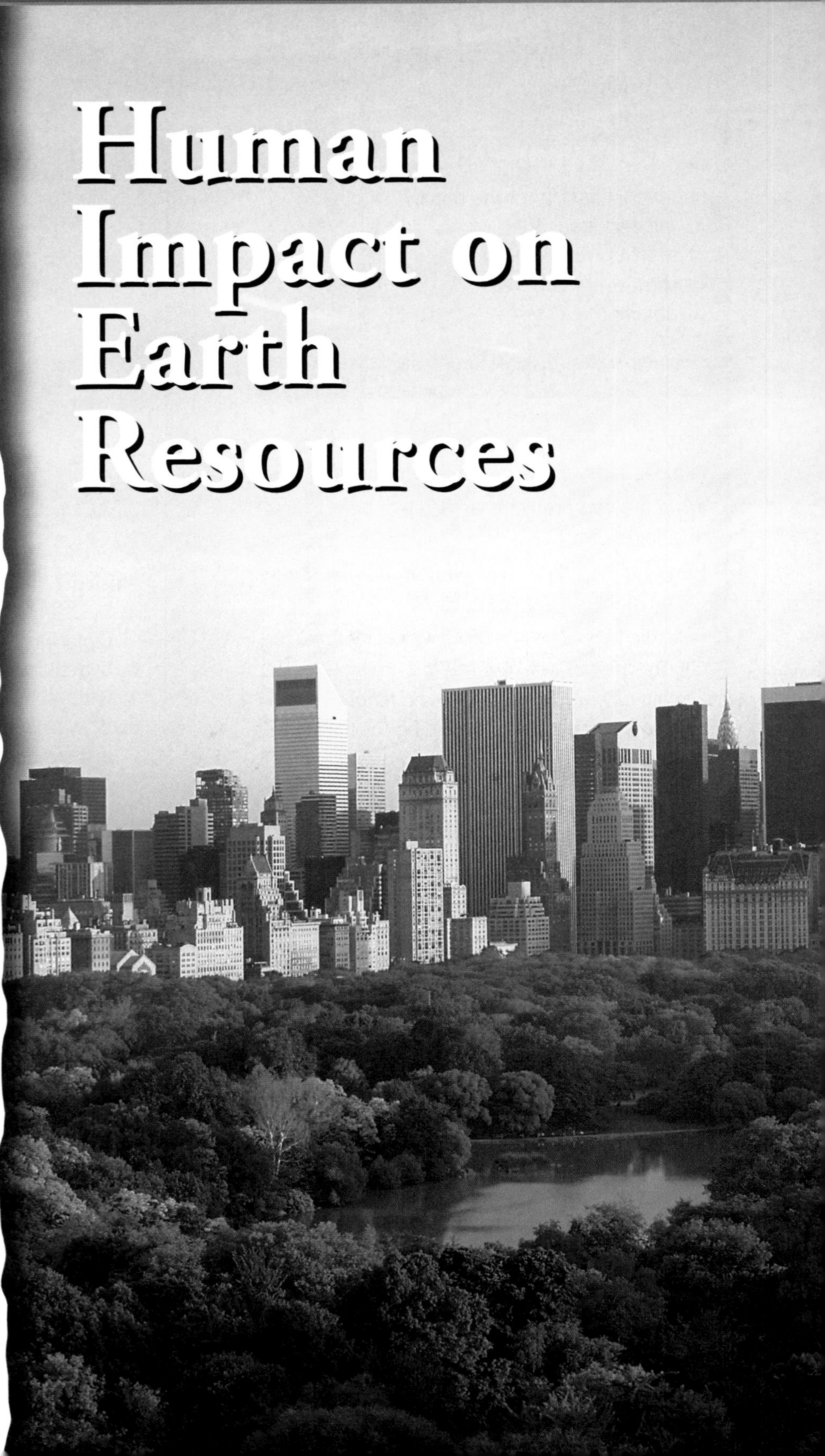

What You'll Learn

- Why all populations require natural resources to exist.
- How the use of natural resources affects Earth's land, air, and water.

Why It's Important

As a result of increasing human population growth, natural resources are being used at increasing rates. The development and use of natural resources may have impacts upon the environment. Humans need to find ways to extract and use natural resources that minimize these impacts.

To learn more about human impact on natural resources, visit the Earth Science Web Site at earthgeu.com

Discovery Lab A Pocketful of Resources

Your everyday activities require resources. In this activity, you will explore the types of resources that you use and some of the global impacts of using them.

1. Make a pile of 10 to 15 items that you brought into the classroom.
2. Combine your pile with those of two or three other students so that you have about 30 different items.
3. In your science journal, make a data table for your items.
4. For each item, record as much of the following information as you can.

- What resources were used to make the item?
- Are the resources renewable or nonrenewable?
- Where was the item made?

CAUTION: Always wear an apron and safety goggles in the lab.

Observe How many different resources are represented by the items in your collection? What are the percentages of renewable and nonrenewable resources? Where were each of the objects made?

SECTION 27.1 *Populations and the Use of Natural Resources*

OBJECTIVES

- **Summarize** *the typical pattern of the population growth of organisms.*
- **Describe** *what happens to populations when they reach carrying capacity.*
- **Identify** *environmental factors that affect population growth.*

VOCABULARY

exponential growth
carrying capacity
density-independent factor
density-dependent factor

Suppose you and some of your friends plan to spend a day at the park. Some of your friends start a volleyball game, while others unpack the cooler and light the barbecue. Later on, maybe all of you will go for a bicycle ride. Did you know that in each of these activities you and your friends would be using natural resources?

RESOURCES AND ORGANISMS

Take a deep breath right now, then let it out again. When you inhale, you take oxygen gas out of the air and into the cells of the blood vessels in your lungs. When you exhale, you release oxygen and carbon dioxide gas and water vapor back into the air. Oxygen, carbon dioxide, and water are examples of natural resources that you, as well as many other living things, use every day.

Like all organisms, you need particular natural resources to maintain life, grow, and reproduce. Among the resources that organisms require are air, food, water, and, in some cases, shelter. To meet their

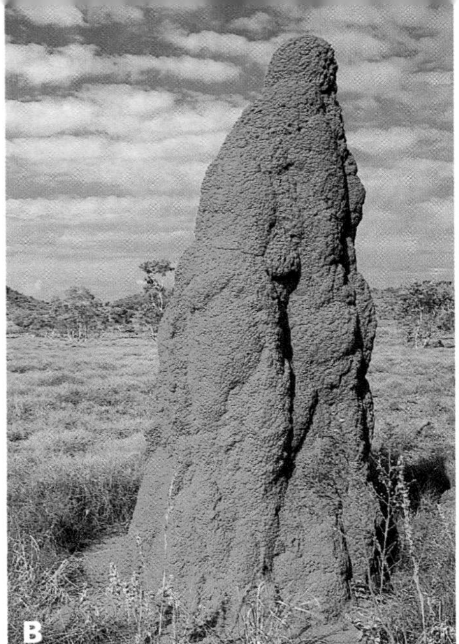

Figure 27-1 Some organisms alter their environment to better provide for their needs. In the middle of their newly created ponds, beavers build domed lodges where they live, nest, and raise their young **(A).** Tropical termites are highly social organisms that live in colonies where each termite has an assigned role that helps the society survive **(B).**

basic needs, most organisms are adapted to their immediate environment; they live in a balance with the natural resources their environment provides. For example, songbirds live in grassy meadows, forage for grass seeds to eat, weave nests out of dried grasses and twigs, and drink water from ponds or streams nearby. Painted turtles live in ponds, swamps, and slow-moving streams, where they eat aquatic plants, insects, and other small animals.

Other organisms, however, alter their environment to better meet their needs. For example, beavers build dams across streams to create ponds where none previously existed. Such alteration of the environment has both positive and negative impacts: it kills some trees and displaces both aquatic and terrestrial organisms, but at the same time, it creates a new wetland environment for other organisms. Termites in tropical areas also create environments that are favorable to themselves by building tall mounds. Both a beaver dam and a termite mound, shown in ***Figure 27-1,*** alter the environment to provide for the basic needs of the organisms that build them. Many other organisms alter their environments to improve their chances of survival. For example, corals build huge, underwater reefs that provide homes for all kinds of marine organisms. Of all organisms, however, humans have an unequaled capacity to modify their environments. This capacity allows us to live in every terrestrial environment on Earth. As a result, humans also have the greatest impact on Earth's natural resources.

Resources and Population Growth

Any type of organism can have an impact on its environment if its population becomes large enough. For example, your consumption of oxygen and release of carbon dioxide usually has little effect on your

Figure 27-2 A large elephant population can damage natural resources, such as forests and riverbeds, which other populations depend upon for survival.

immediate environment. However, if you were in a closed room crowded with people, there would soon be less oxygen and more carbon dioxide than the amount people are used to. If no additional fresh air entered the room, everyone eventually would become uncomfortable and would leave the room. As any population increases, its demand for natural resources increases as well. ***Figure 27-2*** illustrates what happens when an elephant population exceeds the ability of the environment to provide the necessary natural resources.

Population growth is defined as an increase in the size of a population over time. A graph of a growing population resembles a J-shaped curve at first. Whether the population is one of dandelions in a lawn, squirrels in a city park, or herring gulls on an isolated island, the initial increase in population is small because the number of adults capable of reproducing is low. As the number of reproducing adults increases, however, the rate of population growth increases rapidly. The population then experiences **exponential growth,** a pattern of growth in which a population grows faster as it increases in size. As shown in ***Figure 27-3,*** exponential growth results in a population explosion.

Figure 27-3 This graph shows a rapid increase over time in the population of bacteria after just a few individuals were added to a petri dish full of necessary nutrients.

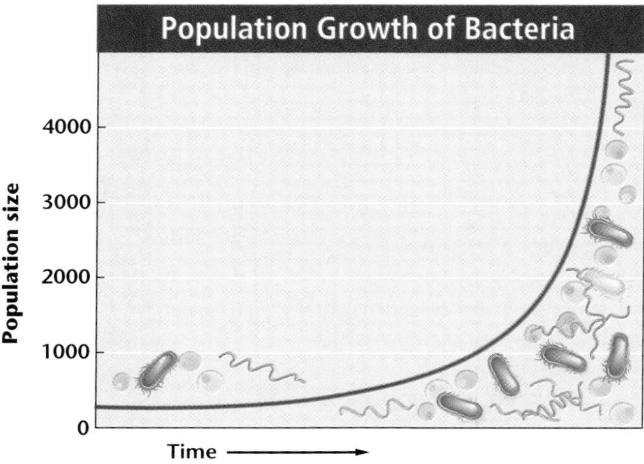

Limits to Population Growth

If the graph in ***Figure 27-3*** was extended for a longer period of time, what do you think would happen to the size of the population? Would it continue to grow exponentially? Many of Earth's natural resources are in limited supply, and therefore, most populations cannot continue to grow forever. Eventually, one or more *limiting factors,* such as the availability of food, water, or clean air, will cause a population to stop increasing. This leveling-off of population size results in an S-shaped curve, similar to the one in ***Figure 27-4.***

The number of organisms that any given environment can support is its **carrying capacity.** When population size has not yet reached the carrying capacity of a particular environment, there will continue to be more births than deaths. If the population size exceeds the carrying capacity temporarily, the number of deaths will increase, or the number of births will decrease, until the population size returns to the carrying capacity. A population that is at the carrying capacity for its environment is in equilibrium. It continues to fluctuate around the carrying capacity as long as natural resources remain available.

Environmental Limits Have you ever seen television or newspaper coverage of the aftermath of a tornado or other violent storm? Storms are environmental factors that limit population growth. Environmental factors that affect population growth, such as storms, extreme changes in temperatures, droughts, floods, and pollution, are **density-independent factors.** These factors affect all populations that they come in contact with, regardless of population size. A flood affects not only the humans whose homes are destroyed, but also trees, birds, and many other populations of organisms.

Other environmental factors that affect population growth, such as disease, parasites, and lack of food, are called **density-dependent factors.** Density-dependent factors increasingly affect a population as the population's size increases. For example, in a large population members may live close together. This enables disease organisms to spread quickly from one member of the population to another. If a population is very dense, disease may wipe out the entire population.

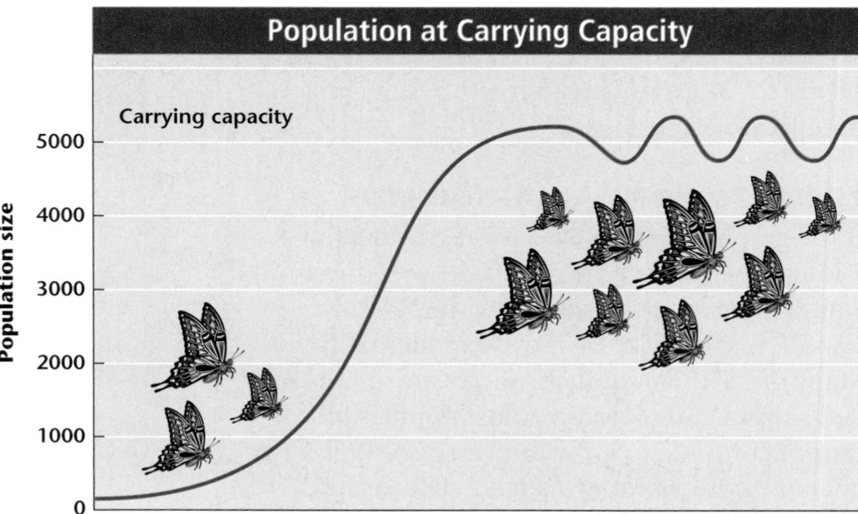

Figure 27-4 Under normal conditions, populations usually reach an equilibrium with the resources available in the environment. From that point on, the population size will fluctuate around the carrying capacity.

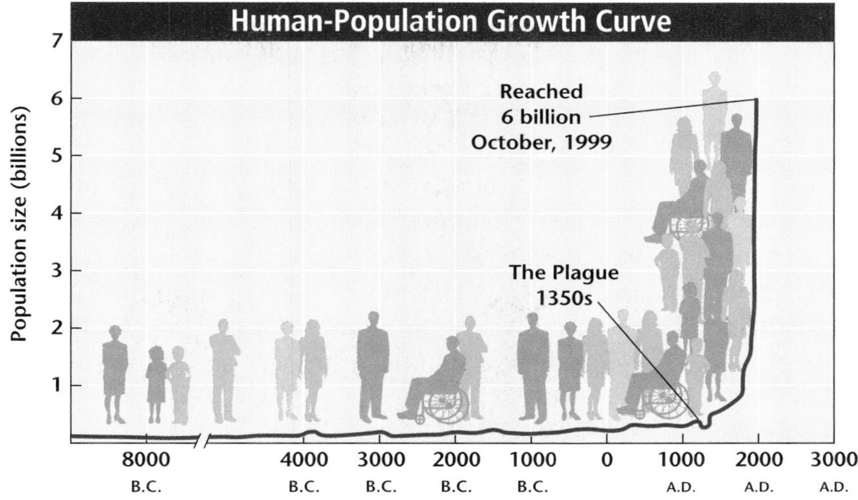

Figure 27-5 A graph of human-population growth still shows a J-shaped curve. ***What does this graph tell you about the growth of the human population on Earth?***

Human Population Growth

No matter where you live, you probably have seen an increase during your lifetime in the number of cars, houses, and roads. The human population on Earth is still growing. As shown in ***Figure 27-5,*** the growth curve of the human population is still in the J-shaped stage. The human population is expected to continue to grow for at least another 50 years.

Although the human population has not yet reached the carrying capacity of Earth, the current rate of growth clearly cannot continue forever. As the population continues to increase, human demand for natural resources also will continue to increase steadily. Although humans are not the most abundant species on Earth, our use of natural resources has impacted the environment on a global scale.

Section Assessment

1. Why do populations stop increasing when they reach carrying capacity?
2. Explain why a hurricane is a density-independent factor that limits population growth.
3. Suppose that a small population of bacteria is placed in a petri dish with limited nutrients. Predict how the population will change over time.
4. **Thinking Critically** What might happen if a human population temporarily exceeded the carrying capacity of an area?

Skill Review

5. **Making Graphs** Make a graph that extends the human-population growth curve shown in ***Figure 27-5.*** Base your graph on the assumption that the carrying capacity of Earth for humans will be reached in the year 2150. For more help, refer to the *Skill Handbook.*

Human Impact on Land Resources

OBJECTIVES

- **Describe** *the environmental impact of mineral extraction.*
- **Discuss** *the environmental problems created by agriculture and forestry, and their solutions.*
- **Explain** *the environmental impact of urban development.*

VOCABULARY

reclamation
biodiversity
monoculture
deforestation
bioremediation

How much land do you think is necessary to grow the food and provide the other materials that you consume and use? Each year, a typical person in North America consumes resources equal to the renewable yield from approximately 5 ha (about 12.35 acres) of forest and farmland. Through our use of mineral resources, food, lumber, and living space, humans have a significant impact on Earth's surface.

EXTRACTION OF MINERAL RESOURCES

Do you spend much time talking on the telephone? Perhaps you use a microwave oven to heat up after-school snacks. Many of the materials in telephones and microwave ovens are derived from land. Modern societies require huge amounts of land resources, including iron, aluminum, copper, sand, gravel, and limestone. Unfortunately, the extraction of these resources often disturbs large areas of Earth's surface. Finding a balance between the need for mineral resources and controlling the environmental change caused by extraction can be difficult.

Surface Mining Mineral and ore deposits found just beneath Earth's surface, such as iron, bauxite (aluminum ore), copper, coal, and gold, can be extracted through mining techniques that involve removing huge amounts of overlying soil and rock, as shown in ***Figure 27-6.*** Unfortunately, extracting land resources in this way completely changes the landscape.

Figure 27-6 Surface-mining techniques include strip-mining and open-pit mining. In a coal strip-mine, surface material up to 60-m deep may be removed to expose the coal seam below **(A).** In an open-pit mine, the mineral resources are extracted from the surface downward, and a gaping hole is gouged into Earth's surface **(B).**

B

Underground Mining Underground mining, also called subsurface mining, is used where mineral resources lie deep under the ground. Underground mining is less disruptive to Earth's surface than surface mining, but it still has impacts on the environment. For example, although the underground mine in ***Figure 27-7*** cannot be seen, the mountains of waste rock dug from under the ground are clearly visible. Rainwater seeping through these piles of mining waste can dissolve toxic metals as well as other chemicals and move them into nearby streams and rivers, where they will cause water pollution. Although many mining companies build large holding ponds to contain polluted water until it can be treated, these ponds sometimes leak.

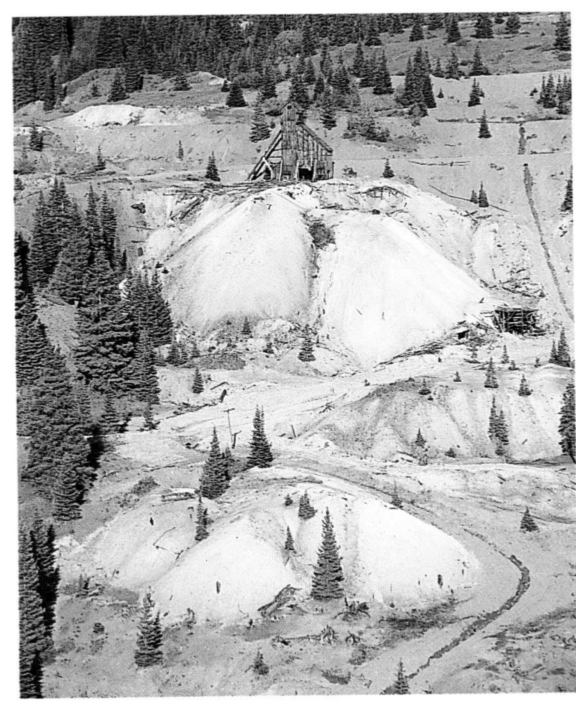

Figure 27-7 Much of the material brought to the surface of underground mines is waste material that must be removed before miners can extract the mineral resource.

Responsible mining companies make efforts to protect the land during mining operations. In the United States, the Surface Mining Control and Reclamation Act of 1977 requires mining companies to restore the land to its original contours and to replant vegetation in the process of **reclamation.** *Figure 27-8* shows a strip-mined area both before and after reclamation. Although reclamation repairs much of the damage that surface mining causes, it can be extremely difficult to restore land to its original contours, as you will discover in the *MiniLab* on the following page.

Figure 27-8 Layers of soil and rock have been stripped away to expose the coal in this strip mine **(A).** During reclamation, soil and rock are replaced and the area is replanted **(B).**

MiniLab

Reclamation

Model the procedure used by mining companies to reclaim an area after strip-mining.

Procedure

1. Using a plastic knife, make four or five cuts across the icing of one cream-filled, iced cupcake. Remove the pieces of icing.
2. Make four or five cuts down into the cake until you reach the cream filling. Cut horizontally just above the cream filling and remove the cake pieces.
3. Remove the cream filling. To restore the area, use the pieces of cake that you cut out to fill in the hole left when you removed the cream filling. Make the surface of the cupcake as level as possible.
4. Replace the icing so that it covers the surface of the restored cupcake.

Analyze and Conclude

1. On the restored cupcake, what does the icing represent? What does the cream filling represent?
2. Does the reclaimed cupcake resemble the original, untouched cupcake?
3. Can reclamation of an area that has been strip-mined restore the land to its original contours? Explain your answer.

AGRICULTURE

In natural ecosystems, such as a forest, many species of organisms interact with one another and with their environment to create a stable ecosystem. For example, scientists have identified as many as 300 species of trees on just 1 ha of land in a tropical rain forest. Even city parks can have a wide variety of different species, called **biodiversity.** In a recent study of a park in Hartford, Connecticut, scientists found and identified a total of 1369 species of organisms in just one 24-hour period. Ecosystems that have high biodiversity are more stable than those with fewer species because they are able to recover more quickly from harmful events such as disease and drought.

Monoculture When land is cleared for food production, a biologically diverse ecosystem is often replaced with a single plant species, such as corn or wheat. The planting of just one species in a field is called **monoculture.** Growing a monoculture crop makes it easier for a farmer to sow, fertilize, and harvest a crop, but this efficiency also brings risks. For example, in a monoculture of corn, illustrated in ***Figure 27-9,*** a fungus or a parasite that destroys corn can spread rapidly and destroy the entire crop. In contrast, in a field that contains several species of crops, disease organisms cannot spread as quickly because they have a more difficult time finding the target species. Even if the entire corn crop in such a field is eventually lost, the farmer can still harvest the other crops growing in the field.

Figure 27-9 Many farmers in the United States grow just one plant species in their fields.

Pesticides A variety of pesticides, including fungicides and insecticides, have played an important role in boosting food production worldwide by eliminating organisms that destroy crops. However, the use of pesticides has drawbacks. Some pesticides remain in the environment for long periods of time. As they slowly accumulate in the food chain, they may harm beneficial organisms, such as fishes and birds. Some pesticides also kill beneficial insect predators along with destructive insects. When pesticides kill decomposers, such as worms, the overall fertility of topsoil deteriorates. In addition, insect populations can quickly develop resistance to an insecticide, causing some farmers to use ever-increasing amounts in an attempt to control pests. Further problems are created when wind and rain carry pesticides away from a farm and cause pollution in nearby waterways.

Topsoil It can take thousands of years for topsoil to form, and thus, once it is lost, it is hard to replace. Erosion of topsoil occurs when forests or grasslands are cleared for the first time, but even established farms can suffer from the loss of topsoil. As shown in ***Figure 27-10,*** whenever fields are plowed and the plants whose roots hold the soil in place are removed, topsoil becomes vulnerable to erosion by wind and water. The addition of fertilizers helps replace some of the nutrients that are depleted by topsoil erosion, but there are other substances in topsoil that fertilizers cannot provide. Topsoil contains trace minerals as well as organisms such as earthworms, which aerate the soil and provide space for plant roots to grow, and nitrogen-fixing bacteria, which take nitrogen out of the air and make it available to plants. Topsoil also has an abundance of organic matter, including fecal material from organisms that live in the soil and dead and dying organisms such as grasses and insects. As organic matter decomposes, it releases nutrients back into the soil.

Figure 27-10 Plowing a field in preparation for planting can result in the loss of topsoil.

Figure 27-11 Good farming practices can help conserve topsoil. In no-till farming, the crop residue is left on the field after harvest to hold soil in place. In the spring, seeds are planted right through the crop residue **(A).** To stabilize the soil and replenish nutrients in a field used for monoculture, farmers plant cover crops such as clover every few years **(B).**

To maintain the fertility of their land, many farmers use methods that help preserve topsoil, such as those shown in ***Figure 27-11.*** When fertilizers are necessary, responsible farmers carefully monitor their use to prevent runoff into streams. Because fertilizers are expensive, farmers also save money by using only as much as the plants require. Methods used by farmers to selectively apply fertilizers where they will provide the greatest benefit include soil analysis, careful mapping of fields, and monitoring of plant growth.

Figure 27-12 In selective logging, foresters mark trees that will be cut down as well as those that will be left standing.

FORESTRY

The clearing of forested land is another way in which topsoil is lost. Worldwide, thousands of hectares of forests are cut down annually to meet the demand for firewood, charcoal, paper, and lumber. In many parts of the world, the clearing of forested land results in **deforestation,** which is the removal of trees from a forested area without adequate replanting. Deforestation often involves clear-cutting, the complete removal of all the trees in an area. Clear-cutting may result in the loss of topsoil through erosion and in the clogging of nearby streams with excess sediment. Fortunately, the negative environmental impacts of deforestation can be minimized through the practices of selective logging, as shown in ***Figure 27-12,*** and the retention of buffer zones of trees along streambeds. In selective logging, workers remove only designated trees rather than clear-cutting an entire forest. This practice reduces the amount of ground left bare and thus helps prevent erosion. In the United States, new logging laws require that buffer zones of trees be left along the banks of streams. Buffer zones of trees slow runoff by catching the sediment that has been eroded from bare ground before it reaches streams.

URBAN DEVELOPMENT

Do you live in the country, or in a town or city? As the human population continues to increase, more and more people live in cities and towns. For example, 70 percent of the population in North America now lives in urban and suburban areas, and an estimated 5 billion people worldwide will be living in cities and towns by the year 2025.

The development of land for the growth of urban areas has many impacts on the environment. When towns and cities expand into rural areas, natural habitats are lost as forests are cleared and wetlands are filled to provide land for roads, houses, and other buildings. When land is prepared for construction, erosion of topsoil often increases until new landscaping can be established.

Development also takes land away from agricultural use, which puts pressure on the remaining farmland for increased production. Other problems are created when concrete and asphalt cover large land areas; because there are fewer opportunities for rainwater to soak into the ground, groundwater supplies are not recharged and flooding increases during heavy rains. Increasing urbanization also produces large volumes of solid waste, as illustrated in ***Figure 27-13.*** Each person in the United States generates an average of 1.5 kg of solid waste per day. Where does it all go? Much of it is buried in landfills. People once thought that because buried waste was out of sight, it was no longer a problem. Many old landfills, however, are creating pollution problems as dangerous chemicals leak out and contaminate water supplies.

Additional contamination occurs as a result of industrial processes. Heavy metals, such as lead and mercury, and poisonous chemicals, such as arsenic, are by-products of many industrial processes and can

Finding Percentages
If 61 million metric tons of the solid waste generated in the United States each year are reclaimed, and if this represents 28 percent of the total solid waste generated, how much total solid waste is generated in the United States each year?

Solid Waste Generated Each Year

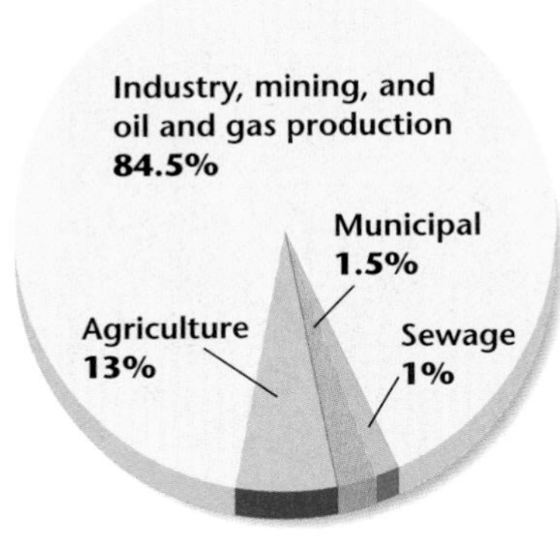

Figure 27-13 This graph shows the percentage of solid waste generated by various users of natural resources in the United States annually. Municipal wastes are those wastes contributed by homes and businesses.

pollute the soil and groundwater in urban areas. Some of this type of contamination was caused by industries that operated before the dangers of improper waste disposal were known. However, accidental spills and illegal dumping continue to be sources of contamination. Even though it is possible to clean up contaminated sites, the processes involved in doing so are difficult and extremely expensive.

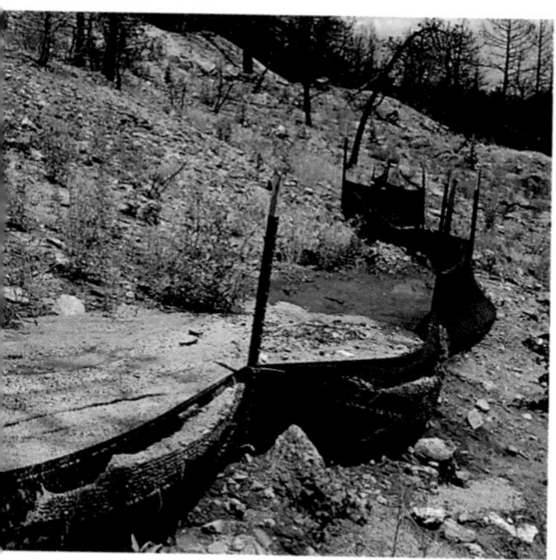

Figure 27-14 During construction, barriers such as these slow water flow and prevent erosion from the site.

Solutions Although urban development can create many environmental problems, most of these problems can be solved. People are becoming aware of the need to protect the environment, and communities are making increased efforts to do so. For example, developers are often required to place barriers, such as those shown in ***Figure 27-14,*** around construction sites to catch sediment from increased erosion. In the United States, wetlands are now recognized as valuable ecosystems and are protected from development. In some cases, if developers destroy a wetland area, they are required to build wetlands somewhere else in return.

Problems associated with waste disposal are more difficult, primarily because the volume of trash is so large. Modern landfills, however, are very different from the dumps of the past; they are carefully designed to minimize leakage of toxic liquids. Impermeable clay or plastic layers are placed beneath a landfill, and each day's trash is compacted by huge machines and buried under a layer of dirt to reduce volume and eliminate wind-blown trash. Ventilation pipes in landfills release methane and other gases that are generated as the garbage decomposes. A diagram of a modern landfill is shown in ***Figure 27-15.***

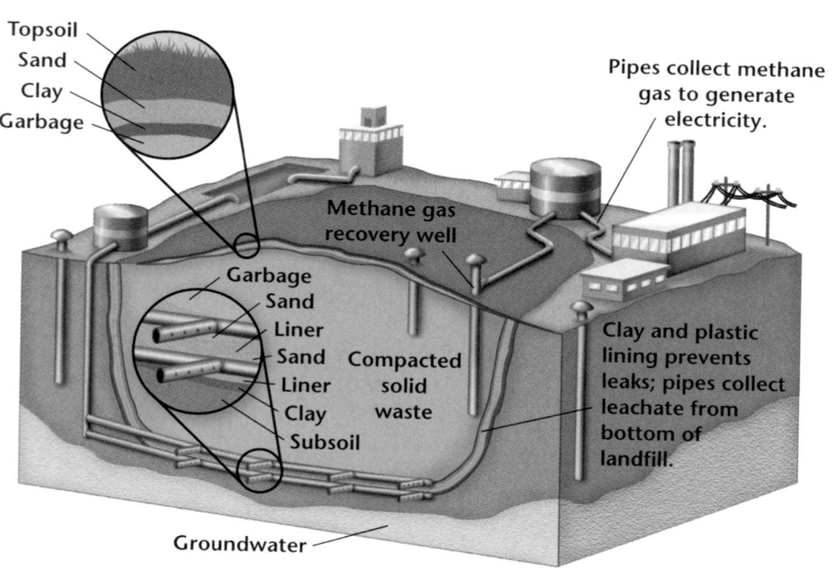

Figure 27-15 Modern landfills are designed to have the least possible impact on the environment.

Figure 27-16 In one type of bioremediation, plants that contain bacterial genes are grown in soils contaminated with toxic compounds. The bacterial genes enable the plants to convert these compounds into nontoxic substances, which the plants then release.

Several methods are available for cleaning up industrial toxic-waste sites. In one method, all the contaminated soil is removed and incinerated at temperatures high enough to destroy the toxic chemicals. The drawbacks to this method are that it can be very expensive when large volumes of soil are involved and it also produces toxic ash. Another method that shows great promise is **bioremediation,** the use of organisms to clean up toxic wastes, illustrated in ***Figure 27-16.*** In some cases, naturally occurring bacteria can be found that eat toxic materials and convert them to less-harmful substances. This technique has been especially useful for contamination caused by spilled gasoline and oil.

SECTION ASSESSMENT

1. What are some of the ways in which mining activities affect the land surface?
2. Compare the positive and negative aspects of pesticide use.
3. What are some of the ways in which urbanization affects the local land environment?
4. **Thinking Critically** If fertilizers cause environmental damage and are expensive, why do farmers use them?

SKILL REVIEW

5. **Concept Mapping** Use the following terms to fill in the concept map below to organize the major ideas in this section: erosion, topsoil loss, water pollution, waste rock, and mineral extraction. For more help, refer to the *Skill Handbook.*

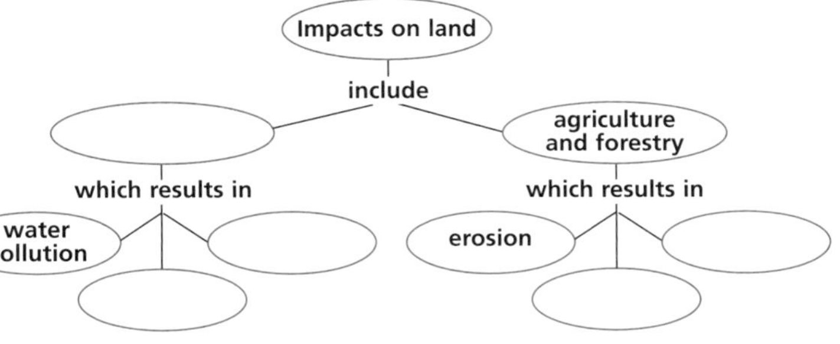

SECTION 27.3

Human Impact on Air Resources

Earth Sciences 4.c, 4.d, 7.b, 8.b, 8.c

OBJECTIVES

- **Describe** *the types and sources of air pollution.*
- **Differentiate** *between the greenhouse effect and global warming.*
- **Sequence** *the reactions that occur as CFCs cause ozone depletion.*
- **Identify** *the causes and effects of acid precipitation.*

VOCABULARY

smog
ozone
global warming
acid precipitation

Can you see the haze that seems to hover above the buildings in ***Figure 27-17?*** This yellow-brown haze is a type of air pollution called **smog,** which is a photochemical haze caused by the action of solar radiation on an atmosphere polluted with hydrocarbons and nitrogen oxides mostly from automobile exhaust systems. When smog occurs in a city, the air becomes harmful to breathe, especially for those who already have some difficulty breathing. The major chemical in smog is **ozone** (O_3), a gas molecule made up of three oxygen atoms. Recall that in the upper atmosphere, solar radiation converts oxygen gas into ozone. Ozone in the upper atmosphere is beneficial because it absorbs and filters out harmful ultraviolet (UV) radiation. However, ground-level ozone is produced when combinations of air pollutants, including nitrogen oxides, carbon monoxide, and hydrocarbons, are exposed to sunlight. Ozone irritates the eyes, noses, throats, and lungs of humans, and it also has harmful effects on plants.

Air pollution also occurs in the form of particulate matter. The solid particles of such materials as ash, dust, pollen, and asbestos fibers range in size from microscopic bits to large grains. When humans breathe in particulates, they can lodge in lung tissues, disrupt normal functions, and cause breathing difficulties and lung disease.

Earth Sciences

8.c Students know the location of the ozone layer in the upper atmosphere, its role in absorbing ultraviolet radiation, and the way in which this layer varies both naturally and in response to human activities.

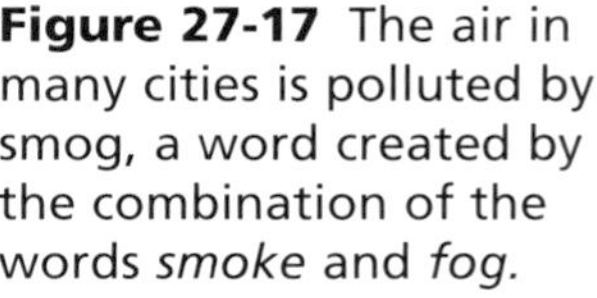

Figure 27-17 The air in many cities is polluted by smog, a word created by the combination of the words *smoke* and *fog.*

GLOBAL IMPACTS OF AIR POLLUTION

Recently, it has become clear that human activities can affect Earth on a global scale. The global atmospheric effects of air pollution include global warming, ozone depletion, and acid precipitation.

Global Warming Recall from Chapter 14 that the greenhouse effect is a natural phenomenon in which Earth's atmosphere traps heat in the troposphere to warm Earth. A phenomenon related to the greenhouse effect is **global warming,** which is the increase in Earth's average surface temperature. Whereas the greenhouse effect is a natural phenomenon, global warming is partly caused by humans. Human activities, especially the burning of fossil fuels by automobiles, are largely responsible for increased levels of carbon dioxide, which is the main greenhouse gas that causes global warming. Fossil fuels contain carbon, and when they are burned, the carbon combines with oxygen to form carbon dioxide. Since the beginning of the industrial revolution, around 1850, humans have been burning fossil fuels at an ever-increasing rate. ***Figure 27-18*** shows how atmospheric carbon dioxide has increased over the past 250 years.

Studies indicate that Earth's mean surface temperature has risen about 0.5°C in the last century. Some scientists hypothesize that this warming trend is the result of global warming and predict that global warming could raise average temperatures by 1 to 3.5°C in the next 100 years. Although this may not seem like much of a temperature change, the consequences could be extreme. Wind and rainfall patterns might

Update To find out more about global warming, visit the Earth Science Web Site at earthgeu.com

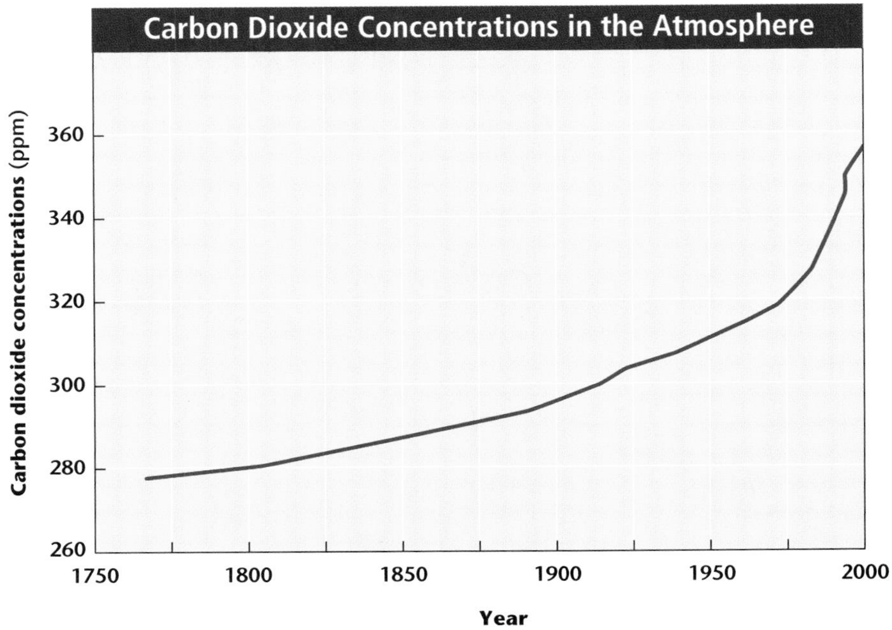

Figure 27-18 The amount of atmospheric carbon dioxide has increased greatly since the 1850s.

Earth Sciences

4.c Students know the different atmospheric gases that absorb the Earth's thermal radiation and the mechanism and significance of the greenhouse effect.

4.d Students know the differing greenhouse condition on Earth, Mars, and Venus; the origins of those conditions; and the climatic consequences of each.

change and affect the major agricultural belts. If climate patterns change too rapidly, plant and animal species may be unable to adapt and may become extinct. Glaciers and ice caps could melt, raising the sea level and flooding low-lying areas. Other scientists, however, assert that humans have not kept weather records long enough to tell whether the present rate of global warming is an artificial or a natural phenomenon. They argue that the increase in Earth's temperature could be part of a natural pattern of climatic change.

Earth Sciences

7.b Students know the global carbon cycle: the different physical and chemical forms of carbon in the atmosphere, oceans, biomass, fossil fuels, and the movement of carbon among these reservoirs.

8.b Students know how the composition of Earth's atmosphere has evolved over geologic time and know the effect of outgassing, the variations of carbon dioxide concentration, and the origin of atmospheric oxygen.

Ozone Depletion Another global change that is a result of human activity involves the ozone layer in the stratosphere. The ozone layer serves as a protective shield as it absorbs and filters out harmful UV radiation, which has been linked to human eye damage and skin cancer, as well as reduced crop yields. In the early 1970s, scientists first suggested that chlorofluorocarbons (CFCs) could destroy ozone in the upper atmosphere. Although CFCs are stable and harmless near Earth's surface, scientists now know that they destroy ozone molecules when they migrate into the upper atmosphere, as shown in ***Figure 27-19.*** Since the mid-1980s, atmospheric studies have detected a thinning of the ozone layer, including an extremely thin area over Antarctica that was publicized in the news media as an "ozone hole." Because all of the CFCs in the atmosphere were released from old refrigerators, cleaning agents, and propellants in aerosol cans, this ozone depletion is entirely a result of human activity.

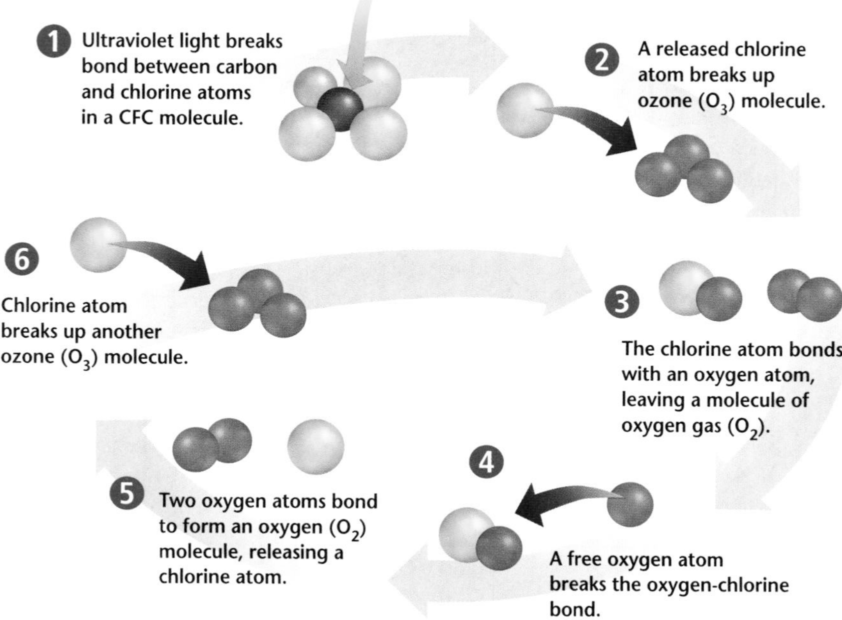

Figure 27-19 Just a few chlorine atoms from CFCs can destroy many ozone molecules.

Global Impacts of Air Pollution

Recently, it has become clear that human activities can affect Earth on a global scale. The global atmospheric effects of air pollution include global warming, ozone depletion, and acid precipitation.

Global Warming Recall from Chapter 14 that the greenhouse effect is a natural phenomenon in which Earth's atmosphere traps heat in the troposphere to warm Earth. A phenomenon related to the greenhouse effect is **global warming,** which is the increase in Earth's average surface temperature. Whereas the greenhouse effect is a natural phenomenon, global warming is partly caused by humans. Human activities, especially the burning of fossil fuels by automobiles, are largely responsible for increased levels of carbon dioxide, which is the main greenhouse gas that causes global warming. Fossil fuels contain carbon, and when they are burned, the carbon combines with oxygen to form carbon dioxide. Since the beginning of the industrial revolution, around 1850, humans have been burning fossil fuels at an ever-increasing rate. ***Figure 27-18*** shows how atmospheric carbon dioxide has increased over the past 250 years.

Studies indicate that Earth's mean surface temperature has risen about 0.5°C in the last century. Some scientists hypothesize that this warming trend is the result of global warming and predict that global warming could raise average temperatures by 1 to 3.5°C in the next 100 years. Although this may not seem like much of a temperature change, the consequences could be extreme. Wind and rainfall patterns might

Update To find out more about global warming, visit the Earth Science Web Site at earthgeu.com

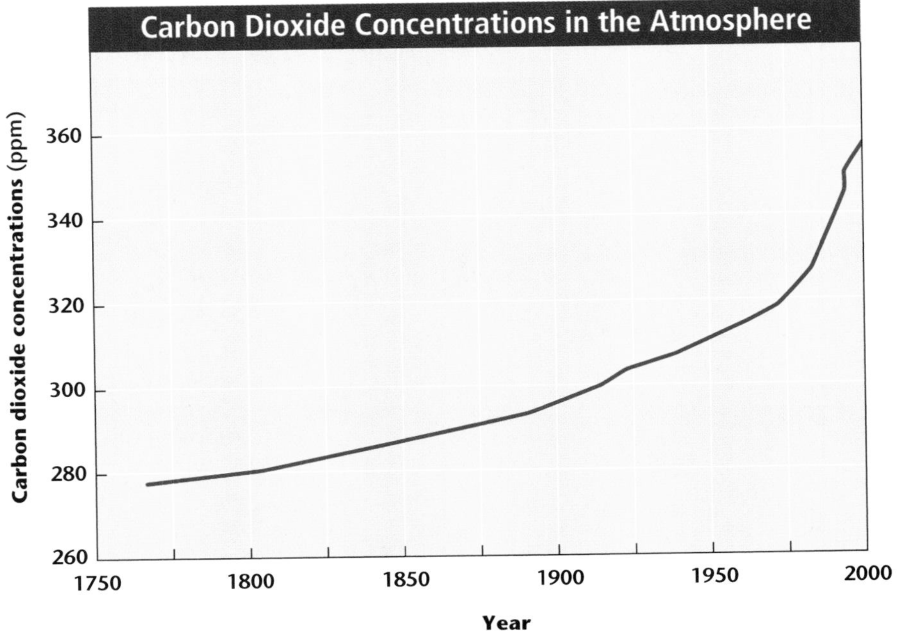

Figure 27-18 The amount of atmospheric carbon dioxide has increased greatly since the 1850s.

Earth Sciences

4.c Students know the different atmospheric gases that absorb the Earth's thermal radiation and the mechanism and significance of the greenhouse effect.

4.d Students know the differing greenhouse condition on Earth, Mars, and Venus; the origins of those conditions; and the climatic consequences of each.

change and affect the major agricultural belts. If climate patterns change too rapidly, plant and animal species may be unable to adapt and may become extinct. Glaciers and ice caps could melt, raising the sea level and flooding low-lying areas. Other scientists, however, assert that humans have not kept weather records long enough to tell whether the present rate of global warming is an artificial or a natural phenomenon. They argue that the increase in Earth's temperature could be part of a natural pattern of climatic change.

Earth Sciences

7.b Students know the global carbon cycle: the different physical and chemical forms of carbon in the atmosphere, oceans, biomass, fossil fuels, and the movement of carbon among these reservoirs.

8.b Students know how the composition of Earth's atmosphere has evolved over geologic time and know the effect of outgassing, the variations of carbon dioxide concentration, and the origin of atmospheric oxygen.

Ozone Depletion Another global change that is a result of human activity involves the ozone layer in the stratosphere. The ozone layer serves as a protective shield as it absorbs and filters out harmful UV radiation, which has been linked to human eye damage and skin cancer, as well as reduced crop yields. In the early 1970s, scientists first suggested that chlorofluorocarbons (CFCs) could destroy ozone in the upper atmosphere. Although CFCs are stable and harmless near Earth's surface, scientists now know that they destroy ozone molecules when they migrate into the upper atmosphere, as shown in ***Figure 27-19.*** Since the mid-1980s, atmospheric studies have detected a thinning of the ozone layer, including an extremely thin area over Antarctica that was publicized in the news media as an "ozone hole." Because all of the CFCs in the atmosphere were released from old refrigerators, cleaning agents, and propellants in aerosol cans, this ozone depletion is entirely a result of human activity.

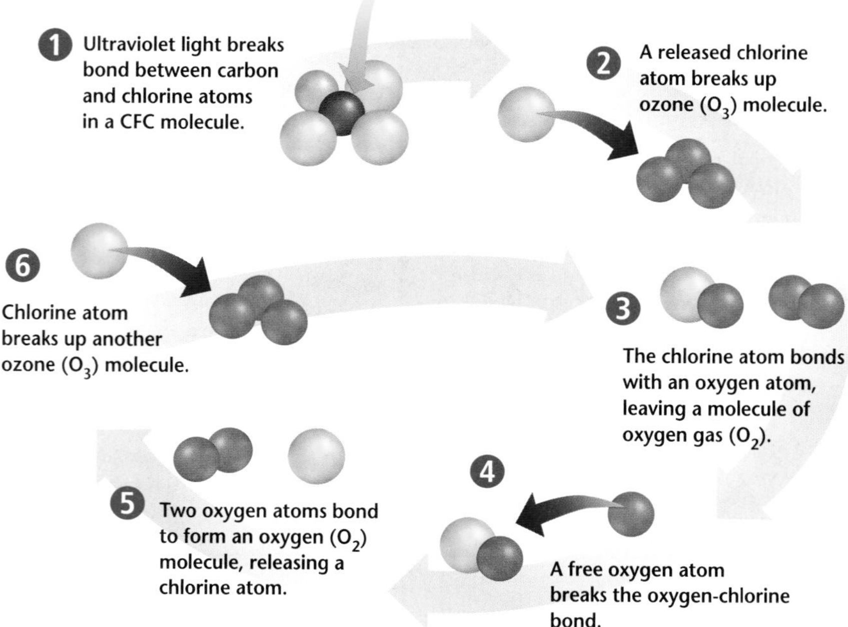

Figure 27-19 Just a few chlorine atoms from CFCs can destroy many ozone molecules.

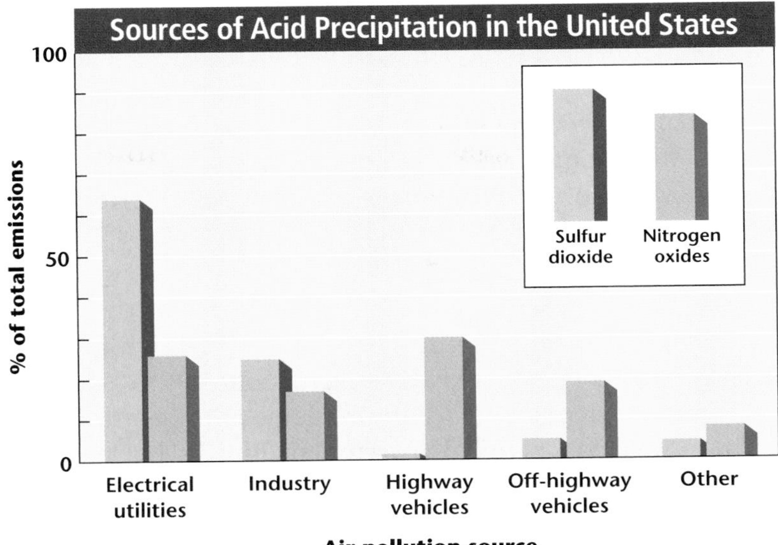

Figure 27-20 Most of the nitrogen oxides that cause acid precipitation come from automobiles. ***What is the greatest source of the sulfur dioxide that causes acid precipitation?***

Acid Precipitation Another major air pollution problem is **acid precipitation,** which is defined as precipitation with a pH of less than 5.0. Recall from Chapter 3 that pH is a measure of the acidity of a substance on a scale of 0 to 14, with 7 being neutral. Natural precipitation has a pH of about 5.0 to 5.6, which is slightly acidic. Acid precipitation forms when sulfur dioxide and nitrogen oxides combine with atmospheric moisture to create sulfuric acid and nitric acid. Acid precipitation includes acidic rain, snow, fog, mist, gas, and dust. Although volcanoes and marshes add sulfur gases to the atmosphere, 90 percent of the sulfur emissions in eastern North America are of human origin. ***Figure 27-20*** shows a comparison of the different sources of acid precipitation.

The type of acid precipitation that has received the most attention is caused by coal-burning power plants in the midwestern United States. These plants burn coal that contains significant amounts of the mineral pyrite (FeS_2) and other sulfur-bearing compounds. When sulfur-rich coal is burned, large amounts of sulfur dioxide are released. The sulfur dioxide generated by midwestern power plants rises high into the air and is carried by winds towards the eastern coast of the United States and Canada. When acids are carried into wet weather, they become part of the rain, snow, or fog that falls to the ground in areas far from their source.

When acid precipitation makes its way into surface waters, such as lakes, streams, ponds, and rivers, it causes damage to aquatic ecosystems and vegetation. Acid precipitation also affects plants and soil. Although trees in areas affected by acid precipitation usually aren't killed outright, acid precipitation weakens them so that they become

Figure 27-21 Buildings and outdoor artwork are damaged by acid precipitation.

more susceptible to damage from insect pests and disease. In addition, acid precipitation depletes the soil of some nutrients needed by plants.

Acid precipitation damages stone buildings and statues, as shown in ***Figure 27-21,*** especially those made of limestone, by accelerating the rate of weathering. It also can corrode metal structures such as bridges, thereby shortening their life spans and increasing maintenance costs.

Reducing Air Pollution

Air pollution is difficult to control because it travels with the wind; pollution produced in one area travels across borders to neighboring regions. Thus, solving air pollution problems requires the cooperation of both state and national governments. In the last decade, the governments of many nations have met several times in an attempt to reduce global air pollution, especially that caused by carbon dioxide

Problem-Solving Lab

Using Graphs

Identify changes in air pollutants The Clean Air Act of 1972 was an attempt to reduce the amount of air pollution in the United States. The emission rates of six major air pollutants are graphed to the right. Use this graph to answer the following questions.

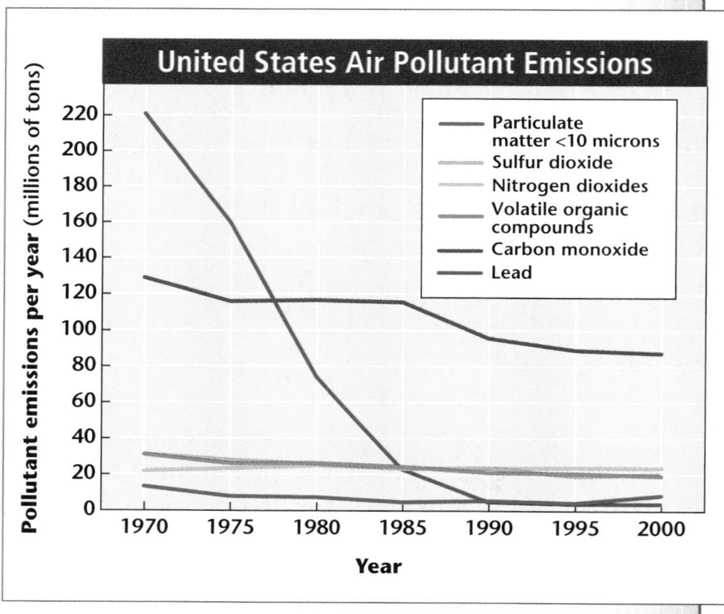

Analysis

1. In 1970, the main source of lead (Pb) air pollution was leaded gasoline. What has happened to the amount of lead emitted into the atmosphere since 1970? What do you think may have caused this change?
2. Modern cars emit significantly fewer air pollutants, such as carbon monoxide and nitrogen dioxides, than cars from the 1970s and earlier. How have the emission rates of these two air pollutants changed since 1970?

Thinking Critically

3. If modern cars are more efficient and less polluting, how would you explain the small amount of change in the levels of nitrogen dioxides over the years?

and CFCs. In the United States, Congress has passed laws to reduce air pollution. For example, the 1990 Clean Air Act set specific reduction goals and enforcement policies for many types of air pollution. This act called for the United States to reduce its sulfur dioxide emissions to 50 percent of their 1980 levels by the year 2000, and to reduce emissions of nitrogen oxides as well. You will find out how the amount of air pollutants in the United States has changed since the 1970s in the *Problem-Solving Lab* on the previous page.

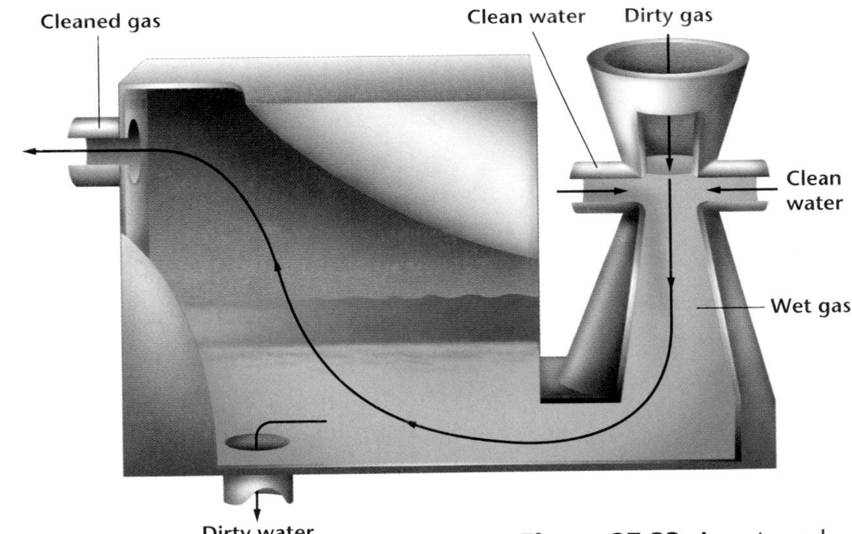

Figure 27-22 A wet scrubber installed on a smokestack of a coal-burning power plant reduces sulfur-dioxide emissions.

Many coal-burning power plants have installed a device such as the wet scrubber illustrated in ***Figure 27-22*** to reduce emissions of particulate matter and sulfur dioxide. In North America and Western Europe, the use of low-sulfur coal and natural gas have helped to reduce such emissions. However, scientists agree that the most effective way to reduce air pollution is to remove older, highly polluting vehicles from roadways. It is estimated that just 10 percent of the motor vehicles in operation produce 50 to 60 percent of the air pollution generated by gasoline-powered engines. Switching to newer cars with more efficient engines could significantly reduce air pollution throughout the world.

Section Assessment

1. Name two forms of pollutants found in air. What are some of the natural and human sources of these pollutants?
2. How is global warming related to the greenhouse effect?
3. How do CFCs cause ozone depletion?
4. What are the effects of acid precipitation on ecosystems?
5. **Thinking Critically** At some point, humans will run out of inexpensive coal resources to burn for fuel. What impact might this have on global warming?

Skill Review

6. **Predicting** The atmosphere of Venus is 90 percent carbon dioxide. Based on this information, what could you infer about the average surface temperature of Venus? Explain your answer. For more help, refer to the *Skill Handbook.*

SECTION 27.4

Human Impact on Water Resources

Earth Sciences 9.c I&E 1.m

OBJECTIVES

- **Summarize** *the types and sources of water pollution.*
- **Describe** *some methods of controlling water pollution.*
- **Identify** *ways to conserve water.*

VOCABULARY

point source
nonpoint source

Humans depend on water in many ways. In 1995, the United States consumed 378 billion L of water per day. Since 1960, freshwater use has nearly doubled, and the demand is expected to continue to increase. Most people use freshwater in their homes for bathing, drinking, cooking, and washing. The irrigation of crops also requires water, but much of it is wasted because it often evaporates or seeps into the ground before it can be used by crops. Still, the greatest demand on water supplies comes from industry, including power plants that use water for cooling purposes. ***Figure 27-23*** shows how water supplies in the United States are distributed among users.

Because water supplies are not distributed evenly on Earth, some areas have less water than is needed. When water supplies are limited, conflicts occur between the needs of people and the needs of other users, including wildlife.

WATER POLLUTION

Pollution is another area in which humans have an impact on water supplies. Some supplies of water have been polluted by human activities and are no longer usable. Water-pollution sources are grouped into two main types. **Point sources** generate pollution from a single point of origin, such as a sewage-treatment plant or an industrial site, while **nonpoint sources** generate pollution from widely spread areas.

Use of Water in the United States

Figure 27-23 This graph illustrates the percentage of total water supplies used for various purposes in the United States.

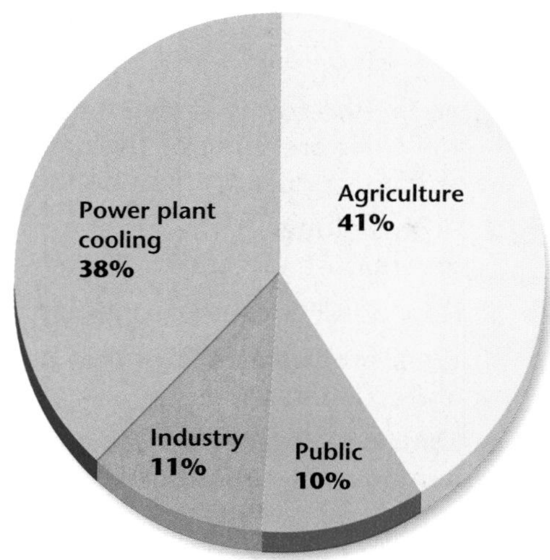

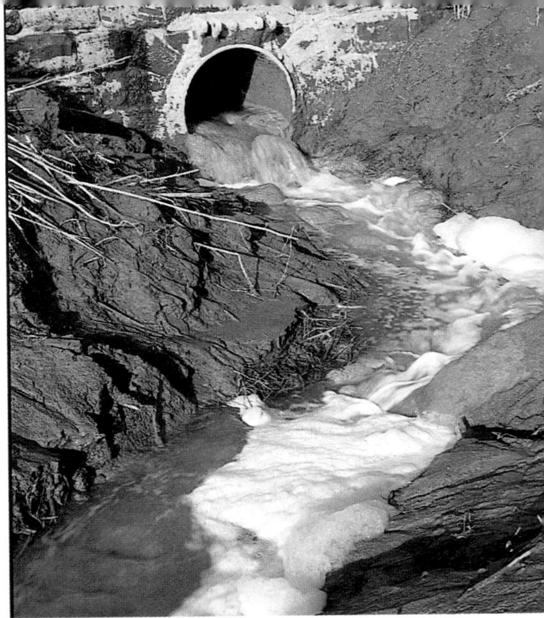

Figure 27-24 Point-source pollution such as this can be identified and cleaned up more easily than nonpoint source pollution.

Point Sources Common point sources of water pollution include bacteria and viruses that enter water systems through improper disposal of sewage, and toxic wastes that enter streams from both illegal dumping and accidental spills. In addition, industries that divert water from streams to use in manufacturing processes may return polluted water to the streams, as shown in ***Figure 27-24.***

Nonpoint Sources Rainwater, a nonpoint source of water pollution, absorbs air pollutants and may become acidic itself. Rainwater also dissolves pesticides and fertilizers and carries them into streams as it drains from farms and lawns. Runoff from roads and parking lots that includes oil, gasoline, and other chemicals is another nonpoint source of water pollution.

Pollution of Groundwater Leaking chemical-storage barrels, underground gasoline-storage tanks, landfills, road salts, nitrates from fertilizers, sewage from septic systems, and other pollutants can seep into the ground and foul underground water supplies. Polluted groundwater may find its way into the drinking-water supplies of people who rely on wells. Once groundwater is contaminated, the pollutants can be very difficult to remove.

Pollution in the Oceans Although human activities have the greatest impact on freshwater supplies, pollution of ocean waters is also a concern. For years, it was thought that the oceans were so big that human activities could not affect them. This may be true for the oceans as a whole, but it is not true for near-shore regions. Nearly 50 percent of the U.S. population lives near coastlines in areas such as San Francisco, shown in ***Figure 27-25.*** Pollutants from such cities often end up in estuaries and other near-shore regions.

Sewage water is a major source of near-shore pollution around coastal areas. Even after treatment, human wastewater contains high levels of nitrogen and phosphorus. These nutrients can create blooms of cyanobacteria that later die and use up the oxygen in the water as they decompose. Some coastal cities dispose of their untreated sewage by pumping it through pipelines that run along the ocean floor and extend far out into the ocean. This practice can create large dead zones on the ocean floor where there are no living organisms. You will identify possible types and sources of pollution along a hypothetical coastline in the *Mapping GeoLab* at the end of this chapter.

Figure 27-25 The area around San Francisco Bay has a high population density, which can lead to pollution of the ocean nearby.

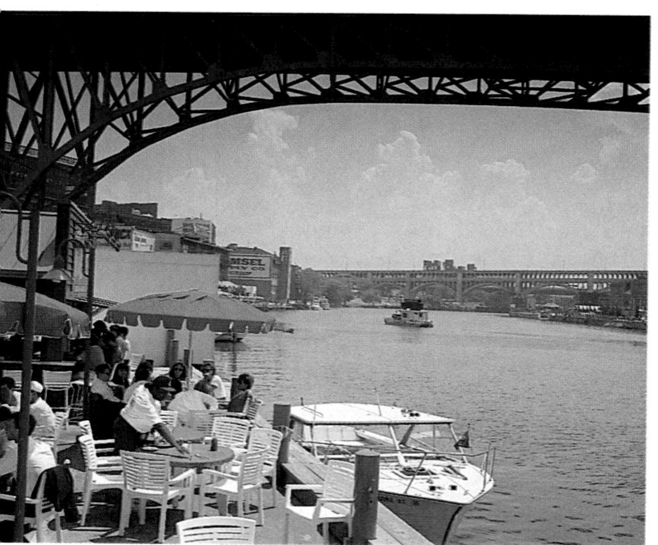

Figure 27-26 In the late 1970s, Ohio's Cuyahoga River was one of the most severely polluted rivers in the world. Today, the river again provides recreational activities for the residents of Akron and Cleveland.

Reducing Water Pollution

In recent decades, many steps have been taken to prevent and reduce water pollution as people have found that it is much cheaper and more efficient to prevent pollution than it is to clean it up later. Two major laws have been passed in the United States to combat water pollution: the Safe Drinking Water Act and the Clean Water Act. The Safe Drinking Water Act of 1974 was designed to ensure that everyone in the United States has access to safe drinking water. Progress is being made, but many water supplies still do not meet the standards consistently. In 1998, 20 percent of public water supplies were in violation of the act at least once in a one-year period. The goal of the Safe Drinking Water Act is to reduce this number to less than 5 percent by the year 2005.

The Clean Water Act of 1972 is the primary federal law that protects our nation's waters. The act was amended in 1977, 1981, and again in 1987. The two main goals of the Clean Water Act are to eliminate discharge of pollutants into rivers, streams, lakes, and wetlands, and to restore water quality to levels that allow for recreational uses of waters, including fishing and swimming. One positive result of this act is shown in ***Figure 27-26.***

Is the Clean Water Act working? Since 1972, the number of people served by sewage-treatment plants has increased from 85 million to 190 million. During that same time period, the annual rate of wetland losses has *decreased* from 146 000 ha to about 32 000 ha. Two-thirds of the nation's waters are now safe for swimming and fishing, compared to only one-third in 1972. However, more improvements must be made; in 1998, 35 percent of U.S. rivers and streams were still in violation of established water-quality levels at some point during that year.

To learn more about controlling water pollution, go to the **National Geographic Expedition** on page 870.

Water Conservation

When there is not enough water to go around, populations have two choices: decrease demand or develop new supplies. In many cases, new supplies are not readily available or may be too expensive to develop. Therefore, water conservation is the most common solution to excessive demand. Because irrigation can waste so much water, efficient irrigation practices can greatly reduce the demand for water. Landscaping with plants that require less water, as illustrated in ***Figure 27-27,*** monitoring soil moisture, improving delivery systems, and raising water prices have all been effective in minimizing the amount of water used for irrigation. Industries can also conserve

Figure 27-27 In arid and semiarid areas, landscaping may include plants adapted to a dry climate. This form of landscaping is called xeriscaping, for the Greek word *xeros,* meaning "dry."

water; many are developing ways to recycle cooling water and wastewater, especially when they are charged high rates for water usage. Manufacturing processes often can use recycled water, or they can be redesigned to conserve water.

Is there a leaky faucet in your home? In the United States alone, 20 to 35 percent of the water taken from public water supplies is lost through leaky pipes, toilets, bathtubs, and faucets. Some cities don't even have water meters to measure, and thus charge for, the public water that households use. Not surprisingly, when water meters were introduced in Boulder, Colorado, the use of water was reduced by more than 30 percent. People tend to fix leaks and conserve water when they have to pay for it. If every person used a little less water, the water conserved would add up to a large volume. Installing more efficient showerheads, as shown in ***Figure 27-28,*** and toilets is just one way to decrease personal water consumption. Consider how you use water. What are some of the ways you might conserve water in your everyday life?

Figure 27-28 Many communities offer low-flow showerheads like these to their residents at low cost to help conserve water.

Section Assessment

1. What are some of the ways in which surface water can be polluted?
2. What are some ways to minimize the need for irrigation?
3. How might residents of a city reduce water consumption?
4. What are some of the positive impacts of the Clean Water Act?
5. **Thinking Critically** Which type of pollution is easier to eliminate, point source or nonpoint source? Give an example of each type and explain how it might be controlled.

Skill Review

6. **Interpreting Graphs** Based on the graph in ***Figure 27-23,*** what percentage of water supplies in the United States is used to provide you with food and the electricity needed to cook it? For more help, refer to the *Skill Handbook.*

Mapping GeoLab

Pinpointing a Source of Pollution

Earth Sciences

9.c Students know the importance of water to society, the origins of California's fresh water, and the relationship between supply and need.

Iris City and the surrounding region are shown in the map on the facing page. Iris City is a medium-sized city of 100 000. It is experiencing many types of environmental impacts. Study the map and the information given to identify these problems and possible solutions.

Preparation

Problem
How can the residents of Iris City identify the source of local water pollution?

Materials
metric ruler
pencil

Procedure

1. Iris City obtains its drinking water from Opal Lake. Studies of the lake have detected increased levels of nitrogen, phosphorus, hydrocarbons, sewage, and silt. The northwest end of Opal Lake is experiencing increased development, while the remainder of the watershed is a mix of forest and logging clear-cuts.
2. Last spring, blooms of cyanobacteria choked parts of the Vista Estuary Nature Preserve. Commercial shellfish beds in Iris Bay have been closed because of sewage contamination.
3. A natural-gas power plant has been proposed for location A, near the Vista Cutoff, an abandoned channel of the Vista River. The plant would provide jobs as well as generate electricity. The company plans to divert 25 percent of the Vista River down the Vista Cutoff.
4. The Lucky Mine was abandoned 60 years ago. A mining company has applied for permits to reopen the mine. An estimated 1 million ounces of gold can be recovered using modern techniques.

Analyze

1. What are some possible sources of water pollution in Opal Lake? What steps might the residents of Iris City take to protect their drinking water?
2. How are the blooms of cyanobacteria and the closing of the shellfish beds in Iris Bay related?
3. What are the positive and negative aspects of diverting water from the Vista River through the Vista Cutoff?
4. If the Lucky Mine is reopened, what effects might it have on the populations of Carlton, Vista, and Iris City?

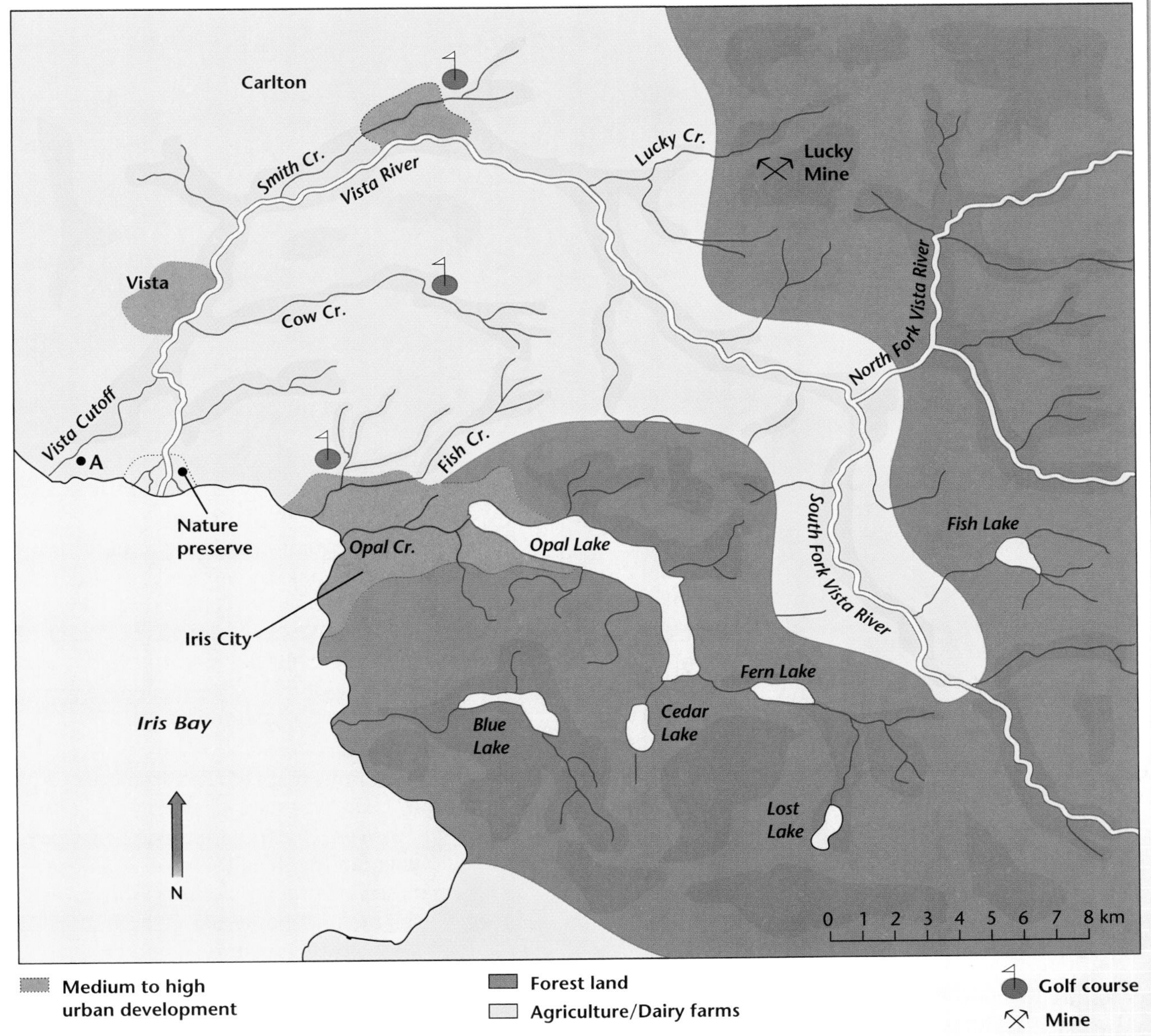

Conclude & Apply

1. Identify the sources of water pollution in Iris Bay. Are these point or nonpoint sources of pollution?
2. How could you identify the source of pollution causing cyanobacteria blooms in the Vista Estuary?
3. If the Vista Cutoff is used to divert water from the Vista River, how will the aquatic habitats of the river be affected?
4. If the Lucky Mine is reopened, what could the mining company do to minimize negative environmental impacts?

Science & the Environment

Methane Hydrates: Hope or Hype?

On April 4, 2000, the United States Congress passed a bill providing nearly $50 million for the research of a new fuel. This new fuel has more than twice the energy potential of all other fossil fuels combined. In addition, the new fuel is plentiful throughout the world. This new fuel could allow many countries to reach energy independence in the new century.

This new fuel is called methane hydrate. Methane hydrates are molecules of frozen methane gas captured inside crystals of regular ice. Many methane hydrate deposits are found deep within the permafrost in arctic regions. Methane gas released when dead plants and animals decay becomes frozen within the ice. Larger deposits of methane hydrates are found under the oceans, particularly along the continental slopes. Heat generated by tectonic processes is thought to release methane gas. The gas first rises, then freezes deep beneath the oceanic crust.

Something Old

Geologists have known about methane hydrates for a long time. The crystal structure of methane hydrates is strong, but not stable. When a hydrate bed is disturbed, the crystals break down and the methane is released as a gas. The rapid breakdown of the hydrate bed causes areas above the bed to slump. This phenomenon has been a major problem for offshore oil drilling platforms, which need a stable ocean floor to rest upon.

Something New

What's new is the idea of mining the hydrates from permafrost or from beneath the ocean floor. Fluctuations in fuel prices have led to an interest in harvesting the energy from this source. Research is underway in Japan, India, Canada, and the United States to develop ways to tap into the methane hydrate beds and remove useful amounts of the fuel. The instability of the hydrates is the biggest stumbling block. The hydrates must either be kept stable and frozen all the way to the processing area to avoid unwanted releases of methane, or the released gas must somehow be captured on-site.

What are the drawbacks to methane hydrates? First, removing hydrates from the seafloor may cause massive slumping along the continental shelves. Second, methane hydrates are fossil fuels, and burning them would cause the same environmental problems as the burning of other fossil fuels. Third, a disruption of the methane hydrate beds might release a large bubble of methane. Scientists have discovered that just such a "big burp" of methane, released 55 million years ago, led to climatic changes and a major extinction of deep-sea species.

Activity

Research and report on the potential advantages and disadvantages of the use of methane hydrates as fuel. Why are scientists interested in using methane hydrates? How might slumping impact the coastal environment?

CHAPTER 27 Study Guide

Summary

SECTION 27.1

Populations and the Use of Natural Resources

Main Ideas

- All organisms use resources to maintain their existence. The use of these resources has an impact on the environment.
- As populations increase, the demand for resources also increases. Because resources are limited, populations eventually will reach the carrying capacity of the environment and stop growing.
- At early stages, populations grow exponentially. Earth is currently experiencing a human population explosion.

Vocabulary

carrying capacity (p. 714)
density-dependent factor (p. 714)
density-independent factor (p. 714)
exponential growth (p. 713)

SECTION 27.2

Human Impact on Land Resources

Main Ideas

- Modern societies require large amounts of land resources. The extraction of land resources can disrupt Earth's surface.
- Growing populations increase the demand for food. Food production can cause habitat loss, erosion, and water pollution.
- Urban development causes habitat loss, increased erosion, and pollution of nearby areas.
- The impact of using land resources can be minimized through the use of modern techniques.

Vocabulary

biodiversity (p. 718)
bioremediation (p. 723)
deforestation (p. 720)
monoculture (p. 718)
reclamation (p. 717)

SECTION 27.3

Human Impact on Air Resources

Main Ideas

- Many human activities create air pollution. Air pollution can cause human health problems.
- Humans have affected Earth's atmosphere on a global scale. Acid precipitation, ozone depletion, and global warming are all caused by human activities.

Vocabulary

acid precipitation (p. 727)
global warming (p. 725)
ozone (p. 724)
smog (p. 724)

SECTION 27.4

Human Impact on Water Resources

Main Ideas

- Humans require freshwater to live. In some regions, there is not enough freshwater to meet the demand. Conservation of existing supplies is the best way to stretch supplies of freshwater.
- Human activities can pollute freshwater supplies and render them unusable.
- The Safe Drinking Water Act and the Clean Water Act are two sets of laws designed to decrease water pollution in the United States.

Vocabulary

nonpoint source (p. 730)
point source (p. 730)

earthgeu.com/vocabulary_puzzlemaker

CHAPTER 27

Assessment

Understanding Main Ideas

1. What is the rapid growth rate of a population after a period of slow growth called?
a. exponential growth **c.** a point source
b. overconsumption **d.** the carrying capacity

2. Which of the following consumes the largest amount of freshwater in the United States?
a. sewage-waste disposal **c.** drinking water
b. irrigation **d.** industrial uses

3. What is the use of organisms to help clean up pollution called?
a. the greenhouse effect **c.** recycling
b. reclamation **d.** bioremediation

4. What type of pollution comes from multiple places?
a. point source **c.** irrigation
b. nonpoint source **d.** reclamation

5. What is the process in which a mining company restores the land after mineral extraction?
a. bioremediation **c.** pollution
b. reclamation **d.** open-pit mining

6. Which of the following is a gas molecule composed of three oxygen atoms?
a. nitrogen oxide **c.** sulfur dioxide
b. ozone **d.** smog

7. Which of the following is the primary source of carbon-monoxide air pollution?
a. motor vehicles **c.** urban development
b. volcanoes **d.** power plants

8. Which of the following gases is (are) responsible for ozone depletion in the upper atmosphere?
a. chlorofluorocarbons **c.** sulfur dioxide
b. carbon dioxide **d.** carbon monoxide

9. What is the variety of species in a habitat called?
a. biodiversity **c.** bioremediation
b. monoculture **d.** biology

Use the following graph to answer questions 10 through 13.

Solid Wastes Generated Annually in the U.S.

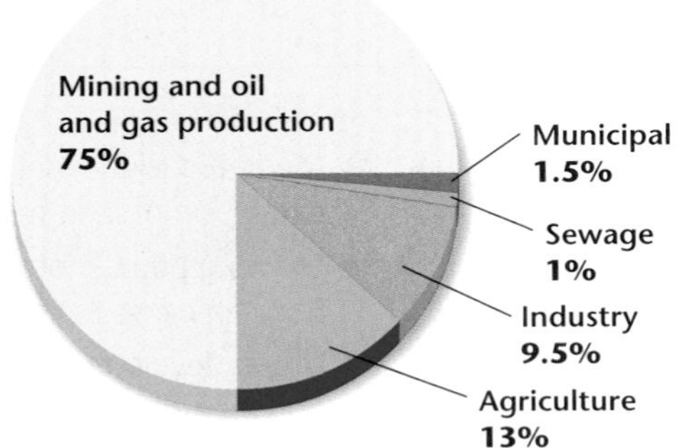

10. What activity is the largest producer of solid wastes in the United States each year?

11. What source produces the smallest amount of solid wastes in the United States each year?

12. What is the total percentage of solid waste produced by industry and resource extraction activities in the United States each year?

13. If you wanted to reduce the total amount of solid wastes produced each year, in what area would you concentrate your efforts? Why?

Test-Taking Tip

Root Words Use roots to learn. Roots can help you group words together as you learn them. If you learn that *bene-* means "good" as in beneficial, you can then group new words such as benefit, benefactor, and benevolent with beneficial, a word you already know.

Assessment

Applying Main Ideas

14. What causes populations to experience periods of explosive growth?
15. What will happen to human population numbers as carrying capacity is approached?
16. How do mining companies protect the environment when they extract mineral resources?
17. What are some of the risks of planting a monoculture?
18. What steps are taken in modern landfills to reduce environmental pollution?
19. Explain why a hurricane is a density-independent factor in limiting population size, whereas a new disease such as the Ebola virus is density-dependent.
20. Why do farmers who use chemical pesticides to control pests have to increase the amount applied to their fields each year?

Thinking Critically

21. What are some of the environmental costs involved in developing a new gold mine, and how might this affect the decision to open the mine?
22. Why is ground-level ozone worse on a sunny weekday than on a sunny weekend?
23. If midwestern power plants generate most of the air pollution, why does most acid precipitation fall in the northeastern United States?
24. How does urban development impact coastal waters?
25. What would be some of the positive environmental effects of reducing consumption of electricity?

Standardized Test Practice

INTERPRETING DATA Use the graphic below to answer the following questions.

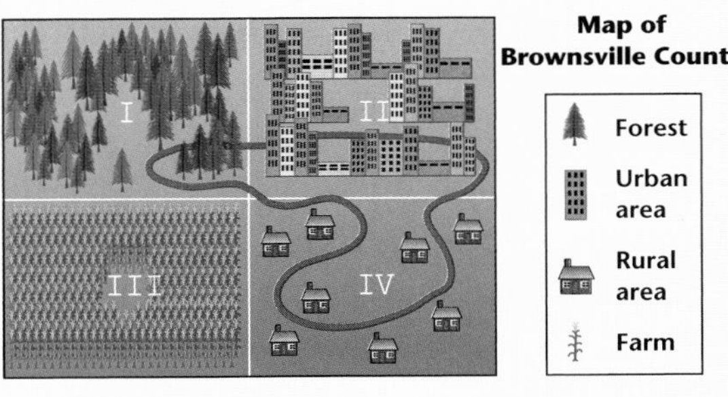

1. Which area of Brownsville is most likely to have problems with flooding during heavy rains?
   a. I **b.** II **c.** III **d.** IV
2. Which area of Brownsville is most likely to have problems with nonpoint source pollution from fertilizer runoff?
   a. I **b.** II **c.** III **d.** IV
3. If Brownsville County decided to clear area I in order to expand area III, Brownsville might develop problems with topsoil erosion and pesticide pollution. What might be one way to minimize harmful effects?
   a. deforestation **c.** monoculture
   b. clear-cutting **d.** selective logging
4. What will happen if the size of Brownsville's human population reaches the carrying capacity for its environment?
   a. There will be more births than deaths.
   b. The death rate will increase and the birth rate will increase.
   c. The population will reach equilibrium.
   d. There will be no more births until the death rate increases.

UNIT 7

GeoDigest

For a **preview** of resources and the environment, study this GeoDigest before you read the chapters. After you have studied the topic, you can use the GeoDigest to **review** the unit.

Resources and the Environment

Earth Resources

Resources Natural resources, which include air, water, land, organisms, rocks, minerals, and nutrients, are resources that Earth provides. Geochemical cycles that move substances through Earth's hydrosphere, lithosphere, biosphere, and atmosphere are also natural resources. Renewable resources, which include living things, surface water, groundwater, fertile soil, air, solar energy, and elements that cycle such as carbon and nitrogen, are replaced by natural processes at a rate that is at least equal to the rate at which they are used. Nonrenewable resources exist in fixed amounts in Earth's crust and can be replaced only by geological, physical, or chemical processes that take hundreds of millions of years. Nonrenewable resources include fossil fuels, such as coal and petroleum, and elements such as gold, copper, and silver. Natural resources are not distributed evenly on Earth.

Land Resources Land resources include topsoil, rocks, minerals, and space for agriculture, housing, roadways, and protected areas such as wildlife refuges and national parks. In the United States, 42 percent of the land is protected from some uses. Topsoil is a mixture of decaying organic matter, eroded rock, minerals, nutrients, oxygen, and water. It takes thousands of years for topsoil to form, yet poor farming practices and erosion can lead to its rapid loss and even to the formation of deserts. Bedrock is unweathered parent rock, which is used

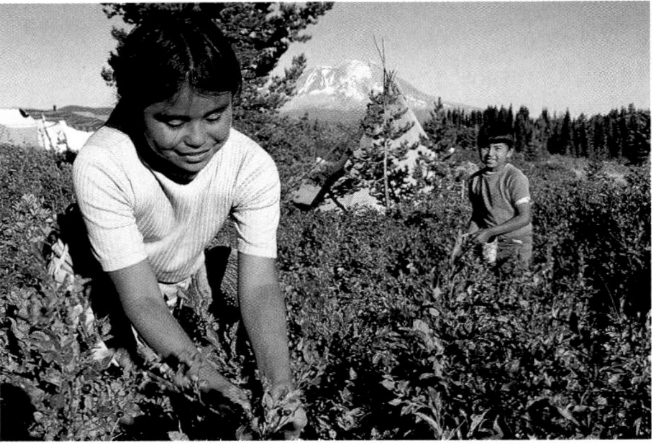

in buildings and monuments. Aggregates such as sand, gravel, and crushed stone, are used in construction. An ore is a natural resource that can be mined at a profit. Ores may be associated with igneous rocks or formed by processes at Earth's surface. Some uses of land resources may have negative effects on the environment.

Air Resources Earth's atmosphere is mostly composed of nitrogen and oxygen, with small amounts of other gases. Early in Earth's history, its atmosphere had no oxygen; over time, oxygen was provided by photosynthetic organisms. When human activities disrupt the balance of the geochemical cycles in Earth's atmosphere, pollution results. Both indoor and outdoor pollution are harmful. Outdoors, pollutants can be transported, diluted, transformed, or removed. Indoor air pollutants remain trapped, resulting in "sick" buildings.

Water Resources Only three percent of the water on Earth is freshwater and it is continually recycled through the water cycle. Water has unique properties that allow life to exist on Earth. It is a liquid over a wide range of temperatures, has high heat-storage capacity, dissolves many substances, and expands when it freezes. Because water is not evenly distributed on Earth, water management techniques and plans are developed to ensure a continuing supply. Management methods include building dams and reservoirs, transporting surface water, tapping groundwater, and desalinating seawater.

Energy Resources

The Sun is Earth's primary energy source. The Sun's energy is transferred from photosynthetic organisms to all other living things. Materials from living things have been used as fuels throughout history. Renewable biomass fuels release energy when they are burned, decomposed, or digested, and thus produce heat and electricity. Wood still serves as a fuel for over half of the world's population. Field crops such as hay, corn, straw, and sugar cane are also used as fuels. Fossil fuels, such as coal and petroleum, formed from organisms that lived millions of years ago. The burning of fossil fuels releases sulfur into the atmosphere and causes air pollution. As ancient organisms died and settled to the bottom of swamps, their remains partially decayed and formed peat. When the peat was subjected to high temperatures and pressure, coal formed. Types of coal include lignite, bituminous coal, and anthracite.

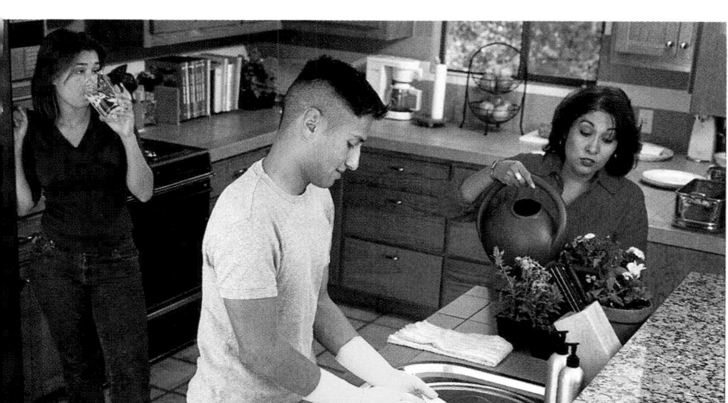

Natural gas and petroleum formed from accumulations of ancient organic material, primarily plankton, in shallow seas. Refined crude oil yields gasoline, kerosene, fertilizer, plastics, lubricants, and medicines.

Alternative Energy Resources

Alternative energy resources are those other than fossil fuels. Solar energy is unlimited, but advances in technology are needed to find better ways to collect and store it. Both passive and active solar techniques are used. Hydroelectric power is derived from the energy of falling or moving water and is commonly used in the production of electricity. In coastal areas, tidal power can be used to produce electricity. Geothermal energy is available in certain areas and is a product of Earth's internal heat. In areas with consistently strong winds, wind is a source of energy. Nuclear energy results when atoms of radioactive elements emit particles in the process known as fission. Technological advances are needed to dispose of or render harmless the dangerous waste products of nuclear energy. Biomass energy comes from the burning, decomposition, or digestion of organic materials such as wood, crops, and animal-waste material. Alcohols that form from biomass can be mixed with fuels such as gasoline to produce gasohol. Biogas, a mixture of methane and carbon dioxide, forms from the decomposition of animal wastes.

Conservation of Energy Resources

Energy resources will last longer if conservation and energy efficiency are further developed, so that fewer resources are used to provide more energy. Making vehicles more fuel efficient and improving efficiency in industry and homes will reduce the rate at which natural resources are used. Achieving sustainable energy use will ensure that current and future energy needs are met while guarding against the degradation of our environment.

Humans and Resources

Populations All organisms use resources to exist and the uses of these resources have an impact on the environment. As populations increase, the demand for resources also increases. At first, populations grow exponentially. But because resources are limited, populations eventually reach the carrying capacity of the environment and stop growing. Earth's current human population explosion has a huge impact on resources and environments.

Impact on Land, Air, and Water The extraction of the mineral resources that modern societies need can disrupt land surfaces and create toxic chemical wastes. Growing populations may increase the demand for food production, which causes habitat loss, erosion, and water pollution. Urban development also causes habitat loss, erosion, and pollution of nearby areas. Many cities are taking steps to preserve habitats, control erosion at construction sites, and clean up pollution. Air pollution causes human health problems. Ground-level ozone is the major component of smog, a type of air pollution common in cities. Humans affect Earth's atmosphere on a global scale, as demonstrated by acid precipitation, ozone depletion, and global warming. Controlling air pollution requires government help because wind blows air pollution across state and national boundaries. Conservation of existing supplies is the best way to ensure that freshwater is available in areas where there is not enough to meet human needs. Pollution of freshwater supplies comes from sewage, pesticide and fertilizer runoff, and chemical and oil spills. The Clean Air, Safe Drinking Water, and Clean Water Acts are laws designed to decrease pollution in the United States.

A Sustainable Society When the total environmental cost of developing and using a resource is considered, some resources may be uneconomical to develop. Resources will run out if the current rate of consumption is not decreased. For a society to be sustainable, it must manage its resources to minimize demand and environmental impact. Reducing, reusing, and recycling materials decreases the demand for resources.

Vital Statistics

Most Populous Cities: 2003	Population
1. Mumbai (Bombay), India	12 383 100
2. Buenos Aires, Argentina	12 116 400
3. Karachi, Pakistan	10 537 200
4. Manila, Philippines	10 232 900
5. Delhi, India	10 203 700
6. São Paulo, Brazil	10 195 000
7. Seoul, South Korea	9 630 600
8. Istanbul, Turkey	9 419 000

FOCUS ON CAREERS

Urban Planner

An urban planner develops ideas for how a city will develop and use its land—now and in the future. He or she works with city leaders to make a plan that meets the needs of all citizens. The urban planner examines the need for greenbelts and parks, and determines how traffic will move through an area. Urban planners usually have bachelor's degrees, and an advanced degree is sometimes necessary.

GeoDigest

ASSESSMENT

Understanding Main Ideas

1. What are resources called that are replaced by natural processes at a rate that is at least equal to the rate at which they are used?
   a. nonrenewable resources
   b. fossil fuels
   c. renewable resources
   d. minerals

2. What type of resources are topsoil, rocks, minerals, and space for agriculture, housing, and roadways?
   a. land
   b. air
   c. water
   d. solar

3. What causes a building to become "sick"?
   a. outdoor air pollution
   b. thinning of the ozone
   c. indoor air pollution
   d. acid precipitation

4. What types of methods are building dams and reservoirs, tapping groundwater, and desalinating seawater?
   a. soil conservation
   b. water management
   c. transportation
   d. cleaning up pollution

5. Which of the following is not a type of coal?
   a. lignite
   b. bituminous coal
   c. anthracite
   d. peat

6. What is the ultimate source of most forms of energy on Earth?
   a. fossil fuels
   b. the Sun
   c. biomass
   d. tidal power

7. Which of these is not a property of water?
   a. liquid over wide temperature range
   b. high heat storage capacity
   c. contracts when it freezes
   d. dissolves many substances

8. What happens when populations reach the carrying capacity of the environment?
   a. They stop growing.
   b. They explode.
   c. They grow exponentially.
   d. They outgrow cities.

9. Acid precipitation, ozone depletion, and global warming demonstrate what type of effects by humans on the atmosphere?
   a. local
   b. regional
   c. national
   d. global

10. Which of these is a land resource that can be mined at a profit?
   a. bedrock
   b. aggregate
   c. ore
   d. topsoil

Thinking Critically

1. Compare and contrast renewable and non-renewable resources.
2. Describe how gasoline is derived from living things.
3. Explain why governments get involved in controlling pollution.

Bauxite

Unit 8

Beyond Earth

The nighttime sky appears to contain only stars, which belong to the Milky Way (right). However, a variety of objects, such as planets, stars, nebulae, and galaxies can be found. Despite their distance, each of these objects impacts our existence here on Earth. Galaxies provide a place for stars to develop. Nebulae provide the materials to form stars. Stars, like the Sun, provide energy and create elements. Star formation often results in the formation of planets.

Unit Contents

Go to the National Geographic Expedition on page 902 to learn more about topics that are connected to this unit.

Milky Way Galaxy

Chapter 28

The Sun-Earth-Moon System

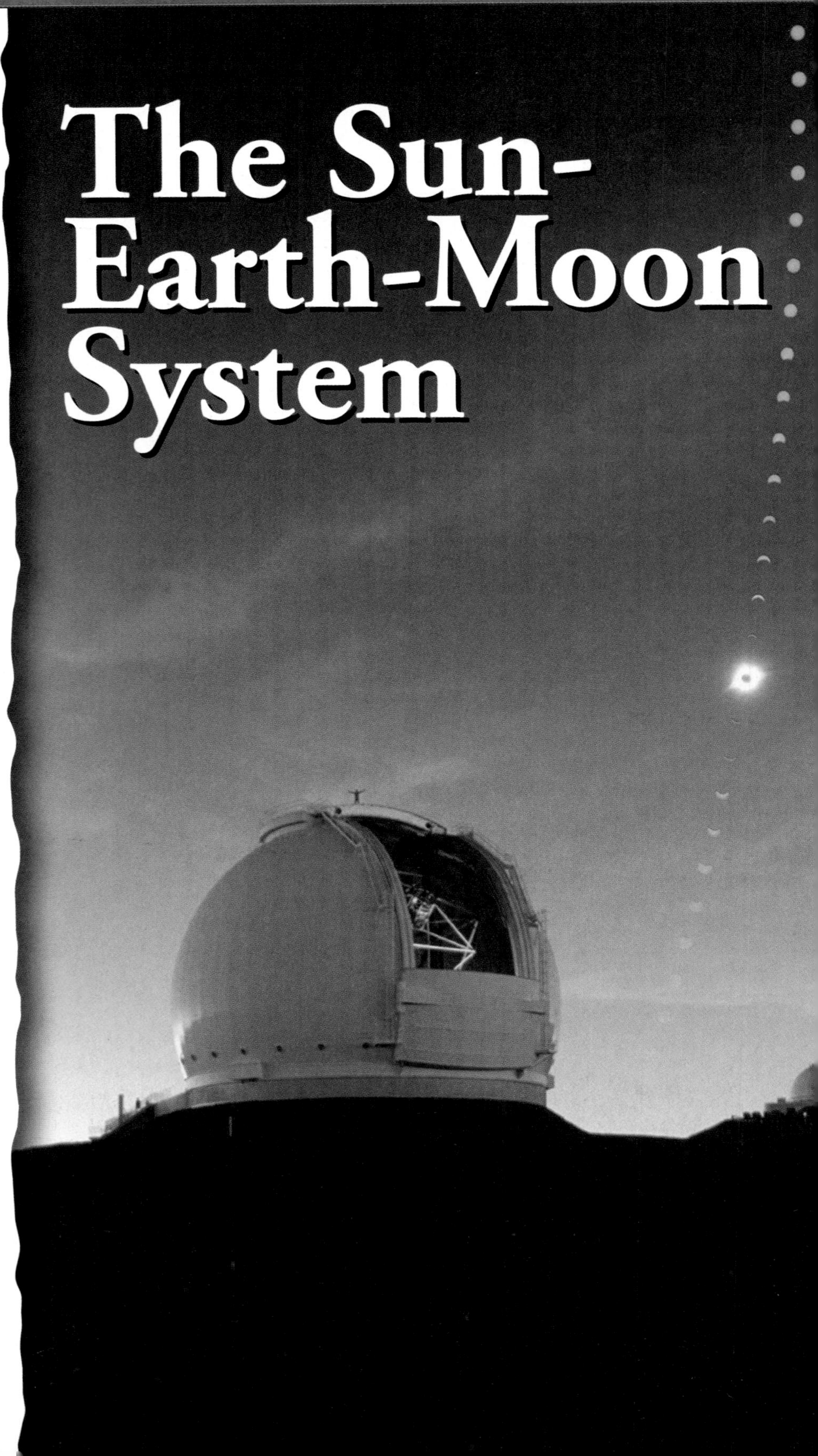

What You'll Learn

- How light and telescopes are used to explore the sky.
- How to identify features on the Moon.
- What theories are used to describe the Moon's origin.
- How to analyze the motions of the Sun, Earth, and the Moon.

Why It's Important

The motions of the Sun-Earth-Moon system affect Earth physically, as well as play an important role in our timekeeping system.

Earth Science Online

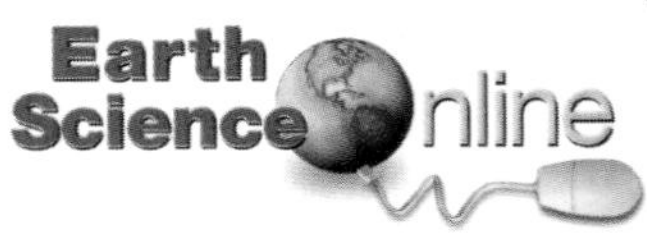

To find out more about the Sun-Earth-Moon system, visit the Earth Science Web Site at earthgeu.com

Discovery Lab Make a Scale Model

The Sun is about 109 times larger in diameter than Earth, and Earth is about 3.7 times larger in diameter than the Moon. The Moon is 30 times farther from Earth than Earth's diameter, and the Sun is 400 times farther away from Earth than is the Moon. In this activity, you will compare relative sizes and distances within the Sun-Earth-Moon system.

1. Calculate the diameters of Earth and the Sun using a scale in which the Moon's diameter is equal to 1 cm.
2. Using your calculations in step 1, calculate the distance between Earth and the Moon and the distance between Earth and the Sun.
3. Cut out circles to represent your scaled Earth and Moon, and place them at the scaled distance apart.

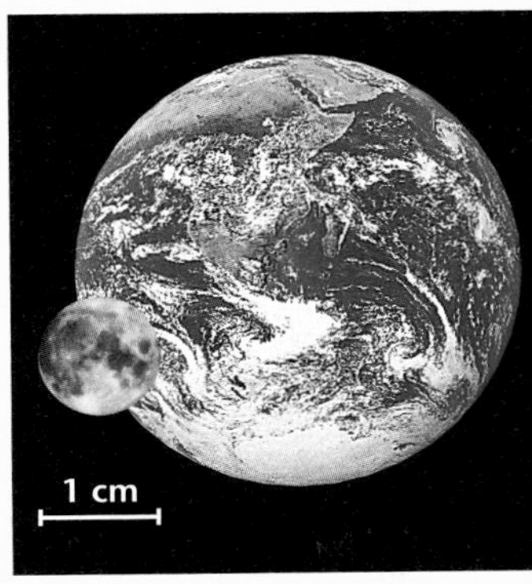

*CAUTION: **Always handle sharp objects with care.***

Observe In your science journal, describe the sizes of your cut-out Earth and Moon compared to the distance between them. Infer why you were not instructed to cut out a scaled Sun and place it at the scaled distance. How would you change this model so that it would fit in your classroom?

SECTION 28.1 *Tools of Astronomy*

OBJECTIVES

- **Describe** *electromagnetic radiation.*
- **Explain** *how telescopes work.*
- **Describe** *space exploration.*

VOCABULARY

refracting telescope
reflecting telescope
interferometry
spinoff

The best tool, and in most cases the only tool, that astronomers can use to learn about the universe is the light that comes to Earth from distant objects. Apart from a few solar-system objects that have been sampled by direct probes and particles and fragments that have made their way into Earth's atmosphere or to Earth's surface, there is no other way to study the cosmos except to analyze the radiation emitted from it. Therefore, it is necessary to understand this radiation.

RADIATION

The radiation from the cosmos that scientists study is electromagnetic radiation. Electromagnetic radiation consists of electric and magnetic disturbances, traveling through space as waves. The human eye can sense only a limited range of all the various wavelengths of electromagnetic radiation. This range is called visible light. Electromagnetic radiation includes not just visible light, but also infrared and ultraviolet radiation, radio waves, microwaves, X rays, and gamma rays.

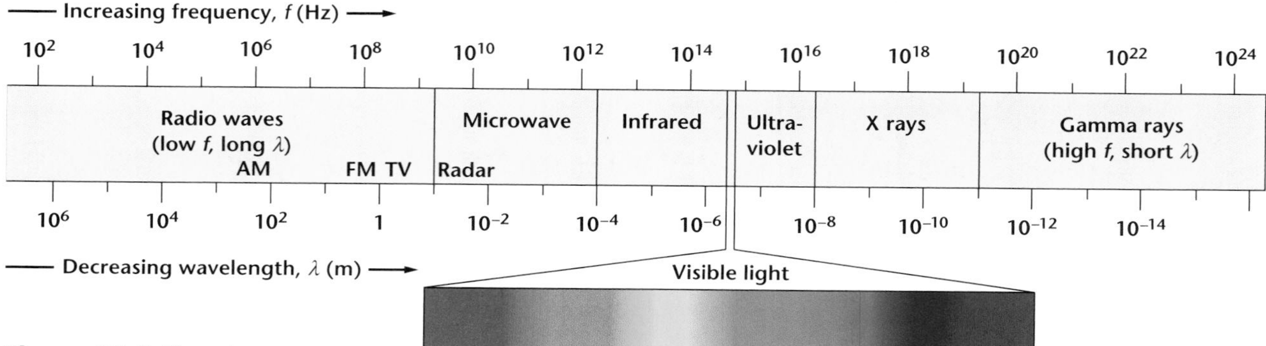

Figure 28-1 The electromagnetic spectrum ranges from radio waves to gamma rays. Wavelength and frequency are related by $c = \lambda f$. Notice how small the range of visible light is compared to the rest of the spectrum.

You may be familiar with some forms of electromagnetic radiation. For example, ultraviolet radiation causes sunburn, and X rays help doctors observe internal injuries and diagnose bone diseases. All the types of electromagnetic radiation, arranged according to wavelength and frequency, form the electromagnetic spectrum, illustrated in ***Figure 28-1.***

Electromagnetic radiation is classified by its wavelengths. Wavelength is the distance between peaks on a wave. You can see in ***Figure 28-1*** that red light has longer wavelengths than blue light, and radio waves have much longer wavelengths than gamma rays. Electromagnetic radiation also can be classified according to frequency, which is the number of waves or oscillations occurring per second. The visible light portion of the spectrum has frequencies ranging from 4.3×10^{14} to 7.5×10^{14} Hz. Frequency is related to wavelength by the mathematical relationship $c = \lambda f$, where c is the speed of light (3.0×10^8 m/s), λ is the wavelength, and f is the frequency. Note that all types of electromagnetic radiation travel at the same speed, c.

Using Math

Using Numbers For a telescope with a circular collector, the collecting area is πr^2, where r is the telescope's radius. If one telescope is twice as large as another, it will collect four times as much light. How much more visible light will a visible-light telescope with a radius of 5 m collect than a human eye that has a pupil with a 1 mm radius?

TELESCOPES

Objects in space emit radiation in all portions of the electromagnetic spectrum. The ability to modify telescopes with different detectors and mirror shapes to observe all wavelengths, especially those the human eye cannot detect, is just one of the benefits of using a telescope. Another benefit is that a telescope collects electromagnetic radiation from a distant object and focuses it at a point where the image of the object can be studied or recorded. The human eye does the same thing with visible light, but the eye is much more limited. A typical human-eye pupil has a diameter of up to 7 mm when it is adapted to darkness, whereas a telescope might be as large as 10 m in diameter. The area of the opening through which electromagnetic radiation enters determines the collecting power of a telescope. The larger the opening, the more electromagnetic radiation that can be gathered. A telescope's ability to collect a large amount of electromagnetic radiation allows astronomers to observe faint or weakly emitting objects.

A third benefit of telescopes is that they allow astronomers to use specialized equipment. A photometer, for example, is used to measure the intensity of visible light. A fourth benefit is that telescopes can be used to make time exposures with the aid of cameras or other imaging devices. In time exposures, electromagnetic radiation is collected over a long period of time. With visible light, the human eye "photographs" what it sees about 10 times per second, so an object too dim to be perceived in one-tenth of a second cannot be seen. This is why telescopes are able to detect objects that are too faint for the human eye to see.

Refracting and Reflecting Telescopes Two different types of telescopes are used to focus visible light. The first telescopes, invented around the year 1600, used lenses to bring visible light to a focus and are called **refracting telescopes,** or refractors. The largest lens on such a telescope is called the objective lens. ***Figure 28-2A*** illustrates how a simple refracting telescope works. In 1668, a new telescope was designed that used mirrors. Telescopes that bring visible light to a focus with mirrors are called **reflecting telescopes,** or reflectors. ***Figure 28-2B*** illustrates how a simple reflecting telescope works.

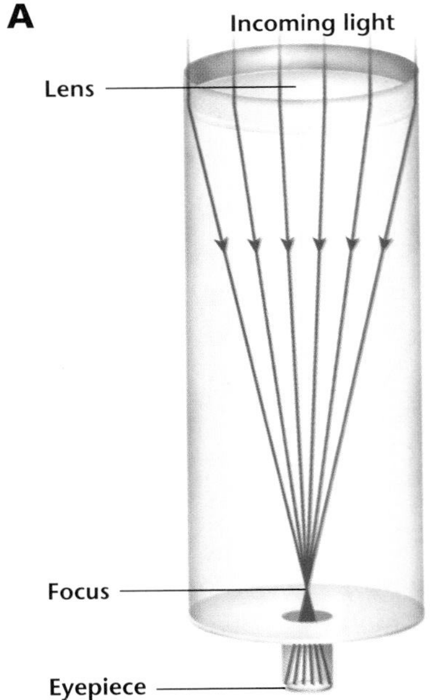

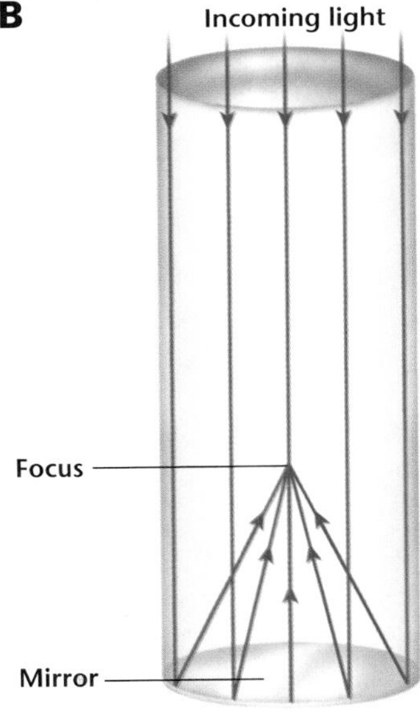

Figure 28-2 A refracting telescope **(A)** uses a lens to bring light to a focus. The largest lens is called the objective lens. A reflecting telescope **(B)** uses a mirror to bring light to a focus. The largest mirror is called the primary mirror.

Although both refracting and reflecting telescopes are still in use today, the majority are reflectors. Most telescopes used for scientific study are located at observatories far from city lights, usually at high elevations where there is less atmosphere overhead to blur images. Some of the best observatory sites in the world are located high atop mountains in the southwestern United States, along the peaks of the Andes Mountains in Chile, and on the summit of Mauna Kea, the gigantic volcano on the island of Hawaii.

To learn more about the *Hubble Space Telescope,* go to the **National Geographic Expedition** on page 902.

Telescopes at Other Wavelengths In addition to using visible-light telescopes, astronomers observe the universe at wavelengths that the human eye cannot detect. For all telescopes, the goal is to bring as much electromagnetic radiation as possible to a focus. Infrared and ultraviolet radiation can be focused by mirrors in much the same way as visible light. X rays cannot be focused by normal mirrors, and thus, special designs must be used. Because gamma rays cannot be focused, telescopes designed to detect the extremely short wavelengths of this type of radiation can determine little more than the general direction from which the rays come.

Figure 28-3 shows a radio telescope consisting of a large dish, or antenna, which resembles a satellite TV dish. The dish plays the same role as the primary mirror in a reflecting telescope, by reflecting radio waves to a focus above the dish. There, a receiver converts the radio waves into electrical signals that can be stored in a computer for analysis. A process called interferometry, which has been used with radio telescopes for a number of years, is now being applied to other telescopes as well. **Interferometry** is the process of linking separate telescopes together so that they act as one telescope. The detail in the images that they produce improves as the distance between the telescopes increases. One of the best-known examples of this technology is the Very Large Array near Socorro, New Mexico.

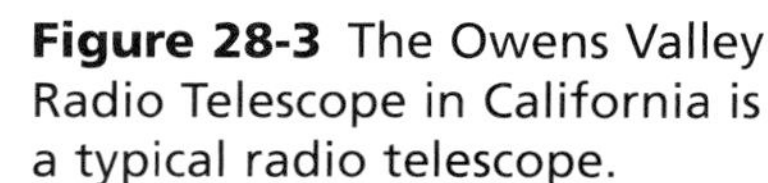

Figure 28-3 The Owens Valley Radio Telescope in California is a typical radio telescope.

SATELLITES, PROBES, AND SPACE-BASED ASTRONOMY

Astronomers often have to send their instruments into space to collect the information they seek. One reason for this is that Earth's atmosphere blocks infrared radiation, ultraviolet radiation, X rays, and gamma rays. In addition, when Earth's atmosphere does allow certain wavelengths to pass through, the images are blurred. Another reason for sending instruments into space is to make close-up observations and even obtain samples from nearby objects in the solar system. Since the late 1960s, American, European, Soviet (later, Russian), and Japanese space programs have launched many space-based observatories to collect data in different wavelengths.

Figure 28-4 On April 25, 1990, the *Hubble Space Telescope* was released from the shuttle *Discovery* during mission STS-31.

One of the best-known space-based observatories, shown in ***Figure 28-4,*** is the *Hubble Space Telescope (HST),* which was launched in 1990 and is scheduled to operate until 2010. *HST* was designed to obtain sharp visible-light images without atmospheric interference, and also to make observations in infrared and ultraviolet wavelengths. Other space-based telescopes, such as the *Far Ultraviolet Spectroscopic Explorer,* the *Chandra X-Ray Observatory,* and the *Spitzer Space Telescope,* are used to observe other wavelengths that are blocked by Earth's atmosphere.

Spacecraft In addition to making observations from above Earth's atmosphere, space-based exploration can be achieved by sending spacecraft directly to the bodies being observed. Robotic probes make close-up observations and sometimes land to collect information directly. Probes are practical only for objects within our solar system, because the stars are much too far away. The robot *Sojourner,* part of the *Pathfinder* probe, explored Mars for almost 3 months in 1997. More recently, the twin robots *Spirit* and *Opportunity* conducted scientific experiments on Mars in 2004–2005 (***Figure 28-5***).

Figure 28-5 One of the twin robots, *Spirit* and *Opportunity,* is shown on a simulated Martian surface on Earth during one of its tests.

Figure 28-6 This photo shows the partially completed *International Space Station* as it orbits Earth.

Human Spaceflight Exploring objects in space has been a top priority for scientists, but they have also been very interested in exploring the effects of space, such as weightlessness. The most recent human explorations and studies have been accomplished with the space shuttle program, which began in 1981. The space shuttle provides an environment for scientists to study the effects of weightlessness on humans, plants, the growth of crystals, and other phenomena. However, because shuttle missions last a maximum of just 17 days, long-term effects must be studied in space stations.

A multi-country space station called the *International Space Station*, shown in ***Figure 28-6,*** is the ideal environment to study the long-term effects of space. Human habitation and research aboard the *International Space Station* began in 2000.

Spinoffs Space-exploration programs have benefited our society far beyond our increased understanding of space. Many technologies that were originally developed for use in space programs are now used by people all over the world. Did you know that the technology for the space shuttle's fuel pumps led to the development of pumps used in artificial hearts? Or that the Apollo program led to the development of cordless tools? In fact, more than 1400 different NASA technologies have been passed on to commercial industries for common use, and are called **spinoffs.** Each year, new technologies are developed that not only benefit astronomers and space exploration, but society also.

Section Assessment

1. How do the various types of electromagnetic radiation differ from each other?
2. What are the advantages of using a telescope compared to making observations with the unaided eye?
3. What is interferometry, and how does it affect the images that are produced?
4. Why do astronomers send telescopes and probes into space?
5. How are space stations beneficial?
6. **Thinking Critically** How would humans' lives and our perceptions of the universe be different without space-based technology and exploration?

Skill Review

7. **Comparing and Contrasting** Compare and contrast refracting telescopes and reflecting telescopes. For more help, refer to the *Skill Handbook*.

earthgeu.com/self_check_quiz

SECTION 28.2 The Moon

Earth Sciences 1.f I&E 1.e

The Moon is a familiar object in the night sky. Despite its proximity to Earth, however, the origins and nature of the Moon have been elusive. Only with advances in telescope and spacecraft technology over the past 100 years have people begun to understand the Moon.

OBJECTIVES

- **Describe** *the development of exploration of the Moon.*
- **Identify** *features on the Moon.*
- **Explain** *the theories about how the Moon formed.*

VOCABULARY

albedo
highland
mare
impact crater
ejecta
ray
rille
regolith

REACHING FOR THE MOON

Astronomers have learned much about the Moon from telescopic observations. However, most of our knowledge of the Moon comes from explorations by space probes, such as *Lunar Prospector* and *Clementine,* and astronauts. Plans for a crewed lunar expedition began in the late 1950s. The first step was taken in 1957 with the launch of the first satellite, *Sputnik I,* by the Soviet Union. Shortly thereafter, in 1961, Soviet cosmonaut Yuri A. Gagarin became the first human in space.

The United States' Project Mercury launched the first American, Alan B. Shepard Jr., shown in ***Figure 28-7,*** into space on May 5, 1961. Project Gemini launched two-person crews into space, and on July 20, 1969, the Apollo program landed Neil Armstrong and Buzz Aldrin on the Moon, during *Apollo 11.*

Lunar Properties Earth's moon is unique among all the moons in the solar system. It is one of the largest moons, especially compared to the size of the planet it orbits. The Moon's radius is about 27 percent of Earth's radius, and its mass is more than 1 percent of Earth's mass, as shown in ***Table 28-1.*** Most moons are much smaller than this in relation to the size of the planets they orbit.

The orbit of the Moon is also unusual in that the Moon is relatively farther from Earth than most moons are from the planets they orbit. Earth's moon is a solid, rocky body, in contrast to the icy composition of the moons of the outer planets Jupiter, Saturn, Uranus, Neptune, and Pluto. Also, Earth's moon is the only large moon among the inner planets. Mercury and Venus have no moons at all, and the moons of Mars are just two tiny chunks of rock.

Table 28-1 The Moon and Earth

	The Moon	Earth
Mass (kg)	7.349×10^{22}	5.9736×10^{24}
Radius (km)	1737.4	6378.1
Volume (km^3)	2.1968×10^{10}	1.08321×10^{12}
Density (kg/m^3)	3340	5515

Figure 28-7 American astronaut Alan B. Shepard Jr., in the *Mercury 7* capsule, prepares for launch.

Earth Sciences

1.f Students know the evidence for the dramatic effects that asteroid impacts have had in shaping the surface of planets and their moons and in mass extinctions of life on Earth.

The Lunar Surface Although the Moon is the brightest object in our nighttime sky, the lunar surface is actually quite dark. The **albedo** of the Moon, the amount of sunlight that its surface reflects, is very small—only about 0.07 (7 percent). In contrast, Earth has an average albedo of nearly 0.31 (31 percent). The sunlight that is absorbed by the surface of the Moon is responsible for the extreme differences in temperatures on its surface. Because the Moon has no atmosphere, sunlight can heat the Moon's surface to temperatures as high as 400 K (127°C). During the absence of sunlight, the Moon's surface temperature can drop to a chilly 100 K (−173°C).

The physical surface of the Moon is very different from that of Earth. There is no erosion on the Moon—except for surface creep and wear caused by recent impacts—because it has no atmosphere or flowing water. The surface of the Moon consists of several features. Regions called **highlands,** shown in ***Figure 28-8A,*** are light in color, mountainous, and heavily covered with craters. Regions called **maria** (*singular,* mare), shown in ***Figure 28-8A,*** are dark, smooth plains, which on average are 3 km lower in elevation than the highlands.

All of the craters on the Moon are **impact craters,** formed when objects from space crashed into the lunar surface. The material blasted out during these impacts fell back to the surface as **ejecta.** Some craters have long trails of ejecta, called **rays,** that radiate outward. Rays are visible as light-colored streaks, as shown in ***Figure 28-8B.***

In contrast to the crater-covered highlands, the surfaces within maria are quite smooth. However, the maria do have a few scattered craters and **rilles,** which are meandering, valleylike structures, as illustrated in ***Figure 28-8D.*** In addition, around some of the maria are mountain ranges, shown in ***Figure 28-8C.***

Why does the Moon have many craters, while Earth has few? Early in the formation of the solar system, Earth was bombarded just as heavily as the Moon, but erosion on Earth has eliminated traces of all but the youngest craters. On the Moon, craters are preserved until one impact covers another.

B

C

D

Figure 28-8 Maria are dark, plains areas on the Moon, while the highlands are very mountainous and heavily cratered **(A).** A relatively recent crater on the Moon has very light ejecta **(B).** This is a mountain range on the surface of the Moon **(C).** Humbolt Crater has a network of rilles surrounding it **(D).**

Composition The Moon is made up of minerals similar to those of Earth—mostly silicates. The highlands, which cover most of the lunar surface, are predominately lunar breccias, which are rocks formed by the fusing together of smaller pieces of rock during impacts. Unlike sedimentary breccias on Earth, most of the lunar breccias are composed of plagioclase feldspar, a silicate containing high quantities of calcium and aluminum but low quantities of iron. The maria are predominately basalts that differ from those on Earth in that they contain no water.

HISTORY OF THE MOON

The entire lunar surface is very old. Radiometric dating of lunar rocks from the highlands indicates an age between 3.8 and 4.6 billion years. Based on the ages of the highlands and the frequency of the impact craters that cover them, scientists theorize that the Moon was heavily bombarded during its first 800 million years, which resulted in the breaking and heating of rocks on the surface of the Moon. This formed a layer of loose, ground-up rock, called **regolith,** on the surface of the Moon. The regolith averages several meters in thickness, but it varies considerably depending on location.

The maria, only slightly younger than the highlands, are between 3.1 and 3.8 billion years old. After the period of intense bombardment in which the highlands formed, lava welled up from the Moon's interior and filled in the large impact basins to form the maria. The maria have remained relatively free of craters because fewer impacts have occurred on the Moon since the time when they formed. However, flowing lava in the maria scarred their surfaces with rilles, which are much like lava tubes found on Earth. During the formation of the maria, the lava often did not fill the basins completely. Instead, the rims of the basins remained above the lava and formed the mountain ranges that now exist around many of the maria. As shown in ***Figure 28-9,*** there are virtually no maria on the far side of the Moon, which is covered almost completely with highlands. Scientists hypothesize that this is because the crust is twice as thick on the far side, which would have made it increasingly difficult for lava to reach the lunar surface. You will determine the relative ages of the Moon's surface features in the *Mapping GeoLab* at the end of this chapter.

Figure 28-9 This photo of the far side of the Moon, shows the heavily cratered surface of the highlands.

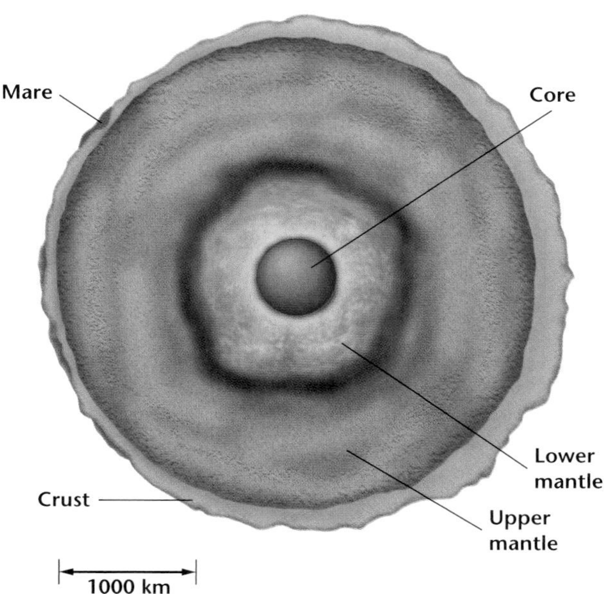

Figure 28-10 The Moon has a layered structure similar to Earth's.

Tectonics on the Moon? Mountain ranges around maria were formed by impacts, not tectonically, as mountain ranges on Earth are. But is that enough evidence to conclude that the Moon is not tectonically active? Scientists infer from seismometer data that the Moon, like Earth, has a layered structure, which consists of the crust, the upper mantle, the lower mantle, and the core, as illustrated in ***Figure 28-10.*** The crust varies in thickness and is thickest on the far side. The Moon's upper mantle is solid, its lower mantle is partially molten, and its core is made of solid iron. Seismometers also measure moonquake strength and their frequency. Although the Moon experiences a moonquake that would be strong enough to cause dishes to fall out of a cupboard approximately once a year, scientists theorize that the Moon is not tectonically active. The fact that the Moon has no active volcanoes and no significant magnetic field supports scientists' theory that tectonics are not occurring on the Moon.

Formation Theories Several theories have been proposed to explain the Moon's unique properties. One of these is the capture theory, which proposes that as the solar system was forming, a large object ventured too near to the forming Earth, became trapped in its gravitational pull, and formed into what is now the Moon. One problem with this theory is that something would have had to slow down the passing object for it to become trapped instead of continuing on its original path. Another problem with the capture theory is that Earth and the Moon are composed of very similar elements. If the Moon had been captured, we would expect the crusts of the Moon and Earth to have different compositions, rather than similar ones.

Another theory, called the simultaneous formation theory, accounts for the problems with the capture theory. According to this theory, the Moon and Earth formed at the same time and in the same general area, and thus the materials from which they formed were essentially the same. Also, because they formed in the same general area, the Moon did not have to be slowed down to become gravitationally trapped. This theory does not account for the different amounts of iron on Earth and on the Moon, however. The Moon is iron poor, while on Earth, iron is relatively abundant.

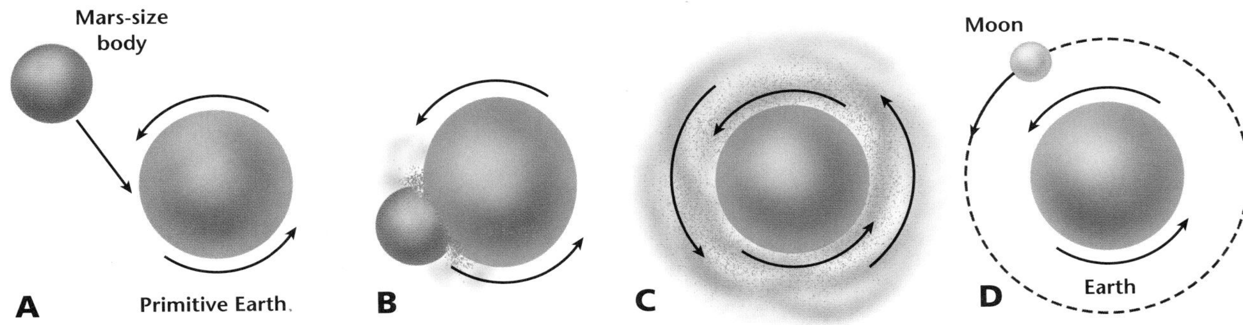

Figure 28-11 The impact theory suggests that a Mars-sized body **(A)** collided with Earth. The impact **(B)** threw material from the body and Earth into space **(C).** This material eventually merged together to form the Moon **(D).** (Not to scale)

The most commonly accepted theory of how the Moon formed, the impact theory, can explain astronomers' observations as a whole. Computer models indicate that the Moon formed as the result of a gigantic collision between Earth and a Mars-sized object about 4.5 billion years ago, when the solar system was forming. As a result of the collision, materials from the incoming body and from Earth's outer layers were ejected into space, where they then merged together to form the Moon, as illustrated by ***Figure 28-11.*** This model accounts for why the Moon is so similar to Earth in chemical composition. If this model is correct, then the Moon is made up of material that was originally part of Earth's iron-deficient crust as well as material that was once part of Earth's mantle. Heat produced by the impact would have evaporated any water that was present and resulted in lunar minerals lacking water. Despite scientists' uncertainty about how the Moon formed, we do know that it plays a vital role in the Sun-Earth-Moon system, as you will learn in the following section.

Section Assessment

1. How is Earth's moon different from the moons of other planets?
2. Why are there many visible craters on the Moon, but few on Earth?
3. Why do scientists believe that tectonic activity is not occurring on the Moon?
4. What is the most accepted theory of how the Moon formed, and what are the problems with the other theories?
5. **Thinking Critically** How would the surface of the Moon look different if the crust on the far side were the same thickness as the crust on the near side?

Skill Review

6. **Concept Mapping** Use the following terms to construct a concept map to organize the major ideas in this section. For more help, refer to the *Skill Handbook.*

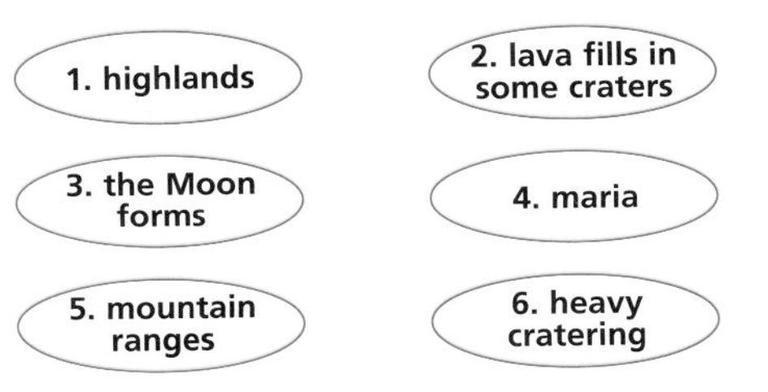

SECTION 28.3 The Sun-Earth-Moon System

OBJECTIVES

- **Identify** *the relative positions and motions of Earth, the Sun, and the Moon.*
- **Describe** *the phases of the Moon.*
- **Explain** *eclipses of the Sun and Moon.*

VOCABULARY

ecliptic
summer solstice
winter solstice
autumnal equinox
vernal equinox
synchronous rotation
solar eclipse
perigee
apogee
lunar eclipse

The relationships between the Sun, Moon, and Earth are important to us in many ways. The Sun provides light and warmth, and it is the source of most of the energy that fuels our society. Additionally, the Moon raises tides in our oceans and illuminates our sky with its monthly cycle of phases. Every society from ancient times to the present has based its calendar and its timekeeping system on the apparent motions of the Sun and Moon.

DAILY MOTIONS

The most obvious pattern of motion in the sky is the daily rising and setting of the Sun, the Moon, the stars, and everything else that is visible in the sky. The Sun rises in the east and sets in the west, as do the Moon, planets, and stars. Today, we understand that these daily motions result from Earth's rotation. The Sun, Moon, planets, and stars do not orbit around Earth every day. It only appears that way to us because we observe the sky from a planet that rotates once every day, or 15° per hour. But how do we know that Earth is rotating?

Earth's Rotation There are two relatively simple ways to demonstrate that Earth is rotating. One is to use a pendulum, which is a weight on a string or wire that is suspended from a support and can swing freely. A Foucault pendulum, which has a long wire and a heavy weight, will swing in a constant direction. But as Earth turns, it appears from our point of view that the pendulum gradually shifts its orientation. With a Foucault pendulum, pegs are often placed on the floor in a circle so that as Earth turns, the pendulum, shown in ***Figure 28-12,*** eventually knocks over each of the pegs. The second method of demonstrating that Earth rotates makes use of the fact that flowing air and water on Earth are diverted from a north-south direction to an east-west direction as a result of Earth's rotation. This diversion of direction is called the Coriolis effect, which you learned about in Chapter 12.

Figure 28-12 A Foucault pendulum, such as this one at the Griffith Observatory in Los Angeles, California, demonstrates that Earth is rotating.

The length of a day as we observe it is a little longer than the time it takes Earth to rotate once on its axis. This is because as Earth rotates, it also moves along in its orbit and has to turn a little farther. The time period from one sunrise or sunset to the next is called a solar day. Our timekeeping system is based on the solar day.

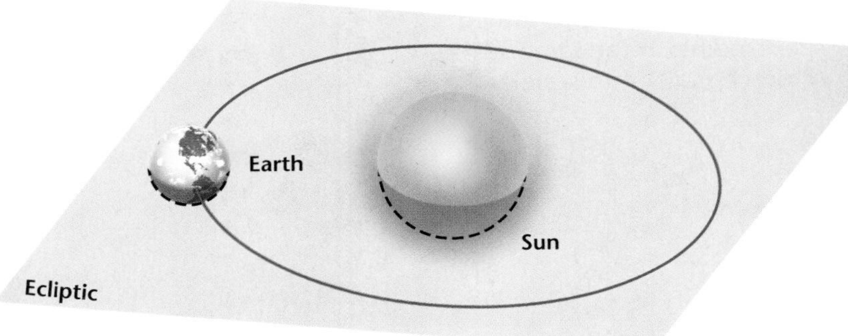

Figure 28-13 The ecliptic is the plane that contains Earth's orbit around the Sun. (Not to scale)

Annual Motions

As you know, the weather changes throughout the year. The length of days varies, and temperatures may range from cold to hot, depending on the latitude where you live. These annual changes are the result of Earth's orbital motion about the Sun. The plane in which Earth orbits about the Sun is called the **ecliptic,** as illustrated in ***Figure 28-13.***

The Effects of Earth's Tilt Earth's axis is tilted relative to the ecliptic at approximately 23.5°. As Earth orbits the Sun, the orientation of Earth's axis remains fixed in space, so that, at one point, the northern hemisphere of Earth is tilted toward the Sun, while at another point, six months later, the northern hemisphere is tipped away from the Sun. Our seasons, as discussed in Chapter 14, are created by this tilt and by Earth's orbital motion around the Sun.

As a result of the tilt of Earth's axis and Earth's motion around the Sun, the Sun changes its altitude in the sky. The way in which altitude of the Sun is measured is illustrated in ***Figure 28-14.***

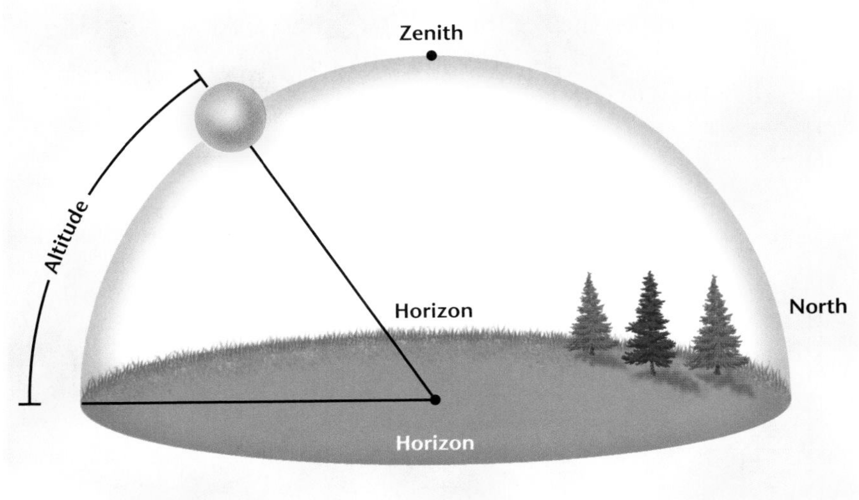

Figure 28-14 Altitude is measured in degrees from the observer's horizon to the object. There are 90 degrees from the horizon to the point directly overhead, called the zenith of the observer.

Figure 28-15 As Earth orbits the Sun, Earth's tilted axis points in the same direction. (Not to scale)

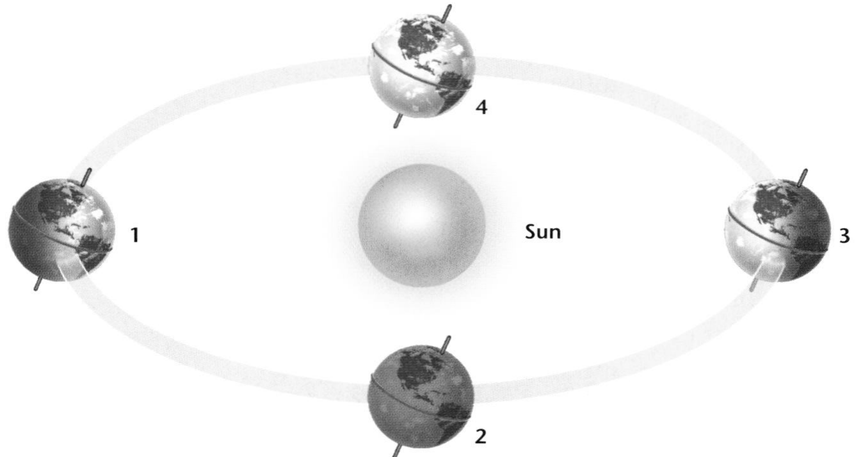

You've probably noticed the change in altitude of the Sun during the northern hemisphere's summer, when the Sun appears higher in the sky than it does during the northern hemisphere's winter. This change occurs gradually throughout Earth's orbit in a cyclic pattern.

Solstices Earth's varying position in its orbit around the Sun and the tilt of Earth's axis are illustrated in ***Figure 28-15.*** As Earth moves from position 1, through position 2, to position 3, the altitude of the Sun decreases in the northern hemisphere. Once Earth is at position 3, the Sun's altitude starts to increase as Earth moves through position 4 and back to position 1. Position 1 corresponds to the Sun's maximum altitude in the sky in the northern hemisphere. At this position, called the **summer solstice,** the Sun is directly overhead at the Tropic of Cancer, which is at 23.5° north latitude, as illustrated in ***Figure 28-16A.*** On the summer solstice, which occurs around June 21 each year, the number of daylight hours for the northern hemisphere is at its maximum, while it is at its minimum for the southern hemisphere. During the summer solstice, the Sun does not set in the region within the arctic circle, and it does not rise in the region within the antarctic circle.

Figure 28-16 The Sun's rays are vertical at the Tropic of Cancer during the summer solstice **(A),** at the Tropic of Capricorn during the winter solstice **(B),** and at the equator during the autumnal equinox **(C)** and the vernal equinox **(D).**

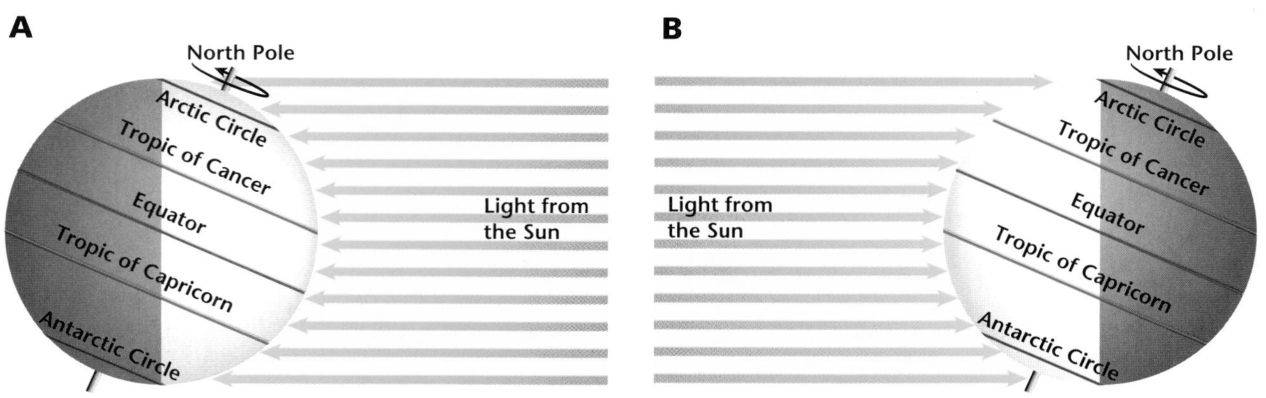

Conversely, when Earth is in position 3 and the northern hemisphere is tilted away from the Sun, the Sun has reached its lowest altitude in the sky. At this position, called the **winter solstice,** the Sun is directly overhead at the Tropic of Capricorn at 23.5° south latitude, as illustrated in ***Figure 28-16B.*** On the winter solstice, which occurs around December 21 each year, the number of daylight hours in the northern hemisphere is at its minimum, while it is at its maximum for the southern hemisphere. During the winter solstice, the Sun never rises in the region within the arctic circle, and it never sets in the region within the antarctic circle. You will model the Sun's position as seen from your location during the summer solstice in the *MiniLab* on this page.

Equinoxes At positions 2 and 4 in ***Figure 28-15,*** Earth's axis is not pointed at the Sun. As a result, both hemispheres receive equal amounts of sunlight, and the Sun is directly overhead at the equator. Thus, the lengths of day and night are equal for both the northern and southern hemispheres when Earth is at position 2, called the **autumnal equinox,** illustrated in ***Figure 28-16C,*** and position 4, called the **vernal equinox,** illustrated in ***Figure 28-16D.*** The term *equinox* means "equal nights."

MiniLab

The Sun's Position

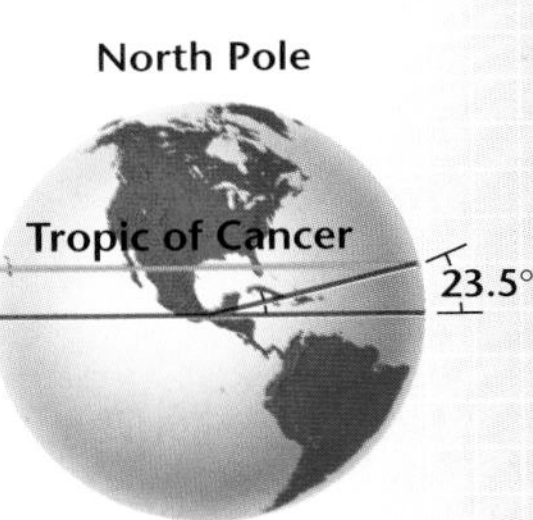

Model the overhead position of the Sun at various latitudes during the summer solstice.

Procedure

1. Draw a circle to represent Earth. Also draw the equator.
2. Use a protractor to find the location of the Tropic of Cancer. Draw a line from Earth's center to the Tropic of Cancer.
3. Using a map, locate that latitude at which you live. With the protractor, mark that latitude on your diagram. Draw a line from Earth's center to this location.
4. Measure the angle between the line to the Tropic of Cancer and the line to your location.
5. Choose two different latitudes, then repeat steps 3 and 4 for these latitudes.

Analyze and Conclude

1. How does the angle vary with latitude?
2. At what southern latitude would you not see the Sun above the horizon?
3. How would the angle change if you used the Tropic of Capricorn?

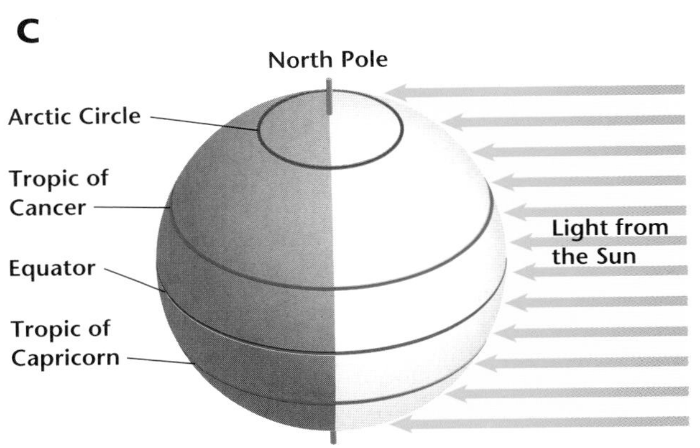

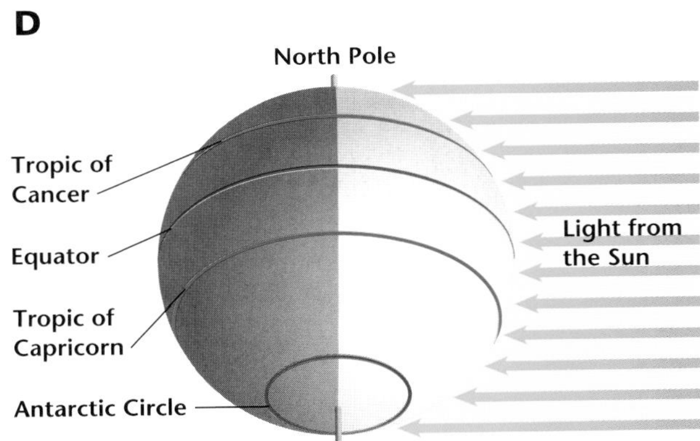

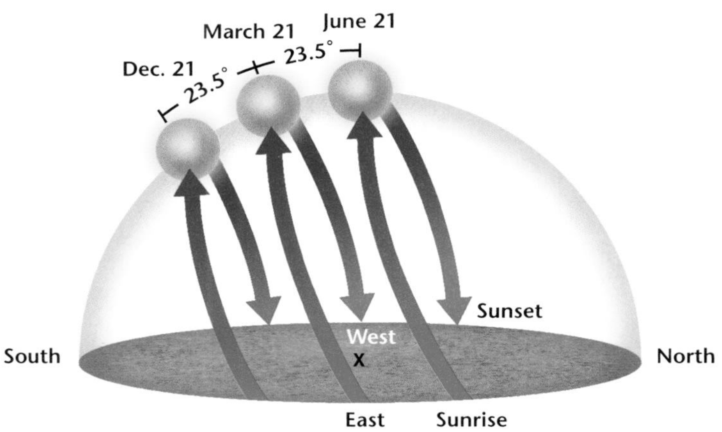

Figure 28-17 For a person standing at the *x* at 23.5° north latitude, the Sun would appear in these positions on the winter solstice, the vernal equinox, and the summer solstice. On the autumnal equinox, the Sun would be at the same altitude as on the vernal equinox.

At the Tropic of Cancer or Tropic of Capricorn, the Sun is 23.5° from the point directly overhead during the equinoxes. In the *Science & Math* feature at the end of this chapter, you will learn how Eratosthenes used the Sun's position and shadows to calculate the circumference and radius of Earth.

Figure 28-17 illustrates how the Sun would appear in the sky to a person at 23.5° north latitude during the solstices and the equinoxes. As you can see, the position of the Sun affects how directly sunlight strikes Earth. When the Sun is at a lower altitude, the sunlight that strikes Earth is spread out over a larger area.

Phases of the Moon

Just as the Sun appears to change its position in the sky, so, too, does the Moon. This is a result of the movement of the Moon around Earth and of our changing viewpoint on Earth relative to the Sun. The sequential changes in the appearance of the Moon are called lunar phases, shown in ***Figure 28-18.***

You have learned that the Moon does not emit visible light. Instead, we see the Moon's reflection of the Sun's light. When the Moon is between Earth and the Sun, however, we cannot see the Moon because the sunlit side is facing away from us. This dark Moon positioned between Earth and the Sun is called a new moon.

As the Moon moves along in its orbit, as illustrated in ***Figure 28-19,*** the amount of reflected sunlight that we can see increases. The increase in the portion of the sunlit side of the Moon that we see is called waxing. When we can see less than half of the sunlit portion of the Moon during this increase, it is called a waxing crescent. When we can see more than half of the sunlit portion of the Moon during this increase, it is called a waxing gibbous. Between these phases, the

Figure 28-18 These photos show the phases of the Moon, except the new moon phase, in which no portion of the Moon's illuminated surface is visible from Earth. The photo starts on the left with a waxing crescent, and ends on the right with a waning crescent. ***What are the phases in between called?***

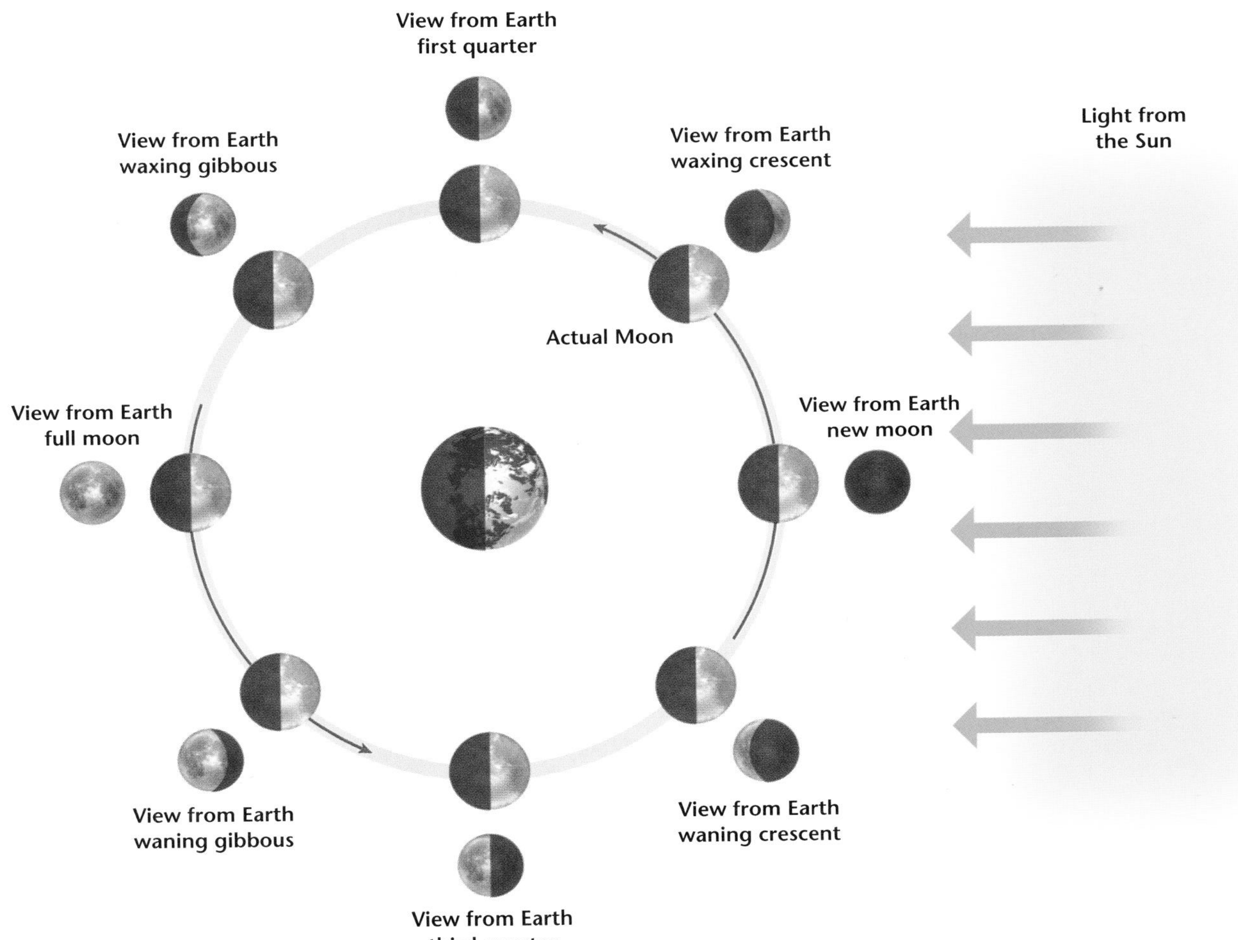

Figure 28-19 As the Moon orbits Earth, the portion of the illuminated side of the Moon that we see from Earth changes, thus creating phases. (Not to scale)

Moon reaches a point in its orbit when we see half of the sunlit side. This is called the first quarter. As the Moon continues farther in its orbit, it moves to a position where it is once again aligned with the Sun. This time, Earth is between the Moon and Sun, and we are able to see the entire sunlit side of the Moon. This is known as a full moon.

Once a full moon is reached, the portion of the sunlit side that we see begins to decrease as the Moon moves back toward the new-moon position. The decrease in the amount of the sunlit side of the Moon that we see is called waning. As in the waxing phases, there is a period during the waning phases when we can see more than half of the sunlit portion of the Moon, as well as a period when we can see less than half of the sunlit portion. These phases are called waning gibbous and waning crescent, respectively. In the middle of the waning phases, the Moon is in a position in its orbit where we can see half of the sunlit portion. This is called the third quarter.

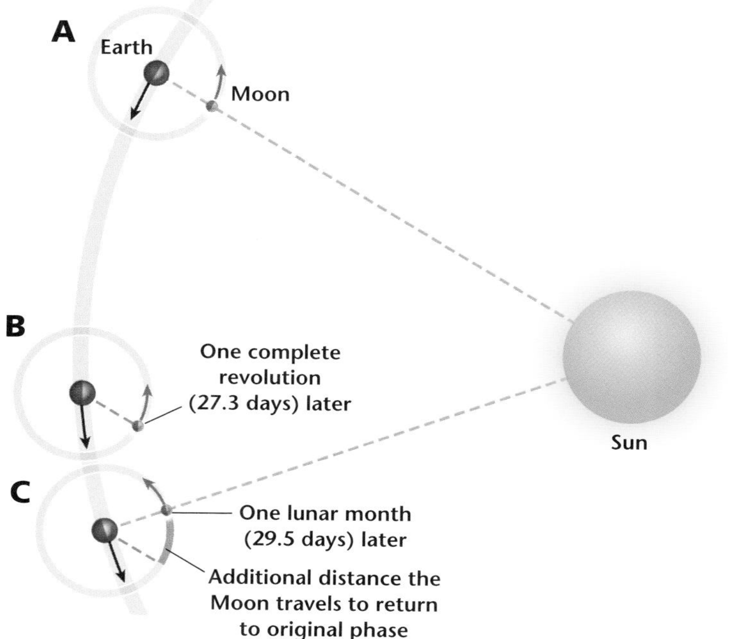

Figure 28-20 As the Moon moves from position **A,** where the Moon is in new moon phase as seen from Earth, to position **B,** it completes one revolution and is in the waning crescent phase as seen from Earth. At position **C,** the Moon has traveled for another 2.2 days and is back to the new moon phase, completing a lunar month. (Not to scale)

Synchronous Rotation You might have noticed that the illuminated surface of the Moon always looks the same. As the Moon orbits Earth, the same side faces Earth at all times. This is because the Moon is rotating with a period equal to its orbital period, so it spins exactly once each time it goes around Earth. This is not a coincidence. Scientists theorize that Earth's gravity slowed the Moon's original spin until the Moon reached **synchronous rotation,** the state at which its orbital and rotational periods are equal.

Motions of the Moon

The length of time it takes for the Moon to go through a complete cycle of phases, for example, from one full moon to the next, is called a lunar month. The length of a lunar month is about 29.5 days, which is longer than the 27.3 days it takes for one revolution, or orbit, around Earth, as illustrated in ***Figure 28-20.*** The Moon also rises and sets 50 minutes later each day because the Moon has moved 13° in its orbit over a 24-hour period, and Earth has to turn an additional 13° for the Moon to rise.

Tides One of the Moon's effects on Earth is the formation of tides. The Moon's gravity pulls on Earth along an imaginary line connecting Earth and the Moon, and this creates bulges of ocean water on both the near and far sides of Earth. Earth's rotation also contributes to the formation of tides, as you learned in Chapter 15. As Earth rotates, these bulges remain aligned with the Moon, so that a person at a shoreline on Earth's surface would observe that the ocean level rises and falls every 12 hours.

The Sun's gravitational effect on the formation of tides is about half that of the Moon's, because the Sun is farther away. However, when the Sun and Moon are aligned along the same direction, the effects of the Sun and Moon combine, and tides are higher than normal. These tides, called spring tides, are especially high when the Moon is nearest Earth and Earth is nearest the Sun in their slightly noncircular orbits. When the Moon is at a right angle to the Sun-Earth line, the result is lower-than-normal tides, called neap tides.

Figure 28-21 This multiple-exposure photograph, taken July 11, 1991, in California, shows a total solar eclipse in the middle of the sequence.

SOLAR ECLIPSES

A **solar eclipse** occurs when the Moon passes directly between the Sun and Earth and blocks our view of the Sun. Although the Sun is much larger than the Moon, it is much farther away, which causes the Sun and Moon to appear to be the same size when viewed from Earth. When the Moon perfectly blocks the Sun's disk, we see only the dim, outer gaseous layers of the Sun. This spectacular sight, shown in ***Figure 28-21,*** is called a total solar eclipse. A partial solar eclipse is seen when the Moon blocks only a portion of the Sun's disk.

The difference between a partial and a total solar eclipse can be explained by the fact that the Moon casts a shadow on Earth. This shadow consists of two regions, as illustrated in ***Figure 28-22.*** The inner portion, which does not receive direct sunlight, is called the umbra. People who witness an eclipse from the umbra see a total solar eclipse. People in the outer portion of this shadow, where some of the Sun's light reaches, are in the penumbra. They see a partial solar eclipse where part of the Sun's disk is still visible. Typically, the umbral shadow is never wider than 270 km, so a total solar eclipse is visible from a very small portion of Earth, whereas a partial solar eclipse is visible from a much larger portion.

Topic: Next Solar Eclipse
To find out more about solar eclipses, visit the Earth Science Web Site at earthgeu.com

Activity: Research future solar eclipses. When will the next solar eclipse be visible in your area?

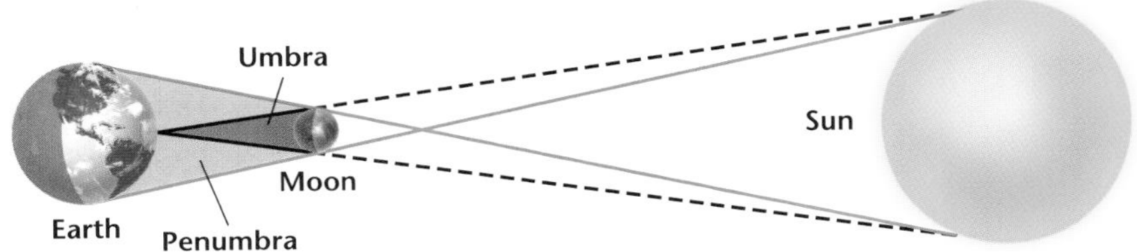

Figure 28-22 During a solar eclipse, the Moon passes between the Sun and Earth. People within the umbral shadow witness a total solar eclipse, while people within the penumbral shadow witness a partial solar eclipse. (Not to scale)

Figure 28-23 This annular eclipse, partly obscured by clouds, was photographed in San Diego, California, in January, 1992.

The Effects of Orbits You might wonder why a solar eclipse does not occur every month, as the Moon passes between the Sun and Earth during the new moon phase. This does not happen because the Moon's orbit is tilted 5° relative to the ecliptic. Usually, the Moon passes north or south of the Sun as seen from Earth, so no solar eclipse takes place. Only when the Moon crosses the ecliptic is it possible for the proper alignment for a solar eclipse to occur, but even that is not enough to guarantee a solar eclipse. The plane of the Moon's orbit also rotates slowly around Earth, and a solar eclipse occurs only when the intersection of the Moon and the ecliptic is in a line with the Sun and Earth. Hence, the proper alignment for solar eclipses does not occur every month with each new moon.

Not only does the Moon move above and below the plane of Earth and the Sun, but also, the Moon's distance from Earth increases and decreases as the Moon moves in its elliptical orbit around Earth. The closet point in the Moon's orbit to Earth is called **perigee,** and the farthest point is called **apogee.** When the Moon is near apogee, it appears smaller as seen from Earth, and thus it does not completely block the disk of the Sun during an eclipse. This is called an annular eclipse because from Earth, a ring of the Sun called an annulus is visible around the dark Moon, as shown in ***Figure 28-23.*** You'll experiment with the different types of solar eclipses in the *Problem-Solving Lab* on this page.

Problem-Solving Lab

Interpreting Scientific Illustrations

Predict how a solar eclipse will look Depending on an observer's location, a solar eclipse can look different.

Analysis

1. Make a drawing of how the solar eclipse would appear to an observer at each labeled location in the illustration.

Thinking Critically

2. Design a data table showing your drawings of how the eclipse would appear at each location.
3. What type of eclipse does each of your drawings represent? Include this information in your data table.

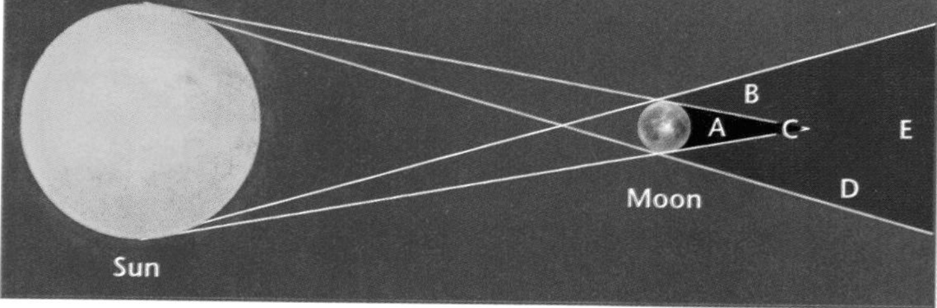

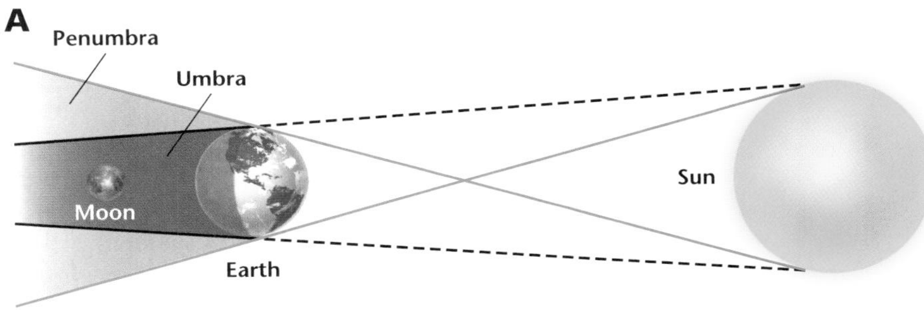

Figure 28-24 When the Moon is completely within the umbra of Earth's shadow **(A),** we observe a total lunar eclipse **(B).** (Illustration not to scale)

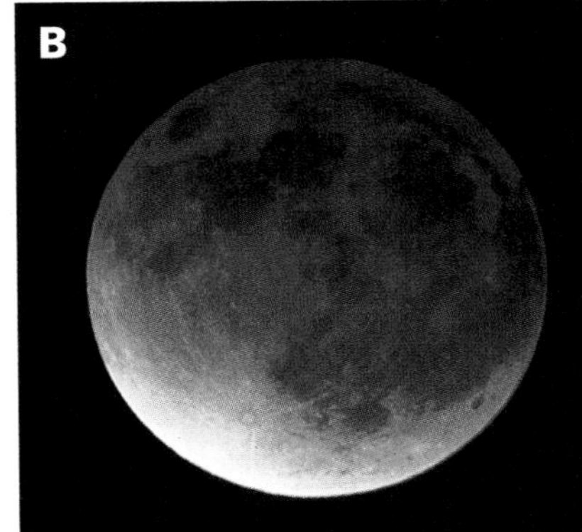

Lunar Eclipses

A **lunar eclipse** occurs when the Moon passes through Earth's shadow. As illustrated in ***Figure 28-24A,*** this can happen only at the time of a full moon, when the Moon is in the opposite direction from the Sun. The shadow of Earth has umbral and penumbral portions, just as the Moon's shadow does. A total lunar eclipse occurs when the entire Moon is within Earth's umbra, and totality lasts for approximately two hours. During a total lunar eclipse, the Moon is faintly visible, as shown in ***Figure 28-24B,*** because sunlight that has passed near Earth has been refracted by Earth's atmosphere. This light can give the eclipsed Moon a reddish color as Earth's atmosphere bends the red light into the umbra, much like a lens. Like solar eclipses, lunar eclipses do not occur every full moon because the Moon in its orbit usually passes above or below the Sun as seen from Earth.

Solar and lunar eclipses occur in almost equal numbers, with slightly more lunar eclipses. The maximum number of eclipses, solar and lunar combined, that can be seen in a year is seven. The last time this occurred was in 1982, and it won't happen again until 2038.

Section Assessment

1. What are the causes of Earth's seasons?
2. What would our seasons be like if Earth's axis were not tilted?
3. Explain why the Moon goes through phases as seen from Earth.
4. Describe solar and lunar eclipses.
5. **Thinking Critically** If Earth's axis were tilted 45°, at what latitudes would the Sun be directly overhead on the solstices and the equinoxes?

Skill Review

6. **Formulating Models** If you were to observe Earth from the Moon, you would see that it goes through phases. Draw a diagram illustrating these phases and the positions of the Sun, Earth, and the Moon. For more help, refer to the *Skill Handbook.*

Mapping GeoLab

Relative Ages of Lunar Features

It is possible to use the principle of cross-cutting relationships, discussed in Chapter 21, to determine the relative ages of surface features on the Moon. By observing which features cross-cut others, you can infer which is older.

Preparation

Problem

How can you use images of the Moon to interpret relative ages of lunar features?

Materials

metric ruler
pencil

Procedure

1. Observe photos I and II. Use the letters to identify the oldest feature in each photo using the principle of cross-cutting relationships. List the other features in order of their relative ages.
2. Observe photo III. List the mare, rille, and craters in order of their relative ages.
3. Observe photo IV. Use the principle of cross-cutting relationships, along with your knowledge of lunar history, to identify the features and list them in order of their relative ages.

Analyze

1. What problems did you encounter?
2. Based on information from all the photos, what features are usually the oldest? The youngest?
3. Could scientists use the process you did to determine the exact age difference between two overlapping craters? Why or why not?
4. If the small crater in photo II, labeled A, is 44 km across, what is the scale for that photo? What is the size of the large crater, labeled D?

Conclude & Apply

1. Which would be older, a crater that had rays crossing it, or the crater that caused the rays? Explain.
2. Is there some type of relative-age dating that scientists can use to analyze craters on Earth? Explain.
3. What do you think caused the chain of craters in photo I? If the crater labeled A is approximately 17 km across, how long is the chain of craters?

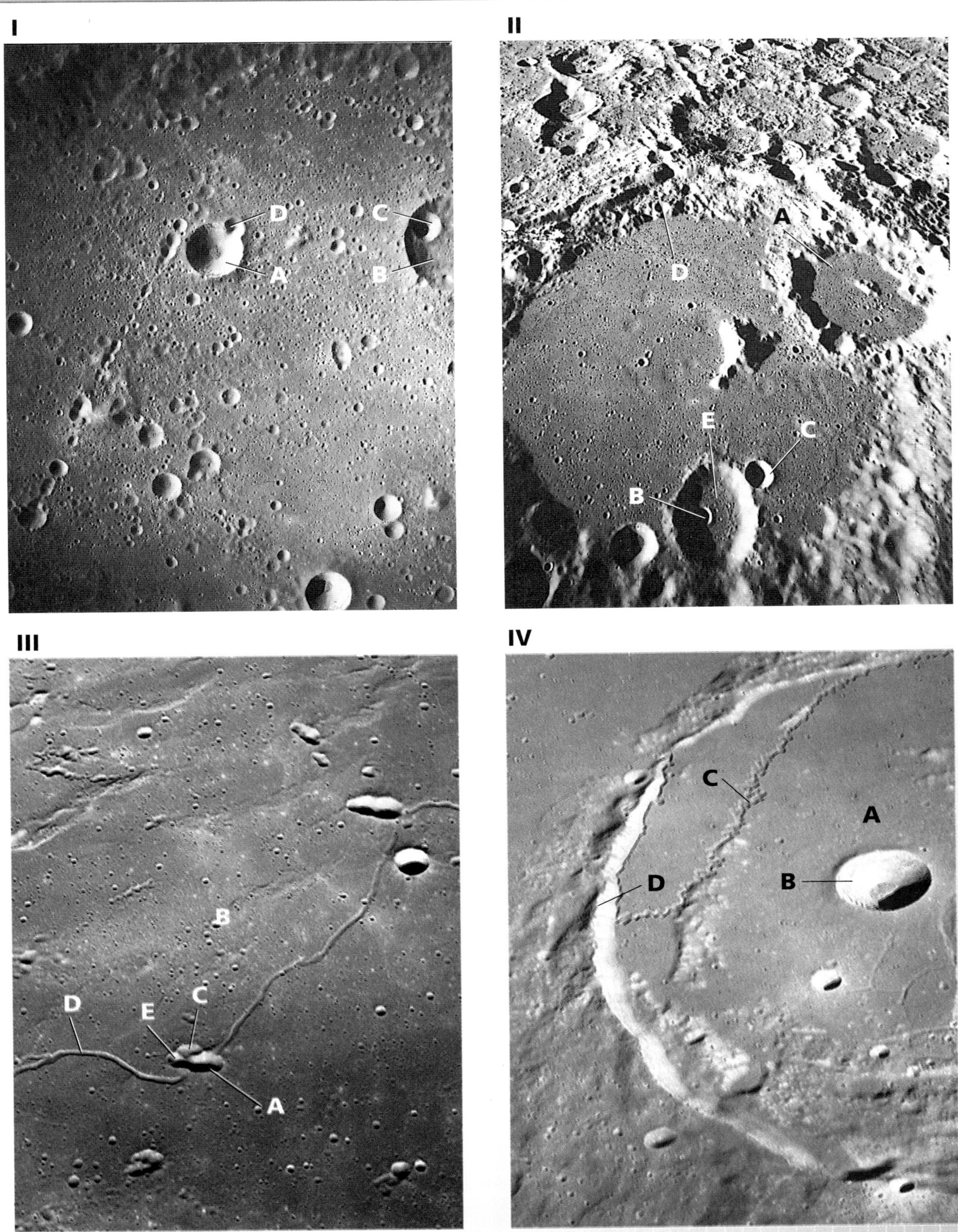
I
D
C
A
B
II
A
D
E
C
B
III
B
D
E
C
A
IV
C
A
D
B

Science & Math

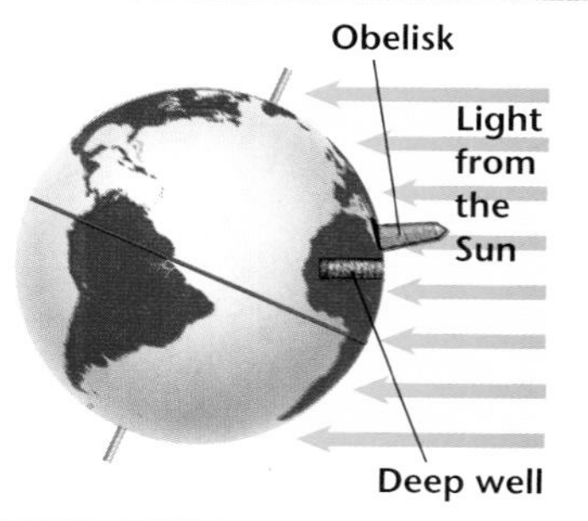

The Size of Earth

We know that Earth is round, but how do we know how large it really is? Have you ever wondered how we measure such a large object? Long ago, before the development of high-tech computers and space shuttles, one man used his knowledge of geometry to determine the circumference of Earth.

Using Geometry

An ancient Greek mathematician, Eratosthenes (276–194 B.C.), was the first to develop a method for determining the circumference of Earth. It was known during his time that at noon on the summer solstice, when the Sun was directly overhead in Syene, Egypt, sunlight reached the bottom of a local well. However, to the north, in Alexandria, the Sun cast a shadow off an obelisk on the same day and at the same time.

Eratosthenes knew that the distance between the two cities was approximately 4900 stadia, an ancient form of measurement equivalent to 770 km by today's estimate. He measured the height of the obelisk and the length of the shadow. Then, by using the relationship

$$\arctan\left(\frac{\text{length of shadow}}{\text{height of obelisk}}\right)$$

he calculated that the Sun was 7° lower than directly overhead. Knowing that Earth was round, and that round objects have a total of 360°, Eratosthenes determined that the difference in latitude of the two cities was 7°. Because sunlight could be seen at the bottom of the well in Syene on the summer solstice, Eratosthenes determined that Syene was at latitude 23.5°N and that Alexandria was at latitude 30.5°N.

Procedure

1. Using a compass and a sheet of paper, draw a diagram of Earth. Mark the equator.
2. Using a protractor, locate Syene at latitude 23.5°N and Alexandria at latitude 30.5°N.
3. Knowing that the difference in latitudes of the two cities is 7° and that a circle has 360°, you can determine what portion of a circle is 7°. This ratio of 7° to 360° can be represented by
$$\frac{d}{C} = \frac{7°}{360°}$$
where d is the distance between Alexandria and Syene, and C is Earth's circumference. Given that $d = 770$ km (4900 stadia), solve the equation for C. Then find Earth's radius using $C = 2\pi r$.
4. Use your answers in step 3 to determine Earth's diameter.

Challenge

1. Earth's radius is actually 6378.1 km. How do your measurements compare to this?
2. What is the percent deviation of your measurement?

$$\text{Percent deviation} = \frac{\text{difference from accepted value}}{\text{accepted value}} \times 100$$

To learn more about Eratosthenes' contributions to science and math, visit the Earth Science Web Site at earthgeu.com

CHAPTER 28

Study Guide

Summary

SECTION 28.1

Tools of Astronomy

Main Ideas

- Visible light, radio waves, infared and ultraviolet radiation, X rays, and gamma rays are types of electromagnetic radiation.
- A telescope collects light over a large area, makes time exposures, and can use other instruments to analyze light.
- Visible-light telescopes can be made using lenses, as in refracting telescopes, or mirrors, as in reflecting telescopes.
- Space is explored by telescopes, satellites, probes, and humans.

Vocabulary

interferometry (p. 750)
reflecting telescope (p. 749)
refracting telescope (p. 749)
spinoff (p. 752)

SECTION 28.2

The Moon

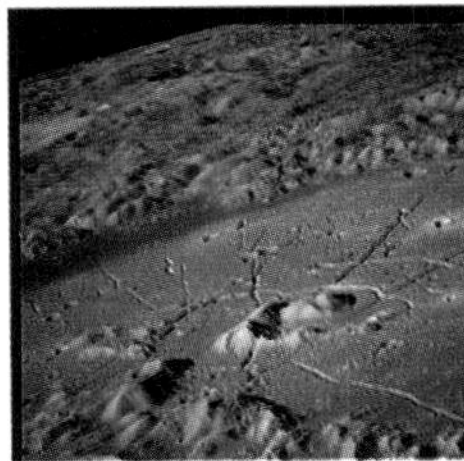

Main Ideas

- The first step toward exploration of the Moon was the launch of the Soviet satellite *Sputnik 1.* The American spacecraft *Apollo 11* was the first crewed exploration of the Moon.
- The Moon's surface has many features that are not present on Earth because the Moon lacks an atmosphere and therefore its surface does not undergo erosion.
- Scientists have three theories on how the Moon formed—simultaneous formation with Earth, a passing object captured by Earth's gravity, or as the result of an object colliding with Earth. The collision theory is the most widely accepted.

Vocabulary

albedo (p. 754)
ejecta (p. 754)
highland (p. 754)
impact crater (p. 754)
mare (p. 754)
ray (p. 754)
regolith (p. 755)
rille (p. 754)

SECTION 28.3

The Sun-Earth-Moon System

Main Ideas

- The entire sky appears to rotate daily because we observe it from a rotating Earth. Our timekeeping system is based on the solar day, the length of day as observed from Earth.
- Our view of the Sun's position changes throughout the year as Earth moves in its orbit about the Sun. Seasons occur on Earth because Earth's axis is tilted.
- The Moon goes through a cycle of phases each lunar month that correspond to our changing view from Earth of the sunlit side of the Moon.
- Tides are caused by the gravitational attraction of the Moon, and to a lesser extent, the gravitational attraction of the Sun.
- A solar eclipse occurs when the Moon lies directly between Earth and the Sun. A lunar eclipse occurs when the Moon passes through Earth's shadow.

Vocabulary

apogee (p. 766)
autumnal equinox (p. 761)
ecliptic (p. 759)
lunar eclipse (p. 767)
perigee (p. 766)
solar eclipse (p. 765)
summer solstice (p. 760)
synchronous rotation (p. 764)
vernal equinox (p. 761)
winter solstice (p. 761)

CHAPTER 28

Assessment

Understanding Main Ideas

1. On what does the light-collecting power of a telescope depend?
   a. the type of telescope
   b. the area of the opening through which light enters
   c. the location of the telescope
   d. the distance from the telescope to the object being observed

2. What is the same for all types of electromagnetic radiation?
   a. frequency
   b. wavelength
   c. color
   d. speed

3. What type of radiation does not have to be observed above Earth's atmosphere?
   a. visible light
   b. X rays
   c. gamma rays
   d. ultraviolet radiation

4. During which of the following is the Sun directly overhead at 23.5° north latitude?
   a. summer solstice
   b. vernal equinox
   c. winter solstice
   d. autumnal equinox

5. Which of the following provides evidence that Earth is rotating?
   a. The Sun rises and sets.
   b. The plane of a Foucault pendulum appears to shift its orientation.
   c. The Moon goes through phases.
   d. The same side of the Moon always faces Earth.

6. Which of the following is in the correct order?
   a. waning crescent, third quarter, waning gibbous, new moon
   b. waxing gibbous, full moon, waning gibbous, third quarter
   c. new moon, waning gibbous, first quarter, waning crescent
   d. waxing crescent, new moon, waning crescent, first quarter

7. List the various forms of electromagnetic radiation according to wavelength, from shortest to longest.

8. Why must some telescopes be launched into space?

Use the diagrams below to answer question 9.

9. List the types of shadows as well as the types of eclipses that will be seen by an observer on the unlit side of Earth.

10. What is electromagnetic radiation?

11. How was the lunar regolith formed?

12. Describe how a lunar month is defined. How long is it?

13. Of all types of electromagnetic radiation, which can the human eye detect?

Test-Taking Tip

CROSSING OUT Cross out choices you've eliminated. If you can't write in the test booklet, list the answer choice letters on the scratch paper and cross them out there. You'll save time and stop yourself from choosing an answer you've mentally eliminated.

earthgeu.com/chapter_test

CHAPTER 28
Assessment

14. How did the mountain ranges around the maria on the Moon form?
15. Why are the temperature fluctuations on the surface of the Moon so extreme compared to those on Earth?

Applying Main Ideas

16. What are the Moon's positions relative to the Sun and Earth when we observe a full moon and a new moon?
17. Why does the Sun's altitude in the sky change throughout the year?
18. If the Moon rotated twice on its axis for every one time it orbited Earth, would it be in synchronous rotation? Explain.
19. Suppose the Moon's orbital plane were exactly aligned with Earth's orbital plane. How often would eclipses occur?
20. Why is it best to get away from city lights to view the nighttime sky?

Thinking Critically

21. How would Earth's surface look if Earth did not have an atmosphere?
22. Why did one-half of the Moon's surface remain hidden from human sight until the era of space probes, which started in 1959?
23. When observers on Earth can see a total lunar eclipse, what kind of eclipse would be seen by an observer on the Moon?
24. In some maria, there are craters. Which are younger, the maria or the craters?
25. How would the topography of the Moon be different if the Moon had an atmosphere?

Standardized Test Practice

1. What is debris from an impact that falls back to the surface of the Moon called?
   a. rilles **c.** ejecta
   b. maria **d.** albedo

2. In December, the South Pole is tilted farther toward the Sun than at any other time of the year, and the North Pole is tilted its farthest away from the Sun. What is the northern hemisphere experiencing at that time?
   a. the winter solstice
   b. the summer solstice
   c. the vernal equinox
   d. the autumnal equinox

INTERPRETING SCIENTIFIC ILLUSTRATIONS Use the diagram below to answer questions 3 and 4.

3. What results on Earth when the Sun and the Moon are aligned along the same direction, as in the diagram?
   a. spring tides **c.** the autumnal equinox
   b. neap tides **d.** the summer solstice

4. If the Moon in this diagram were passing directly between the Sun and Earth, thereby blocking our view of the Sun, what would we be experiencing on Earth?
   a. a lunar eclipse **c.** umbra
   b. a solar eclipse **d.** penumbra

Chapter 29

Our Solar System

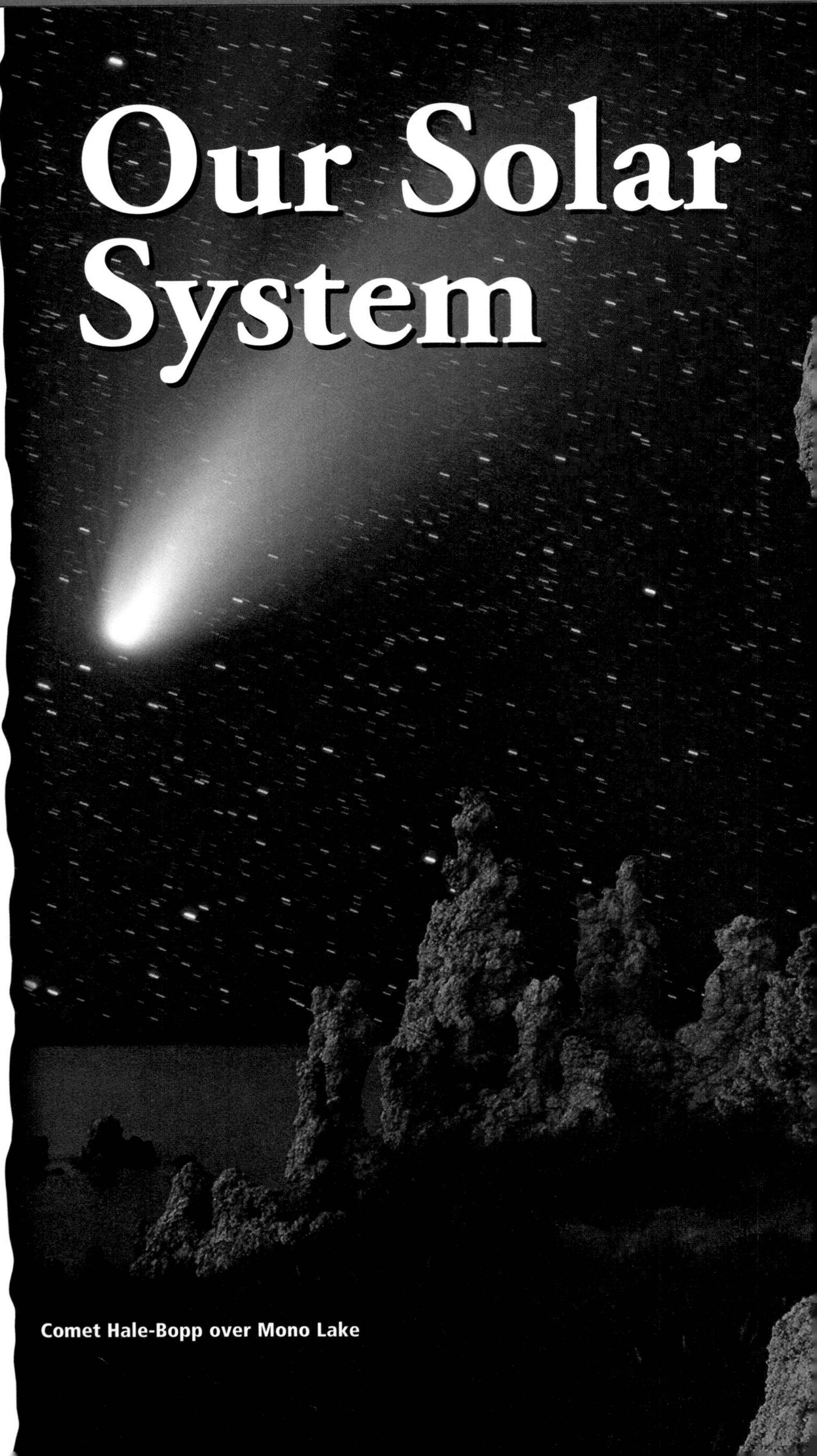

Comet Hale-Bopp over Mono Lake

What You'll Learn

- How gravity and orbits are related.
- The characteristics of planets and interplanetary bodies.
- What theory is used to describe the formation of the solar system.

Why It's Important

The laws of motion and universal gravitation explain how gravity governs the motions of the planets and other planetary bodies. Scientists base the model of our solar system on observations of the organization and nature of the planets and interplanetary bodies.

Earth Science Online

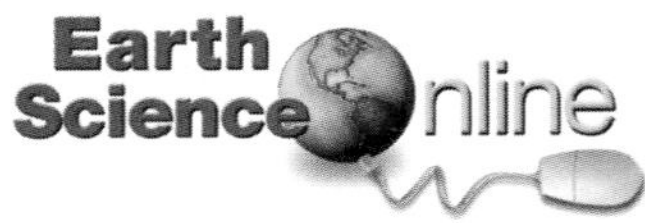

To find out more about our solar system, visit the Earth Science Web Site at earthgeu.com

Discovery Lab

Exploring Our Solar System

All nine planets in our solar system have been explored by uncrewed space probes, or soon will be. You can learn about these missions and their discoveries by finding information on the Web. The agency that sponsors a mission, the scientists involved, or both, usually create extensive Web sites full of information about the design, operation, and scientific goals of the mission.

1. Find at least one Web site for missions to four different planets. Or go to earthgeu.com and follow the links.
2. Make a list of some of the key aspects of each mission.

Summarize Make an outline for each mission. Include the type of mission (flyby, lander, or orbiter), the scientific goals, the launch date and the date of arrival at the planet, and a summary of what was learned, or what scientists hope will be learned.

SECTION 29.1 *Overview of Our Solar System*

Earth Sciences 1.d **I&E** 1.k, 1.n

OBJECTIVES

- **Describe** *early models of our solar system.*
- **Examine** *the modern heliocentric model of our solar system.*
- **Relate** *gravity to the motions of celestial bodies.*

VOCABULARY

retrograde motion
astronomical unit
perihelion
aphelion
eccentricity

Earth is one of nine planets revolving around, or orbiting, the Sun. All the planets, as well as most of their moons, also called satellites, orbit the Sun in the same direction, and all their orbits, except Pluto's, lie near the same plane. The planets of our solar system have various sizes, surface conditions, and internal structures. Scientists have gathered much information about our solar system through the use of technologies developed in the twentieth century, but human beings have been watching the sky for thousands of years, and early ideas about the solar system were developed solely on the basis of Earth-based observations of the sky.

EARLY IDEAS

When viewed from Earth, the planets slowly change position each night relative to the position of the stars. Therefore, ancient astronomers could recognize the difference between stars and planets. These astronomers assumed that the Sun, planets, and stars orbited a stationary Earth in what is now known as a geocentric model, meaning "Earth centered."

Figure 29-1 Mars appears to move from east to west (positions 3 and 4) for a short time **(A)** during its retrograde motion. The heliocentric model **(B)** explains retrograde motion. Follow the lines from Earth's orbital positions to Mars's orbital positions, and then to Mars's position in the sky. Retrograde motion is similar to passing a slower car on the freeway. It appears that the slower car is moving backwards relative to the background. (not to scale)

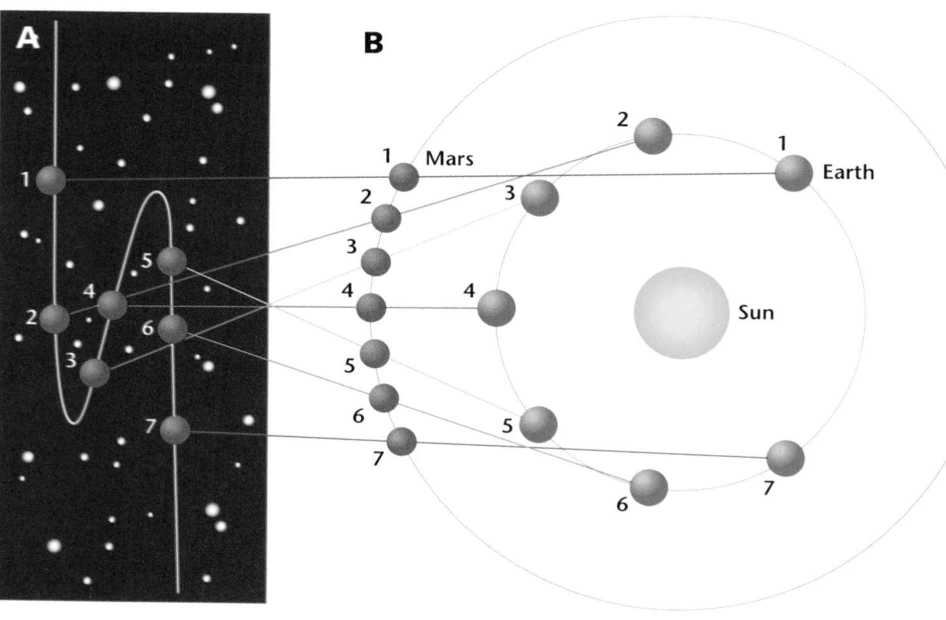

Earth Sciences

1.d Students know the evidence indicating that the planets are much closer to Earth than the stars are.

Also covers: 1.k, 1.n

However, some aspects of planetary motion were difficult to explain with a geocentric model. For example, the normal direction of motion for all planets, as observed from Earth, is toward the east. Occasionally, however, a planet will move in the opposite direction across the sky in what is called **retrograde motion.** The retrograde motion of Mars is shown in ***Figure 29-1A.*** The search for a simple scientific explanation for retrograde motion motivated early astronomers to keep searching for a better model of our solar system.

In 1543, Polish scientist Nicolaus Copernicus suggested that the Sun was the center of the solar system. In this sun-centered, or heliocentric, model, Earth and the other planets orbit the Sun. This model provided a simple explanation of retrograde motion. In a Sun-centered model, the inner planets move faster in their orbits than the outer planets do. As Earth bypasses a slower-moving outer planet, as illustrated in ***Figure 29-1B,*** it appears that the outer planet temporarily moves backward in the sky.

Figure 29-2 The major axis passes through the foci of an ellipse, while the semimajor axis is half of the major axis.

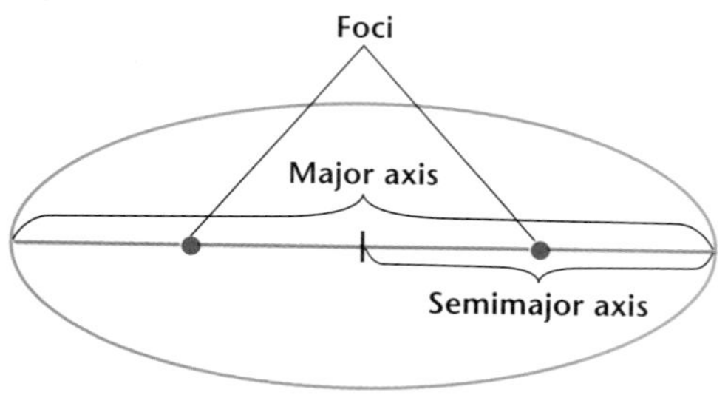

Kepler's First Law The ideas of Copernicus were not initially accepted by the scientific community, but within a century, other astronomers were finding evidence that supported the heliocentric model. For example, from 1576–1601, a Danish astronomer, Tycho Brahe, made many accurate observations of planetary positions. Using Brahe's data, Johannes Kepler demonstrated that each planet orbits the Sun in a shape called an ellipse, rather than in a circle. This is known as Kepler's first law.

An ellipse is an oval shape that is centered on two points instead of a single point, as in a circle. The two points are called the foci (*sing.* focus). The major axis is the line that runs through both foci; it is the maximum diameter of the ellipse, as illustrated in ***Figure 29-2.*** You will experiment with the foci and shapes of ellipses in the *MiniLab* on this page.

Each planet's elliptical orbit is a different shape and size, and the Sun is always at one focus. For each Sun-planet pair, half of the length of the major axis is called the semimajor axis. It is the average distance between the Sun and the planet. For the Sun and Earth, it is 1.496×10^8 km, or 1 **astronomical unit** (AU). The average distances between the Sun and each planet are measured in astronomical units, and therefore these distances are relative to Earth's average distance from the Sun.

Eccentricity A planet in an elliptical orbit is not at a constant distance from the Sun. When a planet is closest to the Sun in its orbit, it is at **perihelion,** and when it is farthest away, it is at **aphelion,** as shown in ***Figure 29-3.*** The shape of a planet's elliptical orbit is defined by **eccentricity,** which is the ratio of the distance between the foci to the length of the major axis. Eccentricity values range from 0 to 1. An eccentricity of 0 is a perfect circle, and an eccentricity of nearly 1 is a very elongated oval. An eccentricity equal to 1 is a parabola. Most of the planets have orbits that are not very eccentric, as shown in *Appendix J,* and are thus close to being circles. The length of time it takes for a planet or other body to travel a complete elliptical orbit around the Sun is called the orbital period.

MiniLab

Eccentricity

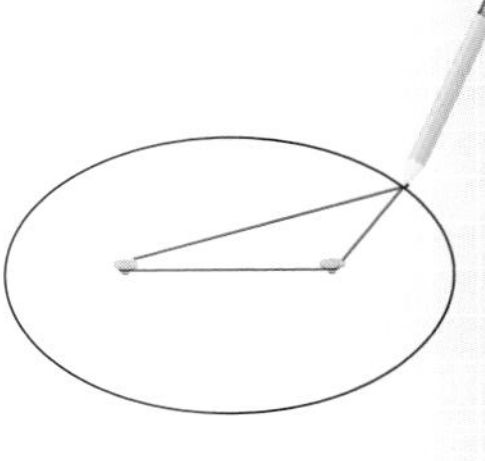

Measure the eccentricity of different ellipses. Eccentricity is the ratio of the distance between the foci to the length of the major axis.

Procedure

1. Tie a piece of string into a loop that fits on a piece of cardboard when it is laid out in a circle.
2. Place a sheet of paper on the cardboard.
3. Stick two pins through the paper close to the center but separated from each other by a few centimeters. Use caution when using sharp objects.
4. Loop the string over the pins and use the pencil to trace around them. Keep the string taut.
5. Measure the major axis and the distance between the pins. Calculate the eccentricity.
6. Repeat steps 3–5 for different separations of the pins.

Analyze and Conclude

1. What do the two pins represent?
2. How does the eccentricity change as the distance between the pins changes?
3. What kind of figure would you form if the two pins were at the same location? What would its eccentricity be?

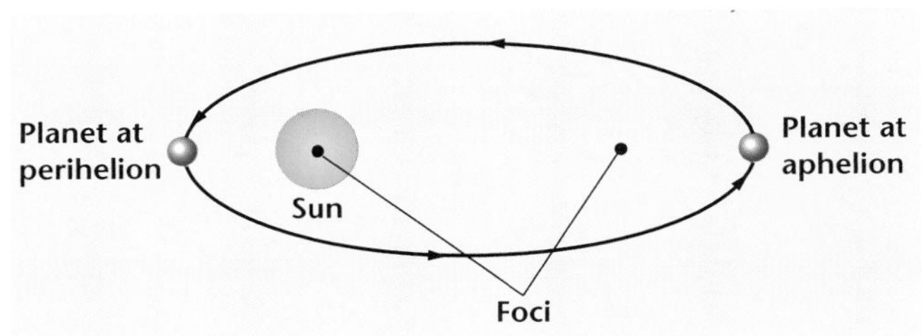

Figure 29-3 A planet is at perihelion when it is closest to the Sun in its orbit and at aphelion when it is farthest. (not to scale)

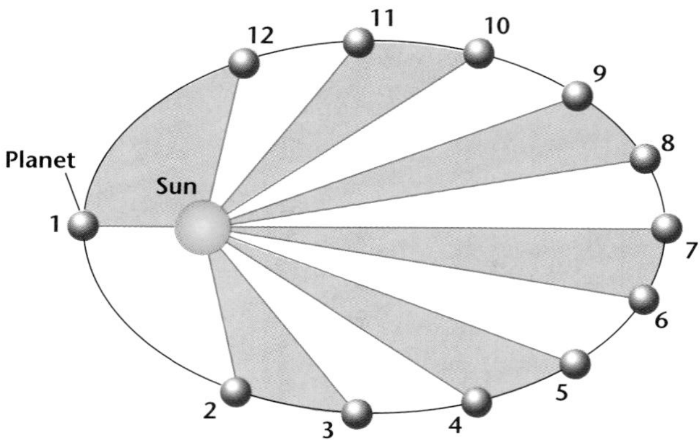

Figure 29-4 Because a planet moves fastest when close to the Sun and slowest when far from the Sun, equal areas are swept out in equal amounts of time, which is Kepler's second law. (not to scale)

Kepler's Second and Third Laws In addition to discovering the true shapes of planetary orbits, Kepler found that an imaginary line between the Sun and a planet sweeps out equal amounts of area in equal amounts of time, as illustrated in ***Figure 29-4.*** This is known as Kepler's second law. Kepler also derived a mathematical relationship between the size of a planet's ellipse and its orbital period. He found that the square of the orbital period (P) equals the cube of the semimajor axis of the orbital ellipse (a). This relationship, called Kepler's third law, is $P^2 = a^3$, where P is a unit of time measured in Earth years, and a is a unit of length measured in astronomical units. You will apply Kepler's third law to each planet in our solar system in the *Problem-Solving Lab* later in this chapter.

While Kepler was developing his ideas, Italian scientist Galileo Galilei became the first person to use a telescope to observe the sky. Galileo made many important discoveries that supported Copernicus's idea that the planets, including Earth, orbit the Sun. The most famous of Galileo's discoveries was that four moons orbit the planet Jupiter. This observation proved that not all celestial bodies orbit Earth, and therefore, Earth is not necessarily the center of the solar system. The underlying explanation for the heliocentric model still remained unknown, however, until 1684, when English scientist Isaac Newton published a mathematical and physical explanation of the motions of celestial bodies. Newton's concepts included the law of universal gravitation, which provided an explanation of how the Sun governs the motions of the planets.

Using Numbers
Newton's law of universal gravitation, found on the next page, can be applied to any two objects that have mass. If one student has a mass of 50.0 kg and is 12.0 m away from another student that has a mass of 65.0 kg, what is the force of gravity between them?

Gravity and Orbits

Newton developed an understanding of gravity by observing the Moon's motion, the orbits of the planets, and the acceleration of falling objects on Earth. He realized that any two bodies attract each other with a force that depends on their masses and the distance between the two bodies. The force grows stronger in proportion to the product of the two masses, but diminishes as the square of the distance between them. For example, if the distance between Earth and the Moon were twice as great, the gravitational force between them would be only one-fourth as strong. At their normal distance apart, if the mass of the Moon were doubled and the mass of Earth were tripled, the force would be greater by a factor of 6.

Gravity Newton's statement of the relationship among the masses of two bodies and the force and distance between them is known as the law of universal gravitation. This law can be stated as follows: Every pair of bodies in the universe attract each other with a force that is proportional to the product of their masses and inversely proportional to the square of the distance between them, or

$$F = G\frac{m_1 m_2}{r^2}.$$

F is the force measured in newtons, and G is the universal gravitation constant, or 6.6726×10^{-11} meters cubed per kilogram per second squared. m_1 and m_2 are the masses of the bodies measured in kilograms, and r is the distance between the two bodies measured in meters.

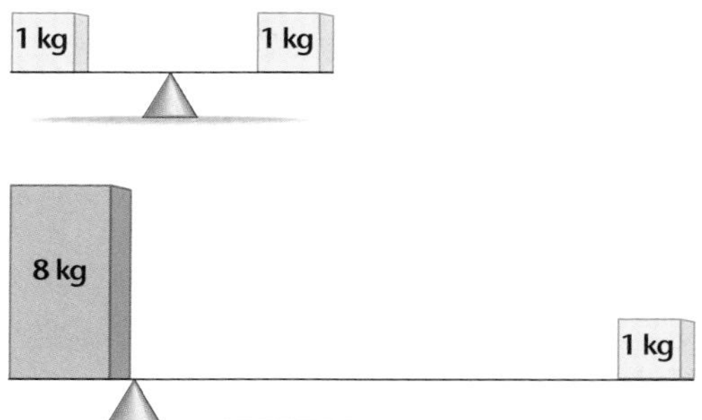

Figure 29-5 The center of mass is halfway between two equal mass objects, but closer to the heavier object when the objects are of unequal mass, similar to the pivot of a see-saw.

Center of Mass Newton also determined that each planet orbits a point between it and the Sun called the center of mass. The center of mass is the balance point between two orbiting bodies, similar to the pivot point on a see-saw. If one person on a see-saw is much heavier than the other, the balance point is closer to the heavier person, as shown in ***Figure 29-5.*** In space, the same is true. If one of two bodies orbiting each other is more massive than the other, the center of mass is closer to the more massive body. If the two bodies have similar masses, the center of mass is near the middle position between them. For any planet and the Sun, the center of mass is just above the surface of the Sun, or within the Sun, because the Sun is much more massive than any planet, as you will learn in the following sections.

Section Assessment

1. Why is retrograde motion an apparent motion?
2. What were the contributions of Copernicus, Kepler, and Galileo in developing the Sun-centered model of the solar system?
3. Describe how the force between two bodies depends on their masses and the distance between them.
4. **Thinking Critically** Your weight is the gravitational force between you and Earth, and the separation between you and Earth is equal to Earth's radius. How would your weight be different if Earth's radius were larger or smaller than it is but Earth's mass remained the same?

Skill Review

5. **Comparing and Contrasting** Compare and contrast the geocentric and the heliocentric models of the solar system. For more help, refer to the *Skill Handbook.*

SECTION 29.2

The Terrestrial Planets

Earth Sciences 1.f, 4.d

OBJECTIVES

- **Describe** *the properties of the terrestrial planets.*
- **Compare** *Earth with the other terrestrial planets.*

VOCABULARY

terrestrial planet
gas giant planet
precession

The nine planets of our solar system can be grouped into two main categories according to their basic properties. The inner four planets, called **terrestrial planets,** are close to the size of Earth and have solid, rocky surfaces. The terrestrial planets are Mercury, Venus, Earth, and Mars, in order from closest to farthest from the Sun. The next four planets from the Sun, called **gas giant planets,** are much larger, more gaseous, and lack solid surfaces. The gas giants include Jupiter, Saturn, Uranus, and Neptune. Pluto, the ninth planet from the Sun, has a solid surface, but it does not fit into either category. On the next several pages, we'll discuss each planet.

MERCURY

Mercury is the closest planet to the Sun and has no moons. Mercury is about one-third the size of Earth and has a smaller mass and radius, as shown in *Appendix J.* Radio observations in the 1960s revealed that Mercury has a slow spin of 1407.6 hours. After Mercury completes one orbit around the Sun, it has rotated one and a half times, and the opposite side of the planet faces the Sun. In two orbits, Mercury spins three times and the side originally facing the Sun faces the Sun again. Thus, in two of Mercury's years, three of Mercury's days have passed.

Atmosphere Unlike Earth, Mercury has essentially no atmosphere, and what little does exist is composed primarily of oxygen and sodium, as illustrated in ***Figure 29-6.*** The daytime surface temperature on Mercury is 700 K (427°C), while temperatures at night fall to 100 K (−173°C). This is the largest day-night temperature difference of all the planets in our solar system.

Figure 29-6 The major components in Mercury's atmosphere are oxygen, sodium, hydrogen, helium, and potassium.

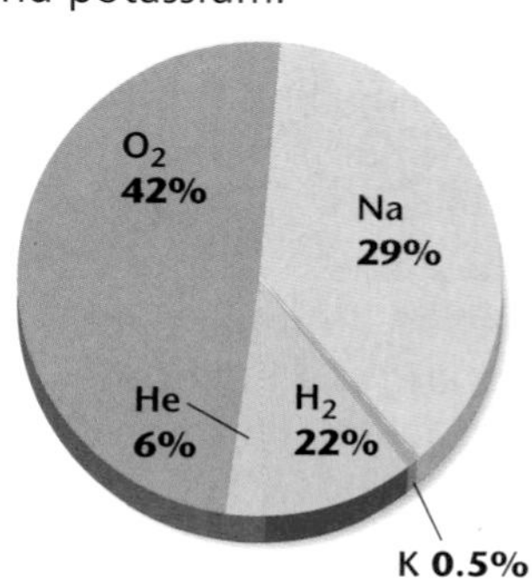

Surface Most of what we know about Mercury is based on radio observations and images from a United States space probe mission, called *Mariner 10,* which passed close to Mercury three times in 1974 and 1975. Images from *Mariner 10* show that Mercury's surface, similar to the Moon's surface, is covered with craters and plains, as shown in ***Figure 29-7A.*** The plains of Mercury's surface are smooth and relatively crater free. It is thought that the plains were formed from lava flows that covered cratered terrain, much like the maria on the Moon. The surface gravity of Mercury is much greater than that of the Moon, and thus crater walls and peaks are lower and ejecta are shorter in length than those on the Moon. Mercury has a planetwide system of cliffs, called scarps, as shown in

Figure 29-7B. Scientists hypothesize that the scarps developed as Mercury's crust shrank and fractured early in the planet's geological history.

Interior Although scientists have no seismic data with which to analyze the interior of Mercury, the high density of the planet suggests that Mercury has an extensive nickel-iron core, filling about 42 percent of Mercury's volume. The detectable magnetic field, only 1 percent of Earth's magnetic field strength, suggests that Mercury has a molten zone in its interior. Mercury's small size, high density, and probable molten interior zone resemble what Earth might be like if its crust and mantle were removed. These observations suggest that Mercury was originally much larger, with a mantle and crust similar to Earth's, and that the outer layers may have been lost in a collision with another celestial body early in its history.

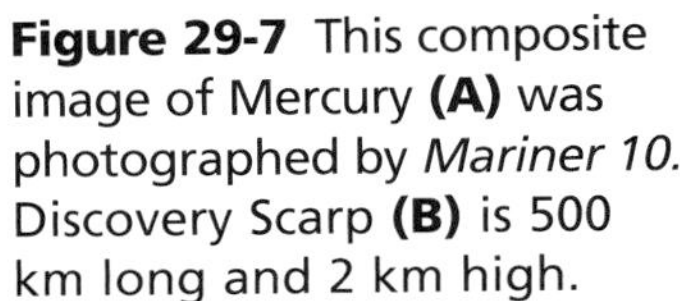

Figure 29-7 This composite image of Mercury **(A)** was photographed by *Mariner 10.* Discovery Scarp **(B)** is 500 km long and 2 km high.

VENUS

Venus and Mercury are the only two planets closer to the Sun than Earth. Like Mercury, Venus has no moons. Venus is the brightest planet in Earth's nighttime sky because it is close and because its albedo is 0.75. The albedo of Venus is the highest of any planet.

The thick clouds that are present in Venus's atmosphere would prevent an observer on the surface of Venus from seeing the stars. These clouds also prevent astronomers from directly observing the surface, except at radio wavelengths. In the 1960s, radar measurements showed that the surface of Venus is very hot, and that Venus is rotating slowly. One day on Venus is 243 Earth days in length. Radar measurements also revealed Venus's clockwise spin, which is opposite the spin of most planets. The backward spin exhibited by Venus, called retrograde rotation, means that for an observer on Venus, the Sun would rise in the west and set in the east. Astronomers hypothesize that the retrograde rotation of Venus may have been caused by a collision between Venus and another body early in our solar system's history.

Today, astronomers know a lot about Venus as a result of close-up observations made by several United States and Soviet (later Russian) spacecraft. Some probes landed on the surface of the planet and made observations there, while others flew by. The 1978 *Pioneer-Venus* and the 1989 *Magellan* missions of the United States used radar to map up to 98 percent of the surface of Venus.

Earth Sciences

1.f Students know the evidence for the dramatic effects that asteroid impacts have had in shaping the surface of planets and their moons and in mass extinctions of life on Earth.

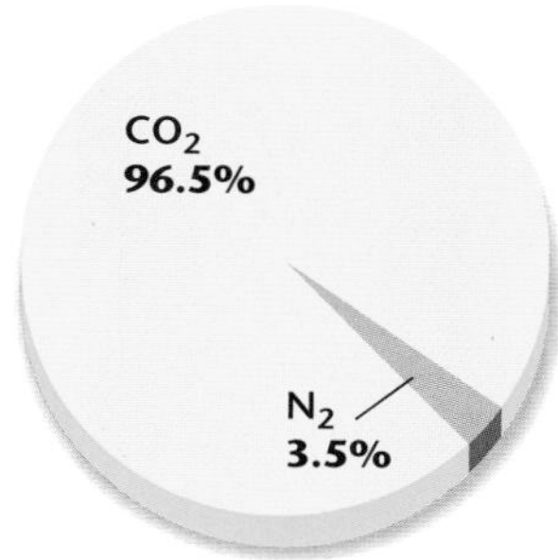

Figure 29-8 The major components of Venus's atmosphere are carbon dioxide and nitrogen.

Earth Sciences

4.d Students know the differing greenhouse conditions on Earth, Mars, and Venus; the origins of those conditions; and the climatic consequences of each.

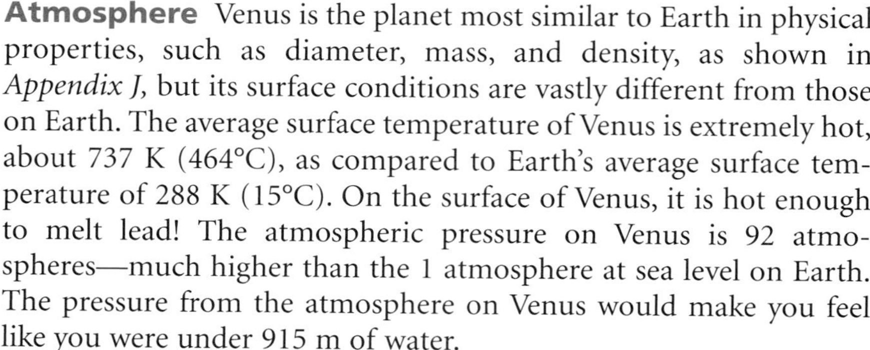

Atmosphere Venus is the planet most similar to Earth in physical properties, such as diameter, mass, and density, as shown in *Appendix J,* but its surface conditions are vastly different from those on Earth. The average surface temperature of Venus is extremely hot, about 737 K (464°C), as compared to Earth's average surface temperature of 288 K (15°C). On the surface of Venus, it is hot enough to melt lead! The atmospheric pressure on Venus is 92 atmospheres—much higher than the 1 atmosphere at sea level on Earth. The pressure from the atmosphere on Venus would make you feel like you were under 915 m of water.

The atmosphere of Venus is primarily carbon dioxide and nitrogen, as illustrated in ***Figure 29-8.*** How does this compare to Earth's atmospheric composition? Similar to Earth, Venus has clouds. But instead of being composed of water vapor and ice, clouds on Venus, shown in ***Figure 29-9A,*** are made of sulfuric acid and are 35 km thick. If it were to rain on Venus, the rain would be sulfuric acid.

Venus has a greenhouse effect, like Earth, but Venus's is more efficient. As you learned in Chapter 14, greenhouse gases in Earth's atmosphere trap infrared radiation and keep Earth at higher temperatures than those that would exist if there were no atmosphere. Carbon dioxide, one of Earth's greenhouse gases, has a high concentration in Venus's atmosphere, which prevents infrared radiation from escaping and keeps the surface extremely hot. In fact, it is so hot that liquid water can't exist. Venus is the hottest planet, even though it is not the closest to the Sun.

Surface The *Magellan* orbiter used radar reflection measurements to map the surface of Venus in fine detail. The surface has been smoothed by volcanic lava flows, as shown in ***Figure 29-9B,*** and it has only a few impact craters. The most recent global episode of volcanic activity took place about 500 million years ago, and therefore the surface of Venus is relatively young. Unlike Earth, there is little evidence of current tectonic activity on Venus, and there is no well-defined system of crustal plates.

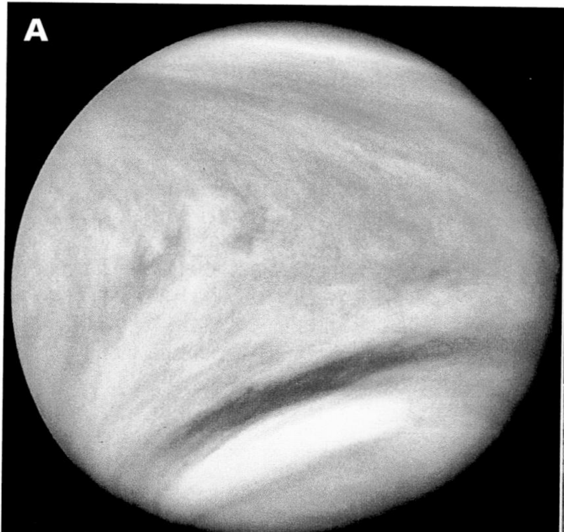

Figure 29-9 The clouds in the atmosphere of Venus **(A)** obscure the surface. By using the radar of *Magellan,* astronomers have been able to map the surface and discover features like the volcano Maat Mons and the surrounding surface smoothed by volcanic lava flows **(B).**

Interior The size and density of Venus are similar to Earth, so the internal structure is most likely similar. However, astronomers have no seismic data with which to prove this. It is theorized that Venus has a liquid metal core that extends halfway to the surface. There is no measurable magnetic field despite this liquid core, which is probably due to Venus's slow rotation rate.

EARTH

Earth, shown in ***Figure 29-10,*** is the third planet from the Sun. Earth has many unique properties when compared with the other planets. Its distance from the Sun and its nearly circular orbit allow liquid water to exist on its surface in all three states: solid, liquid, and gas. Liquid water is required for life, and Earth's vast abundance of liquid water has been important for the development and existence of life on our planet. In addition, our planet's moderately dense atmosphere, which is composed primarily of 78 percent nitrogen and 21 percent oxygen, and a mild greenhouse effect also support conditions suitable for life.

Figure 29-10 This shows Earth as seen by the Apollo astronauts orbiting the Moon.

Precession Earth's axis is tilted, as you have learned, and this tilt creates our seasons. As you learned in Chapter 14, Earth's axis is wobbling, like a toy top that wobbles if you give it a small sideways push while it is spinning. This wobble in Earth's rotational axis, shown in ***Figure 29-11,*** is called **precession.** It takes Earth's rotational axis about 26 000 years to go through one cycle of precession. The sideways pull that causes precession comes from the Moon's gravitational force on Earth, as well as the Sun's gravitational force. However, the Sun's gravitational force plays a lesser part in Earth's precession.

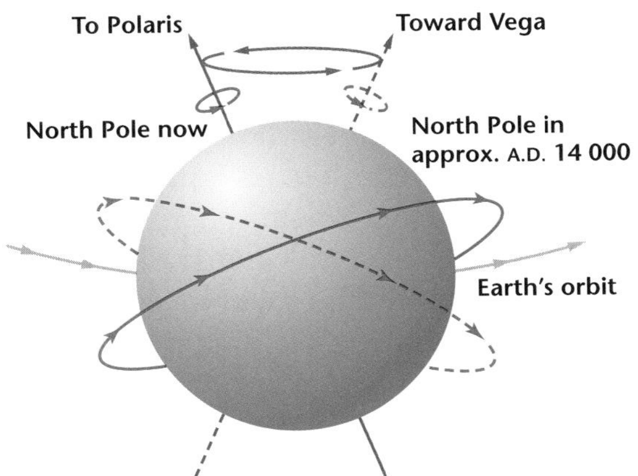

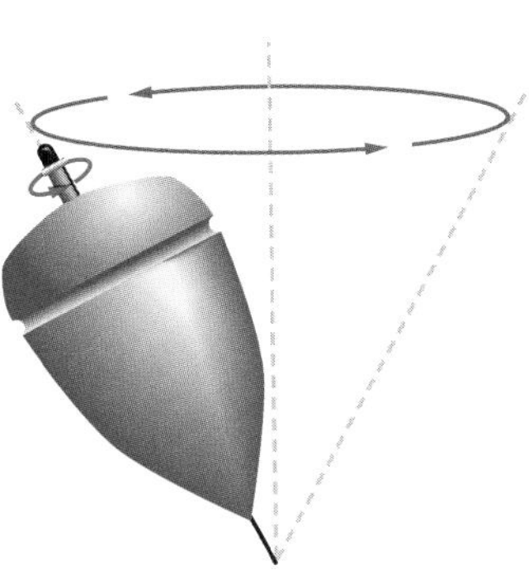

Figure 29-11 Earth precesses, or wobbles, on its axis, much like a toy top. In 12 000 years, our new north star will be Vega.

MARS

Figure 29-12 This view of Mars, taken by the *Hubble Space Telescope (HST),* shows its surface, its atmosphere, and one of its polar caps.

Mars is the fourth planet from the Sun and the outermost of the terrestrial planets. It is often referred to as the red planet because of its reddish surface color, shown in ***Figure 29-12,*** which is caused by a high iron content in the soil. Mars is smaller and less dense than Earth, as shown in *Appendix J,* and has two irregularly-shaped moons, Phobos and Deimos, which are most likely captured asteroids.

Mars has been explored by telescopes on Earth and with probes that have flown by, orbited, or landed. In the 1960s, the United States sent a spacecraft, *Mariner 4,* to explore Mars. Later missions included *Mariner 9,* in 1971, and two *Viking* landers in 1976. More recent missions have been the *Mars Climate Orbiter* in 1998 and the *Mars Exploration Rover Mission* in 2004–2005, which are all part of a long-term NASA plan to explore Mars for evidence of preexisting life.

Atmosphere The composition of Mars's atmosphere, shown in ***Figure 29-13,*** is similar to Venus's atmosphere, but the density and pressure in Mars's atmosphere are much lower, and therefore Mars does not have a strong greenhouse effect, as Venus does. Although the atmosphere is thin, it is turbulent, so there is a constant wind on Mars. Dust storms in the atmosphere may last for weeks at a time.

Figure 29-13 The major components of Mars's atmosphere are carbon dioxide, nitrogen, argon, and oxygen.

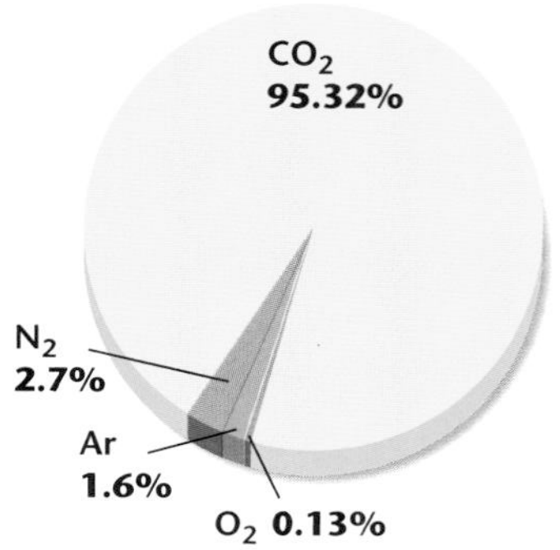

Surface The southern and northern hemispheres of Mars have different types of surfaces. The southern hemisphere is a heavily cratered, highland region, resembling the highlands of the Moon. The northern hemisphere is dominated by plains that are sparsely cratered. Scientists theorize that great lava flows covered the once-cratered terrain of the northern hemisphere. Four gigantic shield volcanoes are located in the northern hemisphere, near a region called the Tharsis Plateau. The largest volcano is Olympus Mons, which is also the largest mountain in the solar system. The base of Olympus Mons would cover the state of Colorado; it is three times higher than Mt. Everest. An enormous canyon, Valles Marineris, shown in ***Figure 29-14,*** lies on the Martian equator and splits the Tharsis Plateau. This canyon seems to have formed as a fracture when the Tharsis Plateau was uplifted more than 3 billion years ago.

Other Martian surface features include dried river and lake beds, outflow channels, and runoff channels. These are all erosional features that suggest that liquid water once existed on the surface of Mars. Astronomers hypothesize that the atmosphere must have once been much warmer, thicker, and richer in carbon dioxide, which would have allowed water to exist on Mars. Although there is a relatively small amount of ice at the poles, astronomers are still searching for water at other locations on the Martian surface today.

Figure 29-14 Valles Marineris stretches for more than 4000 km. A few large volcanoes can be seen in the upper left.

Mars has polar ice caps covering both poles. The caps grow and shrink with the seasons on Mars. Martian seasons are caused by a combination of a tilted axis and a highly elongated orbit. Both caps are made of carbon dioxide ice, which you may know as "dry ice." Water ice lies beneath the carbon dioxide ice in the northern cap, shown in ***Figure 29-12,*** and is exposed during the northern hemisphere's summer when the carbon dioxide ice evaporates. There may also be water ice beneath the southern cap, but the carbon dioxide ice never completely evaporates to expose the water ice.

Interior Astronomers are unsure about the internal structure of Mars. They hypothesize that there is a core of iron and nickel, and possibly sulfur, that extends somewhere between 1200 km and 2400 km from the center of the planet. Because Mars has no magnetic field, astronomers hypothesize that the core is probably solid. Above the solid core is a mantle. There is no evidence of current tectonic activity or tectonic plates on the surface of the crust.

SECTION ASSESSMENT

1. List the similarities and differences between Mercury and the Moon, and between Mercury and Earth.
2. Explain why surface conditions on Venus and Earth are so different.
3. Why do astronomers hypothesize that the southern polar cap of Mars has water ice under the carbon dioxide ice?
4. What evidence do astronomers use to support the hypothesis that there was once tectonic activity on Mercury, Venus, and Mars?
5. **Thinking Critically** What do you think the terrestrial planets would be like today if major impacts had not played a role in their formation and evolution?

SKILL REVIEW

6. **Making Graphs** Using *Appendix J,* create a graph showing the distance from the Sun for each terrestrial planet on the *x*-axis and the orbital period in Earth days on the *y*-axis. For more help, refer to the *Skill Handbook.*

earthgeu.com/self_check_quiz

SECTION 29.3 The Gas Giant Planets

OBJECTIVES

- **Describe** *the properties of the gas giant planets.*
- **Identify** *the unique nature of the planet Pluto.*

VOCABULARY

liquid metallic hydrogen
belt
zone

The interiors of the gas giant planets are composed of fluids, either gaseous or liquid, and possibly small, solid cores. They are composed primarily of lightweight elements such as hydrogen, helium, carbon, nitrogen, and oxygen, and they are very cold at their surfaces. The gas giants have many satellites as well as ring systems, and, as their name implies, they are all very large, ranging from 15 to more than 300 times the mass of Earth, and from about 4 to more than 10 times Earth's diameter. You will compare the relative sizes and distances from the Sun of the nine planets in the *Design Your Own GeoLab* at the end of the chapter.

JUPITER

Jupiter is the largest planet and the fifth planet from the Sun. The diameter of Jupiter is 11 times larger than Earth's, and only 10 times smaller than the Sun's. Jupiter's mass, shown in *Appendix J,* makes up 70 percent of all planetary matter in our solar system. From Earth, Jupiter appears quite bright because of its albedo of 0.343. Telescopic observations reveal that Jupiter has a banded appearance, as shown in ***Figure 29-15A,*** as a result of flow patterns in its atmosphere. Jupiter has four major satellites, which were discovered by Galileo, in addition to many smaller ones, which were discovered by flyby space probes and recent observations.

Jupiter has been explored by several United States space probes. The first were the *Pioneer 10* and *Pioneer 11* missions, which arrived at Jupiter in 1973 and 1974, respectively. In 1979, the *Voyager 1* and *Voyager 2* missions discovered several new satellites and a thin, dim ring around Jupiter. They also detected volcanic activity on Jupiter's closest major moon, Io. In 1995, the United States spacecraft *Galileo* arrived at Jupiter and dropped a probe into the clouds while the main spacecraft orbited the planet for five years and made observations.

Figure 29-15 Jupiter **(A)**, as photographed by *Voyager 1*, has a banded appearance. *Voyager 1* also photographed Jupiter's Great Red Spot **(B)**.

Atmosphere Jupiter has a low density, 1326 kg/m^3, for its huge size because it is composed of lightweight elements. Hydrogen and helium make up the majority of Jupiter's atmospheric gas, as illustrated in ***Figure 29-16.*** In Jupiter's atmosphere, these elements remain in a gas or liquid form. Below the liquid hydrogen, there is a layer of **liquid metallic hydrogen,** a form of hydrogen that has properties of both a liquid and a metal, which can exist only under conditions of very high pressure. Electric currents exist within the layer of liquid metallic hydrogen and generate Jupiter's magnetic field. Theoretical models of Jupiter suggest that Jupiter might have an Earth-sized solid core made of heavier elements that have sunk to the center of the planet.

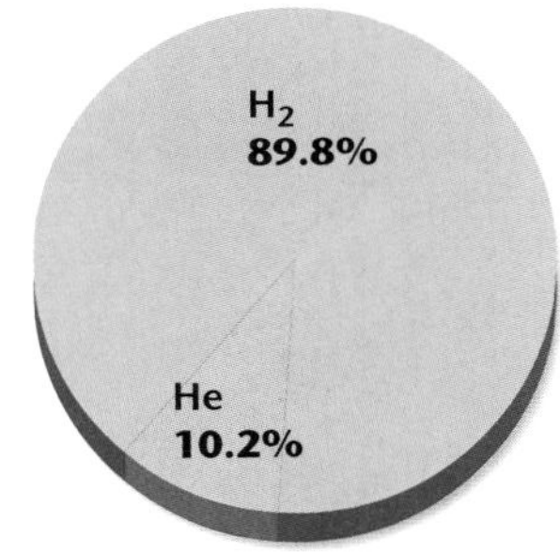

Figure 29-16 The major components of Jupiter's atmosphere are hydrogen and helium.

The rotation of Jupiter is extremely rapid for its huge size. Jupiter spins on its axis in a little less than 10 hours, making it the shortest day in the solar system. This rapid rotation distorts the shape of the planet so that the diameter through its equatorial plane is 7 percent larger than the diameter through its poles. Jupiter's rapid rotation causes its clouds to flow rapidly as well, in alternating cloud types called belts and zones. **Belts** are low, warm, dark-colored clouds that sink, and **zones** are high, cool, light-colored clouds that rise. These are similar to the rotation-driven flows in Earth's atmosphere. ***Figure 29-15B*** shows Jupiter's Great Red Spot, which is an atmospheric storm that has been rotating around Jupiter for more than 300 years.

Figure 29-17 Io is the closest major moon to Jupiter **(A).** The next closest is Europa **(B).**

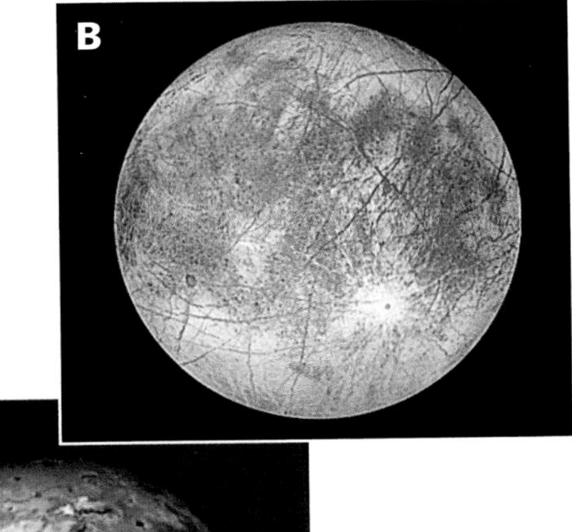

Moons and Rings Jupiter's four largest moons, Io, Europa, Ganymede, and Callisto, are called Galilean satellites, after their discoverer. All but one of them are bigger than Earth's moon, and all are larger than Pluto. These four moons are composed of ice and rock mixtures. The ice content is lower in Io and Europa, shown in ***Figure 29-17,*** because they have been squeezed and heated to a greater extent by Jupiter's gravitational force than the outer moons. In fact, Io has been heated to the point of becoming almost completely molten inside and undergoes constant volcanic eruptions. Gravitational heating has melted the ice in Europa, at least in the past, and astronomers hypothesize that this moon still has a subsurface ocean of liquid water.

The ring of Jupiter was discovered in images from the *Voyager 1* mission. Jupiter's ring is 6400 km wide. Its discovery proved that Saturn is not the only gas giant that has rings. In fact, it is now known that all four of the gas giant planets have rings.

Figure 29-18 This image of Saturn was captured by *HST,* on December 1, 1994.

Saturn

Saturn is the sixth planet from the Sun and the second-largest planet in the solar system. Saturn is shown in ***Figure 29-18.*** Four space probes have visited Saturn, including *Pioneer 10, Pioneer 11,* and *Voyager 1* and *2.* In 2004, the United States *Cassini* mission, launched in 1997, arrived at Saturn and entered orbit around the planet. In 2005, *Cassini* released the *Huygens* probe into the atmosphere of Titan, Saturn's largest moon, to explore surface conditions there.

Atmosphere Saturn is not quite as large as Jupiter and has an average density, shown in *Appendix J,* that is actually lower than that of water. Similar to Jupiter, Saturn rotates rapidly for its size and has flowing belts and zones. Saturn's atmosphere is dominated by hydrogen and helium, as illustrated in ***Figure 29-19,*** but it also includes ammonia ice near the top of the clouds. The internal structure of Saturn is also probably similar to Jupiter. It is most likely fluid throughout with a small, solid core and a magnetic field that is 1000 times stronger than Earth's. Saturn's magnetic field is aligned with its rotational axis, which is unusual among the planets.

Moons and Rings The most striking feature of Saturn is its ring system, shown in ***Figure 29-20,*** which has much broader and brighter rings than those of the other gas giant planets. Saturn's rings are composed of pieces of rock and ice that range from microscopic to the size of houses. There are seven major rings, but each ring is actually made up of narrower rings, called ringlets, and many open gaps. These ringlets and gaps are caused by the gravitational effects of the many moons of Saturn. The rings are very thin, less than 200 m thick, because rotational forces keep all the particle orbits confined to Saturn's equatorial plane. The ring particles have not combined to form a large satellite because Saturn's gravity prevents particles very close to the planet from sticking together. This is why the major moons of the gas giant planets are always found farther out than the rings.

Until recently, astronomers hypothesized that the ring particles were simply left over from the time when Saturn and its moons formed. Now, however, many astronomers hypothesize it more likely that the ring particles are debris left over when a moon was destroyed by a collision with an asteroid or other object, or was ripped apart by Saturn's gravity. Some astronomers hypothesize that Saturn and the other gas giant planets may form new ring systems from time to time as collisions or gravitational effects occasionally destroy their moons.

Figure 29-19 The major components of Saturn's atmosphere are hydrogen and helium.

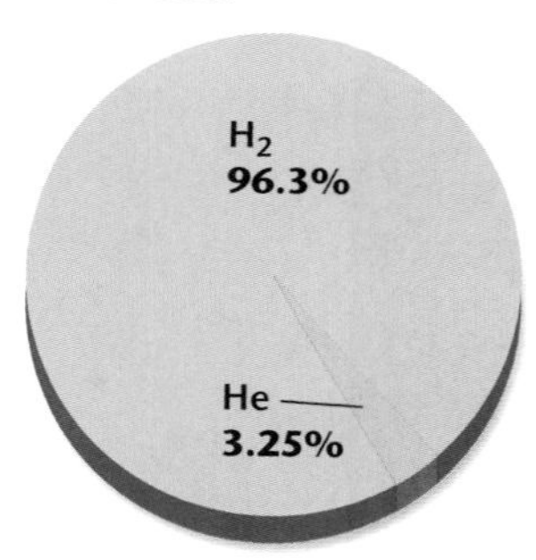

Saturn's many satellites include the giant Titan, seven intermediate-sized moons, and a number of small moons. Titan is larger than Earth's moon, and its atmosphere is made of nitrogen and methane. Methane may exist as a gas, a liquid, and ice on Titan's surface, similar to the three phases of water on Earth's surface.

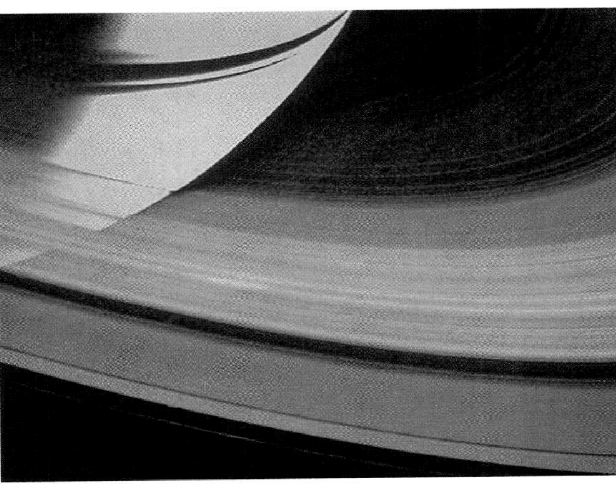

Figure 29-20 This image taken by *Voyager 1* shows the many ringlets that comprise the rings of Saturn.

URANUS

The seventh planet from the Sun, Uranus, was discovered accidentally in 1781. A bluish object was spotted through a telescope, and after tracking it for a couple of days, it was found that the object moved relative to the stars. It was a planet. In 1787, two of Uranus's larger moons, Titania and Oberon, were discovered. Today, we know that Uranus has many moons and 10 rings. In 1986, the United States *Voyager 2* mission visited Uranus and provided detailed information about the planet, including the existence of new moons and rings.

Atmosphere Uranus is 4 times as large and 15 times as massive as Earth, as shown in *Appendix J.* It has a blue, velvety appearance, as shown in ***Figure 29-21A,*** which is caused by its atmospheric composition. The methane gas in Uranus's atmosphere reflects blue light back into space, although most of the atmosphere is composed of helium and hydrogen, as illustrated in ***Figure 29-21B.*** There are very few clouds present, and they differ little in brightness and color from the surrounding atmosphere, making them difficult to detect. In addition, there are no distinct belts or zones like those observed on Jupiter and Saturn, which contributes to Uranus's featureless appearance. The internal structure of Uranus is similar to Jupiter and Saturn; it is completely fluid except for a small, solid core. Uranus also has a strong magnetic field.

The rotational axis of Uranus is tipped over so far that the north pole almost lies in its orbital plane. Astronomers hypothesize that Uranus was knocked sideways by a massive collision with a passing object, such as a very large asteroid, early in the solar system's history. Each pole on Uranus spends 42 Earth years in darkness and 42 Earth years in sunlight due to this tilt and Uranus's long trip around the Sun. Uranus's atmosphere keeps the planet at a temperature of 58 K (–215°C).

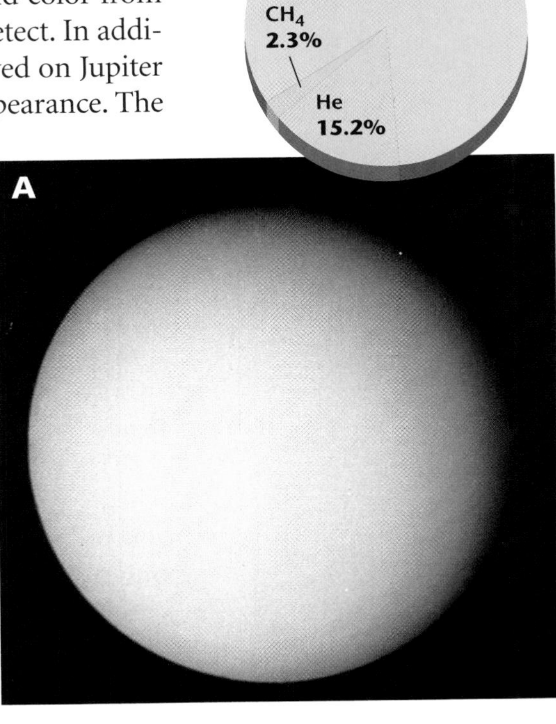

Figure 29-21 Uranus **(A),** photographed by *Voyager 2,* appears to be blue and featureless because of the methane (CH_4) in its atmosphere **(B).**

Figure 29-22 The rings of Uranus, photographed by *Voyager 2,* are very dark. Astronomers hypothesize that they are made of carbon compounds.

Moons and Rings The known moons and rings of Uranus orbit in the planet's equatorial plane. New moons are frequently being discovered, so Uranus's moon counts, like Jupiter's and Saturn's, are always changing. Uranus's rings, shown in ***Figure 29-22,*** are very dark—almost black. This is why they weren't discovered until the brightness of a star behind the rings dipped as Uranus moved in its orbit and the rings blocked the light.

NEPTUNE

The existence of Neptune was predicted before it was discovered. The prediction was based on small deviations in the motion of Uranus and the application of Newton's universal law of gravitation. In 1846, Neptune was discovered where astronomers had predicted it. Few details can be observed on Neptune with an Earth-based telescope, but the *Voyager 2* probe flew past this planet in 1989 and sent back new data and images of the planet.

Atmosphere Neptune is slightly smaller and denser than Uranus, but it is still about four times as large as Earth, as shown in *Appendix J.* Other similarities between Neptune and Uranus include their bluish color caused by methane in the atmosphere, atmospheric compositions, as illustrated in ***Figure 29-23,*** temperatures, magnetic fields, interiors, and particle belts. Unlike Uranus, however, Neptune, shown in ***Figure 29-24A,*** has distinctive clouds and atmospheric belts and zones similar to those of Jupiter and Saturn. In fact, Neptune had a persistent storm, the Great Dark Spot, with characteristics similar to Jupiter's Great Red Spot. The storm disappeared from Neptune in 1994.

Figure 29-23 Like Uranus, the major components of Neptune's atmosphere are hydrogen, helium, and methane.

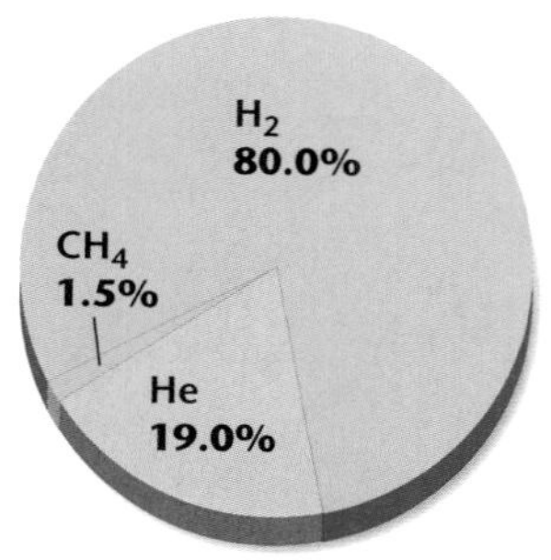

Moons and Rings Neptune has many moons, the largest being Triton. Triton, shown in ***Figure 29-24B*** has a retrograde orbit, which means that it orbits backward, unlike virtually every other large satellite in the solar system. Triton also has a thin atmosphere and nitrogen geysers. The geysers are caused by nitrogen gas below the surface in Triton's south polar ice cap expanding and erupting when heated by the Sun.

The *Voyager 2* flyby increased our knowledge of Neptune's rings, which previously had been only indirectly observed. The six rings are composed of microscopic-sized dust particles. Some parts of the outermost ring appear much brighter than other parts because of the clumping of material. Scientists theorize that these clumps do not spread out evenly in the ring because of the gravitational effects of Neptune's moons.

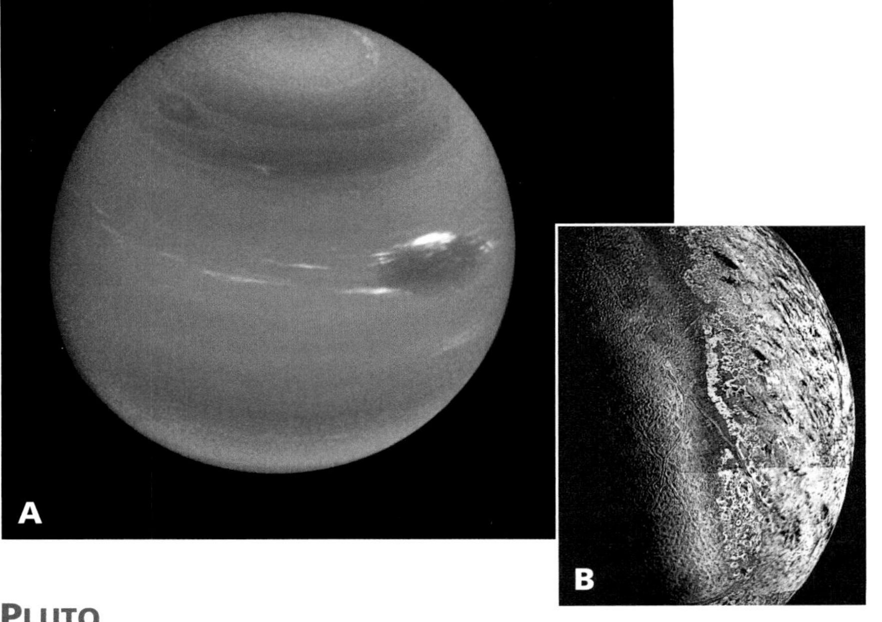

Figure 29-24 Neptune **(A)** has bands in its atmosphere. This image, taken before 1994, shows the Great Dark Spot right of center. This image of Triton **(B),** taken by *Voyager 2,* shows the sooty material of the nitrogen geysers that have been blown downwind by Triton's thin atmosphere.

PLUTO

The ninth planet in our solar system, Pluto, was discovered in 1930. Pluto is very different from the other eight planets of our solar system. Even though it has a solid surface, Pluto is not classified as a terrestrial planet because of its low density and small size. With its solid surface, Pluto does not have properties characteristic of the gas giant planets either. The density of Pluto indicates that it is made of half ice and half rock, and it is smaller than Earth's moon. The atmosphere is composed of methane and nitrogen, but in unknown quantities.

Problem-Solving Lab

Using Numbers

Test Kepler's third law For the six planets closest to the Sun, Kepler observed that $P^2 = a^3$, where P is the orbital period in years and a is the semimajor axis in AU.

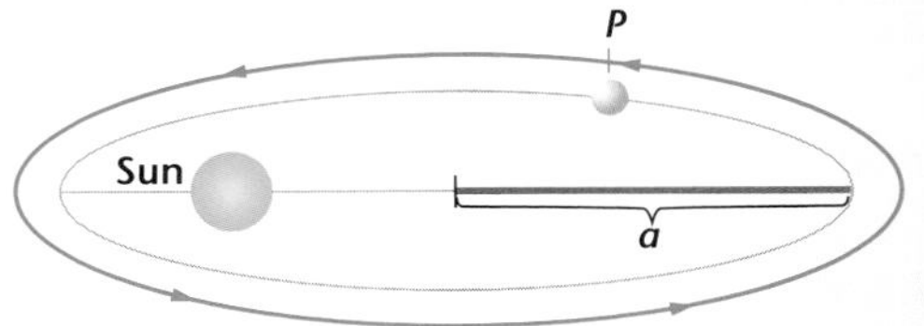

Analysis

1. Use data from *Appendix J* to calculate P^2 and a^3 for each of the nine planets.
2. Compare P^2 to a^3 for each planet.

Thinking Critically

3. Does Kepler's third law fit each of the planets?
4. If Uranus, Neptune, and Pluto had been discovered in Kepler's time, do you think he would have still believed in his law? Explain.
5. What would be the orbital period of an asteroid orbiting the Sun at 2.5 AU?
6. What is the semimajor axis for comet Halley, which has an orbital period of 76 years?

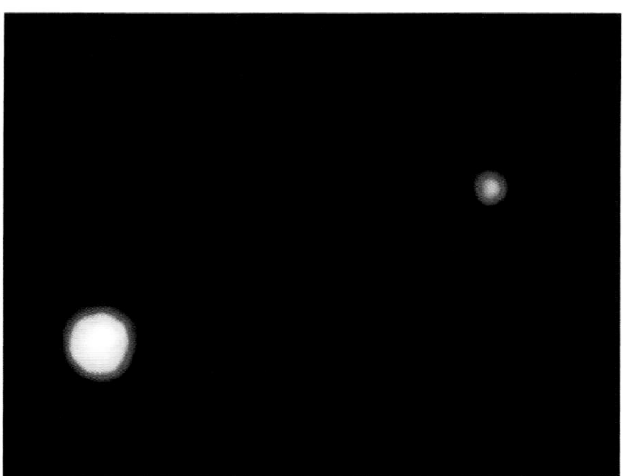

Figure 29-25 This image of Pluto and Charon, taken by *HST* on February 21, 1994, is the best image to date of the pair. Pluto and Charon are separated by a distance of 19 640 km.

The orbit of Pluto is so eccentric that at aphelion, it is 50 AU from the Sun, and at perihelion, it is almost 30 AU from the Sun. While at perihelion, Pluto is closer to the Sun than Neptune is. This happened last between 1978 and 1998. No space probes have traveled out to Pluto, although NASA is planning a flyby mission of Pluto.

Pluto's rotational axis is tipped so far over that its north pole actually points south of its orbital plane. Pluto has a satellite, called Charon, which orbits in Pluto's equatorial plane. Pluto and Charon, shown in ***Figure 29-25,*** have masses that are the most similar of any planet-satellite pair in the solar system. They are in synchronous rotation with each other, which means that each one keeps the same side facing the other. If you visited Pluto at a location on the side facing Charon, you would always see Charon in the sky overhead, throughout day and night.

Many of Pluto's properties, shown in *Appendix J,* are more similar to those of the gas giants' large moons than they are to those of any other planet. One theory suggests that Pluto was once a satellite of Neptune that escaped as a result of a near-collision with Triton. This would help explain Pluto's highly eccentric, tilted orbit and the unusual tilt of its rotational axis, as well as Triton's backward orbital motion. According to another theory, Pluto's composition and eccentric orbit suggest that it is related to a comet.

Section Assessment

1. In what two ways does Jupiter's rapid rotation affect the planet?
2. Describe the rings of Saturn and how they are thought to have formed.
3. What are the similarities between Uranus and Neptune?
4. Describe properties of Pluto that exclude it from being classified as either a terrestrial planet or a gas giant planet.
5. **Thinking Critically** Pluto and Neptune's orbits are arranged in such a way that Pluto is sometimes within Neptune's orbit. Despite this, collisions between the two are not possible. What are some possible reasons for this?

Skill Review

6. **Using Tables** Using values from *Appendix J,* compare the mass of Jupiter with the total mass of the other eight planets. Also, compare the total mass of all the planets, including Jupiter, with the mass of the Sun. For more help, refer to the *Skill Handbook.*

SECTION 29.4 Formation of Our Solar System

Earth Sciences 1.a, 1.b **I&E** 1.g, 1.i

Now that you know some of the characteristics of the solar system and the nature of the celestial bodies that occupy the Sun's vicinity, you are prepared to think about how the solar system formed. Astronomers use Earth-based observations and data from probes to derive theories about how our solar system formed. The significant observations related to our solar system's formation include the shape of our solar system, the differences among the planets, and the oldest planetary surfaces, asteroids, meteorites, and comets.

OBJECTIVES

- **Summarize** *the properties of the solar system that support the theory of the solar system's formation.*
- **Describe** *how the planets formed from a disk surrounding the young Sun.*
- **Explore** *remnants of solar system formation.*

VOCABULARY

planetesimal
asteroid
meteoroid
meteor
meteorite
comet
coma
nucleus
meteor shower

A Collapsing Interstellar Cloud

Stars and planets form from clouds of gas and dust, called interstellar clouds, which exist in space between the stars. The interstellar clouds consist mostly of gas, especially hydrogen and helium. Dust makes interstellar clouds look dark because the dust blocks the light from stars within or behind the clouds. The dust can be thought of as a kind of interstellar smog. Conversely, light from stars reflects off the dust and partially illuminates the clouds. The clouds can also be heated by stars, which can cause them to glow on their own. This is why the interstellar clouds often appear as blotches of light and dark, as shown in ***Figure 29-26.***

Many interstellar clouds can be observed along the Milky Way in regions that have relatively high concentrations of interstellar gas and dust. The density of interstellar gas is very low, much lower than even the best laboratory vacuums created by scientists. However, an interstellar cloud can start to condense as a result of gravity and become concentrated enough to form a star and possibly planets. Astronomers hypothesize that our solar system began in this way.

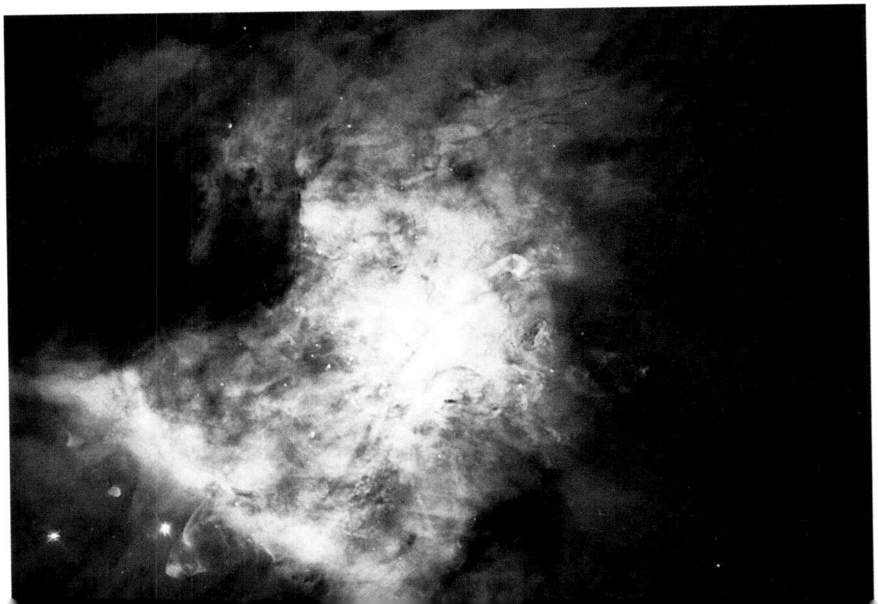

Figure 29-26 This mosaic of images of the Orion Nebula was taken by *HST.* The Orion Nebula is home to newly forming stars.

At first, the collapse is slow, but it accelerates and the cloud soon becomes much denser at its center. If a cloud was rotating at all to begin with, it will spin faster and faster as it contracts, for the same reason that ice-skaters spin faster as they pull their arms close to their bodies. As the collapsing cloud spins, the rotation slows the collapse in the equatorial plane, and the cloud becomes flattened. The cloud eventually becomes a rotating disk with a dense concentration at the center, as shown in ***Figure 29-27.***

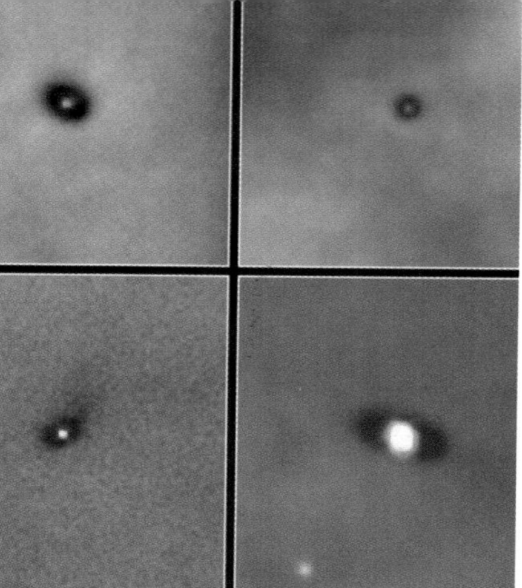

Figure 29-27 These images, taken by *HST* on November 20, 1995, show four young stars with disks of gas and dust around them. Astronomers often use radio and infrared telescopes which can penetrate the interstellar clouds to find these young stars.

Earth Sciences

1.a Students know how the differences and similarities among the sun, the terrestrial planets, and the gas planets may have been established during the formation of the solar system.

Sun and Planet Formation

The disk of dust and gas that formed the Sun and planets is known as the solar nebula. The dense concentration of gas at the center of this rotating disk eventually became the Sun. You will learn more about the Sun's formation as well as the formation of other stars in Chapter 30.

In the disk surrounding the young Sun, the temperature varied greatly with location. It was hottest close to the Sun, possibly as hot as 2000 K (1726°C), and coolest near the edge of the disk, far from the Sun. As the disk began to cool, different substances were able to condense into a liquid or solid form. One of the first elements to form would have been tungsten, because of its high condensing temperature. As the disk cooled further, more elements and compounds, such as aluminum oxide, iron, and silicates, were able to condense.

Eventually, the condensation of materials into liquid and solid forms slowed. The area closest to the Sun was still warm because of the Sun's proximity and energy, while at the outer edge of the disk, it was cold because the Sun was so distant. Thus, different elements and compounds were able to condense depending on their distance from the Sun, as illustrated in ***Figure 29-28,*** which impacted the compositions of the forming planets.

To learn more about The *Hubble Space Telescope,* go to the **National Geographic Expedition** on page 902.

The Growth of Objects Once the condensing slowed, the tiny grains of condensed material started to accumulate and merge together to form larger bodies. These solid particles gradually built up in size as grains collided and stuck together, and as gas particles collected on the surfaces of the grains. As the solid bodies continued to grow, they eventually reached hundreds of kilometers in diameter. These objects are called **planetesimals.** Further growth continued through collisions and mergers of planetesimals. These events were violent and sometimes destroyed the planetesimals. However, the overall result was a smaller number of larger bodies: the planets.

Merging into Planets In the outer solar system, the first large planet to develop was Jupiter. As Jupiter increased in size through mergers of icy planetesimals, its gravity began to attract additional

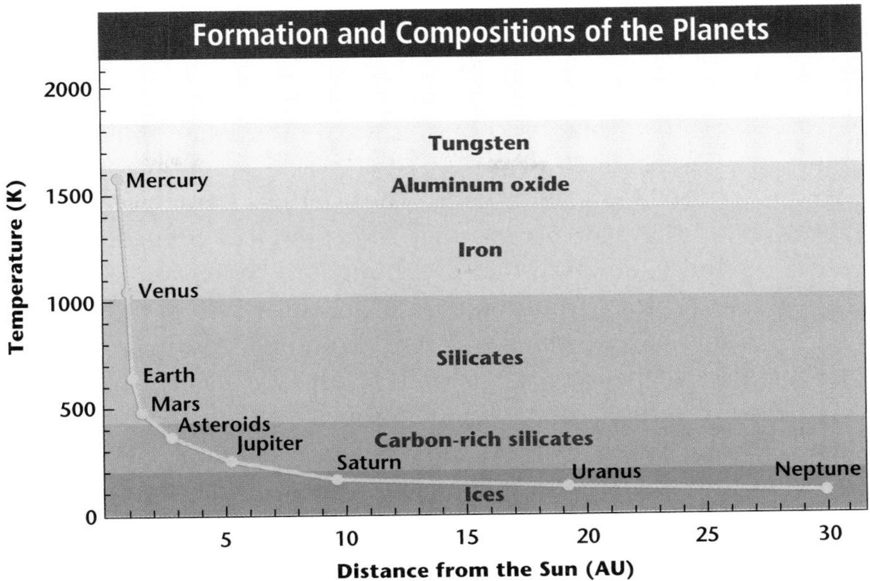

Figure 29-28 Elements and compounds that were able to condense close to the Sun, where it was warm, are called refractory elements, and far from the Sun, where it was cool, volatile elements could condense. Refractory elements, such as iron, comprise the terrestrial planets, which are close to the Sun. Volatile elements, such as ices and gases like hydrogen, comprise the planets farther from the Sun, where it is cool.

Earth Sciences

1.b Students know the evidence from Earth and moon rocks indicates that the solar system was formed from a nebular cloud of dust and gas approximately 4.6 billion years ago.

Also covers: I&E 1.i

gas, dust, and planetesimals, so Jupiter grew even larger. Saturn and the other gas giants formed similarly, but they could not become as large because Jupiter had collected so much of the material in the vicinity. As each gas giant acquired material from its surroundings, a disk formed in its equatorial plane, much like the disk of the early solar system. In the disk, matter coalesced to form satellites.

In the inner part of the main disk, near the young Sun, planets also formed by the merging of planetesimals. These planetesimals, however, were composed primarily of refractory elements, so the inner planets are rocky and dense, in contrast to the gaseous outer planets. Also, the Sun's gravitational force is theorized to have swept up much of the gas in the area of the inner planets and prevented them from acquiring much additional material from their surroundings. Thus, the inner planets initially ended up with no satellites.

Update For an online update on our solar system, visit the Earth Science Web Site at earthgeu.com

Debris Eventually, the amount of interplanetary debris thinned out as it crashed into planets or was diverted out of the solar system. However, not all debris was ejected out of the solar system. The planetesimals in the area between Jupiter and Mars, known as the asteroid belt, remained there because Jupiter's gravitational force prevented them from merging to form a planet.

Asteroids

There are thousands and thousands of bodies that orbit the Sun within the planetary orbits. They are leftovers from the formation of the solar system and are called **asteroids.** Asteroids range from a

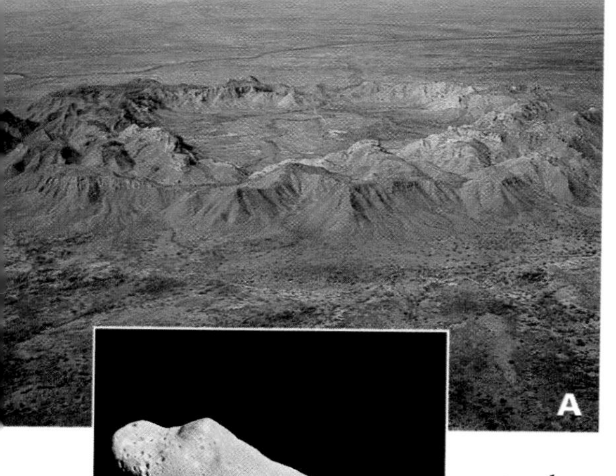

Figure 29-29 Gosses Bluff in central Australia **(A)** is 6 km in diameter and has a rim that is 200 m high. The asteroid Ida **(B)**, photographed by *Galileo,* is 56 km in length and has its own moon, Dactyl.

few kilometers to about 1000 km in diameter and have pitted, irregular surfaces, like the surface of Ida, shown in ***Figure 29-29B.*** Most asteroids are located between the orbits of Mars and Jupiter within the asteroid belt.

Pieces of Asteroids The asteroids were once thought to represent planets that somehow had been destroyed. Today, however, they are thought to be leftover planetesimal pieces from the time of the solar system's formation that never formed planets. Astronomers estimate that the total mass of all the asteroids is only about 0.08 percent of Earth's mass. As the asteroids orbit, they occasionally collide and break into fragments. When this, or any, interplanetary material falls toward Earth and enters Earth's atmosphere, it is called a **meteoroid.** When a meteoroid falls toward Earth, it burns up in Earth's atmosphere and produces a streak of light called a **meteor.** If the meteoroid does not completely burn up, part of it will collide with the ground, and it is then called a **meteorite.** If the meteorite is large, it will cause an impact crater when it collides with the ground. There is evidence of impact craters on Earth, such as Meteor Crater in Arizona and Gosses Bluff in Central Australia, shown in ***Figure 29-29A.*** Any craters visible on Earth must be relatively young because otherwise they would have been erased by erosion.

COMETS

Other remnants from solar system formation are comets. **Comets** are small, icy bodies that have highly eccentric orbits around the Sun. Comets are made of ice and rock, and they range from 1 to 10 km in diameter. There are two clusters, or clouds, of comets: the Kuiper belt and the Oort cloud. The Kuiper belt is close to Pluto and is between 30 and 50 AU from the Sun. The Oort cloud lies more than 100 000 AU from the Sun. Occasionally, a comet is disturbed by the gravity of another object and is thrown into the inner solar system from one of these clusters.

Figure 29-30 Comet Hale-Bopp was visible in 1997, and it will not be visible again until the year 4397.

The Orbits of Comets Cometary orbits are highly eccentric. Some stretch far beyond the orbit of Pluto at aphelion, while others come very close to the Sun at perihelion. When a comet is within 3 AU of the Sun, it begins to evaporate, becomes much brighter, and forms a head and one or more tails. You may have seen the head and tails of comet Hale-Bopp, shown in ***Figure 29-30,*** when it lit up our night skies in the spring of 1997.

The head of a comet consists of the **coma** (KOH muh), an extended volume of glowing gas, and the **nucleus,** the small solid core, as shown in ***Figure 29-31.*** When the nucleus is heated, it releases gases and dust particles that form the coma and tails. The tails are pushed away from the coma by particles and ions coming from the Sun, as well as by the pressure of radiation from the Sun. This is why the tails of comets point away from the Sun, no matter what direction the comet is moving.

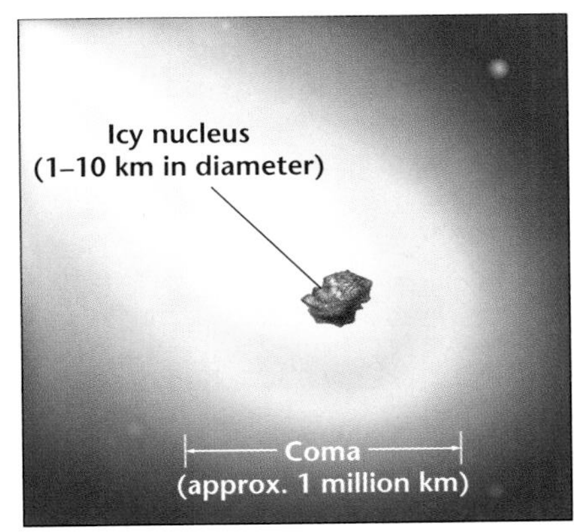

Figure 29-31 A comet consists of a nucleus, a coma, and tails pointing away from the Sun (not to scale).

Periodic Comets Comets that repeatedly orbit into the inner solar system are known as periodic comets. For example, Comet Halley is a well-known short-period comet with a 76-year period. Halley last appeared in 1985–1986, and it is expected again in 2061. Each time a periodic comet comes near the Sun, the Sun vaporizes some of the comet's ice, and the comet loses some of its matter. Eventually, it may break apart completely as the remaining ice evaporates.

When Earth intersects a cometary orbit, we experience a **meteor shower** as particles from the comet burn up upon entering Earth's upper atmosphere. Most meteors are caused by dust particles from comets, while most meteorites, the solid chunks of rock or metal that reach Earth's surface, are fragments of asteroids.

Astronomers theorize that solar system formation occurs commonly among stars. Thus, this has driven the search for planets and systems of planets orbiting other stars. You will learn more about these planets and their detection in the *Science in the News* feature at the end of this chapter.

SECTION ASSESSMENT

1. Explain why the material surrounding a young star forms a disk.
2. Describe why the inner planets are dense and rocky, while the outer planets have low densities and no solid surfaces.
3. Why is it theorized that the asteroids did not merge to form a planet?
4. For what reason did the inner planets not collect gas like the gas giants did?
5. **Thinking Critically** Why are there no comets in an orbit that is always close to the Sun?

SKILL REVIEW

6. **Concept Mapping** Use the following terms to construct a concept map to organize the correct sequence of events. For more help, refer to the *Skill Handbook.*

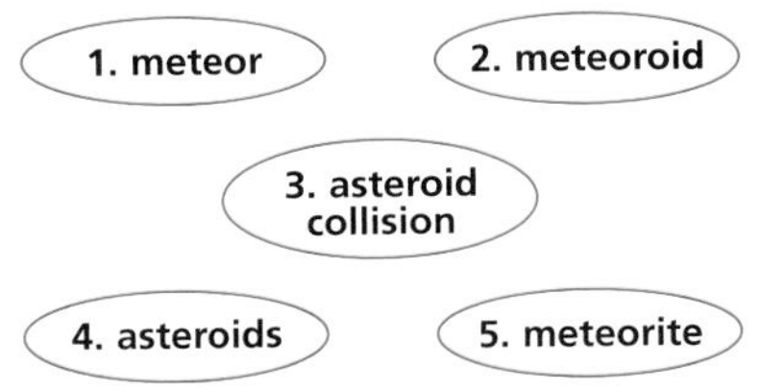

DESIGN YOUR OWN GeoLab

Scaling the Solar System

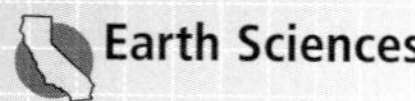

Earth Sciences

1.g Recognize the usefulness and limitations of models and theories as scientific representations of reality.

Astronomers are familiar with both the small, such as interstellar dust particles, and the large, such as the solar system. In order to understand the variety of sizes in the solar system, astronomers use models. These models can be as simple as putting people or objects in marked places or as complex as elaborate computer simulations. The most difficult task with many models is choosing a scale that will display all the information needed, such as distance, rotation rates, and size.

Preparation

Problem

How can the size of the solar system be converted to a scale that will easily demonstrate relative distances between objects in the solar system? Is distance the only measurement that can be demonstrated in a scale model?

Possible Materials

calculator
meterstick
marker
variety of sizes of common round objects
tape measure
stopwatches
masking tape

Hypothesis

Brainstorm about possible models and data needed to create them. Determine where these data are available, and collect them to use as a reference for your model. To have your model fit in your chosen area, make a hypothesis about the appropriate scale to use for distance from the Sun to each planet. Hypothesize how additional solar system measurements can be included in your model. For example, think about including planet diameters, rotation rates, etc.

Objectives

In this GeoLab, you will:

- **Calculate** the distance from the Sun for each planet based on your scale.
- **Determine** how to incorporate additional solar system measurements into your model, or design another model to show that information.
- **Interpret** your results based on your scale, and decide if your scale was an appropriate one based on the problems that may have resulted.

Plan the Experiment

1. As a group, make a list of possible ways you might test your hypotheses. Keep the available materials in mind as you plan your procedure.
2. Be sure your scale is appropriate for the information you are representing. Remember that a model should have the same scale throughout. You may have to try more than one scale before you are successful.
3. Record your procedure and list every step. Determine what materials are needed and the amounts of each.
4. Design and construct a data table for recording your original data and your scaled data.
5. Check the plan. Make sure your teacher has approved your plan before you proceed with your experiment.
6. Carry out your plan.

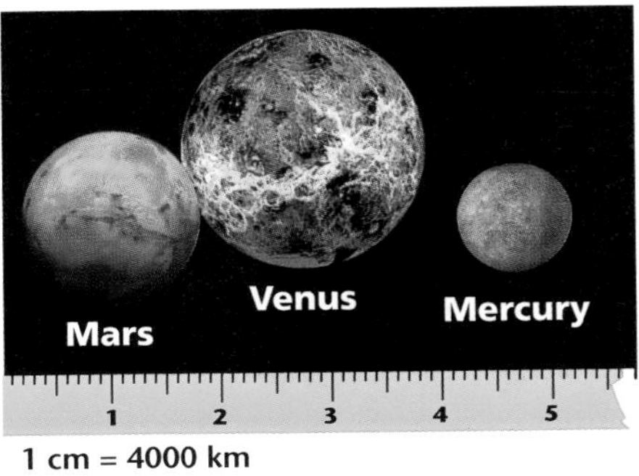

Analyze

1. **Checking Your Hypothesis** Which scale worked the best for your model? Explain.
2. **Interpreting Observations** What problems did you have in finding a scale? Explain how you corrected the problems.
3. **Calculating Results** List and explain the conversions that you used to create your scale model. If multiple steps were necessary to convert to your scale units, how could they be combined to make the process simpler?
4. **Observing and Inferring** What possible problems could result from using a very large scale? A small scale? Explain why depicting a scale model of the solar system on a sheet of notebook paper is extremely difficult.
5. **Compare and Contrast** Compare and contrast your model with one of your classmates'. What were the advantages of the scale you used? What were the disadvantages? How would you improve your model?
6. **Thinking Critically** Suppose that the outer planets are three times farther away than they are now. How would this affect your model? What scale would you choose now? Explain.

Conclude & Apply

1. Proxima Centauri, the closest star to our Sun, is about 4.01×10^{13} km from the Sun. Based on your scale, how far would Proxima Centauri be from the Sun in your model? If you were to fit the distance between the Sun and Proxima Centauri into your classroom, how small would the scaled distance between Pluto and the Sun be?
2. An interstellar dust particle is 1.0×10^{-6} m in length. Convert this measurement to your scale. How many dust particles could fit in the distance between the Sun and Jupiter? Between Mars and Uranus?

Science in the News

Earth Sciences 1.g

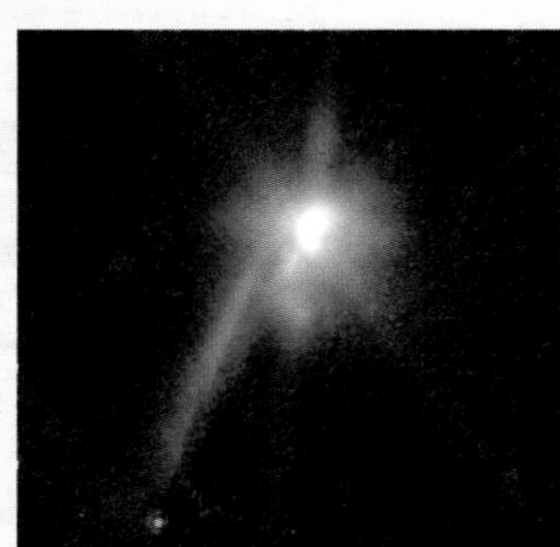

Discovering New Planets

Have you ever looked up at the night sky and wondered whether there are other Earth-like planets? We know that there are other stars similar to the Sun in the galaxy. In 1995 the first evidence of a planet was discovered.

In 1995, after a year of careful measurements, Swiss astronomers announced the discovery of a planet orbiting around the star 51 Pegasi. A planet found outside of our solar system is called an extrasolar planet. Shortly after the discovery of the first extrasolar planet, several more were discovered. More than 100 extrasolar planets have been detected, with the number of discoveries increasing with every passing month.

Detecting Planets

The presence of these extrasolar planets was inferred, not directly detected. The gravitational attraction of a star and a planet causes both bodies to orbit around a center of mass. Because the star is much larger and brighter than the planet, astronomers actually detect the small movement of the star around the center of mass, rather than directly observing the planet. Then, by using Newton's law of universal gravitation, astronomers can determine information such as the planet's minimum mass and orbital period.

In addition to single planets orbiting around stars, astronomers have found more than one planet orbiting around a single star. On April 15, 1999, astronomers announced the discovery of three planets orbiting around the star Upsilon Andromedae. This discovery was significant because scientists then had evidence of a multiple-planet system, similar to our solar system with its nine planets. On March 29, 2000, two planets smaller than Saturn were discovered. Until this time, only larger planets close to the size of Jupiter had been discovered.

Direct Evidence?

In the fall of 1999, astronomers using Earth-based telescopes recorded the distinct dimming of light from a star in the constellation Pegasus. Previously, a planet had been inferred from gravitational effects, but astronomers did not have direct evidence. However, the dimming occurred where the gravitational effects predicted the planet to be. This dimming was caused by the light from the star being blocked out by an extrasolar planet passing in front of the star. This provided direct evidence of an extrasolar planet.

No Earth-sized planets have yet been discovered, but scientists theorize that they exist. Two NASA missions, the *Space Interferometry Mission (SIM)* and the *Terrestrial Planet Finder Mission (TPF),* will be used to study the extrasolar planets. *SIM* will have special telescopes that will be able to detect Earth-sized extrasolar planets. Later, the *TPF* will be used to study the compositions of the atmospheres and surfaces of the planets.

Activity

Visit your library or the Earth Science Web Site at earthgeu.com to research the most recent extrasolar planet discoveries. Present your findings to the class on a poster or in an oral report.

CHAPTER 29

Study Guide

Summary

SECTION 29.1

Overview of Our Solar System

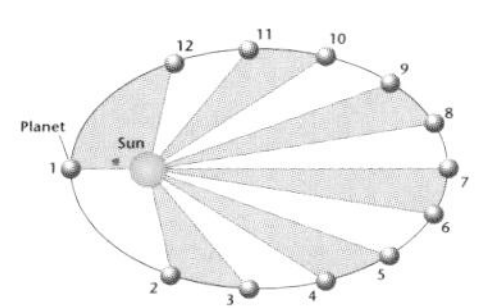

Main Ideas

- Early astronomers explained the motions of the planets with geocentric models, including epicycles.
- Copernicus, Brahe, Kepler, and Galileo developed evidence supporting a heliocentric solar system model.
- Newton developed a law of gravitation that was used to demonstrate the validity of the heliocentric model.

Vocabulary

aphelion (p. 777)
astronomical unit (p. 777)
eccentricity (p. 777)
perihelion (p. 777)
retrograde motion (p. 776)

SECTION 29.2

The Terrestrial Planets

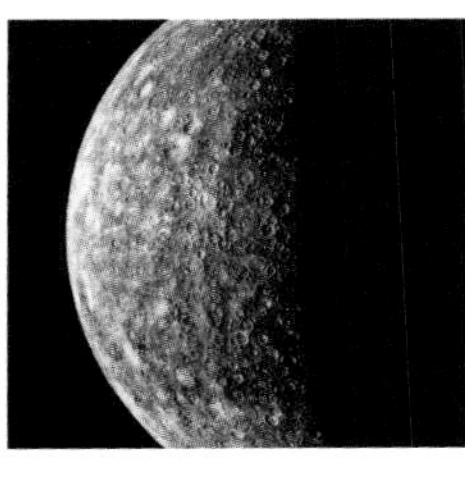

Main Ideas

- The terrestrial planets include the four planets closest to the Sun. They are relatively small and dense, and they have rocky surfaces.
- Mercury has a surface similar to the Moon's, but a very different interior.
- Venus has an extremely hot surface as a result of greenhouse heating, but is similar to Earth in other properties.
- Earth is suitable for life because of its unique orbital position that allows water to exist in all three phases on the surface.
- Mars shows signs of having once had tectonic activity.

Vocabulary

gas giant planet (p. 780)
precession (p. 783)
terrestrial planet (p. 780)

SECTION 29.3

The Gas Giant Planets

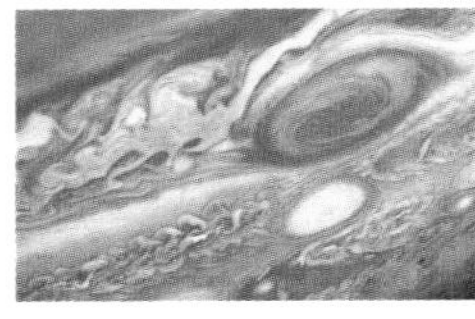

Main Ideas

- The gas giant planets are very large and have low densities, no solid surfaces, ring systems, and many moons.
- Jupiter is the largest of the planets. It has a fluid interior, except for a small rocky core, and several moons. Saturn is slightly smaller than Jupiter and has a more extensive ring system.
- Uranus and Neptune are very similar in size and composition.
- Pluto is not classified as a gas giant or a terrestrial planet.

Vocabulary

belt (p. 787)
liquid metallic hydrogen (p. 787)
zone (p. 787)

SECTION 29.4

Formation of Our Solar System

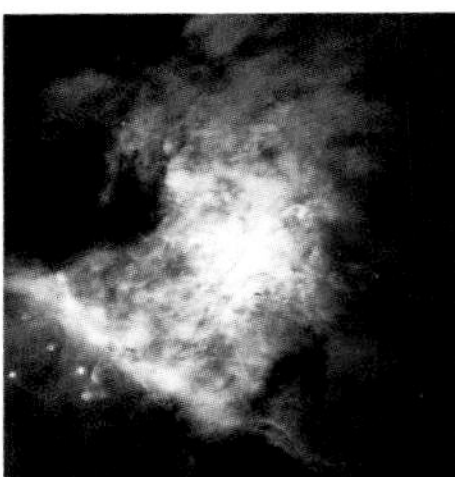

Main Ideas

- The solar system formed from a collapsing interstellar cloud that flattened into a disk from which the planets formed.
- Terrestrial planets formed from refractory materials in the hot inner disk, and gas giants formed from volatile elements in the cold outer disk.
- Asteroids are rocky remnants of the early solar system. Most of them orbit the Sun between Mars and Jupiter.
- Comets have highly eccentric orbits and are made of rock and ice. When they are close to the Sun, they glow brightly and have a head and tails of gas and dust.

Vocabulary

asteroid (p. 795)
coma (p. 797)
comet (p. 796)
meteor (p. 796)
meteor shower (p. 797)
meteorite (p. 796)
meteoroid (p. 796)
nucleus (p. 797)
planetesimal (p. 794)

earthgeu.com/vocabulary_puzzlemaker

CHAPTER 29
Assessment

Understanding Main Ideas

1. What is a planet's backward motion in the sky called?
   a. revolving
   b. retrograde motion
   c. planetary spin
   d. geocentric motion

2. What point in a planet's orbit is closest to the Sun?
   a. focus
   b. semimajor axis
   c. aphelion
   d. perihelion

3. What model of our solar system did Copernicus propose?
   a. a heliocentric model
   b. a retrograde model
   c. a geocentric model
   d. a nested-sphere model

4. A planet's average distance from the Sun is also what part of the orbital ellipse?
   a. the distance between the foci
   b. the eccentricity
   c. the semimajor axis
   d. the circumference

5. How were the plains on Mercury hypothesized to have formed?
   a. from large oceans smoothing out the surface
   b. from lava flows
   c. from wind eroding high points
   d. from landslides on the scarps

6. Why does the Sun rise in the west and set in east when viewed from Venus?
   a. Venus is the closest planet to Earth.
   b. Venus has a retrograde rotation.
   c. Venus's orbital period is 224 Earth days.
   d. The semimajor axis of Venus is only 0.723 AU from the Sun.

7. What two planets have similar appearances, atmospheric compositions, interiors, and particle belts?
   a. Neptune and Uranus
   b. Mars and Venus
   c. Saturn and Neptune
   d. Saturn and Uranus

8. Which condensed into solid form closest to the Sun?
   a. refractory elements
   b. volatile elements
   c. gas giants
   d. the Oort cloud

9. What does Kepler's third law state?

10. Which planet is not classified as a gas giant or a terrestrial planet? Why?

11. Why is the surface temperature of Venus 750 K (476°C)?

Use the diagram below to answer question 12.

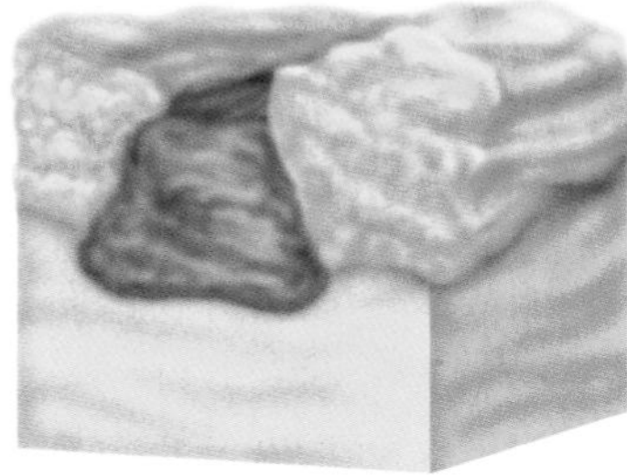

12. Identify a belt and a zone. What is the direction of movement of each?

Test-Taking Tip

MNEMONICS Use memory devices to help you remember terms and orders. For example, My Very Elegant Monkey Just Sat Upon Nancy's Petunias can help you remember the order of the planets in our solar system.

Assessment

13. If the Martian atmosphere is 95 percent carbon dioxide, why doesn't Mars have a strong greenhouse effect?
14. Where are asteroids hypothesized to have originated?
15. What new discovery was made from the *Voyager 1* and *Voyager 2* visits to Jupiter?
16. What does the density of Pluto indicate that it is made of?

Applying Main Ideas

17. What were some of the advantages of the Sun-centered model proposed by Copernicus?
18. How are Venus and Earth similar, and how are they different?
19. How has Jupiter's gravitational force affected the geology of its four major moons?
20. What are the differences between asteroids and comets?
21. How did collisions during the formation of the planets affect Earth, Mercury, Venus, and Uranus?
22. Would it be possible for a gas giant to form close to the Sun? Explain.
23. How are the rings of the gas giants theorized to have formed?

Thinking Critically

24. Why are the volcanoes on Mars so much larger than those on Earth?
25. Why are the rings of each gas giant planet closer to the planet than its large moons are?
26. Should Pluto be considered a planet or a planetesimal? Explain.
27. How would a circular orbit affect Kepler's second and third laws? Restate these laws using the radius of a circle.

Standardized Test Practice

INTERPRETING SCIENTIFIC ILLUSTRATIONS Use the diagram below to answer questions 1 and 2.

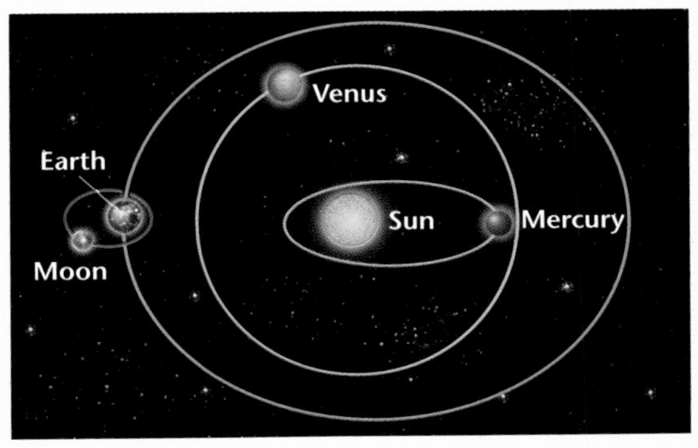

1. Which planet is located at aphelion in its orbit?
   a. Mercury
   b. Venus
   c. Earth
   d. none of the above
2. Which orbit shown has an eccentricity that is closest to 0?
   a. Mercury
   b. Venus
   c. Earth
   d. the Moon
3. How are Mercury and the Moon similar?
   a. Both are covered with craters and plains.
   b. Both have the same night-day temperature difference.
   c. They have the same strength of surface gravity.
   d. Both have an extensive nickel-iron core.
4. What is a piece of interplanetary material that burns up in Earth's atmosphere called?
   a. a meteorite
   b. an asteroid
   c. a meteor
   d. a meteoroid

Chapter 30

Stars

What You'll Learn

- What the structure and processes of the Sun are.
- What properties are used to observe and measure stars.
- How stars change during their lives and what is left when they die.

Why It's Important

The Sun is vital to life on Earth. To understand the Sun, which is a star, it is necessary to understand how all stars function and evolve. Stars are also the building blocks of our galaxy and the source of most elements in the universe.

Earth Science Online

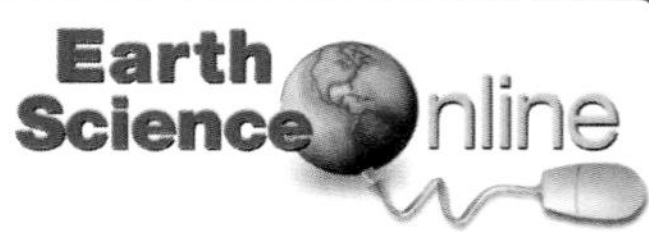

To find out more about the Sun and other stars, visit the Earth Science Web Site at earthgeu.com

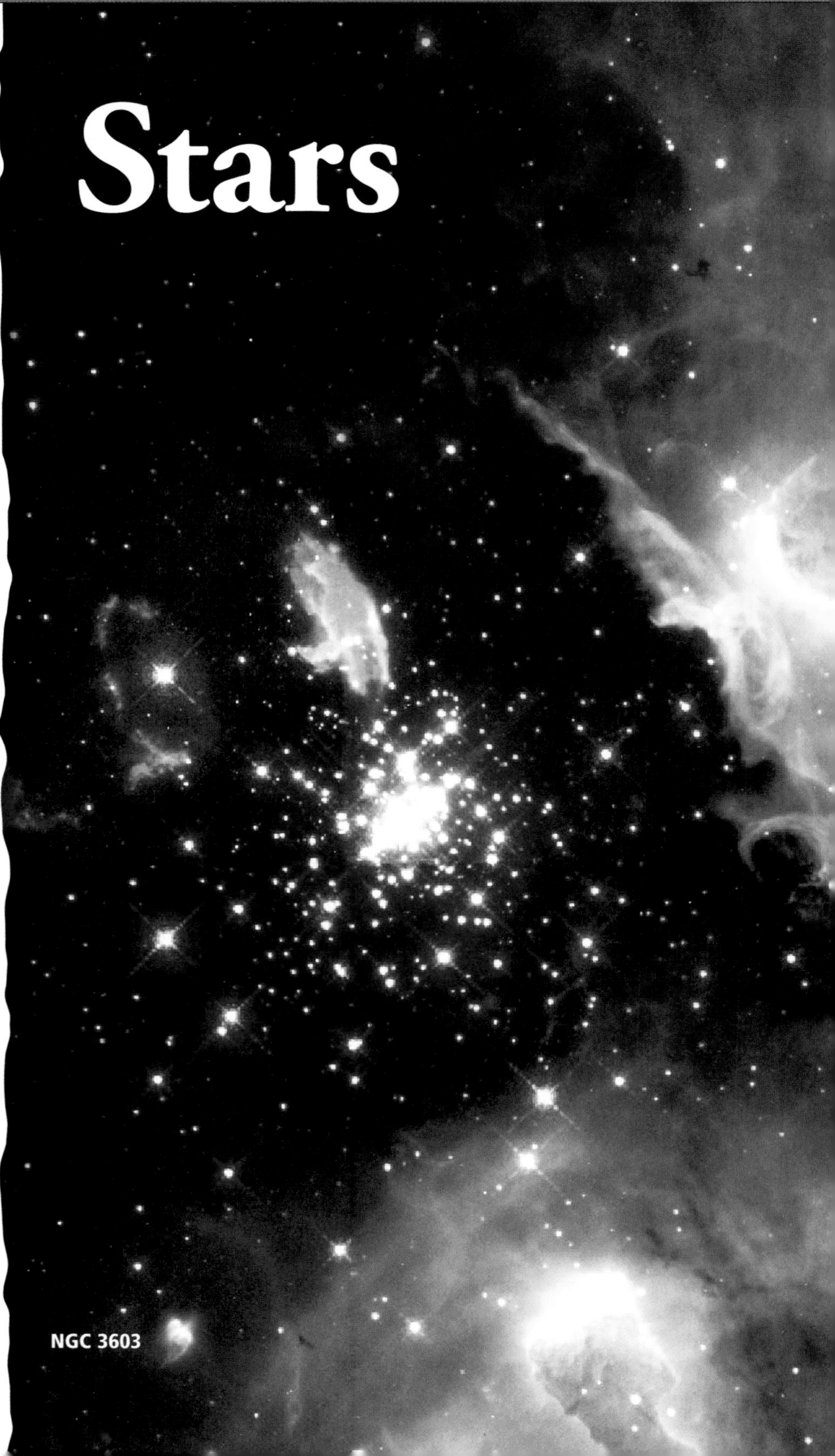

NGC 3603

Discovery Lab Monitoring the Sun

Although the Sun is an average star, it has many complex processes. We get a glimpse of these processes through the solar activity cycle. The activity cycle of the Sun refers to how active its surface is. During the peak of the activity cycle, the surface is violent and eruptive and has numerous dark spots. The activity cycle varies with a period of approximately 11 years and last peaked in late 2000.

1. Find sites on the Web that provide current images of the Sun and observe how the Sun looks today when viewed at different wavelengths. Or, go to earthgeu.com and follow the links.
2. Make notes of features that you observe and the wavelengths from which you observe them.
3. Find out where the Sun currently is in its activity cycle, and compare what you actually observe with what you would expect for this time in the cycle.

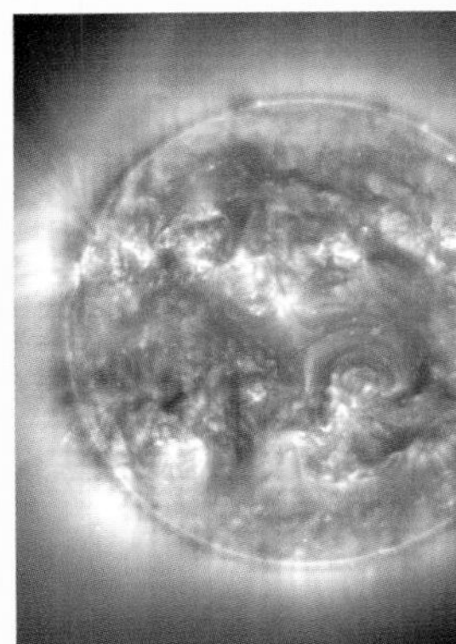

Observe Is the Sun near the peak of its activity cycle? Do the solar images that you observed fit with your expectation of where the Sun is in its activity cycle? Compare and contrast the images obtained at different wavelengths.

SECTION 30.1 The Sun

Earth Sciences 1.e

OBJECTIVES

- **Explore** *the structure of the Sun.*
- **Describe** *the solar activity cycle and how the Sun affects Earth.*
- **Compare** *the different types of spectra.*

VOCABULARY

photosphere
chromosphere
corona
solar wind
sunspot
solar flare
prominence
fusion
fission
spectrum

Humans have probably always been aware of the Sun. However, it has been only recently that astronomers have begun to understand its nature. Through observations and probes such as the *Solar Heliospheric Observatory* (SOHO) and the *Ulysses* mission, astronomers have begun to unravel the mysteries of the Sun. Astronomers still rely on computer models for an explanation of the interior of the Sun because the interior cannot be directly observed.

PROPERTIES OF THE SUN

The Sun is the largest object in the solar system, in both size and mass. It would take 109 Earths lined up edge to edge to fit across the Sun, or almost 10 Jupiters. The Sun is about 330 000 times as massive as Earth and 1048 times the mass of Jupiter. In fact, the Sun contains more than 99 percent of all the mass in the solar system. It should therefore come as no surprise that the Sun's mass controls the motions of the planets and other objects.

Table 30-1 Properties of the Sun			
	Sun	Earth	Jupiter
diameter (km)	1.392×10^6	1.2756×10^4	1.4298×10^5
mass (kg)	1.99×10^{30}	5.9736×10^{24}	1.8986×10^{27}
density (kg/m^3)	1.408×10^3	5.515×10^3	1.326×10^3

The Sun's average density is similar to the densities of the gas giant planets, represented by Jupiter in ***Table 30-1.*** Astronomers can deduce densities at specific points inside the Sun, as well as other information, only by using computer models that explain the observations that they make. These models show that the density in the center of the Sun is about 1.50×10^5 kg/m^3, which is about thirteen times the density of lead! A pair of dice having this density would weigh about two pounds. However, unlike lead, which is a solid, the solar interior is gaseous throughout because of its high temperature—about 1×10^7 K in the center. At this high temperature many of the gases are completely ionized, meaning that they are composed only of atomic nuclei and electrons. This state of matter is known as plasma. The outer layers of the Sun are not quite hot enough to be plasma.

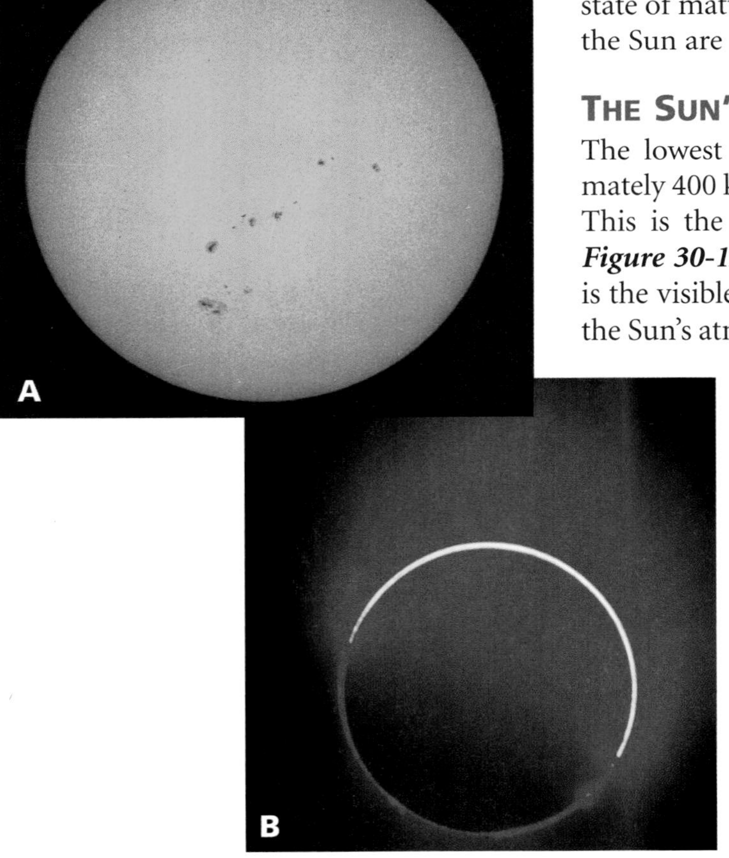

Figure 30-1 The photosphere **(A)** is the visible surface of the Sun. This annular eclipse **(B)** on May 30, 1994 shows the red chromosphere of the Sun in the lower half of the image.

The Sun's Atmosphere

The lowest layer of the Sun's atmosphere, approximately 400 km in thickness, is called the **photosphere.** This is the visible surface of the Sun, as shown in ***Figure 30-1A.*** You may wonder why the photosphere is the visible surface when it is also the lowest layer of the Sun's atmosphere. This is because most of the light emitted by the Sun comes from this layer. The two layers above are transparent at most wavelengths of visible light. Additionally, the top two layers are very dim in the wavelengths that they do emit. The average temperature of the photosphere is about 5800 K.

Above the photosphere is the **chromosphere,** which is approximately 2500 km in thickness and has a temperature of nearly 30 000 K at the top. Normally, the chromosphere is visible only during a solar eclipse, when the photosphere is blocked.

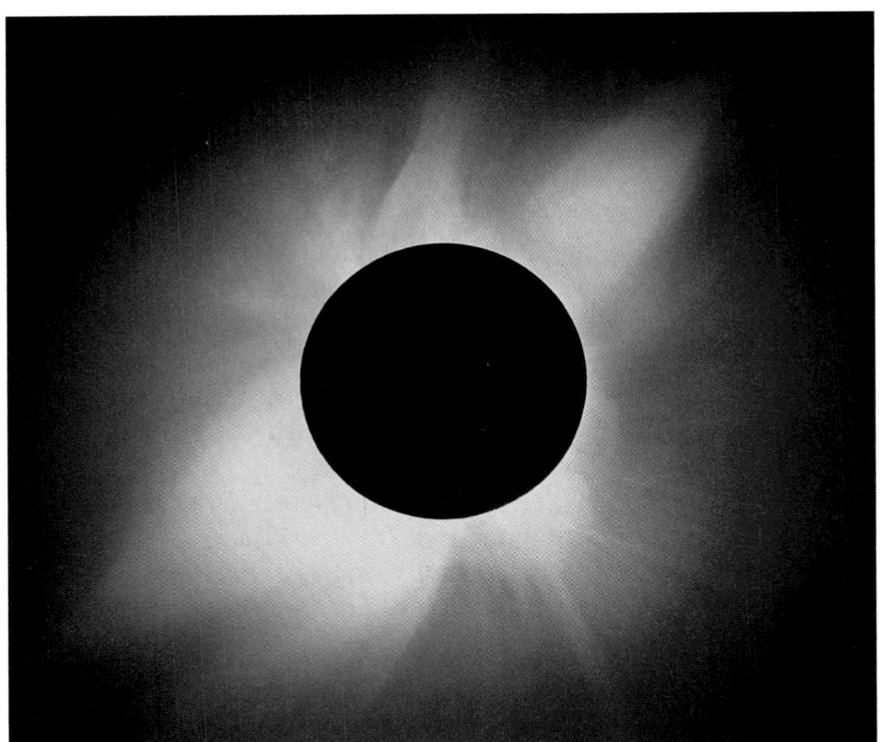

Figure 30-2 The faint corona was a fabulous sight during the eclipse on July 11, 1991 in Baja, California.

However, astronomers can use special filters to observe the chromosphere when the Sun is not eclipsed. The chromosphere appears red, as shown in ***Figure 30-1B,*** because it emits most strongly in a narrow band of red wavelengths.

The top layer of the Sun's atmosphere, called the **corona,** extends several million kilometers from the top of the chromosphere and has a temperature range of 1 million to 2 million K. The density of the gas in the corona is very low, which explains why the corona is so dim that it can be seen only when the photosphere is blocked by either special instruments, as in a coronagraph, or by the Moon during an eclipse, as in ***Figure 30-2.***

Solar Wind The corona of the Sun does not end abruptly. Instead, gas flows outward from the corona at high speeds and forms the **solar wind.** As this wind of charged particles, or ions, flows outward through the entire solar system, it bathes each planet in a flood of particles. At 1 AU, Earth's distance from the Sun, the solar wind flows at a speed of about 400 km/s. The charged particles are deflected by Earth's magnetic field and are trapped in two huge rings in Earth's magnetic field called the Van Allen belts. The high-energy particles in these belts collide with gases in Earth's atmosphere and cause the gases to give off light that we see as the aurora, shown in ***Figure 30-3.***

Figure 30-3 The aurora is the result of the particles from the Sun colliding with gases in Earth's atmosphere. It is most easily viewed from regions around the poles of Earth.

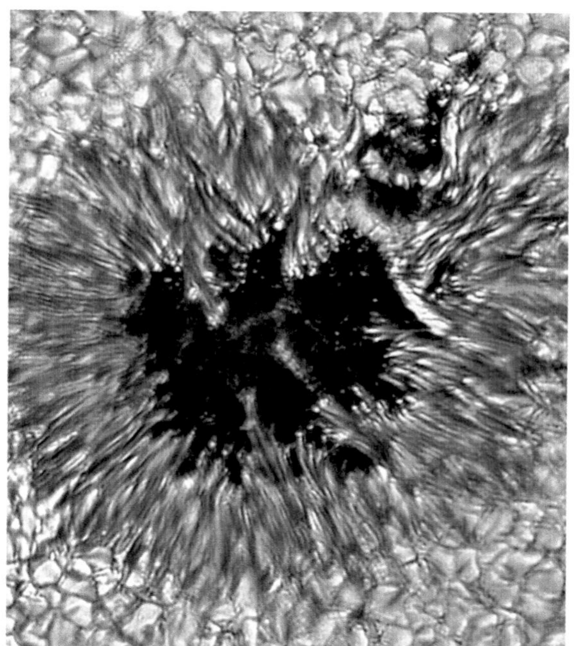

Figure 30-4 Sunspots consist of two regions: the lighter outer ring called the penumbra, and the dark inner ring called the umbra.

Solar Activity

While the solar wind and layers of the Sun's atmosphere are permanent features, other features on the Sun change over time in a process called solar activity. The Sun's magnetic field disturbs the solar atmosphere periodically and causes new features to appear. The most obvious features are **sunspots,** shown in ***Figure 30-4,*** which are dark spots on the surface of the photosphere. Sunspots are actually very bright, but they appear darker than the surrounding areas on the Sun because they are cooler. They are located in regions where the Sun's intense magnetic fields poke through the photosphere. These magnetic fields prevent hot gases inside the Sun from rising to the surface and heating the spots. Sunspots typically last two months. They occur in pairs with opposite magnetic polarities—with a north and a south pole like a bar magnet.

Solar Activity Cycle Astronomers have observed that the number of sunspots changes regularly and on average, reaches a maximum number every 11.2 years on average. Scientists therefore hypothesized that the solar activity cycle is 11.2 years in length. However, when the polarity of the Sun's magnetic field is taken into account, the length of the cycle doubles to 22.4 years. The Sun's magnetic field reverses, so that the north magnetic pole becomes the south magnetic pole and vice versa. When the polarities of the Sun's magnetic poles reverse, the polarities of pairs of sunspots also reverse, because sunspots are caused by magnetic fields. Thus, the solar activity cycle starts with minimum spots and progresses to maximum spots. Then the magnetic field reverses in polarity, and the spots start at a minimum number and progress to a maximum number again. The magnetic field then switches back to the original polarity and completes the solar activity cycle.

Figure 30-5 The dark regions in this X-ray image are coronal holes.

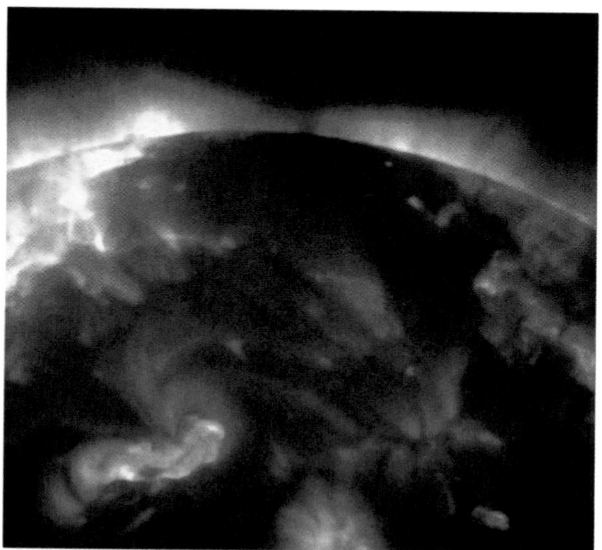

Other Solar Features Coronal holes, shown in ***Figure 30-5,*** are often located over sunspot groups. Coronal holes are areas of low density in the gas of the corona. They are the main regions from which the particles that comprise the solar wind escape. These holes are visible in the X ray region of the electromagnetic spectrum.

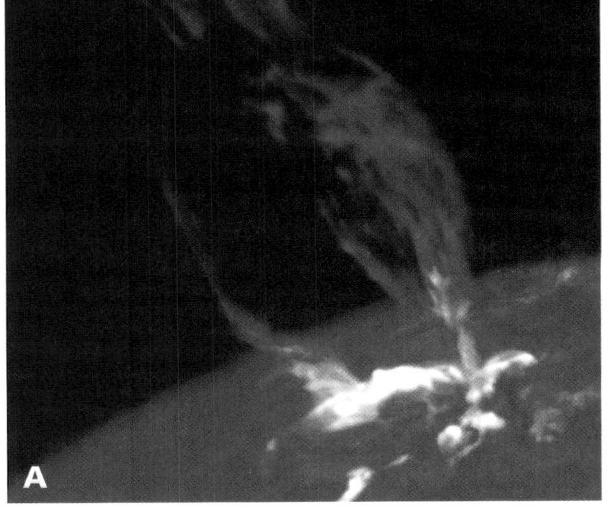

Figure 30-6 This solar flare **(A)** was observed by the National Solar Observatory, and this solar prominence **(B)** was observed by the Big Bear Solar Observatory of the New Jersey Institute of Technology.

Highly active solar flares also are associated with sunspots. **Solar flares** are violent eruptions of particles and radiation from the surface of the Sun, as shown in ***Figure 30-6A.*** Often, the released particles escape the surface of the Sun in the solar wind and Earth gets bombarded with the particles a few days later. The largest solar flare on record, which occurred in April 2001, hurled particles from the Sun's surface at 7.2 million km/hr.

Another active feature, sometimes associated with flares, is a **prominence,** shown in ***Figure 30-6B,*** which is an arc of gas that is ejected from the chromosphere, or gas that condenses in the inner corona and rains back to the surface. Prominences can reach temperatures greater than 50 000 K and can last from a few hours to a few months. Like flares, prominences also are associated with sunspots, and hence, occurrences of both vary with the solar activity cycle.

 **Earth Sciences**

1.e Students know the Sun is a typical star and is powered by nuclear reactions, primarily the fusion of hydrogen to form helium.

Impact on Earth There is evidence that the solar activity cycle affects climates on Earth. For example, some scientists have found evidence of subtle climate variations within 11-year periods. Also, there were severe weather changes on Earth during the latter half of the 1600s when the solar activity cycle stopped and there were no sunspots for nearly 60 years. No one knows why the Sun's cycle stopped. Those 60 years were known as the "Little Ice Age" because the weather was very cold in Europe and North America during those years.

ENVIRONMENTAL CONNECTION

The Solar Interior

You may be wondering where all the energy that causes solar activity and light comes from. Within the core of the Sun, where the pressure and temperature are extremely high, fusion occurs. **Fusion** is the combining of lightweight nuclei, such as hydrogen, into heavier nuclei. This is the opposite of the process of **fission,** which is the splitting of heavy atomic nuclei into smaller, lighter atomic nuclei.

In the core of the Sun, helium is a product of the process in which hydrogen nuclei fuse. The mass of the helium nucleus is less than the combined mass of the hydrogen nuclei, which means that mass is being lost during the process somehow. Albert Einstein's theory of special relativity showed that mass and energy are equivalent, and that matter can be converted into energy and vice versa. This relationship can be expressed as $E = mc^2$, where E is energy measured in joules, m is the quantity of mass that is converted to energy measured in kilograms, and c is the speed of light measured in m/s. This theory explains that the mass lost in the fusion of hydrogen to helium is converted to energy, which powers the Sun. At the Sun's rate of hydrogen fusing, it is about halfway through its lifetime, with another 5 billion years or so left.

Earth Science Online

Topic: Farthest Star
To find out more about the Sun and other stars, visit the Earth Science Web Site at earthgeu.com

Activity: Research the most recent discoveries of stars. Which star is the most distant from Earth? How many light years or AUs is the star from the Earth?

Energy from the Sun The quantity of energy that arrives on Earth every day from the Sun is enormous. Above Earth's atmosphere, 1354 J of energy is received in 1 m^2 per second (1354 W/m^2). In other words, thirteen 100-W lightbulbs could be operated with the solar energy that strikes a 1-m^2 area. However, not all of this energy reaches the ground because some is absorbed and scattered by the atmosphere. You will learn how energy from the Sun can be converted to electricity with solar panels in the *Problem-Solving Lab* on this page.

Problem-Solving Lab

Using Numbers

Calculate energy output from solar panels The energy from the Sun striking the top of Earth's atmosphere is 1354 W/m^2, about half of which reaches the ground. The energy output is measured in watts, because it is energy received per second. Assume that solar panels can convert about 15 percent of the Sun's energy into electricity.

Analysis

1. How much energy from the Sun reaches the ground per square meter per second?
2. How much of the Sun's energy would a 1-m^2 solar panel receive per second?
3. After the 1-m^2 solar panel converts the energy to electricity, how much electricity is available per second?

Thinking Critically

4. How many 1-m^2 solar panels would be needed to produce 1000 W of electricity for a house?
5. How would a cloudy region affect the energy production of solar panels? Is there anything that could be done to counteract this? Explain.

Solar Zones If the energy of the Sun is produced in the core, how does it get to the surface before it travels to Earth? The answer lies in the two zones in the solar interior. Above the core is a region called the radiative zone, which extends approximately 86 percent of the way to the photosphere. In this zone, energy is transferred from particle to particle by radiation, as atoms continually absorb energy and then re-emit it. Above the radiative zone is the convective zone. In this zone, moving volumes of gas carry the energy the rest of the way to the Sun's surface through convection, which you learned about in Chapter 11. The radiative and convective zones are illustrated in ***Figure 30-7.***

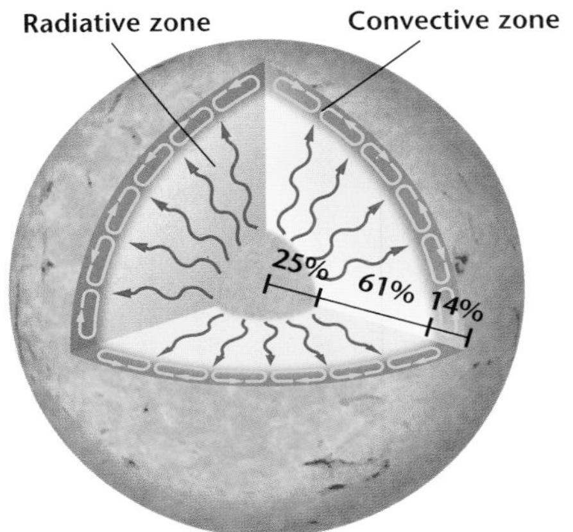

Figure 30-7 It takes about 170 000 years to transfer energy to the surface of the Sun through these zones.

SPECTRA

You are probably familiar with the rainbow that appears when white light is shined through a prism. This rainbow is a **spectrum,** which is visible light arranged according to wavelengths. There are three types of spectra: continuous, emission, and absorption, as shown in ***Figure 30-8.*** All three types will be discussed on the next few pages. A spectrum that has no breaks in it, such as the one produced when light from an ordinary bulb is shined though a prism, is called a continuous spectrum. A continuous spectrum also can be produced by a glowing solid or liquid, or by a highly compressed, glowing gas. However, if you were to observe a spectrum coming from a non-compressed gas, you would see bright lines at certain wavelengths. This is called an emission spectrum, and the lines are called emission lines. The wavelengths of the lines you see depend on the element being observed, because each element has its own characteristic emission spectrum.

Figure 30-8 A continuous spectrum is produced by a hot solid, liquid, or dense gas. When a cloud of gas is in front of this hot source, an absorption spectrum is produced. A cloud of gas without a hot source behind it will produce an emission spectrum **(A).** The Sun's spectrum **(B)** is an absorption spectrum.

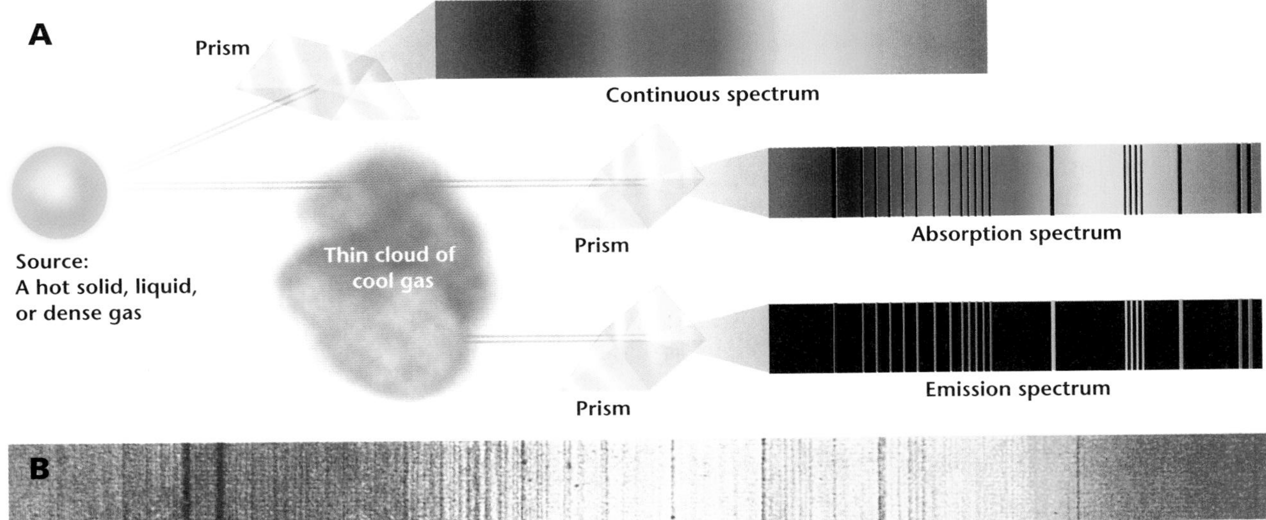

Composition of the Sun by Mass

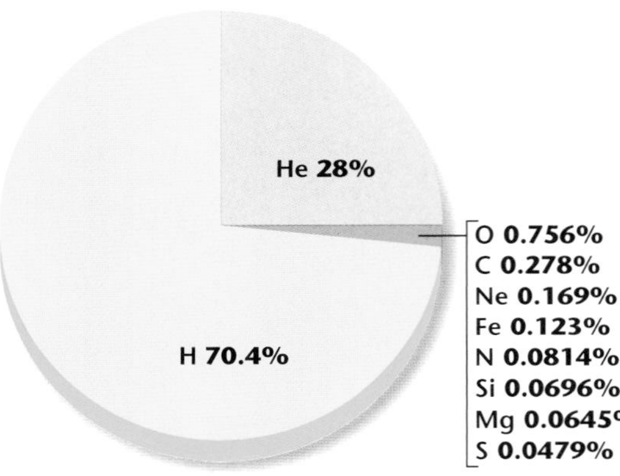

Figure 30-9 The Sun is primarily composed of hydrogen and helium. Trace elements in the Sun that are not listed are Li, Be, B, F, Al, P, Cl, Ar, K, Ca, Sc, Tr, V, Cr, Mn, Co, Ni, Cu, and Zn.

Conversely, if you observe the light from the Sun in a spectrum, you will see a series of dark bands. These dark spectral lines are caused by different chemical elements that absorb light at specific wavelengths. This is called an absorption spectrum, and the lines are called absorption lines. Absorption is caused by a cooler gas in front of a source that emits a continuous spectrum. The absorption lines caused by the element in the gas are in the exact same location as the emission lines made by the same element. Thus, by comparing laboratory spectra of different gases with the dark lines in the solar spectrum, it is possible to identify the elements that make up the Sun's outer layers. You will experiment with identifying spectral lines in the *GeoLab* at the end of this chapter.

SOLAR COMPOSITION

The Sun consists of hydrogen, about 70.4 percent by mass, and helium, 28 percent, as well as a small amount of other elements, as illustrated in ***Figure 30-9.*** This composition is very similar to that of the gas giant planets, which suggests that the Sun and the gas giants represent the composition of the interstellar cloud from which the solar system formed, while the terrestrial planets have lost most of the lightweight gases, as you learned in Chapter 29. The Sun's composition represents that of the galaxy as a whole. Most stars have proportions of the elements similar to the Sun. Hydrogen and helium are the predominant gases in stars, as well as in the entire universe. All other elements are in very small proportions compared to hydrogen and helium.

SECTION ASSESSMENT

1. How do astronomers know what conditions exist inside the Sun?
2. Describe the layers of gas above the Sun's visible surface.
3. How does energy produced in the core of the Sun reach the surface? How long does it take?
4. How are the Sun's magnetic field and its activity cycle related?
5. How are the different types of spectra created?
6. **Thinking Critically** How would the Sun affect Earth if Earth did not have a magnetic field?

SKILL REVIEW

7. **Comparing and Contrasting** Compare and contrast solar flares, prominences, and sunspots. For more help, refer to the *Skill Handbook*.

SECTION 30.2 Measuring the Stars

Earth Sciences 1.d, 2.d, 2.f

When you look up at the sky at night, you can often see the brightest stars, even in the city. These stars appear to be fairly isolated from each other. However, away from city lights, you would notice many more stars grouped together.

OBJECTIVES

- **Describe** *star distribution and distance.*
- **Classify** *the types of stars.*
- **Summarize** *the interrelated properties of stars.*

VOCABULARY

constellation
binary star
parallax
apparent magnitude
absolute magnitude
luminosity
Hertzsprung-Russell diagram
main sequence

Groups of Stars

Long ago, many civilizations looked at the brightest stars and named groups of them after animals, mythological characters, or everyday objects. These groups of stars are called **constellations.** Today, we group stars by the 88 constellations named by ancient peoples. Some constellations can be seen all year long, depending on the observer's location. In the northern hemisphere, you can see constellations that appear to move around the north pole of Earth. These constellations are called circumpolar constellations. Ursa Major, also known as the Big Dipper, is a circumpolar constellation for the northern hemisphere.

Unlike circumpolar constellations, the other constellations can be seen only at certain times of the year because of Earth's changing position in its orbit around the Sun, as illustrated in ***Figure 30-10.*** For example, the constellation Orion can be seen only in the northern hemisphere's winter, and the constellation Hercules can be seen only in the northern hemisphere's summer. This is why constellations are classified as summer, fall, winter, and spring constellations. For maps of the constellations, see *Appendix K.*

Star Clusters Although the stars in constellations appear to be close to each other, very few are gravitationally bound to one other. The reason that they appear to be close together is that human eyes can't distinguish how far or near stars actually are. Two stars could

Figure 30-10 Depending on the time of year, only certain constellations are visible. (not to scale)

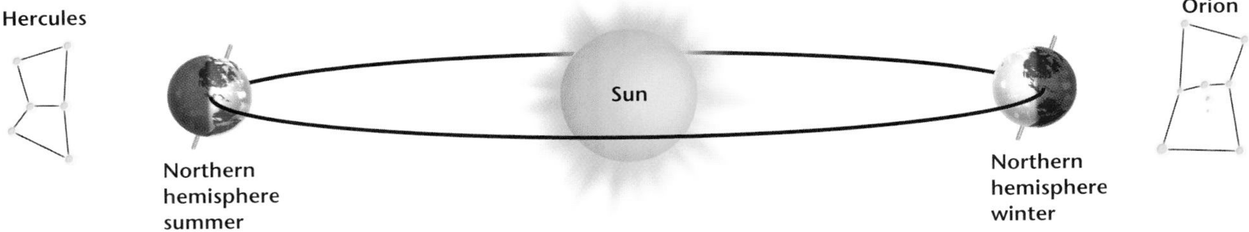

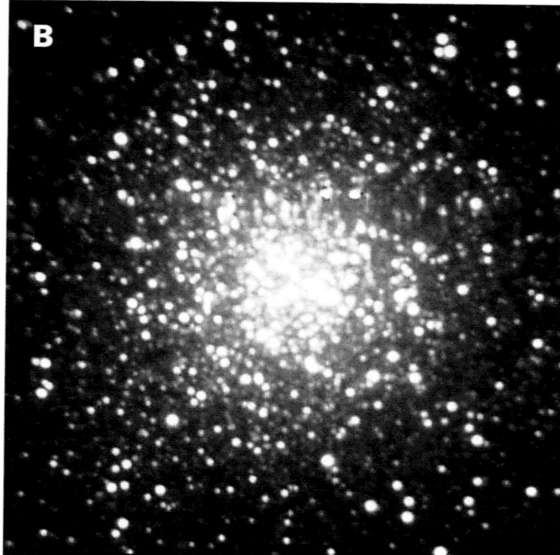

Figure 30-11 The Pleiades **(A)** is an open cluster, and M13 **(B)** is a densely packed globular cluster.

appear to be right next to each other, almost touching, but one might be 1 trillion km away from Earth, and the other might be 2 trillion km away from Earth. However, by measuring distances to stars and observing how they interact with each other, scientists can determine which stars are gravitationally bound to each other. A group of stars that are gravitationally bound to each other is called a cluster. The Pleiades, in the constellation Taurus, shown in ***Figure 30-11A,*** is an open cluster because the stars are not densely packed. In contrast, a globular cluster is a group of stars that are densely packed into a spherical shape, such as M13, in the constellation Hercules, shown in ***Figure 30-11B.***

Figure 30-12 The bright star Sirius and its companion white dwarf, on the left, are a binary system.

Binaries When only two stars are gravitationally bound together and orbit a common center of mass, they are called a **binary star.** More than half of the stars in the sky are either binary stars or members of multiple-star systems. The bright star Sirius is actually a binary system, as shown in ***Figure 30-12.*** Most binary stars appear to be single stars to the human eye, even with a telescope. The two stars are usually too close together to appear separately and one of the two is often much brighter than the other. Astronomers are able to identify binary stars through several methods. For example, even if only one star is visible, accurate measurements can show that its position shifts back and forth as it orbits the center of mass between it and the unseen companion star. Also, the orbital plane of a binary system can sometimes be seen edge-on from Earth. In such cases, the two stars alternately block each other out and cause the total brightness of the two-star system to dip each time one star eclipses the other. This type of binary star is called an eclipsing binary.

STELLAR POSITIONS AND DISTANCES

Astronomers use two units of measure for long distances. One, which you are probably familiar with, is a light-year (ly). A light-year is the distance that light travels in one year, equal to 9.461×10^{12} km. Astronomers often use a larger unit than a light-year, a parsec. A parsec (pc) is equal to 3.26 ly, or 3.086×10^{13} km.

Precise position measurements are an important tool for finding distances to stars. To estimate the distance of stars from Earth, astronomers make use of the fact that nearby stars shift in position as observed from Earth. This apparent shift in position caused by the motion of the observer is called **parallax.** In this case, the motion of the observer is the change in position of Earth as it orbits the Sun. As Earth moves from one side of its orbit to the opposite side, a nearby star appears to be shifting back and forth, as illustrated in ***Figure 30-13.*** The closer the star, the larger the shift. The distance to a star can be estimated from its parallax shift. In fact, a parsec is defined as the distance at which an object has a parallax of 1 arcsecond. Using the parallax technique, astronomers only could find accurate distances to stars up to about 100 pc, or approximately 300 ly, away until recently. Now with advancements in technology, such as the *Hipparcos* satellite, astronomers can find accurate distances up to 500 pc by using parallax. In the *MiniLab* later in this section, you will experiment with distance and how it affects parallax shifts.

Earth Sciences

2.d Students know that stars differ in their life cycles and that visual, radio, and X-ray telescopes may be used to collect data that reveal those differences.

1.d Students know the evidence indicating that the planets are much closer to Earth than the stars are.

Basic Properties of Stars

The basic properties of stars include diameter, mass, brightness, energy output (power), surface temperature, and composition. The diameters of stars range from as little as 0.1 times the Sun's diameter to hundreds of times larger, while their masses vary from a little less than 0.01 to 20 or more times the Sun's mass. The most massive stars can be as massive as 50 to 100 Suns but are extremely rare.

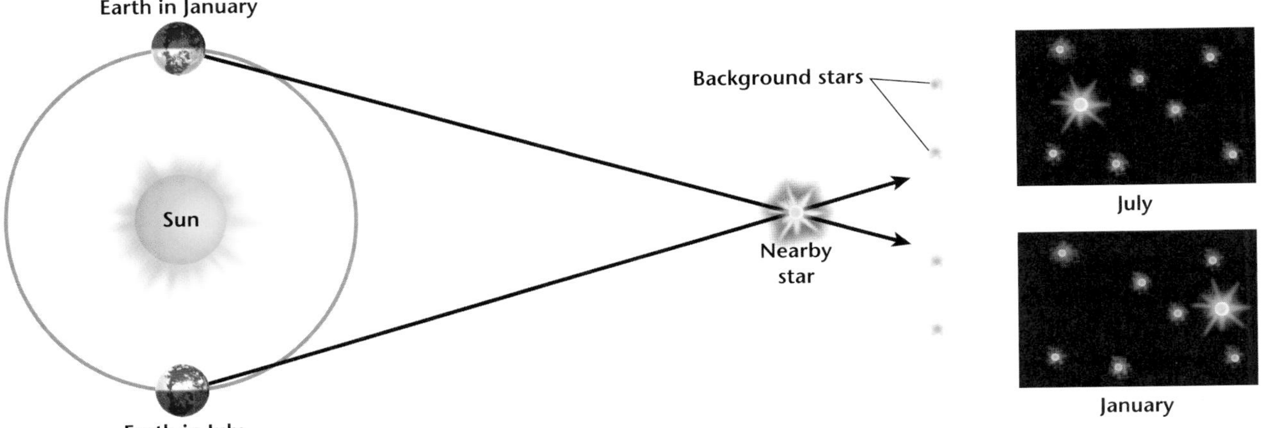

Figure 30-13 The shift in position of a star as viewed from opposite sides of Earth's orbit around the Sun is called parallax.

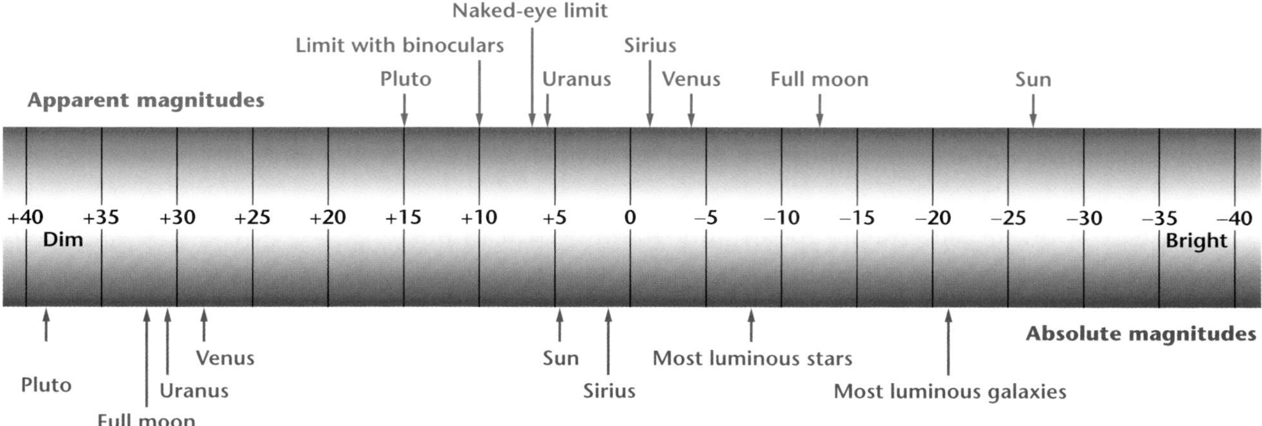

Figure 30-14 This shows the apparent and absolute magnitudes of some familiar celestial objects.

Magnitude One of the most basic observable properties of a star is how bright it appears. The ancient Greeks established a classification system based on the brightnesses of stars. The brightest stars were given a ranking of +1, the next brightest +2, and so on. Today's astronomers still use this system, but they have refined it.

Using Numbers The difference in brightness between a magnitude +12 star and a magnitude +9 star is $2.512 \times 2.512 \times 2.512 = 2.512^3 = 15.85$. What is the difference in brightness between a magnitude +21 star and a magnitude +14 star?

Apparent Magnitude Astronomers have defined the ancient Greek system of classification as **apparent magnitude,** or how bright a star appears to be. In this system, a difference of 5 magnitudes corresponds to a factor of 100 in brightness. Thus, a magnitude +1 star is 100 times brighter than a magnitude +6 star. A difference of 1 magnitude corresponds to a factor of 2.512 in brightness. The modern magnitude system extends to objects that are both brighter and fainter than those that were included in the ancient Greek system. For objects brighter than magnitude +1, such as the Sun, the Moon, Venus, and some of the very brightest stars, negative numbers are assigned. The apparent magnitudes of several objects are shown in ***Figure 30-14.***

Absolute Magnitude Apparent magnitude does not actually indicate how bright a star is, because it does not take distance into account. A faint star can appear to be very bright because it is relatively close to Earth, while a bright star can appear to be faint because it is far away. To account for this phenomenon, astronomers have developed another classification system for brightness. **Absolute magnitude** is the brightness an object would have if it were placed at a distance of 10 pc. The classification of stars by absolute magnitude allows comparisons that are based on how bright the stars would appear at equal distances from an observer. The absolute magnitudes for several objects are shown in ***Figure 30-14.*** The disadvantage of absolute magnitude is that it can be calculated only when the actual distance to a star is known.

Luminosity Apparent magnitudes do not give an actual measure of energy output. To measure the energy output from the surface of a star per second, called its **luminosity,** an astronomer must know both the star's apparent magnitude and how far away it is. The brightness we observe for a star depends on both its luminosity and its distance, and because brightness diminishes with the square of the distance, a correction must be made for distance. Luminosity is measured in units of energy emitted per second, or watts. The Sun's luminosity is about 3.85×10^{26} W. This is equivalent to 3.85×10^{24} 100-W lightbulbs! The values for other stars vary widely, from about 0.0001 to more than a million times the Sun's luminosity. No other stellar property varies by so much.

Spectra of Stars

You have learned that the Sun has dark absorption lines at specific wavelengths in its spectrum. Other stars also have dark absorption lines in their spectra and are classified according to their patterns of absorption lines.

Classification Stars are assigned spectral types in the following order: O, B, A, F, G, K, and M. Each class is subdivided into more specific divisions with numbers from 0 to 9. For example, a star may be classified as being a type A4 or A5. The classes were originally based only on the pattern of spectral lines, but astronomers later discovered that the classes correspond to stellar temperatures, with the O stars being the hottest and the M stars being the coolest. Thus, by examination of a star's spectrum, it is possible to estimate its temperature. The Sun is a type G2 star, which corresponds to a surface temperature of about 5800 K. Surface temperatures range from about 50 000 K for the hottest O stars to as low as 2000 K for the coolest M stars.

MiniLab

Parallax in the Classroom

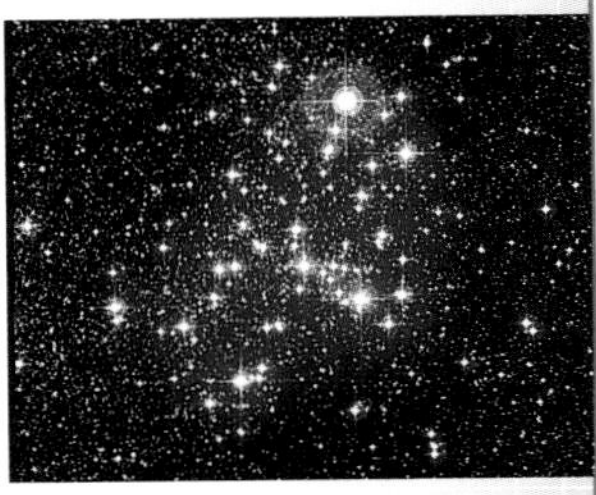

Model stellar parallax and the change in parallax angle with distance.

Procedure

1. Place a meterstick at a fixed position and attach a 4-m piece of string to each end.
2. Stand away from the meterstick and hold the two strings together to form a triangle. Be sure to hold the strings taut. Measure your distance from the meterstick. Record your measurement.
3. Measure the angle between the two pieces of string with a protractor. Record your measurement of the angle.
4. Repeat steps 2 and 3 for different distances from the meterstick by shortening or lengthening the string.
5. Make a graph of the angles versus their distance from the meterstick.

Analyze and Conclude

1. What does the length of the meterstick represent? The angle?
2. What does the graph show? How does parallax angle depend on distance?
3. Are the angles that you measured similar to actual stellar parallax angles? Explain.

Figure 30-15 These are the typical black and white absorption spectra of a class B5 star **(A),** a class F5 star **(B),** a class K5 star **(C),** and a class M5 star **(D).**

All stars, including the Sun, have nearly identical compositions, despite the differences in their spectra, shown in ***Figure 30-15.*** The differences in the appearance of their spectra are almost entirely a result of temperature effects. Hotter stars have fairly simple spectra while cooler stars have spectra with more lines. The coolest stars have bands in their spectra due to molecules, such as titanium oxide, in their atmospheres. Typically, about 73 percent of a star's mass is hydrogen, about 25 percent is helium, and the remaining 2 percent is composed of all the other elements. While there are some variations in the composition of stars, particularly in that final 2 percent, all stars have this general composition.

Earth Sciences

2.f Students know the evidence indicating that the color, brightness, and evolution of a star are determined by a balance between gravitational collapse and nuclear fusion.

Wavelength Shifts Spectral lines provide other information about stars in addition to composition and temperature. Spectral lines are shifted in wavelength by motion between the source of light and the observer. The shifts in spectral lines are an example of the Doppler effect.

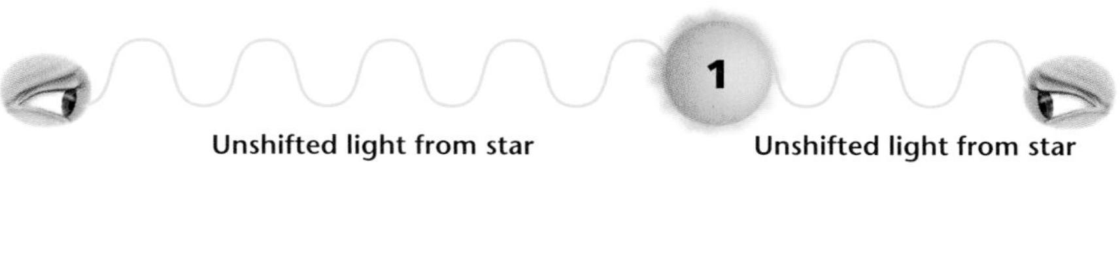

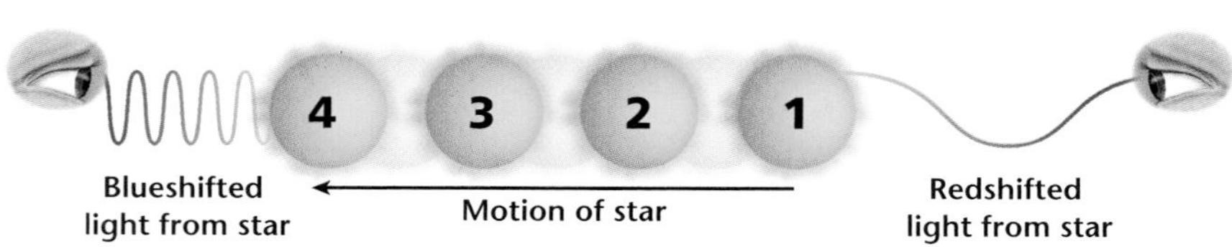

Figure 30-16 As a star moves towards or away from an observer, light is blueshifted or redshifted, respectively.

If a star is moving toward the observer, the spectral lines are shifted toward shorter wavelengths, or blueshifted. However, if the star is moving away, the wavelengths become longer, or redshifted, as illustrated in ***Figure 30-16.*** The higher the speed, the larger the shift, and thus careful measurements of spectral line wavelengths can be used to determine the speed of a star's motion.

Because there is no Doppler shift for motion that is sideways to the line of sight, astronomers can learn only about the portion of a star's motion that is directed toward or away from Earth. The Doppler shift in spectral lines can be used to detect binary stars as they move about their center of mass towards and away from Earth with each orbit.

H-R Diagrams The properties of mass, luminosity, temperature, and diameter are closely related. Each class of star has a specific mass, luminosity, magnitude, temperature, and diameter. These relationships can be demonstrated on a graph called the **Hertzsprung-Russell diagram** (H-R diagram) with absolute magnitude plotted on the vertical axis and temperature or spectral type plotted on the horizontal axis, as shown in ***Figure 30-17.*** This graph was first plotted in the early twentieth century. An H-R diagram with luminosity plotted on the vertical axis looks very similar to the one in ***Figure 30-17.***

About 90 percent of stars, including the Sun, fall along a broad strip of the H-R diagram called the **main sequence,** which runs diagonally from the upper-left corner, where hot, luminous stars are represented, to the lower-right corner, where cool, dim stars are represented. The interrelatedness of the properties of these stars indicates that all these stars have similar internal structures and functions.

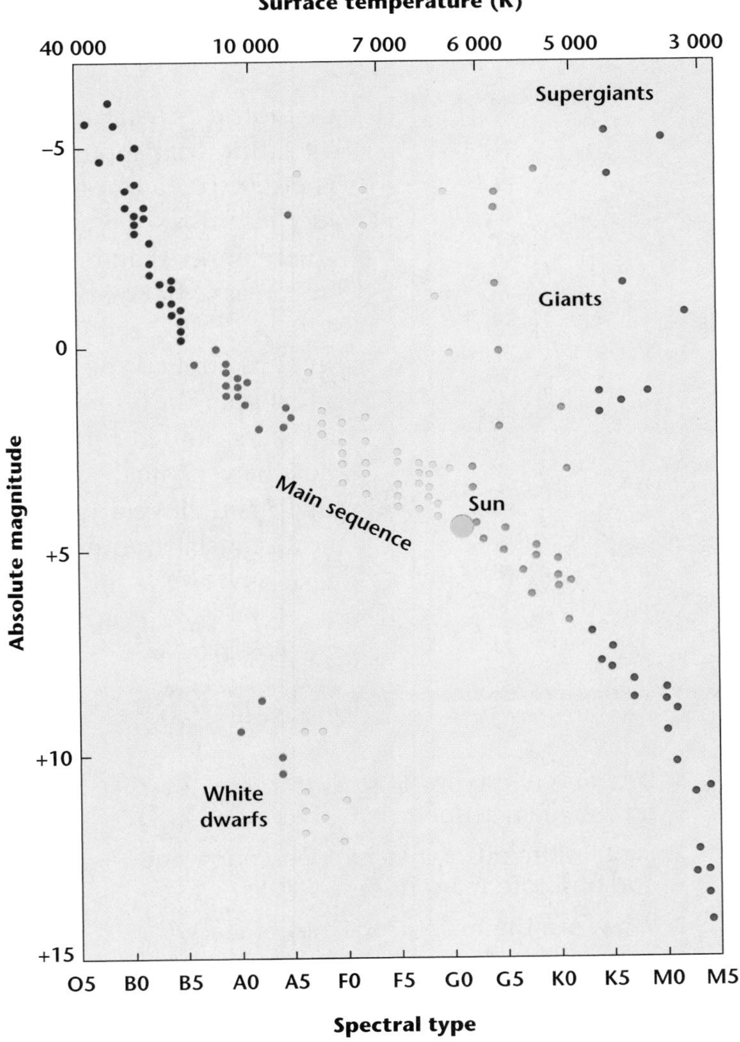

Figure 30-17 This H-R diagram shows the relationship between absolute magnitude, surface temperature, and spectral type. Mass decreases from left to right, while luminosity increases from bottom to top.

Table 30-2 Properties of Main-Sequence Stars

Spectral Type	Mass*	Surface Temperature (K)	Luminosity*	Radius*
O5	40.0	40 000	5×10^5	18.0
B5	6.5	15 500	800	3.8
A5	2.1	8500	20	1.7
F5	1.3	6580	2.5	1.2
G5	0.9	5520	0.8	0.9
K5	0.7	4130	0.2	0.7
M5	0.2	2800	0.008	0.3

*These properties are given relative to the Sun. For example, an O5 star has a mass 40 times that of the Sun, or $40 \times 1.99 \times 10^{30}$ kg = 7.96×10^{31} kg. The mass, luminosity, and radius of the Sun are 1.99×10^{30} kg, 3.846×10^{26} W, and 6.96×10^5 km, respectively.

Table 30-2 summarizes the basic properties of main sequence stars. But what about the stars that do not lie on the main sequence? The stars plotted at the upper right of the H-R diagram are cool, yet very luminous. Because cool surfaces emit much less radiation per square meter than hot ones do, these cool stars must have large surface areas to be so bright. For this reason, these larger, cool, luminous stars are called red giants. Red giants are so large—more than 100 times as large as the Sun in some cases—that Earth would be swallowed up if the Sun were to become a red giant! Conversely, the dim, hot stars plotted in the lower-lefthand corner of the H-R diagram must be very small, or else they would be far more luminous. These small, dim, hot stars are called white dwarfs. A white dwarf is about the size of Earth but has a mass about as large as the Sun's. You will learn how all the different stars are formed in the following section.

Section Assessment

1. Describe two types of stars that are not on the main sequence.
2. Explain the difference between apparent and absolute magnitudes.
3. What are the main properties of stars?
4. How do astronomers know that some stars are binary stars?
5. **Thinking Critically** How do the Sun's properties compare with those of other stars?

Skill Review

6. **Interpreting Diagrams** Use the H-R diagram in ***Figure 30-17*** to describe the properties of an A and an M star on the main sequence. For more help, refer to the *Skill Handbook*.

SECTION 30.3 Stellar Evolution

Earth Sciences 1.e, 2.c, 2.d I&E 1.c

Using observations as their guide, astronomers have developed models of stars that successfully explain the properties that you have just learned about. These models, like the solar model, are based on equations describing physical processes that occur inside stars. However, these models are accepted only when they reproduce the external properties that have been observed.

OBJECTIVES

- **Explain** *how astronomers learn about the internal structure of stars.*
- **Describe** *how the Sun will change during its lifetime and how it will end up.*
- **Compare** *the evolutions of stars of different masses.*

VOCABULARY

nebula
protostar
neutron star
supernova
black hole

BASIC STRUCTURE OF STARS

Mass governs a star's temperature, luminosity, and diameter. In fact, astronomers have discovered that the mass and the composition of a star determine nearly all its other properties. The more massive a star is, the greater the gravity pressing inward, and the hotter and denser the star must be inside to balance gravity. The temperature inside a star governs the rate of nuclear reactions, which in turn determines the star's energy output, or luminosity. The balance between gravity squeezing inward and pressure from nuclear fusion and radiation pushing outward, called hydrostatic equilibrium, must hold for any stable star; otherwise, the star would expand or contract, as illustrated in ***Figure 30-18.*** This balance is governed by the mass of a star.

Fusion Inside a star, conditions vary in much the same way that they do inside the Sun. The density and temperature increase toward the center, where energy is generated by nuclear fusion. Stars on the main sequence all produce energy by fusing hydrogen into helium, as the Sun does. Stars that are not on the main sequence either fuse different elements in their cores or do not undergo fusion at all.

Fusion reactions involving elements other than hydrogen can occur. Once a star's core has been converted into helium, the helium may fuse to form carbon if the temperature is high enough. At even higher temperatures, carbon can react with helium to form oxygen, then neon, then magnesium, and then silicon. Other types of reactions can produce even heavier elements, but few heavier than iron. Each of these reactions produces energy according to the equation $E = mc^2$ as a small fraction of mass is converted into energy. This energy stabilizes a star by producing the pressure needed to counteract gravity.

Figure 30-18 This star is stable and will not expand or contract.

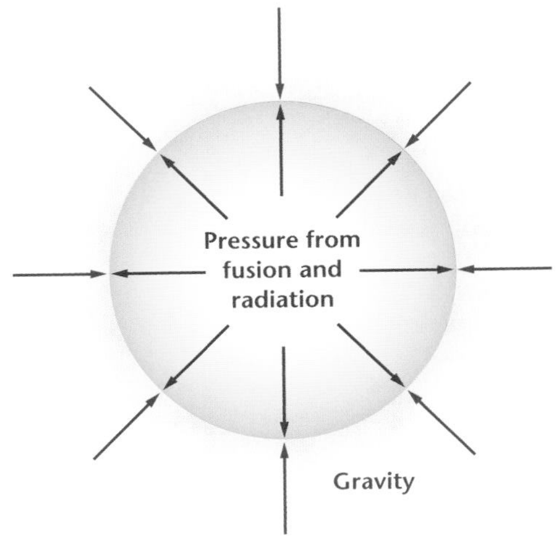

Figure 30-19 A protostar, formed from a disk of gas and dust **(A)**, will become a star when fusion begins. The Triffid Nebula **(B)** is illuminated by new stars, as shown by the *Hubble Space Telescope.*

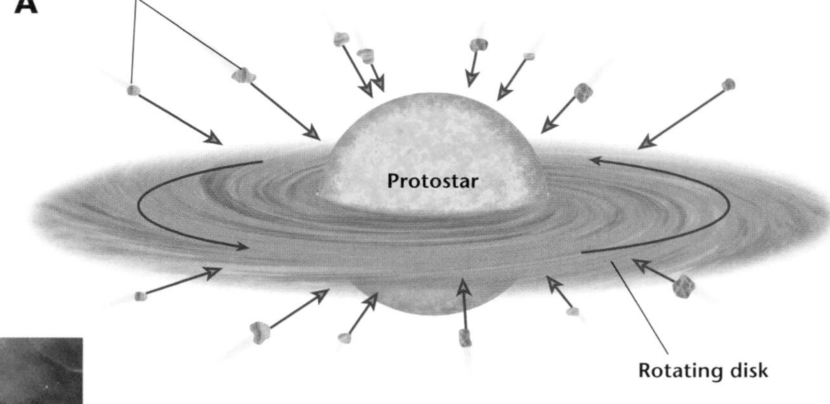

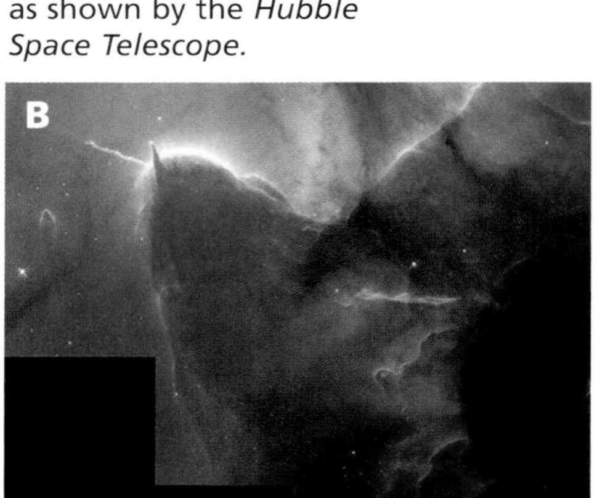

STELLAR EVOLUTION AND LIFE CYCLES

A star changes as it ages because its internal composition changes as nuclear fusion reactions in the star's core convert one element into another. As a star's core composition changes, its density increases, its temperature rises, and its luminosity increases. Eventually, the nuclear fuel runs out. Then the star's internal structure and mechanism for producing pressure must change to counteract gravity.

Earth Sciences

1.e Students know the Sun is a typical star and is powered by nuclear reactions, primarily the fusion of hydrogen to form helium.

Star Formation All stars form in much the same manner as the Sun did. The formation of a star begins with a cloud of interstellar gas and dust, called a **nebula** (*pl.* nebulae), which collapses on itself as a result of its own gravity. As the cloud contracts, its rotation forces it into a disk shape with a hot condensed object at the center, called a **protostar,** as illustrated in ***Figure 30-19A.*** The condensed object will become a new star. A protostar is brightest at infrared wavelengths.

To learn more about the *Hubble Space Telescope*, go to the **National Geographic Expedition** on page 902.

Fusion Begins Eventually, the temperature inside a protostar becomes hot enough for nuclear fusion reactions to begin. The first reaction to ignite is always the conversion of hydrogen to helium. Once this reaction begins, the star becomes stable because it then has sufficient internal heat to produce the pressure needed to balance gravity. The object is then truly a star and takes its place on the main sequence according to its mass. A new star often illuminates the gas and dust surrounding it, as shown in ***Figure 30-19B.***

THE SUN'S LIFE CYCLE

What happens during a star's life cycle depends on its mass. For example, as a star like the Sun converts hydrogen into helium in its core, it gradually becomes more luminous because the core density and temperature rise slowly and increase the reaction rate. It takes about 10 billion years for a star with the mass of the Sun to convert all of the hydrogen in its core into helium. Thus, such a star has a main sequence lifetime of 10 billion years.

Only the innermost 10 percent or so of a star's mass can undergo reactions because temperatures outside of this core never get hot enough for reactions to occur. Thus, when the hydrogen in its core is gone, a star has a helium center and outer layers made of hydrogen-dominated gas. Some hydrogen continues to react in a thin layer at the outer edge of the helium core, as illustrated in ***Figure 30-20.*** The energy produced in this layer forces the outer layers of the star to expand and cool. The star then becomes a red giant, because its luminosity increases while its surface temperature decreases due to the expansion.

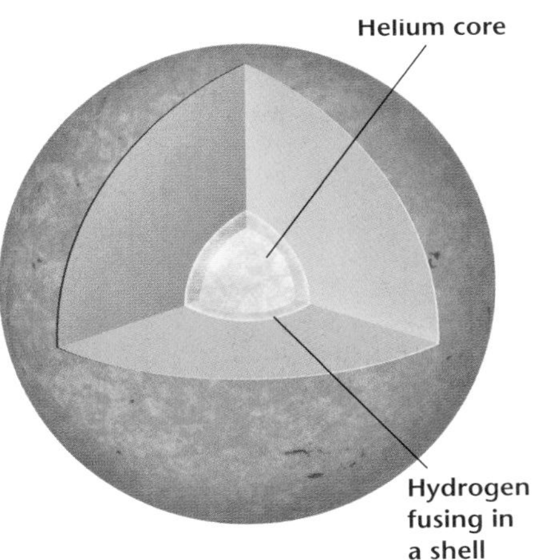

Figure 30-20 After hydrogen fusing is done in the core, it continues in a shell. (not to scale)

While the star is a red giant, it loses gas from its outer layers. The star is so large that its surface gravity is very low and thus the outer layers can be driven away by small expansions and contractions or pulsations, of the star due to instability. Meanwhile, the core of the star becomes hot enough, at 100 million K, for helium to react and form carbon. The star contracts back to a more normal size, where it again becomes stable for awhile. The helium-reaction phase lasts only about one-tenth as long as the earlier hydrogen-burning phase. Afterwards, when the helium in the core is all used up, the star is left with a core made of carbon.

Earth Sciences

2.c Students know the evidence indicating that all elements with an atomic number greater than that of lithium have been formed by nuclear fusion in stars.

A Nebula Once Again A star of the Sun's mass never becomes hot enough for carbon to react, so the star's energy production ends at this point. The outer layers expand once again and are driven off entirely by pulsations that develop in the outer layers. This shell of gas is called a planetary nebula. It has nothing to do with planets, despite its name. In the center of a planetary nebula, shown in ***Figure 30-21,*** the core of the star becomes exposed as a small, hot object about the size of Earth. The star is then a white dwarf made of carbon.

Pressure in White Dwarfs A white dwarf is stable despite the lack of nuclear reactions because it is supported by the resistance of electrons being squeezed close together, and does not require a source of heat to be maintained. This pressure counteracts gravity and can support the core as long as the mass of the remaining core is less than about 1.4 times the mass of the Sun.

A star that has less mass than that of the Sun has a similar life cycle, except that helium may never form carbon in the core, and the star ends as a white dwarf made of helium. The main sequence lifetime of such a star is much longer, however, because low-mass stars are dim and do not use up their nuclear fuel very rapidly.

Figure 30-21 NGC 6751 is a planetary nebula. This image was taken by the *Hubble Space Telescope.* The white dwarf is the white object at the center of the nebula.

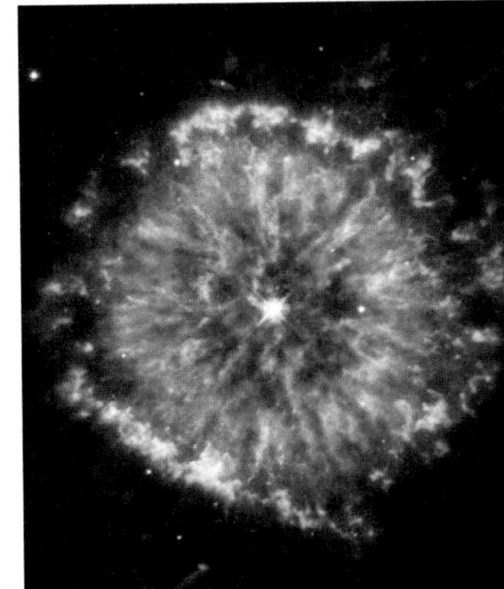

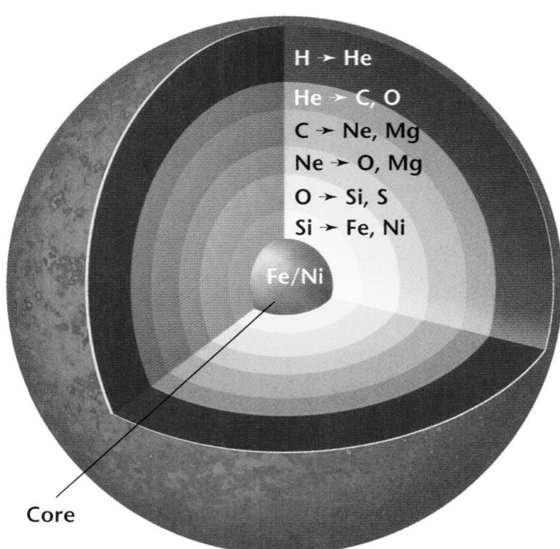

Figure 30-22 A massive star can have many shells fusing different elements. Massive stars are responsible for producing heavier elements. (not to scale)

Life Cycles of Massive Stars

For stars more massive than the Sun, evolution is very different. A more-massive star begins its life in the same way, but much higher on the main sequence, with hydrogen being converted to helium. However, the star's lifetime in this phase is short, because the star is very luminous and uses up its fuel quickly.

A massive star undergoes many more reaction phases and thus produces a rich stew of many elements in its interior. The star becomes a red giant several times as it expands following the end of each reaction stage. As more shells are formed by the fusion of different elements, illustrated in ***Figure 30-22,*** the star expands to a larger size and becomes a supergiant, such as Betelgeuse in the Orion constellation.

A massive star loses much of its mass during its lifetime as gas drifts from its outer layers, or is driven away by a stellar wind. A star that begins with as many as 8 times the Sun's mass may end up as a white dwarf with a final mass less than 1.4 times the Sun's mass. The composition of a white dwarf is determined by how many reaction phases the star went through before it stopped reacting altogether. Thus, there can be white dwarfs made of oxygen, white dwarfs made of neon, and so on.

Earth Sciences

2.d Students know that stars differ in their life cycles and that visual, radio, and X-ray telescopes may be used to collect data that reveal those differences.

Supernovae Some stars do not lose enough mass to become white dwarfs. A star that begins with a mass between about 8 and 20 times the Sun's mass will end up with a core that is too massive to be supported by electron pressure. Such a star comes to a very violent end. Once reactions in the core of the star have created iron, no further energy-producing reactions can occur, and the core of the star violently collapses in on itself, as illustrated in ***Figure 30-23A.*** As it does so, protons and electrons in the core merge to form

Figure 30-23 The core of the star collapses **(A)** forming a neutron star **(B),** on which infalling material bounces off and causes a supernova **(C).** (not to scale)

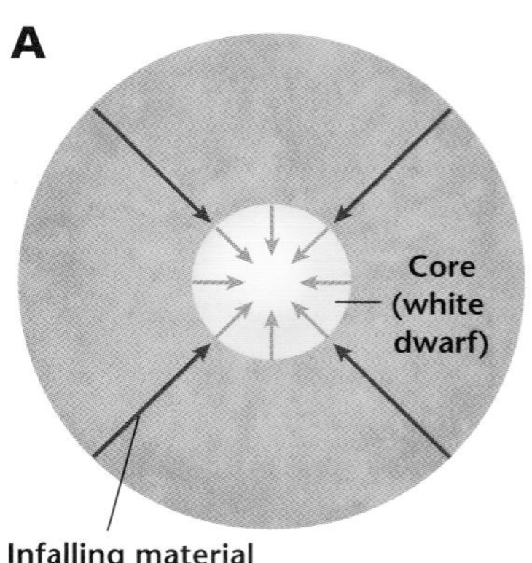

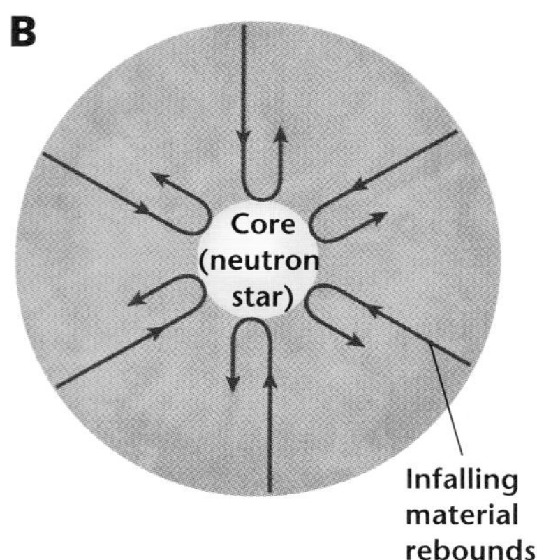

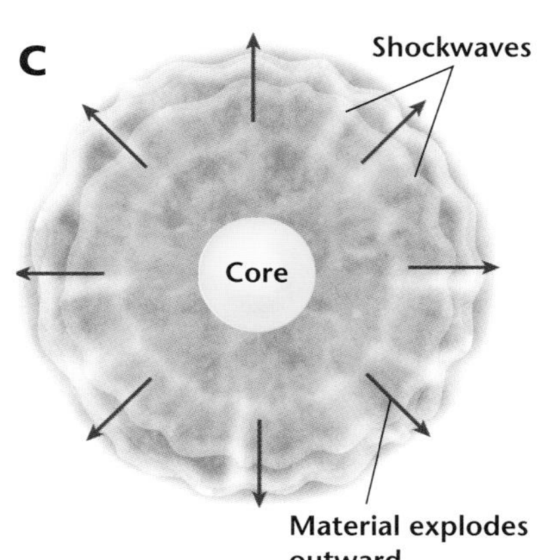

neutrons. Like electrons, neutrons can't be squeezed too closely together. Their resistance to being squeezed creates a pressure that halts the collapse of the core, and the core becomes a **neutron star,** as illustrated in ***Figure 30-23B.*** A neutron star has a mass of 1.5 to 3 times the Sun's mass but a radius of only about 10 km! The density is incredibly high—about 100 trillion times more dense than water—and is comparable to that of an atomic nucleus.

A neutron star forms quickly while the outer layers of the star are still falling inward. This infalling gas rebounds when it strikes the hard surface of the neutron star as illustrated in ***Figure 30-23B,*** and explodes outward, as illustrated in ***Figure 30-23C.*** The entire outer portion of the star is blown off in a massive explosion called a **supernova** (*pl.* supernovae), shown in ***Figure 30-24.*** This explosion creates elements that are heavier than iron and enriches the universe.

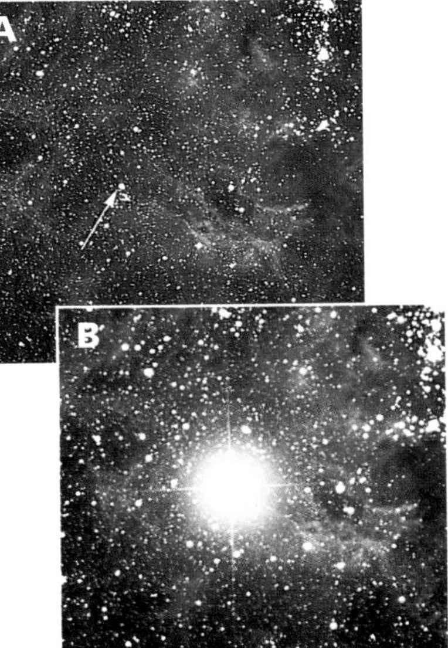

Figure 30-24 The top image **(A),** shows the region before Supernova 1987A in the Large Magellanic Cloud, while the bottom image **(B)** shows the supernova in full bloom.

Black Holes Some stars are too massive even to form neutron stars. The pressure from the resistance of neutrons being squeezed together cannot support the core of a star if the star's mass is greater than about 3 times the mass of the Sun. A star that begins with more than about 20 times the Sun's mass will end up above this mass limit, and it cannot form a neutron star. The resistance of neutrons to being squeezed is not great enough to stop the collapse and the core of the star simply continues to collapse forever, compacting matter into a smaller and smaller volume. The small, but extremely dense, object that remains is called a **black hole** because its gravity is so immense that nothing, not even light, can escape it. You will learn about the search for black holes in the *Science in the News* feature at the end of this chapter.

SECTION ASSESSMENT

1. How do astronomers learn about the internal structure and evolution of stars?
2. How does a new star form?
3. What causes a supernova to occur? Explain.
4. Is the lifetime of a massive star shorter or longer than a star like the Sun? Why?
5. In what ways is the evolution of a massive star similar to the evolution of the Sun, and in what ways is it different?
6. **Thinking Critically** How would the universe be different if massive stars did not explode at the ends of their lives?

SKILL REVIEW

7. **Comparing and Contrasting** Compare and contrast how pressure and gravity are balanced or not balanced in main-sequence stars, white dwarfs, neutron stars, and black holes. For more help, refer to the *Skill Handbook.*

Identifying Stellar Spectral Lines

An astronomer studying a star or other type of celestial object often starts by identifying the lines in the object's spectrum. The identity of the spectral lines gives astronomers information about the chemical composition of the distant object, along with data on its temperature and other properties.

 **I&E**

1.c Identify possible reasons for inconsistent results, such as sources of error or uncontrolled conditions.

Preparation

Problem
Identify stellar spectral lines based on two previously identified lines.

Materials
ruler

Objectives
In this Geolab, you will:
- **Develop** a scale based on the separation between two previously identified spectral lines.
- **Measure** wavelengths of spectral lines.
- **Compare** measured wavelengths to known wavelengths of elements to determine composition.

Procedure

1. Measure the distance between the two identified spectral lines on star 1. Be sure to use units that are small enough to get accurate measurements.
2. Calculate the difference in wavelengths between the two identified spectral lines.
3. Set up your scale by dividing the difference in wavelengths by the measured distance between the two identified spectral lines. This will allow you to measure wavelengths based on your distance measurement unit. For example, 1 mm = 12 nm.
4. Measure the distance to spectral lines from one of the two previously identified spectral lines.
5. Convert your distances to wavelengths using your scale. You have measured the difference in wavelength. This difference must be added or subtracted to the wavelength of the line you measured from. If the line you measured from is to the right of the line you are identifying, then you must subtract. Otherwise, you add.
6. Compare your wavelength measurements to the table of wavelengths emitted by elements, and identify the elements in the spectrum.
7. Repeat this procedure for star 2.

Star 1

Star 2

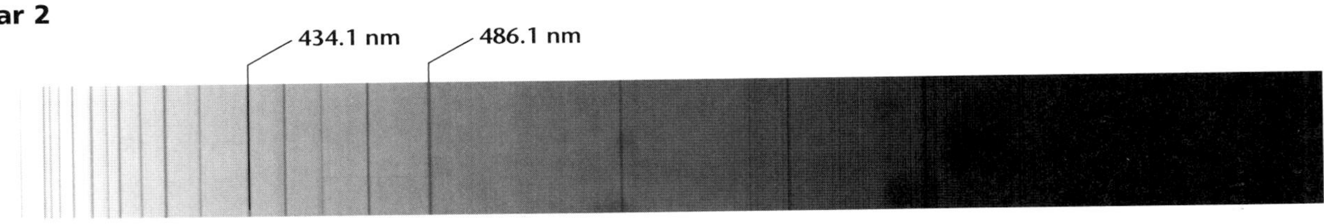

Analyze

1. What elements are present in the stars?
2. How does your list of elements compare with the list of elements seen in the periodic table in *Appendix G?*
3. Can you see any clues in the star's spectrum about which elements are most common in the stars? Explain.

Conclude & Apply

1. Do both stars contain the same lines for all the elements in the table?
2. You should notice that some absorption lines are wider than others. What are some possible explanations for this?
3. How do the thicker absorption lines of some elements in a star's spectrum effect the accuracy of your measurements? Is there a way to improve your measurements? Explain.
4. Using the following formula, calculate the percent deviation for 5 of your measured lines.

$$\text{Percent deviation} = \frac{\text{difference from accepted value}}{\text{accepted value}} \times 100$$

Is there a value that has a high percent deviation? If so, what are some possible explanations for this?

Possible Elements and Wavelengths

Element/Ion	Wavelengths (nm)
H	383.5, 388.9, 397.0, 410.2, 434.1, 486.1, 656.3
He	402.6, 447.1, 492.2, 587.6, 686.7
He^{+}	420.0, 454.1, 468.6, 541.2, 656.0
Na	475.2, 498.3, 589.0, 589.6
Ca^{+}	393.4, 480.0, 530.7

Science in the News

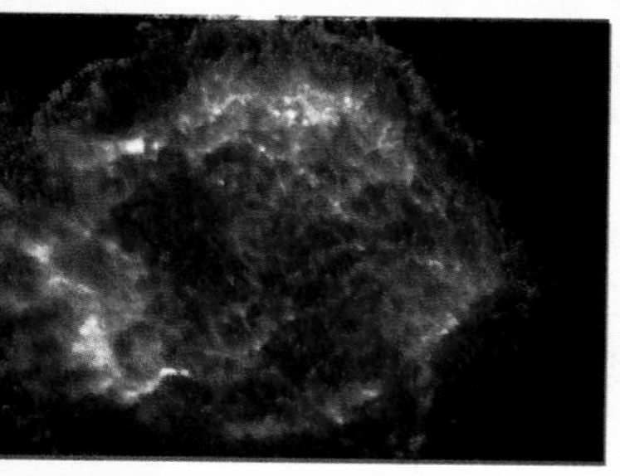

Supernova remnant
Cassiopeia A

Chandra—An Eye on the Universe

Some of the hottest action in the universe—action that takes place in black holes, exploding stars, and colliding galaxies—can't be observed with the naked eye. With the 1999 launch of the Chandra X-ray Observatory, however, scientists now have an unprecedented view.

Chandra is the third of NASA's Great Observatories—space telescopes designed to capture images beyond the reach of Earth-based telescopes. Unlike the Hubble Space Telescope and the Compton Gamma-Ray Observatory, which take pictures created by visible light and gamma rays, Chandra studies X rays. These rays are absorbed by Earth's atmosphere. Thus, they are best studied from a position high above our planet.

The Keen Eye

X-ray telescopes are not new. However, Chandra represents a vast improvement in technology. According to NASA, Chandra is a billion times more powerful than the first X-ray telescope, built just decades ago. "Chandra's resolving power," a NASA document explains, "is equivalent to the ability to read a 1- cm newspaper headline at the distance of a half-mile."

The $2.8 billion observatory is composed of three separate elements. The spacecraft system contains computers, data recorders, and communication equipment to transmit information back to Earth. The telescope system includes an assembly of high-resolution mirrors, the largest and most polished of their kind. The science instrument system is equipped with a high-resolution camera to record X-ray images of turbulent, high-temperature events, such as supernovae. Armed with this space-age technology, Chandra is in a unique position to help scientists solve some of astronomy's most baffling puzzles.

Black Holes

Chandra has turned its eye toward black holes—high-gravity objects that can suck in entire stars. Black holes are thought to exist at the center of galaxies. They may be the source of the astounding amounts of energy and radiation emitted by galactic centers. Because nothing, not even light, can escape from black holes, they cannot be observed directly. However, when matter is pulled into a black hole, it is heated to incredibly high temperatures. The matter becomes so hot that it emits X rays. Chandra is studying these X-ray emissions, documenting particles of matter up until the very millisecond before they disappear into the black hole. Using this information, scientists hope to learn more about the nature of black holes and the energy they produce. The information Chandra gathers promises to change our views of the universe.

Activity

Chandra was originally called the Advanced X-ray Astrophysics Facility. Go to earthgeu.com to find links to more information on how the observatory received its new name. Who was it named for? Why?

Study Guide

Summary

SECTION 30.1

The Sun

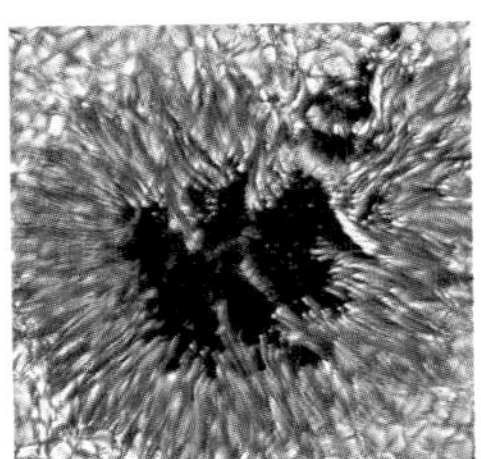

Main Ideas

- The Sun contains most of the mass in the solar system and is made up primarily of hydrogen and helium.
- Astronomers learn about conditions inside the Sun by a combination of observation and theoretical models.
- The Sun's atmosphere consists of the photosphere, the chromosphere, and the corona.
- The Sun has a 22-year activity cycle caused by reversals in its magnetic field polarities.
- Sunspots, solar flares, and prominences are active features of the Sun.
- The solar interior consists of the core, where fusion of hydrogen into helium occurs, and the radiative and convective zones.

Vocabulary

chromosphere (p. 806)
corona (p. 807)
fission (p. 809)
fusion (p. 809)
photosphere (p. 806)
prominence (p. 809)
solar flare (p. 809)
solar wind (p. 807)
spectrum (p. 811)
sunspot (p. 808)

SECTION 30.2

Measuring the Stars

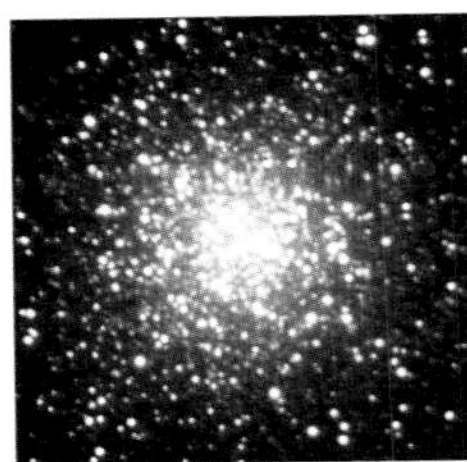

Main Ideas

- Positional measurements of the stars are important for measuring distances through stellar parallax shifts.
- Stellar brightnesses are expressed in the systems of apparent and absolute magnitude.
- Stars are classified according to the appearance of their spectra, which indicate the surface temperatures of stars.
- The H-R diagram relates the basic properties of stars: class, mass, temperature, and luminosity.

Vocabulary

absolute magnitude (p. 816)
apparent magnitude (p. 816)
binary star (p. 814)
constellation (p. 813)
Hertzsprung-Russell diagram (p. 819)
luminosity (p. 817)
main sequence (p. 819)
parallax (p. 815)

SECTION 30.3

Stellar Evolution

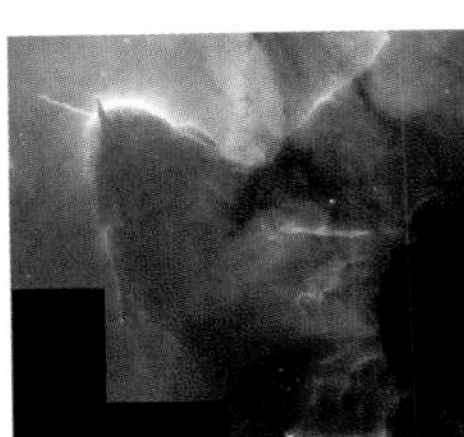

Main Ideas

- The mass of a star determines its internal structure and its other properties.
- Gravity and pressure balance each other in a star.
- If the temperature in the core of a star becomes high enough, elements heavier than hydrogen but lighter than iron can fuse together.
- Stars such as the Sun end up as white dwarfs. Stars up to about 8 times the Sun's mass also form white dwarfs after losing mass. Stars with masses between 8 and 20 times the Sun's mass end as neutron stars, and more massive stars end as black holes.
- A supernova occurs when the outer layers of the star bounce off the neutron star core, and explode outward.

Vocabulary

black hole (p. 825)
nebula (p. 822)
neutron star (p. 825)
protostar (p. 822)
supernova (p. 825)

earthgeu.com/vocabulary_puzzlemaker

CHAPTER 30

Assessment

Understanding Main Ideas

1. Which of the following is not a part of the Sun's atmosphere?
a. the corona
b. the chromosphere
c. the solar wind
d. the photosphere

2. Which of the following is not created by the Sun's magnetic field?
a. the radiative zone
b. prominences
c. solar flares
d. sunspots

3. Which type of spectrum, if any, does the Sun emit?
a. an absorption spectrum
b. an emission spectrum
c. a continuous spectrum
d. no spectrum

4. Where does nuclear fusion in the Sun occur?
a. in the convective zone
b. in the radiative zone
c. in the photosphere
d. in the core

5. If a star begins its evolution with 10 times the mass of the Sun, but ends with 2 times the mass of the Sun, what type of object does it form?
a. a white dwarf
b. a nebula
c. a neutron star
d. a black hole

6. How would you calculate how much brighter a magnitude +4 star is than a magnitude +7 star?
a. 2.512×2.512
b. $2.512 \times 2.512 \times 2.512$
c. $2.512 \times 2.512 \times 2.512 \times 2.512$
d. $2.512 \times 2.512 \times 2.512 \times 2.512 \times 2.512$

7. Which of the following is the correct order of units from largest to smallest?
a. pc, ly, AU, km
b. ly, pc, km, AU
c. km, pc, ly, AU
d. km, AU, pc, ly

8. What is the difference between nuclear fusion reactions and nuclear fission reactions?

9. Why do we say that the Sun's activity cycle is 22 years long, when the number of sunspots follows an 11-year pattern?

10. Why do sunspots appear to be darker than their surroundings?

11. What is the energy source for all main-sequence stars?

12. Does the Doppler shift provide complete information about the motion of a star? Explain.

Use the diagram below to answer question 13.

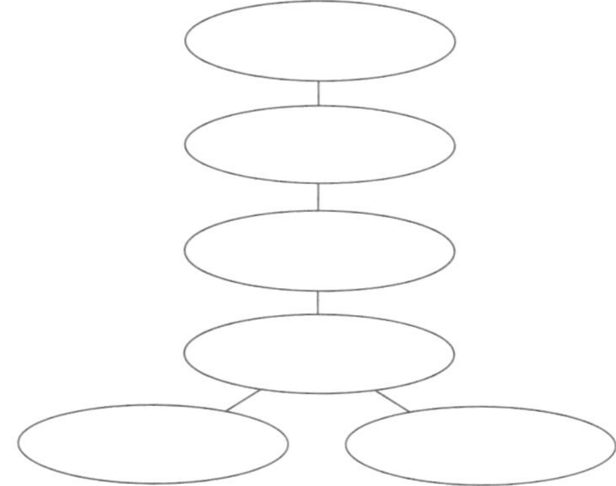

13. Use the following terms to fill in the concept map of the evolution of a star like the Sun.

star	nebula	red giant
protostar	white dwarf	planetary nebula

Test-Taking Tip

Slow Down Read the questions and answer choices carefully. Remember that doing most of the problems and getting them right is always preferable to doing all the problems and getting many of them wrong.

CHAPTER 30

Assessment

14. Why can the Sun's chromosphere and corona be observed without special instruments only during times of total solar eclipses?

Applying Main Ideas

15. Why do some stars explode as supernovae?
16. What happens to a star when the hydrogen in its core has all been converted into helium?
17. If a white dwarf can exist only if its mass is less than 1.4 times the Sun's mass, how can a star having 5 times the Sun's mass end as a white dwarf?
18. Is a magnitude –3 star brighter or dimmer than a magnitude 0 star? By how much?
19. What can astronomers learn about a star from its spectral lines?
20. How would the appearance of the Sun be different if it did not have a magnetic field?

Thinking Critically

21. Why does a star change over time?
22. Why do the most massive stars end up as black holes instead of neutron stars?
23. Why do the Sun and the gas giant planets have similar compositions? Why are the terrestrial planets different?
24. How would astronomers know if the theory of hydrostatic equilibrium inside the Sun were not correct?
25. How would an increase in mass affect the basic properties of a star? You may want to use an H-R diagram to help you.
26. Is a star's eventual fate determined by its initial mass or by the mass it has at the end of its life? Explain.

Standardized Test Practice

1. What causes sunspots on the Sun?
   a. intense magnetic fields poking through the photosphere
   b. charged particles flowing outward into the solar system
   c. spots on the surface of the photosphere which are hotter than the surrounding areas
   d. areas of low density in the gas of the Sun's corona

INTERPRETING DATA Use the table below to answer questions 2–4.

Stellar Magnitudes

Star	Apparent Magnitude	Absolute Magnitude
Procyon	+0.38	+2.66
Altair	+0.77	+2.22
Becrux	+1.25	−3.92
Bellatrix	+1.64	−1.29
Denebola	+2.14	+1.54

2. Which is the brightest star as seen from Earth?
   a. Procyon
   b. Becrux
   c. Bellatrix
   d. Denebola
3. Which is the brightest star as seen from 10 parsecs?
   a. Procyon
   b. Becrux
   c. Bellatrix
   d. Denebola
4. Which is the dimmest star as seen from 10 parsecs?
   a. Bellatrix
   b. Altair
   c. Procyon
   d. Becrux

Chapter 31

Galaxies and the Universe

What You'll Learn

- What the Milky Way Galaxy is like.
- How galaxies are distributed and what their characteristics are.
- What astronomers know about the origin and history of the universe.

Why It's Important

The study of galaxies and the structure of the universe helps scientists to better understand the origin of our solar system as well as Earth's origin.

To find out more about galaxies and the universe, visit the Earth Science Web Site at earthgeu.com

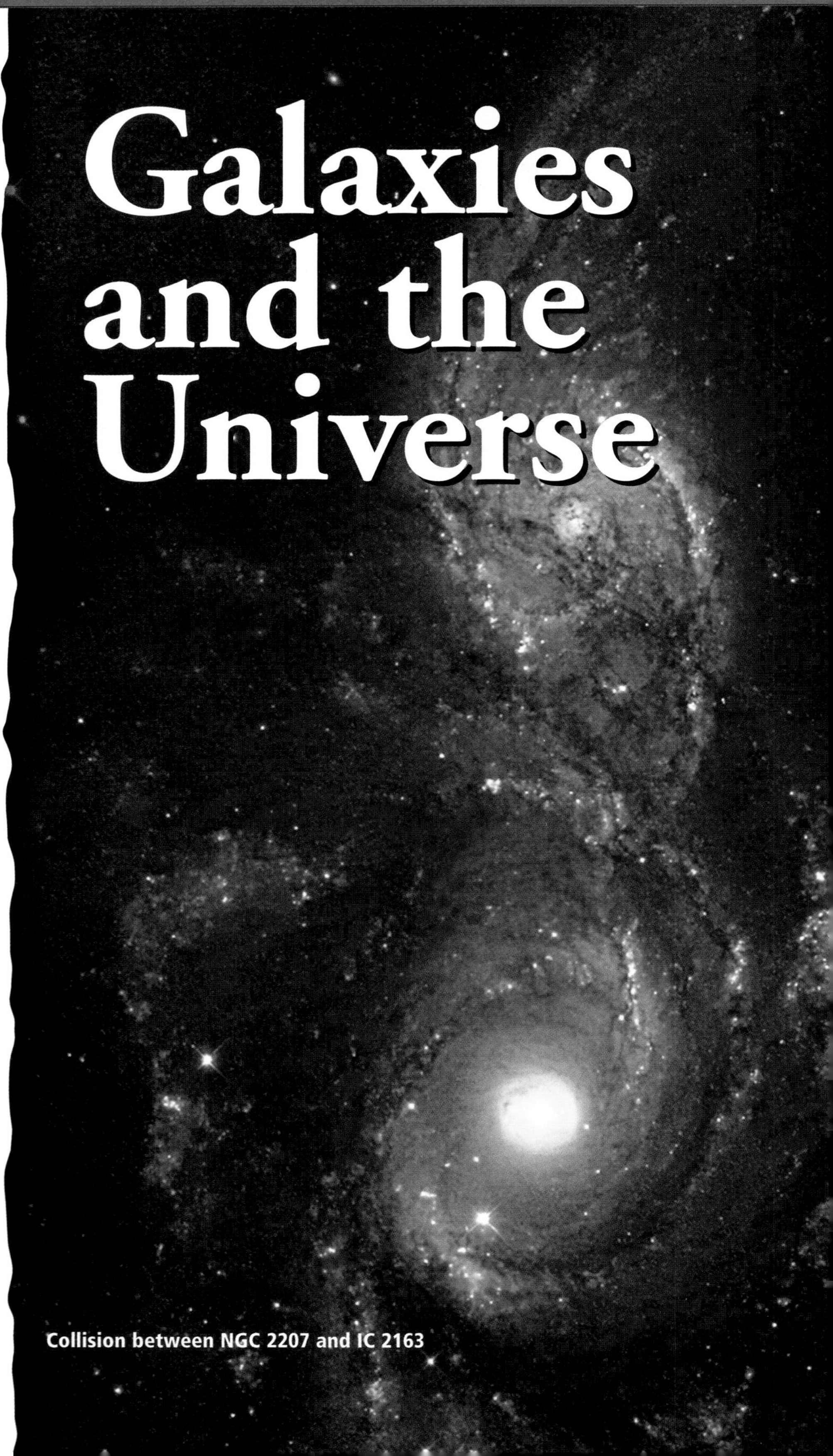

Collision between NGC 2207 and IC 2163

Discovery Lab Model The Milky Way

Our solar system seems large when we compare it to the size of Earth. However, the Milky Way dwarfs the size of our solar system.

1. The Milky Way has a diameter of approximately 8.25×10^9 AU. Given that 206 265 AU = 3.26 ly, what is the diameter of the Milky Way in light-years?
2. Given that the orbit of Pluto has a diameter of 80 AU, what is the diameter of Pluto's orbit in ly?
3. If you were to apply the scale 1 mm = 1 ly, how large would the Milky Way be?
4. The Sun is located 28 000 ly from the center of the Milky Way. Based on the scale that you used in question 3, how many millimeters away from the center of the Milky Way would the Sun be?
5. If you included Pluto's orbit in your model, how many millimeters across would its orbit be?

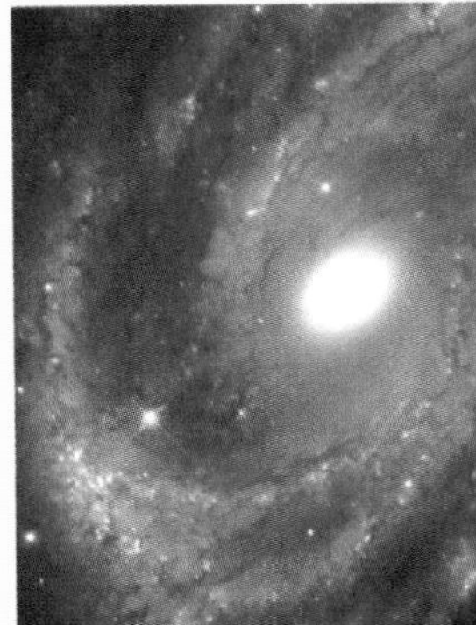

Observe In your science journal, describe what your model of the Milky Way would look like if you actually built it. Explain why it would be a problem to show the size of our solar system in comparison to the Milky Way. Explain how you would change your model to include the size of Earth.

Section 31.1 *The Milky Way Galaxy*

Earth Sciences 2.a, 2.b

OBJECTIVES

- **Determine** *the size and shape of the Milky Way, as well as Earth's location within it.*
- **Describe** *how the Milky Way formed.*

VOCABULARY

variable star
RR Lyrae variable
Cepheid variable
halo
spiral density wave

When you see the hazy band of light across the sky known as the Milky Way, you are looking at our galaxy from the inside. It is an awesome sight, but one that you can see only if you get away from city lights on a moonless night. The Milky Way is a great disk made of stars orbiting a central point in the disk. Our Sun is just one of perhaps 100 billion stars that make up the Milky Way.

Discovering the Milky Way

When we look at the Milky Way, it is difficult to see its size and shape because not only are we too close, but we are also inside the galaxy. We see the band of stars stretching across the sky, and it's not hard to imagine that this is an edge-on view of a disk. However, we can't tell how big it is, where its center is, or what Earth's location is within it. These were some of the mysteries that astronomers were interested in solving. In fact, astronomers are still refining their measurements.

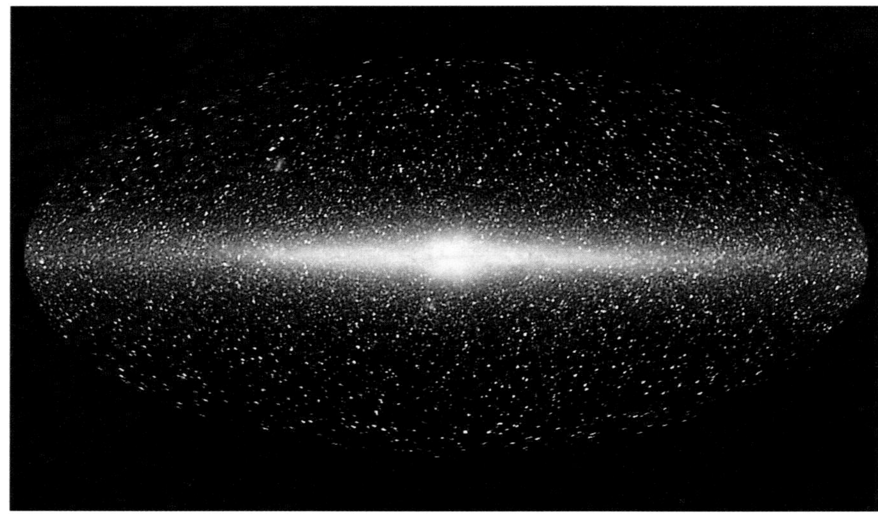

Figure 31-1 Globular clusters are located above and below the plane of the disk of the Milky Way. The Milky Way is the white band across the middle of this photo.

Earth Sciences

2.a Students know the solar system is located in an outer edge of the disc-shaped Milky Way galaxy, which spans 100,000 light years.

2.b Students know galaxies are made of billions of stars and comprise most of the visible mass of the universe.

Variable Stars In the 1920s, astronomers mapped out the locations of globular clusters. These huge, spherical star clusters are located above or below the plane of the galactic disk, shown in ***Figure 31-1.*** Astronomers estimated the distances to the clusters by identifying variable stars in them. **Variable stars** are stars in the giant branch of the Hertzsprung-Russell diagram, discussed in Chapter 30, that pulsate in brightness because of the expansion and contraction of their outer layers. For certain types of variable stars, there is a relationship between a star's luminosity and its pulsation period, which is the time between brightenings. For example, **RR Lyrae variables** have periods of pulsation between 1.5 hours and 1 day, and on average, they have the same luminosity. **Cepheid variables,** however, have pulsation periods between 1 and 100 days. The longer the period of pulsation, the greater the luminosity of the star. By measuring a star's period of pulsation, astronomers can determine the star's luminosity. This, in turn, allows them to compare the star's luminosity, or absolute magnitude, to its apparent magnitude and calculate how far away the star must be to appear as dim or bright as it does.

The Galactic Center Astronomers used RR Lyrae variables to determine the distances to the globular clusters. They discovered that these clusters are located very far from our solar system, and that their distribution in space is centered on a distant point, 28 000 ly away. Astronomers reasoned that the globular clusters were orbiting the center of the Milky Way. The center is a region of very high star density, as shown in ***Figure 31-2,*** much of which is obscured by interstellar gas and dust. The direction of the galactic center is toward the constellation Sagittarius.

The Shape of the Milky Way

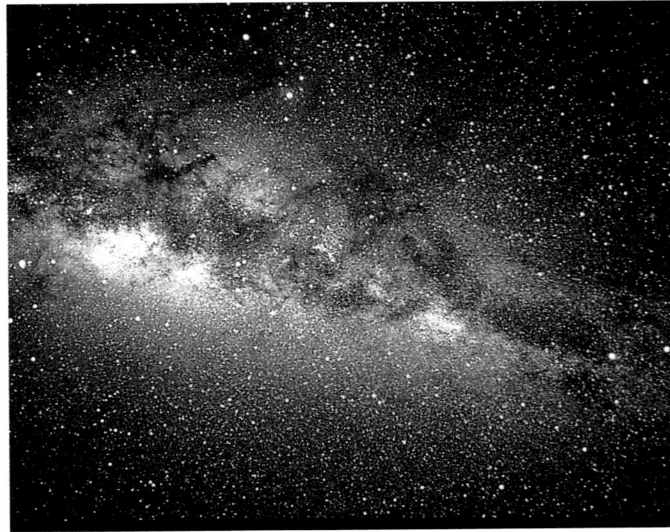

Figure 31-2 The center of the Milky Way is densely populated by stars, many of which are obscured by dust.

Only by mapping the galaxy with radio waves have astronomers been able to determine its shape. This is because radio waves can penetrate the interstellar gas and dust without being scattered or absorbed. Astronomers have discovered, by measuring radio waves as well as infrared radiation, that the galactic center, also called the nucleus, is surrounded by a nuclear bulge, which sticks out of the galactic disk much like the yolk in a fried egg. Around the nuclear bulge and disk is the **halo,** a spherical region where globular clusters are located, as illustrated in ***Figure 31-3.***

Spiral Arms Knowing that our galaxy has a disklike shape with a central bulge, astronomers speculated that it might also have spiral arms, as many other galaxies do. This was very difficult to prove, however, because astronomers have no way to get outside of the galaxy and look down on the disk. Astronomers have used hydrogen atoms to discover the spiral arms.

Hydrogen atoms in a very low-density gas can emit radiation at a wavelength of 21 cm. The use of this hydrogen emission for mapping the spiral arms of our galaxy has several advantages: first, hydrogen is the most abundant element in space; second, the interstellar gas, composed mostly of hydrogen, is concentrated in the spiral arms; and third, the 21-cm wavelength of hydrogen emission can penetrate the interstellar gas and dust and be detected all the way across the galactic disk. Using the 21-cm hydrogen emission as a guide, astronomers have identified four major spiral arms and numerous minor arms in the Milky Way. The Sun is located in the minor arm Orion at a distance of about 28 000 ly from the galactic center. The Sun's orbital speed is about 220 km/s, and thus its orbital period is about 240 million years. It is hypothesized that in its 5 billion-year life, the Sun has orbited the galaxy approximately 20 times.

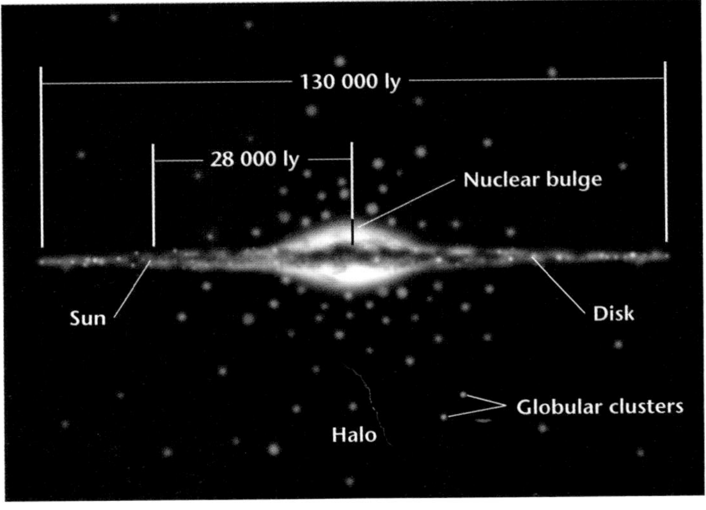

Figure 31-3 The Milky Way consists of a nuclear bulge in the center of a disk. The disk and bulge are surrounded by a spherical region called the halo. (not to scale)

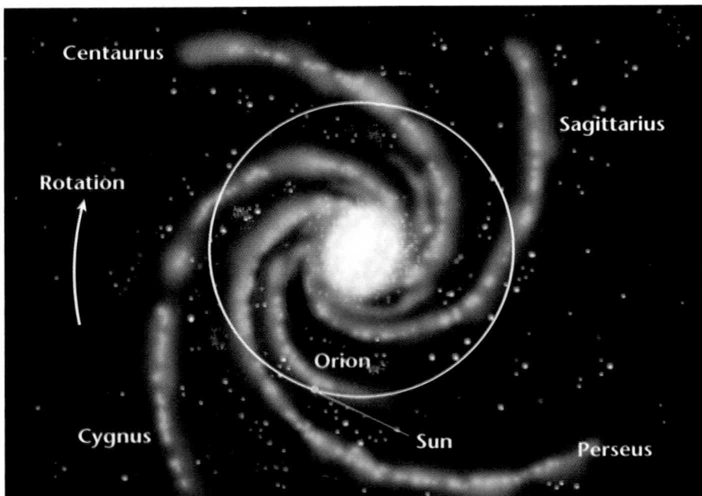

Figure 31-4 The Sun is located on the minor arm Orion and follows an orbital path around the nuclear bulge as shown. (not to scale)

Mass of the Milky Way

The mass located within the circle of the Sun's orbit through the galaxy, outlined in ***Figure 31-4,*** is about 100 billion times the mass of the Sun. Because the Sun is about average in mass, astronomers have concluded that the galaxy contains about 100 billion stars within its disk.

However, astronomers have found evidence that much more mass exists outside the disk of the galaxy. The stars and gas clouds that orbit in the outer disk are moving faster than they would if the galaxy's mass were concentrated near the center of the disk. Evidence indicates that as much as 90 percent of the galaxy's mass is contained in the halo. This mass is not observed in the form of normal stars, however. Some of this unseen matter is probably in the form of dim stellar remnants such as white dwarfs, neutron stars, or black holes, but the nature of the remainder of this mass is a mystery. As you'll discover, the problem of unseen matter, usually called dark matter, extends to other galaxies and to the universe as a whole.

A Galactic Black Hole Careful studies of the motions of the stars that orbit close to Sagittarius A*, the center of the galaxy, indicate that this area has about 2.6 million times the mass of the Sun but is smaller than our solar system. Data gathered by the *Chandra X-Ray Observatory* reveal intense X-ray emissions as well. Astronomers believe that Sagittarius A* is a supermassive black hole that glows brightly because of the hot gas surrounding it and spiraling into it. This black hole probably formed early in the history of the galaxy, at the time when the galaxy's disk was forming. Gas clouds and stars within the disk probably collided and merged to form a single, massive object that collapsed to form a black hole.

Stars in the Milky Way

The halo of the Milky Way contains the oldest-known objects in the galaxy: globular clusters. These clusters are estimated to be as old as 12 to 14 billion years. Stars in the globular clusters have extremely small amounts of elements that are heavier than hydrogen and helium. All stars contain small amounts of these heavy elements, but in globular clusters, the amounts are mere traces. Stars like the Sun are composed of about 98 percent hydrogen and helium, whereas in globular cluster stars, this proportion can be as high as 99.9 percent. This indicates their extreme age. The nuclear bulge of the galaxy also contains stars with compositions like those of globular cluster stars.

Stellar Populations Most of the young stars in the galaxy are located in the spiral arms of the disk, where the interstellar gas and dust are concentrated. Most star formation takes place in the arms. There is little interstellar material and very few stars currently forming in the halo or the nuclear bulge of the galaxy. In fact, the galaxy could be divided into two components: the round part made up of the halo and bulge, where the stars are old and contain only traces of heavy elements; and the disk, especially the spiral arms, where stars are still forming, as illustrated in ***Figure 31-5.*** Astronomers divide stars in these two regions into two classes. Population I stars are those in the disk and arms and have small amounts of heavy elements. Population II stars are those in the halo and bulge and contain only traces of heavy elements. The Sun is a Population I star.

Formation and Evolution of the Milky Way

The fact that the halo and bulge are made exclusively of old stars suggests that these parts of the galaxy formed first, before the disk that contains only younger stars. Astronomers therefore hypothesize that the galaxy began as a spherical cloud in space. The first stars formed while this cloud was round. This explains why the halo, which contains the oldest stars, is spherical. The nuclear bulge, which is also round, represents the inner portion of the original cloud. The cloud eventually collapsed under the force of its own gravity, and rotation forced it into a disklike shape. Stars that formed after this time have orbits lying in the plane of the disk.

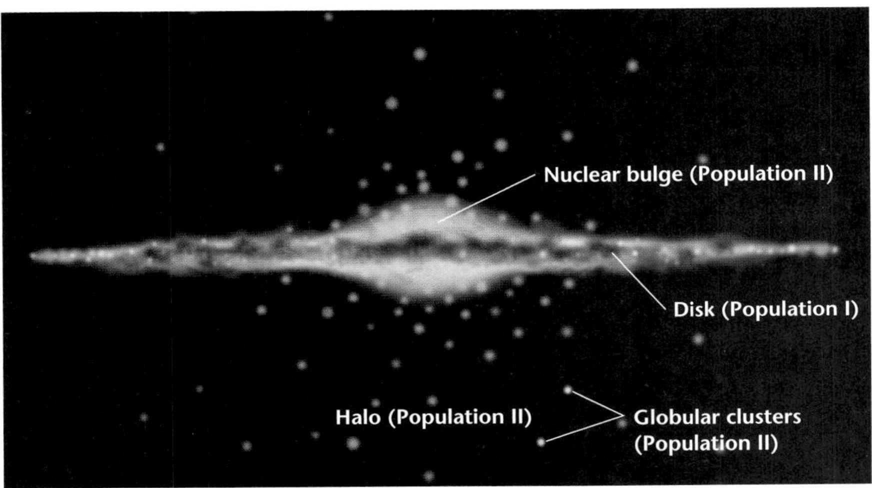

Figure 31-5 Globular clusters and the nuclear bulge contain old stars poor in heavy elements. The disk contains young stars that have a higher heavy element content. (not to scale)

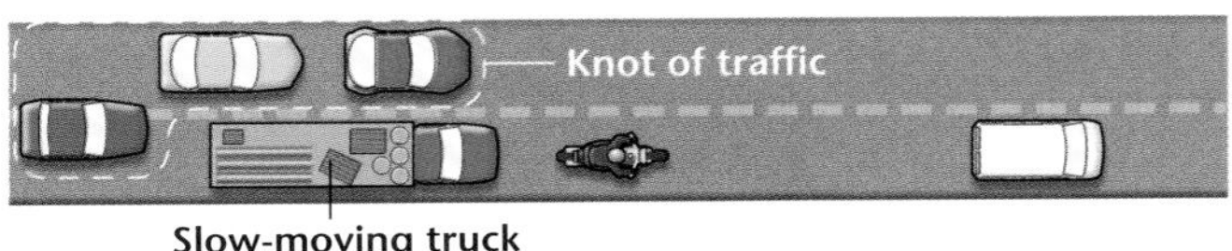

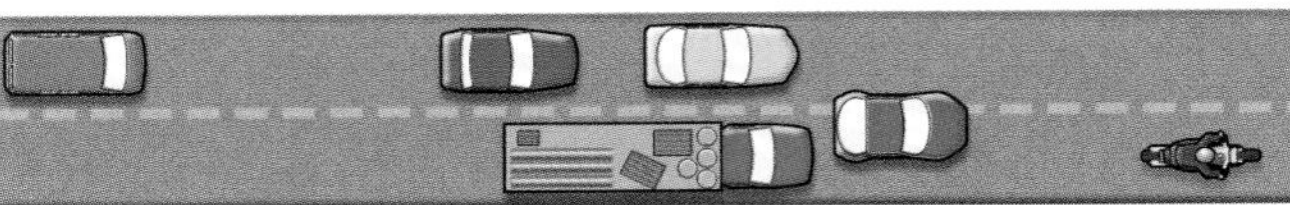

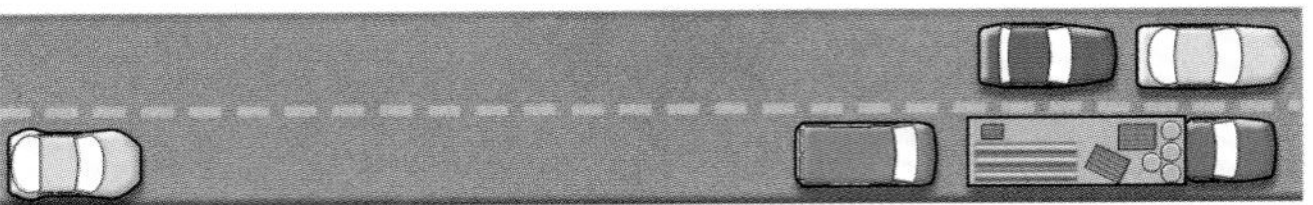

Figure 31-6 A slow truck on a highway causing a buildup of cars around it is similar to how spiral density waves create spiral arms in a galaxy.

They also contain greater quantities of heavy elements because they formed from gas that had been enriched by previous generations of massive stars.

Maintaining Spiral Arms

Most of the main features of the galaxy are clearly understood by astronomers, except for the way in which the spiral arms are maintained. The Milky Way is subject to gravitational tugs by neighboring galaxies and is periodically disturbed by supernovae explosions, both of which can create spiral arms. There are two different theories about how these arms are maintained. One is that a kind of wave called a spiral density wave is responsible. You can think of a **spiral density wave** as alternating dense and less-dense regions that are frozen in place and rotate as a rigid pattern. As the wave moves through gas and dust, it causes a temporary buildup of material, like a slow truck on the highway that causes a buildup of cars, illustrated in ***Figure 31-6.*** The other theory is that the spiral arms are not permanent structures but instead are continually forming as a result of disturbances such as supernovae explosions. The Milky Way has a broken spiral-arm pattern, which most astronomers think fits this second model best. However, some galaxies have a prominent 2-armed pattern, almost certainly created by density waves.

Section Assessment

1. How did astronomers determine where Earth is located within the Milky Way?
2. What do measurements of the mass of the Milky Way indicate?
3. How are Population I stars and Population II stars different?
4. How can variable stars be used to determine the distance to globular clusters?
5. **Thinking Critically** If our solar system were slightly above the disk of the Milky Way, why would astronomers still have difficulty determining the shape of the galaxy?

Skill Review

6. **Concept Mapping** Use the following terms to construct a concept map to organize the major ideas in this section. For more help, refer to the *Skill Handbook.*

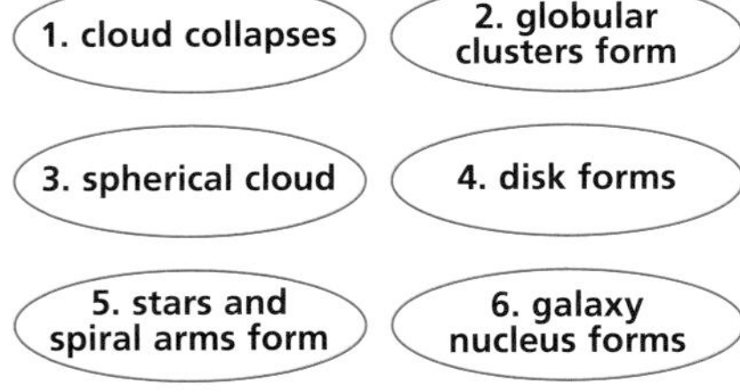

SECTION 31.2 Other Galaxies in the Universe

Earth Sciences 2.b, 2.g

Our galaxy is just one of billions of galaxies in the universe. We live in a cosmos of galaxies that have a wide variety of sizes and shapes. By observing those galaxies that are farthest away, astronomers get an idea of how the universe looks as a whole. Because it takes so long for light to reach us from remote galaxies, these far away galaxies also provide an idea of what the universe was like long ago.

OBJECTIVES

- **Describe** *how astronomers classify galaxies.*
- **Identify** *how galaxies are organized into clusters and superclusters.*
- **Describe** *the expansion of the universe.*

VOCABULARY

supercluster
Hubble constant
radio galaxy
active galactic nucleus
quasar

DISCOVERING OTHER GALAXIES

Astronomers were aware of galaxies outside the Milky Way long before they knew what these objects were. Many objects had been observed scattered throughout the sky, some with spiral shapes, but astronomers disagreed about their nature. Some astronomers hypothesized that these objects were nebulae or star clusters within the Milky Way. Others hypothesized that they were distant galaxies, as large as the Milky Way.

The question of what these objects were was answered by Edwin Hubble in 1924, when he discovered Cepheid variable stars in the Great Nebula in the Andromeda constellation. Using these stars to measure the distance to the nebula, Hubble showed that they were much too far away to be located in our own galaxy. The Andromeda Nebula then became known as the Andromeda Galaxy, shown in ***Figure 31-7.***

Classification of Galaxies Hubble went on to study galaxies and sort them into categories according to their shapes. The disklike galaxies with spiral arms were called spiral galaxies. These were divided into two subclasses: normal spirals and barred spirals. Barred spirals have

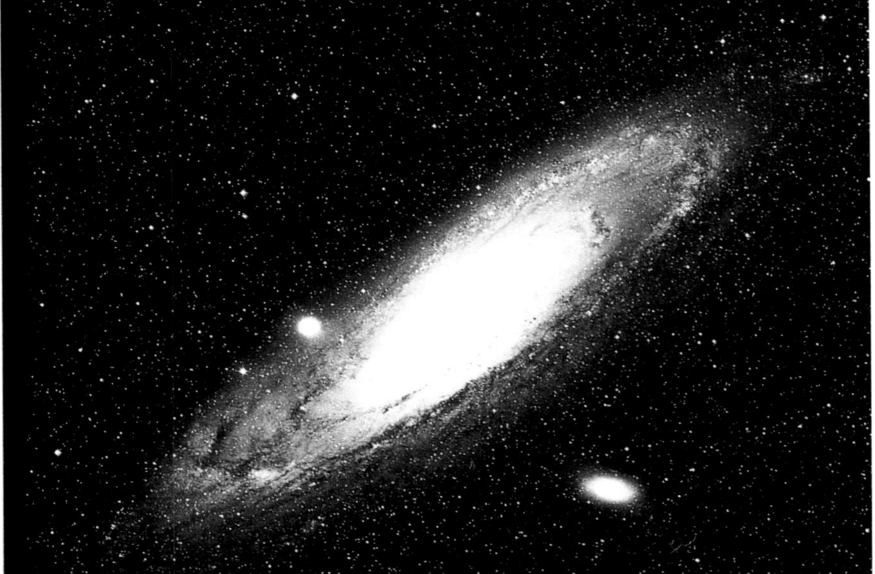

Figure 31-7 The Andromeda Galaxy is a spiral galaxy like the Milky Way. The two bright elliptical objects are small galaxies orbiting the Andromeda Galaxy.

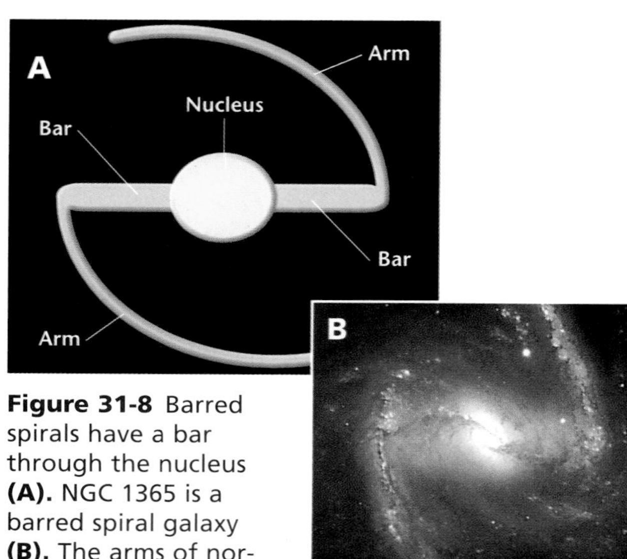

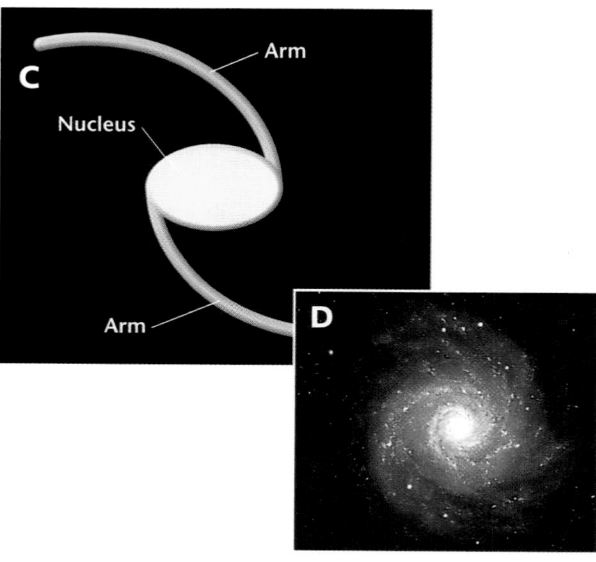

Figure 31-8 Barred spirals have a bar through the nucleus **(A).** NGC 1365 is a barred spiral galaxy **(B).** The arms of normal spirals extend directly from the nucleus **(C).** NGC 628 is a normal spiral galaxy **(D).**

an elongated central region, or a bar, from which the spiral arms extend, as shown in ***Figures 31-8A*** and ***B.*** Normal spirals do not have bars, as shown in ***Figures 31-8C*** and ***D.*** A normal spiral is denoted by the letter *S,* and a barred spiral is denoted by *SB.* Normal and barred spirals are further subdivided based on how tightly the spiral arms are wound, as well as the size and brightness of the nucleus. The letter *a* represents tightly wound arms and a large, bright nucleus. The letter *c* represents loosely wound arms and a small, dim nucleus. The letter *b* represents characteristics between those of *a* and *c.* Thus, a normal spiral with class *a* arms and nucleus is denoted *Sa,* while a barred spiral with class *a* arms and nucleus is denoted *SBa.* Galaxies with flat disks, that do not have spiral arms are denoted *S0.* When an *S0* galaxy is observed edge-on, it often has a band of dark dust crossing the nuclear bulge.

In addition to spiral galaxies, there are galaxies that are not flattened into disks and do not have spiral arms, as shown in ***Figure 31-9.*** These are called elliptical galaxies. Ellipticals are divided into subclasses based on the apparent ratio of their major and minor axes. Round ellipticals are classified as *E0,* while very elongated ellipticals are classified as *E7.* Others are denoted by the letter *E* followed by a numeral *1* through *6.* The classification of both spiral and elliptical galaxies can be summarized by Hubble's tuning-fork diagram, illustrated in ***Figure 31-10.*** You will learn more about galaxy classification in the *Internet GeoLab* at the end of this chapter.

Some galaxies do not have distinct shapes, and thus do not fit into either the spiral or elliptical classification. These are called irregular galaxies and are denoted by *Irr.* The Large and Small Magellanic Clouds are irregular galaxies that are close to the Milky Way.

Figure 31-9 This galaxy, M32, is an elliptical galaxy. Notice the uniform distribution of stars throughout the galaxy. Despite their flat appearance, ellipticals are three-dimensional, and are similar in shape to a football.

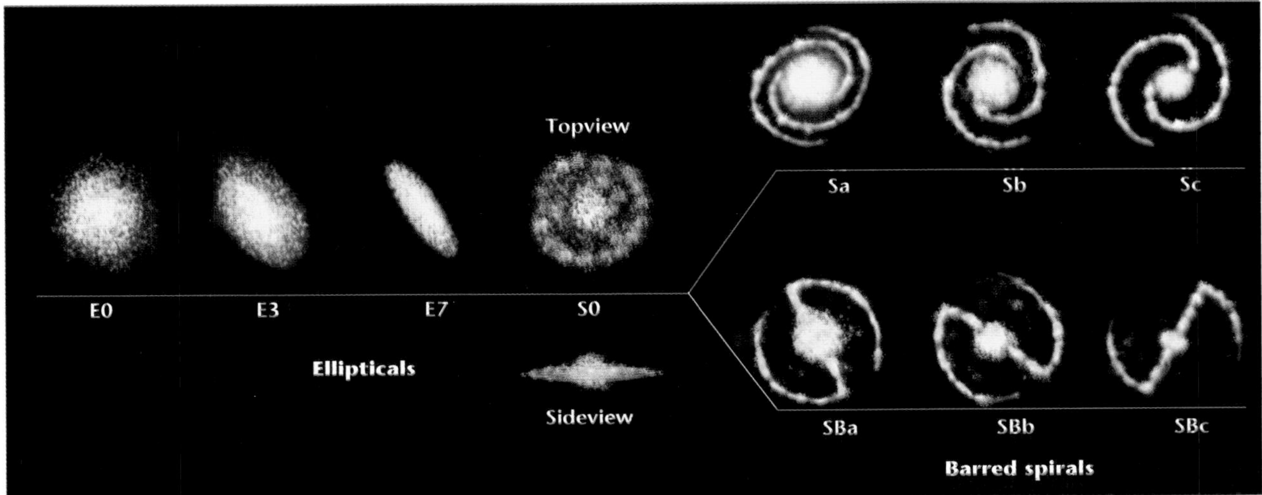

Figure 31-10 The Hubble tuning fork diagram summarizes the Hubble classification scheme for non-unusual galaxies. An S0 galaxy is related to both spirals and ellipticals.

Masses of Galaxies Masses of galaxies range from the dwarf ellipticals, which have masses of perhaps one million Suns; to large spirals, such as the Milky Way, with masses of around 100 billion Suns; to the largest galaxies, called giant ellipticals, which have masses as high as 100 trillion times the Sun's mass. Measurements of the masses of many galaxies indicate that they have extensive halos containing more mass than is visible, just as the Milky Way does.

Groups and Clusters of Galaxies

Most galaxies are located in groups, rather than being spread uniformly throughout the universe. The Milky Way belongs to a small cluster of galaxies called the Local Group. The diameter of the Local Group is roughly 2 million ly. There are about 35 known members, most of which are dwarf ellipticals. The Milky Way and Andromeda Galaxies are the largest galaxies in the group.

Clusters larger than the Local Group may have hundreds of members and diameters in the range of about 5 to 30 million ly. The Virgo cluster is shown in ***Figure 31-11.*** Most of the galaxies in the inner region of a large cluster are ellipticals, while there is a more even mix of ellipticals and spirals in the outer portions.

In regions where galaxies are as close together as they are in large clusters, gravitational interactions among galaxies have many important effects. Galaxies often collide and form strangely shaped galaxies,

Figure 31-11 Galaxies are located in groups, or clusters, like the Virgo cluster.

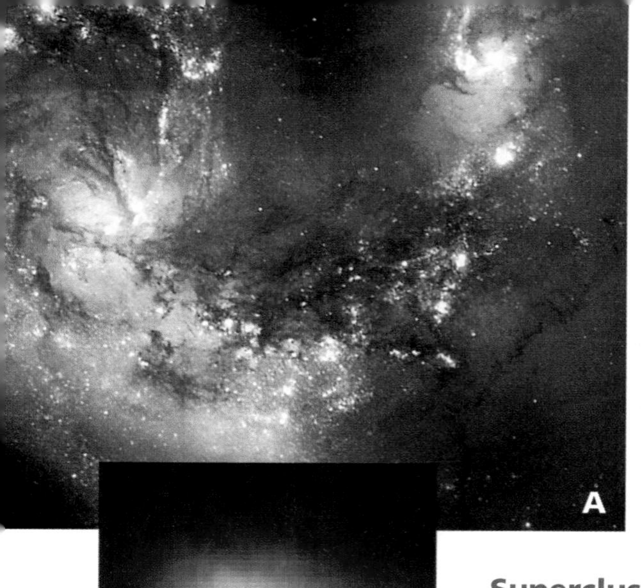

such as those shown in ***Figure 31-12A,*** or galaxies with more than one nucleus, such as the Andromeda Galaxy, shown in ***Figure 31-12B.***

Masses of Clusters For clusters of galaxies, the mass determined by analyzing the motion of member galaxies is always much larger than the sum of the visible masses of the galaxies. This suggests that most of the mass in a cluster of galaxies is invisible, which provides astronomers with the strongest evidence that the universe contains a great amount of dark matter. The nature of dark matter is still unknown.

Figure 31-12 Strangely shaped galaxies, such as the Antennae Galaxies (NGC 4038 and NGC 4039) **(A),** are often the result of a collision. At the center of the Andromeda Galaxy, the *Hubble Space Telescope* uncovered two nuclei **(B),** which are probably a result of one galaxy being consumed by another.

Superclusters Clusters of galaxies are organized into even larger groups called **superclusters.** These gigantic formations, hundreds of millions of light-years in size, can be observed only when astronomers map out the locations of many galaxies ranging over huge distances. These superclusters appear in sheetlike and threadlike shapes, giving the appearance of a gigantic bubble bath with galaxies located on the surfaces of the bubbles, and the inner air pockets void of galaxies.

THE EXPANDING UNIVERSE

Edwin Hubble made yet another dramatic discovery, this time in 1929. It was known at the time that most galaxies have redshifts in their spectra, indicating that all galaxies are moving away from Earth. Hubble measured the redshifts and distances of many galaxies and found that the redshift of a galaxy depends on its distance from Earth. The farther away a galaxy is, the faster it is moving away. The universe is expanding.

You might infer from this that Earth is at the center of the universe. Actually, this is not the case. An observer located in any galaxy, at any place in the universe, will observe the same thing we do: all other galaxies move away with speeds that depend on their distances. In a medium that is uniformly expanding, all points are moving away from all other points, and no point has to be at the center. You will experiment with uniform expansion in the *MiniLab* later in this section.

Earth Sciences

2.b Students know galaxies are made of billions of stars and comprise most of the visible mass of the universe.

2.g Students know how the red shift from different galaxies and the cosmic background radiation provide evidence for the "big bang" model that suggests that the universe has been expanding for 10 to 20 billion years.

Hubble's Law Hubble determined that the universe is expanding by making a graph of the speed at which a galaxy is moving versus its distance. The result is a straight line, which can be expressed as a simple equation, $v = Hd$, where v is the speed at which a galaxy is moving away measured in kilometers per second; d is the distance to the galaxy measured in megaparsecs (Mpc), where 1 Mpc $= 1 \times 10^6$ pc; and H is a number called the **Hubble constant,** that is measured in kilometers per second per megaparsec. *H* represents the slope of the line. You will plot Hubble's law in the *Problem-Solving Lab* on this page.

To measure *H* requires finding distances and speeds for many galaxies, out to the largest possible distance, and constructing a graph to find the slope. This is a difficult task, because it is hard to measure accurate distances to the most remote galaxies. Hubble himself could obtain only a crude value for *H.* Measuring an accurate value for *H* was one of the key goals of astronomers who designed the *Hubble Space Telescope.* It took nearly ten years after the *Hubble Space Telescope*'s launch to gather enough data to pinpoint the value of *H* to the satisfaction of most astronomers. Currently, the best measurements indicate a value of approximately 70 kilometers per second per megaparsec.

Once the value of *H* is known, it can be used to find distances to far away galaxies. By measuring the speed at which a galaxy is moving, astronomers can locate that speed on the graph and can then determine the corresponding distance of the galaxy on the graph. This method works for the most remote galaxies that can be observed, and it allows astronomers to measure distances to the edge of the observable universe.

To learn more about the *Hubble Space Telescope,* go to the **National Geographic Expedition** on page 902.

Problem-Solving Lab

Making and Using Graphs

The Hubble constant The table lists distances and speeds for a number of galaxies.

Analysis

1. Use the data to construct a graph. Plot the distance on the *x*-axis and the speed on the *y*-axis.
2. Use a ruler to draw a straight line through the center of the band of points on the graph, so that about as many points lie above the line as below it. Make sure your line starts at the origin.
3. Measure the slope by choosing a point on the line and dividing the speed at that point by the distance.

Thinking Critically

4. What does the slope represent?
5. How accurate do you think your value of *H* is? Explain.
6. How would an astronomer improve this measurement of *H?*

Galaxy Data

Distance (Mpc)	Speed (km/s)	Distance (Mpc)	Speed (km/s)
3.0	210	26.5	2087
8.3	450	33.7	2813
10.9	972	36.8	2697
16.2	1383	38.7	3177
17.0	1202	43.9	3835
20.4	1685	45.1	3470
21.9	1594	47.6	3784

Figure 31-13 M87 is a large radio galaxy **(A).** In addition to radio lobes, M87 has a jet of gas that emits visible light **(B).**

Active Galaxies

Radio-telescope surveys of the sky have revealed a number of galaxies that are extremely bright. These galaxies, called **radio galaxies,** are often giant elliptical galaxies that emit as much as or more energy in radio wavelengths than they do in wavelengths of visible light. Radio galaxies have many unusual properties. The radio emission usually comes from two huge lobes of very hot gas located on opposite sides of the visible galaxy. These lobes are linked to the galaxy by jets of very hot gas. The type of emission that comes from these regions indicates that the gas is ionized, and that electrons in the gas jets are traveling nearly at the speed of light. Many radio galaxies have jets that can be observed only at radio wavelengths. One of the brightest of the radio galaxies, a giant elliptical called M87, shown in ***Figure 31-13,*** also has a jet of gas that emits visible light extending from the galactic center out toward one of the radio-emitting lobes.

In some unusual galaxies, some sort of highly energetic object or activity exists in the core. This object or activity emits as much or more energy than the rest of the galaxy. The output of this energy often varies over time, sometimes as little as a few days. The cores of galaxies where these highly energetic objects or activities are located are called **active galactic nuclei,** or AGNs.

Quasars

In the 1960s, astronomers discovered another new type of object. These objects looked like ordinary stars, but some were strong radio emitters. Most stars are not. The spectra of these new objects were completely different from the spectra of normal stars. Whereas most stars have spectra with absorption lines, these new objects had mostly

Figure 31-14 From a ground-based telescope, a quasar (right of center) looks much like a star **(A).** From the *Hubble Space Telescope,* a quasar can be seen in the nucleus of a dim galaxy **(B).**

emission lines in their spectra. These starlike objects with emission lines in their spectra are called **quasars.** A quasar is shown in ***Figure 31-14A.*** At first, astronomers could not identify the emission lines in the spectra of quasars. Finally, they realized that the emission lines were spectral lines of common elements, such as hydrogen, shifted very far toward longer wavelengths. Soon, astronomers also discovered that many quasars vary in brightness over a period of a few days. Once astronomers had identified the large spectral-line shifts of quasars, they wondered whether these could be redshifts caused by the expansion of the universe.

The redshifts of quasars were much larger than any that had been observed in galaxies up to that time, which would mean that the quasars were much farther away than any known galaxy. Some astronomers at first doubted that quasars were far away, but in the decades since quasars were discovered, more evidence supports the hypothesis that quasars are indeed far away. One bit of supporting evidence is that some quasars are associated with clusters of galaxies with the same redshift, verifying that they are the same distance away. Another, more important discovery is that most quasars are nuclei of very dim galaxies, as shown in ***Figure 31-14B.*** The quasars appear to be extra-bright active galactic nuclei—so much brighter than their surrounding galaxies that astronomers could not even see the galaxies at first.

Looking Back in Time Because many quasars are far away, it takes their light a long time to reach Earth. For this reason, astronomers observing a quasar are seeing it as it was a long time ago. For example, it takes light from the Sun approximately eight minutes to reach Earth. When we observe that light, we are seeing the Sun as it

MiniLab

Measuring Redshifts

Model uniform expansion of the universe and the redshifts of galaxies that result from expansion.

Procedure

1. Use a felt tip marking pen to make four dots in a row, each separated by 1 cm, on the surface of an uninflated balloon. Label the dots 1, 2, 3, and 4.
2. Partially inflate the balloon. Using a piece of string and a meterstick, measure the distance from dot 1 to each of the other dots. Record your measurements.
3. Inflate the balloon further, and again measure the distance from dot 1 to each of the other dots. Record your measurements.
4. Repeat step 3 with the balloon fully inflated.

Analyze and Conclude

1. Are the dots still separated from each other by equal distances? Explain.
2. How far did each dot move away from dot 1 after each inflation?
3. What would be the result if you had measured the distances from dot 4 instead of dot 1? From dot 2?
4. How does this activity illustrate uniform expansion of the universe and redshifts of galaxies?

was eight minutes earlier. When we observe the Andromeda Galaxy, we see the way it looked two million years earlier. The most remote quasars are several billion light-years away, which indicates that they existed billions of years ago. If quasars are extra-bright galactic nuclei, then the many distant ones are nuclei of galaxies as they existed when the universe was young. This suggests that many galaxies went through a quasar stage when they were young. In that case, today's active galactic nuclei might be former quasars that are not quite as energetic as they were long ago.

Source of Power The AGNs and quasars emit far more energy than ordinary galaxies, but they are as small as solar systems. This suggests that all of these objects are supermassive black holes. Recall that our own galaxy is hypothesized to contain such an object within its core, and that this object has a mass of about one million Suns. The black holes in the cores of AGNs and quasars are much more massive, up to hundreds of millions of times the mass of the Sun. The beams of charged particles that stream out of the cores of radio galaxies and form jets are probably created by magnetic forces. As material falls into a black hole, the magnetic forces push the charged particles out into jets. There is evidence that similar beams or jets occur in other types of AGNs and in quasars. In fact, radio-lobed quasars have jets that are essentially related to radio galaxies.

The role of AGNs and especially quasars as early stages in the evolution of galaxies provide important clues for astronomers as they study the origin and evolution of the universe itself. With the discovery of new and more distant objects, astronomers are getting closer to unraveling the story of the universe.

Earth Science Online

Update For an online update on quasars, visit the Earth Science Web Site at earthgeu.com

SECTION ASSESSMENT

1. How did astronomers discover that there are other galaxies beyond the Milky Way?
2. Why do astronomers theorize that most of the matter in galaxies and clusters of galaxies is dark matter?
3. Why is it difficult for astronomers to accurately measure a value for the Hubble constant, *H*? Once a value is determined, what can it be used for?
4. Explain the differences among normal spiral, barred spiral, elliptical, and irregular galaxies.
5. **Thinking Critically** How would the nighttime sky look from Earth if we lived in an elliptical galaxy?

SKILL REVIEW

6. **Using Numbers** Suppose that a distant galaxy is moving away from Earth at a velocity of 20 000 km/s. Assuming that the value of the Hubble constant is 70 kilometers per second per megaparsec, how far away is this galaxy? For more help, refer to the *Skill Handbook*.

SECTION 31.3 Cosmology

Earth Sciences 2.e, 2.g

You've learned about Earth, the planets, the Sun, stars, and galaxies. Now you are ready to consider the universe as a whole. The study of the universe, its current nature, and its origin and evolution, is called **cosmology.**

As in other areas of science, astronomers use a combination of observations and theoretical models in cosmology. One difference between cosmology and other areas of study is that in cosmology, objects that have a range of properties cannot be compared, as they can be in the study of stars, for example. Astronomers have only one universe to consider and nothing to compare it with. However astronomers can use particle accelerators to learn about the universe, as you'll learn in the *Science & Technology* feature at the end of this chapter.

OBJECTIVES

- **Explain** *the different theories about the formation of the universe.*
- **Describe** *the possible outcomes of universal expansion.*

VOCABULARY

cosmology
Big Bang theory
steady-state theory
cosmic background radiation
inflationary universe

MODELS OF THE UNIVERSE

The fact that the universe is expanding implies that it had a beginning. The theory that the universe began as a point and has been expanding ever since is called the **Big Bang theory.** The Big Bang theory is not an explosion into space, but instead is an expansion of space with matter going along for the ride. Not all astronomers agree that the universe had a beginning, which led to an alternative theory. The **steady-state theory** proposes that the universe looks the same on large scales to all observers, and that it has always looked that way. The steady-state universe doesn't change with time. However, if the universe is expanding, how can this be true? Supporters of the steady-state theory propose that new matter is created and added to the universe as it expands, and thus, the overall density of the universe does not change, as illustrated in ***Figure 31-15.***

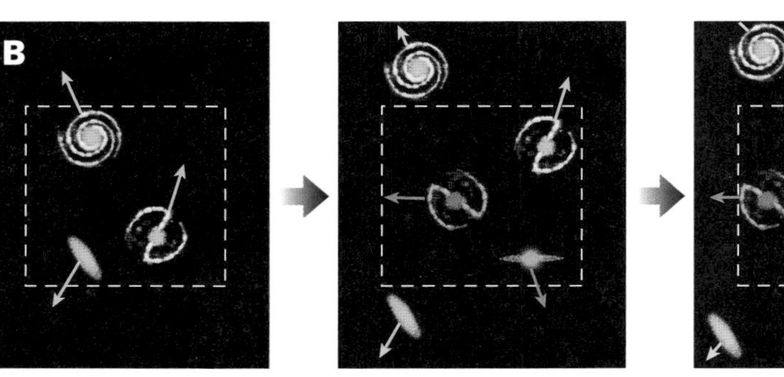

Figure 31-15 Without the creation of new matter, the area within the dotted box would not contain 3 galaxies after a time **(A).** The steady-state theory requires new matter to be added so that the area within the dotted box always contains 3 galaxies **(B).**

Figure 31-16 The cosmic background radiation was discovered accidentally with this radio antenna at Bell Labs, in Holmdel, New Jersey.

Earth Sciences

2.g Students know how the red shift from different galaxies and the cosmic background radiation provide evidence for the "big bang" model that suggests that the universe has been expanding for 10 to 20 billion years.

Many observational tests can be conducted to provide support for either the Big Bang or the steady-state theory. Each theory predicts what the universe should look like. By observing the universe, astronomers can determine which predictions are actually met. This in turn provides support for one theory or the other. Today, the evidence weighs in favor of the Big Bang.

Cosmic Background Radiation If the universe began in a highly compressed state, as the Big Bang theory suggests, it would have been very hot, and the high temperatures would have filled it with radiation. If radiation filled the universe back then, that same radiation should still fill the universe. Originally, when the radiation was able to escape, it would have been in the form of visible light and infrared radiation, which have high energy and short wavelengths. However, as the universe expanded and cooled, the radiation would have been Doppler shifted to lower energies and longer wavelengths.

The existence of such radiation in the universe today would not support the steady-state theory of the universe, which does not predict such radiation. In 1965, some scientists discovered a persistent background noise in their radio antenna, shown in ***Figure 31-16.*** This noise was caused by weak radiation, called the **cosmic background radiation,** that appeared to come from all directions in space and corresponded to an emitting object having a temperature of about 2.735 K (–270°C). This was very close to the temperature predicted by the Big Bang theory, and the radiation was interpreted to be from the beginning of the Big Bang.

Figure 31-17 The map of the cosmic background radiation made by *WMAP* shows slight temperature differences of only a few millionths of a degree.

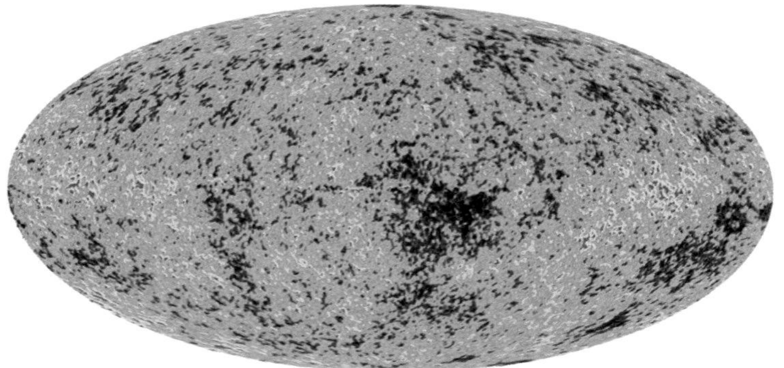

Mapping the Radiation Since the discovery of the cosmic background radiation, extensive observations have confirmed that it matches the properties of the predicted leftover radiation from the early, hot phase in the expansion of the universe. Earth's atmosphere blocks much of the radiation, so it is best observed from high-altitude balloons or satellites. An orbiting observatory called the *Wilkinson Microwave Anisotropy Probe (WMAP),* launched by NASA in 2001, mapped the radiation in greater detail, as shown in ***Figure 31-17.*** The peak of the radiation it measured has a wavelength of approximately 1 mm; thus, it is microwave radiation in the radio portion of the electromagnetic spectrum.

Proponents of the steady-state universe theory have not succeeded in explaining the cosmic background radiation. This is one of the major reasons that most astronomers do not accept this model. Other tests that can be used to validate either the steady-state theory or the Big Bang theory are whether or not each theory can explain redshifts of galaxies, the evolution of galaxies, and quasars.

THE BIG BANG MODEL

In the Big Bang model, there is a competition between the outward momentum of the expansion of the universe and the inward force of gravity as the matter in the universe acts to slow the expansion, as illustrated in ***Figure 31-18A.*** What ultimately will happen depends on which of these two forces is stronger. There are three possible outcomes for the universe: an open universe, in which the expansion will never stop; a closed universe, in which the expansion will stop and turn into a contraction; and a flat universe, in which the expansion will slow to a halt in an infinite amount of time—but it will never contract. All three cases, illustrated in ***Figure 31-18B,*** are based on the premise that the rate of expansion has slowed down since the beginning of the universe. The question would then appear to be, How much further will it slow down?

The Critical Density The total amount of matter in the universe is one of the factors that will determine whether or not the expansion will stop. The total amount of matter in the universe is expressed in terms of the average density of matter, something that astronomers can observe. The *critical density,* about 10^{-26} kg/m^3, is the dividing point between a closed or open universe. If the average density is higher than the critical density, the universe is closed, but if the average density is lower than the critical density, the universe is open. If the density equals the critical density, the universe is flat. By observing the average density of matter, astronomers can predict how the universe will change on a large scale. Observations of visible galaxies reveal an average density much less than the critical density. However, as you've learned, there is evidence that a great amount of dark matter exists in the universe, which impacts the average density.

Figure 31-18 Gravity competes with the momentum of the expansion **(A).** The universe could be open, flat, or closed **(B).** The flat universe line will gradually become a horizontal line, while the open universe line will keep increasing.

A

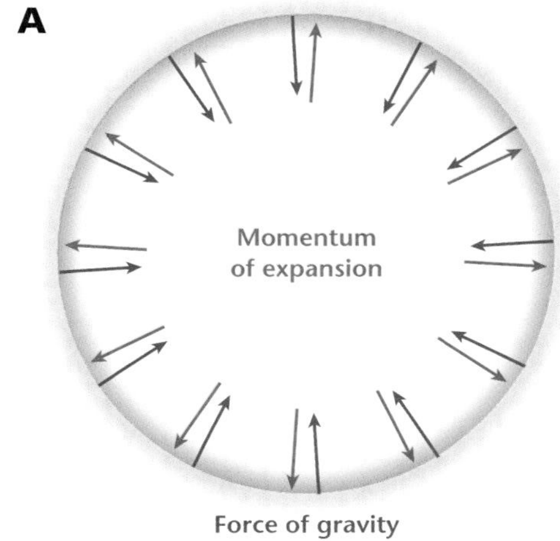

B

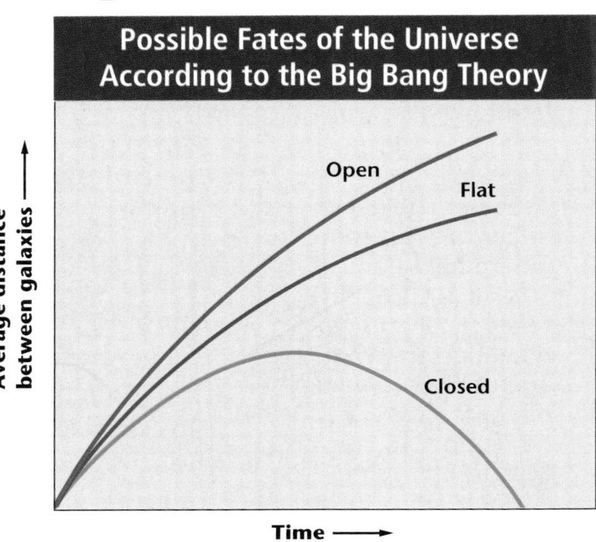

Figure 31-19 This distant cluster of galaxies, Abell 2218, allows astronomers to look back into time.

Expansion Rate Another approach to determining the fate of the universe is to measure how much slowing has occurred so far in its expansion. This will indicate how much further it will slow down. The slowing of the expansion of the universe can be measured by comparing the expansion rate today with the rate long ago. Recall that when astronomers observe the most distant galaxies, such as the ones shown in ***Figure 31-19,*** they are observing the galaxies as they existed far back in time. By measuring the redshifts of the most remote galaxies, it is possible for astronomers to determine the expansion rate long ago. This also requires finding the distances to those very far away galaxies.

The most recent observations of the change in the expansion rate of the universe have been startling. Astronomers have found that the rate of expansion is speeding up! It appears that the rate of expansion slowed for a while but is now gathering speed. The only explanation offered so far is that a previously unknown force is acting to push the galaxies apart. According to this explanation, when the universe was smaller and denser, gravity was strong enough to slow the expansion, but some time ago this unknown force began to dominate over gravity.

Using Numbers The time since the expansion of the universe began can be calculated by

$$t = \frac{1}{H}.$$

Astronomers hypothesize that H is somewhere between 50 and 100 kilometers per second per megaparsec. What age does each value of H predict? Remember to convert Mpc to km, using 1 Mpc = 3.1×10^{19} km.

The Inflationary Model The combination of the observed density of the universe, including an allowance for dark matter, and the apparent acceleration of the expansion, fit together into a model in which the universe is flat. A flat universe was predicted in the 1980s by a version of the Big Bang model called the inflationary universe. In the **inflationary universe** model, the universe began as a fluctuation in a vacuum and expanded very rapidly for a fraction of a second, as illustrated in ***Figure 31-20,*** before settling into a more orderly expansion. The initial inflationary step is similar to what can happen in water that is supersaturated with soap: if the water is disturbed, it forms a sea of bubbles almost instantaneously.

When the rate of expansion of the universe is known, it is possible to calculate the time since the expansion started, or the age of the universe. When the distance to a galaxy and the rate at which it is moving away from Earth are known, it is simple to calculate how long ago that galaxy and the Milky Way were together. In astronomical terms,

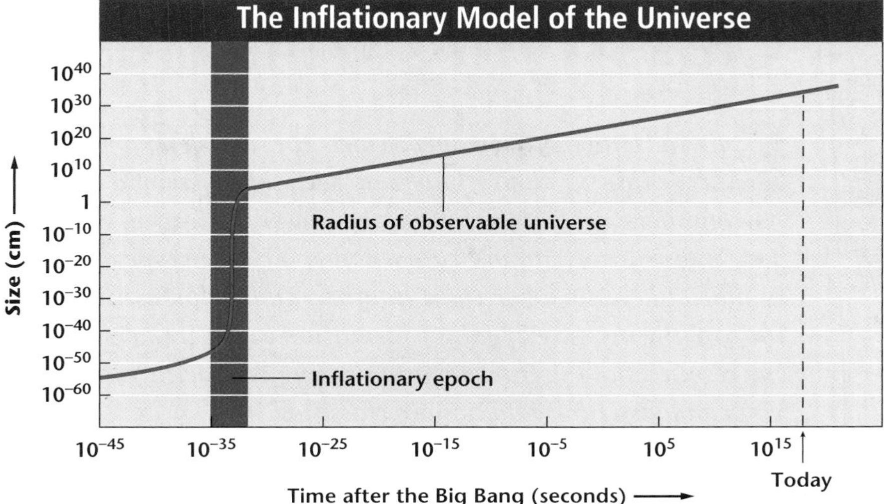

Figure 31-20 According to the inflationary model, the universe expanded very quickly in the early portion of the Big Bang. The time in which inflation occurred is the shaded portion of the graph, where the size of the universe increases drastically.

if the value of *H*, the expansion constant, is known, then the age of the universe can be determined. A correction is needed to allow for the fact that the expansion was more rapid at the beginning, and now it seems that this correction must also take into account the acceleration of the expansion that is occurring now. Based on the best value for *H* that has been calculated from *Hubble Space Telescope* data, as well as the appropriate corrections, the age of the universe is hypothesized to be about 13.5 billion years. This fits with what astronomers know about the age of our own galaxy, which is estimated to be between 12 and 14 billion years, based on the ages of the oldest star clusters. However, refinements to these measurements are still being made.

Section Assessment

1. What are the differences between the steady-state model of the universe and the Big Bang model of the universe?
2. Describe the inflationary model and its predicted impact on the structure of the universe.
3. Why is dark matter important in determining the density of matter in the universe?
4. Why was the cosmic background radiation an important discovery?
5. **Thinking Critically** How would a large value of *H*, the Hubble constant, affect the age of the universe? What about a small value of *H*?

Skill Review

6. **Comparing and Contrasting** Compare and contrast an open universe, a closed universe, and a flat universe. For more help, refer to the *Skill Handbook*.

Classifying Galaxies

Edwin Hubble developed rules for classifying galaxies according to their shapes as seen in telescopic images. Astronomers are interested in the classification of galaxies. This information can indicate whether a certain type of galaxy is more likely to form than another and helps astronomers unravel the mystery of galaxy formation in the universe. Using the resources of the Internet and sharing data with your peers, you can learn how galaxies are classified.

Preparation

Problem

How can different galaxies be classified?

Hypothesis

How might galaxies be classified using Hubble's classification system? Are there absolute classifications based solely on shape? **Form a hypothesis** about how you can apply Hubble's galaxy classification system to galaxy images on the Internet.

Objectives

- **Gather** and **communicate** details about galaxy images on the Internet.
- **Form conclusions** about the classification of different galaxies.
- **Reconstruct** the tuning-fork diagram with images that you find.

Data Sources

Go to the Earth Science Web Site at earthgeu.com to find links to galaxy images on the Internet. You can also visit a local library or observatory to gather images of galaxies and information about them.

Plan the Experiment

1. Find a resource with multiple images of galaxies and, if possible, names or catalog numbers for the galaxies. The Earth Science Web Site lists sites that have galaxy images.
2. Choose one of the following types of galaxies to start your classification: spirals, ellipticals, or irregular galaxies.
3. Gather images and information, such as catalog numbers and names of galaxies, from the links on the Earth Science Web Site or the library.
4. Sort the images by basic types: spirals, ellipticals, or irregular galaxies.

5. For each basic type, compare the galaxies to each other and decide which galaxy best represents each class and subclass of Hubble's galaxy classification system: Sa, Sb, Sc, SBa, SBb, SBc, S0, E0–E7, and Irr. Try to find at least one galaxy for each subclass.
6. Arrange the galaxy images to construct a tuning-fork diagram like Hubble's.

GALAXY DATA			
Galaxy Name	**Sketch of Galaxy**	**Classification**	**Notes**
NGC 3486		Sc	

Procedure

1. Complete the data table. Add any additional information that you think is important.
2. Go to the Earth Science Web Site at earthgeu.com to post your data.
3. Visit sites listed on the Earth Science Web Site for information about other galaxies.

Conclude & Apply

Sharing Your Data Find this Internet GeoLab on the Earth Science Web Site at earthgeu.com. Post your data in the table provided for this activity. Use the additional data from other students to complete your chart and answer the Conclude & Apply questions.

1. Were there any galaxy classes or subclasses that were difficult to find images for? If so, which ones?
2. How many of each type of galaxy—normal spiral, barred spiral, elliptical, and irregular—did you find?
3. Calculate the percentages of the total number of galaxies that each type represents. Do you think this reflects the actual percentage of each type of galaxy in the universe? Explain.
4. Were there any galaxy images that you found that didn't fit your classification scheme? If so, why?
5. Was it difficult to distinguish between a normal spiral and a barred spiral in some cases? Explain your method.
6. What problems did you have with galaxies that are edge-on as seen from Earth?
7. Ellipticals are usually a difficult type of galaxy to classify. Why?

Science & Technology

Earth Sciences 2.e

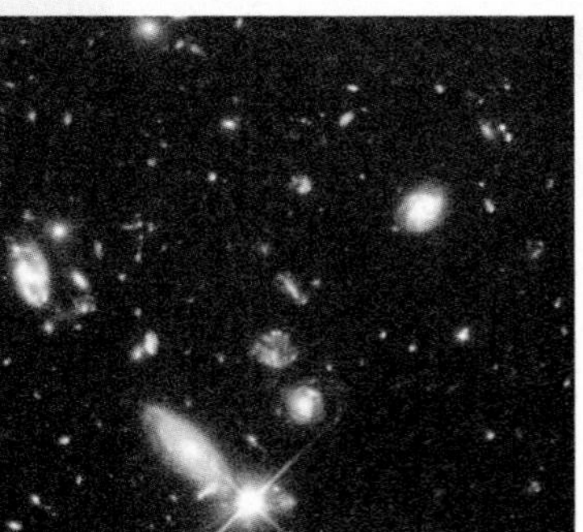

The Early Universe

Imagine looking back 10 billion years to a time when the universe was young and galaxies were just beginning to form. The Hubble Deep Field (HDF), pictured at left, is a snapshot of the early universe by the Hubble Space Telescope. *The HDF shows at least 1500 galaxies in various stages of evolution. Some are so far away that we see them now as they looked less than 1 billion years after the Big Bang.*

Back in Time

Images such as the HDF offer tantalizing clues to questions about conditions in the early universe. Did it resemble the universe we know today? How did the universe begin? Most astronomers hypothesize that the universe was formed by the Big Bang. This violent event left particles rushing away from each other at tremendous speeds. The energy created billions of years ago still propels galaxies away from each other as the universe continues to expand.

Accelerated Particles

Scientists learn about the world by making observations about events in progress, or by creating specific conditions that will cause an event to occur. Because the Big Bang can neither be observed nor recreated, other methods must be used to study it. One tool that scientists use to replicate conditions in the early universe is the particle accelerator, a machine that smashes charged particles into various targets at speeds close to the speed of light. The particle accelerator allows scientists to test models of how matter and energy behaved as soon as 1×10^{-11} s after the Big Bang.

All particle accelerators are based on the principle that electrically charged objects exert a force on each other. Intense magnetic fields are applied to charged particles moving through tubes. Particles gain speed until they crash into some target. High-speed collisions between particles often result in strange particles that scientists have never before observed.

Limitations

In 1989, workers began digging miles of tunnels in a geologically stable area in Texas where the Supercolliding Superconductor (SCC) was to be built. A 54-mile (86.9-km) oval, the SCC would have been the world's largest, most powerful particle accelerator, allowing astronomers to probe deeper into the early conditions of the universe. Despite the efforts of physicists and astronomers worldwide, however, Congress cut off funding for the project in 1993. Until larger, more powerful particle accelerators are built, scientists are limited in what they can learn about the early universe.

Activity

The Fermilab Tevatron is the world's most powerful accelerator. Go to the Earth Science Web Site at earthgeu.com to research Fermilab. Do you think particle accelerators should be built despite their large cost? Why?

CHAPTER 31

Study Guide

Summary

SECTION 31.1

The Milky Way Galaxy

Main Ideas

- Because our solar system is inside the Milky Way galaxy, it was difficult at first for astronomers to determine the size and shape of our galaxy, and where Earth is located within it.
- The Milky Way consists of a nuclear bulge, a disk, and a halo. Much of the mass of the galaxy is not visible.
- It is hypothesized that the Milky Way began as a spherical cloud of gas that eventually collapsed into a disk.
- Population I stars contain small amounts of heavy elements and are located in the arms and disk. Population II stars are located in the bulge and halo, and contain only trace amounts of heavy elements.

Vocabulary

Cepheid variable (p. 834)
halo (p. 835)
RR Lyrae variable (p. 834)
spiral density wave (p. 838)
variable star (p. 834)

SECTION 31.2

Other Galaxies in the Universe

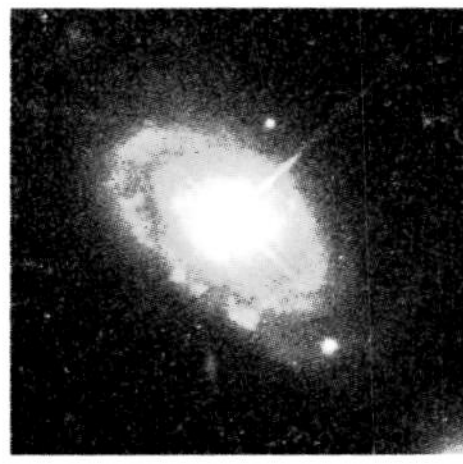

Main Ideas

- Galaxies are classified according to their shapes as normal spirals, barred spirals, ellipticals, irregulars, or dwarf ellipticals.
- Most galaxies occur in clusters, which are further organized into superclusters.
- The universe is expanding, and the Hubble constant, H, measures the rate of expansion.
- Because it takes the light from very distant galaxies so long to reach Earth, when astronomers observe these galaxies, they are looking back in time.
- Many galaxies have energetic objects or activities at their cores called active galactic nuclei.

Vocabulary

active galactic nucleus (p. 844)
Hubble constant (p. 843)
quasar (p. 845)
radio galaxy (p. 844)
supercluster (p. 842)

SECTION 31.3

Cosmology

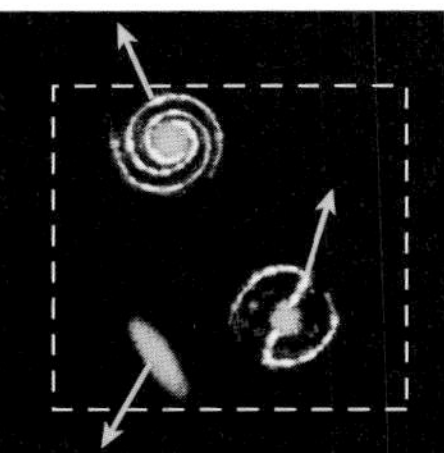

Main Ideas

- The Big Bang theory proposes that the universe began as a single point and has been expanding ever since.
- The steady-state theory proposes that the universe is the same as it always has been.
- The universe is filled with cosmic background radiation that is left over from the early, hot stages in the Big Bang expansion of the universe.
- In the Big Bang model, the universe could be open, closed, or flat.
- The inflationary model can explain the walls and voids of the distribution of galaxies. It also predicts that the universe is flat, which is supported by observations.
- Current observations indicate that an unknown force is accelerating the expansion of the universe.

Vocabulary

Big Bang theory (p. 847)
cosmic background radiation (p. 848)
cosmology (p. 847)
inflationary universe (p. 850)
steady-state theory (p. 847)

CHAPTER 31

Assessment

Understanding Main Ideas

1. What are the oldest objects in the Milky Way?
   a. globular clusters
   b. Cepheid variables
   c. spiral arms
   d. Population I stars

2. Where in the Milky Way are new stars being formed?
   a. in the nuclear bulge
   b. in globular clusters
   c. in the spiral arms
   d. in the halo

3. Where does the energy emitted by AGNs and quasars most probably come from?
   a. material falling into a supermassive black hole
   b. a neutron star
   c. a supernova explosion
   d. a pulsar

4. What theory proposes that the universe does not change with time?
   a. the Big Bang
   b. the cosmological constant
   c. the heliocentric universe theory
   d. the steady-state universe theory

5. What is the origin of the cosmic background radiation?
   a. It is emitted by stars.
   b. It is a remnant of the Big Bang.
   c. It is emitted by radio galaxies.
   d. It comes from the steady-state universe.

6. In the Big Bang model, which of the following describes a universe that will stop expanding and begin to contract?
   a. open
   b. flat
   c. closed
   d. inflationary

7. Why are pulsating variable stars useful for finding distances to globular clusters?

8. How do astronomers observe the spiral structure of our galaxy?

9. Why do astronomers think that there is a great amount of mass in the halo of the Milky Way?

10. What possible forms might the dark matter in the halo of the Milky Way be in?

11. Why do astronomers hypothesize that there is a supermassive black hole at the center of the Milky Way?

Use the diagram below to answer question 12.

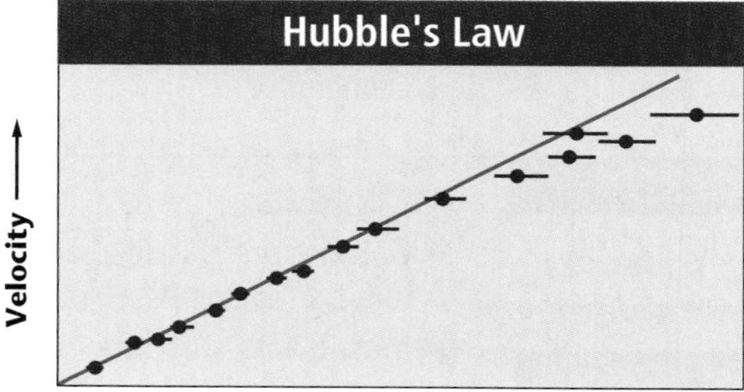

12. The straight line represents Hubble's law, $v = Hd$, and the points are actual data for galaxies. Why don't the points agree with the straight line of Hubble's law?

13. Why are the stars in globular clusters classified as Population II stars?

14. Does the fact that all galaxies are receding from Earth suggest that it is located at the center of the universe? Explain.

15. What is the significance of the Hubble constant?

16. What evidence indicates that many quasars existed long ago?

Test-Taking Tip

MISTAKES Your mistakes can teach you. The mistakes you make before the test are helpful because they show you the areas in which you need more work.

earthgeu.com/chapter_test

CHAPTER 31
Assessment

17. What are the main differences between the steady-state model and the Big Bang model of the universe?
18. What would be the different outcomes of the open, flat, and closed universe models?

Applying Main Ideas

19. Why are astronomers actually looking into the past when they observe objects that are very far away?
20. What evidence indicates that much of the mass in the universe is in an invisible form called dark matter?
21. Why is it important for astronomers to determine the average density of the universe?
22. How did Edwin Hubble reach the conclusion that the universe is expanding?
23. Why do astronomers theorize that many quasars are active nuclei of young galaxies?

Thinking Critically

24. How would a star that forms in our galaxy a few billion years in the future compare with the Sun?
25. How is the formation of our galaxy like the formation of a star?
26. A quasar and three galaxies are observed, and the ranking of their redshifts is as follows: galaxy 1, largest redshift; galaxy 2, second-largest redshift; the quasar, third-largest redshift; and galaxy 3, fourth-largest redshift. List these objects in order from the nearest to the farthest from Earth.
27. How do the properties of quasars suggest that the steady-state model of the universe is not accurate?

Standardized Test Practice

USING GRAPHS Use the graph below to answer question 1.

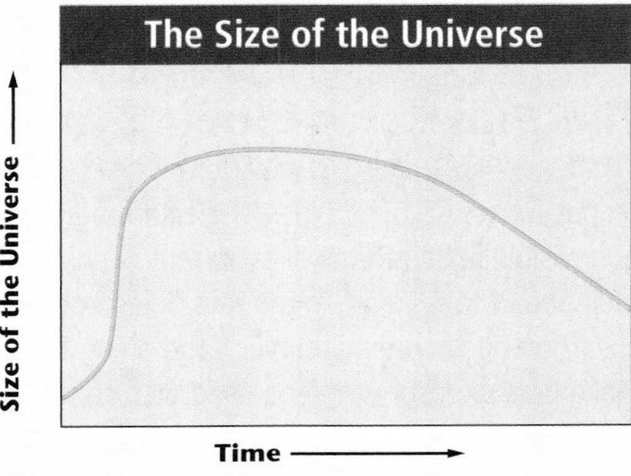

1. What possible fate of the universe does this represent?
   a. flat
   b. open
   c. closed
   d. steady
2. Without doing any calculations, what can astronomers determine from a variable star's period of pulsation?
   a. distance
   b. apparent magnitude
   c. luminosity
   d. age
3. What two measurements are required to determine the Hubble constant?
   a. distance and speed
   b. distance and absolute magnitude
   c. apparent magnitude and speed
   d. apparent and absolute magnitudes
4. Which of the following does the existence of the cosmic background radiation support?
   a. critical density
   b. the steady-state theory
   c. the inflationary model
   d. the Big Bang theory

UNIT 8

GeoDigest

For a **preview** of the universe beyond Earth, study this GeoDigest before you read the chapters. After you have studied these chapters, you can use the GeoDigest to **review** the unit.

Beyond Earth

The Sun-Earth-Moon System

Tools of Astronomy Different types of telescopes make use of all types of electromagnetic radiation: visible light, radio waves, infrared and ultraviolet radiation, X rays, and gamma rays. Telescopes collect electromagnetic radiation over large areas. They can make time exposures and be used with cameras and other equipment to analyze and measure electromagnetic radiation. Refracting telescopes use lenses to bring light to a focus, whereas reflecting telescopes use mirrors. Some forms of radiation are blocked by Earth's atmosphere, and thus telescopes that collect these forms of radiation must be placed in space. Space probes make observations from space or land and take samples for analysis.

John Glenn with suit technicians, 1998

The Moon The Moon has surface features that survive due to its lack of an atmosphere and erosion. Earth's moon is larger and more distant, relative to the size of its planet, than any other satellite in the solar system. These characteristics may be a result of the Moon's being formed in a collision between Earth and a Mars-sized body. The Moon is covered with highlands and maria.

The Sun-Earth-Moon System From a rotating Earth, the sky appears to rotate daily. Our timekeeping system is based on the solar day as observed from Earth. On Earth, we see the Moon go through phases that complete a lunar month. The phases correspond to our changing view of the sunlit side of the Moon. A lunar eclipse occurs when the Moon passes through Earth's shadow; a solar eclipse occurs when the Moon passes between Earth and the Sun, and the Moon's shadow intersects the surface of Earth. Earth's tilt on its axis and its changing position in orbit causes seasons. Tides are caused by the gravitational pull of the Moon and the Sun on the water on Earth's surface.

The Solar System

Motion and Formation Early astronomers explained the motions of the planets with geocentric models. Copernicus, Brahe, Kepler, and Galileo collected data that supported a heliocentric model of the solar system. Newton's law of gravitation demonstrated the validity of the heliocentric model. The solar system formed from a collapsing

GeoDigest

Earth Sciences 2.c, 2.g

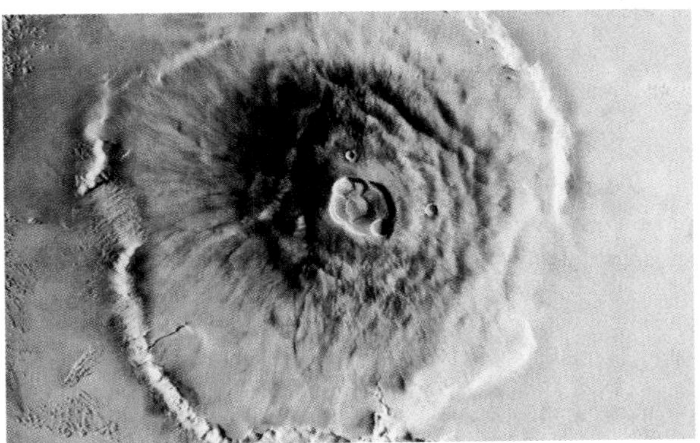

Olympus Mons, Mars

interstellar cloud that flattened into a disk. The terrestrial planets formed from refractory materials in the hot, inner disk; the gas giants formed from volatile elements in the cold, outer disk. Planets orbiting other stars have recently been discovered.

The Planets The terrestrial planets are the four planets closest to the Sun. Mercury has a surface similar to the Moon's. Venus's surface is heated to extreme temperatures through the greenhouse effect. Liquid water exists on Earth, making it suitable for life. Mars may have once had active tectonics. Between Mars and Jupiter is a belt of asteroids, which are rocky remnants of the early solar system. The huge gas giant planets have low densities, no solid surfaces, ring systems, and many moons. Jupiter, the largest and most massive planet, has a fluid (gas or liquid) interior, a small, rocky core, and many moons. Saturn is slightly smaller than Jupiter and has an extensive ring system. Uranus and Neptune are similar in size, composition, and the presence of moons and rings. Pluto is neither a gas giant nor a terrestrial planet. Its moon is close to its own size. Comets are made of rock and ice, and they have highly eccentric orbits. When comets are close to the Sun, they glow brightly and have tails of gas and dust that point away from the Sun.

Stars

The Sun Astronomers learn about the Sun through observations and models. They have found that the Sun contains most of the mass in the solar system and is made primarily of hydrogen and helium. The layers of its atmosphere are the photosphere, chromosphere, and corona. The interior of the Sun consists of radiative and convective zones that transfer energy. The nuclear fusion of hydrogen into helium within the core powers the Sun. Surface features of the Sun include sunspots, solar flares, and prominences. The Sun has an 11-year sunspot cycle and a 22-year activity cycle.

Measurement of Stars Positional measurements of the stars are important for measuring distances through apparent shifts in position, called parallax. Stellar brightness is expressed by apparent and absolute magnitude. The appearance of stars' spectra classifies them by their surface temperatures. Stellar characteristics—class, mass, temperature, magnitude, diameter, and luminosity—are related on a diagram called the Hertzsprung-Russell diagram. The mass of a star determines its internal structure and other properties; gravity and pressure are balanced in a star.

Evolution of Stars If the temperature in the core of a star becomes high enough, elements heavier than hydrogen but lighter than iron can fuse together. When the hydrogen in a star is used up, the star may go through several red-giant phases as reaction stages begin and end. Stars like the Sun end as white dwarfs. Stars up to about 8 times the Sun's mass also become white dwarfs after losing mass. Stars between 8 and 20 times the Sun's mass become supernovae and end as neutron stars. Stars more than 20 times as massive as the Sun end as black holes.

Galaxies

The Milky Way The Milky Way consists of a nuclear bulge, a disk, and a halo; much of its mass is not visible. In the galaxy, Population I stars contain small amounts of heavy elements and are located in the arms and disk. Population II stars are located in the bulge and halo, and contain only trace amounts of heavy elements. Because Earth is inside the Milky Way, it has been difficult to determine the size and shape of the galaxy, and also where Earth is located within it.

Star trails around Polaris

Other Galaxies Most galaxies occur in clusters that are further organized into superclusters. Galaxies are classified according to their shapes as normal spirals, barred spirals, ellipticals, or irregulars. Some galaxies have energetic objects or activities at their cores called active galactic nuclei. The universe is expanding, as measured by the Hubble constant. When astronomers observe distant galaxies, they are looking back in time because the light takes so long to reach Earth.

Cosmology Cosmology is the study of the universe as a whole. The Big Bang model proposes that the universe began as a single point and has been expanding ever since. The universe is filled with cosmic background radiation that is left over from the early, hot stages in the Big Bang's expansion. According to the Big Bang model, the universe could be open, closed, or flat. The steady state theory of cosmology, which proposes that the universe is and always will be the same, is not accepted by most astronomers. The inflationary theory explains the walls and voids in the distribution of galaxies. Current observations indicate that the expansion of the universe is accelerating.

Vital Statistics

The Sun's Nearest Neighbors

Star	Distance from the Sun
1. Proxima Centauri	4.24 ly
2. Alpha Centauri	4.34 ly
3. Barnard's Star	5.97 ly
4. Wolf 359	7.80 ly
5. Lalande 21185	8.19 ly
6. Luyten 726-8	8.55 ly
7. Sirius	8.68 ly
8. Ross 154	9.52 ly
9. Ross 248	10.37 ly

FOCUS ON CAREERS

Cosmologist

Cosmologists study the universe as a whole in an attempt to discover how it formed and how it will end. They usually obtain doctoral degrees in physics, astronomy, or both and have a strong background in math and computer analysis. Cosmologists usually work and teach at universities, or other research institutions.

ASSESSMENT

Understanding Main Ideas

1. A telescope that uses mirrors rather than lenses to bring light to a focus is what kind of telescope?
   a. refracting
   b. reflecting
   c. electromagnetic
   d. probing

2. What is one theory about how the Moon formed?
   a. Numerous asteroids were pulled together.
   b. Earth and Mars collided.
   c. Earth and a Mars-sized body collided.
   d. Earth captured an existing moon from Mars.

3. The planets formed from a collapsing interstellar cloud that flattened into a disk. From which part of the disk did terrestrial planets form?
   a. the inner section of the disk
   b. the outer section of the disk
   c. the exact center of the disk
   d. sections throughout the entire disk

4. Which planets have low densities, no solid surfaces, ring systems, and many moons?
   a. the terrestrial planets
   b. the gas giants
   c. Pluto and Mercury
   d. Mars and Earth

5. The atmosphere of which object in the solar system is made up of the photosphere, the chromosphere, and the corona?
   a. Earth **c.** Venus
   b. Jupiter **d.** the Sun

6. What tool of astronomy relates the class, mass, temperature, magnitude, diameter, and luminosity of stars?
   a. parallax
   b. absolute magnitude
   c. absolute brightness
   d. the Hertzsrpung-Russell diagram

7. Stars that are 8 to 20 times more massive than the Sun end up as what kind of stars?
   a. red giants **c.** neutron stars
   b. white dwarfs **d.** black holes

8. Where are Population II stars located in the Milky Way?
   a. in the bulge and halo
   b. in the arms
   c. in the disk
   d. in the heavy elements

9. How are galaxies classified?
   a. by size
   b. by closeness to Earth
   c. by shape
   d. by the Hubble constant

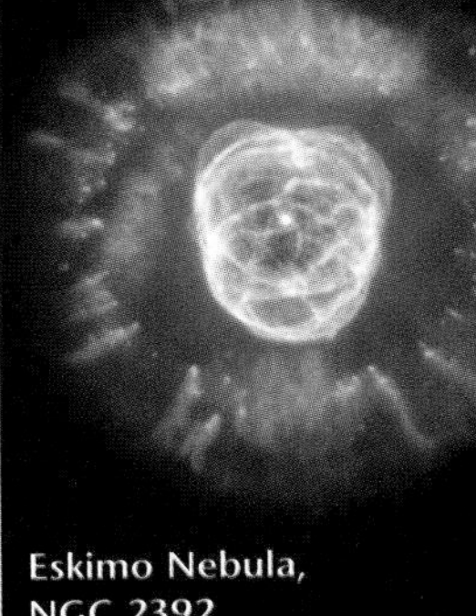
Eskimo Nebula, NGC 2392

10. What is the study of the universe as a whole?
    a. astronomy **c.** physics
    b. cosmology **d.** astrometry

Thinking Critically

1. Compare lunar and solar eclipses.
2. Sequence the following star outcomes in order of the mass of the original star, with the smallest mass first: neutron star, white dwarf, and black hole.
3. Explain the Big Bang model of the formation of the universe.

Explore the *Titanic*...

Tame the mighty Colorado River...

Dig for dinosaurs...

What is it like to map the highest mountain on Earth? Or to explore the ghostly wreck of the "unsinkable" *Titanic?* The **National Geographic Expeditions** allow you to share in the excitement and adventures of explorers, scientists, and lovers of the environment as they delve into the unknown. You can explore the sparkling beauty of a cave in a river of ice. Or watch the explosion of a dying star. Each Expedition will take you on a journey that reaches from the distant galaxies to the depths of the oceans.

As you learn about our dynamic planet, you can use the **National Geographic Expeditions** to extend your knowledge and challenge yourself. To learn more about the Expeditions, go to the **Earth Science Web Site** (earthgeu.com) and click on the **National Geographic Expeditions** link. Or read the original **National Geographic Society** magazine articles to find out how geologists work to discover how a volcano erupts and try to predict future eruptions. The goal is to save lives! It's explorers and scientists like these that discovered the fascinating topics you are studying in your Glencoe **Earth Science: Geology, the Environment, and the Universe** textbook.

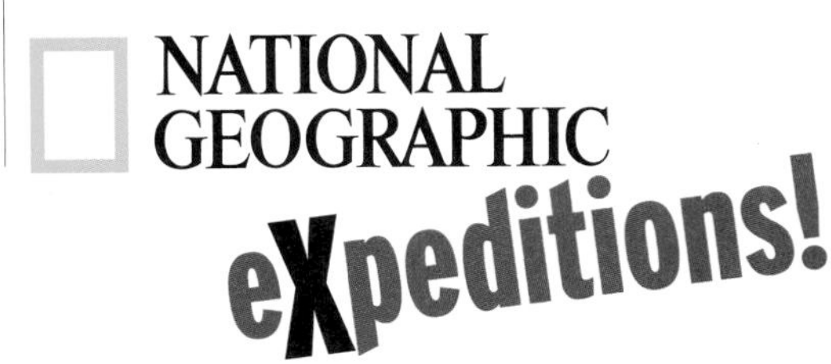

National Geographic Expeditions! are referenced within the chapters at point of use, to support or extend chapter content.

Table of Contents

Taking Cartography to New Heights

Mapping Mount Everest

No one could accuse Bradford Washburn of rushing into his plan to map Mount Everest. Before embarking on such an enormous undertaking, the climber-cartographer thought things over for a while. For 48 years, in fact.

What got Washburn thinking was a conversation at National Geographic Society headquarters in Washington, D.C. It was January 1936, and Washburn was working on a report about an expedition he'd led to the Yukon. At the National Geographic Society he ran into an old friend and hero, Captain Albert W. Stevens, who was an aviator and photographer.

ABOVE: The earliest maps of Mount Everest, compiled in 1921, relied on the work of British surveyors Henry Morshead (right) and Oliver Wheeler. "They were the giants on whose shoulders we stood," said Bradford Washburn, who remapped Everest more recently.

LEFT: The highest point on Earth, Everest proved to be a tough subject for remapping. In 1984 a jet plane with sophisticated photographic equipment provided the basis for more accurate depictions than the cartographers of the 1920s were able to achieve.

Not-so-Casual Conversation

Stevens was writing about his trip aboard the *Explorer II,* a hot-air balloon that had reached the record height of 22 080 meters (72 395 feet). The flight had given Stevens an idea: Why not fly over Everest? Aerial photographs would be an invaluable tool for mapping Earth's tallest peak. "It's a terrific project," he exulted.

Washburn never forgot those words. They took on new force in 1980, when he retired after decades as director of Boston's Museum of Science. "As I looked around to see what a 70-year-old cartographer with a keen interest in mountains could do," wrote Washburn, "I remembered that long-ago conversation with Steve." The project still sounded terrific.

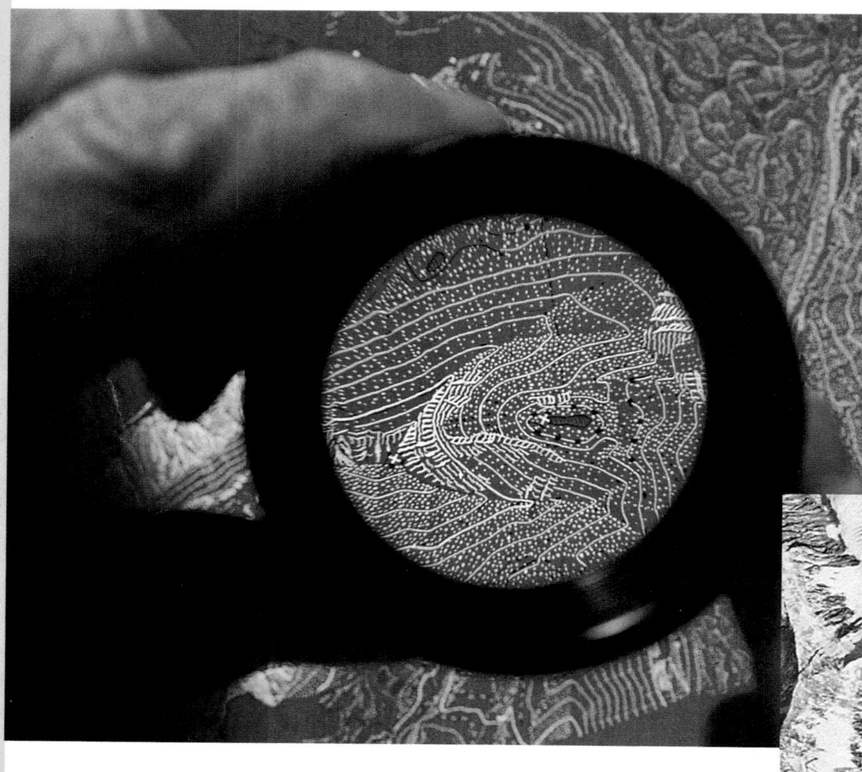

LEFT: In a process that took two years, stereophotographs of Mount Everest and its environs were converted into a conventional contour map, which showed the dramatic lay of the land in remarkable detail.

ABOVE: A portrait of Mount Everest from directly above is one of 160 images obtained in the jet's overflight. The mountain is discernable as a huge pyramid (center left) flanked by glaciers.

Politics and Planning

Washburn began with a goal: To map 777 square kilometers (300 square miles) of the world's most rugged mountain wilderness. The vision was simple; bringing it to life would not be. In fact, it took several years.

First, Washburn had to chart a political landscape because Mount Everest straddles the China-Nepal border. Nepal approved the project after China's consent. "We visited China," wrote Washburn, "and when we returned we carried with us a treasure that no one had ever brought out of China before—permission to map Everest from the air."

Next, Washburn needed money. Half of the flight costs came from his former employer, Boston's Museum of Science. National Geographic provided the rest.

Finally the team had to schedule the flight carefully. "To make an accurate map," explained Washburn, "we needed not only cloudless weather but also a mountainscape that was free of loose snow." December looked like the best time for the expedition. The gales that hit Everest late each fall would sweep the peak clean, and the skies should be clear all winter.

In 1984, after almost half a century, Bradford Washburn was ready to map Mount Everest.

To the Brink and Back

He wasn't ready for what happened next. The Washburns had barely arrived in Nepal for the long-awaited aerial expedition when Barbara Washburn's temperature began soaring. Her doctors could find no cause, and the fever raged on.

Washburn took his wife to Bangkok, Thailand, for specialized medical labs. Thai doctors delivered grim news. Barbara Washburn was dying of cancer and needed to return home immediately.

Frantic, Bradford Washburn recruited Barry Bishop to lead the team. Bishop, an Everest veteran, was vice chairman of National Geographic's Committee for Research and Exploration. Expert pilot Werner Altherr, who worked for Swissair Photo + Surveys of Zurich, agreed to oversee the aerial photography. The expedition would go on—without the Washburns.

They were back in Boston, where doctors gladly lifted Barbara Washburn's death sentence. She did not have cancer; she had a rare blood disease. They had caught it just in time.

ABOVE: Leader of the project to create a new map of Mount Everest, Bradford Washburn, at left, examines the map (see detail at left) with surveying experts in Berne, Switzerland.

LEFT: An artist uses an airbrush to create the stark relief on the most accurate and detailed aerial survey of Mount Everest ever accomplished.

Santa With a Swiss Accent

Barbara Washburn was home from the hospital in time for Christmas. The Washburn's holiday began with a phone call. Werner Altherr was calling from Switzerland.

"I've just arrived at Zurich," he said, "In my briefcase are 160 gorgeous negatives of Everest. This news and my love are a Christmas present to you both from the team in Kathmandu!"

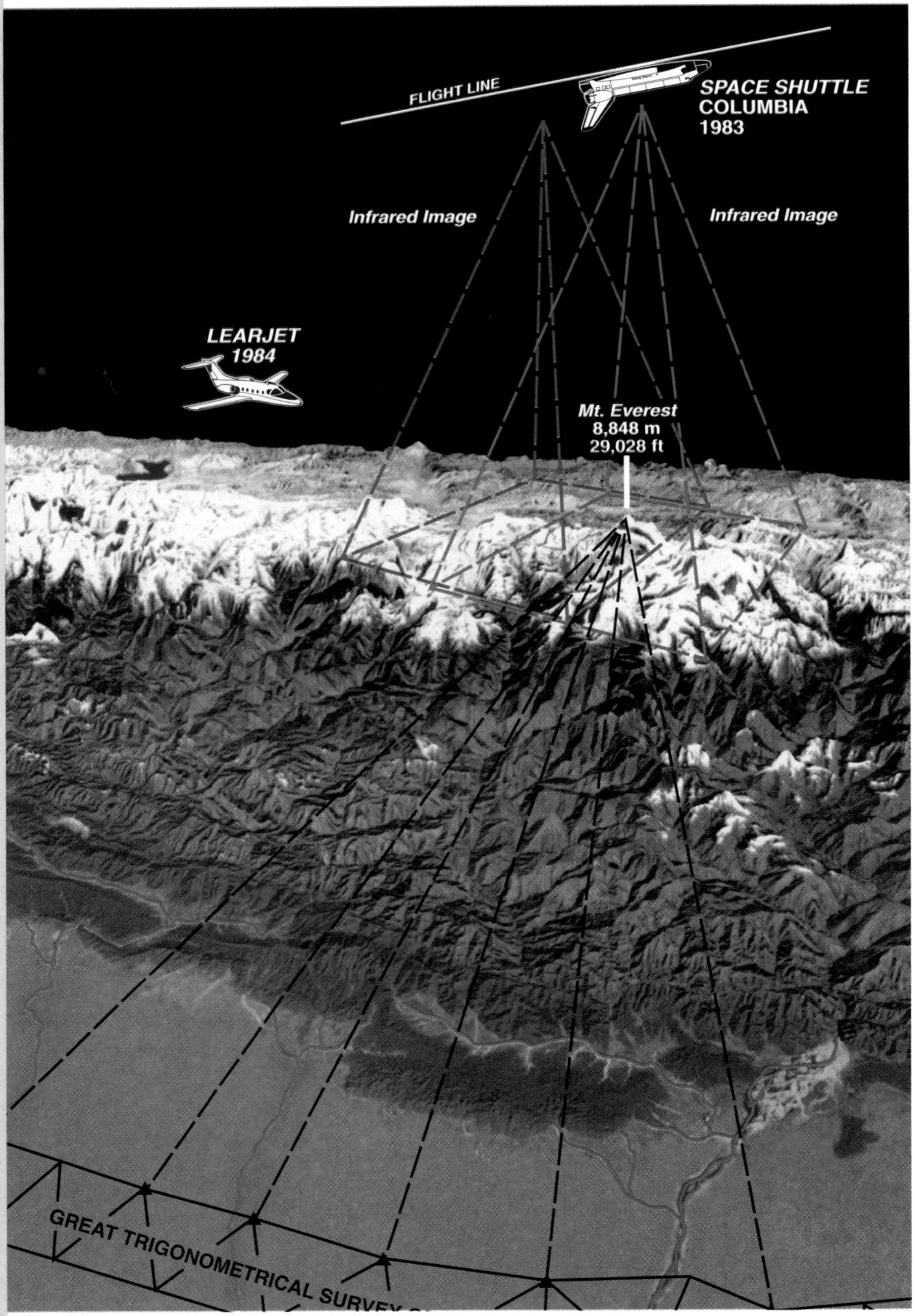

The flight had gone perfectly, he reported. Aboard a Learjet, the team spent three and a half hours photographing Everest and its Himalayan neighbors. Three hours later, the images were already developed. "It all seemed so very fast," mused Washburn, "so very easy at the end of so many years of dreaming, hoping, planning."

BELOW: The view of Everest from 12 300 kilometers (40 000 feet) aboard a Learjet was "utterly hazeless, like the surface of the moon," said aerial photographer Kurt Keller.

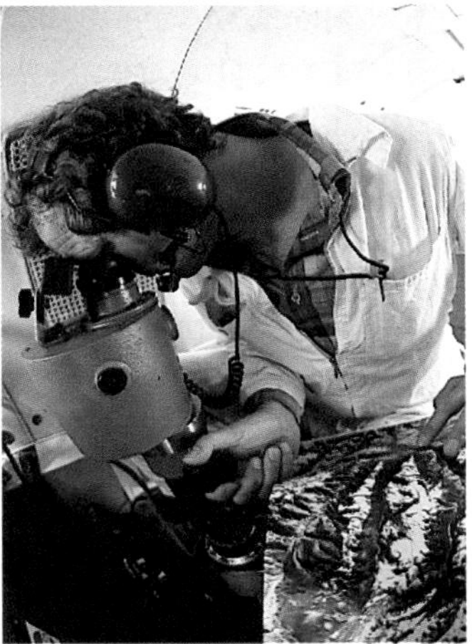

LEFT: Mapping of Mount Everest in the 1980s was preceded by surveys almost 150 years older. British surveyors in the 1840s took sightings of Himalayan peaks from six stations in northern India (bottom of diagram). In 1852, the position and elevation of the peak were determined. It took until 1954, however, to establish the height of Everest at 8848 meters (29 028 feet). The new map in the 1980s made use of this information, plus photos from the Learjet and stereophotographs taken from space by equipment on the space shuttle *Columbia* in 1983. Subsequent investigations, in the late 1990s, established Everest's peak two meters (7 feet) higher, at 8850 meters (29 035 feet).

This text is based on "Mount Everest, Surveying the Third Pole," in the November 1988 issue of NATIONAL GEOGRAPHIC. *Use resources to read about a 1999 expedition that sought to measure Mount Everest.*

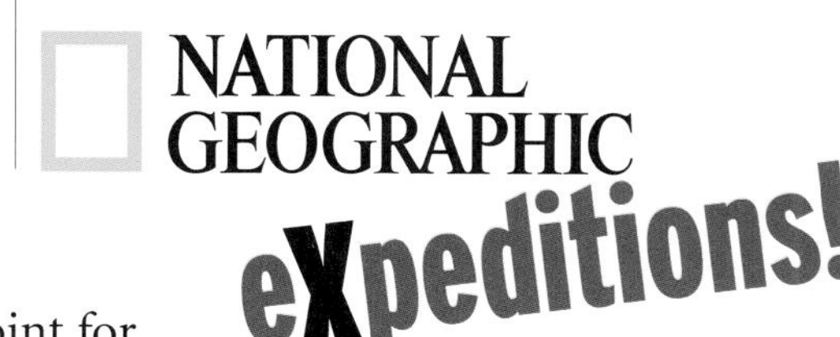

Fast Flight, Painstaking Processing

The next stage of the project was not fast. Valuable though they were, the aerial images were just a starting point for cartographers. The photographs had to be studied carefully and compared with a variety of other resources, including a computer database of old British, Chinese, and Austrian maps of the region.

NASA provided infrared images taken when the space shuttle *Columbia* orbited 250 kilometers (156 miles) above Everest in December 1983. The crystal-clear shots helped Washburn and his team confirm the location of key points. "These are as important to the map," stressed Washburn, "as a high-quality steel framework is to a big building."

Once the team was satisfied with the data, mapmakers in Switzerland went to work. Portraying the convoluted terrain took thousands of patient hours. Some days a cartographer might complete only a single square inch. Meanwhile, National Geographic researchers worked with experts in Nepal and China to ensure the accuracy of each place-name.

The final product—blue, gray, and beautiful—accompanied the November 1988 issue of National Geographic. It was, indeed, a triumph.

LEFT: Space shuttle photographs were useful in evaluating the accuracy of earlier maps of the Himalaya. Because clouds obscured part of the range in this 1984 image made from *Challenger* as it passed over Everest, the mapping team relied on photographs from the 1983 *Columbia* flight.

Expeditions Activity

Despite the accuracy of high-tech equipment for surveying mountain systems and measuring the height of peaks, the figures that scientists have obtained are subject to change even today. Discuss factors that might lead to new measurements and more accurate images of Everest. What are the possibilities for geological changes that would actually alter Everest's vital statistics?

Grand Managed Canyon

The Remaking of a River

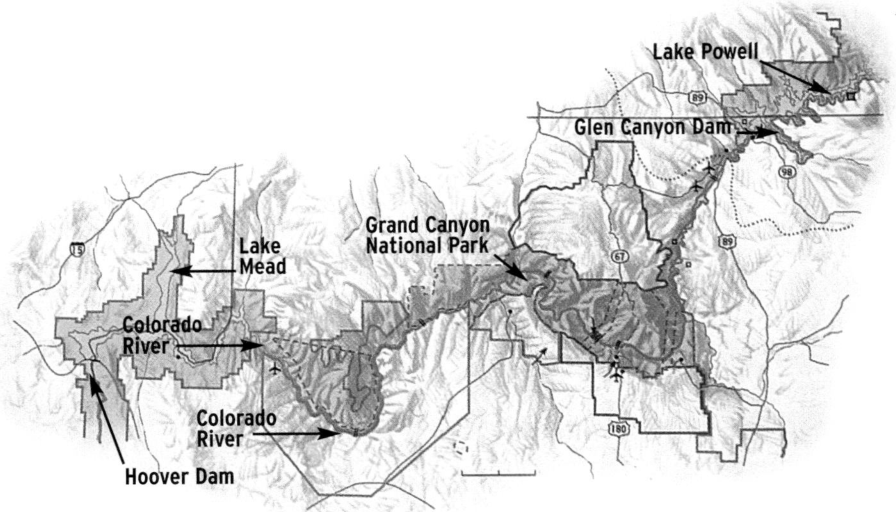

"You're not going to write about this, are you?" boatman Martin Litton joked to writer Mike Long. The two men, along with several other passengers from a dory called the *Sequoia*, were drying off after capsizing in the Colorado River, the mighty stream that carved the Grand Canyon.

Long had come to the Colorado to study the effects of Glen Canyon Dam, built in 1963 to provide a reservoir for the thirsty Southwest and to generate hydroelectric power. As part of his fieldwork for NATIONAL GEOGRAPHIC, he embarked on a 451-kilometer (280-mile) ride downstream. The trip started well. Long saw incredible scenery and gathered tales about river life.

Then came Crystal Rapids—"the centerpiece," wrote Long, "of one of the most intimidating half miles of white water in North America." To make matters worse, an oar escaped from Litton's left hand, and the passengers aboard *Sequoia* found themselves heading straight into a two-story-high wave. "I see wavelets frothing on the curling crest," Long wrote, "and instantly we are flipped like frying eggs, sunny-side down, into the green, wet womb of the Colorado." No one suffered more than scrapes or bruised pride, but Long's baptism made one thing clear—rivers may be dammed, but not necessarily tamed.

ABOVE: The Colorado River, which used to rage with spring floods, is restricted now by Glen Canyon Dam. Today the river rises and falls according to the demand for electric power.

LEFT: Water rocketed from outlet tubes at the base of Glen Canyon Dam for seven days in the spring of 1996. The artificial flood was designed to gather sediment from the riverbed to rebuild downstream beaches long eroded by the dam's fluctuating outflows.

ABOVE: On the Colorado River, boaters travel an ecosystem transformed upriver by Glen Canyon Dam. Now available to the masses, Rainbow Bridge is only a half-mile stroll after a comfortable two-hour boat ride from Wahweap Marina on Lake Powell. In predam days, before Lake Powell existed, a mere handful of visitors trudged here. Today more than 300 000 people make the trip each year.

Water Wars

Not that people haven't tried to tame rivers. Americans at the start of the twentieth century largely saw nature as an obstacle to overcome. Minor details such as a desert climate were no reason not to settle in Arizona, Nevada, or southern California. American ingenuity would get water there somehow.

"Somehow" turned out to be more complicated than anyone expected, especially for states along the Colorado River. The very question of who owned the water sparked controversy, and the Colorado, according to a *Denver Post* columnist, "earned the reputation as the most legislated, litigated, and debated river in the world."

In 1922, seven western states signed the Colorado River Compact, which determined each state's share of the water. Two decades later, the U.S. guaranteed a portion of the flow to Mexico. Enforcing all these agreements meant wresting control of the river from Mother Nature.

Dams, Dams, Everywhere!

With scores of reservoirs and diversion dams, hundreds of miles of aqueducts and tunnels, dozens of pumping stations, thousands of miles of canals, and more than 30 hydroelectric plants, the Colorado River basin contains one of the world's most controlled river systems. Among the jewels in this technological crown is Glen Canyon Dam, authorized by the U.S. Congress in 1956. Located near Page, Arizona,

This text is based on "Grand Managed Canyon" in the July 1997 issue of NATIONAL GEOGRAPHIC.

the dam is 92 meters (300 feet) thick at its base. It stands 217 meters (710 feet) high, creating an upstream reservoir known as Lake Powell.

The dam's influence also extends downstream, for nearly 483 kilometers (300 miles). Decades ago, the Colorado raged fierce and wild seasonally through the Grand Canyon. Now it runs year-round like water in a pipe, controlled not by nature but by human need.

New, Improved Nature?

"The predam Colorado, muddy and fierce," wrote Long, "rose and fell with the seasons. Carrying millions of tons of sediment, its floods scoured streamside vegetation while depositing enormous sandbars at the river's edge. A mud-caked cadre of river runners mucked it up in the world-class rapids during the few months of the strong flow, which ebbed to a chilled trickle in winter."

That all changed once the dam was built. Seasonal shifts gave way to a steady flow of clear, cool water. Depending on the amount pumped through Glen Canyon's hydroelectric plants, the river level rises and falls daily, sometimes by about 4 meters (13 feet). The Colorado now carries much more debris, which alters the dynamics of the river and creates new rapids—including the white water that soaked Mike Long.

Whose Ecosystem Is It, Anyway?

"One ecosystem was destroyed and replaced by another," said Ron Arnberger, superintendent of Grand Canyon National Park. Willow and tamarisk plants now crowd the Colorado's banks. The foliage sustains insects and migratory birds. Yet other birds—cliff swallows and endangered Bell's vireos, for instance—have declined. Three fish species have vanished. A fourth, the humpback chub, is endangered.

Experts disagree on what's best for the Colorado River ecosystem, and the human thirst for water and electricity further complicates the issue. According to former Secretary of the Interior Bruce Babbitt, "The more we look into our relation to the land, the more likely we will reaffirm natural visions. But final decisions will be made by the American people."

ABOVE: A Bewick's wren brings its brood a meal of insect larva in a tamarisk tree.

LOWER LEFT: Tamarisks stand on the shoreline at Eminence Break. Brought to the United States from the Middle East nearly 200 years ago, they now crowd many western waterways. Here they offer food and shelter to birds.

Expeditions Activity

Draining Lake Powell and abandoning the Glen Canyon Dam is a subject of much debate. Develop arguments both for and against this possibility. A good place to start might be the Web sites of the Glen Canyon Institute and of the Friends of Lake Powell. You can also search the Environmental News Network for recent developments: Click on their link at earthgeu.com

Really Cool Caves

Some of the toughest spelunkers—people who explore caves—don't see a single rock on their slow, dangerous, exhilarating journeys. That's because these cavers, a relatively new breed, navigate caves of ice.

Icy labyrinths twist and turn within the glaciers that cover 10 percent of the planet's land surface. The determined adventurers who brave those frigid mazes find them at least as challenging and diverse as the more familiar caves of rock that lie underground.

ABOVE: Specially designed screws for securing one's steps in ice enable Stefan Geissler to reach this point, some 15 meters (50 feet) above the floor of an ice cave in southern Germany.

LEFT: An ice shaft more than 91 meters (300 feet) deep surrounds rappelling adventurers in the Greenland ice sheet. A new breed of explorer is discovering that cave systems in glaciers can be just as challenging as caves underground.

Birth of a Glacier

Glaciers, huge masses of ice that inch over Earth's surface, form when snow piles up faster than it can melt. Each new snowfall compresses the layers beneath until what were once snowflakes fuse into solid ice.

As the ice forms, growing thicker and thicker and heavier and heavier, it begins to creep downward, driven by the force of gravity. As soon as a mass of ice starts moving, it becomes a glacier.

Glaciers come in two varieties. Alpine glaciers, which form on mountainsides, resemble frozen rivers. Continental glaciers, also called ice sheets, have a wide mound in the center and spread out in all directions, like pancake batter on a griddle. Ice sheets blanket most of Greenland and Antarctica.

Glaciers on the Go

There's a reason people complain when something moves at a "glacial pace." Driven by gravity, glaciers generally creep ever so slowly over the landscape, some advancing by only a few centimeters a day. Alpine glaciers drift downhill, slicing their way through valleys. Continental glaciers spread outward slowly from the center.

The layers of a glacier move at different rates. The bottom trudges, slowed by friction with soil and rocks. Icy layers in the middle move more swiftly, and the brittle crust at the top moves fastest of all—fast, that is, for a glacier.

Those varying speeds of movement create enormous pressures within glaciers. As the ice flows toward the margin of the glacier, it may encounter obstacles that cause the ice to slow down. The ice may also speed up if the slope of the glacier becomes steeper. The resulting stresses within the glaciers may cause the surface to fracture into deep cracks known as crevasses or produce shifting spaces that form glacial caves deep within.

Chameleon Caves

"They are the chameleons of caves," says German photographer-spelunker Carsten Peter of the glacial caves he has explored in Europe

ABOVE: In search of the perfect ice cave, a team of Italian cavers carefully crosses Moreno Glacier, on Lake Argentino, in the Andes mountains of Argentina. They have to watch for hidden, deep crevasses and for ice that may thinly cover a cave.

RIGHT: Rising heat from geothermal springs melts an Iceland glacier from the inside out, creating a huge hollow in the ice. Layers of ice, marked by bands of volcanic ash, form a monstrous "eye" that casts its frozen stare on caver Arne Kaiser.

and Greenland. "They change all the time."

Summer warmth often melts some of the ice and snow atop a glacier, creating torrents that both form and erase caves. Slithering through tiny fractures, meltwater can drip down into caves, where frigid temperatures turn the water back into ice. The result: enormous icicles that dangle like bizarre crystal chandeliers. They are an ice cave's version of stalactites.

In addition, geothermal springs within Earth give off heat that melts glaciers from the inside out. The melting makes and endlessly remakes large bowl-like chambers within the ice. Iceland is particularly noted for its geothermal and volcanic activity.

Each fall, as the deluge of meltwater recedes, cavers return to Greenland and other glacier sites to explore the latest batch of creations. It's an opportunity traditional spelunkers can only dream about. "As often as I return," says Carsten Peter, "I have never entered the same glacial cave twice."

Bleak Beauty

During their short lives, glacial caves can be hauntingly gorgeous. Depending on how thick the ice is, refracted sunlight may reach deep into a cave, causing walls to glow deep blue, or slick cave walls may gleam like mirrors.

Caves near a glacier's surface may even have "skylights"—thin, clear sheets of ice that offer glimpses of Sun and clouds.

ABOVE: With his helmet light casting a warm glow, Daniel Jehle uses ice axes to climb huge ice stalactites in a cave in Germany. The stalactites formed when meltwater from the surface dripped through tiny fractures and into the cave, where cold air froze it into huge icicles.

Into an Icy World

Glacial caves lure spelunkers willing to face discomfort and danger in order to enjoy the fragile beauty. To do so, ice cavers use some of the same tools that traditional cavers employ: ropes, lamps, helmets, and more. Ice cavers also bring tools specially designed for their frozen world. These include screws that dig deep into icy walls, boots with long spikes that keep spelunkers from slipping, and ice axes.

In 1995, an Italian expedition took a new approach to exploring glacial caves in Argentina. They suited up like scuba divers and plunged into a flooded maze. With the water barely a degree above freezing, the spelunkers could not stay under for more than a few minutes. Twice, divers found that the air valves on their breathing devices had frozen. Despite these obstacles, the expedition located one of the longest glacial caves yet recorded—1037 meters (3400 feet).

The team also faced a challenge that might surprise traditional

ABOVE: Tracing the path of a stream over ice was a difficult chore, so the Italian cavers in Argentina used fluorescent dye to mark the water, making following its path much easier.

RIGHT: Some ice caves form extremely close to the surface of a glacier, as this one did in southern Argentina. Only a fragile skin of ice covered the tubular cave—until it broke.

This text is based on "Into the Heart of Glaciers" in the February 1996 issue of National Geographic.

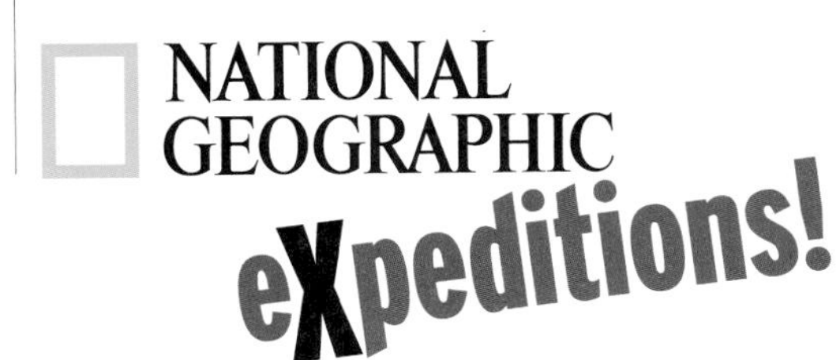

spelunkers—distinguishing the water from the walls. "The ice is blue, the water is blue," explained diver Antonio De Vivo. "It is very difficult to tell the difference between the ice and the water." Their creative solution? Add fluorescent yellow dye to the water.

A season later, of course, that notable cave might have melted away, but the turmoil beneath the surface of Earth's glaciers always promises replacements—brief, blue, and beautiful.

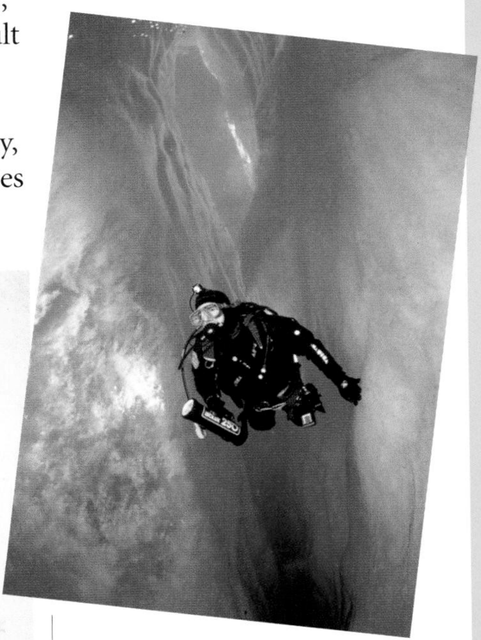

ABOVE: Swimming in a flooded ice cave presents its own peculiar difficulty—distinguishing the blue water from the blue ice. Diver Matteo Diana gets help in finding the icy wall with a flashlight.

Expeditions Activity

The article describes perils that glacial spelunkers face due to the nature of the ice and water. What advantages might there be to caving in glaciers? Research typical spelunking and compare the dangers that explorers of rock caves face with the dangers faced by those who probe into glaciers.

Finding the *Titanic*

Plunging Into History

"God himself could not sink this ship," boasted a crewman aboard the R.M.S. *Titanic*. Famous last words. Many aboard the sumptuous liner—the largest and most expensive ship of her era—shared the sailor's optimism. As the ship left England on her maiden voyage, she seemed the very emblem of technological triumph.

Probably most people in America know the unhappy ending. The *Titanic* hit an iceberg at 11:40 P.M. on April 14, 1912. She sank about two hours later, killing 1522 people. (The ship's lifeboats saved 705 others.) The "unsinkable" ship didn't survive a single Atlantic crossing.

ABOVE: A proud and allegedly unsinkable R.M.S. *Titanic* departs her berth in Southampton, England, in April 1912, on her fateful maiden voyage to the North Atlantic.

LEFT: The ghostly hulk of the *Titanic* lies in her grave more than two miles deep, her bow festooned by decades of rust and sediment.

Yet by sinking, the *Titanic* achieved gruesome immortality. When scientists found the shipwreck in 1985, they earned attention seldom given to marine exploration.

Deadly Decision

Fred Fleet was on lookout duty the night the *Titanic* sank. From his perch in the crow's nest, he spotted an iceberg directly in the ship's path, only a quarter of a mile away. He phoned the bridge with a terse, terrifying warning: "Iceberg right ahead!"

The officer-in-charge tried to reverse engines while turning hard to starboard (to the right)—a combination that proved fatal. Because the ship was in reverse, the change in steering actually moved her to port (to the left), and the *Titanic*'s starboard side hit the iceberg. Six out of 16 watertight compartments flooded, pulling the bow (front) into frigid water.

Ironically, the *Titanic* might have survived a crash head-on. Two or three watertight compartments might have flooded, but that wouldn't have been enough to sink the ship.

Out of Sight, Much in Mind

Shivering in icy terror, survivors watched the *Titanic* disappear shortly after 2 A.M. on April 15. "I saw that ship sink," recalled Eva Hart. "I saw all the horror of its sinking. And I heard, even more dreadful, the cries of drowning people." Eventually, the screams gave way to utter silence: No one remaining in the -2°C (28°F) North Atlantic waters was left alive.

Hart and 704 others were the last people to see the *Titanic* for 73 years. Various expeditions searched for the wreck, but they were defeated by the unpredictable North Atlantic weather, the enormous depth to which the ship sank, and conflicting accounts of her final hours.

Meanwhile, books and movies—notably, *A Night to Remember*, by Walter Lord—helped keep the story alive. The Titanic Historical Society gathered memorabilia, documents, and survivors' stories. In addition, a *Titanic* memorial by the Potomac River was included with the many monuments of Washington, D.C.

ABOVE: A storage area for *Titanic* lifeboats (top) serves as a landing platform for the manned submersible *Alvin*, resting on the deck at the far right. This scene was photographed by the remotely operated robot *Jason Jr.*, which is taking a closer look at the same spot (below), peering into a first-class cabin.

High Tech on the High Seas

On June 28, 1985, a French-American expedition set out to search for the *Titanic* with the most advanced tools available: side-scan sonar, sonic beacons, submersible video cameras, computers, and much more. Led by Robert D. Ballard and Jean-Louis Michel, the team examined a 386-square-kilometer (150-square-mile) target zone about 644 kilometers (400 miles) south of Newfoundland. Like gardeners mowing a lawn, the expedition's ships patiently went back and forth, surveying strips of the seafloor.

Their perseverance paid off at 1:05 A.M. on September 1, 1985. Video cameras aboard the *Argo,* a submersible research vessel, focused on one of the *Titanic's* enormous boilers. It lay some 4000 meters (13 000 feet) beneath the surface.

Over the next four days, the expedition shot more than 20 000 pictures of the well-preserved wreck. Marine organisms had devoured much of the woodwork—and any human remains—but the bulk of the ship looked achingly familiar.

ABOVE: The research vessel *Knorr*, from the Woods Hole Oceanographic Institution in Woods Hole, Massachusetts, floats more than two miles above *Titanic*. Serving as a support ship for the submersible search vessel *Argo, Knorr* is specially equipped to allow it to maneuver sideways as well as forward and backward.

Face-to-Face With the Ship

On a second expedition, in 1986, Ballard and two colleagues became the first people to actually visit the wreck. Scrunched up in the research submersible *Alvin,* they had a mishap-filled descent to the bottom. Sonar died, salt water seeped into the sub's batteries, alarms wailed. The navigator, back at the surface in a support ship, was having difficulty directing them.

Hollywood could have scripted the next scene. Just as the pilot was ready to abort the mission, Ballard spotted "an endless slab of black steel rising out of the bottom." *Titanic!* Ballard truly was the first

LEFT: Robert Ballard and other members of the joint French-American search team monitor video screens during a descent by *Argo*.

High-tech Explorer Plumbs New Depths

ALVIN

1 Still camera with strobe **2** Manipulator arm with lights and still and video cameras **3** Cable winch and reel **4** Scanning sonar **5** Hatch **6** Titanium sphere **7** Viewing port **8** Jettisonable iron ballast **9** Acoustic telephone **10** Hoisting bitt **11** Tanks for air and variable seawater ballast **12** Batteries **13** Pressure housings for electric controls **14** Thrusters **15** Emergency tether cutter **16** Down-looking, low-light-level, black-and-white TV camera **17** Forward-looking, low-light-level, black-and-white TV camera

PAINTING BY WILLIAM H. BOND, NATIONAL GEOGRAPHIC ARTIST

person to see the wreck—a moment he'd spent a lifetime dreaming about and working for.

Ballard and his team made 11 more dives. They also dispatched a "swimming eyeball"—a robotic submersible called *Jason Jr.*—to explore parts of the wreck too small or too dangerous for humans. Such robots, Ballard has argued, represent an essential advance for marine exploration.

ABOVE: The three-person submersible *Alvin* carries a tethered robot named *Jason Jr.* in a so-called garage at the front of the vehicle. *Alvin* can park and let *Jason Jr.* explore areas that are too isolated or dangerous for the manned vehicle to enter.

LEFT: Above the wreck site, divers stand atop the submersible *Alvin* to secure safety lines that hold *Jason Jr.*'s garage as the vehicles are lifted out of the water to the deck of the support ship. Both vehicles are operated by the Woods Hole Oceanographic Institution in Woods Hole, Mass.

Surprises on the Seafloor

Many *Titanic* experts had believed the doomed ship went down in one piece. But the wreckage suggests a different story. The bow (front) and stern (rear) lie some 400 meters (1200 feet) apart. Ballard theorized that the ship tore apart at the surface. The bow descended first. The stern plunged afterwards and may have rotated en route. "In short," said Ballard, "the bow landed, the stern crashed."

This text is based on four NATIONAL GEOGRAPHIC *articles: "How We Found* Titanic*" (December 1985), "A Long Last look at* Titanic*" (December 1986)," "Epilogue for* Titanic*" (October 1987), and "*Titanic*: Tragedy in Three Dimensions" (August 1998).*

Even more surprising was the starboard bow. *Titanic* accounts had always assumed that the iceberg slashed open the hull, but the expedition found no such wound. Either the fatal gash was lower on the bow, an area that lay buried in mud, or it never existed. Ballard noted, "*Titanic*'s massive steel plates probably would have been been bent or forced apart rather than ripped open by ice."

Tragedy for Sale

Ballard hoped that only "gentle exploration" would befall the wreck he found: "It would be wrong to attempt recovery of any of the remains." Not all agreed, and an American entrepreneur secured salvage rights to the ship. *Titanic* buffs can now see artifacts on exhibit or even buy pieces of coal meant to power the ship.

Late in the 1990s, the *Titanic* inspired a Broadway musical and a Hollywood movie—the first film to gross a billion dollars. The unsinkable ship that sank still sails on in millions of imaginations.

ABOVE: A grenadier fish eyes *Alvin*'s mechanical arm as it pauses above a china cup on the seafloor. The cup still bears the emblem of the White Star Line, a white star on a red flag.

LEFT: A brass-and-crystal light fixture dangles, decorated by a feathery sea pen. Handsome features abounded in areas such as the skylighted grand staircase (lower) connecting first-class decks.

Expeditions Activity

Robert Ballard has discovered the hulks of other famous ships from the 20th century, as well as ancient vessels lost at sea. Which other ships has he discovered? Were the techniques used to find these more recently found vessels similar to the techniques used to find *Titanic?* Research Robert Ballard—like him, you'll start to discover some amazing information.

NATIONAL GEOGRAPHIC
eXpeditions!

Mount Pinatubo's Eruption

One Hot Topic

ABOVE: Much of the damage inflicted by Mount Pinatubo's eruption in the Philippines in June 1991 was not immediate. The weight of accumulated ash on roofs, particularly after it mixed with torrential rains, eventually brought down thousands of sheds and homes.

LEFT: At full blast, Mount Pinatubo lies completely hidden by turbulent clouds of superheated ash that surge from the volcano.

Someone was screaming. Yowls of terror woke NATIONAL GEOGRAPHIC's Noel Grove, who was on the Philippine island of Luzon researching an article about volcanoes. He'd borrowed a bed in a small, crowded boardinghouse about 19 kilometers (12 miles) from Mount Pinatubo. The creaking structure didn't inspire confidence. "I worry about the house," said the owner. "It isn't solid."

Seismic waves, rolling through the Earth, struck that night, jolting the house and its edgy inhabitants. Then came those screams, and the veteran writer noticed he was shaking. So was his bed. So was his room. Only then did Grove realize who was yelling—he was.

A Day of Geologic Drama

Noel Grove, and anyone else near Pinatubo in June 1991, had good reason to be nervous. In the first hours of June 15, the 1760-meter (5770-foot) volcano roared awake after a six-century sleep. Pinatubo belched clouds of gas and ash known as pyroclastic flows. Their temperature: 816°C (1500°F). Streams of ash and sulfur dioxide rocketed 40 kilometers (25 miles) into the stratosphere.

Another blast at dawn blew away the side of the mountain. So much ash and pumice choked the air that the sky grew black by

Pinatubo: Ready to erupt

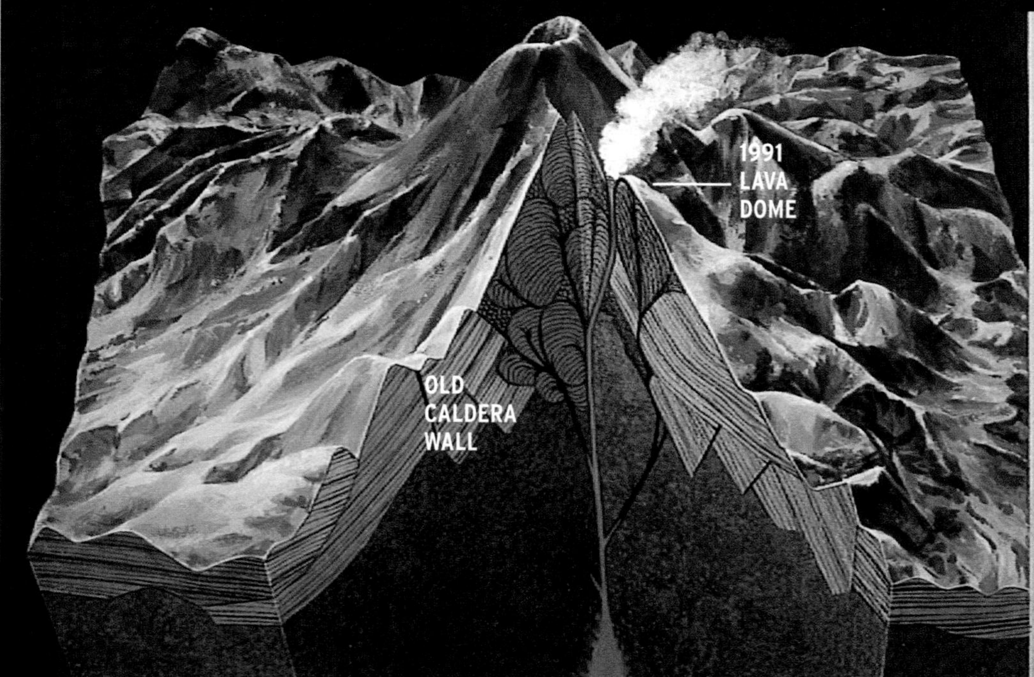

Eve of eruption

Pinatubo sits in the center of a five-kilometer-wide (three-mile) caldera, a depression from an earlier eruption that had made the volcano collapse in upon itself (tilted blocks at left). A new cone formed—a dome of thick clumps of dacite rock overlain by layers of ash-flow deposits. Geothermal vents on the north side of the mountain gave the only clue that the volcano was still active. On April 2, 1991, steam and ash erupted from the site. Monitoring instruments were quickly installed.

afternoon, and chunks of volcanic rock fell like dirty hail. Meanwhile, new pyroclastic flows swept down Pinatubo's remaining slopes, filling canyons and ravaging farmland. Some flows reached as far as 18 kilometers (11 miles) from the volcano.

Evening brought new trauma—earthquakes. Pinatubo's eruption had created an underground cavern that finally caved in on itself, creating tremors. The remorseless assaults of June 15 now came from below as well as above. It was, said a woman in Manila, "like the end of the world."

Just Add Water

July 1991 brought new suffering, as typhoons struck Luzon, where huge amounts of ash covered the landscape. Pinatubo had ejected enough ash to cover Washington, D.C., 15 stories high.

Water and ash blended into a cement-like slurry that soon charged over the terrain in streams called lahars. They snapped power-line poles like twigs and moved cars like cardboard stage props. Lahars also wrecked bridges and waterworks, and obliterated fields and villages. In fact, the deadly torrents destroyed twice as many homes as did the original eruption.

ABOVE: Quiet for 600 years, Mount Pinatubo was an innocent-looking peak in central Luzon. Then, in April 1991, vigorous steam eruptions, swarms of shallow earthquakes, increased sulfur dioxide emissions, and rapid growth of a lava dome heralded a powerful eruption. Scientists from the Philippine Institute of Volcanology and Seismology, later joined by a team from the U.S. Geological Survey, monitored the vents and called for evacuation of nearby villages. A series of minor explosions began on June 12.

This text is based on the article "Crucibles of Creation: Volcanoes," in the December 1992 National Geographic.

Making matters even worse, lahars reappeared each year during the rainy season. The recurring floods of sludge defeated efforts to rebuild around Pinatubo and left more than a million people homeless. "In terms of magnitude, Mount Pinatubo is unprecedented," said volcanologist Kelvin Rodolfo. "No lahar flows of this magnitude and duration have ever been recorded [before]," he said.

Pinatubo's Price Tag

Nine hundred people perished during the eruption. Five hundred others died of disease or exposure in the rudimentary camps that sheltered some 200 000 evacuees. Volcanic fallout destroyed 42 000 homes, and lahars washed away 110 000 more.

Ash laid waste to 40 500 hectares (100 000 acres) of farmland, and that was just one aspect of the devastation that left 650 000 Filipinos with no way to make a living. Damages were estimated conservatively at $440 million. Rebuilding would cost billions.

The Science of Survival

Stunned, homeless, and jobless, the 200 000 people evacuated from the area around Pinatubo before the eruption probably did not feel all that lucky. Yet only a generation earlier, the volcano might have killed

BELOW: A small dam built to restrain lahars, which are floods of rain-soaked volcanic ash, was undercut by the lahars that followed Pinatubo's eruption. They scoured out a channel under the dam 18 meters (60 feet) deep.

LEFT: Masked against the choking gloom of volcanic ash, jobless young men seek assistance—a few of the 650 000 Filipinos who were left without a source of income.

them. The difference resulted from scientists' increasing ability to detect volcanic activity early and predict its severity.

The first signs that Pinatubo's long nap was ending came on April 2, 1991, when villagers spotted plumes of steam rising from vents in the mountainside. Filipino scientists raced to the scene, and colleagues came from the United States to pitch in.

The volcanologists deployed seismometers to measure movements beneath Pinatubo and tiltmeters to detect any swelling in the mountainsides. "There are no magic new machines," seismologist Paul Okubo told Grove, "only steady improvement in using what we have."

Airborne instruments monitored gases escaping from the volcano. By May, sulfur dioxide levels had increased tenfold—a warning that new and active magma was rising toward the surface.

BELOW: Steaming vents called fumaroles were still active eight months after Pinatubo's violent eruption. Rainwater had collected into a lake in the base of the massive crater.

The scientists also conducted field studies and examined aerial photographs for geological evidence of the extent of previous eruptions. The team found volcanic debris 16 kilometers (10 miles) from the mountain, a sign that Pinatubo had a seriously violent past. The next blast was likely to be dramatic and deadly.

By June 11, Pinatubo was beginning to emit ash, and the mountain's lava dome had doubled in size. Explosion was imminent. Based on the scientists' findings, Filipino authorities evacuated everyone within 24 kilometers (15 miles) of the volcano. When villagers balked at leaving home, the savvy officials didn't waste time on lengthy arguments. Instead they showed a video of pyroclastic flows and lahars. Viewers packed quickly.

Earth Never Sleeps

Mount Pinatubo's 1991 eruption was extraordinary—perhaps the mightiest of the entire twentieth century. Yet it was also oddly ordinary, part of Earth's long history of violent change. The planet, after all, is home to some 550 known active volcanoes, and a dozen or two are erupting at any given moment.

For humans in harm's way, such eruptions amount to unspeakable tragedy. Yet for Earth itself, represent an essential, vibrant cycle.

Expeditions Activity

A great many people live close to volcanoes that could erupt in their lifetime. What kinds of precautions should they take? Break up into groups. Each group should act as if it is responsible for emergency planning for a community near a volcano. Consider specific plans for detecting an eruption and for organizing a mass evacuation. Compare plans among the groups.

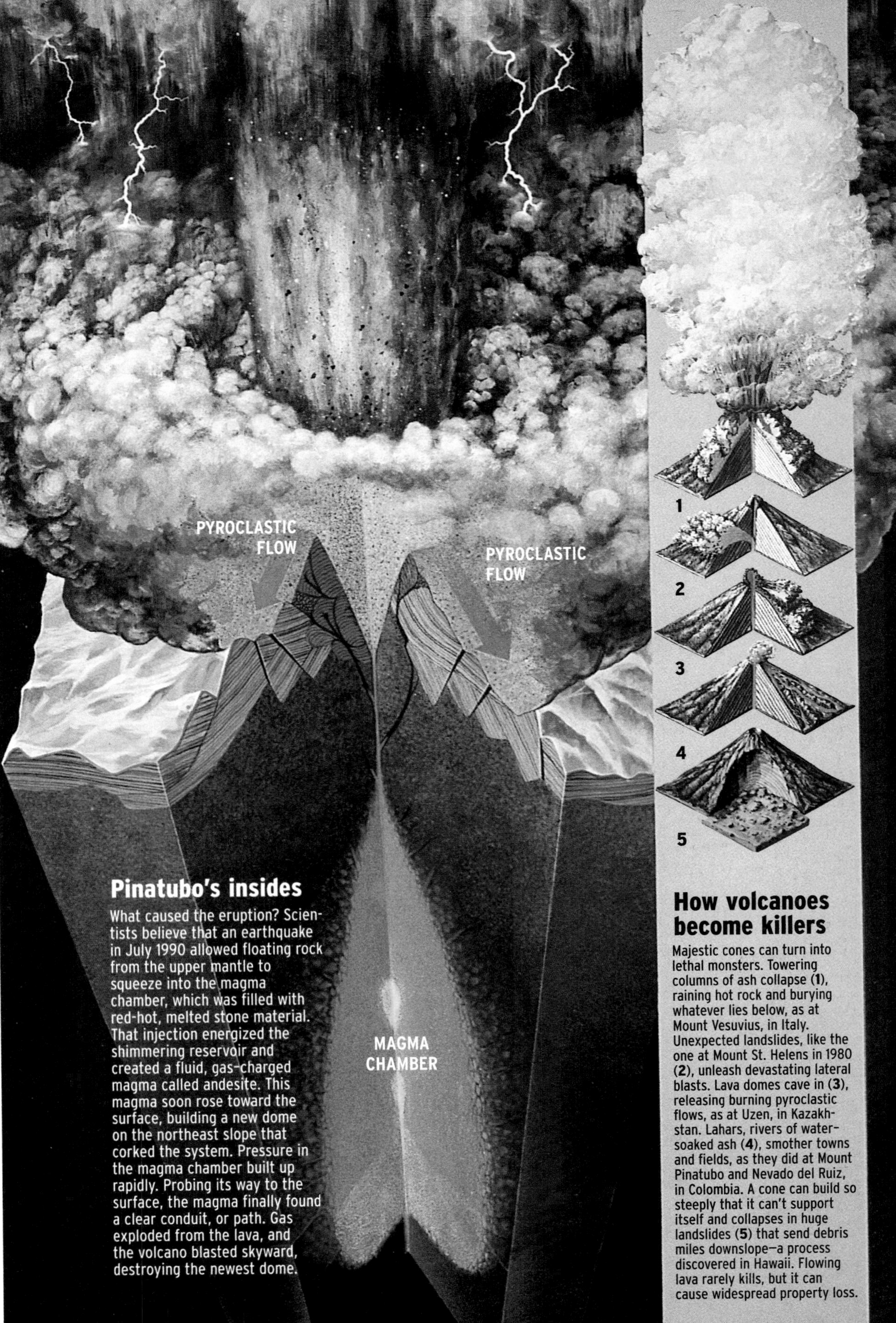

Pinatubo's insides

What caused the eruption? Scientists believe that an earthquake in July 1990 allowed floating rock from the upper mantle to squeeze into the magma chamber, which was filled with red-hot, melted stone material. That injection energized the shimmering reservoir and created a fluid, gas-charged magma called andesite. This magma soon rose toward the surface, building a new dome on the northeast slope that corked the system. Pressure in the magma chamber built up rapidly. Probing its way to the surface, the magma finally found a clear conduit, or path. Gas exploded from the lava, and the volcano blasted skyward, destroying the newest dome.

How volcanoes become killers

Majestic cones can turn into lethal monsters. Towering columns of ash collapse (**1**), raining hot rock and burying whatever lies below, as at Mount Vesuvius, in Italy. Unexpected landslides, like the one at Mount St. Helens in 1980 (**2**), unleash devastating lateral blasts. Lava domes cave in (**3**), releasing burning pyroclastic flows, as at Uzen, in Kazakhstan. Lahars, rivers of water-soaked ash (**4**), smother towns and fields, as they did at Mount Pinatubo and Nevado del Ruiz, in Colombia. A cone can build so steeply that it can't support itself and collapses in huge landslides (**5**) that send debris miles downslope—a process discovered in Hawaii. Flowing lava rarely kills, but it can cause widespread property loss.

Fossils With Feathers

Dinosaurs Take Wing

Ji Qiang gasped aloud as he opened the green silk box. As director of the National Geological Museum in Beijing, China, Ji was used to seeing fossils. But this one, sent to Ji in 1996 by a dealer in the far northeastern province of Liaoning, was extraordinary.

The more than 120-million-year-old remains belonged to a creature about the size of a chicken (see model at left). The large skull held needle-sharp teeth, and the bony tail was nearly twice as long as the rest of the body. The creature had short, stout forelimbs.

Ji and his fellow scientists named the new creature *Sinosauropteryx prima,* which means "first Chinese dragon feather." They classified it as a theropod, or carnivorous dinosaur. (*Sinosauropteryx*'s gut actually held remains from its last meal, a small mammal.) "The fossil looked a lot like a small meat-eating dinosaur," Ji told National Geographic, "except for one thing."

ABOVE: A *Sinosauropteryx* model peers down with arms extended, ready to pounce on a lizard or small mammal. If its body covering first evolved to preserve heat rather than for display, it would suggest that *Sinosauropteryx* was warm-blooded. Scientists have long debated the question of dinosaur metabolism.

LEFT: Sealed at death more than 120 million years ago, the eye of a juvenile *Sinosauropteryx* glints with black carbon that crystalized during fossilization. If growths on the head and neck were primitive plumage, they may have served to trap body heat or as courtship display. *Sinosauropteryx* has been classified as a theropod, a meat-eating dinosaur.

"One Thing" That Stirred Scientists

The "one thing" was a thin, dark ridge along *Sinosauropteryx*'s back. Looking closer, Ji realized the ridge consisted of tiny fibers—thousands of them (see photograph, opposite). They appeared to be the incredibly well-preserved remains of protofeathers, precursors of the feathers that cover modern-day birds. The find reheated a smoldering debate about the relationship between birds and dinosaurs. Most experts today believe that birds evolved from dinosaurs, a view first espoused by naturalist Thomas Henry Huxley in the 1860s. Hux-

ley's critics argued, however, that both birds and dinosaurs descended from a common ancestor, and their theory held for decades.

Huxley's hypothesis found a new champion in the 1970s—paleontologist John Ostrom. Ostrom painstakingly compared newly excavated dinosaur fossils to the remains of the earliest known bird, *Archaeopteryx,* and was overwhelmed by the similarities. "Dinosaurs did not become extinct," he proclaimed; they became birds.

But scientists still needed a missing link. *Sinosauropteryx*—a dinosaur that seemed to have feathers—came closer than any other fossil Ji had seen.

More Fossils, More Feathers

Soon after *Sinosauropteryx* was discovered, Liaoning Province yielded two more feathered dinosaurs. The region is a paleontologist's dream, thanks to volcanic ash that smothered the landscape more than 120 million years ago. The fallout killed myriad plants and animals, then preserved them beautifully. No place on Earth has provided more information on the origin of birds.

Another Liaoning discovery—*Protarchaeopteryx robusta*—looked like a larger, more primitive version of *Archaeopteryx,* the first known bird. Scientists who have examined *Protarchaeopteryx*'s feathers doubt that the creature could fly. (*Archaeopteryx,* in contrast, was probably able to fly somewhat.)

Posed in courtship display, a model of *Caudipteryx zoui* depicts a creature nearly a meter (3 feet) long that stunned paleontologists when reconstructed from its fossil—a dinosaur with feathers! It is more than 120 million years old. Months before *Caudipteryx* was found, farmer Li Yin Fang (top, second from right) found the feathered dinosaur *Sinosauropteryx prima* (previous page).

Studying *Protarchaeopteryx* specimens, Ji and Canadian paleontologist Phillip Currie were startled by inconsistencies. "Were we looking at males and females?" Currie wrote. "Was *Protarchaeopteryx* a creature with much variation?" The two scientists watched eagerly as technician Kevin Aulenback used a microscope and dental tools to scrape excess stone from the fossil.

Slowly it became clear that the three men were looking at an entirely new creature. They named it *Caudipteryx zoui* (*Caudipteryx* means "tail feather") for the plumes that the dinosaur probably fanned out when it wooed mates or intimidated enemies (see model at left). A speedy runner, *Caudipteryx* did not fly.

Caudipteryx did, however, share a striking trait with modern birds—a gizzard full of small stones. Such pebbles, called gastroliths, ground food for the dinosaur, which had only a few teeth. Birds today use gastroliths for the same purpose.

"*Caudipteryx* and *Protarchaeopteryx*," wrote Currie, "make the dividing line between dinosaurs and birds even less distinct and strengthen the theory that birds evolved from small, carnivorous, ground-dwelling dinosaurs."

Early Birds

Hundreds of Liaoning fossils have given scientists an extraordinary look at yet another early bird—*Confuciusornis sanctus* ("sacred Confucius bird"). The species differed from the feathered dinosaurs in a crucial way—it could fly. "*Confuciusornis* is the earliest bird we know of that could fly for any distance," said fossil specialist Hou Lianhai, who named the creature. "It had wings nearly as primitive as those of *Archaeopteryx*, but it had other, more modern features—lighter bones and a shorter tail."

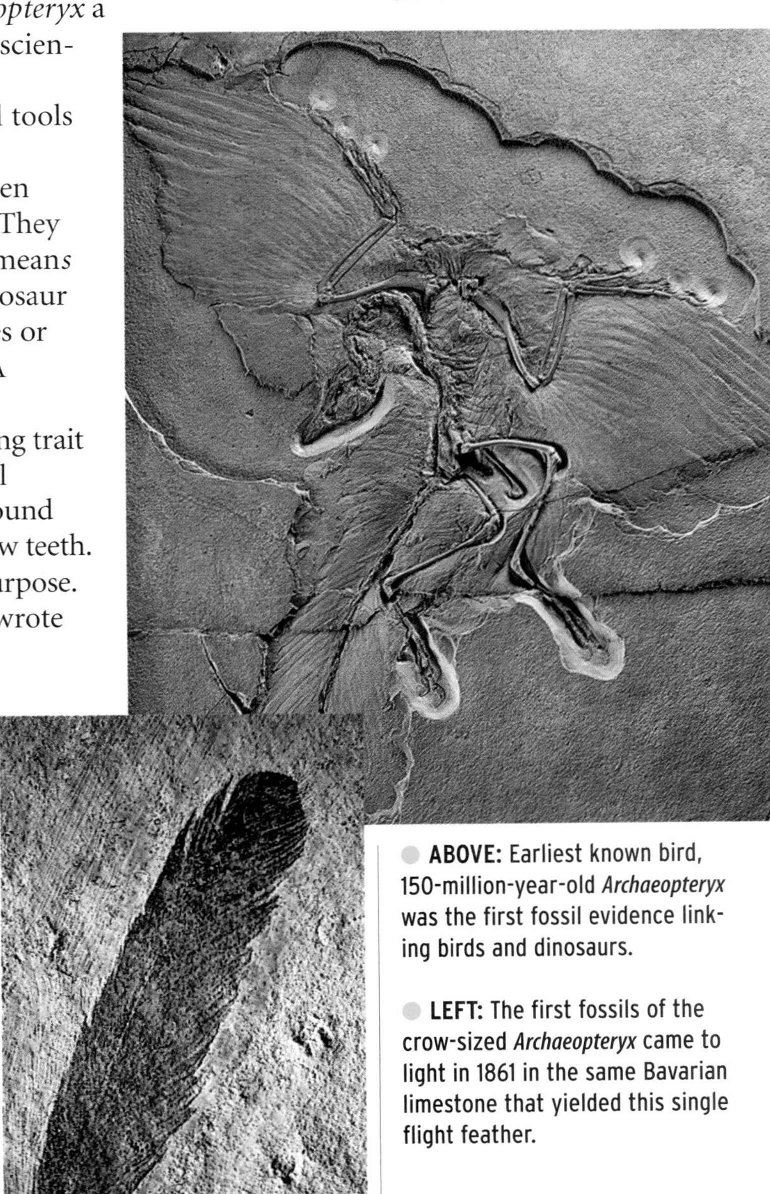

ABOVE: Earliest known bird, 150-million-year-old *Archaeopteryx* was the first fossil evidence linking birds and dinosaurs.

LEFT: The first fossils of the crow-sized *Archaeopteryx* came to light in 1861 in the same Bavarian limestone that yielded this single flight feather.

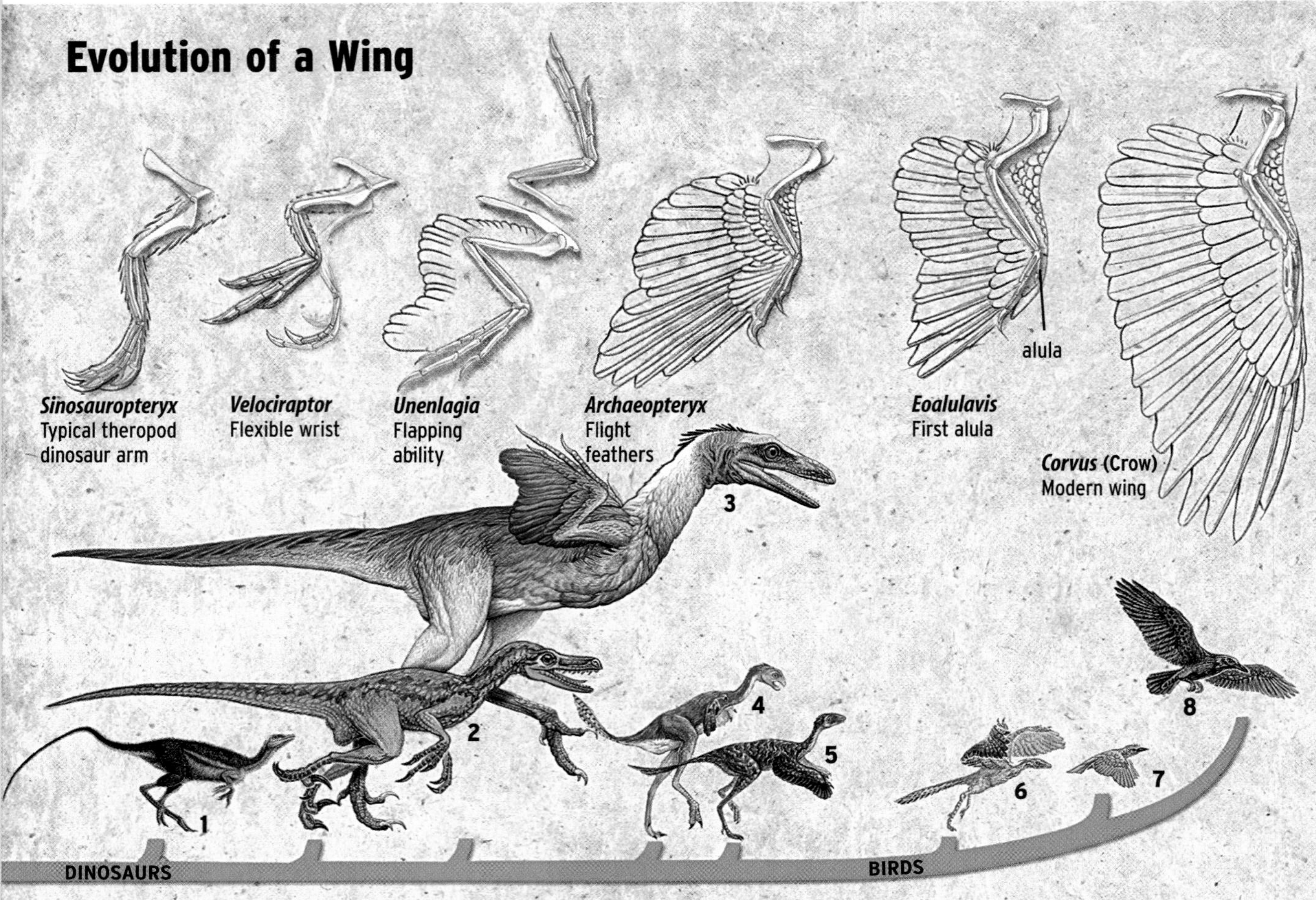

1 *Sinosauropteryx*
Covered with filaments that may have evolved for insulation or display, *Sinosauropteryx* was a ground-dwelling runner with short arms and three-fingered hands.

2 *Velociraptor*
This predatory theropod, whose fossils were found in Mongolia, was endowed with a wrist bone that permitted the animal's grasping hands to swivel, helping it capture prey. A flexible wrist is required for powered flight.

3 *Unenlagia*
Found in Patagonia, this flightless, 2.5-meter-long (8-foot) creature could move its arms up and down much as a person on a surfboard moves the arms for balance. A precursor to flapping, this action is critical to the flight stroke.

4 *Caudipteryx*
Straddling the realms of dinosaurs and birds, *Caudipteryx* is but one in a series of sensational fossil finds in China. A speedy runner, it was covered with primitive feathers that lacked the aerodynamic quality necessary for flight.

5 *Protarchaeopteryx*
Another discovery in China, *Protarchaeopteryx* resembles *Archaeopteryx* in many ways but is more primitive. The symmetrical feathers on its arms and tail appear longer than those of *Caudipteryx*, but it probably could not have achieved true powered flight.

6 *Archaeopteryx*
The feathers of this bird's celebrated fossils are asymmetrical, with the leading edge narrower and more streamlined than the trailing edge. This enabled the wings to slice through the air, permitting at least rudimentary flight.

7 *Eoalulavis*
Found in Spain, this bird exhibits the earliest known alula, a tuft of feathers attached to the thumb. By altering airflow, the alula permits good maneuverability and control at low flying speeds, crucial for takeoffs and landings.

8 *Corvus*
At the peak of forelimb evolution, the crow and other modern avian species, with their shortened tailbones and expansive wing surfaces, represent the full flowering of the ability to fly.

Unlike *Archaeopteryx,* which had teeth, *Confuciusornis* had a toothless beak similar to those of modern birds. The discovery indicated that beaks appeared 70 million years earlier than previously believed. Like birds themselves, scientists' knowledge was evolving.

Consensus and a Tough Question

By 2000, science had largely circled back to supporting Thomas Huxley's contention that birds were the descendants of dinosaurs. "The anatomical similarities are overwhelming," said Mark Norell of the American Museum of Natural History, in New York City. Birds and dinosaurs, he pointed out, share more than a hundred features—including wishbones.

Yet critics point out that *Archaeopteryx* is 30 million years older than the feathered dinosaurs from Liaoning. A true missing link, the critics stress, should predate the first bird. Until it appears, they're tracing birds back to a pre-dinosaur reptile.

Even advocates of the dinosaur–bird link concede the need for more, and older, fossils. "We don't have the evidence to support any kind of direct lineal descent," said Ostrom. "We're in the business of connecting dots scattered in time and space."

This text is based on the article "Dinosaurs Take Wing" in the July 1998 National Geographic.

FAR LEFT: Just as humans are mammals, birds belong in the clan of dinosaurs in the view of most paleontologists. The idea that the defining traits of birds first arose in dinosaurs dates from the early arguments of evolutionary theory.

This family tree is not a chronological progression, but an illustration of how the traits of the modern wing evolved in different creatures in different places at different times.

Feathers that evolved to keep a ground-dweller warm, for example, might give it the ability to glide and better catch prey, leading to more feather development.

LEFT: Hopes for another big discovery draw scientists to the dig at Sihetun, in China, where layers of rock entomb birds, dinosaurs, plant life, and multitudes of fish. Dating from more than 120 million years ago, the lake bed formation is 31 meters (100 feet) thick and covers 52 square kilometers (20 square miles).

Expeditions Activity

Have you ever stopped to ponder the features that make birds so different from ground-dwelling animals? Think about weight, shape, skeleton structure, and feather design. Now list specific traits, such as webbed feet or bald heads, that various birds have developed that allow them to exercise specific behaviors.

NATIONAL GEOGRAPHIC
eXpeditions!

River Restoration

Streams of Hope

ABOVE: Knee-high with muck, a Boys Club member totes trash pulled from a New Hampshire section of the Nashua River.

LEFT: Rank and lifeless by the 1960s (inset), the Nashua River in Massachusetts was a toxic stew of sewage, running red with dye from paper mills. Today it's a haven for anglers and canoeists and a model for communities striving to clean the waters they have fouled.

One day the river ran red. Then green. Then white, or orange, or maybe even blue. Sound like sci-fi or Dr. Seuss? Alas, no. It's an actual part of American environmental history. In the 1960s the Nashua River, which flows quietly for 90 kilometers (56 miles) in Massachusetts and New Hampshire, was one of the most polluted streams in the entire United States.

Mills along the Nashua produced lumber, shoes, paper, and textiles. For more than a century these industries dumped their waste into the water. Dyes in the wastewater turned the river into such a chameleon that locals used to take bets on what color it would be the next day.

People could laugh—darkly—about the changing colors. But there was nothing funny about river fumes that blackened paint on nearby buildings, or about kids being rushed to the doctor for tetanus shots when they got too near the water. It was mortifying that environmental authorities classified the Nashua as unfit for sewage disposal.

Perhaps the river was red because it was dying of embarrassment. It was certainly dying of something.

One Hero, Many Hands

Marion Stoddart didn't plan on letting the Nashua die quietly. When she explored the watershed in the 1960s, she was appalled. "The highest form of life in the North Nashua was sludge worms," she said. She began a campaign to bring the river back to life.

Stoddart knew that she'd need help—lots of it. She reached out to everyone, including labor leaders and businesspeople. "Marion Stoddart didn't go out there as a lone person in the wilderness," said Geographic reporter Erla Zwingle. "She did it by making key allies."

"What we were working to do," said Stoddart, "was to get everyone on every level working together." Toward that end, she learned how to present her case in economic as well as environmental terms. Stoddart knew the value of gestures. She presented bottles of polluted river water—emblems of the cost of inaction—to key officials.

"Marion Stoddart qualifies as a hero," said Zwingle, "but the Nashua shows that heroes don't accomplish things on their own. Everything worked because everybody worked."

Possible Dream

So what did this hero and her helpers actually do? Stoddart began by setting an "impossible" goal—making the Nashua safe for swimming. When she presented this demand at a water-pollution hearing, almost everyone was flabbergasted. "Come on, be realistic," they responded. But Stoddart stood her ground.

The people who live near and depend on the Nashua then set about doing the impossible. Mill owners began treating their waste, and citizens pitched in to monitor water quality. With help from the

ABOVE: On the leaf-flecked Nashua River, student oarsmen are too young to recall the days when capsized boaters raced to a clinic for tetanus shots because the water was so polluted. Although parts of the Nashua River have been successfully cleaned up, much remains to be done in the watershed to improve water quality.

This text is based on the article "Restoration: New Ideas, New Understanding, New Hope" in National Geographic*'s special issue on water, published in November 1993.*

federal government, officials built and upgraded water-treatment plants. There are now eight along the river.

Builders, conservationists, and zoning authorities collaborated on creating a "greenway" along half the Nashua and two of its major tributaries. Development is limited in the greenway, which lessens the amount of pollution entering the river.

Today you can swim in the Nashua River, and you can bet that its water will be clear again tomorrow.

ONE DOWN, THOUSANDS TO GO

Magnificent as the Nashua's rebirth is, it's just one small stream in a nation with two dozen major watersheds. Making all of them healthy requires a daunting, complex interplay of science, economics, and politics. And it can mean serious costs.

Farmers and developers, for instance, may balk at setting fields and houses farther from a stream bank to lessen runoff. Engineers are not often eager to change bridge and highway designs so that a river can meander naturally. Even eco-minded towns gasp at the price tag for separating storm drains from sewer lines. These are but a few examples of the challenges in repairing rivers.

"I'm optimistic about what we can do," said river advocate John Cairns, Jr. "I'm pessimistic about what we will do." Another water expert, pondering the need for cross-border cooperation, asked at a conference, "How do you get people to give up power in order to survive?"

TAKING WATER TO HEART

Perhaps the seed of an answer lies in Marion Stoddart's defiant proposal—a place to swim. When her neighbors looked at the Nashua and imagined themselves immersed in red (or green or white or orange) water, the river suddenly wasn't so distant. They were polluting part of their own lives.

Cleaning up the Nashua didn't magically get cheaper or easier, of course, but a clearer connection to the river may have helped people take hard steps on its behalf. Many river advocates stress the importance of such connections, of seeing streams as more than just water. "Water is a living thing," said Gwichin Indian Chief Clarence Alexander. "We treat it like it's got a soul of its own."

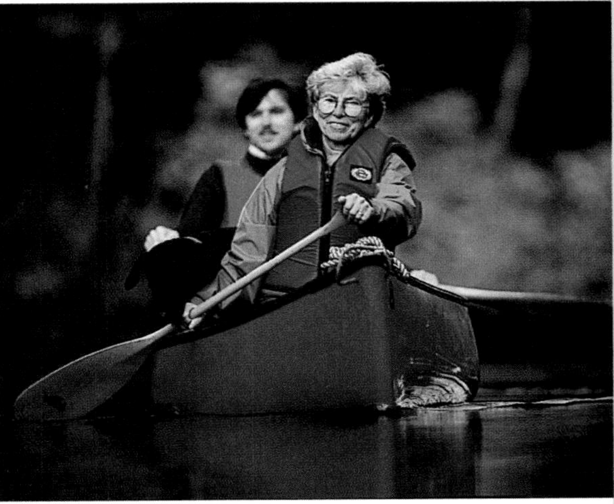

ABOVE: When Marion Stoddard explored the Nashua Basin in the 1960s, almost no form of life survived in the river's polluted waters. Outraged, she has spent 30 years getting industries to treat waste, citizens to monitor water quality, and builders to limit riverbank development. Today the activist canoes the waters she helped revive.

Expeditions Activity

If you had to clean up a river, what measures would you take? What activities would you try to stop in order to reduce pollution? What kinds of plants or animals could you introduce to help clean the river? How would protecting or restoring wetlands affect your project? Investigate these issues and report on the cleanup of a specific river, such as the Cuyahoga, in Ohio.

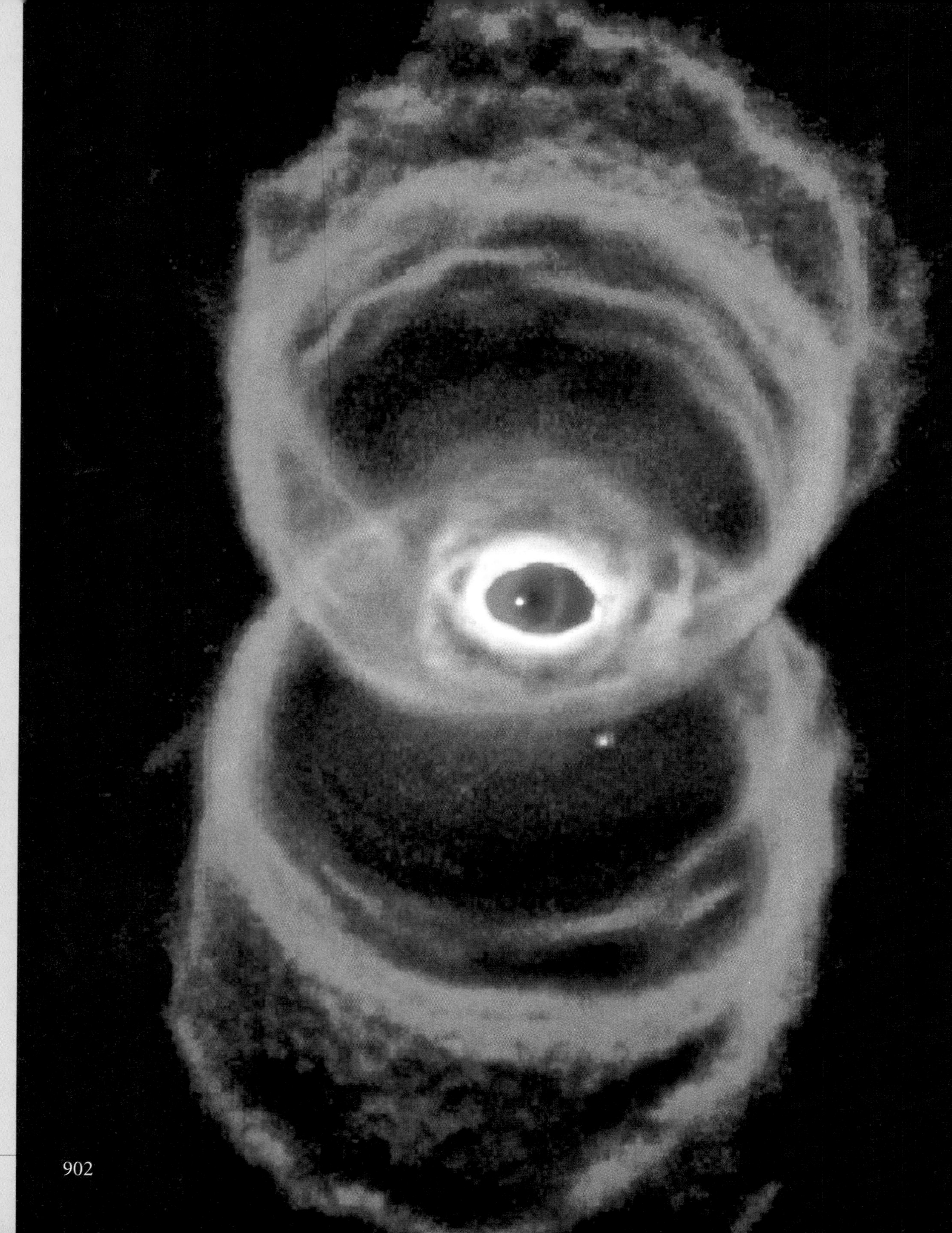

Hubble Space Telescope

NASA's Eye in the Sky

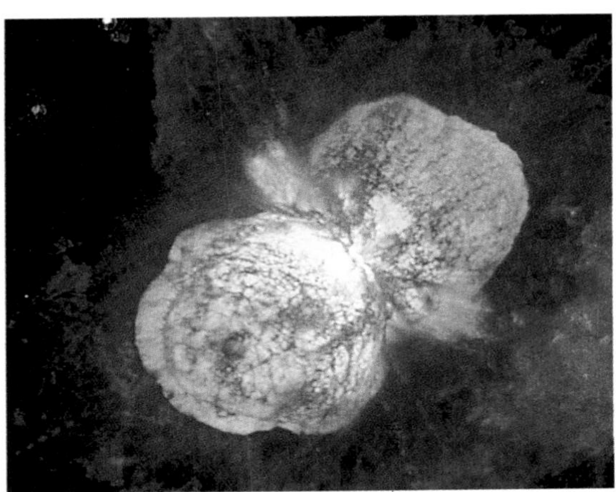

ABOVE: Clouds of gas and dust pour out from the exploding star **Eta Carinae** at 2.4 million kilometers (1.5 million miles) per hour. Still burning five million times brighter than our Sun, Eta Carinae illuminates the clouds from the inside, like frosted Christmas bulbs.

LEFT: The **Etched Hourglass Nebula** is a shell of gas expanding from a dying star.

Hermann Oberth was frustrated. The German scientist yearned to explore space, but even the best telescopes available to him—to anyone—were hampered by Earth's atmosphere. The gases that make the planet habitable also distort our views of the heavens. Oberth's frustration sparked a bold idea: Why not design a telescope that could orbit the planet? It could offer an unparalleled view of the universe. But would it ever exist?

Oberth conceived his orbiting telescope in 1923. Air travel was still new; space travel was still science fiction. As the saying goes, he was ahead of his time—by decades. NASA launched two small space telescopes in the 1960s and 1970s, but neither came close to being the marvel that Oberth and others yearned for.

Astronomy's Dream Machine

Five, four, three, two, one—blastoff! In April 1990, four months after Hermann Oberth's death, the shuttle *Discovery* set the *Hubble Space Telescope (HST)* into orbit 595 kilometers (370 miles) above Earth's surface. Named for American astronomer Edwin Hubble, the telescope would finally offer scientists the undistorted images for which they had waited so long. *Hubble* promised to be, as headlines and hype put it, a "window on the universe."

The new telescope could create images that were ten times sharper than anything seen from Earth-based observatories. That's comparable to distinguishing the headlights on a car that is 4800 kilometers (3000 miles) away. *Hubble*'s sharp instruments are sensitive to incredibly faint light; the telescope detects objects that would have to be 50 times brighter before they showed up on a ground-based telescope. Being above the atmosphere, moreover, *Hubble* can observe ultraviolet wavelengths that do not reach the ground.

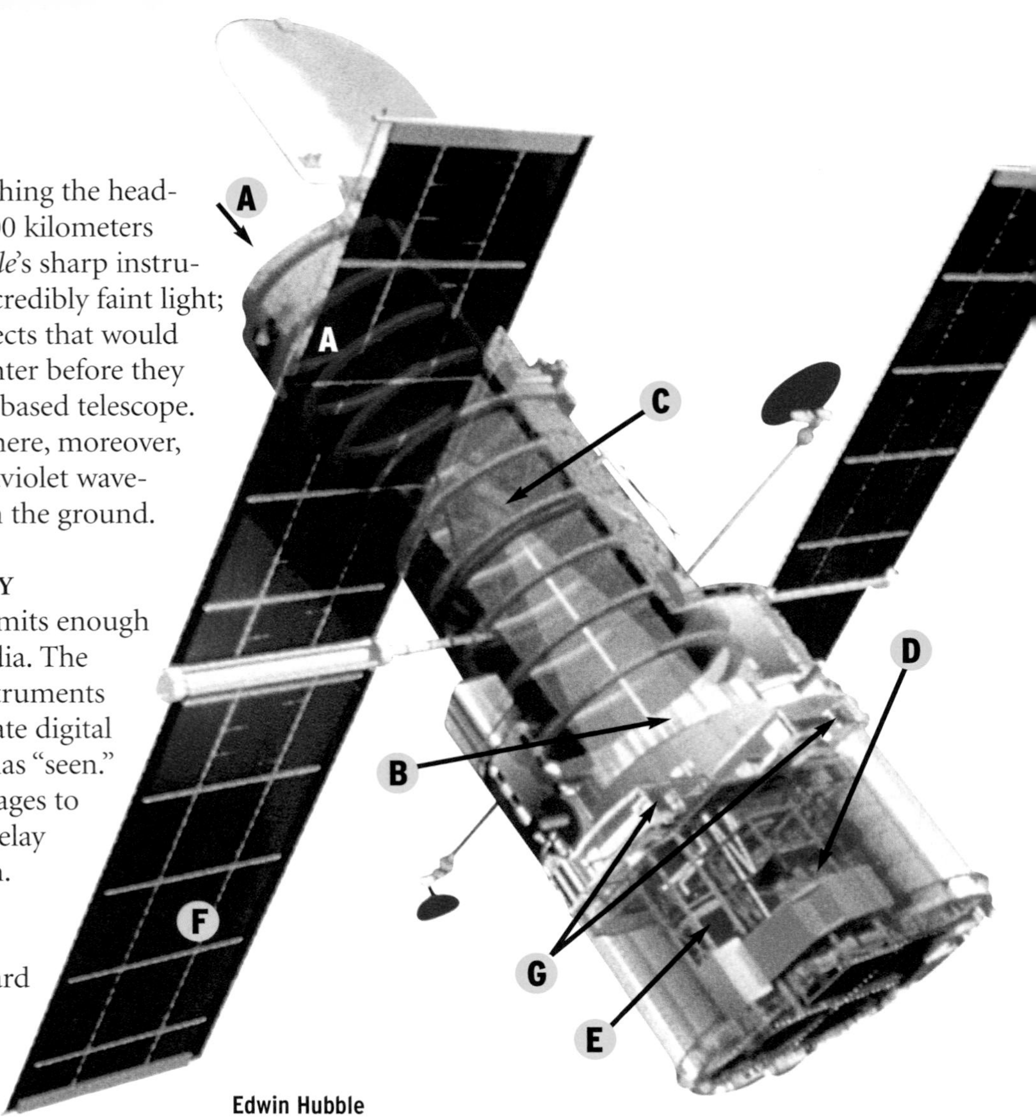

Information Odyssey

In a day, *Hubble* transmits enough data to fill an encyclopedia. The process begins when instruments aboard the telescope create digital images of what *Hubble* has "seen." *Hubble* transmits the images to the Tracking and Data Relay Satellites that orbit Earth.

The communications satellites route the data stream to NASA's Goddard Space Flight Center in Greenbelt, Maryland. Goddard serves as mission control for *Hubble*. Whenever the telescope is directed to point toward a new target or alter the way it gathers data, the command comes from Greenbelt. Goddard dispatches *Hubble* data, by telephone links, to the Space Telescope Science Institute in Baltimore, Maryland.

Edwin Hubble

"Eye Doctors" in Orbit

In 1990 *Hubble* was aloft, and the mechanisms for getting data to scientists were in place. But *Hubble* hype turned to horror after the telescope began transmitting images. Where were the crystal-clear shots everyone had expected? Taxpayers roared, and NASA officials cringed. *Hubble*'s blurry pictures looked as if someone had stretched fine gauze over the lens of a camera.

DIAGRAM (ABOVE): *Hubble*'s "front door" is the **Aperture (A),** a long tube that collects light particles called photons. The hinged cover closes when the light is too bright. Once inside, photons bounce from the **Primary Mirror (B)** to the **Secondary Mirror (C)** to a **Spectrograph (D)** and a **Spectrometer (E)** that collect data. The probe's **Solar Panels (F)** collect the minimal energy—only about 150 watts—needed to operate the telescope. The solar panels move constantly to stay aligned with the Sun. *Hubble*'s **Fine Guidance Sensors (G)** aim the satellite with ten times the accuracy of ground-based telescopes.

NATIONAL GEOGRAPHIC
eXpeditions!

TOP LEFT: Astronauts from the Space Shuttle *Discovery* make repairs to the *Hubble Space Telescope.* During their week in space in December 1999, the shuttle crew overhauled the telescope for another decade of observation and discovery. Another servicing mission will take place early in the new century.

BOTTOM LEFT: The miracle of star birth unfolds in the **Eagle Nebula.** Deep inside pillars of dense, cool gas and dust—the tallest one here measures three light-years in height—molecular hydrogen and dust condense into lumps that contract and ignite under their own gravity to become stars.

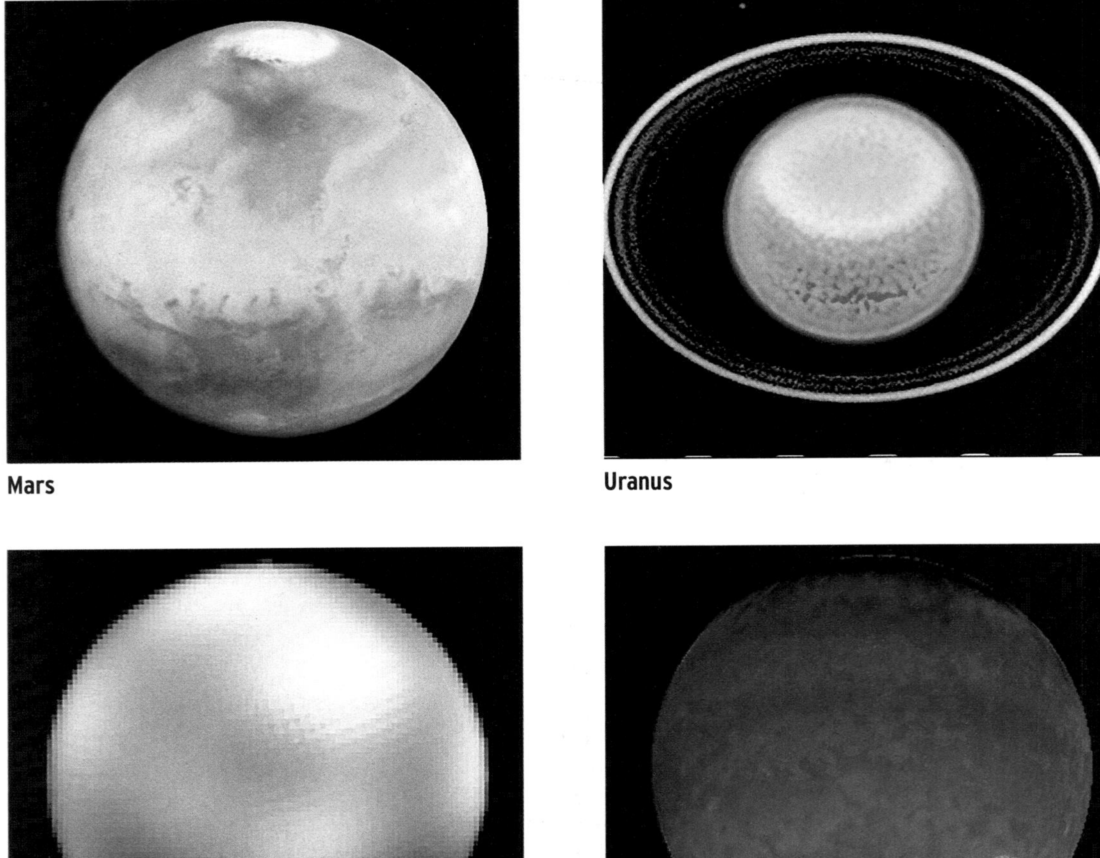
Mars

Uranus

Pluto

Neptune

NASA quickly diagnosed the problem: The primary mirror was slightly defective. As a result, *Hubble*'s images weren't focused properly. The eye in the sky needed glasses. Responding to a 911 call from space takes time. The shuttle *Endeavour* reached the telescope in December 1993. Replacing the primary mirror was impossible, so the astronauts installed supplementary mirrors that compensated for the flaw. At last the telescope could see clearly.

Keeping the Telescope in Shape

Hubble's creators had expected occasional repairs and upgrades, so they made sure astronauts could get to the telescope's key systems. The telescope's parts were made to be removed and replaced easily. Thus, the shuttle crew can repair the *Hubble Space Telescope* in orbit, rather than haul it back to Earth.

ABOVE AND RIGHT: Even the solar system unveiled secrets to *Hubble*'s powerful eye. No one had clearly seen Pluto's face before, and no telescope had captured such a sharp picture of Mars. *Hubble* also provided new looks at clouds on Jupiter and Saturn and at the charcoal black rings of Uranus.

This text is based on "Time Exposures" in the April 1997 issue of NATIONAL GEOGRAPHIC.

Saturn

Where *Hubble* Has Looked

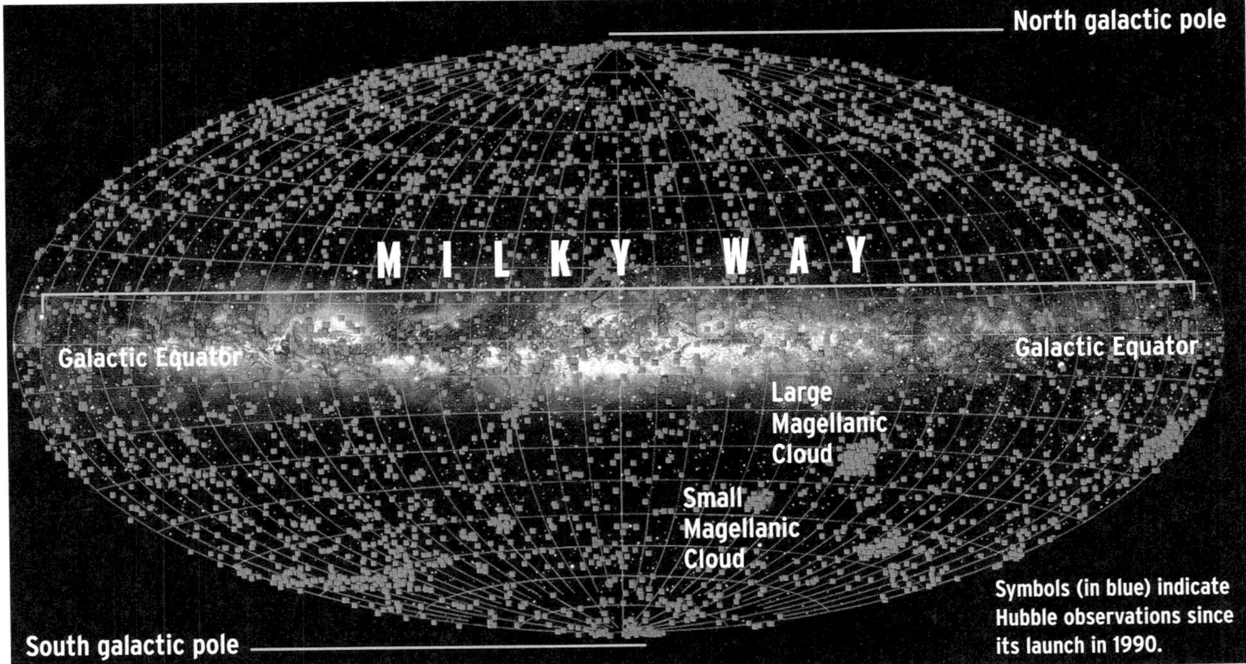

In 1997 the shuttle *Discovery* made the second house call to *Hubble.* In addition to routine maintenance, the astronauts replaced the telescope's scientific instruments with newer, more powerful devices. They also lifted *Hubble* about 15 kilometers (9 miles) higher above the planet's surface.

Discovery visited *Hubble* again in 1999. This third servicing mission replaced the gyroscopes that aim the telescope at targets and installed a new computer that works 20 times faster than its predecessor.

NASA originally planned to retire *Hubble* in 2005, but the telescope's incredible output—120 000 images of 10 000 objects during the first decade of operation—inspired NASA to think again. The agency announced in 1997 that *Hubble* would keep its powerful eyes open until 2010. Despite its problems, the *Hubble Space Telescope* has lived up to—even surpassed—Hermann Oberth's dream.

Space will never look the same again.

Expeditions Activity

Do you want to know more about *Hubble Space Telescope*? Consult the April 1997 NATIONAL GEOGRAPHIC article, "Time Exposures"—or explore science.glencoe.com to compare and contrast photos taken by *Hubble* with earlier images from satellites and space probes.

Appendix

Contents

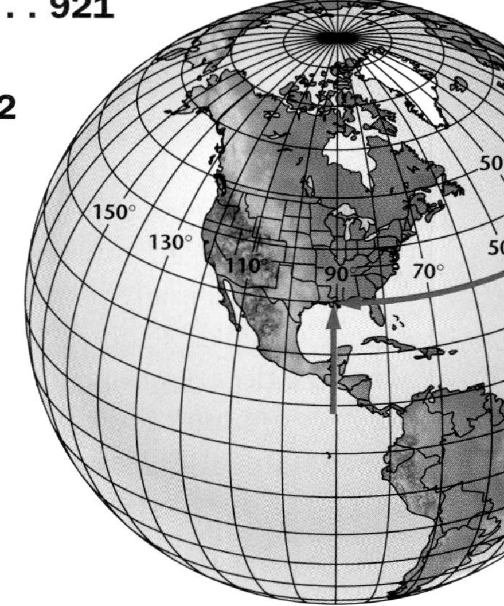

Appendix A International System of Units

The International System of Units (SI) is accepted as the standard for measurement throughout most of the world. Three base units in SI are the meter, the kilogram, and the second. Frequently used SI units are listed below.

Temperature measurements in SI are often made in degrees Celsius. A Celsius degree (°C) is a supplementary unit derived from the base unit Kelvin. The Celsius scale has 100 equal graduations between the freezing temperature (0°C) and the boiling temperature (100°C) of water. The following relationship exists between the Celsius and Kelvin temperature scales:

$$K = °C + 273$$

Several other supplementary SI units are listed in Table A-2.

If you want to convert measurements in SI/Metric units to English units, use Table A-3.

Table A-1 Frequently Used SI Units

Length	1 millimeter (mm) = 1000 micrometers (μm)
	1 centimeter (cm) = 10 millimeters (mm)
	1 meter (m) = 100 centimeters (cm)
	1 kilometer (km) = 1000 meters (m)
	1 light-year (ly) = 9 460 000 000 000 kilometers (km)
Area	1 square meter (m^2) = 10 000 square centimeters (cm^2)
	1 square kilometer (km^2) = 1 000 000 square meters (m^2)
Volume	1 milliliter (mL) = 1 cubic centimeter (cm^3)
	1 liter (L) = 1000 milliliters (mL)
Mass	1 gram (g) = 1000 milligrams (mg)
	1 kilogram (kg) = 1000 grams (g)
	1 metric ton = 1000 kilograms (kg)
Time	1 s = 1 second
	1 h = 1 hour, 60 seconds

Table A-2 Supplementary SI Units

Measurement	Unit	Symbol	Expressed in Base Units
Energy	Joule	J	$kg \cdot m^2/s^2$ or $N \cdot m$
Force	Newton	N	$kg \cdot m/s^2$
Power	Watt	W	$kg \cdot m^2/s^3$ or J/s
Pressure	Pascal	Pa	$kg/(m \cdot s^2)$ or $N \cdot m$

Table A-3 SI/Metric-to-English Conversions

	When You Want to Convert:	Multiply By:	To Find:
Length	Inches	2.54	centimeters
	Centimeters	0.39	inches
	Feet	0.30	meters
	Meters	3.28	feet
	Yards	0.91	meters
	Meters	1.09	yards
	Miles	1.61	kilometers
	Kilometers	0.62	miles
Mass and Weight*	Ounces	28.35	grams
	Grams	0.04	ounces
	Pounds	0.45	kilograms
	Kilograms	2.20	pounds
	Tons	0.91	metric tons
	Metric tons	1.10	tons
	Pounds	4.45	newtons
	Newtons	0.23	pounds
Volume	Cubic inches	16.39	cubic centimeters
	Cubic centimeters	0.06	cubic inches
	Cubic feet	0.03	cubic meters
	Cubic meters	35.31	cubic feet
	Liters	1.06	quarts
	Liters	0.26	gallons
	Gallons	3.78	liters
Area	Square inches	6.45	square centimeters
	Square centimeters	0.16	square inches
	Square feet	0.09	square meters
	Square meters	10.76	square feet
	Square miles	2.59	square kilometers
	Square kilometers	0.39	square miles
	Hectares	2.47	acres
	Acres	0.40	hectares
Temperature	Fahrenheit	5/9 (°F – 32)	Celsius
	Celsius	9/5 °C + 32	Fahrenheit

*Weight as measured in standard Earth gravity

Appendix B Safety in the Laboratory

The Earth Science laboratory is a safe place to work if you are aware of important safety rules and if you are careful. You must be responsible for your own safety and for the safety of others. The safety rules given here will protect you and others from harm in the laboratory. While carrying out procedures in any of the activities or *GeoLabs,* take note of the safety symbols and caution statements. The safety symbols are listed and explained in Table B-2.

Safety Rules

1. Always obtain your teacher's permission to begin an investigation.
2. Study the procedure outline in the text. If you have questions, ask your teacher. Make sure that you understand all safety symbols shown on the page.
3. Use the safety equipment provided for you. Safety goggles and an apron should be worn during all investigations that involve the use of chemicals.
4. When heating test tubes, always slant them away from yourself and others.
5. Never eat or drink in the lab, and never use lab glassware as food or drink containers. Never inhale chemicals. Do not taste any substances or draw any material into a test tube with your mouth.
6. If you spill any chemical, wash it off immediately with water. Report the spill immediately to your teacher.
7. Know the location and proper use of the fire extinguisher, eye wash, safety shower, fire blanket, first-aid kit, and fire alarm.
8. Keep materials away from flames. Tie back hair and loose clothing when you are working with flames.
9. If a fire should break out in the lab, or if your clothing should catch fire, smother it with the fire blanket or a coat, get under a safety shower, or use the fire department's recommendation for putting out a fire on your clothing: **Stop, Drop, and Roll. NEVER RUN.**
10. Report any accident or injury, no matter how small, to your teacher.

Clean-up Procedures

1. Turn off the water and gas. Disconnect electrical devices.
2. Return all materials to their proper places.
3. Dispose of chemicals and other materials as directed by your teacher. Place broken glass and solid substances in the proper containers. Never discard materials in the sink.
4. Clean your work area.
5. Wash your hands thoroughly after working in the laboratory.

Table B-1 First Aid in the Science Laboratory

Injury	Safe Response
Burns	Apply cold water. Call your teacher immediately.
Cuts and bruises	Stop any bleeding by applying direct pressure. Cover cuts with a clean dressing. Apply cold compresses to bruises. Call your teacher immediately.
Fainting	Leave the person lying down. Loosen any tight clothing and keep crowds away. Call your teacher immediately.
Foreign matter in eye	Flush with plenty of water. Use an eyewash bottle or fountain.
Poisoning	Note the suspected poisoning agent and call your teacher immediately.
Any spills on skin	Flush with large amounts of water or use safety shower. Call your teacher immediately.

Safety symbols in the following table are used in the lab activities to indicate possible hazards. Learn the meaning of each symbol.

We recommend that you wear safety goggles and apron at all times in the lab. This may be required in your school district.

Appendix

Table B-2 Safety Symbols

SAFETY SYMBOLS	HAZARD	EXAMPLES	PRECAUTION	REMEDY
DISPOSAL	Special disposal procedures need to be followed.	certain chemicals, living organisms	Do not dispose of these materials in the sink or trash can.	Dispose of wastes as directed by your teacher.
BIOLOGICAL	Organisms or other biological materials that might be harmful to humans	bacteria, fungi, blood, unpreserved tissues, plant materials	Avoid skin contact with these materials. Wear mask or gloves.	Notify your teacher if you suspect contact with material. Wash hands thoroughly.
EXTREME TEMPERATURE	Objects that can burn skin by being too cold or too hot	boiling liquids, hot plates, dry ice, liquid nitrogen	Use proper protection when handling.	Go to your teacher for first aid.
SHARP OBJECT	Use of tools or glassware that can easily puncture or slice skin	razor blades, pins, scalpels, pointed tools, dissecting probes, broken glass	Practice common-sense behavior and follow guidelines for use of the tool.	Go to your teacher for first aid.
FUME	Possible danger to respiratory tract from fumes	ammonia, acetone, nail polish remover, heated sulfur, moth balls	Make sure there is good ventilation. Never smell fumes directly. Wear a mask.	Leave foul area and notify your teacher immediately.
ELECTRICAL	Possible danger from electrical shock or burn	improper grounding, liquid spills, short circuits, exposed wires	Double-check setup with teacher. Check condition of wires and apparatus.	Do not attempt to fix electrical problems. Notify your teacher immediately.
IRRITANT	Substances that can irritate the skin or mucous membranes of the respiratory tract	pollen, moth balls, steel wool, fiberglass, potassium permanganate	Wear dust mask and gloves. Practice extra care when handling these materials.	Go to your teacher for first aid.
CHEMICAL	Chemicals can react with and destroy tissue and other materials	bleaches such as hydrogen peroxide; acids such as sulfuric acid, hydrochloric acid; bases such as ammonia, sodium hydroxide	Wear goggles, gloves, and an apron.	Immediately flush the affected area with water and notify your teacher.
TOXIC	Substance may be poisonous if touched, inhaled, or swallowed.	mercury, many metal compounds, iodine, poinsettia plant parts	Follow your teacher's instructions.	Always wash hands thoroughly after use. Go to your teacher for first aid.
FLAMMABLE	Flammable chemicals may be ignited by open flame, spark, or exposed heat.	alcohol, kerosene, potassium permanganate	Avoid open flames and heat when using flammable chemicals.	Notify your teacher immediately. Use fire safety equipment if applicable.
OPEN FLAME	Open flame in use, may cause fire.	hair, clothing, paper, synthetic materials	Tie back hair and loose clothing. Follow teacher's instruction on lighting and extinguishing flames.	Notify your teacher immediately. Use fire safety equipment if applicable.

Eye Safety
Proper eye protection should be worn at all times by anyone performing or observing science activities.

Clothing Protection
This symbol appears when substances could stain or burn clothing.

Animal Safety
This symbol appears when safety of animals and students must be ensured.

Handwashing
After the lab, wash hands with soap and water before removing goggles.

Appendix C Physiographic Map of Earth

CHUKCHI CAP
CANADA ABYSSAL PLAIN
ELLESMERE ISLAND
MELVILLE ISLAND
LAPTEV SEA
NEW SIBERIAN ISLANDS
WRANGEL ISLAND
CHUKCHI SEA
BANKS I.
VICTORIA ISLAND
BAFFIN BAY
BAFFIN ISLAND
BROOKS RANGE
ALASKA
BERING STRAIT
BERING SEA
Anchorage
VERKHOYANSK MTS.
CHERSKIY MTS.
KOLYMSKOYE MTS.
STANOVOY MTS.
OKHOTSK SEA
SAKHALIN ISLAND
KAMCHATKA
HUDSON BAY
LABRADOR
NORTH AMERICA
ROCKY MOUNTAINS
Vancouver
ALEUTIAN TRENCH
ALEUTIAN ABYSSAL PLAIN
KURIL TRENCH
EMPEROR SEAMOUNTS
JAPAN SEA
YELLOW SEA
Seoul
Tokyo
Shanghai
TAIWAN
MENDOCINO FRACTURE ZONE
MURRAY FRACTURE ZONE
MOLOKAI FRACTURE ZONE
CLARION FRACTURE ZONE
CLIPPERTON FRACTURE ZONE
Chicago
Montreal
New York
Washington
New Orleans
MARIANA TRENCH
NEW GUINEA
Darwin
GIBSON DESERT
SIMPSON DESERT
AUSTRALIA
NULLARBOR PLAIN
Perth
Sydney
Melbourne
TASMANIA
NEW ZEALAND
SOUTH AMERICA
AMAZON
ANDES
MATO GROSSO
Quito
Lima
La Paz
Santiago
Buenos Aires
Montevideo
FALKLAND ISLANDS
BELLINGSHAUSEN ABYSSAL PLAIN
ANTARCTICA

Appendix D Topographic Map Symbols

Roads and Railroads

Primary highway, hard surface

Secondary highway, hard surface

Light-duty road, hard or improved surface

Unimproved road

Railroad: single track and multiple track

Railroads in juxtaposition

Buildings and Structures

Buildings

School, church, and cemetery — cem

Barn and warehouse

Wells, not water (with labels) — oil gas

Tanks: oil, water, etc. (labeled if water) — water

Open-pit mine, quarry, or prospect

Tunnel

Benchmark — BM Δ 293

Bridge

Campsite

Habitats

Marsh (swamp)

Wooded marsh

Woods or brushwood

Vineyard

Submerged marsh

Mangrove

Coral reef, rocks

Orchard

Urban area

Perennial streams

Elevated aqueduct

Water well and spring

Small rapids

Large rapids

Intermittent lake

Intermittent stream

Glacier

Large falls

Dry lake bed

Surface Elevations

Spot elevation — **x 7369**

Water elevation — **670**

Index contour — 100

Intermediate contour

Depression contour

Boundaries

National

State

County, parish, municipal

Civil township, precinct, town, barrio

Incorporated city, village, town, hamlet

Reservation, national or state

Small park, cemetery, airport, etc.

Land grant

Township or range line, United States land survey

Township or range line, approximate location

Appendix E Weather Map Symbols

Sample Plotted Report at Each Station

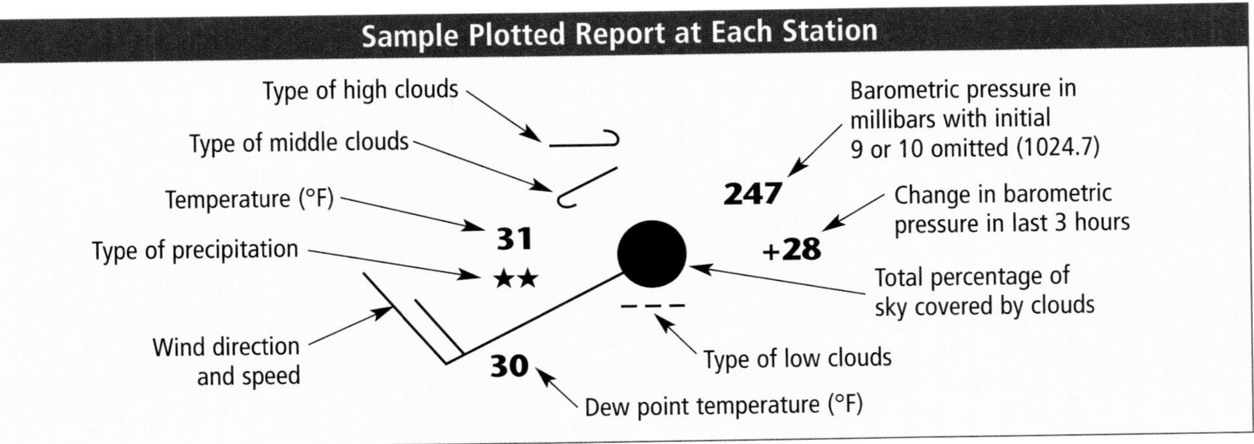

Symbols Used in Plotting Report

Precipitation	Wind Direction and Speed	Sky Coverage	Fronts and Pressure Systems	
Fog	0 calm	No cover	(H) or High	Center of high- or
Snow	1–2 knots	1/10 or less	(L) or Low	low-pressure system
Rain	3–7 knots	2/10 to 3/10		Cold front
Thunderstorm	8–12 knots	4/10		Warm front
Drizzle	13–17 knots	1/2		Occluded front
Showers	18–22 knots	6/10		Stationary front
	23–27 knots	7/10		
	48–52 knots	Overcast with openings		
	1 knot = 1.852 km/h	Completely overcast		

Clouds

Some Types of High Clouds	Some Types of Middle Clouds	Some Types of Low Clouds
Scattered cirrus	Thin altostratus layer	Cumulus of fair weather
Dense cirrus in patches	Thick altostratus layer	Stratocumulus
Veil of cirrus covering entire sky	Thin altostratus in patches	Fractocumulus of bad weather
Cirrus not covering entire sky	Thin altostratus in bands	Stratus of fair weather

Appendix F Relative Humidity

Relative Humidity %										
Dry-Bulb Temperature	Dry-Bulb Temperature Minus Wet-Bulb Temperature, °C									
	1	2	3	4	5	6	7	8	9	10
0°C	81	64	46	29	13					
1°C	83	66	49	33	18					
2°C	84	68	52	37	22	7				
3°C	84	69	55	40	25	12				
4°C	85	71	57	43	29	16				
5°C	85	72	58	45	32	20				
6°C	86	73	60	48	35	24	11			
7°C	86	74	61	49	38	26	15			
8°C	87	75	63	51	40	29	19	8		
9°C	87	76	65	53	42	32	21	12		
10°C	88	77	66	55	44	34	24	15	6	
11°C	89	78	67	56	46	36	27	18	9	
12°C	89	78	68	58	48	39	29	21	12	
13°C	89	79	69	59	50	41	32	22	15	7
14°C	90	79	70	60	51	42	34	26	18	10
15°C	90	80	71	61	53	44	36	27	20	13
16°C	90	81	71	63	54	46	38	30	23	15
17°C	90	81	72	64	55	47	40	32	25	18
18°C	91	82	73	65	57	49	41	34	27	20
19°C	91	82	74	65	58	50	43	36	29	22
20°C	91	83	74	66	59	51	44	37	31	24
21°C	91	83	75	67	60	53	46	39	32	26
22°C	92	83	76	68	61	54	47	40	34	28
23°C	92	84	76	69	62	55	48	42	36	30
24°C	92	84	77	69	62	56	49	43	37	31
25°C	92	84	77	70	63	57	50	44	39	33
26°C	92	85	78	71	64	58	51	46	40	34
27°C	92	85	78	71	65	58	52	47	41	36
28°C	93	85	78	72	65	59	53	48	42	37
29°C	93	86	79	72	66	60	54	49	43	38
30°C	93	86	79	73	67	61	55	50	44	39
31°C	93	86	80	73	67	62	56	50	45	40
32°C	93	86	80	74	68	62	57	51	46	41

Appendix G Periodic Table of the Elements

PERIODIC TABLE OF THE ELEMENTS

Key: Element — Hydrogen; Atomic number — 1; Symbol — H; Atomic mass — 1.008; State of matter

Metal / Metalloid / Nonmetal / Recently discovered

Gas / Liquid / Solid / Synthetic elements

	1A 1	2A 2	3B 3	4B 4	5B 5	6B 6	7B 7	8B 8	9	10	1B 11	2B 12	3A 13	4A 14	5A 15	6A 16	7A 17	8A 18
1	Hydrogen 1 H 1.008																	Helium 2 He 4.003
2	Lithium 3 Li 6.941	Beryllium 4 Be 9.012											Boron 5 B 10.811	Carbon 6 C 12.011	Nitrogen 7 N 14.007	Oxygen 8 O 15.999	Fluorine 9 F 18.998	Neon 10 Ne 20.180
3	Sodium 11 Na 22.990	Magnesium 12 Mg 24.305											Aluminum 13 Al 26.982	Silicon 14 Si 28.086	Phosphorus 15 P 30.974	Sulfur 16 S 32.066	Chlorine 17 Cl 35.453	Argon 18 Ar 39.948
4	Potassium 19 K 39.098	Calcium 20 Ca 40.078	Scandium 21 Sc 44.956	Titanium 22 Ti 47.88	Vanadium 23 V 50.942	Chromium 24 Cr 51.996	Manganese 25 Mn 54.938	Iron 26 Fe 55.847	Cobalt 27 Co 58.933	Nickel 28 Ni 58.693	Copper 29 Cu 63.546	Zinc 30 Zn 65.39	Gallium 31 Ga 69.723	Germanium 32 Ge 72.61	Arsenic 33 As 74.922	Selenium 34 Se 78.96	Bromine 35 Br 79.904	Krypton 36 Kr 83.80
5	Rubidium 37 Rb 85.468	Strontium 38 Sr 87.62	Yttrium 39 Y 88.906	Zirconium 40 Zr 91.224	Niobium 41 Nb 92.906	Molybdenum 42 Mo 95.94	Technetium 43 Tc 97.907	Ruthenium 44 Ru 101.07	Rhodium 45 Rh 102.906	Palladium 46 Pd 106.42	Silver 47 Ag 107.868	Cadmium 48 Cd 112.411	Indium 49 In 114.82	Tin 50 Sn 118.710	Antimony 51 Sb 121.757	Tellurium 52 Te 127.60	Iodine 53 I 126.904	Xenon 54 Xe 131.290
6	Cesium 55 Cs 132.905	Barium 56 Ba 137.327	Lanthanum 57 La 138.906	Hafnium 72 Hf 178.49	Tantalum 73 Ta 180.948	Tungsten 74 W 183.84	Rhenium 75 Re 186.207	Osmium 76 Os 190.2	Iridium 77 Ir 192.22	Platinum 78 Pt 195.08	Gold 79 Au 196.967	Mercury 80 Hg 200.59	Thallium 81 Tl 204.383	Lead 82 Pb 207.2	Bismuth 83 Bi 208.980	Polonium 84 Po 208.982	Astatine 85 At 209.987	Radon 86 Rn 222.018
7	Francium 87 Fr 223.020	Radium 88 Ra 226.025	Actinium 89 Ac 227.028	Rutherfordium 104 Rf (261)	Dubnium 105 Db (262)	Seaborgium 106 Sg (263)	Bohrium 107 Bh (262)	Hassium 108 Hs (265)	Meitnerium 109 Mt (266)	Darmstadtium 110 Ds (269)	Roentgenium 111 Rg (272)	Ununbium * 112 Uub (277)		Ununquadium * 114 Uuq (285)				

The number in parentheses is the mass number of the longest lived isotope for that element.

*Names not officially assigned. Discovery of element 114 recently reported. Further information not yet available.

Lanthanide series	Cerium 58 Ce 140.115	Praseodymium 59 Pr 140.908	Neodymium 60 Nd 144.24	Promethium 61 Pm 144.913	Samarium 62 Sm 150.36	Europium 63 Eu 151.965	Gadolinium 64 Gd 157.25	Terbium 65 Tb 158.925	Dysprosium 66 Dy 162.50	Holmium 67 Ho 164.930	Erbium 68 Er 167.26	Thulium 69 Tm 168.934	Ytterbium 70 Yb 173.04	Lutetium 71 Lu 174.967
Actinide series	Thorium 90 Th 232.038	Protactinium 91 Pa 231.036	Uranium 92 U 238.029	Neptunium 93 Np 237.048	Plutonium 94 Pu 244.064	Americium 95 Am 243.061	Curium 96 Cm 247.070	Berkelium 97 Bk 247.070	Californium 98 Cf 251.080	Einsteinium 99 Es 252.083	Fermium 100 Fm 257.095	Mendelevium 101 Md 258.099	Nobelium 102 No 259.101	Lawrencium 103 Lr 260.105

Appendix

Appendix H Minerals

Table H-1 Minerals with Metallic Luster

Mineral (Formula)	Color	Streak	Hardness	Specific Gravity	Crystal System	Breakage Pattern	Uses and Other Properties
Bornite (Cu_5FeS_4)	bronze, tarnishes to dark blue purple	gray-black	3	4.9–5.4	tetragonal	uneven fracture	source of copper called "peacock ore" because of the purple shine when it tarnishes
Chalcopyrite ($CuFeS_2$)	brassy to golden yellow	greenish black	3.5–4	4.2	tetragonal	uneven fracture	main ore of copper
Chromite ($FeCr_2O_4$)	black or brown	brown to black	5.5	4.6	cubic	irregular fracture	ore of chromium, stainless steel, metallurgical bricks
Copper (Cu)	copper red	copper red	3	8.5–9	cubic	hackly	coins, pipes, gutters, wire, cooking utensils, jewelry, decorative plaques; malleable and ductile
Galena (PbS)	gray	gray to black	2.5	7.5	cubic	cubic cleavage perfect	source of lead, used in pipes, shields for X rays, fishing equipment sinkers
Gold (Au)	pale to golden yellow	yellow	2.5–3	19.3	cubic	hackly	jewelry, money, gold leaf, fillings for teeth, medicines; does not tarnish
Graphite (C)	black to gray	black to gray	1–2	2.3	hexagonal	basal cleavage (scales)	pencil lead, lubricants for locks, rods to control some small nuclear reactions, battery poles
Hematite (specular) (Fe_2O_3)	black or reddish brown	red or reddish brown	6	5.3	hexagonal	irregular fracture	source of iron; roasted in a blast furnace, converted to "pig" iron, made into steel
Magnetite (Fe_3O_4)	black	black	6	5.2	cubic	conchoidal fracture	source of iron, naturally magnetic, called lodestone
Pyrite (FeS_2)	light, brassy yellow	greenish black	6.5	5.0	cubic	uneven fracture	source of iron, "fool's gold," alters to limonite
Pyrrhotite ($Fe_{1-x}S$)* *contains one less atom of Fe than S	bronze	gray-black	4	4.6	hexagonal	uneven fracture	an ore of iron and sulfur; may be magnetic
Silver (Ag)	silvery white, tarnishes to black	light gray to silver	2.5	10–12	cubic	hackly	coin, fillings for teeth, jewelry, silverplate, wires; malleable and ductile

Table H-2 Minerals with Nonmetallic Luster

Mineral (Formula)	Color	Streak	Hardness	Specific Gravity	Crystal System	Breakage Pattern	Uses and Other Properties
Augite ((Ca, Na) (Mg, Fe, Al) (Al, Si)$_2$O$_6$)	black	colorless	6	3.3	monoclinic	2-directional cleavage	square or 8-sided cross section
Corundum (Al$_2$O$_3$)	colorless, blue, brown green, white pink, red	colorless	9	4.0	hexagonal	fracture	gemstones: ruby is red, sapphire is blue; industrial abrasive
Feldspar (orthoclase) (KAlSi$_3$O$_8$)	colorless white to gray, green, and yellow	colorless	6	2.5	monoclinic	two cleavage planes meet at 90° angle	insoluble in acids; used in the manufacture of porcelain
Feldspar (plagioclase) (NaAlSi$_3$O$_8$) (CaAl$_2$Si$_3$O$_8$)	gray, green, white	colorless	6	2.5	triclinic	two cleavage planes meet at 86° angle	used in ceramics; striations present on some faces
Fluorite (CaF$_2$)	colorless, white, blue, green, red, yellow, purple	colorless	4	3–3.2	cubic	cleavage	used in the manufacture of optical equipment; glows under ultraviolet light
Garnet (Mg, Fe, Ca, Mn)$_3$, (Al, Fe, Cr)$_2$, (SiO$_4$)$_3$	deep yellow-red, green, black	colorless	7.5	3.5	cubic	conchoidal fracture	used in jewelry; also used as an abrasive
Hornblende Ca$_2$Na(Mg, Fe^2)$_4$, (Al, Fe^3, Ti)$_3$, Si$_8$O$_{22}$(O, OH)$_2$	green to black	gray to white	5–6	3.4	monoclinic	cleavage in two directions	will transmit light on thin edges; 6-sided cross section
Limonite (hydrous iron oxides)	yellow, brown, black	yellow, brown	5.5	2.74–4.3	—	conchoidal fracture	source of iron; weathers easily, coloring matter of soils
Olivine ((Mg, Fe)$_2$ SiO$_4$)	olive green	colorless	6.5	3.5	orthorhombic	conchoidal fracture	gemstones, refractory sand
Quartz (SiO$_2$)	colorless, various colors	colorless	7	2.6	hexagonal	conchoidal fracture	used in glass manufacture, electronic equipment, radios, computers, watches, gemstones
Topaz (Al$_2$SiO$_4$ (F, OH)$_2$)	white, pink, yellow, pale blue, colorless	colorless	8	3.5	orthorhombic	basal cleavage	valuable gemstone

Appendix I Rocks

Rock Types		
Rock Type	**Rock Name**	**Characteristics**
Igneous (intrusive)	Granite	Large mineral grains of quartz, feldspar, hornblende, and mica. Usually light in color.
	Diorite	Large mineral grains of feldspar, hornblende, and mica. Less quartz than granite. Intermediate in color.
	Gabbro	Large mineral grains of feldspar, hornblende, augite, olivine, and mica. No quartz. Dark in color.
Igneous (extrusive)	Rhyolite	Small or no visible grains of quartz, feldspar, hornblende,and mica. Light in color.
	Andesite	Small or no visible grains of quartz, feldspar, hornblende, and mica. Less quartz than rhyolite. Intermediate in color.
	Basalt	Small or no visible grains of feldspar, hornblende, augite, olivine, and mica. No quartz. Dark in color. Vessicles may be present.
	Obsidian	Glassy texture. No visible grains. Volcanic glass. Fracture is conchoidal. Color is usually black, but may be red-brown or black with white flecks.
	Pumice	Frothy texture. Floats. Usually light in color.
Sedimentary (clastic)	Conglomerate	Coarse-grained. Gravel- or pebble-sized grains.
	Sandstone	Sand-sized grains 1/16 to 2 mm in size. Varies in color.
	Siltstone	Grains are smaller than sand but larger than clay. Varies in color.
	Shale	Smallest grains. Usually dark in color.
Sedimentary (chemical or biochemical)	Limestone	Major mineral is calcite. Usually forms in oceans, lakes, rivers, and caves. Often contains fossils. Effervesces in dilute HCl.
	Coal	Occurs in swampy, low-lying areas. Compacted layers of organic material, mainly plant remains.
Sedimentary (chemical)	Rock salt	Commonly forms by the evaporation of seawater.
Metamorphic	Gneiss	Well-developed banding because of alternating layers of different minerals, usually of different colors. Common parent rock is granite.
	Schist	Well-developed parallel arrangement of flat, sheetlike minerals, mainly micas. Common parent rocks are shale and phyllite.
	Phyllite	Shiny or silky appearance. May look wrinkled. Common parent rocks are shale and slate.
	Slate	Harder, denser, and shinier than shale. Common parent rock is shale.
Metamorphic (nonfoliated)	Marble	Interlocking calcite or dolomite crystals. Common parent rock is limestone.
	Soapstone	Composed mainly of the mineral talc. Soft with a greasy feel.
	Quartzite	Hard and well cemented with interlocking quartz crystals. Common parent rock is sandstone.

Appendix J Solar System Charts

The Planets

	Mercury	Venus	Earth	Mars	Jupiter	Saturn	Uranus	Neptune	Pluto
Mass (kg)	3.302×10^{23}	4.8685×10^{24}	5.9736×10^{24}	6.4185×10^{23}	1.8986×10^{27}	5.6846×10^{26}	8.6832×10^{25}	1.0243×10^{26}	1.25×10^{22}
Equatorial radius (km)	2439.7	6051.8	6378.1	3397	71 492	60 268	25 559	24 764	1195
Mean density (kg/m^3)	5427	5243	5515	3933	1326	687	1270	1638	1750
Albedo	0.056	0.750	0.306	0.250	0.343	0.342	0.300	0.290	0.145
Semimajor axis (km)	5.791×10^{7}	1.0821×10^{8}	1.4960×10^{8}	2.2792×10^{8}	7.7857×10^{8}	1.43353×10^{9}	2.87246×10^{9}	4.49506×10^{9}	5.86966×10^{9}
Orbital period (Earth days)	87.969	224.701	365.256	686.980	4332.589	10 759.22	30 685.4	60 189	90 465
Orbital inclination (degrees)	7.00	3.39	0.00	1.850	1.304	2.485	0.772	1.769	17.16
Orbital eccentricity	0.2056	0.0067	0.0167	0.0935	0.0489	0.0565	0.0457	0.0113	0.2444
Rotational period (hours)	1407.6	5832.5[R]	23.9345	24.6229	9.9250	10.656	17.24[R]	16.11	153.2928[R]
Axial tilt (degrees)	0.01	177.36	23.45	25.19	3.13	26.73	97.77	28.32	122.53
Average surface temperature (K)	440	737	288	210	129	97	58	58	50
Number of known moons	0	0	1	2	61	31	25*	13	1

*Number as of November 2003.

[R] indicates retrograde rotation.

The Moon

Mass (kg)	7.349×10^{22}
Equatorial radius (km)	1737.4
Mean density (kg/m^3)	3340
Albedo	0.067
Semimajor axis (km)	3.844×10^{5}
Orbital period (Earth days)	27.3217
Lunar period (Earth days)	29.53
Orbital inclination (degrees)	5.145
Orbital eccentricity	0.0549
Rotational period (hours)	655.728

The Sun

Mass (kg)	1.99×10^{30}
Equatorial radius (km)	6.96×10^{5}
Mean density (kg/m^3)	1408
Absolute magnitude	+4.83
Luminosity (W)	384.6
Spectral type	G2
Rotational period (hours)	609.12
Average temperature (K)	5778

Appendix K Star Charts

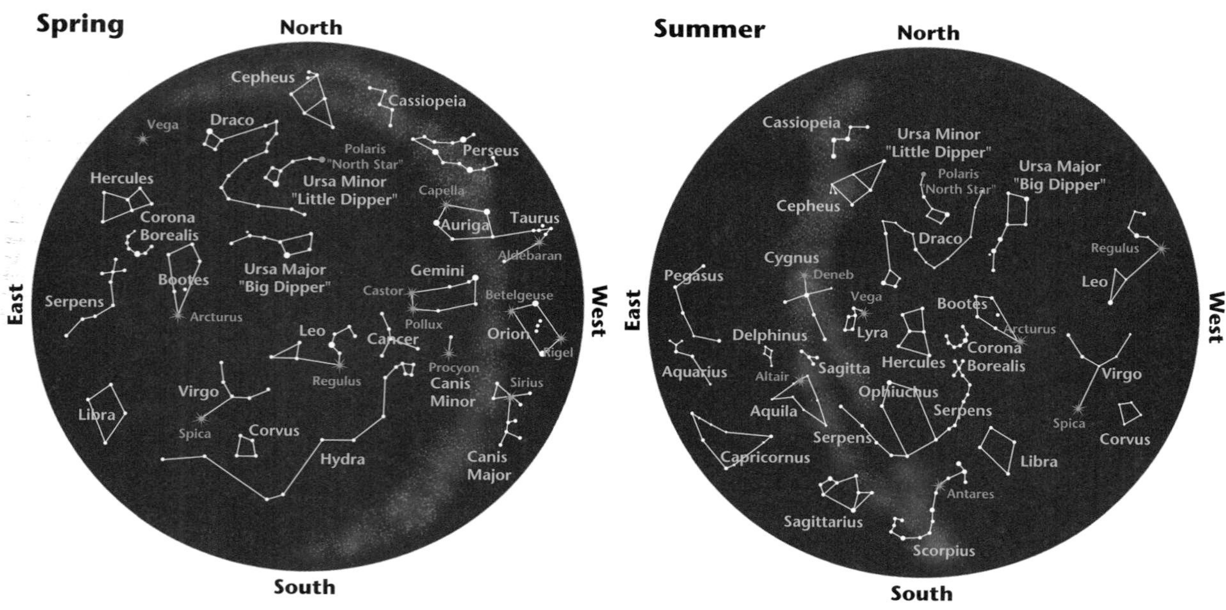

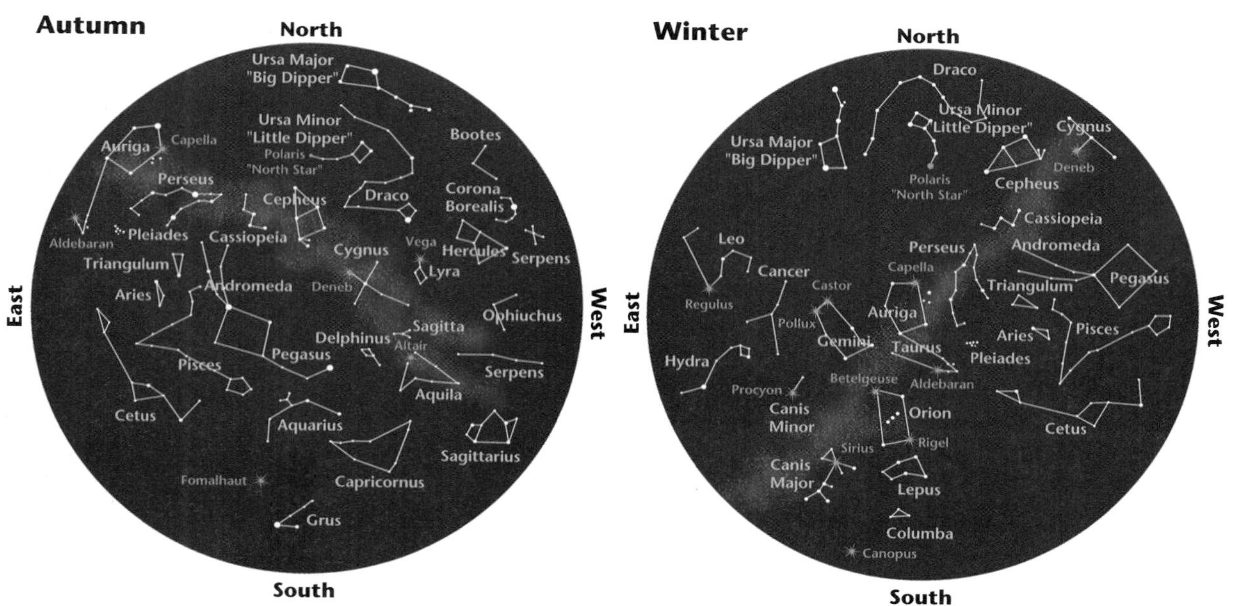

Skill Handbook

Contents

Thinking Critically

Practicing Scientific Methods

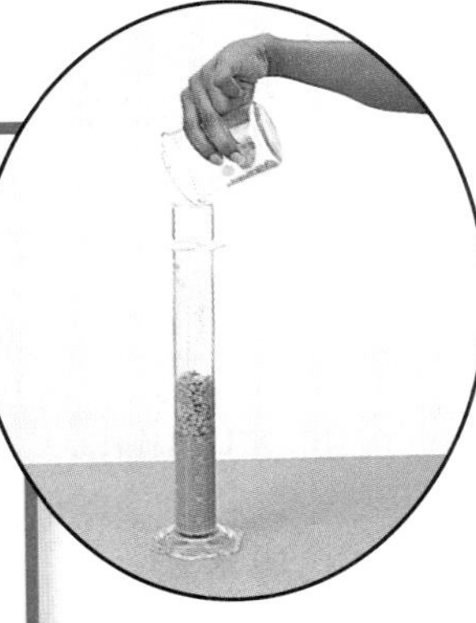

Organizing Information

Thinking Critically

Observing and Inferring

The process of science often begins with an observation. An observation can be as simple as noticing that the sky is blue, or that seawater is salty. Scientists try to make careful and accurate observations. Some observations describe an event or object using only words. Such observations are called qualitative observations. If you were making qualitative observations of clouds, for example, you might use words such as *puffy, curly, feathery, thin,* or *flat.*

Other observations describe how much of something there is. These are quantitative observations, in which numbers are used as well as words. Quantitative observations of a cloud might include that the cloud's base is at a height of 2000 m and that the cloud extends to a height of 6000 m. When it is possible, scientists use instruments, such as microscopes, thermometers, pan balances, and tape recorders, to make quantitative observations. Measurements provide numerical data, information that can be checked and reproduced. Data collection allows scientists to determine the how and why of their observations.

When you make observations in science, you will find it helpful to first examine an entire object or event. Use your senses of sight, touch, smell, and hearing to examine the object or event in detail. Write down everything that you observe. For example, suppose that you have been given a piece of a quartz crystal. What kinds of observations can you make about that crystal? You might note that it has a specific shape and size. You might also describe how the crystal feels and whether light can pass through it. All of these observations can help you to identify the crystal.

Scientists often use their observations to make inferences. An inference is an attempt to explain, interpret, or determine what caused an observation. For example, if you observed a fist-sized, heavy, round rock called a geode, you might infer that the rock is solid all the way through. However, the rock might be hollow inside, or it might contain beautiful mineral crystals. Or, the rock might be a conglomerate that has a smooth surface. The only way to be sure that your inference is correct is to investigate further. How could you investigate the composition of a rock further?

Before you make an inference, be certain that you have made accurate observations and recorded them carefully. Then, based on everything you know, try to explain or interpret what you have observed. For example, what can you infer from observing the outside of a piece of granite? If possible, investigate further to determine whether your inference is correct.

Thinking Critically

Comparing and Contrasting

You can analyze and then organize your observations by noting the similarities and differences between two or more objects or events. When you examine objects or events to determine their similarities, you are comparing them. When you examine similar objects or events to determine their differences, you are contrasting them.

Suppose you are asked to compare and contrast the minerals halite and quartz. You could begin by making observations of each type of mineral. Then, you could divide a piece of paper into two columns and list the ways in which the two minerals are similar in one column and the ways in which they are different in the other column. After completing your list, you could organize your findings in a data table or a graph.

Similarities you might point out are that both minerals are solids that occur as crystals, and both are inorganic compounds. Differences might include that halite has a cubic crystal structure, whereas quartz has a hexagonal crystal structure. You could further investigate these two minerals by testing for hardness, luster, color, streak, and cleavage or fracture.

Recognizing Cause and Effect

Have you ever seen something happen, and then tried to figure out why or how it happened? If so, you observed an event and inferred a reason for the event. The event or result of an action is an effect, and the reason for the event is the cause.

Suppose that you take a 2-L bottle of soda pop out of the refrigerator and accidentally drop it on the floor. You pick the bottle up, then unscrew the cap and take it off. Immediately, the soda pop fizzes up and spills out of the bottle all over the counter. What was the effect, and what would you infer was the cause? The effect was the soda pop fizzing up and spilling out of the open bottle.

Thinking Critically

You might infer that the cause was your dropping the soda-pop bottle on the floor. In determining the cause and effect, you made a logical inference based on your observations.

Perhaps the soda pop fizzed up and spilled out of the bottle because your little brother or sister shook the bottle before he or she placed it in the refrigerator, or perhaps the soda had been in the refrigerator only for a short time, and it was still warm. When scientists are unsure of the cause of a certain event, they often design controlled experiments to determine what caused the event that they observed. Although you may have made a sound judgment about the cause of the soda pop fizzing and spilling, you would have to perform an experiment to be certain that it was your dropping the bottle on the floor that caused the effect you observed.

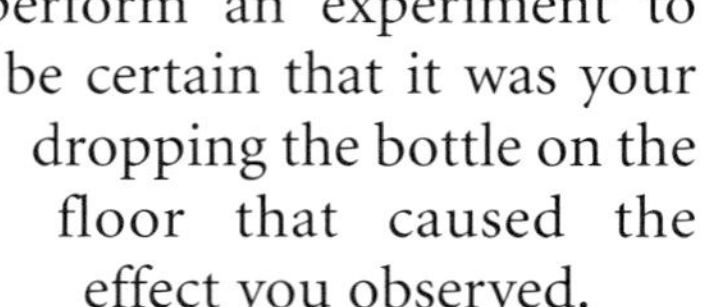

Interpreting Scientific Illustrations

Illustrations are included in your textbook to help you understand, interpret, and remember what you read. Whenever you encounter an illustration, examine it carefully and read the caption. The caption explains or identifies the illustration.

Scientific illustrations often are simplified to enable you to understand a certain portion of a much more complicated object or event. Often, illustrations are simple models of objects or events. Sometimes, scientific illustrations use enhanced or exaggerated figures to help you visualize objects or events.

Some illustrations are designed to show you things that you cannot observe easily on your own. Look at the illustration of a water molecule on the next page. A water molecule, H_2O, is made up of two hydrogen atoms and one oxygen atom. Although you can see water, you cannot see that it is made up of these two types of atoms. A scientific illustration such as the one shown allows you to learn about something that is too small to be seen with the unaided eye. Other illustrations help you to understand something that is too big to be seen easily. A map of the ocean basins, for example, helps you to see the features of the seafloor, something you could not observe yourself.

Some illustrations are designed to show how the internal parts of a structure are arranged. Look at the first illustration of a volcano. The volcano is represented as having been cut lengthwise to show a section that runs along the height of the volcano. An illustration

Thinking Critically

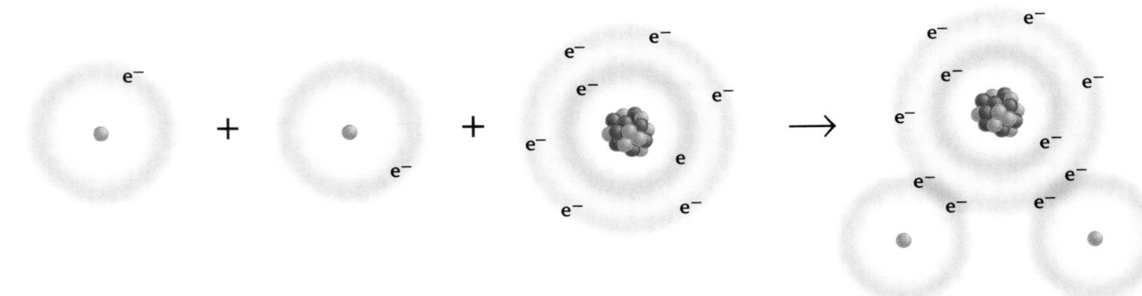

that is represented as having been cut along the length or height of a structure is called a longitudinal section. An illustration that is represented as having been cut crosswise at right angles to the length is a cross section. The second illustration is a cross section of a volcano.

Some scientific illustrations are a series of photographs or diagrams that describe processes. The formation of soil, for example, can best be explained in a series of diagrams that show how rock begins to fracture and break down, then weathers into smaller fragments, then is mixed with organic matter, and finally becomes fertile soil as organic matter decays. Many other processes, such as the formation of sedimentary rocks, also are best shown in a series of diagrams.

Scientific illustrations also can be used to show events that cannot be observed in a human lifetime. For example, illustrations of tectonic plate movements can show where Earth's landmasses were located millions of years ago. The formation of fossil fuels from organic matter, a process that takes millions of years, can also be described in illustrations.

Symmetry refers to a similarity or likeness of parts. Many organisms and objects have symmetry. When something can be divided into two similar parts lengthwise, it has bilateral symmetry. Look at the photograph of the

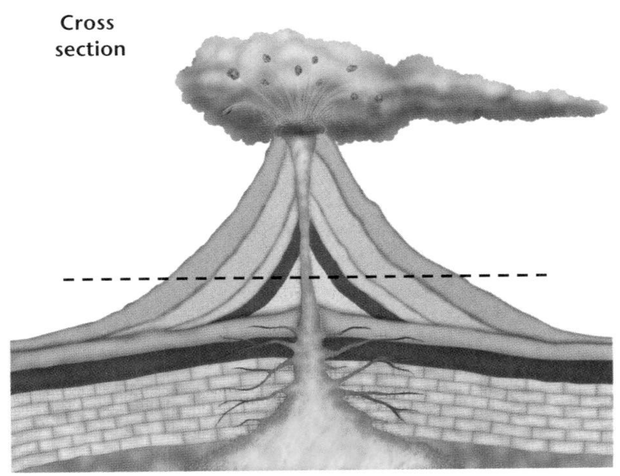

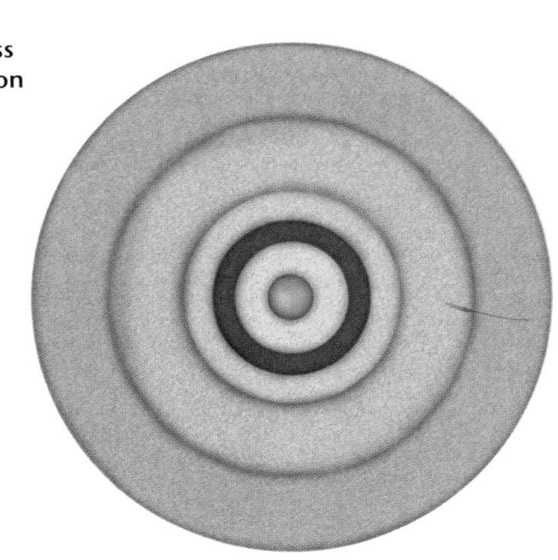

dragonfly. If you drew a line lengthwise through the dragonfly, you would see that the right side looks just the same as the left side. The dragonfly therefore has bilateral symmetry. Other organisms and objects have radial symmetry. Radial symmetry exists when similar parts are arranged around a central point in a circular pattern. The sand dollar in the photograph has radial symmetry because it can be divided anywhere through its center into equal parts. Some organisms and objects cannot be divided into two similar parts. When an organism or an object cannot be divided into two similar parts, it is asymmetrical. Study the amethyst crystal. No matter how you try to divide the crystal, you cannot draw a line that divides it into two equal parts.

Practicing Scientific Methods

The work of a scientist may be described as solving problems. Scientists generally use experiments to solve problems and answer questions. An experiment is a method in which a scientist uses an organized process to attempt to solve a problem or answer a question.

Experimentation involves defining a problem and then formulating and testing a hypothesis, which is a proposed solution to the problem. The proposed solution is tested during an experiment, which includes making careful observations and collecting data. After the collected data have been analyzed, a conclusion is formed and compared to the hypothesis.

Forming a Hypothesis

Suppose you want to earn a perfect score on an algebra test. You think of several possible ways to accomplish a perfect score. You base these possibilities on your own past experiences and on your observations of your friends' results. All of the following are hypotheses you might consider that could explain how it would be possible to score 100 percent on your test.

If the test is easy, then I will score 100 percent.

If I am intelligent, then I will score 100 percent.

If I study hard, then I will score 100 percent.

Practicing Scientific Methods

Scientists use hypotheses that they can test to explain the observations that they have made. Perhaps a scientist has observed that acid precipitation is damaging the statues in a pioneer cemetery, and also that plants growing nearby are not healthy. The scientist wonders how acid precipitation affects plants. The scientist may form a hypothesis that states: If plants are exposed to increasing levels of acid precipitation, they will become sick and eventually die.

Designing an Experiment

Once you have stated a hypothesis, you probably want to find out whether or not it explains an event or an observation. This requires a test. To be valid, a hypothesis must be testable by experimentation. How could you conduct an experiment to test the hypothesis about the effect of acid precipitation on plants?

First, you would obtain identical, young plants in small pots from a garden center. If you were a scientist conducting an experiment, you would use many plants in each experimental group; you would repeat your experiment many times to generate large amounts of data; and you would average your data at the end of all of your experiments. In the science classroom, however, time and resources are more limited, so it is likely that you will only be able to obtain the same number of plants as the number of acidity levels that you want to test, plus one more to serve as the control plant. Perhaps you can obtain six plants for your experiment. Label each plant A, B, C, and so on until each plant has a label. Make a data table that lists each plant by letter, and record the height and width of each plant. In your data table, draw a picture of each plant, noting any areas that are discolored or damaged. Then, collect small spray bottles and label them as well. Pour 200 mL of distilled water (pH 7, neutral) into spray bottle A. This bottle represents your control. In the spray bottle labeled B, pour 200 mL of white vinegar (pH 2, strong acid). This represents the worst case of acid precipitation. In the remaining spray bottles, add vinegar to distilled water to create acid solutions with varying pH values. You can determine the pH values with wide-range indicator paper.

Next, spray each plant, and the soil in each pot, with the corresponding solutions in each

Practicing Scientific Methods

spray bottle once a day for seven days. Each day, record how the plants look. Continue to make daily drawings of each plant and identify any discolored or damaged areas.

From the data you recorded, you will draw a conclusion and make a statement about your results. If your conclusion supports your hypothesis, then you can say that your hypothesis is reliable. If your conclusion does not support your hypothesis, then you will have to make new observations and state a new hypothesis, one that you can also test. Do the data you collected support the hypothesis that acid precipitation causes plants to sicken and die?

Separating and Controlling Variables

When scientists perform experiments, they must be careful to manipulate or change only one condition and keep all other conditions in the experiment the same. A condition that changes is called a *variable*. A condition that does not change is called a *constant*.

Independent Variable How does a scientist determine if an observation is the result of a certain change in condition? Scientists determine this by setting up an experiment in which only one condition changes, then observing the results. The condition that changes is called the independent variable because it is not influenced by the experimental procedure. The independent variable is the condition that a scientist manipulates in order to observe its effect. In the acid precipitation experiment, the independent variable is the acidity of the solutions sprayed on the plants and the soil.

Constant A constant is a condition that does not change during the course of an experiment. In the acid precipitation experiment, the constants are using plants of the same species, size, and age; using the same amounts of solutions to spray the plants each day; and keeping all the plants in the same conditions of sunlight and temperature at all times during the experiment. For scientists to determine that only the independent variable in an experiment caused any observed changes, they must keep all other factors the same.

Dependent Variable What are these changes observed in an experiment? They are the dependent variables. They are called dependent variables because changes in them depend upon the values of the independent variables. The dependent variables are any changes that result from manipulating the independent variables. In the case of the acid precipitation experiment, the dependent variables are the changes in the health of the plants over seven days as measured by their growth.

Control Scientists also use a control to be certain that the observed changes resulted only from the manipulation of one independent variable. A control is a sample that is treated exactly like each experimental group except that the independent variable is not applied to the control. Controls allow scientists to observe the effect of the independent variable because the assumption is made that whatever changes occur in the control group are not due to the independent variable. The control group provides a means of comparison. The control in the acid precipitation experiment is distilled water, which was sprayed on plant A each day for the same number of times as the other plants in the experiment were sprayed with other solutions. After the experiment, if there has been any change in the dependent variable of the control sample, such as height of the plant, it can be compared to changes in the experimental group. For example, if the control plants had grown several centimeters in seven days, but the other plants had not, you could assume that the lack of growth may be associated with the acidity of the solutions in each spray bottle.

Why is it important to find out what levels of acid precipitation plants can and cannot tolerate? Normal precipitation is slightly acidic, at pH 5.6, and all plants can tolerate normal precipitation. But the use of fossil fuels, such as gasoline in cars and coal in power plants, results in increasing levels of acid precipitation. Knowing how much acidity plants can tolerate can help us determine how much pollution should be allowed.

Measuring in SI

The metric system is a uniform system of measurement developed in 1795 by a group of scientists. The development of the metric system helped scientists worldwide avoid problems resulting from the use of different units of measurement by providing an international standard of comparison for measurements. A modern form of the metric system called the International System, or SI, was adopted for worldwide use in 1960. You will find that this text uses metric units, with a few exceptions.

Base Metric Units The metric system is easy to use because it has a systematic method of naming units and a decimal base. For example, the meter is the base unit for measuring length, the gram is the base unit for measuring mass, and the liter is the base unit for measuring volume. Unit sizes vary by multiples of 10. When converting from smaller units to larger ones, you divide by a multiple of 10. When converting from larger units to smaller ones, you multiply by a multiple of 10. Prefixes are used to name larger and smaller units. The following table provides some common metric prefixes and their meanings.

Metric Prefixes

Prefix	Symbol	Meaning
kilo-	k	1000 (thousand)
hecto-	h	100 (hundred)
deka-	da	10 (ten)
deci-	d	0.1 (tenth)
centi-	c	0.01 (hundredth)
milli-	m	0.001 (thousandth)

Do you see how the prefix *kilo-* attached to the unit *gram* is *kilogram,* meaning 1000 grams? The prefix *deci-* attached to the unit *meter* is *decimeter,* that is, one-tenth (0.1) of a meter. These prefixes can be attached to any unit of measure in the metric system to form another unit that is a multiple of 10.

Units of Length The meter is the SI unit used to measure length. To visualize the length of a meter, think of a baseball bat, which is about 1 m long. For the measurement of smaller distances, the meter is divided into smaller units called centimeters and millimeters.

Practicing Scientific Methods

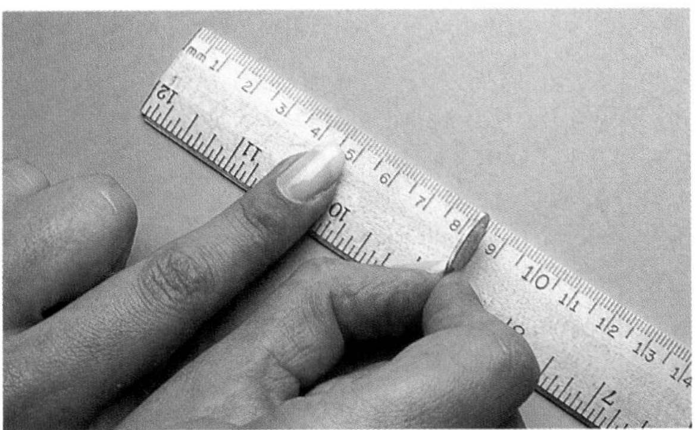

A centimeter is one-hundredth (0.01) of a meter, which is about the width of the fingernail on your index finger. A millimeter is one-thousandth (0.001) of a meter, about the thickness of a dime. The photograph shows this comparison.

Most metric rulers have lines indicating centimeters and millimeters. The centimeter lines are the longer, numbered lines, and the millimeter lines are shorter lines between the centimeter lines. As you might guess, there are 10 millimeters to a centimeter.

Units of length also are used to measure surface area. The standard unit of area is the square meter, m^2, which is a square 1 m long on each side. Similarly, a square centimeter, cm^2, is a square 1 cm long on each side. To find surface area, you multiply the number of units in length times the number of units in width. If you have a surface that is 4 cm long and 5 cm wide, the surface area is 4 cm $\times$ 5 cm = 20 cm^2. Suppose you want to find the surface area of a cubic crystal that is 3 cm long and 3 cm wide. The surface area of a side would be found by multiplying 3 cm $\times$ 3 cm. But because a cube has six sides, you would multiply the surface area of one side by 6 to find the surface area of the cube. The surface area of this cube would be $3 \times 3 \times 6 = 54$ cm^2.

Units of Volume The cubic meter (m^3) is the standard SI unit of volume. A cubic meter is a cube 1 m long on each side. You can determine the volume of rectangular solids by multiplying length by width by height.

Liquid volume is measured using a unit called a liter. A liter has a volume of 1000 cubic centimeters. Recall that there are 1000 milliliters in a liter. Thus, a milliliter equals one cubic centimeter. One milliliter of liquid would completely fill a cube measuring 1 cm on each side.

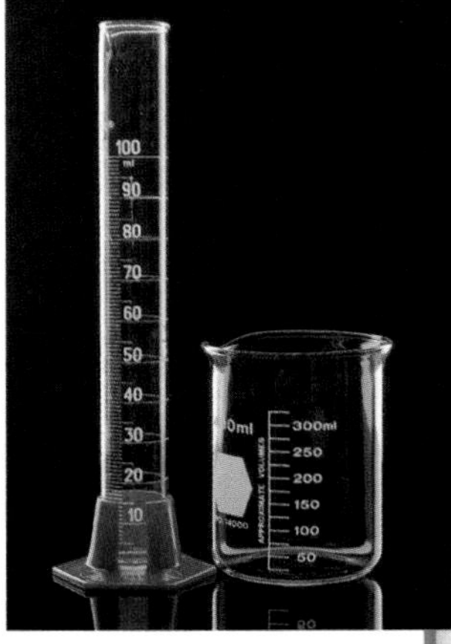

You will measure liquids using beakers and graduated cylinders marked in milliliters. A graduated cylinder is marked with lines from bottom to top. Each graduation represents one milliliter.

Units of Mass You will likely use a beam balance when you want to find the masses of objects. Often, you will measure mass in grams. On one side of the beam balance is a pan, and on the other side is a set of beams. Each beam has an object of a known mass called a rider that slides on the beam.

Before you find the mass of an object, you must set the balance to zero by sliding all the riders back to the zero point. Check the

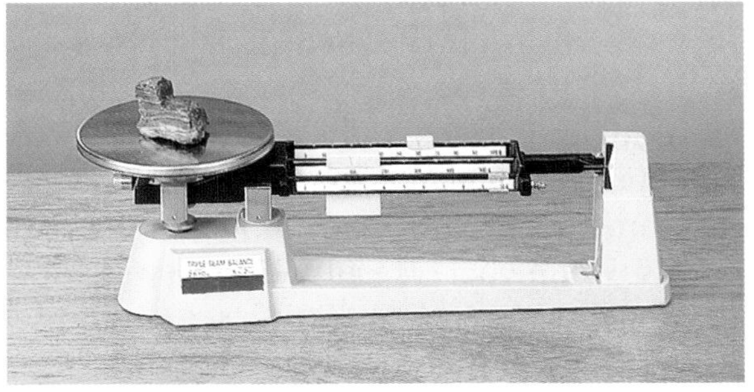

pointer to make sure it swings an equal distance above and below the zero point on the scale. If the swing is unequal, find and turn the adjusting screw until the swing is equal. Then wait until the pointer comes to rest exactly at the zero point.

You can use the balance to find the mass of an object. Place an object on the pan. Slide the rider with the largest mass along the beam until the pointer drops below the zero point. Then move the rider back one notch. Repeat the process for each beam until the pointer comes to a stop on the zero point. Read the masses indicated on the beams. The sum of the masses is the mass of the object.

Never place a hot object or pour chemicals directly on the pan of the balance. Instead, find the mass of a clean container, such as a beaker or glass jar. Place the object or chemical that you want to measure in the container. Next, find the combined mass of the container and the object or chemical. Calculate the mass of the object or chemical by subtracting the mass of the empty container from the combined mass.

Measuring Temperature Most measurements of temperature in SI are made using the Celsius scale. You can easily convert temperatures from the Fahrenheit scale to Celsius using the following formulas.

Conversion of Fahrenheit to Celsius

$$°C = 5/9\ (°F - 32)$$

Conversion of Celsius to Fahrenheit

$$°F = (9/5 + °C) + 32$$

Organizing Information

CLASSIFYING

Classifying is grouping objects or events based on common features. When you classify, you first make careful observations of the group of items to be classified. Then, you select one feature that is shared by some items in the group but not by others. Place the items that share this feature in a subgroup. Ideally, the items in the second subgroup will have some feature in common with one another. After you decide on the first feature that separates the items into subgroups, examine the items in each subgroup for other features that some items possess but others don't. In this way, form further subgroups until the items no longer can be distinguished enough to identify them as belonging to another distinct subgroup.

How would you classify a collection of minerals? Classify the collection first based on observable features. You might begin classifying minerals by dividing them into two groups: one for minerals with metallic luster, and one for minerals with nonmetallic luster. Within the first group, you could classify minerals by the type of crystal system, by the breakage patterns, and by color. Note that

Organizing Information

Classifying Minerals with Metallic Luster

Mineral	Color	Crystal System	Breakage Pattern
Copper	copper red	cubic	hackly
Graphite	black to gray	hexagonal	basal cleavage
Hematite	black or reddish brown	hexagonal	irregular fracture
Pyrite	light, brassy yellow	cubic	uneven fracture
Silver	silvery white	cubic	hackly

each mineral can only belong to one group under each type of classification. For example, copper has a cubic crystal system, has a hackly breakage pattern, and is copper red. Continue to select features until all the members of the collection have been classified. The table shows one method of classification.

Remember, when you classify, you are grouping objects or events for a purpose. The purpose could be general, such as how each rock or mineral formed. The classification of your collection might be different, however, if you are interested in the economic value of each specimen or how suitable each would be for use in a piece of jewelry.

Sequencing

A sequence is an arrangement of things or events in a particular order. A common sequence with which you may be familiar is the order of steps you must follow to make an omelette. Certain steps of preparation have to be followed for the omelette to be cooked properly.

When you are asked to sequence things or events, you must identify what comes first. You then decide what should come second. Continue to choose things or events until they are all in order. Then go back over the sequence to make sure that each thing or event logically leads to the next.

Suppose you wanted to watch a movie that just came out on videotape. What sequence of events would you have to follow to watch the movie? You would first turn the television set to the channel that receives images from the videotape player. Then you would turn on the videotape player and insert the videotape. Once the tape began playing, you might adjust the sound and picture. When the movie was over, you would rewind the tape and return it to the video store. What would happen if you did things out of order, such as adjusting the sound before you put in the tape?

Concept Mapping

If you were taking an automobile trip, you would probably take along a road map. The road map would show your location, your destination, and other places along the way. By examining the map, you would understand where you were in relation to other locations on the map.

A concept map is similar to a road map, except that a concept map shows the relationships among ideas or concepts rather than places. A concept map is a diagram that visually shows how concepts are related. Because a concept map shows the relationships among ideas or concepts, it can clarify their meanings and help you to more clearly understand what you are studying. There is usually not a single correct way to construct a concept map. As you are constructing a concept map, you might discover that an arrangement that is different from the one you began with shows the

Organizing Information

relationships more accurately. There are three common types of concept maps: network trees, events chains, and cycle concept maps.

Network Trees Note that some words in the concept map below are circled. The circled words are science concepts. The lines in the map show related concepts and the words written on the lines describe relationships between the concepts. This map is an example of a network-tree concept map showing the relationships among various types of rocks.

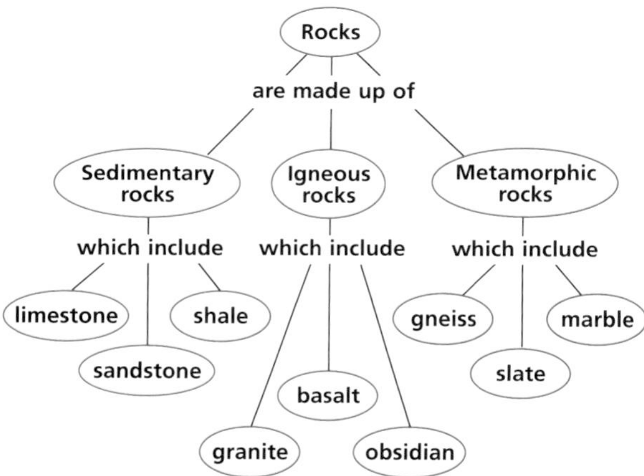

When you begin to construct a network tree, first state the topic and identify the major concepts. Then, identify related concepts and put them in order from most general to most specific. Branch the related concepts from the major concept, and describe each relationship on the lines. Continue in this way with the more specific concepts. Write the relationships between the concepts on the lines until all the concepts are mapped. Examine the concept map for relationships that cross branches, and add them to the concept map as well.

Events Chains An events-chain concept map is used to describe ideas in order. In science, an events-chain concept map can be used to describe a sequence of events, the steps in a procedure, or the stages of a process.

When making an events-chain concept map, you first must identify the one event that starts the chain. This event is called the initiating event. You then identify the next event in the chain, and continue until you reach an outcome. Suppose you wanted to make a concept map of the steps involved in making a fossil. An events-chain map might look like the one shown here. Note that connecting words may not be necessary in this kind of concept map.

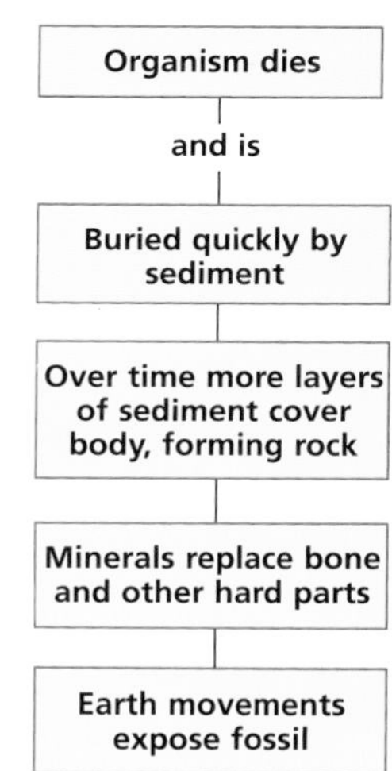

Cycle Concept Maps A cycle concept map is a special type of events-chain map. In a cycle concept map, the series of events does not produce a final outcome. The last event in the chain relates back to the initiating event. Because there is no outcome and the last event relates back to the initiating event, the cycle repeats itself. Follow the stages shown in the concept map of the water cycle.

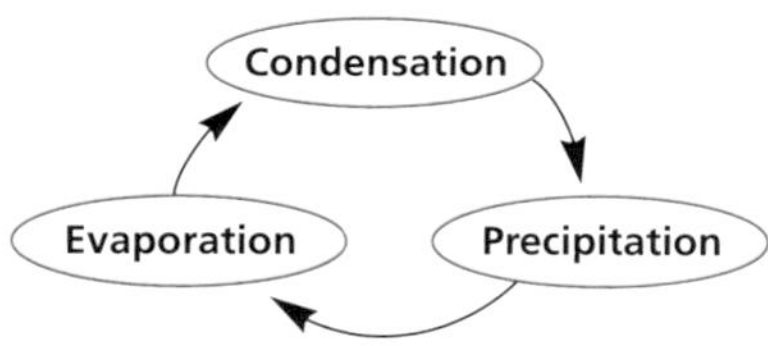

Organizing Information

Making and Using Tables

Browse through your textbook, and you will notice many data tables both in the text and in the labs. The tables in the text arrange information in such a way that it is easier for you to understand. Also, many labs in your text have tables for you to complete as you do the labs. Lab tables help you organize the data that you collect during a lab so that they can be interpreted more easily.

A data table has a title that describes what is being presented. The table itself is divided into columns and rows. The column headings list the items that are being compared. The row headings list the specific characteristics that are being compared among those items. Within the grid of the table, the collected data are recorded.

Making and Using Graphs

After scientists organize data in tables, they often manipulate and then display the data in graphs. A graph is a diagram that shows a comparison between variables. Because graphs show pictures of collected data, they make interpretation and analysis of the data easier. The three basic types of graphs are line graphs, bar graphs, and circle graphs.

Line Graphs A line graph is used to show the relationship between two variables. The variables being compared go on two axes of the graph. The independent variable always goes on the horizontal axis, called the *x*-axis. The independent variable is the one being manipulated. The dependent variable always goes on the vertical axis, the *y*-axis. The dependent variable is any change that results from manipulating the independent variable.

Suppose your class wanted to collect data about humidity. You could make a graph of the amount of water vapor that air can hold at various temperatures. The temperature of the air is the independent variable and therefore should be plotted on the *x*-axis of your graph. The amount of water vapor found per cubic meter of air is the dependent variable and thus it is plotted on the *y*-axis.

You can use unlined paper or graph paper to construct a graph. After drawing your axes, label each axis with a scale. The *x*-axis simply lists the temperatures for which data were collected. To construct the *y*-axis, use the data in the data table below. Because the lowest amount of water vapor in air is 5 g/m^3 of air, you know that you will have to start numbering from at least 5 and continue to number through 80. You might decide to start numbering at 0 and number by 5s spaced at equal distances through 90.

Amount of Water Vapor in Air at Various Temperatures

Air Temperature (°C)	grams/cubic meter of air
0	5
5	8
10	10
15	13
20	18
25	23
30	31
35	39
40	50
45	66
50	80

Organizing Information

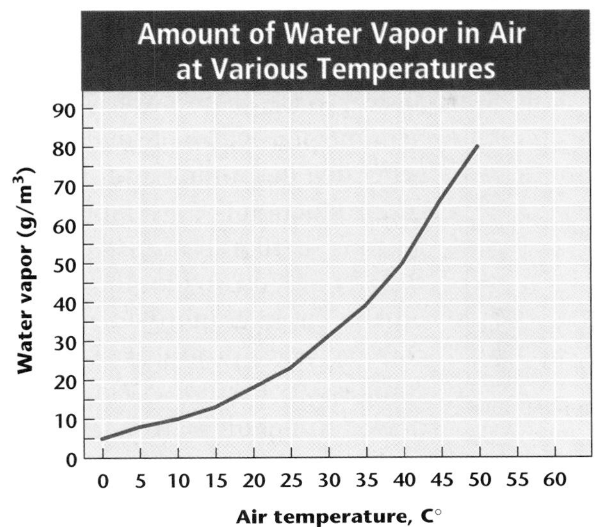

Next you plot the data points. The first pair of data that you want to plot is the temperature 0°C and the amount 5 g of water vapor in one cubic meter of air. Locate 0°C on the *x*-axis and 5 on the *y*-axis. Where an imaginary vertical line from each axis would meet, place the first data point. Place the other data points in the same way. You can then see if the data have any direction. That is, you can see if water vapor increases or decreases with air temperature. According to the graph, does the amount of water vapor in air increase or decrease with air temperature? Scientists often draw a line that appears to connect the data points on a line graph. Actually, scientists don't draw a line that connects individual data points. Instead, they draw a line that best fits the direction that the data points indicate. Because no one can ever collect enough data to plot every single point on a line graph, lines drawn in a graph come close to clusters of data points and show the general direction of the data points. Although the line graphs used in this book appear to have straight lines drawn between data points, remember that these are not really continuous lines but rather the best fit for the data points that have been plotted. For the purposes of this book, you may connect the data points plotted on your graph with a smooth line.

What if you wanted to compare the data about humidity collected by your class with similar data collected a year ago by a different class? The data of the other class can be plotted on the same graph to make the comparison. In this case, you would include a key indicating that the two different sets of symbols each indicate a different set of data.

Bar Graphs Bar graphs are similar to line graphs, except that they are used to show comparisons between different sets of data or to display data that are not continuous. In a bar graph, bars rather than data points show the relationships among data.

To make a bar graph, set up the *x*-axis and the *y*-axis as you would for a line graph. The data are plotted by drawing bars from the *x*-axis up to an imaginary point where the *y*-axis would intersect the bar if it were extended.

Look at the bar graph on the next page comparing the net energy efficiency of various ways to heat an enclosed space, such as a house. The independent variable is the way to heat an enclosed space, and the dependent variable is the net energy efficiency of each. The net energy efficiencies of various ways to heat an enclosed space are being compared. Can you see that the heating methods being compared do not have any influence on each

Organizing Information

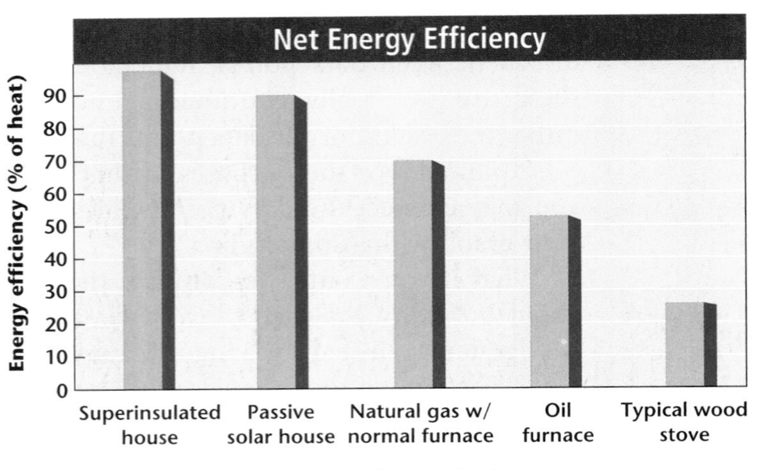

other? That is, the bar graph is illustrating data that are not continuous.

Circle Graphs A circle graph, often called a pie graph, uses a circle divided into sections to display data. Each section represents a part of the whole. When all the sections are added together, they equal 100 percent of the whole.

Suppose you wanted to make a circle graph that shows the percentage of solid waste generated by various industries in the United States each year. The total amount of solid waste generated each year is estimated at 10 billion metric tons. The whole circle graph will therefore represent this amount of solid waste.

Mining and oil and gas production are responsible for approximately 7.5 billion metric tons of solid waste each year. To determine what percentage of the circle graph this represents, divide the amount of solid waste generated by mining and oil and gas production by the total amount of solid waste generated, 10 billion metric tons. To find the number of degrees on a circle this represents, multiply the answer by 360, the number of degrees in a circle. Round the answer to the nearest whole number. The percentage of solid waste generated by mining and oil and gas production is: 7.5 billion metric tons $\div$ 10 billion metric tons $\times$ 360° = 270°. To plot these data on the circle graph, you need a compass and a protractor. Use the compass to draw a circle. Then draw a straight line from the center to the edge of the circle. Place your protractor on this line and use it to mark a point on the edge of the circle at 270°. Connect this point with a straight line to the center of the circle. This is the section that represents the percent of solid waste generated by mining and oil and gas production each year. Now, try to perform the same operation for the other data to find the number of degrees of the circle that each represents, and draw them in as well: agriculture, 1.3 billion metric tons; industry, 0.95 billion metric tons; municipal, 0.15 billion metric tons; and sewage sludge, 0.1 billion metric tons. Complete your graph by labeling the sections of the graph and giving the graph a title.

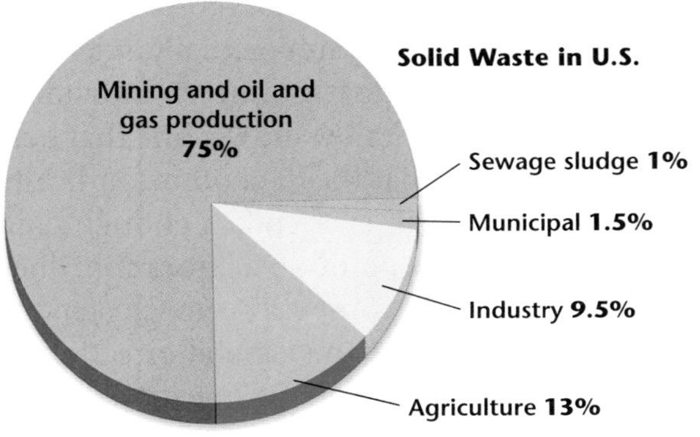

Glossary/ Glosario

A multilingual science glossary at **earthgeu.com** includes Arabic, Bengali, Chinese, English, Haitian Creole, Hmong, Korean, Portuguese, Russian, Spanish, Tagalog, Urdu, and Vietnamese.

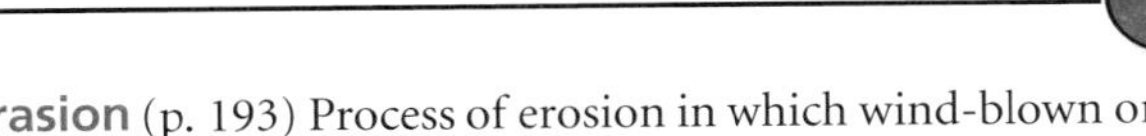

A

abrasion (p. 193) Process of erosion in which wind-blown or waterborne particles, such as sand, scrape against rock surfaces or other materials and wear them away.

absolute magnitude (p. 816) Brightness an object would have if it were placed at a distance of 10 pc; classification system for stellar brightness that can be calculated only when the actual distance to a star is known.

abyssal plain (p. 426) Smooth, flat part of the seafloor covered with muddy sediments and sedimentary rocks that extends seaward from the continental margin.

Acadian Orogeny (p. 608) Tectonic event in which Avalonia collided with the southeastern margin of Laurasia and added faults, folds, and igneous intrusions to the rocks of the Taconic Orogeny.

acid (p. 65) Solution that contains hydrogen ions (H^+).

acid precipitation (p. 727) Any precipitation with a pH of less than 5.0 that forms when sulfur dioxide and nitrogen oxides combine with moisture in the atmosphere to produce sulfuric acid and nitric acid.

active galactic nucleus (AGN) (p. 844) A galaxy's core in which highly energetic objects or activities are located.

aggregate (p. 660) Naturally accumulating mixture of sand, gravel, and crushed stone found in floodplains, alluvial fans, or glacial deposits.

air mass (p. 301) Large body of air that takes on the characteristics of the area over which it forms; can be described by its stability, temperature, and humidity.

air mass modification (p. 304) Exchange of heat or moisture with the surface over which an air mass moves.

air-mass thunderstorm (p. 331) Type of thunderstorm in which air rises because of unequal heating of Earth's surface within a single air mass and is most common during the afternoon and evening.

abrasion/abrasión (pág. 193) Proceso erosivo en que las partículas arrastradas por el viento o en el agua, como la arena, raspan contra las superficies de piedra las rocas u otros materiales y los desgastan.

absolute magnitude/magnitud absoluta (pág. 816) Brillo que tendría un objeto si estuviera a una distancia de 10 pc; sistema de clasificación para el brillo estelar que puede calcularse sólo cuando se sabe la distancia verdadera hasta una estrella.

abyssal plain/llanura abisal (pág. 426) Parte plana y nivelada del fondo del mar cubierto con sedimentos fangosos y piedras sedimentarias que se extiende desde el margen continental hacia el mar.

Acadian Orogeny/orogenia arcadiana (pág. 608) Acontecimiento tectónico en que Avalonia chocó contra el margen del sudeste de Laurasia y agregó fallas, dobleces e intrusiones ígneas a las rocas de la orogenia Tectónica.

acid/ácido (pág. 65) Solución que contiene iones hidrógeno (H^+).

acid precipitation/precipitación ácida (pág. 727) Cualquier precipitación con un pH de menos de 5.0, que se forma cuando se combinan el dióxido de azufre y óxidos los de nitrógeno con la humedad en la atmósfera para producir ácido sulfúrico y ácido nítrico.

active galactic nucleus (AGN)/núcleo galáctico activo (NGA) (pág. 844) centro de la galaxia en que se localizan cuerpos u ocurren actividades sumamente energéticos.

aggregate/agregado (pág. 660) Mezcla de arena, grava y piedra molida acumulada naturalmente, que se encuentra en llanuras aluviales, abanicos aluviales o depósitos glaciales.

air mass/masa de aire (pág. 301) Cantidad enorme de aire que toma las características del área sobre la cual se forma; puede describirse según su estabilidad, su temperatura y su humedad.

air mass modification/modificación de masa de aire (pág. 304) Intercambio de calor o humedad con la superficie sobre la que se mueve una masa de aire.

air-mass thunderstorm/tormenta eléctrica de masa de aire (pág. 331) Tipo de tormenta en que el aire se eleva debido al calentamiento desigual de la superficie terrestre dentro de una masa de aire y el cual es muy común durante la tarde y el crepúsculo.

air pollution (p. 665) Occurs when air contains harmful levels of pollutants; can be caused by natural phenomena, such as forest fires or volcanic eruptions, or by human activities, such as burning of fossil fuels.

air pollution/contaminación del aire (pág. 665) Ocurre cuando el aire contiene niveles perjudiciales de contaminantes; puede ser causada por fenómenos naturales, como los incendios forestales o erupciones las volcánicas, o por actividades humanas, como la combustión de combustibles fósiles.

albedo (p. 754) Amount of sunlight that is reflected by the surface of a planet or a satellite, such as the Moon.

albedo/albedo (pág. 754) Cantidad de luz solar que refleja la superficie de un planeta o un satélite, como por ejemplo, la Luna.

Alleghenian Orogeny (p. 615) Mountain-building event that occurred when the part of Gondwana that is now Africa began to collide with Laurasia during the Late Pennsylvanian.

Alleghenian Orogeny/orogenia alegeniana (pág. 615) Acontecimiento formador de montañas que ocurrió cuando la parte de Gondwana que es ahora África empezó a chocar contra Laurasia durante el Pennsylvaniano tardío.

altered hard parts (p. 567) Fossils whose organic material has been removed and whose hard parts have been changed by recrystallization or mineral replacement.

altered hard parts/partes duras alteradas (pág. 567) Fósiles cuya materia orgánica ha desaparecido y cuyas partes duras han sido transformadas por recristalización o por reemplazo mineral.

amino acids (p. 590) Building blocks of proteins that were likely abundant on Earth during the Archean.

amino acids/aminoácidos (pág. 590) Componentes de las proteínas que probablemente eran abundantes en la Tierra durante el Arqueano.

amniote egg (p. 616) Egg with a shell, providing a complete environment to a developing embryo, that allowed reptiles to colonize land during the Late Mississippian.

amniote egg/huevo amniótico (pág. 616) Huevo con cascarón que provee un ambiente completo para un embrión en desarrollo y el cual permitió que los reptiles colonizaran la tierra firme durante el Mississippiano Tardío.

analog forecast (p. 320) Weather forecast that compares current weather patterns to patterns that occurred in the past.

analog forecast/pronóstico análogo (pág. 320) Pronóstico del tiempo que compara las pautas actuales del clima con pautas que ocurrieron en el pasado.

Ancestral Rockies (p. 614) Mountain range that formed from the collision between Gondwana and Laurasia during the Late Paleozoic.

Ancestral Rockies/Rocosas antiguas (pág. 614) Cordillera de montañas que se formó debido a la colisión entre Gondwana y Laurasia durante el Paleozoico Tardío.

anemometer (p. 313) Weather instrument used to measure wind speed.

anemometer/anemómetro (pág. 313) Instrumento meteorológico que se utiliza para medir la velocidad de viento.

angiosperms (p. 630) Seed-bearing, flowering plants that evolved during the Cretaceous.

angiosperms/angiospermas (pág. 630) Plantas con flores, portadoras de semillas que evolucionaron durante el Cretáceo.

Antler Orogeny (p. 608) Collisional tectonic event that impacted the passive western margin of Laurentia.

Antler Orogeny/orogenia Antler (pág. 608) Evento de colisión tectónica que impactó el margen occidental pasivo de Laurentia.

aphelion (p. 777) Point in a planet's orbit where it is farthest from the Sun.

aphelion/afelio (pág. 777) Punto en la órbita de un planeta cuando se encuentra más alejado del Sol.

apogee (p. 766) Farthest point in the Moon's elliptical orbit to Earth.

apogee/apogeo (pág. 766) Punto de la órbita elíptica de la Luna en que ésta se encuentra más alejada de la Tierra.

apparent magnitude (p. 816) Classification system based on how bright a star appears to be; does not take distance into account so cannot indicate how bright a star actually is.

apparent magnitude/magnitud aparente (pág. 816) Sistema de clasificación que se basa en el grado de brillo aparente de una estrella; no toma en cuenta la distancia y por lo tanto no puede indicar el brillo real de la estrella.

aquifer (p. 243) Permeable underground layer through which groundwater flows relatively easily.

aquifer/acuífero (pág. 243) Capa subterránea permeable por la cual el agua subterránea fluye de manera relativamente fácil.

artesian well (p. 253) Fountain of water that spurts above the land surface when a well taps a deep, confined aquifer containing water under pressure.

asteroid (p. 578) Metallic or silica-rich object, 1 km to 950 km in diameter, that bombarded early Earth, generating heat energy; (p. 795) rocky remnant of the early solar system found mostly between the orbits of Mars and Jupiter in the asteroid belt.

asthenosphere (p. 8) Partially molten, plasticlike, flowing layer located below the solid part of Earth's mantle.

astronomical unit (AU) (p. 777) Average distance between Earth and the Sun; approximately 150 million km.

astronomy (p. 6) Study of objects beyond Earth's atmosphere.

atmosphere (p. 9) Blanket of gases surrounding Earth that contains about 78 percent nitrogen, 21 percent oxygen, and 1 percent other gases such as argon, carbon dioxide, and water vapor.

atom (p. 54) Smallest particle of an element, having all the characteristics of that element; the basic building block of matter, consisting of protons, neutrons, and electrons.

atomic mass (p. 58) Average of the mass numbers of the isotopes of an element.

atomic number (p. 54) Number of protons contained in an atom's nucleus.

autumnal equinox (p. 761) Occurs when the Sun is directly overhead at the equator and results in day and night of equal length for both northern and southern hemispheres.

avalanche (p. 187) Landslide that occurs in a mountainous area when snow falls on an icy crust, becomes heavy, slips off, and slides swiftly down a mountainside.

artesian well/pozo artesiano (pág. 253) Fuente de agua que brota hacia la superficie terrestre cuando un pozo conecta con un acuífero profundo y confinado con agua bajo presión.

asteroid/asteroide (pág. 578) Cuerpo metálico o rico en sílice, de 1 a 950 km de diámetro, que bombardeó la Tierra primitiva generando energía térmica; (pág. 795) resto rocoso del sistema solar primitivo que se encuentra principalmente entre las órbitas de Marte y Júpiter en el cinturón de asteroides.

asthenosphere/astenosfera (pág. 8) Capa fluida, parcialmente fundida, tipo plástico, localizada debajo de la parte sólida del manto de la Tierra.

astronomical unit (AU)/unidad astronómica (UA) (pág. 777) La distancia promedio del sol y Tierra; aproximadamente 150 milliones de km.

astronomy/astronomía (pág. 6) Estudio de los cuerpos que se encuentran más allá de la atmósfera de la Tierra.

atmosphere/atmósfera (pág. 9) Manto de gases que rodea la Tierra y que contiene aproximadamente un 78 por ciento de nitrógeno, un 21 por ciento de oxígeno y 1 por ciento de otros gases como el argón, el dióxido de carbono y el vapor del agua.

atom/átomo (pág. 54) La partícula más pequeña de un elemento, la cual posee todas las características de ese elemento; el componente básico de la materia que consiste en protones, neutrones y electrones.

atomic mass/masa atómica (pág. 58) Promedio de los números de masa de los isótopos de un elemento.

atomic number/número atómico (pág. 54) Número de protones que contiene el núcleo de un átomo.

autumnal equinox/equinoccio otoñal (pág. 761) Ocurre cuando el Sol está directamente por encima del ecuador y resulta en que el día y la noche son de igual duración, tanto para el hemisferio norte como para el hemisferio sur.

avalanche/avalancha (pág. 187) Deslizamiento de tierra que ocurre en un área montañosa cuando la nieve cae sobre la corteza helada, se vuelve pesada, se desprende y se resbala rápidamente montaña abajo.

B

banded iron formations (p. 586) Unique, shallow marine deposits made up of alternating bands of chert and iron oxides that formed due to locally high oxygen levels produced by stromatolites.

barometer (p. 312) Weather instrument used to measure air pressure.

banded iron formations/formaciones ferrosas bandeadas (pág. 586) Depósitos marinos, superficiales y poco comunes compuestos de bandas alternadas de óxidos ferrosos y cuarzo criptocristalino que se formaron debido a la cercanía de los altos niveles de oxígeno producidos por estromatolitos.

barometer/barómetro (pág. 312) Instrumento meteorológico que se utiliza para medir la presión atmosférica.

barrier island (p. 418) Long ridge of sand or other sediment deposited or shaped by longshore currents that is separated from the mainland and can be up to tens of kilometers long.

base (p. 66) Solution that contains hydroxide ions (OH^-).

Basin and Range Province (p. 637) Area in the southwestern United States and North-Central Mexico where extensional tectonism resulted in the formation of mountain ranges separated by long valleys.

batholith (p. 476) Coarse-grained, irregularly shaped, igneous rock mass that covers at least 100 km^2, generally forms 10-30 km below Earth's surface, and is common in the interior of major mountain chains.

beach (p. 415) Sloping band of loose sediments, such as sand, pebbles, or mud, deposited along a shoreline.

bedding (p. 126) Horizontal layering in sedimentary rock that can range from a millimeter thick to several meters thick.

bed load (p. 217) Describes sediments that are too heavy or large to be kept in suspension or solution and are pushed or rolled along the bottom of a streambed.

bedrock (p. 660) Unweathered, solid parent rock that may consist of limestone, marble, granite, or other quarried rock.

belt (p. 787) Low, warm, dark-colored cloud that sinks and flows rapidly in the Jovian atmosphere.

Big Bang theory (p. 847) Proposes that the universe began as a single point and has been expanding ever since.

binary star (p. 814) Describes two stars that are bound together by gravity and orbit a common center of mass.

biodiversity (p. 718) Biological diversity of an ecosystem, which is determined by the variety of species of plants and animals.

biogas (p. 696) Biomass fuel that is a mixture of gases, primarily methane, produced when plant and animal wastes are acted on by anaerobic bacteria in a digester.

bioremediation (p. 723) Use of organisms to clean up toxic waste.

biosphere (p. 9) All of Earth's organisms and the environments in which they live.

barrier island/barrera de islas (pág. 418) Grandes ondas de arena u otro sedimento depositado o formado por las corrientes litorales y separadas del continente; pueden medir hasta decenas de kilómetros de largo.

base/base (pág. 66) Solución que contiene iones hidróxido (OH^-).

Basin and Range Province/Provincia Basin and Range (pág. 637) Área en el suroeste de Estados Unidos y el norte y centro de México en donde la tectónica extensional resultó en la formación de cordilleras de montañas separadas por grandes valles.

batholith/batolito (pág. 476) Masa rocosa ígnea de grano grueso irregularmente formada que cubre por lo menos 100 km^2 y que se forma generalmente de 10 a 30 km bajo de la superficie terrestre y frecuentemente se encuentra en el interior de las principales cadenas montañosas.

beach/playa (pág. 415) Banda inclinada de sedimentos sueltos, como la arena, los guijarros o el barro, depositado a lo largo de una costa.

bedding/estratificación (pág. 126) Capa horizontal de roca sedimentaria que puede medir de un milímetro a varios metros de grueso.

bed load/carga de cauce (pág. 217) Término que describe los sedimentos que son demasiado pesados o grandes para ser mantenidos en suspensión o solución y los cuales son empujados o arrastrados a lo largo del fondo del lecho de una corriente.

bedrock/lecho rocoso (pág. 660) Roca madre sólida no meteorizada que puede consistir en piedra caliza, mármol, granito u otra piedra de cantera.

belt/cinturón (pág. 787) Nube baja, tibia y oscura que desciende y fluye rápidamente en la atmósfera joviana.

Big Bang theory/teoría de la Gran Explosión (pág. 847) Propone que el universo empezó en un solo punto y se ha estado expandiendo desde entonces.

binary star/estrella binaria (pág. 814) Describe dos estrellas unidas por la gravedad, las cuales giran alrededor de un centro común de masa.

biodiversity/biodiversidad (pág. 718) Diversidad biológica de un ecosistema, que está determinada por la variedad de especies de plantas y animales.

biogas/biogas (pág. 696) Combustible de la biomasa que es una mezcla de gases, principalmente metano, que se produce cuando las bacterias anaerobias actúan sobre los desechos de plantas y animales en un aparato digestivo.

bioremediation/biorremediación (pág. 723) Uso de organismos para limpiar desechos tóxicos.

biosphere/biosfera (pág. 9) Todos los organismos de la Tierra y los ambientes en que éstos viven.

black hole (p. 825) Small, extremely dense remnant of a star whose gravity is so immense that not even light can escape its gravity field.

Bowen's reaction series (p. 103) Sequential, predictable, dual-branched pattern in which minerals crystallize from cooling magma.

breaker (p. 400) Collapsing wave that forms when a wave reaches shallow water and is slowed by friction with the ocean bottom.

Burgess Shale (p. 604) Canadian fossil formation that contains Cambrian soft-bodied organisms as well as organisms with hard parts.

black hole/agujero negro (pág. 825) Restos de una estrella, muy densos y pequeños cuya gravedad es tan grande que ni la luz puede escapar de su campo de gravedad.

Bowen's reaction series/serie de reacción de Bowen (pág. 103) Patrón doblemente ramificado, predecible y secuencial en el cual se cristalizan los minerales a partir de magma que se enfría.

breaker/cachón (pág. 400) Ola en estado de colapso que se forma cuando una ola alcanza aguas poco profundas y la cual decelera debido a la fricción contra el fondo del océano.

Burgess Shale/esquisto de Burguess (pág. 604) Formación de fósiles canadiense que contiene tanto organismos Cámbricos de cuerpo blando como organismos con partes duras.

C

caldera (p. 481) Large crater, up to 50 km in diameter, that can form when the summit or side of a volcano collapses into the magma chamber during or after an eruption.

Caledonian Orogeny (p. 608) Collisional tectonic event that closed the ocean between Laurentia and Baltica, forming the larger continent, Laurasia.

Canadian shield (p. 581) Name given to the Precambrian shield in North America because much of it is exposed in Canada.

carrying capacity (p. 714) Number of organisms that a specific environment can support.

cartography (p. 27) Science of mapmaking.

cast (p. 568) Fossil formed when an earlier fossil of a plant or animal leaves a cavity that becomes filled with minerals or sediment.

cave (p. 245) Underground opening connected to Earth's surface, usually formed when groundwater dissolves limestone.

ceilometer (p. 313) Weather instrument used to measure the height of cloud layers; estimates the amount of sky that is covered by clouds.

cementation (p. 125) Process of sedimentary rock formation that occurs when dissolved minerals precipitate out of groundwater and either a new mineral grows between the sediment grains or the same mineral grows between and over the grains.

Cepheid variable (p. 834) Star with pulsation periods ranging from 1 to 100 days and luminosity that increases with the length of the pulsation period.

caldera/caldera (pág. 481) cráter grande, de hasta 50 km de diámetro, que puede formarse cuando la cumbre o el lado de un volcán se desploman en la cámara de magma, durante o después de una erupción.

Caledonian Orogeny/orogenia Caledoniana (pág. 608) Acontecimiento tectónico de colisión que cerró el océano entre Laurencia y Báltica, formando el continente más grande llamado Laurasia.

Canadian shield/escudo canadiense (pág. 581) Nombre dado al escudo precámbrico en Norteamérica porque la mayor parte está expuesta en Canadá.

carrying capacity/capacidad de carga (pág. 714) Número de organismos que un ambiente específico puede sustentar en un momento dado.

cartography/cartografía (pág. 27) Ciencia de elaboración de mapas.

cast/molde (pág. 568) Fósil que se forma cuando un fósil precedente de una planta o un animal deja una cavidad que se llena con minerales o sedimentos.

cave/cueva (pág. 245) Abertura subterránea conectada a la superficie terrestre, generalmente se forma cuando el agua subterránea disuelve la piedra caliza.

ceilometer/cielómetro (pág. 313) Instrumento meteorológico para medir la altura de las capas de nubes; estima la porción del cielo que está cubierta por nubes.

cementation/cementación (pág. 125) Proceso de formación de roca sedimentaria que ocurre cuando los minerales disueltos se precipitan del agua subterránea y un mineral nuevo crece entre los granos de sedimento o el mismo mineral crece entre los granos y sobre ellos.

Cepheid variable/variable cefeida (pág. 834) Estrella con períodos de pulsación que van desde 1 hasta 100 días y cuya luminosidad aumenta con la longitud del período de la pulsación.

chemical bond (p. 60) Force that holds the atoms of elements together in a compound.

chemical reaction (p. 64) Change of one or more substances into other substances.

chemical weathering (p. 155) Process by which rocks and minerals undergo changes in their composition due to chemical reactions with agents such as acids, water, oxygen, and carbon dioxide.

chromosphere (p. 806) Layer of the Sun's atmosphere above the photosphere and below the corona that is about 2500 km thick and has a temperature around 30 000 K at its top.

cinder-cone volcano (p. 482) Steep-sided, generally small volcano that is built by the accumulation of tephra around the vent.

cirque (p. 201) Deep depression scooped out by a valley glacier.

clastic (p. 122) Describes rock and mineral fragments produced by weathering and erosion and classified according to particle size and shape.

clastic sedimentary rock (p. 128) Most common type of sedimentary rock, such as breccia, sandstone, and shale, formed by lithification of clastic sediments.

cleavage (p. 86) Ability of a mineral to break easily and evenly along one or more flat planes.

climate (p. 300) Average weather of a particular area over a long period of time; (p. 359) includes annual variations in temperature and precipitation, which are influenced by latitude, closeness of lakes and oceans, topography, wind patterns, and air masses.

climatology (p. 359) Study of Earth's climate in order to understand and predict climatic change, based on past and present variations in temperature, precipitation, wind, and other weather variables.

coalescence (p. 289) Process that occurs when cloud droplets collide and form larger droplets, which eventually become too heavy to remain aloft and can fall to Earth as precipitation.

cogeneration (p. 700) Production of two usable forms of energy at the same time from the same process, which can conserve resources and generate income.

cold wave (p. 349) Extended period of lower-than-normal temperatures caused by large, high-pressure systems of continental polar or arctic origin.

chemical bond/enlace químico (pág. 60) Fuerza que mantiene unidos los átomos de los elementos en un compuesto.

chemical reaction/reacción química (pág. 64) Cambio de una o más sustancias en otras sustancias.

chemical weathering/meteorización química (pág. 155) Proceso mediante el cual las rocas y los minerales experimentan cambios en su composición debido a las reacciones químicas con agentes como los ácidos, el agua, el oxígeno y el dióxido de carbono.

chromosphere/cromosfera (pág. 806) Capa de la atmósfera del Sol por encima de la fotosfera y debajo de la corona, la cual mide aproximadamente 2500 km de ancho y tiene una temperatura máxima de alrededor de 30 000 K.

cinder-cone volcano/volcán de cono de carbonilla (pág. 482) Volcán generalmente pequeño de lados empinados que se forma debido a la acumulación de tefrita alrededor de la abertura.

cirque/circo (pág. 201) Depresión profunda formada por un glaciar de valle.

clastic/clástica (pág. 122) Describe los fragmentos de roca y de mineral producidos por la meteorización y la erosión y los cuales se clasifican según su tamaño de partícula y forma.

clastic sedimentary rock/roca sedimentaria clástica (pág. 128) Tipo más común de roca sedimentaria, que incluye rocas como la brecha, la arenisca y el esquisto y la cual se forma por litificación de sedimentos clásticos.

cleavage/crucero (pág. 86) Capacidad de un mineral para romperse fácil y uniformemente a lo largo de uno o más planos lisos.

climate/clima (pág. 300) Tiempo promedio de cierta área a lo largo de un período de tiempo considerable; (pág. 359) incluye las variaciones anuales en temperatura y precipitación y se ve influenciado por la latitud, la cercanía de lagos y océanos, la topografía, los patrones de viento y las masas de aire.

climatology/climatología (pág. 359) Estudio del clima de la Tierra para entender y pronosticar los cambios meteorológicos; se basa en variaciones pasadas y presentes de temperatura, precipitación, viento y otras variables meteorológicas.

coalescence/coalescencia (pág. 289) Proceso que ocurre cuando las gotas de nube chocan y forman gotas más grandes, que a la larga se vuelven demasiado pesadas para permanecer suspendidas y pueden caer a la Tierra como precipitación.

cogeneration/cogeneración (pág. 700) Producción simultánea de dos formas utiles de energía a partir del mismo proceso y la cual puede conservar recursos y ser lucrativa.

cold wave/onda fría (pág. 349) Período extendido de temperaturas más bajas que lo normal causado por grandes sistemas de alta presión de origen polar continental o ártico.

coma (p. 797) Extended volume of glowing gas that forms when a comet's nucleus is heated.

comet (p. 796) Small, eccentrically orbiting body made of rock and ice that consists of a nucleus, a coma, and one or more tails that point away from the Sun.

composite volcano (p. 482) Large, sloping volcano built by violent eruptions of volcanic fragments and lava that accumulate in alternating layers.

compound (p. 60) Substance composed of atoms of two or more different elements that are chemically combined.

condensation (p. 69) Process by which a cooling gas changes into a liquid and releases thermal energy; (p. 279) change of matter from a gas to a liquid.

condensation nuclei (p. 285) Small particles in the atmosphere around which cloud droplets can form.

conduction (p. 276) Energy transfer that occurs when molecules collide; takes place only when substances are in contact with each other.

conic projection (p. 32) Map that is highly accurate for small areas, made by projecting points and lines from a globe onto a cone.

constellation (p. 813) Group of stars that forms a pattern in the sky that resembles an animal, mythological character, or everyday object.

contact metamorphism (p. 135) Local effect that occurs when molten rock meets solid rock.

continental drift (p. 444) Wegener's hypothesis that Earth's continents were joined as a single landmass, called Pangaea, that broke apart about 200 million years ago and slowly moved to their present positions.

continental glacier (p. 200) Glacier that forms over a broad, continent-sized area of land and usually spreads out from its center.

continental margin (p. 422) Submerged part of a continent and shallowest part of the ocean that consists of the continental shelf, the continental slope, and the continental rise.

continental rise (p. 425) Gently sloping accumulation of sediments deposited by a turbidity current at the foot of a continental margin.

coma/cabellera (pág. 797) Volumen extendido de gas resplandeciente que se forma cuando se calienta el núcleo de un cometa.

comet/cometa (pág. 796) Cuerpo pequeño de órbita excéntrica compuesto de roca y hielo y que consta de un núcleo, una cabellera y una o más colas que apuntan hacia el lado contrario del Sol.

composite volcano/volcán compuesto (pág. 482) Volcán grande e inclinado que se forma debido a erupciones violentas de fragmentos y lava volcánicos, los cuales se acumulan en capas alternantes.

compound/compuesto (pág. 60) Sustancia compuesta de átomos de dos o más elementos diferentes que se combinan químicamente.

condensation/condensación (pág. 69) Proceso por el cual un gas enfriador se transforma en un líquido y libera energía térmica; (pág. 279) el cambio de la materia gaseosa a líquida.

condensation nuclei/núcleos de condensación (pág. 285) Partículas pequeñas en la atmósfera alrededor de las cuales se pueden formar las gotas de nubes.

conduction/conducción (pág. 276) Transferencia de energía que ocurre cuando las moléculas chocan; sólo sucede cuando las sustancias están en contacto.

conic projection/proyección cónica (pág. 32) Mapa que es sumamente exacto para áreas pequeñas y el cual se elabora mediante la proyección de puntos y líneas de un globo a un cono.

constellation/constelación (pág. 813) Grupo de estrellas que forman en el firmamento un patrón que parece un animal, un personaje mitológico o un objeto cotidiano.

contact metamorphism/metamorfismo de contacto (pág. 135) Efecto local que ocurre cuando la roca fundida se encuentra con roca sólida.

continental drift/deriva continental (pág. 444) Hipótesis de Wegener que propone que los continentes de la Tierra estaban unidos en una sola masa terrestre, llamada Pangaea, la cual se separó hace aproximadamente 200 millones de años y que los fragmentos se movieron lentamente a sus ubicaciones actuales.

continental glacier/glaciar continental (pág. 200) Glaciar que se forma sobre un área amplia de terreno del tamaño de un continente y que generalmente se extiende a partir de su centro.

continental margin/margen continental (pág. 422) Parte sumergida de un continente y la parte menos profunda del océano formada por la plataforma continental, el talud continental y el declive continental.

continental rise/declive continental (pág. 425) Acumulación de sedimentos levemente inclinados, depositados por una corriente de turbidez al pie de un margen continental.

continental shelf (p. 423) Shallowest part of a continental margin, with an average depth of 130 m and an average width of 60 km, that extends into the ocean from the shore and provides a nutrient-rich home to large numbers of fish.

continental slope (p. 424) Sloping oceanic region found beyond the continental shelf that generally marks the edge of the continental crust and may be cut by submarine canyons.

contour interval (p. 34) Difference in elevation between two side-by-side contour lines on a topographic map.

contour line (p. 33) A line on a topographic map that connects points of equal elevation.

control (p. 12) Standard for comparison in an experiment.

convection (p. 277) Transfer of energy by the flow of a heated substance.

convergent boundary (p. 457) Place where two of Earth's tectonic plates are moving toward each other; is associated with trenches, islands arcs, and folded mountains.

Cordillera (p. 626) Mountain ranges formed in western North America when active subduction along North America's western coast continued through the Middle Triassic.

Coriolis effect (p. 305) Deflects moving particles such as air to the right above the equator and to the left below the equator; caused by Earth's rotation and combines with the heat imbalance found on Earth to create the trade winds, polar easterlies, and prevailing westerlies.

corona (p. 807) Top layer of the Sun's atmosphere that extends from the top of the chromosphere and ranges in temperature from 1 million to 2 million K.

correlation (p. 561) Matching of rock outcrops of one geographic region to another.

cosmic background radiation (p. 848) Weak radiation that is left over from the early, hot stages of the Big Bang expansion of the universe.

cosmology (p. 847) Study of the universe, including its current nature, origin, and evolution, based on observation and the use of theoretical models.

covalent bond (p. 60) Attraction of two atoms for a shared pair of electrons that holds the atoms together.

continental shelf/plataforma continental (pág. 423) Parte más superficial de un margen continental, con una profundidad promedio es de 130 m y un ancho promedio de 60 km que se extiende hacia el océano desde la costa y provee un vivero rico en nutrientes a un gran número de peces.

continental slope/talud continental (pág. 424) Región oceánica inclinada que se encuentra más allá de la plataforma continental, que generalmente marca la orilla de la corteza continental y que puede estar seccionada por cañones submarinos.

contour interval/intervalo entre curvas de nivel (pág. 34) Diferencia en la elevación entre dos curvas de nivel seguidas en un mapa topográfico.

contour line/curva de nivel (pág. 33) Curva en un mapa topográfico que conecta puntos de igual elevación.

control/control (pág. 12) Estándar de comparación en un experimento.

convection/convección (pág. 277) Transferencia de energía debido al flujo de una sustancia calentada.

convergent boundary/límite convergente (pág. 457) Lugar donde dos placas tectónicas terrestres se mueven una hacia la otra. Está asociado con fosas abismales, arcos de islas y montañas plegadas.

Cordillera/cordillera (pág. 626) Conjunto de montañas formado en Norteamérica occidental cuando una subducción activa, a lo largo de la costa occidental de Norteamérica, continuó durante el Triásico Medio.

Coriolis effect/efecto de Coriolis (pág. 305) Desviación de partículas en movimiento, como el aire, hacia la derecha al norte del ecuador y hacia la izquierda al sur del ecuador; lo causa la rotación de la Tierra y se combina con el desequilibrio térmico presente en la Tierra para crear los vientos alisios, los vientos polares del este y los vientos ponientes prevalecientes.

corona/corona (pág. 807) Capa superior de la atmósfera del Sol que se extiende desde la parte superior de la cromosfera y tiene un rango de temperatura de 1 a 2 millones K.

correlation/correlación (pág. 561) Apareamiento de afloramientos rocosos de una región geográfica a otra.

cosmic background radiation/radiación cósmica de fondo (pág. 848) Radiación residual débil de las calientes etapas iniciales de la expansión de la Gran Explosión del universo.

cosmology/cosmología (pág. 847) Estudio del universo que incluye su naturaleza actual, su origen y evolución y el cual se basa en la observación y el uso de modelos teóricos.

covalent bond/enlace covalente (pág. 60) Atracción de dos átomos hacia un par compartido de electrones que mantiene unidos los átomos.

crater (p. 480) Bowl-shaped depression, usually less than 1 km in diameter, that forms around the central vent at the summit of a volcano.

creep (p. 184) Slow, steady downhill movement of loose weathered Earth materials, especially soils, causing objects on a slope to tilt.

crest (p. 399) Highest point of a wave.

cross-bedding (p. 126) Depositional feature of sedimentary rock that forms as inclined layers of sediment are carried forward across a horizontal surface.

cross-cutting relationships (p. 559) Principle stating that a fault or intrusion is younger than the rock it cuts across.

crystal (p. 79) Solid in which atoms are arranged in repeating patterns.

crystalline structure (p. 67) Regular geometric pattern of particles in most solids, giving a solid a definite shape and volume.

cyanobacteria (p. 585) Microscopic, photosynthetic prokaryotes that formed stromatolites and changed early Earth's atmosphere by generating oxygen.

cyclothems (p. 613) Repeating cyclic patterns of stacked sediments that represent cycles of glacial-interglacial periods; from base to top are generally sandstone-shale-coal-shale-limestone-shale.

crater/cráter (pág. 480) Depresión en forma de tazón, generalmente, de menos de 1 km de diámetro que se forma alrededor de la abertura central en la cumbre de un volcán.

creep/corrimiento (pág. 184) Movimiento cuesta abajo constante y lento de materias erosionadas sueltas de la Tierra, especialmente los suelos y la cual hace que se inclinen los objetos en una cuesta.

crest/cresta (pág. 399) Punto más alto de una ola.

cross-bedding/estratificación cruzada (pág. 126) Característica de la depositación de rocas sedimentarias que se forma a medida que las capas inclinadas de sedimento son arrastradas hacia adelante a través de una superficie horizontal.

cross-cutting relationships/relaciones de estratificación cruzada (pág. 559) Principio que establece que una falla o una intrusión es más reciente que la roca que atraviesa.

crystal/cristal (pág. 79) Sólido en el cual los átomos están ordenados en patrones repetitivos.

crystalline structure/estructura cristalina (pág. 67) Patrón geométrico y regular de partículas en la mayoría de los sólidos, la cual le da forma y volumen definidos al sólido.

cyanobacteria/cianobacterias (pág. 585) Organismos procariotas fotosintéticos microscópicos que formaron estromatolitas y cambiaron la atmósfera temprana de la Tierra al producir oxígeno.

cyclothems/ciclotemas (pág. 613) Patrones cíclicos repetitivos de sedimentos amontonados que representan ciclos de períodos glaciales- interglaciales; desde la base hasta la superficie generalmente son: arenisca-esquisto arcilloso–carbón–esquisto arcilloso–piedra caliza-esquisto arcilloso.

D

deep-sea trench (p. 426) Long, relatively narrow depression in the seafloor that can extend for thousands of kilometers, is the deepest part of the ocean basin, and is found primarily in the Pacific Ocean.

deflation (p. 192) Lowering of land surface caused by wind erosion of loose surface particles, often leaving coarse sediments behind.

deforestation (p. 720) Removal of trees from a forested area without adequate replanting, often using clear-cutting, which may result in loss of topsoil and water pollution.

delta (p. 226) Triangular deposit, usually made up of silt and clay particles, that forms where a stream enters a large body of water.

deep-sea trench/fosas abismales submarinas (pág. 426) Depresión larga y relativamente estrecha en el fondo marino que puede extenderse por miles de kilómetros; es la parte más profunda de la cuenca oceánica y se encuentra principalmente en el océano Pacífico.

deflation/deflación (pág. 192) Depresión de la superficie terrestre causada por la erosión eólica de partículas superficiales sueltas, la cual deja a menudo sedimentos gruesos.

deforestation/deforestación (pág. 720) Eliminación de árboles de un área forestal sin replantar adecuadamente, a menudo se utiliza corta a hecho, que puede tener como resultado la pérdida de la capa superficial del suelo y la contaminación del agua.

delta/delta (pág. 226) Depósito triangular, compuesto generalmente de partículas de cieno y arcilla, que se forma en donde una corriente entra a una gran masa de agua.

dendrochronology (p. 564) Science of comparing annual growth rings in trees to date events and environmental changes.

density current (p. 403) Movement of ocean water that occurs in depths too great to be affected by surface winds and is generated by differences in water temperature and salinity.

density-dependent factor (p. 714) Environmental factor, such as disease or lack of food, that increasingly affects a population as the population's size increases.

density-independent factor (p. 714) Environmental factor, such as a storm, flood, or pollution, that limits the growth of all affected populations regardless of size.

dependent variable (p. 12) Factor in an experiment that can change if the independent variable is changed.

deposition (p. 123) Occurs when sediments are laid down on the ground or sink to the bottom of a body of water; (p. 162) final stage of the erosional process in which the movement of transported materials slows and they are dropped in another location.

desalination (p. 674) Process that removes salt from ocean water in order to provide freshwater.

desertification (p. 660) Process by which productive land becomes desert; in arid areas can occur through the loss of topsoil.

dew point (p. 279) Temperature to which air is cooled at a constant pressure to reach saturation, at which point condensation can occur.

differentiation (p. 580) Process in which a planet becomes internally zoned, with the heavy materials sinking toward the center and the lighter materials accumulating near its surface.

digital forecast (p. 319) Weather forecast that uses numerical data to determine how atmospheric variables change over time.

dike (p. 478) Pluton that cuts across preexisting rocks and often forms when magma invades cracks in surrounding rock bodies.

dinosaur (p. 631) Terrestrial reptile that developed an upright posture and dominated the Mesozoic land.

dendrochronology/dendrocronología (pág. 564) Ciencia de la comparación de los anillos de crecimiento anual de los árboles para datar acontecimientos y cambios ambientales.

density current/corriente de densidad (pág. 403) Movimiento de las aguas oceánicas que ocurre a grandes profundidades, no se ve afectado por los vientos superficiales y se genera debido a las diferencias en temperatura y salinidad del agua.

density-dependent factor/factor dependiente de la densidad (pág. 714) Factor ambiental, como las enfermedades o la falta de alimento, que afecta cada vez más una población, a medida que aumenta el tamaño de la población.

density-independent factor/factor independiente de la densidad (pág. 714) Factor ambiental, como las tempestades, las inundaciones o la contaminación, que limita el crecimiento de todas las poblaciones afectadas, sea cual sea el tamaño de la población.

dependent variable/variable dependiente (pág. 12) Factor en un experimento que puede cambiar, si se cambia la variable independiente.

deposition/depositación (pág. 123) Ocurre cuando los sedimentos son depositados en el suelo o se hunden al fondo de una masa de agua; (pág. 162) la etapa final del proceso erosivo en que disminuye el movimiento de materiales transportados y éstos son arrojados en otra ubicación.

desalination/desalinización (pág. 674) Proceso de eliminación de la sal del agua marina para proveer agua dulce.

desertification/desertificación (pág. 660) Proceso mediante el cual la tierra productiva se convierte en desierto; en áreas áridas puede ocurrir debido a la pérdida de la capa superficial del suelo.

dew point/punto de rocío (pág. 279) Temperatura a la cual el aire se enfría a presión constante para alcanzar la saturación y en este punto puede ocurrir la condensación.

differentiation/diferenciación (pág. 580) Proceso en que un planeta se divide internamente en zonas, con el hundimiento de los materiales pesados hacia el centro, mientras que los materiales más ligeros se acumulan cerca de su superficie.

digital forecast/pronóstico digital (pág. 319) Pronóstico del tiempo que utiliza los datos numéricos para determinar cómo cambian las variables atmosféricas con el tiempo.

dike/dique (pág. 478) Plutón que atraviesa las rocas preexistentes y el cual se forma frecuentemente cuando el magma invade las grietas en cuerpos rocosos circundantes.

dinosaur/dinosaurio (pág. 631) Reptil terrestre que desarrolló una postura vertical y dominó la Tierra Mesozoica.

discharge (p. 218) Measure of a volume of stream water that flows over a specific location in a particular amount of time.

divergent boundary (p. 456) Place where two of Earth's tectonic plates are moving apart; is associated with volcanism, earthquakes, and high heat flow, and is found primarily on the seafloor.

divide (p. 215) Elevated land that divides one watershed, or drainage basin, from another.

Doppler effect (p. 315) Change in the wave frequency that occurs in energy when that energy moves toward or away from an observer.

downburst (p. 336) Violent, damaging thunderstorm wind that is concentrated in a local area.

drawdown (p. 252) Difference between the water level in a pumped well and the original water-table level.

drought (p. 347) Extended period of low rainfall, usually caused by shifts in global wind patterns, allowing high-pressure systems to remain for weeks or months over continental areas.

drumlin (p. 202) Elongated landform that results when a glacier moves over an older moraine.

dune (p. 194) Pile of wind-blown sand that develops over time, whose shape depends on sand availability, wind velocity and direction, and amount of vegetation present.

discharge/descarga (pág. 218) Medida de un volumen de agua corriente que fluye sobre una ubicación dada, en cierto lapso de tiempo.

divergent boundary/límite divergente (pág. 456) Lugar donde se alejan dos placas tectónicas terrestres; se asocia con la actividad volcánica, con los terremotos y el flujo de calor alto y se halla principalmente en el fondo marino.

divide/divisoria (pág. 215) Tierra elevada que divide una cuenca hidrográfica de otra.

Doppler effect/efecto Doppler (pág. 315) Cambio en la frecuencia de onda que ocurre en la energía, cuando esa energía se mueve hacia un observador o se aleja de él.

downburst/chaparrón violento (pág. 336) Viento de una tormenta eléctrica violento y dañino, que se concentra en un área local.

drawdown/tasa de agotamiento (pág. 252) Diferencia entre el nivel de agua en un pozo bombeado y el nivel original de la capa freática.

drought/sequía (pág. 347) Período prolongado de poca precipitación, generalmente causado por los cambios en los patrones de vientos globales, que permite que los sistemas de alta presión permanezcan sobre áreas continentales durante semanas o meses.

drumlin/drumlin (pág. 202) Formación alargada de tierra que se forma cuando un glaciar se mueve sobre una morrena más antigua.

dune/duna (pág. 194) Pila de arena arrastrada por el viento, la cual se desarrolla con el tiempo y cuya forma depende de la disponibilidad de la arena, la velocidad y dirección del viento y la cantidad de vegetación presente.

eccentricity (p. 777) Ratio of the distance between the foci to the length of the major axis; defines the shape of a planet's elliptical orbit.

ecliptic (p. 759) Plane that contains Earth's orbit around the Sun.

ectotherm (p. 633) Animal whose body temperature is regulated by the temperature of its surroundings.

Ediacaran fauna (p. 592) Fossils of multicellular, varied organisms lacking a mouth, anus, and gut that were widely distributed in the shallow oceans of the late Proterozoic and flourished between 570 and 670 million years ago.

ejecta (p. 754) Material that falls back to the lunar surface after being blasted out by the impact of a space object.

eccentricity/excentricidad (pág. 777) Proporción de la distancia entre los focos a la longitud del eje mayor; define la forma de la órbita elíptica de un planeta.

ecliptic/eclíptica (pág. 759) Plano que contiene la órbita de la Tierra alrededor del Sol.

ectotherm/poiquilotermo (pág. 633) Animal cuya temperatura corporal es regulada por la temperatura de sus alrededores.

Ediacaran fauna/fauna Ediacarana (pág. 592) Fósiles de organismos multicelulares diversos, que carecen de boca, ano, e intestino, los cuales estaban ampliamente distribuidos en los océanos poco profundos del Proterozoico Tardío y que abundaron hace 570 a 670 millones de años.

ejecta/ejecta (pág. 754) Material que regresa a la superficie lunar después de ser expulsado por el impacto de un objeto del espacio.

El Niño (p. 370) Warm ocean current that develops off the western coast of South America and can cause short-term climatic changes felt worldwide.

electromagnetic spectrum (p. 37) Arrangement of electromagnetic radiation, including gamma rays, X rays, and microwaves.

electron (p. 55) Tiny atomic particle with little mass and a negative electrical charge that surrounds the nucleus in energy levels.

element (p. 53) Natural or artificial substance that cannot be broken down into simpler substances by physical or chemical means.

endotherm (p. 633) Animal that maintains a relatively constant body temperature, regardless of its surroundings.

energy efficiency (p. 698) A type of conservation in which energy resources are used in the most productive ways.

energy level (p. 55) Represents the area in an atom where electrons are most likely to occur.

eon (p. 554) Longest time unit in the geological time scale, measured in billions of years.

epicenter (p. 499) Point on Earth's surface directly above the focus of an earthquake.

epoch (p. 556) Time unit in the geological time scale, smaller than a period, measured in millions of years to tens of millions of years.

equator (p. 27) Imaginary line that lies at 0° latitude and circles Earth midway between the north and south poles, dividing Earth into the northern hemisphere and the southern hemisphere.

era (p. 554) Second-longest time unit in the geological time scale, measured in hundreds of millions of years, and defined by differences in life-forms that are preserved in rocks.

erosion (p. 153) Movement of weathered materials from one location to another by agents such as water, wind, glaciers, and gravity.

esker (p. 202) Long, winding ridge of layered sediments deposited by streams that flow beneath a melting glacier.

estuary (p. 415) Coastal area of brackish water formed where the lower end of a freshwater river or stream enters the ocean; provides an excellent source of food and shelter to commercially important marine organisms.

eukaryote (p. 591) Organism that is composed of cells that contain nuclei and is generally larger and more complex than a prokaryote.

El Niño/El Niño (pág. 370) Corriente oceánica cálida que se desarrolla lejos de la costa occidental de Sudamérica y que puede causar cambios climáticos a corto plazo y cuyos efectos se sienten por todo el mundo.

electromagnetic spectrum/espectro electromagnético (pág. 37) Arreglo de la radiación electromagnética; incluye los rayos gamma, los rayos X y las microondas.

electron/electrón (pág. 55) Partícula atómica diminuta con masa pequeña y carga eléctrica negativa que rodea el núcleo en los niveles energéticos.

element/elemento (pág. 53) Sustancia natural o artificial que no puede separarse en sustancias más simples por medios físicos o químicos.

endotherm/endotermo (pág. 633) Animal que mantiene una temperatura corporal relativamente constante, sea cual sea la temperatura de sus alrededores.

energy efficiency/eficacia energética (pág. 698) Tipo de conservación en el cual los recursos energéticos se utilizan de maneras más productivas.

energy level/nivel de energía (pág. 55) Representa el área en un átomo donde es más probable que se encuentren los electrones.

eon/eon (pág. 554) Unidad más larga de tiempo en la escala de tiempo geológico; se mide en mil millones de años.

epicenter/epicentro (pág. 499) Lugar en la superficie terrestre directamente encima del foco de un terremoto.

epoch/época (pág. 556) Unidad de tiempo en la escala de tiempo geológico, más pequeña que un período; se mide en millones a decenas de millones de años.

equator/ecuador (pág. 27) Línea imaginaria localizada en la latitud 0° que da vuelta a la Tierra a medio camino entre los polos norte y sur, dividiendo la Tierra en dos hemisferios: el norte y el sur.

era/era (pág. 554) Segunda unidad más larga de tiempo en la escala del tiempo geológico; se mide en centenas de millones de años y se define según las diferencias en las formas de vida preservadas en las rocas.

erosion/erosión (pág. 153) Movimiento de materiales erosionados de un lugar a otro por agentes como el agua, el viento, los glaciares y la gravedad.

esker/ésker (pág. 202) Formación larga y sinuosa de sedimentos en capas, depositados por corrientes que fluyen debajo de un glaciar que se derrite.

estuary/estuario (pág. 415) Área costera de agua salobre que se forma en donde el extremo más bajo de un río o corriente de agua dulce entra al océano; provee una fuente excelente de alimento y refugio para organismos marinos comercialmente importantes.

eukaryote/eucariota (pág. 591) Organismo que está compuesto de células nucleadas; generalmente es más grande y más complejo que un procariota.

eutrophication (p. 230) Process by which lakes become rich in nutrients from the surrounding watershed, resulting in a change in the kinds of organisms in the lake.

eutrophication/eutroficación (pág. 230) Proceso mediante el cual los lagos se convierten en lugares ricos en alimentos nutritivos provenientes de las cuencas circundantes, lo que resulta en un cambio en los tipos de organismos en el lago.

evaporation (p. 68) Vaporization—change of state, involving thermal energy; (p. 290) process by which water changes from a liquid to a gas.

evaporation/evaporación (pág. 68) Vaporización: cambio de estado que involucra energía térmica; (pág. 290) proceso mediante el cual el agua cambia de un líquido a un gas.

evaporite (p. 130) Chemical sedimentary rock that forms mainly in restricted ocean basins in areas with high evaporation rates.

evaporite/evaporita (pág. 130) Roca química sedimentaria que se forma principalmente en cuencas oceánicas restringidas en áreas con tasas de evaporación altas.

evolution (p. 566) Adaptation of life-forms to changing environmental conditions.

evolution/evolución (pág. 566) Adaptación de los seres vivos a las condiciones ambientales cambiantes.

exfoliation (p. 155) Mechanical weathering process in which outer rock layers are stripped away, often resulting in dome-shaped formations.

exfoliation/exfoliación (pág. 155) Proceso de meteorización mecánica mediante el cual se eliminan las capas rocosas exteriores, a menudo resultando en formaciones en forma de domo.

exosphere (p. 274) Outermost layer of Earth's atmosphere that is located above the thermosphere and contains light gases such as helium and hydrogen.

exosphere/exosfera (pág. 274) Capa más externa de la atmósfera terrestre localizada por encima de la termosfera y la cual contiene gases livianos como el helio y el hidrógeno.

exponential growth (p. 713) Pattern of growth in which a population of organisms grows faster as it increases in size, resulting in a population explosion.

exponential growth/crecimiento exponencial (pág. 713) Patrón de crecimiento en que una población de organismos crece más rápidamente a medida que aumenta de tamaño, dando como resultado una explosión demográfica.

extrusive (p. 99) Fine-grained igneous rock that is formed when molten rock cools quickly and solidifies on Earth's surface.

extrusive/extrusiva (pág. 99) Roca ígnea de grano fino que se forma cuando la roca fundida se enfría rápidamente y se solidifica en la superficie terrestre.

eye (p. 344) Calm center of a tropical cyclone that develops when the winds around its center reach at least 120 km/h.

eye/ojo (pág. 344) Centro tranquilo de un ciclón tropical que se desarrolla cuando los vientos alrededor de su centro alcanzan por lo menos 120 km/h.

eyewall (p. 344) Band where the strongest winds in a hurricane are usually concentrated, surrounding the eye.

eyewall/pared del ojo de huracán (pág. 344) Banda que rodea el ojo de un huracán, donde generalmente se concentran los vientos más fuertes.

fault (p. 497) Fracture or system of fractures in Earth's crust that occurs when stress is applied too quickly or stress is too great; can form as a result of horizontal compression (reverse fault), horizontal shear (strike-slip fault), or horizontal tension (normal fault).

fault/falla (pág. 497) Fractura o sistema de fracturas en la corteza terrestre que ocurre cuando se aplica tensión rápidamente o cuando la tensión es demasiado grande; se puede formar como resultado de la compresión horizontal (falla invertida), del cizallamiento horizontal (falla de transformación) o de la tensión horizontal (falla normal).

fault-block mountains (p. 538) Mountains that form when large pieces of crust are tilted, uplifted, or dropped downward between large normal faults.

fault-block mountains/montañas de bloque de falla (pág. 538) Montañas que se forman cuando los trozos grandes de corteza se inclinan, se elevan, o se hunden entre fallas normales grandes.

felsic (p. 107) Light-colored igneous rocks that contain quartz and feldspars and have high silica contents.

felsic/félsica (pág. 107) Rocas ígneas de color claro que contienen cuarzo y feldespatos y tienen un contenido alto de sílice.

fission (p. 809) Process in which heavy atomic nuclei split into smaller, lighter nuclei.

fission/fisión (pág. 809) Proceso mediante el cual los núcleos atómicos pesados se parten en núcleos más livianos y pequeños.

flood (p. 219) Potentially devastating natural occurrence in which water spills over the sides of a stream's banks onto adjacent land areas.

flood/inundación (pág. 219) Acontecimiento natural potencialmente devastador en que el agua se desborda de las riberas de una corriente y cubre los terrenos adyacentes.

floodplain (p. 219) Broad, flat, fertile area extending out from a stream's bank that is covered with water during floods.

floodplain/llanura aluvial (pág. 219) Área fértil, plana y ancha que se extiende desde las riberas de una corriente, la cual se cubre con agua durante las inundaciones.

focus (p. 499) Point of the initial fault rupture where an earthquake originates that usually lies at least several kilometers beneath Earth's surface.

focus/foco (pág. 499) Punto de la ruptura de la falla inicial donde se origina un terremoto y el cual generalmente se encuentra por lo menos varios kilómetros debajo de la superficie terrestre.

foliated (p. 136) Metamorphic rock, such as schist or gneiss, whose minerals are squeezed under high pressure and arranged in wavy layers and bands.

foliated/foliada (pág. 136) Roca metamórfica, como el esquisto o el gneis, cuyos minerales se comprimen bajo presiones altas y se ordenan en capas y bandas onduladas.

fossil (p. 566) Remains or evidence of a once-living plant or animal.

fossil/fósil (pág. 566) Restos o pruebas de una planta o un animal que vivió alguna vez.

fossil fuel (p. 686) Nonrenewable energy resource formed over geologic time from the compression and partial decomposition of organisms that lived millions of years ago.

fossil fuel/combustible fósil (pág. 686) Recurso energético no renovable que se forma a lo largo del tiempo geológico a partir de la compresión y descomposición parcial de organismos que vivieron hace millones de años.

fractional crystallization (p. 103) Process in which different minerals crystallize from magma at different temperatures, removing elements from magma.

fractional crystallization/cristalización fraccionaria (pág. 103) Proceso en el cual diferentes minerales se cristalizan a partir del magma a diferentes temperaturas, eliminando elementos del magma.

fracture (p. 87) Ability of a mineral to break into pieces with arclike, rough, or jagged edges.

fracture/fractura (pág. 87) Capacidad de un mineral de romperse en trozos con bordes ásperos, arqueados, o serrados.

frequency (p. 38) Number of waves that pass a given point each second.

frequency/frecuencia (pág. 38) Número de ondas que pasan por un punto dado cada segundo.

front (p. 308) Boundary between two air masses of differing densities; can be cold, warm, stationary, or occluded and can stretch over large areas of Earth's surface.

front/frente (pág. 308) Frontera entre dos masas de aire con diferentes densidades; puede ser frío, cálido, estacionario u ocluido y puede extenderse sobre grandes áreas de la superficie de la Tierra.

frontal thunderstorm (p. 331) Type of thunderstorm usually produced by an advancing cold front, which can result in a line of thunderstorms hundreds of kilometers long, or, more rarely, an advancing warm front, which can result in a relatively mild thunderstorm.

frontal thunderstorm/tormenta eléctrica frontal (pág. 331) Tipo de tormenta que generalmente produce un frente frío en avance y el cual puede resultar en una línea de tormentas de cientos de kilómetros de largo, o con menor frecuencia, en un frente cálido que avanza, el cual puede ocasionar una tormenta relativamente ligera.

frost wedging (p. 154) Mechanical weathering process that occurs when water repeatedly freezes and thaws in the cracks of rocks, often resulting in rocks splitting.

frost wedging/grietas causadas por el hielo (pág. 154) Proceso mecánico de meteorización que ocurre cuando el agua se congela y se descongela en repetidas ocasiones en las grietas de rocas, frecuentemente dando como resultado el rompimiento de las rocas.

fuel (p. 684) Material, such as wood, peat, or coal, burned to provide power or heat.

fuel/combustible (pág. 684) Materiales, incluyendo la leña, la turba o el carbón, que se queman para proporcionar potencia o calor.

Fujita tornado intensity scale (p. 338) Classifies tornados according to their wind speed, duration, and path of destruction on a scale ranging from F0 to F5.

fusion (p. 809) Process in a star's core in which lightweight hydrogen nuclei combine into heavier helium nuclei.

Fujita tornado intensity scale/escala Fujita de intensidad de tornados (pág. 338) Clasifica los tornados según la velocidad de viento, la duración y el daño causado a su paso, en una escala que va de F0 a F5.

fusion/fusión (pág. 809) Proceso en el centro de una estrella en el cual los núcleos livianos de hidrógeno se combinan para formar núcleos más pesados de helio.

G

gangue (p. 663) Material left after ore is extracted that may release harmful chemicals into surface water or groundwater.

gas giant planets (p. 780) Large, gaseous planets that are very cold at their surfaces, have ring systems, many moons, and lack solid surfaces—Jupiter, Saturn, Uranus, and Neptune.

gasohol (p. 696) Biomass fuel that is a mixture of gasoline and ethanol and can be used in conventional gasoline engines.

gem (p. 90) Rare, precious, highly prized mineral that can be cut, polished, and used for jewelry.

geologic time scale (p. 554) Record of Earth's history from its origin to the present used to correlate geologic events, environmental changes, and development of life-forms that are preserved in rocks.

geology (p. 6) Study of materials that make up Earth and the processes that form and change these materials.

geothermal energy (p. 693) Energy produced from Earth's own internal steam and hot water.

geyser (p. 251) Explosive hot spring that erupts regularly.

glacier (p. 198) Large, moving mass of ice that forms near Earth's poles and in mountainous regions at high elevations.

glass (p. 68) Solid that consists of densely packed atoms with a random arrangement and lacks crystals or has crystals that are not visible.

Global Positioning System (GPS) (p. 40) Satellite-based navigation system that permits a user to pinpoint his or her exact location on Earth.

global warming (p. 376) Rise in global temperatures, which may be due to increases in atmospheric CO_2 from deforestation and burning of fossil fuels; (p. 725) a phenomenon related to the greenhouse effect.

gangue/ganga (pág. 663) Material que queda después de que se extrae un mineral y el cual puede liberar sustancias químicas perjudiciales en el agua superficial o subterránea.

gas giant planets/planetas gigantes gaseosos (pág. 780) Planetas grandes y gaseosos con superficies muy frías, sistemas de anillos, muchas lunas y que carecen de superficies sólidas: Júpiter, Saturno, Urano y Neptuno.

gasohol/gasohol (pág. 696) Combustible de biomasa que es una mezcla de gasolina y etanol y que se puede utilizar en motores de gasolina convencionales.

gem/gema (pág. 90) Mineral sumamente valioso, precioso y escaso que se puede cortar, pulir y utilizar en joyería.

geologic time scale/escala del tiempo geológico (pág. 554) Registro de la historia de la Tierra desde su origen hasta el presente; se utiliza para correlacionar los acontecimientos geológicos, los cambios ambientales y el desarrollo de las formas de vida preservadas en las rocas.

geology/geología (pág. 6) Estudio de los materiales que componen la Tierra y de los procesos que forman y cambian estos materiales.

geothermal energy/energía geotérmica (pág. 693) Energía que se produce a partir del vapor interno y del agua caliente que posee la Tierra.

geyser/géiser (pág. 251) Manantial termal explosivo que hace erupción regularmente.

glacier/glaciar (pág. 198) Enorme masa móvil de hielo que se forma cerca de los polos de la Tierra y en regiones montañosas a grandes elevaciones.

glass/vidrio (pág. 68) Sólido que consiste en átomos densamente empacados con un arreglo aleatorio; carece de cristales o tiene cristales que no son visibles.

Global Positioning System (GPS)/Sistema de Posicionamiento Global (GPS) (pág. 40) Sistema de navegación con satélite que le permite a un usuario localizar, con toda precisión, su ubicación exacta sobre la Tierra.

global warming/calentamiento global (pág. 376) Aumento en las temperaturas globales, que puede ser producto de los aumentos en el CO_2 atmosférico debido a la deforestación y la quema de combustibles fósiles; (pág. 725) fenómeno relacionado con el efecto de invernadero.

gnomonic projection (p. 33) Map useful in plotting long-distance trips by boat or plane that is made by projecting points and lines from a globe onto a piece of paper that touches the globe at a single point.

gnomonic projection/proyección gnomónica (pág. 33) Mapa útil en el trazado de viajes de distancias largas por barco o por avión; se elabora proyectando los puntos y las líneas de un globo sobre una hoja de papel que toca el globo en un solo punto.

Gondwana (p. 612) Large continent in the southern hemisphere that formed as a result of collisional events during the Late Paleozoic.

Gondwana/Gondwana (pág. 612) Inmenso continente en el hemisferio sur que se formó como resultado de las colisiones durante el Paleozoico Tardío.

graded bedding (p. 126) Depositional feature of sedimentary rock in which particles are progressively heavier and coarser toward the bottom layers of bedding.

graded bedding/estratificación graduada (pág. 126) Característica de la depositación de rocas sedimentarias en la cual las partículas son progresivamente más pesadas y toscas hacia las capas inferiores de la estratificación.

greenhouse effect (p. 375) Natural heating of Earth's surface by certain atmospheric gases, which helps keep Earth warm enough to sustain life.

greenhouse effect/efecto de invernadero (pág. 375) Calentamiento natural de la superficie terrestre por ciertos gases atmosféricos; ayuda a mantener la Tierra lo suficiente cálida para sustentar vida.

gully erosion (p. 163) Erosion that occurs when a rill channel widens and deepens.

gully erosion/erosión de barrancos (pág. 163) Erosión que ocurre cuando el cauce de un arroyuelo se ensancha y se profundiza.

half-life (p. 563) Period of time it takes for a radioactive isotope, such as carbon-14, to decay to one-half of its original amount.

half-life/media vida (pág. 563) Período de tiempo que demora un isótopo radiactivo, como el carbono 14, en desintegrarse a la mitad de su cantidad radiactiva original.

halo (p. 835) Spherical region that surrounds the Milky Way's nuclear bulge and disk.

halo/halo (pág. 835) Región esférica que rodea la protuberancia y el disco nucleares de la Vía Láctea.

hardness (p. 86) Measure of how easily a mineral can be scratched, which is determined by the arrangement of a mineral's atoms.

hardness/dureza (pág. 86) Medida de la facilidad de un mineral para rayarse, lo cual se determina por el arreglo de los átomos del mineral.

heat (p. 278) Energy transfer that occurs because of a difference in temperature between substances and flows from an object of higher temperature to an object of lower temperature.

heat/calor (pág. 278) Transferencia de energía que ocurre debido a una diferencia de temperatura entre las sustancias y que fluye de un objeto de temperatura más alta a un objeto de temperatura más baja.

heat island (p. 368) Urban area where climate is warmer than in the surrounding countryside due to factors such as numerous concrete buildings and large expanses of asphalt.

heat island/isla de calor (pág. 368) Área urbana donde el clima es más caliente que en el área rural circundante debido a factores como los numerosos edificios de concreto y las grandes extensiones de asfalto.

heat wave (p. 348) Extended period of higher-than-normal temperatures caused by large, high-pressure systems that warm by compression and block cooler air masses.

heat wave/ola de calor (pág. 348) Período extenso con temperaturas más altas de lo normal; lo causan los grandes sistemas de alta presión que se calientan por compresión y bloquean las masas de aire más frescas.

Hertzsprung-Russell diagram (H-R diagram) (p. 819) Graph that relates stellar characteristics—class, mass, temperature, magnitude, diameter, and luminosity.

Hertzsprung-Russell diagram/diagrama de Hertzsprung-Russell (diagrama H-R) (pág. 819) Gráfica que relaciona las características estelares, incluyendo la clase, la masa, la temperatura, la magnitud, el diámetro y la luminosidad.

highlands (p. 754) Light-colored, mountainous, heavily cratered areas of the Moon, composed mostly of lunar breccias.

highlands/tierras altas (pág. 754) Áreas en la Luna de color claro, con muchos cráteres y montañosas, compuestas en su mayor parte de brechas lunares.

hominids (p. 640) Group of upright, bipedal primates, including Homo sapiens.

hominoids (p. 640) Group of primates, including hominids and the great apes.

Homo sapiens (p. 641) Species to which humans belong.

hot spot (p. 486) Unusually hot area in Earth's mantle that is stationary for long periods of time, where high-temperature plumes of mantle material rise toward the surface.

hot spring (p. 251) Thermal spring with temperatures higher than that of the human body.

Hubble constant (p. 843) Value (H) used to calculate the rate that the universe is expanding; measured in kilometers per second per megaparsec.

humidity (p. 283) Amount of water vapor in the air.

hydrolysis (p. 156) Chemical reaction of water with other substances.

hydrosphere (p. 8) All the water in Earth's oceans, lakes, seas, rivers, and glaciers plus all the water in the atmosphere.

hydrothermal metamorphism (p. 135) Occurs when very hot water reacts with rock, altering its mineralogy and chemistry.

hydrothermal vent (p. 591) Hot-water, deep-sea vent that has the energy and nutrients needed for the beginnings of life.

hygrometer (p. 313) Weather instrument used to measure relative humidity.

hypothesis (p. 11) A suggested explanation for an observation often stated in the form of a question that can be answered by the results of an experiment.

hominids/homínidos (pág. 640) Grupo de primates erguidos y bípedos que incluye *Homo sapiens.*

hominoids/hominoideos (pág. 640) Grupo de primates que incluye los homínidos y los grandes simios.

Homo sapiens/Homo sapiens (pág. 641) Especie a la cual pertenecen los seres humanos.

hot spot/foco caliente (pág. 486) Área muy caliente del manto de la Tierra, estacionaria durante largos períodos de tiempo, en donde las plumas de alta temperatura de material del manto suben hacia la superficie.

hot spring/fuente termal (pág. 251) Manantial termal con temperaturas más altas que las del cuerpo humano.

Hubble constant/constante de Hubble (pág. 843) Valor (*H*) utilizado para calcular la velocidad a la que se expande el universo; se mide en kilómetros por segundo por megaparsec.

humidity/humedad (pág. 283) Cantidad de vapor de agua en el aire.

hydrolysis/hidrólisis (pág. 156) Reacción química del agua con otras sustancias.

hydrosphere/hidrosfera (pág. 8) Toda el agua en los océanos, los lagos, los mares, los ríos y los glaciares de la Tierra, además de toda el agua en la atmósfera.

hydrothermal metamorphism/metamorfismo hidrotérmico (pág. 135) Ocurre cuando agua muy caliente reacciona con la roca, alterando su mineralogía y su química.

hydrothermal vent/chimenea hidrotérmica (pág. 591) abertura de agua caliente en las profundidades marinas que contiene la energía y los nutrientes necesarios para el comienzo de la vida.

hygrometer/higrómetro (pág. 313) Instrumento meteorológico que se usa para medir la humedad relativa.

hypothesis/hipótesis (pág. 11) Explicación sugerida para una observación; a menudo se formula en forma de pregunta y puede ser contestada por los resultados de un experimento.

ice age (p. 369) Period of extensive glacial coverage, producing long-term climatic changes.

igneous rock (p. 99) Intrusive or extrusive rock formed from the cooling and crystallization of magma; can be classified as felsic, mafic, intermediate, and ultramafic, according to its mineral composition.

impact crater (p. 754) Moon crater formed when space material impacted on the lunar surface.

ice age/glaciación (pág. 369) Período de amplia cobertura glacial que produce cambios climáticos de largo plazo.

igneous rock/roca ígnea (pág. 99) Roca intrusiva o extrusiva formada por el enfriamiento y cristalización del magma; pueden clasificarse en félsica, máfica, intermedia y ultramáfica, según su composición mineral.

impact crater/cráter de impacto (pág. 754) cráter lunar que se forma cuando el material proveniente del espacio choca contra la superficie lunar.

independent variable (p. 12) Factor that is manipulated by the experimenter in an experiment.

index fossils (p. 568) Remains of plants or animals that were abundant, widely distributed, and existed briefly that can be used by geologists to correlate or date rock layers.

infiltration (p. 240) Process by which precipitation that has fallen on land surfaces enters the ground and becomes groundwater.

inflationary universe (p. 850) Model predicting that the universe is flat; can explain the walls and voids in the distribution of galaxies.

interferometry (p. 750) Process that links separate telescopes so they act as one telescope, producing more detailed images as the distance between them increases.

International Date Line (p. 31) The 180° meridian, which serves as the transition line for calendar days.

intrusive (p. 99) Coarse-grained igneous rock that is formed when molten rock cools slowly and solidifies inside Earth's crust.

ionic bond (p. 63) Attractive force between two ions with opposite charge.

ions (p. 62) Electrically charged atoms or groups of atoms.

iridium (p. 633) Metal that is rare in rocks at Earth's surface but is relatively common in meteorites and asteroids.

isochron (p. 452) Line on a map that connects points of the same age.

isopleth (p. 317) Line drawn on a weather map that connects points of equal or constant values.

isostasy (p. 525) Condition of equilibrium that describes the displacement of Earth's mantle by Earth's continental and oceanic crust.

isostatic rebound (p. 527) Slow process of Earth's crust rising as the result of the removal of mass from the crust.

isotopes (p. 58) Atoms of the same element that have different mass numbers and the same chemical properties.

independent variable/variable independiente (pág. 12) Factor que manipula el investigador en un experimento.

index fossils/fósiles guías (pág. 568) Restos de plantas o animales que fueron abundantes, extensamente distribuidos y de breve existencia, los cuales pueden utilizar los geólogos para correlacionar o para datar las capas rocosas.

infiltration/infiltración (pág. 240) Proceso mediante el cual la precipitación que cae sobre la superficie terrestre entra al suelo y se convierte en agua subterránea.

inflationary universe/universo inflacionario (pág. 850) Modelo que predice que el universo es plano; puede explicar las paredes y los vacíos en la distribución de las galaxias.

interferometry/interferometría (pág. 750) Proceso que une telescopios separados de tal manera que actúan como un solo telescopio, produciendo imágenes más detalladas al aumentar la distancia entre ellos.

International Date Line/Línea Internacional de Cambio de Fecha (pág. 31) El meridiano 180°, que sirve como la línea de transición para los días del calendario.

intrusive/intrusiva (pág. 99) Roca ígnea de grano grueso que se forma cuando roca fundida se enfría lentamente y se solidifica en el interior de la corteza de la Tierra.

ionic bond/enlace iónico (pág. 63) Fuerza atractiva entre dos iones con cargas opuestas.

ions/iones (pág. 62) Átomos o grupos de átomos cargados eléctricamente.

iridium/iridio (pág. 633) Metal escaso en las rocas de la superficie terrestre, pero relativamente común en los meteoritos y los asteroides.

isochron/isocrona (pág. 452) Línea en un mapa que conecta puntos con la misma antigüedad.

isopleth/isopleta (pág. 317) Línea en un mapa meteorológico que conecta los puntos de valores iguales o constantes.

isostasy/isostasia (pág. 525) Condición de equilibrio que describe el desplazamiento del manto terrestre por la corteza continental y la oceánica de la Tierra.

isostatic rebound/rebote isostático (pág. 527) Proceso lento de elevación de la corteza terrestre como resultado de la eliminación de masa de la corteza.

isotopes/isótopos (pág. 58) Átomos del mismo elemento que tienen diferentes números de masa, pero las mismas propiedades químicas.

jet stream (p. 307) High-altitude, narrow, westerly wind band that occurs above large temperature contrasts and can flow as fast as 185 km/h.

jet stream/corriente de chorro (pág. 307) Banda de vientos ponientes, estrecha y de gran altitud situada por encima de los grandes contrastes de temperatura y que puede fluir tan rápido como 185 km/h.

karst topography (p. 246) Irregular topography with sinkholes, sinks, and sinking streams caused by groundwater dissolution of limestone.

key bed (p. 565) Sediment layer that serves as a time marker in the rock record and results from volcanic ash or meteorite-impact debris that spread out and covered large areas of Earth.

kimberlite (p. 113) Rare, ultramafic rock that can contain diamonds and other minerals formed only under very high pressures.

Köeppen classification system (p. 364) Divides climates into five basic types, based on the mean monthly values of temperature and precipitation and types of vegetation.

karst topography/topografía cárstica (pág. 246) Topografía irregular con sumideros, hundimientos y corrientes en declive causados por la disolución de la piedra caliza por el agua subterránea.

key bed/capa clave (pág. 565) Capa de sedimento que sirve como un marcador de tiempo en el registro de las rocas; se origina de las cenizas volcánicas o de los escombros de los impactos de meteoritos que se esparcen y cubren grandes áreas de la Tierra.

kimberlite/kimberlita (pág. 113) Roca ultramáfica escasa que puede contener diamantes y otros minerales; sólo se forma bajo presiones muy altas.

Köeppen classification system/sistema de clasificación de Köeppen (pág. 364) Divide los climas en cinco tipos básicos, basándose en los valores mensuales promedio de temperatura y precipitación y en los tipos de vegetación.

laccolith (p. 477) Relatively small, mushroom-shaped pluton that forms when magma intrudes into parallel rock layers close to Earth's surface.

lake (p. 228) Natural or human-made body of water that can form when a depression on land fills with water.

Landsat satellite (p. 38) Information-gathering satellite that uses visible light and infrared radiation to map Earth's surface.

landslide (p. 186) Rapid downslope movement of a mass of loose soil, rock, or debris that has separated from the bedrock; can be triggered by an earthquake.

latent heat (p. 286) Stored energy in water vapor that is not released to warm the atmosphere until condensation takes place.

latitude (p. 27) Distance in degrees north and south of the equator.

Laurentia (p. 583) Ancient continent formed during the Proterozoic that is the core of modern-day North America.

lava (p. 99) Magma that flows out onto Earth's surface.

law (p. 19) Basic fact that describes the behavior of a natural phenomenon.

laccolith/lacolito (pág. 477) Plutón relativamente pequeño con forma de champiñón que se forma cuando se introduce el magma en las capas rocosas paralelas cerca de la superficie terrestre.

lake/lago (pág. 228) Masa de agua natural o hecha por el hombre que se forma cuando una depresión de la tierra se llena de agua.

Landsat satellite/satélite Landsat (pág. 38) Satélite que recoge información y que utiliza la luz visible y la radiación infrarroja para trazar la superficie terrestre.

landslide/deslizamiento (pág. 186) Rápido desplazamiento cuesta abajo de una masa de tierra, rocas o escombros sueltos que se ha separado de del lecho rocoso; puede ser causado por un terremoto.

latent heat/calor latente (pág. 286) Energía almacenada en el vapor del agua que se libera para calentar la atmósfera hasta que ocurre la condensación.

latitude/latitud (pág. 27) Distancia en grados al norte y al sur del ecuador.

Laurentia/Laurencia (pág. 583) Antiguo continente que se formó durante el Proterozoico y que en la actualidad corresponde al corazón de Norteamérica.

lava/lava (pág. 99) Magma que fluye por la superficie terrestre.

law/ley (pág. 19) Acontecimiento básico que describe el comportamiento de un fenómeno natural.

Le Système International D'Unités (SI) (p. 14) Modern version of the metric system based on a decimal system using the number 10 as the base unit; includes the meter (m), liter (L), second (s), and kilogram (kg).

lifted condensation level (LCL) (p. 279) Height at which condensation occurs, which often corresponds with the base of clouds.

liquid metallic hydrogen (p. 787) Form of hydrogen with both liquid and metallic properties that exists as a layer in the Jovian atmosphere.

lithification (p. 124) Transformation of sediments into rock through compaction and cementation.

lithosphere (p. 8) Earth's rigid outer shell, including the crust and the solid, uppermost part of the mantle.

loess (p. 197) Thick, wind-blown, fertile deposit of silt that contains high levels of nutrients and minerals.

longitude (p. 29) Distance in degrees east and west of the prime meridian.

longshore bar (p. 416) Submerged sandbar located in the surf zone of most beaches.

longshore current (p. 416) Current that flows parallel to the shore, moves large amounts of sediments, and is formed when incoming breakers spill over a longshore bar.

luminosity (p. 817) Energy output from the surface of a star per second; measured in watts.

lunar eclipse (p. 767) Occurs only during a full moon, when Earth passes between the Sun and the Moon, and Earth's shadow falls on the Moon.

luster (p. 84) Describes the metallic or nonmetallic way that a mineral reflects light from its surface.

Le Système International D'Unités/Sistema Internacional de Unidades (SI) (pág. 14) Versión moderna del sistema métrico basado en un sistema decimal que utiliza el número 10 como unidad base; incluye el metro (m), el litro (L), el segundo (s) y el kilogramo (kg).

lifted condensation level (LCL)/nivel de condensación por ascenso (NCA) (pág. 279) Altura a la cual ocurre la condensación, que a menudo corresponde con la base de nubes.

liquid metallic hydrogen/hidrógeno metálico líquido (pág. 787) Forma del hidrógeno con propiedades tanto líquidas como metálicas que existe en forma de capa en la atmósfera joviana.

lithification/litificación (pág. 124) Transformación de los sedimentos en roca debido a la compactación y la cementación.

lithosphere/litosfera (pág. 8) Parte exterior rígida de la Tierra, que incluye la corteza y la parte sólida más externa del manto.

loess/loes (pág. 197) grueso depósito fértil de cieno, soplado por el viento y el cual contiene niveles altos de nutrientes y minerales.

longitude/longitud (pág. 29) Distancia en grados al este y oeste del primer meridiano.

longshore bar/barra costera (pág. 416) Barra de arena sumergida, localizada en la zona del oleaje de la mayoría de las playas.

longshore current/corriente costera (pág. 416) Corriente que fluye paralela a la costa, transporta grandes cantidades de sedimentos y se forma cuando entran las olas rompientes y se desbordan sobre una barra costera.

luminosity/luminosidad (pág. 817) Energía que irradia la superficie de una estrella por segundo; se mide en vatios.

lunar eclipse/eclipse lunar (pág. 767) Ocurre sólo durante la luna llena, cuando la Tierra pasa entre el Sol y la Luna y la sombra de la Tierra cae sobre la Luna.

luster/lustre (pág. 84) Describe la manera en que un mineral refleja la luz de su superficie; puede ser metálico o no metálico.

mafic (p. 107) Dark-colored igneous rocks that are rich in iron and magnesium.

magma (p. 80) Molten material found beneath Earth's crust that forms minerals with large crystals when it cools slowly and forms minerals with small crystals when it cools rapidly.

mafic/máfica (pág. 107) Rocas ígneas oscuras, ricas en hierro y magnesio.

magma/magma (pág. 80) Material fundido que se encuentra bajo la corteza terrestre; forma minerales con cristales grandes cuando se enfría lentamente y forma minerales con cristales pequeños cuando se enfría rápidamente.

magnetic reversals (p. 451) Changes in Earth's magnetic field over geologic time, recorded in ocean-floor rocks and continental basalt flows.

magnetometer (p. 448) Device used to map the ocean floor that detects small changes in magnetic fields.

magnitude (p. 505) Measure of the energy released during an earthquake, which can be described using the Richter scale.

main sequence (p. 819) In an H-R diagram, the broad, diagonal band that includes about 90 percent of all stars and runs from hot, luminous stars in the upper-left corner to cool, dim stars in the lower-right corner.

map legend (p. 35) Key that explains what the symbols on a map represent.

map scale (p. 35) Ratio between the distances shown on a map and the actual distances on Earth's surface.

mare (p. 754) Dark-colored, basaltic, smooth plain on the Moon, between 3.1 and 3.8 billion years old, that formed from lava welling up and filling in large impact basins.

mass extinction (p. 611) Occurs when an unusually large number of organisms become extinct over a relatively short period of geologic time.

mass movement (p. 181) Downslope movement of Earth materials, due to gravity, that can occur suddenly or very slowly, depending on the weight of the material, its resistance to sliding, and whether a trigger, such as an earthquake, is involved.

mass number (p. 54) Combined number of protons and neutrons in the nucleus of an atom.

Maunder minimum (p. 372) Period of very low sunspot activity that occurred between 1645 and 1716 and closely corresponded with a cold climatic episode known as the "Little Ice Age."

meander (p. 224) Curve or bend in a stream formed when a stream's slope decreases, water builds up in the stream channel, and moving water erodes away the sides of the streambed.

mechanical weathering (p. 154) Process that breaks down rocks and minerals into smaller pieces but does not involve any change in their composition.

magnetic reversals/inversiones magnéticas (pág. 451) Cambios en el campo magnético de la Tierra a lo largo del tiempo geológico, registrado en las rocas del fondo del océano y en los flujos de basalto continentales.

magnetometer/magnetómetro (pág. 448) Dispositivo que se utiliza para hacer mapas del fondo oceánico, el cual detecta los cambios pequeños en los campos magnéticos.

magnitude/magnitud (pág. 505) Medida de la energía que se libera durante un terremoto; se puede describir utilizando la escala de Richter.

main sequence/secuencia principal (pág. 819) En un diagrama H-R, es la banda diagonal ancha que incluye cerca del 90 por ciento de todas las estrellas y que va desde las estrellas calientes y luminosas, en la esquina superior izquierda, hasta las estrellas frías de brillo débil, en la esquina inferior derecha.

map legend/leyenda del mapa (pág. 35) Clave que explica los símbolos en un mapa.

map scale/escala del mapa (pág. 35) Proporción entre las distancias que se muestran en un mapa y las distancias reales en la superficie terrestre.

mare/mar (pág. 754) Planicie lunar lisa, basáltica, de color oscuro, que data entre 3.1 y 3.8 billones de años de antigüedad, que se formó cuando brotó la lava y llenó las grandes cuencas de impacto.

mass extinction/extinción en masa (pág. 611) Ocurre cuando se extingue un número inusualmente grande de organismos, durante un período relativamente corto de tiempo geológico.

mass movement/movimiento de masa (pág. 181) Movimiento cuesta abajo de materiales terrestres debido a la gravedad; puede ocurrir repentina o muy lentamente, dependiendo del peso del material, su resistencia a deslizarse y si está implicado un acontecimiento iniciador, como un terremoto.

mass number/número de masa (pág. 54) Número combinado de protones y neutrones en el núcleo de un átomo.

Maunder minimum/mínimo de Maunder (pág. 372) Período de muy baja actividad de manchas solares que ocurrió entre 1645 y 1716 y correspondió aproximadamente con un episodio climático frío conocido como la "Pequeña Glaciación."

meander/meandro (pág. 224) Curva o recoveco en una corriente que se forma cuando disminuye la inclinación de la corriente, el agua se acumula en el canal y su movimiento erosiona los costados del lecho de la corriente.

mechanical weathering/meteorización mecánica (pág. 154) Proceso que rompe las rocas y los minerales en trozos más pequeños, pero que no implica ningún cambio en la composición del material.

Mercator projection (p. 32) Map with parallel lines of latitude and longitude that shows true direction and the correct shapes of landmasses but distorts areas near the poles.

mesosphere (p. 274) Layer of Earth's atmosphere above the stratopause.

meteor (p. 796) Streak of light produced when a meteoroid falls toward Earth and burns up in Earth's atmosphere.

meteorite (p. 579) Small asteroid or asteroid fragment that fell on early Earth, generating heat; (p. 796) meteoroid that does not completely burn up in Earth's atmosphere and strikes Earth's surface, sometimes causing an impact crater.

meteoroid (p. 796) Piece of interplanetary material that falls toward Earth and enters its atmosphere.

meteorology (p. 6) Branch of Earth science that studies the air that surrounds our planet; (p. 299) study of atmospheric phenomena such as fog, clouds, snow, rain, and lightning.

meteor shower (p. 797) Occurs when Earth intersects a cometary orbit and comet particles burn up as they enter Earth's upper atmosphere.

microclimate (p. 367) Localized climate that differs from the surrounding regional climate.

microcontinents (p. 582) Small pieces of continental crust that collided with one another throughout the Proterozoic, forming the cores of the continents.

mid-ocean ridges (p. 426) Chains of underwater mountains that run throughout the ocean basins, have a total length over 65 000 km, and contain countless active and extinct volcanoes.

mineral (p. 77) Naturally occurring, inorganic solid with a specific chemical composition and a definite crystalline structure that forms from magma or from supersaturated solution.

modern fauna (p. 628) New marine organisms that evolved during the Mesozoic, including crabs, shrimps, sponges, sea urchins, modern corals, snails, and clams, plus major vertebrate groups such as bony fishes, aquatic reptiles, sharks, and aquatic mammals.

modified Mercalli scale (p. 506) Measures earthquake intensity on a scale from I to XII; the higher the number, the greater the damage the earthquake has caused.

Mercator projection/proyección de Mercator (pág. 32) Mapa con las líneas de latitud y de longitud paralelas que muestra la dirección verdadera y las formas correctas de las masas terrestres, pero el cual distorsiona las áreas cercanas a los polos.

mesosphere/mesosfera (pág. 274) Capa de la atmósfera terrestre por encima de la estratopausa.

meteor/meteoro (pág. 796) Rayo luminoso que se produce cuando un meteorito cae a la Tierra y se quema en su atmósfera.

meteorite/meteorito (pág. 579) Asteroide pequeño o fragmento de asteroide que cayó en la Tierra temprana produciendo calor; (pág. 796) meteoroide que no se quema completamente en la atmósfera terrestre y llega a la superficie terrestre, en ocasiones causando un cráter de impacto.

meteoroid/meteoroide (pág. 796) Trozo de material interplanetario que cae a la Tierra y entra a su atmósfera.

meteorology/meteorología (pág. 6) Rama de las ciencias terrestres que estudia el aire que rodea nuestro planeta; (pág. 299) estudio de los fenómenos atmosféricos, como la niebla, las nubes, la nieve, la lluvia y los relámpagos.

meteor shower/lluvia de meteoros (pág. 797) Ocurre cuando la Tierra interseca la órbita de un cometa y las partículas de éste se queman al entrar a la atmósfera superior de la Tierra.

microclimate/microclima (pág. 367) Clima localizado que difiere del clima regional circundante.

microcontinents/microcontinentes (pág. 582) Trozos pequeños de corteza continental que chocaron unos con otros a lo largo del Proterozoico, formando los núcleos de los continentes.

mid-ocean ridges/dorsales mediooceánicas (pág. 426) Cadenas montañosas submarinas que se extienden a través de las cuencas oceánicas, tienen una longitud total de más de 65 000 km y contienen innumerables volcanes activos y extintos.

mineral/mineral (pág. 77) sólido inorgánico de ocurrencia natural con una composición química específica y una estructura cristalina definida que se forma a partir del magma o de una solución sobresaturada.

modern fauna/fauna moderna (pág. 628) Nuevos organismos marinos que evolucionaron durante el Mesozoico; incluyen los cangrejos, los camarones, las esponjas, los erizos de mar, los corales modernos, los caracoles y las almejas, además de grupos de vertebrados principales como los peces óseos, los reptiles acuáticos, los tiburones y los mamíferos acuáticos.

modified Mercalli scale/escala de Mercalli modificada (pág. 506) Mide la intensidad de un terremoto en una escala de I a XII; mientras más alto sea el número, mayor es el daño causado por el terremoto.

mold (p. 568) Fossil that can form when a shelled organism decays in sedimentary rock and is weathered away, leaving a hollowed-out impression.

molecule (p. 61) Combination of two or more atoms joined by covalent bonds.

moment magnitude scale (p. 506) Scale used to measure earthquake magnitude—taking into account the size of the fault rupture, the rocks' stiffness, and amount of movement along the fault—using values that can be estimated from the size of several types of seismic waves.

monoculture (p. 718) Planting of a single plant species, such as corn or wheat, in a field.

moraine (p. 202) Ridge of mixed debris deposited by a melting glacier.

mudflow (p. 185) Rapidly flowing, often destructive mixture of mud and water that may be triggered by an earthquake, intense rainstorm, or volcanic eruption.

mold/molde (pág. 568) Fósil que se puede formar cuando un organismo con concha se desintegra en roca sedimentaria, se meteoriza y queda una impresión hueca.

molecule/molécula (pág. 61) Combinación de dos o más átomos unidos por un enlace covalente.

moment magnitude scale/escala de magnitud momentánea (pág. 506) Escala que se utiliza para medir la intensidad de terremoto (tomando en cuenta el tamaño de la ruptura de la falla, la rigidez de la roca y la cantidad del movimiento a lo largo de la falla) usando valores que pueden estimarse a partir del tamaño de varios tipos de ondas sísmicas.

monoculture/monocultivo (pág. 718) Sembrado de una sola especie de planta, como el maíz o el trigo, en un campo.

moraine/morrena (pág. 202) Acumulación de escombros mezclados que deposita un glaciar que se está derritiendo.

mudflow/flujo o corriente de lodo (pág. 185) Mezcla de lodo y agua que fluye rápidamente y que a menudo es destructiva; puede ser causada por un terremoto, una lluvia intensa o una erupción volcánica.

natural resources (p. 655) Resources provided by Earth, including air, water, land, all living organisms, nutrients, rocks, and minerals.

nebula (p. 822) Large cloud of interstellar gas and dust that collapses on itself, due to its own gravity, and forms a hot, condensed object that will become a new star.

neutron (p. 54) Tiny atomic particle that is electrically neutral and has about the same mass as a proton.

neutron star (p. 825) Collapsed, dense core of a star that forms quickly while its outer layers are falling inward, has a radius of about 10 km, a mass 1.5 to 3 times that of the Sun, and contains only neutrons.

nonfoliated (p. 136) Metamorphic rock that lacks mineral grains with long axes in one direction.

nonpoint source (p. 730) Water-pollution source that generates pollution from widely spread areas, such as runoff from roads.

nonrenewable resource (p. 657) Resource that exists in Earth's crust in a fixed amount and can be replaced only by geological, physical, or chemical processes that take hundreds of millions of years.

natural resources/recursos naturales (pág. 655) Recursos que provee la Tierra; incluyen el aire, el agua, la tierra, todos los organismos vivos, los nutrientes, las rocas y los minerales.

nebula/nebulosa (pág. 822) Nube extensa de gas y polvo interestelares que se hunde sobre sí misma debido a su propia gravedad; forma un cuerpo condensado caliente, que se convertirá en una estrella nueva.

neutron/neutrón (pág. 54) Partícula atómica diminuta, eléctricamente neutra; posee aproximadamente la misma masa de un protón.

neutron star/estrella de neutrones (pág. 825) Centro denso y colapsado de una estrella que se forma rápidamente, mientras sus capas exteriores caen hacia adentro; tiene un radio de aproximadamente 10 km, una masa de 1.5 a 3 veces la del Sol y contiene sólo neutrones.

nonfoliated/no foliada (pág. 136) Roca metamórfica que carece de granos minerales con ejes largos en una dirección.

nonpoint source/fuente no localizada (pág. 730) Fuente de contaminación del agua que genera contaminación a partir de áreas muy separadas, como la escorrentía de los caminos.

nonrenewable resource/recurso no renovable (pág. 657) Recurso que existe en la corteza terrestre en una cantidad fija y sólo puede ser reemplazado por procesos geológicos, físicos o químicos que demoran centenas de millones de años.

normals (p. 360) Standard values for a location, including rainfall, wind speed, and temperatures, based on meteorological records compiled for at least 30 years.

nucleus (p. 54) Positively charged center of an atom, made up of protons and neutrons and surrounded by electrons in energy levels; (p. 797) small, solid core of a comet.

normals/normales (pág. 360) Valores estándares para una ubicación; incluyen la lluvia, la velocidad del viento y las temperaturas y se basan en los registros meteorológicos recopilados durante por lo menos 30 años.

nucleus/núcleo (pág. 54) Centro atómico positivamente cargado, compuesto de protones y neutrones y rodeado por electrones en niveles de energía; (pág. 797) centro pequeño y sólido de un cometa.

O

oceanography (p. 6) Study of Earth's oceans including the creatures that inhabit its waters, its physical and chemical properties, and the effects of human activities; (p. 385) discipline usually considered to have begun with the Challenger.

ore (p. 89) Mineral that contains a valuable substance that can be mined at a profit; (p. 661) natural resource that may be associated with igneous rocks or formed by processes at Earth's surface.

original horizontality (p. 558) Principle stating that sedimentary rocks are deposited in horizontal or nearly horizontal layers.

original preservation (p. 566) Describes a fossil with soft and hard parts that have not undergone any change since the organism's death.

Ornithischia (p. 632) "Bird-hipped," herbivorous dinosaurs.

orogeny (p. 528) Cycle of processes that form all mountain ranges, resulting in broad mountain belts, most of which are associated with plate boundaries.

orographic lifting (p. 285) Cloud formation that occurs when warm moist air is forced to rise up the side of a mountain.

Ouachita Orogeny (p. 614) Mountain-building event that occurred during the Late Paleozoic when Gondwana collided with the southeastern margin of Laurasia.

outwash plain (p. 202) Area at the leading edge of a glacier, where outwash is deposited by meltwater streams.

oxidation (p. 156) Chemical reaction of oxygen with other substances.

ozone (O3) (p. 273) A gas, formed by the addition of a third oxygen atom to an oxygen molecule, that exists in a layer in the atmosphere and absorbs ultraviolet radiation from the Sun; (p. 724) the major chemical in smog.

oceanography/oceanografía (pág. 6) Estudio de los océanos de la Tierra, incluyendo los seres que habitan en sus aguas, sus propiedades físicas y químicas y los efectos de las actividades humanas sobre ellos; (pág. 385) disciplina que generalmente se considera haber comenzado con las exploraciones del *Challenger*.

ore/mena (pág. 89) Mineral que contiene una sustancia valiosa que se puede extraer con fines de lucro; (pág. 661) recurso natural que puede estar asociado con las rocas ígneas o que puede haberse formado mediante procesos en la superficie terrestre.

original horizontality/horizontalidad original (pág. 558) Principio que establece que rocas sedimentarias se depositan en capas horizontales o casi horizontales.

original preservation/conservación original (pág. 566) Describe un fósil con partes blandas y duras que no han experimentado ningún cambio desde la muerte del organismo.

Ornithischia/Ornithischia (pág. 632) Dinosaurios herbívoros con "caderas de ave".

orogeny/orogenia (pág. 528) Ciclo de procesos que forman todas las cadenas montañosas, dando como resultado cadenas montañosas anchas, la mayor parte de ellos asociados con los límites de las placas.

orographic lifting/levantamiento orográfico (pág. 285) Formación de nubes que ocurre cuando aire húmedo caliente es forzado a elevarse por el costado de una montaña.

Ouachita Orogeny/orogenia Ouachita (pág. 614) Acontecimiento de formación de montañas que ocurrió durante el Paleozoico Tardío, cuando Gondwana chocó contra el margen sureste de Laurasia.

outwash plain/planicie derrubiada (pág. 202) Área en el borde frontal de un glaciar, donde las corrientes de agua de nieve derretida depositan los derrubios.

oxidation/oxidación (pág. 156) Reacción química del oxígeno con otras sustancias.

ozone/ozono (O_3) (pág. 273) Gas que se forma al añadir un tercer átomo de oxígeno a una molécula del oxígeno; existe en una capa atmosférica y absorbe la radiación ultravioleta del Sol; (pág. 724) sustancia química principal en el smog.

P

paleogeography (p. 601) Ancient geographic setting of an area.

paleomagnetism (p. 451) Study of Earth's magnetic record using data gathered from iron-bearing minerals in rocks that have recorded the orientation of Earth's magnetic field at the time of their formation.

Paleozoic fauna (p. 609) Collective name for the animals that dominated the Middle Paleozoic seas.

Pangaea (p. 444) Ancient landmass made up of all the continents that began to break apart about 200 million years ago.

parallax (p. 815) Apparent positional shift of an object caused by the motion of the observer.

partial melting (p. 102) Process in which different minerals melt into magma at different temperatures, changing its composition.

passive margin (p. 602) Edge of a continent along which there is no tectonic activity.

peat (p. 686) Light, spongy, organic fossil fuel derived from moss and other bog plants.

pegmatites (p. 112) Vein deposits of extremely large-grained minerals that can contain rare ores such as lithium and beryllium.

perigee (p. 766) Closest point in the Moon's elliptical orbit to Earth.

perihelion (p. 777) Point in a planet's orbit where it is closest to the Sun.

period (p. 555) Third-longest time unit in the geological time scale, measured in tens of millions of years to hundreds of millions of years, and defined by life-forms that were abundant or became extinct.

permeability (p. 242) Ability of a material to let water pass through, which is high in material with large, well-connected pores and low in material with few pores or small pores.

permineralization (p. 567) Process in which pore spaces in a fossil are filled in with mineral substances.

photosphere (p. 806) Lowest layer of the Sun's atmosphere that is also its visible surface, has an average temperature of 5800 K, and is about 400 km thick.

paleogeography/paleogeografía (pág. 601) Ubicación geográfica antigua de un área.

paleomagnetism/paleomagnetismo (pág. 451) Estudio del registro magnético de la Tierra; utiliza la información recogida de los minerales ferrosos en las rocas, los cuales registraron la orientación del campo magnético de la Tierra en el momento en que se formaron las rocas.

Paleozoic fauna/fauna del Paleozoico (pág. 609) Nombre colectivo para los animales que dominaron los mares del Paleozoico Medio.

Pangaea/Pangaea (pág. 444) Antigua masa terrestre compuesta de todos los continentes, los cuales empezaron a separarse hace aproximadamente 200 millones de años.

parallax/paralaje (pág. 815) Cambio de la posición aparente de un objeto causado por el movimiento del observador.

partial melting/fundición parcial (pág. 102) Proceso en el cual los diferentes minerales se funden en el magma a diferentes temperaturas, cambiando su composición.

passive margin/margen pasivo (pág. 602) Perifería de un continente a lo largo de la cual no ocurre actividad tectónica.

peat/turba (pág. 686) Combustible fósil liviano, esponjoso y orgánico derivado del musgo y otras plantas de ciénaga.

pegmatites/pegmatitas (pág. 112) Vetas de minerales de grano extremadamente grueso que pueden contener minerales raros, como el litio y el berilio.

perigee/perigeo (pág. 766) Punto más cercano a la Tierra en la órbita elíptica de la Luna.

perihelion/perihelio (pág. 777) Punto en la órbita de un planeta cuando está más cercano al Sol.

period/período (pág. 555) Tercera unidad de tiempo más grande en la escala del tiempo geológico; se mide en decenas de millones a centenares de millones de años y se define según las formas de vida que abundaron o que se extinguieron.

permeability/permeabilidad (pág. 242) Capacidad de un material de permitir que el agua lo atraviece, la cual es grande en materiales con poros grandes y bien conectados y baja en materiales con pocos poros o poros pequeños.

permineralization/permineralización (pág. 567) Proceso en que los espacios porosos en un fósil se llenan con minerales.

photosphere/fotosfera (pág. 806) Capa más baja de la atmósfera solar, la cual es su superficie visible. Tiene una temperatura promedio de 5800 K y mide aproximadamente 400 km de ancho.

photovoltaic cells (p. 691) Thin, transparent wafers that convert sunlight into electrical energy and are made up of layers of boron- and phosphorus-enriched silicon.

pillow basalt (p. 536) Igneous rock formed when magma pushes through dikes and erupts onto the seafloor.

planetesimal (p. 794) Space object built of solid particles that can form planets through collisions and mergers.

plasma (p. 69) Hot, highly ionized, electrically conducting gas.

plutons (p. 476) Intrusive igneous rock bodies, including batholiths, stocks, sills, and dikes, formed through mountain-building processes and oceanic-oceanic collisions; can be exposed at Earth's surface due to uplift and erosion.

point source (p. 730) Water-pollution source that generates pollution from a single point of origin, such as an industrial site.

polar easterlies (p. 307) Global wind system that lies between 60∞ latitude and the poles and is characterized by cold air.

polar zones (p. 362) Areas of Earth where solar radiation strikes at a low angle, resulting in temperatures that are nearly always cold; extend from 66.5° north and south of the equator to the poles.

pollutant (p. 665) Substance that disrupts Earth's geochemical cycles and can harm the health of living things or adversely affect their activities.

porosity (p. 129) Percentage of open spaces between grains in a rock; (p. 241) is highest in well-sorted sediments.

porphyritic (p. 110) Rock texture characterized by large, well-formed crystals surrounded by finer-grained crystals of the same mineral.

porphyroblasts (p. 136) Large crystals that form in solid rock by the reorganization of atoms during metamorphism.

Precambrian shield (p. 581) Continental core of Archean and Proterozoic rock that may be exposed at the surface or buried by sedimentary rocks.

precession (p. 783) Wobble in Earth's rotational axis.

photovoltaic cells/celdas fotovoltaicas (pág. 691) Láminas delgadas y transparentes que convierten la luz solar en energía eléctrica y que están compuestas de capas de silicio enriquecido con boro y fósforo.

pillow basalt/almohada de basalto (pág. 536) Roca ígnea que se forma cuando el magma atraviesa diques y hace erupción en el fondo del mar.

planetesimal/planetesimal (pág. 794) Cuerpo espacial formado por partículas sólidas y los cuales pueden formar planetas mediante choques y fusiones.

plasma/plasma (pág. 69) Gas caliente, altamente ionizado y conductor de electricidad.

plutons/plutones (pág. 476) Cuerpos rocosos ígneos intrusivos, que incluyen los batolitos, los troncos, las intrusiones y los diques, que se formaron mediante los procesos de creación de montañas y las colisiones oceánicas-oceánicas; pueden estar expuestos a la superficie terrestre debido al levantamiento y la erosión.

point source/fuente localizada (pág. 730) Fuente de contaminación de agua que genera contaminación a partir de un solo punto de origen, como por ejemplo, una ubicación industrial.

polar easterlies/vientos polares del este (pág. 307) Sistema de vientos globales que se encuentra entre los 60° de latitud y los polos y lo caracteriza el aire frío.

polar zones/zonas polares (pág. 362) Áreas de la Tierra donde la radiación solar llega a un ángulo bajo, dando como resultado temperaturas que son casi siempre frías; se extienden desde los 66.5° al norte y al sur del ecuador hasta los polos.

pollutant/contaminante (pág. 665) Sustancia que perturba los ciclos geoquímicos de la Tierra y puede causar daños de salud a los seres vivos o afectar adversamente sus actividades.

porosity/porosidad (pág. 129) Porcentaje de espacios abiertos entre los granos de una roca; (pág. 241) es más alta en sedimentos bien clasificados.

porphyritic/porfirítica (pág. 110) Textura rocosa caracterizada por grandes cristales bien formados y rodeados por cristales del mismo mineral de grano más fino.

porphyroblasts/porfiroblastos (pág. 136) Cristales grandes que se forman en la roca sólida por la reorganización de átomos durante el metamorfismo.

Precambrian shield/escudo precámbrico (pág. 581) Núcleo continental de rocas Arqueanas y Proterozoicas que puede estar expuesto en la superficie o sepultado por rocas sedimentarias.

precession/precesión (pág. 783) Temblor en el eje rotacional de la Tierra.

precipitation (p. 289) All solid and liquid forms of water—including rain, snow, sleet, and hail—that fall from clouds.

prevailing westerlies (p. 306) Global wind system that lies between 30∞ and 60∞ north and south latitude, where surface air moves toward the poles in an easterly direction.

primary wave (p. 498) Seismic wave that squeezes and pulls rocks in the same direction that the wave travels, causing rock particles to move back and forth.

primate (p. 640) Mammal that developed specialized traits, such as opposable thumbs and two eyes directed forward, primarily because of arboreal ancestry.

prime meridian (p. 29) Imaginary line representing 0∞ longitude, running from the north pole, through Greenwich, England, to the south pole.

prokaryote (p. 591) Single-celled organism that lacks a nucleus, is generally small, belongs to the Kingdom Monera, and existed before the Proterozoic.

prominence (p. 809) Arc of gas ejected from the chromosphere, or gas that condenses in the Sun's inner corona and rains back to the surface, that can reach temperatures over 50 000 K and is associated with sunspots.

proton (p. 54) Tiny atomic particle that has mass and a positive electrical charge.

protostar (p. 822) Hot, condensed object at the center of a nebula that will become a new star when nuclear fusion reactions begin.

pyroclastic flow (p. 484) Swift-moving, potentially deadly clouds of gas, ash, and other volcanic material produced by a violent eruption.

precipitation/precipitación (pág. 289) Todas las formas líquidas y sólidas de agua (lluvia, nieve, aguanieve y granizo) que cae de las nubes.

prevailing westerlies/vientos ponientes prevalecientes (pág. 306) Sistema de vientos globales que se localiza entre los 30° y 60° de latitud norte y sur, en donde el aire superficial se mueve hacia los polos en dirección este.

primary wave/onda primaria (pág. 498) Onda sísmica que comprime y hala las rocas en la misma dirección en que se mueve la onda, haciendo que las partículas de la roca se muevan de adelante hacia atrás.

primate/primate (pág. 640) Mamífero que desarrolló rasgos especializados, como los pulgares oponibles y dos ojos dirigidos hacia el frente, principalmente a causa de su ascendencia arbórea.

prime meridian/primer meridiano (pág. 29) Línea imaginaria que representa la longitud 0 y la cual va desde el polo norte, pasa por Greenwich, Inglaterra, hasta el polo sur.

prokaryote/procariota (pág. 591) Organismo unicelular que carece de núcleo, es generalmente pequeño, pertenece al Reino Monera y existió desde antes del Proterozoico.

prominence/prominencia (pág. 809) Arco de gas expulsado de la cromosfera, o gas que se condensa en la corona interna del Sol y que se precipita de nuevo sobre su superficie; puede alcanzar temperaturas mayores a los 50 000 K y se asocia con las manchas solares.

proton/protón (pág. 54) Partícula atómica diminuta que tiene masa y una carga eléctrica positiva.

protostar/protoestrella (pág. 822) Cuerpo condensado caliente en el centro de una nebulosa que se convertirá en una estrella nueva cuando empiecen las reacciones de fusión nuclear.

pyroclastic flow/flujo piroclástico (pág. 484) Nubes de gas, cenizas y otros materiales volcánicos, muy rápidas y potencialmente mortales, producidas por una erupción violenta.

Q

quasars (p. 845) Starlike, very bright, extremely distant objects with emission lines in their spectra.

quasars/cuásares (pág. 845) Cuerpos semejantes a estrellas, muy brillantes y extremadamente lejanos con líneas de emisión en sus espectros.

R

radiation (p. 275) Energy transfer through space by visible light, ultraviolet radiation, and other forms of electromagnetic waves.

radiation/radiación (pág. 275) Transferencia de energía por el espacio a través de luz visible, radiación ultravioleta y otras formas de ondas electromagnéticas.

radioactive decay (p. 562) Emission of atomic particles at a constant rate from a radioactive substance and its resulting change into other elements over time.

radioactivity (p. 58) Spontaneous process of decay, or breaking apart, through which unstable nuclei emit radiation.

radio galaxy (p. 844) Very bright, often giant, elliptical galaxy that emits as much or more energy in the form of radio wavelengths as it does wavelengths of visible light.

radiometric dating (p. 562) Process used to determine the absolute age of a rock or fossil by determining the ratio of parent nuclei to daughter nuclei within a given sample.

radiosonde (p. 314) Balloon-borne weather instrument whose sensors measure air pressure, humidity, temperature, wind speed, and wind direction.

ray (p. 754) Long trail of ejecta that radiates outward from a Moon crater.

recharge (p. 252) Process by which water from precipitation and runoff is added to the zone of saturation.

reclamation (p. 717) Process in which a mining company restores land used during mining operations to its original contours and replants vegetation.

red beds (p. 587) Sedimentary rock deposits that contain oxidized iron, providing evidence that free oxygen existed in the atmosphere during the Proterozoic.

reflecting telescope (p. 749) A type of telescope that uses mirrors to focus visible light.

refracting telescope (p. 749) A type of telescope that uses lenses to focus visible light.

regional metamorphism (p. 134) Process that affects large areas of Earth's crust, producing belts classified as low, medium, or high grade, depending on pressure on the rocks, temperature, and depth below the surface.

regoligh (p. 755) Layer of loose, ground-up rock on the lunar surface.

regression (p. 604) Occurs when sea level falls, causing the shoreline to move seaward, and results in shallower-water deposits overlying deeper-water deposits, with adjacent sediment types overlying each other in vertical succession.

radiactive decay/desintegración radiactiva (pág. 562) Emisión de partículas atómicas de una sustancia radiactiva, a una tasa constante; con el tiempo la sustancia se convierte en otros elementos.

radiactivity/radiactividad (pág. 58) Proceso de desintegración o ruptura espontánea mediante el cual emiten radiación los núcleos inestables.

radio galaxy/radiogalaxia (pág. 844) Galaxia elíptica muy brillante, a menudo gigantesca, que emite tanta o más energía en forma de longitudes de ondas radiales, como lo hace en longitudes de ondas de luz visible.

radiometric dating/datación radiométrica (pág. 562) Proceso que se utiliza para establecer la edad absoluta de una roca o fósil determinando la proporción de núcleos originales a núcleos derivados dentro de una muestra dada.

radiosonde/radiosonda (pág. 314) Instrumento meteorológico que se monta en un globo y cuyos sensores miden la presión atmosférica, la humedad, la temperatura, la velocidad y dirección del viento.

ray/rayo (pág. 754) Largo rastro de ejecta que irradia hacia afuera de un cráter lunar.

recharge/recarga (pág. 252) Proceso mediante el cual el agua de la precipitación y de la escorrentía se añade a la zona de saturación.

reclamation/reclamación (pág. 717) Proceso en que una compañía minera restablece la tierra utilizada durante las operaciones de minería y planta nueva vegetación.

red beds/lechos rojos (pág. 587) Depósitos de roca sedimentaria que contienen hierro oxidado, lo cual proporciona pruebas de que existió oxígeno libre en la atmósfera durante el Proterozoico.

reflecting telescope/telescopio reflector (pág. 749) Tipo de telescopio que utiliza espejos para enfocar la luz visible.

refracting telescope/telescopio refractor (pág. 749) Tipo de telescopio que utiliza lentes para enfocar la luz visible.

regional metamorphism/metamorfismo regional (pág. 134) Proceso que afecta grandes áreas de la corteza terrestre; produce cinturones de bajo, de medio o de alto grado, dependiendo de la presión sobre las rocas, la temperatura y la profundidad debajo de la superficie.

regoligh/regolito (pág. 755) Capa de roca molida suelta en la superficie lunar.

regression/regresión (pág. 604) Ocurre cuando baja el nivel del mar, haciendo que la costa se mueva hacia el mar y resulta en que los depósitos de agua más superficiales cubren los depósitos de aguas profundas con tipos de sedimentos adyacentes que se acumulan uno encima del otro en sucesión vertical.

rejuvenation (p. 227) Process during which a stream resumes downcutting toward its base level, increasing its rate of flow.

relative humidity (p. 283) Ratio of water vapor contained in a specific volume of air compared with how much water vapor that amount of air actually can hold.

remote sensing (p. 37) Process of gathering data about Earth from far above the planet's surface.

renewable resource (p. 656) Natural resource, such as fresh air and most groundwater, than can be used indefinitely without causing a reduction in the available supply.

residual soil (p. 168) Soil located above its parent bedrock.

retrograde motion (p. 776) A planet's backward motion in the sky.

Richter scale (p. 505) Numerical scale used to measure the magnitude of an earthquake, using values based on the size of the earthquake's largest seismic waves.

ridge push (p. 461) Tectonic process associated with convection currents in Earth's mantle that occurs when the weight of an elevated ridge pushes an oceanic plate toward a subduction zone.

rift valley (p. 456) Long, narrow depression that forms when continental crust begins to separate at a divergent boundary.

rill erosion (p. 163) Erosion in which water running down the side of a slope carves a small stream channel.

rille (p. 754) Valleylike structure that meanders across some regions of the Moon's maria.

rock cycle (p. 138) Continuous, dynamic set of processes by which rocks are changed into other types of rock.

RR Lyrae variable (p. 834) Stars with pulsation periods ranging from 1.5 hours to 1 day, generally having the same luminosity, regardless of pulsation period length.

runoff (p. 212) Water that flows downslope on Earth's surface and may enter a stream, river, or lake; its rate is influenced by the angle of the slope, vegetation, rate of precipitation, and soil composition.

rejuvenation/rejuvenecimiento (pág. 227) Proceso en que una corriente reanuda la erosión hacia su nivel base y aumenta su velocidad de flujo.

relative humidity/humedad relativa (pág. 283) Proporción del vapor de agua que contiene un volumen específico de aire comparado con la cantidad de vapor de agua que en realidad puede contener esa cantidad de aire.

remote sensing/percepción remota (pág. 37) Proceso para reunir datos acerca de la Tierra desde muy por encima de la superficie del planeta.

renewable resource/recurso renovable (pág. 656) Recurso natural, como el aire fresco y la mayoría de las aguas subterráneas, que se puede utilizar indefinidamente sin causar una reducción en el suministro disponible.

residual soil/suelo residual (pág. 168) Tierra localizada encima de su lecho rocoso madre.

retrograde motion/movimiento retrógrado (pág. 776) Movimiento hacia atrás de un planeta en el firmamento.

Richter scale/escala de Richter (pág. 505) Escala numérica que se usa para medir la intensidad de un terremoto; emplea valores basados en el tamaño de las ondas sísmicas más grandes del terremoto.

ridge push/empuje de las cordilleras (pág. 461) Proceso tectónico asociado con las corrientes de convección en el manto de la Tierra que ocurre cuando el peso de una cordillera elevada empuja una placa oceánica hacia una zona de subducción.

rift valley/valle de fosas tectónicas (pág. 456) Depresión larga y estrecha que se forma cuando la corteza continental se empieza a separar en un límite divergente.

rill erosion/erosión por surcos (pág. 163) Erosión en la cual el agua que corre cuesta abajo forma un canal pequeño.

rille/surco (pág. 754) Estructura tipo valle que serpentea a través de algunas regiones de los mares lunares.

rock cycle/ciclo de las rocas (pág. 138) Conjunto de procesos continuos y dinámicos a través de los cuales las rocas se transforman en otros tipos de roca.

RR Lyrae variable/variable tipo RR Lyrae (pág. 834) Estrellas con períodos de pulsación que van de 1.5 horas a 1 día; en general tienen la misma luminosidad, a pesar de la longitud del período de la pulsación.

runoff/escorrentía (pág. 212) Agua que fluye cuesta abajo sobre la superficie terrestre y que puede incorporarse a una corriente, río o lago; su velocidad está influenciada por el ángulo de la pendiente, la vegetación, la tasa de precipitación y la composición del suelo.

S

Saffir-Simpson hurricane scale (p. 344) Classifies hurricanes according to air pressure in the center, wind speed, and property damage potential on a scale ranging from Category 1 to Category 5.

salinity (p. 392) Measure of the amount of salts dissolved in seawater, which is 35 ppt, on average.

Saurischia (p. 632) "Lizard-hipped" dinosaurs—quadrupedal, herbivorous sauropods and bipedal, carnivorous theropods.

scientific notation (p. 16) A type of shorthand used by scientists in which a number is expressed as a multiplier and a power of 10.

sea-breeze thunderstorm (p. 331) Local air-mass thunderstorm that commonly occurs along a coastal area during the summer.

seafloor spreading (p. 453) Hess's theory that new ocean crust is formed at mid-ocean ridges and destroyed at deep-sea trenches; occurs in a continuous cycle of magma intrusion and spreading.

sea level (p. 388) Level of the oceans' surfaces, which is presently rising 1 to 2 mm per year due to melting glaciers

seamount (p. 428) Basaltic, submerged volcano on the seafloor that is more than 1 km high.

season (p. 370) Short-term period of climatic change caused by regular variations in temperature, hours of daylight, and weather patterns that are due to the tilt of Earth's axis as it revolves around the Sun, causing different areas of Earth to receive different amounts of solar radiation.

secondary wave (p. 498) Seismic wave that causes rock particles to move at right angles to the direction of the wave.

sediments (p. 121) Solid particles deposited on Earth's surface that can form sedimentary rocks by processes such as weathering, erosion, deposition, and lithification.

seismic gap (p. 515) Place along an active fault that has not experienced an earthquake for a long time.

seismogram (p. 500) Record produced by a seismometer that can provide individual tracking of each type of seismic wave.

Saffir-Simpson hurricane scale/escala de huracanes de Saffir-Simpson (pág. 344) Clasifica los huracanes según la presión del aire en el centro, la velocidad del viento y el potencial de daño a la propiedad en una escala que va desde la Categoría 1 hasta la Categoría 5.

salinity/salinidad (pág. 392) Medida de la cantidad de sales disueltas en el agua de mar, que en promedio es de 35 ppt.

Saurischia/Saurischia (pág. 632) dinosaurios de "cadera de lagarto"; cuadrúpedos, saurópodos herbívoros y terópodos carnívoros bípedos.

scientific notation/notación científica (pág. 16) Tipo de escritura simplificada que usan los científicos en que un número se expresa como un multiplicador y una potencia de 10.

sea-breeze thunderstorm/tormenta eléctrica de brisa marina (pág. 331) Tormenta de masa de aire local que ocurre comúnmente a lo largo de un área costera durante el verano.

seafloor spreading/expansión del suelo marino (pág. 453) Teoría de Hess que dice que la nueva corteza oceánica se forma en las dorsales mediooceánicas y es destruida en las fosas submarinas profundas; ocurre en un ciclo continuo de intrusión y esparcimiento de magma.

sea level/nivel del mar (pág. 388) Nivel de las superficies de los océanos que actualmente sube de 1 a 2 mm por año, debido al deshielo de los glaciares

seamount/montaña submarina (pág. 428) Volcán basáltico sumergido en el fondo marino y el cual es de más de 1 km de alto.

season/temporada (pág. 370) Período de cambio climático de corto plazo debido a las variaciones regulares en temperatura, horas de luz solar y patrones meteorológicos debido a la inclinación del eje de la Tierra cuando gira alrededor del Sol, lo cual hace que las diferentes áreas de la Tierra reciban diferentes cantidades de radiación solar.

secondary wave/onda secundaria (pág. 498) Onda sísmica que hace que las partículas de las rocas se muevan formando ángulos rectos a la dirección de la onda.

sediments/sedimentos (pág. 121) Partículas sólidas depositadas en la superficie terrestre que puede formar rocas sedimentarias mediante procesos como la metorización, la erosión, la deposición y la litificación.

seismic gap/brecha sísmica (pág. 515) Lugar a lo largo de una falla activa que no ha sufrido un terremoto durante mucho tiempo.

seismogram/sismograma (pág. 500) Registro que produce un sismógrafo y el cual puede proporcionar un registro individual de cada tipo de onda sísmica.

seismometer (p. 500) Instrument used to measure horizontal or vertical motion during an earthquake.

seismometer/sismógrafo (pág. 500) Instrumento que se usa para medir los movimientos horizontales o verticales durante un terremoto.

shield volcano (p. 481) Broad volcano with gently sloping sides built by nonexplosive eruptions of basaltic lava that accumulates in layers.

shield volcano/volcán de escudo (pág. 481) Volcán ancho con laderas levemente inclinadas formado por las erupciones no explosivas de lava basáltica que se acumula en capas.

side-scan sonar (p. 386) Technique that directs sound waves at an angle to the seafloor or deep-lake floor, allowing underwater topographic features to be mapped.

side-scan sonar/sonar de escaneo lateral (pág. 386) Técnica que dirige las ondas sonoras a un ángulo hacia el fondo marino o de un lago profundo, lo que permite trazar el relieve topográfico submarino.

silicate (p. 81) Mineral that contains silicon (Si) and oxygen (O) and usually one or more other elements.

silicate/silicato (pág. 81) Mineral que contiene silicio (Si) y oxígeno (O) y uno o más de otros elementos, generalmente.

sill (p. 477) Pluton that forms when magma intrudes parallel rock layers.

sill/intrusión (pág. 477) Plutón que se forma cuando el magma penetra las capas paralelas de roca.

sinkhole (p. 246) Depression in Earth's surface formed when a cave collapses or bedrock is dissolved by acidic rain or moist soil.

sinkhole/sumidero (pág. 246) Depresión en la superficie terrestre que se forma cuando una caverna se desploma o la lluvia ácida o la tierra húmeda disuelven el lecho rocoso.

slab pull (p. 462) Tectonic process associated with convection currents in Earth's mantle that occurs as the weight of the subducting plate pulls the trailing lithosphere into a subduction zone.

slab pull/tracción de placa (pág. 462) Proceso tectónico asociado con las corrientes de convección del manto de la Tierra que ocurre cuando el peso de la placa subductora hala la litosfera hacia una zona de subducción.

slump (p. 187) Mass movement that occurs when Earth materials in a landslide rotate and slide along a curved surface, leaving a crescent-shaped scar on a slope.

slump/desprendimiento (pág. 187) Movimiento de masa que ocurre cuando los materiales terrestres de un deslizamiento giran y se deslizan por una superficie curva y dejan una cicatriz con forma de medialuna en una pendiente.

smog (p. 724) Yellow-brown, photochemical haze that occurs when solar radiation acts on air polluted with hydrocarbons and nitrogen oxides from automobile exhausts.

smog/smog (pág. 724) Neblina fotoquímica ocre que se presenta cuando la radiación solar actúa sobre el aire contaminado con óxidos de nitrógeno e hidrocarburos provenientes de los escapes de automóviles.

soil (p. 167) Loose covering of weathered rock and decayed organic matter overlying Earth's bedrock that is characterized by texture, fertility, and color and whose composition is determined by its parent rock and environmental conditions.

soil/suelo (pág. 167) Cubierta suelta de roca erosionada y materia orgánica degradada que cubre el lecho rocoso terrestre; se caracteriza por su textura, fertilidad y color; la roca original y las condiciones ambientales determinan su composición.

soil horizon (p. 169) A distinct layer within a soil profile.

soil horizon/horizonte del suelo (pág. 169) Capa distintiva dentro de un perfil del suelo.

soil profile (p. 168) Vertical sequence of soil layers, containing horizon A (topsoil), horizon B (subsoil), and horizon C (weathered parent material).

soil profile/perfil del suelo (pág. 168) Sucesión vertical de capas de tierra que comprende los horizontes A (capa superficial del suelo), B (subsuelo) y C (material original erosionado).

solar eclipse (p. 765) Occurs when the Moon passes between Earth and the Sun and the Moon casts a shadow on Earth, blocking Earth's view of the Sun; can be partial or total.

solar eclipse/eclipse solar (pág. 765) Ocurre cuando la Luna pasa entre la Tierra y el Sol y la Luna proyecta su sombra sobre la Tierra y bloquea el Sol; puede ser parcial o total.

solar flare (p. 809) Violent eruption of radiation and particles from the Sun's surface that is associated with sunspots.

solar flare/erupción solar (pág. 809) Violenta erupción de radiación y partículas desde la superficie del Sol que se asocia con manchas solares.

solar wind (p. 807) Wind of charged particles that flows throughout the solar system and begins as gas flowing outward from the Sun's corona at high speeds.

solar wind/viento solar (pág. 807) Viento de partículas cargadas que fluye a través del sistema solar y comienza como un gas que fluye de la corona del Sol hacia el exterior a altas velocidades.

solution (p. 65) Homogeneous mixture whose components cannot be distinguished and can be classified as liquid, gaseous, solid, or a combination; (p. 215) in a stream, is created when materials, such as silica (SiO_2), calcium (Ca), and sodium (Na), dissolve in the stream's water.

solution/solución (pág. 65) Mezcla homogénea cuyos componentes no se pueden distinguir; puede clasificarse como líquida, gaseosa, sólida o una combinación de éstas; (pág. 215) en una corriente de agua, se crea cuando materiales como el sílice (SiO_2), el calcio (Ca) y el sodio (Na) se disuelven en la corriente de agua.

sonar (p. 40) Use of sound waves to detect and measure objects underwater.

sonar/sonar (pág. 40) Uso de las ondas sonoras para detectar y medir objetos submarinos.

specific gravity (p. 87) Ratio of the weight of a substance to the weight of an equal volume of H_2O at 4°C.

specific gravity/gravedad específica (pág. 87) Proporción del peso de una sustancia al peso de un volumen igual de H_2O a 4°C.

spectrum (p. 811) Arrangement of visible light ordered according to wavelength.

spectrum/espectro (pág. 811) Arreglo de la luz visible ordenada según su longitud de onda.

spinoff (p. 752) NASA technology that has been passed to commercial industries for common use.

spinoff/tecnología derivada (pág. 752) Tecnología de la NASA que se les ha pasado a las industrias comerciales para el uso común.

spiral density wave (p. 838) Type of wave that creates spiral arms in a galaxy and is composed of alternating regions with variable density that rotate in a fixed pattern.

spiral density wave/onda de densidad espiral (pág. 838) Tipo de onda que crea brazos en espiral en una galaxia y la cual se compone de regiones alternadas con densidad variable que giran en un patrón fijo.

spring (p. 249) Natural discharge of groundwater at Earth's surface where an aquifer and an aquiclude come in contact.

spring/manantial (pág. 249) Descarga natural de agua subterránea en la superficie terrestre donde un acuífero y un acuicludo entran el contacto.

stability (p. 286) Ability of an air mass to resist rising.

stability/estabilidad (pág. 286) Capacidad de una masa de aire para resistirse a ascender.

stalactite (p. 248) Cone-shaped or cylindrical dripstone deposit of calcium carbonate that hangs like an icicle from a cave's ceiling.

stalactite/estalactita (pág. 248) Depósito rocoso de carbonato de calcio, de forma cónica o cilíndrica, que se forma por goteo y que cuelga como un carámbano del techo de una caverna.

stalagmite (p. 248) Mound-shaped dripstone deposit of calcium carbonate that forms on a cave's floor beneath a stalactite.

stalagmite/estalagmita (pág. 248) Depósito de carbonato de calcio, con forma de montículo, que se forma por goteo en el piso de una caverna, debajo de una estalactita.

station model (p. 317) Record of weather data for a specific place at a specific time, using meteorological symbols.

station model/código meteorológico (pág. 317) Registro de los datos del tiempo para un lugar específico en un tiempo dado, el cual utiliza símbolos meteorológicos.

steady-state theory (p. 847) Proposes that the universe is the same as it has always been.

steady-state theory/teoría del estado estacionario (pág. 847) Propone que el universo es igual a lo que siempre ha sido.

stock (p. 477) Irregularly shaped pluton that is similar to a batholith but smaller, generally forms 10-30 km beneath Earth's surface, and cuts across older rocks.

stock/tronco (pág. 477) Plutón de forma irregular que parece un batolito pero más pequeño, generalmente se forma de 10 a 30 km bajo la superficie terrestre y atraviesa rocas más antiguas.

storm surge (p. 345) Occurs when powerful, hurricane-force winds drive a mound of ocean water toward shore, where it washes over the land, often causing enormous damage.

storm surge/oleaje de tempestad (pág. 345) Ocurre cuando los vientos poderosos con fuerza huracanada arrojan una gran cantidad de agua del océano hacia la costa y la lanzan sobre la tierra, causando a menudo un daño enorme.

strain (p. 496) Deformation of materials in response to stress.

strain/tensión (pág. 496) Deformación de materiales en respuesta a un estrés.

stratosphere (p. 274) Layer of Earth's atmosphere that is located above the tropopause and is made up primarily of concentrated ozone.

stratosphere/estratosfera (pág. 274) Capa de la atmósfera terrestre ubicada por encima de la tropopausa y la cual está compuesta principalmente de ozono concentrado.

Glossary/Glosario

streak (p. 85) Color a mineral leaves when it is rubbed across an unglazed porcelain plate or when it is broken up and powdered.

stream bank (p. 222) Ground bordering each side of a stream that keeps the moving water confined.

stream channel (p. 222) Narrow pathway carved into sediment or rock by the movement of surface water.

stress (p. 495) Forces per unit area that act on a material—compression, tension, and shear.

stromatolites (p. 585) Large mats and mounds composed of billions of photosynthesizing cyanobacteria that dominated the Proterozoic's shallow oceans.

subduction (p. 457) Process by which one tectonic plate slips beneath another tectonic plate.

sublimation (p. 68) Process by which a solid slowly changes to a gas without first entering a liquid state.

summer solstice (p. 760) Occurs when the Sun is directly overhead at 23.5∞ north latitude, around June 21, and results in the maximum number of daylight hours for the northern hemisphere and the minimum number for the southern hemisphere.

sunspot (p. 808) Dark spot on the surface of the photosphere that typically lasts two months, occurs in pairs, and has a penumbra and an umbra.

supercell (p. 334) Extremely powerful, self-sustaining thunderstorm characterized by intense, rotating updrafts.

supercluster (p. 842) Gigantic threadlike or sheetlike cluster of galaxies that is hundreds of millions of light-years in size.

supernova (p. 825) Massive explosion that occurs when the outer layers of a star are blown off.

superposition (p. 558) Principle stating that in an undisturbed rock sequence, the oldest rock layers are at the bottom and each successive layer is younger than the layer beneath.

surface current (p. 403) Wind-driven movement of ocean water that primarily affects the upper few hundred meters of the ocean.

surface wave (p. 498) Seismic wave that moves in two directions as it passes through rocks, causing the ground to move both up and down and from side to side.

streak/veta (pág. 85) Color que deja un mineral cuando se frota contra un plato de porcelana sin barnizar o cuando se rompe y se pulveriza.

stream bank/cauce de corriente de agua (pág. 222) Suelo limítrofe a cada lado de una corriente que mantiene restringida el agua en movimiento.

stream channel/canal de corriente de agua (pág. 222) Sendero estrecho labrado en el sedimento o en la roca por el movimiento del agua superficial.

stress/estrés (pág. 495) Fuerzas por unidad de área que actúan sobre un material; compresión, tensión y cizallamiento.

stromatolites/estromatolitos (pág. 585) Montículos grandes compuestos de miles de millones de cianobacterias fotosintéticas que dominaron los océanos superficiales del Proterozoico.

subduction/subducción (pág. 457) Proceso en que una placa tectónica se desliza por debajo de otra.

sublimation/sublimación (pág. 68) Proceso en que un sólido cambia lentamente a un gas, sin pasar primero por el estado líquido.

summer solstice/solsticio de verano (pág. 760) Ocurre cuando el Sol está directamente sobre los 23.5° de latitud norte, alrededor del 21 de junio y tiene como resultado el número máximo de horas de luz de día, para el hemisferio norte y el número mínimo, para el hemisferio sur.

sunspot/mancha solar (pág. 808) Sitio oscuro en la superficie de la fotosfera que dura típicamente dos meses, ocurre en pares y tiene una penumbra y una umbra.

supercell/supercelda (pág. 334) Tormenta autosostenible extremadamente poderosa, caracterizada por intensas corrientes ascendentes giratorias.

supercluster/supercúmulo (pág. 842) Cúmulo gigantesco de galaxias con forma de filamento o lámina que mide centenares de millones de años luz.

supernova/supernova (pág. 825) Enorme explosión que ocurre cuando estallan las capas exteriores de una estrella.

superposition/superposición (pág. 558) Principio que dice que en una sucesión rocosa no alterada, las capas de roca más antiguas están en el fondo y cada capa sucesiva es más reciente que la capa debajo.

surface current/corriente superficial (pág. 403) Movimiento de las aguas del océano producido por el viento y el cual afecta principalmente los primeros cientos de metros superiores del océano.

surface wave/onda superficial (pág. 498) Onda sísmica que se mueve en dos direcciones al pasar a través de las rocas; hace que el suelo se mueva de arriba para abajo y de lado a lado.

suspension (p. 216) State in which small particles, such as silt or sand, are held up and carried along by the turbulence of a stream's moving water.

suspension/suspensión (pág. 216) Estado en el cual las partículas pequeñas, como el cieno o la arena, no se precipitan y son arrastradas por la turbulencia de una corriente de agua en movimiento.

sustainable energy (p. 703) Involves global management of Earth's natural resources to ensure that current and future energy needs will be met without harming the environment.

sustainable energy/energía sostenible (pág. 703) Involucra la administración global de los recursos naturales de la Tierra para asegurar que se satisfagan las necesidades energéticas actuales y futuras sin causar daños al ambiente.

sustainable yield (p. 657) Replacement of renewable resources at the same rate at which they are consumed.

sustainable yield/rendimiento sostenible (pág. 657) Reemplazo de recursos renovables a la misma tasa en que se consumen.

synchronous rotation (p. 764) Describes the state at which the Moon's orbital and rotational periods are equal.

synchronous rotation/rotación sincronizada (pág. 764) Describe el estado en que los períodos de la órbita y de la rotación de la Luna son iguales.

Taconic Orogeny (p. 607) Mountain-building event that occurred during the Middle Ordovician.

Taconic Orogeny/orogenia Tacónica (pág. 607) Acontecimiento de formación de montañas que ocurrió durante el Ordoviciense Medio.

temperate zones (p. 362) Areas of Earth that extend between 23.5° and 66.5° north and south of the equator and have moderate temperatures.

temperate zones/zonas templadas (pág. 362) Áreas de la Tierra que se extienden entre los 23.5° al norte y los 66.5° al sur del ecuador y las cuales experimentan temperaturas moderadas.

temperature (p. 278) Measurement of how quickly or how slowly molecules move around, which can be measured in degrees Fahrenheit (°F), degrees Celsius (°C), or kelvins (K).

temperature/temperatura (pág. 278) Medida de la rapidez con que se mueven las moléculas; puede medirse en grados Fahrenheit (°F), grados centígrados (°C) o Kelvin (K).

temperature inversion (p. 281) Increase in temperature with height in an atmospheric level, which inverts the temperature-altitude relationship and can worsen air-pollution problems.

temperature inversion/inversión de la temperatura (pág. 281) Aumento de temperatura con la altura en un nivel atmosférico, el cual invierte la relación de la altitud con la temperatura y puede empeorar los problemas de contaminación del aire.

temperature profile (p. 396) Plots changing ocean water temperatures with depth, which varies, depending on location and season.

temperature profile/perfil de temperatura (pág. 396) Diagramas que analizan cómo cambia la temperatura del agua oceánica con la profundidad, la cual varía según la ubicación y la temporada.

tephra (p. 483) Rock fragments, classified by size, that are thrown into the air during a volcanic eruption and fall to the ground.

tephra/tefrita (pág. 483) Fragmentos rocosos que se clasifican por tamaño, son lanzados por el aire durante una erupción volcánica y luego caen al suelo.

terrestrial planets (p. 780) Rocky-surfaced, relatively small, dense inner planets closest to the Sun—Mercury, Venus, Earth, and Mars.

terrestrial planets/planetas terrestres (pág. 780) Planetas internos densos, relativamente pequeños con superficie rocosa y cercanos al Sol: Mercurio, Venus, la Tierra y Marte.

Tethys Sea (p. 638) Narrow sea with a strong westward current that moved organisms great distances and, following the breakup of Pangaea, separated Africa and Eurasia before they collided.

Tethys Sea/mar de Tetis (pág. 638) Mar estrecho con una fuerte corriente hacia el oeste que movió los organismos por grandes distancias y, después del rompimiento de Pangaea, separó África y Eurasia antes de que chocaran estos continentes.

theory (p. 19) An explanation based on many observations during repeated experiments that is valid only if it is consistent with observations, makes predictions that can be tested, and is the simplest explanation of observations.

theory of plate tectonics (p. 455) States that Earth's crust and upper mantle are broken into plates, which are huge rock slabs that move in different directions and at different rates over Earth's surface.

thermocline (p. 396) Transitional ocean layer that lies between the relatively warm, sunlit surface layer and the colder, dark, dense bottom layer and is characterized by temperatures that decrease rapidly with depth.

thermometer (p. 312) Weather instrument used to measure temperature that contains a column of mercury or alcohol that expands and rises when heated and contracts and falls when cooled.

thermosphere (p. 274) Layer of Earth's atmosphere that is located above the mesopause and contains only a minute portion of the atmosphere's mass.

tide (p. 400) Periodic rise and fall of sea level caused by the gravitational attraction among Earth, the Moon, and the Sun.

***Topex/Poseidon* satellite** (p. 39) Data-gathering satellite that uses radar to map features on the ocean floor.

topographic map (p. 33) Map that uses contour lines, symbols, and color to show changes in the elevation of Earth's surface and features such as mountains, bridges, and rivers.

tornado (p. 338) Violent, whirling column of air in contact with the ground that forms when wind direction and speed suddenly change with height, is often associated with a supercell, and can be extremely damaging.

trade winds (p. 305) Global wind system that flows at 30∞ north and south latitude, where air sinks, warms, and returns to the equator in a westerly direction.

Transcontinental Arch (p. 601) Series of islands dividing southern Laurentia from north to south during the Cambrian Period, when Laurentia was mostly covered by a shallow, tropical sea.

transform boundary (p. 459) Place where two tectonic plates slide horizontally past each other that is characterized by long faults and shallow earthquakes.

theory/teoría (pág. 19) Una explicación basada en muchas observaciones hechas durante experimentos repetidos que es válida sólo si es consistente con las observaciones, hace predicciones que pueden probarse y es la explicación más sencilla de las observaciones.

theory of plate tectonics/teoría de las placas tectónicas (pág. 455) Establece que la corteza terrestre y su manto superior están separados en placas, que son trozos inmensos de roca que se mueven en diferentes direcciones y a distintas velocidades sobre la superficie terrestre.

thermocline/termoclina (pág. 396) Capa de transición del océano que se encuentra entre la capa superficial iluminada por el Sol y relativamente tibia y la capa inferior, densa, oscura y más fría; se caracteriza por la disminución rápida de las temperaturas con la profundidad.

thermometer/termómetro (pág. 312) Instrumento meteorológico que se utiliza para medir la temperatura que contiene una columna de mercurio o de alcohol, la cual se expande y se eleva cuando se calienta y se contrae y desciende cuando se enfría.

thermosphere/termosfera (pág. 274) Capa de la atmósfera terrestre ubicada por encima de la mesopausa y la cual representa sólo una porción diminuta de la masa de la atmósfera.

tide/marea (pág. 400) Ascenso y descenso periódicos del nivel del mar causados por la atracción gravitatoria entre la Tierra, la Luna y el Sol.

Topex/Poseidon* satellite/satélite *Topex/Poseidon (pág. 39) Satélite de recolección de datos que usa un radar para trazar el relieve del fondo del océano.

topographic map/mapa topográfico (pág. 33) Mapa que usa curvas de nivel, símbolos y colores para mostrar los cambios en la elevación de la superficie terrestre y rasgos como las montañas, los puentes y los ríos.

tornado/tornado (pág. 338) Violenta columna giratoria de aire en contacto con el suelo, la cual se forma cuando la dirección y la velocidad del viento cambian repentinamente con la altura; se le asocia a menudo con una supercelda y puede ser extremadamente dañino.

trade winds/vientos alisios (pág. 305) Sistema de vientos globales que fluye a 30° de latitud norte y sur, en donde el aire se hunde, se calienta y regresa al ecuador con dirección oeste.

Transcontinental Arch/arco trascontinental (pág. 601) Serie de islas que dividieron el sur de Laurencia de norte a sur durante el Período Cámbrico, cuando Laurencia estaba cubierta, en su mayor parte, por un mar tropical poco profundo.

transform boundary/límite transformante (pág. 459) Lugar donde dos placas tectónicas se deslizan horizontalmente una sobre otra; se caracteriza por fallas grandes y terremotos superficiales.

transgression (p. 603) Occurs when sea level rises and causes the shoreline to move inland, resulting in deeper-water deposits overlying shallower-water deposits with adjacent depositional facies overlying each other in vertical succession.

transgression/transgresión (pág. 603) Ocurre cuando el nivel del mar se eleva y hace que la costa se mueva hacia el interior, resultando en depósitos de agua más profunda que cubren los depósitos de agua menos profunda con biofacies depositados adyacentemente unos sobre otros, en sucesión vertical.

transported soil (p. 168) Soil that has been moved away from its parent material by water, wind, or a glacier.

transported soil/suelo transportado (pág. 168) Tierra que ha sido retirada de su material de origen por el agua, el viento o un glaciar.

travertine (p. 248) A type of limestone found in dripstone formations.

travertine/travertina (pág. 248) Tipo de piedra caliza que se encuentra en en las formaciones rocosas por goteo.

tropical cyclone (p. 341) Large, low-pressure, rotating storm that gets its energy from the evaporation of warm ocean water and the release of heat.

tropical cyclone/ciclón tropical (pág. 341) Tormenta giratoria grande de baja presión, la cual obtiene su energía de la evaporación del agua oceánica tibia y la liberación de calor.

tropics (p. 362) Area of Earth that receives the most solar radiation, is generally warm year-round, and extends between 23.5° south and 23.5° north of the equator.

tropics/trópicos (pág. 362) Área de la Tierra que recibe la mayor cantidad de radiación solar, generalmente es caliente todo el año y se extiende entre 23.5° sur y 23.5° norte del ecuador.

troposphere (p. 274) Layer of the atmosphere closest to Earth's surface, where most of the mass of the atmosphere is found and in which most weather takes place and air pollution collects.

troposphere/troposfera (pág. 274) Capa de la atmósfera más cercana a la superficie terrestre donde se encuentra gran parte de la masa atmosférica y en la cual ocurre la mayor parte del tiempo y se acumula la contaminación del aire.

trough (p. 399) Lowest point of a wave.

trough/seno (pág. 399) Punto más bajo de una ola.

tsunami (p. 513) Large, powerful, ocean wave generated by the vertical motions of the seafloor during an earthquake; in shallow water, can form huge, fast-moving breakers exceeding 30 m in height that can damage coastal areas.

tsunami/tsunami (pág. 513) Enorme y poderosa ola marina generada por los movimientos verticales del fondo del mar durante un terremoto; en aguas superficiales, puede formar inmensas olas rompientes muy rápidas que exceden 30 m de altura y pueden causar daños a las áreas costeras.

turbidity current (p. 424) Rapidly flowing ocean current that can cut deep-sea canyons in continental slopes and deposit the sediments in the form of a continental rise.

turbidity current/corriente de turbidez (pág. 424) Corriente oceánica de rápido flujo que puede cortar cañones en los taludes continentales y depositar los sedimentos en forma de una elevación continental.

ultramafic (p. 108) Igneous rocks with low silica content and very high levels of magnesium and iron.

ultramafic/ultramáfica (pág. 108) Rocas ígneas con bajo contenido de sílice y niveles muy altos de magnesio y hierro.

unconformity (p. 560) Gap in the rock record caused by erosion or weathering.

unconformity/disconformidad (pág. 560) Brecha en el registro rocoso causado por la erosión o la meteorización.

Uniformitarianism (p. 557) States that processes such as mountain building, erosion, and sea-level changes that are occurring today have been occurring since Earth formed.

uniformitarianism/uniformitarianismo (pág. 557) Establece que los procesos como la formación de montañas, la erosión y los cambios del nivel del mar que ocurren en nuestros días han estado ocurriendo desde que se formó la Tierra.

uplifted mountains (p. 537) Mountains that form when large regions of Earth are forced slowly upward without much deformation.

uplifted mountains/montañas levantadas (pág. 537) Montañas que se forman cuando grandes regiones de la Tierra son forzadas a moverse lentamente hacia arriba sin mucha deformación.

upwelling (p. 405) Upward movement of ocean water that occurs when winds push surface water aside and it is replaced with cold, deep water that originates on the ocean bottom.

upwelling/corriente resurgente (pág. 405) Movimiento ascendente de las aguas del océano que ocurre cuando los vientos empujan hacia un lado el agua superficial y ésta es reemplazada con agua fría profunda que se origina en el fondo de océano.

valence electron (p. 57) Electron in an atom's outermost energy level.

valley glacier (p. 199) Glacier that forms in a valley in a mountainous area and widens V-shaped stream valleys into U-shaped glacial valleys as it moves downslope.

Varangian Glaciation (p. 592) Widespread glaciation event in which ice reached nearly to the equator between 700 and 800 million years ago.

variable star (p. 834) Star in the giant branch of the Hertzsprung-Russell diagram that pulsates in brightness due to its outer layers expanding and contracting.

varves (p. 565) Alternating light-colored and dark-colored sedimentary layers of sand, clay, and silt deposited in a lake that can be used to date cyclic events and changes in the environment.

vascular plants (p. 609) Land plants with the ability to move water through their stems and stalks and which, during the Early Silurian, were small and leafless and grew along the ground.

vent (p. 480) Opening in Earth's crust through which lava erupts and flows out onto the surface.

ventifact (p. 193) Rock shaped by wind-blown sediments.

vernal equinox (p. 761) Occurs when the Sun is directly overhead at the equator and results in day and night of equal length for both northern and southern hemispheres.

viscosity (p. 474) A substance's internal resistance to flow.

valence electron/electrón de valencia (pág. 57) Electrón en el nivel más externo de energía de un átomo.

valley glacier/glaciar de valle (pág. 199) Glaciar que se forma en un valle de un área montañosa; ensancha los valles de corrientes con forma en V generando valles glaciales con forma de U, al moverse cuesta abajo.

Varangian Glaciation/glaciación Varangiana (pág. 592) Amplio acontecimiento glacial en que el hielo casi alcanzó a llegar al ecuador, hace 700 a 800 millones de años.

variable star/estrella variable (pág. 834) Estrella en la rama gigantesca del diagrama de Hertzsprung-Russell cuya luminosidad tiene lugar a impulsos debido a la expansión y contracción de sus capas exteriores.

varves/varves (pág. 565) Capas sedimentarias de colores claros y oscuros alternados, compuestas de arena, arcilla y limo depositados en un lago, las cuales pueden utilizarse para datar acontecimientos cíclicos y cambios en el ambiente.

vascular plants/plantas vasculares (pág. 609) Plantas de tierra con la capacidad de mover agua por sus pedúnculos y tallos y las cuales eran pequeñas y sin hojas y crecían a lo largo del suelo durante el Siluriano Temprano.

vent/chimenea (pág. 480) Abertura en la corteza terrestre por la cual sale lava que fluye hacia la superficie.

ventifact/ventifacto (pág. 193) Roca moldeada por sedimentos arrastrados por el viento.

vernal equinox/equinoccio de invierno (pág. 761) Ocurre cuando el Sol está directamente arriba del ecuador y resulta en que el día y la noche son de igual longitud para los hemisferios norte y sur.

viscosity/viscosidad (pág. 474) Resistencia interna a fluir de una sustancia.

water cycle (p. 290) Continual movement of water between Earth's surface and the atmosphere through evaporation, condensation, and precipitation.

watershed (p. 215) Land area drained by a stream system.

water table (p. 241) Upper boundary of the zone of saturation that rises during wet seasons and drops during dry periods.

water cycle/ciclo del agua (pág. 290) Movimiento continuo del agua entre la superficie terrestre y la atmósfera a través de la evaporación, la condensación y la precipitación.

watershed/cuenca (pág. 215) Área de terreno drenada por un sistema de corrientes de agua.

water table/nivel freático (pág. 241) Límite superior de la zona de saturación que sube durante la temporada de lluvias y disminuye durante los períodos de sequía.

wave (p. 399) Rhythmic movement that carries energy through matter or space and, in oceans, is generated mainly by wind moving over the surface of the water.

wave refraction (p. 414) Process in which waves advancing toward shore slow when they encounter shallower water, causing the initially straight wave crests to bend toward the headlands.

weather (p. 300) Current state of the atmosphere, including short-term variations such as temperature and precipitation.

weathering (p. 153) Chemical or mechanical process that breaks down and changes rocks on or near Earth's surface and whose rate is influenced by factors such as precipitation and temperature.

well (p. 252) Deep hole drilled or dug into the ground to reach a reservoir of groundwater.

wetland (p. 230) Low-lying land area, such as a bog or marsh, that is covered in water a large part of the year and supports specific plant species.

wind-chill factor (p. 350) Phenomenon measured by the wind-chill index, which estimates the heat loss from human skin caused by a combination of wind and cold air.

winter solstice (p. 761) Occurs when the Sun is directly overhead at 23.5∞ south latitude, around December 21, and results in the minimum number of daylight hours for the northern hemisphere and the maximum number for the southern hemisphere.

wave/onda u ola (pág. 399) Movimiento rítmico que conduce energía a través de la materia o el espacio y, en los océanos, se genera principalmente por el movimiento del viento sobre la superficie del agua.

wave refraction/refracción de onda (pág. 414) Proceso en que las olas avanzan hacia la costa y deceleran cuando encuentran agua menos profunda, lo que hace que sus crestas, inicialmente rectas, se doblen hacia los cabos.

weather/tiempo (pág. 300) Estado actual de la atmósfera que incluye las variaciones a corto plazo, como la temperatura y la precipitación.

weathering/meteorización (pág. 153) Proceso químico o mecánico que rompe y cambia las rocas sobre o cercanas de la superficie terrestre y cuya velocidad se ve influenciada por factores como la precipitación y la temperatura.

well/pozo (pág. 252) Hoyo profundo taladrado o excavado en el suelo para alcanzar un depósito de agua subterránea.

wetland/humedal (pág. 230) Área de tierra baja, como una ciénaga o un pantano, que se encuentra cubierta de agua gran parte del año y la cual alberga especies específicas de plantas.

wind-chill factor/sensación térmica (pág. 350) Fenómeno que mide el índice de sensación térmica y el cual estima la pérdida de calor de la piel humana, causada por una combinación del viento y del aire frío.

winter solstice/solsticio de invierno (pág. 761) Ocurre cuando el Sol está directamente arriba de los 23.5° de latitud sur, alrededor del 21 de diciembre; resulta en el número mínimo de horas de luz solar para el hemisferio norte y el número máximo para el hemisferio sur.

Z

zircon (p. 578) Very stable mineral commonly found in granite that can provide evidence of Earth's crust being at least 4.1 to 4.2 billion years old through radiometric dating.

zone (p. 787) High, cool, light-colored cloud that rises and flows rapidly in the Jovian atmosphere.

zone of saturation (p. 241) Depth below Earth's surface where all the pores of a material are completely filled with groundwater.

zircon/circón (pág. 578) Mineral sumamente estable que se encuentra comúnmente en el granito y que puede proporcionar prueba de que edad de la corteza terrestre es por lo menos de 4.1 a 4.2 mil millones de años por datación radiométrica.

zone/zona (pág. 787) Nube alta, fresca y levemente colorida que se eleva y fluye rápidamente en la atmósfera joviana.

zone of saturation/zona de saturación (pág. 241) Profundidad debajo de la superficie terrestre en donde todos los poros de un material están completamente llenos de agua subterránea.

Index

Page number in boldface indicates the page where the entry is defined.

C

Index

Index

Index

Index

Index

I

Index

Index

Index

Q

R

Index

Index

Index

Index

U

Index

Index

Art Credits

Morgan-Cain and Associates: **xvii, 11, 18, 25, 30**(b), **31, 34, 35**(t), **38-39, 41, 45**(c), **47**(l), **78, 81**(t), **59, 66, 100-104, 106-107, 110, 122, 125, 130, 133, 134**(b), **135-136, 139, 145, 157-158, 160-161, 173, 179, 200, 209, 212-214, 217, 219-220, 223, 225, 233, 236, 241-243, 246, 250, 252, 254-256, 261, 263, 272, 274-275, 278-279, 283, 296-297, 301, 305-311, 317, 320, 326, 331, 332-335, 338, 343, 345, 356, 361-364, 367, 369-370, 372-373, 375, 383, 388-389, 395-403, 405-406, 411, 414, 416-417, 422-423, 435, 448, 450-452, 454, 456-457, 459, 461, 465-466, 468-469, 472, 476, 493, 496-504, 509, 519-520, 525-526, 529-531, 533, 535-538, 541**(b), **544-545, 554, 558-562, 567-569, 575, 578, 580-581, 595, 597, 599, 602**(t), **603, 606, 610, 613, 622, 623, 626, 630**(b), **631, 632**(t), **635, 646, 658, 665, 667, 672, 681, 689, 698, 705, 708-709, 714-715, 721-722, 725, 727, 730, 738-739, 748-749, 756-757, 759-767, 770, 772-773, 776-780, 782-784, 787-791, 795, 797, 801-803, 811-813, 815-816, 818-819, 821-824, 827, 835-838, 840-841, 847, 849, 851, 855-857, 914-915, 922, 927**(b), **937-938**

Barbara Hoopes Ambler: **604, 609, 615, 629, 630**(t), **632**(b), **640-641**

Michael Woods: **167, 169, 592**

Glencoe: **56-58, 61-63, 67, 73-74, 81**(b), **82, 726, 917**

Precision Graphics: **28, 30**(t), **32-33, 35**(b), **45**(t), **47**(r), **126, 134**(t), **159, 170, 182-184, 186-187, 191-193,195-198, 201, 205, 215, 251, 273, 276, 285-288, 291, 303, 315, 319, 323, 330, 342, 353, 390, 393, 404, 415, 434, 443-445, 455, 473, 477, 481-483, 485, 487, 508, 510, 514, 517, 524, 528, 582-583, 602**(b), **605, 607-608, 612, 633, 637-638, 671, 735, 908, 927**(t)

Photo Credits

Cover Galen Rowell/Mountain Light Photography; **vii** EIT/SOHO/NASA; **ix** (t)Geoff Tompkinson/Science Photo Library/Photo Researchers, (b)Poulet/Liaison; **x** (t)M.I. Walker/Photo Researchers, (b)Galen Rowell/CORBIS; **xi** Joe Sohm/The Stock Market; **xii** Jack Dykinga, **xiii** NASA; **xiv** Matt Meadows; **xv** James L. Amos/Photo Researchers; **xvi** Warren Faidley/ Weatherstock; **xviii** Wolfgang Kaehler/CORBIS; **xx** AP/Wide World Photos; **xxi** CXC/SAO/NASA; **xxii** Ed Viesturs; **xxiii** John Noel Photographic Collection; **2-3** Jack Dykinga; **4** Larry Ulrich; **5** Matt Meadows; **6** (t)Roger Ressmeyer/ CORBIS, (b)Simon Fraser/Science Photo Library/Photo Researchers; **7** (tl)Everett Johnson/Stone, (tr)Jeremy Woodhouse/DRK Photo, (cl)Paul A. Souders/CORBIS, (cr)Chip Clark, (bl)Laurence Parent, (br)Herb Lingl/Black Star; **8** Flip Nicklin/ Minden Pictures; **9** Jeremy Woodhouse/DRK Photo; **10** Larry Lefever/Grant Heilman Photography; **12** Matt Meadows; **14** NASA; **15** (t)Matt Meadows, (b)Runk/Schoenberger/Grant Heilman Photography; **16** David Malin/Anglo-Australian Observatory; **17** Matt Meadows; **19** (l)Victoria & Albert Museum, London/Art Resource, NY, (r)Bettmann/CORBIS; **20** Matt Meadows; **22** Ira Block/National Geographic Image Collection; **23** (t)Jeremy Woodhouse/DRK Photo, (c,b)Matt Meadows; **26** CORBIS; **27** Aaron Haupt; **35** USGS; **36** National Geographic Maps; **37** Boeing; **38** NASA/AP/Wide World Photos; **40** Doug Martin; **43** USGS; **44** Geostock/PhotoDisc; **45** Boeing; **48** Wayne Newton/PhotoEdit; **49** USGS/NASA; **50-51** Roger Du Buisson/The Stock Market; **52** James L. Stanfield/National Geographic Image Collection; **53** Doug Martin; **54** (l)Richard Megna/Fundamental Photographs, (c)L.S. Stepanowicz/ Visuals Unlimited, (r)Kaj R. Svenson/Science Photo Library/Photo Researchers; **55** Matt Meadows; **60** Richard Megna/Fundamental Photographs; **64** (t)Bruce Iverson, (b)PhotoDisc; **65** (t)Matt Meadows, (b)John Evans; **66** (l to r) StudiOhio, Mark Burnett, StudiOhio, Matt Meadows, Amanita Pictures, StudiOhio, Aaron Haupt; **67** (l)Andrew Syred/Science Photo Library/Photo Researchers, (r)Chip Clark; **68** (t)Aaron Haupt, (c)John Evans, (b)Tony Freeman/PhotoEdit; **69** EIT/SOHO/NASA; **70 71** Matt Meadows; **72** (l)Andrew Syred/Science Photo Library/Photo Researchers, (c)Dr. Jeremy Burgess/Science Photo Library/Photo Researchers, (r)SCIMAT/Photo Researchers; **73** (t)Bruce Iverson, (b)Tony Freeman/PhotoEdit; **76** Chip Clark; **77** Matt Meadows; **78** (t)M. Claye Jacana/Photo Researchers, (b, l to r)Charles D. Winters/Photo Researchers, Mark A. Schneider/Visuals Unlimited, Mark A. Schneider/Visuals Unlimited, Biophoto Associates/Photo Researchers, Runk/Schoenberger/Grant Heilman Photography, Doug Martin; **79** (l)Robert De Gugliemo/Science Library/Photo Researchers, (r)Barry L. Runk/Grant Heilman Photography; **80** Martin Miller/Visuals Unlimited; **83** Chip Clark; **84** (l)Mark A. Schneider/Visuals Unlimited, (c)Charles D. Winters/Photo Researchers, (r)Runk/Schoenberger/Grant Heilman Photography; **85** (tl)Mark A. Schneider/Visuals Unlimited, (tr)Runk/Schoenberger/Grant Heilman Photography, (b)Fundamental Photographs; **87** (l)Tom Ives/The Stock Market, (c)Runk/Schoenberger/Grant Heilman Photography, (r)Doug Martin; **88** Paul Silverman/Fundamental Photographs; **89** (t,br)Paul Silverman/Fundamental Photographs, (bl)AP/Wide World Photos; **90** AFP/CORBIS; **91** (l)Biophoto Associates/Photo Researchers, (r)Chip Clark; **92** Matt Meadows; **94** Rick Gayle/The Stock Market; **95** (t)Biophoto Associates/Photo Researchers, (b)Tom Ives/The Stock Market;
97 Photo Researchers; **98** Alfred Pasieka/Science Photo Library/Photo Researchers; **99** Matt Meadows; **100** (l)Runk/Schoenberger/Grant Heilman Photography, (r)Phillip Hayson/Photo Researchers; **101** Robert Garvey/CORBIS; **104** A.J. Copley/Visuals Unlimited; **105** Craig A. Cox; **108** (l)Paul Silverman/Fundamental Photographs, (r)Christena M. Cox; **109** (tl)Andrew J. Martinez/Photo Researchers, (tr)Doug Martin, (b)Andrew J. Copley/Visuals Unlimited; **110** (l)Barry L. Runk/Grant Heilman Photography, (r)Doug Martin; **111** (t)Aaron Haupt, (bl)Jeremy Woodhouse/DRK Photo, (br)AFP/CORBIS; **112** (t)Tom McHugh/Photo Researchers, (bl)Doug Martin, (br)University of Houston; **113** (l)Chip Clark, (r)Jeremy Woodhouse/DRK Photo; **115 116** Matt Meadows; **117** (t)A.J. Copley/Visuals Unlimited, (b)Jeremy Woodhouse/DRK Photo; **120** Tim Fitzharris/Minden Pictures; **121** Matt Meadows; **122** William E. Ferguson; **123** (tl)T.A. Wiewandt/DRK Photo, (tr)Kim Heacox Photography/DRK Photo, (bl)William E. Ferguson, (br)Grant Heilman/Grant Heilman Photography; **124** (l)Kazuyoshi Nomachi/Photo Researchers, (r)Anthony Bannister/Photo Researchers; **125** Christena M. Cox; **126** (t)John Cancalosi/Peter Arnold, Inc.; (b)Jim Steinberg/Photo Researchers; **127** (t)Marc Epstein/Visuals Unlimited, (b)Adrienne T. Gibson/Earth Scenes; **129** (t,c)Doug Martin, (b)John Sohlden/Visuals Unlimited; **130** Scott T. Smith/CORBIS; **131** (l)Mark Burnett/Photo Researchers, (r)M.I. Walker/Photo Researchers; **135** Farley Lewis/Photo Researchers; **136** (t)Alfred Pasieka/Science Photo Library/Photo Researchers, (bl,bcl) William E. Ferguson, (bcr)Doug Martin, (br)Andrew J. Martinez/Photo Researchers; **137** (tl)A.J. Copley/Visuals Unlimited, (tr)Doug Martin, (b)Breck P. Kent/Earth Scenes; **140** Matt Meadows; **142** Dr. Jeremy Burgess/Science Photo Library/Photo Researchers; **143** (t)Marc Epstein/Visuals Unlimited, (c)Scott T. Smith/CORBIS, (b)Breck P. Kent/Earth Scenes; **146** Yoav Levy/Phototake; **147** Mark A. Schneider/Photo Researchers; **148** Craig Blacklock/Larry Ulrich Stock; **150-151** Jack Dykinga/Stone; **152** Jack Dykinga; **153** Matt Meadows; **154** (t)Catherine Ursillo/Photo Researchers, (b)William E. Ferguson; **155** (tl)Susan Rayfield/Photo Researchers, (tc)Cliff Leight, (tr)Bud Lehnhausen/Photo Researchers, (b)J. Serrao/Photo Researchers; **156** (t)Scott T. Smith/CORBIS, (b)Wilbur E. Garrett/National Geographic Image Collection; **160** Jack Dykinga; **162** Gary Braasch; **163** (l)Jack Dykinga, (r)William E. Ferguson; **164** (t)NASA, (c)Arthur Morris/Visuals Unlimited, (b)John Shelton; **165** (t)Cliff Leight, (b)Jeff Foott/DRK Photo; **166** Vince Streano/CORBIS; **168** Kevin Fleming/CORBIS; **169** Doug Martin; **171** Fletcher & Baylis/Photo Researchers; **174** Matt Meadows; **176** Bob Jordan/AP/Wide World Photos; **177** (t)Susan Rayfield/Photo Researchers, (c)William E. Ferguson, (b)Kevin Fleming/CORBIS; **180** Andres Leighton/AP/Wide World Photos; **182** Daniel D. Lamoreux/Visuals Unlimited; **185** (l)Gary Braasch, (r)Giuseppe Mastullo/Grazia Neri/Sygma; **186** Michael Gallacher/Missoulian/Liaison Agency; **187** Chuck Place/Stock Boston; **188** (t)Gilbert Vogt/Liaison Agnecy, (bl)Don & Pat Valenti/DRK Photo, (br)Stouffer Productions/Earth Scenes; **189** Ricardo Mazalan/AP/Wide World Photos; **190** (l)Michael Habicht/Earth Scenes, (r)Jim Steinberg/Photo Researchers; **191** Tony Freeman/PhotoEdit; **192** David Muench/Corbis; **193** (t)Jeff Foott/DRK Photo, (b)Galen Rowell/CORBIS; **194** Tony Wilson-Bligh/CORBIS; **196** Scott W. Smith/Earth Scenes; **199** (t)Kim Heacox Photography/DRK Photos, (b)John Gerlach/Earth Scenes; **201** Tom Bean/DRK Photo; **202** Ken M. Johns/Photo Researchers; **203** (l)Tom Bean/DRK Photo, (r)Cliff Leight; **206** Johnny Johnson/Stone; **207** (t)Michael Gallacher/Missoulian/Liaison Agnecy, (c)Tony Wilson-Bligh/CORBIS, (b)Kim Heacox Photography/DRK Photo; **210** Tom Bean/Stone; **211** Matt Meadows; **214** John Shelton; **217** Jack Dykinga; **218**

Tom Bean/DRK Photo; **219** Geoff Tompkinson/Science Photo Library/Photo Researchers; **220** John Eastcott & Yva Momatiuk/Photo Researchers; **221** Cliff Leight; **222** Jack Dykinga; **223** Norbert Rosing/Earth Scenes; **224** (t)Bob Krist/CORBIS, (b)Brian Milne/Earth Scenes; **225** Jeff Lapore/Photo Researchers; **226** (t)Michael Collier/DRK Photo, (b)NASA/TSADO/Tom Stack & Associates; **227** Jack Dykinga; **228** Scott T. Smith/CORBIS; **229** Cliff Leight; **230** (t)W. Banaszewski/Visuals Unlimited, (b)Farrell Grehan/Photo Researchers; **231** Gordon & Cathy Illg/Animals Animals; **232** Matt Meadows; **234** Konrad Wothe/Minden Pictures; **235** (t)Jack Dykinga, (c)Jeff Lapore/Photo Researchers, (b)Gordon & Cathy Illg/Animals Animals; **238** Peter & Ann Bosted/Tom Stack & Associates; **239** Doug Martin; **244** Keren Su/CORBIS; **245** John W. Bova/Photo Researchers; **246** John Shelton; **247** Sheila Terry/Science Photo Library/Photo Researchers; **248** Richard Thom/Visuals Unlimited; **249** A.J. Copley/Visuals Unlimited; **251** Jeff Foott/Tom Stack & Associates; **259** USGS; **260** T.A. Wiewandt/DRK Photo; **261** (t)Richard Thom/Visuals Unlimited, (b)A.J. Copley/Visuals Unlimited; **264** Cliff Leight; **265** (l)Larry Ulrich, (r)Theo Allofs/Stone; **267** Oldrich Karasek/Stone; **268-269** Laurence Parent; **270** Art Wolfe; **271** Matt Meadows; **277** (l)Chuck Savage/The Stock Market, (r)Chris Bartlett/FPG; **281** Tom Prettyman/PhotoEdit; **282** Bernhard Edmaier/Science Photo Library/Photo Researchers; **284** CORBIS; **284** (t)Chris Bartlett/FPG, (c)Bernhard Edmaier/Science Photo Library/Photo Researchers, (b)Michael Collier/DRK Photo; **289** Michael Collier/DRK Photo; **290** NCAR/Tom Stack & Associates; **292** Matt Meadows; **294** TOMS/NASA; **298** Larry Ulrich; **299** Matt Meadows; **300** (l)Nuridsany Et Perennou/Science Source/Photo Researchers, (c)Warren Faidley/Weatherstock, (r)Adam Jones/Photo Researchers; **307** NASA; **312** (tl)Greg Vaughn/Tom Stack & Associates, (tr,br)Doug Martin, (bl)Leonard Lessin, FBPA/Photo Researchers; **313** Randy Trine; **314** United Nations; **316** TSADO/NCDC/NOAA/Tom Stack & Associates; **319** Bob Daemmrich/Stock Boston; **321** Science Visualization Studio/The SeaWiFs Project/Goddard Space Flight Center/NASA; **324** William L. Wantland/Tom Stack & Associates; **325** (t to b)Adam Jones/Photo Researchers, NASA, Randy Trine, Bob Daemmrich/Stock Boston; **327** United Nations; **328** Warren Faidley/Weatherstock; **329** Matt Meadows; **333** CORBIS; **334** Warren Faidley/Weatherstock; **337** (tl)CORBIS, (tr)Gene Moore/Phototake/PictureQuest, (b)Warren Faidley/Weatherstock; **339** (t)H. Baker/Weatherstock, (c)Keith Brewster/Weatherstock, (b)W. Balzer/Weatherstock; **341** NASA/TSADO/Tom Stack & Associates; **343** NASA; **345** Mark Allen Stack/Tom Stack & Associates; **347** CORBIS; **351** (l)Dick Blume, Syracuse Newspapers/AP/Wide World Photos, (r)Gary Walts, Syracuse Newspapers/AP/Wide World Photos; **354** CORBIS; **355** (t to b)CORBIS, W. Balzer/Weatherstock, NASA/TSADO/Tom Stack & Associates, Gary Walts Syracuse Newspapers /AP/Wide World Photos; **358** David Barnes/Stone; **363** Francois Gohier/Photo Researchers; **365** Frans Lanting/Minden Pictures; **366** Baron Wolman/Stone; **367** (l)Tui De Roy/Minden Pictures, (r)Peter B. Kaplan/Photo Researchers; **368** SVS/Goddard Space Flight Center/NASA; **371** (l)Mitch Reardon/Photo Researchers, (r)Doug Sokell/Tom Stack & Associates; **374** NASA/Roger Ressmeyer/CORBIS; **377** (t)Howard Buffett/Grant Heilman Photography, (b)Bill Bachman/PhotoEdit; **378** Tom Stack; **380** Jacques Jangoux/Stone; **381** (t to b)Francois Gohier/Photo Researchers, Frans Lanting/Minden Pictures, Doug Sokell/Tom Stack & Associates, Howard Buffett/Grant Heilman Photography; **384** Kim Heacox/Stone; **385** Matt Meadows; **386** (l)AP/Wide World Photos, (r)USGS/AP/Wide World Photos; **387** (t)NASA/Science Photo Library/Photo Researchers, (b)S. Nielsen/DRK Photo; **389** Los Almos National Laboratory/Science Photo Library/Photo Researchers; **390** Perry Conway/Tom Stack & Associates; **391** TSADO/NASA/Tom Stack & Associates; **392** Jose Manuel Sanchis Calvete/CORBIS; **393** CORBIS; **394** Randy Morse/Tom Stack & Associates; **396** Peter David/Photo Researchers; **400** Paul Berger/Stone; **408** Underwood & Underwood/CORBIS; **409** (t)NASA/Science Photo Library/Photo Researchers, (c)Randy Morse/Tom Stack & Associates, (b)Paul Berger/Stone; **412** Jack Dykinga; **413** Geoff Butler; **415** Wolfgang Kaehler/CORBIS; **418** (l)Melvin B. Zucker/Visuals Unlimited, (r)Philip Gould/CORBIS; **419** Tom Bean/DRK Photo; **420** Tom Stack/Tom Stack & Associates; **421** John Shelton; **424** Official U.S. Navy Photo by R.F. Diel; **425 426** Marie Tharp; **427** Dr. Ken MacDonald/Science Photo Library/Photo Researchers; **428** Doug Martin; **429** Science VU/Visuals Unlimted; **431** USGS; **432** Peter David/Photo Researchers; **433** (t)Wolfgang Kaehler/CORBIS, (b)Science VU/Visuals Unlimited; **436** William L. Wantland/Tom Stack & Associates; **437** StockTrek/The Stock Market; **438** Larry Ulrich; **439** (t)Jeff Foott/DRK Photo, (b)Andrew Syred/Science Photo Library/Photo Researchers; **440-441** David Ball/Stone; **442** Emory Kristoff/National Geographic Society Image Collection; **446** Betty Crowell/Faraway Places; **449** Dr. Ken MacDonald/Science Photo Library/Photo Researchers; **453** NOAA/NGDC; **458** (t)Harold E. Wilson/Earth Scenes, (b)Marie Tharp; **460** Richard Megna/Fundamental Photographs; **462** Stephen P. Grand/Rob D. van der Hilst/Sri Widiyantoro/*GSA Today* 4/97; **463** NASA/Tom Stack & Associates; **467** (t to b)Betty Crowell/Faraway Places, NOAA/NGDC, Harold E. Wilson/Earth Scenes, NASA/Tom Stack & Associates; **470** Richard A. Cooke III/Stone; **471** Aaron Haupt; **474** G. Brad Lewis/Stone; **475** J.D. Griggs/CORBIS; **478** (l)Art Attack/Photo Researchers, (r)Tom Bean/DRK Photo; **479** Larry Ulrich; **480** Michael Collier/DRK Photo; **481** C.C. Lockwood/DRK Photo; **482** David Muench Photography; **483** (l)C. Sharp/Earth Scenes, (r)Pat & Tom Leeson/Photo Researchers; **484** Morris J. Elsing/National Geographic Image Collection; **486** Yann Arthus-Bertrand/CORBIS; **488** PhotoDisc; **490** NASA; **491** (t)G. Brad Lewis/Stone, (c)Larry Ulrich, (b)Pat & Tom Leeson/Photo Researchers; **494** Wally Santana/AP/Wide World Photos; **495** Bob Daemmrich; **497** John Shelton; **505** Winfield Parks/National Geographic Image Collection; **506** Barry Sweet/AP/Wide World; **511** Nik Wheeler/CORBIS; **512** William H. Mullins/Photo Researchers; **513** James Mori/Institute of Geophysics, National Central University, Taiwan, and Kyoto University, Japan; **515** David Parker/Science Photo Library/Photo Researchers; **518** (l)Earl & Nazima Kowall/CORBIS, (r)Shahpari Sohaie/CORBIS; **519** (t)John Shelton, (c)Winfield Parks/National Geographic Image Collection, (b)David Parker/Science Photo Library/Photo Researchers; **522** Barbara Rowell/Mountain Light Photography; **523** Doug Martin; **527** Warren Morgan/CORBIS; **529** Roger Ressmeyer/CORBIS; **530** Betty Crowell/Faraway Places; **531** Galen Rowell/CORBIS; **532** William Felger/Grant Heilman Photography; **534** Kenneth Murray/Photo Researchers; **536** B. Murton/Southampton Oceanography Centre/Science Photo Library/Photo Researchers; **537** Lynn Gerig Photography/Tom Stack & Associates; **538** A.J. Copley/Visuals Unlimited; **539** Richard A. Cooke/CORBIS; **541** USGS; **542** AP/Wide World Photos; **543** (t)Warren Morgan/CORBIS, (c)Roger Ressmeyer/CORBIS, (b)A.J. Copley/Visuals Unlimited; **546** Yann Arthus-Bertrand/CORBIS; **547** Sharna Balfour/Gallo Images/CORBIS; **548** Wilbur Garrett/National Geographic Image Collection; **549** Cliff Leight; **550-551** John M. Roberts/ The Stock Market; **552** Larry Ulrich; **553** Matt Meadows; **555** Tom Bean/DRK Photo; **556** Laynne Kennedy/CORBIS; **557** Tom Bean/DRK Photo; **559** William E. Ferguson; **561** Christena M. Cox; **564** Tom Till/DRK Photo; **565** Charlie Ott/Photo Researchers; **566** C. Munoz-Yague/Phototake; **567** (tl)Dr. Dennis Kunkel/Phototake, (tr)Tom Bean/DRK Photo, (b)James L. Amos/CORBIS; **568** (t)Mark A. Schneider/Photo Researchers, (b)Tom Bean/DRK Photo; **569** (tl)Fred Bruemmer/Peter Arnold, Inc., (tr)Ray Nelson/Phototake, (b)Matt Meadows; **572** Francis Latreille; **573** (t to b)Tom Bean/DRK Photo, William E. Ferguson, Charlie Ott/Photo Researchers, Tom Bean/DRK Photo; **576** Chip Clark; **577** Matt Meadows; **579** Keith Bedford/AP/Wide World Photos; **584** Erwin and Peggy Bauer/Tom Stack & Associates; **585** (tl)Sinclair Stammers/Science Photo Library/ Photo Researchers, (tr)M. Abbey/Photo Researchers, (bl)Roland Seitre/Peter Arnold, Inc., (br)Sinclair Stammers/Science Photo Library/Photo Researchers; **586** (l)John Cancalosi/Peter Arnold, Inc., (r)Jim Wark/Peter Arnold, Inc.; **587** Jack Dykinga; **589** Jim Sugar Photography/CORBIS; **590** (l)Prof. Oscar L. Miller/Science Photo Library/Photo Researchers, (r)Science Source/Photo Researchers; **591** Fred McConnaughey/Photo Researchers; **596** NASA/Science Photo Library/Photo Researchers; **597** (t)Keith Bedford/AP/Wide World Photos, (c)John Cancalosi/Peter Arnold, Inc., (b)Fred McConnaughey/Photo Researchers; **600** Adam Jones/Photo Researchers; **601** Doug Martin; **603** Tom Bean/DRK Photo; **604** Chip Clark; **606** David Cavagnaro/DRK Photo; **607** Michael P. Gadomski/Photo Researchers; **608** Don Duckson/Visuals Unlimited; **609** (tl)William E. Ferguson, (tr)Doug Martin, (b)Dick Keen/Visuals Unlimited; **610** (l)Gregory K. Scott/Photo Researchers, (c)John Cancalosi/DRK Photo, (r)John Cancaloci/DRK Photo; **614** (t)Laurence Parent, (b)ProFiles West/Index Stock Photography; **615** (l)Kaj R. Svensson/Science Photo Library/Photo Researchers, (r)Fred McConnaughey/Photo Researchers; **618** Doug Martin; **620** Sinclair Stammers/Science Photo Library/Photo Researchers; **621** (t)Tom Bean/DRK Photo, (c)Doug Martin, (b)Kaj R. Svensson/Science Photo Library/Photo Researchers; **624** Francois Gohier/Photo Researchers; **625** (t)Dee Breger/Photo Researchers, (b)Juergen Berger/Max-Planck Institute/Science Photo Library/Photo Researchers; **626** Christena M. Cox; **627** Jack Dykinga; **628** M.I. Walker/Photo Researchers; **629** Sinclair Stammers/Science Photo Library/Photo Researchers; **631** Ken Lucas/Visuals Unlimited; **636** Dominique Braud/Tom Stack & Associates; **639** John Shaw/Tom Stack & Associates; **642** Ken Lucas Photo/Visuals Unlimited; **643** Stephen J. Krasemann/DRK Photo; **644** Mark Burnett; **645** (t to b)Dee Breger/Photo Researchers, Sinclair Stammers/Science

Photo Library/Photo Researchers, Dominique Braud/Tom Stack & Associates, John Shaw/Tom Stack & Associates; **648** Jonathan Blair/CORBIS; **649** Carolina Biological Supply Company/Phototake; **650** Jonathan Blair/CORBIS; **651** David M. Dennis; **652-653** ©1995 Hallmark Cards, Inc. Photography by John Perryman/courtesy of THE WILDS; **654** Joseph Sohm/ChromoSohm Inc./CORBIS; **655** Geoff Butler; **656** Larry Ulrich/Stone; **657** (t)Martin Bond/Science Photo Library/Photo Researchers, (b)Matt Meadows; **659** Joel Sartore/Grant Heilman Photography; **660** (t)Wolfgang Kaehler, (b)Grilly Bernard/Stone; **661** (t)Cliff Leight, (b)John D. Cunningham/Visuals Unlimited; **662** Volkmar K. Wentzel/National Geographic Image Collection; **663** Michael K. Nichols/National Geographic Image Collection; **664** Jacques Jangoux; **666** (t)Bettmann/CORBIS, (c)Wolfgang Kaehler, (b)Conor Caffrey/Science Photo Library/Photo Researchers; **668** Blair Seitz/Photo Researchers; **669** PLI/Science Photo Library/Photo Researchers; **670** (t)Dan Bosler/Stone, (b)Roine Magnusson/Stone; **672** AFP/CORBIS; **673** (l)Robert Fried/Tom Stack & Associates, (r)John Livsey/Stone; **675** Bob Rowan/Progressive Image/CORBIS; **676** Matt Meadows; **677** Courtesy Water Group System, Inc., San Diego, CA 92154; **678** The Newark Museum/Art Resource, NY; **679** (t to b)Larry Ulrich/Stone, Wolfgang Kaehler, Conor Caffrey/Science Photo Library/Photo Researchers, Roine Magnusson/Stone; **682** Wolfgang Kaehler/CORBIS; **683** Matt Meadows; **684** (tl)Grant Heilman/Grant Heilman Photography, (tr)Doug Martin, (b)Grafton Marshall Smith/The Stock Market; **685** Enzo & Paolo Ragazzini/CORBIS; **686** Dominique Braud/Tom Stack & Associates; **687** Breck P. Kent/Earth Scenes; **688** (t)Matt Meadows, (b)William E. Ferguson; **690** Aaron Haupt; **691** (t)Michael Bisceglie/Earth Scenes, (b)Paolo Koch/Photo Researchers; **692** (t)Lowell Georgia/Photo Researchers, (b)Thomas Kitchin/Tom Stack & Associates; **693** Oddur Sigurdsson/Visuals Unlimited; **694** John Mead/Science Photo Library/Photo Researchers; **695** W. Cody/CORBIS; **696** Colin Garratt/Milepost 92_/CORBIS; **697** Bailey, Kristof/National Geographic Image Collection; **700** Stephen J. Krasemann/DRK Photo; **701** (t)Grantpix/Photo Researchers, (b)David Young-Wolff/PhotoEdit; **702** Matt Meadows; **703** Aaron Haupt; **706** (l)Wolfgang Kaehler, (r)Konrad Wothe/Minden Pictures; **707** (t)Dominique Braud/Tom Stack & Associates, (c)John Mead/Science Photo Library/Photo Researchers, (b)Grantpix/Photo Researchers; **710** Lawrence Manning/CORBIS; **711** Matt Meadows; **712** (l)Gregory K. Scott/Photo Researchers, (r)Bill Bachman/Photo Researchers; **713** M. Harvey/DRK Photo; **716** (l)Michael Collier/DRK Photo, (r)Leonard L.T. Rhodes/Earth Scenes; **717** (t)Tom Bean/DRK Photo, (bl)Marty Cordano/DRK Photo, (br)C. P. Hickman/Visuals Unlimited; **718** Aaron Haupt; **718** CORBIS; **719** R. & E. Thane/Earth Scenes; **720** (tl)Jim Strawser/Grant Heilman Photography, (tr)Rod Planck/Photo Researchers, (b)Owen Franken/CORBIS; **722** Renee Lynn/Photo Researchers; **723** Courtesy Edenspace Systems Corporation; **724** Jonathan Blair/National Geographic Image Collection; **727** Gary Milburn/Tom Stack & Associates; **731** (t)David Woodfall/DRK Photo, (b)Jim Corwin/Photo Researchers; **732** Mark Burnett; **733** (t)Tom Bean/DRK Photo, (b)Geoff Butler; **736** USGS; **737** (t to b)M. Harvey/DRK Photo, Leonard L.T. Rhodes/Earth Scenes, Jonathan S. Blair/National Geographic, David Woodfall/DRK Photo; **740** David Hiser/Stone; **741** Bob Daemmrich; **743** Doug Martin; **744-745** Frank Zullo/Photo Researchers; **746** Roger Ressmeyer/CORBIS; **747** (l)NASA, (r)NASA/Tom Stack & Associates; **750** M.L. Sinibaldt/The Stock Market; **751** (t)NASA/Science Photo Library/Photo Researchers, (b)JPL/NASA; **752 753** NASA; **754** (t)NASA/Science Source/Photo Researchers, (c)NASA/CORBIS, (bl)NASA, (br)NASA/National Geographic Society Image Collection; **755** PhotoDisc; **758** Anthony Cook/Griffith Observatory; **762** file photo; **765** George Post/Science Photo Library/Photo Researchers; **766** Tom Brewster/PhotoTake; **767** Dennis di Cicco; **769** NASA; **771** (t)M.L. Sinibaldt/The Stock Market, (c)NASA/National Geographic Society Image Collection, (b)Dennis di Cicco; **774** William James Warren/CORBIS; **775** CORBIS; **781** (t)JPL/TSADO/Tom Stack & Associates, (b)A.S.P./Science Source/Photo Researchers; **782** (l)JPL/NASA, (r)JPL/TSADO/Tom Stack & Associates; **783** NASA; **784** HST/NASA; **785** NASA/Roger Ressmeyer/CORBIS; **786** (l)JPL/CORBIS, (r)JPL/NASA; **787** (t)NASA/JPL/Photo Researchers, (b)JPL/NASA; **788** HST/NASA; **789** (t)NASA/JPL/TSADO/Tom Stack & Associates, (b)JPL/NASA; **790** NASA/Roger Ressmeyer/CORBIS; **791** (l)JPL/CORBIS, (r)JPL/NASA; **792** NASA/Science Photo Library/Photo Researchers; **793** NASA; **794** HST/NASA/CORBIS; **796** (t)G.R. Roberts, (c)NASA/JPL/Photo Researchers, (b)Bill & Sally Fletcher/Tom Stack & Associates; **799** (l)NASA, (c)NASA/Mark Marten/Science Source/Photo Researchers, (r)USGS/Science Photo Library/Photo Researchers; **800** S. Tereby/Extrasolar Research Corp./NASA; **801** (t)JPL/TSADO/Tom Stack & Associates, (c)JPL/NASA, (b)NASA; **804** NASA; **805** EIT/NASA; **806** (t)NASA, (b)Jerry Lodriguss/Photo Researchers, **807** (t)Rev. Ronald Royer/Science Photo Library/Photo Researchers, (b)Pekka Parviainen Photo Library/Photo Researchers; **808** (t)TSADO/NASA/Tom Stack & Associates, (b)SAO/IBM Research/TASDO/Tom Stack & Associates; **809** (l)National Solar Observatory/AP/Wide World Photos, (r)Big Bear Solar Observatory/California Institute of Technology; **810** Russell Illig/PhotoDisc; **811** Deutsches Museum/Hansen Planetarium; **814** (tl)Celestial Image Co./Science Photo Library/Photo Researchers, (tr)Dr. Rudolph Schild/Science Photo Library/Photo Researchers, (b)UCO/Lick Observatory; **817** Celestial Image Co./Science Photo Library/Photo Researchers; **818** Spectral Images created by R. Pogge (OSU)/spectra from Jacoby, G.H., Hunter, D.A., & Christian, C.A. 1984, *A Library of Stellar Spectra,* Astrophysical Journal Supplement Series, vol. 56, p. 257-281.; **822** Jeff Hester/Arizona State University/NASA; **823** STScI/AURA/NASA; **825** David Malin/Anglo-Australian Observatory; **827** Spectral Images created by R. Pogge (OSU)/spectra from Jacoby, G.H., Hunter, D.A., & Christian, C.A. 1984, *A Library of Stellar Spectra,* Astrophysical Journal Supplement Series, vol. 56, p. 257-281.; **828** CXC/SAO/NASA; **829** (t)TSADO/NASA/Tom Stack & Associates, (c)Dr. Rudolph Schild/Science Photo Library/Photo Researchers, (b)Jeff Hester/Arizona State University/NASA; **832** Reuters NewMedia Inc./CORBIS; **833** AFP/CORBIS; **834** TSADO/COBE/NASA/Tom Stack & Associates; **835** Ronald Royer/Science Photo Library/ Photo Researchers; **839** Celestial Image Co./Science Photo Library/Photo Researchers; **840** (tl)TSADO/ESO/Tom Stack & Associates, (tr)Jean-Charles Cuillandre/Canada-France-Hawaii Telescope/Science Photo Library/Photo, (b)USNO/TSADO/Tom Stack & Associates; **841** Celestial Image Co./Science Photo Library/Photo Researchers; **842** (t)Space Telescope Science Institute/NASA/Science Photo Library/Photo Researchers, (b)HST/NASA/CORBIS; **844** (t)David Malin/Anglo-Australian Observatory, (c)AFP/CORBIS, (bl)Royal Greenwich Observatory/Science Photo Library/Photo Researchers, (br)Space Telescope Science Institute/NASA/Scienec Photo Library/Photo Researchers; **845** Matt Meadows; **848** (t)Bettmann/CORBIS, (b)COBE/NASA; **850** HST/A. Fruchter and the ERO Team (STScI, ST-ECF)/NASA; **853** Jean-Charles Cuillandre/Canada-France-Hawaii Telescope/Science Photo Library/Photo Researchers **854** HST/NASA/CORBIS; **855** (t)Ronald Royer/Science Photo Library/ Photo Researchers, (b)Space Telescope Science Institute/NASA/Science Photo Library/Photo Researchers; **858 859** NASA; **860** Dr. Fred Espenak/Science Photo Library/Photo Researchers; **861** NASA; **864** Ed Viesturs; **865** John Noel Photographic Collection; **866** (l)Adam Woolfitt, (r)Swissair Photo & Surveys, Zurich; **867** Adam Woolfitt; **868** (l)NGS Cartographic Division/computer-generated image supplied by Dynamic Graphics, Inc., KRS Remote Sensing, a Kodak Company, & JPL/Alex Tait, Equator Graphics, (r)William Thompson; **869** NASA; **870** Peter Essick; **871** ©National Geographic Maps; **872 873** Peter Essick; **874 875** Carsten Peter; **876** Sandro Irsara/La Venta; **877** Carsten Peter; **878** Sandro Irsara/La Venta; **879** Roberto Rinaldi; **880** ©Robert D. Ballard & Martin Bowen, Woods Hole Oceanographic Institution; **881** Courtesy The Mariners' Museum, Newport News, VA; **882** (t)©Robert D. Ballard & Martin Bowen, Woods Hole Oceanographic Institution, (b)©Woods Hole Oceanographic Institution; **883** Emory Kristof; **884** (t)William H. Bond, (b)Perry Thorsvik; **885** (t,c)©Robert D. Ballard & Martin Bowen, Woods Hole Oceanographic Institution, (b)From *The Shipbuilder,* Midsummer 1911, Special Number, courtesy John P. Eaton and Charles A. Haas Collection; **886** Alberto Garcia, Saba Press Photos; **887** Shawn G. Henry; **888** William H. Bond; **889** (l)Van Cappellen/Saba Press Photos, (r)Roger Ressmeyer/CORBIS; **890** Roger Ressmeyer/CORBIS; **891** (l)William H. Bond, (r)Christopher A. Klein; **892 893 894 895** O. Louis Mazzatenta; **896** Portia Rollings/*Sinosauropteryx* by Michael Skrepnick—*Unenlagia* by Aldo Chiappe based on fossil interpretation by Fernando Novas; **897** O. Louis Mazzatenta; **898** George Steinmetz, (inset)Courtesy Nashua River Watershed Association; **899 900 901** George Steinmetz; **902** Raghvendra Sahai, John Trauger and NASA; **903** Jon Morse, Kris Davidson, and NASA; **904** (t)Don Foley, (b)Courtesy The Huntington Library, San Marino, CA; **905** (t)NASA, (b)Jeff Hester, Paul Scowen, and NASA; **906** (tl)Philip James, Steven Lee, and NASA, (tr)Erich Karkoschka and NASA, (bl)Alan Stern, Marc Buie, ESA/NASA, (br)Lawrence Sromovsky and NASA; **907** (t)Reta Beebe, Diane Gilmore, Louis Bergeron, and NASA, (b)National Geographic Maps, Source: Zoltan G. Levay, Space Telescope Science Institute; **908** University of Houston; **912-913** Marie Tharp; **923** Doug Martin; **924** Matt Meadows; **925** (t)Matt Meadows, (b)Doug Martin; **926** Doug Martin; **928** (l)L. West/Photo Researchers, (c)Chip Clark, (r)Jeffrey L. Rotman/CORBIS; **929** Matt Meadows; **932** (l)Doug Martin, (r)Matt Meadows; **933** Matt Meadows.